International Deep-Drawing Research Group (IDDRG 2020)

IOP Conference Series: Materials Science and Engineering
Volume 967

Seoul, South Korea
26 - 30 October 2020

Editors:

Myoung-Gyu Lee
Daeyong Kim
Jung Han Song
Ji Hoon Kim

ISBN: 978-1-7138-4788-5
ISSN: 1757-8981

Printed from e-media with permission by:

Curran Associates, Inc.
57 Morehouse Lane
Red Hook, NY 12571

Some format issues inherent in the e-media version may also appear in this print version.

Institute of Physics
Dirac House, Temple Back
Bristol BS1 6BE UK

Phone: 44 1 17 929 7481
Fax: 44 1 17 920 0979

techtracking@iop.org

Additional copies of this publication are available from:

Curran Associates, Inc.
57 Morehouse Lane
Red Hook, NY 12571 USA
Phone: 845-758-0400
Fax: 845-758-2633
Email: curran@proceedings.com
Web: www.proceedings.com

TABLE OF CONTENTS

IOP Conf. Series: Materials Science and Engineering **967** (2020) 011001 doi:10.1088/1757-899X/967/1/011001

39th International Deep Drawing Research Group Conference

Be more flexible: extending flexibility of sheet metal forming

Seoul, Republic of Korea

International Deep Drawing Research Group annual conference 2020 was virtually held on 26–30 October in Seoul, Republic of Korea. The conference was organized by the Korean Society for Technology of Plasticity. The conference chairs were Prof. Myoung-Gyu Lee, Seoul National University, Dr. Daeyong Kim, Korea Institute of Materials Science, Dr. Jung Han Song, Korea Institute of Industrial Technology, and Prof. Ji Hoon Kim, Pusan National University.

The IDDRG 2020 was originally planned to be held on 7–11 June at Busan, Republic of Korea. However, the conference was postponed to October because of safety precautions over the COVID-19 outbreak. Later, the executive committee of IDDRG decided to hold the conference virtually considering the difficulties in international travel.

During IDDRG 2020, all engineering aspects of sheet metal forming were discussed, such as behavior of workpiece materials, formability, tools, tribology, process robustness, simulation methods and experiments. Apart from the general topics, the theme of IDDRG 2020 was *Be More Flexible: Extending Flexibility of Sheet Metal Forming*

International Deep Drawing Research Group

The International Deep Drawing Research Group (IDDRG) was started in the late 1950s as an organization of national groups devoted to the study of sheet metal forming including forming processes, materials, formability issues, tooling, tribology and many other interesting aspects of sheet metal forming research and industrial practices.

At the IDDRG, the leading sheet metal specialists and researchers from all over the world gather for fruitful discussions on challenging technical topics in forming technology. The conference is designed to provide an inspiring environment for knowledge development, discussions and interactions.

Virtual Conference

IOP Conf. Series: Materials Science and Engineering **967** (2020) 011001 doi:10.1088/1757-899X/967/1/011001

As a virtual conference platform, the specially designed online platform was constructed apart from the original conference website. The outline of the virtual conference platform is shown below.

[Main page]

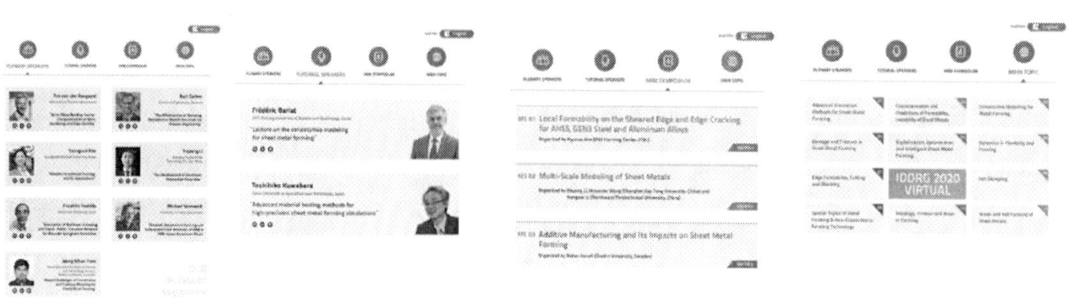

[Technical sessions]

IOP Conf. Series: Materials Science and Engineering **967** (2020) 011001 doi:10.1088/1757-899X/967/1/011001

[Sponsorships and resources]

In the virtual conference, all talks were pre-recorded and uploaded before the conference. The registered audiences could participate in the conference and access all the presentations without restrictions of place and time during the conference period.

The lengths of presentation for plenary lecture and regular talk were 40 and 20 minutes, respectively. Also, to take full advantage of the virtual conference platform, we invited two distinguished professors in the field of plasticity theory and experiment and they provided 1-2 hour tutorial sessions.

In the technical session, instead of conventional Q&A session, the virtual conference used the text or message based Q&A. For audiences who had questions to the speakers, they left notes in the specially prepared message window. Then, the speaker could respond to the audience. Alternatively, they can contact the speaker directly by email.

IOP Conf. Series: Materials Science and Engineering **967** (2020) 011001 doi:10.1088/1757-899X/967/1/011001

The virtual conference had many sponsors from industry and Korean government. Since no physical exhibition booth is available, the advertisement for the sponsored organization could be made virtually as well. There was a sponsor section in which the companies could upload their advertisement in the form of movie or slides.

Topics

The theme of IDDRG 2020 conference was *Be More Flexible: Extending Flexibility of Sheet Metal Forming*. Besides the main theme of the conference, topical sessions with the focus on the specific aspects, current trends and challenges in the field of sheet metal forming was organized by internationally renowned experts. These topics are:

- Formability of sheet metals and characterization methodologies
- Material constitutive modelling
- Necking, instability and formability prediction
- In-line control in sheet metal forming
- Digitalization in tooling and intelligent tools for sheet metal forming
- Advanced simulation methods for sheet metal forming
- Damage and fracture in sheet metal forming
- Tribology, friction and wear in forming
- Shear cutting and blanking
- Sheared edge formability
- Local versus global formability
- AHSS and Gen-3 steels
- Warm and hot forming of light metals
- Hot stamping
- Interaction of forming and crashworthiness

Furthermore, three mini-symposia were organized:

- Local formability on the sheared edge and edge cracking for AHSS, GEN3 steel, and aluminum alloys (Dr. Hyunok Kim, EWI, USA)
- Multi-scale modeling of sheet metals (Dr. Dayong Li and Dr. Huamiao Wang, Shanghai Jiao Tong University, China, and Dr. Hongwei Li, Northwest Polytechnical

University, China)

- Additive manufacturing and its impacts on sheet metal forming (Dr. Nader Asnafi, Örebro University, Sweden)

Plenary speakers

Seven keynote speeches were given by academic and industrial experts:

Prof. Ton van den Boogaard (University of Twente, Netherland)
An in-plane bending test for characterization of work hardening and edge ductility

Dr. Bart Carleer (Autoform, Germany)
The effective use of stamping simulation to stretch the limits for process engineering

Prof. Youngsuk Kim (Kyungpook National University, Korea)
Flexible incremental forming and its applications

Dr. Yuqiang Li (Shanghai Superior Die Technology Co., Ltd., China)
Aluminum automotive panel stamping die developing technology

Prof. Fusahito Yoshida (Hiroshima University, Japan)
Description of nonlinear unloading and elastic-plastic transition behavior for accurate springback simulation

Prof. Michael Worswick (University of Waterloo, Canada)
Warm and hot forming of high strength aluminum alloys

Prof. Jeong Whan Yoon (KAIST, Korea / Deakin University, Australia)
Recent challenges of constitutive and fracture modeling for sheet metal forming

Tutorial sessions

In order to draw more attentions from students and young researchers for the virtual conference, two tutorial courses were delivered.

IOP Conf. Series: Materials Science and Engineering **967** (2020) 011001 doi:10.1088/1757-899X/967/1/011001

Prof. Frédéric Barlat (GIFT, Pohang University of Science and Technology, Korea)
Lecture on the constitutive modeling for sheet metal forming

Prof. Toshihiko Kuwabara (Tokyo University of Agriculture and Technology, Japan)
Advanced material testing methods for high-precision sheet metal forming simulations

Participants & Authors

In total 260 participants from 25 countries attended IDDRG 2020. Chart below shows the number of participants by country. From these, approximately over 30% of participants had industrial affiliation and 70% were from academic institutes.

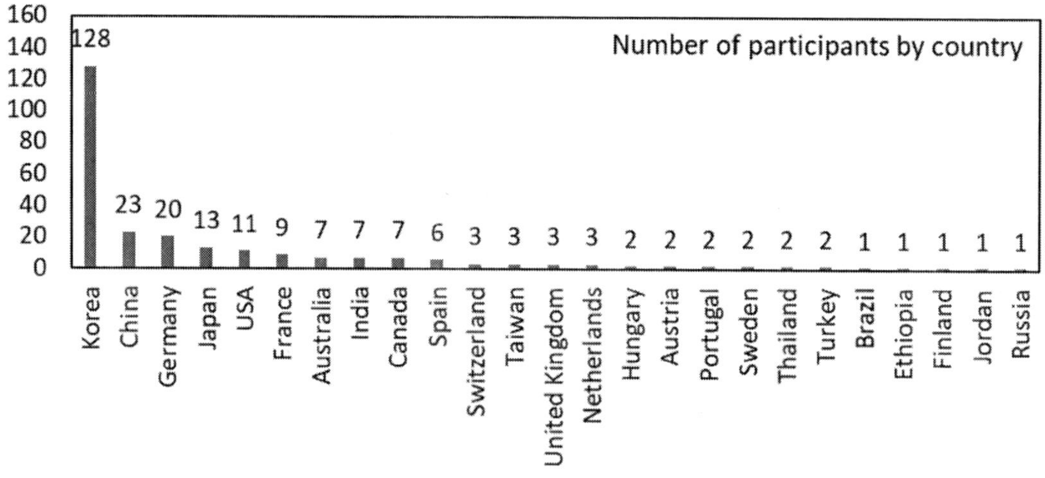

In overall, after pre-selection of 183 abstracts from 26 countries, 112 full paper submissions were made to IDDRG 2020 from 25 countries. Of these submissions, 91 manuscripts were selected based on 240 reviews in total by the external reviewers and the scientific committee. Chart below shows the number of authors by country.

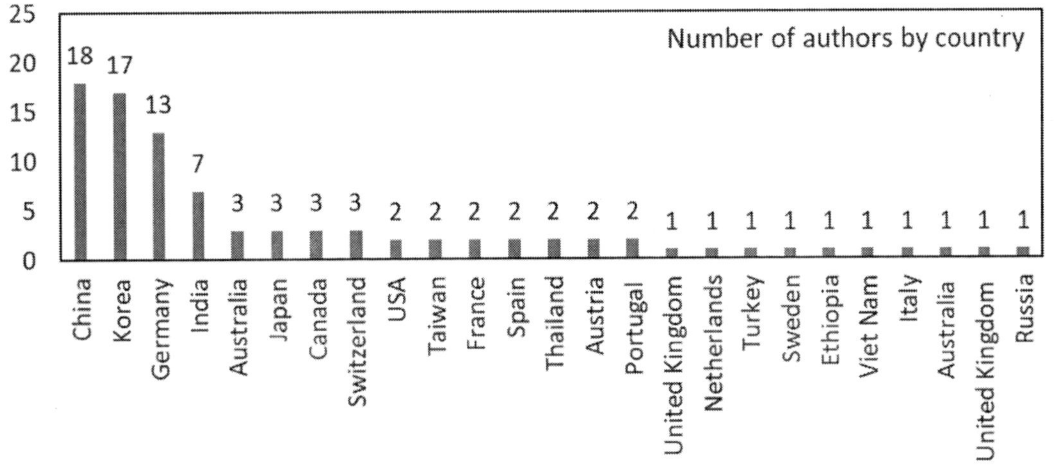

The presentation format of the virtual event was pre-recording basis. Total 146 presentations including 7 plenary talks were provided during the conference period. Chart below shows the number of presented technical talks by country.

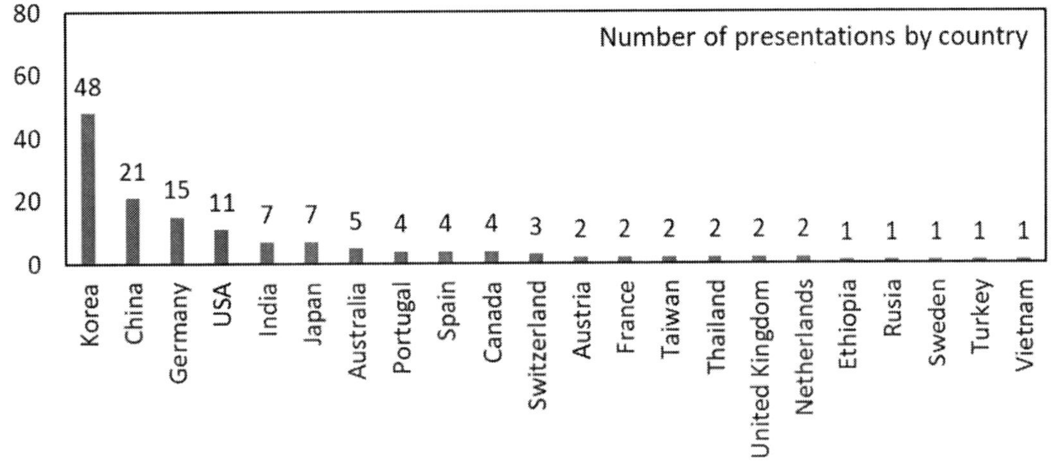

Scientific Committee

B. Rolfe (Australia)
D. Green (Canada)
K. Inal (Canada)
M. Worswick (Canada)
D. Li (China)
M. Zhan (China)
S. H. Zhang (China)
H. Huh (Korea)
B. S. Kang (Korea)
J. Yoon (Korea)
L. Zhang (Korea)
N. Langerak (Netherlands)
van den Boogaard (Netherlands)
A. Santos (Portugal)

J. Yoon Hanyang University

Advisory Board

F. Barlat Pohang University of Science and Technology

Y. S. Kim Kyungpook National University

Y. H. Moon Pusan National University

J. W. Yoon Korea Advanced Institute of Science and Technology

Acknowledgement

The conference organizing committee would like to express their gratitude to all who supported IDDRG 2020, including the participants, reviewers, keynote speakers, scientific committee members, local organizing committee members and the conference sponsors and exhibitors.

Sponsors and Exhibitors

The main sponsor of the conference was AutoForm, and all sponsors are listed below.

Main Partner of the IDDRG 2020-Virtual

In addition to the sponsorships from industry, the conference was supported by Korean Federation of Science & Technology Societies (KOFST), SNU Materials Education/Research Division for Creative Global Leaders.

Editors

Myoung-Gyu Lee (Seoul National University)

Daeyong Kim (Korea Institute of Materials Science)

Jung Han Song (Korea Institute of Industrial Technology)

Ji Hoon Kim (Pusan National University)

IOP Conf. Series: Materials Science and Engineering **967** (2020) 011002 doi:10.1088/1757-899X/967/1/011002

Peer review declaration

All conference organisers/editors are required to declare details about their peer review. Therefore, please provide the following information:

- **Type of peer review: Single-blind**
- **Conference submission management system:** Easychair(https://easychair.org)
- **Number of submissions received: 112**
- **Number of submissions sent for review: 109**
- **Number of submissions accepted: 91**
- **Acceptance Rate (Number of Submissions Accepted / Number of Submissions Received X 100): 81.2**
- **Average number of reviews per paper: 2**
- **Total number of reviewers involved: 70**
- **Any additional info on review process: We notified all authors the information on the publication and publishing licence in the review process with the following statement, "Please note that authors must agree to the terms of the IOP Proceedings License and by submitting a paper for publication it is assumed all authors of the paper agree, in full, to the terms of the license. The license terms and conditions are also available in the author guidelines".**
- **Contact person for queries:**

Prof. Ji-Hoon Kim, Pusan National University (Email: jhkim@pusan.ac.kr)

IOP Conf. Series: Materials Science and Engineering **967** (2020) 012001 doi:10.1088/1757-899X/967/1/012001

Semi – automatic approach for the creation of non - proportional load paths

R. Norz[1*], W. Volk[1]

[1]Chair of Metal Forming and Casting, Technical University Munich, Walther-Meissner Strasse 4, 85748 Garching, Germany

*roman.norz@utg.de

Abstract. Many industrial sheet metal parts experience a non – proportional loading history. The determination of such non – proportional loading paths on a laboratory scale is still very challenging and time - consuming. Most tests require different machines, tools and measurement techniques. Norz and Volk [1] have presented a new approach to create arbitrary non – proportional loading paths by using a cruciform specimen in combination with a draw bead tool. This simple experimental setup lead to a significant reduction in the experimental effort. It allows the generation of arbitrary loading paths by using only one machine, one tool and one specimen geometry without any further processing of the specimen between the forming steps. Nevertheless, the strain path could only be controlled manually after a certain forming step. In this paper, an in-line strain measurement was used to observe the real – time strain path during the experiment. This information was used to create a semi-automatic approach in which the draw bead height is adapted in order to create a prescribed strain path.

1. Introduction

The development of sheet-metal parts is often done by Finite element analysis (FEA) to reduce the try-out procedures. The accuracy of those FEA is dependent on reliable material parameters including the strength, anisotropy and formability. Following this, the characterization of the formability is done by Nakajima- or Marciniak tests according to ISO12004-2 [2]. These determined Forming Limit Curves (FLC) are only valid for proportional strain paths. Experimental studies of Bergström and Ölund [3] and Graf and Hoshford [4] have shown that the strain path history has a great influence on the formability of different materials. Therefore, different models have been introduced to predict this behaviour, for example the enhanced modified maximum force criterion (eMMFC) by Hora et al. [5], the polar effective plastic strain (PEPS) by Stoughton and Yoon [6] or the Generalized Forming Limit Concept (GFLC) by Volk and Suh [7]. The GFLC-model, as a purely phenomenological approach, needs a solid database consisting of different non-proportional experiments. Those experiments are often complex and time-consuming and not yet standardized. Hence, many different experimental approaches have been published. Jocham et al. [8] used laser cutted pre-formed Nakajima specimens to create non-proportional load paths. With this proposed method it is not possible to create arbitrary strain paths as the specimen width can only be reduced. To solve this problem, a modified Marciniak tool for the pre-straining of specimens can be used, see [9]. An area with a homogenous pre-forming allows the extraction of all widths of Nakajima specimen. Nevertheless, this approach needs several tools and optical strain measurement systems. Due to this, cruciform specimen were developed. One approach is the use of in-plane biaxial tension machines to determine the formability, see Leoting et al. [10]. Jocham

IOP Conf. Series: Materials Science and Engineering **967** (2020) 012001 doi:10.1088/1757-899X/967/1/012001

et al. [11] used a draw bead tool consisting of four independent draw beads on a conventional Nakajima testing machine in combination with a cruciform specimen. The height of the draw beads influences the retention forces and hence the strain path. A reduction of thickness in the centre of the specimen is needed in all presented approaches to ensure a crack occurrence in the evaluation area. Norz and Volk [1] have used a laser-welded cruciform specimen out of three single sheets in combination with the draw bead tool published by Jocham et al. [11]. This specimen, shown in Figure 1, was validated by conventional Nakajima tests and is capable to create non-proportional load paths.

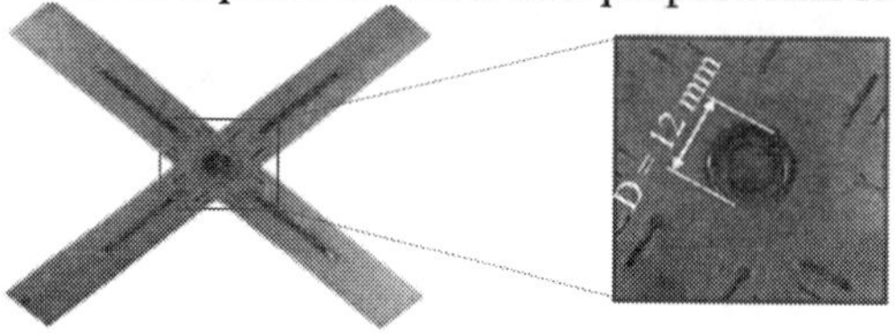

Figure 1. Laser-welded cruciform specimen according to Norz and Volk [1].

The strains of the pre-formed specimens in the Marciniak tool or the draw bead tool can only be determined after the forming process. To create certain given strains specimens have to be formed and evaluated. It is necessary to know which draw depth leads to which strain. To overcome this problem, in-line measurement of the strains is necessary. In the recent years the real-time digital image correlation (DIC) has been further developed. Tian et al. [12] have improved a video extensometer to measure the strains in real-time for small strains. Wang et al. [13] and Huang et al. [14] have shown, that the real-time strain measurement requires high-performance computers in order to obtain a satisfying frame rate and a sufficient size of the evaluation area. The bigger the evaluation area, the lower the possible frame rate is. Blug et al. [15] have used the real-time strain measurement for controlling low cycle fatigue tests. A frame rate of 3 Hz was reached. This frame rate and the low latency of only 2 milliseconds allowed the replacement of the mechanical extensometer. The information obtained by the DIC-system can also be used for closed loop control. Fischer et al. [16] used a closed loop control in the production of sinks. They used information about the draw-in of the material to adjust the blank holder force. This approach lead to a stable production process. By using 3D-image correlation in combination with a closed control loop, Hao and Duncan [17] reduced the geometrical error in incremental sheet forming.

1.2 Experimental approach

The creation of non-proportional loading paths is a challenging task as described before. In this paper the real-time strain measurement is used to semi-automatically create arbitrary non-proportional load paths. Therefore, the draw bead tool of Jocham et al. [11] is used in combination with a control loop. The control loop compares a given strain increment to the measured strains in real-time. The flow chart of the control loop is shown in Figure 2.

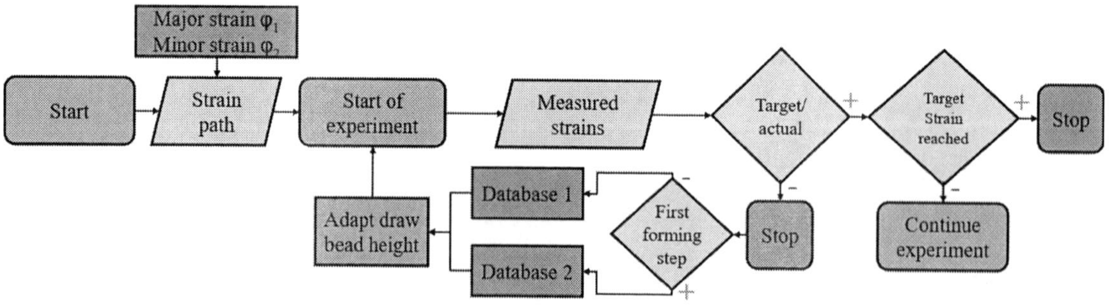

Figure 2. Flow chart of the LabView tool.

The controlling of the draw bead tool is done by the software LabView™. Measured real-time strains from ARAMIS™ are transferred to a LabView™ tool. These strains are compared to a given strain increment. If there are deviations to the target strain increment or the target strain has been reached, the experiment is stopped and the draw bead height is adjusted. This adjustment is done by the testXpert™

IOP Conf. Series: Materials Science and Engineering **967** (2020) 012001 doi:10.1088/1757-899X/967/1/012001

software. The testXpert software to control the machine is also connected to a LabView™ tool which controls the draw bead heights. Nevertheless, the clamping of the specimens as well as the start of the experiment have to be done manually due to safety reasons. The initial draw bead height is set according to the first forming step. This approach reduces the experimental effort and allows an accurate creation of non-proportional load paths with defined strain increments.

2. Experimental setup

The draw bead tool is used in combination with a Zwick Roell™ BUP1000. This sheet metal resting machine allows punch speeds from 0.02 – 50 mm/s and has a clamping and punch force of up to 1000 kN. For the strain measurement the optical measurement system GOM ARAMIS™ 4M is used. The measurement frequency is 10 Hz according to ISO 12004-2. The onset of necking is determined by the Time Dependent Evaluation Method (TDEM) proposed by Volk and Hora [18]. The used material for the experimental investigations is an uncoated micro-alloyed steel HC340LA as in [1]. The experimental setup can be found in Figure 3. For controlling the strain path the mean major and mean minor strain inside the evaluation area are used. The deviation of the strains inside the evaluation area is neglected. It is also possible to control the strain path by measuring the strains only on single facets, but as sometimes deviations occur, the use of the mean values is chosen. The communication between the different computers the TCP/IP–protocol is used. Both the LabView™ and the testXpert™ software have a TCP/IP interface which can be used.

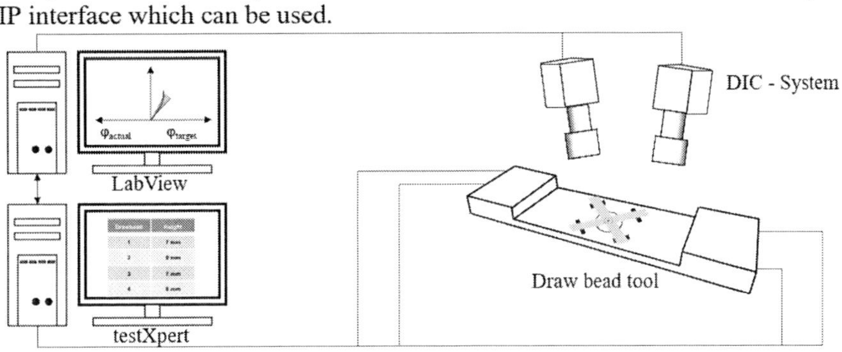

Figure 3. Experimental setup with strain measurement and draw bead height adjustment.

3. Results and discussion

The database needed for the adjustment of the draw beads was created by conducting various experiments. Several different draw bead heights were tested in order to determine which draw bead height leads to which strain ratio $\beta = \varphi_2/\varphi_1$, see Figure 4. Also, bi-linear experiments were conducted to quantify the influence of the biaxial pre-forming as well as the influence of the strain history on a consecutive forming step.

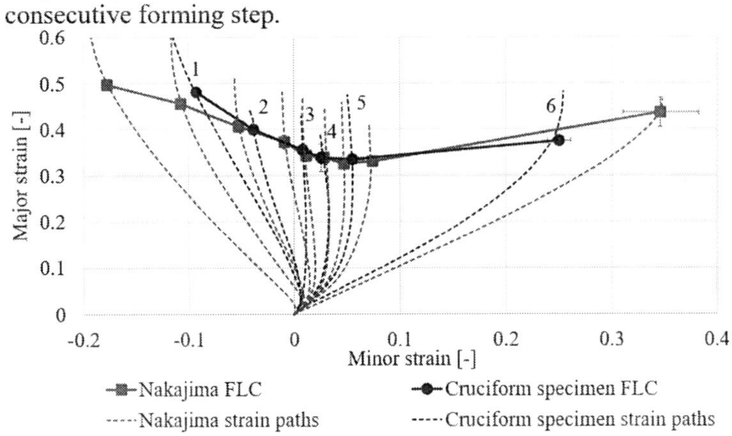

Number	Draw bead 1 & 3	Draw bead 2 & 4
1	7 mm	0 mm
2	7 mm	1 mm
3	7 mm	2.5 mm
4	7 mm	3.5 mm
5	7 mm	4.5 mm
6*	0 mm	0 mm

* Specimen is clamped only

Figure 4. Different strain paths for different draw bead heights.

IOP Conf. Series: Materials Science and Engineering **967** (2020) 012001 doi:10.1088/1757-899X/967/1/012001

In the experimental setup, a hemispherical Nakajima punch is used. As a result of this setup, a biaxial pre-strain at the beginning of the experiment occurs which is clearly visible in Figure 4. This pre-strain affects the strain paths with a negative strain ratio the most. The biaxial pre-forming has to be taken into account for the first forming step. All following forming steps do not show a biaxial pre-forming. Therefore, lower strain ratios are possible in the consecutive forming steps shown in Figure 5. The shown database in Figure 4 is hence not valid for bi- or multi-linear experiments. A different database for bi-linear experiments has to be implemented in the LabView™ tool. The strain paths with a positive strain ratio β are not significantly affected as the biaxial pre-forming does not affect the strain paths as much as negative strain ratios.

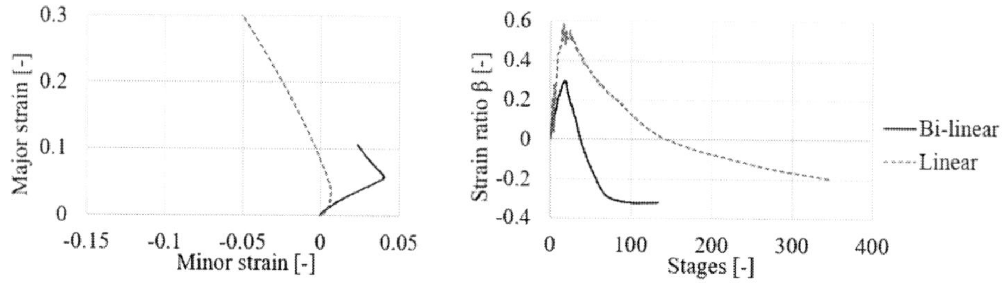

Figure 5. Influence of a consecutive forming step on the strain ratio in the uniaxial load case.

The draw bead height adjustment can only be done between the forming steps when the clamping force is removed. Removing the clamping force and the heading back of the punch leads to a relaxation of the specimen. The following reduction of the major and minor strain can be clearly seen. The resulting springback is approximately 0.01 in major strain, see Figure 6. This springback value is almost constant for all strain ratios β. Hence, a constant correction value is taken into account in the control loop.

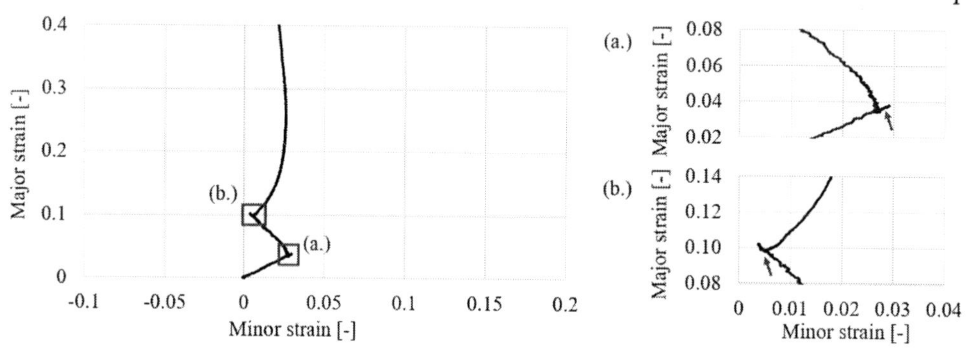

Figure 6. Strain path for a three-step experiment. Arrows indicate the start of the next forming step.

For some strain path combinations, the draw bead height has to be reduced. For example, a plane strain increment with a draw bead height of 7 mm and 3.5 mm respectively is followed by an uniaxial strain increment with a draw bead height of 7 mm and 0 mm respectively. Therefore, the specimen is rotated by 45° in the tool, see Figure 7 (a). The movement of the tool which applies the clamping force is used to remove the draw beads (Figure 7 (b)). This procedure does affect the material in the draw bead zone. Nevertheless, during the conducted experiments with and without multiple strain path changes, no failure of the specimens in this area has occurred.

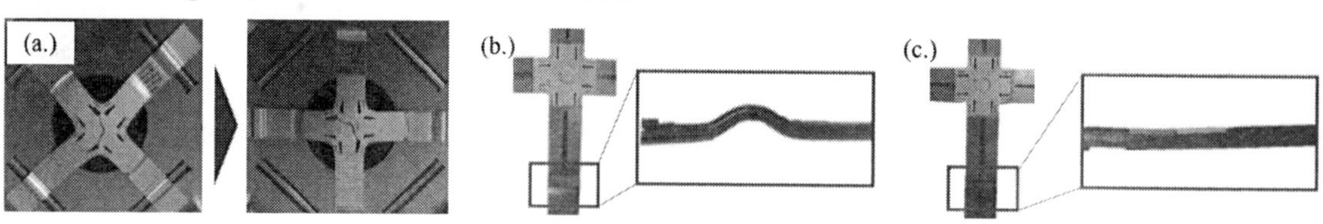

Figure 7. Removing of the draw beads, (a) Rotating the specimen by 45°, (b) draw bead height before removing, (c) draw bead height after removing.

IOP Conf. Series: Materials Science and Engineering **967** (2020) 012001 doi:10.1088/1757-899X/967/1/012001

To ensure a sufficient measuring frequency of the strains a limited evaluation area has to be chosen. The laser–welded cruciform specimen has a small area with a reduced thickness (d = 12 mm), see Figure 1. If the in-line strain measurement and calculation is done in this area, a frequency of 8 Hz can be reached. As this frequency is close to the suggested measuring frequency of 10 Hz in the DIN ISO 12004-2, a sufficient measuring frequency is ensured.
The control loop allows therefore the creation of accurate non-proportional load paths, see Figure 8. A big advantage of the presented approach is, that in comparison to the used method in [1] also small strain increments are feasible. This is possible due to the fact that with the new approach the strains are measured in-line during the experiment and the relaxation of the specimen is taken into account. This relaxation leads to a decrease in the dome height of the formed specimen. Also, the strain increase with increasing punch height is not linear. At higher strains, only a small increase in punch height can lead to a significant increased strain. Thus, the punch height is not a sufficient parameter to create accurate strain increments. As the measured and calculated strains during the experiment are directly transferred to the LabView™ tool, the strain path of the experiment can directly be extracted from the software. For the last forming step before failure no in-line measurement is used to ensure a frame rate of 10 frames per second as stated in the ISO12004-2.

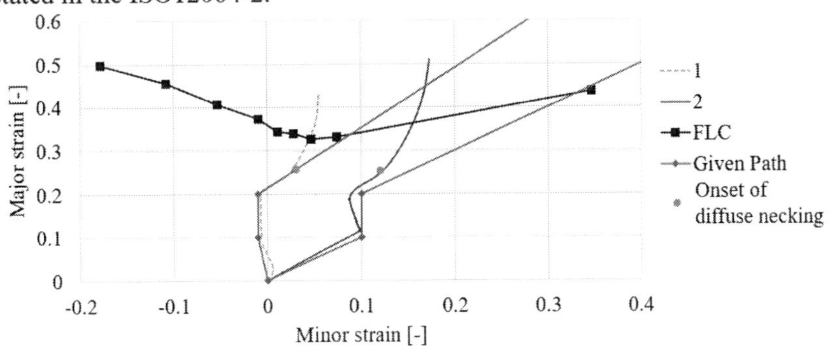

Figure 8. Comparison of given strain path to the measured strain path

At the beginning of the experiment a deviation between the given and the measured strain path is detectable. This is due to the biaxial pre-forming of the specimen. An adaption of the draw bead height does not lead to a more accurate strain path. As for the last forming step, no adaption of the strain path is done, the strain path shows a significant deviation which is also due to the non-linearity of the strain path after the onset of diffuse necking. This non-linearity is also occurring for linear strain paths, see Figure 4. Nevertheless, the creation of defined non-proportional load paths is possible. The onset of necking for the conducted multi-linear experiments can accurately be predicted by the Generalized Forming Limit Concept (GFLC) as proposed by Volk and Suh [7], shown in Figure 9.

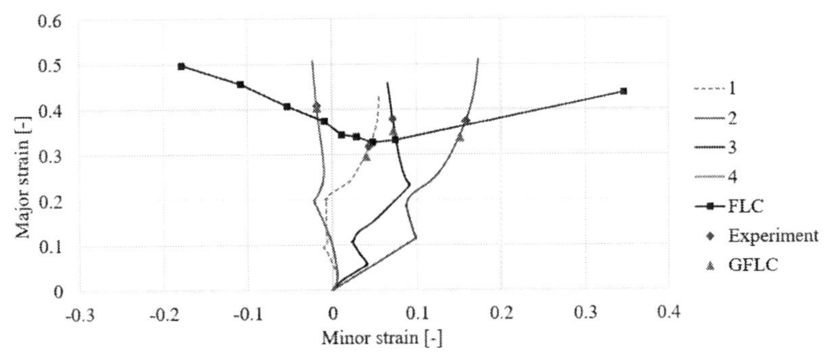

Figure 9. Different created strain paths by the semi-automatic approach and the predicted failure by the GFLC.

4. Conclusions

With the presented in-line measurement almost arbitrary non-proportional loading paths can be created. Due to the limited evaluation area of the cruciform specimen a sufficient measuring frequency is possible. This allows to react on occurring deviations from a given strain path. Due to the hemispherical punch used in the experimental setup, a biaxial pre-strain occurs at the beginning of the experiments. The biaxial pre-forming does not occur in the following forming steps. This leads to the need of a second database for bi- and multi-linear strain paths. To be able to adjust the draw bead height after a strain increment, the clamping force has to be removed. This leads to a relaxation of the cruciform specimen. All those effects are taken into account by the presented semi-automatic approach. Nevertheless, this experimental setup presents an easy way to create non-proportional load paths. To predict the onset of necking for the conducted non-proportional load paths with several different forming increments, the GFLC-model is used. It is shown, that even strain paths with four different strain increments are predicted accurately.
Due to the use of the experimental setup using a passive draw bead tool to create the retention forces, some limitations are given. For certain strain increment combinations, for which the draw bead height has to be lowered, the beadings have to be removed and formed again. By using active actors this problem can be solved. Also by using active actors, the specimens would not have to be relaxed between the forming steps while adjusting the retention forces.

Acknowledgments
The authors would like to thank the German Research Foundation (DFG) for the financial support under the grant number VO1487/34-1.

References

[1] Norz R. and Volk W. 2019 Investigation of non-proportional load paths by using a cruciform specimen in a conventional Nakajima test. *IOP Conf. Ser.: Mater. Sci. Eng.* **651**
[2] Deutsches Institut für Normung e.V. 2009) *Metallische Werkstoffe – Bleche und Bänder - Bestimmung der Grenzformänderungskurve – Teil 2: Bestimmung von Grenzformänderungskurven im Labor* (Berlin: Beuth Verlag GmbH)
[3] Bergström Y. and Ölund S. 1982 The forming limit diagram of sheet metals and effects of strain path changes on formability: a dislocation treatment. *Materials Science and Engineering* **56**:47–61
[4] Graf A. and Hosford W. 1994 The influence of strain-path changes on forming limit diagrams of Al 6111 T4. *International Journal of Mechanical Sciences* **36**:897–910
[5] Hora P., Tong L. and Berisha B. 2013 Modified maximum force criterion, a model for the theoretical prediction of forming limit curves. *Int J Mater Form* **6**:267–279
[6] Stoughton T.B. and Yoon J.W. 2012 Path independent forming limits in strain and stress spaces. *International Journal of Solids and Structures* **49**:3616–3625
[7] Volk W. and Suh J. 2013 Prediction of formability for non-linear deformation history using generalized forming limit concept (GFLC) (AIP) pp 556–61
[8] Jocham D., Gaber C., Böttcher O. and Volk W. 29. - 30. June, 2015 *Prediction of Formability for multi-linear strain paths* (Zürich)
[9] Weinschenk A. and Volk W. FEA-based development of a new tool for systematic experimental validation of nonlinear strain paths and design of test specimens:1–6
[10] Leotoing L., Guines D., Zhang S.Y. and Ragneau E. 2013 A Cruciform Shape to Study the Influence of Strain Paths on Forming Limit Curves. *Key Engineering Materials:*41–46
[11] Jocham D., Gaber C., Böttcher O., Wiedemann P. and Volk W. 2017 Experimental prediction of sheet metal formability of AW-5754 for non-linear strain paths by using a cruciform specimen and a blank holder with adjustable draw beads on a sheet metal testing machine. *Int J Mater Form* **10**:597–605
[12] Tian L., Yu L. and Pan B. 2018 Accuracy enhancement of a video extensometer by real-time error compensation. *Optics and Lasers in Engineering* **110**:272–278

IOP Conf. Series: Materials Science and Engineering **967** (2020) 012001 doi:10.1088/1757-899X/967/1/012001

[13] Wang T., Kemao Q., Seah H.S. and Lin F. 2018 A flexible heterogeneous real-time digital image correlation system. *Optics and Lasers in Engineering* **110:**7–17

[14] Huang J.W., Zhang L.Q., Jiang Z.Y., Dong S., Chen W., Liu Y.P., Liu Z.J., Zhou L.C. and Tang L.Q. 2018 Heterogeneous parallel computing accelerated iterative subpixel digital image correlation. *Sci. China Technol. Sci.* **61:**74–85

[15] Blug A., Regina D.J., Eckmann S., Senn M., Bertz A., Carl D. and Eberl C. 2019 Real-Time GPU-Based Digital Image Correlation Sensor for Marker-Free Strain-Controlled Fatigue Testing. *Applied Sciences* **9:**2025

[16] Fischer P., Heingärtner J., Renkci Y. and Hora P. 2018 Experiences with inline feedback control and data acquisition in deep drawing. *Procedia Manufacturing* **15:**949–954

[17] Hao W. and Duncan S.R. *IEEE International Conference on Automation Science and Engineering* pp 779–84

[18] Volk W. and Hora P. 2011 New algorithm for a robust user-independent evaluation of beginning instability for the experimental FLC determination. *Int J Mater Form* **4:**339–346

IOP Conf. Series: Materials Science and Engineering **967** (2020) 012002 doi:10.1088/1757-899X/967/1/012002

Finite Element Study on Pattern of Pins Supporting an Elastic Blankholder Used for Stamping Irregular Shaped Sheet Metal Parts

K J Fann[1] and M Liewald[2]

[1]Department of Mechanical Engineering, National Chung Hsing University, 250, Kuokuang Road, Taichung 40227, Taiwan, R.O.C
[2]Institute of Metal Forming Technology, University of Stuttgart, 17, Holzgarten-Str., 70174 Stuttgart, Germany

E-mail: kjfann@nchu.edu.tw

Abstract. Working with individual in-process controllable elements to support an elastic blankholder is the state-of-the-art for stamping irregular shaped sheet metal parts without wrinkles and cracks. It is thus needed to carefully determine the position for each expensive device. This study is therefore aimed to appraise a feasible approach with finite element analysis, in that the blankholder is first modelled as a rigid object instead of an elastic one to evaluate the contact pressure of the blankholder to the blank during the forming process. The sites showing a locally relative high contact pressure, which are usually induced by local thickening of material, can be regarded as the locations of the supporting elements to intently form a local high pressure to control the material flow and thus to eliminate the local thickening there. As a result, since the blankholder force can be efficiently delivered from the supporting elements directly to the blankholder as well as to the blank, it is not only applied to the rectangular box-shaped part but also to a fender-like irregular sheet metal part. Furthermore, the initiated contact pressure can be maintained at locations during the whole stamping process. This is also experimentally validated by stamping a sound fender-like irregular sheet metal part with an elastic blankholder supported by elements located accordingly at those sites suggested by this study.

1. Introduction

In the last few decades, a tremendous amount of research has been involved in overcoming the failures, such as wrinkles and cracks, presented on irregular shaped sheet metal parts during the stamping process [1]. Especially when working with an elastic blankholder supported by individual in-process force-controllable cylinders, which is based on the monitoring of part wall stresses, a sound stamped part can be easily carried out [2]. However, this achievement is investment-intensive, because lots of expensive hydraulic proportional valves are needed to realize such robust stamping process. Therefore, it has been alternatively proposed by [3] to transfer the achievement to the press shops, where the presses are usually equipped with a die cushion, by implementing length-adjustable thermal pins as die cushion pins, with which no more expensive hydraulic valve is needed. However, it is still unclear whether the available pin pattern is appropriate for forming irregular shaped parts using an elastic blankholder. If not, it is still required to figure out another approach to support the elastic blankholder in properly building up the localized contact pressure to control the material flow. Based on this idea, this study is thus initiated to

IOP Conf. Series: Materials Science and Engineering **967** (2020) 012002 doi:10.1088/1757-899X/967/1/012002

investigate how well the layout of the cushion pins works for locating the thermal pins to support the elastic blankholder, as well as to further figure out a preliminary layout of the thermal pin supports if the pin pattern is not appropriate to build up a localized contact pressure on the elastic blankholder. Actually, even in the phase working with individual cylinders for stamping irregular shaped sheet metal parts, it is still required to specify the locations of the individual cylinders.

On this basis, this study starts with the process of forming a rectangular box-shaped sheet metal part with round corners, which can be considered as a basic irregular shape, to investigate how well the accessible pattern of the cushion pins works with an elastic blankholder to stamp the box, as shown in Figure 1 (a). If it does not work well, the layout of supporting elements, which stand on an intermediate plate raised by the cushion pins and then support the elastic blankholder to deliver the cushion force to the blankholder as shown in Figure 1 (b), is then investigated. At the same time, an approach should be proposed to preliminarily specify a layout to individually position each supporting element. Once the layout of the supporting elements is specified for forming irregular shaped sheet metal part, the optimization of the process parameters can then be further conducted. In comparison to direct using robust design optimization approaches, it can benefit the engineers by eliminating the design parameters, such as number and locations of the supporting pins, and thus reducing lots of trials while developing the process by using the approach proposed by this study. The approach suggested by this study will be further validated in this study by forming a fender-like irregular sheet metal part stamped with an elastic blankholder supported by such elements.

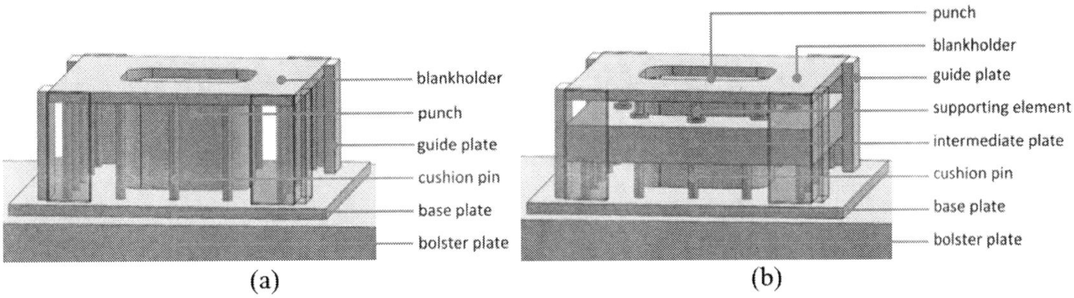

Figure 1. Illustration of layout for an elastic blankholder supported (a) directly by the cushion pins and (b) by the supporting elements standing on the intermediate plate raised by the cushion pins

2. Deep Drawing of Rectangular Box-Shaped Sheet Metal Part

A rectangular sheet with a width of 210 mm and a length of 260 mm is approximately determined for drawing a rectangular box-shaped sheet metal part having dimensions of 100 mm wide, 150 mm long, and 50 mm high according to the handbook [4]. The punch for stamping has a corner radius of 20 mm and a top radius of 7.5 mm, while the die has a radius of 5 mm.

2.1. Finite element analysis with a rigid blankholder

A stamping process of sheet metal parts can be efficiently and robustly simulated with an explicit finite element code [5]. In this study, the finite element analysis for stamping is then carried out by the commercial explicit code LS-DYNA. Normally, the steel tools around the sheet metal blank can be modelled as rigid parts in shell elements for further efficient computation [6]. Thus, the drawing process of this rectangular box-shaped sheet metal part is firstly checked with a finite element model, in which the blankholder is regarded as rigid. Figure 2 (a) depicts the thickness distribution on the sheet metal parts stamped with a blankholder force of 75 kN, which is determined according to the handbook [4] with the blankholder pressure of 2.0 MPa for a deep-drawing steel DC04 used in this study. It can be observed that the whole sheet metal part has at least a thickness of 0.84 mm from the initial thickness of 1 mm. However, a bulge can be found on the long side wall of the part, as shown the dark blue fringed area in Figure 2 (b). This can be attributed to that the part suffers a compression stress in the length (Y) direction there as shown in Figure 2 (b), which should be reduced in a reasonably acceptable range.

2.2. Finite element analysis with an elastic blankholder directly carried by cushion pins

To minimize the bulge on the side wall of the part, which might be considered as a possible drawback of the part, an elastic steel blankholder is used in this study to provide localized contact pressure to depress the potential origin, the compressive stress in the length direction. The blankholder used in the common design has a dimension of 600 mm each in length and width to enclose the four guide pins and bushings at the corners as well as 8 cushion pins around the die cavity. The 8 cushion pins are modelled as rigid parts, which have a diameter of 50 mm each and are positioned according to the pin hole pattern centered on the press bed with a distance of 150 mm to each other in the directions of front-back and left-right. The blankholder has a thickness of 25 mm, which might be reasonably considered as a compliant one to sensibly bring a specific contact pressure to the blank. Most of the keywords used for finite element modelling the stamping process with the elastic blankholder are the same as with the rigid blankholder, except that the elastic steel blankholder is modelled with the selective reduced integrated solid elements. Either solid elements or shell elements that are used for the blankholder have an average mesh size of 8 mm, while the other rigid tools modelled in shell elements have a finer mesh with at least 11 elements around the 90° corners to keep the curvature detail there. The plastic behaviour of the DC04 sheet is formulated with three parameter Barlat material model. The penalty surface-to-surface contact was taken between the blank and the tools as well as between the cushion pins and the blankholder. The coefficient of friction is set as 0.125 to reduce the potential of wrinkling formations.

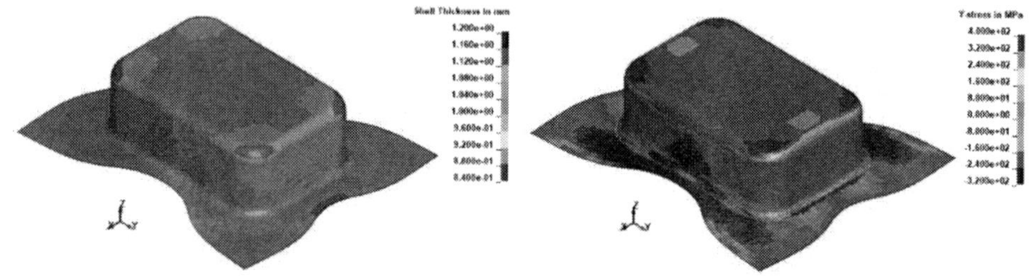

(a) (b)

Figure 2. (a) Thickness and (b) stress in the length (Y) direction on the sheet part stamped with a blankholder force of 75 kN

In comparison to the results as shown in Figure 2, the thickness and the stress obtained with the elastic blankholder shown in Figure 3 depict no significant difference to those computed with the rigid blankholder. It means that an elastic blankholder directly supported by the cushion pins might not provide localized contact pressure to adequately depress the bulge on the side wall of the part. It needs thus to explore another measurement.

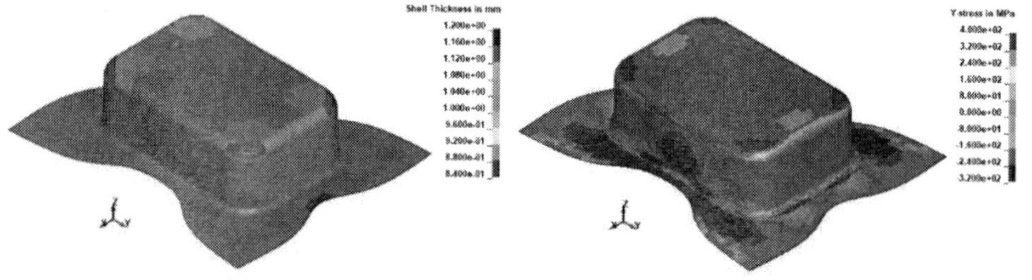

(a) (b)

Figure 3. (a) Thickness and (b) stress in the length (Y) direction on the sheet part stamped with an elastic blankholder having a thickness of 25 mm supported by 8 cushion pins under a force of totally 75 kN

IOP Conf. Series: Materials Science and Engineering **967** (2020) 012002 doi:10.1088/1757-899X/967/1/012002

Because there is no reaction force provided for the results computed with the rigid blankholder via the current available postprocessor LS-PrePost, to expose potential inducement, the contact pressure of the blank to the blankholder is shown alternatively in Figure 4 and 5 for the results computed with each kind of blankholder. It can be observed in Figure 4 (a) that even with the rigid blankholder the pressure is not homogenously distributed. There already exist sites having high pressure far away from the designed one of 2.0 MPa that are concentrated around the middle of each side of the outer edge of the blank at the very beginning of the forming process as the die stroke just reaches 10 mm. Actually, the material at these sites accumulates and becomes thicker there. It means that, other than the typical sites in deep drawing around the corners, the material there suffers a compressive stress in the circumferential direction along the edge of the die cavity as well. Especially, the contact pressure around the corners is measured to be significantly lower than the expected value. Around the edge of the die cavity, a contact pressure about the designed value is present. This might be attributed to that the material there is bent and tilts a little bit over the die corner radius. For the further die stroke to 40 mm, the contact pressure around the outer edge middles and around the corners of the blank stays high and closer to the die cavity as shown in Figure 4 (b), because the thickness there becomes thicker. Around the die cavity, small amount of contact pressure still can be found due to bending over the die radius.

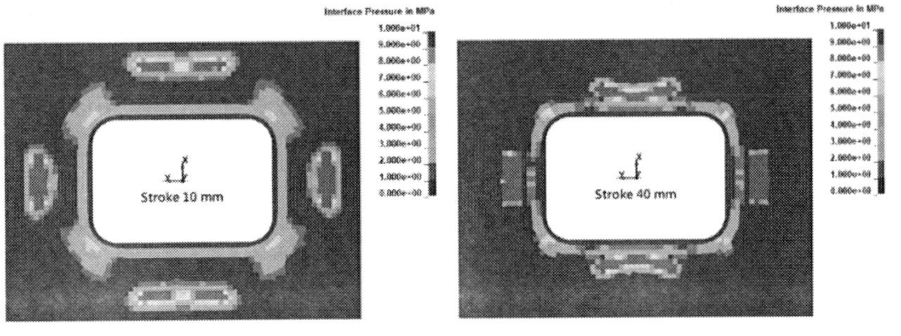

(a) (b)

Figure 4. Resultant contact pressure of the blank on the blankholder under a force of totally 75 kN computed with a rigid blankholder at the stroke (a) 10 mm and (b) 40 mm of the die

The contour of the blank that reveals a relative high contact pressure to the blankholder computed in elastic model at the die stroke of 10 mm can somehow reflect the rectangular shape of the blank as shown in Figure 5 (a), which is complete different to that computed in rigid model as shown in Figure 4 (a). Only the area of the blankholder underneath the die projection is shown in Figure 5, because the area out of the projection of the die face has no contact to the blank and it is thus ignored for easy comparison to the results computed with the rigid blankholder. Part of the 8 yellow rings shown in Figure 5 is the outer contour of the 8 cushion pins used to demonstrate where the 8 cushion pins are located. The highest contact pressure is present in front of the short side wall, where there two cushion pins are located. It means that the force delivered from the cushion can directly act on the blankholder as shown in Figure 6 (pin #4 and #8). The pressure is thus dominant there. Furthermore, even the four cushion pins located at the corners do not contribute any of the contact pressure on the blankholder, because there is no resultant force on them as shown in Figure 6 (the remaining pins). However, the high contact pressure presents around the corners can be attributed to the material becoming thicker there because of the compressive stress existing in the circumferential direction. In contrast to the contact pressure around the corners, there is no significant pressure appearing on the area in front of the long side wall, even there is a resultant force on the cushion pins there as shown in Figure 6 (pin #2 and #6). In general, the elastic blankholder having the thickness of 25 mm can efficiently deliver the cushion force as the localized contact pressure of the blankholder, if the thickness of the blank on the blankholder is relatively homogeneous. However, as the die increases its stroke, the thickness of the blank becomes inhomogeneous and more material accumulates around the outer edge middles of the blank and around the corners of the die cavity, the elastic blankholder losses its function to build up the localized contact pressure on the area above the cushion pins. It can be observed that the contact pressure shown in Figure 5 (b) is more or less similar to those shown in Figure 4 (b).

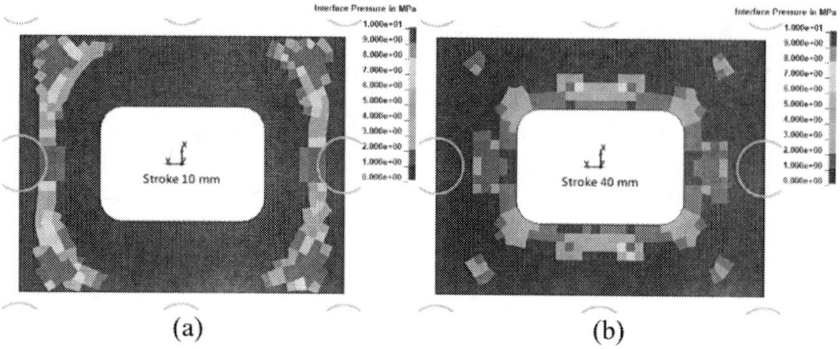

Figure 5. Resultant contact pressure of the blank on the blankholder under a force of totally 75 kN computed with an elastic blankholder having a thickness of 25 mm supported by 8 cushion pins at the stroke (a) 10 mm and (b) 40 mm of the die

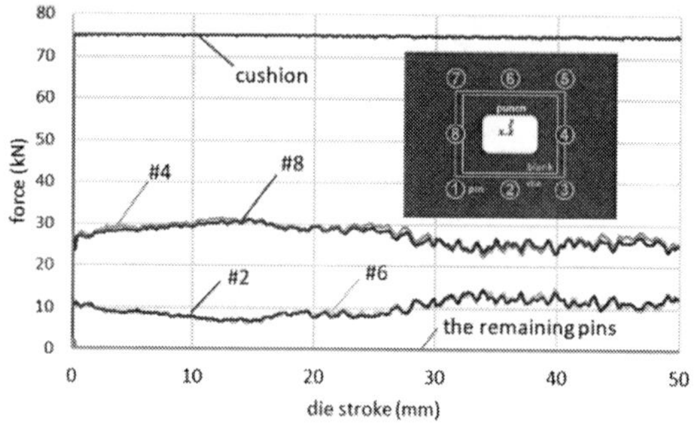

Figure 6. Resultant force acting on the 8 cushion pins supporting the elastic blankholder having a thickness of 25 mm in comparison to the cushion force over the die stroke

2.3. Finite element analysis with an elastic blankholder carried by supporting elements

Based on the above earned results that the localized contact pressure of the blankholder to the blank can be built up directly by the supporting elements and can be maintained to the end of the forming process, it is thus meaningful to select the sites that have the relative higher contact pressure at the beginning as well as at the end of the forming process as the locations for the supporting elements. It is easily done by executing a simulation with a rigid modelled blankholder. For the above investigated rectangular box-shaped sheet metal part, eight sites are accordingly chosen, four at the corners and four at the side middles, to position the supporting elements. Each of the supporting elements is a cylindrical rod having a diameter of 50 mm standing on a disc base. All the elements are located such that there is a small gap of 3.5 mm to the die cavity of the blankholder and that they are completely underneath the blank, as indicated by the gold rings shown in Figure 7. Figure 7 depicts the contact pressure of the blankholder to the blank. Around the location of the supporting elements a relative higher pressure is maintained as intended through the process. The resultant force on each supporting element is then plotted in Figure 8. In comparison to the results with the blankholder supported by the cushion pins as shown in Figure 6, all the elements can provide the blankholder force, which actually is the main idea behind using the elastic blankholder. With this layout of the supporting elements, it is easily to have another resultant force diagram for each element for each length. Figure 9 depicts the results for the setup, in that the elements closest to the long side walls (pin #2 and #6) have a length only 0.01 mm longer than the

IOP Conf. Series: Materials Science and Engineering **967** (2020) 012002 doi:10.1088/1757-899X/967/1/012002

remaining elements. The resultant force on those elements is raised about 14.5 kN. Therefore, an optimization of the process can be carried out to eliminate the bulge on the long side wall by this approach, which is not conducted in this study.

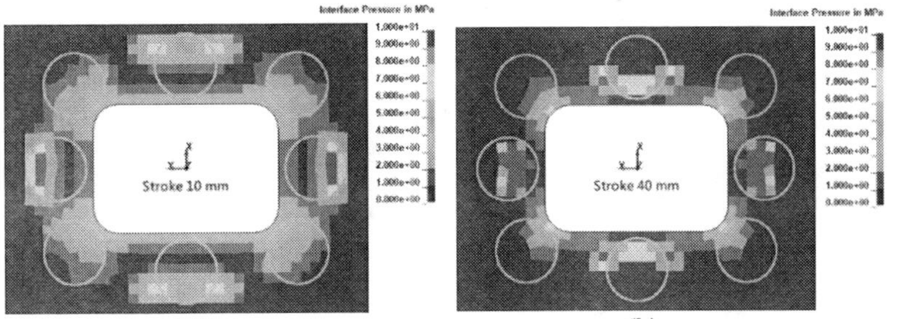

(a) (b)

Figure 7. Resultant contact pressure of the blankholder to the blank under a force of totally 75 kN computed with an elastic blankholder having a thickness of 25 mm supported by 8 elements at the stroke (a) 10 mm and (b) 40 mm of the die

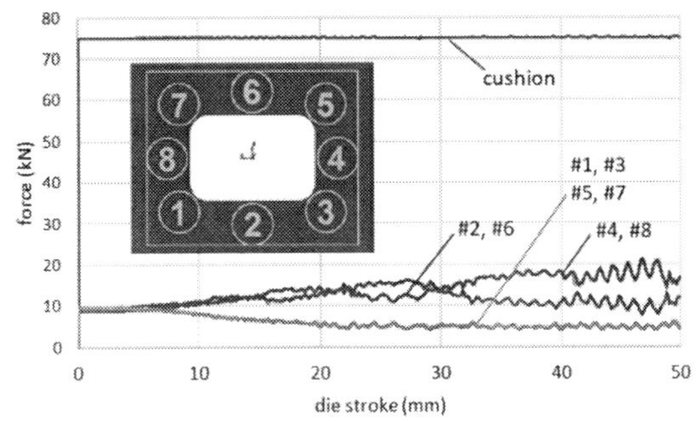

Figure 8. Resultant force acting on the 8 elements directly underneath the blank to support the elastic blankholder

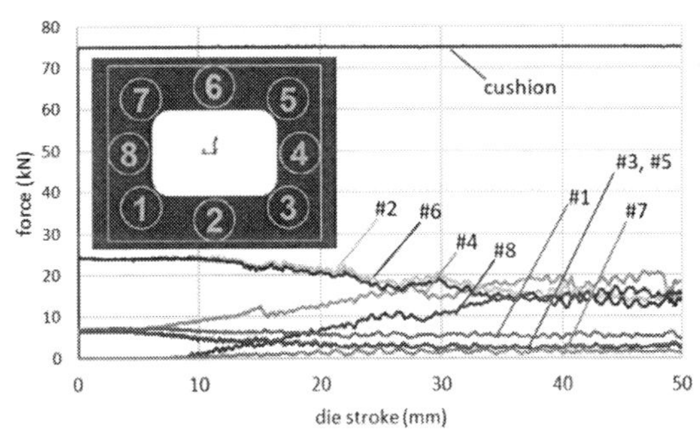

Figure 9. Resultant force acting on the 8 supporting elements, in that element #2 and #6 are longer than the others

3. Deep Drawing of Fender-Like Sheet Metal Part

To validate the procedure proposed by this paper, a fender-like irregular sheet metal part is used, which is based on [7] and has a slender right trapezoid on its top face and also on its bottom projection with different radii at the corners. There exists an inclined wall on each side, in that the radii and size on the bottom projection are larger than those on the top face. Figure 10 depicts the thickness of the sheet metal

IOP Conf. Series: Materials Science and Engineering **967** (2020) 012002 doi:10.1088/1757-899X/967/1/012002

part stamped to the die stroke of 70 mm under a cushion force of 650 kN with an initial DC04 blank in octagonal shape having a thickness of 1 mm and the contact pressure of the blank to the blankholder modelled as a rigid part. It confirms the above mentioned approval that the material at the sites having a relative high contact pressure accumulates and becomes thicker there. Because the existing layout of 20 cushion pins for the die set to stamp such an irregularly shaped part [8] cannot provide a promised contact pressure when the blankholder is modelled as an elastic part [9], the ten sites shown in Figure 10 (b) are thus accordingly applied to set up the layout of the elements supporting the blankholder.

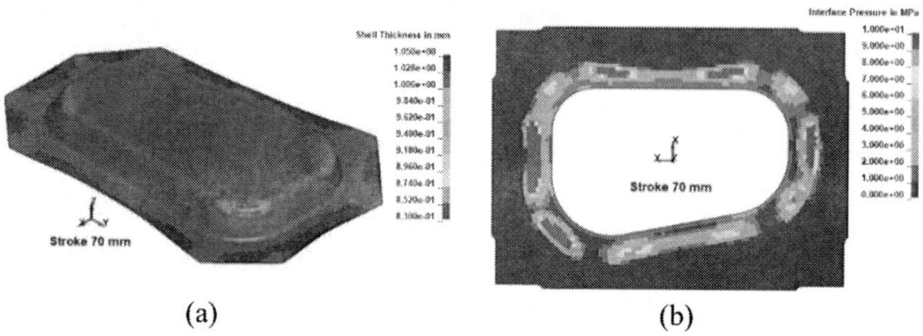

Figure 10. (a) Thickness of the fender-like sheet metal part stamped with a blankholder force of 650 kN and (b) contact pressure on the blankholder modelled as a rigid part

Figure 11 depicts the contact pressure of the blankholder to the blank with the layout of the supporting elements, which are suggested by this study for stamping the fender-like sheet metal part and shown as black rings. There is a gap of 20 mm between each supporting element and the die cavity. It can be observed that around the location of the supporting elements a relative higher pressure could be maintained through the stamping process. This layout is actually similar to the 10 hydraulic individual cylinders as the multipoint cushion system built by [10] to apply the blankholder force on the blankholder for forming such fender-like irregular sheet metal parts. The blankholder force for stamping the part with the multipoint cushion system has been optimized by the previous study [7]. Figure 12 depicts the fender-like part having a height of 70 mm stamped from a DC04 sheet in thickness of 1 mm with the above-developed setup under the approximation of the optimized blankholder force by adding the height of the supporting element with feeler gauges, which was determined in [9], and with the compensation of the backlash presenting in the die set, which is detailed in [11].

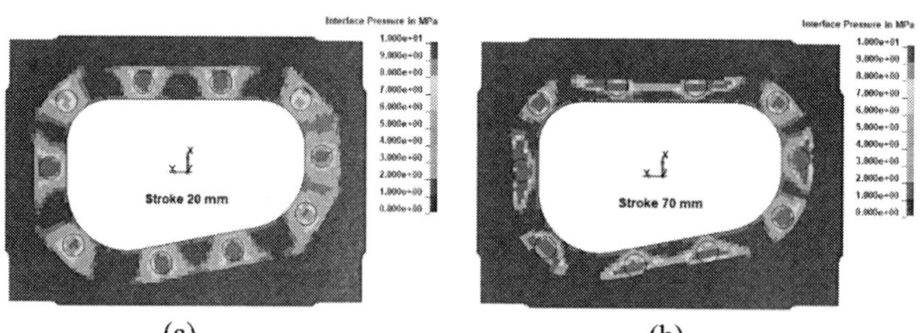

Figure 11. Resultant contact pressure of the elastic-modelled blankholder to the blank stamped with a blankholder force of 650 kN supported by 10 elements at the die stroke of (a) 20 mm and (b) 70 mm

4. Conclusion and Prospects

Based on the fact that the blankholder directly supported by the available cushion pins might not work properly for stamping irregular shaped sheet metal parts, using separate supporting elements instead of directly using the cushion pins is one of the ways to build up a localized contact pressure of an elastic blankholder to the blank in order to efficiently control the material flowing into the die cavity and thus produce a sound part. To effortlessly find a suitable layout of the supporting elements out of unlimited combinations, this study proposed an easy approach to determine the layout with finite element analysis, in which all the tools involved in the stamping process are modelled rigid. The sites, where a relative higher contact pressure appears on the blankholder to the blank at the beginning as well as at the end of the stamping process, can be considered as candidate locations for the supporting elements. In the study, a rectangular box-shaped sheet metal part, which can be regarded as a basic irregular shape, is used to demonstrate this procedure. As a result, it has been proven that the blankholder force can be more efficiently delivered from the supporting elements directly to the blankholder as well as to the blank, if the elements are located at the sites suggested by this study. A fender-like irregular sheet metal part is further used to demonstrate and validate the procedure to layout the supporting elements with experiment as well. It seems that the procedure proposed by this study works well for stamping irregular shaped sheet metal part, in this study, a quadrilateral box-shaped part with round corners. Thus, for the future works, it is worth to conduct further studies involving different materials, different anisotropy, different shape and dimension of initial blank and so on, so that the procedure proposed by this study can benefit most of applications.

Figure 12. The stamped fender-like sheet metal part on the blankholder supported by 10 elements

Acknowledgements
The per diem for the first author working on this research was supported by the Ministry of Science and Technology of the Republic of China with the grant coded MOST 108-2918-I-005-008.

References

[1] Siegert K and Doege E 1993 CNC hydraulic multipoint blankholder systems for sheet metal forming presses *CIRP Ann-Manuf Techn* **42** pp 319–322

[2] Blaich C, Liewald M and Wurster K 2010 Optimization of blank holder force distribution for deep drawing by the use of wrinkle criteria *ANSYS Conference & 28th CADFEM Users' Meeting 2010*

[3] Liewald M and Barthau M 2018 Adaptive control strategies for deep drawing of high performance sheet metal materials *37th International Deep-Drawing Research Group Conference 2018*

[4] Suchy I 2006 *Handbook of Die Design*, 2nd Ed., (New York: McGraw-Hill) chapter 9 pp 415–424

[5] Mattiasson K 2000 On finite element simulation of sheet metal forming processes in industry *European Congress on Computational Methods in Applied Science and Engineering*

[6] Roll K 2008 Simulation of sheet metal forming – necessary developments in the future *LS-DYNA Anwenderforum*

[7] Liewald M and Blaich C 2009 Approaches for closed-loop control and optimization of deep drawing processes *ANSYS Conference & 27th CADFEM Users' Meeting 2009*

[8] Radonjic R, Liewald M and Han F 2015 Influence of blank holder stiffness on part surface quality in deep drawing process *12th International Conference on Accomplishments in Electrical and Mechanical Engineering and Information Technology*

[9] Fann K-J, Liewald M, Riedmüller M, Radonjic R and Karadogan C 2019 Finite element study on sheet metal forming with an elastic blankholder supported by individual length-adjustable pins *2nd Asian Pacific Symposium on Technology of Plasticity*

[10] Häusermann M 2000 Multipoint – cushion – technology (advances and die design) *New Developments in Sheet Metal Forming* ed K Siegert (Frankfurt am Main: MAT-Info) pp 341–366

[11] Fann K-J, Liewald M and Riedmüller K M 2020 Localized contact pressure of blankholder for stamping irregular sheet parts by configuring the height of the supporting elements *13th International Conference on the Technology of Plasticity* (submitted)

IOP Conf. Series: Materials Science and Engineering **967** (2020) 012003 doi:10.1088/1757-899X/967/1/012003

Improved coining force calculations through incorporation of key process parameters

D Cotton[1,3], A Maillard[2], and J Kaufmann[2]

[1] Arts et Métiers Institute of Technology, LABOMAP, HESAM University, 71250 Cluny, France
[2] CETIM – Technical Centre for Mechanical Industries – France
[3] Author to whom any correspondence should be addressed

E-mail address: dominique.cotton@ensam.eu

Abstract. Among the sheet forming processes, coining is a specific operation which makes it possible to correct shape defects or to perform thickness reductions in the parts. This operation often requires a very high force which is likely to have an impact on the functioning of the tool or the press. It is therefore important that the coining forces be evaluated accurately, but this is not allowed by the existing analytical calculation formulas. The objective of this study is to improve the accuracy of the calculation of the coining forces via the adaptation of an existing formula. Such adaptation was carried out based on a study of the influence of the flat coining parameters. To that end, a series of experimental measurements was performed. An instrumented force measuring setup made it possible to measure the maximum coining force. Several parameters were analysed, including the coining ratio and the sheet thickness. Numerical simulations were carried out at the same time, in order to understand the influence of certain parameters on the coining force. 2D numerical models were developed using the Forge NXT2 software. The simulations and the experimental tests were analysed and the results revealed the influential phenomena which have to be taken into account in the analytical formula, in particular the coining surface and the friction. In order to study the friction more thoroughly, ring compression tests were performed so as to determine the friction coefficients, based on the Coulomb's law limited to Tresca. This type of test is representative of the stresses undergone by the metal during the coining operation. Finally, a new calculation formula is proposed, in order to integrate the coining surface and friction more accurately.

1. Introduction

Making parts through sheet metal forming usually requires blanking, drawing and bending operations. However, in certain cases, specific operations are carried out, such as coining, an operation during which the sheet metal material is "compressed" along its thickness by the action of a punch and a die (Figure 1). Coining may be required to flatten the surface of a part, locally reduce a thickness, form a specific surface relief (Figure 2) or correct geometrical defects caused by springback (Figure 3). Coining generates significant forces that are much higher than those generated in the other forming operations. This can therefore have repercussions on the necessary force of the presses, the stiffness and endurance of the tools and the actual accuracy of the tools which undergo these very high forces.

IOP Conf. Series: Materials Science and Engineering **967** (2020) 012003 doi:10.1088/1757-899X/967/1/012003

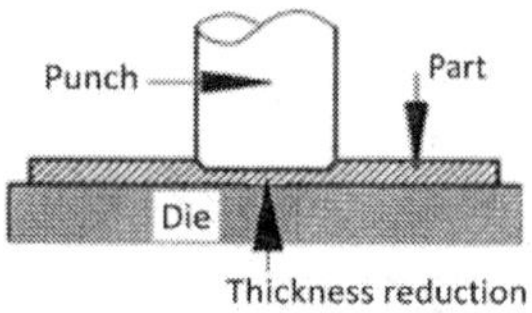

Figure 1. Principle of the coining operation

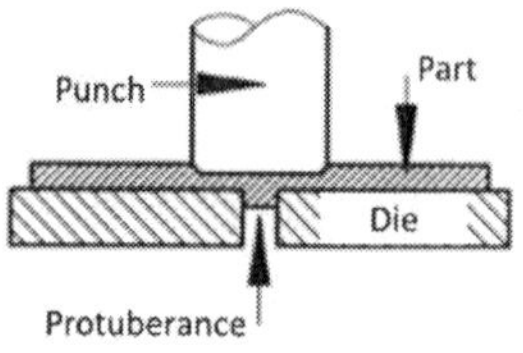

Figure 2. Protuberance coining

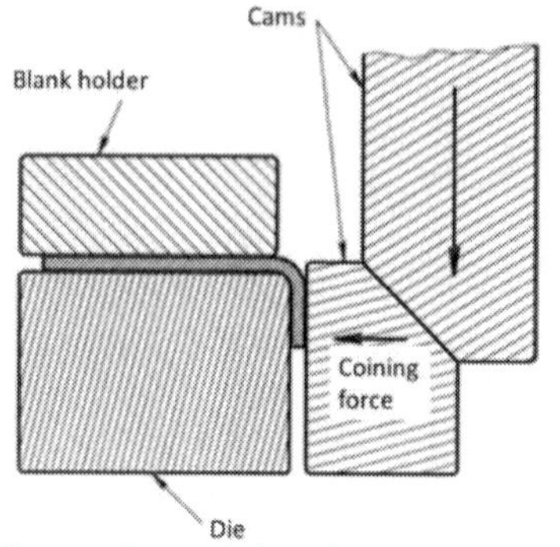

Figure 3. Coining for springback correction

This is why these forces need to be properly estimated right from the design of the tools. With regard to thin sheet metal forming, some practical methods to calculate the coining force are commonly used, but remain inaccurate, such as the rough estimation of the bending force during coining [1], [2] which is still employed today, or the use of abacuses or calculation principles based on the knowledge of the compression stress and the coined area [3], [4]. The latter method remains the basis for force calculation in forging [4] with the adaptation of calculation formulae based on the shape of the part, the material flow modes and the friction [5].
In the presented study, the coining shape considered (Figure 4) features a constant compressed cross-section. In this case, a simplified form of the force calculation formula may allow the coining force to be calculated with an acceptable level of accuracy [4], [5].
The simplified force calculation form selected at the beginning of the study is described in (1):

$$F = 1.3\,\sigma.S_p \quad (1) \qquad \sigma = K\left[\ln\left(1+\frac{t_0-t_1}{t_0}\right)\right]^n \quad (2)$$

$$S_p = (lp - 2R + 2S).w_s \quad (3) \qquad S = R.\sin\left\{\cos^{-1}\left[\frac{R-(t_0-t_1)}{R}\right]\right\} \quad (4)$$

Where K is the strengthen coefficient and n is the strain hardening exponent. The strain ε is integrated into (2) via the expression: $\ln\left(1+\frac{t_0-t_1}{t_0}\right)$. t_0 and t_1 are, respectively, the initial thickness and post-coining thickness of the sheet. The projected area below the punch S_p is calculated using (3) and (4), with l_p being the coining width, R the radius of the coining punch, S the calculated contact length (4) of the punch radius and w_s the width of the coined sheet (Figure 4). 1.3 is a coefficient widely used in forging; it takes into account a constant effect of the friction between the sheet and the tool.

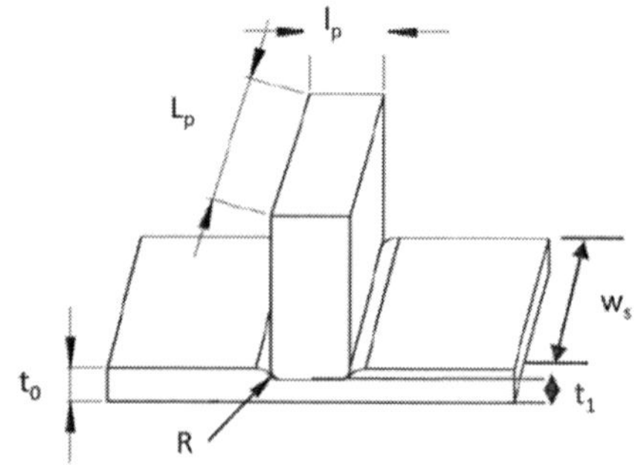

Figure 4. Type of coining under study

The objective of the presented study is to improve the proposed calculation formula with an experimental and numerical study of the influence of coining parameters.

2. Experimental procedure

Two sheet metal grades were used in this study: X6Cr17 stainless steel with thicknesses of 1 and 1.5 mm and HC380LA high strength low alloy steel with thicknesses of 0.8 and 1.5 mm. The samples used in the coining tests are 40 x 100 mm in size. The tests were carried out in dry conditions.

2.1. Tool used, force measurements and characterisation of the coined area

The tool used in the study is comprised of the main components listed below (Figure 5):

- A flat punch with a radius of 1.5 mm and a width of 7.5 mm, made of X153CrMoV12 steel, treated to 60 HRC
- A flat die, made of X153CrMoV12 steel, treated to 60 HRC
- Clamps to hold the coined samples
- A die with ball guide pillars

The setup is installed on a COFMO KI 2128 mechanical press with a maximum capacity of 63 tonnes.

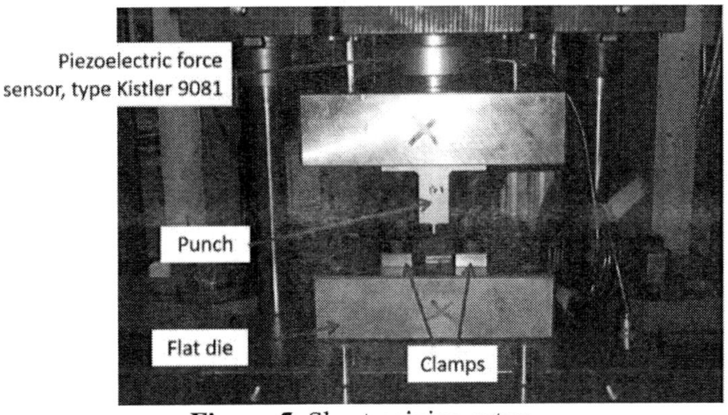

Figure 5. Sheet coining setup

The coining forces are measured by a piezoelectric force sensor located above the punch. Thickness measurements were carried out in the coined areas of the sheets, using a dial indicator, before and after coining. The coined areas were also examined using a confocal microscope. Finally, the contact areas were measured on the upper faces.

2.2. Ring compression test

Since cold forging and sheet metal coining are similar operations, we used the ring compression test, a test widely used in forging to estimate the values of the friction coefficient at the interface between the sheet and the tool. These values will then be used in the coining simulation software.

This test, developed by Male and Cockroft [6], consists in deforming metal rings between two flat dies. When the ring is compressed, the variation of the inside diameter is extremely susceptible to the friction between the ring and the tools. A series of curves is obtained by numerical simulation of the ring compression test and describes the change in the inside diameter as a result of the change in height, for a given friction coefficient. The experimental data is obtained with a series of ring compression tests, with different compression heights.

The rings feature the following dimensions: 3.6 mm for the inside diameter and 10.5 mm for the outside diameter. The thickness of the sheet used determines the thickness of the ring.

3. Numerical model

We used the Forge NXT2 software to perform the various simulations, namely simulations of the coining operation in order to better understand the evolution of the contact between the metal sheet and the tool, and simulations of the ring compression test.

3.1. Geometries

The flat coining numerical model is presented below (Figure 6). The simulations were carried out based on a 2D planar model (plane strain condition). The element type is here 2D and triangular. The tools (flat die, clamps, and the punch) are considered rigid bodies. The flat die and the clamps are fixed. A 0.5 mm gap is imposed between clamps and the coined sample. The mesh size in the coined sample has been reduced under the punch to 0.2 mm.

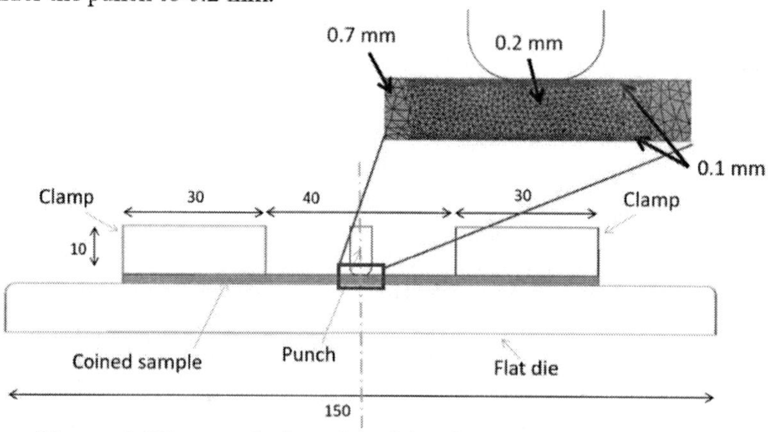

Figure 6. 2D numerical model of the sheet and the coining tools

3.2. Behaviour laws

We used an elastic-plastic model in the coining simulations. The plastic behaviour data of the coined metals were determined with tensile tests. Here, the behaviour is assumed to be isotropic. The tests were carried out at 0°, 45° and 90° with respect to the rolling direction, with a tensile test speed of 5 mm/min. The behaviour coefficients were determined by adjustment on the 3 tensile stress curves. The behaviour law used in the numerical simulations is of the Hollomon type (5).

$$\sigma = K\varepsilon^n \quad (5)$$

The behaviour coefficients stemming from the tensile tests are described in the Table 1:

Table 1. Behaviour coefficients used in the 2D models of the sheet metal coining operations

	K (in MPa)	n
HC380LA	727	0.155
X6Cr17	717	0.194

The Young's modulus of elasticity is considered equal to 200 GPa. In this law, the initial yield strength of the material is neglected and the plastic state of the material is, in our assumption, mainly responsible for the stress taken into account in the calculation.

IOP Conf. Series: Materials Science and Engineering **967** (2020) 012003 doi:10.1088/1757-899X/967/1/012003

3.3. Friction

Several laws exist to describe the friction which occurs between two bodies. In our sheet metal coining case, the contact pressures are significant. With the slip, the asperities present on the surfaces of the two bodies are gradually crushed (Coulomb friction) until the contact area between the two bodies no longer increases. Friction then occurs as a result of the deformation of the surface of the softer body (Tresca friction). Thus, we propose the use of a Coulomb friction law with a Tresca law type limitation (6), with the latter integrating a yield criterion. This law was introduced into the sheet metal coining simulations and into the simulations used to define the ring compression test abacuses.

$$\tau_c = Min(\mu\sigma_n, \bar{m}\frac{\sigma_0}{\sqrt{3}}) \quad (6)$$

Where τ_c is the critical shear stress; μ is the Coulomb friction coefficient; σ_n is the normal stress; σ_0 is the Von Mises stress; and $\bar{m}$ is the Tresca friction coefficient

As a result, the simulated curves of the ring compression tests served to determine two friction coefficients simultaneously. The studies described by Zhang et al. [7] show that a relationship $\bar{m} = 2\ \mu$ could be applied.

4. Results and discussions

4.1. Ring compression tests

The results of the ring compression test are given in Figure 7. The relative change in the inside diameter of the ring (ΔDi/Di) is presented as a function of the coining ratio (t_1/t_0). Note that, for the HC380LA grade, the experimental points are correctly positioned with respect to the calculated curve for $\bar{m} = 0.3$ and $\mu = 0.15$. However, for the X6Cr17 grade, it is more difficult to position the experimental points over a single curve. As a matter of fact, for a coining ratio ranging between 0.9 and 0.7, the friction coefficients are rather close to $\bar{m} = 0.6$ and $\mu = 0.3$ and, for coining ratios below 0.7, the friction coefficients are higher: $\bar{m} = 1$ and $\mu = 0.5$. These last values, which are better correlated with the experimental measurements, were selected for the X6Cr17 grade.

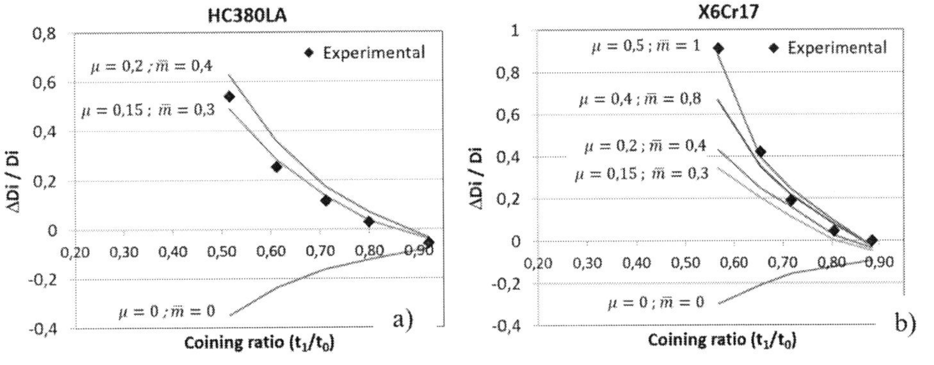

Figure 7. Results of the ring compression tests on HC 380 LA (a) and X6Cr17 (b)

The friction coefficients are extremely high for the X6Cr17 grade. This can be explained by a magnetic attraction phenomenon between the flat die and the X6Cr17 work pieces.

4.2. Analysis of the coined areas

The coined areas were examined in order to determine whether or not they were correctly estimated by the force calculation formula (1). Figures 8 a), b) and c) illustrate the top and bottom sides of a non-lubricated X6Cr17 work piece after coining.

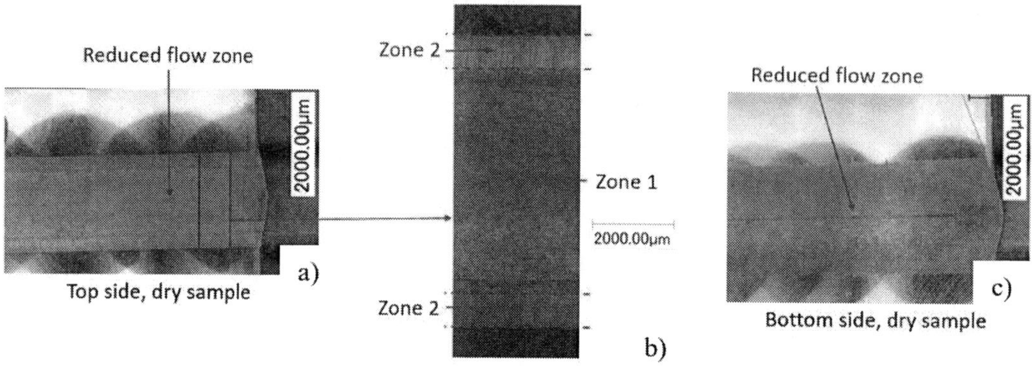

Figures 8. Examples of coined areas on a X6Cr17 work piece – Punch/sheet dry contact (a and b), die/sheet dry contact (c)

The coined area features different zones: a zone 1 marked with a black line and located in the centre, the simulation of which shows that the sheet/tool sliding velocities are almost equal to zero, and a zone 2, more visible on Figures 8 b), which is characterised by two lines running lengthwise in the coined area and limiting said zone on its sides. During the coining operation, the surface asperities are compressed by the punch. Zone 2 still exhibits some visible asperities, which means that there is very little or even no contact with the punch in this zone at the end of the coining operation. The asperities run along the direction of the material flow and correspond to a surface which was in contact with the punch during the coining operation and which flowed outwards. To conclude, we can say that the contact area at the end of the coining operation is located in zone 1 and that it is therefore smaller than the one calculated with the force calculation formula.

The contact area was quantified through numerical simulation (Figure 9) and through measurements with a confocal microscope. The results show that the contact area on the radius of the punch is not effective and that the areas calculated with the current force estimation formula are overestimated by 23.5% on average when compared with the measured areas. If the punch radii are not taken into account in the area calculation (S = 0 (4)), then the calculated area is markedly more accurate and barely underestimated, by 8.3% on average.

These results show that it is advisable to disregard the area below the radius of the punch.

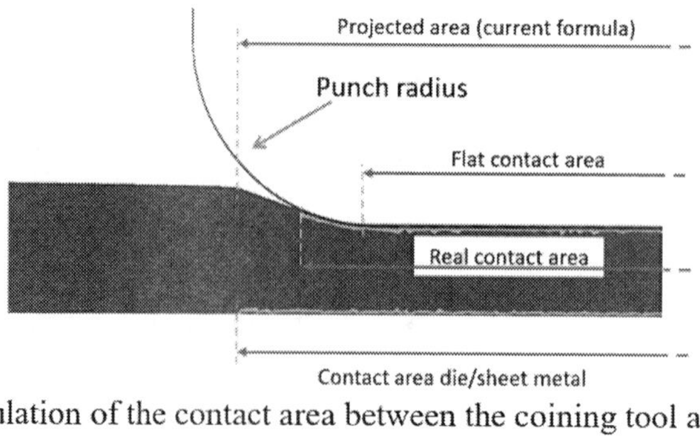

Figure 9. 2D simulation of the contact area between the coining tool and the coined sheet

4.3. Adjustment and validation of the force estimation formula

Two major changes were made to the coining force calculation. The first change is the omission of the punch radii in the calculation of the projected area (S = 0). The other major change is a more accurate integration of friction in the force calculation. The slice/slab method [8-9] is an analytical calculation method serving to estimate the hot compression forces applied to cylindrical slugs. This analytical calculation integrates the friction coefficient $\overline{m}$. The integration of the friction coefficient into the calculation formula using the slice/slab method allows us to propose a new force estimation formula (7):

$$F_{slices} = K.\left[ln\left(1+\frac{t_0-t_1}{t_0}\right)\right]^n.\left(1+\frac{1}{3\sqrt{3}}\overline{m}\frac{2\left(\frac{S_p}{\pi}\right)^{1/2}}{t_1}\right).S_p \tag{7}$$

The slice/slab method does not take the Coulomb friction into account, therefore only the Tresca coefficient is integrated into the new formula.

The forces measured experimentally and calculated with the current formula and with the new formula are presented in Figure 10. This figure also shows the results of the numerical simulation. These results are given for the two material grades, two coining ratios and different sheet thicknesses.

The results reveal that the current calculation formula considerably underestimates the coining forces. The difference in the forces is greater for the X6Cr17 grade (-70.9% maximum) than for the HC380LA grade; this can be related to the friction coefficients found in the ring compression tests and to the need to accurately integrate this factor. The force is much better estimated with the new formula, with a maximum difference of 21% with respect to the measured values. In the cases with a coining ratio of 0.7, the difference between the measured forces and the forces calculated with the new formula amounts to 14% on average. The simulated forces are also very close to the measurements, except for the X6Cr17 coining case with a coining ratio of 0.9. In this latter case, the simulation and the new calculation formula overestimated the forces. This difference may be explained by the friction coefficient, the value of which can be better adjusted.

The results obtained seem to substantiate the validity of the new force calculation formula. Nevertheless, the accuracy of the new calculation formula appears to be strongly dependent on the accuracy of the value of the friction coefficient, which can be complex to measure or estimate.

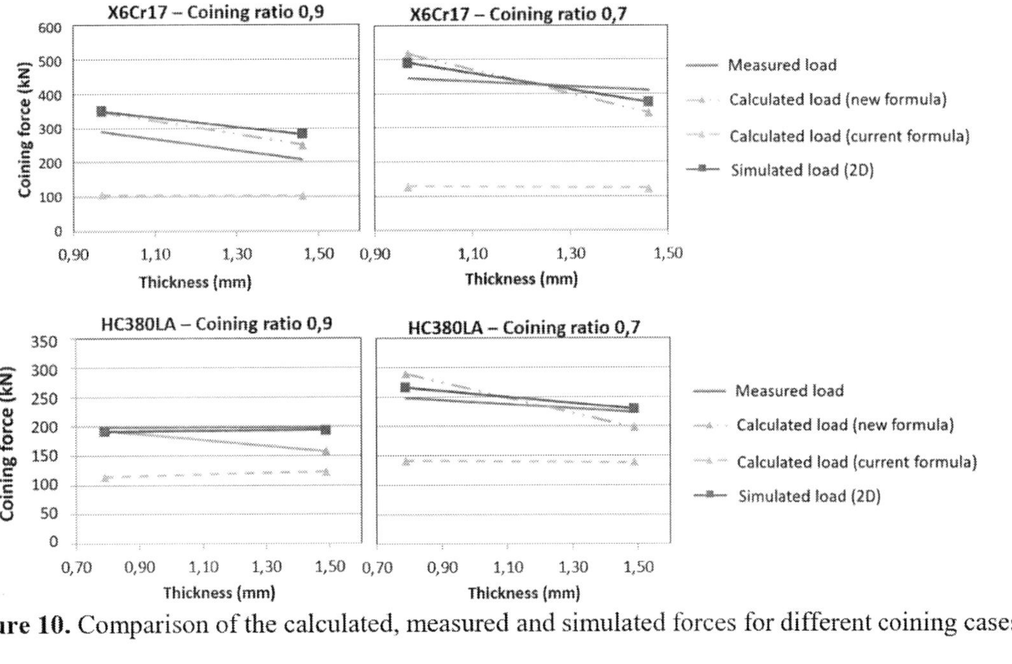

Figure 10. Comparison of the calculated, measured and simulated forces for different coining cases

IOP Conf. Series: Materials Science and Engineering **967** (2020) 012003 doi:10.1088/1757-899X/967/1/012003

5. Conclusion

The objective of this study was to propose an improved empirical calculation formula for sheet metal coining, based on stress and coined area. For the two sheet metal grades considered, the comparison of the coined area obtained with the current calculation formula and the coined area obtained after simulation analysis showed that the former was overestimated by approximately 23%. The study also highlighted the influence of the friction, which is not taken into account in the current formula. Thus a Coulomb friction coefficient limited to Tresca was proposed and the values were determined from the ring compression test. Then this friction was integrated into a new force calculation formula derived from the slice/slab method.

The comparison with the measured forces revealed that the new calculation formula gave better results with regard to the estimation of the coining forces. However, as friction has a strong influence in the estimation of the forces, the friction coefficients must be accurately measured, for all coining ratios. In order to be fully validated, this new formula also needs to be tested on other materials. In addition, while the results are much improved for this particular geometry, whether this is the case for all applications is not demonstrated, e.g., much large R, much large lp, etc. Also, varying frictional conditions with various lubricants should be investigated.

References

[1] Bliss Power Press Handbook 1950 E W Bliss Co Toledo Ohio
[2] Aberwald TS 1996 Manufacturing strategies Volume 6 PMA Symposium Proceeding Metalform, Nashville, Tenessee
[3] Sachs J 1966 Principles and methods of Sheet metal Fabricating Second edition Chapter Impact intrusion and coining Reynold Publishing Corporation New York, 10022
[4] Smith D 1990 Die Design Handbook Third edition Chapter Compression dies ISBN n° 0-87263-374-6
[5] Schultes T, Sevenler K, Altan T 1981 Prediction of Forging Load and Stresses Using a Programmable Calculator Topical Report No. 4, Battelle Columbus laboratories
[6] Male AT and Cockroft MG 1964 *J. I. Met.* **93** 38
[7] Zhang DW and Ou H 2016 *Tribol. Int.* **95** 13
[8] Boujut JF and Tichkiewitch S 1992 *J. Mater. Process. Tech.* **34** 163
[9] Marin PH, Boujut JF and Tichkiewitch S 1994 *J. Mater. Process. Tech.* **45** 43

The study on the relationship between the friction characteristic and the formability of the automotive steel sheet during the stamping process

J Y Kim[1,2], S C Yoon[1], J S Hyun[1], M.-G. Lee[2]
[1] Products Development Assistance Team, Hyundai-steel, Dangjin City, Korea
[2] Department of Materials Science and Engineering, Seoul National University, Korea

Abstract. This study aims to analyse the formability characteristics by friction characteristics of automotive steel sheets to improve sheet forming during the stamping process. During the stamping operation, the steel sheet is pushed into the tool surface, a complex phenomenon occurs, and many parameters interact. Therefore, automotive steel researchers want to optimize the friction of automotive steel sheets to improve the forming process, such as drawing ratio limit, dome height limit, cup drawing test, etc. In addition, in this study, the finite element method is used in combination with a numerical solution to present the friction behavior of high strength steel sheets in automobiles, taking into account the operating deformation mechanism during the formability test.

1. Introduction

Recently, many studies have been conducted to overcome concerns about the future environment and maximize energy efficiency. Various materials development and optimal method research are being conducted for this purpose, and in the automobile field, many efforts are being made to contribute to industrial development based on low carbon-green growth by achieving maximum energy efficiency through light weight of the vehicle body. Above all, many studies are being conducted on the development and application of high-strength steel for automobiles to reduce the weight of the vehicle and at the same time improve the strength, which has the advantages of light weight effect and passenger safety. As a result, interest in environmental regulations and passenger safety, which are discussed internationally, is increasing, and the application rate of high tensile strength steel is rapidly increasing.[1~8)]. Above all, the steel sheet applied to the car body for light weight should be designed and properly developed and applied in consideration of the characteristics required for each component. However, in the case of metal materials in general, strength and elongation have an inverse relationship, so we face the problem of overcoming the plasticity of high-strength steel for complex structural component parts of cars, and a lot of research is being done in various ways to solve this. Various high-strength steel sheets applied to the car body are manufactured through sheet metal forming processes and applied to actual parts. Since there are many process variables in the manufacturing process, whether to use them is determined by how to control them, and thus an optimization process is required. Generally, there are various process variables such as blank size, blank holding force, shape, sheet thickness and speed in the sheet metal forming process.[9~17)]

In general, friction, wear, and lubrication should always be considered for all mechanical elements in contact relative motion, and many studies have been conducted to improve the characteristics at the interface through solid lubricants, surface coatings, and high performance lubricants. In this regard, it is possible to improve the formability of high-strength steel sheet because it is possible to induce high

IOP Conf. Series: Materials Science and Engineering **967** (2020) 012004 doi:10.1088/1757-899X/967/1/012004

formability by improving the mold settling characteristics of automotive steel sheet. Specific application techniques include Advanced Zinc Coating, Lubrication Sheet, and Surface Texturing.
However, due to the low reproducibility of the experimental values, the friction coefficient of the material is generally used only for relative comparison, which makes it difficult to reflect the friction coefficient in the FEM. This characteristic is due to the fact that the coefficient of friction (COF) is not the inherent properties of the material, but the value of the system properties that are affected by the surrounding environment and test conditions.
Therefore, in this study, in order to investigate the forming efficiency and deformation behaviour of automobile steel sheet through the change of friction characteristics, which have been recognized as uncontrolled factors, firstly, the reproducibility of friction coefficient value was verified by controlling the test conditions of friction coefficient measurement test. Second, we tried to evaluate the effect of the friction characteristics of the steel sheet on the formability by reflecting the COF as the main factor in the FEM.

2. Friction Test

In this study, reproducibility of friction coefficient test equipment was first evaluated to determine whether the friction factor values are reflected in the FEM(Finite Element Method). The friction coefficient test equipment used in this study was a single-sided sliding type pneumatic friction test equipment with a contact pressure of 20 MPa and 78.4 MPa, with a drawing distance of 0.1 m, a drawing speed of 0.2 m / min, and anti-rust oil applied after ethanol degreasing.
The equipment schematic is shown in Figure 1, and the test conditions are shown in Table 1.
In this experiment, SGAFC590 steel sheets were measured three times at 10-day intervals to verify the reproducibility of friction coefficient measurements. The yield strength of SGAFC590 specimen is 323 MPa, tensile strength is 577MPa, elongation is 28% and thickness is 0.7mm. The coefficient of friction value is expressed as the mean and standard deviation of the 30 measurements.

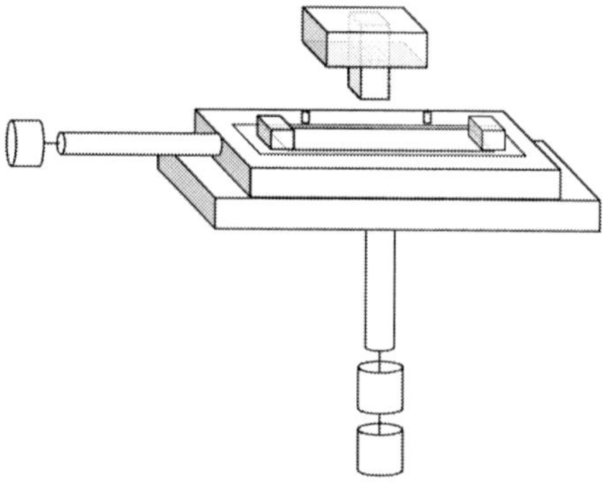

Figure 1. Schematic of sliding type friction equipment.

Table 1. Test condition

Test factors	Test condition
Contact pressure	20MPa, 78.4MPa
Drawing speed	0.2m/min
Travel distance	100 mm
Working temperature	Room temperature
Degreasing	Ethanol
Lubrication	Rust prevention oil

A reproducibility test result of friction test was shown in the Table 2.
In addition, in order to ensure the reproducibility of the test results, the vertical load was applied uniformly by adopting a method of loading from bottom to top. In order to minimize specimen tearing or slip, process factors such as correction of counter-edge curvature and checking of jamming load were checked in advance. As a result of the test, specimen slip did not occur until the contact pressure was 160MPa.

Table 2. Reproducibility test result of friction test

Contact Pressure	1st		2nd		3rd		Average	
	COF	SD	COF	SD	COF	SD	COF	SD
20MPa	0.1405	0.00121	0.1430	0.00077	0.1375	0.00071	0.1403	0.0024
78.4MPa	0.1432	0.00187	0.1436	0.00210	0.1374	0.00127	0.1414	0.0033

Under the conditions of contact pressure of 20MPa and 78.4MPa, the values of the friction coefficient were measured at 0.1403, 0.1414, and the standard deviation was 0.0024 and 0.0033, respectively.

3. FEM analysis

In order to evaluate the influence of factors affecting the formability of automotive steel sheets, the shape of the Dash Pot was selected, and the deformation behaviour was investigated using the Fractional Factorial Design method. Dash pot was used to evaluate the effect of the main factors affecting each deformation behaviour. In order to induce high formability, the variable corresponding to the processing efficiency was designed to have the maximum value.
Figure 2 shows the shape of the dash pot and the forming analysis conditions that are commonly used to examine this formability, and to analyse the effects of the main factors, the major factors are set in the range of -1 to 1.
Fractional Factorial Design was used for friction, work hardening index, material strength factor, and R-value in each direction. As a result, the Fractional Factorial Design was conducted through the results of wrinkles, safety, and fracture. Marked results correspond to the area obtained from the FEM results. In Figure 3, the optimal conditions are set for the case where the wrinkles are less than 2%, 0% for the fracture and 98% for the rest of the range is safe, and the resulting satisfaction is obtained. In the graph showing safety results, the larger the slope, the greater the sensitivity of the value. In the case of friction, it can be seen that it has a much larger value than the slope of other factors.
Comparing comprehensively, the smaller the friction, the larger the work hardening index, the smaller the material strength factor, and the larger the R value in each direction, the more favourable the formability. It can be seen that the large sensitivity factor is the effect of friction.
Based on these results, we tried to check the impact of the center pillar, which is the main part of the automobile. Figure 4 shows the forming analysis conditions of the commonly used Single Action type of Center Pillar. At this time, only the coefficient of friction coefficient was change, and the degree of

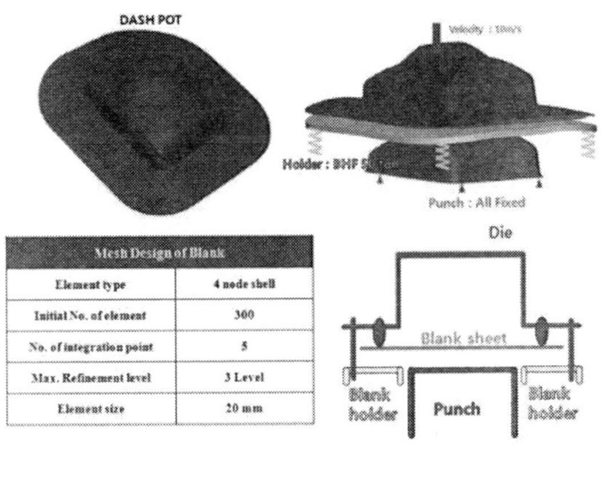

Figure 2. Forming condition of Dash pot

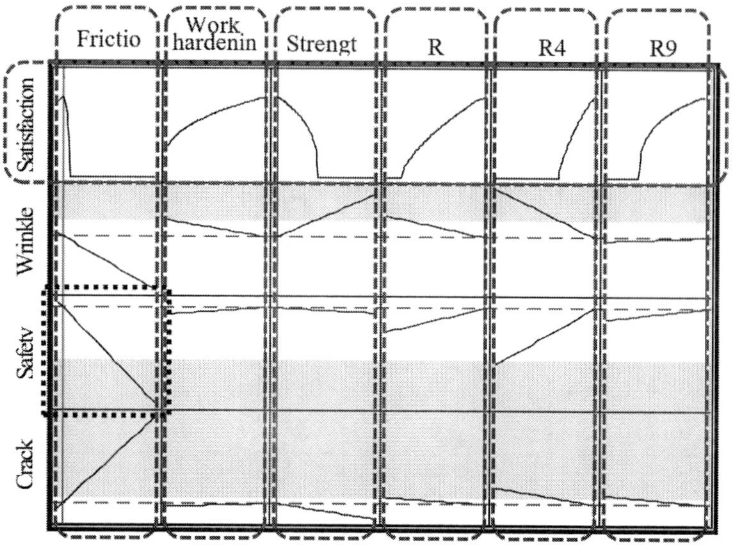

Figure 3. Satisfaction Analysis for Major Factors

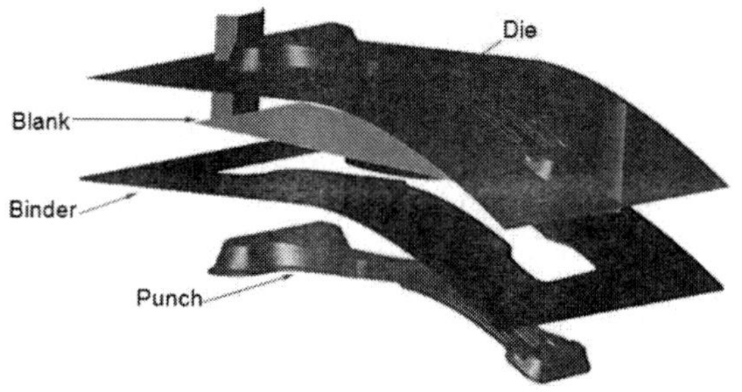

Figure 4. FEM condition of Center Pillar

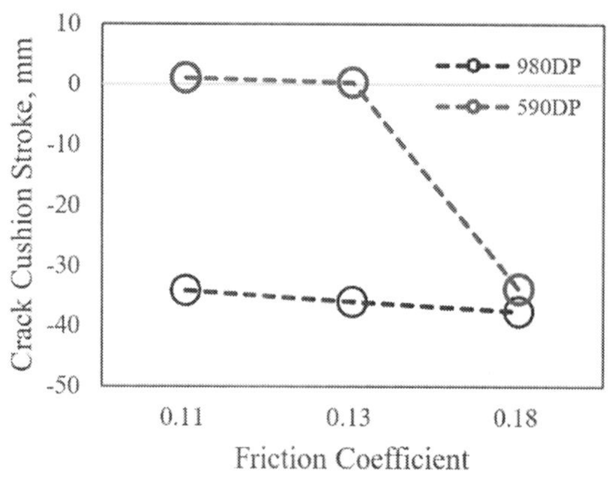

Figure 5. Crack cushion stroke due to changes in COF

improvement in formability was analysed by marking the fracture and possible cushion stroke at the forming limit curve.

In the case of 980DP steel sheets, the initial fracture occurs gradually decreasing as COF decreases.

On the other hand, in the case of 590DP, when the COF was 0.13 and 0.105, no breakage occurred and the forming was safely completed.

Figure 5 shows the point of the Cushion Stroke for each type of steel in which the initial fracture occurs. Even though the COF decreases for 980DP, the FEM result shows that the initial fracture is not significantly reduced, whereas for 590DP, the initial fracture stroke changes significantly as the friction decreases.

Based on this, assuming that the standard deviation of the COF for fracture and crease is about 0.005, and the overall tendency follows the normal distribution, the difference in the degree of occurrence of crease and fracture according to the change of the COF can be confirmed. When the COF decreased from 0.18 to 0.13, the wrinkles showed a difference of about 1.72%, and the fracture occurred about 0.8%.

In addition, if the coefficient of friction decreases from 0.13 to 0.105, there will be a difference of about 0.7% in wrinkles and about 0.32% in fractures. Through this, it is considered that the change of the friction coefficient affects the formability of automobile steel sheet.

4. Conclusion

In this study, to maximize the formability of automotive steel sheets, we tried to improve the efficiency by controlling the friction characteristics that were considered as the uncontrolled factors. The effect of friction was investigated through the FEM and result analysis of the center pillar.

(1) By controlling the test environment of the friction test, the Coefficient Of Friction(COF) within the deviation 0.004 was obtained. Through this, it was confirmed that the COF the steel sheets can be reflected as a major factor in the FEM.
(2) A Fractional Factorial Design method was used using the Dash pot.
(3) The analysis of friction, work hardening index, material strength factor and R value in each direction showed that the sensitivity of friction coefficient had the greatest influence.
(4) The FEM was performed by changing the coefficient of friction from 0.18 to 0.105 for the center pillar, which is the main part of the vehicle body. For 590DP, the fracture occurred only when the COF was about 0.18.
(5) Assuming that the standard deviation of the friction coefficient for the occurrence of fracture and creases is approximately 0.005 and that the overall distribution of distributions follows the pattern of normal distribution, it can be found that there is a difference of up to 2.42% for creases and an increased probability of occurrence up to 1.12% for fractures.

References

[1] Bayraktar E, Isac E and Arnold G 2005 *J. Mater. Pro. Tech.* **162–163** 471–6
[2] Perlade A, Bouaziz O and Furnémont Q 2003 *Mater. Sci. Eng.* A **356** 145–52
[3] Hamada A S, Karjalainen L P and Somani M C 2007 *Mater. Sci. Eng.* A **467** 114–24
[4] Bouaziz O and Guelton N 2001 *Mater. Sci. Eng.* A **391–321** 246-9
[5] Vercammen S, Blanpain B, De Cooman B C and Wollants P 2004 *Acta. Mater.* **52** 2005–12
[6] Rana R, Bleck W, Singh S B and Mohanty O N 2007 *Mater. Lett.* **61** 2919–22
[7] Keeler S P 1994 *J. Mater. Pro. Tech.* **46** 443–54
[8] Merklein M and Lechler J 2006 *J. Mater. Pro. Tech.* **177** 452–5
[9] Choudhury I A, Lai O H and Wong L T 2006 *Simul. Model. Pract. Theory* **14** 71–81
[10] Voevodin A A and Zabinski J S 2006 *Wear* **261**, 1285–92
[11] Borghi A, Gualtieri E, Marchetto D, Moretti L and Valeri S 2008 *Wear* **265** 1046–51
[12] Andersson P, Korkinen J, Varjus S, Gerbig Y, Haefke H, Georgiou S, Zhmud B and Buss W 2007 *Wear* **262** 369–79
[13] Sellgren U, Björklund S and Andersson S 2003 *Wear* **254** 1180–8
[14] Ryk G and Etsion I 2006 *Wear* **261** 792–6
[15] Etsion I and Sher E 2009 *Tribol. Inter.* **42** 542–7
[16] Rapoport L, Moshkovich A, Perfilyev V, Gedanken A, Koltypin Y, Sominski E, Halperin G and Etsion I 2009 *Wear* **267** 1203–7
[17] Verma R K and Chandra S 2006 *J. Mater. Pro. Tech.* **172** 218–24
[18] Dieter D E 1976 *Mechanical Metallurgy*
[19] Yang T S 2010 *Tribol. Inter.* **43** 1104–12

IOP Conf. Series: Materials Science and Engineering **967** (2020) 012005 doi:10.1088/1757-899X/967/1/012005

Investigation on water as lubricant in combination with a structured tool surface in micro metal forming

L Rathmann[a,*], T Czotscher[a], T Radel[a] and F Vollertsen[a,b]

[a] BIAS - Bremer Institut für angewandte Strahltechnik GmbH, Klagenfurter Straße 5, 28359 Bremen, Germany
[b] University of Bremen, Bibliothekstraße 1, 28359 Bremen, Germany

* rathmann@bias.de

Abstract. The use of lubricants in forming processes is common because they reduce friction between the tools and workpiece during the forming process. Today, mainly mineral oils are used as conventional lubricant. In order to cut their amount in the future, the friction between tools and workpiece is investigated by using water as lubricant in this work. Therefore, the tools' surfaces are structured with laser induced periodic surface structures (LIPSS) to make them hydrophilic and strip drawing tests are carried out at different traverse velocities to evaluate the resulting friction force. The results show that at a traverse velocity of v = 10 mm/s the maximum friction force is lowest for the structured tool with water (F = 7.0 N) in comparison with unstructured tools in combination with a conventional lubricant. Consequently, the potential of water as lubricant combined with a hydrophilic surface structure is highlighted.

1. Introduction

One of the main tasks of the forming-oriented research areas is to reduce energy consumption and environmental pollution [1]. In sheet metal forming, however, mainly liquid lubricants based on mineral oil are used [2] and thus contribute to environmental pollution. These lubricants also contaminate the surface of the workpiece and therefore make a cleaning necessary. This leads to a reduction of the process efficiency due to additional steps in the process chain [3].

As a result, in recent years the industry has increasingly used aqueous solutions to which additives have been added in order to improve the adhesion of the solution to the tools. The proportion of additives in these solutions can be held under 5 % as investigations show making these solutions capable for deep drawing processes and internal high pressure forming [1]. If water is applied to a friction test, the friction coefficient is halved from $\mu = 0.30$ to $\mu = 0.15$ [4]. In addition to reducing the coefficient of friction by using water as a lubricant, studies by Jia et al. [5] show that there is also a reduction in wear. For this purpose, friction tests were carried out using a bronze-graphite material and stainless steel (1Cr18Ni9Ti) under water at a speed of v = 0.53 m/s.

The composition of the water also plays a role regarding the friction behavior of two contact partners. The dropwise use of deionized water in a ball-on-plate test with friction partners made of rolling bearing steel (GCr15) at a load of F = 2 N leads to friction coefficients greater than $\mu = 0.5$. The cause is attributed to the relatively large contact angle of deionized water on steel of $\Theta = 80°$. This results in metal-to-metal contact and thus, high friction coefficients as well as high abrasion.

During micro forming, the viscosity of the lubricant used also has a significant effect on the occurring process forces [7]. It is shown that the lubricant with the lowest viscosity (AK3080, $\nu = 52$ mm²/s at T = 20 °C) has the highest punch forces in the process, while the lubricant with the highest viscosity (HBO 947/11, $\nu = 400$ mm²/s at T = 40 °C) leads to the lowest punch forces during a strip drawing test with double deflection. The higher process forces can be explained by the lubrication pocket model and the resulting increase in friction during forming processes in the micro range. The lubrication pocket model is based on the size effect of the microstructure, which is described in [8]. The lower the viscosity of the lubricant used, the sooner it flows out of the cavities between the surfaces and prevents effective separation of the workpiece from the tool. This results in a solid contact with correspondingly high friction coefficients, which are reflected in the increased punch force. Vollertsen and Hu experimentally investigated the influence of the punch speed and the contact pressure on the friction function [9]. The friction function describes the coefficient of friction in dependency of the acting contact pressure between the blank and the tools during the forming process. For this purpose, strip drawing tests with double deflection were carried out, whereby tribological size effects could be displayed. These effects reveal in the change of the course of the friction function in dependency of the punch velocity. It was discovered, that increasing punch velocities caused decreasing coefficients of friction [10].

Due to the tendency of the water on untreated and smooth surfaces to move with the contact zone of two contact partners, the surface must be modified to prevent the water from moving. Such surface treatments result, for example, in laser induced periodic surface structures (LIPSS) which were described by Sipe et al. [11]. These structures enable the functionalization of surfaces and prevent or provide them from wetting with liquids. The degree of hydrophilicity or hydrophobia caused by LIPSS can be adjusted via the pulse overlap. Eckert's work shows that the surfaces made of aluminum (EN AW-1050A) become more hydrophilic with increasing pulse overlap [12].

As pointed out, water can improve the contact behavior between tools and worksheet during sheet metal forming, if the process conditions are selected accordingly. These include high punch velocities and the modification of the tool's surfaces. For this reason, strip drawing tests were carried out in this work on a tensile testing machine using structured tools and water as lubricant to investigate their influence as a combination on the friction force in comparison with a conventional lubricant.

2. Methodology

2.1. Tool structuring

The structuring of the strip drawing jaws is made with a pico-second laser TrueMicro 5050 by Trumpf with the parameters shown in Table 1.

Table 1. Parameters of the pico-second laser to generate the LIPSS on the strip drawing jaw's surface.

wavelength λ	laser power p_L	spot size	scanning velocity v_f
1030 nm	2.5 W	46 µm	500 mm/s

The strip drawing jaw consists of an aluminium bronze CW307G and the contact surface has an area of $A_C = 40$ mm². To create the LIPSS in this area, a zigzag trajectory with a line spacing of $d_h = 34$ µm was chosen. The trajectory was scanned once while Argon was used as shielding gas.

2.2. Strip drawing

The strip drawing tests were carried out on a tensile testing machine made by Zwick Roell. The machine is equipped with the experimental setup shown in Figure 1. It is built up from two strip drawing jaws which are framed and screwed down in tool holders. These holders are mounted on a linear guidance to allow only a horizontal movement of the strip drawing jaws. One of these jaws is in contact with a force sensor which detects the normal load F_N. To apply this load, the second jaw is pushed against a metal strip and therefore against the other one and the force sensor using a built-in micrometer screw. The screw enables the precise adjustment of the normal load.

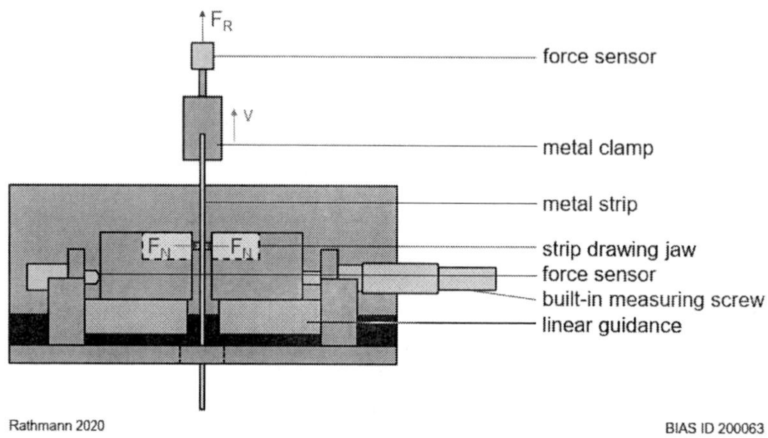

Figure 1. Experimental setup for strip drawing.

The metal strip is made from 1.4301 (X5CrNi18 - 10). A second force sensor that links the strip to the vertical axis of the tensile testing machine detects the friction force F_R. Using Coulomb's law of friction, the coefficient of friction was calculated via:

$$\mu = F_R / F_N \tag{1}$$

In this equation F_R represents the friction force while μ is the coefficient of friction and F_N is the normal load. For the measurements, the normal force was set at $F_N = 15$ N. The drawing speed varies between v = 1 mm/s, v = 5 mm/s and v = 10 mm/s. Demineralized water in combination with a laser-induced periodic surface structure serves as lubricant. In order to be able to draw a comparison with conventional lubricants, the ester oil-based lubricant Wisura LS240A was used. Its properties are shown in Table 2. In this case, the tools were not structured. In addition, drawing tests were carried out with unstructured tools in conjunction with demineralized water and dry tools for further comparisons.

Table 2. Properties of the used lubricant during strip drawing.

base oil	viscosity at 20 °C	density at 20 °C	dissociation temperature
ester oil	52 mm^2/s	842 kg/m^3	170 °C

3. Results

The structuring of the tool surfaces with the laser parameters listed in Table 1 leads to the structured surface shown in Fig. 2. The resulting LIPSS are parallel to the tension direction of the metal strip and show a homogeneous spread over the contact area.

The friction coefficient resulting from the combination of the structure presented in Fig. 2 and demineralized water in simple strip drawing tests are shown in Fig. 3. For the comparison, additional experiments with unstructured tools and different lubricants as well as the dry condition of the unstructured< tools were carried out and shown in Fig. 3. As it is presented, the coefficient of friction decreases with increasing traverse velocity in case of a structured tool surface in conjunction with demineralized water from $\mu = 0.26$ to $\mu = 0.18$. This contrasts the increase in the coefficient with increasing forming speed from $\mu = 0.16$ to $\mu = 0.21$ when an unstructured tool with conventional lubricant is used. An unstructured tool in combination with demineralized water and in a dry condition show the highest maximum friction forces at a traverse velocity of v = 10 mm/s. In addition, Fig. 3 shows that the process gets more stable in case of the structured tool with LIPSS and demineralized water since the deviations of the force values decrease with increasing traverse velocities. On the other hand, a dry and unstructured tool shows the highest deviations and therefore is the most unstable process.

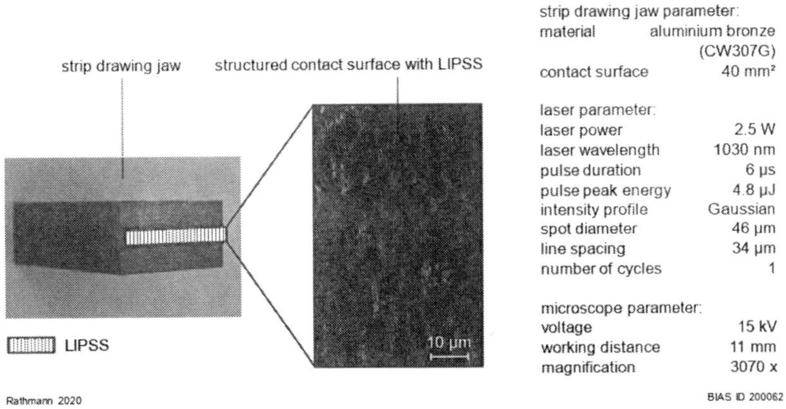

Figure 2. Representation of the LIPSS structures on the contact surface of the strip drawing jaws.

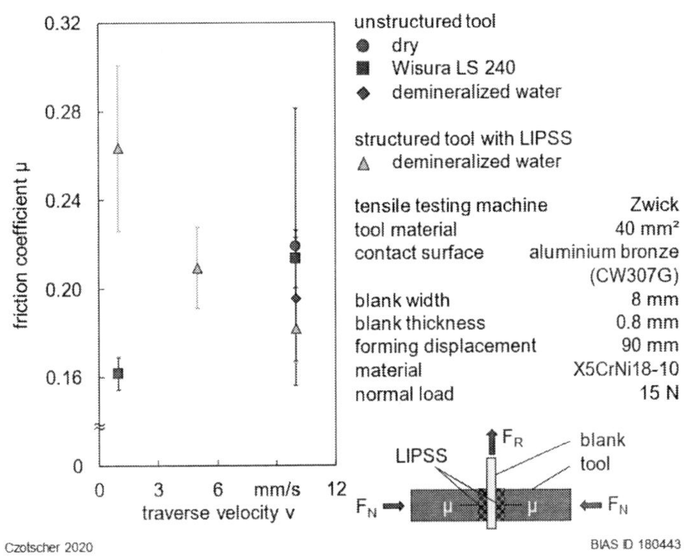

Figure 3. Friction coefficient µ in dependency of the traverse velocity v for different regimes of tool surfaces and lubricant.

It is in fact surprising that the coefficient of friction at a drawing speed of v = 10 mm/s is lowest for the structured tools and water as lubricant. This might be attributed to the low contact pressure (0.375 MPa) applied in the experiments and will be evaluated further.

4. Discussion

The given conditions of the strip drawing jaws, unstructured and dry, unstructured and water as lubricant, unstructured and conventional lubricant as well as structured and water as lubricant, lead in this order to decreasing friction coefficients at the highest traverse velocity. As Fig. 4 shows, the use of water as lubricant reduces the friction coefficient regardless of whether the strip drawing jaw's surface is structured or not. This observation corresponds to those made by Kim et al. where the coefficient of friction was halved from $\mu = 0.30$ to $\mu = 0.15$ [4]. In this work the friction coefficient was reduced from $\mu = 0.22$ with a dry tool via $\mu = 0.20$ using demineralized water as lubricant to $\mu = 0.18$ with a structured strip drawing jaw's surface and water as lubricant at a traverse velocity of v = 10 mm/s.

However, using a traverse velocity of v = 1 mm/s a structured surface in combination with water shows a higher coefficient of friction than an unstructured tool with a conventional lubricant. Due to the difference in the viscosity between water ($\nu = 1$ mm²/s at 20 °C) and Wisura LS240 ($\nu = 52$ mm²/s at 20 °C), this phenomenon can be addressed to the lubricant pocket model, which was also used to describe the observations in [7]. According to the model described in [13], the high friction force and therefore the high coefficient of friction is a result of the proportion of open to closed lubricant pockets. The more open lubricant pockets exist in comparison to closed ones, the higher the friction force and the coefficient of friction are. If a lubricant with a low viscosity and a low traverse velocity are used, the lubricant has the possibility to escape from a lubricant pocket. In contrast, a lubricant with a high viscosity remains in the lubricant pocket. The former favors the number of open lubrication pockets, particularly at the edge of the contact between the sheet metal strip and strip drawing jaws. This could explain the increased coefficient of friction at low speeds for structured tools with water as lubricant in accordance with [7].

Work by [14] in the field of deep drawing in the macro range in turn shows that structured surfaces can reduce friction without the addition of a lubricant. Such observations are investigated especially in the field of dry forming. At this point, it cannot be excluded that a reduction in friction can be achieved by applying structures to tools in the micro range alone. Therefore, further investigations are needed to determine the role of water in the application of structures.

5. Conclusion

It can be concluded that the use of water in combination with a hydrophilic surface structure as well as increased traverse velocities enables the reduction of the coefficient of friction during strip drawing tests at least at low contact pressures. This opens up possibilities to place the LIPSS in regions like the drawing edge where there are high frictional stresses in order to reduce them and to deliver an alternative to conventional lubricants.

Acknowledgement

Funding by the Deutsche Forschungsgemeinschaft DFG (project 416530419 "Wasser als Schmierstoff für die Hochgeschwindigkeitsumformung mithilfe von LIPSS") is gratefully acknowledged.

References

[1] Bay, N., Azushima, A., Groche, P. (2010) Environmentally benign tribo-systems for metal forming. CIRP Ann. - Manufacturing Technology 59, 760-780.

[2] Schuler GmbH (Hrsg.): Handbuch der Umformtechnik. 1. Auflage, Springer Verlag, Heidelberg 1996, ISBN 3-540-61099-5.

[3] Vollertsen, F., Schmidt, F. (2014) Dry metal forming: Definition, chances and challanges, Int. Journal of Prec. Eng. And Man. - Green Tech. 1, 59-62.

[4] Kim, H.J., Dong, G.S., Dae, E.K. (2016) Friction behaviour between silicon and steel coated with graphene oxide in dray sliding and water lubrication conditions, Int. Journal of Precision Engineering and Manufacturing - Green Tech. 3, 91-97.

[5] Jia, J.H., Jian, M.C., Hui, D.Z., Jing, B.W., Hua, Z. (2004) Friction and wear properties of bronze-graphite composite under water lubrication, Tribology International 37, 423-429.

[6] Liang, S., Zhigang, S., Min, Y., Lei, L., Xiaojing, Z., Shulin, M. (2016) In-situ exfoliated graphene for high performance water-based lubricants, Carbon 96, 1181-1190.

[7] Tetzel, H., Bohlen, A., Vollertsen, F. (2015) Influence of lubricant viscosity on punch force in strip drawing test, Proceedings of the 4M/ICOMM2015 Conference (ICOMM2015), eds.: Annoni, M., Fassi, I., Wiens, G., Dimov, S., Research Publishing Milano, 223-226.

[8] Vollertsen, F. (2008) Categories of size effects, Prod. Eng. 2, 377-383.

IOP Conf. Series: Materials Science and Engineering **967** (2020) 012005 doi:10.1088/1757-899X/967/1/012005

[9] Vollertsen, F., Hu, Z. (2006) Tribological size effects in sheet metal forming measured by a strip dawing test. CIRP Ann. 55/1, 291-294.
[10] Vollertsen, F., Hu, Z. (2010) Analysis of punch velocity dependent process window in micro deep drawing, Prod. Eng. 4, 553-559.
[11] Sipe, J.E., Young, J.F., Preston, J.S., Van Driel, H.M. (1983) Laser-induced periodic surface structure. I. Theory, Phys. Rev. B 27, 1141-1154.
[12] Eckert, S., Köhnsen, A., Vollertsen, F. (2017) Surface finish using laser-thermochemical machining, Lasers in Manufacturing (LIM17) eds.: L. Overmeyer, U. Reisgen, A. Ostendorf, M. Schmidt.
[13] Geiger M., Kleiner, M., Eckstein, R., Tiesler, N., Engel, U. (2001) Microforming. CIRP Annals – Manufacturing Technology 50, 445-462.
[14] Brosius, A., Mousavi, A. (2016) Lubricant free deep drawing process by macro structrured tools. CIRP Ann. – Manufacturing Tech. 65, 253-256.

IOP Conf. Series: Materials Science and Engineering **967** (2020) 012006 doi:10.1088/1757-899X/967/1/012006

Influence of anisotropy evolution on strain path independent failure limits

K Bandyopadhyay[1], **S Basak**[2] **and M G Lee**[2]
[1]Department of Mechanical Engineering, IIT Bhilai, Raipur 492015, India

[2]Department of Material Science and Engineering, Seoul National University, Seoul 08826, Korea

email: kaushik@iitbhilai.ac.in

Abstract. Stress based forming limit diagram (SFLD) and polar effective plastic strain (PEPS FLD) diagram are two techniques reported to be independent of strain path change. Thus, these techniques are capable of predicting failure in pre-strained sheet metal samples. Therefore, in this study evolution in anisotropy properties of sheet metals were considered in terms of evolution in yield shape during conversion of conventional strain based forming limit diagram (FLD) to SFLD and PEPS-FLD. The evolution in anisotropic properties were modelled as a function of effective plastic strain depending on which the anisotropic coefficients of non-quadratic yield model Yld2000-2d were calculated. Further, to analyse the strain path change effect, metal sheets of extra deep drawing quality steels were first deformed under biaxial condition in Marciniak in-plane stretch forming setup. The pre-strained samples were subsequently deformed with hemispherical punch till the necking appears. For failure prediction, finite element modelling of the complete procedure was performed considering evolution in the yield shape. Finally, both the failure criteria were implemented for failure prediction. Implementation of the evolution in yield function in finite element analysis and failure criteria provided better results.

1. Introduction

Sheet metal forming is an integral part of automotive industries due to manufacturing of various auto mobile body parts like car fender, A, B, C pillars, roof, bonnet etc. The increasing aesthetic requirements associated with car design also increasingly making the sheet metal forming complex to obtain the desired final product without any failure. Concern over car emission, fuel efficiency also led to application of different light weight metallic alloys. Many times, the newly introduced lightweight materials are not easy to form and specific arrangements are required. Hence, recent research activities in the field of sheet metal forming is directed toward either development of more flexible forming techniques or to enhance theoretical understanding of the material behaviour during forming to increase the extent of formability. For the theoretical investigation numerical simulation is a reliable and effective tool which reduces the requirement of large number of experimentation and thus saves resources and human effort. In case of finite element (FE) simulation, the information related to specific material is modelled based on the mechanical properties of the material such as elastic modulus, Poisson's ratio, density, yield function and hardening parameters. The yield function defines the elastic limit of the material and normally serves as the plastic potential. In general, the coefficients of yield function are

estimated based on plastic anisotropy parameters, and the yield stress is treated to be constant throughout the deformation process. Further, size of the yield function is assumed to be increased proportionally with strain hardening while keeping its shape constant with the assumption of isotropic hardening. However, recent development in the field of metal plasticity revealed that anisotropy properties of the material changes with plastic deformation. Zamiri et al. [1] incorporated change in R-values due to deformation was into FE simulation to improve prediction accuracy. Wang et al. [2] incorporated changes in stress ratio along different anisotropic direction into FE simulation by expressing yield function in terms of effective plastic strain. A fourth order polynomial function was utilised by Safaei et al. [3] to demonstrate the evolution in directional stress ratio. However, not only the evolution in the anisotropic properties but also proper failure criteria need to be utilised for accurate prediction of the forming behaviour. Mostly strain based forming limit diagram (FLD), estimated either from experimentation or by theoretical approach, is implemented as the failure criterion. Requirement of obtaining complex parts needs multiple forming step. It has been well established that strain based FLD depends on strain history of the blank material. Therefore, after each forming operation a new FLD must be calculated for prediction of the formability in subsequent step. Hence, in case of theoretical estimation of FLD, strain path change need to be considered. Considering the huge experimental effort or the complexity in the calculation to make strain FLD applicable for multistep forming, several other theoretical approaches were explored by different researchers. Among them stress based FLD and polar effective plastic strain (PEPS) based FLD diagrams are the most popular approaches. The concept of stress based FLD (SFLD) was popularised by Stoughton [4]. It was reported that SFLD does not depend on deformation history if the amount of pre-strain remain relatively lower (<0.1) [5]. The path independent nature of SFLD was challenged by Yoshida et al. [6] and it was found that SFLD remains path independent if unloading takes place between subsequent deformation step. The effectiveness of PEPS FLD diagrams was established by Yoon and Stoughton [7].
In the present work, a representative study is presented where influence of evolution in anisotropy parameters are incorporated into prediction of SFLD and PEPS FLD. Evolution in the yield shape is also incorporated into FE simulation for enhancing the accuracy in failure prediction of the material. Both SFLDs and PEPS FLDs are used to predict the formability of pre-strained samples and a comparative study is presented here to highlight the effectiveness of incorporating the evolution in anisotropic parameters of material in both FE simulation and failure criterion as both criteria is path independent in the domain of the study.

2. Methodologies

2.1. Evaluation of mechanical properties of the material

The Uniaxial tensile testing on extra deep drawing quality (EDD) steel sheet (1.2 mm thickness) were performed along three material directions i.e. rolling (RD), transverse (TD) and diagonal direction (DD) to obtain the mechanical properties as shown in Table 1. The Lankford anisotropy parameters (R-value) along different anisotropic directions were also calculated using the relation $R\ (=\frac{\varepsilon_w}{\varepsilon_t})$ where, ε_w and ε_t are the true width and thickness strain, respectively. Stack compression tests were performed to assess the stress-strain relationship and R-value like parameter, $R_{BB}\ (=\frac{d\varepsilon_{TD}}{d\varepsilon_{RD}})$, of the material in balanced biaxial condition. It is mostly observed that the stress ratios along different directions varies with increase in plastic strain. However, variations in R-values and strain rate sensitivity of the material were neglected.

Table 1: Mechanical properties of the selected materials

Direction	Yield strength (MPa)	Ultimate tensile strength (MPa)	Uniform elongation (%)	R value	K value (MPa)	n	ε_0
					$\bar{\sigma} = K(\varepsilon_0 + \varepsilon_p)^n$		
RD	262.7	414.9	21.25	1.56	626.5	0.22	0.005
DD				1.47	622.8	0.21	0.006
TD				1.71	631.2	0.21	0.005
BB				1.11	618.2	0.19	0.006

2.2. Two-step forming

Square blank with dimension of 200 mm × 200 mm were trimmed out from the as received EDD sheets and the blanks were deformed in an in-plane stretch forming set up. In case of in-plane stretch forming, a flat bottom punch of diameter 101.6 mm was used to stretch the blank clamped under draw bead with a diameter of 150 mm. The blank was deformed until it showed an average strain (principal strain) of 5% on the surface. The average strain at the bottom of the deformed blank was measured by grid marking analysis. The actual punch displacement to obtain 5% strain was calibrated after rigorous FE simulation. Subsequent to the in-plane stretching, a circular blank with a diameter of 90 mm was taken out from the flat portion of the stretched blank for further deformation under a hemispherical punch. In this second stage of deformation, punch diameter was kept 50 mm with a die opening of 54 mm and the blank holding force was 132 kN. Detailed discussion on the two-stage forming set is available in Basak and Panda [8]. Different strain path was followed in the second stage of deformation and corresponding sample geometry was fabricated from the 90 mm blank. The deformed blank after each stage of deformation is shown in Figure 1.

Figure 1: Deformed pre-strained specimens

2.3. Stress based forming limit diagram

The stress based FLD or SFLD maps each point in the strain based FLD to the stress space. The pair of limiting strain is selected from strain based FLD following a particular strain path and strain ratio is calculated ($\rho = \frac{d\varepsilon_2}{d\varepsilon_1}$). The strain ratio can be related to the stress ratio $\left(\alpha = \frac{\sigma_2}{\sigma_1}\right)$ through the selected yield function. A parameter ξ is defined as the ratio of the effective stress to the major stress ($\xi = \frac{\bar{\sigma}}{\sigma_1}$) and can be expressed using yield function. For plane-stress condition, the parameters α, ξ and ρ can be correlated with effective strain following the plastic work equivalence.

$$\bar{\varepsilon} = {}^{\varepsilon_1}/_{\xi}\,(1 + \rho\alpha). \quad (1)$$

With effective plastic strain determined from the equation, effective stress can also be estimated using any isotropic hardening law. Therefore, the major and minor stress can be estimated using parameters α, and ξ. However, if evolution in anisotropy is considered then the complete set of equations need to be calculated iteratively as yield function itself becomes a function of effective plastic strain.

2.4. Polar effective plastic strain diagram

Effective strain falls under the category of stress metric as it is uniquely related to stress tensors, thus independent of the path change effect. Hence, Stoughton and Yoon [9] proposed a limiting curve to be plotted in a polar diagram and termed as polar effective plastic strain (PEPS) based FLD. The x-axis of the diagram was defined as $(\bar{\varepsilon} \times sin\theta)$ and y-axis was defined as $(\bar{\varepsilon} \times cos\theta)$ where, $\theta = \tan^{-1}\frac{\varepsilon_2}{\varepsilon_1}$. In the present study, effective plastic strain data were collected for each strain path of the FLD during the conversion of FLD data to SFLD. The angle θ is also obtained from the strain ratio of a particular strain path and PEPS FLD diagrams were plotted. As the estimation of PEPS FLD relied upon calculation of effective plastic strain as demonstrated above, evolution in anisotropy was already implemented.

2.5. Finite element simulation

Commercial explicit FE software package LSDYNA was utilized in the study. Quarter symmetric models were prepared for each step of the two-step forming operation. The tools were modeled as rigid body as per the dimensions of the original tools. The blanks were modeled with Belytschko–Lin-Tsay shell elements. Yld2000-2d was used as the yield function to model the material along with Swift's hardening law. In the study, evolution in anisotropy parameters were implemented by treating the coefficient of the yield function as functions of effective plastic strain. Therefore, with every increment in effective plastic strain the directional stress ratio was recalculated based on which yield coefficients were evaluated. Hence, instead of constant yield coefficient evolving set of yield coefficients were used. The FE simulation performed considering evolution in anisotropy parameters is termed as EVO_FE and the FE simulation the evolution in anisotropy not incorporated in yield function is termed as NONEVO_FE.

3. Results and Discussion

3.1. Application of stress based FLD

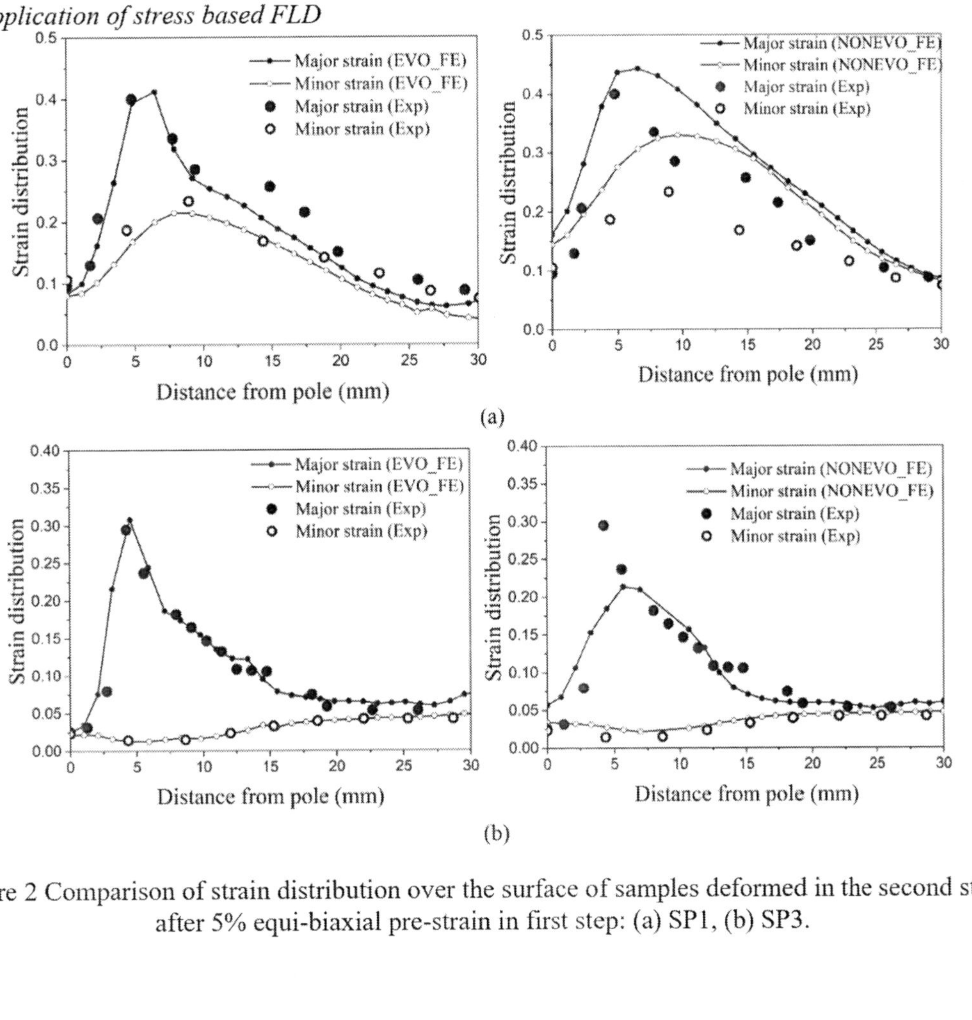

Figure 2 Comparison of strain distribution over the surface of samples deformed in the second stage after 5% equi-biaxial pre-strain in first step: (a) SP1, (b) SP3.

The major and minor strain distributions, over the deformed dome surfaces of pre-strained EDD sheet materials, were predicted from the FE simulation for the similar LDH. Strain distributions predicted by both EVO_FE and NONEVO_FE were correspondingly validated with the experimental results in Figure 2. These strain distributions were plotted with respect to the deformed dome center referred as a pole. It was found that the numerical value of the peak strain (major strain) was highest for SP1 specimen. From the minor strain distribution, it was evident that the minimum minor strain was positive for all the cases including the SP3 specimen. This is attributed to the biaxial pre-strain of the samples. It was also found that the maximum strain was induced at a certain distance from the pole where the punch lost contact with the sheet metal during deformation. Over the punch friction restricted thinning of the material. However, at the punch detaching point the material got stretched, and hence more strain was observed. The excellent correlation between the experimental and EVO_FE predicted strain distributions indicates the reliability of the developed FE model in the present work. The strain distribution predicted by NONEVO_FE could not follow the experimentally observed trend as compared in Figure 2. For the simulations with NONEVO_FE, both major and minor strain distributions could not be predicted for the SP1, while the major strain was considerably underestimated at the distance of 5 mm from the pole for the SP3. The enhanced accuracy of EVO_FE is attributed to the fact that in can accommodate the evolving shape of yield surface with increase in plastic strain. It is observed that yield surface becomes blunter at the balanced biaxial condition delaying the change of strain state to plane strain condition. As EVO_FE resulted better prediction in strain distribution over the deformed surface of the pre-strained samples, in the next sections discussions were kept limited to EVO_FE only.

3.2. Application of stress based FLD

In this study, an experimentally obtained strain based FLD of EDD steel reported by Basak and Panda [8] was converted to SFLD according to the methods described in section 2.3 incorporating anisotropic evolution in yield function. To differentiate the estimated SFLD from SFLDs calculated without evolution in yield surface, the SFLD was termed as EVO_SFLD. The EVO_SFLD was implemented for failure prediction of pre-strained samples. In case of implementation of SFLD in failure prediction, element stress data were collected from the simulation and superimposed on the SFLD curves (Figure 3a) as SFLD cannot be directly incorporated in LS-DYNA. The predicted LDHs are shown in Figure 3b. It is shown that EVO_SFLD rather under-predicted the LDH in case of lubricated biaxial specimen (SP1) though LDH is predicted accurately for other two sample geometries.

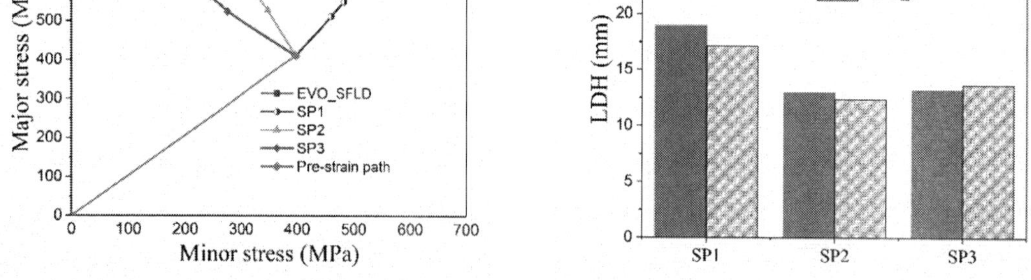

(a) (b)

Figure 3 a) stress path of pre-strained specimen in SFLD, b) Comparison of LDH predicted with SFLD

3.3. PEPS FLD

The PEPS FLD was also implemented to predict the failure during the second step deformation of the blanks under hemispherical punch. The critical elements were identified and the corresponding major and minor strain data were collected with regular interval. The major and minor strain data were then converted to points at PEPS space. Figure 4a depicts the strain path of the blanks in second stage of deformation superimposed within a PEPS FLD. It was observed that LDH was predicted for the SP1 and SP3 samples reasonably well (Figure 4b). However, it was observed that LDH was slightly over predicted for the case of SP2.

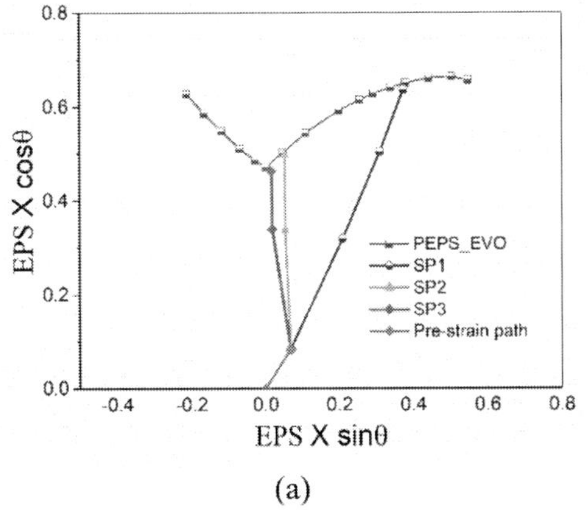

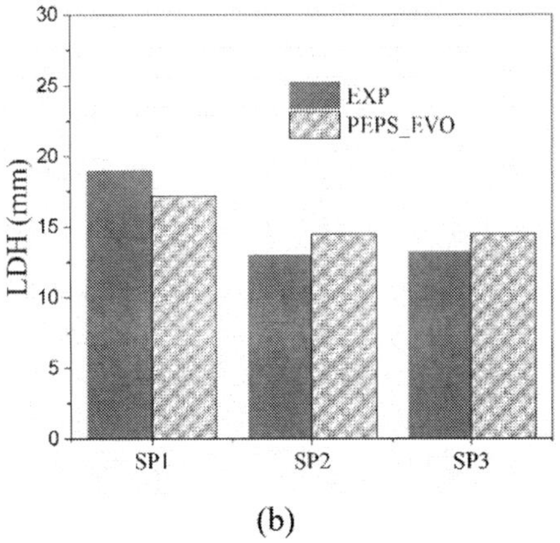

Figure 4 a) stress path of pre-strained specimen in SFLD, b) Comparison of LDH predicted with SFLD

4. Conclusion

In this study evolution in yield function was extended to estimate stress based FLD and polar effective plastic strain based FLD. Both of these failure criteria do not depend on the strain path history and robust in nature to predict formability in multi-stage forming. Hence, stress based FLD and PEPS FLD incorporating evolution in yield function were implemented to predict the formability in terms of LDH in multistage forming. The following conclusion can be inferred.

1. Evolution in yield function was successfully incorporated in estimation of SFLD and PEPS FLD. Both SFLD and PEPS FLD were successfully implemented to predict failure in two step forming.

2. The LDH predicted with SFLD corroborated well with experimental results for all three specimens SP1, SP2, SP3. However, PEPS FLD slightly over predicted the LDH in plane strain condition for specimen SP2.

3. The predicted strain distribution over the deformed surface by EVO_FE followed the experimental trend properly.

Bibliography

[1] A. Zamiri, F. Pourboghrat, F. Barlat, Int. J. Plast. 23 (2007) 1126–1147. doi:10.1016/j.ijplas.2006.10.012.

[2] H. Wang, M. Wan, X. Wu, Y. Yan, Comput. Mater. Sci. 47 (2009) 12–22. doi:10.1016/j.commatsci.2009.06.008.

[3] M. Safaei, M. Lee, S. Zang, W. De Waele, Comput. Mater. Sci. 81 (2014) 15–29. doi:10.1016/j.commatsci.2013.05.035.

[4] T.B. Stoughton, J. Eng. Mater. Technol. 123 (2001) 417. doi:10.1115/1.1398083.

[5] P.D. Wu, A. Graf, S.R. MacEwen, D.J. Lioyd, M. Jain, K.W. Neale, Int. J. Solids Struct. 42 (2005) 2225–2241.

[6] K. Yoshida, T. Kuwabara, M. Kuroda, Int. J. Plast. 23 (2007) 361–384. doi:10.1016/j.ijplas.2006.05.005.

[7] J.W. Yoon, T.B. Stoughton, ICAA13 Pittsburgh, Springer International Publishing, Cham, 2016: pp. 723–730.

[8] S. Basak, S.K. Panda, F Int. J. Mech. Sci. 151 (2019) 356–374. doi:10.1016/j.ijmecsci.2018.10.065.

[9] T.B. Stoughton, J. Yoon, Int. J. Solids Struct. 49 (2012) 3616–3625. doi:10.1016/j.ijsolstr.2012.08.004.

IOP Conf. Series: Materials Science and Engineering **967** (2020) 012007 doi:10.1088/1757-899X/967/1/012007

A novel experimental set-up for *in-situ* microstructural characterization during continuous strain path change

S Dhara[1]*, S Taylor[1], Ł Figiel[1], B Shollock[2] and S Hazra[1]

[1] *WMG, University of Warwick, Coventry, CV4 7AL, UK*

[2] *Department of Engineering, King's College London, London, WC2R 2LS, UK*

* *Email: Sisir.Dhara@warwick.ac.uk*

Abstract

Strain path change is a typical phenomenon during continuous stamping operations of sheet metal for a variety of applications including automotive body parts. During stamping, a punch continuously deforms a metal sheet to produce a desired geometry while following various strain path transitions depending on overall design of the stamping process. The strain path change can potentially alter the expected forming limit of the material. Previous researchers investigated the effect of changing strain path by loading sample in two distinct steps. Typically, between the steps the sample is unloaded before being re-loaded in the new strain path. This practice reflects the key challenge in elucidating this strain path dependent deformation, which is the ability to control the strain path change in a single deformation stage in an experimental set-up. In this work, a novel testing rig and specimen geometry that is capable of changing the strain path of a sample continuously without unloading the specimen were conceptualised, modelled and subsequently manufactured. Using this apparatus, the specimen was deformed in the uniaxial strain path in the first step before being deformed biaxially without unloading in between the steps. Thus, the apparatus ensures that the sample undergoes a continuous strain path change without unloading between the steps. The size of this mechanical test rig permits it to be placed inside a scanning electron microscope (SEM) chamber in order to study strain path transition *in-situ* to highlight strain localization and related microstructural changes in real time. Utilizing this test set-up, strain path change and corresponding strain values along each strain path were evaluated. The changes in material microstructure were concurrently investigated using *in-situ* SEM and electron back scattered diffraction (EBSD) analysis.

Keywords: Strain Path Change, Finite Element Modelling, Microstructural Characterization, *In-situ* SEM and EBSD.

Corresponding author

Sisir Dhara

WMG, University of Warwick

Coventry, CV4 7AL, UK

Email: Sisir.Dhara@warwick.ac.uk

IOP Conf. Series: Materials Science and Engineering **967** (2020) 012007 doi:10.1088/1757-899X/967/1/012007

1 Introduction

The automotive sector and its suppliers are one of the largest manufacturers of stamped components. Stamping is a multi-stage process where a sheet material is drawn in the first stage and then redrawn, flanged and pierced in subsequent stages. It is common for the material to undergo two types of strain path change: a continuous strain path change in the draw stage and a discontinuous strain path change when the drawn shell is re-drawn and flanged in subsequent stages. In the first mechanism, the strain path change occurs when material is drawn and stretched over a typically complex tool geometry. In the second mechanism, the strain path change occurs when the pre-strained drawn shell is unloaded from the draw tools and subsequently re-drawn or flanged along a different strain path. The effect of strain path changes have been shown by Graf and Hosford [1], [2]. Their experiments replicated the discontinuous mechanism of strain path change. They studied the effect of uniaxial, biaxial and plain strain pre-strain on the forming limit of Al 2008 T4 aluminium alloy. It was found that the biaxial pre-strain lowered the forming limits. However, the pre-straining in uniaxial tension raised the forming limits when the direction of pre-strain was normal to rolling direction. The pre-straining near plane strain produced a slight increase in forming limits when the direction of pre-strain was normal to rolling direction. Dhara et al. [3] expanded on this work by using a Zwick tensile machine to pre-strain AA 5754-O aluminium alloy in uniaxial tension and then re-loaded their samples in the biaxial strain path using a Nakajima test with a 50 mm diameter punch. They found that the forming limit strains of the pre-strained materials were increased when the pre-strain direction was along transverse direction. However, the limiting strains decreased when the rolling direction was the pre-strain direction. Collins et al. [4] uniaxially pre-strained low carbon ferritic steel samples in a stamping press and then re-loaded their samples in the biaxial strain path using a purpose-built biaxial tester. They found that a uniaxial pre-strain followed by biaxial deformation provided a significant gain in strain to failure. A common characteristic of these experiments into discontinuous strain path change is that the two stages were carried out on different experimental setups and the samples may have experienced dis-similar loading strain rates during both stages.

The effect of changing strain path on formability is thought to be due to the way it changes the texture and work hardening of the material. To investigate the relationship between external loads and microstructure, Ghadbeigi et al. [5] and Celotto et al. [6] used a recently developed experimental technique where mechanical tests were carried out in the chamber of an SEM using a miniaturised test rig with the aim of imaging the deformation of the sample *in-situ*. They proportionally loaded their samples in the uniaxial strain path using a test rig that was compact enough to fit in the chamber of a scanning electron microscope (SEM). They imaged their samples periodically during the test and post-processed using the digital image correlation (DIC) technique to analyse the full-field strain of the samples at the meso-scale. At this scale, they were able to analyse the effect of the external loading on the strain distribution in the ferritic and martensitic phases of their dual-phase 1000 (DP1000) steel and transformation induced plasticity 800 (TRIP800) steel materials respectively. Subsequently, they observed the micro-void nucleation and fracture mechanism. Caër and Pesci [7] contributed to the design of sample for the *in-situ* deformation technique. The designed sample ensured the strain localisation at the centre of the sample and using the sample in a micromechanical test rig, they examined grain rotation and misorientation development in an annealed AISI 304 stainless steel *in-situ* during biaxial loading. Similarly to Ghadbeigi et al. and Celotto et al., they placed their test rig in an SEM to measure the evolution of texture by measuring electron back scattered diffraction (EBSD) data *in-situ* from the centre of the biaxial specimen periodically during the loading process. They explained that the loading path had a potential to orient the grains along a particular direction. Kubo et al. [8] carried out a biaxial tensile test of interstitial free (IF) steel using a biaxial specimen with the notch at the four corners of the central part in an SEM chamber and studied inverse pole figure (IPF) maps, grain average misorientation (GAM) maps and Taylor factor (TF) maps to characterise microstructural changes during biaxial tension. Conclude that this *in-situ* technique can be used to measure the evolution of strain and texture during plastic deformation and that it is a useful tool in understanding the formability of material that undergo strain path changes. The gap is the lack of a method to change strain path during a mechanical test.

The aim of this work is to develop an experimental method to investigate the effect of the continuous strain path transition that occurs in the drawing stage on material microstructure. A Micromecha biaxial testing rig was used with a cruciform sample design based on the Caër and Pesci's design. The design was modified to enable the strain path change to occur at a set point during the experiment. The strain evolution of the samples were observed at the macroscopic and microscopic scales using DIC while the texture evolution was observed with EBSD. Initial results showed that the modified sample design was effective at altering the applied strain path and that the material experienced observable changes to its strain path and texture.

2 Development of the methodology for continuous strain path change

A Micromecha Proxima test rig was used for the strain path transition experiment. The test method was developed in two steps. In the first step, the strain evolution of a cruciform sample was analysed along a proportionate biaxial load path. In the second step, the cruciform sample was redesigned to allow a change in the strain path during loading.

The rig and the specimen used in the first step is presented in Figure 1. The horizontal (inline grips) and vertical (transverse grips) grips of the rig are connected through a 45° wedge (Figure 2b). A motor drives one end of the horizontal grip while the other end is securely fastened to the body of the rig. When the motor moves the horizontal grips, its motion is translated into a vertical motion via the wedge, thus loading the sample vertically at the same time. The inline grip was moved along X-axis and the transverse grips were moved along Y-axis during the experiments. The external dimension of the cruciform sample was 30 mm by 30 mm and its thickness was 1.2 mm (Figure 2a). The central region was reduced in thickness (0.25 mm) to concentrate the strain in that region and to prevent overloading the rig and the load cell. DP600 steel was chosen for the specimen material, as it is a high strength material used commonly in automotive applications. The material properties and chemical compositions are detailed in Table 1 and Table 2 respectively. The rig and sample were modelled using the finite element (FE) method. The validated model was used as the basis for the second step. The model consisted of the horizontal grips, vertical grips and the sample (Figure 2b). The grips and the gripping areas of the sample were modelled as rigid bodies while the sample was modelled as a deformable body. All bodies were meshed with solid elements of an appropriate size that optimised speed and accuracy. Yielding was modelled with the von Mises model and power-law hardening was assumed. The yield model and hardening law were chosen the simplest one as the finite element modelling was done to optimize the design of sample and overall assembly. The anisotropy coefficient of the sample material was nearly 1 and hence, isotropic material model was considered to simplify the model. More advanced material model such as crystal plasticity based model with anisotropic behaviour and hardening law were beyond the scope of current research and would be studied in future. To model the clamping of the sample to the grips, the nodes in the sample gripping areas were joined to the nodes of the grips as constrained rigid bodies. Displacement (1.5 mm) was applied to in-line grip 1 along X-axis (Figure 2b) to represent the motor in the physical rig. Inline grip 2 was fixed. The motion of transverse grips 1 and 2 was restricted to the Y-axis, to reflect the physical rig. The model was solved using LS-Dyna explicit v.10.1.0. The X-axis (ε_x) and Y-axis (ε_y) strain contour shows the strain concentration and uniformity of the strain distribution at the central region (Figure 3a-b). The strain path at the centre of the specimen showed that the specimen was undergone biaxial strain path (Figure 3c).

The model was validated against physical trials. During trial experiments with the rig and the biaxial sample, the strain evolution at the centre of the specimen needs to be calculated in order to confirm whether the specimen is suitable enough for performing biaxial elongation and suitable for *in-situ* experiment. The strain evolution of the sample during the experiment was measured with the GOM 12M macro-DIC system. The DIC system consisted of two 12MP cameras fitted with lenses to capture the strains in the arms and central region of the specimens. Measurement frame rate was 1 frame/second and samples were loaded at 0.3 mm/min. The experimental ε_x and ε_y strain contour of the specimen is presented in Figure 4a-b. The strain path at the centre of the specimen was analysed from the experiment and depicted in Figure 4c. There was some slack at the start of the experiment as the loading mechanism fully loaded the specimen. After ε_x=0.02, the strain path was similar to that predicted by the FE model (slope of 1.2 experimental vs 1.1 predicted).

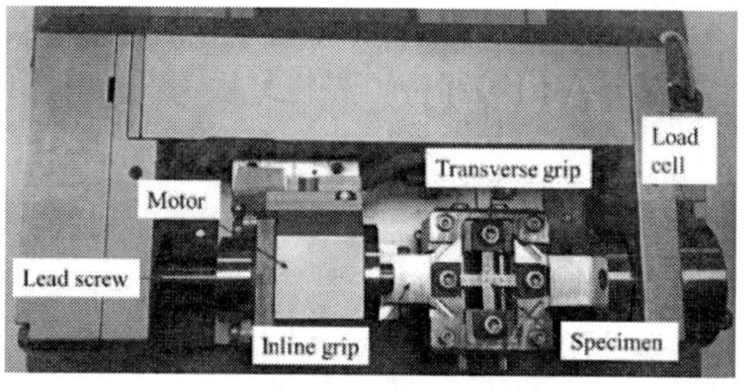

Figure 1: Experimental set-up consisting of the rig and the specimen

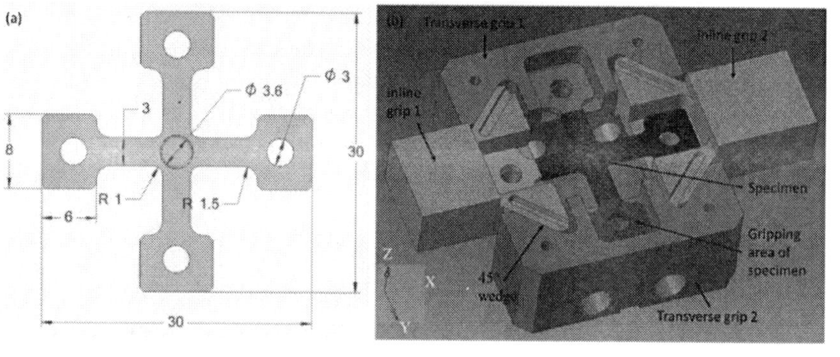

Figure 2: (a) Biaxial stretching specimen and (b) FE model of the assembly consists of the rig and the specimen

Table 1: Material properties of DP600 steel used in this work

Material		DP600
Gauge (mm)		1.2
Young modulus (GPa)		208
Yield strength (MPa)		407
Ultimate tensile strength (MPa)		665
% Elongation		22
Anisotropy parameters	r_0	0.96
	r_{45}	0.93
	r_{90}	1.05
Hardening parameters according to Hollomon law	Strength coefficient, K (MPa)	1125
	Hardening exponent, n	0.21
Density (g/cm^3)		7.87
Poisson's ratio		0.3

IOP Conf. Series: Materials Science and Engineering **967** (2020) 012007 doi:10.1088/1757-899X/967/1/012007

Table 2: Chemical composition (wt %) of DP600 steel used in this work

Material	C	Mn	Si	Al	Mo	Cr	Cu	S	P	Fe
DP600	0.09	1.84	0.36	0.05	0.01	0.02	0.03	0.005	0.01	Balance

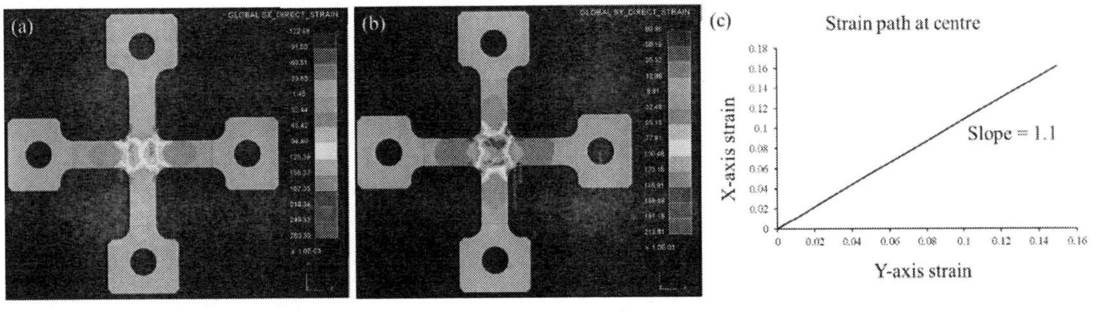

Figure 3: (a) X-axis and (b) Y-axis strain contours of the specimen predicted from FE modelling, and (c) predicted strain path at the centre of the specimen during biaxial stretching

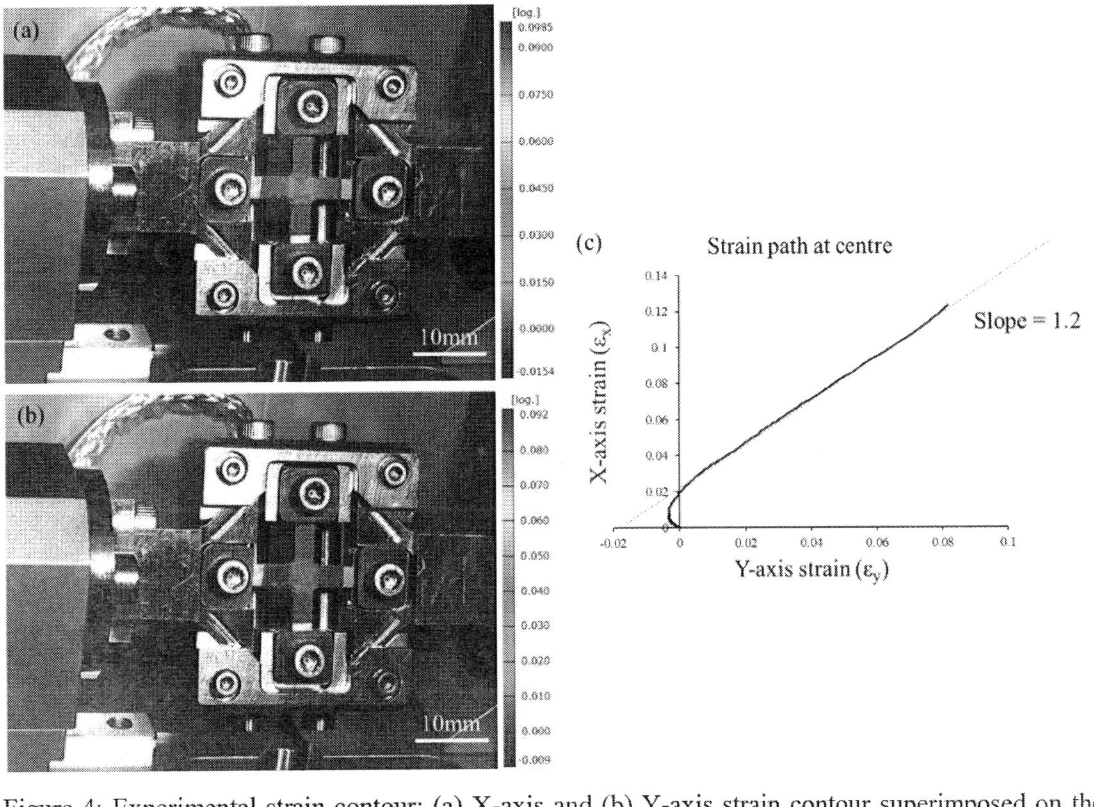

Figure 4: Experimental strain contour: (a) X-axis and (b) Y-axis strain contour superimposed on the specimen, and (c) strain path at the centre of the specimen during biaxial stretching

IOP Conf. Series: Materials Science and Engineering **967** (2020) 012007 doi:10.1088/1757-899X/967/1/012007

To change the strain path during testing, the cruciform specimen geometry was modified to include slotted holes in the vertical arms (Figure 5a). At the start of the test, the slots allowed the bolts of the vertical (transverse) grips to translate without loading the sample. After a specified elongation that is determined by the length of the slots (0.2 mm), the transverse grips engage and the specimen is deformed biaxially. Hence, the material is deformed uniaxially at the start of the test before the strain path changes to biaxial strain path. The test in step 1 was repeated with this modified sample geometry. It ensured that the specimen was loosely held in the transverse grips to allow the grips to travel freely in the 0.2 mm slot in the first stage of the travel (Figure 5b). The experimental ε_x and ε_y contour of the specimen before and after the strain path transition was evaluated and is presented in Figure 6a-d. The ε_x and ε_y strain evolution at the centre of the specimen showed that the sample was deformed uniaxially initially and then biaxially towards the end of the test (Figure 6e).

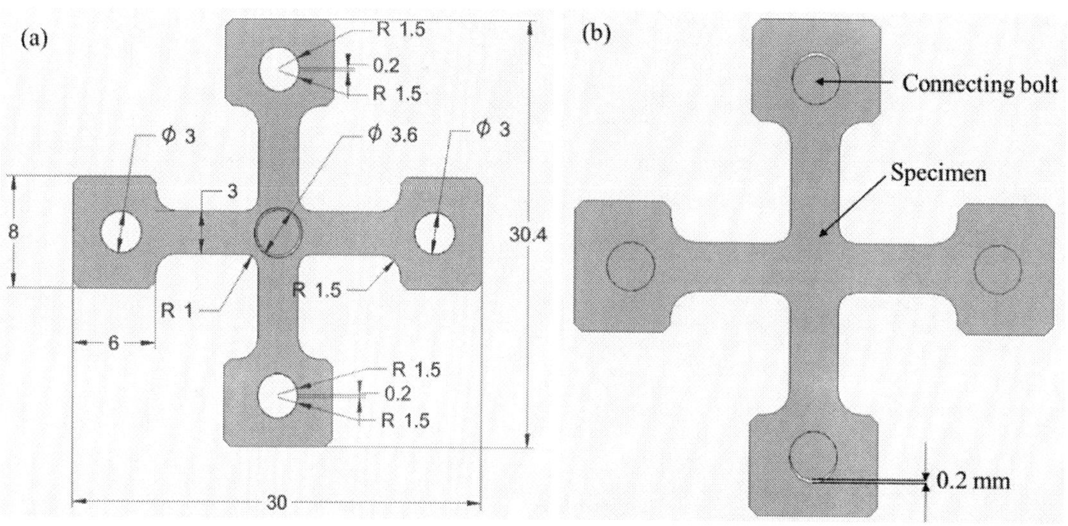

Figure 5: (a) Design of strain path change specimen and (b) assembly design showing 0.2 mm travel of connecting bolt during strain path change

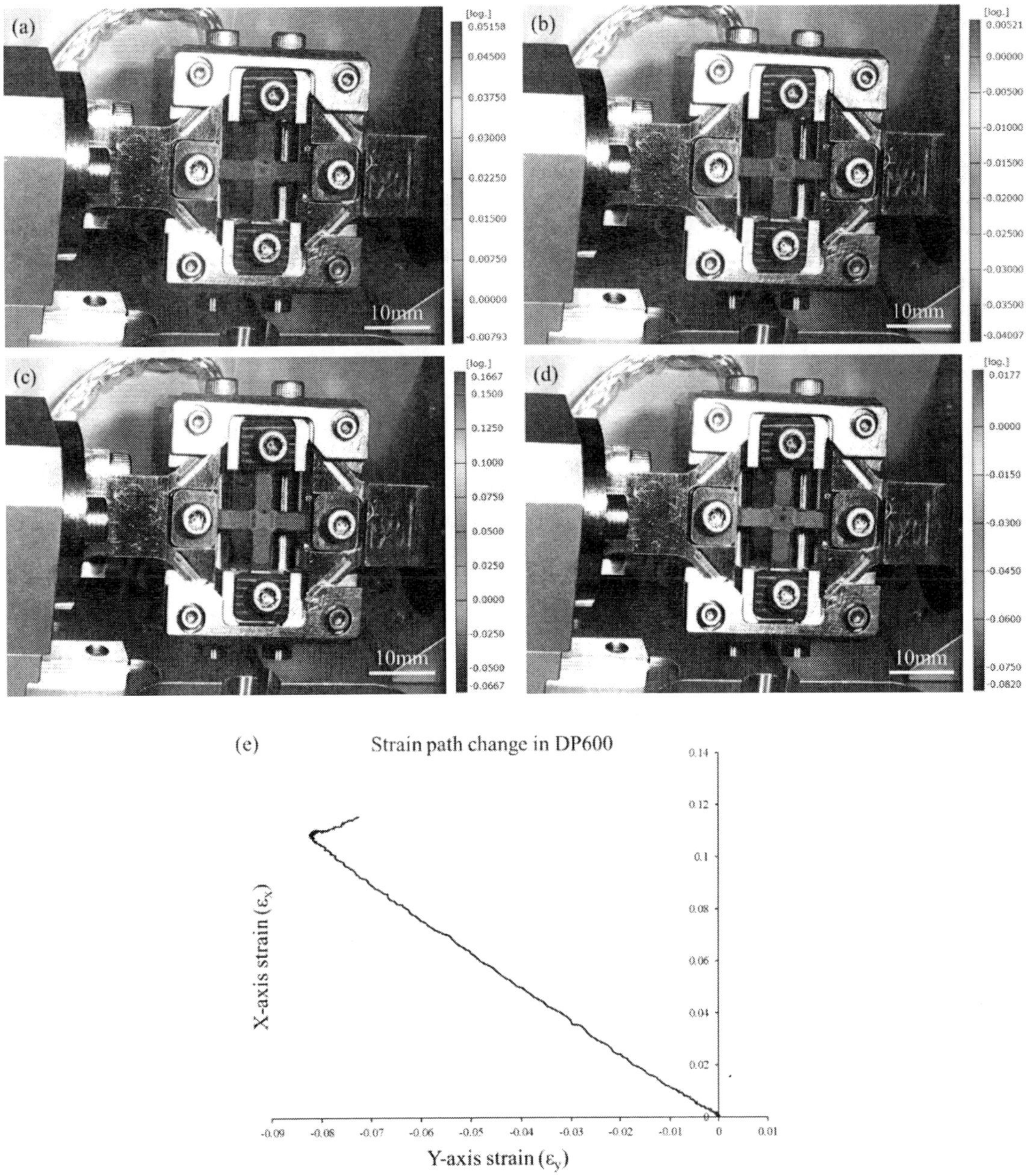

Figure 6: Experimental strain contour: (a) X-axis and (b) Y-axis strain contour superimposed on the specimen before strain path transition, (c) X-axis and (d) Y-axis strain contour superimposed on the specimen after strain path transition, and (e) strain path change at the centre of the specimen

3 Measuring microstructural strains and texture evolution during continuous strain path change

The specimens for strain path change were prepared from DP600 steel sheet and were polished by using silicon carbide (SiC) paper, diamond suspension solutions and finally 0.05 µm colloidal silica solution. Samples for microstructural strain evaluation were etched with 2% Nital solution for 10 seconds. The etched sample was loaded in the rig. To carry out the microstructural measurements, the rig was placed

IOP Conf. Series: Materials Science and Engineering **967** (2020) 012007 doi:10.1088/1757-899X/967/1/012007

in the chamber of a ZEISS SIGMA field emission gun scanning electron microscope (FEG-SEM). The rig was controlled using the POROS 2 software supplied with the Micromecha rig. The inline grip was elongated at a rate of 5 μm/sec. The elongation was arrested every 0.2 mm travel to perform SEM and EBSD scans for strain and texture analysis respectively. The microstructural observations consisted of micro (grain level) DIC analysis and EBSD analysis. For micro DIC analysis, the SEM scan was done by using secondary electrons (SE2) with 20 keV voltage. The captured SEM images were post-processed with Lavision's digital image correlation software (Davis 10) to obtain the distribution of strain in the microstructure. The subset size was used as 25 μm and step size was used as 8 μm during the analysis. The test was repeated to obtain texture measurements with the SEM's EBSD detector. During EBSD scanning, the step size was kept at 0.5 μm. The EBSD results were analysed in HKL Channel 5 software. These *in-situ* measurements allowed the detailed analysis of microstructural deformation as a result of the continuous strain path change.

4 Results and discussion

The ε_x and ε_y strain distributions calculated from micro DIC analysis show that the strain distributions are heterogeneous with a wide spread of strains (Figure 7 and 8). This is because the dual phase steel is made up of a hard martensite phase and a softer ferrite phase. Lower strains are approximately associated with the martensite phase. During the initial stage of the experiment when the sample was loaded uniaxially, ε_x increased while ε_y decreased. Based on the average strains in the data, the $\varepsilon_y/\varepsilon_x$ ratio was found to be -0.75, which is in between uniaxial tension (ratio -0.5) and pure shear deformation (ratio -1.0). This may be due to the extent of the strain variance as a result of the heterogeneity of the microstructural strain field. After 1 mm elongation of inline grip, the ε_y started to increase confirming that the strain path change (uniaxial to biaxial) started after 1 mm elongation of the inline grip. The banding of ε_x and ε_y strain (indicated by black line in Figure 7 and 8) align approximately 45° to the initial loading direction. When the strain path change occurs, the angle of these lines appears to rotate. The rotation is more prominent in ε_x strain distribution than ε_y strain distribution. Further analysis on microstructural changes was performed by carrying out EBSD mapping of around 250 grains at the centre of the specimen. The EBSD maps are shown in Figure 9 along with the inverse pole figure (IPF) Z colouring. Evidence of deformation was clear within the grain structure, with grain elongation and rotation occurring with increased elongation of the sample. During the strain path change, some of the grains (indicated by black dotted circle in Figure 9) underwent rotation and their crystallographic direction changed from [101] towards [001]. The grain rotation was further studied with the help of pole figures as shown in Figure 10. There was a difference in colour contour in the pole figures in {100} and {111} plane before and after the strain path transition with a noticeable increase in multiples of uniform distribution (MUD) values immediately before failure.

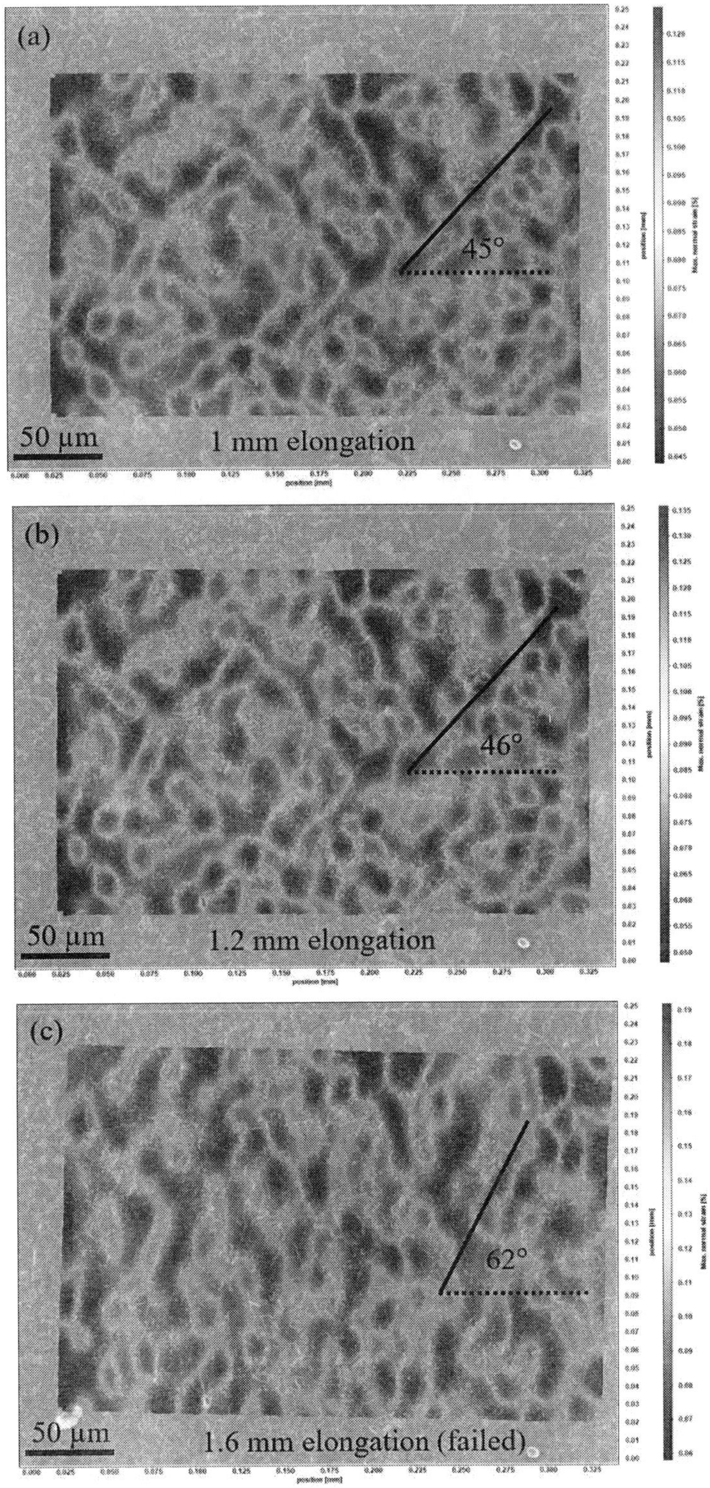

Figure 7: ε_x strain distribution at the centre of the specimen calculated from micro DIC analysis after (a) 1 mm, (b) 1.2 mm and (c) 1.6 mm elongation of the specimen along X-axis during strain path change

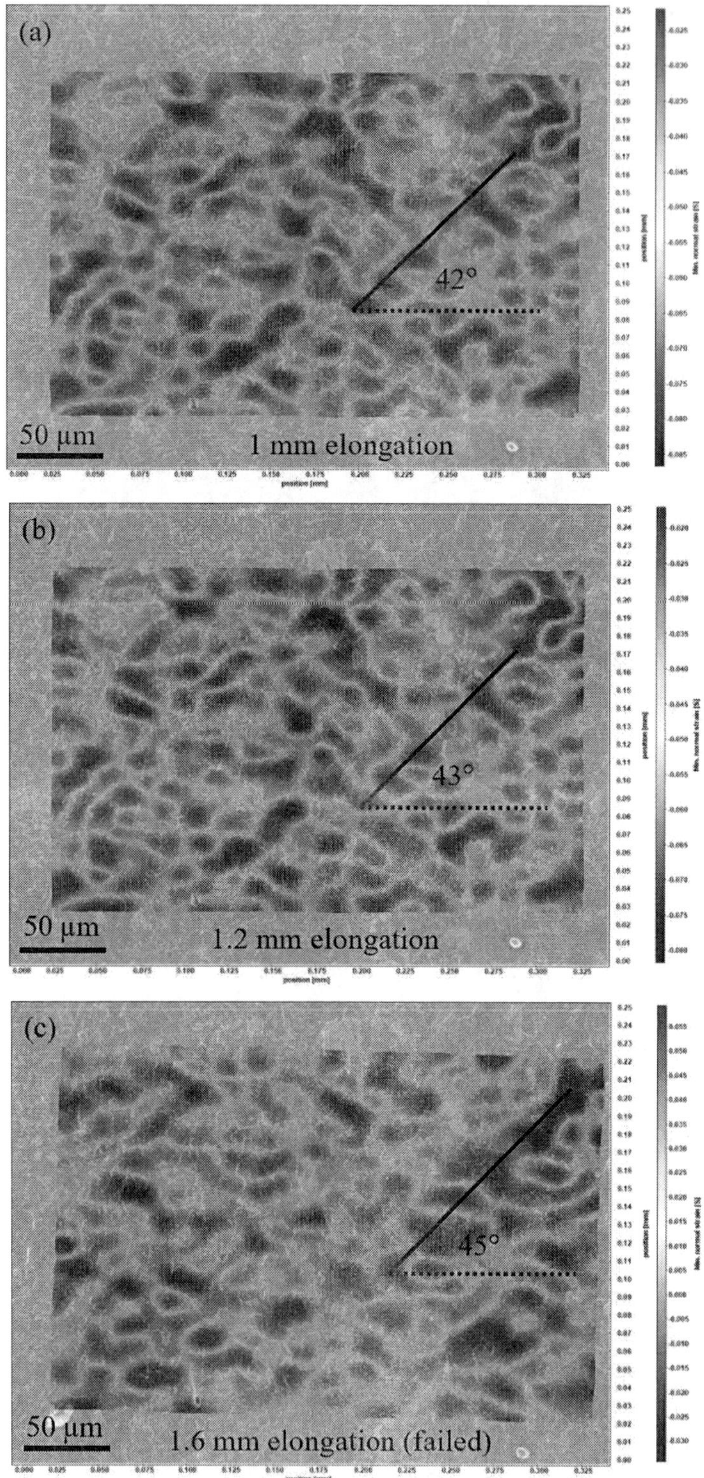

Figure 8: ε_y strain distribution at the centre of the specimen calculated from micro DIC analysis after (a) 1 mm, (b) 1.2 mm and (c) 1.6 mm elongation of the specimen along X-axis during strain path change

IOP Conf. Series: Materials Science and Engineering **967** (2020) 012007 doi:10.1088/1757-899X/967/1/012007

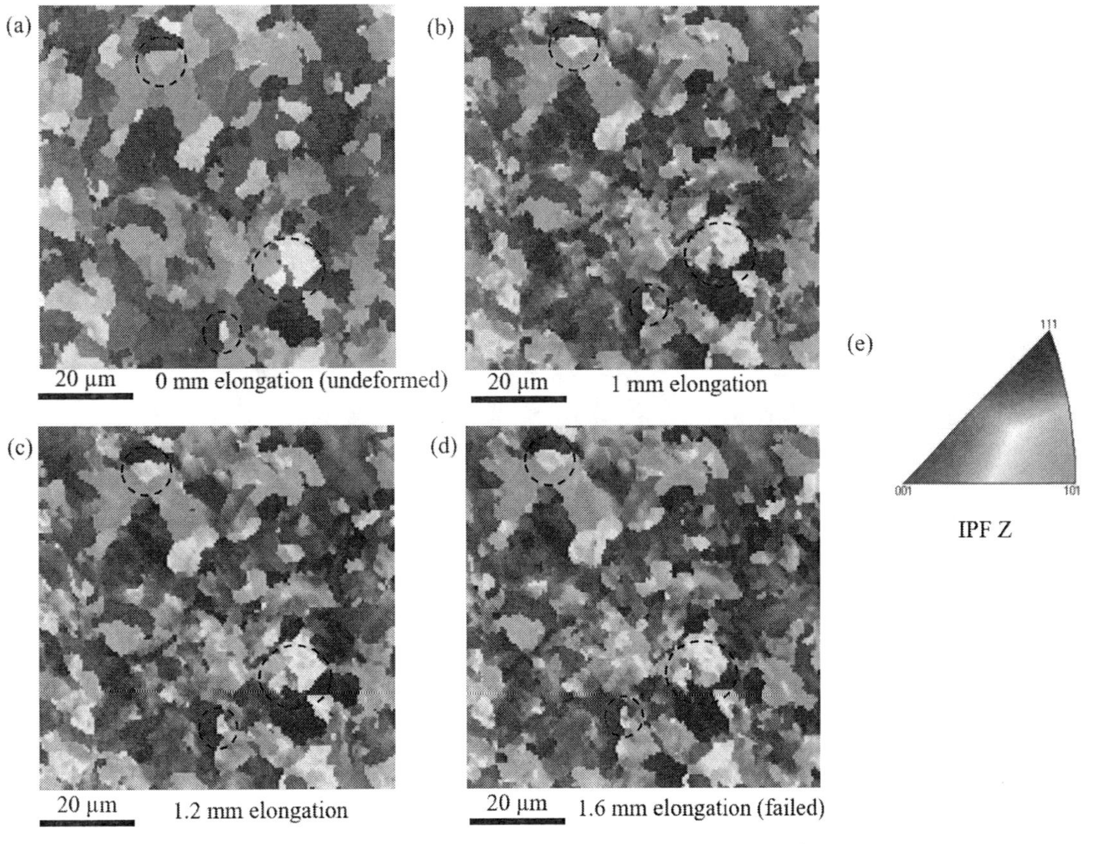

Figure 9: EBSD mapping of the grains at the centre of the specimen after (a) 0 mm, (b) 1 mm, (c) 1.2 mm and (d) 1.6 mm elongation of the specimen along X-axis during strain path change, and (e) IPF Z colouring

IOP Conf. Series: Materials Science and Engineering **967** (2020) 012007 doi:10.1088/1757-899X/967/1/012007

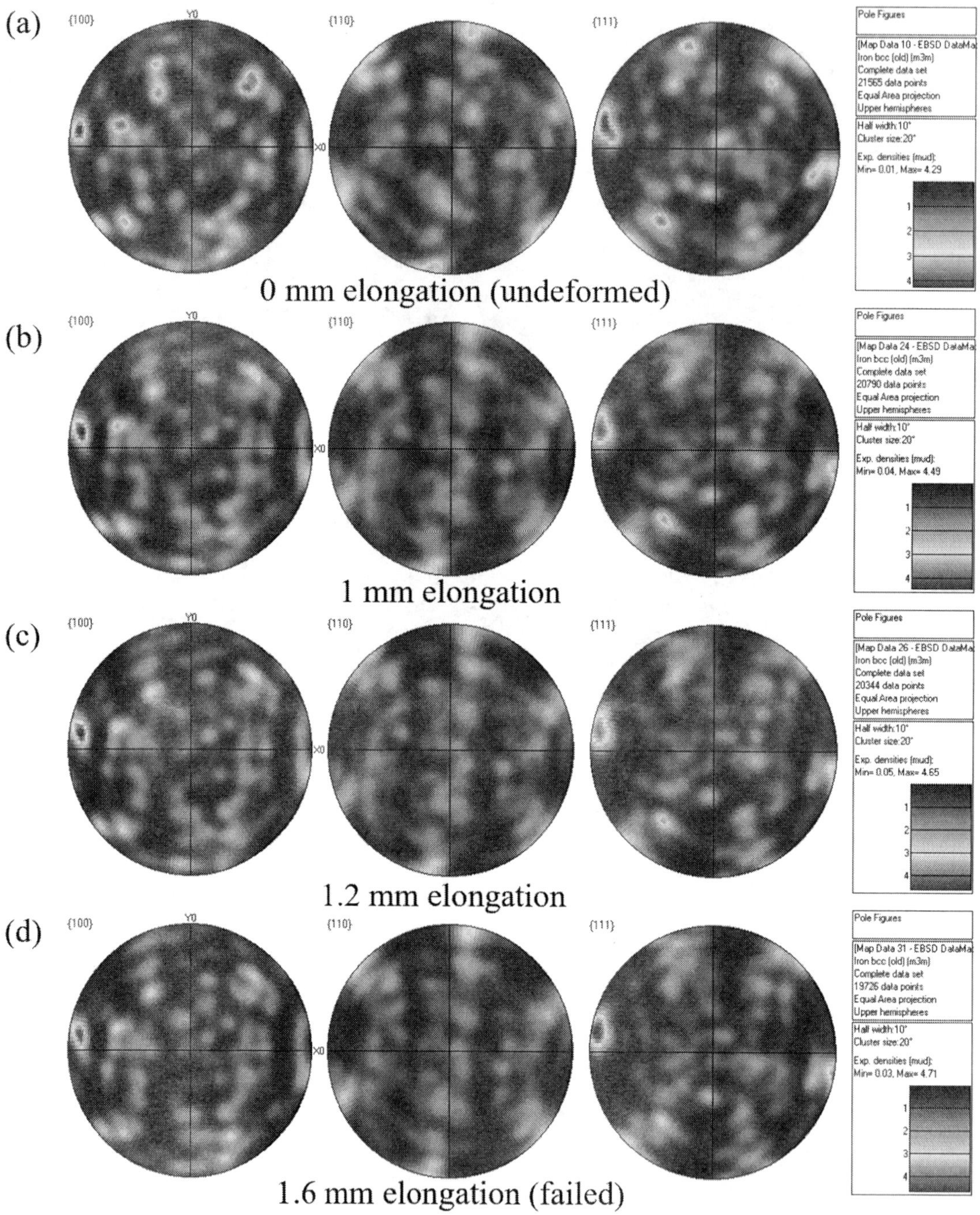

Figure 10: Pole figures of the grains at the centre of the specimen after (a) 0 mm, (b) 1 mm, (c) 1.2 mm and (d) 1.6 mm elongation of the specimen along X-axis during strain path change

5 Conclusions

An experimental technique was developed to test cruciform samples along (a) a proportional strain path and (b) a non-proportional strain path. Suitable specimen for proportional and non-proportional strain

IOP Conf. Series: Materials Science and Engineering **967** (2020) 012007 doi:10.1088/1757-899X/967/1/012007

path change was also designed. The mechanism developed for strain path change has the potential to perform strain path transition continuously without withdrawal of the applied load from the specimen. DIC analysis at macroscale showed that the DP600 steel specimen underwent biaxial strain path in proportionate test and uniaxial to biaxial strain path transition in the non-proportionate test. Hence, the mechanism is ideal to study the effect of strain path change. Moreover, as the overall set-up can easily be placed inside an SEM chamber, performing microstructural analysis was extremely simple during strain path change. DIC analysis at microscale revealed the heterogeneous strain banding on the sample surface and the angle of the bands was rotated as the strain path transition occurred. The EBSD and pole figure results confirmed the evidence of grain rotation during strain path change. Thus, DP600 steel is susceptible to grain rotation during strain path transition. The microscale DIC and EBSD analysis confirmed that the microstructural evolution could easily be tracked during *in-situ* strain path change using the developed mechanism. In future, a detailed study on the microstructural changes will be carried out during strain path change.

References

[1] A. F. Graf and W. F. Hosford, "Calculations of forming limit," *Metall. Trans. A*, vol. 24, no. 11, pp. 2497–2501, 1993.

[2] A. Graf and W. Hosford, "The influence of strain-path changes on forming limit diagrams of Al 6111 T4," *Int. J. Mech. Sci.*, vol. 36, no. 10, pp. 897–910, 1994.

[3] S. Dhara, S. Basak, S. K. Panda, S. Hazra, B. Shollock, and R. Dashwood, "Formability analysis of pre-strained AA5754-O sheet metal using Yld96 plasticity theory: Role of amount and direction of uni-axial pre-strain," *J. Manuf. Process.*, vol. 24, pp. 270–282, 2016.

[4] D. M. Collins *et al.*, "A synchrotron X-ray diffraction study of non-proportional strain-path effects," *Acta Mater.*, vol. 124, pp. 290–304, 2017.

[5] H. Ghadbeigi, C. Pinna, S. Celotto, and J. R. Yates, "Local plastic strain evolution in a high strength dual-phase steel," *Mater. Sci. Eng. A*, vol. 527, no. 18, pp. 5026–5032, 2010.

[6] S. Celotto, H. Ghadbeigi, C. Pinna, B. A. Shollock, and P. Efthymiadis, "Deformation-Induced Microstructural Banding in TRIP Steels," *Metall. Mater. Trans. A*, vol. 49, no. 7, pp. 2893–2906, 2018.

[7] C. Caër and R. Pesci, "Local behavior of an AISI 304 stainless steel submitted to in situ biaxial loading in SEM," *Mater. Sci. Eng. A*, vol. 690, pp. 44–51, 2017.

[8] M. Kubo *et al.*, "Development of Biaxial Tensile Test System for *In-situ* Scanning Electron Microscope and Electron Backscatter Diffraction Analysis," *ISIJ Int.*, vol. 56, no. 4, pp. 669–677, 2016.

IOP Conf. Series: Materials Science and Engineering **967** (2020) 012008 doi:10.1088/1757-899X/967/1/012008

Influence of a drawbead passage in deep drawing processes on surface values and the tribological system

Harald Schmid[a*] and Marion Merklein[a]

[a]Institute of Manufacturing Technology LFT, Friedrich-Alexander-Universität Erlangen-Nürnberg, Egerlandstr. 13, 91058 Erlangen, Germany

*Email: harald.schmid@fau.de

Abstract. In recent times, lightweight design and functional integration in metal forming and especially in deep drawing operations lead to complex geometries. These geometries can cause variable defects in production. Especially with modern materials like AHSS or UHSS, deep drawing becomes more challenging. One common way to meet these challenges and manufacture sheet metal parts without defects is the application of drawbeads in the forming tool. Drawbeads are used for an exact control and forecast of the material flow during the deep drawing process. It was already shown, that different material parameters like the tensile strength and the fracture strain are changed significantly after running through a drawbead. In addition, there are various indications that the tribological system in a drawbead passage is also changed significantly and that this has further influence on the ongoing drawing process.

This assumption is examined with an experimental setup derived from industrial deep drawing processes. Therefore, different sheet metal materials are drawn through a drawbead geometry under the same conditions like the blank holder force, the drawing velocity or lubrication. After the drawbead passage, 3D surface parameters are measured and compared to the values in its initial state. Following, conventional strip-drawing tests with preloaded strips are carried out. The variation of the friction coefficient after a drawbead passage is analysed and compared to the initial state. The results are correlated with the surface values. For these investigations, three different materials are used: a conventional deep drawing steel because of its wide application as well as an advanced high strength steel AHSS and an aluminium alloy of the 6xxx series as representatives for modern lightweight materials.

1. Introduction

In modern forming processes like deep drawing of metal parts, functional integration and complex design are more present than ever before [1]. This leads to demanding and challenging geometries which can cause multiple defects and failure in production series. Especially for the new developed high strength steel classes and modern aluminium alloys, deep drawing becomes more challenging. One option to address these challenges is the accurate control of the material flow during the forming process. Next to changes in the lubrication system and different tool coatings also an adjustment of the blank holder force can be done according to Doege [2]. Very often, high retention forces are needed in the different forming areas and drawbeads are the method of choice as they can be implemented directly in the tools. Next to the retention forces, they also affect other areas in the forming process. It could already be shown by the authors in [3], that mechanical properties such as the yield stress or the uniform elongation ε_u are changed after a drawbead passage and that this affects the forming process. There are also optical indications, that the tribological system after a drawbead is changed and this may influence the metal forming.

In 1902, Stribeck [4] found a connection between the velocity and the friction coefficient which is still known as the "Stribeck curve" today. It was also investigated, that the friction coefficient is a function of many more parameters [4]. Nowadays, it is well known, that the contact pressure in the friction area affects the tribological system, which is for example analysed by Zöller in [5]. In addition, it is found that the velocity, the lubrication system and the tool and sheet metal parameters influence the tribological conditions in general [6].

Filzek investigated in 2004 [7], that in the drawbead geometry the contact pressure and other parameters are changed significantly in comparison to the rest of the forming area, which indicates a distinct change of the friction coefficient. Also Christiany and Groche analysed in a strip drawing test with drawbead and the use of Advanced High Strength Steels (AHSS), that different tool materials lead to different wear resistance and therefore also to different tribological conditions [8]. Ludwig [9] and Leocata [10] investigated the friction coefficient only for one material deep drawing steel and aluminium), but not under same conditions and did not consider the different sheet metal sides. In general, this is not considered in simulations by using one global friction coefficient. Hol and Wiebenga implemented a commercial system to forecast friction coefficients under different conditions by using a multi-scale model [11]. This could prove a good accuracy to standard friction tests [12]. Summarized, a change of the tribological system in a drawbead passage is assumed and should be analysed. Therefore a modified strip drawing test with drawbead geometry, which setup was also presented by the authors in [3], is used to preload steel and aluminium sheet metal in a drawbead passage. Afterwards, conventional strip drawing tests are done with the preloaded material from the drawbead to investigate the friction behaviour after a drawbead treatment and especially to analyse the influence of the changed surface properties. The transformations of the friction coefficients are compared to the measurements in the initial state. For a widespread investigation, 3D confocal analysis of the sheets are done before and after the drawbead pass to correlate the profile result and the tribological conditions.

2. Materials

In modern deep drawing processes, many different material grades are used to fulfil the multiple challenges in forming processes. To provide an overview on the different sheet materials, three varying sheets are used for this investigation. First, the aluminium alloy EN AW AA6014 according to EN 573-1:2005-02 is investigated in T4, which means the material was annealed and naturally aged to a stable state. This alloy is widely used in bodyshell applications and for structural components in the automotive industry [13]. Apparently, aluminium is also considered as a lightweight material because of its low density and its high specific strength compared to steel sheets. Additionally, a conventional mild steel CR 3 GI 50/50-U according to VDA 239-100 [14] is investigated. It is hot dip zinc coated with 50 g/m^2 on each side and is considered because of its wide application in the industry. It is also widely known as DC04. Finally, the AHSS CR 440Y780T-DP GI40/40 [14], is analysed because of its application as a high strength steel in modern lightweight design, especially in the automotive industry and will be called DP800 in this publication. All materials are used in the initial thickness of $t_0 = 1.0$ mm to make them comparable.

As it can be seen in Table 1, results from the optical 3D uniaxial tensile tests are differing between the sheets. As expected the AHSS DP800 has a way higher tensile strength TS compared to the aluminium alloy while DC04 and AA6014 have a higher uniform elongation ε_u in comparison.

Additionally, 3D surface parameters are measured by a confocal microscope (Nanofocus) in the initial state according to [15] and were averaged over the two different sides. Pfestorf and Staeves already found in 1997 different area roughness parameters of metal sheets [16] to be important for the lubrication system in deep drawing processes. In this work, two parameters are chosen: the mean arithmetic height of the area S_a to quantify the surface globally and the closed void volume V_{Cl} as an attribute to quantify the oil storage ability of the sheet metal. Analysing the surface parameters in Table 1, a correlation between the values S_a and V_{Cl} is evident. Aluminium has a quite smooth surface testified by S_a of 0.8 μm with a layer of aluminium oxide resulting from oxidation processes. Next to this, the DP800 steel with the zinc coating has a mean arithmetic value of S_a of 1.18 μm and is comparable. The mild steel DC04 has a higher roughness of S_a=2.45 μm.

Table 1. Material parameters in uniaxial tensile test and 3D surface parameters with corresponding standard deviations in the initial state before preloading

	Yield Strength YS [MPa]	**Tensile strength TS [MPa]**	**Uniform elongation ε_u [%]**	**Mean arithmetical height S_a [μm]**	**Closed void volume V_{cl} [mm^{-1}]**
AA6014	141.9 ± 1.1	249.6 ± 1.7	20.0 ± 0.2	0.81± 0.07	436.2 ± 19.7
DC04	165.0 ± 0.5	303.7 ± 0.3	25.7 ± 0.6	2.45 ± 0.22	1576.8 ± 235.1
DP800	505.4 ± 1.9	808.2 ± 3.4	12.9 ± 0.3	1.18 ± 0.10	597.3 ± 19.8

All parameters are used to compare the different sheet metal materials in the initial state and after a preloading by a drawbead.

3. Experimental setup and applied methodology

For the preloading with a drawbead, a conventional strip drawing test setup with a drawbead geometry is used. The methodology for the test setup is shown in Figure 1.

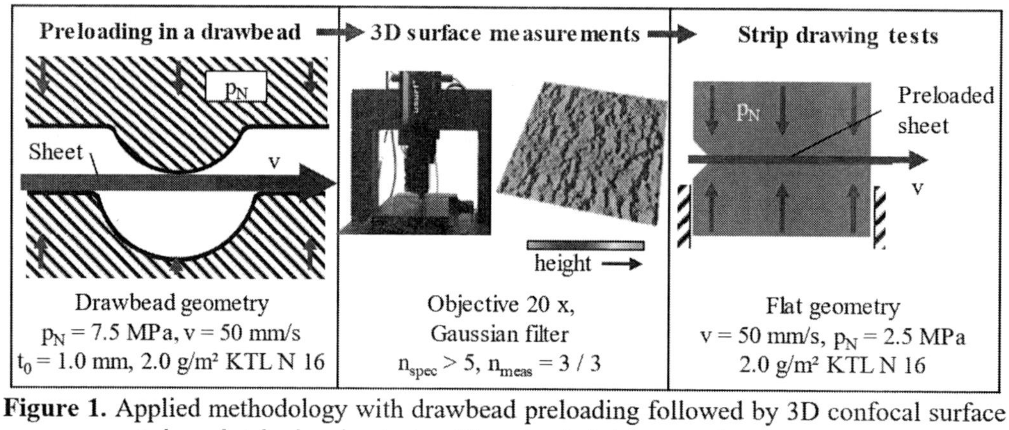

Figure 1. Applied methodology with drawbead preloading followed by 3D confocal surface measurements in and strip drawing tests with preloaded sheet metals.

The used drawbead has a radius of 6.5 mm and a height of 5.0 mm according to the Numisheet Benchmark 2008 [17]. In the first step, a strip with the geometry of 40 mm x 750 mm for steel materials and 58 mm x 750 mm for aluminium is cleaned with acetone and afterwards lubricated to 2.0 ± 0.1 g/m² by using the spot oil KTL N 16 (Zeller+Gmelin GmbH & Co. KG). The amount of lubrication is controlled by an oil film gauge Infralytic NG2 (Infralytic GmbH) [18]. The strip is then clamped into the drawing unit and the tool is closed with a blank holder force according to the pressure of 7.5 MPa to the projected parallel area. Afterwards, the strip is pulled through the drawbead with a velocity of v = 50 mm/s and the normal force is controlled. The remaining oil is measured on both sides of the preloaded sheet. Afterwards, the preloaded specimen is cut in two parts by a laser cutting. This is done to avoid any damages or scratches by the following strip drawing test on the measurement area. This can also be seen in Figure 2 a).

The one part of the preloaded specimen is examined in the confocal microscope NanoFocus µSurf (Nanofocus AG) with a 20 x objective and a Gaussian filter is applied onto the data. Thus, on the inner and outer side of the drawbead three measurements for each side $n_m = 3 / 3$ with a size of 3 mm x 3 mm are done and averaged for each side. This is applied to testify that there are no deviations over the strip width. This procedure is repeated for at least five specimen $n_s = 5$ and the surface values S_a and V_{Cl} are averaged and can then be compared to the initial state in section 2.

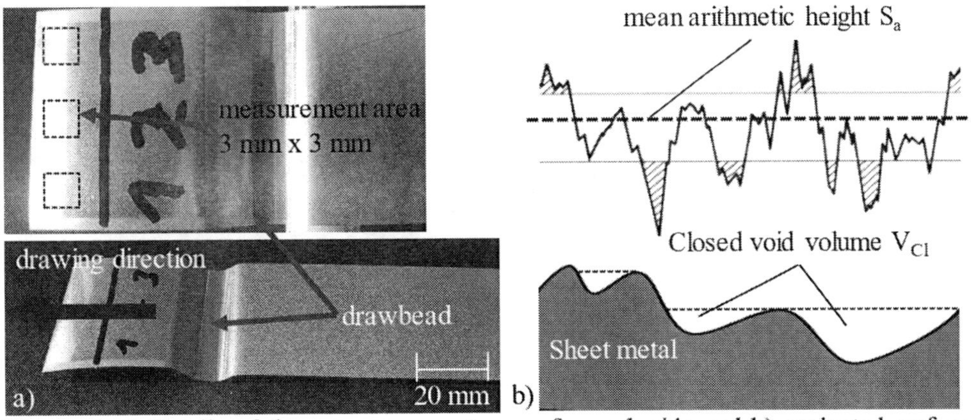

Figure 2. a) Details of three measurement areas for each side and b) evaluated surface parameters mean arithmetic height S_a and closed void volume V_{Cl} according to [16]

Afterwards, conventional flat strip drawing tests are performed to analyse the friction coefficient of the preloaded sheets. This is tested on tool made of 1.2379 with a tool size of 55 mm x 100 mm and the sheet is fully covered by the tool surface. In preliminary studies, it was investigated that for preloaded aluminium heavy abrasion takes place on the edges and the friction coefficient could not be determined. It was decided to laser cut the aluminium strip to a width of w = 40 mm from 58 mm before. In this step, all preloaded material strips have the same width of w = 40 mm. To analyse the influence of the surface changes and separate this from the lubrication system and different remains on the sheets after a drawbead, the strips are cleaned with acetone and re-lubricated with KTL N 16 oil to 2.0 ± 0.1 g/m² as before. This is done with the intention to separate the influence of the lubrication from the surface changes in a drawbead, which is shown in section 4.2. Afterwards, $n_s = 5$ specimens are drawn through the flat tool and the friction coefficient is evaluated according to a test setup following VDA 230-213 [19]. The corresponding blank holder pressure for friction tests is set to p_N = 2.5 MPa to avoid any tendencies to abrasion. The drawing velocity is defined to v = 50 mm/s taking into account the preloading with the same velocity. To avoid further influences of the preloaded edges, they were deburred manually before testing.

4. Results and analysis

In this section, the results of the confocal measurements and the strip drawing tests with preloaded specimen are described. As could be also shown by the authors in [3], the mechanical properties and the retention forces for the identical test setup can be assumed as constant.

4.1. Confocal measurements of sheets preloaded in a drawbead passage

In Figure 3, the analysis of the chosen surface parameters are presented. On the left hand side, Figure 2 a), the arithmetic mean height S_a can be seen in the initial state and after preloading for the in- and outside of the drawbead. In the initial state, a difference between the different materials is visible as noted in Table 1. Also, a smoothing for the steel materials after the drawbead can be seen directly. This smoothing effect is more significant on the inside of the sheet metal that passed the drawbead. For the aluminium alloy AA6014, a constant value of S_a or a kind of roughing of the outside is detected. This is in contrast to the results of the steel materials used.

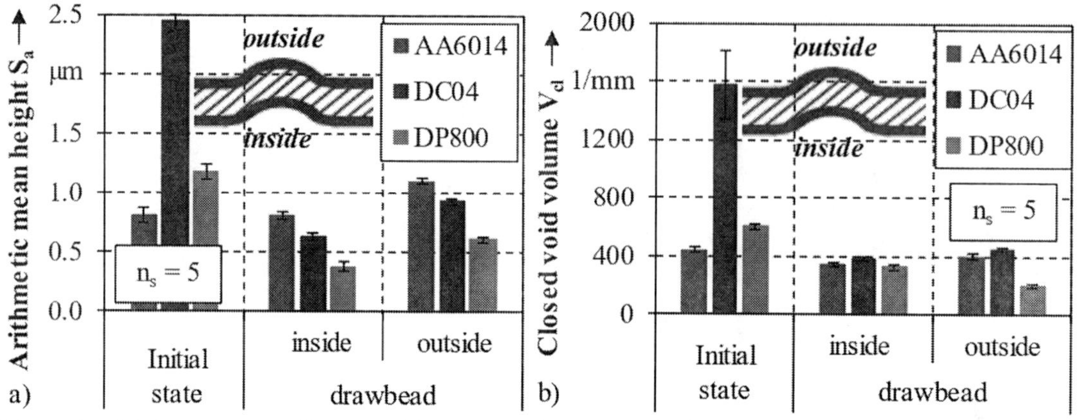

Figure 3. a) Measurement of mean arithmetic height S_a and b) closed void volume V_{Cl} in the original state and after a drawbead preloading

As S_a only shows a global overview of the surface characteristic, in Figure 2 b) the closed void volume V_{Cl} can be compared in initial state and after preloading. This value reflects the ability of sheet metal surfaces to store lubrication. The original state can be correlated to the initial state of S_a, but the development after preloading the sheets is differing. While the S_a value of AA6014 is rising on the outside, the closed void volume V_{Cl} stays nearly constant or is even reduced on the inside. This indicates a loss of the lubrication storage ability with a significant roughening of the global value S_a at the same time.

For the steel materials DC04 and DP800, a significant reduction of the closed void volume V_{Cl} is visible, especially for the AHSS DP800. The oil storage ability becomes lower, in particular on the outside. To give a qualitative comparison in Figure 4, the 3D profile height of the different sheet metal materials in the initial state and after the drawbead preloading is shown.

Especially for the aluminium AA6014, it is obvious that the surface got valleys or fine cracks in the area perpendicular to the rolling and drawing direction. The outside of the drawbead here is affected more, which is also analyzed in the results of S_a and V_{Cl} in Figure 3. The surface profiles for DC04 and DP800 in the 2nd and 3rd row show a significant smoothing, that also certifies the findings before. Also here, a difference between the inside and outside is determined. As conclusion for the confocal microscopy of sheet metals preloaded in a drawbead, it can be stated that AA6014 is roughened and fine valleys appear perpendicular to the drawing direction. For the steel sheets, a smoothing is visible and the lubrication storage ability is for example limited down to 25 % for DC04. This illustrates the surface changes happening in a drawbead.

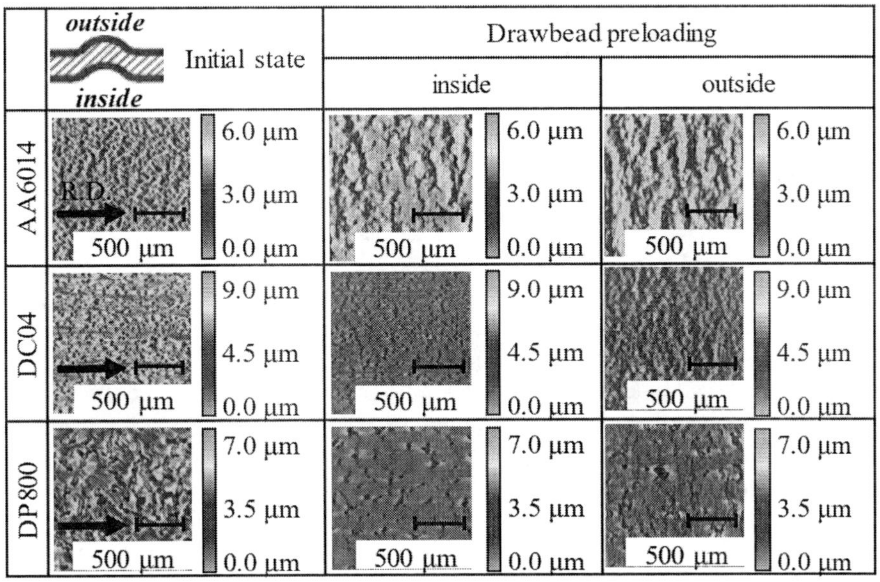

Figure 4. Profile height results of confocal microscopy in original state and after drawbead preloading with respect to rolling direction R.D.

4.2. Strip drawing tests with preloaded sheet metal

Next to the confocal microscopy, strip drawing tests with preloaded specimens after a drawbead passage are performed. Thus, the preloaded specimen are cut to a suitable length. In Figure 5, the corresponding results of the friction coefficient in the initial state and after a drawbead preloading with the recorded test parameters are presented

Drawbead preloading:
$n_s > 5$
$p_N = 7.5$ MPa,
v = 50 mm/s,
Numisheet Benchmark 2008
2.0 g/m² KTL N 16

Strip drawing test
$n_s > 5$
$p_N = 2.5$ MPa, v = 50 mm/s,
w = 40 mm
2,0 g/m² KTL N 16

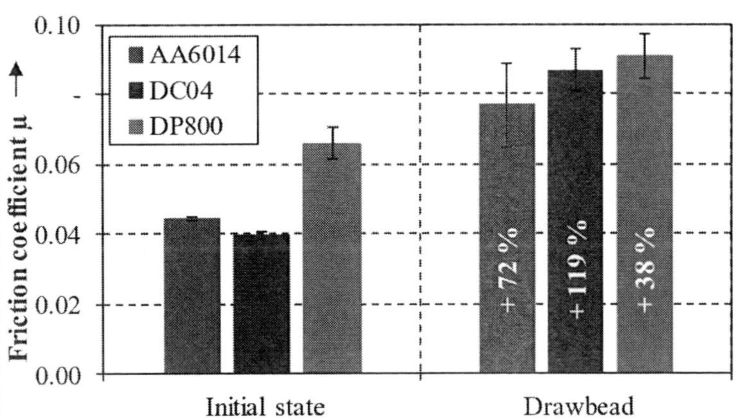

Figure 5. Friction coefficient tested in strip drawing test in the initial state and after a drawbead preloading for three different sheet metals (cleaning an re-lubrication with 2.0 g/m²)

In the initial state on the left side of the diagram, AA6014 and DC04 show nearly the same friction coefficient with about $\mu = 0.04$, while the value of DP800 is located around $\mu = 0.06$. The results for the preloaded specimen finally all show significantly higher values. The standard deviations in between the variations is also increased, which could be a consequence of the whole process with many steps, which can influence the results. In comparison to the initial state, the friction coefficient of AA6014 rises about 72 %. The preloaded mild steel DC04 shows the most significant change and the value is more than doubled to nearly 120 %. Compared to the other materials, DP800 has the smallest increase of the friction coefficient with only 38 %. Anyway, all friction coefficients tested in the strip drawing test after preloading in a drawbead are notably higher than before, although the lubrication was renewed after the drawbead passage. It can be shown that carrying out strip drawing tests without changing the lubrication condition after the drawbead test leads to even greater values of the friction coefficient.

5. Discussion

In this section, a detailed examination of the results presented in the chapters before will be done and possible explanations for the material behaviour shall be found. Comparing the surface parameters in Figure 3 and the friction coefficients after preloading in Figure 5, a correlation between the reduction of the closed void volume V_{Cl} and the modification of the friction coefficients μ is displayed as a function of the ratio parallel normal pressure p_N to the Yield Stress YS.

An important finding is, that the most distinctive reduction of the oil storage ability for the sheet metal DC04 is accompanied with the most significant increase of the friction coefficient. As the lubrication is renewed to 2.0 g/m², the influence of the surface change is isolated. In comparison, the surface parameters of DP800 are not reduced in the same amount. This can also be seen in the lower increase of the friction coefficient after the drawbead passage. It needs to be mentioned, that both steel materials have a zinc coating with 50 g/m² for DC04 or respectively 40 g/m² for DP800. This also has influences on the forming process, which is also mentioned by Murakawa et al. in [20].

Aluminium alloys have an oxide layer, and show under tensing with bending a different behaviour on the in- and outside, which was analysed by Azushima et al. [21]. This behaviour can also be responsible for the roughening of AA6014 surfaces after the drawbead passage.

As the blank holder pressure is responsible for the contact pressure and the tool distance in the drawbead area, a different behavior of the three materials at the same load was expected. As the YS is a characteristic to determine the onset of yielding, a connection between the blank holder pressure, the YS and the material behavior in a drawbead is assumed. This means, the same blank holder pressure p_N is expected to have higher impact on materials with a lower YS, for example in comparison between DC04 and DP800. This assumption was analyzed with the existing data. In Figure 6, the friction coefficient μ is displayed over the ratio p_N / YS (blank holder pressure p_N and yield strength YS).

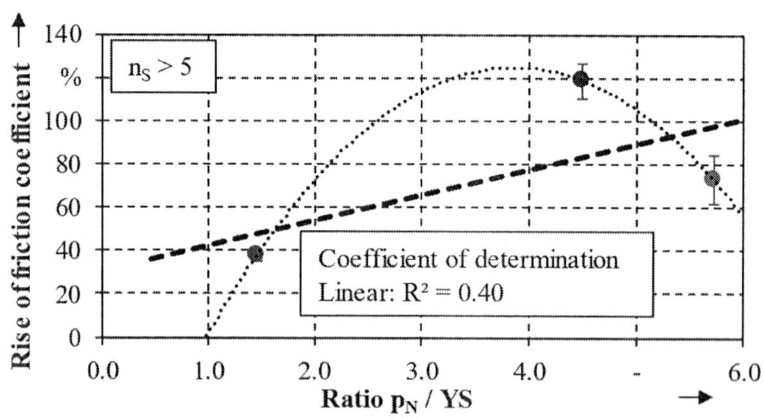

Figure 6. Rise of friction coefficient after drawbead vs. the ratio p_N / YS (blank holder pressure p_N and yield strength YS)

As explained, this ratio is an individual index for the loading for each material in the drawbead. A linear correlation is drawn on the diagram, but shows a poor coefficient of determination with $R^2 = 0.40$ whereas a polynomial in three points, of course, can be fitted perfectly.

There are different explanations for this. First of all, the materials have different coatings and therefore frictional behaviour. The aluminium alloy has an oxide layer and shows a roughening compared to the steel materials with a zinc coating. One more aspect is the combination between the applied blank holder pressure p_N and the resulting gap between the die and blank holder (see also Figure 1). Depending on the sheet metal material, a constant blank holder pressure leads to a corresponding distance between the tools. This, of course is followed by different contact pressure in the radii areas, which are mainly responsible for smoothing or roughening of the materials. On the inside peak of the drawbead, the contact pressure for example is quite high, which explains the higher smoothing for most materials on the inside. Next to this, it needs to be mentioned that the drawing oil was renewed in the strip drawing tests. The assumption of a linear correlation could not be proved here, more experimental data is necessary.

One aspect that was not considered in this evaluation yet is the removal of oil during the forming process in the drawbead. By oil film thickness measurements after the drawbead passage on each side with help of the Infralytic NG 2, it is testified that a significant reduction of lubrication is happening as is displayed in Figure 7. As already mentioned in section 3, the initial lubrication is set to 2.0 g/m² and is measured with a low standard deviation for all materials on the in- and outside before the preloading by a drawbead. For the aluminium alloy AA6014, a reduction of less than 0.2 g/m² was found on the inside. The value on the outside is even a little bit lower. In comparison, the oil film thickness for DC04 is analysed to 1.0 g/m² on the inside and 0.6 g/m² on the outside. For DP800, the same trend occurs with a reduction on the inside to 1.0 g/m² and lower than 0.5 g/m² on the outside. This distribution seems to be quite the opposite of the development in Figure 3, where the smoothing on the inside is higher whereas the oil reduction is more evaluable on the outside. This fact need to be checked with respect to the original oil film thickness after the drawbead passage. Thus, a process without re-lubrication needs to be developed.

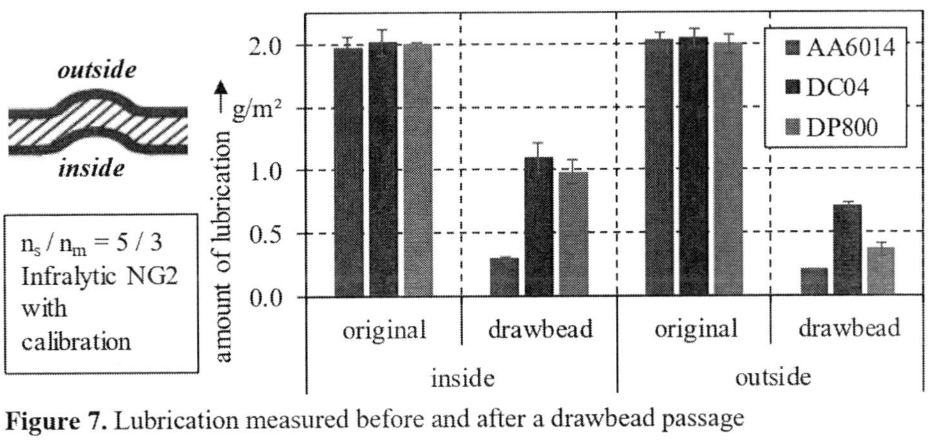

Figure 7. Lubrication measured before and after a drawbead passage

6. Summary and outlook

In this publication, strip drawing tests with drawbead geometry were performed with three different sheet metal materials. Thus, a drawbead according to Numisheet Benchmark 2008 and a height of h = 5 mm was used with a velocity of v = 50 mm/s and a blank holder pressure of p_N = 7.5 MPa. In the next step, the preloaded area of the specimen was examined by confocal microscopy. The surface parameters S_a and V_{Cl} were analysed in the initial state and after preloading in a drawbead. Following

IOP Conf. Series: Materials Science and Engineering **967** (2020) 012008 doi:10.1088/1757-899X/967/1/012008

this, strip drawing tests with preloaded specimen were performed with a re-lubrication to 2.0 g/m² to separate the influence of the surface changes. This process is also schematically shown in Figure 1.

The most distinctive findings in this paper are:

- Steel materials DC04 and DP800 are smoothed after a drawbead passage, especially on the inside. This can be testified by the evaluation of the arithmetic mean height S_a and the closed void volume V_{Cl} in confocal microscopy.
- In contrast, the aluminium alloy AA6014 is roughened by an alternating bending process.
- For all sheet metals, the friction coefficient is increased after a drawbead passage between 39 % and 119 %, although the sheet metal is cleaned and re-lubricated to 2.0 g/m² again.
- This shows the influence of the drawbead preloading on the friction coefficient, which is also emphasized by the surface changes analysed.
- A correlation between the blank holder pressure p_N in combination with the yield strength YS and the friction coefficient μ shows a linear correlation.
- After the drawbead, the lubrication is reduced heavily and is also depending on the side.

It is shown, that a drawbead passage has significant influences on the 3D profile of preloaded surfaces. By separating the lubrication system, influence of the surface variation was shown in subsequently performed strip drawing tests and the evaluation of the friction coefficients. For further investigations, an evaluation with the present oil film thickness after a drawbead is indicated. Thus, the process needs to be re-developed to investigate preloaded specimen with the same conditions. It would be also interesting to analyse variations of the blank holder pressure p_N, the drawbead geometry like the radius or the drawbead height or the drawing velocity. Also, the strip drawing test for friction coefficients can be done at different pressure or velocities to develop more complex friction models. These results could also be implemented in numerical models with drawbeads.

In this paper, the investigations were done with a with a constant pressure p_N or blank holder force F_N, which lead to different gaps between blank holder and die for each material. According to this, a comparison must take this into account. The most interesting finding is the behaviour of each single material compared to its initial state. Another comparison could be done, when the drawbead passage is processed with a constant distance between the tools that guarantees the same geometrical conditions for all material, although this might have other influences. A further analysis of surface parameters and the friction coefficient would help to understand the tribological processes that occur in a drawbead in detail.

Acknowledgement

For the support in the research project EFB 05/217 (AiF 20015N) *"Analysis of the influence of tribological properties on the deep drawing process of sheet metals after passing a drawbead"*, the authors would like to thank the European Research Association for Sheet Metal Working e.V. (EFB) as well as the German Federation of Industrial Research Associations „Otto von Guericke" e.V. (AiF).

The authors would also like to show their gratitude to Mr. Fabian Schmitt, who participated in this experimental study within the scope of his Master's thesis.

7. References

[1] Barthau M, Liewald M and Christian H 2017 *J. Phys.: Conf. Ser.* **896** 12040

[2] Doege E and Behrens B-A 2010 *Handbuch Umformtechnik: Grundlagen, Technologien, Maschinen (VDI-Buch)* 2nd edn (s.l.: Springer-Verlag)

[3] Schmid H and Merklein M 2019 *Manufacturing Rev.* **6** 14

[4] Jacobson B 2003 *Tribology International* **36** 781–9

[5] Zöller F, Sturm V and Merklein M 2015 *KEM* **639** 403–10

[6] Hol J, Meinders V T, Geijselaers HJM and van den Boogaard A H 2015 *Tribology International* **85** 10–25

[7] Filzek J 2004 *Kombinierte Prüfmethode für das Reib-, Verschleiß- und Abriebverhalten beim Tief- und Streckziehen (Berichte aus Produktion und Umformtechnik* vol 62*)* (Aachen: Shaker)

[8] Christiany M and Groche P 2014 *AMR* **1018** 293–300

[9] Ludwig M 2017 *Bewertung von Einflussgrößen auf die Reibverhältnisse in der Blechumformung und deren Korrelation mit numerisch ermittelten Belastungsverläufen (Berichte aus Produktion und Umformtechnik* Band 106*)* (Aachen: Shaker Verlag)

[10] Leocata S, Senner T, Reith H and Brosius A 2020 *Int J Adv Manuf Technol* **286-287** 66

[11] Hol J Multi-scale friction modeling for sheet metal forming *PhD Thesis - Research UT, graduation UT* University of Twente

[12] Hol J, Wiebenga J H, Hörning M, Dietrich F and Dane C 2016 *J. Phys.: Conf. Ser.* **734** 32092

[13] Bloeck M 2012 Aluminium sheet for automotive applications *Advanced Materials in Automotive Engineering* (Elsevier) pp 85–108

[14] VDA Verband der Autoindustrie - German Association of the Automotive Industry 2016 Flacherzeugnisse aus Stahl zur Kaltumformung *(*VDA 239-100*)* (VDA)

[15] DIN Deutsches Institut für Normung e. V. 2012 Geometrical product specifications (GPS) – Surface texture: Areal – Part 2: Terms, definitions and surface texture parameters *(*DIN EN ISO 25178-2, 2012-09,*)* (Berlin: Beuth Verlag GmbH)

[16] Pfestorf M, Staeves J and Wagner K *Stahl und Eisen* **117** 89–94

[17] Hora P (ed) 2008 *Numisheet 2008: International Conference and Workshop on Numerical Simulation of 3D Sheet Metal Forming Processes ; September 1 - 5, 2008, Interlaken, Switzerland* (Zurich: ETH)

[18] Infralytic GmbH 2015 *Bedienungsanleitung Ölauflagensensor NG2: Version 2.1.3de , http://www.infralytic.de*

[19] VDA Verband der Autoindustrie - German Association of the Automotive Industry Prüfverfahren für die Produktklassen Prelube, Prelube 2, Hotmelt, Spot lubricant VDA 230-213

[20] Murakawa M, Koga N, Watanabe S and Takeuchi S 1998 *Surface and Coatings Technology* **108-109** 425–30

[21] Azushima A and Sakuramoto M 2006 *CIRP Annals - Manufacturing Technology* **55** 303–6

Determination of the properties of semi-finished parts in blanking processes

S Wiesenmayer, P Frey, M Lechner and M Merklein

Institute of Manufacturing Technology (LFT), Friedrich-Alexander University Erlangen-Nürnberg (FAU), Egerlandstraße 13, 91058 Erlangen, Germany

sebastian.wiesenmayer@fau.de

Abstract. Deviations of the properties of semi-finished sheet metal parts affect the finished part's geometry and its properties. With knowledge of the deviations, forming processes can be adjusted to a certain degree in order to maintain the part quality. Typically, in mass production, the sheets are blanked before deep drawing. Within the blanking processes, the necessary cutting force and cutting stroke are both influenced by the properties of the material. Thus, by the analysis of the process data, which is obtained in blanking processes, deviations of the properties of semi-finished parts could be determined. Within the scope of the work, it is analyzed to which degree process data, which is gained during blanking processes, can be used for the determination of the properties of semi-finished parts and their deviations. For this purpose, a blanking process is investigated for two different batches of the dual phase steel DP600 with a nominal sheet thickness of 2.0 mm and varying mechanical properties. In addition, the sheet thickness of the specimens is reduced by 5 and 10 % for one batch. The recorded process data from the blanking process is correlated with the properties of the sheet materials, which are determined with uniaxial tensile tests.

1. Introduction

The properties of semi-finished parts are subject to deviations, which influence forming processes as well as previous and subsequent manufacturing steps. Thus, they determine the properties and the quality of finished components. With regard of the increasing complex geometries and thinner wall thicknesses of car body parts, process windows become smaller and forming processes more challenging [1]. Knowledge of the deviations and their influence on the processes and part properties allows the use of adaptive tool systems for the compensation in order to maintain the part quality and avoid rejection. Requirements are the identification of the deviations and the data exchange between the different manufacturing steps. However, by common destructive characterization methods, the properties of semi-finished parts can only be determined for random samples and extrapolated, based on statistics, as the properties cannot be measured continuously in mass production. An approach for the non-destructive inline determination of mechanical properties is the use of eddy current [2]. Since the microstructure of the material influences its electromagnetic and mechanical properties, these characteristics can be correlated [3]. However, next to additional measurement equipment, mathematical models are needed for the correlation of the eddy current and the mechanical properties [2]. Moreover, the calibration of the setup is necessary for each material. Another approach is the use of data, which is generated during the production process. Until now, processes are mostly monitored individually, while the obtained data remains unused [4]. In [5] it was principally shown how information on the tensile strength of the sheet materials can be gained during blanking and bending processes and used for the adaption of the blank

holder force in a deep drawing operation. However, only maximum forces from the blanking and the bending process were considered in order to calculate the tensile strength of the used materials. Yet, for a full process control, knowledge of the flow behavior and the ductility as well as the sheet thickness of the used materials is also important.
Blanking the initial shape of the semi-finished parts from a coil is often the first step in the manufacturing chain of formed sheet metal components. Thus, within the scope of this work, it is analyzed to which degree data, which is gained during blanking processes, can be used for the determination of the properties of semi-finished parts and their deviations. For this purpose, a round hole punching process is investigated, which is typical for the production of holes for locating pins in progressive dies. The mechanical properties of the used material DP600, which is common within the automotive industry for deep drawn complex structural parts, are determined with uniaxial tensile tests. Afterwards, the properties are correlated with the force-displacement curves, which are obtained during the blanking process. While process parameters like the cutting clearance are held constant, the material properties and sheet thicknesses are varied in order to determine the influence of deviations of the properties.

2. Materials and setup

Within the investigation, a round-hole punching process is analyzed. The experiments are conducted on a universal testing machine RM400 from Schenck-Trebel with a punch velocity of 100 mm/min. The blanking tool consists of a punch, which has a diameter of 10 mm, a blank holder and a die with an inner diameter of 10.4 mm. Therefore, a cutting clearance of 0.2 mm is set, which equals a relative cutting clearance of 10 % for the investigated sheet thickness of 2.0 mm. The blank holder is connected via springs with the inner punch. The springs have a stiffness of 2200 N/mm and a pre-load of circa 0.7 kN. The specimens are quadratic and have a width of 45 mm.
A force-displacement curve, which is gained in the round-hole punching process, is depicted in figure 1. In the initial phase (1), the sheet is only elastically deformed. The force, which is necessary for the plastic deformation of the material depends on its properties and the geometry of the cutting line. Due to work hardening, the force increases during the actual cutting phase (2). In this phase, the maximum of the process force F_{max} is reached and cracks are initiated. With further movement of the punch, the force slightly decreases as the cracks propagate, until the remaining cross section breaks and the force instantly drops. Afterwards, the cut out slug is pushed out from the die (3).

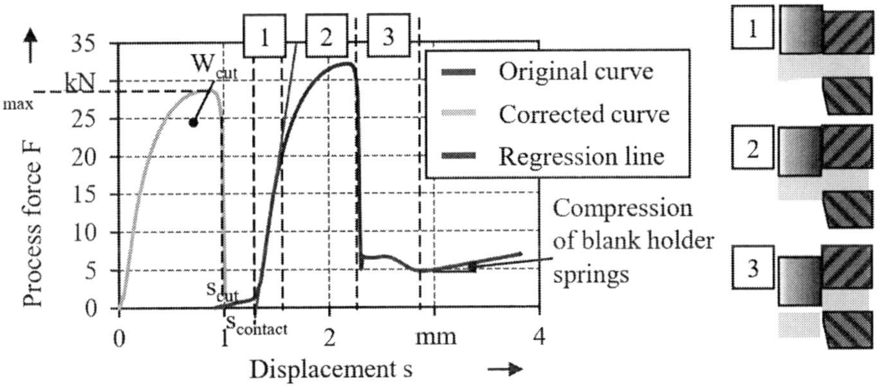

Figure 1. Force-displacement curve of a round hole punching process with blank holder

Due to the tool setup, the applied blank holder force also contributes to the measured process force. In order to extract the actual cutting force and work, the force-displacement curves are corrected. The linear increase of the force, which is caused by the compression of the blank holder springs as well as the pre-load of the springs are subtracted from the curve. Moreover, the curves are only considered until the actual cut, which is indicated by the steep decline of the force. The phase, in which the slug is pushed through the die, is not taken into account for the calculation of the cutting work W_{cut}. Next to the maximum cutting force $F_{cut\ max}$, the displacement at the cut s_{cut} and the cutting work W_{cut}, the punch displacement $s_{contact}$, which marks the initial contact between the punch and the sheet is also taken into account for the correlation. In order to determine $s_{contact}$ from the original, uncorrected force-displacement curves, a regression line is defined for the initial elastic deformation in a section, where the process force is between 4 and 12 kN and linear behavior can be assumed. Therefore, influences from the initial compression of the blank holder springs before the contact of the punch and the sheet and the plastic deformation of the specimen for higher forces are avoided. The intercept of the line and the x-axis is defined as $s_{contact}$ and marks the starting point of the displacement for the corrected curves. The correlation of the maximum cutting force $F_{cut\ max}$ and the properties of the semi-finished parts is based on equation (1), which is commonly used for the calculation of blanking processes [6]. The tensile strength TS, the length of the cutting line l as well as the thickness t are considered with a linear relationship in the equation. The equation is based on empirical observations and shows a good correlation, although material properties from the tensile tests are obtained under a uniaxial stress state, while blanking is dominated by a multiaxial stress state. The shearing constant c accounts for the dissimilar forming behavior of different materials and has to be determined experimentally in order to increase accuracy. Beside the influence of the material, the shearing constant c is mixed with influences of the sheet thickness, the cutting clearance and tool wear [6]. It is therefore not possible to separate those influences and determine a general shearing constant c for each material without considering the circumstances of the blanking process. However, for the estimation of cutting forces a shear factor c in the range of 0.6 is applied for harder materials whereas 0.8 is used for ductile materials [6].

$$F_{cut\,max}=t \cdot l \cdot c \cdot TS \tag{1}$$

For the investigation, the cold rolled dual phase steel DP600 with a nominal sheet thickness of 2.0 mm is used. For the analysis of the influence of deviations of the mechanical properties of the semi-finished components on the blanking process, two different batches of DP600 with zinc coating are used. In order to simulate the variation of the sheet thickness while maintaining the mechanical properties, specimen from batch A with an original thickness of 2.0 mm were grinded to 1.9 mm and 1.8 mm. Thus, deviations of the mechanical properties due to different material batches can be avoided. With this method it is intended to separate influences of the sheet thickness and the material batch on the blanking process. However, due to grinding the zinc coating was removed on both sides from the specimens of batch A with a thickness of 1.9 mm and 1.8 mm. The true stress-true strain curves in figure 2 and the mechanical properties of the materials, which are displayed in table 1, were determined with uniaxial tensile tests according to DIN EN ISO 6892. The samples from batch A have a similar yield strength and elongation at break. However, despite the almost equal uniform elongation, the grinded specimen with a thickness of 1.9 mm and 1.8 mm have a slightly higher tensile strength and hardening behavior than the original material with a thickness of 2.0 mm which is attributed to the removed zinc coating and favorable surface conditions of the grinded specimen. Considering the standard deviation, the properties of the specimens with a thickness of 1.9 mm and 1.8 mm are equal. Batch B has a distinctly higher strength, a higher work hardening and a lower ductility than batch A.

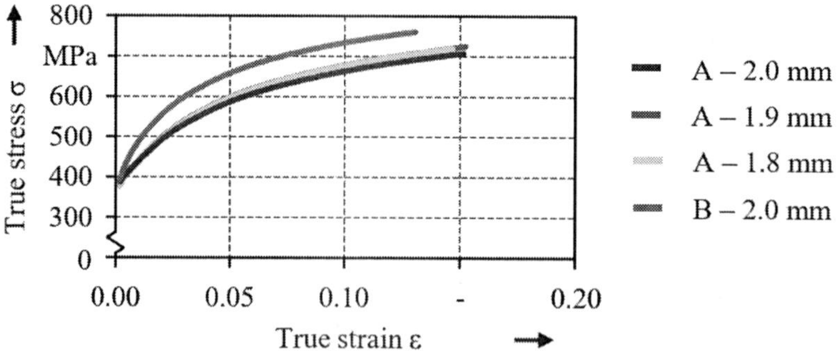

Figure 2. Exemplary true stress-true strain curves of used materials

IOP Conf. Series: Materials Science and Engineering **967** (2020) 012009 doi:10.1088/1757-899X/967/1/012009

Table 1. Mechanical properties of used materials (n = 5)

Batch	Sheet thickness (mm)	Yield strength (MPa)	Tensile strength (MPa)	Uniform elongation (%)	Elongation at break (%)
A	2.0	379 ± 3	604 ± 4	16.5 ± 0.3	25.0 ± 0.5
A	1.9 (grinded)	378 ± 1	623 ± 2	16.2 ± 0.2	24.5 ± 0.7
A	1.8 (grinded)	378 ± 3	622 ± 4	16.1 ± 0.2	24.6 ± 0.3
B	2.0	391 ± 2	666 ± 2	13.9 ± 0.1	21.7 ± 0.2

3. Results and discussion

In figure 3 the corrected force-displacement curves of all conducted experiments are shown. For the variation of the sheet thickness as well as for the different batches, there is no significant difference in the elastic deformation behavior in the beginning of the process. However, higher forces are necessary for the plastic deformation of sheets with a higher thickness due to the larger shear area. In addition, specimen with smaller sheet thicknesses are cut at lower punch displacements. When the mechanical properties are considered, there is also no significant difference in the initial elastic behavior. However, for batch B, which has a higher initial flow stress and work hardening behavior, higher forces are necessary for the plastic deformation of the sample. Due to the reduced uniform elongation and elongation at break of batch B, the specimens are cut at lower punch displacements. The extracted information from the force-displacement curves regarding the properties of the semi-finished parts and the limits of the accuracy are discussed in the following.

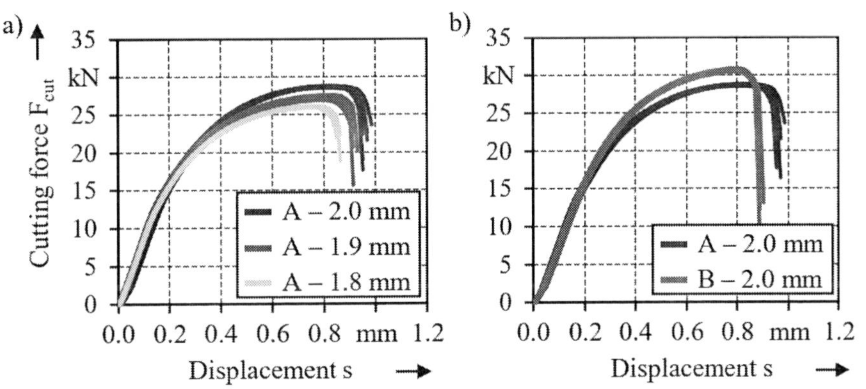

Figure 3. Force-displacement curves in dependence of the sheet thickness (a) and different mechanical properties (b)

In figure 4 a detailed analysis of the influence of the properties of the semi-finished parts on the maximum process force is given. The maximum process force increases almost linearly with the sheet thickness with a coefficient of determination of R^2 = 97.7 % which is in good accordance with the linear relationship postulated in equation (1). Since changes in thickness and mechanical properties occur gradually in rolled sheet metal coils it is appropriate to compare the mean values of a sample. Due to the compensation of random scatter, the differences are more robust compared to single values. In this study, a sample size of n = 10 is used. Therefore, the sheet thickness and its deviations can be correlated with the maximum process force. Deviations from this linear correlation can be attributed to a change in mechanical properties which is shown in figure 4 b).

The maximum force increases from 28.7 ± 0.1 kN for batch A, which has a tensile strength of 604 ± 4 MPa to 30.7 ± 0.2 kN for batch B, which has a tensile strength of 666 ± 2 MPa. Therefore, deviations of the mechanical properties, in particular of the tensile strength, also lead to a change of the maximum process force, which allows the correlation of the tensile strength and its deviations and the maximum process force. The mechanical properties were determined with uniaxial tensile tests with a representative sample size of n = 5 due to the low scattering of the experiments. For the maximum cutting force a significant difference of more than 0.235 kN at a reliability level of 99.9 % can be detected for a sample size of n = 10. However, the cutting constant c varies in dependence of thickness and other conditions, thus no satisfying linear correlation can be derived.

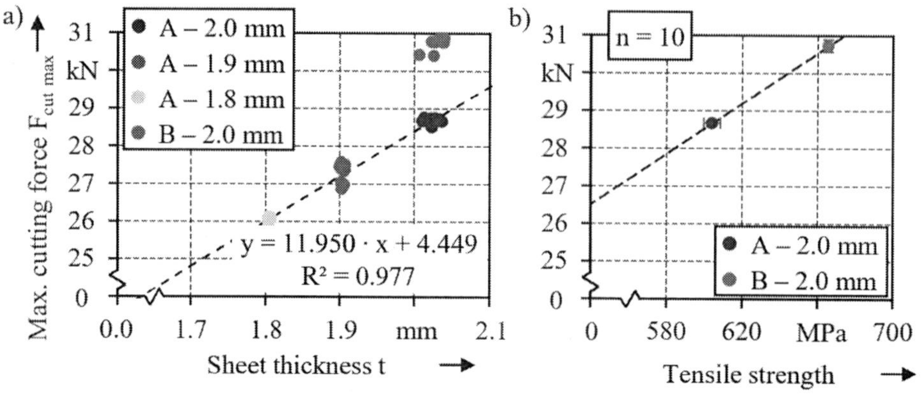

Figure 4. Maximum process force in dependence of sheet thickness (a) and the tensile strength (b)

Another factor, which is influenced by the properties of the semi-finished parts, is the punch displacement at the actual cut. It therefore also allows the correlation of the gained data with the properties of the semi-finished parts and their deviations. Figure 5 displays the cutting stroke s_{cut} in dependence of the semi-finished part's properties, which gives an estimate of the fracture behavior. A longer cutting stroke indicates higher formability.

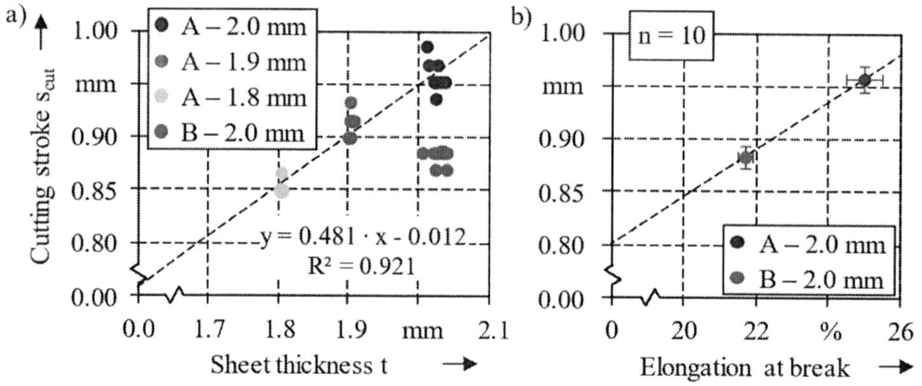

Figure 5. Punch displacement at cut in dependence of sheet thickness (a) the elongation at break (b)

The necessary cutting stroke s_{cut} increases almost linearly with the thickness of the sheet, with a mean standard deviation of ± 0.0105 mm. Therefore, the cutting stroke s_{cut} also allows the determination of the sheet thickness and its deviations through correlation with the obtained process data. However, in case of the cutting stroke, a distinct effect can be detected for a difference of 0.039 mm at a reliability level of 99.9 %. Requirement for the determination of the sheet thickness is also the knowledge on the mechanical properties of the semi-finished part, since the cutting stroke is also influenced by the elongation at break of the material. Therefore, for batch B, which has a 13 % lower elongation at break, the cutting stroke is reduced by 8 % in comparison to batch A. The low scattering of the data allows the correlation of the cutting stroke and the ductility of the material. The third factor, which can be deducted from the force-displacement curve, is the cutting work W, which equals the integral of the process force F over the punch displacement s. The comparison in figure 6 a) for the different sheet thicknesses from batch A, shows a linear increase of the cutting work with the sheet thickness, which allows the correlation of the factor with the sheet thickness and its deviations. When the specimen from batch A and B with a thickness of 2.0 mm are compared, there is no significant difference between the batches regarding the cutting work. Higher forces are necessary in order to cut the samples from batch B. However, due to the lower elongation at break, the specimen are cut at lower displacements, which results in an almost equal work for both batches in figure 6 b). Therefore, no significant information can be derived for different mechanical properties.

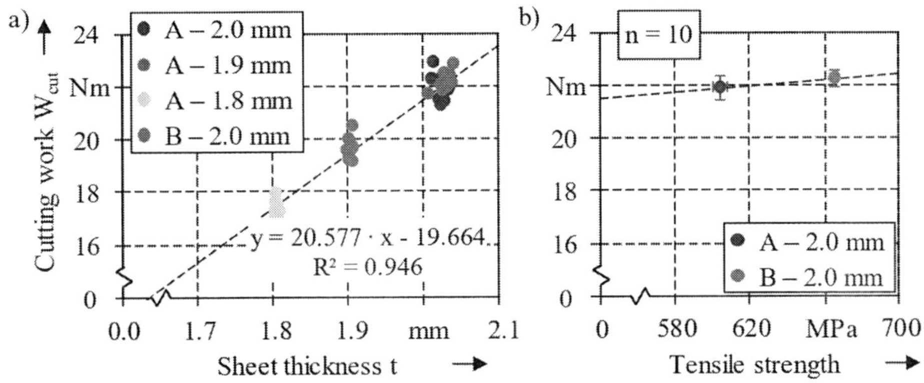

Figure 6. Cutting work in dependence of sheet thickness (a) and the tensile strength (b)

Since the factors maximum cutting force, cutting stroke and cutting work depend on the sheet thickness and the mechanical properties of the material, knowledge on either of the properties is required in order to determine the other. A simple approach to determine deviations of the sheet thickness is the consideration of punch displacement at the initial contact of the punch and the sheet $s_{contact}$. For the different materials, the displacement at the contact was determined with the original, uncorrected curves and the presented approach (figure 7). While for the specimen of batch A with a nominal sheet thickness of 2.0 mm a displacement at the contact of 1.304 ± 0.005 mm was measured in average, this value increases to 1.413 ± 0.002 for the specimen with 1.9 mm thickness and to 1.503 ± 0.003 mm for the samples with 1.8 mm thickness. Therefore, the difference in the punch displacement at $s_{contact}$ correlates with the difference of the sheet thicknesses. For the specimen of batch B an average $s_{contact}$ of 1.275 ± 0.006 mm was measured. When the displacement $s_{contact}$ is calibrated for one sheet thickness, the actual sheet thickness can be recalculated for each specimen. Given the variance of the recalculated sheet thickness, deviation of more than 0.008 mm of the mean value of a sample size of n = 10 can be detected with a reliability level of 99.9 %.

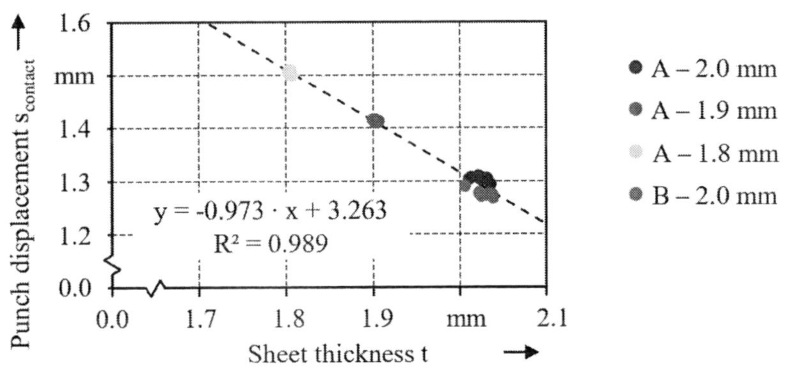

Figure 7. Punch displacement at contact with the sheet in dependence of the sheet thickness

With knowledge of the sheet thickness, deviations of the mechanical properties can be determined. Therefore, the process force is normalized with the cutting area, which equals the product of the length of the cutting line l and the sheet thickness t in figure 8 a). For this purpose equation (1) is rearranged, resulting in equation (2), which allows the calculation of the cutting resistance, which equals the product of the shear constant c and the tensile strength TS.

$$c \cdot TS = \frac{F_{cut\,max}}{t \cdot l} \tag{2}$$

For the normalization, equation (2) is also used for the currently prevailing process force and not only for the maximum $F_{cut\ max}$. Following the approach in uniaxial tensile tests, the force is related to the initial cutting area. The deformation induced reduction of the cross section is not considered. Also, analogous to the determination of the yield strength in uniaxial tensile tests, the end of the first phase and therefore the beginning of the plastic deformation of the specimens is determined with the intercept of the corrected force-displacement curve and a regression line for the elastic deformation, which was parallel shifted. For the shift a value of 0.01 mm was chosen.

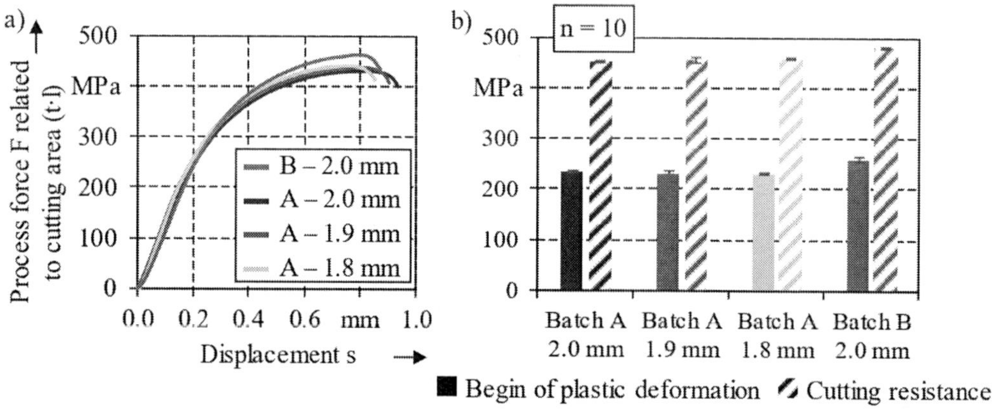

Figure 8. Normalized force-displacement curve (a) and properties of the semi-finished part (b)

For the beginning of the plastic deformation of the sheets, there is almost no difference between the different specimens from batch A. All three sheet thicknesses are plastically deformed at normalized forces of 232 to 234 MPa, which is in good accordance with the tensile tests, which have shown, that despite the grinding process, all sheet thicknesses of batch A have a similar yield strength. In contrast, for specimens from batch B, which has a higher yield strength, an average stress of 260 ± 5 MPa is necessary for the plastic deformation of the sheet. Therefore, there is a correlation between the beginning of the plastic deformation of the sheets in the blanking process and the yield strength. However, the deviation of the flow stress between the batches cannot be quantified reliably, as for batch B, a 12 MPa higher yield stress leads to a 27 MPa higher begin of the plastic deformation in the blanking process, which contradicts the calculated relation of the normalized process force and the yield strength of circa 232 MPa / 378 MPa = 0.62 to 260 MPa / 391 MPa = 0.66. The cutting resistances, which equal the maximum cutting force $F_{cut\ max}$ related to the cutting area, differ slightly for the different sheet thicknesses from batch A. This is in good accordance with the findings of [6], as a smaller sheet thickness leads to an increase of the shear constant c. For batch B, which has a 60 MPa higher tensile strength, a 31 MPa higher cutting resistance of 482 ± 2 MPa was determined. The reason is that the shear constant c is also influenced by the ductility of the material, which is lower for batch B. Therefore, with knowledge of the sheet thickness deviations of the mechanical properties can be determined. However, the accuracy is also limited, as the shear factor c is influenced by the properties of the semi-finished part and therefore not constant.

4. Conclusion

Within the scope of this work it was analyzed to which degree information on the properties of semi-finished parts can be obtained from process data, which is gained during a blanking process. Considering the information, which was obtained from the force-displacement curves, the following statements on the correlation of the data with the sheet thicknesses as well as with the mechanical properties of the semi-finished parts and their deviations can be made:

- The sheet thickness and its deviations indicates a linear correlation with the maximum cutting force, the cutting stroke and the cutting work. These factors are interdependent with the mechanical properties which can lead to deviation of the linear correlation, which indicates a variation of the properties. However, for the cutting work no a significant deviation is observed neither for the linear correlation nor for the direct comparison, thus rendering this parameter inappropriate for the conclusion of mechanical properties.
- An independent factor without influences of the mechanical properties is the displacement at contact of punch and sheet, which is based on the increase of force at contact. The derived displacement at contact can be used to recalculate the sheet thickness. However, this requires the calibration of the system based on known sheet thicknesses of one sample group. Without calibration at least deviations of the sheet thickness can be obtained with high accuracy.
- Deviations of the ductility of the material can be correlated with the cutting stroke and changes of the strength with the process force. In comparison to the sheet thickness, the sensitivity is reduced.
- With knowledge of the sheet thickness, the flow behavior of the sheet material can be estimated from the normalized process force-displacement curve.
- The accuracy of the determination of deviations and their quantification are limited due to the interdependence of the sheet thickness and the mechanical properties.
- Due to the gradual variation of properties of sheet metal, a moving average is suitable in order to increase robustness of the measured effects and to discover small variations.

Since the force-displacement curve is influenced by the mechanical properties and the sheet thickness, knowledge of one of the characteristics is required in order to determine the other reliably. A simple approach for the determination of the sheet thickness and the mechanical properties was shown. In future investigations the presented approach for the determination of the properties of semi-finished sheet metal parts and their deviations will be investigated for further materials and sheet thicknesses.

Acknowledgements
The authors gratefully acknowledge the support of the European Research Association for Sheet Metal Working (EFB) and the project-related committee for funding the project "Konzeption einer adaptiven Prozesskette für das mechanische Fügen - Mechanisches Fügen 4.0" (EFB 02/217).

References
[1] Hora P, Heingärtner J, Manupolo N, Tong L, Hortig D, Neumann A and Roll K 2011 On the Way from an Ideal Virtual Process to the Modelling of the Real Stochastic, *4th Forming Technology Forum 2011* (Zurich: ETH Zürich) pp 3-14

[2] Heingärtner J, Born M and Hora P 2010 Online Acquisition of Mechanical Material Properties of Sheet Metal for the Prediction of Product Quality by Eddy Current *10th European Conf. on Non-Destructive Testing (ECNDT 2010)* vol 3 (Red Hook: Curran)

[3] Ružovič M 2004 Die zerstörungsfreie Ermittlung von genauen Zugversuchsdaten mit dem Wirbelstromverfahren (Zurich: ETH Zürich)

[4] Hagenah H, Schulte R, Vogel M, Hermann J, Scharrer H, Lechner M and Merklein M 2018 4.0 in metal forming – questions and challenges *Procedia CIRP 12th CIRP Conference on Intelligent Computation in Manufacturing Engineering* ed R Teti and D M D'Adonna (Gulf of Naples) vol 79

[5] Schulte R, Frey P, Hildenbrand P, Vogel M, Betz C, Lechner M and Merklein M 2017 *J. Phys.: Conf. Ser.* **896** 012037

[6] Spur G and Stöferle T 1985 *Handbuch der Fertigungstechnik Band 2/3 Umformen und Zerteilen* ed G Spur (Munich, Vienna: Carl Hanser Verlag) chapter 9.2 pp 1384-96

IOP Conf. Series: Materials Science and Engineering **967** (2020) 012010 doi:10.1088/1757-899X/967/1/012010

The effect of friction and lubrication modelling in stamping simulations of the Ford Transit hood inner panel: a numerical and experimental study

S Berahmani[1], C Bilgili[2], G Erol[2], J Hol[1] and B Carleer[3]

[1] TriboForm Engineering, Hengelosestraat 500, 7521 AN, Enschede, the Netherlands
[2] Ford Otosan-Gölcük, Akpınar Mahallesi Hasan Basri Caddesi No:2 Sancaktepe İstanbul, Turkey
[3] AutoForm Engineering Deutschland GmbH, Joseph-von-Fraunhofer-Straße 13a, D-44227 Dortmund, Germany

E-mail: s.berahmani@triboform.com

Abstract. Reducing weight of Body in white (BIW) is a well-known solution to tackle environmental issues in automotive industry. Replacing steel by lighter aluminium sheet material is an accepted alternative. However, this trend brings challenges like the forming of Aluminum into complex car body parts. These challenges signify the necessity of more advanced tools to obtain defect-free parts. Nowadays, it is a common practice to implement forming simulations at the early stage of part design in order to identify critical features and to optimize the production process. Although, it is known that friction and lubrication conditions are one of the most influential sources affecting product quality in aluminium sheet metal forming, it is currently not considered in detail in stamping simulations. The current application case focuses on the role of friction and lubrication modelling in stamping simulations of an aluminium hood inner part of the new Ford Transit. A comparison was made between different friction models to identify and resolve split issues in the forming simulations. Simulation results were used to guide try-out towards efficient and successful elimination of these issues on the physical panels. Finally, different possible scenarios corresponding to production conditions were investigated to define a robust process window.

1. Introduction

At present, the automotive industry is facing several trends, particularly the increased use of aluminium and thinner sheet materials. The trends are geared towards light weighting, and they bring challenges, including the fact that forming thinner sheet materials and aluminium is difficult. These challenges signify the necessity for more advanced tools to obtain defect-free parts. Nowadays, it is a common practice to implement forming simulations at the early stage of part design before try-out to identify critical features in the design and to optimize the forming process. Simulations are used to support and get through the try-out phase without issues. Series production will be initiated only if the try-out phase is successful.

Although it is known [1] that friction and lubrication conditions are two of the most influential sources affecting product quality in aluminium sheet metal forming, it is not considered in detail in stamping simulations. The current industrial standard is to use a constant coefficient of friction. This

limits the overall accuracy of the forming-process simulation. It is also known that the friction conditions are not only affected by the properties of the sheet metal and tooling, but also by the process conditions, such as sliding velocity, contact pressure, temperature and lubrication type and amount.
This paper describes a selection of results of an application case by Ford Otosan examining split issues observed in the try-out phase of the Ford Otosan Transit hood inner panel. Figure 1 shows the all new Ford Transit. The respective part and stamping process is described in detail in Section 2. The simulation approach followed to predict the split issues is described in Section 3, followed by a sensitivity study to identify a robust process window in reality. In Section 4, it is explained how split issues were tackled in simulations and subsequently in try-out together with a summary of the sensitivity analysis. Finally, a conclusion and points of future work are given in Section 5.

Figure 1. The all new Ford Transit

2. Stamping process of the Ford Transit hood inner panel

2.1. Tribology system: sheet material, lubricant and die material

The combination of the utilized sheet material, lubricant type and die material is referred to as the tribology system. The hood inner part used for this study is shown in Figure 2. The sheet material used is an Aluminum 6xxx series. The sheet material is pre-lubricated with a hotmelt lubrication. An average pre-applied lubrication amount of 1.2 g/m^2 was considered. The utilized tooling material is GGG70L. The die surfaces are polished and hardened. It is assumed that the surface properties of the blank resemble a commonly used AA6xxx series. The surface topography of the sheet was captured using 3D confocal microscopy measurements and was seen to have a surface roughness of 1.0 μm. An average tooling roughness of Sa = 0.4 μm was assumed using a common surface topography from the same tooling family.

Figure 2. Hood inner part with critical regions showing splits (red boxes)

2.2. Stamping process and observed try-out issues
This highly complex part was stamped using a single action process in the Ford Otosan press shop in Kocaeli, Turkey. The formed hood inner part is shown in Figure 2, whereby the critical regions are indicated in the red boxes. At first it was assumed that these splits must be a tryout issue, and that the problems had arisen from the die spotting level etc. However, after several trials to solve the issue by checking the dies at the press, the same result continued over a couple of weeks period.

3. Stamping simulations

3.1. Sheet metal forming simulation approach
This section describes the simulation approach followed in this work, as visualized in Figure 3. The sheet metal forming simulations in this study were all performed with AutoForm plus R7.0.3. After observing the split issues, Ford performed a full investigation to detect the reason of discrepancy. In their evaluation, material properties, blank size, punch force, and die spotting level were compared with reality, and were confirmed to show good agreement between reality and simulation. When additional oil was applied to the sheet some sensitivity was observed which highlighted the influence of tribological behavior on the aluminum part quality.

3.2. Simulation of friction and lubrication conditions
Initial simulations were performed with a constant Coulomb coefficient of friction of $\mu = 0.12$, which is commonly used in industry to describe frictional behavior of parts from Aluminum. that In the second step, a pressure-velocity dependent friction model, referred to as P-V dependent friction model, was adopted. It was out of scope of this study to develop a P-V dependent friction model based on the experimental data so the settings of the P-V dependent friction model were based on the default settings provided by AutoForm with a base friction coefficient of 0.12 [2]. The model takes into account the decrease of friction coefficient due to an increase in contact pressure and increase in relative sliding velocity between the sheet and the contacting tools. The effective coefficient of friction μ_{eff} is calculated according to the formula:

$$\mu_{eff} = \mu\left(\frac{p}{p_{ref}}\right)^{e-1} - a\, ln\frac{\max\left(v_{rel}, v_{ref}\right)}{v_{ref}}$$

Where

- μ - base friction coefficient; in this case 0.12
- p - contact pressure
- p_{ref} – reference pressure (Default value of 4.0 MPa)
- e – pressure exponent (Default value of 0.9)
- a – velocity factor (Default value of 0.015)
- v_{rel} – velocity of the sheet relative to the tool in contact
- v_{ref} – reference velocity (Default value of 10 mm/s)

Finally, to more accurately account for friction and lubrication conditions in sheet metal forming simulations, the TriboForm software was used. See Figure 3.

The TriboForm software allows for multi-scale modeling of a time and spatially varying friction coefficient under a wide range of process conditions [3, 4]. The tribology system specification as described in the Section 2.1 (sheet material: Aluminum 6xxx series; hotmelt lubrication; tooling material: Cast iron, GGG70L), combined with hardening behavior of AA6xxx and viscosity data of the lubricant, enables the creation of a TriboForm Library which includes the friction conditions for the considered tribology system according to the procedure described in [5]. The resulting TriboForm friction models are a function of local contact pressure, straining of the sheet material, relative sliding velocity and interface temperature. The friction model can be imported into the AutoForm software using the TriboForm FEM Plug-In, replacing the constant coefficient of friction. A more detailed description of this simulation approach can be found in [3, 4, 5].

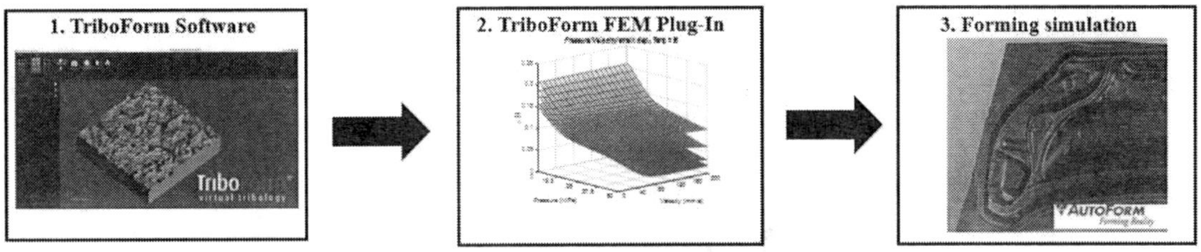

Figure 3. Schematic of simulation approach for friction and lubrication modelling in sheet metal forming simulations

3.3. Parameter variations as encountered in press-shop reality

A sensitivity analysis was performed on the final part, whereby the part has been simulated with different levels of lubrication amounts and tool roughness values. By setting these different tribological conditions in TriboForm, and performing AutoForm simulations with the resulting friction models, it is possible to simulate the forming process including variations that can occur in press-shop reality. The first TriboForm model was defined to simulate low friction conditions, i.e. generating a friction model with a high lubrication amount (2 g/m^2) and a low tool roughness (Sa = 0.2 μm), See Figure 4. Also, a TriboForm friction model with a low lubrication amount (0.5 g/m^2) and a high tool roughness (Sa = 0.6 μm) was generated, corresponding to a friction model with high friction conditions (see Figure 4).

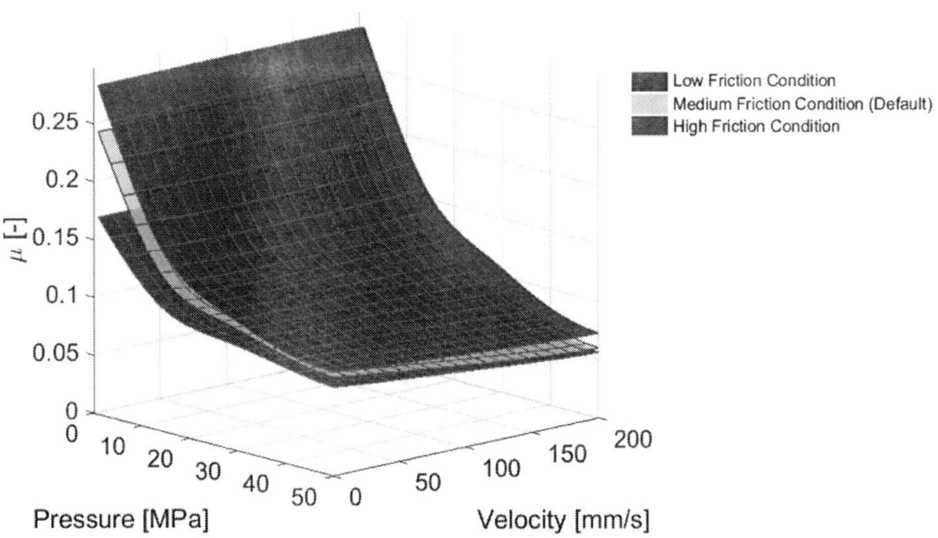

Figure 4. Three TriboForm friction models correspond to low, medium and high friction conditions including variations that can occur in press-shop reality.

4. Results and discussion

In this section, a comparison is made between simulation and try-out results. AutoForm simulations were performed using a Coulomb coefficient of friction of μ = 0.12, the P-V dependent friction model and the TriboForm friction model respectively. In addition, it is described how split issues were resolved in simulations and subsequently in try-out. Finally, the results of the sensitivity study are discussed.

4.1. Try-out vs. simulation results: formability

IOP Conf. Series: Materials Science and Engineering **967** (2020) 012010 doi:10.1088/1757-899X/967/1/012010

Figure 5 shows the formability results using Coulomb friction which fails in predicting the critical regions in the product. The corresponding region with high risk of splits from try-out is also indicated in the forming limit diagram (FLD). It can be seen that these points are far below the forming limit curve, and therefore "safe". In the next step, the P-V dependent friction model was used. The P-V dependent friction model is an enhanced version of the Coulomb friction model, however, also with this model the splits were not predicted in the part (see Figure 6). In the third simulation, the results using the TriboForm friction model are shown which predicts the splits in the exact same location as they occurred during try-out (Figure 7).

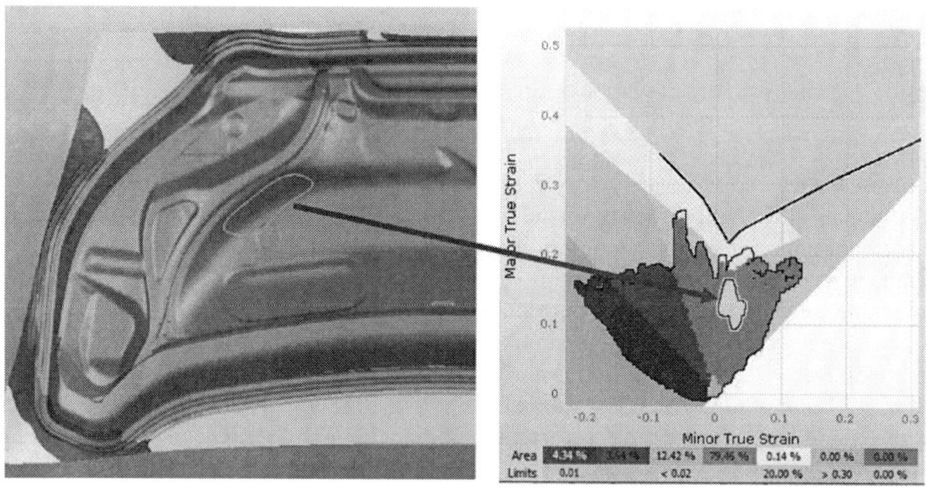

Figure 5. Initial simulation results using coulomb 0.12 showing no risk of splits

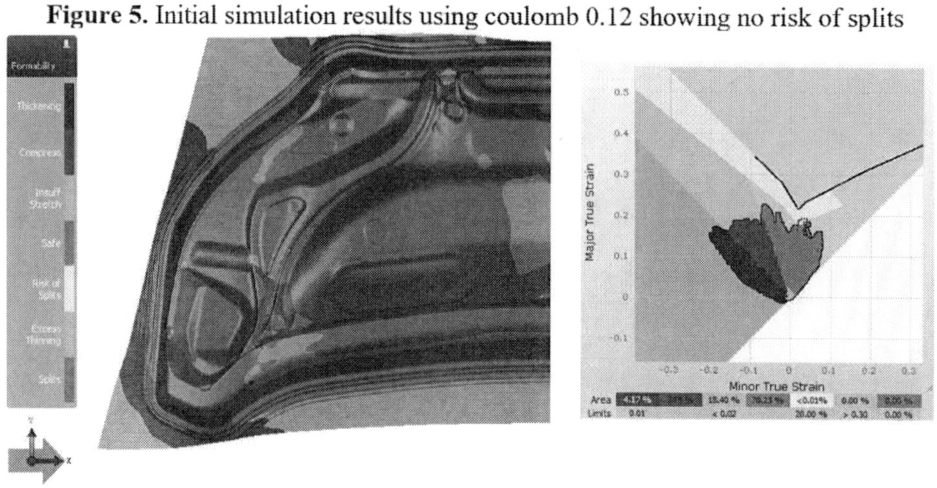

Figure 6. Simulation results using P-V friction model also showing no risk of splits

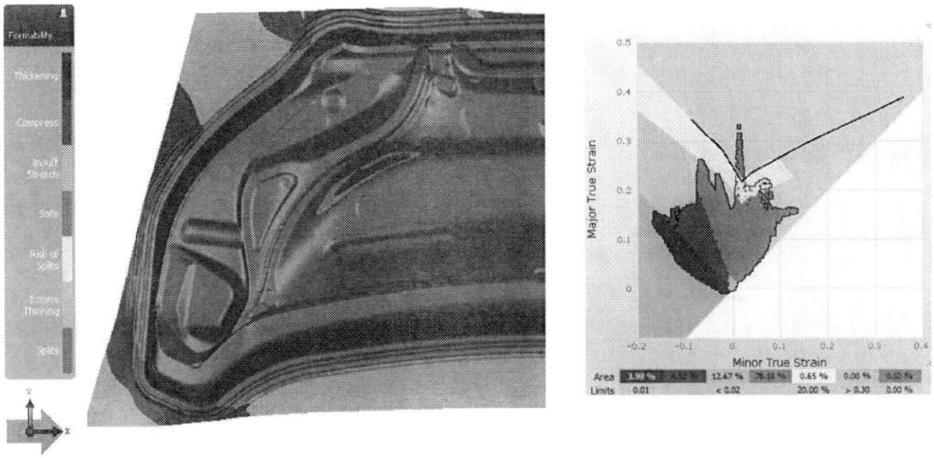

Figure 7. Simulations using the TriboForm friction model predicting splits in the same location as in try-out

4.2. Reverse engineering to resolve the split issue

Using the improved stamping simulation that included the TriboForm friction model, Ford started to change the simulation model to solve the split issues in try-out. After varying different parameters, it was found that modifying the 2nd drawbead in the simulation, as shown in Figure 7-left, resulted in a green simulation with no risk of splits. Subsequently, Ford verified this solution by die modification in reality. Figure 8-right shows that indeed the split issues were solved in the part according to the modified drawbead geometry.

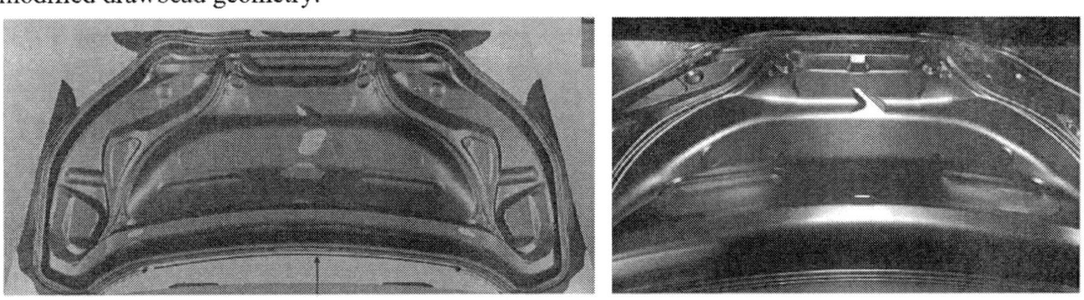

Figure 8. Modifying the outer drawbead in simulation resolved the split issues; this was also later verified in try-out.

4.3. Sensitivity study

Evaluating the effect of lubrication amount and tool roughness on the formability of the part shows that the part is not affected when friction conditions are decreased (lower friction, see Figure 9- left). However, increasing friction conditions (higher friction) can result in critical regions which are located at different locations compared to the default simulation (medium friction conditions, see Figure 9-right). These results demonstrate the robustness of the part with respect to tribological conditions in terms of formability. It is important to understand that lowering the tool roughness, compared to an average roughness value of 0.4μm, means additional polishing of the tools which is a costly procedure in terms of both man power and time. At the other hand, the tools should not exceed the average tool roughness of 0.4μm too much, as this can result again to split issues. Therefore, based on this study, additional polishing of the tools is not necessary, but it may increase the safety margin to possible account for other variations encountered in reality such as blank holder force and press velocity.

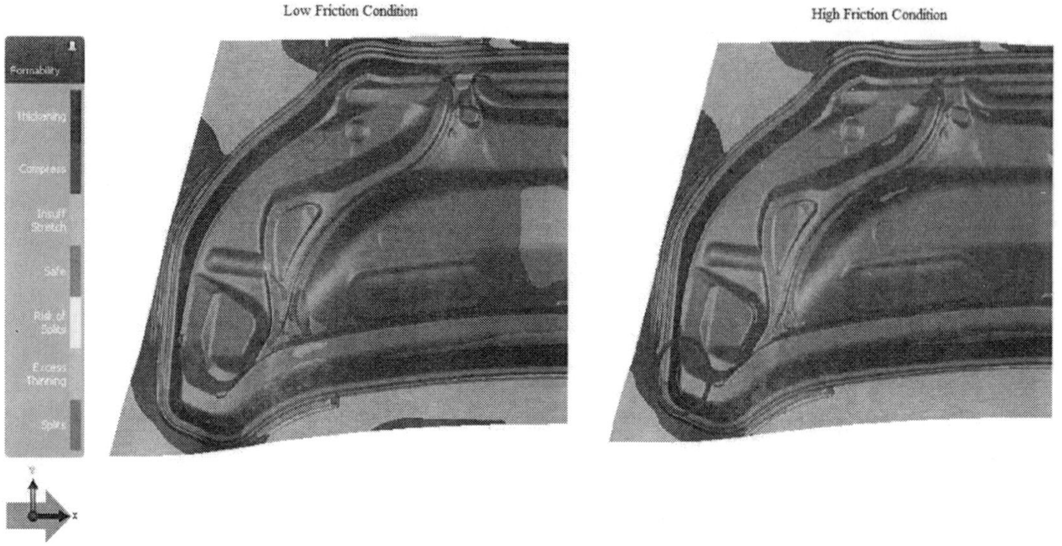

Figure 9. Low friction conditions (left) resulted in part with no forming issues while the high friction conditions (right) caused splits in the part.

5. Conclusions and future work

This work clearly demonstrates the strong influence of friction and lubrication on part quality of the Ford Otosan Transit hood inner panel. More importantly, it was demonstrated that the observed try-out issues could be simulated accurately by accounting for an advanced friction model in sheet metal forming simulations. This subsequently enabled Ford to first virtually predict production issues and subsequently improve the forming process to produce a defect free product. This can now be done by controlling, adjusting and optimizing the tribological conditions both in simulations and production.
Future work of Ford Otosan will focus on continuing the implementation of TriboForm's advanced friction and lubrication modeling technology into AutoForm's sheet metal forming simulations. The ultimate goal is to further enhance the prediction accuracy of forming processes to improve the prediction of part quality. Specifically, the use of such advanced tribological models for aluminum parts will be considered, as this is considered as indispensable for virtual design of lightweight car bodies.

References

[1] Tatipala S, Pilthammar J, Sigvant M, Wall J, Johansson CM, Introductory study of sheet metal forming simulations to evaluata process robustness, *37th Internatopnal Deep Drawing Research Group Conference*, June 3-7, 2018, Waterloo, Canada.
[2] AutoFormplus R7.0.3 user manual, 2017, AutoForm Engineering GmbH.
[3] Hol J, Meinders VT, deRooij MB, vandenBoogaard AH, Multi-scale friction modeling for sheet metal forming: The boundary lubrication regime, *Tribology International*, Volume 81, 2015, Pages 112-128.
[4] Hol J, Meinders VT, Geijselaers HJM, vandenBoogaard AH, Multi-scale friction modeling for sheet metal forming: The mixed lubrication regime, *Tribology International*, Volume 85, 2015, Pages 10-25.
[5] Sigvant M, Pilthammar J, Hol J, Wiebenga JH, Chezan T, Carleer B, Boogaard T van den, Friction in sheet metal forming: simulations of the Volvo XC90 inner door, *Forming Technolgy Forum*, September 12-13, 2016, Munchen, Germany
[6] Bolay C, Wied J, Naegele P, and Hol J, Advanced friction modelling in sheet metal forming simulation and their effect on drawbead models, *NEBU 2018 – Neuere Entwicklungen in der Blechumformung*, 15-16 May 2018, Stuttgart, Germany.

Experimental Investigation of Multi-stage Deep Drawing of IN718 Sheet at Elevated Temperature

Kanhu C Nayak[1], Prashant P Date[1]

[1]Department of Mechanical Engineering, Indian Institute of Technology Bombay, Mumbai 400076, India

E-mail: nayakkanhu83@gmail.com

Abstract. IN718 is a nickel-based super alloy that retains its mechanical properties at elevated temperatures and hence has numerous applications in the aerospace industry. In this study, deep drawing of IN718 sheet at elevated temperatures (700-1000°C) has been investigated experimentally. Initially, stress-strain response with failure strain at different temperatures was evaluated by uniaxial tensile tests along different orientations. Subsequently, hot deep drawing and re-drawing tools were designed and fabricated. A circle-grid marked circular blank of 3mm thickness was used in the first draw to form a circular cup followed by re-drawing (2nd draw). The major and minor diameter of the circular grid on the blank after the deep drawing was used to examine the strain distribution in the drawn cup and further used to calculate the thickness strain. These deep drawing operations were carried out at different temperatures with a constant blank holding force. Further, the deep drawing force at different temperatures was examined. The effect of the test temperatures during hot deep drawing and the process parameters on the variation of the thickness strain over the surface of the drawn cup has been studied. Finally, the major and minor strains are plotted in Keeler and Arcelor V9 forming limit curve to examine the formability of IN718 sheets at elevated temperature.

Keywords: Forming limit diagram; hot deep drawing; Inconel 718; thickness strain

1. Introduction

Inconel 718 is widely used in high-temperature applications such as in aerospace industry, oil, and chemical industries, and also nuclear power plants due to its high strength, creep resistance, oxidation, corrosion properties and structural stability [1-3]. Different manufacturing processes are involved to produce Inconel alloy based mechanical components. Sheet metal forming such as, deep drawing and stamping are among the important manufacturing processes for aerospace and nuclear industry. However, it is difficult to deep draw the components from Inconel super alloy sheets at room temperature in comparison to other aerospace alloys [4-6]. Therefore, hot forming of these sheets at elevated temperature is a promising processing technique. Many researchers have investigated to improve the formability with reduced forming load for different alloys [4,7]. Hot deep drawing process has advantages to produce drawn parts without springback from high strength materials like steel, Inconel and lightweight materials like Ti alloy, Al alloy etc. In addition, the deep drawing at elevated temperature reduces the load and hence high strength materials can be easily drawn [8].

Recently the formability of solution treated IN718 material has been investigated [9-11] at room temperature. Further, the formability of In718 has been studied at elevated temperatures by conducting the tensile test, compression test and numerical modelling [12,13]. There is limited literature related to the hot drawability of IN718 thick sheet.

In the present work, a hot deep drawing die has been designed and fabricated to investigate the drawability of the IN718 sheet at elevated temperature. In addition, the hot tensile test has been performed in 0 and 90° to the rolling direction. The variation of deep drawing load with punch travel at different drawing temperatures has been studied. The thickness strains are calculated from the measured diameters of deformed grid marks on the deep drawn and re-drawn cups. The earing on these cups has been examined. Finally, the major and minor strains are plotted in a forming limit curve based on Keeler and Arcelor V9 and the drawability was discussed.

2. Materials and methods

In the present work, commercially available IN718 sheets of 3mm thickness were used for the experiment. Tensile as well as samples for plastic strain ratio, were cut from these sheets. The tensile properties in two orientations (90° and 0° to the rolling direction of sheet) were measured at three different temperatures of 700, 850 and 1000°C. Similarly, the plastic strain ratio was calculated at 0, 45 and 90° to the rolling direction of the sheet. Thereafter, circular blanks of 114mm diameter were cut from 3mm thick of IN718 sheet. Circular grid pattern (2(+0.02 to 0.04)mm diameter and center to center distance is 2.5((+0.02 to 0.04)) was laser marked with a depth of impression 70 microns on one side of the blanks (Figure 1). In-plane, major and minor diameters of deformed grid circles were measured using Camscope followed by analysis using the ImageJ software. Further, these measured diameters were used for the calculation of major and minor strains. Then, thickness strains were estimated by the principle of volume constancy.

The hot deep-drawing set up is schematically illustrated in Figure 2(a), whereas the experimental set up is shown in Figure 2(b). Diameter of first stage deep drawing die and re-drawing die is 69mm and 50mm respectively. The hot deep drawing was carried out in a hydraulic press of 200 tonnes capacity. The die and blank holder were heated to 250°C using cartridge heater. A thermocouple was connected to the die for measuring the temperature (Figure 2(b)). The grid circle marked circular blank was heated separately in a muffle furnace, which was kept near to the hydraulic press. This blank was maintained at a temperature of 900°C in one set of drawing experiments and 1000°C for another, using a muffle furnace. Re-drawing operations were carried out at 1000°C. There is a transportation (bringing blanks from furnace to deep drawing station) temperature loss of approximatly15-20°C over approximately 2-3 seconds. Nickel-based anti-seize compound (NTC, solid lubricant up to 1400°C) was used as a lubricant on the die surface. First, the furnace is heated up to the required temperature, and then the blank was put in the furnace for 5 minutes of soaking time for uniform heating. After that, the heated blank was placed centrally on the die and immediately, the spring-loaded blank holding force (4.75kN) was applied to the blank and then the drawing process started. The punch speed was kept 14mm/sec during the deep-drawing process. The entire drawing process took 3-4 seconds.

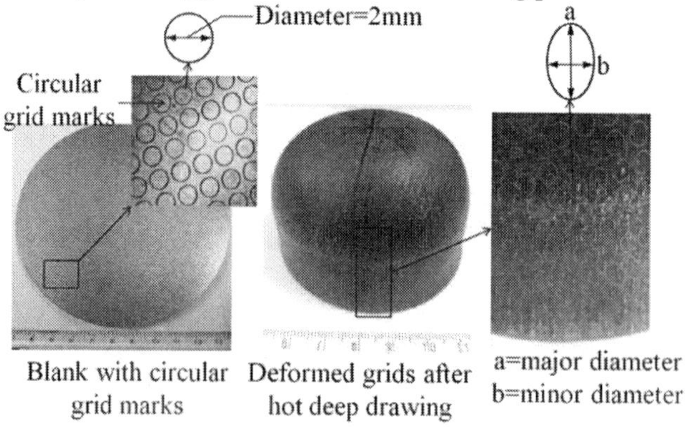

Figure 1. Circular blank with laser marked circular grid patterns

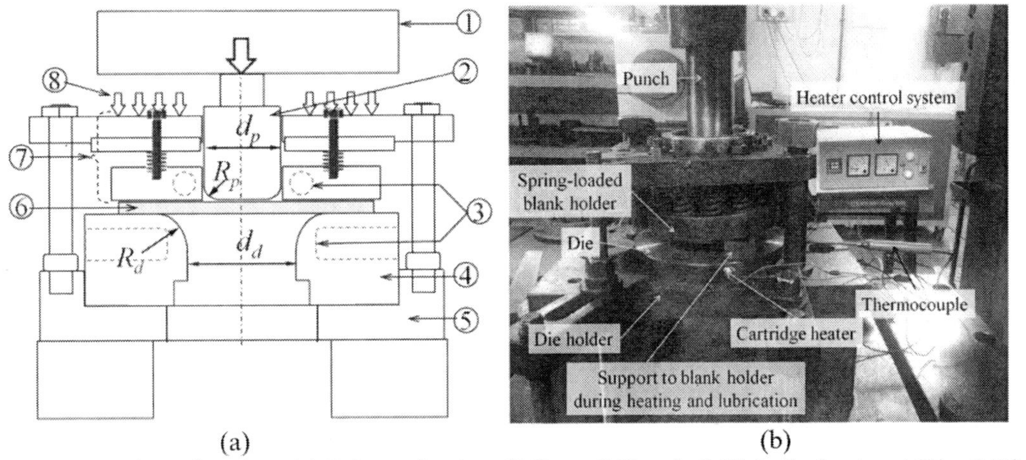

Figure 2. Hot deep-drawing, (a) Schematic view (1-Ram, 2-Punch, 3-Hole for heater, 4-Die, 5-Die holder, 6-Blank, 7-Spring loaded blank holding and 7-Blank holding force) and (b) experimental set up

3. Results and discussion

3.1. Tensile properties

The engineering stress-strain curves of IN718 Inconel at different temperatures are shown in Figure 3(a) and (b). It is observed that the ultimate and yield strength of samples along the rolling direction is approximately 50MPa smaller than the samples that are perpendicular to rolling direction up to the test temperature of 850°C. The stress-strain curve at a temperature of 700°C shows a hardening behaviour irrespective of the rolling direction. This behaviour has been studied using the Hollomon equation [14]. The strength coefficient and strain hardening exponent of IN718 sheet along the rolling direction is 1305.9MPa and 0.340 respectively, whereas along transverse direction, these values are 1443.7MPa and 0.345 respectively. The percentage of elongation of IN718 sheet at 1000°C is higher than at a temperature of 850°C as per the fracture strain. At 1000°C, the yield strength of this sheet is low, and elongation is more. Therefore, in this study for low deep drawing load and more drawability, the deep drawing temperature was maintained at 900 and 1000°C, whereas re-drawing temperature was kept at 1000°C. The material was found to be close to isotropic with an R value of 1.071 at the drawing temperature. The ΔR value is -0.414. The nature of fracture and length of tested tensile samples are shown in Figure 3(c). The trends of the overall length of the tested samples shown in Figure 3(c) are in good agreement with the engineering strain, as observed in Figure 3(a) and (b). There was no distortion of the sample at higher temperatures in the pin region where it was fixed with the tensile fixture.

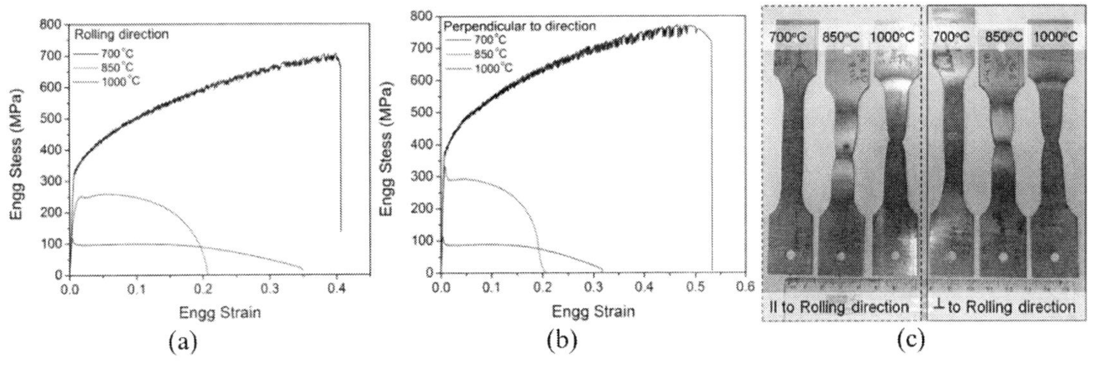

IOP Conf. Series: Materials Science and Engineering **967** (2020) 012011 doi:10.1088/1757-899X/967/1/012011

Figure 3. Engineering stress-strain of IN718, (a) rolling direction, (b) perpendicular to the rolling direction and (c) samples after tensile test

3.2. High temperature deep drawing

3.2.1 Hot deep drawn cups and forming load

The cylindrical cups were drawn from a circular blank of 114mm diameter and 3mm thick IN718 Inconel sheets. The deep drawn (first drawn) and re-drawn cups are presented in Figure 4(a). The variations in dimensions of the circular grid patterns on deep drawn cups were compared with the re-drawn cup in Figure 4(a). The deformed grid patterns on the curvilinear surface and wall of a re-drawn cup are illustrated in Figure 4(b). The grid circles are deformed to shape of an ellipse. This deformed grid circles further gives the strain distribution on the surface of the deep drawn and re-drawn cups that have been discussed in the successive section.

The deep drawing load at a drawing temperature of 900°C is compared with the re-drawing load in Figure **5**(a), whereas Figure **5**(b) shows the comparison of deep drawing load between first deep drawn cups at 1000°C with re-drawn cup at 1000°C. There is an effect of temperature on the load required for the deep drawing process. Deep drawing load decreased with increasing operating temperature and vice versa for an isothermal condition. At a constant operating temperature re-drawn cup required larger load compared to first deep drawn cups as seen in Figure **5**(b). It is because of the accumulated strain hardening in the first drawn cup results the increase in drawing load during re-drawing. Further, the load vs. punch travel curve can be divided into three zones such as initial punch travel, actual deep drawing operation and end of deep drawing. During deep drawing process, load gradually increased and attained a peak load.

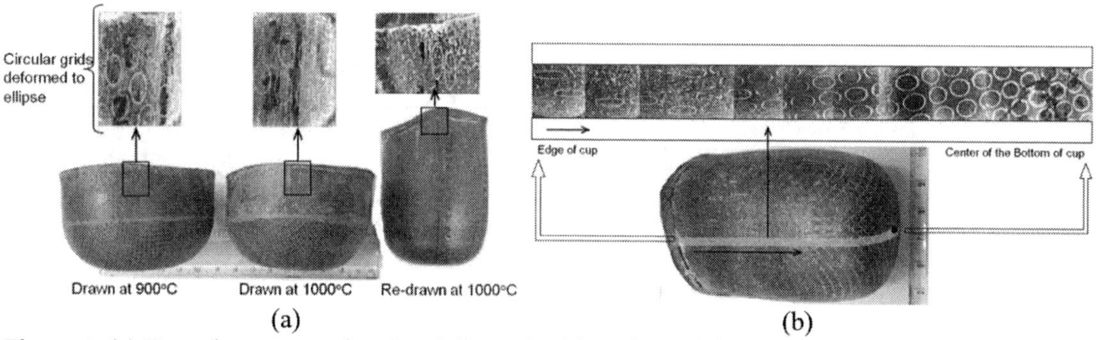

Figure 4. (a) Deep drawn cups showing deformed grid marks and (b) the evolution of deformation of circular grids during re-drawing at 1000°C

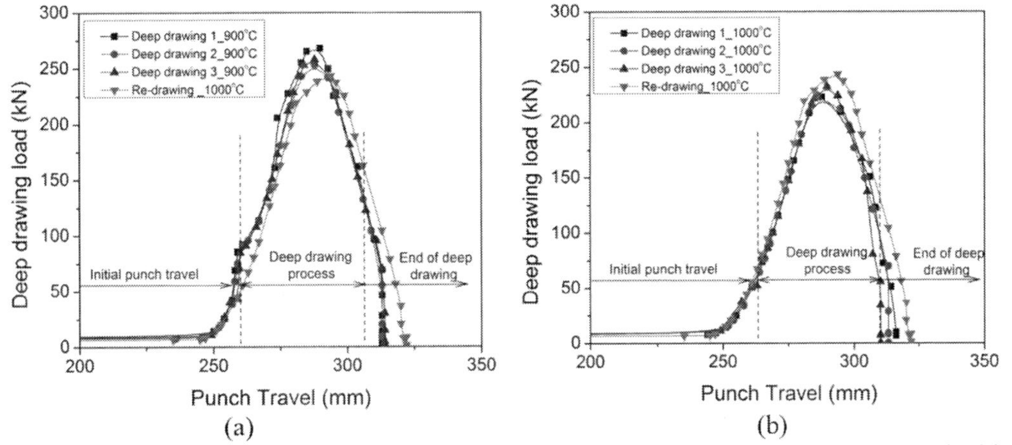

Figure 5. Deep drawing load with punch travel; (a) deep drawing load at 900°C is compared with re-drawing load at 1000°C and (b) deep drawing load at 1000°C is compared with re-drawing load at 1000°C

3.2.2 Thickness strain and earing

In sheet metal forming, thickening and thinning are the two major issues. The final thickness distribution of deep drawn cup and re-drawn cups can be measured directly or by indirect calculation from the thickness strains (ε_3). The thickness strains are obtained using the principle of volume consistency.The variation of thickness strain with curvilinear distance (Distance from one end of the cup to the other end in a longitudinal sectional plane) of the cup along 0, 45 and 90° to the rolling direction are presented in Figure 6(a)-(c). It has been seen that thinning occurs at the corner of the cup and thickening occurs at the edge of the cup. The grid circles at the center of the bottom of the cup do not show any variation in diameter for the first deep drawn cups. Zero thickness strain results in a flat land, as is evident from Figure 6(a) and (b) for the first drawn cups. However, in the case of re-drawing this land becomes narrow even if it has a flat bottom of 10mm diameter. It is because of increase in grid circle diameter due to deformation at a higher temperature (here 1000°C). Therefore, one cannot expect a flat land in Figure 6(c) as in the case of deep drawing at room temperature on the bottom of the cup. Hence there is an effect of temperature on dimensions of grid circles resulting in additional strain. Moreover, Maximum thickness strain occurs at corner radius of cup in both deep drawing operation and it is higher for re-drawn cups. However, the thickness strain distribution is independent of the rolling direction of the sheet as well as deep drawing temperature. The small variation in thickness strain at the corner of the cylindrical cup as shown in Figure 6(a)-(c) is due to distortion of grid circle dimension at higher temperature. Similar observation found at the edge of the re-drawn cylindrical cup.

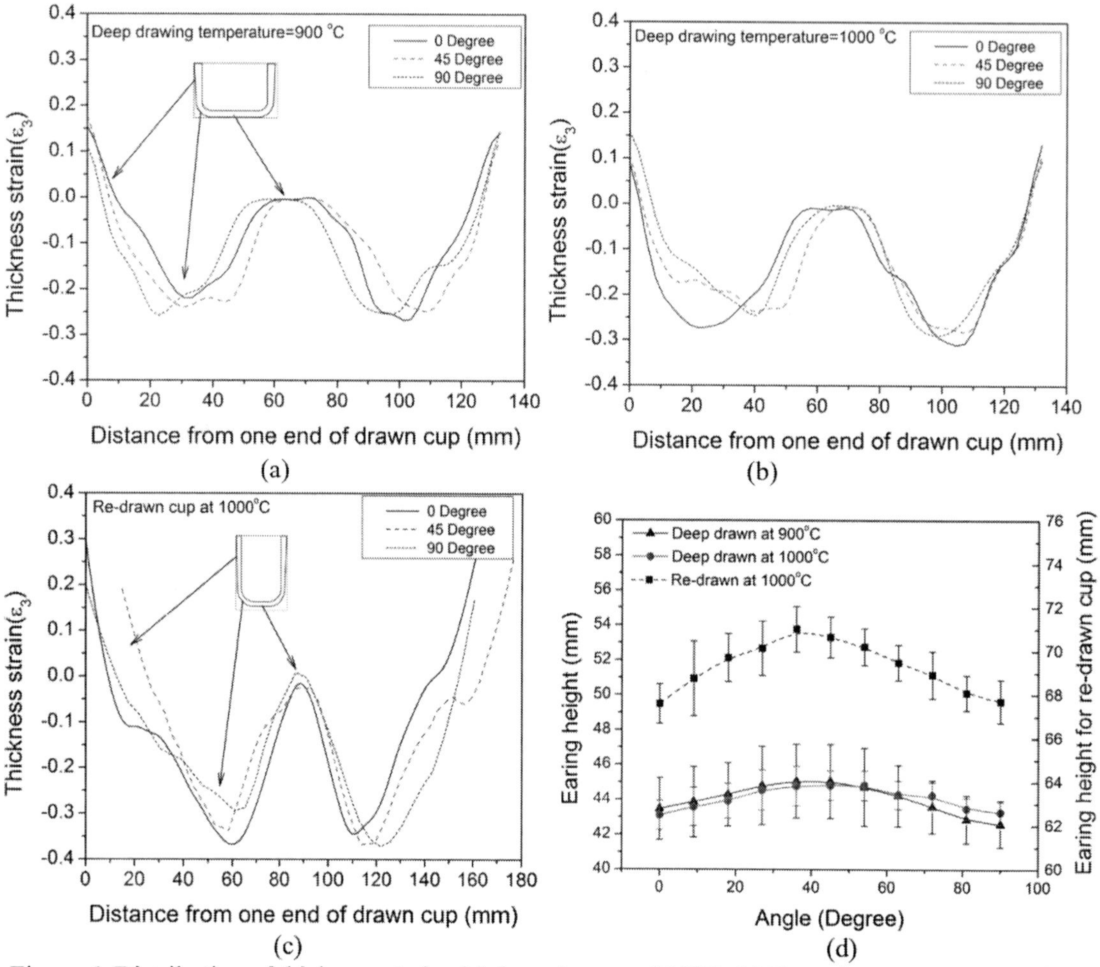

Figure 6. Distribution of thickness strain; (a) deep drawn at 900°C, (b) Deep drawn at 1000°C, (c) re-drawn at 1000°C and (d) earing of different drawn cups

Figure 6 (d) illustrates the earing that occurs during the deep drawing process. It is one of the major defects observed during the deep drawing process due to the anisotropic nature of the sheet. There is no significant earing in the first draw. However, higher earing occurs for re-drawn cups compared to first drawn cup.

3.2.3 Formability analysis

The major and minor strain distribution in three different directions (0, 45 and 90 degree) to rolling for cylindrical cups, drawn from as received IN718 Inconel sheet is shown in Figure 7. Both the first drawn cups and the re-drawn cup give five modes of deformation namely uniaxial mode, biaxial mode, pure shear mode, equi-biaxial mode and plain strain mode as seen in Figure 7. It is observed that major strain and minor strain distribution is also independent of rolling direction of sheet. To check the formability of IN718 sheet at higher temperature, two forming limit curves have been used. One is Keeler's criterion [15] and the other is Arcelor's V9 criterion [16]. These two forming limit curves are obtained only using the various parameters from uniaxial tensile test in the mathematical model according to Keeler and Arcelor. The Keeler's forming limit curve (FLC) is used because of its simplicity where tensile test data (strain hardening exponent=0.345, strength coefficient=1305MPa and strength=640MPa) at 90° to rolling direction and thickness data are required to generate the FLC. For the Arcelor V9 FLC model, ultimate tensile strength, percentage of deformation, yield strength, strain hardening exponent, strength coefficient, normal anisotropy and thickness of the sheet data is required to generate the FLC. In this study, the mechanical properties obtained from the hot tensile test are used to produce both the FLC. The options are available in the AutoForm to generate the Keeler and Arcelor V9 FLC which is explained elsewhere [16]. FLD and fracture FLD for commercially available IN718 of 1mm thick at different temperatures (max 700°C) have been developed by Mahalle et al. [17], using limit dome height test. It is observed, the forming limit curve in tension-compression region is similar to Keeler's FLC. By increasing the forming temperature, the major safe strain increases. The major and minor strains are in safe zone with respect to these two FLCs. However, thinning at the corner of cylindrical cup occurs without fracture. Also thicker sheet have more volume and hence can be stretched to a greater extent, increasing the thinning without causing fracture.

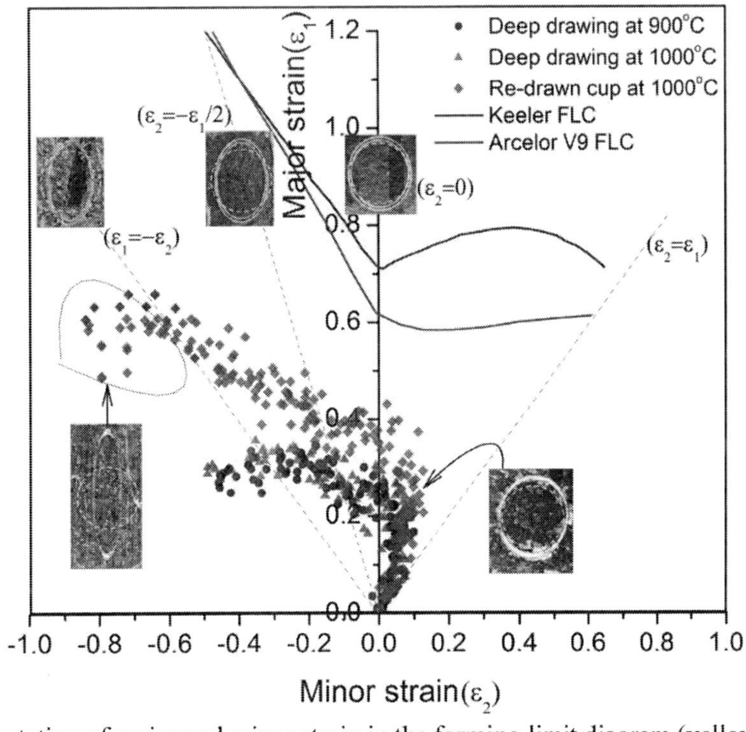

Figure 7. Presentation of major and minor strain in the forming limit diagram (yellow colored dash line indicates the initial grid circle)

4. Conclusions

In this research work, IN718 sheets were used for hot deep drawing process. Mechanical properties of this sheet were investigated at high temperature. Cylindrical cups were drawn from circular blanks cut from sheets and strain distribution was observed along different directions inclined to rolling directions. The conclusions drawn from the results can be summarized as follows:

- Deep drawing load decreased with increase in the drawing temperature.
- First deep drawing cups of 40mm height and re-drawing cups of 67mm height cylindrical cups are successfully drawn without necking from circular blanks.
- Both the first drawn and re-drawn cups undergo five modes of deformation namely uniaxial mode, biaxial mode, pure shear mode, equi-biaxial mode and plain strain mode.
- Thickness strain, major strain and minor strain distribution is independent of rolling direction of sheet.
- Thinning occurs at the corner of deep drawn and re-drawn cylindrical cups without causing fracture and it is temperature dependent.

References

[1] Yuan H and Liu W C 2005 *Mater. Sci. Eng. A* **408** 281–289
[2] Prasad K S, Panda S K, Kar S K, Sen M, Murty S N and Sharma S C 2017 *J. Mater. Eng. Perform.* **26(4)** 1513-30
[3] El-Bagoury N, Hessien M M, Mohammed A, Mahmoud M H H, Alanazi A K, Alshanbari N A 2019 *Metall. Microstructure, and Analysis* **8** 642–655
[4] Toros S, Ozturk F and Kacar I 2008 *Journal of Materials Processing Technology* **207** 1–12
[5] Date P P and Padmanabhan K A 2001 *Journal of Materials Processing Technology* **112** 68-77
[6] Kesharwani R K, Panda S K and Pal S K 2015 *Journal of Materials Engineering and Performance* **24** 1038–1049
[7] Lee Min-sik, Baeck Seung-cheol and Kang Chung-gil 2012 *Journal of Engineering Manufacture* **226(5)** 898–908
[8] Mori K, Bariani P F, Behrens B A, Brosius A, Bruschi S, Maeno T, Merklein M and Yanagimoto J 2017 *CIRP Annals - Manufacturing Technology* **66** 755–777
[9] Prasad K S, Panda S K, Kar S K, Sen M, Murty S N and Sharma S C 2017 *Journal of Materials Engineering and Performance* **26** 1513-1530
[10] Prasad K S, Panda S K, Kar S K, Murty S N and Sharma S C 2018 *Materials Science & Engineering A* **733** 393–407
[11] Jinhui D, Xudong L, Qun D and Luo Y 2011 *Adv. Mater. Res.* **197–198** 1125–1128
[12] Yuan H and Liu W C 2005 *Mater. Sci. Eng. A* **408** 281–289
[13] Prasad K S, Panda S K, Kar S K, Singh S K, Murty S N and Sharma S C 2018 *IOP Conf. Series: Materials Science and Engineering* **418** 012055
[14] Kleemola H J and Nineminen M A 1974 *Metall. Trans.* **5** 1863–1866.
[15] Keeler S P 1978 *Forming Limit Criteria-Sheets* (In: Burke J.J., Weiss V. (eds) Advances in Deformation Processing, Springer, Boston, MA)
[16] Jadhav S, Schoiswohl M and Buchmayr B 2018 *BHM* **163 (3)** 109–118
[17] Mahalle G, Morchhale A, Kotkunde N, Gupta, A K, Singh S K and Lin Y C 2020 Journal of Manufacturing Processes **56** 482-499

IOP Conf. Series: Materials Science and Engineering **967** (2020) 012012 doi:10.1088/1757-899X/967/1/012012

Comparative Investigation on Aluminum Material Modeling with applications to Springback Prediction

Z Deng

Aleris North American Automotive Innovation Center, Madison Heights, MI 48071, USA

Email: Zhi.Deng@aleris.com

Abstract. A comparative investigation on springback prediction of aluminum stampings has been completed using an industry-type "shotgun" die as the benchmark case. 5xxx aluminum alloy at a thickness of 1.5 mm was used for the shotgun die tryout. After stamping tryout, the formed panel was scanned to capture its sprung geometry. The studied aluminum alloy was characterized by means of uniaxial tensile test and hydraulic bulge test to determine the material isotropic hardening behavior and planar anisotropy. In order to determine the material kinematic hardening behavior, the exclusive Yoshida model fitting test was conducted in tension-compression (T-C) and compression-tension (C-T) loading modes with the one-cycle and three-cycle loading schemes. The shotgun benchmark case was simulated using FEA software LS-DYNA®. By comparing the simulation results with the tryout measurements, the accuracy of springback prediction was examined to evaluate the influence of material modeling, loading mode and loading cycle in the Yoshida model fitting test. Based on the current benchmark study, the best practice in springback prediction and compensation of aluminum stampings and in conducting the Yoshida model fitting test of aluminum alloys were summarized for industrial applications.

1. Introduction

Springback prediction and compensation are a challenging task for stamping die design and engineering, which is also one of the controlling factors for the die manufacturing cost. This is particularly true for automotive aluminum stampings, since aluminum alloys tend to have large springback resulting from the lower elastic rigidity. For aluminum autobody closures, springback phenomenon is far beyond simple open-angle springback and is more representative of the side wall curl distortions and twist springback. Springback complexity on aluminum stampings has become a major issue in maintaining a consistent part geometric performance out of the press lines. Springback is an inherent material response during a reverse deformation process, which is extremely affected by the Bauschinger effect. Apparently, considering the Bauschinger effect in springback analysis is a must. Up to date, the most widely accepted material model with the Bauschinger effect included in the stamping simulation software is the one developed by Yoshida and Uemori (Yoshida model) [1, 2]. This nonlinear kinematic hardening model provides a framework which can be incorporated with any types of anisotropic yield functions and has been proven to be able to significantly improve the accuracy of springback prediction. However, most of these published investigations targeted the sheet metals other than aluminum alloys [3-5]. The previous experimental study showed that aluminum alloys have a strong cyclic strain hardening behavior, but comparatively weak Bauschinger effect [6]. So far, it is unclear to what extent the Yoshida model can improve the accuracy of springback prediction for aluminum stampings. With the increasing use of aluminum alloys in the automotive industry, it is of utmost importance to find a way leading to accurate springback prediction and thus reliable die compensation for aluminum stampings.

The current work was aimed to address those outstanding demands, based on the experimental investigations, physical die tryout and extensive springback simulations. In the experimental investigations, the targeted material of an AA 5000 alloy at a thickness of 1.5 mm was characterized by uniaxial tensile test and hydraulic bulge test for its isotropic hardening and planar anisotropic behavior. The same material was also examined with the Yoshida model fitting test to determine its kinematic hardening behavior under different loading conditions. In the physical die tryout, the studied material was stamped with the industry-type "shotgun" die [7] as the benchmark case to investigate the relationship between material modeling and the accuracy of springback prediction. Based on the current benchmark study, the best practice in springback prediction and compensation of aluminum stampings and in conducting the Yoshida model fitting test of aluminum alloys were summarized for industrial applications.

2. Material property testing

Two kinds of material property testing have been performed to characterize the material isotropic and kinematic hardening behavior as well as the material planar anisotropy. Uniaxial tensile test and hydraulic bulge test were carried out to determine the basic material property parameters, material anisotropic yielding and isotropic hardening behavior. The material kinematic hardening behavior was examined by using the Yoshida model fitting test. In Yoshida model fitting test, the sample material is continuously deformed under cyclic loading and reverse loading condition, providing the essential testing data to determine the Yoshida model parameters for springback simulations.

2.1 Uniaxial tensile and hydraulic bulge tests

The material isotropic strain hardening and anisotropic yielding behavior have been determined with these two tests for the studied 5000 aluminum alloy. Uniaxial tensile test was conducted by following the ISO 6892 standard with the type 2 sample geometry at room temperature. The testing samples were cut along the 0^0, 45^0 and 90^0 relative to the rolling direction, in order to identify the planar anisotropy. Hydraulic bulge test was used to explore the material isotropic hardening behavior in a higher strain range, following the ISO 16808 standard. Digital image correlation (DIC) techniques were used for all the tests to record the deformation process.

2.2 Yoshida model fitting test

The test was performed with a sophisticated testing apparatus featuring a customized anti-buckling mechanism. By maintaining a real-time controlled side force, the testing specimens can be continuously

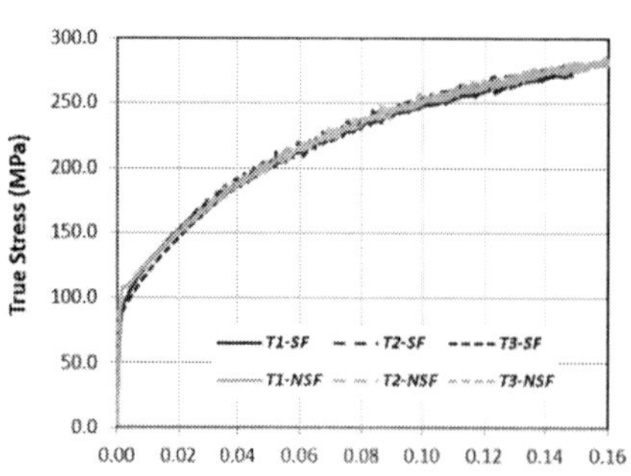

Figure 1. The true stress-true strain curves measured with and without a side force applied.

deformed under either tension-compression (T-C) or compression-tension (C-T) loading cycles. Figure 1 compares the true stress-true strain curves measured with and without a side force applied. It is clear from Figure 1 that the applied side force has little effect on the testing measurements. For all the Yoshida model fitting tests, 3D DIC techniques were used in a real time mode for strain measurements. All the testing samples were cut along the sheet rolling direction and tested at a constant strain rate of 0.002/sec.

In order to evaluate the influence of the loading mode and loading cycle on the determination of the Yoshida model parameters and the resultant springback prediction accuracy, the tests were conducted under the following conditions: (1) one-cycle T-C test, (2) three-cycle T-C test, and (3) three-cycle C-T test. Each of these tests was repeated three times. For the one-cycle T-C test, the loading path followed (1) tension to 5% strain, (2) compression to -5% strain, and (3) tension to 0%. For the three-cycle T-C test, the loading path followed (1) tension to 2.5% strain, (2) compression to -2.5% strain, (3) tension

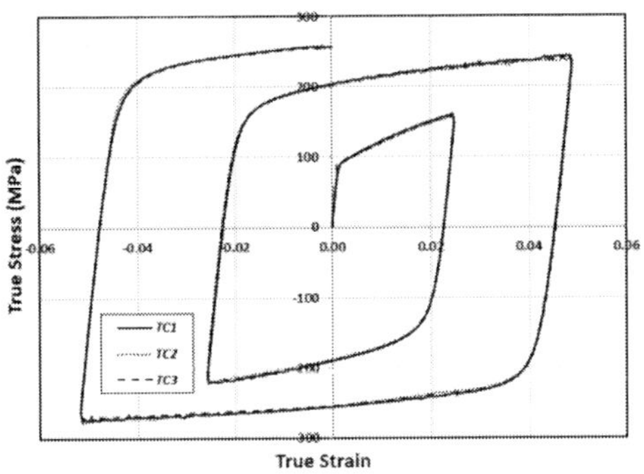

Figure 2. The true stress-true strain curves measured from the T-C test.

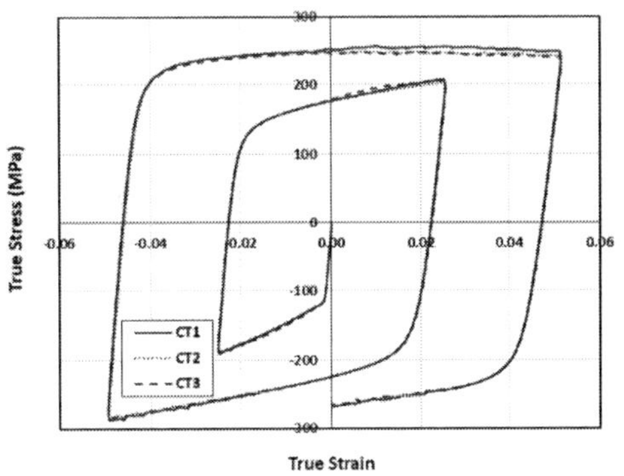

Figure 3. True stress-true strain curves measured from the C-T test.

to 5% strain, (4) compression to -5% strain and (5) tension to 0% strain. For the three-cycle C-T test, the loading path followed (1) compression to -2.5% strain, (2) tension to 2.5% strain, (3) compression to -5% strain, (4) tension to 5% strain and (5) compression to 0% strain. Figure 2 and 3 compare the true stress-true strain curves from the three-cycle T-C and C-T tests respectively. With these testing measurements, the Yoshida model parameters can be determined for the simulations.

3. Shotgun Panel Tryout

As detailed in [7], the industry-type "shotgun" draw die was specifically designed with the stake beads to provide a post-stretch force over the formed blank near the end of the drawing process. Figure 4 clearly shows that the use of the stake beads can greatly reduce the springback on the formed part.

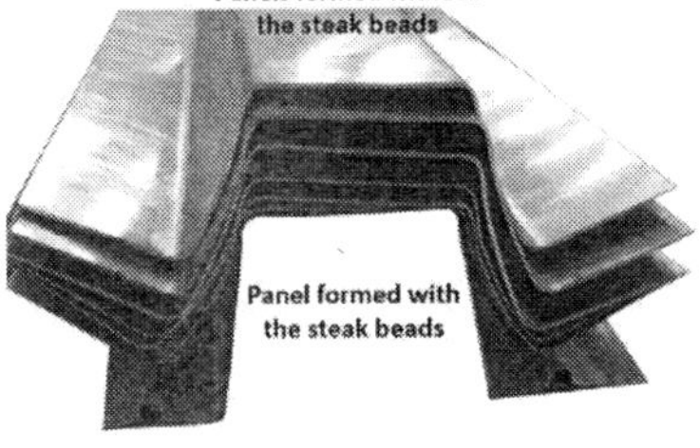

Figure 4. The tryout panels formed with and without the steak beads.

Figure 5. The scanned part geometry.

The details of the shotgun tryout die and its stake bead geometry can be found in [7]. For the current tryout, the B4H8 stake bead was adapted which has the bead radii of 2 mm and the bead height of 8 mm. A total of ten blanks were laser cut along the sheet rolling direction for the tryout. After tryout, two panels were selected for white light scanning to capture their springback geometries as shown in Figure 5. The scanned panel geometries also provided the blank draw-in boundaries for the simulations to match.

4. Springback Analysis and Evaluation

The forming simulation and springback analysis of the shotgun part were conducted with the commercial software LS-DYNA. Figure 6 depicts the tooling and blank setup in the simulations. The tooling surfaces used in the simulations were rebuilt from the scanned data of the real tryout tools. This practice can significantly reduce the CPU time and also improve the simulation accuracy. The blank was modelled with the fully integrated shell elements with 7 integration points along the sheet thickness direction. The binder clamping force and forming velocity were defined according to the tryout measurements.

The material models used for the current comparative study were MAT36: 3-PARAMETER_ BARLAT, MAT226: MAT_KINEMATIC_HARDENING_BARLAT89, MAT37: TRANSVERSELY _ANISOTROPIC_ELASTIC_PLASTIC and MAT_125: MAT_KINEMATIC_HARDENING_ TRANSVERSELY_ANISOTROPIC. MAT36 is the material model incorporating the Barlat89 planar anisotropic yield function, and MAT37 uses the Hill48 transverse anisotropic yield criterion. Both MAT36 and MAT37 are widely used in stamping simulations, and even in springback analysis when the testing data for using the Yoshida model are not available. MAT226 and MAT125 are the advanced material models of MAT36 and MAT37 respectively, being coupled with the Yoshida nonlinear kinematic hardening model. Based on the material property testing data detailed in the Section 2, all the parameters in these four models can be determined for forming simulations and springback analyses.

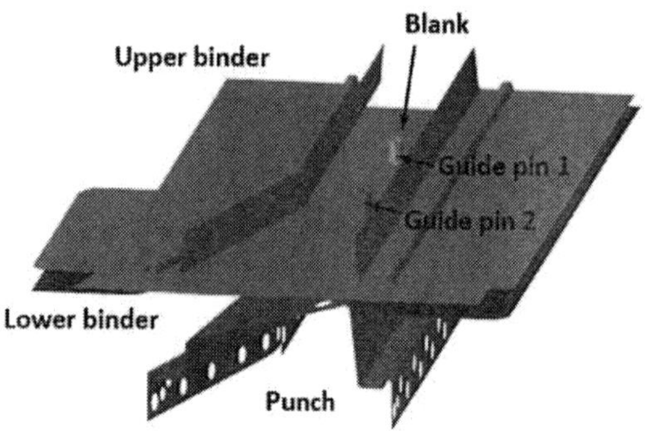

Figure 6. Tooling setup in simulations.

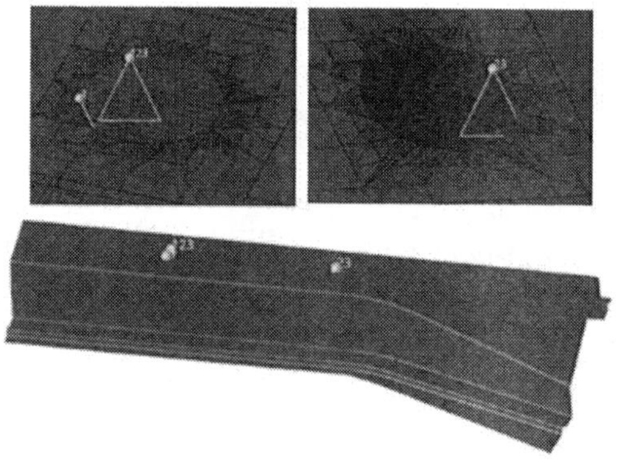

Figure 7. Constraints applied for springback evaluation.

Springback prediction was evaluated by comparing the part geometries from the simulations with the scanned tryout part geometries. The tryout parts were scanned on their outer surfaces, while the middle surfaces of the blank were used in simulations. In order to have a reasonable comparison, the sprung middle surfaces of the blank in the simulations were offset with a half thickness. In all the springback calculations, the 3-2-1 constraints on the formed parts were defined along the edges of the blank location holes as shown in Figure 7. This would ensure the real constraining condition in the tryout be applied in the simulations.

5. Springback Results and Discussions

The shotgun part has three distinguished geometric regions (Figure 8). The material in the uniform channel region has a balanced stretch-bend deformation along the side walls during the forming process. The springback in this region is mainly attributed to the open-angle springback. In the transition region, the material undergoes a slightly unbalanced stretch-bend deformation at both the sides with the extra stretch at the top and also along the right-side wall curvature. Since the extra stretch is not significant, the springback in this region is still dominated by the open-angle springback plus a little side wall curl springback. The material in the aggressive region of the part has a strong unbalanced stretch-bend deformation at both the sides. The part cross sections are progressively enlarged toward the end of the part, resulting in an aggressively enhanced stretch forming at the top and along the right-side wall of the part. The springback in this region is dominated by the side-wall curl and twist springback. It is worthy to state that the springback difference in these three regions is mainly applied to the material at the right-side of the first blank-tooling contact line A-B (Figure 8). At the left side of the first contact line A-B, the open-angle springback is the major component due to the straight and uniform part geometry. In order to reflect the springback complexity on the part, the springback evaluation was performed based on the cross-section cuts in each of those regions as shown in Figure 9.

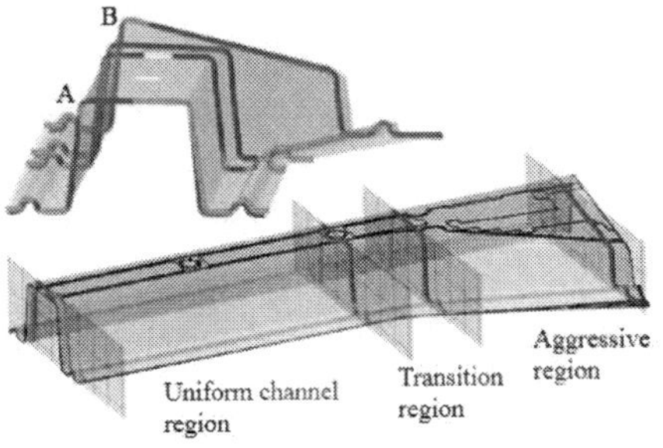

Figure 8. The geometric and springback characteristics of the shotgun part.

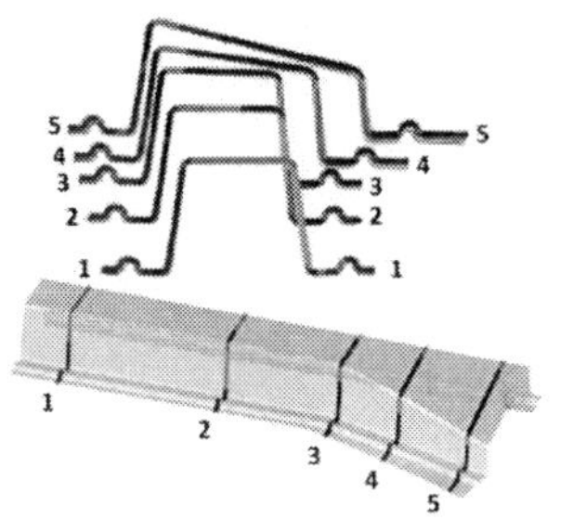

Figure 9. The cross section cut scheme for springback comparison.

5.1 Influence of material kinematic hardening

To evaluate the material kinematic hardening effect, the five sprung cross-section profiles predicted using MAT226 and MAT125 are compared with the profiles predicted using MAT36 and MAT37, and also with the scanned profiles (Figure 10). It is found from Figure 10 that at the left side of the contact line A-B where the open-angle springback is dominated, the springback predictions using MAT125, MAT36 and MAT37 are very close to each other. They are also in excellent agreement with the scanned data, especially along the side wall and inside the drawbead area. In contrast, the material model MAT226 predicts the worst springback results among the studied four material models. The derivations between the predicted springback profiles using MAT226 and the scanned profiles are in the range of 2.0 to 4.0 mm with the maximum derivation of 3.9 mm at the cross-section cut #4. Overall, the material model MAT125 gives the best springback prediction at the left side of the contact line A-B.

At the right side of the contact line A-B where a complex springback is observed, the accuracy of springback prediction varies from the region to region. As shown in Figure 10(a), the predicted springback profiles using these four models are fairly close to each other and also in good agreement with the scanned profiles in the uniform channel region. In this region, MAT226 gives the improved springback results. Compared to the scanned data, the maximum derivation of MAT226 decreases from 1.9 mm at the cross-section cut #1 to 1.3 mm at the cross-section cut #2. Contrastingly, both MAT36 and MAT37 predict the worst springback results at the cross-section cut #2 with the maximum derivation of 1.9 mm. The improvement in springback prediction using MAT226 is highlighted in Figure 10(b), where the predicted sprung cross-section profiles are in excellent agreement with the

IOP Conf. Series: Materials Science and Engineering **967** (2020) 012012 doi:10.1088/1757-899X/967/1/012012

scanned profiles for both the transition and aggressive regions. For both MAT36 and MAT37, the difference between the predicted and scanned sprung cross-section profiles increases with the increase in the side-wall curl and twist springback. The maximum derivation reaches 5.8 mm and 7.1 mm inside and outside the drawbead area respectively at the cross-section cut #5.

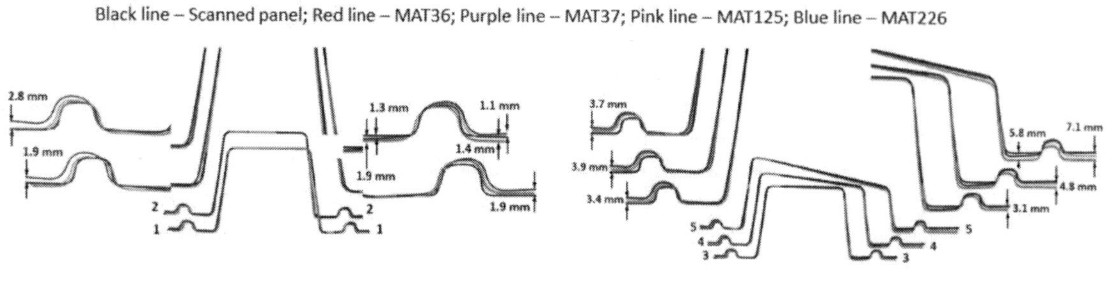

(a) Sections in the uniform channel region (b) Sections in the transition and aggressive regions

Figure 10. Springback comparison between the material models with and without kinematic hardening.

The testing data in Figure 2 and 3 indicate that aluminum alloys have weak Bauschinger effect and less kinematic hardening, as compared with other materials. The comparisons discussed above still demonstrate a moderate improvement in springback prediction using the advanced material models with the consideration of the Bauschinger effect and material kinematic hardening behaviour. In particular, a significant improvement has been observed in the aggressive region of the shotgun part where sophisticated deformation creates a complicated springback phenomenon. For aluminum autobody closures, springback is mainly dominated by twist springback. Based on the findings above, it can be concluded that the advanced material models should be used in order to obtain more accurate springback prediction and thus more reliable springback compensation.

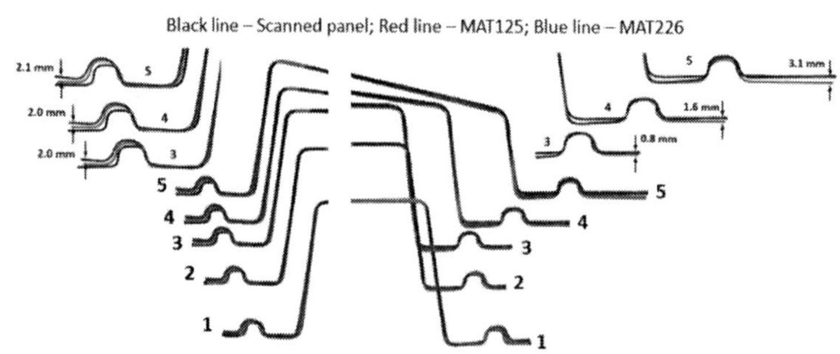

Figure 11. Comparison in springback simulation results between MAT125 and MAT226.

5.2 Influence of yielding criterion

In aluminum stamping simulations, part formability evaluation and springback prediction are always associated with each other. Depending on testing data availability, the advanced material models of MAT125 and MAT226 may be used interchangeably. Even though both the models include the Yoshida nonlinear kinematic hardening rule, the yield criterion incorporated in the models are different. The influence of the Barlat89 and Hill48 models on springback prediction was investigated with the Yoshida model parameters derived from the three-cycle T-C tests. The associated springback results are compared with the scanned data as shown in Figure 11. Overall, the predicted cross-section profiles using these two models are close to each other and also to the scanned profiles. The difference in springback prediction between these two models is underlined in the aggressive region where the open-angle, side-wall cur and twist springback are all encountered. Figure 11 clearly reveals that MAT125 has better springback predictions at the left-side of the first contact line A-B, while predicts the worse results at the right side of the line A-B. Since MAT226 can provide more accurate results in forming simulations, it is therefore recommended to use it for springback analysis of aluminum stampings having complex geometries.

5.3 Influence of loading mode in Yoshida model fitting test

Yoshida model fitting test is usually carried out in a cyclic T-C loading mode. Technically, the test can also be conducted in a cyclic C-T loading mode. This reverse loading mode imposes the Bauschinger effect in the tension with early re-yielding, permanent stress softening, and work-hardening stagnation as demonstrated in Figure 3. In real stamping operations, both T-C and C-T loading modes may simultaneously exist on a part at different locations. It is of great interest to investigate the influence of the loading mode on the determination of the Yoshida model parameters and further on springback simulation results. In the current study, the three-cycle T-C and C-T testing data were used to calibrate the Yoshida model parameters for springback simulations. The corresponding parameter values using these two sets of testing data are compared in Table 1. Apparently, the loading mode applied in the test has significant effect on calibrating the Yoshida model parameters.

Table 1 Yoshida model parameters calibrated with the testing data from T-C and C-T loading

Loading mode	CB (MPa)	Y (MPa)	SCI	K (MPa)	RSAT	SB (MPa)	H	C1	C2
T-C	106.0	68.0	950.0	19.0	160.0	16.0	0.3	0.003	0.22
C-T	132.4	85.7	977.0	16.4	136.0	22.5	0.2	0.006	0.20

Figure 12 (a) and (b) compares the predicted sprung cross-section profiles with the tryout results for MAT125 and MAT226 respectively. It can be seen from Figure 12 that for the advanced material models of MAT125 and MAT226, little difference in springback prediction is observed with those two sets of the Yoshida model parameters. The notable difference may be found at the cross-section cut #5 for MAT125, where the maximum deviation in the predicted cross-section profiles between the T-C and C-T modes is 1.3 mm. Figure 12 (a) and (b) also proves that the predicted cross-section profiles using MAT125 and MAT226 with those two sets of the Yoshida model parameters are in good agreement with the scanned profiles. It can therefore be concluded that even though the loading mode in Yoshida model fitting test has big influence on the Yoshida model parameters, it has little effect on springback prediction.

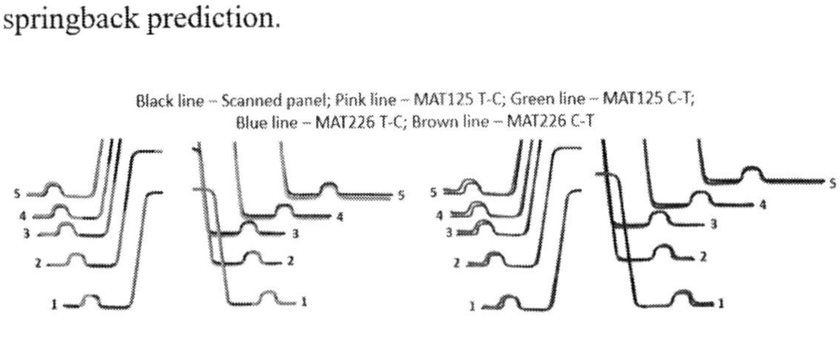

(a) MAT125 T-C vs. C-T (b) MAT226 T-C vs. C-T

Figure 12. Springback comparison between the Yoshida model parameters calibrated with the T-C and C-T testing data

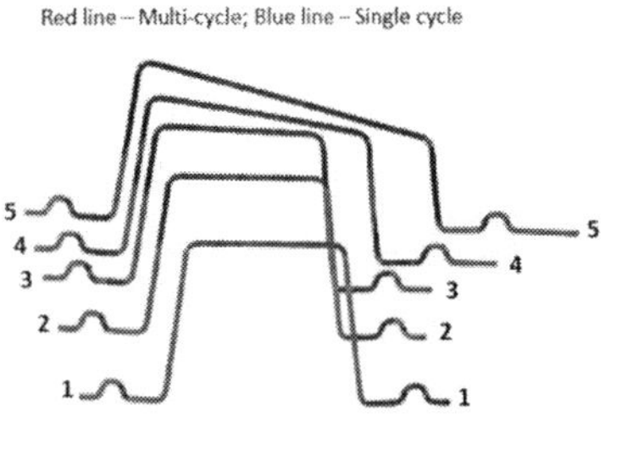

Figure 13. Influence of loading cycles on springback prediction (MAT125)

5.4 Influence of testing data from single and multiple loading cycles
Yoshida model fitting test may be performed at single or multiple T-C loading cycles. For this study, the Yoshida model parameters derived from the single cycle and multi-cycle T-C tests are compared in Table 2. Figure 13 illustrates the difference in the springback simulation profiles using the parameters

Table 2 Yoshida model parameters calibrated with the testing data from single cycle and multi-cycle T-C tests

Loading cycle	CB (MPa)	Y (MPa)	SCI	K (MPa)	RSAT	SB (MPa)	H	C1	C2
Single	101.25	90.0	911.0	19.5	146.7	28.4	0.3	0.01	0.32
Multiple	106.0	68.0	950.0	19.0	160.0	16.0	0.3	0.003	0.22

listed in Table 2 for MAT125. The comparison in Figure 13 clearly shows that there is no notable difference in springback prediction for all the profiles examined, even though a notable difference in the parameter values can be observed in Table 2. Since this benchmark part has open-angle, side-wall curl and twist springback altogether, it implies that a single cycle T-C test is good enough to fully characterize material kinematic hardening behavior for complex springback simulations.

6. Conclusions
The current benchmark study demonstrates that for aluminum autobody closures, a higher accuracy of springback prediction leading to better springback compensation can be obtained using the advanced material models considering the Bauschinger effect and material kinematic hardening behavior. It is highly recommended to use the advanced material model of MAT226 for forming simulation and springback analysis of aluminum stampings being subjected to complex deformation and springback. For simple shape parts having dominated open-angle springback, MAT125 is able to give accurate springback prediction. The investigations with the shotgun part proved that the Yoshida model parameters determined from the tension-compression test or compression-tension test have little influence on springback prediction. It was also observed from the study that a single cycle of tension-compression test can provide sufficient data to calibrate the Yoshida model parameters for springback analysis of aluminum stampings.

Acknowledgements
The author would like to express great appreciation to Dr. DJ Zhou of FCA for help in conducting the shotgun die tryout and providing the scanned data, to Dr. Fadi Abu-Farha of FADI-AMT LLC for performing the Yoshida model fitting tests, and also to Dr. XH Zhu of ANSYS for his technical support in using LS-DYNA for simulations.

References
[1] Yoshida F and Uemori T 2002 *Int. J. Plasticity* **18** p 661
[2] Yoshida F and Uemori T 2003 *Int. J. Mech Sci* **45** p 1687
[3] Chen XM, Shi MF, Zhu XH, Du CQ, Xie ZC, Xu SG and Wang CT 2009 *SAE Tech Paper,* **2009-01-0981**
[4] Shi MF, Zhang L and Zhu XH 2016 *J. of Phys.: Conf. Ser.* **734** 032118
[5] Aryanpour A and Green DE 2012 *12th Int. LS-DYNA Users Conf.*
[6] Tamura S, Sumikawa S, Uemori T, Hamasaki H and Yoshida F 2011 *Mater Trans* **52** p 868
[7] Zhou DJ, Du CQ, Hsiung CK, Schmid K, Ren F and Liasi E 2016 *in: Proc. of IDDRG* (Stahlwelt Linz) p 623

IOP Conf. Series: Materials Science and Engineering **967** (2020) 012013 doi:10.1088/1757-899X/967/1/012013

Crystal Plasticity Simulation of the AZ31B Alloy Sheet under Uniaxial Deformation Considering Grain Boundary Sliding

Zihan Li[1], Dayong Li[1,2,*], Weiqin Tang[1], Huamiao Wang[1] and Yinghong Peng[1]

[1] State Key Laboratory of Mechanical Systems and Vibration, Shanghai Jiao Tong University, Shanghai 200240, China

[2] Materials Genome Initiative Center, Shanghai Jiao Tong University, Shanghai 200240, China

* dyli@sjtu.edu.cn

Abstract. The aim of present work is to develop a crystal plasticity modeling approach to integrate slip, dynamic recrystallization (DRX) and grain boundary sliding (GBS) for simulating the deformation and texture evolution of magnesium alloys deformed at the high temperature. A GBS model is developed to evaluate strain and grains' rotation induced by GBS, and implemented into the polycrystal plasticity framework VPSC. The VPSC-DRX-GBS model can well reproduce the stress-strain curves and texture evolution. The calculated GBS contribution ratio in tension is obviously higher than in compression due to easier cavity nucleation on grain boundaries under tensile creep-aging conditions. The significantly different texture development in tension and compression deformation due to GBS is well reproduced by the proposed model.

1. Introduction

Magnesium (Mg) alloys have attracted much attention due to their light weight potential especially in the transportation industry in the past two decades. Mg alloy sheets generally exhibit poor ductility at room temperature due to their hexagonal close packed (HCP) crystal structure with limited slip systems. Their ductility can be enhanced at high temperatures which is partially attributed to the dynamic recrystallization (DRX) and grain boundary sliding (GBS) [1].

According to previous experimental research, the GBS involved deformation of Mg alloys usually is accompanied by DRX and includes two stages. In the early stage, the deformation is mainly accommodated by dislocation slip and DRX, through which the grains are refined; in the second stage, GBS plays an important or even dominant role by taking advantage of the already refined grains [2]. The plastic deformation and corresponding texture evolution depend on the combined action of three different mechanisms: dislocation slip, DRX and GBS. Currently, there is no comprehensive modeling method to incorporate the three mechanisms, which can realize the integrated simulation of texture evolution and mechanical response in GBS deformation. The aim of present work is to develop a crystal plasticity modeling approach to integrate slip, DRX and GBS for the first time in order to simulate the deformation and texture evolution of magnesium alloys deformed at the high temperature.

2. Polycrystal plasticity modeling

A commercial AZ31B-H24 alloy sheet was used in the present work, which was previously studied by Zhou et al. [3, 4]. The uniaxial tension and compression tests at 300°C along rolling direction (RD) at a strain rate of 0.0018/s, as well as subsequent optical microscopy (OM) and X-ray diffraction (XRD) observation have been introduced in detail elsewhere [4]. In this paper, GBS mechanism is incorporated into the visco-plastic self-consistent (VPSC)-DRX framework [5] to develop a polycrystal plasticity based method for simulating the deformation of the AZ31B sheet at 300°C.

2.1. Grain rotation mediated by GBS

In the GBS regime, initial basal texture of the rolled and annealed magnesium alloys became more random with the increase of deformation, and strain rate and GBS contribution are critical factors for grains rotation. In the present work, a model to introduce GBS induced grain rotation is proposed as below.

When GBS takes effect, grains affected by GBS are randomly selected to rotate in addition to the rotation induced by dislocation slip. The number of the picked grains is proportional to the ratio of GBS strain rate to total strain rate. Therefore, the expectation of rotated grains z_{GBS} is determined by

$$z_{GBS} = \sum_{i=1}^{n_g} \eta_g \xi_{GBS,i} \tag{1}$$

where η_g is a fitting parameter; n_g is the total grain number; $\xi_{GBS,i}$ is the GBS contribution ratio for grain i.

$$\xi_{GBS,i} = \frac{\dot{\varepsilon}_{GBS,i}}{\dot{\varepsilon}_{total}} \tag{2}$$

where $\dot{\varepsilon}_{GBS,i}$ is the GBS strain rate for the grain i, $\dot{\varepsilon}_{total}$ is the total strain rate. The extra random rotation degree of each selected grain should be within 45°.

Under GBS dominated deformation, the compatible conditions (five slip systems needed for compatible and homogeneous deformation under plastic deformation) are not necessary to be fulfilled, and grain rotation will randomize the deformation texture. In uniaxial tension of the present AZ31B sheet, the c-axis will rotate toward the sheet normal by basal slip, while prismatic slip rotates the grain around c-axis to the balance position. After the onset of GBS, the grain will rotate about c-axis by a random angle with no specific balance direction.

2.2. VPSC-GBS-DRX model

The GBS method is implemented into the VPSC-DRX framework. Figure 1 shows the flowchart of the proposed VPSC-DRX-GBS approach. The computational procedures are as follows:

1. If strain is less than the critical strain for initiating GBS, ε_{GBS} (0.2 for the present AZ31B alloy), the status of a given aggregate, including stress, strain and orientation, is updated by the VPSC model, and the total strain rate $\dot{\boldsymbol{\varepsilon}}_{total,i}$ for a grain i is equal to the plastic strain rate $\dot{\boldsymbol{\varepsilon}}_i$.
2. When the critical dislocation density for DDRX initiation is achieved, new DRX grains will nucleate, followed by growth through boundary migration.
3. If strain reaches ε_{GBS}, the GBS strain rate for the grain i is calculated by

$$\dot{\varepsilon}_{GBS,i} = \frac{6b\upsilon_D}{d_i}\sinh\left(\nu\frac{\sigma_i}{K_B T(1-D_{c,i})}\right)\exp\left(\frac{-\Delta F}{RT}\right) \tag{3}$$

where υ_D is a typical lattice vibrational frequency, and $\upsilon_D \approx 10^{13}$; d_i is the grain size; ν is the activation volume, taken as $\bar{b}^3$; σ_i is the effective stress of grain i; K_B is the Boltzmann's constant (1.38×10^{-23} J/K); T is the temperature; ΔF is the activation energy for grain boundary diffusion (92 kJ/mol for Mg alloys); R is the gas constant (8.314 J/(mol·K)).

The cavitation damage parameter D_c is introduced to characterize the process of void nucleation, growth and coalescence during high-temperature deformation. For a grain i, $D_{c,i}$ is determined by the rupture strain ε_f and GBS strain rate as below, considering the difference in tension and compression.

$$\begin{cases} D_{c,i} = 0, \sigma_{m,i} < 0 \\ \dot{D}_{c,i} = \dfrac{1}{3\varepsilon_f}\dot{\varepsilon}_{GBS,i}, \sigma_{m,i} \geq 0 \end{cases} \tag{4}$$

where $D_{c,i}$ is permanently set to zero in compression, and the hydrostatic pressure $\sigma_{m,i} = \frac{1}{3}(\sigma_{1,i} + \sigma_{2,i} + \sigma_{3,i})$.

4. Update the macroscopic plastic strain rate $\dot{\boldsymbol{\varepsilon}}$, in which the weighted average of the plastic strain rate over the aggregate has to coincide with the macroscopic counterpart, i.e.

$$\dot{\boldsymbol{\varepsilon}} = \langle \dot{\boldsymbol{\varepsilon}}_i \rangle = \langle \dot{\boldsymbol{\varepsilon}}_{total,i} - \dot{\boldsymbol{\varepsilon}}_{GBS,i} \rangle \tag{5}$$

where $\dot{\boldsymbol{\varepsilon}}_{GBS,i}$ and $\dot{\boldsymbol{\varepsilon}}_i$ represent strain rates of grain i due to GBS and dislocation slip respectively and the bracket "$\langle\ \rangle$" denotes the average over grains, weighted by associated volume fractions.

5. Calculate stress and strain by using the macroscopic plastic strain rate in the framework of VPSC, and update effective stress σ_i for grains.

6. Calculate rotation of grains by Eqs. (1) and (2).

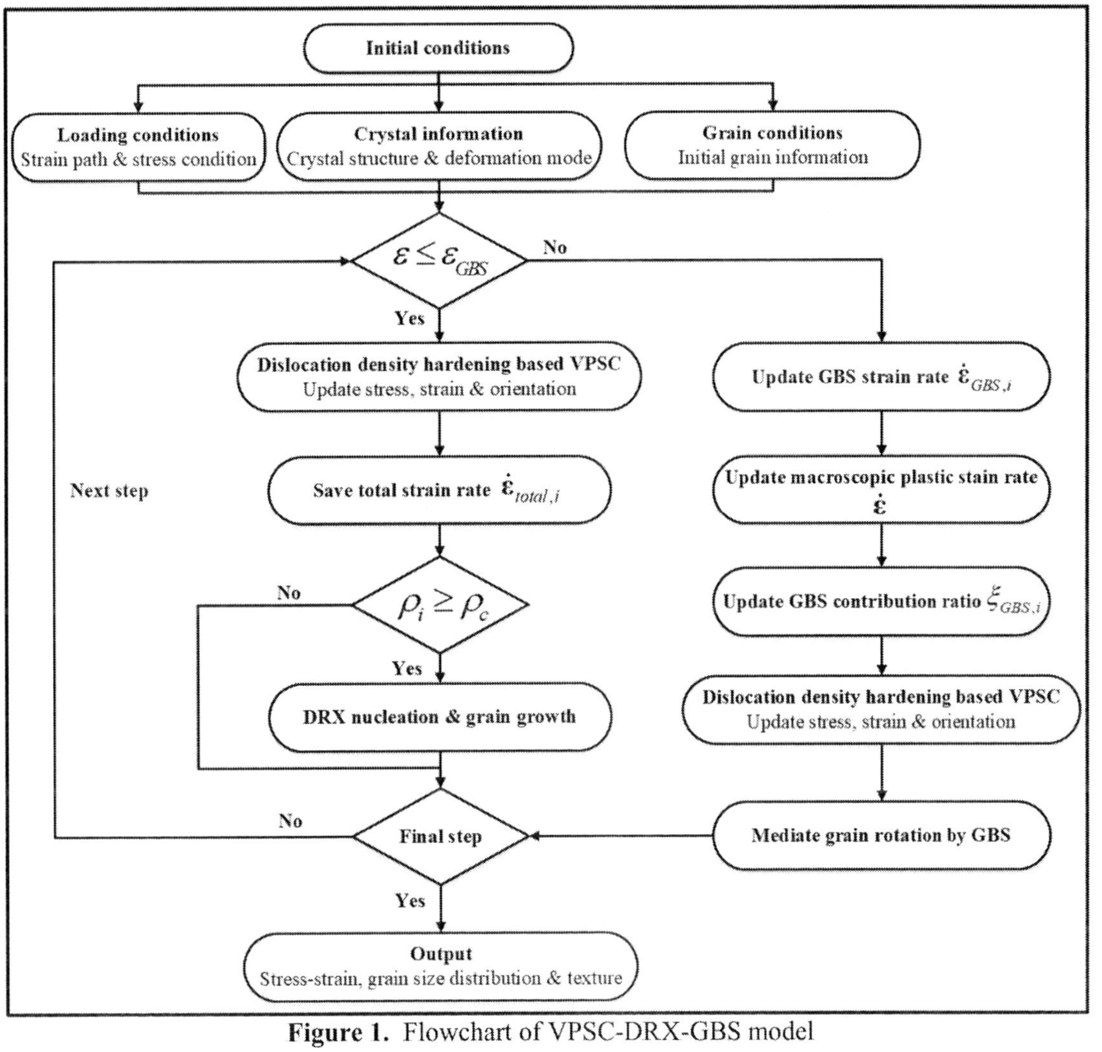

Figure 1. Flowchart of VPSC-DRX-GBS model

3. Simulation results

3.1. Flow stress

Figure 2 shows the comparison between the experimental and fitted stress-strain curves. The calculated results by the VPSC-DRX-GBS model agree well with the experimental results of RD tension and RD compression. For comparison, the stress-strain curves without considering GBS are also calculated and shown in Figure 2, by turning the GBS module off. Without consideration of GBS, the stress-strain curve without GBS fits the experimental result for RD compression, while the calculation obviously overestimates the stress after 0.2 strain for RD tension, and cannot reproduce the softening behavior. GBS gives rise to more softening in tension than in compression, because contribution of GBS to deformation is significantly reduced in compression.

Generally, there is no evident difference between tension and compression for the softening by DRX, and the DRX will lead to the grain refinement. In contrast, due to the greater stress softening in tension and stable grain size distribution after 0.2 strain for the AZ31B Mg sheet [4], the softening behavior cannot be attributed to DRX, but GBS after 0.2 strain.

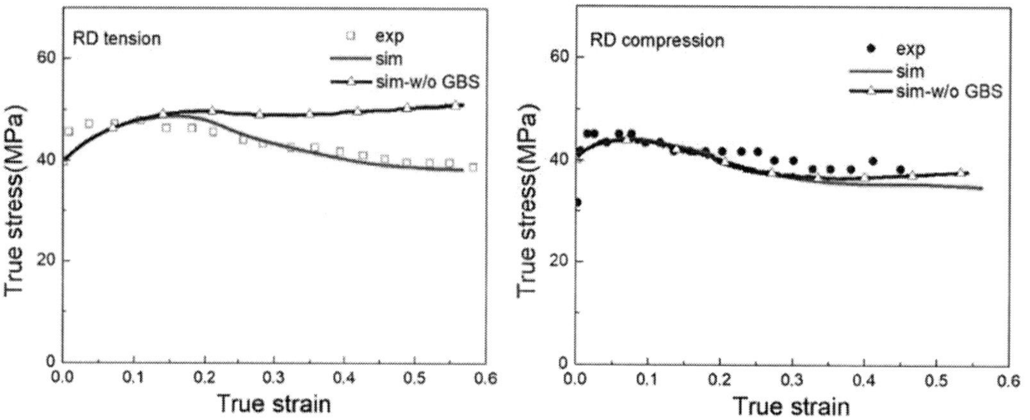

Figure 2. The experimental and fitted stress-strain results of the AZ31B sheets at 300°C. ("w/o GBS" represents "without GBS".)

3.2. Texture analysis

The calculated texture at 0.4 strain in the form of pole figures is compared with experimental results in RD tension and RD compression, as shown in Figure 3. The experimental texture was tested from XRD measurement, and more details about the experimental texture development will be exhibited elsewhere. The predicted texture in RD tension considering GBS is consistent with measurements, and texture intensity with GBS module is evidently reduced compared to that without considering GBS. In RD tension, the dominantly active basal <a> and prismatic <a> slip modes cause the suppression of basal texture in the (0002) pole figure and formation of hexagonal maxima distribution in the (10-10) pole figure. With the enhancement of GBS, random rotation of grains reduces overall texture intensity, which results in more disperse distribution texture in the (0002) pole figure, and hexagonal maxima distribution in the (10-10) pole figure disappears.

The predicted texture in RD compression is also consistent with measurements, while texture intensity is slightly reduced with GBS module. In compression, the texture distributions with and without considering GBS are similar due to less activation of GBS. In RD compression, the activation of the basal slip <a> and prismatic <a> slip modes split the basal texture in the (0002) pole figure and results in two maxima distribution in the (10-10) pole figure [3].

Therefore, RD tension and RD compression show the obvious difference between tension and compression in terms of texture development, which can be attributed to less activation of GBS in compression.

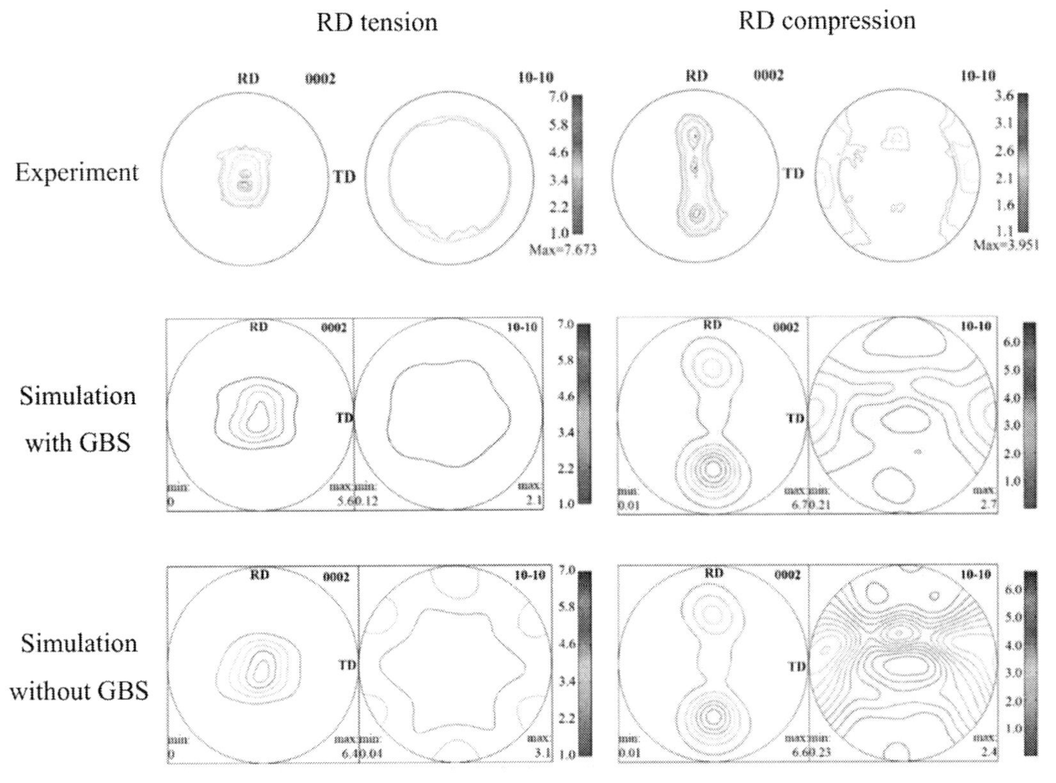

Figure 3. The comparison of texture between simulations and experiments at 0.4 strain: (a) RD tension and (b) RD compression.

3.3. GBS contribution

The calculated mean GBS contribution ratios in RD tension and RD compression, with the proceeding of deformation after 0.2 strain, are presented in Figure 4. The GBS contribution ratio in RD tension is obviously higher than in RD compression. The evolution trends of GBS contribution ratio vary with the loading modes. For RD tension, the value of GBS contribution ratio rises steadily, with increasing strain. For RD compression, the GBS contribution ratio decreases slowly from 0.16 to 0.14, resulting from the stress softening.

The significant reduction of the GBS contribution ratio in the compression mode is calculated by the current VPSC-DRX-GBS model. Much higher tension GBS contribution ratio than compression can be attributed to the easier cavity nucleation on grain boundaries under tensile creep conditions.

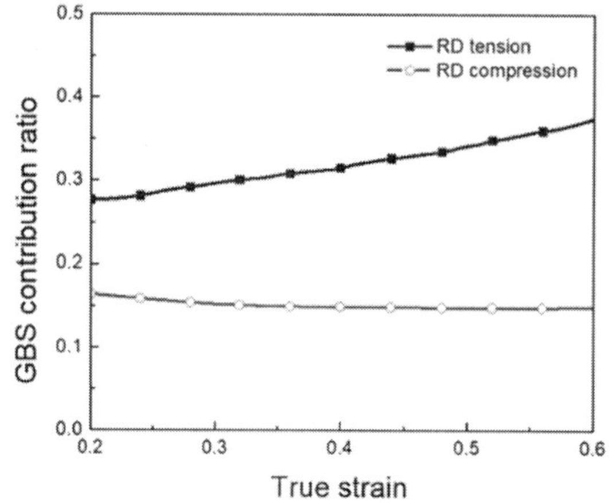

Figure 4. The evolution of mean GBS contribution ratio to overall strain rate

4. Conclusions

In this work, a VPSC-DRX-GBS model is developed by introducing GBS model into the VPSC-DRX framework to simulate the deformation and texture evolution of the Mg alloy. The following conclusions can be drawn.

1) A polycrystal plasticity based integration modeling method incorporating GBS is developed, and the stress-strain curves and texture evolution can be simultaneously calculated.
2) By introducing the cavitation damage parameter D_c into GBS model, the polycrystal plasticity model can reproduce the tension-compression difference induced by GBS.
3) Due to GBS, tension and compression deformation results in significant different texture development. The proposed model for GBS mediated grain rotation can well reproduce the texture development. In RD tension, the random grains' rotation induced by GBS produces more disperse distribution texture in the (0002) pole figure, and hexagonal maxima distribution disappears. In the compression modes, GBS effect is much weaker, which hardly influences texture distribution.

Acknowledgments

The authors would like to acknowledge the support of the National Natural Science Foundation of China (No. 51675331, 51775337), Major Projects of the Ministry of Education (No. 311017).

References

[1] Somekawa H, Kinoshita A, Washio K and Kato A 2016 *Mater. Sci. Eng. A* **676** 427–33
[2] Panicker R, Chokshi A H, Mishra R K, Verma R, Krajewski P E, 2009 *Acta Mater.* **57** 3683–93
[3] Zhou G, Li Z, Li D, Peng Y, Wang H, Wu P 2018 *Mater. Sci. Eng. A.* **730** 438–56
[4] Zhou G, Jain M K, Wu P, Shao Y, Li D, Peng Y 2016 *Int. J. Plast.* **79** 19–47
[5] Zhou G, Li Z, Li D, Peng Y, Zurob H S, Wu P 2017 *Int. J. Plast.* **91** 48–76

IOP Conf. Series: Materials Science and Engineering **967** (2020) 012014 doi:10.1088/1757-899X/967/1/012014

Reverse-loading coefficients identification for updated homogeneous anisotropic hardening model

Shin-Yeong Lee*, Jin-Hwan Kim, Frédéric Barlat

Graduate Institute of Ferrous Technology, Pohang University of Science and Technology, 77 Cheongam-ro, Nam-gu, Pohang, Gyeongbuk 790-784, Republic of Korea

E-mail: sylee413@postech.ac.kr, jinkim@postech.ac.kr, f.barlat@postech.ac.kr

Abstract. The homogeneous anisotropic hardening (HAH) model formulation, which is solely based on distortional plasticity, was improved recently. In this study, a robust coefficient identification scheme was developed for this updated model. Tension-compression tests were conducted on a mild steel sheet sample to assess the hardening behavior in the forward-reverse loading mode. An inverse identification procedure was employed to calibrate the coefficients of the model. The Nelder-Mead simplex and genetic optimization methods were investigated and the results regarding the accuracy of the calibration were compared. As a result, both optimization methods led to reasonable coefficients for the new model.

1. Introduction

The effect of a strain path change is one of the important factors to consider for the prediction of accurate sheet metal forming simulation results. For example, many researchers have shown that the modeling of the Bauschinger and related effects after load reversal is essential for an accurate prediction of springback [1]. Although kinematic hardening is a classical approach to capture the Bauschinger effect, a recently proposed distortional plasticity concept [2] was shown to describe strain path change effects reasonably well. The homogeneous anisotropic hardening model, so-called HAH, is formulated based on pure distortional plasticity without back-stress. The application of the HAH model in finite element simulations of forming processes has led to a good prediction of springback in the U-draw bending test [3,4]. However, some researchers found out that the HAH model exhibits a singularity for strain path changes close to cross-loading [5]. In addition, the permanent softening behavior under cross-loading conditions could not be predicted satisfactorily in the original HAH model.

Recently, Barlat et el [6] proposed a revised version of the HAH model to mitigate these issues. The updated HAH model, called HAH_{20}, showed indeed a much-improved capability to predict the material behavior after strain path change between pure reverse loading and pure cross-loading. In addition, the singularity problem was virtually eliminated by updating the evolution equations of the so-called microstructure deviator, a tensorial state variable.

However, a robust calibration method should be established in order to proliferate the use of this updated HAH model adequately. Some researchers have already developed various calibration methods for the original HAH model [7,8] but the robustness of which was not demonstrated. The calibration of the coefficients is an important aspect of material modeling because an unsuitable coefficient set could lead to problems in numerical simulations [4].

The aim of this work is to investigate optimization methods for the calibration of the HAH_{20} model. Since five coefficients are available to describe the reverse loading effects, including permanent softening, the possibility to end up the calibration at the local minima of the objective function is likely.

Therefore, an adequate optimization scheme is necessary to avoid these traps. Two optimization methods, the Nelder-Mead simplex and genetic algorithm are considered in this work.

2. Experiments

2.1. Tension-compression test

A custom-made tension-compression testing machine was used to characterize the material behavior under reverse loading cycles. A 1.2 mm thick EDDQ (Extra Deep Drawing Quality) mild steel was investigated. Details about the specimen size and testing equipment are given elsewhere [9]. The holding force preventing buckling was set to 1.96 kN and the displacement rate of the moving grip to 0.03 mm/s. The stress state was reversed when the engineering strain reached the values of 2, -2, 4, -4 and 4 %. Therefore, the stress-strain curve was composed of 2 full cycles after a pre-strain. The final accumulated strain was approximately 0.28, which is considered large enough for various sheet metal forming simulations. After an experiment, a correction scaling factor was multiplicatively applied to the measured stress to compensate for the friction effect. The biaxial stress effect was neglected as the normal force was sufficiently small. Figure 1 shows the true stress-true strain and absolute true stress-accumulated true strain curves.

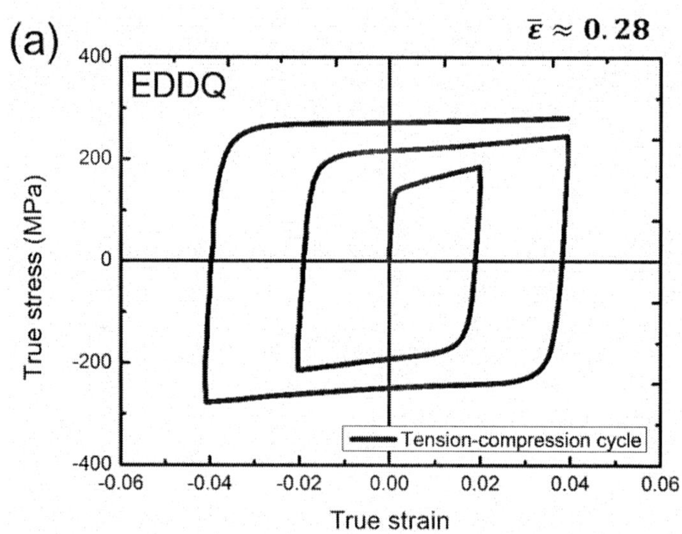

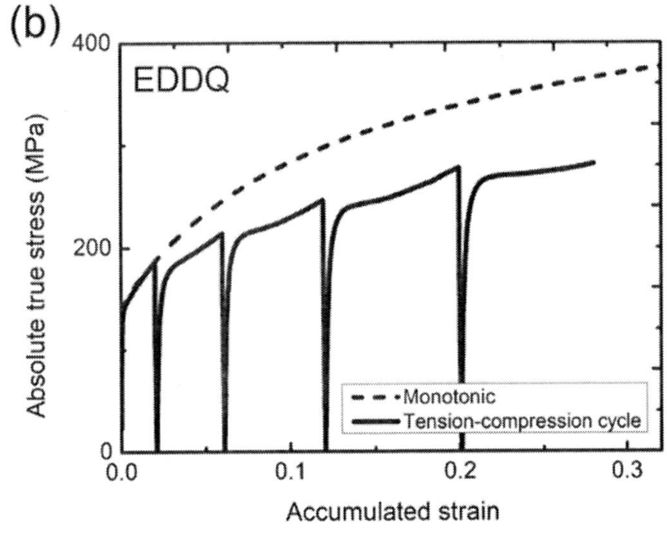

Figure 1. (a) True stress-strain curve and (b) Absolute true stress-Accumulated strain curve in the tension-compression test for EDDQ steel

3. Inverse identification

Two constitutive models are required before the identification of the HAH_{20} model. First, an isotropic hardening equation that represents monotonic loading is calibrated using the combination of uniaxial tension and bulge test data. The combination curve was made by the same method in reference [10]. This experimental data was able to cover an effective strain range of up to 0.9. The Swift-modified Voce hardening law, defined as

$$\bar{\sigma} = K(\varepsilon_0 + \bar{\varepsilon})^n + \rho\bar{\varepsilon} + \sigma_y + \sigma_b(1 - exp(-\eta\bar{\varepsilon})) \tag{1}$$

was selected for the isotropic hardening model. The inverse method [11] was adopted to determine the Swift-modified Voce coefficients using the Nelder-Mead simplex optimization method. Table 1 shows the coefficients of the isotropic hardening model. The other is yield function that could represent material anisotropy. The von-Mises isotropic yield function is selected in this study.

Table 1. Coefficients of Swift-modified Voce equation

Swift-modified Voce	K	ε_0	n	ρ	σ_y	σ_b	η
	326.2	0.013	0.313	7.526	55.14	89.45	12

The new model, HAH_{20}, employs enhanced evolution equations for the state variables, including a second microstructure deviator which delays the evolution of the original deviator. In addition, the influence of the hydrostatic pressure on plasticity was incorporated in the yield condition, allowing the occurrence of a strength-differential (SD) effect between tension and compression. In this work, the pressure coefficient was set to the constant value of 20^{-1} TPa, as recommended in [12], which corresponds to flow stress about 5% higher in compression compared to tension.

The formulations of the new model, HAH_{20}, are not described in this article but the reader is referred to Reference [6] for details. The five reverse loading coefficients, which include permanent softening, are denoted k_1, k_2, k_3, k_4 and k_5 as in the original paper. The role of these five coefficients is similar to that of the former model [2], although the state variable evolution equations were modified. k_1 controls the transient change of the hardening rate just after reversal. k_2 and k_3 controls the reloading stress after unloading and reversal. k_4 and k_5 are related to the rate and limit of permanent softening. In this work, these five coefficients were calibrated using inverse identification. The experimental tension-compression curve was compared with the simulated curve obtained from a stand-alone FORTRAN code containing the constitutive model to evaluate an error function. This error (or objective) function was minimized by an optimization method.

Two optimization methods were adopted to use inverse identification [11]. One is the Nelder-Mead simplex method (NM) and the other is the genetic algorithm (GA) both of which built-in MATLAB software. Both NM and GA algorithms require a search domain, which is set by selecting a large range of suitable coefficients. In contrast, only the NM algorithm requires an initial set of coefficients, which corresponds to the middle of the range for each coefficient. Table 2 lists the search ranges and initial values. Figure 2 shows the experimental and optimized curves from both optimization methods. Table 2 provides detailed values of the coefficients, mean error and computation time. Fig. 2 shows that the approximated curves using both optimization methods are in excellent agreement with the experimental data for this EDDQ steel sheet sample although both optimization methods lead to somewhat different coefficients, especially k_2. A sensitivity study is needed to assess the validity of this result. Table 2 indicates that the error percentage obtained between these methods is very similar but the computation time is much larger for the GA algorithm. This is not a significant issue because the computation time is still acceptable for calibration and, more importantly, because the GA algorithm is expected to converge towards the global minimum of the objective function.

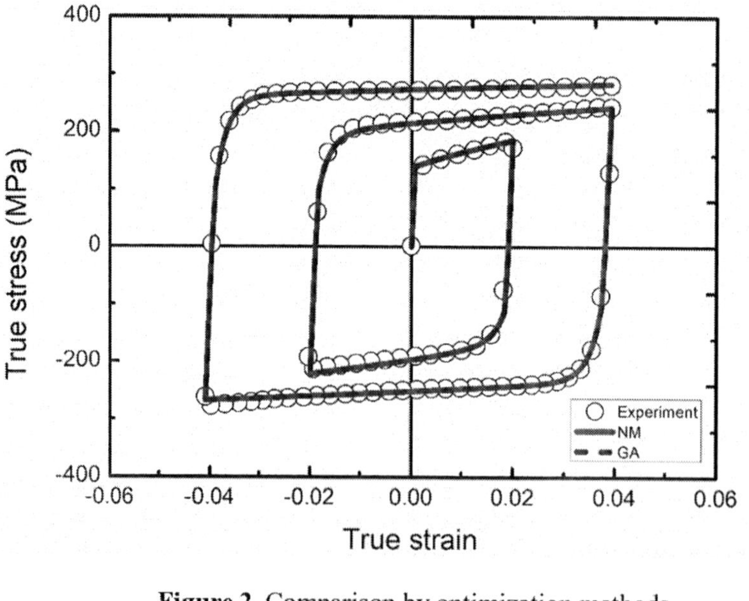

Figure 2. Comparison by optimization methods

Table 2. Coefficients by optimization methods

Tension-compression in EDDQ steel	NM	k_1	k_2	k_3	k_4	k_5	Error	Time
		399.99	100.15	0.69	0.78	28.1	1.68 %	10 min
	Genetic	k_1	k_2	k_3	k_4	k_5	Error	Time
		391.12	158.87	0.75	0.78	27.9	1.52 %	312 min
	Search range (Initial value)	k_1	k_2	k_3	k_4	k_5		
		50~400 (225)	50~200 (125)	0.1~0.8 (0.45)	0.7~1 (0.85)	5~30 (17.5)		

4. Conclusion

The coefficients of the updated HAH model, HAH_{20}, related to reverse loading were calibrated from a tension-compression test data obtained on EDDQ mild steel. Two optimization methods, the Nelder-Mead simplex and genetic algorithms were used to calibrate the coefficient. Both optimizations led to similar curves, in excellent agreement with the experiment, but one coefficient, k_2 was significantly different from one model to the other. Additional research is required to understand this result.

Acknowledgments
The authors gratefully acknowledge the support of the POSCO.

References

[1] Wagoner R H, Lim H and Lee M-G 2013 Advanced Issues in springback *Int. J. Plast.* **45** 3–20
[2] Barlat F, Vincze G, Grácio J J, Lee M G, Rauch E F and Tomé C N 2014 Enhancements of homogenous anisotropic hardening model and application to mild and dual-phase steels *Int. J. Plast.* **58** 201–18
[3] Lee J-Y, Lee J-W and Lee M-G 2012 An application of homogeneous anisotropic hardening to springback prediction in pre-strained U-draw/bending *Int. J. Solids Struct.* **49** 3562–72
[4] Choi J, Lee J, Bong H J and Lee M-G 2018 Advanced constitutive modeling of advanced high strength steel sheets for springback prediction after double stage U-draw bending *Int. J. Solids Struct.* **151** 152–64
[5] Qin J, Holmedal B and Hopperstad O S 2019 Experimental characterization and modeling of aluminum alloy AA3103 for complex single and double strain-path changes *Int. J. Plast.* **112** 158–71
[6] Barlat F, Yoon S-Y, Lee S-Y, Wi M-S and Kim J-H 2020 HAH_{20} distortional plasticity framework including pressure effect *Int. J. Solids Struct. subimitted*
[7] Ha J, Lee M G and Barlat F 2013 Strain hardening response and modeling of EDDQ and DP780 steel sheet under non-linear strain path *Mech. Mater.* **64** 11–26
[8] Zaman S Bin, Barlat F and Kim J H 2018 Deformation-induced anisotropy of uniaxially prestrained steel sheets *Int. J. Solids Struct.* **134** 20–9
[9] Lee S, Kim J-H and Barlat F 2019 Multi-Coefficient Optimization of Homogeneous Anisotropic Hardening Model for Ahss *IOP Conf. Ser. Mater. Sci. Eng.* **651** 012018

[10] Park N, Stoughton T B and Yoon J W 2020 A new approach for fracture prediction considering general anisotropy of metal sheets *Int. J. Plast.* **124** 199–225
[11] Pottier T, Vacher P, Toussaint F, Louche H and Coudert T 2012 Out-of-plane Testing Procedure for Inverse Identification Purpose: Application in Sheet Metal Plasticity *Exp. Mech.* **52** 951–63
[12] Spitzig W A and Richmond O 1984 The effect of pressure on the flow stress of metals *Acta Metall.* **32** 457–63

Soft sensing force measurement approach in sheet metal forming facilities

Jon Olaizola[a], Aitzol Iturrospe[a], Jose Manuel Abete[a], Lander Galdos[a] and Eneko Sáenz de Argandoña[a]*

[a] *Mondragon Unibertsitatea, Loramendi 4, Mondragon 20500, Gipuzkoa, Spain*
* *esaenzdeargan@mondragon.edu*

Keywords: Stamping force, soft sensors, real time monitoring

Abstract. Process stability and facility security are the two main purposes for measuring the force in sheet metal forming facilities. These two purposes look for guaranteeing the good health of the facility avoiding overloads and at the same time look for variations in the process force that could lead to products out of the demanded quality. Traditionally there have been different approaches to measure the force exerted by the press. Piezoelectric sensors and strain gages are the most widely used technologies. However there are some common limitations that reduce the success in the measurement of the force by the aforementioned sensors: force sensors tend to have a signal drift and the introduction of the sensors together with the electronics represents an extra cost that in some cases can be high. As an alternative to the implementation of force sensors, soft sensing approaches take advantage of the already available signals in the facility to estimate non measured variables, the force in this case. The present work aims at predicting the force curve that a servo press exerts in forming processes reading the internal servomotor signals. The approach has been evaluated with several forming processes achieving an error lower than 3% for all the cases.

1. Introduction

Monitoring plays a crucial role in modern industrial processes and machines. It provides useful information about the condition of the machine and the process quality. Furthermore, monitoring can detect signs of malfunction that may cause the machine's failure and its subsequent downtime. In the specific case of a servo press, a press that is driven by a servo motor, the process force has a high impact in the quality of the manufactured part. The monitoring of the force can reveal aspects related to the quality of the produced part, such as information about the profile of the applied force that it was subjected to during a stroke. The industry's classical approach for monitoring press force is to place sensors on machine components. Due to ease of installation and the little wear they suffer compared with other locations, such sensors are commonly installed on the press frame and the connecting rods [1]. However, force sensors located in forming tools provide more precise information, since they are closer to the process. When the tool is replaced to produce a new component, these sensors are usually replaced, which increases the cost of monitoring [1]. These sensors also experience higher levels of wear than those mounted on the frame or connecting rods. It is also well known that force sensors tend to lose their initial calibration over the working cycles, and they need periodic maintenance for recalibration [2].

IOP Conf. Series: Materials Science and Engineering **967** (2020) 012015 doi:10.1088/1757-899X/967/1/012015

To address the increased costs and the recalibration need associated with physical sensors, other solutions, such as model-based soft(ware) sensors (MBSS), can be adopted, which take advantage of the already existing signals and a model of the targeted machine. A MBSS is a software-based sensor that uses a mathematical representation (model) of the machine and/or process and some measurements to provide estimations [3]. Models are described by state variables (also known as states, as an example the ram position or velocity) that determine the behaviour of the system. At the same time, the system can be driven by external inputs. Those inputs can be either known or unknown depending on whether they are measured or unmeasured. In the case of servo presses, the press force is modelled as an unmeasured input. The signals utilized by the control of the system (in this case torque and angular position of the servomotor) can be used by soft sensors to estimate states and the unknown input (unmeasured force) of the servo press.

In this paper the dual particle filter algorithm presented by [4] has been employed to monitor different metal forming processes. The dual particle filter is an algorithm able to estimate in real time both the states of the system as well as the unknown inputs. Therefore, it takes advantage of the already available signals of the system provided by the installed sensors, avoiding the installation of new ones. The rest of the paper is organized as follows. Section II presents the model of the servo press. Section III briefly describes the proposed dPF-based approach. Section IV shows the experimental results of three different experiments conducted for three different metal forming operations. Finally, Section V provides the conclusions that were extracted from the developed work.

2. Servo press model

A servo press is formed by mechanical and electric components so that it can perform the forming of workpieces. Its operation is powered by a servomotor which creates a rotary motion that is translated into the linear motion of the ram by means of the kinematic chain that is in between. Permanent magnets synchronous motors (PMSP) are the most widely used servomotors to feed servo presses. PMSMs are conventionally controlled by the field oriented control scheme, which takes advantage of the servomotor's angular position and electric signals such as voltages and currents of the different phases that are present in the servomotor.

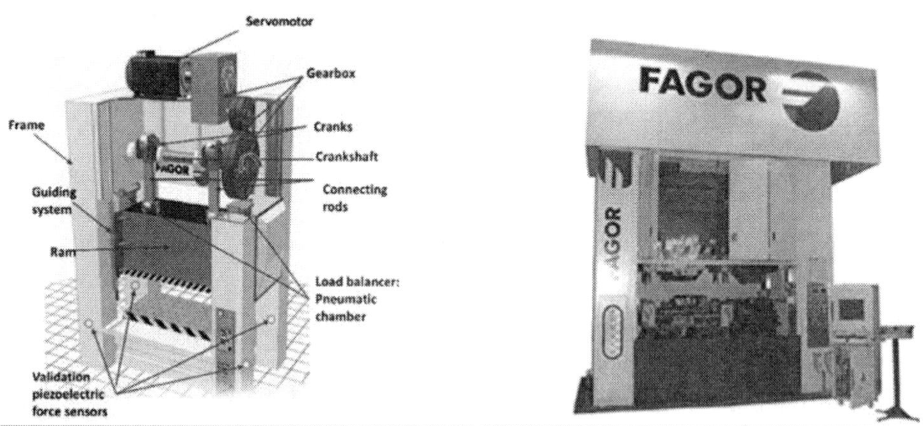

Press capacity (kN)	4,000 at 20 mm from the BDC
Number of connection rods	2
Working torque Maximum / Nominal (N.m)	5,5000 / 3,000
Max stroke (mm)	400
Die height (mm)	1,000 to 1,2000 (str 400mm)
Stroke adjustment (mm)	200

Table size (mm x mm)	2,400 x 1,200
Max number of stroke (spm)	40
Die cushion capacity (kN)	400
Die cushion stroke (mm)	150
Main motor power Maximum / Nominal (kW)	450 / 250

Figure 1. Components and characteristics of Fagor Arrasate's SDM2-400-2400-1200 servo press.

The mechanical part of the servo press is based on a ram-crank mechanism, as illustrated in Figure 1. Its transmission chain is comprised of a gearbox, a crankshaft, two connecting rods and a ram, which transforms the electric torque τ_e generated by the servomotor into a linear force. A load balancer is also included in the model, since it actuates in the real system compensating the weight of the ram. The model also integrates a friction model that explains some of the friction phenomena affecting the performance of the servo press. The components and characteristics of the servo press used at the present research work are shown in Figure 1. The schematic representation of the mechanical components of the servo press and the complete modelling of the servo press are shown in Figure 2. The modelling of a ram-crank type servo press was carried out by the Lagrange method presented in [5] and [6]. All the components considered in the model are shown in left side of Figure 2 and its representation in the model is shown in right side of Figure 2. All symbols in the modelling equation can be found in [4].

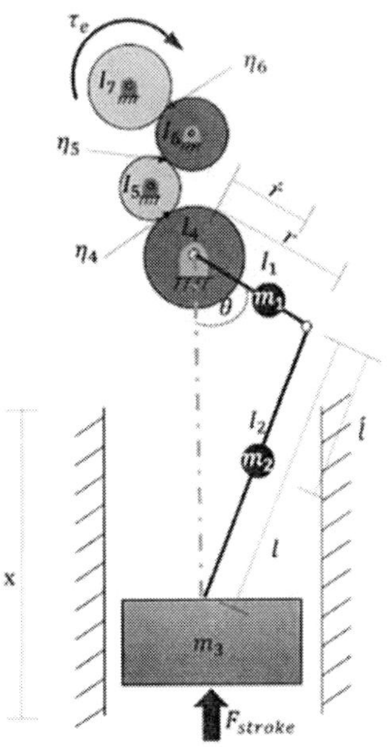

$$
\begin{aligned}
&\ddot{\theta}\Bigg(I_1 + m_1\acute{r}^2 + \frac{I_2 r^2 \cos^2\theta}{c^2} + \frac{(l-\acute{l})^2}{l^2} m_2 r^2 \cos^2\theta \\
&+ \sin^2\theta\left(m_2 r^2 \frac{(r\acute{l}\cos\theta + lc)^2}{l^2c^2} + m_3 r^2 \frac{(r\cos\theta + c)^2}{c^2}\right) + I_4 \\
&+ I_5\eta_4^2 + I_6(\eta_4\eta_5)^2 + I_7(\eta_4\eta_5\eta_6)^2\Bigg) \\
&+\frac{1}{2}\dot{\theta}^2\Bigg(\frac{2I_2 r^4 \cos^3\theta \sin\theta}{c^4} - \frac{2I_2 r^2 \cos\theta \sin\theta}{c^2} \\
&-\frac{2m_2 r^2 \cos\theta \sin\theta\,(l-\hat{l})^2}{l^2} \\
&+ 2\sin\theta\cos\theta\left(m_2 r^2 \frac{(r\acute{l}\cos\theta + lc)^2}{l^2c^2} + m_3 r^2 \frac{(r\cos\theta + c)^2}{c^2}\right) \\
&+ 2\sin^2\theta\Bigg(\frac{m_2 r^3 \hat{l} \sin\theta\,(r\acute{l}\cos\theta + lc)(r\cos\theta - c)(r\cos\theta + c)}{l^2c^4} \\
&+\frac{m_3 r^3 \sin\theta\,(r\cos\theta - c)^2(r\cos\theta + c)^2}{c^4}\Bigg)\Bigg) \\
&+ g\left(\sin\theta\,(m_1\acute{r} + m_2 r + m_3 r) + \sin\theta\cos\theta\left(\frac{m_2 r^2 \hat{l}}{lc} + \frac{m_3 r^2}{c}\right)\right) \\
&= \tau_e\eta - \tau_{fric} - (F_{stroke} + F_{lb} + F_{fric}) r \sin\theta\left(1 + \frac{r\cos\theta}{c}\right)
\end{aligned}
$$

Figure 2. Schematic representation of the mechanical components of the servo press (left side) and the complete model of the servo press (right side).

The servo press model is used to generate the predictions of the magnitudes of interest of the system so that the soft sensor can make inferences from those predictions and obtain the desired estimations of the dynamic response of the servo press and the process force.

3. The soft sensor: the dual particle filter

The employed algorithm, the dual particle filter (dPF) is based on the conventional particle filter (PF) algorithm which was originally formulated for the estimation of the system states that explain the dynamic response of the system, such as the angular position, angular speed and acceleration of the crankshaft of the servo press. The dual scheme of the dPF integrates two PFs, making the algorithm able to estimate not only the states of the system but also an unknown input such as the process force of the servo press.

The workflow of the algorithm is shown in Figure 3. Each PF of the dPF performs four steps to carry out the estimation of states: the Prediction & Update step, the Normalization step, the Resampling step and the Averaging step, as described in [4].

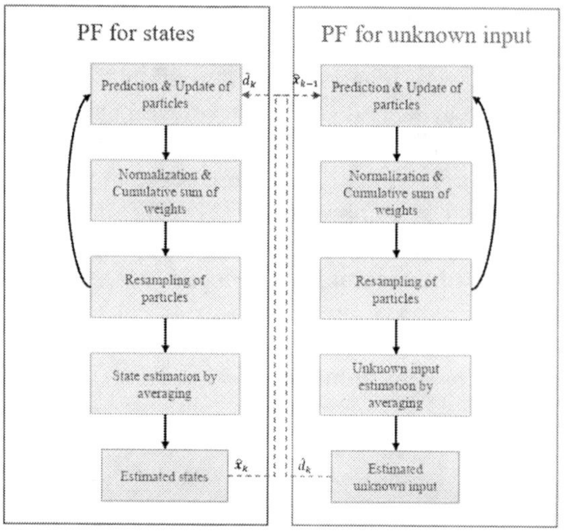

Figure 3. Workflow of the employed dPF.

4. Experimental results

The dPF was evaluated under three different industrial processes. Two measurements were employed to feed the model: the three phase currents and the angular position of the servomotor. The estimated process force was compared with actual force signals measured by two piezoelectric sensors installed in drilled holes of the two connecting rods of the servo press. The evaluated industrial processes are shown next.

4.1 Single strokes against rigid cylinders

In this process, the ram applies the force against two rigid cylinders that are placed on the ram right under the connection rods of the servo press. The main purpose of the experiments was to generate a parabolic force curve that is not very complex in shape but changes very rapidly in time. This way, the response of the prediction algorithm to sudden force changes wanted to be analyzed. The servo press was working at a cadence of 6 strokes per minute in these tests.

The maximum load that the servo press can perform is 400 t. In order to evaluate the algorithm at several force levels, four different set-ups were analysed. In order to modify the press force the ram position with respect the connection rods was changed at each of the set-ups. Several tests were carried out in order to evaluate the repeatability of the prediction algorithm finding similar results between them.

Table 1 gathers the methodological set-up to carry out the four different force levels in the experiments.

Experiment	Maximum force measured	Adjustment of ram (height)	Angular speed during force profile
STR1	88.65 t	848.0 mm	36 °/s
STR2	184.27 t	847.6 mm	36 °/s
STR3	232.19 t	847.3 mm	36 °/s
STR4	343.11 t	846.6 mm	36 °/s

IOP Conf. Series: Materials Science and Engineering **967** (2020) 012015 doi:10.1088/1757-899X/967/1/012015

Figure 4 shows four measured and estimated process forces of four different experiments with their corresponding instantaneous normalised absolute deviations. The instantaneous absolute normalised deviation is obtained by means of the absolute value of the difference between the estimated and measured signals, normalised with a factor of 400 t, which is the maximum load capacity of the servo press. The maximum deviation shown in the highest force point is given in the fourth experiment, 2.97%. Apart from that some fluctuation emerge around the 0° angular position of the crankshaft. This is due to control's activity to correct the speed deviation that occurs as the structure recovers from its elastic deformation. The control system of the servomotor was not included in the model due to the lack of information.

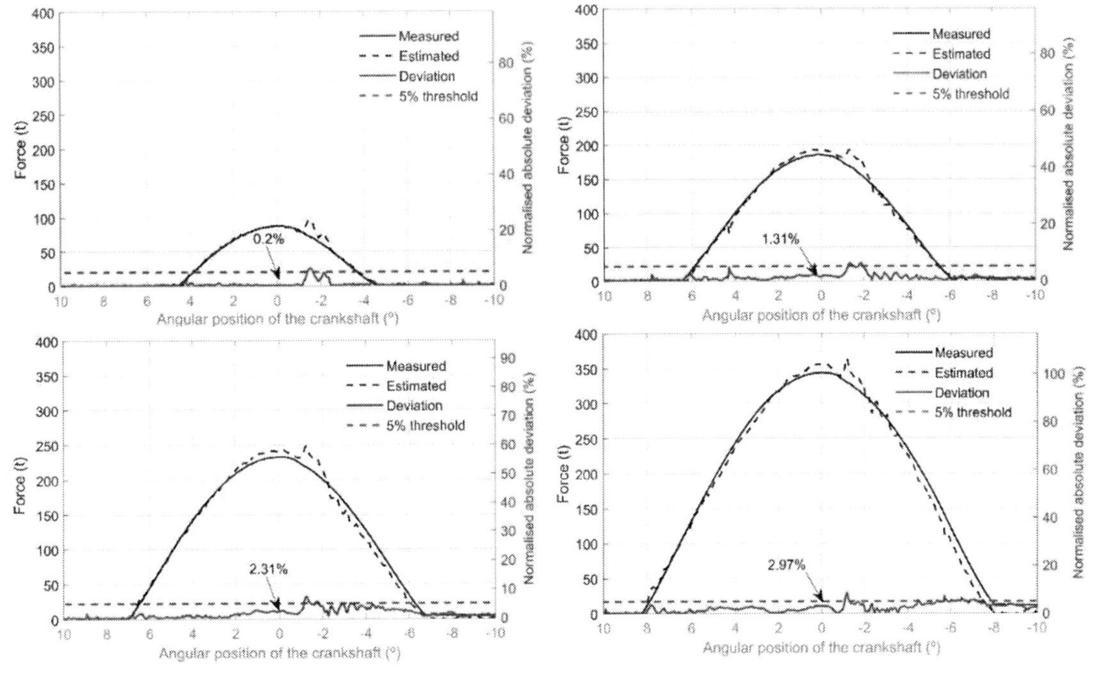

Figure 4. Measured and estimated process forces of the four experiments and the instantaneous normalised absolute deviation.

4.2 Deep drawing processes

The second type of process that was carried out to evaluate the dPF was the deep drawing process. Three different deep drawing processes were analyzed giving as a result different force. Compared to the stroke against two rigid steel cylinders, the deep drawing process is characterized by prolonged force profiles that begin in an earlier and end in a later angular position of the crankshaft. Furthermore the evolution of the force is not as homogeneous as it was in the case of the stroke against two rigid steel cylinders. The main purpose of these experiments was to evaluate if the prediction algorithm was able to calculate the force in real forming processes. The cadence of the servo press during these experiments was 20 strokes per minute for DD1 and DD2 and 6 strokes per minute for DD3.

Table 2 gathers the methodological set-up to carry out the three different deep drawing experiments.

Table 2. Experimental set-up of the deep drawing processes

Experiment	Maximum force measured	Adjustment of ram (height)	Angular speed during force profile	Sheet material
DD1	79.52 t	850.2 mm	120 °/s	DP1000
DD2	184.40 t	850.2 mm	120 °/s	DP1000
DD3	49.66 t	800.0 mm	36 °/s	DP1000

The estimation results of the three types of deep drawing processes are shown in Figure 5. Regarding the deviations at the maximum force position in the three experiments, the obtained deviations are 0.9254%, 2.1381% and 1.1715% for DD1, DD2 and DD3 respectively. Then, some fluctuation can be seen again in experiments DD1 and DD2, which are attributed again to the controls activity.

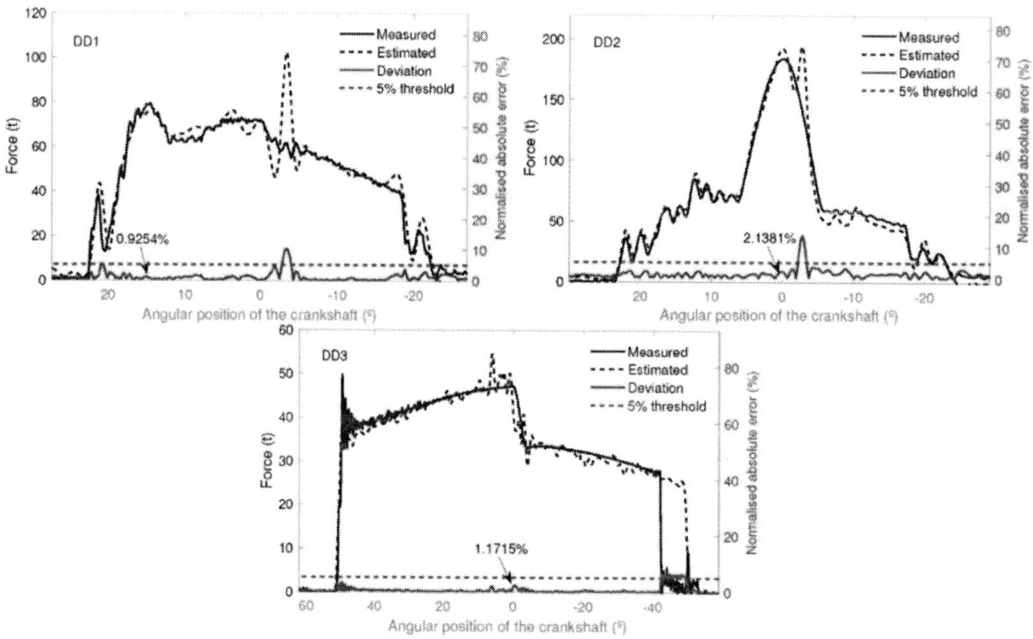

Figure 5. Measured and estimated process forces of the three deep drawing experiments and the instantaneous normalised absolute deviation.

4.3 Semi solid forging

The third process analysed at the present research work was the semi solid forging where a workpiece that is in a semisolid state is formed in a closed die. In this process the force increases until reaching the maximum value at the bottom dead centre (BDC) (lowest position of the ram), and then the ram is stopped at that position during a determined period of time forming the workpiece. The main purpose of this analysis was to evaluate the prediction capacity of the algorithm under static conditions close to the BDC of the servo press.

Table 3 gathers the methodological set-up to carry out the two different semi solid forging experiments.

Table 3. Experimental set-up of the semi solid forging processes

Experiment	Maximum force measured	Adjustment of ram (height)	Maximum force position	Angular speed during force profile	Workpiece material
SSF1	417.72 t	960.0 mm	19.12°	90 °/s	42CrMo4
SSF2	297.60 t	960.0 mm	3°	0 °/s	S48C

The results of semi solid forging experiments are shown in Figure 6. In SSF1 the maximum force of the servo press was applied before reaching the BDC, so the safety mechanism of the press switched off the servomotor around 18° angular position before reaching the BDC and therefore the estimations are not representative from that point onwards. Otherwise the normalized absolute deviation remains under the 5% threshold. Regarding the second experiment, a complete SSF operation was performed, where the ram was stopped for 6 seconds at the BDC. A significant peak is illustrated between seconds 11 and 12, which is due to control's activity to correct the speed deviation that occurs as the structure recovers from its elastic deformation.

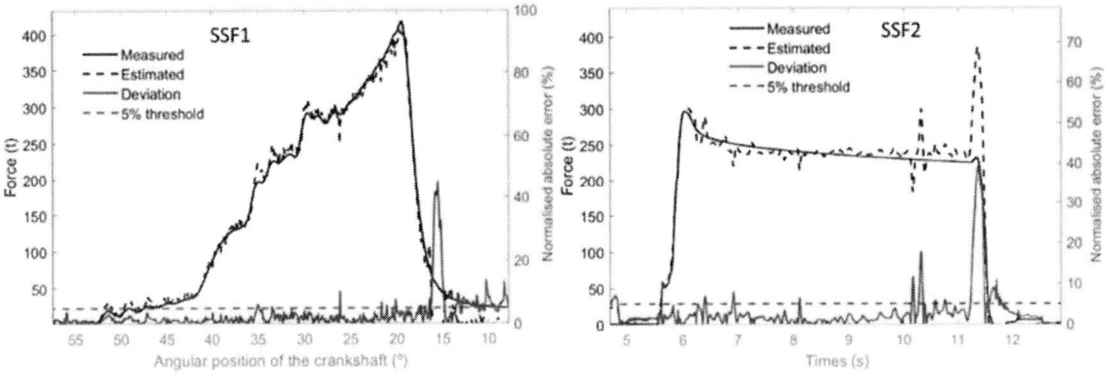

Figure 6. Measured and estimated process forces of the two semi solid forging experiments and the instantaneous normalised absolute deviation.

Based on the previous results, the proposed work could give rise to control strategies that act on the processes to improve production and the quality of the formed parts. Furthermore, the estimation of the press forces and states can also reveal information about the condition of the machine, thus allowing machine failures and downtime to be prevented. Moreover, the developed algorithm and its hardware implementation can be ported to other machines and processes by modifying the underlying model while keeping the rest of the infrastructure the same.

5. Conclusions

This paper presents experimental results of the servo press's process force obtained under three different metal forming operations applying the dPF soft sensor. The employed algorithm takes advantage of the already available signals used by the CNC, thereby avoiding the need to introduce new sensors to the system. The results show that the estimated process forces approximate well the measured force signals being the maximum error 2.97% of the measured force.

6. Acknowledgements

The technical support of Andoitz Aranburu from Fagor Arrasate S. Coop is gratefully acknowledged.

7. References

[1] T. Altan and A. E. Tekkaya, Sheet metal forming: fundamentals. Asm International, 2012

[2] R. Kumme, O. Mack, B. Bill, C. Gossweiler, H. Haab, and K. I. AG, “Dynamic properties and investigations of piezoelectric force measuring devices,” VDI Berichte, vol. 1685, pp. 161–172, 2002.

[3] R. Doraiswami and L. Cheded, “Robust model-based soft sensor: design and application,” IFAC Proceedings Volumes, vol. 47 DOI 10.3182/20140824-6-ZA-1003.00245, no. 3, pp.5491–5496, 2014

[4] J. Olaizola, C. Bouganis, E. S. de Argandoña, A. Iturrospe and J. M. Abete, "Real-Time Servo Press Force Estimation Based on Dual Particle Filter," in IEEE Transactions on Industrial Electronics, vol. 67, no. 5, pp. 4088-4097, May 2020.

[5] R. Halicioglu, L. C. Dulger, and A. T. Bozdana, “Modelling, design, and implementation of a servo press for metal-forming application,” The International Journal of Advanced Manufacturing Technology, DOI 10.1007/s00170-016-9947-8, pp. 1–12, 2016.

[6] J. Olaizola, J. M. Abete, A. Iturrospe, and E. Saenz de Argandoña, “Modelling, simulation and validation of a ram-crank servo driven press,” in 19th Scientific Convention on Engineering and Architecture, pp. 112–123, 2018.

IOP Conf. Series: Materials Science and Engineering **967** (2020) 012016 doi:10.1088/1757-899X/967/1/012016

Prediction of Mechanical Behaviors of the Q&P980 Steel Under Different Temperatures and Strain Rates Using Crystal Plasticity Method

Hao Yang[a], Huamiao Wang[a], Dayong Li[a,b]*, Yinghong Peng[a], Peidong Wu[c]

[a]State Key Laboratory of Mechanical System and Vibration, Shanghai Jiao Tong University, Shanghai, 200240, China
[b]Materials Genome Initiative Center, Shanghai Jiao Tong University, 200240, China
[c]Department of Mechanical Engineering, McMaster University, Hamilton, Ontario, L8S 4G7, Canada

*Corresponding author: dyli@sjtu.edu.cn

Abstract. The martensitic transformation in quenching and partitioning (Q&P) steels is greatly influenced by loading conditions such as environmental temperature and loading speed, and thus impacts the macroscopic mechanical properties during the deformation process. Within this work, an elastic-visco-plastic self-consistent (EVPSC) framework coupling with phase transformation model is used to simulate the stress strain responses as well as the microstructure evolution of the multi-phases Q&P980 steel under uniaxial tension process. A temperature and strain rate dependent transformation kinetics is incorporated into the model and phase transformation behaviors of the Q&P980 steel under different temperatures (25℃~100℃) and strain rates (0.0002s-1~2s-1) are successfully characterized. The corresponding stress strain responses under different loading conditions are predicted and compared with the experimental data.

1. Introduction

As a representative of the 3rd generation advanced high strength steels (3G-AHSS), Q&P steels possess good combination of strength and ductility and offers great potential in forming light-weight vehicle structures [1]. The Q&P concept was originally proposed by Speer et al. [2] and has been explored for years [3]. The Q&P980 steel has been commercialized for fabrication of automotive structural and safety parts such as cross members, longitudinal beams, B-pillar reinforcements, sills, and bumper reinforcements [4].

There are three phases in Q&P980 steel at the initial state: retained austenite (RA), ferrite (F) and tempered martensite (TM). During deformation, phase transformation occurs from RA to newly formed martensite (NM), enhancing the ductility and strength. Under press forming process of sheet metal, the strain rate is in the range of 0.1~10 s-1 and the temperature rising due to plastic work is below 100 ℃. These variant environmental conditions affect the transformation behaviors and impact the mechanical behaviors.

An EVPSC framework coupling with phase transformation model is used to study phase transformation behaviors of the Q&P980 steel under different temperatures and strain rates, and predict the macroscopic flow stress behaviors. The microstructure evolution is also characterized by the model and compared with experimental results.

2. Experiment

Experimental results from Hu et al. [5] and Zou et al. [6] were used in this study. The lattice strain evolution of each phase during tension is from Hu et al. [5]. Phase transformation behaviors and stress strain responses under different temperatures and strain rates are from Zou et al. [6]. Both of the Q&P980 steels in the literature are from BAOSTEEL. The chemical composition, microstructure characterization and quasi-static uniaxial tension stress strain curves of the two steels were compared and little difference was found. Therefore, the experimental data could be used together.

3. Model description

A crystallographic phase transformation model was implemented into an EVPSC [7, 8] framework, which can be summarized as follows:

For each crystal, the total strain rate is divided by elastic strain ($\dot{\boldsymbol{\varepsilon}}^{e}$) and plastic strain. The plastic strain consists of crystal slip ($\dot{\boldsymbol{\varepsilon}}^{p}$) and transformation strain ($\dot{\boldsymbol{\varepsilon}}^{pt}$).

$$\dot{\boldsymbol{\varepsilon}}^{p} = \dot{\gamma}_0 \sum_{\alpha} \mathbf{P}^{\alpha} \left| \frac{\tau^{\alpha}}{\tau_{cr}^{\alpha}} \right|^{\frac{1}{m}-1} \frac{\tau^{\alpha}}{\tau_{cr}^{\alpha}} \tag{1}$$

where $\dot{\gamma}_0$ is the reference value for the slip rate; m the rate sensitivity, $\mathbf{P}^{\alpha}$ the Schmid tensor. τ^{α} and τ_{cr}^{α} are the resolved shear stress (RSS), and the critical resolved shear stress (CRSS), respectively. τ_{cr}^{α} is characterized in the form of modified Voce hardening law [9]:

$$\tau_{cr}^{\alpha} = \tau_{0}^{\alpha} + (\tau_{1}^{\alpha} + \theta_{1}^{\alpha}\Gamma)(1 - \exp(-\Gamma\left|\frac{\theta_{0}^{\alpha}}{\tau_{1}^{\alpha}}\right|)) \tag{2}$$

where $\Gamma = \sum_{\alpha} \gamma^{\alpha}$ is the accumulated shear in the grain, τ_{0}^{α} the initial CRSS, θ_{0}^{α} the initial hardening rate, θ_{1}^{α} the asymptotic hardening rate, and $(\tau_{0}^{\alpha} + \tau_{1}^{\alpha})$ the back-extrapolated CRSS, respectively.

For the ferrite and martensite grains, there is no transformation strain ($\dot{\varepsilon}^{pt}=0$). For the austenite grains, strain-induced martensitic transformation (SIMT) can occur and induce transformation strain when the accumulated shear exceeds a critical value.

The phase transformation model can be summarized as follows:

(1) For an austenite grain that has not experienced phase transformation, if the accumulated shear of the grain reaches a critical value Γ_c, then the transformation of the grain will initiate.

(2) There are 24 possibly formed martensite variants, according to the K-S orientation relationship, but only the one with the maximum transformation energy potential ($U_i = \sigma : \varepsilon_i^{pt}, i = 1,...,24$) is chosen as the nucleation of NM.

(3) If a transformation martensite grain is nucleated, its initial stress state is set to be equal to the current parent austenite grain considering the stress balance, and its initial shape is set to be a plate-like with axis ratio of 10:10:1, based on the experimental observation [10].

(4) If one variant is selected, other variants will be restricted. This assumption is not only based on experimental results [11] but also on common sense that growing up on an existing nucleus is much easier than nucleation again.

(5) A modified phenomenological transformation kinetics equation is used to describe the volumetric evolution of an austenite grain.

$$f = 1 - \exp\{-\beta[1 - \exp(\alpha\Gamma)]^2\} \tag{3}$$

where $\alpha(T,\dot{\varepsilon})$ and $\beta(T,\dot{\varepsilon})$ are transformation parameters.

(6) The transformation associated dilatational and shear strains are about 3.7% and 22%, respectively. The Eigen transformation strain tensor ε^{PT} of the variant with $(011)_\alpha//(111)_\gamma$ and $[11\bar{1}]_\alpha//[10\bar{1}]_\gamma$ is given by (referred to austenitic axes)

$$\varepsilon^{PT} = \frac{(F^T F - I)}{2} = \begin{bmatrix} -0.0061 & 0 & -0.0242 \\ 0 & 0.131 & 0 \\ -0.0242 & 0 & -0.0958 \end{bmatrix} \tag{4}$$

The transformation strain tensors of other variants are calculated by multiplication of ε^{PT} with associated crystal symmetry operation matrices.

A uniform macroscopic strain is imposed on a material point. An aggregate of 10000 grains is considered for the Q&P980 alloy, and consists of four phases: RA, F, TM and NM. According to Zou et al. [6], the volume fractions of the RA, F, TM and NM are taken to be respectively 10.6% (1060 grains), 37.5% (3750 grains), 51.9% (5190 grains) and zero at the initial state. The initial texture measured by electron backscatter diffraction (EBSD) is used as input for each phase. For the RA, 12 {111}<110> slip systems are used without considering the twinning mechanism. For the F, TM and NM, the 12 {110}<111> and 12 {112}<111> slip systems are used in this study. The single-crystal elastic constants for the γ (RA) and α (F, TM and NM) phases are set to be: $C_{11}^{\gamma} = 209\text{GPa}$, $C_{12}^{\gamma} = 133\text{GPa}$,

$C_{44}^{\gamma} = 121\text{GPa}$ [12] and $C_{11}^{\alpha} = 222\text{GPa}$, $C_{12}^{\alpha} = 143\text{GPa}$, $C_{44}^{\alpha} = 114\text{GPa}$ [13], respectively. The calibrated model parameters of each phase are listed in Table 1.

Table 1 Model parameters

phase	Slip mode	τ_0	τ_1	θ_0	θ_1
RA	{111}<110>	380	1	800	20
F	{110}<111>	190	150	1200	150
	{112}<111>	190	150	1200	150
TM	{110}<111>	380	50	600	50
	{112}<111>	380	50	600	50
NM	{110}<111>	450	40	2500	2200
	{112}<111>	450	40	2500	2200

4. Results

The stress strain responses as well as the volumetric evolution of RA is shown in Figure 1a. The volume fraction of RA continues decreasing to 6% at ε=0.15. The lattice strain of each phase during deformation is shown in Figure 1b. After yielding, the F and NM show obvious hardening behavior, while the lattice strains of RA and TM almost do not change. The texture of each phase after 15% tension is shown in Figure 2. After deformation, the <111> fiber texture of RA and the <110> fiber texture of F and TM enhanced. The current model is able to reproduce the macroscopic mechanical properties as well as microstructure evolutions of the multi-phase Q&P980 steel.

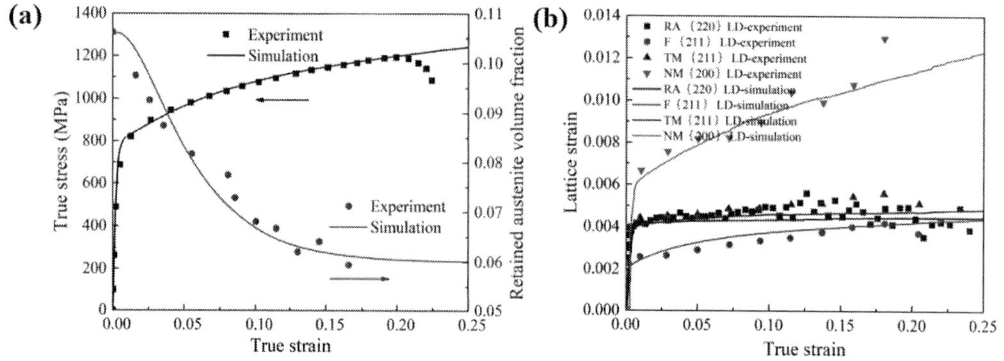

Figure 1. (a) experimental and simulated flow stress and volumetric evolution of RA, (b) experimental and simulated lattice strain evolution of each phase

	Experiment	Simulation

IOP Conf. Series: Materials Science and Engineering **967** (2020) 012016 doi:10.1088/1757-899X/967/1/012016

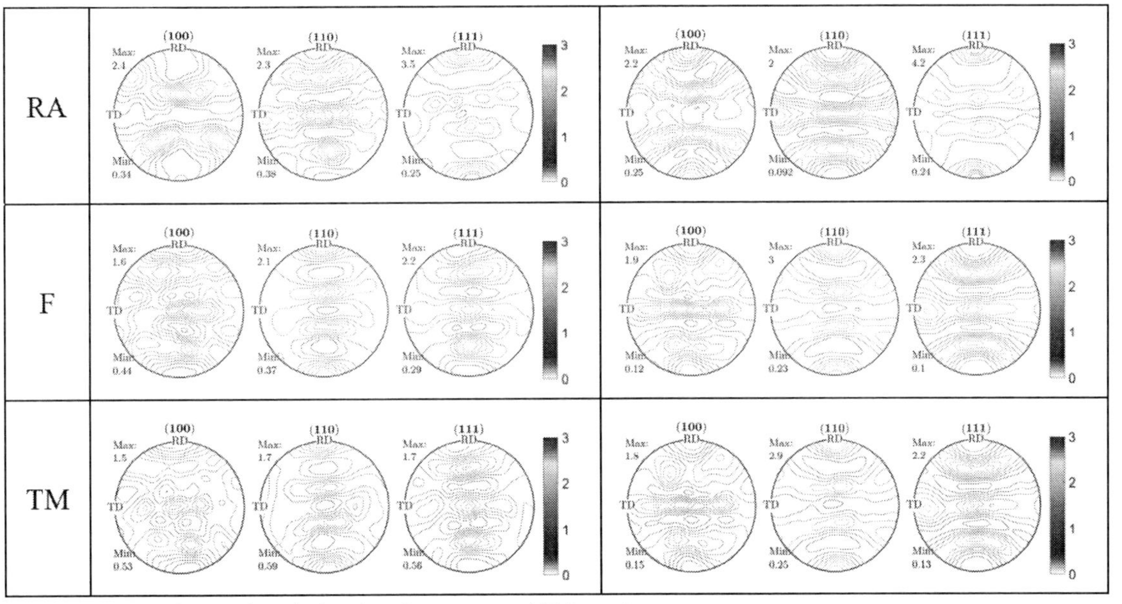

Figure 2. experimental and simulated texture at 15% strain

The relationship between transformation parameters (α, β) and loading conditions (T, $\dot{\varepsilon}$) is given by:

$$\alpha = 2.2 f_1 f_2 - 12.7956(f_1 + f_2) + 80.0012 \tag{5}$$

where $f_1 = -0.005T + 6.36$,

$$f_2 = -0.0003(\ln\dot{\varepsilon})^4 + 0.0035(\ln\dot{\varepsilon})^3 + 0.17(\ln\dot{\varepsilon})^2 + 1.31\ln\dot{\varepsilon} + 8.8$$

$$\beta = 0.81 g_1 g_2 \text{-} 0.8132(g_1 + g_2) + 0.4585 \tag{6}$$

where $g_1 = -0.0047T + 0.65$,

$$g_2 = 0.000546(\ln\dot{\varepsilon})^4 + 0.0107(\ln\dot{\varepsilon})^3 + 0.0757(\ln\dot{\varepsilon})^2 + 0.1988\ln\dot{\varepsilon} + 0.52222$$

The phase transformation behaviors and stress strain responses under different temperatures (where $\dot{\varepsilon}$=0.0002s^{-1}) are shown in Figure 3a and Figure 3b, respectively. With temperature increasing, phase transformation is constrained and the strength is weakened. The phase transformation behaviors and stress strain responses under different strain rates (where T=25°C) are shown in Figure 4a and Figure 4b, respectively. With strain rate increasing, phase transformation is constrained firstly but enhanced after $\dot{\varepsilon}$>0.1s^{-1}. Therefore, the strength is not enhanced with strain rate monotonously and the flow stress crosses with each other at low strain rate ($\dot{\varepsilon}$<0.1s^{-1}).

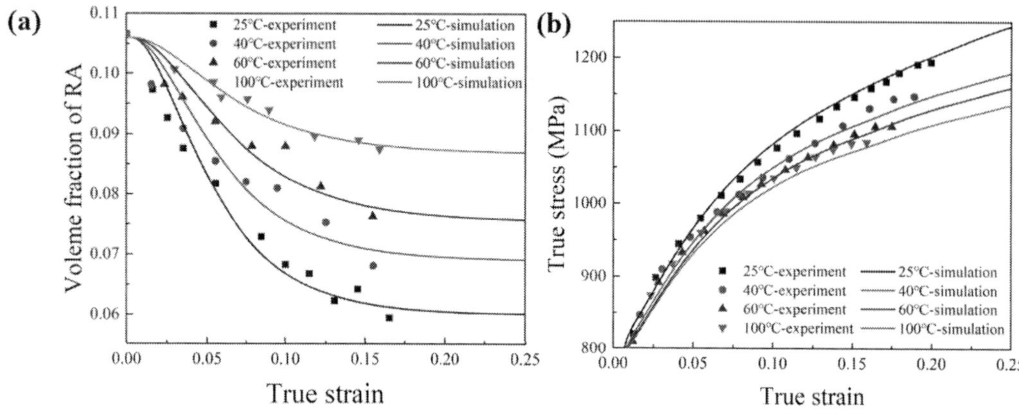

Figure 3. (a) measured and simulated volume fraction of RA under different temperatures, (b) measured and simulated stress strain responses under different temperatures

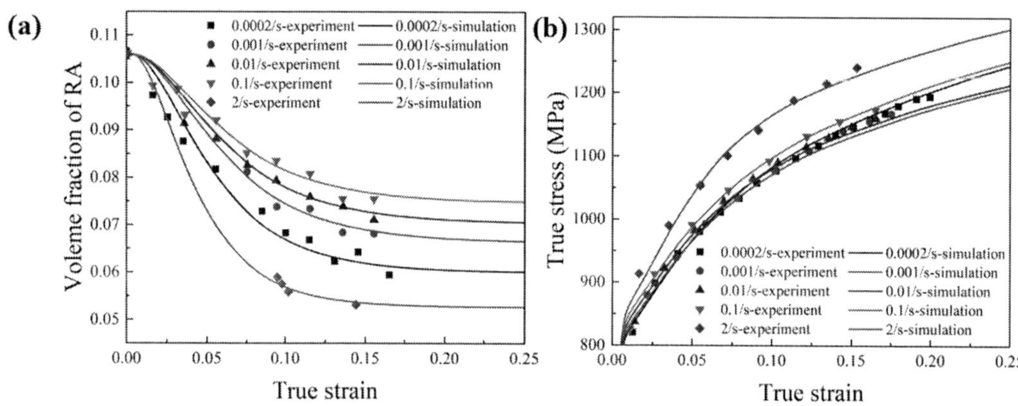

Figure 4. (a) measured and simulated volume fraction of RA under different strain rates, (b) measured and simulated stress strain responses under different strain rates

5. Conclusions

An EVPSC framework was combined with crystallographic phase transformation model to capture the macroscopic mechanical behavior as well as microstructure evolutions of the Q&P980 steel. The developed model was able to characterize the phase transformation behaviors and predict the stress strain responses under different temperatures and strain rates.

Acknowledgements

DL acknowledges the support by NSFC [Grant number U1860110]. The original EBSD data was supplied by D. Q. Zou et al.

References

[1] D.K. Matlock, Speer, J.G., Third generation of AHSS: microstructure design concepts, microstructure and texture in steels, 2009.

[2] J. Speer, D.K. Matlock, B.C. De Cooman, J.G. Schroth, Carbon partitioning into austenite after martensite transformation, Acta Materialia 51(9) (2003) 2611-2622.

[3] L. Wang, W. Feng, Development and Application of Q&P Sheet Steels, 2011.

[4] L. Wang, J.G. Speer, Quenching and Partitioning Steel Heat Treatment, Metallography, Microstructure, and Analysis 2(4) (2013) 268-281.

[5] X.H. Hu, X. Sun, L.G. Hector, Y. Ren, Individual phase constitutive properties of a TRIP-assisted QP980 steel from a combined synchrotron X-ray diffraction and crystal plasticity approach, Acta Materialia 232 (2017) 230-244.

[6] D.Q. Zou, S.H. Li, J. He, Temperature and strain rate dependent deformation induced martensitic transformation and flow behavior of quenching and partitioning steels, Materials Science and Engineering: A 680 (2017) 54-63.

[7] H. Wang, P.D. Wu, C.N. Tomé, Y. Huang, A finite strain elastic–viscoplastic self-consistent model for polycrystalline materials, Journal of the Mechanics and Physics of Solids 58(4) (2010) 594-612.

[8] H. Wang, Y. Jeong, B. Clausen, Y. Liu, R.J. McCabe, F. Barlat, C.N. Tomé, Effect of martensitic phase transformation on the behavior of 304 austenitic stainless steel under tension, Materials Science and Engineering: A 649 (2016) 174-183.

[9] C. Tome, G.R. Canova, U.F. Kocks, N. Christodoulou, J.J. Jonas, The relation between macroscopic and microscopic strain hardening in F.C.C. polycrystals, Acta Metallurgica 32(10) (1984) 1637-1653.

[10] S. Turteltaub, A.S.J. Suiker, Transformation-induced plasticity in ferrous alloys, Journal of the Mechanics and Physics of Solids 53(8) (2005) 1747-1788.

[11] N. Gey, B. Petit, M. Humbert, Electron backscattered diffraction study of ε/α' martensitic variants induced by plastic deformation in 304 stainless steel, Metallurgical & Materials Transactions A 36(12) (2005) 3291-3299.

[12] H.M. Ledbetter, Predicted monocrystal elastic constants of 304-type stainless steel, Physica B & C 128B+C(1) (1985) 1-4.

[13] M.R. Fellinger, L.G. Hector, D.R. Trinkle, Effect of solutes on the lattice parameters and elastic stiffness coefficients of body-centered tetragonal Fe, Computational Materials Science 152 (2018) 308-323.

IOP Conf. Series: Materials Science and Engineering **967** (2020) 012017 doi:10.1088/1757-899X/967/1/012017

Mechanical properties and formability of EN AW-7075 in cold forming processes

B-A Behrens[1], S Hübner[1], H Vogt[1], O Golovko[2], S Behrens[2] and F Nürnberger[2]

[1] Leibniz Universität Hannover, Institut für Umformtechnik und Umformmaschinen, An der Universität 2, 30823 Garbsen, Germany

[2] Leibniz Universität Hannover, Institut für Werkstoffkunde, An der Universität 2, 30823 Garbsen, Germany

E-Mail: vogt@ifum.uni-hannover.de

Abstract. Due to a low density and high tensile strength, the aluminum alloy EN AW 7075 T6 offers a high lightweight potential for structural components. Since its formability is limited at room temperature in the T6 temper state, the potential of this alloy for automotive bodies is only utilizable by adapted deep drawing processes. In recent years, process chains suited for warm and hot forming have been researched and developed. However, warm and hot forming solutions require additional process steps and a complex tooling system in comparison to cold forming processes. Alternatively, the forming of such blanks at room temperature in the W temper state is favorable since conventional tools can be used. The W temper state is a heat treatment condition achieved after solution heat treatment and subsequent quenching, which is characterized by an increased ductility. However, this condition is unstable, due to the onset of natural ageing. With increasing time after the quenching step, the strength of the material increases, which leads to a reduction of formability. Another phenomenon that occurs after quenching is the Portevin Le-Chatelier effect. This effect causes the formation of flow lines during cold forming and results in a decrease of ductility. Hence, the objective of the investigations was to determine the formability of EN AW 7075 as a function of the natural ageing time after solution heat treatment and quenching. Therefore, tensile tests of various aged samples were carried out. The results show a relation of the formability to the natural ageing time and a dependency on the quenching rate. Furthermore, a heat treatment strategy for EN AW-7075 was developed, that considers manufacturing processes like the cathodic dip coating. The influence of the quenching rate, ageing time and temperature as well as the influence of temperature of the paint baking process after the cathodic dip coating were considered. Therefore, a design of experiments and tensile tests were carried out. Thus, the deep drawing of EN AW-7075 at room temperature is particularly promoted.

1. Introduction

In car bodies aluminium is mainly used as lightweight material. The highest strength and therefore the highest lightweight potential belong to the 7000 series aluminum alloys. These materials have the same specific strength like hot-stamped steels and additionally feature an increased elongation at break [1]. In particular, the substitution of hot-stamped steel components by parts manufactured with 7000 aluminum alloys can lead to a significant reduction in body weight. However, these alloys have so far rarely been used in car body construction. The reason for this is the low formability of the materials at room temperature in the highest strength state. Therefore, various approaches are currently being investigated to improve the formability of the alloys.

One approach is the hot forming process route. Therefore, the aluminum blanks are heated to temperatures above the recrystallization temperature. The sheet material is subsequently transferred into a water-cooled press tool and simultaneously formed and quenched [2]. Argandoña et al. [3] have used this process route to form defect-free B-pillars. Another approach is the warm forming of the high strength material after quenching and artificial ageing. The improved formability of this process is based on the dissolution of η' precipitations and dynamic recovery effects at elevated temperatures [4]. Sotirov et al. [1] and Behrens et al. [5] have determined the increased formability for this process route for the alloy EN AW-7075. Noder et al. have also investigated a process variant for non-isothermal forming [6]. However, warm and hot forming solutions require additional process steps and a complex tooling system in comparison to cold forming processes. Alternatively, the forming of such blanks at room temperature in the W temper state is favourable since conventional tools can be used. Therefore, in this paper a cold forming process route is investigated.

2. Cold forming and heat treatment process route

To increase the formability of 7000 aluminum alloys, approaches like the W temper process are examined. This process is named after the W heat treatment state. This condition is achieved after solution heat treatment and quenching of the material. The alloying elements remain dissolved and a supersaturated state with a high number of solute atoms and vacancies is formed. Due to its high ductility, this heat treatment state allows cold forming. However, the blanks must be formed as soon as possible after quenching, otherwise the material will age naturally. As a result, the strength of the material increases and thus its formability is reduced. Argandoña [7] investigated the formability of EN AW-7075 at room temperature in the ductile W condition. However, due to the solution heat treatment process, the formed components have no usable mechanical properties. Therefore, an artificial ageing step must be carried out afterwards. The increase in strength bases on the forming of finely dispersed metastable precipitations [2]. The precipitation sequence determined by Degischer [8] and Löffler [9] describes partially coherent η' precipitations as the main hardening phase for 7000 aluminum alloys. The heat treatment can be combined by the artificial ageing process and the cathodic dip coating (cdc) process, see Figure 1. Another influence on the mechanical properties is the quenching rate after the solution heat treatment. For the alloy EN AW-7075, Behrens et al. [10] have investigated a dependency of the mechanical properties on the copper content of the alloy. Thereby, an extension of the W temper process route is possible. By partial quenching of the blanks it is possible to produce components with load-optimised tailored properties. For example, a spray-field can be used to quench certain areas of the blank rapidly, while the remaining areas cool down slowly [11, 12, 13]. Following artificial ageing, high strengths are achieved in the rapidly quenched areas. In the slowly cooled areas the strength is lower, but the ductility increases. However, in order to be able to adjust these partial properties, the heat treatment parameters of the complete process chain have to be adapted to each other.

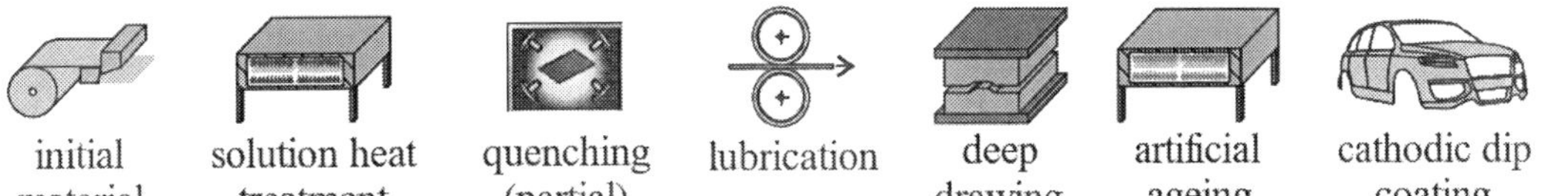

Figure 1. Process route for cold forming and heat treatment of 7000 aluminum alloys.

3. Experimental setup

The material used for the investigations is the copper-containing alloy EN AW-7075. The copper content is essential, due to its significant influence on the quenching sensitivity of this alloy. The chemical composition of the investigated material is listed in Table 1. The delivery state of the material was T6.

Table 1. Chemical composition of the investigated aluminum sheet metal material in ma.-% determined by spark spectroscopy.

Alloy	Zn	Mg	Cu	Mn	Si	Cr	Fe	Ti
EN AW-7075	6.1	2.35	1.63	0.05	0.08	0.19	0.17	0.02

3.1. Influences on tensile properties after quenching

Tensile tests were carried out to determine the influence of the quenching method and the natural ageing time after quenching on the formability of the material. For this purpose, samples of the test material were first heated at a solution heat treatment temperature T_{SHT} of 475 °C for a duration t_{SHT} of 10 min. Subsequently, a share of the specimens were rapidly quenched in a pneumatically operated plate tool. The quenching rate was 250 K/s. The other share of the samples was slowly cooled at air. In this case a quenching rate of 1 K/s was obtained. The quenching rate was determined by a type K thermocouple. After quenching, the specimens were aged naturally for various times. The aging duration was varied between 5 min and 30 d after quenching. However, the aging period was not started until the temperature of the blanks dropped below 50 °C during quenching, so that forming at room temperature could be ensured for the samples cooled at air. The different aging durations were chosen according to data of Ostermann et al. as it shown in Figure 2. This is based on the increasing hardness of the material with increasing natural ageing time.

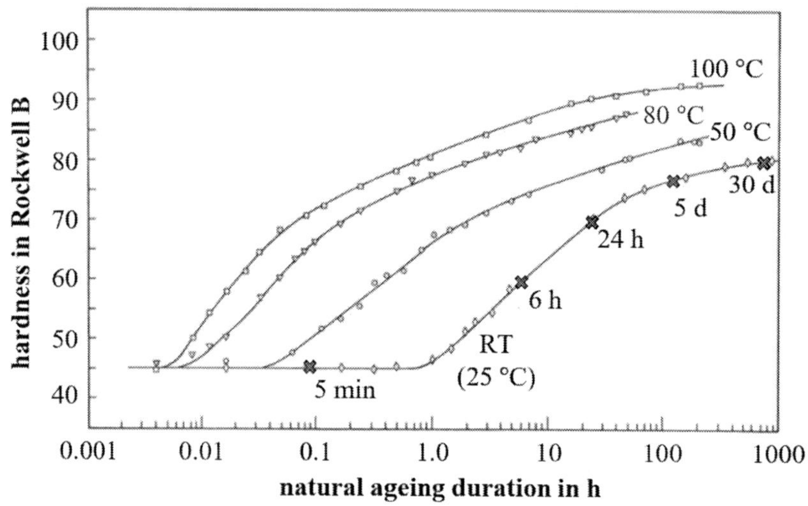

Figure 2. Hardness development in dependence of the natural ageing time according to [14] and selected natural ageing times according to the hardness evolution.

The quenched samples were used subsequently to determine a stress-strain curve for each quenching rate and natural ageing duration. A tensile test machine S100/ZD by DYNA-MESS was used for this purpose. The tests were carried out according to DIN EN ISO 6892-1. The specimens were prepared according to DIN 50125. The shape H with an initial gauge length of 50 mm was used. The investigated alignment of the specimens to the rolling direction were 0°, 45° and 90°.

3.2. Influence of heat treatment parameters on the mechanical properties of the components

The mechanical properties of the components produced by the process route described in chapter 2 depend mainly on four different process parameters. These are, on the one hand, the quenching method selected after solution heat treatment, the artificial ageing temperature and duration and whether cathodic dip coating is carried out at the end of the process or not. The influence and significance of the influence of these parameters on the ultimate tensile strength as well as the elongation at break was tested within a design of experiments. In order to be able to determine the quadratic influence of individual parameters, a response surface design was selected as the experimental plan. Figure 3 shows the structure of the design. The artificial ageing time and temperature were varied in five stages between 2 h and 24 h, respectively 100 °C and 180 °C. For each temperature and time combination three tensile tests were carried out. The central point was tested a total of fifteen times to investigate a scattering of the results.

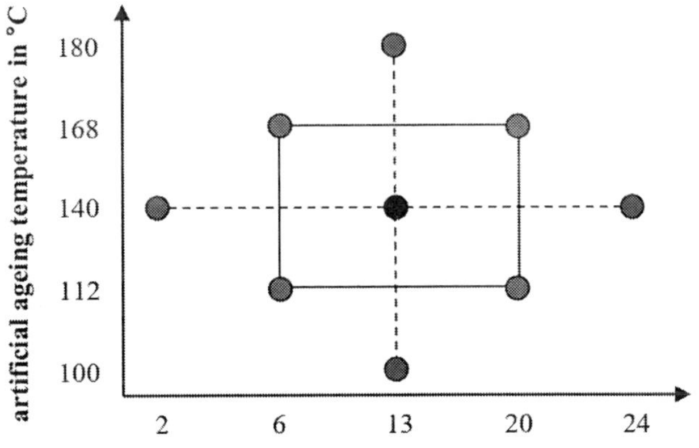

Figure 3. Response surface design to investigate the influence of the process parameters on the mechanical properties.

The influence of the quenching method was investigated in two levels. First a rapid cooling in the plate tool mentioned in chapter 3.1 and second a slow cooling at air. For the investigations regarding the influence of the cathodic dip coating, its heat treatment process was reproduced in a furnace. A two-level factor was therefore also applied in the response surface design. The first level includes a final heat treatment by means of cdc. In the second level, no further heat treatment is carried out after artificial ageing. Thus the response surface design shown in Figure 3 is repeated four times for the different combinations of quenching method and cdc. The complete heat treatment sequence is shown in Figure 4. After quenching in the tool or at air, a five-day natural ageing at room temperature was carried out. The final cdc process is based on Kumar et al. [15]. This five-step heat treatment considers not only the paint baking process but also further drying steps in the cdc-process.

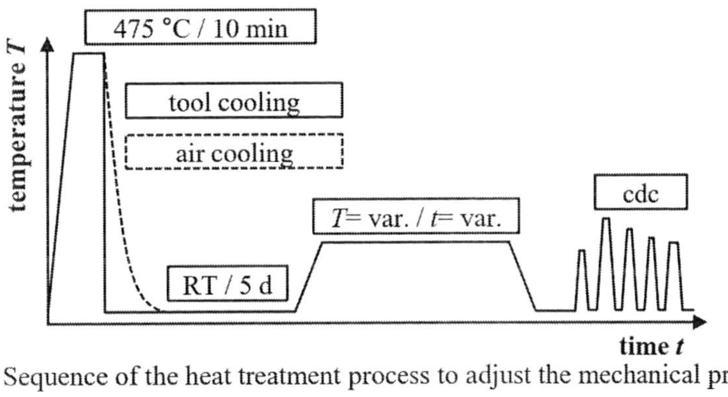

Figure 4. Sequence of the heat treatment process to adjust the mechanical properties.

4. Results of the experiments

4.1. Influences on tensile properties after quenching

IOP Conf. Series: Materials Science and Engineering **967** (2020) 012017 doi:10.1088/1757-899X/967/1/012017

First, the influence of slow cooling at air and the natural ageing duration on the stress-strain curves was determined. The results are shown in Figure 5 for an alignment of the tensile test samples 90° to the rolling direction. As a reference, the stress-strain curve of EN AW-7075 in T6 condition is also shown. As can be seen, after solution heat treatment and air cooling, a significantly lower stress level is achieved. With decreasing natural ageing time, the stress also decreases until a minimum stress level is reached for 24 h. Although an increase of the elongation with decreasing ageing time, was not determined. The same behavior was also investigated for the other rolling directions.

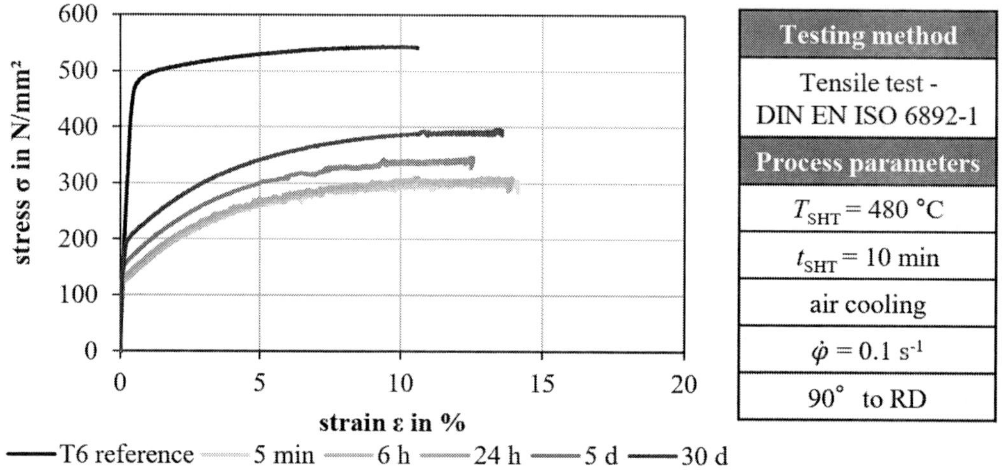

Figure 5. Stress-strain-curves for EN AW-7075 after solution heat treatment and subsequent air cooling depending on the natural ageing time.

A different behavior was determined for the samples quenched rapidly in the plate tool. The results are shown in Figure 6. Due to the high quenching rate, a similar stress level as in the T6 condition was obtained after a natural ageing time of 30 d. The elongation, however, was significantly increased compared to air cooling and the T6 reference. As the natural ageing time decreases, the yield and tensile strength decreases like for the slowly cooled samples. The reachable elongation at break also decreases with decreasing natural ageing time. A further phenomenon can be observed in this case. With decreasing natural ageing time, dynamic strain ageing occurs increasingly. This phenomenon is also known as Portevin-Le Chatelier effect. The effect appears for slowly and rapid quenched samples but is more pronounced in samples quenched quickly and exposed to a short natural ageing time. This can be explained by the higher amount of solute atoms immediately after quenching. With a natural aging time of 5 min, this effect is pronounced to such an extent it causes the stress-strain curve to leave its typical path and break at about an elongation of only 7.5 %.

IOP Conf. Series: Materials Science and Engineering **967** (2020) 012017 doi:10.1088/1757-899X/967/1/012017

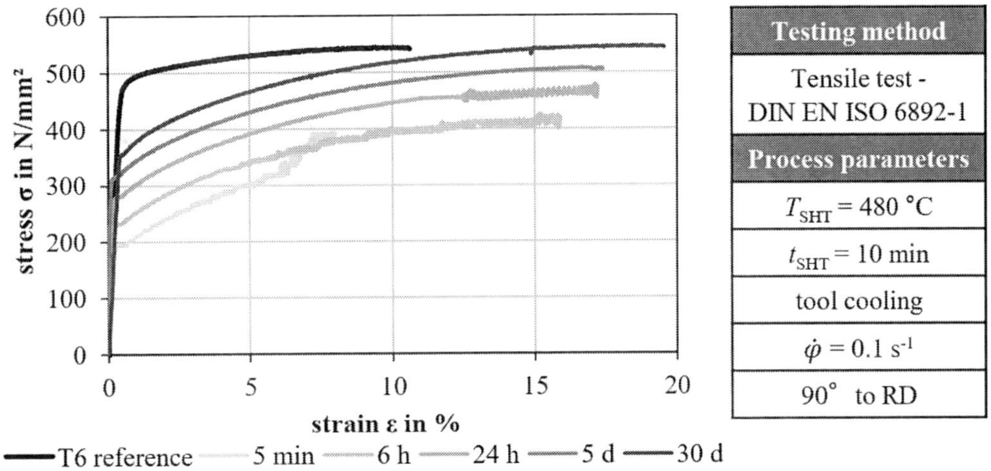

Figure 6. Stress-strain-curves for EN AW 7075 after solution heat treatment and subsequent tool cooling depending on the natural ageing time.

4.2. Influence of heat treatment parameters on the mechanical properties of the components

The results of the tensile tests carried out as a part of the design of experiments are shown in Figure 7. The increasing artificial ageing time has a negative influence on the ultimate tensile strength and elongation at break. The value of both properties decreases with increasing ageing time. The ageing temperature has a similar influence. With increasing temperature, the level of both mechanical properties decreases. In addition, the artificial ageing temperature also has a quadratic influence on the ultimate tensile strength and elongation at break, resulting in a maximum in the temperature range from 100 °C to 120 °C. For the two parameters investigated in two stages - quenching method and cathodic dip coating - only a linear dependency was observed. As expected, the quenching method has a great influence on the mechanical properties. At low quenching rates a low strength and a high elongation is ensured. For the rapid quenching method in the tool the trend is the opposite. The cdc influences only the elongation at break. If a cdc is carried out, it causes an overageing of the material and thus a reduction in elongation at break.

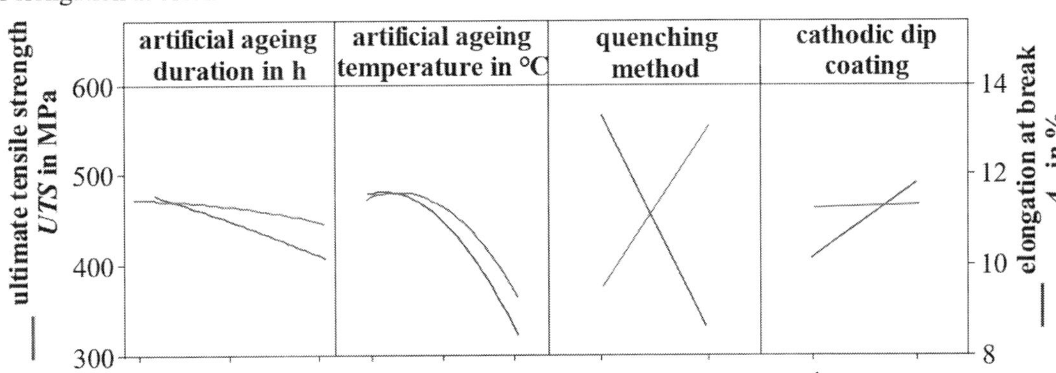

Figure 7. Influence of the investigated process parameters on the ultimate tensile strength and the elongation at break of EN AW-7075.

To determine the significance of the influence of the individual parameters and their interactions with each other, a Pareto-diagram was compiled regarding ultimate tensile strength and elongation at break (s. Figure 8). The significance limit was calculated for both parameters at a standardized effect of 2.02. This means that parameters or interactions of different parameters above this value have a significant effect on the respective property. The quenching method has the highest influence on both investigated mechanical properties. Other significant factors influencing the ultimate tensile strength are the artificial ageing temperature and duration and the interaction between these parameters. The cdc process, as well as the interactions with this parameter, have a high influence on the elongation at break.

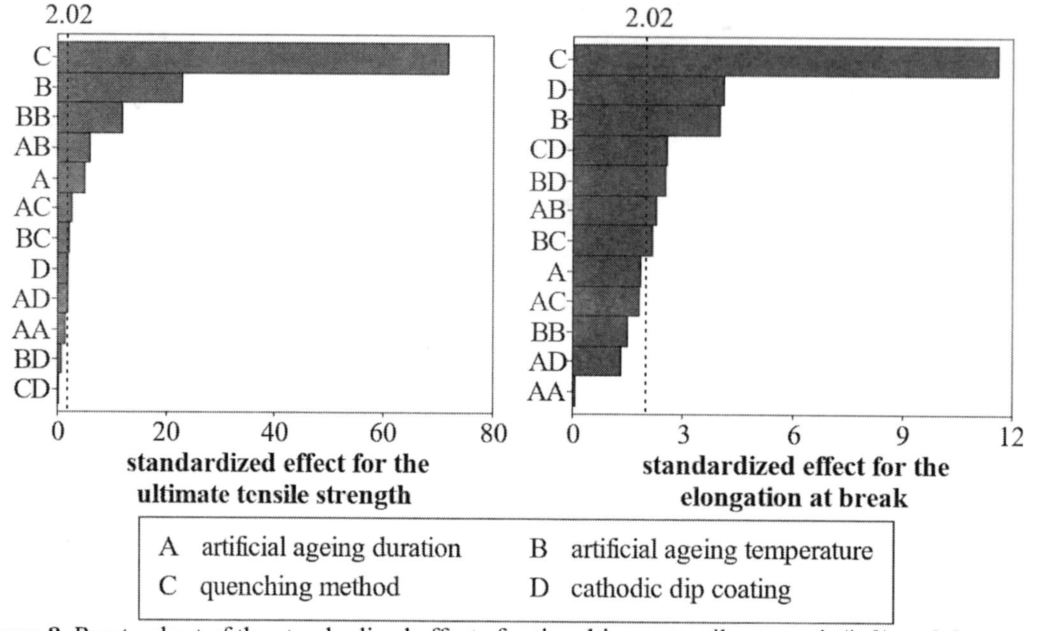

Figure 8. Pareto chart of the standardized effects for the ultimate tensile strength (left) and the elongation at break (right)

5. Summary and Outlook

The influence of the quenching rate and the natural ageing time after solution heat treatment on the formability of the alloy EN AW-7075 were investigated by tensile tests. It was shown that the quenching rate has an influence on the flow stress necessary for deformation and the achievable elongation at break. Low quenching rates tend to result in lower flow stresses compared to rapid rates. However, the ductility is influenced positively at higher quenching rates. In the case of rapid cooling, an increase in ductility compared to the T6 reference state was determined for ageing times above 6 h and an increasing dynamic strain ageing observed. At a short ageing time of only 5 min, the strain ageing is so pronounced that the ductility of the material is reduced significantly. This behavior is uncommon for the alloys of the 7000 series and shall be explored in future examinations by Nakajima tests.

In further experiments the influence of the different heat treatment parameters of the cold forming process route on the mechanical properties was investigated at the example of tensile tests. A significant influence of the quenching rate after solution heat treatment and the artificial ageing temperature was determined. As expected, the strength increases with increasing quenching rates, whereas high elongation at break is achieved with low quenching rates. An increasing ageing temperature above 120 °C has a negative influence on the mechanical properties. It has also been observed that the cdc-process, which is obligatory in the automotive industry, has a negative influence on the elongation at break for the examined material. In future investigations, heat treatment parameters shall be determined which, in the case of a partially quenched component, lead to maximum ductility in the slowly quenched area and to a strength level that is comparable to the T6 condition in the rapidly quenched component area.

According to the obtained results, the presented W temper process route provides an alternative to established forming routes like warm or hot forming of high-strength aluminum alloys. The increased ductility of the alloy EN AW-7075 after quenching suggests an improvement in formability at room temperature. However, these results must be examined and verified in deep-drawing tests. Together with the possibility of manufacturing tailored parts, the W temper process route represents a good opportunity to promote lightweight design in modern car bodies.

Acknowledgements
The project 'Tailored Tempering of 7xxx-Aluminum Alloys' Ref.-No. AiF 20190N was financed and supervised by the European Research Association for Sheet Metal Working (EFB). In the scope of the program to promote Industrial Collective Research it was funded by the German Federation of Industrial Research Associations (AiF) with means of the Federal Ministry of Economic Affairs and Energy (BMWi) on the basis of a decision by the German Bundestag. Furthermore, the authors would like to thank the industrial partners in this research project.

References
[1] Sotirov N, Simon P, Waltenberger T, Uffelmann D and Melzer C 2011 Verbessertes Umformverhalten von höchstfesten 7xxx-Aluminiumblechen durch Halbwarmumformen *HTM J. Heat Treatm. Mat.* **66** Issue 1 pp 37-43
[2] El Fakir O, Wang L, Balint D, Dear J P, Lin J and Dean T A 2014 Numerical study of the solution heat treatment, forming, and in-die quenching (HFQ) process on AA5754 *International Journal of Machine Tools & Manufacture* **87** pp 39-48
[3] Mendiguren J, Saenz de Argandoña E and Galdos L 2016 Hot stamping of AA7075 aluminum sheets. *IOP Conf. Ser.: Mater. Sci. Eng.* **159**
[4] Kumar M, Sotirov N and Chimani C M 2014 Investigations on warm forming of AW-7020-T6 alloy sheet *Journal of Materials Processing Technology* **214** pp 1769-76
[5] Noder J, DiCecco S, Butcher C and Worswick M 2017 Finite Element Simulation of Non-Isothermal Warm Forming of High-Strength Aluminum Alloy Sheet. *AIP Conference* Proceedings **1896**
[6] Behrens B-A, Hübner S and Vogt H 2018 Formability of 7000 aluminum alloys in warm and hot forming condition. *IOP Conf. Ser.: Mater. Sci. Eng.* **418**
[7] Sáenz de Argandoña E, Galdos L, Ortubay R, Mendiguren J and Agirretxe X 2015 Room temperature forming of AA7075 aluminum alloys: W-temper process *Key Engineering Materials* **651-653** pp 199-204
[8] Degischer H P, Lacom W, Zahra A and Zahra C Y 1980 Decomposition process in an Al-5%Zn-1%Mg alloy *Zeitschrift für Metallkunde* **71** pp 231-38
[9] Löffler H, Kovács I and Lendvai J 1983 Review: Decomposition processes in Al-Zn-Mg alloys *Journal of Materials Science* **18** pp 2215-40
[10] Behrens B-A, Nürnberger F, Bonk C, Hübner S, Behrens S and Vogt H 2017 Influences on the formability and mechanical properties of 7000-aluminum alloys in hot and warm forming *Journal of Physics: Conference Series* **896** Issue 1
[11] Nürnberger F, Diekamp M, Moritz J, Wolf L, Hübner S, and Behrens B-A 2014 Spray Cooling of Early Extracted Hot Stamped Parts. Paper presented at the 143rd TMS Annual Meeting, San Diego, California, 16-20 February 2014, 983-90
[12] Wolf L, Diekamp M, Gretzki T, Nürnberger F, Bach F-W, Rodman D, Moritz J, Schrödter J, Hübner S and Behrens B-A 2014 Hot Stamping and Subsequent Spray Cooling: A New Manufacturing Approach. In: Plastic Deformation of Metals. Dnipropetrovsk, UA; Akzent, p 36–55
[13] Diekamp M, Hübner S, Nürnberger F, Schaper M, Behrens B-A and Bach F-W 2011 Optimised press-hardening process using spray cooling - Process integrated heat treatment of 22MnB5 sheet metal *HTM J. Heat Treatm. Mat.* **66** Issue 6 pp 316-22
[14] Ostermann F 2014 *Anwendungstechnologie Aluminium* Berlin Heidelberg Springer Vieweg 3rd edition
[15] Kumar M and Ross N 2015 Warm forming and paint bake response of AW-7xxx sheet in various tempers *Key Engineering Materials* **651-653** pp 47-52

IOP Conf. Series: Materials Science and Engineering **967** (2020) 012018 doi:10.1088/1757-899X/967/1/012018

Effects of Lubrication and Temperature on the Tribological Behavior of a Magnesium Alloy under Contact Sliding

C H Wu[1], L C Zhang[1]*, P L Qu[2], S Q Li [2], Z L Jiang [2] and Y Wang[1]

[1]School of Mechanical and Manufacturing Engineering, The University of New South Wales, NSW 2052, Australia
[2]Baoshan Iron & Steel Co., Ltd., Shanghai 200941, China
*corresponding author's E-mail: Liangchi.Zhang@unsw.edu.au

Abstract. This paper aims to experimentally investigate the effects of lubrication and temperature on the tribology behavior of a magnesium alloy, AZ31B, sliding against 86CrMoV7. The effects of lubricant additives, including oil film improver and extreme pressure (phosphorous donors), on the friction coefficient and surface morphology in the sliding zone were studied. It was found that the relatively lower surface hardness of the material significantly affects the wear of the materials in the sliding zone. The lubricant with oil film improver and extreme pressure can reduce friction. Temperature rise brings about lower friction due to the lubricant additives.

1. Introduction

In the automotive industry, magnesium alloys have been widely used to reduce the vehicle weight because of its characteristics of lightweight and high specific strength. However, the fabrication of magnesium alloy components faces critical challenges due to the limited ductility of the materials at room temperature. As such, the forming process of magnesium alloy products are usually carried out at an elevated temperature, e.g., in a warm rolling condition, which can improve the material's ductility [1-4]. Additionally, the friction within a contact pair involving a magnesium alloy usually leads to poor surface quality [5] and material adhesion [6]. Particularly, the adhesion at the contact interface has been a major problem in a forming process at an elevate temperature and usually occurs due to insufficient lubrication. Thus, it is necessary to develop proper lubricant to reduce the interface friction and hence to improve the surface finish of magnesium alloys.

This paper aims to experimentally investigate the effects of lubrication and temperature on the tribological behavior including the influence of oil film improver and extreme pressure, on the friction coefficient and surface morphology at the sliding zone.

2. Experimental configuration

To measure the interface friction of magnesium alloy AZ31B sliding against 86CrMoV7, a CETR UMT 100 tribometer with a heating chamber and liquid container was used, as shown in Fig. 1. The disc sample with a thickness of 5mm was fixed inside the liquid container while the pin was fixed by the pin holder. Before the tribology test, the surface hardness of AZ31B was measured by indentation on Struers DuraScan. The surface morphology of the contact pair was measured by the 3D optical surface stylus, Zygo NewView 700 Scanning White Light Interferometer. The diameter of pin is 2 mm while the contact radius for the sliding test is 24mm, as shown in Fig. 2

IOP Conf. Series: Materials Science and Engineering **967** (2020) 012018 doi:10.1088/1757-899X/967/1/012018

Fig. 1 CETR UMT 100 tribometer with heat chamber and liquid container

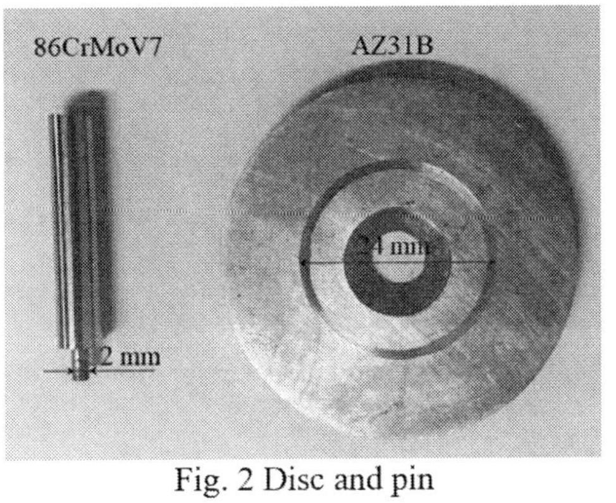

Fig. 2 Disc and pin

To study the effects of lubricant and its additives on the tribological behaviour and surface morphology, four different types of lubricant were prepared [7], as listed in Table 1. The effects of two typical lubricant additives, i.e., oil film improver, extreme pressure (phosphorous donors), used in the production on the friction were investigated.

Table 1 Lubricant samples (wt %)

Samples	Base oil (Sunflower oil)	Oil film improver	Extreme pressure (phosphorous donors)
#1	100%	0%	0%
#2	90%	10%	0%
#3	90%	5%	5%
#4	90%	0%	10%

3. Experimental procedures

The tests were carried out under the same contact sliding conditions. The sliding speed and normal load were fixed at 5 mm/s and 5 N, respectively. The temperature range in the tests varied from 25 °C to 101 °C. Each test ran for 20 minutes and repeated for three times. During a test, the instant friction coefficient of a contact pair was plotted against the running time. The above procedures were repeated for each type of the lubricant listed in Table 1. At the end of the tests for each lubricant, the surface morphology of the disc was measured.

IOP Conf. Series: Materials Science and Engineering **967** (2020) 012018 doi:10.1088/1757-899X/967/1/012018

4. Results and discussion

4.1 Hardness

Before the test, the surface hardness of AZ31B was measured. A mapping method was used to measure the surface hardness of a rectangular area in the surface of AZ31B, as shown in Fig. 3. The obtained surface hardness was about 50 HV, as shown in Fig. 4. For each sliding test, the hardness of the sample was measured again after the elevated temperature of samples completely decreased to the room temperature. It was found that the hardness was not changed on the non-sliding zone (see Fig. 4) because the measured hardness with the error bar was almost in the same magnitude. However, the surface hardness increased significantly inside the sliding-zone at a lower temperature, e.g., 25 °C. Yet, it decreased rapidly as the temperature increased from 25 °C to 200 °C, as shown in Fig. 4. This reveals that the sliding zone of AZ31B may undergo work hardening process due to the contact pressure for a lower temperature. With an increase in the temperature, the mechanical hardening decreases gradually and can be ignored for a higher temperature, e.g., 200°C. This is critical for understanding the tribology behaviour of AZ31B under the condition of elevated temperature.

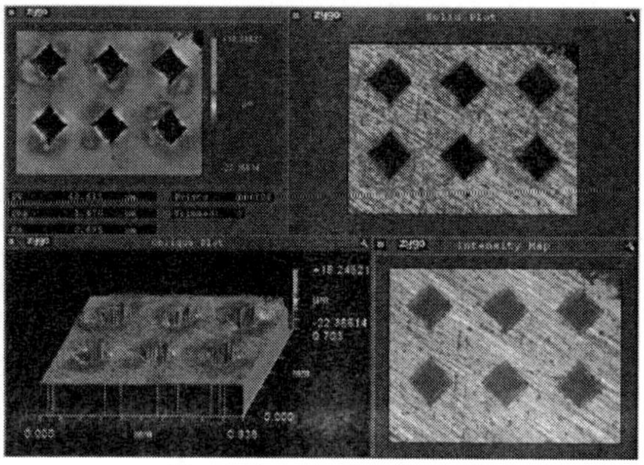

Fig. 3Measured points

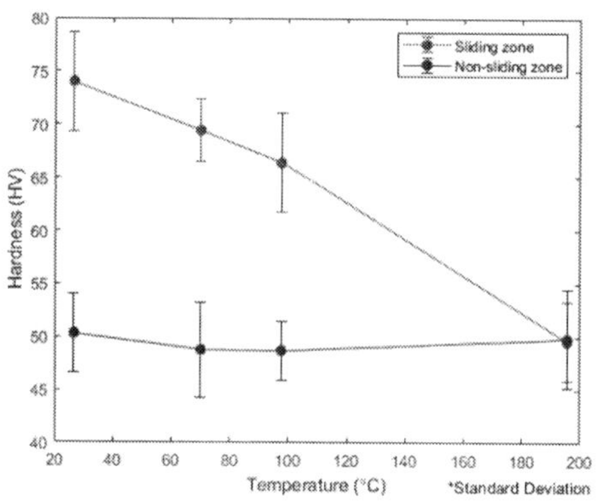

Fig. 4 Measured surface hardness

4.2 Friction coefficient

The friction coefficient of the AZ31B/86CrMoV7 pairs under dry contact sliding are shown in Figs. 5 to 8. The friction coefficient is about 0.3 under a lower temperature, e.g., less than 46 °C. A lower friction coefficient was observed at 72 °C. However, it becomes very unstable when the temperature reaches 101 °C, e.g., a sudden increase from 0.3 to 0.5, because of the surface hardness softness caused by the elevated temperature. Overall, a relatively higher friction coefficient can be expected for the dry surface contact.

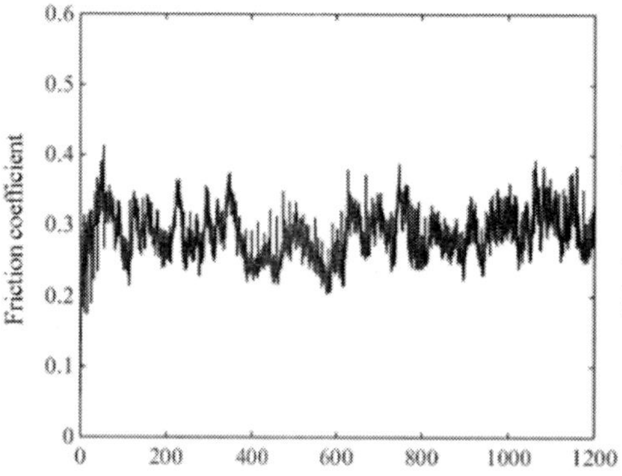

Fig. 5 Dry friction T=25°C

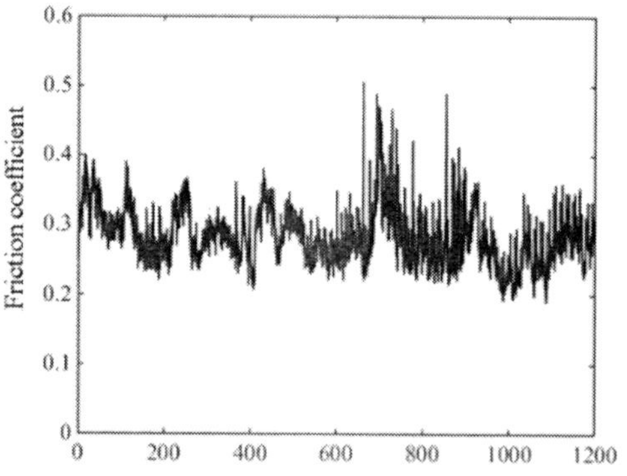

Fig. 6 Dry friction T=46°C

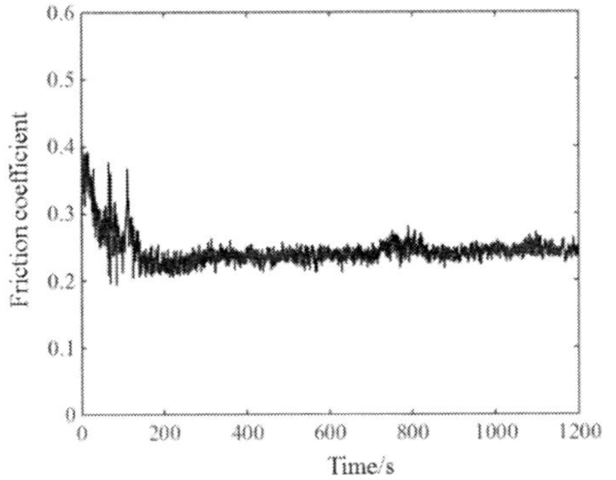

Fig. 7 Dry friction T=72°C

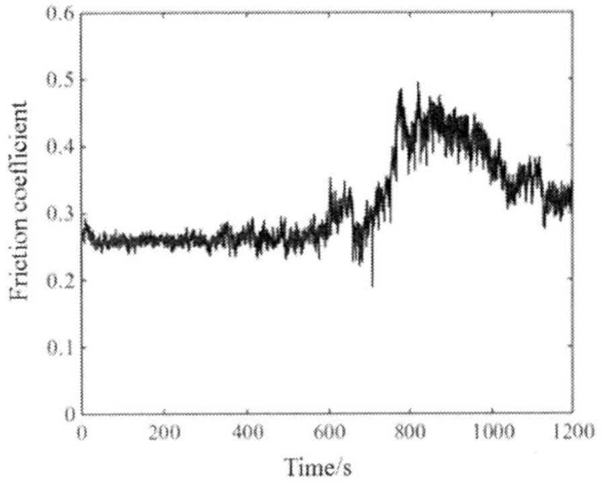

Fig. 8 Dry friction T=101°C

Figs. 9 to 12 show the lubricant and additive effects on friction, using the base oil without any additive. Compared with the results under dry sliding, a much lower and stable friction was observed in the whole temperature range tested (25 to 101 °C).

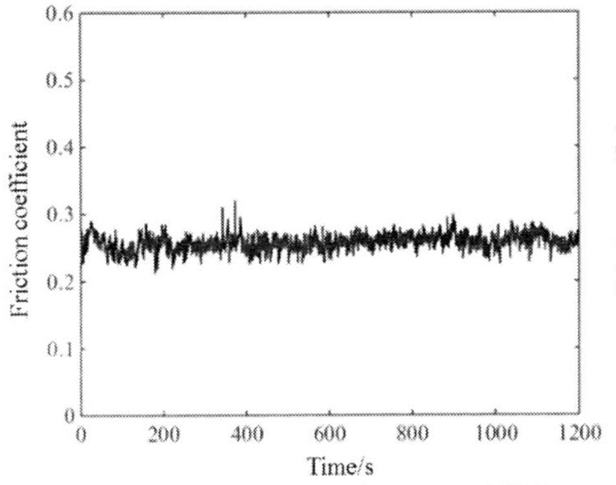

Fig. 9 Sample #1 T=25°C

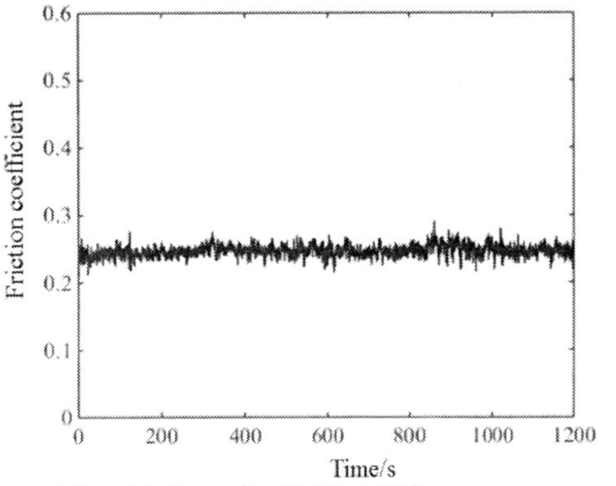

Fig. 10 Sample #1 T=46°C

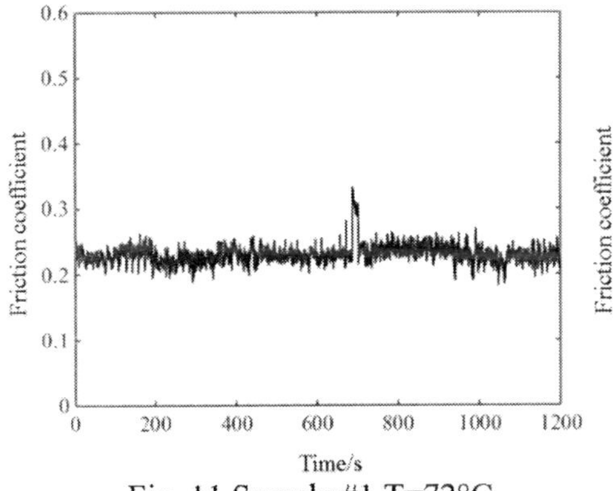

Fig. 11 Sample #1 T=72°C

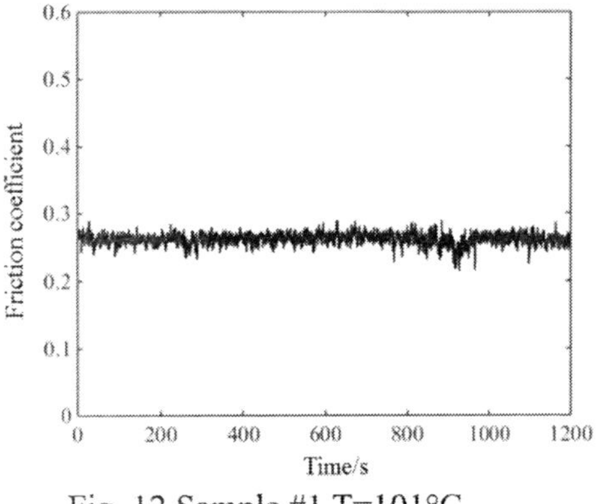

Fig. 12 Sample #1 T=101°C

When the oil film improver was applied, friction coefficient decreases significantly, as shown in Figs. 13 to 16. Moreover, its effect becomes much more sensitive as temperature increases. For instance, the friction coefficient becomes lower than 0.1 when temperature reaches 101°C. Compared with the case using the base oil, the oil film improver can dramatically improve the lubrication performance.

IOP Conf. Series: Materials Science and Engineering **967** (2020) 012018 doi:10.1088/1757-899X/967/1/012018

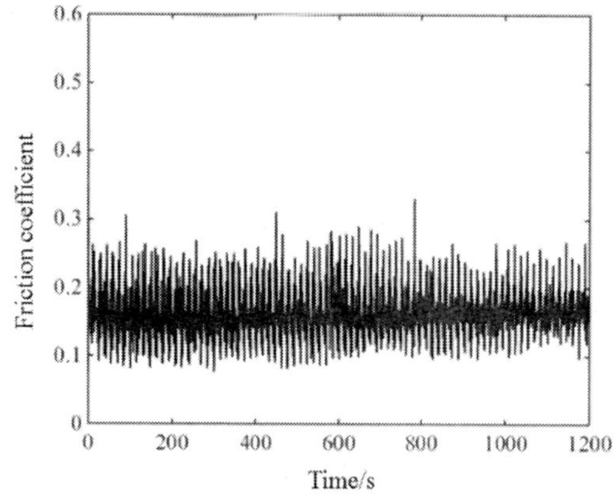

Fig. 13 Sample #2 T=25°C

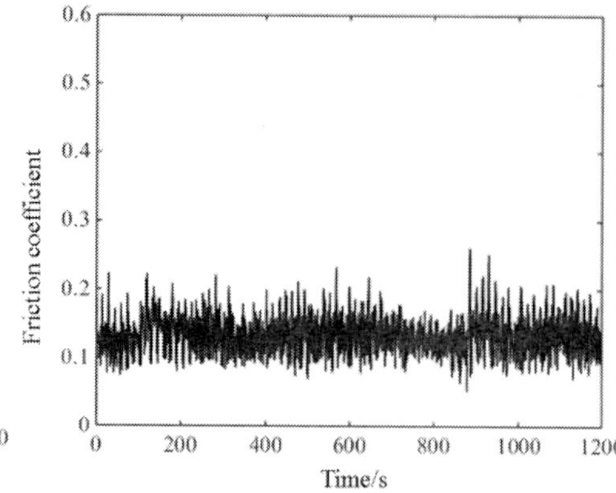

Fig. 14 Sample #2 T=46°C

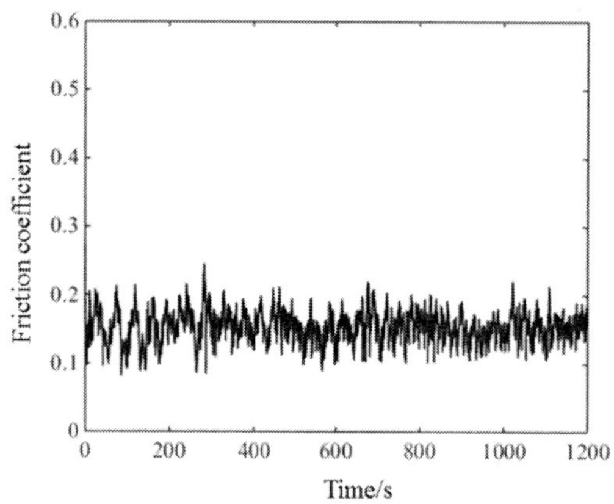

Fig. 15 Sample #2 T=72°C

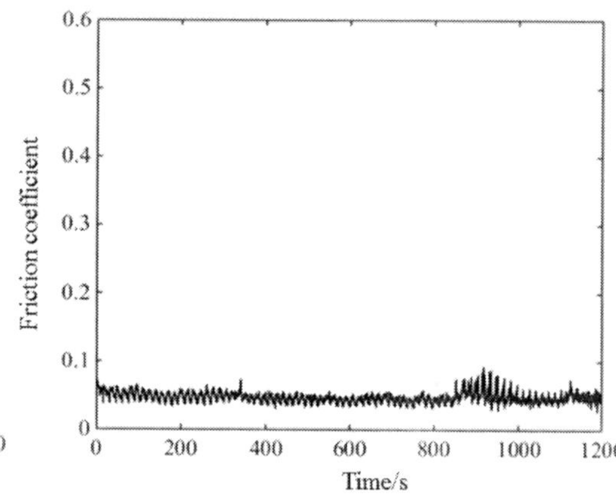

Fig. 16 Sample #2 T=101°C

With pure extreme pressure (10%), the lubricant performance does not seem to be as effective as that with the pure oil film improver at room temperature. However, the extreme pressure can effectively lead to the reduction in friction when the temperature increases, as shown in Figs. 17 to 20. It should be noted that friction becomes less stable compared with the lubricant with pure oil film improver.

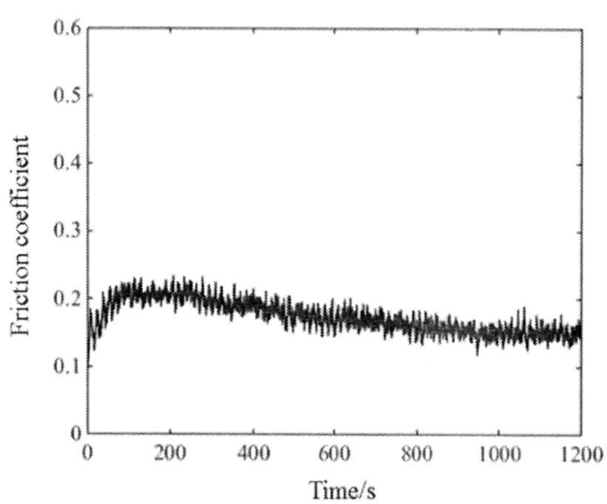

Fig. 17 Sample #4 T=25°C

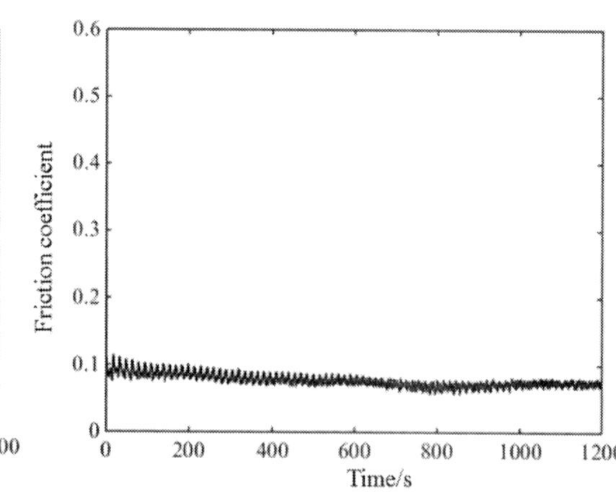

Fig. 18 Sample #4 T=46°C

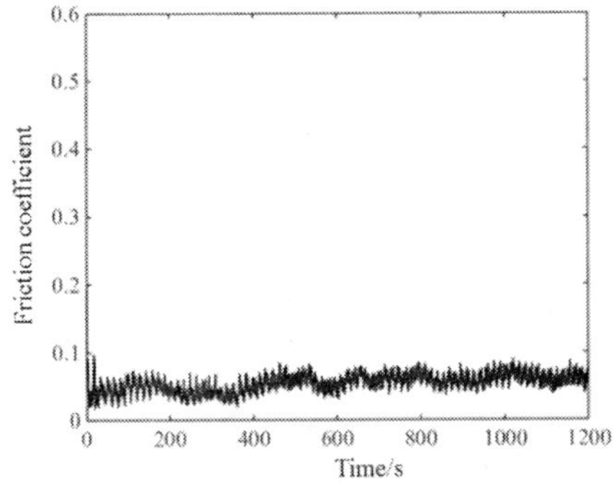

Fig. 19 Sample #4 T=72°C

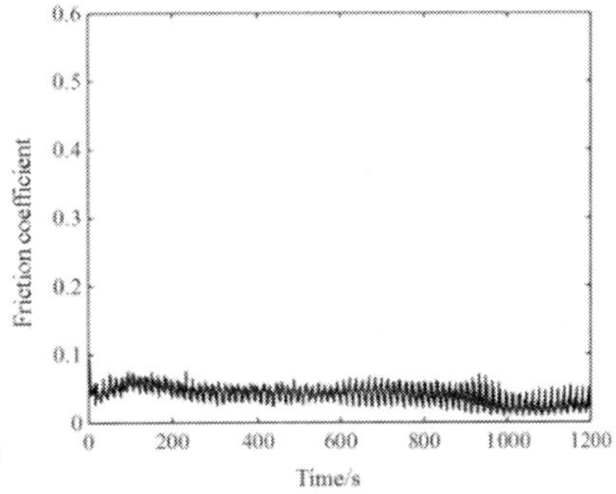

Fig. 20 Sample #4 T=101°C

When oil film improver and extreme pressure were used together, the lubricant performance was excellent, as shown in Figs. 21 to 24. The friction coefficient became less than 0.1 and an even lower friction was observed when temperature reached 101 °C (see Fig. 24). Overall, the lubricant with oil film improver and extreme pressure can significantly reduce the friction of the AZ31B/86CrMoV7 contact pair.

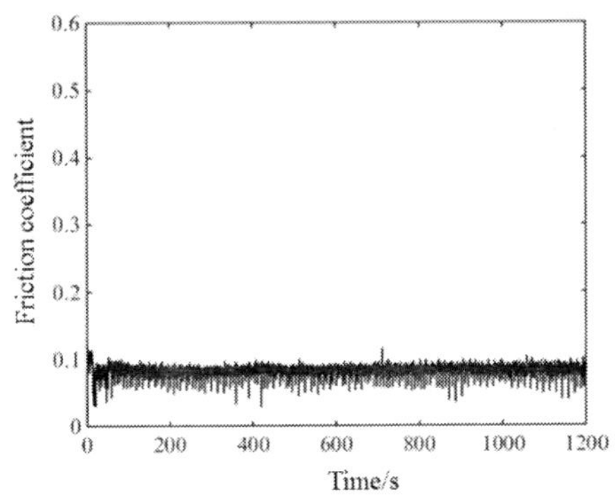

Fig. 21 Sample #3 T=25°C

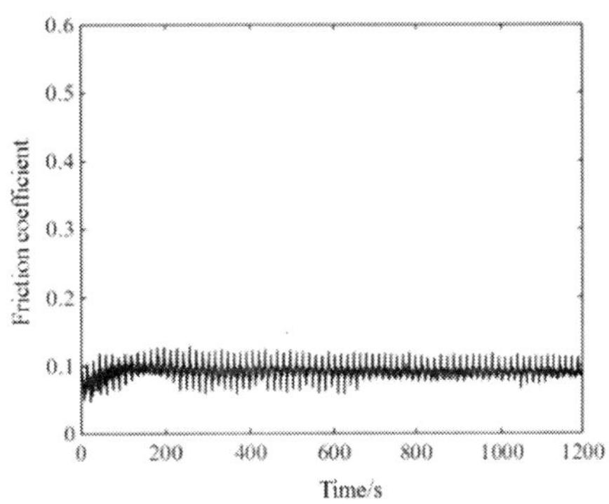

Fig. 22 Sample #3 T=46°C

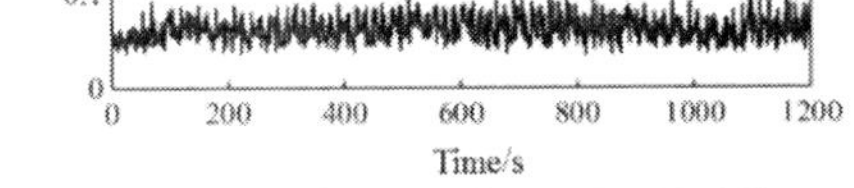

Fig. 23 Sample #3 T=72°C

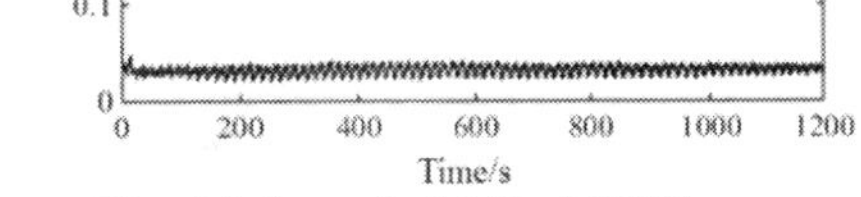

Fig. 24 Sample #3 T=101°C

Generally, the lubrication results in a lower interface friction because of the asperity-lubricant interaction. The lubricant additive can further bring about a friction reduction because of a boundary layer formed by a physical absorption or a chemical reaction process. In fact, the elevated temperature can enhance the chemical reaction process, which results in a decrease in the interface friction [8].

4.3 Surface morphology

Apart from the friction coefficient of the sliding zone, the effects of lubricant and additives on the surface morphology were investigated as well. Fig. 25 shows the measured 3D surface topography of AZ31B under dry sliding. A very rough surface was observed due to the significant wear at the sliding zone.

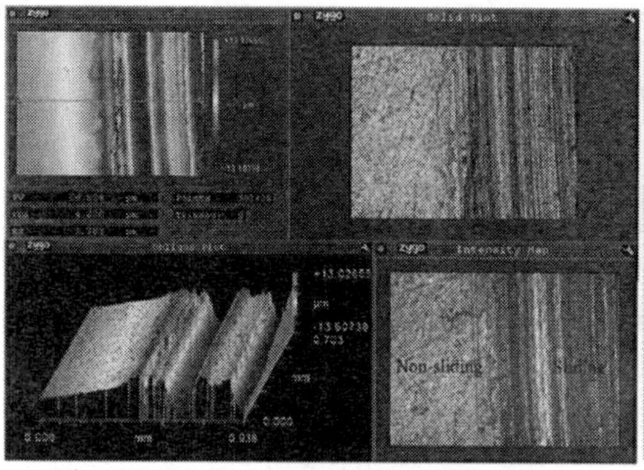

Fig. 25 Surface morphology after dry test

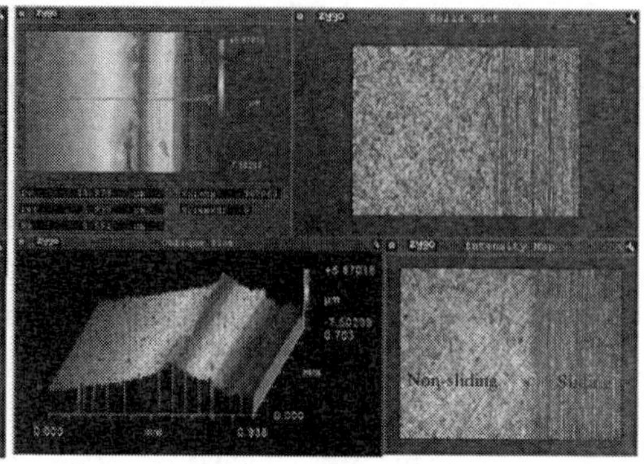

Fig. 26 Surface morphology sample #1

Compared with the dry surface contact, the pure base oil can improve the surface quality inside the sliding zone, as evidenced by Fig. 26. However, pile-up appeared at the edge between the sliding and non-sliding zone. This could be caused by the material's plastic deformation under the contact stresses. The softening of the AZ31B alloy might also occur as temperature increases. Compared with the pure based oil, the application of the oil film improver led to surface oxidation in the sliding zone, as shown in Fig. 27. Moreover, this oxidation surface layer cannot be easily removed after the test. Figs. 28 to 29 demonstrate that the extreme pressure can form a solid lubricant layer on the non-sliding zone. Fig. 30 shows the optical images of the AZ31B surfaces. A shining surface appeared when the lubricant with oil film improver was used, while a dark surface was observed when using of the extreme pressure. Further investigations are necessary to explore the underlying mechanisms.

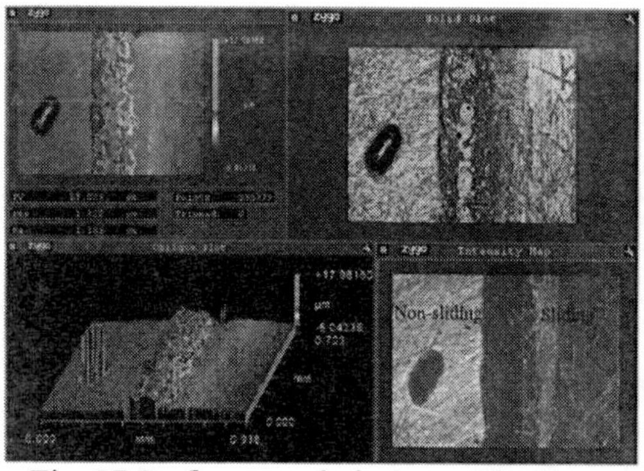

Fig. 27 Surface morphology sample #2

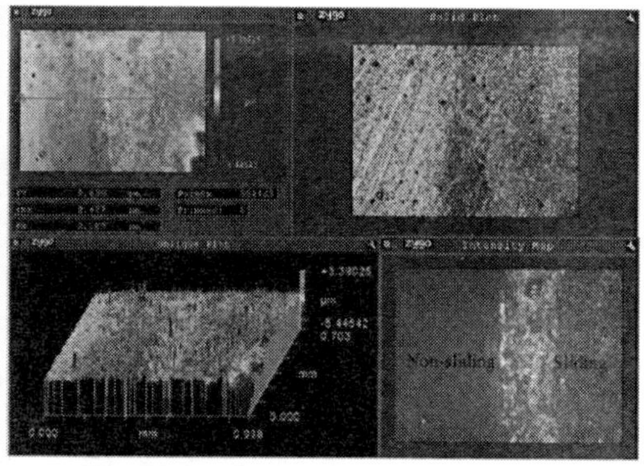

Fig.28 Surface morphology sample #3

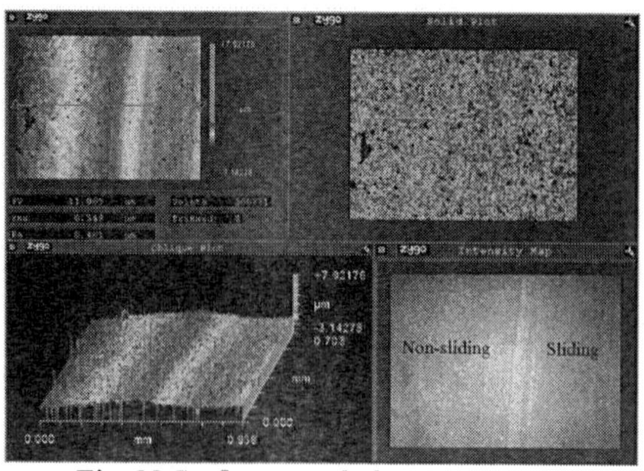

Fig. 29 Surface morphology sample #4

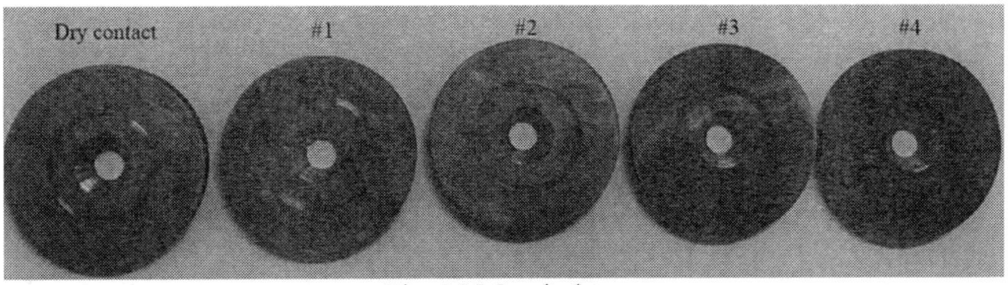

Fig. 30 Morphology

5. Conclusions

This paper has investigated the tribological behaviour of AZ31B sliding against 86CrMoV7 under a range of elevated temperatures. The effect of lubricant with/without additives has also been studied. The result has led to the following conclusions:

(1) The surface hardness of AZ31B was about 50 HV before contact sliding but increased significantly inside the sliding zone for a lower temperature due to the work hardening process. However, the effects of work hardening can be ignored as the temperature increases. The relatively low surface hardness of the AZ31B makes the wear significant, shown by, e.g., the pile-ups at the edge between the sliding and non-sliding zone.

(2) The lubricant with oil film improver and extreme pressure can reduce friction because of a boundary layer formed by a physical absorption or a chemical reaction process. Temperature increase makes the friction smaller due to the effect of lubricant additives.

(3) Compared with the pure base oil, both the oil film improver and extreme pressure can form an oxidation layer on an AZ31B surface. However, the former leads to a layer within only the contact zone while the latter also results in a layer outside the contact zone. The optical images of the AZ31B surfaces demonstrate that the lubricant with oil film improver leads to a shiny surface but the extreme pressure results in a dark surface. Further study is essential for a deeper understanding of the mechanisms.

References

[1] Chen F-K, Huang T-B. Formability of stamping magnesium-alloy AZ31 sheets. J Mater Process Technol. 2003;142:643-647.
[2] Ambrogio G, Filice L, Manco G. Warm incremental forming of magnesium alloy AZ31. CIRP annals. 2008;57:257-260.
[3] Hsu E, Carsley JE, Verma R. Development of forming limit diagrams of aluminum and magnesium sheet alloys at elevated temperatures. Journal of materials engineering and performance. 2008;17:288-296.
[4] Huang G-S, Zhang H, Gao X-Y, Bo S, Zhang L. Forming limit of textured AZ31B magnesium alloy sheet at different temperatures. Transactions of Nonferrous Metals Society of China. 2011;21:836-843.
[5] Habibnejad-Korayem M, Mahmudi R, Ghasemi H, Poole W. Tribological behavior of pure Mg and AZ31 magnesium alloy strengthened by Al2O3 nano-particles. Wear. 2010;268:405-412.
[6] Nguyen Q, Sim Y, Gupta M, Lim C. Tribology characteristics of magnesium alloy AZ31B and its composites. Tribol Int. 2015;82:464-471.
[7] Shen J, Wu C, Zhang L. Effects of sliding speed and lubrication on the tribological behaviour of stainless steel. The International Journal of Advanced Manufacturing Technology. 2018;94:341-350.
[8] Wu C, Qu P, Zhang L, Li S, Jiang Z. A numerical and experimental study on the interface friction of ball-on-disc test under high temperature. Wear. 2017;376:433-442.

Acknowledgments

This research forms part of the Baosteel Australia Research and Development Centre (BAJC) portfolio of projects and has received support through the Centre, project BAJC 18002. The AZ31B and 86CrMoV7 testing samples were provided by Dr. Shiwei Xu (Baoshan Iron & Steel Co., Ltd., China). The authors appreciate the support by Dr. Shiwei Xu's group to this study.

Nondestructive detection of the interfacial failure of steel-polymer sandwich composites during forming process

Jewook Yang[1], Sungjin Han[1] and Woong-Ryul Yu[1]

[1]Seoul National University, 1, Gwanak-ro, Gwanak-Gu, Seoul 08826, Korea

Abstract. Steel-polymer composites have low density, high specific flexural stiffness and vibration damping properties over other metal materials. However, the weak interfaces between steel and polymer sheet are the typical locations where the failure occurs. Furthermore, interfacial failure is a fatal reason of the degradation of mechanical properties. Therefore, predicting interfacial failure of the final product is an important technology in commercialization of steel-polymer composites. In this study, we built forming limit diagram (FLD) based on two different failure modes: failure of the bottom skin steel and interfacial failure between the skin steel and core polymer. Acoustic emission (AE) technique was conducted to observe the interfacial failure moment. Peak frequency was found to be the key feature to distinguish interfacial failure. Peak frequency band that indicates interfacial failure was from 300 kHz to 500 kHz. This peak frequency band made it possible to build FLD considering the interfacial failure.

Keywords: *steel-polymer composites, forming limit diagram, acoustic emission, failure modes*

1. Introduction

Steel-polymer composites are sandwich structure made of top and bottom skin steel and core polymer as shown in Figure 1. Steel-polymer composites are being widely used in the automotive, aerospace and construction fields due to their high specific flexural stiffness and functional properties [1, 2]. Therefore, the forming limit of steel-polymer composites is important. One of the common tools for characterizing the forming limit of sheet material is forming limit diagram (FLD). As such, some researchers have studied FLD of sandwich composites [3]. They focused on the localized necking of the bottom skin material. Also, numerical simulations have been conducted to predict FLD of sandwich material based on the failure of the skin metal sheet [4]. All these researches neglected interfacial failure during the forming process. However, the formability characteristic of steel-polymer composites is more complicated because it consists of materials with different mechanical properties. Furthermore, the interfacial separation between the skin and the core degrades the performance of steel-polymer composites [5]. Therefore, experimental tool for interfacial failure should be studied. Few researches have been carried out on building FLD considering interfacial failure.

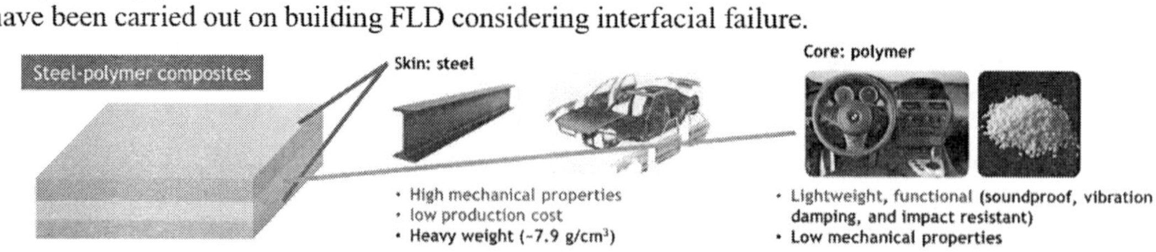

Figure 1. Composition of steel polymer composites

Acoustic emission (AE) technique is widely used for structural health-monitoring system to detect failure of materials in real time. Ultrasonic wave induced by elastic strain energy when material damages is observed by the AE sensor. Many researchers have carried out on classifying failure modes by observing AE signal features: amplitude, rise time, and peak frequency [6].
In this study, key AE signal feature that classifies the failure modes of steel-polymer sandwich composites was investigated. There are three different failure modes: steel fracture, polymer fracture, and interfacial failure. According to the peak frequency of AE signals, only interfacial failure showed unique characteristics. During the formability test, in-situ AE signals were obtained, and finally, FLD considering the interfacial failure can be obtained.

2. Experiments

2.1. Specimen preparation
Electro galvanized (EG) steel was used as skin material. The thickness of skin steels was 0.6 mm. Polyamide-6 (PA6) polymer was used as core material. The thickness of core polymer was 1 mm. Acetyl type adhesive, Loctite 401 (Henkel) was used to adhere the steel and polymer. Mechanical properties of EG steel and PA6 polymer are presented in Table 1.

Table 1. Mechanical properties of EG steel and PA6 polymer

	Elastic modulus (GPa)	Yield strength (MPa)	Tensile strength (MPa)	Elongation
EG steel	177	178.5	302	0.68
PA6	1.13	32.7	54.9	5.43

2.2. Formability test methods
Hemispherical punch tests were conducted to build forming limit diagram of the steel-polymer sandwich composites. Figure 2 shows the punch test setup. Load and displacement of the punch were obtained by the universal testing machine connected to the hemispherical punch. The diameter of the punch is 50 mm and the diameter of hole where the punch moved was 56.5 mm.

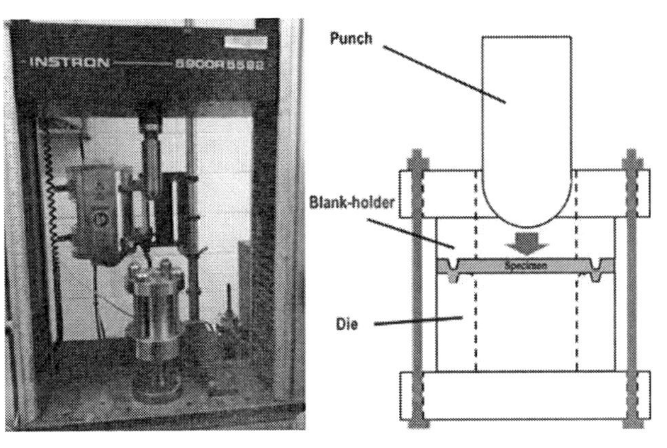

Figure 2. Punch test setup.

Specimen was fabricated with different shape varying from 10 mm to 100 mm width as shown in Figure 3(a). After specimen placed on the die, the blank-holder was clamped by bolts and nuts. The beads helped specimen not to slip. Speed of the hemispherical punch was 10 mm/min. Each specimen showed different strain paths from simple tension to biaxial tension.

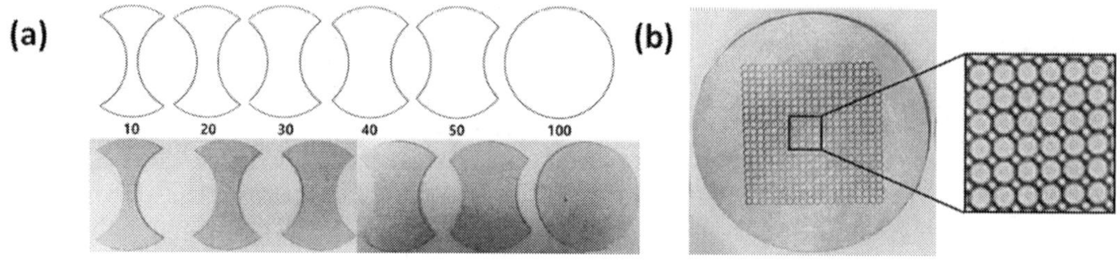

Figure 3. (a) Specimen for the punch test, and (b) specimen with circle grid painted by a stamp

Circle grid with 3 mm diameter was painted on the specimen with a stamp as shown in Figure 3(b). Deformed grids around the failure region was moved to a flat surface using a tape and major strain and minor strain were measured. Forming limit curve was constructed between the safe grids and failed grids

2.3. Acoustic emission test methods

Acoustic emission test was conducted to capture the interfacial failure moment. Figure 4 shows the acoustic emission test setup. M204A wide-broad sensor (Fuji ceramics corporation) was used. Operating frequency of the sensor is from 100 kHz to 600 kHz. To amplify the signal, A1002 preamplifier (Fuji ceramics corporation) was used. Then signals were acquired by DAQ board and analysed by the software (Mistras).

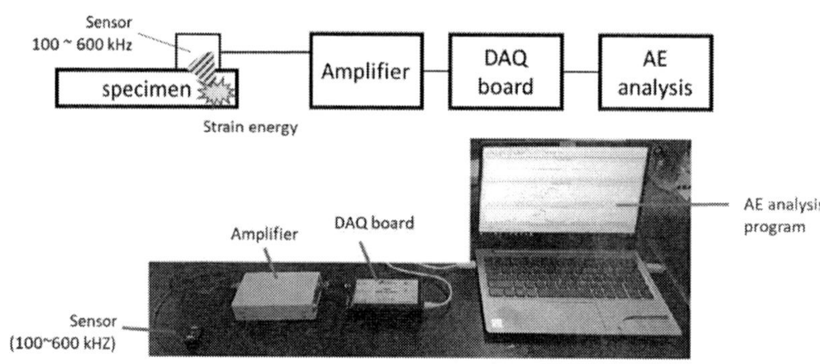

Figure 4. Acoustic emission test setup

2.4. Tests for failure modes

Steel fracture, polymer fracture and interfacial failure can occur during forming process. To distinguish interfacial failure from other failure modes during the punch test, acoustic emission signal features of interfacial failure must be studied. To observe AE signal features of steel fracture and polymer fracture, tensile tests of single steel and single polymer were conducted. For observing AE signal features of interfacial failure, lap shear test and tensile test of steel-polymer composites were conducted. AE sensor was placed on the specimen directly as shown in Figure 5.

Additionally, tensile test of steel-polymer composites was conducted. For one type of specimen there was no adhesive between EG steel and PA6 layer. The other type of specimen was adhered with acrylic adhesive. AE signals from tensile test of two types of specimen were collected to observe AE features of interfacial failure.

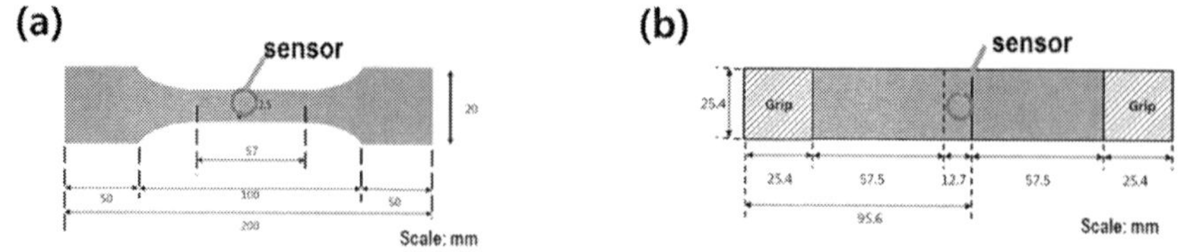

Figure 5. (a) Tensile test specimen and AE sensor, (b) lap shear test specimen.

3. Results and discussion

3.1. FLD of steel-polymer composites with steel fracture

Steel fracture was detected observing load drop during the punch test. Load-displacement curve of the punch test is shown in Figure 6. While displacement of the punch increased, the load increased linearly until the bottom steel broke. Signal of interfacial failure could not be observed from the load-displacement curve. FLDs of single steel and steel-polymer composites without considering interfacial failure were compared as shown in Figure 7. The minimum point of steel-polymer composites was higher than that of steel due to the thickness effect.

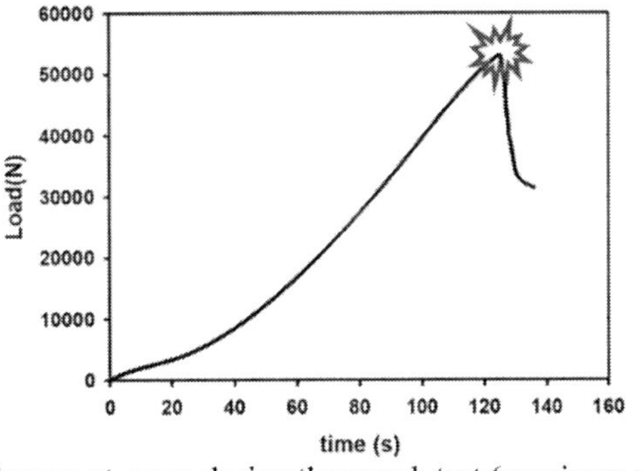

Figure 6. Load-displacement curve during the punch test (specimen with 100 mm width).

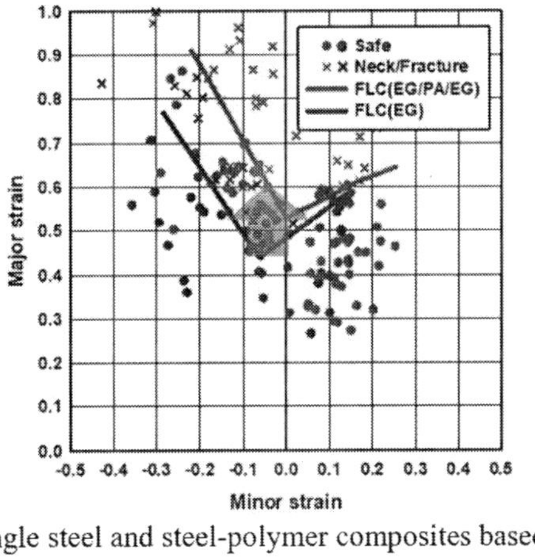

Figure 7. FLDs of single steel and steel-polymer composites based on the steel fracture

3.2. AE signal features

One of the acoustic emission waves measured by AE system during the tensile test of EG steel was shown in Figure 8(a). Peak frequency of the AE wave was calculated using fast Fourier transform as shown in Figure 8(b). The time when AE signals generated and plotted on X axis and Y2 axis of the stress-time curve as shown in Figure 9.

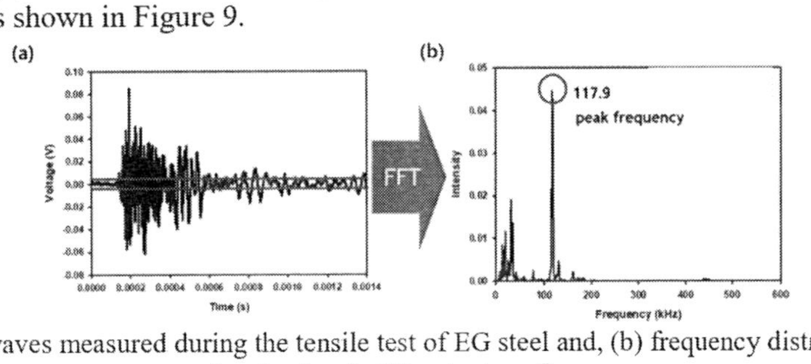

Figure 8. (a) AE waves measured during the tensile test of EG steel and, (b) frequency distribution of the AE signal using fast Fourier transform.

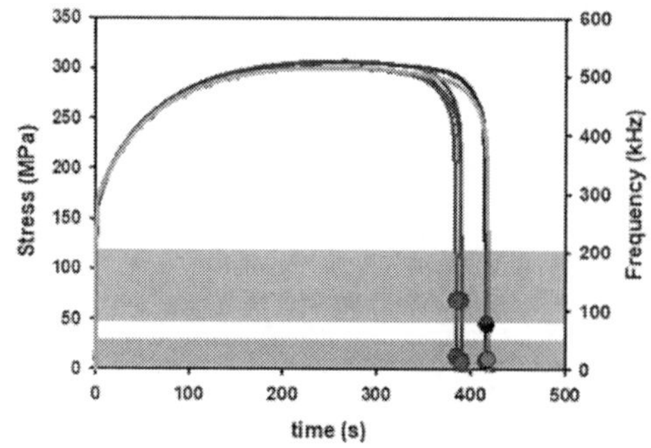

Figure 9. Peak frequency bands of tensile test of single steel.

As shown in Figure 9, AE signals which have peak frequency of 10~100 kHz and 100~200 kHz were observed at the moment of steel fracture. This indicates that strain energy release from steel fracture generates ultrasonic wave with 10~100 kHz and 100~200 kHz peak frequency. Likewise, AE signals from the fracture of the single PA6 showed 30~70 kHz peak frequency. Lap shear test was also conducted and AE signals from interfacial failure had peak frequency of 10~200 kHz and 300~500 kHz. Comparing the results, signal with peak frequency of 300 kHz~500 kHz was only observed at interfacial separation.

Stress-time and frequency-time curves of tensile tests of steel-polymer composites with and without adhesion were shown in Figure 10(a) and 10(b) respectively. Steel-polymer composites specimen with adhesion showed additional frequency band of 300 kHz~500 kHz when compared to the specimen without adhesion. When the specimen with adhesion went into the fracture, the interfacial separation at the side of the specimen occurred due to different necking behavior of EG steel and PA6 as shown in Figure 11. This results confirm that AE signals of 300 kHz~500kHz peak frequency band are key features to distinguish the interfacial failure from the other failure modes.

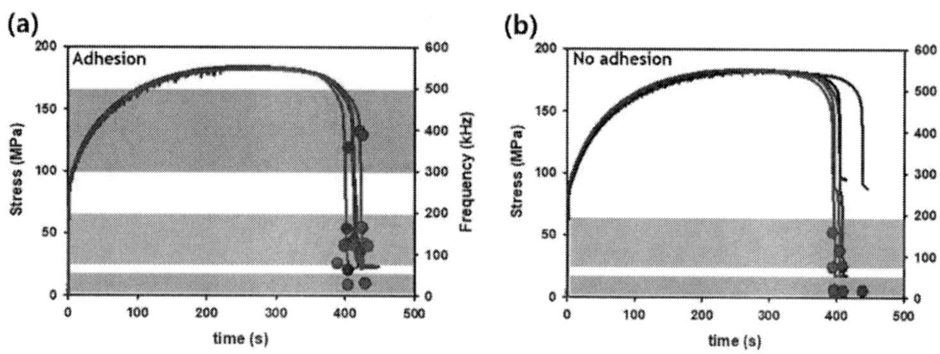

Figure 10. Peak frequency bands of tensile test of steel-polymer sandwich composites (a) with adhesion and (b) without adhesion.

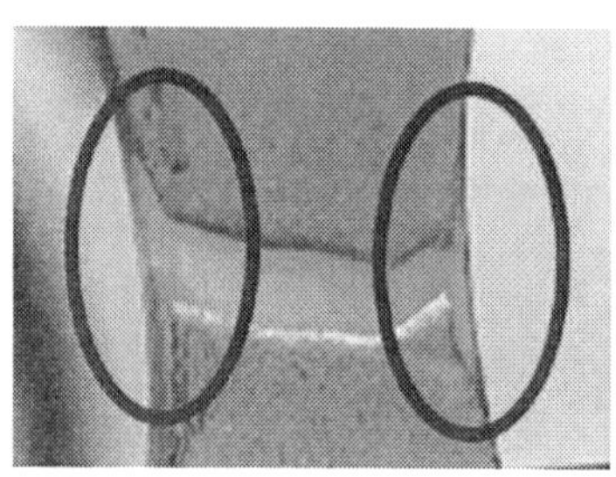

Figure 11. Interfacial separation after the tensile test of steel-polymer composites with adhesion.

4. Conclusions

In this study, the formability of steel-polymer sandwich composites was investigated. Acoustic emission technique was used to define key features of interfacial failure signal. Single steel and polymer tensile tests were compared with lap shear test. Furthermore, tensile test of steel-polymer composites were carried out with and without adhesion. It was shown that peak frequency of 300 kHz~500 kHz was only found in interfacial failure.

5. References

[1] Nakra B. Vibration control in machines and structures using viscoelastic damping. Journal of sound and vibration. 1998;211(3):449-66.
[2] Kari L, Lindgren K, Feng L, Nilsson A. Constrained polymer layers to reduce noise: reality or fiction?—An experimental inquiry into their effectiveness. Polymer testing. 2002;21(8):949-58.
[3] Liu J, Liu W, Xue W. Forming limit diagram prediction of AA5052/polyethylene/AA5052 sandwich sheets. Materials & Design. 2013;46:112-20.
[4] Parsa M, Ettehad M, Matin P. Forming limit diagram determination of Al 3105 sheets and Al 3105/polypropylene/Al 3105 sandwich sheets using numerical calculations and experimental investigations. Journal of engineering materials and technology. 2013;135(3).
[5] Rinker M, John M, Zahlen PC, Schäuble R. Face sheet debonding in CFRP/PMI sandwich structures under quasi-static and fatigue loading considering residual thermal stress. Engineering Fracture Mechanics. 2011;78(17):2835-47.
[6] Gutkin R, Green C, Vangrattanachai S, Pinho S, Robinson P, Curtis P. On acoustic emission for failure investigation in CFRP: Pattern recognition and peak frequency analyses. Mechanical systems and signal processing. 2011;25(4):1393-407.

Acknowledgments
This work was supported by the POSCO (2018Z093).

Formability simulation of steel-polymer sandwiched composites considering the adhesion strength

S Han[1], J Yang[1] and W-R Yu [1]

[1] Seoul National University, 1, Gwanak-ro, Gwanak-Gu, Seoul 08826, South Korea

Abstract. Developed to increase the energy efficiency of automobiles, steel-polymer sandwiched composites are attracting attention due to their lightweight and multi-functionality and are being actively researched recently. In order to apply those composites to industrial fields, their formability must be investigated in order to prevent the degradation of mechanical properties of the composites during the forming process in various shapes. Therefore, experimental studies have recently been conducted to evaluate the formability of the composites. However, there are few studies to predict the formability of the composites considering the mechanical properties of core polymer and interfacial properties. In this study, formability tests were carried out for evaluating the forming limit diagram of the composites. Also, the formability test of the composites was simulated considering the viscoplastic properties of the core polymer using Arruda-Boyce model and the interfacial properties using the cohesive zone model. Finally, the effect of the interfacial adhesion on the formability of the composites was investigated, from which an optimal condition was explored for the interfacial adhesion strengths that can ensure the formability of the composites for specific applications.

1. Introduction

Steels have been widely used in various field due to their good mechanical properties and low production cost. However, they have been require to be replaced because of their heavy weight. For that reason, automotive industries have developed new structural materials with lightweight to enhance energy efficiency. Steel-polymer sandwiched composites have been researched to reduce the weight of structural materials while lightweight maintaining their performance. For example, the sandwiched composites with a volume fraction of polymer of 30% theoretically show that the flexural rigidity remains 97%, but the weight decrease by 75% compared to the steel of the same volume. In addition, these composites have showed improved damping and soundproof properties compared to pure steel structure.

In order to apply the composites to industrial fields, their formability must be investigated. Various experimental studies have recently been conducted to evaluate the formability of the composites [1]. Although there were studies evaluating the formability of the composites using the deep drawing test, there were some disadvantages that the used materials were limited. Also, in the case of the steel-polymer sandwiched composites, most fractures are caused by delamination at the interface. Therefore, formability of steel-polymer sandwiched composites should be investigated considering the interfacial properties. Also, there were studies to evaluate the forming limit of the composites through simulation. However, they did not consider the delamination at the interface and did not match the actual forming limit.

For simulating the formability of the composites, a mechanical model is required to accurately simulate the interfacial properties and the mechanical properties of steel and polymer. For example, a ductile damage criterion was used in simulation for the steel. Also, during forming of the steel-polymer sandwiched composites, the core polymer undergoes the plastic deformation. Generally, the mechanical properties of the polymer are affected by the tensile rate and temperature. Therefore, in order to accurately simulate the mechanical properties of polymer, a mechanical model for the polymer that should reflect the viscoplastic properties of the polymer. In addition, for simulating the interfacial properties, the cohesive zone model is necessary.
In this study, the formability test of the sandwiched composites was simulated considering the viscoplastic properties of the core polymer and the interfacial properties using the cohesive zone model. Among various effective traction-separation law [2], a linear softening relationship was used [3]. Finally, the effects of the adhesion strength on the formability of the composites were investigated, and the optimal conditions for the adhesion strength to ensure the formability of the composites were investigated.

2. Theoretical background

2.1. Formability

The formability is the ability of materials such as metals to undergo the plastic deformation without being damaged. Generally, the formability of the material is affected by its deep drawability, ductility and so on. There are various indices to show the formability of the material, but the forming limit diagram is typically used. The forming limit diagram, also known as the forming limit curve, is used in sheet metals. Through repeated measurement, the shape of the curve could be obtained experimentally. Alternatively, the forming limit diagram could be generated by mapping the shape of a failure criterion into formability limit domain. The example of the forming limit diagram is shown as Figure 1.

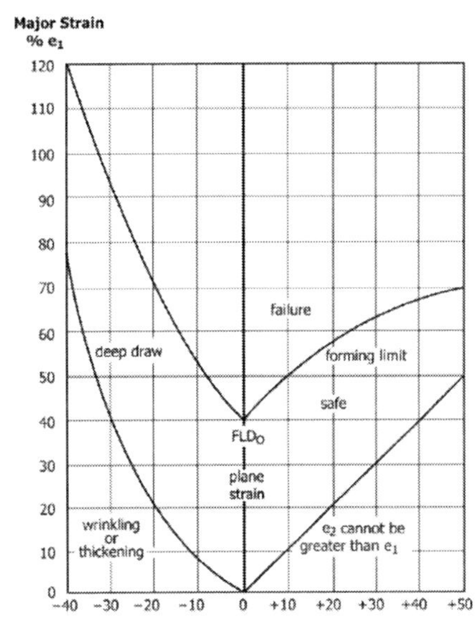

Figure 1. Example of forming limit diagram (ASTM E2218-15)

2.2. Ductile damage model for steel

A ductile criterion is a phenomenological model for predicting the onset of damage due to nucleation, growth, and coalescence of voids [4]. The model assumes that the equivalent plastic strain ($\bar{\varepsilon}_D^{pl}$) at the onset of damage is a function of stress triaxiality (η) and strain rate [5] (see Equation (1)).

$$\bar{\varepsilon}_D^{pl}(\eta) = \frac{\varepsilon_T^+ \sinh[k_0(\eta^- - \eta)] + \varepsilon_T^- \sinh[k_0(\eta - \eta^+)]}{\sinh[k_0(\eta^- - \eta^+)]} \quad (1)$$

Figure 2 illustrates the characteristic stress-strain behaviour of the material undergoing damage. In the context of an elastic-plastic material with isotropic hardening, the damage manifests itself in two forms: softening of the yield stress and degradation of the elasticity. The solid curve in the Figure 1 represents the damaged stress-strain response, while the dashed curve is the response in the absence of damage. In Figure 2, σ_{y0} and $\bar{\varepsilon}_0^{pl}$ are the yield stress and equivalent plastic strain at the onset of damage, and $\bar{\varepsilon}_f^{pl}$ is the equivalent plastic strain at failure. That is, when the overall damage variable reaches the value D=1. The overall damage variable, D, captures the combined effect of all active damage mechanisms and is computed in terms of the individual damage variable. The value of the equivalent plastic strain at failure, $\bar{\varepsilon}_f^{pl}$, depends on the characteristic length of the element and cannot be used as the material parameter for the specification of the damage evolution law. Instead, the damage evolution law is specified in terms of equivalent plastic displacement or in terms of fracture energy dissipation.

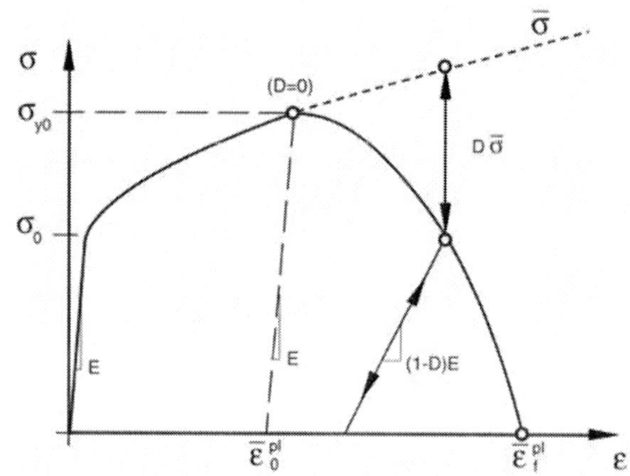

Figure 2. Stress-strain curve with progressive damage degradation [6]

2.3. Viscoplastic model for polymer

Arruda-Boyce model [7] was selected as a model to simulate the viscoplastic behavior of the polymer, and the viscoplastic behavior of the polymer could be simulated by deriving the material constants of the polymer (see Figure 3).

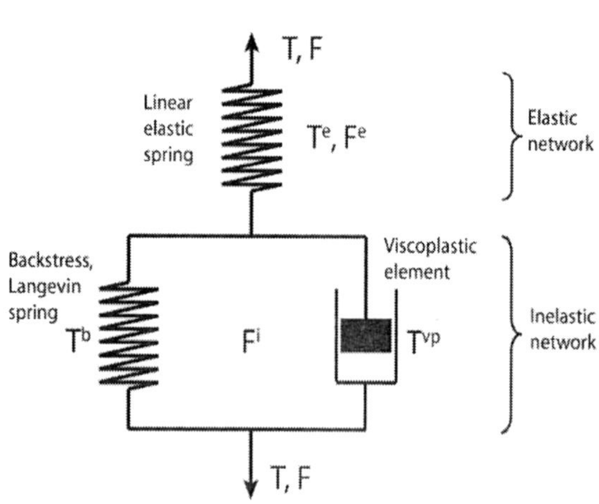

Figure 3. A rheological model of polymer [8]

IOP Conf. Series: Materials Science and Engineering **967** (2020) 012020 doi:10.1088/1757-899X/967/1/012020

2.4. Cohesive zone model for interfacial properties

To describe the delamination behavior of the interface, a cohesive zone model was used. The adhesive layer between the steel and the polymer was represented by cohesive elements, whose delamination behavior was characterized by a traction-separation law in Figure 4.

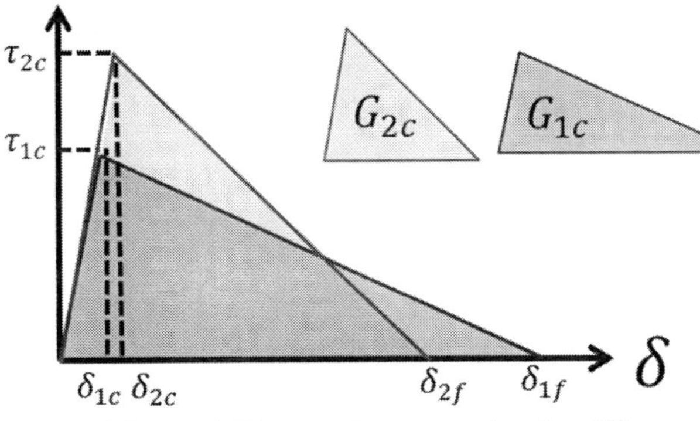

Figure 4. The traction-separation law [3]

3. Experimental

3.1. Mechanical properties

Tensile tests were carried out for evaluating the mechanical properties of steel, polymer, and the hybrid composite using a universal tensile machine. In addition, in order to obtain the material constants of polymer, tensile test and unloading test of polymer were carried out under various strain rate conditions.

3.2. Interfacial properties

For evaluating the interfacial properties of the adhesive layer between steel and polymer, four types of tests were carried out. For evaluating the failure stress of mode I and mode II, the butt joint test and lap shear test were carried out. Also, for evaluating the fracture toughness of mode I and mode II, double cantilever beam test and end-notched flexure test were carried out [9].

3.3. Formability simulation

A commercial finite element analysis software (ABAQUS/explicit, Simulia Inc., USA) was used for simulation. A three-dimensional model was built with the same specifications as the formability test by Nakazima [10]. The forming limit diagram was derived using the strain at the fracture of the specimen. The results obtained from the simulation were verified by comparing with the results obtained through Nakazima tests.

4. Results and discussion

Since the polymer did not fully recover in the unloading test, the plastic properties should be used in simulation. Experiments were conducted to confirm that the recovery of polymer was affected by strain rate. The results were shown in Figure 5. Also, through the tensile tests at various strain rates, the material constants of the polymer were obtained for simulation. Four experiments were performed to consider the properties of the adhesive layer. The failure stress in the each mode served as a criterion for the initiation of interfacial failure. And the fracture toughness in each mode was the basis of the final fracture at the interface.

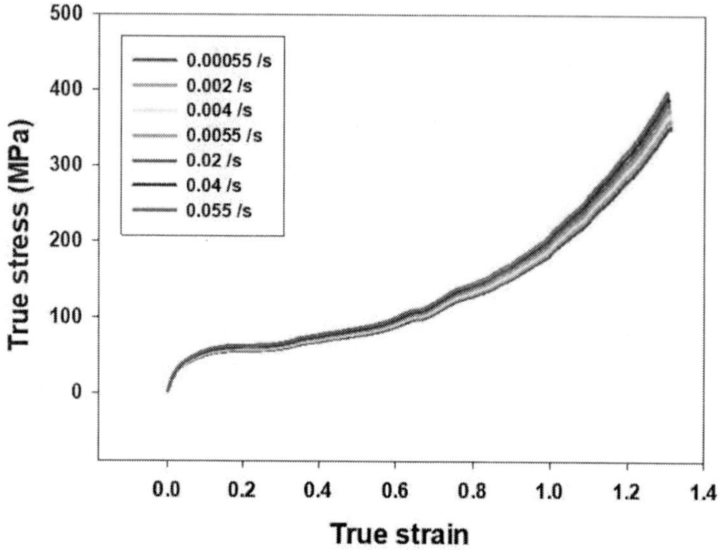

Figure 5. Stress-strain curve for polymer (PA6) according to tensile rate

In this study, 4 kinds of tests such as butt joint test, lap shear test, double cantilever beam test and end-notched flexure test were conducted to measure the interfacial properties between steel and polymer. Through these tests, the failure stress and the fracture toughness in mode I and model II were obtained. The results are shown in the table below.

Table 1. Interfacial properties between steel and polymer.

	Mode I	Mode II
Failure stress (MPa)	1.85	5.36
Fracture toughness (N/mm)	0.36	0.8

The forming limit of the steel/polymer/steel hybrid composites could be predicted through the simulation. The simulation results in Figure 6 showed that the forming limit of the hybrid composites were highly accurate when compared with the experimental results

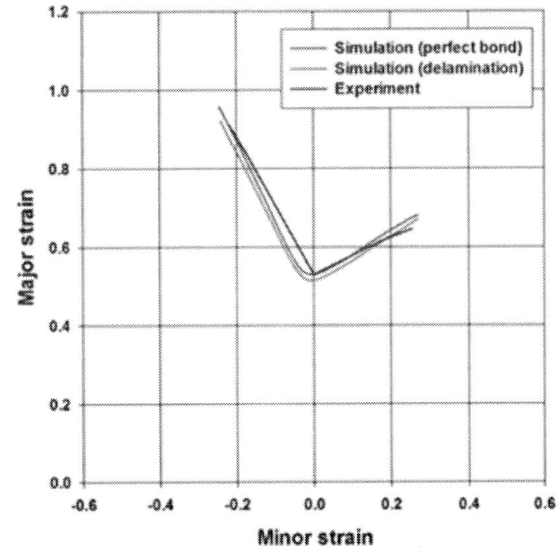

Figure 6. The forming limit diagram of steel-polymer sandwiched composites

Finally, the effect of the interfacial adhesion on the formability of the hybrid composites was investigated in shown as Figure 7, from which an optimal condition was explored for the interfacial adhesion strengths that can ensure the formability of the hybrid composites for specific applications. The forming limit of steel-polymer sandwiched composites increased as the adhesion increased. Under actual experimental conditions, the forming limit of the composites was determined by delamination, and this tendency was more shown as the adhesion was lower. However, assuming that the failure stress at mode I and mode II were 10 times stronger than the actual, the forming limit of the composites was very close to that in a perfect bonding condition where delamination did not occur. Therefore, in order to derive the forming limit of the composites, it is necessary to consider the interfacial properties because most fracture of the specimen occurs. In addition, it is understood that the adhesion at the interface must be strengthened for stable forming of the composites.

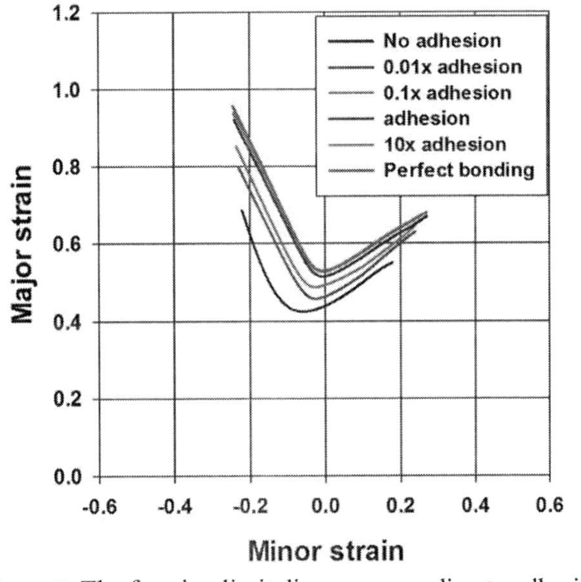

Figure 7. The forming limit diagram according to adhesion

5. Conclusion

In this study, we set the analytic model for steel, polymer, and interface, and used them to confirm the forming limit of steel-polymer sandwiched composites through the simulation. The simulation results verified the accuracy through comparison with the experimental results. When the forming limit of the composites was derived while changing the adhesion, it was confirmed that the forming limit increased as the adhesion increased. In most cases, the fracture at the interface occurred earlier than the fracture of the specimen, and the strain at delamination was further reduced as the adhesion decreased. Thus, for stable forming of the steel-polymer sandwiched composites, it is necessary to strengthen the interfacial properties.

6. References

1. Kim, K., et al., *Formability of AA5182/polypropylene/AA5182 sandwich sheets.* Journal of Materials Processing Technology, 2003. **139**(1-3): p. 1-7.
2. Park, K. and G.H. Paulino, *Cohesive zone models: a critical review of traction-separation relationships across fracture surfaces.* Applied Mechanics Reviews, 2011. **64**(6): p. 060802.
3. Alfano, G., *On the influence of the shape of the interface law on the application of cohesive-zone models.* Composites Science and Technology, 2006. **66**(6): p. 723-730.
4. Hooputra, H., et al., *A comprehensive failure model for crashworthiness simulation of aluminium extrusions.* International Journal of Crashworthiness, 2004. **9**(5): p. 449-464.
5. Andrade, F., J.C. de Sá, and F.A. Pires, *Assessment and comparison of non-local integral models for ductile damage.* International Journal of Damage Mechanics, 2014. **23**(2): p. 261-296.
6. Version, A., *6.13 Documentation (Abaqus), 2013.* Abaqus user's guide.
7. Boyce, M.C., *Large inelastic deformation of glassy polymers.* 1986, Massachusetts Institute of Technology.
8. Tømmernes, V., *Implementation of the Arruda-Boyce Material Model for Polymers in Abaqus.* 2014, Institutt for konstruksjonsteknikk.
9. Borg, R., L. Nilsson, and K. Simonsson, *Simulating DCB, ENF and MMB experiments using shell elements and a cohesive zone model.* Composites Science and Technology, 2004. **64**(2): p. 269-278.
10. Uthaisangsuk, V., et al., *Experimental and numerical failure criterion for formability prediction in sheet metal forming.* Computational materials science, 2008. **43**(1): p. 43-50.

IOP Conf. Series: Materials Science and Engineering **967** (2020) 012021 doi:10.1088/1757-899X/967/1/012021

The Investigation of Zinc Coating Quality with Different Thickness Levels After Forming of Flat Steels

E Tamer[1] and O Yildirim[1]

[1]Borcelik Celik Sanayi Ticaret A.S., R&D Center, Gemlik, 16601, Bursa, Turkey

Abstract. In this study, the deformation performance of galvanized deep drawing steel is investigated depending on coating thickness and forming limit of non-alloy pure zinc coated steel. Regarding to this; 100 and 150 g/m^2 Zn-coated samples are used to see the relevance with finite element analysis. To determine the deformation condition of Zn coating the samples were prepared at different elongation levels, then the samples are examined with surface views and cross-sectional analyses with scanning electron microscope to observe the amount of deformation at the onset of crack in the zinc coating cross-section. The change of zinc coating with elongation at cross-section region of the material will be investigated via microstructural analyses. It is aimed to determine the most suitable zinc coating level depending on the deep drawing amount in interstitial free (IF) steel usage. Furthermore, the performance of zinc coating will be investigated with finite element analyses with the goal of validating experimental results. Depending on the results of this study, the influence of zinc coating for a physical component will be carried out as a new study.

1. Introduction

The quality of sheet metal forming products commonly depends on the tribological conditions especially in deep drawing analysis since the friction affects the deformation condition. Flat steel products' elongation can be determined by using finite element analysis with different zinc coating levels. Moreover, the corrosion protection can be developed by applying different coating layers. Even though forming of galvanized steel the zinc coating undergoes large strain without significant damage, there are two major problems in zinc coating performance during forming operation which are the damage of coating and the effect of coating in friction

The formability of coated flat steel is affected by many factors such as formability behaviour of the uncoated material, existence of an intermetallic layer between the steel substrate and the zinc coating, zinc texture, grain size of zinc grains, as well as quality and quantity of alloying elements in zinc [1]. Different studies in literature including expansion tests for finite element analysis (FEA) have been investigated for galvanized steels and summarized in [2-5].

With the aim of investigating the zinc coating change in thickness at a certain deformation level, different tensile specimens have been prepared. The tests were performed up to different elongations. The change of zinc layer on steel surface a sequence of microstructural analyses was carried out at the cross-section region of the steel. It is aimed to determine the most suitable zinc coating level depending on deformation states. After mechanical characterization tests and microstructural experiments, tensile test simulations with two coating thicknesses (100 and 150 gr/m^2) were executed to compare with experimental results.

2. Experimental Setup

The competency in performance of sheet metal forming simulations remarkably rely on the chosen materials' constitutive models. Thus, material modelling is a field of study which gains increased attention of the scientific community, especially in case of sheet metal forming.

2.1. Mechanical characterization tests

To investigate the performance of zinc coating which will be compared with FEA data, a sequence of tensile tests is performed. To measure mechanical properties DX54D+Z tensile specimens with 100 and 150 gr/m^2 coating thicknesses were fulfilled. The DX54D+Z material, continuous annealed and hot dip galvanized IF steels, were chosen owing to its wide application range in automotive industry. The nominal thickness of the sheet materials was 0.85 mm, in contrast the individually measured thicknesses both for two different zinc layers were in the range of 0.83 - 0.87 mm. The tests were carried out in rolling direction with Euro-norm test speed. The average values with standard deviation of the test results are figured out in table 1. The average true stress versus true strain curves and the tested specimens are shown in figure 1 and figure 2.

Table 1. The mechanical properties of DX54D+Z with different coating thicknesses.

Amount of coating	t (mm)	Yield Stress(MPa)	Tensile Stress (MPa)	A_{80} (%)	n	r
100 gr/m^2	0.85 ± 0.02	175	316	42.4	0.22	1.92
150 gr/m^2	0.85 ± 0.02	167	307	42.4	0.22	1.89

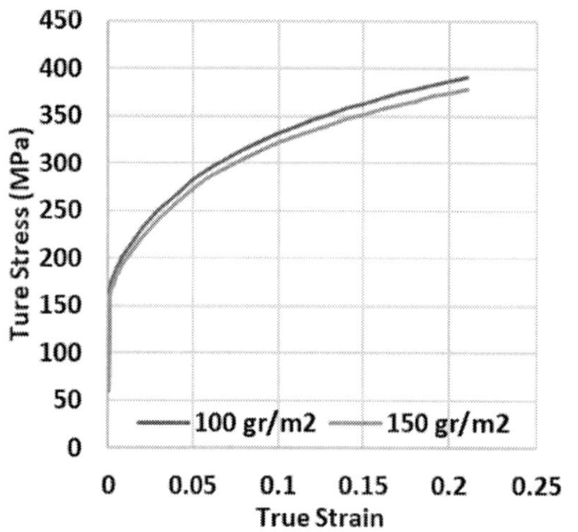

Figure 1. True stress-strain curve of DX54D+Z with the thickness of 0.86 mm.

Figure 2. Tensile testing device in Borcelik and test specimens with different deformation levels.

2.2. Microstructural analyses

The microstructural analyses were performed at Zeiss Axio Imager optical microscope and Zeiss EVO 50 scanning electron microscope (SEM). At the very beginning, specimens were cut to a covering 30 mm area in the middle region of the samples. To demonstrate the zinc-coating distribution an undeformed specimen is measured at SEM with 200X magnification. The shorn tensile test specimens were prepared to be measured under the optical microscope. The SEM images are shown in figure 3 and figure 4 both for 100 gr/m^2 and 150 gr/m^2 with different elongation levels. To illustrate the coating thickness, shorn samples were investigated under optical microscope (figure 5). The results are going to be assessed in section 4.

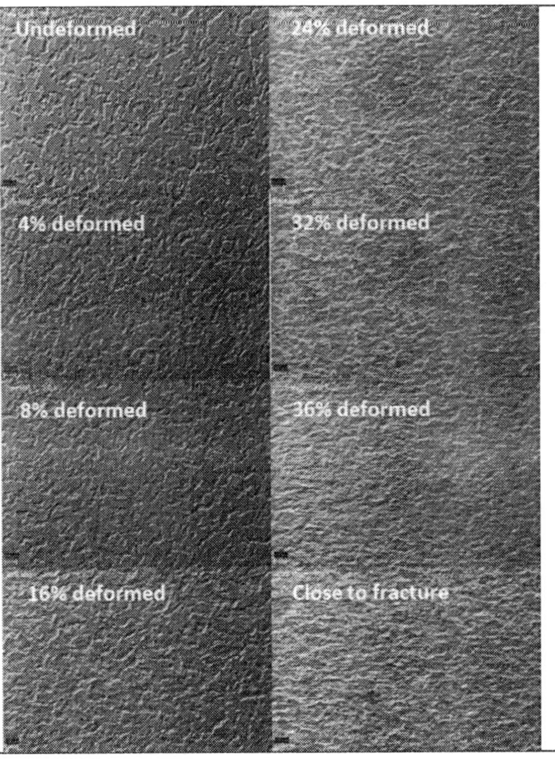

Figure 3. SEM measurements in the middle region of deformed tensile specimens for 100 gr/m^2.

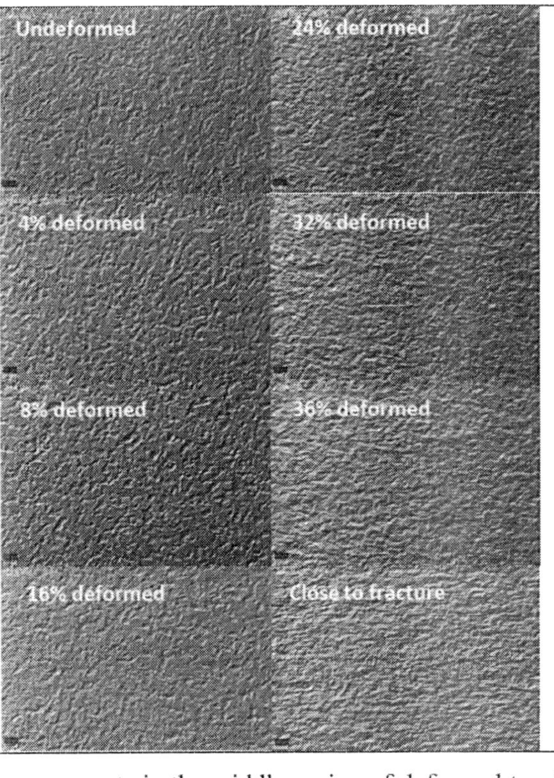

Figure 4. SEM measurements in the middle region of deformed tensile specimens for 150 gr/m^2.

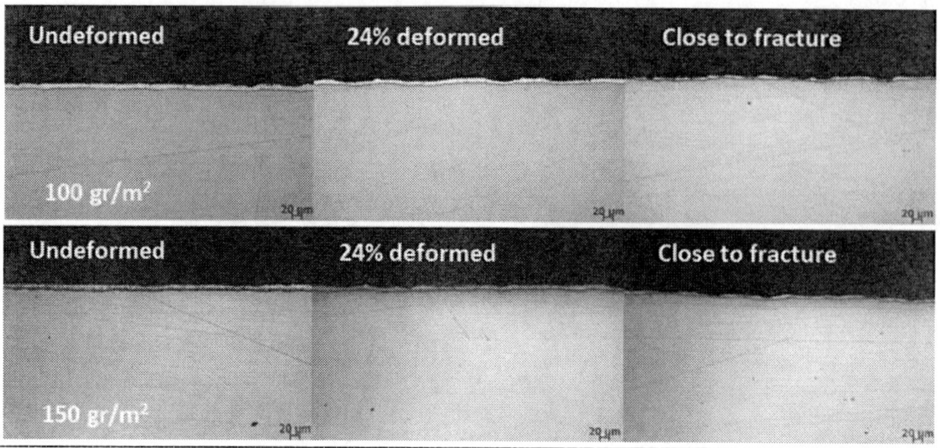

Figure 5. Optical microscope measurements of three deformation levels (undeformed-24% deformed-near fracture) for both zinc coating cases.

3. Numerical Modeling

In this study, tensile test simulation was executed with two coating layers. The analyses were performed in Autoform environment. The finite element analyses were conducted with Borcelik material cards. The initial mesh size was 0.06 mm, with 11 layers and 6 refinement levels. Vegter 2017 material model was used as a yield surface model for the sheet material. The mesh size and deformation result of galvanized sheet strip are demonstrated in figure 6. The coating thicknesses of pure zinc were taken from experimentally measured data as 0.0095 mm for 100 gr/m^2 and 0.015 mm for 150 gr/m^2.

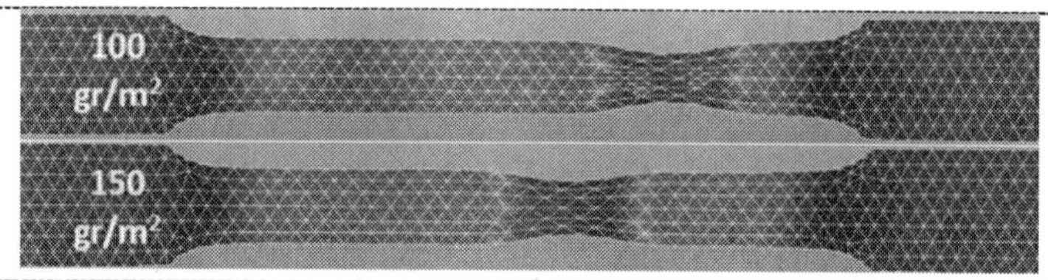

Figure 6. Major strain results at different localized necking regions.

Coating of pure zinc is modelled using tensile properties which were investigated in [6]. The specimens of pure zinc were used from cast samples and tensile tests were performed at room temperature. The test specimens were manufactured by pouring the molten zinc at a temperature of 430 ^{0}C. Mechanical properties used in Autoform simulation are given in table 2.

Table 2. The mechanical properties of pure zinc given in literature [6].

	Yield Stress(MPa)	Tensile Stress (MPa)	Ag (%)	n
Pure Zinc	27.5	30.0	0.788	0.008

4. Comparison of Results

To figure out a correlation between physical and virtual experiments of galvanized sheet metal in terms of deformation performance on different coating levels, the assessments have been made by using an orthogonal X section profiles on each part as shown in figure 7.

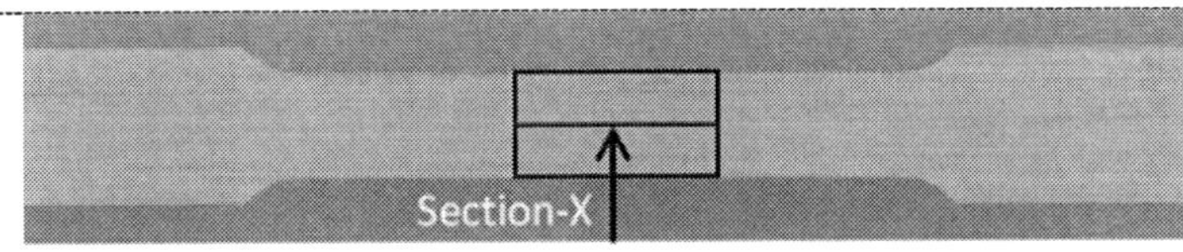

Figure 7. Measurement direction both for microstructural and finite element analyses..

4.1. Results of microstructural analyses

To demonstrate the initial microstructure of zinc coating on sheet material, samples of tensile specimens were investigated in scanning electron microscope. As shown in first picture in figure 8, the surface consists of homogenous zinc layer. According to the figure, as the amount of deformation level increases, the zinc coating on the surface also elongates. However, the zinc coating on the surface does not disappear on material raw surface of steel at fracture.

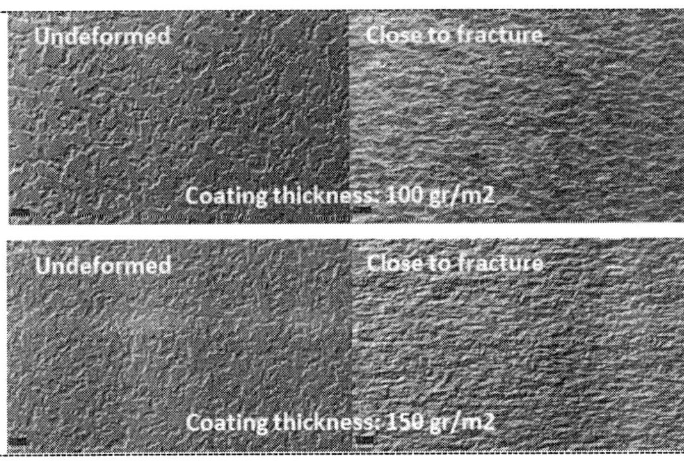

Figure 8. SEM measurements of undeformed and elongated up to fracture tensile strips in the mid-region of the specimen.

Moreover, the coating thickness was measured from cross sectional region of the samples via optical microscope. In figure 9, it can be considered that the thickness of the zinc coating decreases, but there is no tearing on steel surface up to rupture.

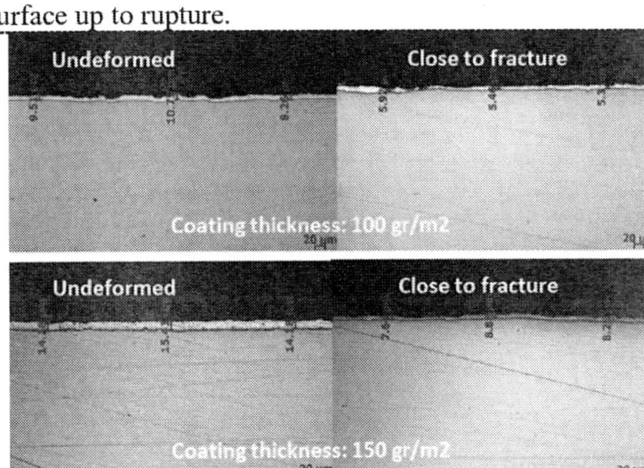

Figure 9. Optical microscope results in the mid-region of the specimen for two deformation levels.

Even though there is no tearing on material surface, the elongation of the zinc layer at 150 gr/m^2 coating is notably higher than the 100 gr/m^2 coated surface.

4.2. Results of FEA

In figure 10, the results of tensile specimen cross sectional measurements are illustrated. The thickness change in coating shows same behaviour as in the experiments.

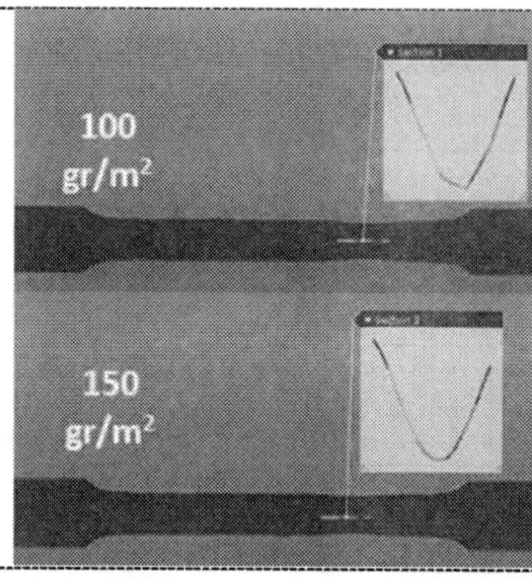

Figure 10. FEA results of two different coating levels near fracture.

FEA enables this study of zinc coating with different thicknesses with accurate material model, assuming no effect of measurement errors and supposing ideal mechanical properties of pure zinc.

4.3. Comparison of the results

As the distribution of the zinc layer, the investigation of the coating thickness is performed at only upper surface. The length of the section is 30 mm and the comparison are done at x direction as it is shown in figure 7. The results are presented in figure 11.

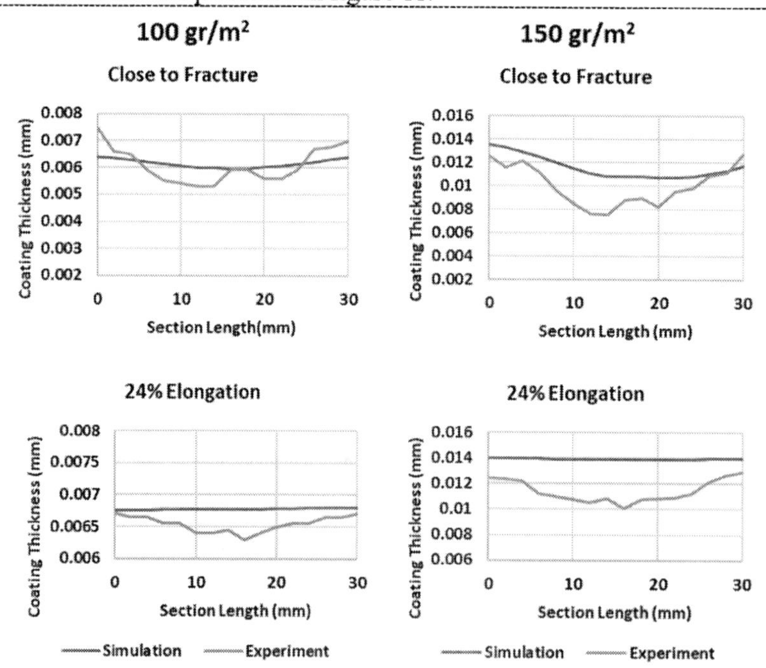

Figure 11. FEA results of two different coating thicknesses with two deformation levels.

One can see from figure 11 the experimental results of maximum elongated specimen agree well with simulation results with a maximum deviation of 3%. On the other hand, the experimental results of 24% elongated specimen differs from simulation results. It should be noted that the thickness of the coating layer is much smaller than steel. Additionally, the difference between the experimental and numerical analysis results might be caused by measurement errors.

5. Conclusion

In this study, the deformation performance of galvanized deep drawing steel is investigated depending on coating thickness of non-alloy pure zinc coated steel. In this context, 100 and 150 g/m^2 Zn-coated samples are used to investigate the consistency with numerical analysis. With the aim of describing the deformation condition of Zn coating, the samples were tested to different elongation levels, then the samples were examined with surface views and cross-sectional analyses with scanning electron microscope to observe the amount of deformation at the onset of crack at the zinc coating cross-section. The difference in terms of elongation at cross-section region of the steel has been investigated via microstructural analyses. It has been concluded that the experimental results have good agreement at higher deformation levels rather than lower deformation states. Therefore, especially in high formable deep drawing operations it is possible to decide which coating thickness could be offered to end-user with the execution of numerical analysis.

6. Futurework

The experimental tests were performed at rolling direction. With a purpose of investigating the coating thickness effect on anisotropy, the authors will carry out tensile tests at transverse and diagonal directions. According to these results, the effect of coating thickness on drawability will be investigated.

7. References

[1] Lazik S Esling C and Wegria J 1995 *Textures and Microstructures* Vol 123 pp 131-147
[2] Parisot R Forest S Gourgues A F Pineau A and Mareuse D 2000 *Computational Materials Science* Vol_19 pp 189-204
[3] Parisot R Forest S Pineau A Grillon F Demonet X and Mataigne J M 2004 *Metallurgical and Materials Transactions A* Vol 35A pp 797-811
[4] Lietzau J Philippe M J Esling C Wegria J Dubois M 1998 *Zinc-based Steel Coating Systems* in F E Goodwin (Ed.) T.M.S pp 207-217
[5] Maeda C Shimomura J Fujisawa H Konishi M 1996 *Scr.Metall.* Vol 35 pp 333-338.
[6] Freeman J R Sillers F Brandt P F 1926 *Pure Zinc at Normal and Elevated Temperatures US Government Printing Office* Vol 1 p 672

IOP Conf. Series: Materials Science and Engineering **967** (2020) 012022 doi:10.1088/1757-899X/967/1/012022

Fuzzy modelling of surface scratching in contact sliding

Wei Li[1], Liangchi Zhang[2*], Xinping Chen[3], Chuhan Wu[1], Zhenxiang Cui[3] and Chao Niu[3]

[1] Laboratory for Precision and Nano Processing Technologies, School of Mechanical and Manufacturing Engineering, The University of New South Wales, NSW 2052, Australia
[2] Department of Mechanics and Aerospace Engineering, Southern University of Science and Technology, Shenzhen, 518055, Guangdong, China
[3] Baoshan Iron & Steel Co., Ltd., Shanghai 200941, China

*corresponding author e-mail: zhanglc@sustech.edu.cn

Abstract. This paper presents a fuzzy modelling technique for the prediction of surface scratching problems encountered in sheet metal forming processes involving contact sliding. To generalize the contact conditions with die-sheet interactions in such processes, ball-on-disk sliding investigations, including modelling and experiment, were carried out with various ball diameters to mimic the die-contact radii effect and a range of contact loads to simulate the effect of contact stress variation experienced in forming processes. Because die-contact radius, surface roughness and contact stress changes cannot be exactly defined in a practical production process, in the modelling of the ball-on-disk configuration, ball diameter, normal load, surface roughness, sliding cycles and extent of scratching depth in a metal sheet surface were treated as fuzzy variables for the prediction. Based on the comparison between predictions with the relevant experimental measurements, it can be seen that the fuzzy modelling technique developed in this paper can predict very well the surface scratching.

1. Introduction

In the deep drawing process, sliding between the metal sheet and die/mould surfaces undergoes high contact stresses. As a result, such contact sliding can cause severe surface wear and subsequent scratching damage of the counterparts. However, it has turned out to be difficult to predict the evolution process of surface scratching damage because of the complexities in deformation and wear mechanisms. Therefore, it is of great importance to develop a feasible and effective method to predict the extent and evolution of surface scratching in the contact sliding process.

To solve this problem, some researchers, e.g. Archard [1], Rhee [2] and Bayer [3], have carried out empirical investigations to study the surface wear in the sliding process and relate the wear volume with the normal load, sliding distance and surface hardness. In these empirical studies, the influence of real contact stress on surface wear was not taken into consideration. For the wear problem in deep drawing process, Ersoy-Nürnberg et al. [4] presented a modified Archard's model, in which the wear coefficient was given as a function of accumulated wear work. Groche et al. [5] combined this method with finite element method and investigated the effects of sliding speed and contact stress. Despite

IOP Conf. Series: Materials Science and Engineering **967** (2020) 012022 doi:10.1088/1757-899X/967/1/012022

these researches and findings, the evolution process of microscopic surface damage during the contact sliding process was not presented in the studies mentioned above.

To investigate the surface damage evolution and wear mechanisms properly, some experimental methods were proposed and utilized. In the research carried out by Gåård et al. [6], a slider-on-flat-surface tribological system was developed to figure out the wear process of sheet metal surfaces. The initiation and evolution process of surface galling was studied and presented. In addition, the acoustic emission technique was also used to identify the initiation and variation process of surface wear in the studies conducted by Skåre and Krantz [7] and Shanbhag et al. [8]. However, it is still challenging to obtain reliable acoustic emission signals during the sheet metal forming process.

In general, these aforementioned approaches are not capable of predicting the surface damage evolution in the contact sliding process. It is preferable to establish appropriate models which can make full use of the available knowledge and provide reliable prediction results about the surface scratching damage. The artificial intelligence techniques have proved to be appropriate choices for this purpose. For example, the fuzzy logic approach, proposed by Zadeh [9], has been identified with good representation capability. It was used by Ali and Zhang to predict the residual stresses [10], surface roughness [11] and surface burn [12] in surface grinding processes.

In this paper, fuzzy modelling technique was used to predict the surface scratching damage induced by contact sliding. The experimental results of ball-on-disk sliding tests were adopted to establish a fuzzy prediction model, in which the ball diameter, normal load, surface roughness of sheet metal, sliding cycles and surface scratching depth were incorporated as fuzzy variables. In addition, a modified quantum-behaved particle swarm optimisation (QPSO) algorithm was developed for further optimization of the fuzzy inference model. The prediction performance of the optimized fuzzy inference model was verified by the comparison between prediction results and experimental measurements.

2. Experimental setup

Ball-on-disk sliding experiments were carried out by using a CETR tribometer, as shown in Figure 1. The materials of sheet metals and balls were advanced high-strength steel QP980 and high carbon chromium bearing steel GCr15, respectively. In the sliding process, the ball slid against the sheet metal surface reciprocally and the instant friction coefficient was given by the CETR tribometer. The sheet metal surfaces had three different levels of surface roughness, including R_a = 0.04 μm, 0.06 μm and 0.08 μm. After each sliding test, a Zygo microscope (NewView 700) was used to measure the sliding track in the sheet metal surface and calculate the maximum scratching depth (h_{max}). In this study, the value of h_{max} was adopted as the output variable to characterise the extent of surface scratching damage. Contact pressure in the ball-on-disc sliding tests can be estimated by employing the Hertz contact theory [13] and its average values in different contact conditions as illustrated in Figure 2. The yield strength (Rp0.2) of the QP980 steel is around 700MPa [14]. Thus, the contact pressure in the current sliding tests is of the same order of magnitude as that in the sheet metal forming process, demonstrating the close correlation between laboratory and industrial conditions.

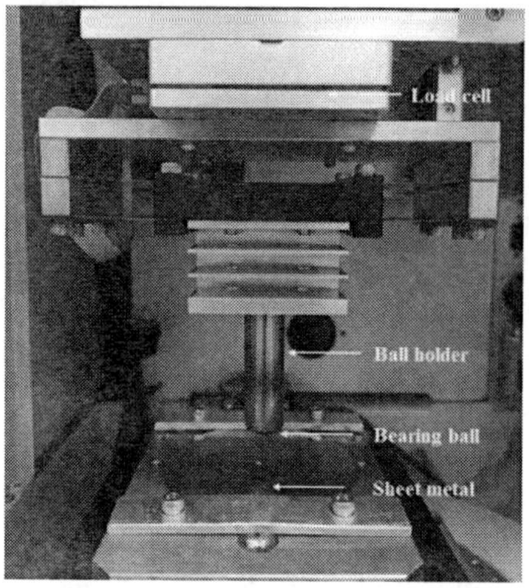

Figure 1. Experimental setup of ball-on-disk sliding test

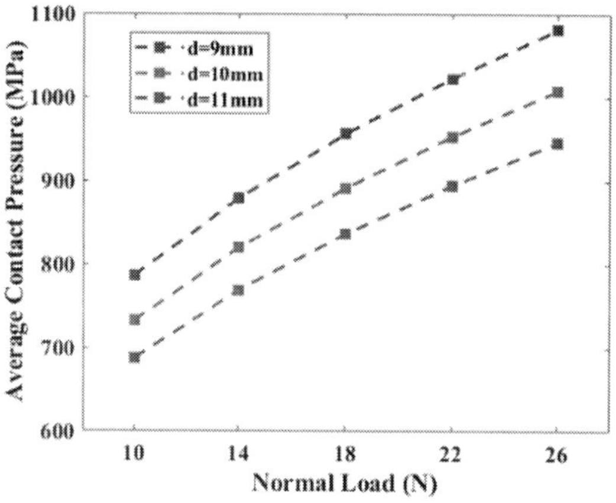

Figure 2. Average contact pressure in ball-on-disc sliding conditions

3. Fuzzy modelling technique

In this study, a rule-based fuzzy inference model was developed by using the fuzzy logic approach. To further improve the feasibility and learning ability of the established fuzzy prediction model, a modified QPSO algorithm was proposed to refine its definition parameters and adjust the fuzzy inference relationships.

3.1. Fuzzy logic approach

The fuzzy logic approach has been widely used to deal with the continuous transition between different states. In fuzzy modelling, the input variables are transformed into their corresponding fuzzy linguistic terms by using membership functions. The general form of a membership function can be given in Figure 3. Its value μ can describe the continuous transitions from "zero-belonging (μ=0)" to "total-belonging (μ=1)" and its accuracy is of great importance to the modelling performance.

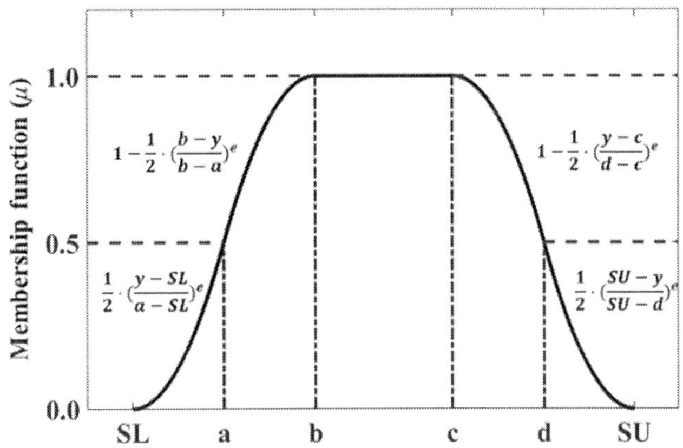

Figure 3. General form of a membership function

In general, a fuzzy modelling consists of three steps: (1) the fuzzification of input and out variables; (2) the establishment of fuzzy rule base and inference engine; (3) the defuzzification of prediction results. In this study, the Mamdani-type [15] fuzzy inference system was adopted, and the weighted-sections method was utilized to produce defuzzification results.

3.2. Development of fuzzy prediction model

In this section, four critical parameters were adopted as the input variables, including ball diameter, normal load, surface roughness of sheet metal and sliding cycles. Besides, the maximum scratching depth h_{max} was incorporated in the fuzzy model as the output variable. Simplified membership functions were adopted in this stage by taking b=c, as presented in Figure 4. By taking the experimental results as the training database, a preliminary fuzzy prediction model was established. To test the prediction performance of the established fuzzy model, the parameter ε, given by Eq. (1), was used to calculate the average prediction error between the predicted results and experimental measurements.

$$e=\frac{1}{M}\times\sum_{j=1}^{M}\frac{abs(R^j-R^{*j})}{R^j} \tag{1}$$

where R^* and R are the results of prediction and experimental measurements, respectively. For the preliminary fuzzy model, it can reproduce the original training database with the ε value of 6.85%. Therefore, the fuzzy model needs further optimization to improve its prediction performance by adaptively tuning the membership functions.

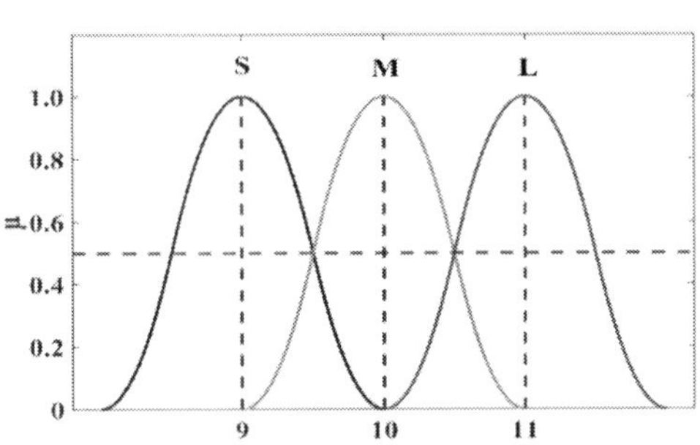

(a) Ball diameter (mm)

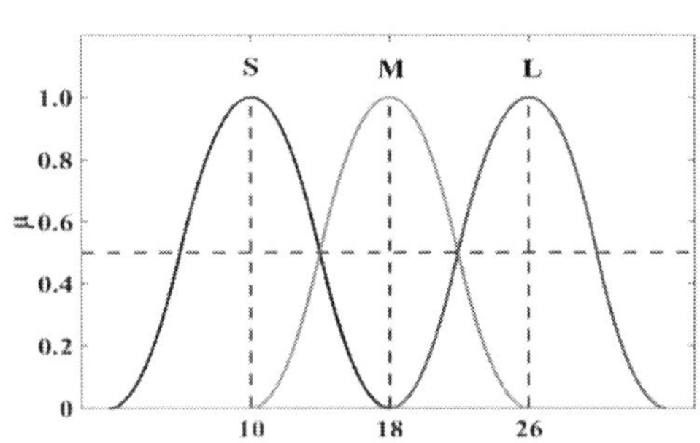

(b) Normal Load (N)

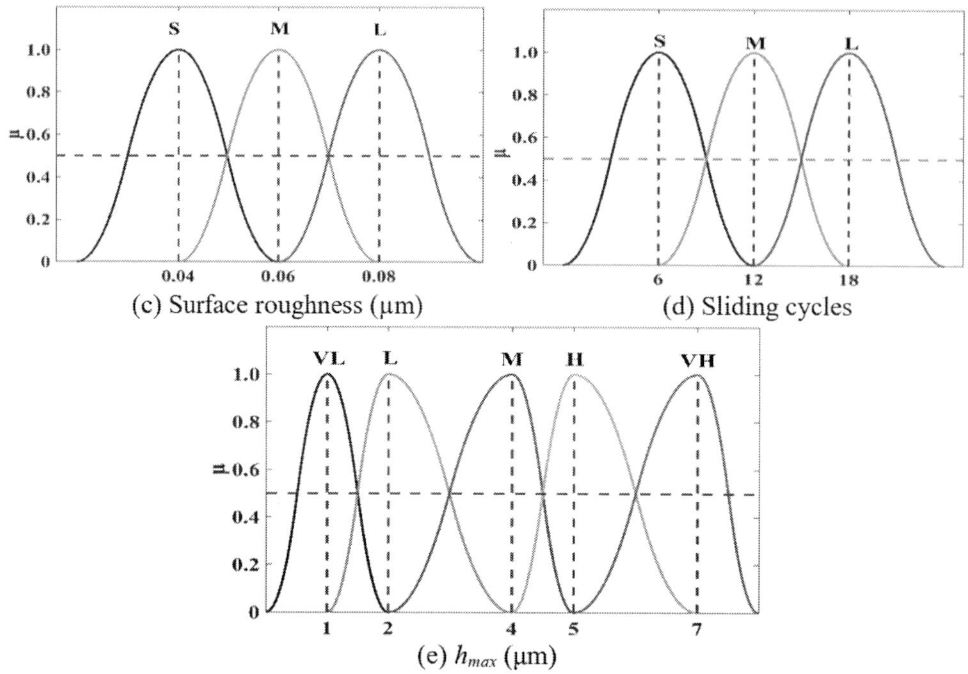

Figure 4. Membership functions of input and output variables
(VL: Very Low, L: Low, M: Medium, L: Large, VL: Very Large)

3.3. Optimization of fuzzy prediction model

In this section, the optimization of the preliminary fuzzy prediction model can be achieved by refining the definition parameters of membership functions to minimise the value of ε. Thus, the fuzzy model optimization can be converted into an N-dimensional optimisation problem with the objective function of ε. In the preliminary fuzzy model, the total number of definition parameters of both input and output variables is 102, which means the value of N will be 102 as well.

For the N-dimensional optimisation problems, particle swarm optimisation (PSO) algorithm [16] and quantum-behaved particle swarm optimisation (QPSO) algorithm [17] have been widely used. In these algorithms, the particle swarm refers to a population of possible solutions and each solution is defined as a particle. However, both optimization algorithms are unable to deal with high-dimensional optimisation processes. Therefore, the QPSO algorithm is further developed in this study by exerting an adaptive crossover operator. When the particle swarm is found to get trapped into a local minimum zone, all the *pbest* particles, which refer to the personal best positions found by each particle, are sorted based on their function values and some lagging *pbest* particles with worse function values will be selected. These selected lagging *pbest* particles will be hybridized with their crossbreeding particles randomly generated in the search space. If a newly-generated particle manages to find a better solution, it will be adopted as the new *pbest* particle. Such selection and crossover operations will be repeated until the termination criterion is reached.

It is found that the modified QPSO algorithm manages to achieve a good balance between convergence speed and search ability. By combining the fuzzy logic approach and the further developed QPSO algorithm, an efficient artificial intelligence method is proposed in this study, which is capable to learn from the training data and refine the definition parameters adaptively.

4. Results and discussions

4.1. Results of model optimization

In this section, the further developed QPSO algorithm was employed to train the preliminary fuzzy prediction model by refining its membership functions. For the model optimization, the average prediction error ε was adopted as the objective function and the population size and maximum iteration times in the QPSO algorithm were set as 40 and 2000, respectively. The optimized membership functions of input and output variables are shown in Figure 5.

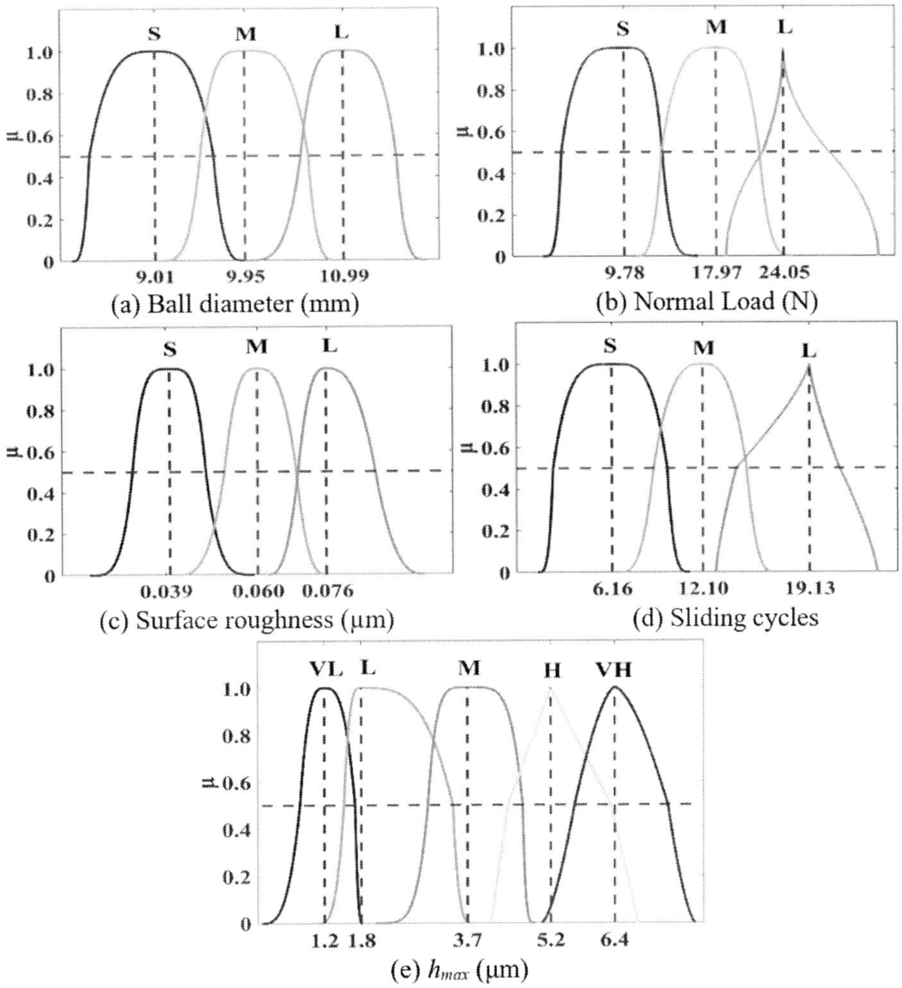

Figure 5. Optimized membership functions of input and output variables

After model optimization, an optimal combination of all the definition parameters is determined. It can be found that the average prediction error in reproducing the original training database was significantly reduced, from 6.85% to 2.89%. In addition, the optimized membership functions in Figure 4 are obviously different from their original ones in Figure 3, which is determined by the inherent nonlinear characteristics of surface damage problem. Overall, the developed fuzzy inference model can learn from the training database and tune its definition parameters adaptively by coupling the advantages of the modified QPSO algorithm.

4.2. Results of model verification

To check the performance of the optimized fuzzy inference model, additional sliding experiments were carried out and the experimental results were adopted as the verification database. For comprehensive comparison and analysis, the combinations of input variables used in the verification experiments were different from those used in the training database. From the comparison results in Figure 6, the average prediction error was 3.49% when the optimized fuzzy inference model was utilized to predict the value of maximum scratching depth h_{max}. However, the average prediction error of the preliminary fuzzy inference model was 8.15%. Clearly, the learning capacity of developed fuzzy inference model and the optimization ability of modified QPSO algorithm were demonstrated by the significant decrease in the average prediction error.

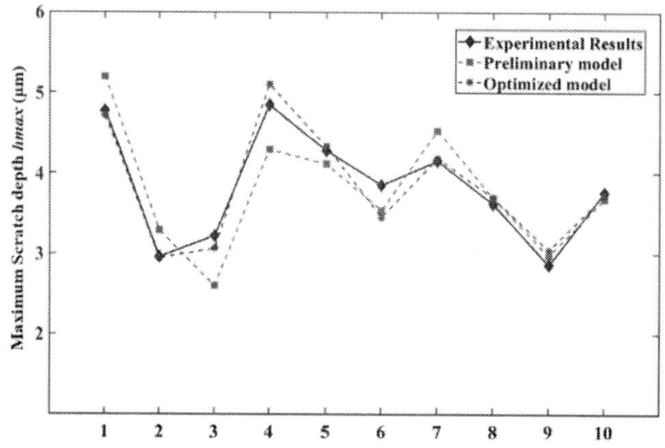

Figure 6. Comparison between prediction and experimental results

5. Conclusions

This paper has managed to develop and optimize a fuzzy inference model, which is capable to predict the extent of maximum scratching depth in the sheet metal surface with high prediction accuracy. Ball-on-disk sliding experiments were carried out to simulate the contact conditions in the sheet metal forming process and the experimental results were adopted as the database. The fuzzy inference model has the ability to learn from the experimental database and then further enhance its prediction performance with the application of the modified QPSO algorithm. The feasibility and reliability of the developed fuzzy prediction model have been demonstrated by the verification results. It can be expected that the fuzzy modelling approach will play an important role in the complex engineering problems where the traditional prediction methods are not applicable.

Acknowledgements

This research forms part of the Baosteel Australia Research and Development Centre (BAJC) portfolio with Project BA17001, as well as part of the ARC Research Hub project IH140100035. The first author is financially supported by China CSC and UNSW TFS scholarships.

References

[1] J. Archard, "Contact and rubbing of flat surfaces," *Journal of applied physics,* vol. 24, no. 8, pp. 981-988, 1953.
[2] S. Rhee, "Wear equation for polymers sliding against metal surfaces," *Wear,* vol. 16, no. 6, pp. 431-445, 1970.
[3] R. Bayer, "A general model for sliding wear in electrical contacts," *Wear,* vol. 162, pp. 913-918, 1993.
[4] K. Ersoy-Nürnberg, G. Nürnberg, M. Golle, and H. Hoffmann, "Simulation of wear on sheet metal forming tools—An energy approach," *Wear,* vol. 265, no. 11-12, pp. 1801-1807, 2008.
[5] P. Groche, N. Moeller, H. Hoffmann, and J. Suh, "Influence of gliding speed and contact pressure on the wear of forming tools," *Wear,* vol. 271, no. 9-10, pp. 2570-2578, 2011.

IOP Conf. Series: Materials Science and Engineering **967** (2020) 012022 doi:10.1088/1757-899X/967/1/012022

[6] A. Gåård, P. Krakhmalev, and J. Bergström, "Wear mechanisms in deep drawing of carbon steel–correlation to laboratory testing," *Tribotest,* vol. 14, no. 1, pp. 1-9, 2008.
[7] T. Skåre and F. Krantz, "Wear and frictional behaviour of high strength steel in stamping monitored by acoustic emission technique," *Wear,* vol. 255, no. 7-12, pp. 1471-1479, 2003.
[8] V. V. Shanbhag, B. F. Rolfe, N. Arunachalam, and M. P. Pereira, "Investigating galling wear behaviour in sheet metal stamping using acoustic emissions," *Wear,* vol. 414, pp. 31-42, 2018.
[9] L. A. Zadeh, "Outline of a new approach to the analysis of complex systems and decision processes," *IEEE Transactions on systems, Man, and Cybernetics,* no. 1, pp. 28-44, 1973.
[10] Y. Ali and L. Zhang, "Estimation of residual stresses induced by grinding using a fuzzy logic approach," *Journal of Materials Processing Technology,* vol. 63, no. 1-3, pp. 875-880, 1997.
[11] Y. Ali and L. Zhang, "Surface roughness prediction of ground components using a fuzzy logic approach," *Journal of materials processing technology,* vol. 89, pp. 561-568, 1999.
[12] Y. Ali and L. Zhang, "A fuzzy model for predicting burns in surface grinding of steel," *International Journal of Machine Tools and Manufacture,* vol. 44, no. 5, pp. 563-571, 2004.
[13] K. L. Johnson and K. L. Johnson, *Contact mechanics.* Cambridge university press, 1987.
[14] X. Chen, C. Niu, C. Lian, and J. Lin, "The evaluation of formability of the 3rd generation advanced high strength steels QP980 based on digital image correlation method," *Procedia engineering,* vol. 207, pp. 556-561, 2017.
[15] E. H. Mamdani and S. Assilian, "An experiment in linguistic synthesis with a fuzzy logic controller," in *Readings in Fuzzy Sets for Intelligent Systems*: Elsevier, 1993, pp. 283-289.
[16] J. Kennedy and R. Eberhart, "Particle swarm optimization," in *Proceedings of ICNN'95-International Conference on Neural Networks*, 1995, vol. 4: IEEE, pp. 1942-1948.
[17] J. Sun, B. Feng, and W. Xu, "Particle swarm optimization with particles having quantum behavior," in *Proceedings of the 2004 congress on evolutionary computation (IEEE Cat. No. 04TH8753)*, 2004, vol. 1: IEEE, pp. 325-331.

IOP Conf. Series: Materials Science and Engineering **967** (2020) 012023 doi:10.1088/1757-899X/967/1/012023

Development of Sheet Metal Forming Analysis Incorporating Phase-transformation Model for Hot Stamping

K Uenishi[1], M Kubo[2], T Suzuki[1] and K Okamura[1]

[1]Steel Research Lab, Nippon Steel Corp., 1-8 Fuso, Amagasaki, Hyogo, 660-0891, Japan

[2]Nagoya R&D Lab., Nippon Steel Corp., 5-3 Tokaimachi, Tokai City, Aichi, 476-8686, Japan

Abstract. To predict the spring back after hot stamping of sheet metals, a user material routine incorporated with phase transformation effects has been developed and implemented in general-purpose commercial finite element software. The mechanical response of the metallographic structure incorporates variation of transformation plastic strain and transformation volume expansion. Additionally, the thermal response incorporates the dependencies of thermal properties and latent heat of phase transformation. The start timing of diffusive transformation was determined using Scheil's law, and the Kolmogorov–Johnson–Mehl–Avrami equation was adopted for the progression of diffusive transformation. The Koistinen and Marburger equation was used for the progression of martensite transformation. Using the developed model, we considered the effects of phase transformation on shape fixability. We conducted experiments that focused on measuring the shape fixability of U-bend hot formed parts. In this study, we show that the spring back of the U-bend parts increases under conditions in which phase transformation occurs before formation has been completed. Furthermore, the spring back could accurately predict by using the developed model.

1. Introduction

In the automotive body parts, both improvement of collision safety performance and weight reduction to improve fuel efficiency are required. Against the background of these social demands, hot stamping (HS) parts made of ultra-high-strength steel with a tensile strength of 1.5 GPa class or more are used for automotive body parts. Furthermore, another feature of HS technology is that it can produce ultra-high strength parts with better shape fixability than cold forming parts. However, it is known that the shape fixability of HS parts is affected by changes in forming start temperature and material cooling conditions. It is important to include transformation plasticity in the constitutive equation to analytically predict the shape fixability, and the implicit stress integration method by the return mapping algorithm for this purpose was developed[1-3]. As a result, this analysis technique has been applied to HS.

Although previous studies reflect the effect of the grain size before forming on the calculation[5], and has also been applied to TWB, most of them use three-dimensional block elements despite plate forming[4-6]. This element is disadvantageous in terms of calculation time because it requires element division in the thickness direction. Additionally, there are reports of calculations using shell elements, and a problem of convergence and accuracy arises because they do not use the return mapping algorithm[7]. Therefore, applying these methods to efficiently verify mass production conditions of stamped parts is problematic.

In this study, we acquired physical property data through experiments and developed the analysis model that incorporated phase transformation and implemented the developed model as a general-purpose solver. The developed model enables analysis using the shell element model via the return mapping algorithm in the plane stress state and achieves both accuracy and calculation time. Additionally, by using the developed model, we considered the effects of phase transformation on shape fixability.

2. Numerical methods and material model

Figure 1 shows a conceptual diagram of the numerical analysis using in the developed model. The developed model implements user subroutines related to elastoplastic response and thermophysical properties that can accurately predict the phase transformation behavior during HS analysis using LS-DYNA. The subroutine *umat* for the mechanical response of a material considers the transformation plastic strain and the expansion or shrinkage due to thermal and/or phase transformation. The subroutine *thumat* for the thermal response considers latent heat and the variation of thermophysical material properties accompanying the phase transformation.
The adaptability of the additive rule of the isothermal transformation in infinitesimal time period is assumed to calculate the diffusion-type phase transformation in continuous cooling. The start of phase transformation is determined by Scheil's law[8], and the transformation rate is calculated by the Kolmogorov–Johnson–Mehl–Avrami (KJMA) equation[9].

$$\xi = 1 - \exp[-a(T)t^{n(T)}], \quad (1)$$

where the functions $a(T)$ and $n(T)$ are polynomials fit to the measured isothermal transformation curve. The volume fraction of the martensite ξ_M is calculated using the Koistinen and Marburger (KM) equation[10].

$$\xi_M = 1 - \exp[-a(T - M_s)], \quad (2)$$

where a is a material constant and M_s is the martensitic start temperature calculated by Kunitake's empirical[11],

$$M_s(°C) = 560.5 - 407.3(\%C) - 7.3(\%Si) - 37.8(\%Mn) - 19.8(\%Cr) - 19.5(\%Ni) - 4.5(\%Mo). \quad (3)$$

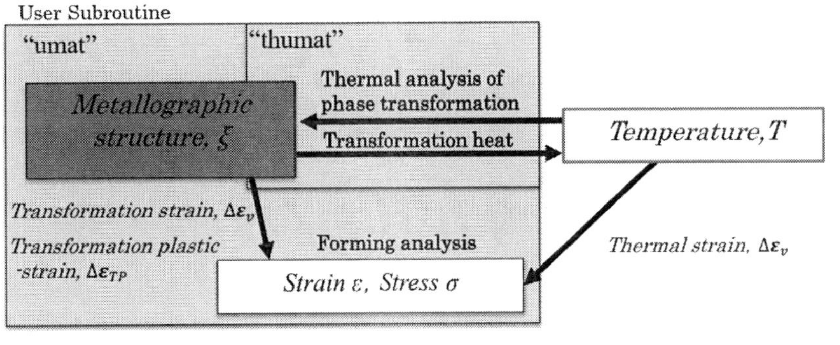

Figure 1 Coupling effects during anisothermal phase transformation.

The total strain increment $\Delta\boldsymbol{\varepsilon}$ comprises elastic $\Delta\boldsymbol{\varepsilon}_e$, plastic $\Delta\boldsymbol{\varepsilon}_p$, thermo-metallurgical $\Delta\boldsymbol{\varepsilon}_v$, and transformation plastic part $\Delta\boldsymbol{\varepsilon}_{TP}$ under the assumption of additive decomposition.

$$\Delta\boldsymbol{\varepsilon} = \Delta\boldsymbol{\varepsilon}_e + \Delta\boldsymbol{\varepsilon}_p + \Delta\boldsymbol{\varepsilon}_v + \Delta\boldsymbol{\varepsilon}_{TP}. \tag{4}$$

The stress at the end of increment $t+\Delta t$ can be expressed as

$${}^{t+\Delta t}\boldsymbol{\sigma} = {}^{t+\Delta t}\mathbf{E} : {}^{t+\Delta t}\boldsymbol{\varepsilon}_e = {}^{t+\Delta t}\mathbf{E} : ({}^{t}\boldsymbol{\varepsilon}_e + \Delta\boldsymbol{\varepsilon}_e) = \boldsymbol{\sigma}^{(T)} - {}^{t+\Delta t}\mathbf{E} : (\Delta\boldsymbol{\varepsilon}_p + \Delta\boldsymbol{\varepsilon}_{TP}), \tag{5}$$

where the subscript on left side denotes time hereafter. Then, ${}^{t+\Delta t}\mathbf{E}$ is the elastic modulus as a function of temperature and metallic structure at time $t+\Delta t$, and

$$\boldsymbol{\sigma}^{(T)} = {}^{t+\Delta t}\mathbf{E} : ({}^{n}\boldsymbol{\varepsilon}_e + \Delta\boldsymbol{\varepsilon} - \Delta\boldsymbol{\varepsilon}_v) \tag{6}$$

is trial stress that comprises variables that are already known because the temperature is calculated before calling *umat*. The increment $\Delta\boldsymbol{\varepsilon}_v$ can be calculated from the change of density and given as

$$\Delta\boldsymbol{\varepsilon}_v = \Delta\varepsilon_v \mathbf{I} = \frac{1}{3}\left(\frac{{}^{t}\rho(T,\zeta)}{{}^{t+\Delta t}\rho(T,\zeta)} - 1\right)\mathbf{I}, \tag{7}$$

where the density can be approximated by linear mixture rule, and the density of each metallic part can refer to Miettinen [12] and Okamura[13].
Unknown variables in the equation (5) are $\Delta\boldsymbol{\varepsilon}_p$ and $\Delta\boldsymbol{\varepsilon}_{TP}$. According to the associated flow rule and using the isotropic yield function $f(\boldsymbol{\sigma})$, the plastic strain increment $\Delta\boldsymbol{\varepsilon}_p$ is given as

$$\Delta\boldsymbol{\varepsilon}_p = \Delta p \frac{\partial f(\boldsymbol{\sigma})}{\partial \boldsymbol{\sigma}} = \Delta p \frac{3}{2}\frac{{}^{t+\Delta t}\mathbf{s}}{{}^{t+\Delta t}\bar{\sigma}}, \tag{8}$$

$$f(\boldsymbol{\sigma}) = \bar{\sigma} - Y(\bar{\varepsilon}_p, T, \xi), \qquad \bar{\sigma} = \left(\frac{3}{2}\mathbf{s} : \mathbf{s}\right)^{\frac{1}{2}}, \tag{9}$$

where Δp is a plastic multiplier that is equal to the equivalent plastic strain increment, $\mathbf{s}$ is deviatoric stress, $\bar{\sigma}$ is equivalent stress, and $\bar{\varepsilon}_p$ is equivalent plastic strain, $Y(\bar{\varepsilon}_p, T, \xi)$ is flow stress approximated using the linear mixture rule. For the $\Delta\varepsilon_{TP}$, the formula by Denis et al.[14] is adopted as

$$\Delta\boldsymbol{\varepsilon}_{TP} = \sum_{I=1}^{N-1} 3K_I \xi_A \Delta\xi_I \, {}^{t+\Delta t}\mathbf{s} = \Delta h \, {}^{t+\Delta t}\mathbf{s}, \tag{10}$$

$$\xi_A = 1 - \sum_{I=1}^{N-1} \xi_I, \quad \Delta h = \sum_{I=1}^{N-1} 3K_I \xi_A \Delta\xi_I, \tag{11}$$

where $K_I (I = 1, \cdots, N-1)$ is the transformation plasticity coefficient from austenite to Ith phase, and ξ_A is the volume fraction of untransformed austenite. The value of Δh is known at the start of increment. Substituting Eqs. (8) and (10) in (5) leads to

$${}^{t+\Delta t}\boldsymbol{\sigma} = \boldsymbol{\sigma}^{(T)} - {}^{t+\Delta t}\mathbf{E} : \left(\frac{3\Delta p}{2\,{}^{t+\Delta t}\bar{\sigma}} + \Delta h\right){}^{t+\Delta t}\mathbf{s}. \tag{12}$$

The normal stress to shell surface must be zero. The elastic modulus in equation (4) must be replaced with the plane stress one $[E']$ and can be rewritten in matrix form as

$$\{\sigma\}=\{\sigma^{(T)}\}-(\frac{3\Delta p}{2\bar{\sigma}}+\Delta h)[E']\{s\}, \tag{13}$$

$$\{\sigma\}^T=(\sigma_{11} \quad \sigma_{22} \quad \tau_{12} \quad \tau_{23} \quad \tau_{31}),\ \{s\}^T=(s_{11} \quad s_{22} \quad 2s_{12} \quad 2s_{23} \quad 2s_{31}). \tag{14}$$

By introducing the deviation operator matrix $[P]$, which works as $\{s\}=[P]\{\sigma\}$, and multiplying the orthogonal operator matrix $[Q]^T$ $([Q][Q]^T=[I])$, Eq.(13) leads to

$$[Q]^T\{\sigma\}=[Q]^T\{\sigma^{(T)}\}-(\frac{3\Delta p}{2\bar{\sigma}}+\Delta h)[Q]^T[E'][P]\{\sigma\}, \tag{15}$$

$$\{\sigma^*\}=\{\sigma^{(T)*}\}-(\frac{3\Delta p}{2\bar{\sigma}}+\Delta h)[Q]^T[E'][Q][Q]^T[P][Q]\{\sigma^*\}, \tag{16}$$

where $\{\sigma^*\}=[Q]^T\{\sigma\}$ and $\{\sigma^{(T)*}\}=[Q]\{\sigma^{(T)}\}$. Because the matrixes $[Q]^T[E'][Q]$ and $[Q]^T[P][Q]$ are orthogonal, we can easily express $\{\sigma^*\}$ by $\{\sigma^{(T)*}\}$. After performing certain calculations, we have

$$\bar{\sigma}^2=\frac{1}{2}\left(\frac{\sigma_{11}^{(T)*}}{1+(\frac{3\Delta p}{2\bar{\sigma}}+\Delta h)\frac{E}{3(1-\nu)}}\right)^2+\frac{3}{2}\left(\frac{\sigma_{22}^{(T)*}}{1+(\frac{3\Delta p}{2\bar{\sigma}}+\Delta h)2G}\right)^2$$
$$+3\left(\frac{\sigma_{12}^{(T)*}}{1+(\frac{3\Delta p}{2\bar{\sigma}}+\Delta h)2G}\right)^2+3\left(\frac{\sigma_{23}^{(T)*}}{1+(\frac{3\Delta p}{2\bar{\sigma}}+\Delta h)2G_c}\right)^2+3\left(\frac{\sigma_{31}^{(T)*}}{1+(\frac{3\Delta p}{2\bar{\sigma}}+\Delta h)2G_c}\right)^2. \tag{17}$$

Then, the plastic strain increment $\Delta\boldsymbol{\varepsilon}_p$ can be determined by solving Δp to satisfy the next equation

$$g(\Delta p)=\bar{\sigma}^2-Y({}^t\bar{\varepsilon}_p+\Delta p,{}^{t+\Delta t}T,{}^{t+\Delta t}\xi)^2=0. \tag{18}$$

The stress $\{{}^{t+\Delta t}\sigma\}$ can be calculated by $\{\sigma\}=[Q]\{\sigma^*\}$ after obtaining $\{\sigma^*\}$ by substituting Δp into Eq.(16).

IOP Conf. Series: Materials Science and Engineering **967** (2020) 012023 doi:10.1088/1757-899X/967/1/012023

3. Experimental methods

3.1 *Verification of phase transition in thermal expansion behavior*

The prediction accuracy of the phase transformation behavior in the developed model was verified using a simple model of the specimen size.

The test material used was steel sheet of 22MnB5 grade for HS. The thermal expansion behavior was experimentally measured by simulating the thermal history of HS. The specimen size was 1.6mm×20mm×200mm. The thermal history of HS was simulated by electric heating the specimen and applying feedback control to the temperature. The heating temperature was 950°C, and the heating time at the heating temperature was 120s. After that, it was cooled to 750°C at 15°C/s to simulate transfer and cooled to room temperature at 50°C/s to simulate die cooling. The thermal expansion behavior during this heat treatment was measured with a contact thermal dilatometer.

Figure 2 shows the analysis model. The analysis model simulating the shape of the specimen was created with a thick shell element. The thermal history measured in the experiment was given as the thermal boundary condition, and the strain change accompanying the temperature change was calculated.

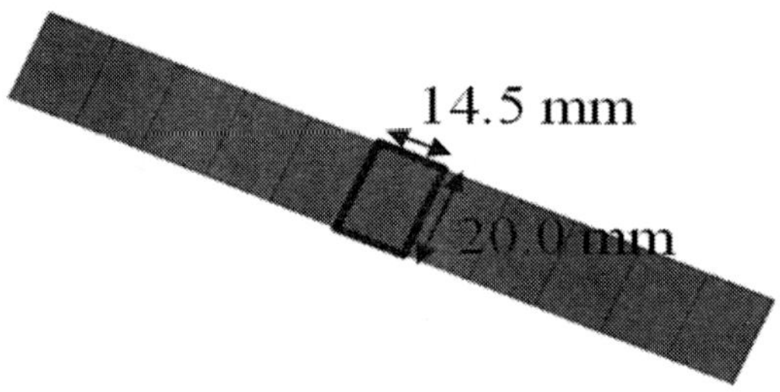

Figure 2 Analysis model.

3.2 *Effect of phase transformation on part shape fixability*

Using the U-bend molding model shown in **Figure 3**, the results of numerical analysis using the developed material model were compared with the experimental results, and the accuracy of the developed model was verified. This U-bend forming part has a shape in which slight changes in stress generated during molding are easily reflected in the amount of spring back, making it suitable for verification of shape fixability. The blank is placed on the pins of mold. Therefore, the temperature of the blank hardly decreases because it does not come into contact with the mold until the start of forming. Also, the effect of diffusion-type phase transformation and martensitic transformation on shape fixability was investigated by changing air cooling rate conditions during transportation. Under low cooling rate conditions, the blanks are air-cooled during transport to achieve a diffusion-type phase transformation. On the other hand, under high cooling rate conditions, the martensite phase transformation can be achieved by forced convection cooling of the blank during transport.

The experimental process of the U-bend forming test is shown in **Figure 4**, and the experimental conditions are shown in **Table 1**. Shape fixability was evaluated by measuring the change in the curvature radius ΔR of the U-bend parts.

Figure 5 shows the analysis process of the developed model. After each analysis process, the development model passes history variables (e.g., temperature, stress, and volume fraction of a phase) to the next analysis process. **Table 2** shows the analysis conditions. The analysis conditions were set to correspond to the experiment.

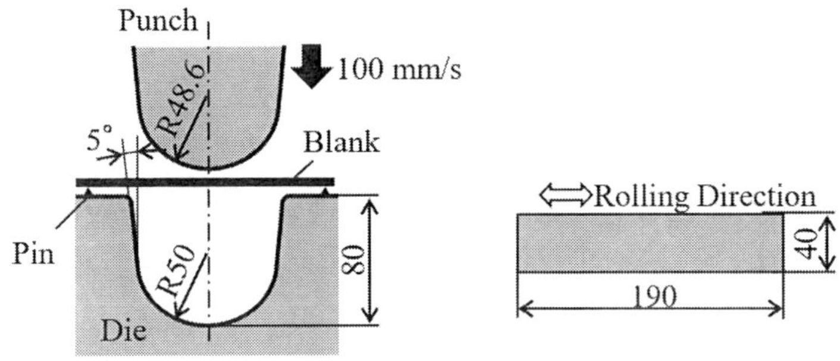

(a) (b)

Figure 3 Schematic of U-bend forming. (a) U-bend molding die. (b) Blank.

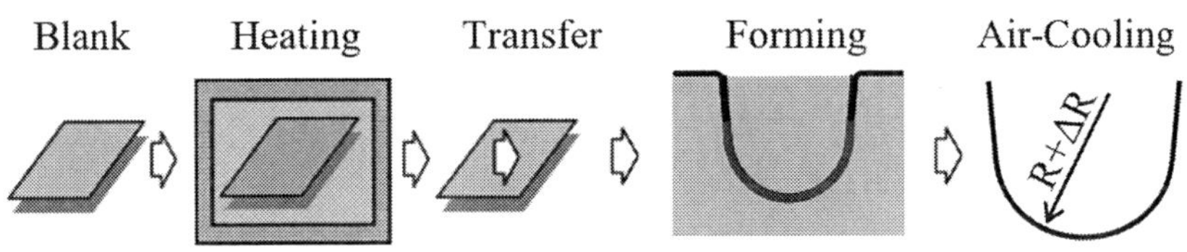

Figure 4 Experimental process.

Table 1 Experimental conditions

Blank material	22MnB5
Blank size (mm)	1.4×190×40
Austenitizing condition	900°C×240s
Cooling rate during transfer condition	(a) Low cooling rate condition: -5°C/s (b) High cooling rate condition: -32°C/s
Forming speed (mm/s)	100
Forming start temp.	(a) Low cooling rate condition: 700–550°C (b) High cooling rate condition: 250–650°C
Closed tool time (s)	15

Transfer the History variables(temperature, Stress) to next process

	Heating	Air cooling	Forming	Die quench	Spring back
Developed model	—	○	○	○	○
Conventional model	—	—	○	○	—

Figure 5 Comparison of conventional and development methods in the analysis process.

IOP Conf. Series: Materials Science and Engineering **967** (2020) 012023 doi:10.1088/1757-899X/967/1/012023

Table 2 Analysis conditions

Element model	Thick Shell element (TSHELL)
Number of integration points in the thickness direction	Temperature : 3 Stress : 7
Mesh size (mm)	1.4
Initial temperature (°C)	900
Cooling rate during transfer condition	(a) Low cooling rate condition: -5°C/s (b) High cooling rate condition: -32°C/s
Forming speed (mm/s)	1000 (time scale =10)
Forming start temperature	(a) Low cooling rate condition: 700–550°C (b) High cooling rate condition: 250–650°C
Holding time at bottom dead center (sec)	0.30 (time scale =100)

4. Results and discussion

4.1 *Verification of phase transition in thermal expansion behavior*

Figure 6 shows a comparison of the experimental and analytical results of the thermal expansion behavior. In addition to the thermal expansion behavior, the prediction of the martensite start temperature is consistent with experiments and analysis, indicating that the phase transformation behavior of HS can be accurately predicted using the developed model.

Figure 7 shows the comparison of the continuous cooling transformation (CCT) diagram between the simulation and experiment. The result shows that the developed model can accurately predict the onset of diffusion-type phase transformation (ferrite and bainite) and martensitic phase transformation could be predicted.

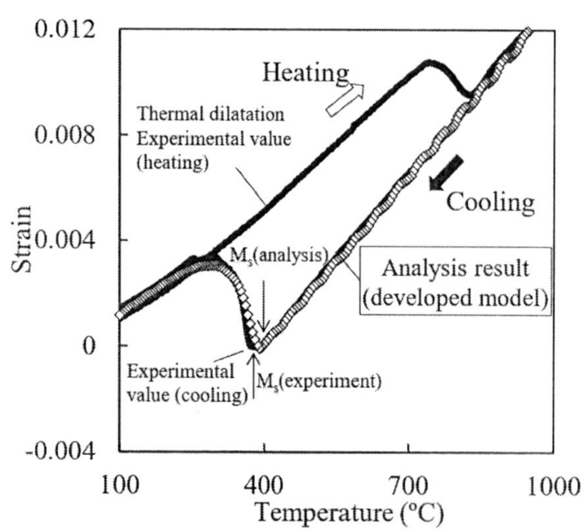

Figure 6 Temperature–strain relation under thermal history simulating the HS process.

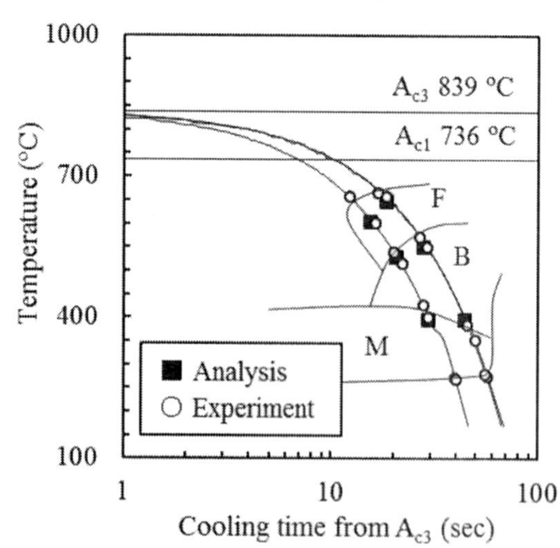

Figure 7 Comparison of the CCT diagram between the simulation and experiment.

4.2 *Effect of phase transformation on part shape fixability*

Figure 8 shows the comparison of forming start temperature–shape fixability relation between simulation and experiment and the relationship between the forming start temperature and the phase fraction obtained from the analysis results. In the experiment of low cooling rate, the change in curvature radius increases as the forming start temperature decreases. Moreover, the results of analysis using the developed model show that the experimental results can be accurately reproduced. According to the analysis results, the forming start temperature at which shape fixability is reduced coincides with the temperature at which diffusion-type phase transformation occurs. On the other hand, under condition of high cooling rate, shape fixability decreases below the martensite start temperature.

Figure 9 shows the effect of fraction of martensite and transformation plasticity on shape fixability. Under the condition of analytically the transformation plasticity coefficient of martensite $K = 0$, the shape fixability is reduced. Therefore, the reason for the decrease in shape fixability is that when processed after phase transformation, residual stress does not decrease due to transformation plasticity, and the deformation resistance of the transformed phase is higher than that of austenite. It is considered necessary to form austenite to ensure the shape fixability of the HS parts.

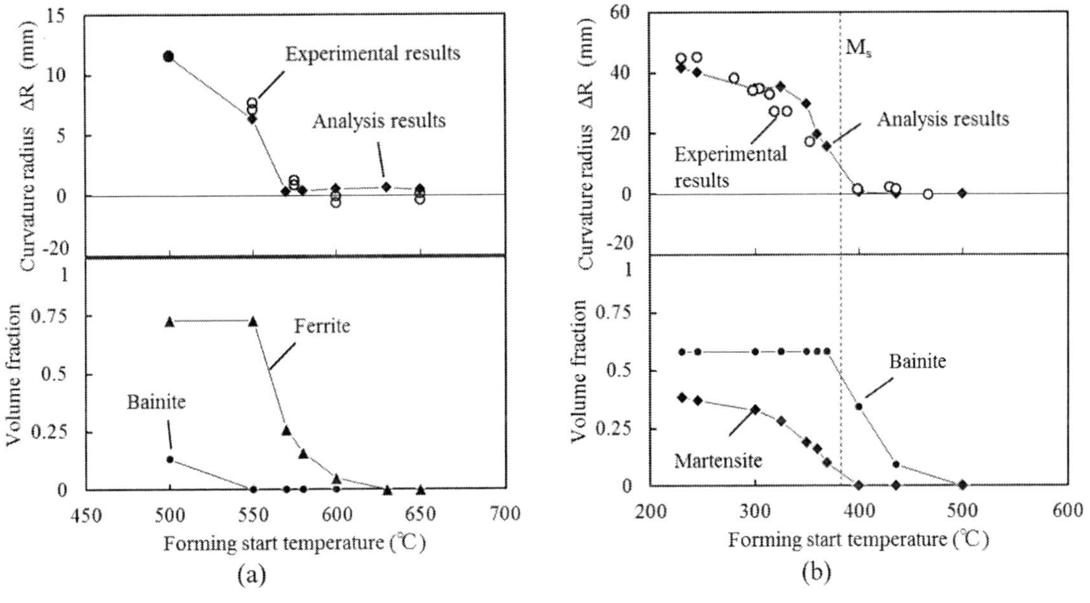

Figure 8 Comparison of the forming start temperature–shape fixability relation between the simulation and experiment, and the relationship between the forming start temperature and the phase fraction.
(a) Cooling rate during transfer is low. (b) Cooling rate during transfer is high.

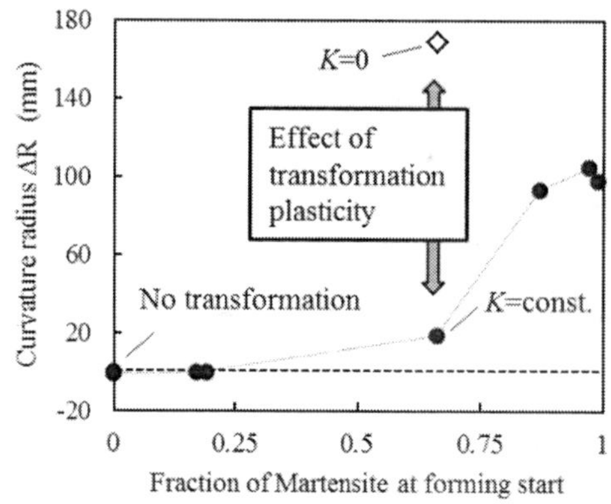

Figure 9 Effect of the fraction of martensite and transformation plasticity on shape fixability.

5. Summary

In this study, sheet metal forming analysis technology incorporating a phase transformation model for HS was developed and verified. The findings obtained from this study are shown below.

(1) Using the developed model, the thermal expansion behavior during the HS process caused by temperature change and phase transition can be accurately predicted.
(2) At low forming start temperatures, the spring back of the part considerably changed after being released from the tools, and the analysis results agreed well with the experimental results. In other words, the suggestion was made that considering the phase transition behavior of the HS process was necessary.
(3) To develop HS parts with good shape fixability, it is necessary to form them in the austenitic phase.

References

[1] Kim J, Im S and Kim H-G 2005 *Int J Plasticity,* **21** 1383-1408
[2] Lee M G, Kim S J, Han H N and Jeong W C 2009 *Int. J. Plasticity* **25** 1726-58
[3] Kawaragi Y, Fukumoto M,and Okamura K 2015 *J. Soc. Mat. Sci. Japan* **64** 258-65
[4] Lee M G, Kim S J, Han H N and Jeong W C 2009 *Int. J. Mechanical Sci.* **51** 888-98
[5] Bok H H, Choi J W, Barlat F, Suh D W and Lee M G 2014 *Int. J. Plasticity* **58** 154-83
[6] Bok H H, Choi J W, Suh D W, Lee M G and Barlat F 2015 *Int. J. Plasticity* **73** 142-70
[7] Bergman G and Oldenburg M 2004 *Int. J. Numer. Meth. Eng.* **59** 1167-86
[8] Scheil E 1935 *Arch. Eisenhüttenwes.* **12** 565-67
[9] Johnson W A and Mehl R F 1939 *Trans. AIME* **135** 416-58
[10] Koistinen D P and Marburger R E 1959 *Acta Metallurgica* **7** 59-60
[11] Kunitake T 2001 *J. Japan Soc. Heat Treatment* **41**164-69
[12] Miettinen J 1997 *Metallurgical and Materials Transactions* B **28** 281-97
[13] Okamura K 2006 *J Society of Materials Science Japan* **55** 529-535
[14] Denis S, Gautier E, Simon A and Beck G 1985 *Materials Science and Technology* **1** 805-14

IOP Conf. Series: Materials Science and Engineering **967** (2020) 012024 doi:10.1088/1757-899X/967/1/012024

Numerical analysis of the bulge test in temperature for the EN AW 6061-T6 sheet

A Boyer[1,2,*], L Germain[1], H Laurent[1], D M Neto[2] and M C Oliveira[2]

[1]Univ. Bretagne Sud, UMR CNRS 6027, IRDL, F-56100, Lorient, France
[2]CEMMPRE, Department of Mechanical Engineering, University of Coimbra, Polo II, Rua Luís Reis Santos, Pinhal de Marrocos, 3030-788 Coimbra, Portugal

[*]E-mail: adrien.boyer@univ-ubs.fr

Abstract. The equibiaxial expansion test allows the evaluation of the mechanical behaviour of materials on a large strain range. In this context, a bulge test device was developed which enables a fast heating stage, a uniform heating of the specimen, and the control of the temperature during the expansion phase. Although during the test, the temperature is uniform at the pole of the cap, the contact with the tools imposes a gradient along its radius. Therefore, a thermo-mechanical model of the bulge test was built in order to improve knowledge about the influence of the temperature gradient in the evaluation of the hardening behaviour at the pole of the cap. Numerical simulations of the bulge test were performed considering both isothermal and anisothermal conditions. The anisothermal conditions considered reproduce the temperature gradient observed experimentally. This enables the analysis of the impact of the thermal gradient in the stress *versus* strain evolution at the pole of the cap and the comparison with experimental results.

1. Introduction

Nowadays, the quest for lightweight vehicles is fundamental to guarantee environmental sustainability. The norms imposed from environmental and safety agencies are particularly strong for the automotive industry, which has led to the development of materials with an improved ratio between strength and weight, or new-forming processes. In this context, aluminium alloys are increasingly used in the automotive industry. However, they are known for being more prone to springback phenomenon than conventional steels and to present lower formability than these steels, at room temperature. Such problems can be overcome using warm forming processes, *i.e.* by performing the deep drawing operation at an intermediate temperature, below the recrystallization temperature. These new forming conditions require either new material characterization methods or the current methods need to be adapted to high temperature conditions.

This study focus on the possibility of performing the characterization of the hardening behaviour of the materials using the equibiaxial expansion test in temperature. The material under analysis is the EN AW 6061-T6 alloy (initial thickness of 0.98 mm), for which experimental uniaxial tensile tests were previously performed on a Gleeble machine [1].

2. Experimental procedure

The setup used to perform the bulge test is presented in figure 1. A circular blank with a diameter of 240 mm is clamped into a blank holder and die, both electrically isolated and with a ceramic drawbead. The specimen is heated using the Joule's effect. Three pairs of electrodes are used with an electrical current intensity that can go up to 6000 A. Each pair is activated in a rotary manner as described in [2]. This device allows a fast heating stage, a uniform heating of the specimen, and the control of the temperature during the expansion phase. The heating is conducted in three phases, where the first one is a fast heating to the testing temperature, followed by a homogenization phase. Then, the temperature of the specimen is maintained during the expansion phase of the experiment, while an argon inert gas is used to deform the specimen. A type K thermocouple, welded on the blank centre by capacitive discharge, is used to control and monitor the temperature of the blank.

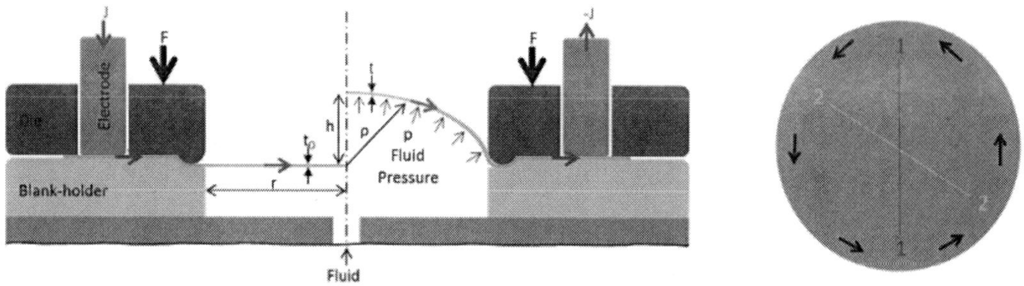

Figure 1. The bulge expansion device used, a) Principle of the expansion device using Joule's effect to heat the blank, b) Rotary activation of the three pairs of electrodes.

The bulge tools are mounted in an Instron 8803 tensile test machine equipped with a 500 kN load cell. The die diameter is 120 mm and its fillet radius is 5 mm, following the recommendations of the standard ISO 16808:2014 [3]. Thus, the device can be used for sheets with an initial thickness up to 1.4 mm. The strain field at the pole is monitored with the Aramis-4M DIC system. A TESCOM ER5000 pressure controller is used as well as a 0-70 bars pressure sensor located inside the cavity to monitor the pressure. Figure 2 presents a scheme of the bulge test highlighting the definition of the bulge pole's height, *h*; the die cavity radius, *a*; the die fillet radius, *df*; the blank initial thickness, t_0; the pole thickness, *t*; and the bulge pole radius of curvature, ρ.

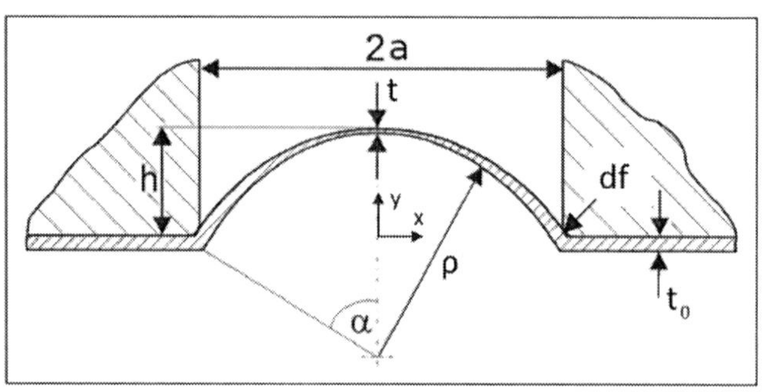

Figure 2. Bulge test scheme highlighting the relevant parameters.

The evaluation of the stress evolution during the test is based on the classical membrane theory, which relates the stress to the thickness (t) and radius of curvature of the blank's apex (ρ) as follows:

$$\sigma_{membrane} = \frac{P\rho}{2t} \quad (1)$$

where P is the pressure applied. When using a DIC system, the thickness strain at every instant can be evaluated using the acquired principal strains at the pole, by applying the volume conservation. The evolution of the radius of curvature is computed by fitting a sphere to the surfaces points using a least square method, which assumes equal radii of curvature along the rolling and the transverse directions. Although some authors report differences in these radii [4], for the EN AW 6061-T6 alloy under analysis the planar anisotropy coefficient is close to zero [1], leading to a negligible difference in the radii of curvature for these directions.

The bulge tests were performed for the EN AW 6061-T6 alloy at 200°C considering five thermocouples, welded at initial positions from the centre of 5, 15, 30, 40 and 50 mm, enabling the acquisition of the temperature evolution during the test as well as the spatial gradient.

3. Finite element model for the bulge test

All numerical simulations were performed with the in-house solver DD3IMP, originally developed to simulate isothermal sheet metal forming processes [5] and that has been continuously updated to enable an improved description of the contact conditions [6][7] and the simulation of warm-forming processes. The thermomechanical problem is solved using the staggered coupled strategy proposed by Martins *et al.* [8]. The experimental setup heats the specimen using the joule's effect. However, in this study, the numerical modelling of the heat generated by electrical current was carried out through an energy rate generation in the volume of the specimen, which was evaluated in each increment to assure the user-prescribed constant heating rate. In the following sections, the details about the modelling of the mechanical behaviour of the EN AW 6061-T6 alloy and the tests conditions considered are given.

3.1. Material Behaviour

In order to describe the work hardening of aluminium alloys, it is normally recommended to use a phenomenological Hockett–Sherby type law, to enable the description of the saturation behaviour. The strain rate dependency is commonly introduced using a power law, as initially proposed by Wagoner *et al.* [9] and used in [10][11]. Under these assumptions, the yield stress, σ_y, evolves with temperature T and strain rate $\dot{\varepsilon}$ as follows:

$$\sigma_y(T,\dot{\varepsilon}) = [\sigma_0 + Q(1-\exp(-b(\bar{\varepsilon}^p)^n))]\left[\frac{\dot{\varepsilon}}{\dot{\varepsilon}_0}\right]^m \quad (2)$$

where $\bar{\varepsilon}^p$ is the equivalent plastic strain, σ_0 is the initial yield stress, Q is the maximum change in size of the yield surface, b denotes the growth rate of the yield surface (or the hardening), n is strain hardening coefficient, $\dot{\varepsilon}_0$ is a constant strain rate normalization factor and m is the strain rate sensitivity coefficient.

The temperature dependence was modelled by taking Q, b, n and m to be functions of the temperature. The strain rate sensitivity index m evolves exponentially with the temperature as follows:

$$m(T) = m_0 \exp\left(m_1 \frac{T}{T_m}\right) \quad (3)$$

where m_0 and m_1 are material parameters and T_m is the material melting temperature, which can be assumed to be equal to 600°C for aluminium alloys. The strain-hardening index n was assumed to evolve linearly with temperature:

$$n(T) = n_0 - n_1\left(\frac{T}{T_m}\right) \quad (4)$$

where n_0 and n_1 are material parameters. The evolution of the size of the yield surface with the temperature is given by the following equation:

$$Q(T) = Q_0 + a_1\left[1-\exp\left(a_2\frac{T}{T_m}\right)\right] \quad (5)$$

where Q_0, a_1 and a_2 are material parameters. In the original formulation of the hardening law, it is assumed that, σ_0 is constant (see equation (1)). However, as shown in figure 3, this assumption is not valid for the EN AW 6061-T6 alloy. Therefore, a quadratic evolution was considered:

$$\sigma_0(T) = \sigma_0 - s_1\left(\frac{T}{T_m}\right) - s_2\left(\frac{T}{T_m}\right)^2 \quad (6)$$

where σ_0, s_1 and s_2 are material parameters. Thus, the identification procedure took into account a total of 12 parameters: $\dot{\varepsilon}_0$, b, m_0, m_1, n_0, n_1, Q_0, a_1, a_2, σ_0, s_1 and s_2. The parameters were identified by minimizing the difference between the experimental [1] and the numerical stress-strain curves, using a Generalized Reduced Gradient (GRG) algorithm and a two-step procedure described in [11]. The parameters identified are presented in table 1. Figure 3 shows the comparison between the experimental and numerical results, highlighting that the modified Hockett-Sherby law enables an accurate description of the material behaviour.

Table 1. Modified Hockett-Sherby hardening law parameters.

Parameter	Value	Parameter	value
$\dot{\varepsilon}_0$	$1.0\times10^{-4}s^{-1}$	n_1	0.63
σ_0	272.1 MPa	Q_0	230.3 MPa
b	3.2	a_1	116.4 MPa
m_0	4.3×10^{-6}	a_2	2.7
m_1	26.1	s_1	0.0 MPa
n_0	0.80	s_2	795.3 MPa

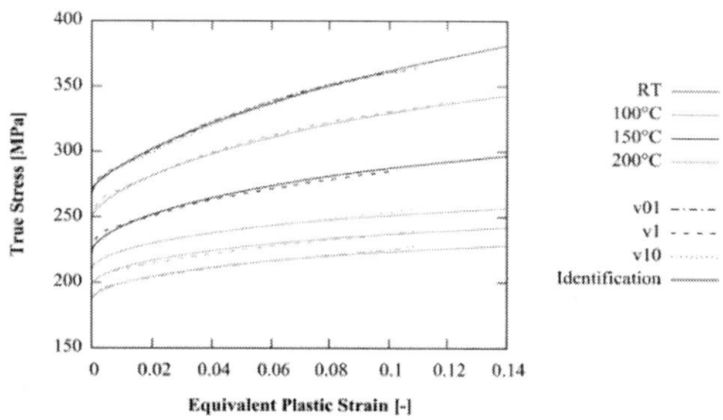

Figure 3. Comparison between experimental [1] and analytical stress-strain curves.

The elastic behaviour was assumed isotropic, described by the Generalized Hooke law. Although previous results show that the Young modulus of aluminium alloys decreases with the temperature increase, this effect was neglected in the numerical model. The other mechanical and thermal properties were also assumed as constant, as shown in table 2. The plastic behaviour was also assumed isotropic, described by the von Mises yield criterion.

Table 2. Mechanical and thermal properties of the EN AW 6061-T6 aluminium alloy.

Parameter	Value
Young Modulus	68900 MPa
Poisson's ratio	0.33
Density	2700 kg/m^3
Specific Heat Capacity	900 J/kg/K
Thermal Conductivity	3 W/m/K

3.2. Numerical procedure

Due to geometrical and material symmetry conditions, it was decided to consider only a quarter of the model as shown in figure 4. The tools are assumed as rigid and, consequently, only the surfaces in contact with the blank are described. Figure 4 shows the coarse discretization adopted for the tools, since Nagata patches are used to assure a smooth interpolation [12]. The blank is discretized with a total of 11523 nodes and 7440 hexahedral linear finite elements. Two layers are considered through the thickness and the central zone is discretized with a relative fine structured mesh, since it corresponds to the zone of interest. The same mesh is used for the mechanical and the thermal problems. However, full integration is adopted in the thermal problem, while the mechanical resorts to the selective reduced integration technique.

The model shown in figure 4 is used for both isothermal and anisothermal conditions. The pressure is applied to the bottom of the blank, considering a linear increase over time. However, for anisothermal conditions, it is also necessary to model the heat losses. The thermal conductance between the sheet and the tools will promote heat losses, as long as the tools present a different temperature than the blank. Based on previous numerical results, regarding the analysis of the heating stage for a boron steel [13], it was decided to keep the tools at room temperature. Note that it is very difficult to evaluate the coefficients that characterize the thermal conductance. From the numerical point of view, this heat loss is usually modelled as convection taking place on the contact interface, where the interfacial heat transfer coefficient (h_c) is the main parameter. In this work it was decided to adopt the same function presented in [14], which states that the heat flux due to thermal conductance is given by:

$$\dot{q}_c = h_c(T_{blank} - T_{tool}) = \left[h_{sup}\exp(-m_c g_n)\right](T_{blank} - T_{tool}), \tag{7}$$

where T_{blank} and T_{tool} correspond to the blank and the tool temperature, respectively. Moreover, h_c depends on the distance (g_n) between the blank and the tools, such that when contact occurs it attains an upper threshold value (h_{sup}) that decreases exponentially with the increase of the gap distance. m_c is the parameter used to control the rate of decrease with g_n. This allows obtaining a smooth variation of the heat transfer coefficient between contact and non-contact (free convection) areas. In this study the values used for these parameters were: h_{sup}=1 W/m^2.K and m = 8.59. Moreover, convection heat losses were also considered between the blank top surface and the surrounding air (h_{air}) and the blank bottom surface and the argon inert gas (h_{gas}). These convection coefficients were determined based on a trial and error approach, until obtaining a proper adjustment between the experimental and numerical temperature evolution during the test, for the five locations of thermocouples previously mentioned. In this study the value used were: h_{air}=0.0195 W/m^2.K and h_{gas}=0.03 W/m^2.K.

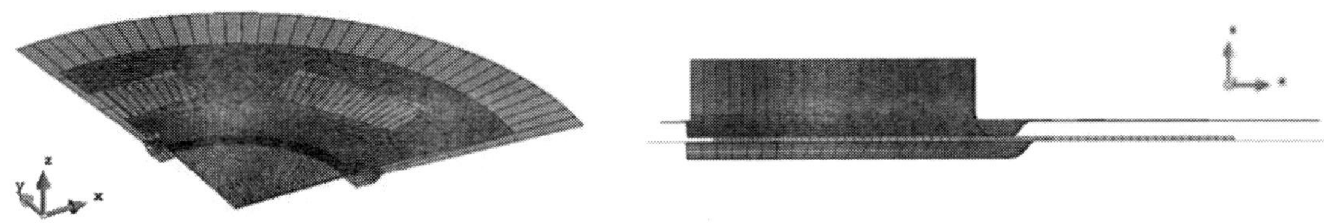

Figure 4. Schematic representation of one quarter of the model: blank (grey) and initial position of the blank holder (green) and die (blue).

The anisothermal test conditions try to replicate the experimental test. Therefore, in the first phase the blank holder is moved against the die until attaining a maximum force of 75kN (one quarter of the total force, due to symmetry conditions). The second phase corresponds to the heating of the blank until the value of 200 °C is attained at the centre of the blank. Then, the third phase corresponds to the temperature homogenization. Finally, the fourth phase corresponds to the forming one, with a linear increase of pressure, until a maximum value of 5 MPa is attained. Nevertheless, it should be pointed out that the cooling and springback phases were not considered in this work. The isothermal test conditions impose that the initial temperature of the blank and the tools is already 200°C. Thus, the test involves only the blank holder clamping phase and the forming one, considering the same maximum blank holder force and pressure values as for the anisothermal test.

3.3. Temperature distribution

Figure 5a) shows the temperature distribution in the blank at the end of the heating phase for the anisothermal test. It is possible to observe that in the centre region the temperature is quite uniform. However, since the tools are electrically isolated, they remain at the room temperature, which generates a temperature gradient mostly due to the heat losses by thermal conductance.

The comparison between the experimental and the numerical temperature evolution during the test is presented in figure 5b). As previously mentioned, the convection coefficients were determined trying to minimize the difference between those results. Therefore, as expected, there is a good correlation between them. However, it should be mentioned that the points located at a distance higher than 15 mm show a decreasing trend at the end of the bulging phase that is not observed in the experimental data. This suggests that the interfacial heat transfer coefficient can be overestimated.

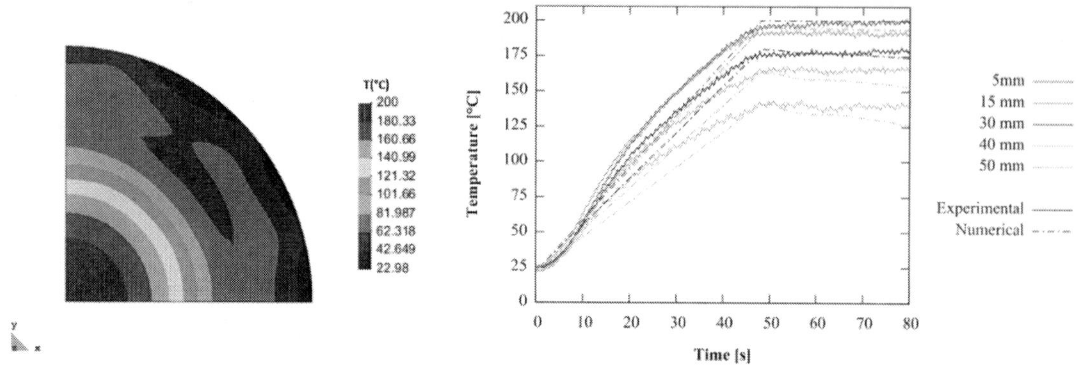

a) b)

Figure 5. Anisothermal test conditions: a) Temperature distribution at the end of the heating phase; b) Comparison between the experimental and the numerical temperature evolution during the test, for nodes located at the same initial positions of the thermocouples used in the experimental test.

4. Results analysis and discussion

Figure 6a) presents the evolution of the pressure in function of the pole's vertical displacement for isothermal, anisothermal and experimental conditions. The pressure value required to attain the same vertical displacement is higher for the anisothermal conditions. This is connected with the fact that the isothermal conditions promote a more homogenous strain field, as the whole blank surface is heated to 200°C. On the other hand, for anisothermal conditions the temperature gradient (see figure 5) induces a higher strain localization at the blank's apex, due to the gradient of material properties (see figure 3). Globally, the anisothermal case reproduces well the experimental pressure evolution, indicating a good description of the material behaviour. Figure 6b) presents the evolution of the thickness at the blank's apex, as predicted by the isothermal and the anisothermal models and the experimental result. Globally, both numerical tests underestimate the thickness reduction when compared to the experimental. This may be connected with the adoption of the von Mises yield criterion, since it was observed that the alloy presents a normal anisotropy coefficient lower than 1 [15]. Nevertheless, the anisothermal conditions predict a trend closer to the experimental one, which may be related with the higher localization of the deformation at the blank's apex.

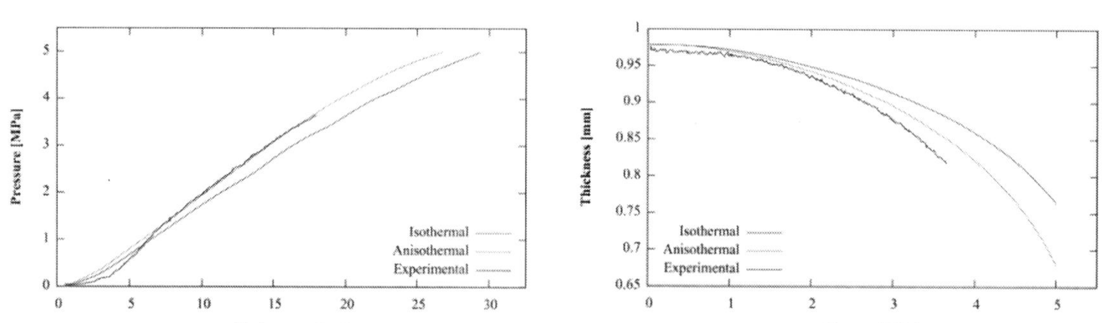

a) b)

Figure 6. Comparison between isothermal, anisothermal and experimental conditions: a) Evolution of the pressure in function of the pole's vertical displacement; b) Evolution of the thickness in the centre of the blank in function of the pressure.

Figure 7a) shows the evolution of the stress in function of the equivalent plastic strain, as predicted for the integration point (Gauss point) located closer to the mentioned distance to the centre. The stress-strain evolution is mainly dictated by the blank's temperature. Therefore, all points of the isothermal test follow a similar curve, while for anisothermal conditions the points located further away from the centre present higher yield stress values (see also figure 3). The vertical lines mark the maximum values attained for the strain, highlighting that, for each distance, the anisothermal condition leads to higher equivalent plastic strain values. This is connected with the different geometry of the blank surface, since a larger height at the apex is attained (see figure 6), which means that there is an increase of surface area along the blank radius to accommodate the strain localization at the pole.

The evolution of the strain-rate in function of the equivalent plastic strain is presented in figure 7b), for the same points. It is observed that the influence of the temperature gradient is marginal, with the strain rate showing a linear increase with equivalent plastic strain in logarithmic scale. The range for the strain-rate is approximately from $1x10^{-4}$ to $1x10^{-1}s^{-1}$, with a monotonous increase resulting in a smooth evolution of the stress curve, as shown in figure 7a) for both conditions.

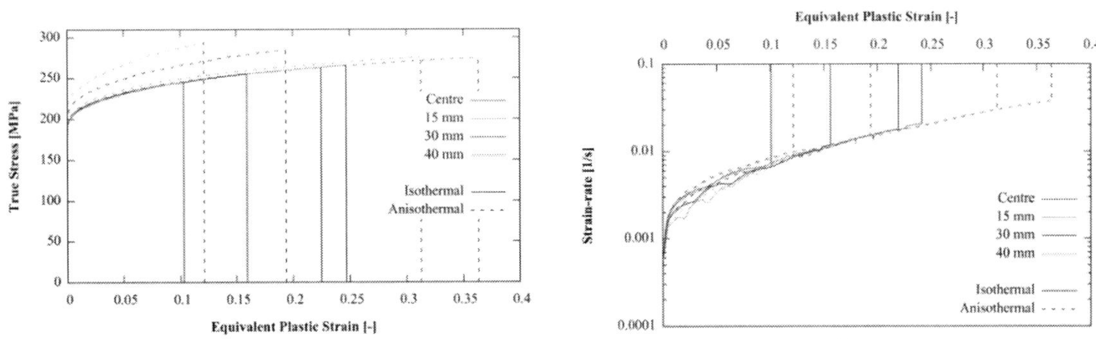

a) b)

Figure 7. Comparison between isothermal and anisothermal conditions: a) Evolution of the stress with the equivalent plastic strain; b) Evolution of the strain-rate with the equivalent plastic strain.

The evolution of the true stress in function of the equivalent plastic strain for the integration point located closer to the specimen apex is presented in figure 8. This figure also shows the yield stress evolution (see equation (2)) for 200°C and the two extreme values of strain rate observed in figure 7b), to highlight the influence of the evolution of the strain rate on the hardening behaviour. Moreover, the experimental results are also presented. The comparison between the experimental and the numerical results indicates that the strain rate increase is higher in the experimental results, which is consistent with the thickness evolution presented in figure 6.

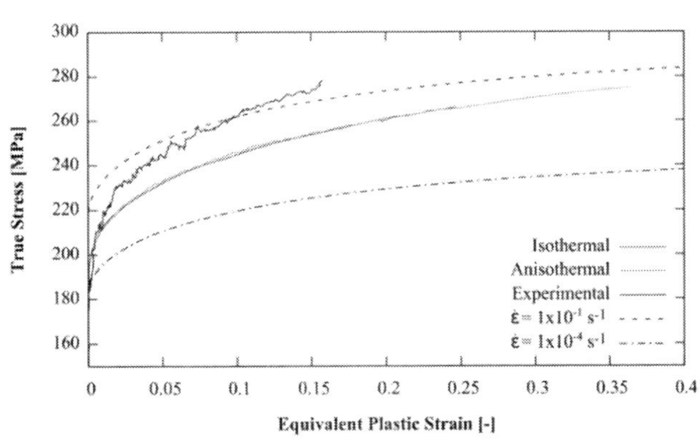

Figure 8. Comparison between numerical, analytical (for 200°C and constant strain rate values of $1x10^{-4}s^{-1}$ and $1x10^{-1}$), and experimental results for the evolution of the true stress in function of the equivalent plastic strain.

5. Conclusion

The thermomechanical behaviour of the EN AW 6061-T6 alloy was described using a modified Hockett-Sherby hardening law, based on experimental results extracted from uniaxial tensile tests, performed in a Gleeble device. The temperature gradient observed in the blank during the experimental bulge test at 200°C was analysed and an anisothermal model was built using the in-house finite-element solver DD3IMP. The same code was used to perform an isothermal numerical analysis of the process.

Globally, the identified hardening law enabled an accurate prediction of the evolution of the specimen geometry during the bulge test, particularly for anisothermal conditions. Both numerical models predict a similar exponential increase of the strain-rate during the bulge test. This evolution of the strain-rate influences the hardening observed for the stress-strain curves, due to the positive strain-rate sensitivity of the material. The knowledge acquired will contribute to improve the analysis of the bulge test at warm temperature for the EN AW 6061-T6 and potentially other materials.

Acknowledgements

The authors would like to acknowledge the funding that sponsored this research work: the national funds from the French Ministry of Higher Education and the Portuguese Foundation for Science and Technology (FCT) via the projects Materiais e Tecnologias Industriais Sustentaveis (MATIS) CENTRO-01-0145-FEDER-000014 and UIDB/00285/2020.

References

[1] V. M. Simões, M. C. Oliveira, H. Laurent, and L. F. Menezes, "The punch speed influence on warm forming and springback of two Al-Mg-Si alloys," *J. Manuf. Process.*, vol. 38, pp. 266–278, Feb. 2019, doi: 10.1016/j.jmapro.2019.01.020.

[2] N. Demazel *et al.*, "Simulations of joule effect heating in a bulge test," in *AIP Conference Proceedings*, 2016, vol. 1769.

[3] "ISO16808:2014 : Metallic materials -- Sheet and strip -- Determination of biaxial stress-strain curve by means of bulge test with optical measuring systems." International Organization for Standardization, Aug. 2014, [Online]. Available: https://www.iso.org/standard/57777.html.

[4] J. Min, T. B. Stoughton, J. Carsley, B. E. Carson, J. Lin, and X. Gao, "Accurate characterization of biaxial stress-strain response of steel metal from bulge testing," *Int J Plast Artic.*, no. 94, pp. 1–22, 2016, doi: https://doi.org/10.1016/j.ijplas.2016.02.005.

[5] L. F. Menezes and C. Teodosiu, "Three-dimensional numerical simulation of the deep-drawing process using solid finite elements," *J. Mater. Process. Technol.*, vol. 97, no. 1–3, pp. 100–106, 2000, doi: 10.1016/s0924-0136(99)00345-3.

[6] M. C. Oliveira, J. L. Alves, and L. F. Menezes, "Algorithms and Strategies for Treatment of Large Deformation Frictional Contact in the Numerical Simulation of Deep Drawing Process," *Arch. Comput. Methods Eng.*, vol. 15, no. 2, pp. 113–162, Jun. 2008, doi: 10.1007/s11831-008-9018-x.

[7] D. M. Neto, M. C. Oliveira, and L. F. Menezes, "Surface smoothing procedures in computational contact mechanics," *Arch. Comput. Methods Eng.*, vol. 24, no. 1, pp. 37–87, 2017, doi: 10.1007/s11831-015-9159-7.

[8] J. M. P. Martins *et al.*, "A new staggered algorithm for thermomechanical coupled problems," *Int. J. Solids Struct.*, vol. 122, pp. 42–58, 2017.

[9] R. H. Wagoner, E. Nakamachi, and Y. Germain, "Analysis of sheet forming operations using the finite element method," Toronto, 1988.

[10] H. Laurent, J. Coër, P.-Y. Manach, M. C. Oliveira, and L. F. Menezes, "Experimental and numerical studies on the warm deep drawing of an Al–Mg alloy," *Int. J. Mech. Sci.*, 2015.

[11] M. C. Oliveira, L. Germain, H. Laurent, D. M. Neto, J. L. Alves, and L. F. Menezes, "A modified Hockett-Sherby law enabling the description of the thermomechanical behaviour of the AA6061-T6," 2020, vol. 47, pp. 896–903, doi: 10.1016/j.promfg.2020.04.277.

[12] D. M. Neto, M. C. Oliveira, L. F. Menezes, and J. L. Alves, "Applying Nagata patches to smooth discretized surfaces used in 3D frictional contact problems," *Comput. Methods Appl. Mech. Eng.*, vol. 271, pp. 296–320, doi: 10.1016/j.cma.2013.12.008.

[13] A. Boyer, "Essais d'expansion à haute température sur un acier trempant," Master's thesis, Université Bretagne Sud, Lorient, France, 2017.

[14] D. M. Neto, M. Pais, M. C. Oliveira, J. L. Alves, and L. F. Menezes, "Thermomechanical analysis of the draw bead test," *Adv. Mater. Process. Technol.*, vol. 5, no. 3, pp. 401–417, 2019.

[15] V. M. Simões, H. Laurent, M. Oliveira, and L. Menezes, "The influence of warm forming in natural aging and springback of Al-Mg-Si alloys," *Int. J. Mater. Form.*, vol. 12, no. 1, pp. 57–68, Jan. 2019, doi: 10.1007/s12289-018-1406-7.

Identification of methods for the in-situ measurement of cutting forces in a tool-bound punching machine

S Niessner[1] and M Liewald[2]

[1] University of Stuttgart, Graduate School of Excellence advanced Manufacturing Engineering, Nobelstr. 12, 70569 Stuttgart, Germany
[2] University of Stuttgart, Institute for Metal Forming Technology, Holzgartenstr. 17, 70174 Stuttgart, Germany

E-Mail: stephan.niessner@gsame.uni-stuttgart.de

Abstract. With the increasing implementation of industry 4.0 technologies in production plants, issues such as control and monitoring as well as feedback control of plants and processes are becoming more and more important in current development trends. Thereby, potential approaches to such issues are generally accompanied by an increased use of sensors in production systems, aimed at gaining operation-, product- and process-specific data. In this respect, this paper focusses on the shear cutting process of a tool-bound punching machine. One of the most important parameters for the evaluation of shear cutting processes is the cutting force required. However, the precise measurement of this highly dynamic process parameter proves to be relatively challenging. The limited installation space within the tools and the loads caused by highly dynamic axis movements decisively restrict the options for sensor-based process monitoring. In order to face these challenges during sensor implementation in such punching machines, issues regarding measuring concept, localization and resulting measurement deviation of the sensors have to be solved simultaneously. In this paper, relevant measurement concepts and their properties are discussed, and the most favourable integration possibilities are derived. Based on this, several measurement concepts are identified that could be suitable for the application considered. These measuring methods include hydraulic pressure sensors, strain gauges or piezoceramics. The major contribution of this paper is an experimentally conducted systematic comparative study of these methods, in which the accuracy of the sensors used is evaluated and the influence of different integration positions on the measurement result is investigated.

1. Introduction

As a result of the increasing globalization of markets and the demand for digitalization concepts regarding autonomous production, flexible sheet metal manufacturing industry is currently facing a multitude of challenges. Here, one challenge arises from the continuous enhancement of the degree of automation of corresponding manufacturing processes in order to ensure economical production. The key to this lies in the metrological acquisition of process data, enabling data-driven monitoring systems and process control systems to be integrated into production processes.

In case of shear cutting processes, efficiency of the processes is particularly influenced by deviations in the cutting geometry caused by built-up edges, rounding and edge breakout [1]. If these deviations exceed a certain level, quality of the cut part is impaired in the area of the cutting surfaces. By means of suitable measuring technology, the influences of the cutting geometry deviations can be detected in-situ via changes in the process data. Here, the process force applied for the cutting process is of essential

importance and is therefore used as an elementary parameter for monitoring and controlling punching machines [2]. The added value of the acquired process signals depends on the signal quality, the measurement concept and the localization [3].
In the past, various methods have already been presented for the signal acquisition of highly dynamic cutting processes in automatic punching presses [3] [4]. The measurement concepts used differ in measurement position and the measurement method. The integration of individual sensor systems based on strain gages, piezoceramics or thin-film technology at different measuring positions is addressed in [5] [6] [7] [8] [9] [10]. However, a comparison of the various measuring methods and the effects of different measuring positions on the quality of the measurement results has not been carried out so far. Therefore, in this paper simultaneously applicable measuring concepts and measuring positions for a tool-bound punching machine TRUPUNCH 5000 are identified and compared by quantifiable quality measurements.

2. Experimental setup

2.1. Machine design

In the investigation reported about in this paper, a hydraulically driven punching machine with a flexible tool system was used. The machine and tooling concept are shown in Figure 1. The force is applied to the punching process via a hydraulic unit that operates at two pressure levels. For the vertical movement of the stroke, pressure is applied to the two differently sized surfaces on the top and bottom side of the ram in the punching head of the machine. Depending on the pressure ratio exerted, the ram will move up or down. To position the ram in starting position a small amount of pressure, is always applied to the bottom surface. The flexible tool concept basically consists of four the elements punch, alignment ring, stripper and die.

Figure 1. a) Punching machine TruPunch 5000; b) Tooling concept

2.2. Applied Material

The test material investigated in this study was the deep-draw steel DC01 (EN-AW 1.0330) having nominal sheet thickness of 2.0 mm. Table 1 shows the chemical composition of the material.

Table 1. Chemical composition of the test material.

Material	C	Si	Mn	P	S	Cr	Al	Ni
EN-AW 1.0330	0.035	0.044	0.209	0.009	0.010	0.026	0.037	0.020

To characterise the mechanical properties of mentioned material, uniaxial tensile tests were carried out with various orientations to rolling direction. The resulting mechanical properties are displayed in table 2. The lowest value of ultimate tensile strength occurs against rolling direction. Diagonally to the rolling direction, the value is significantly higher. However, since only rotationally symmetrical tools were used in the investigations presented here, the deviations for the tested orientations had no effect on the measured values.

Table 2. Mechanical properties of the test material gained by tensile tests.

Material	Rolling direction [°]	E-Modul E [MPa]	Yield strength $R_{p0,2}$ [MPa]	UTS R_m [MPa]	Uniform elongation A_g [%]	Fraction elongation A_{100} [%]
	0	199272	183.702	307.190	23.25	41.83
EN-AW 1.0330	45	216689	200.563	324.311	20.76	36.73
	90	205556	191.198	301.162	22.62	40.41

2.3. Measurement methods and setup

For the investigation, three measuring methods were chosen for comparing at three different integration positions. These measuring methods include hydraulic pressure sensors, strain gauges and piezo-ceramics. In addition to the force measurements, the stroke movement of the punch was measured using a linear measuring system on the ram rod.

2.3.1. Strain gauges. The idealized load case of a cutting punch during a punching operation corresponds to that of a tension/compression rod. For this reason, the cutting force was measured using strain gauge sensors (SG) applied to the outer surface of the cutting punch. Superimposed bending strains as well as thermal expansions were compensated by a suitable bridge circuit for mirror-symmetrical rod cross-sections [11]. In the present study, N2K-06-S5038P-10C/DG/E5 type strain gages from Vishay were applied to the punch. The measuring concept, the applied circuit and the equipped punch are shown in figure 2.

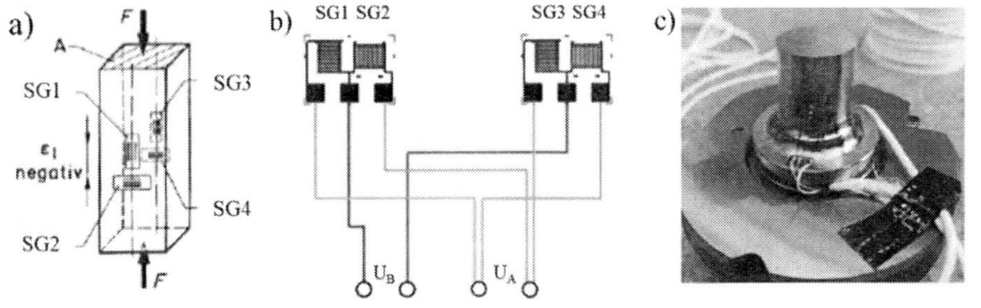

Figure 2. a) Strain measurement on a compression bar [11], b) Circuit diagram of the applied Wheatstone bridge circuit (U_B = bridge supply voltage, U_A = output voltage); c) Punch with applied strain gauges and soldered bridge circuit

2.3.2. Piezoceramics. According to [3], piezoelectric elements show a higher sensitivity than the strain gauges for the same mechanical strains and compressions. Therefore, such piezoelectric elements are superior to strain gages when measuring smaller forces on the same part and can provide more precise measurement results on stiff components showing smaller strains. In the field of application considered here, a concept has been developed for both the punch and the die in order to integrate piezoceramics in the force flow of the cutting elements. The tool construction including the piezoceramics is shown in figure 3. Three rotationally symmetrically arranged load cells (KISTLER type 9021A) were integrated in the die between die holder and cutting unit. For mounting the load cell in the punch, the punch was divided into two parts and the piezoceramic (KISTLER type 9061A) was clamped in between. Both piezoceramic sensors were installed with a preload force of 20 percent of the measuring range.

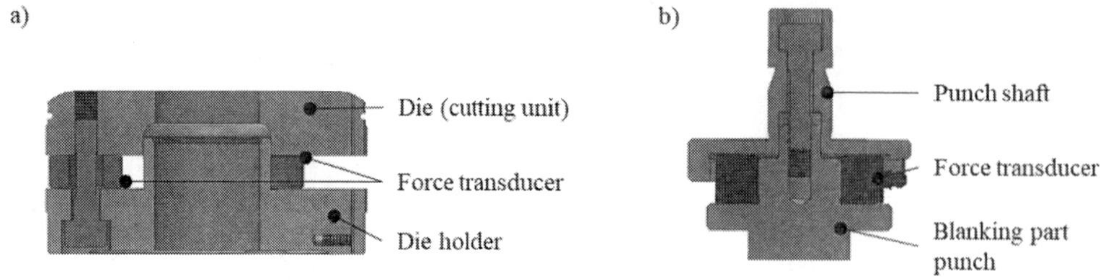

Figure 3. a) Three piezoceramics in die; b) Piezoceramic in punch

2.3.3. Hydraulic pressure. Force measurements were additionally carried out via the hydraulic pressures of the hydraulic press. Therefore, pressure sensors were applied to the upper (Baumer type PDRB E002.S14.B440DE) and lower pressure chamber (Baumer type PDRB E002.S14.B416DE) of the machine ram. Under consideration of the deviating area size of the pressure chambers A_1 and A_2, the resulting force F_{HYD} could subsequently be determined by the following equation

$$F_{HYD} = p_1 A_1 - p_2 A_2 \quad (1)$$

with p_1 = pressure in upper chamber

A_1 = area size upper chamber (9976 mm²)

P_2 = pressure in lower chamber

A_2 = area size lower chamber (4661 mm²).

2.3.4. Setup for the experiments. In order to ensure comparability of the measuring signals detected by the different measurement methods, time synchronicity of the measuring signals had to be guaranteed. For this purpose, imc measuring system PL8 was used, which allows to record all required signals simultaneously. The cutting parameters and combinations of measuring methods examined are summarized in table 3.

Table 3. Setup summery of investigated parameters.

Factors	Settings
Cutting clearance	10%
Punch diameter	8 mm, 16 mm, 32 mm
Cutting edge	No rounding
Stripper diameter	10 mm, 18 mm, 34 mm
Stripper force	0 kN (approx. 0.6 mm above the sheet metal surface)
Sheet metal material	DC01
Sheet metal thickness	2 mm
Sheet metal format	1000 x 1000 mm
Blanking operations	4000 per measurement series
Measurement combination 1	Punch piezoceramic, die piezoceramics, HYD pressure
Measurement combination 2	Punch strain gauges, die piezoceramics, HYD pressure

For the experimental investigation, the two measurement combinations (1 and 2) were compared directly with each other. Therefore, the corresponding experiments were carried out using the same punch diameter (8 mm, 16 mm and 32 mm). This resulted in a total of six test series, each with 4000 cutting operations. The tools applied with strain gauges as well as the tools built up with piezoceramic sensors were separately calibrated in the punching machine using a pre-stressed load cell before the examination.

3. Analyses of the experimental investigation

All 4000 signal curves of each test series are superimposed for evaluation. Due to the large number of individual signals, the superimposition results in an enveloping surface that represents the signal quality of each measurement method. In addition, the mean values of all individual signals are inserted into the diagram. The range of variance of the signals for each measurement method is represented by the distance between the mean values and the border areas of the enveloping surfaces. The signals of the stroke movement are presented in the same way and can be used to explain possible irregularities in the force curves. For an improved understanding of the signal curves the characteristic phases of a cutting process are marked numerically in the evaluation diagrams. The simple numbers represent the digits for the tool integration measurement methods. For hydraulic force determination, the numerical values are marked with an apostrophe. The following nomenclature was used for representing the characteristic phases in the analysis:

0	= start of the vertical stroke movement
1	= penetration of the punch into the sheet, the force increases continuously
2	= transition from elastic to plastic deformation, shearing starts
3	= peak of cutting force
4	= the incipient cracks expatiate and separate sheet and shearing part
5	= the shearing force decreases rapidly
6	= the vertical stroke movement reached the return point

3.1. Comparison between the two measuring method combinations for 8 mm punch diameter

In this first study, the influence of the localization and the measurement methods were investigated for a small load level. Therefore, the punch diameter with 8 mm was chosen and the sheet metal material were kept constant. As shown in Figure 4 the tool-based measuring methods (punch piezoceramic, die piezoceramic) show a very similar course and a rather small variance. In addition, no significant difference between the measurements of the upper and lower tool part can be determined. In the case of hydraulic force measurement, it is to be noted that the initialization of the vertical stroke movement already leads to a signal increase. Furthermore, hydraulic force measurement shows a delayed rise about 0.7 ms in the signal curve and the peak value is approx. 50% above the peak of the piezo-based measuring methods. The hydraulic force signal also shows a larger scattering. It is also noticeable that the force signal of the hydraulic measuring method assumes negative values after reaching the return point of stroke movement.

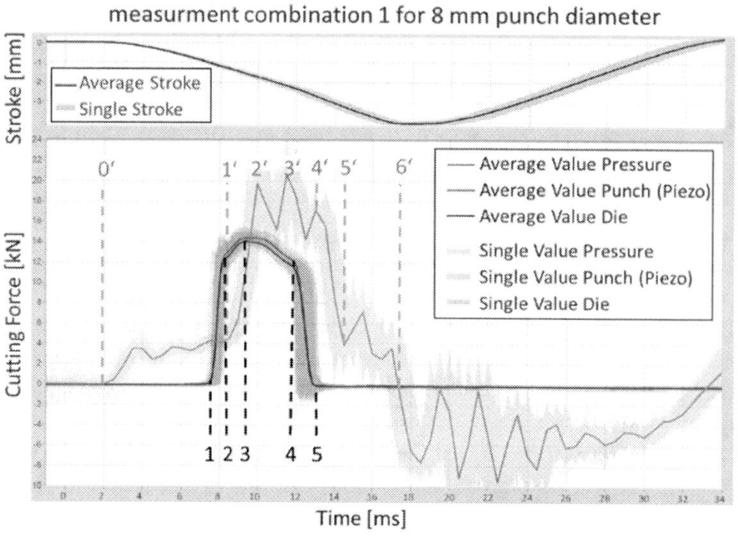

Figure 4. Force signals for measurement combination 1 for 8 mm punch diameter: punch piezoceramic (green), die piezoceramics (red), HYD pressure (yellow).

When comparing the signal curves of measurement combination 2 in figure 5 with the curves of combination 1, no significant difference between the curves measured with piezo elements and those measured with strain gauge can be noted. Both the measured variance and the absolute values correspond for all tool-bound measuring methods in the first investigation. The hydraulic force signal also shows an identical course.

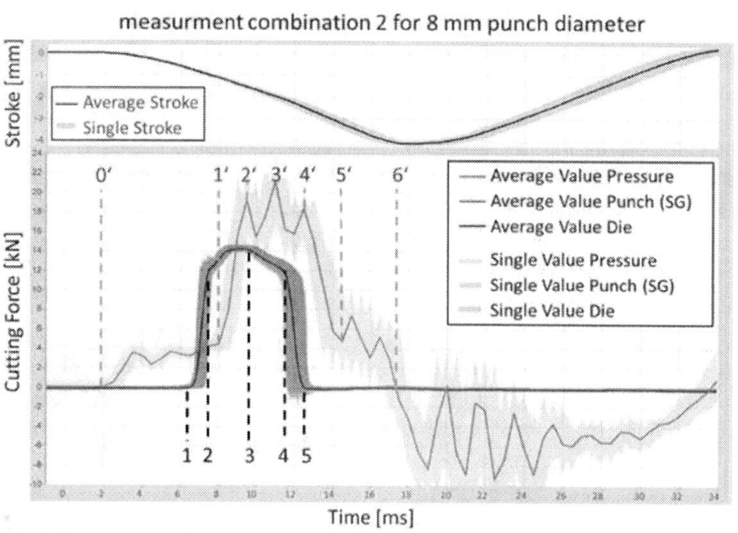

Figure 5. Force signals for measurement combination 2 for 8 mm punch diameter: punch strain gauges (green), die piezoceramics (red), HYD pressure (yellow).

3.2. Comparison between the two measuring method combinations for 16 mm punch diameter

As described in the previous section, the measurement combinations were subjected to a validation study in which larger load conditions were applied. In this investigation, cutting clearance, sheet material and sheet thickness were unchanged. Figure 6 shows the signal characteristics of measuring combination 1 with piezoceramics in die and punch and hydraulic pressure sensors for a punch diameter of 16 mm. Again, the two tool-bound measuring methods show an accurate agreement of their signal characteristics. Only in the area of the maximum force the measurement on the die side shows slightly higher values. For the hydraulic force curves, similar characteristics as in 3.1 can be noted. The premature increase in force due to the downward movement of the punch, the delayed response behaviour and the stronger signal fluctuations are still visible. However, in contrast to the previous investigation, the approach of the hydraulically determined force signals approximates to the tool bound.

The detected difference in the maximum values is less than 20%. In addition, it is notable for all measuring methods that the range of the signals increases strongly after achieving the peak value. Comparing the current stroke movement with the one in 3.1, it is noticeable that it varies considerably after reaching the peak force. As a result, the scattering in the signal curves after the decrease in force increases.

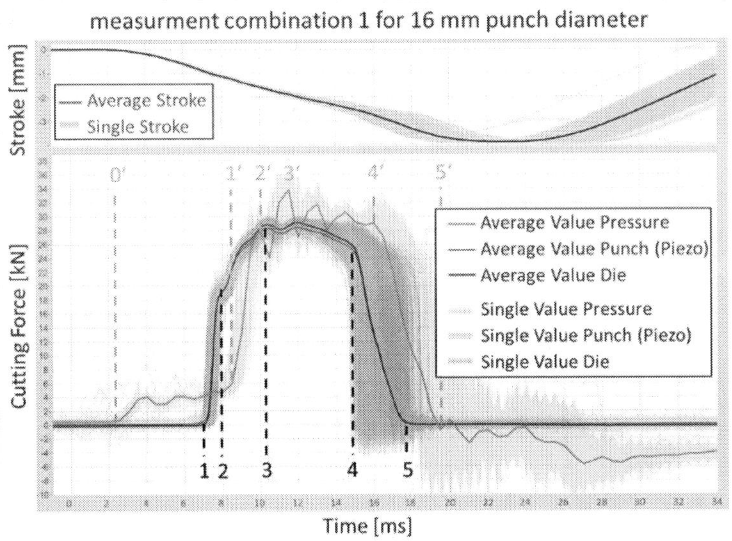

Figure 6. Force signals for measurement combination 1 for 16 mm punch diameter: punch piezoceramic (green), die piezoceramics (red), HYD pressure (yellow).

When comparing the signal curves of measurement combination 2 in figure 7 with the curves of combination 1, no significant difference between the curves measured with piezo elements and those measured with strain gauge can be noted. Also the difference registered in Figure 6 between the use of sensors on punch or on die is no longer visible. Both the measured variance and the absolute values are the same for all tool-related measuring methods, as in the first investigation. The hydraulic force signal also shows an identical course.

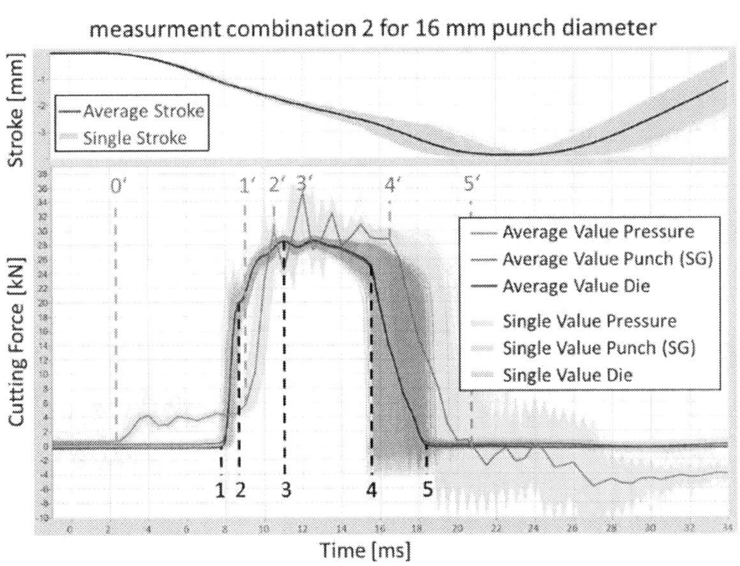

Figure 7. Force signals for measurement combination 2 for 16 mm punch diameter: punch strain gauges (green), die piezoceramics (red), HYD pressure (yellow).

3.3. Comparison between the two measuring method combinations for 32 mm punch diameter

Figure 8 shows the signal characteristics for measurement combination 1 with a punch diameter of 32 mm. In principle, the trends in curve characteristics identified in 3.2 are continuing. The curves of the two measurement methods on tool are in proximity and only show a small deviation in the peak value. The approximation of the hydraulically determined force curves to the tool-bound measuring methods is particularly distinctive. Except for the already known delay in the response behaviour, there is almost no difference to be identified. The collapse of all force signals in the elastic range is due to the pressure switchover. This is also shown by the delayed downward movement of the punch in the stroke movement at this area. Due to the low strains on the 32 mm diameter punch, no suitable results could be gained for measurement method using strain gages under these load conditions. The results of measurement combination 2 were therefore excluded from the evaluation.

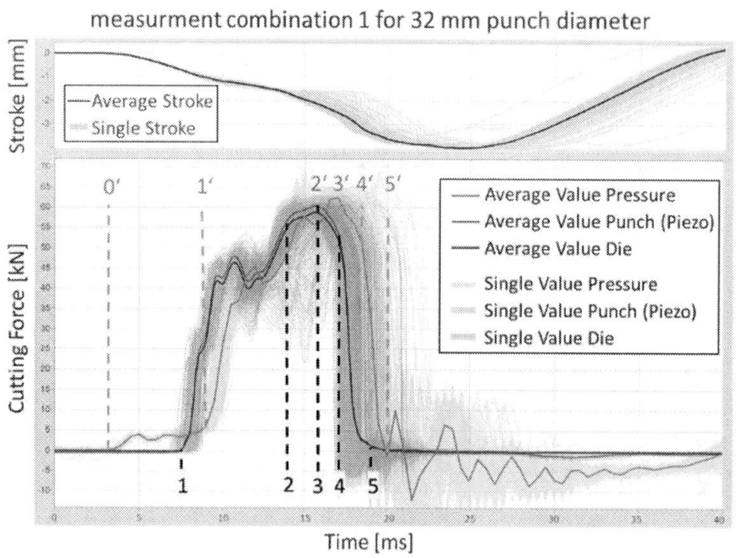

Figure 8. Force signals for measurement combination 1 for 32 mm punch diameter: punch piezoceramic (green), die piezoceramics (red), HYD pressure (yellow).

4. Conclusion and Outlook

In the experimental studies carried out within the scope of this paper various possibilities for measuring the highly dynamic force signals during shear cutting processes in a tool-bound punching machine were investigated. Here, three measurement methods with various localizations were analysed regarding their influence on the resulting measurement deviation. For the experimental comparative study, two possible combinations with alternative measurement concepts were identified, allowing three measurement methods to be integrated into the tool system simultaneously. In parallel, the measurement methods under investigation were validated for different load levels with regard to their continuous signal quality. Based on these results, it could be stated that the proximity of the sensor to the main cutting operation is of major importance for the quality of the measurement signals, especially in lower load conditions, and the resulting deviations become less significant with increasing load. This is particularly visible for hydraulic force measurement, which only achieves close approximation to the signal quality of other force measurement methods at higher load conditions. This is due to the fact, that with increasing load levels, the signal-distorting influence of additional tolerance chains and mass dynamics lose significance and the measured sensor signals increasingly resemble each other. The consistently observed signal delays with the hydraulic force measurement correspond in their extent to the manufacturer's specifications, but can possibly be reduced by using faster responding sensors. A detectable distinction between the force signals using sensors on the punch or die could not be recorded in the study. Likewise, the advantage of the piezoceramic sensors mentioned in 2.3.2 could not be determined in relation to the measuring method with strain gauges. Depending on the specific requirements and load cases, under which the force signals have to be detected, each of the presented measurement methods might produce usable measurement results. However, it is essential to be aware of the signal deviations occurring as a result of the measurement method and localization. Suitable measurement signals are only obtained if the combination of method and localization is suitable to the application. This is also the basic requirement for use in the context of industry 4.0. In field of Industry 4.0, sensors and instrumentation are one of the central driving forces for innovation. Intelligent decisions of complex systems are based on the knowledge of the system as well as ambient conditions and influence factors provided with high accuracy by sensors. Therefore, the importance of sensors, measurement science and smart evaluation, as presented in this paper for a specific application, will gain an increasingly major role for Industry 4.0.

5. References

[1] Brizer F 1996 Feinschneiden und Umformen: Wirtschaftliche Fertigung von Präzisionsteilen aus Blech, *Verlag Moderne Industrie Band 134*

[2] Liewald M and Wagner S 2018 Scheren und Schneiden *In: Grote K-H, Bender B, Göhlich D Dubbel: Taschenbuch für den Maschinenbau* (Springer-Verlag, ISBN 9783662548059)

[3] Kolbe M and Hellwig W 2018 Überwachung von Stanzwerkzeugen. *In: Spanlose Fertigung Stanzen (Springer Vieweg, Wiesbaden, DOI 10.1007/978-3-658-20024-4_16)*

[4] Hoogen M 1999 Einfluß der Werkzeuggeometrie auf das Scherschneiden und Reißen von Aluminiumfeinblech *utg Forschungsberichte Band 6* (ISBN 3-89791-045-4)

[5] Schweitzer M 2001 Prozessspezifische Merkmale des Rotationsschneidens *Dissertation*

[6] Biehl S N and Mintenbeck D 2010 Einsatz neuartiger Dünnschicht-Kraftsensoren zur *Überwachung von Schneidprozessen: Abschlussbericht, AiF/IGF-Vorhaben-Nr.16113 BG*

[7] Kopp C M 2011 Einfluss der Schneidparameter beim Scher- und Keilschneiden von Magnesiumblechwerkstoffen *Dissertation*

[8] Panico T 2011 Beurteilung der Kantenbelastung beim Scherschneiden höchstfeste Stahlbleche mit Hilfe der Finiten-Element-Methode *Dissertation*

[9] Katrin Nothhaft 2014 Scherschneiden höchstfester Blechwerkstoffe im offenen Schnitt *Dissertation*

[10] Mackensen A 2014 Presswerkorientierte Analyse der Umform- und Scherschneideigenschaften von Mehrphasenstählen *Dissertation*

[11] Hoffmann K 1987 Eine Einführung in die Technik des Messens mit Dehnungsmessstreifen *HBM* pp 237-240

IOP Conf. Series: Materials Science and Engineering **967** (2020) 012026 doi:10.1088/1757-899X/967/1/012026

A Method to Incorporate Grain Boundary Strength and its Effects on Plastic Deformation in FCC Polycrystals

W Muhammad[1,2], A P Brahme[1], J Kang[2], E Cyr[1], D S Wilkinson[3] and K Inal[1]

[1] Department of Mechanical and Mechatronics Engineering, University of Waterloo, Waterloo, Ontario, N2L 3G1, Canada
[2] CanmetMATERIALS, 183 Longwood Road South, Hamilton, Ontario, L8P 0A5 Canada
[3] Department of Materials Science and Engineering, McMaster University, 1280 Main Street West, Hamilton, Ontario, L8S 4L7 Canada

Email: kinal@uwaterloo.ca; jidong.kang@canada.ca

Abstract. A three-dimensional (3D) crystal plasticity finite element method (CPFEM) is developed to investigate the effect of grain boundary strength on heterogeneous strain partitioning in FCC polycrystals. The proposed method incorporates electron backscatter diffraction (EBSD) maps into finite element analyses. The numerical analysis accounts for crystallographic texture, its evolution and 3D grain morphologies. Furthermore, grain boundaries are also mapped with a special finite element framework that allows material properties to be assigned to the grain boundaries. The material parameters and the grain boundary strengths are obtained by calibration to experimental uniaxial tension curves for single and polycrystals. Numerical simulations of uniaxial tension are performed and the effects of grain boundary strength on the onset of non-uniform deformation is investigated. The predicted local strain evolution is compared with corresponding experimental results from digital image correlation (DIC) measurements of an AA5754 aluminium sheet. The results showed that an excellent agreement was reached when the grain boundary properties were set so that the hardness was five times that of the average polycrystal response.

1. Introduction

Majority of the engineering materials of interest are polycrystalline in nature, consisting of several grains having different crystallographic orientations. Understanding the role of different contributing factors that are responsible for the overall macroscopic behaviour of these materials is of critical importance. These factors include response of each individual crystals/grain, the effect of grain boundaries as well as the effect on each grain due to the neighbouring grains. The simplest approach is to consider the polycrystal as an aggregate of the individual grains and completely ignore the effect of the neighbours. Classical approaches include enforcing either equal strain on each crystal as the polycrystal [1] or equal stress [2]. Thereafter, several approaches have been proposed. With the increase in the computing power, models that incorporate the crystal plasticity theory into the finite element formulation (CPFEM) have been the choice to capture these and other effects [3-5]. The CPFEM models enforce continuity and compatibility at the elemental level and have been used to successfully predict the mechanical properties as well as texture evolution during deformation for various polycrystalline systems such as FCC, BCC and HCP. A comprehensive review article by Roters et al. [6] provides an excellent overview of this approach.

Crystal plasticity theory accounts for the anisotropic behaviour of each grain based on its crystallographic orientation. This allows for non-uniform stress and strain distributions to be predicted as well as provides new insights on the influence of deformation mechanisms on localized deformation and failure. There have been many cases of crystal plasticity being used to study a range of material instabilities. To name a few: Peirce et al. [3] used crystal plasticity to predict shear bands in tensile single crystals. Wittridge and Knutsen [7] investigated surface roughness in aluminium alloys using a polycrystalline model. Dao and Li [8] and more recently Muhammad et al. [9-10] used crystal plasticity to predict crack initiation and propagation during bending of aluminium polycrystals. Potirniche et al. [11] used crystal plasticity to investigate void growth and coalescence in single crystals. Crystal plasticity has been used to predict the anisotropic behaviour of single crystals very accurately [12]. A set of material parameters can be fit to a single crystal orientation, which in turn can be used to predict the response of other orientations. However, in most cases the material parameters obtained from single crystal data need to be modified to match the polycrystalline response. There have been few works that try to solve this problem. Lim et al. [13] used single crystal continuum constitutive equations with a two scale (grain scale and meso scale) scheme to incorporate the effect of grain boundaries for larger grains. Kluseman et al. [14] used single crystal results to get the material parameters and combined them with size effect due to lack of grain size homogeneity to match deformation behaviour of thin sheets of Fe-3%Si. Cheong and Busso [15] investigated the effect of grain misorientation as well as strain hardening to match the polycrystal behaviour using single crystal material properties. Most of these works use data measured from oligo crystalline samples (samples with several grains covering the gage section) to validate and test their models. These microstructures provide an excellent opportunity to study grain scale plasticity [16-17].

In the literature, most of the existing crystal plasticity models do not have any special calculation or consideration at the grain boundary; they merely represent the boundary by having two neighbouring elements of different orientations [18-21]. Some models have been proposed where cohesive zone elements have been placed along the grain boundaries to account for small amounts of grain boundary slip, fracture, and grain boundary accommodation [22-23]. Lim et al. [13] introduced a model that uses dislocation density-based crystal plasticity in combination with 2D planar elements at the grain boundaries which act as a membrane to control dislocation transfer between neighbouring grains. This approach, in conjunction with the limited slip transfer through the boundary allows the model to predict the dislocation pileup seen in experimental observations and predict the characteristics seen in the grain boundary affected zone. Even though the grain boundary is a planar interface the mismatch of lattice structure at the grain boundary impedes dislocation motion causing dislocation pileup. This dislocation pileup can affect the material properties surrounding the grain boundary. Grain boundary affected zones have been shown to typically extend several micrometres into the grain [24].

The primary goal of this work is to incorporate the effect of grain boundaries into the CPFEM framework that will allow for inclusion of the associated strain inhomogeneity and material strengthening effect. In the first part of this work a method to incorporate the effect of boundaries will be proposed and validated using the DIC data as well as the polycrystalline flow curves provided by Zhu et al. [25] on large grained AA5754 sheets. The grain boundary affected zone will be identified and modelled as 3D elements with the element thickness equal to the grain boundary affected zone. In the next part the effect of boundary roughness on the overall material response will be studied.

2. Experimental procedures

2.1. Large grained AA5754 sample

Zhu et al. [25] created a large grained aluminum sample with two layers of columnar grains. The authors studied the deformation heterogeneity and strain localization and compared strain concentrations and correlated them to Taylor and Schmid factors. The advantage of the columnar structure is that 2D EBSD scans can be used to establish the 3D material microstructure. Since the sample consist of two layers, a scan from each side of the sheet is required and the structure can be made with the assumption that the two layers interface at the center of the sheet thickness. The sample identified as Sample A in Zhu et al. [25] will be reanalyzed and used in this work to generate the 3D material microstructure. The microstructure measured on both sides of the gauge area is shown in Figure 1 (a) has 12 grains and (b)

IOP Conf. Series: Materials Science and Engineering **967** (2020) 012026 doi:10.1088/1757-899X/967/1/012026

has 15 grains. The electron backscatter diffraction (EBSD) measurements were made to obtain the crystallographic orientation of the grains. A step size of 15 μm was used for the measurements and multiple scans were obtained and were stitched together to obtain the overall large microstructure. The area shown in Figure 1 is slightly larger than the gage area. Note that the inverse pole figures presented here are rotated so that all directions are parallel to ND while the figures presented in Zhu et al. [25] have the directions parallel to RD. The EBSD dataset was cleaned using an iteration of neighbor confidence index (CI) of 0.2 to remove bad data points. The reader is referred to Zhu et al. [25] for further experimental details. The microstructure shows large grains with large variation in the grain size and an average grain size of 1.25 mm. These maps will be used to create the mesh that will be used in the CPFEM simulations.

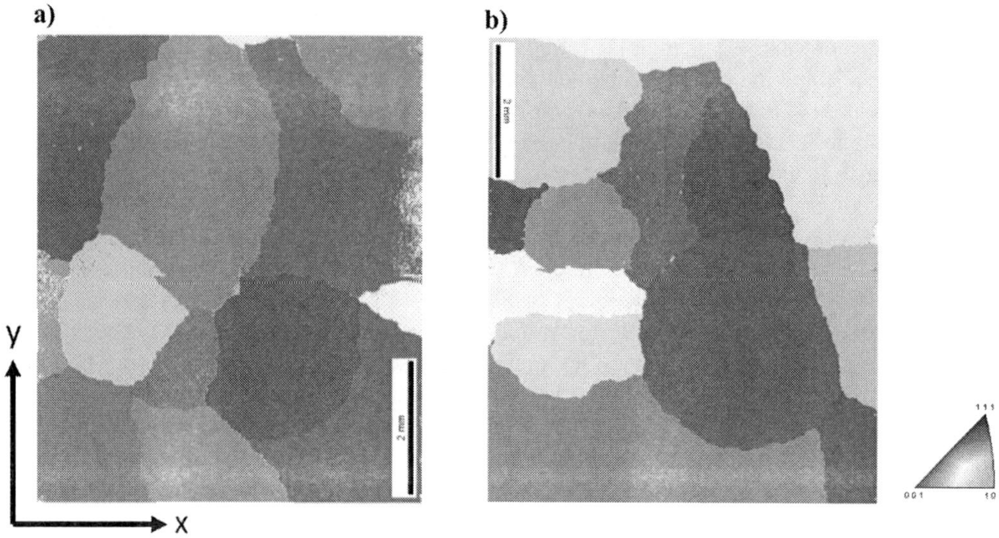

Figure 1: Inverse pole figure for large grained AA5754 aluminum polycrystal (a) Side A, (b) Side B.

The sample was then pulled in tension along the y-direction shown in Figure 1. The tensile tests were performed at a strain rate of $2.7x10^{-3}/s$ and the DIC images were recorded every 2s during the tensile test. The deformed speckled patterns were processed using the Aramis® software package and a regular grid of nodes was established as well as nodal movements throughout the test. As the DIC data is very sensitive to speckle pattern and image quality, there are some regions within images that are processed with lesser degree of confidence than the bulk of the image. These regions of "bad data" are filtered out and removed. The DIC data is further processed to plot the effective strain contour maps as shown in Figure 2. The applied strain at this point is 14%. This data is later used for comparison with simulated results. The tensile stress-strain response was also measured and is shown in Figure 3. The yield stress is approx. 50 MPa and the material shows significant work hardening prior to failure.

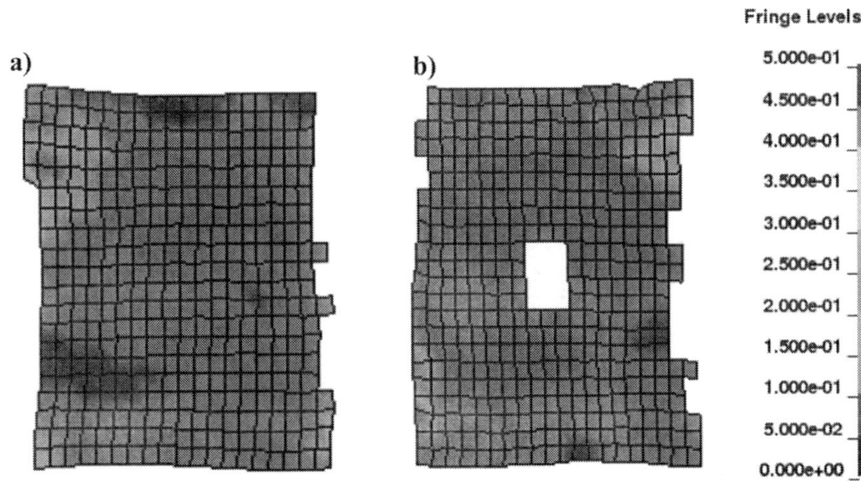

Figure 2: Experimental DIC measured effective strain maps for (a) Side A and (b) Side B of the specimen's gauge region at 14% strain for AA5754 aluminum polycrystal.

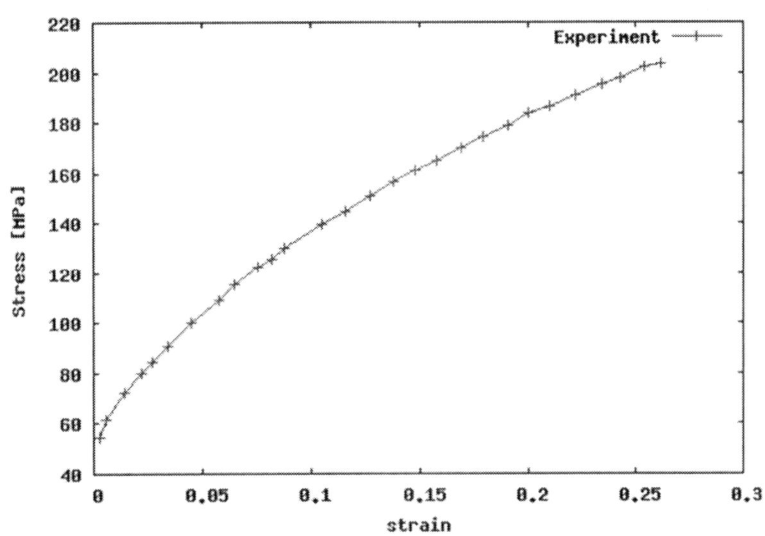

Figure 3: Experimentally measured stress vs. strain response for AA5754 aluminum polycrystal.

2.2. Single crystal data

Inoko et al. [26] presented aluminum single crystal data with tensile axis along ⟨001⟩, ⟨111⟩ and ⟨112⟩ reproduced in Figure 4. This data on single crystals will be used to calibrate the material parameters for CPFEM simulations.

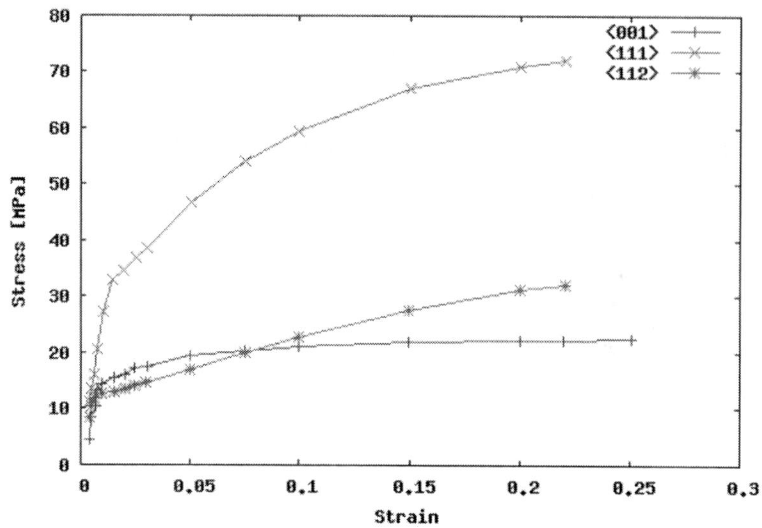

Figure 4: Aluminum single crystal tensile response along three different loading axes.

3. Numerical procedures

Crystal plasticity models are utilized on two levels. At the single crystal level; to identify the material parameters using single crystal constitutive model and at the polycrystal level; employing the Sample data A in Zhu et al. [25]. The single crystal CPFEM simulations employ single elements and are useful to assess the effect of crystallographic texture as the measured properties of single crystal data does not have any grain boundary effect. Two separate polycrystal models were developed to investigate the effect of grain boundaries. These will be discussed in detail in section 3.2.

3.1. Constitutive model

The simulations performed for this paper use a rate dependent crystal plasticity theory formulated by Asaro and Needleman [27] implemented into the commercial FEM software LS-DYNA. For completeness, some important details of the formulation are summarized next.
The velocity gradient $\mathbf{L}^P$ is given by:

$$L^P = \sum_{\alpha=1}^{12} \dot{\gamma}^{\alpha} s^{\alpha} \otimes m^{\alpha} \tag{1}$$

where $\dot{\gamma}^{\alpha}$ is the shear rate on the αth slip system and s^{α} and m^{α} are the slip direction and slip plane normal, respectively. The shear rate is defined as:

$$\dot{\gamma}^{\alpha} = \dot{\gamma}_0^{\alpha} \left|\frac{\tau^{\alpha}}{g^{\alpha}}\right|^{1/m} sgn(\tau^{\alpha}) \tag{2}$$

where $\dot{\gamma}_0^{\alpha}$, τ^{α} and g^{α} are the reference shear rate, resolved shear stress and the hardness of the slip system α and m is the rate sensitivity exponent. The slip system resistance term includes the particular choice of hardening rate $\dot{g}^{\alpha} = \sum h_{\alpha\beta}\dot{\gamma}^{\beta}$; $h_{\alpha\beta} = q_{\alpha\beta}h_{\beta}$ where $h_{\alpha\beta}$ are the hardening moduli, $q_{\alpha\beta}$ is the latent hardening matrix and h_{β} is the single slip hardening rate. The complete description of the constitutive model can be found in Rossiter et al. [28]. The work of Rossiter et al. [28] utilized a power law hardening. The only modification made to the constitutive model is the slip hardening model used for the single crystal material fit. The hardening model was modified to one that can be found in Harren et al. [29]:

$$h_{(\beta)} = h_s + (h_o - h_s)\,\mathrm{sech}^2\left\{\left(\frac{h_o-h_s}{\tau_s-\tau_o}\right)\gamma_a\right\} \tag{3}$$

where h_o is the system's initial hardening rate, h_s is the system's asymptotic hardening rate, τ_o is the critical resolved shear stress of the system, τ_s is the saturation value of the shear stress, and γ_a is the accumulated slip. The reason for implementing this hardening law is that the single crystal experimental data found in Figure 4 displays a saturation hardening characteristic which is not obtainable using the power law hardening model used previously. The model that does not contain grain boundary zones and single crystal material parameters still uses the original power law hardening model from Pierce et al. [30] as described in Rossiter et al. [28].

$$h_{(\beta)} = h_o \left(\frac{h_0 \gamma_a}{\tau_0 n} + 1\right)^{(n-1)} \quad (4)$$

τ_0 is the critical resolved shear stress and n is the hardening exponent.

The constitutive model described above is used for the integration points in the single crystal and the polycrystal model, excluding the elements/integration points within the grain boundary zones. For the grain boundary zones, a simple elasto-plastic Mises plasticity constitutive model (LS-DYNA MAT24) is used. The stress vs. strain curve that is provided to the model is the experimentally measured stress vs. strain curve obtained from the tensile test modified by a scalar, which will be discussed in Section 3.3.2.

3.2. FE mesh generation

To study the effect of grain boundaries two separate meshes were generated. Both meshes were derived from the measured data from Zhu et al. [25]. The difference between the two meshes is the presence of grain boundary zones modeled with ~20 µm elements in one of them. The details of how the meshes are generated from the 2D EBSD maps are given next.

The ESBD data was processed to clean any bad data points and each pixel was assigned a grain number to which it belonged. The map was then analyzed to generate a list of pixel pairs such that each pixel in the pair belonged to two different grains. For every such pair a node representing a boundary point on the grain boundary was placed between them. The maps were processed again where every possible region of 2x2 square pixels was analyzed and regions that contained pixels from 3 or more different grains were identified. A node representing a triple point was placed in the center of each of these 2X2 regions. Once all the grain boundary points and triple points were identified in the map, the mesh for the grain boundary zone was constructed in the following manner. A piece of boundary was considered to be series of nodes separating two grains joining two triple points. Starting with a triple point at one end of the boundary, a series of bar elements was created connecting each boundary nodes along a boundary until the triple point on the opposite end of the boundary. This procedure was repeated for all boundaries. The result was a map of the grain boundaries consisting of 1-D bar elements. Since the sample consists of 2 separate layers of grains, the procedure for mapping the boundaries was conducted for both sides and then overlaid to produce a single map that contained outlines of all the grain boundaries on both layers of grains. The reason for combining the two layers of grain boundaries onto a single mesh is that when the final mesh is created, the elements and nodes at the interface between the two layers of grains need to match completely.

Once the system of bar elements representing the grain boundaries was completed, the 2D region within each grain was then meshed with shell elements. At this point, the addition of grain boundary zones was performed and the two distinct meshes were produced separately. For the mesh that includes the grain boundary zone, the nodes that represent the grain boundary nodes were duplicated, separated by approximately 20 µm, and a quadrilateral shell element was placed in the gap. This resulted in triangular holes in the mesh at each triple point, which were filled with triangular elements.

Both meshes were then extruded to the thickness of the sample. The regions representing each grain were then combined and assigned a unique part designation and the orientations of the grains as measured from EBSD were assigned to the elements. The grain boundary zone elements for the whole structure were treated as a single part and were assigned a conventional plasticity material model.

The first step after building the polycrystal meshes is to determine the material parameters to be used in the simulations. For the model with grain boundaries, the material parameters are fitted using the single crystal data (Section 3.3.1) while for the model without grain boundaries, the material parameters are determined using the data published by Zhu et al. [25]. Once these are determined the model behaviors are compared between the single crystal and polycrystal fits. Next grain boundary zones are assigned properties and the predictions (Section 3.3.2) are compared to the data published by Zhu et al. [25]. Once a satisfactory value for the grain boundary parameters is obtained two independent studies are performed (i) investigating the effect of grain boundary roughness (Section 4.3) and (ii) effect of boundaries in a complex forming operation (Section 4.4).

3.3. Model Parameters

Since the goal of this work is to develop a method of modeling polycrystals that can account for grain boundary effect, first a system which has no grain boundaries (single crystal) should be studied. In the single crystal case, only the grain orientation affects the stress strain response. This allows the single crystal constitutive model to be fitted with published single crystal data. Section 3.3.1 presents the method for determination of single crystal parameters. Once the single crystal constitutive properties have been established, the grain boundary properties can be established using the macroscopic polycrystal response of the sample. For the model that does not contain grain boundary zones, the macroscopic polycrystal response can be used to fit the constitutive model to act as a comparison to evaluate the benefits of modeling grain boundary zones.

3.3.1. Single crystal

In order to identify the appropriate material parameters for the crystal plasticity constitutive model, single crystal simulations were performed and compared with published single crystal aluminum data. The single crystal stress-strain response published by Inoko et al. [26] and reproduced in Figure 4 is used for the parameter determination. A single element simulation of uniaxial tension was setup in LS-Dyna and LS-Opt was used to perform a parametric study to identify the appropriate material parameters. This allows one to find the parameter set that simultaneously fits all the curves represented in Figure 4. Initially for determination of parameters for the single crystal data, the power law slip hardening (see Eqn.4) from Peirce et al. [30] was used. The problem that arose from using this slip hardening model was that it is not capable of reproducing the saturation effect observed in the <001> loading direction seen in Figure 4. In order to obtain a closer fit with the single crystal material data, the slip hardening model was modified to the model found in Harren et al. [29] presented in Eqn. 3. The material parameters obtained by the LS-Opt optimization were then checked in a single integration point, crystal plasticity, material simulator to ensure the predicted uniaxial tension results given by the fitted parameters are in good agreement with the experimental data and are independent of the FE model. The material parameters determined using this procedure are reported in the Table 1.

Table 1: Crystal plasticity constitutive model material parameters.

Parameter	$\dot{\gamma}_0$ (s^{-1})	h_o (MPa)	τ_o (MPa)	h_s (MPa)	τ_s (MPa)	n
With Boundary Zones (single crystal parameters)	0.001	86	1.94	6.4	10	N/A
Without Boundary Zones	0.001	700	15	N/A	N/A	0.4

3.3.2. Polycrystals

Once the single crystal material parameters were identified, the polycrystal meshes developed in Section 3.2 can be used to fit the grain boundary response. Initially, a simulation using these material parameters that does not contain grain boundary zones was performed and the overall stress vs. strain response is given in Figure 5. When compared to the experimental stress vs. strain curve, the model without grain boundary zones far under predicts the material response. This is expected since the hardening due to the grain boundaries is not included in the model. The model containing grain boundary zones was then used where the grain boundary zones were assigned a piecewise phenomenological plasticity material model along with a flow stress curve. To obtain the flow stress curve for the phenomenological model, the overall macroscopic stress vs. strain response from the polycrystal experimental results (see Figure 3) was taken as the base for the curve. A parametric study using LS-Opt was performed where a scalar value was used to modify the flow stress curve given to the grain boundary zones. This scalar was tuned so that it gave the lowest mismatch between the model prediction and the measured flow data. The results from the fitting revealed that when the experimental stress vs. strain curve used in the grain boundary zones is multiplied by a scalar of 5, the overall response of the polycrystal matches the experimental data, as shown in Figure 5. The implication of this result is that a 20 μm zone which encompasses a grain boundary should have a response that is 5 times harder than that of the macroscopic response. Given the existence of dislocation pileups and lattice mismatch at grain boundaries, this finding is in line with expectations. The exact value of this "5 times harder" is subjective since the required hardness applied to the grain boundaries in a model would change depending on the width chosen for the grain boundary zones as well as the exact material flow curve. Another factor that will affect the hardness required in the grain boundary zones would be the single crystal data used for the crystal plasticity model parameters. The data used for this study is pure aluminum single crystal data. The polycrystal used in this work is an AA5754 alloy. The intra-granular material properties might be slightly different than those found for pure aluminum. Nevertheless, the goal of this research is to investigate the benefit of using this modelling approach and its effect on simulated results will be investigated.

In order to compare the modelling approach of using grain boundary zones proposed in this work, an additional polycrystal fit was performed using the previous technique of fitting the single crystal constitutive material parameters to achieve the macroscopic response of the polycrystal. For this model, the mesh that does not contain grain boundary zones was used for the LS-Opt parametric fitting. It was found that the power law slip-hardening model given in equation 2 gave a near perfect fit using the material parameters found in Table 1 for the model without grain boundary zones.

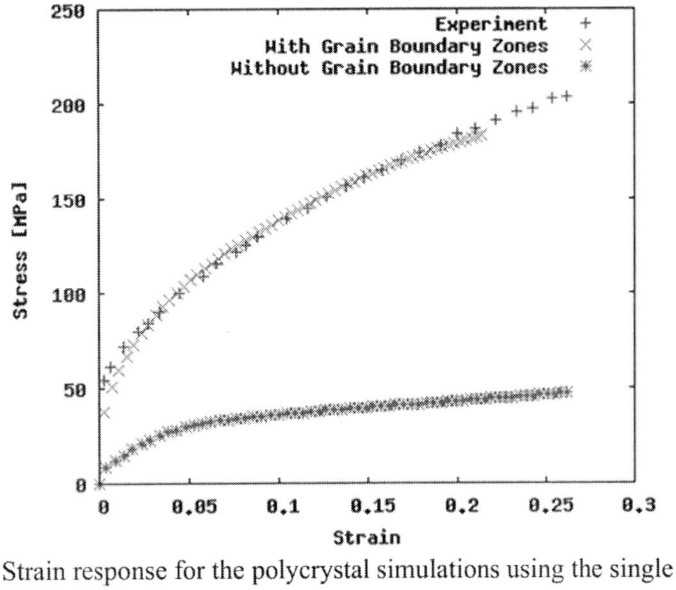

Figure 5: Stress vs Strain response for the polycrystal simulations using the single crystal fitted material response.

4. Results and discussion

4.1. Uniaxial tension loading

Figure 6 (a) shows a deformed mesh at 14% strain for the model that includes grain boundary zones and Figure 6 (b) shows a deformed mesh with the same amount of strain for the model that does not have grain boundary zones (both having the same overall stress-strain behavior). Both these figures show the contours of effective strain and have the same scale. The two models show that the locations of hotspots are similar, but the value of effective strain is different in the two. For example, the highest strain is in the rectangular grain near the top left in the model with grain boundaries while the highest strain is within the equiaxed grain near the middle-right edge for the model without boundary zones. Both models show that the concentration is close to boundary between two grains.

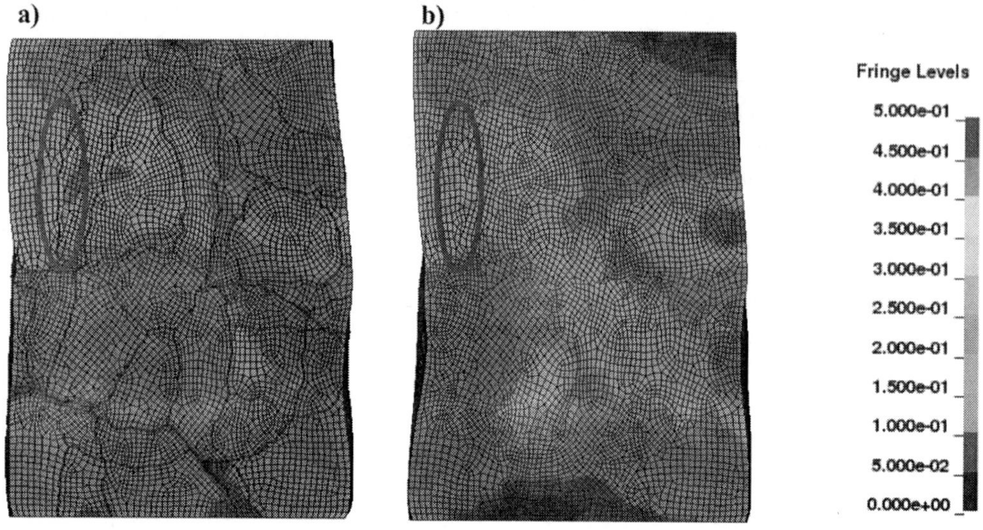

Figure 6: Deformed effective strain maps for the numerical simulations (a) with and (b) without grain boundary zones at 14% applied strain.

A more detailed investigation of the deformed meshes presented in Figure 6 (a) and Figure 6 (b) shows that the hot spots are not exactly at the grain boundaries. Figure 7 provides a magnified view of the regions encircled in Figure 6 (a) and Figure 6 (b). In Figure 7 (a) the grain boundary zones/elements have lower strain than the surrounding region. The model without grain boundary zone has no such abrupt change in strain localization, as shown in Figure 7 (b). This is due to the substantial increase in the hardness that is given to the grain boundary zone in the material properties and is an artifact of the modeling approach. True grain boundaries exist as a gradient of material properties that changes as you get farther from the actual boundary interface. The DIC in situ strain measurement technique used to collect the experimental data is not at a sufficiently high resolution and magnification required to capture any low strain region such as that of a grain boundary zone.

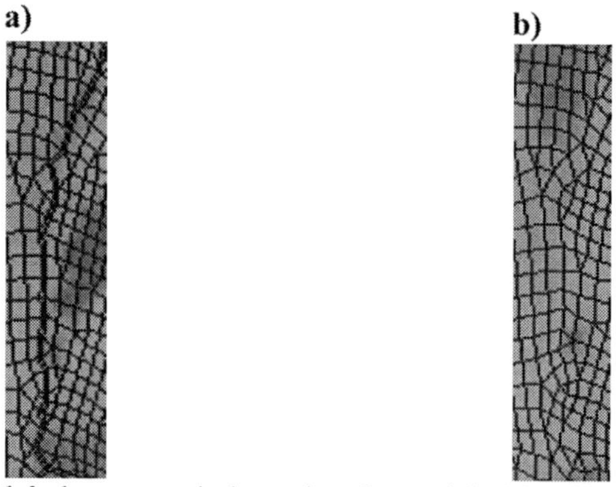

Figure 7: Region of the model close to grain boundary in models (a) with grain boundary zone and (b) without grain boundary zone. These regions are encircled in red in Figure 6.

To validate and quantify the improvements in the model by adding grain boundary zones, the in-situ DIC strain measurements from the experimental tensile test were compared with the strain distributions measured from the simulations. The distributions were evaluated and quantified to establish that the addition of grain boundaries increased the accuracy of the model.

IOP Conf. Series: Materials Science and Engineering **967** (2020) 012026 doi:10.1088/1757-899X/967/1/012026

4.2. Comparison with experimental results

The raw DIC data that was processed and presented in Zhu et al. [25] consists of a regular grid of surface points and their in-plane displacements for each side of the sample. This raw data was processed by creating a finite element mesh consisting of quadrilateral shell elements where the DIC surface points represents the nodes of the finite element mesh and the displacements of the points represent the nodal movements. The displacements were processed using quadrilateral element shape functions to create strain distribution maps. As mentioned in Section 2.1, Figure 2 shows an example of this mesh where the bad data points were at a minimum. Note that Figure 2 represents a smaller area than the one depicted in Figure 6 (a) and Figure 6 (b). Comparing Figure 2 (a) to Figure 6 (a) and Figure 6 (b), the hot spots and low spots of the model containing grain boundary zones are in better agreement with the DIC data. This qualitative comparison implies that the predictions of the model with grain boundary zones are more accurate. In order to quantify differences between the models and measured data, the models were processed to allow a direct comparison with the DIC data. The regular grid of DIC data points was overlaid onto the undeformed finite element mesh of the polycrystal. The surface nodes of the mesh that were the closest to the DIC data points were identified. The simulation results were data mined to extract the surface in-plane nodal movements of these nodes. A new data set was produced that matched the format of the initial DIC data which contained the locations of the mesh nodes closest to the DIC points and tracks their displacements throughout the simulation history. This new data set was then processed in the same manner as the initial DIC data to get a new distribution of strains that have the same resolution and direct element-to-element match with the DIC data.

A pixel-based Root Mean Squared (RMS) error was used to quantify the accuracy of the two models (with and without grain boundary zones).

$$E_{RMS} = \sum_{i=1}^{N} \sum_{j=1}^{9} \left(\left(\varepsilon_{eff}^{sim} \right)^2 - \left(\varepsilon_{eff}^{DIC} \right)^2 \right)^{1/2} \tag{5}$$

where, i sums over all the pixels (N), j sums over immediate neighborhood (first and second nearest neighbors) ε_{eff} is the effective strain. A square 9-pixel window was marched along each of the three data sets (DIC data set and the two models) in 1-pixel steps. For each step, the effective strain found within the window was averaged. For each window location, the average effective strain in the DIC data set was subtracted from the averaged strain in the models. Finally, an RMS error calculation was performed to obtain the total error from all locations. The RMS error is 0.122 for the model with boundary zones and the RMS error is 0.304 for the model without boundary zones. This is a significant improvement on the model results. By including the harder grain boundary zones, 60% of the error in the model has been eliminated.

4.3. Mesh coarsening effects

The results from the previous section demonstrate that modeling the grain boundary zones has a positive effect on the results. Including the grain boundary zones in the model increases the computational time. The small elements that make up the zone elements cause a significant drop in the maximum time step size for the explicit FE calculations. In order to reduce the computation time of the model, it would be beneficial if the overall size of the grain elements could be increased to coarsen the mesh. The current size of the grain mesh is a function of the size of the small features found along the grain boundaries. If these features could be smoothed, then larger elements could be used along the grain boundaries and the overall mesh could be coarser. In order to smooth the grain boundaries, each triple point was identified, and a parabolic curve was fitted between the two triple points separating each section of the grain boundary. The curves were fitted to the grain boundaries using the statistical software MYSTAT to minimize the error between the parabolic curve connecting the triple points and the grain boundary nodes found from the EBSD data. A series of equal length bar elements were placed along the grain boundaries and the length of the bar elements could be varied to adjust the coarseness of the mesh. Once the bar elements were created, the remaining mesh was created using the same procedure as outlined in Section 3.2.

Two additional meshes were created, one where the same element coarseness as the original mesh was preserved (~80 μm elements) and a second where the average element size was increased to ~180 μm. The smoothed meshes along with the original mesh are presented in Figure 8. These meshes were then subjected to tension similar to the original mesh using the same material properties identified previously. The uniaxial tensile stress vs strain response of these smoothed meshes along with the original mesh and the experimental data are shown in Figure 9. It shows that smoothing the grain boundaries did not significantly affect the macroscopic stress strain response of the model, however coarsening the elements did cause a change in the resulting stress vs. strain curve. Figure 10 shows the effective strain distributions for the three simulated meshes. Even though the overall strain distribution is similar when the grain boundaries are smoothed, the intensities of the strain hot spots are different. The coarse and smoothed meshes have the strain hot spots closer to triple junction points. Thus, the smoothening of the grain boundary and coarsening do have a significant effect on the overall predictions of the model.

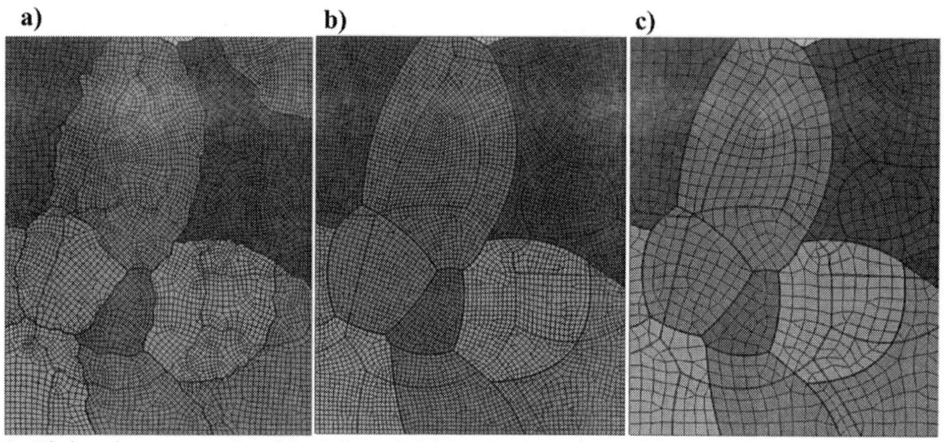

Figure 8: Finite element mesh with a) the original microstructure, (b) smoothed grain boundaries with the original element size (~80 μm), and (c) smoothed grain boundaries with coarsened element size (~180 μm).

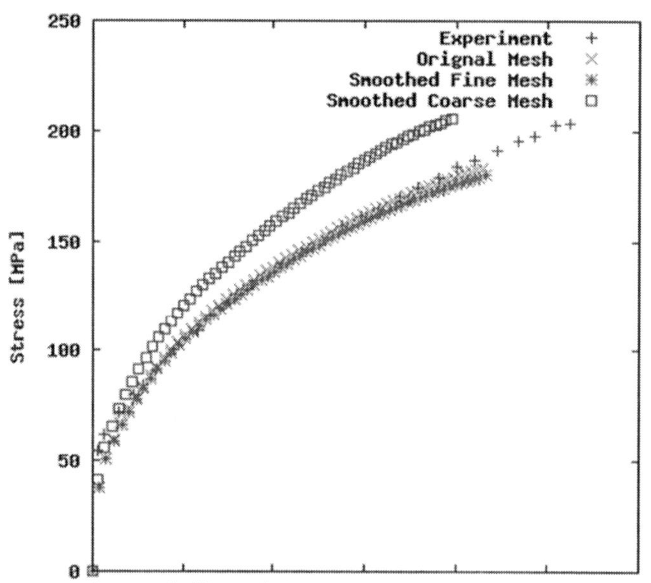

Figure 9: Stress vs Strain response for the polycrystal simulations comparing the effect of smoothing the grain boundaries and coarsening the mesh.

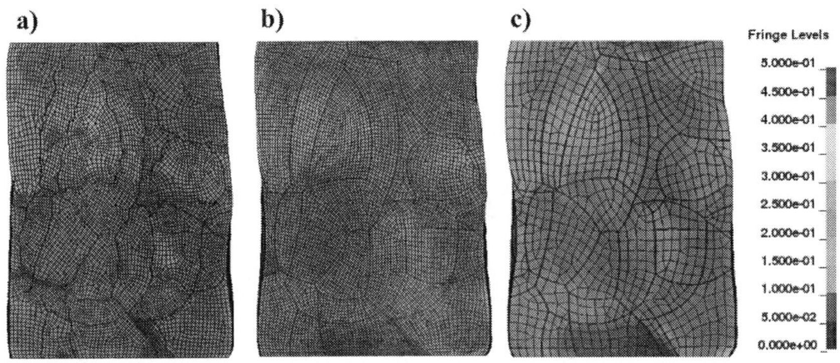

Figure 10: Effective strain distributions for the finite element mesh with a) the original microstructure, (b) smoothed grain boundaries with the original element size (~80 μm), and (c) smoothed grain boundaries with coarsened element size (~180 μm).

In this work, all grain boundary zones were treated as being identical, which might not be reflective as it is expected of boundaries with different grain boundary characters to have different response [24]. Also, this study treats the boundary zones as rectangular region of equal thickness (~20 μm) with hard cut-offs. In reality, this zone exhibits smoother transition, where the influence reduces as one moves farther away from the boundary towards grain centers [31]. Despite these, the models do indicate that the grain boundaries have a significant effect and can be used without the need of other approaches like Hall-Petch etc. to account for the higher hardening of polycrystalline materials as compared to single crystals.

5. Conclusions

A new modeling approach was developed to use single crystal material properties by fitting it to experimental data and translating it to polycrystal models. To achieve this, the grain boundaries were identified, and separate grain boundary zone was created. This zone was assigned separate material parameters. The predictions of the model were validated by comparing predicted results to measured DIC data. Based on the work presented, the following conclusions can be drawn:

- It is feasible to use single crystal constitutive behaviour and extend the predictions to polycrystal model by using separate properties for the grain boundary zone.
- The grain boundary zone has a large impact on the strain partitioning as it is harder than the grain centers and forces the matrix to accommodate more strain.
- Comparison with DIC data shows a 60% reduction in the total error between measured and predicted strain distributions when using the correct grain boundary properties.
- Coarsening of the mesh and smoothening of the grain boundaries had a significant effect on local strain distribution especially in vicinity of the triple junctions. This implies any attempt at speed up of the procedure using these techniques needs to be validated before use.

Acknowledgements

This work was supported by the Natural Sciences and Engineering Research Council of Canada (NSERC) and General Motors of Canada through the Collaborative Research and Development (CRD) grant number CRDPJ401744-10.

References

[1] Taylor G I, 1934 *Proc. Roy. Soc. A.* **165** 362.
[2] Sachs G, 1928 *Z. Ver. Dtsch. Ing.* **12** 134–136.

[3] Peirce D, Asaro R J, and Needleman A, 1982 *Acta Metallurgica* **30** 1087-1119.
[4] Asaro R J, 1983 *In: John W. Hutchinson and Theodore Y. Wu, Editor(s), Advances in Applied Mechanics, Elsevier* **23** 1-115.
[5] Anand L, and Kalidindi S R 1994 *Mechanics of Materials* **17** 223-243.
[6] Roters F, Eisenlohr P, Hantcherli L, Tjahjanto D, Bieler T R, and Raabe D 2010 *Acta Materialia* **58** 1152-1211.
[7] Wittridge N J, and Knutsen R D 1999 *Materials Science and Engineering: A* **269** 205-216.
[8] Dao M, and Li M 2001 *Philosophical Magazine A* **81**(8) 1997-2020.
[9] Muhammad W, Kang J, Brahme A P, Ali U, Hirsch J, Brinkman H, Engler O, Mishra R K and Inal K 2019 *Materials Science and Engineering: A* **753** 179-191.
[10] Muhammad W, Brahme A P, Ali U, Hirsch J, Engler O, Aretz H, Kang J, Mishra R K and Inal K 2019 *Materials Science and Engineering: A* **754** 161-177.
[11] Potirniche G P, Hearndon J L, Horstemeyer, M F, and Ling, X W 2006 *International Journal of Plasticity* **22** 921.
[12] Izadbakhsh A, Inal K, and Mishra R K 2010 *Computational Materials Science* **50** 571-585.
[13] Lim H, Lee M G, Kim J H, Adams B L, and Wagoner R H 2011 *International Journal of Plasticity* **27** 1328-1354.
[14] Klusemann B, Svendsen B, and Vehoff H 2012 *Computational Materials Science* **52**(1) 25-32.
[15] Cheong, S, and Busso E P 2006 *Journal of the Mechanics and Physics of Solids* **54**(4) 671-689.
[16] Raabe D, Sachtleber M, Zhao Z, Roters F, and Zaefferer S 2001 *Acta Materialia* **49**(17) 3433-3441.
[17] Lim H, Carroll J D, Battaile C C, Buchheit T E, Boyce B L, Weinberger C R 2014 *International Journal of Plasticity* **60** 1-18.
[18] Li S, Van Houtte P, and Kalidindi S R 2004 *Modelling and Simulation in Materials Science and Engineering* **12** 845-870.
[19] Delannay L, Jacques P J, Kalidindi S R 2006 *International Journal of Plasticity* **22** 1879-1898.
[20] Erinosho TO, Cocks A C F, and Dunne F P E International Journal of Plasticity **50** 170-192.
[21] Muhammad W, Brahme A, Kang J, Mishra R, and Inal K 2016 *E. Williams (Ed.) Light Metals 2016* 169-174.
[22] Wei Y J, and Anand L 2004 *Journal of the Mechanics and Physics of Solids* **52** 2587 – 2616.
[23] Sfantos G K, and Aliabadi M H 2006 *International Journal for Numerical Methods in Engineering* **69** 1590-1626.
[24] Randle V, Hansen N, and Jensen D J 1996 *Philosophical Magazine A* **73**(2) 265-82.
[25] Zhu G, Hu X, Kang J, Mishra R, and Wilkinson D 2011 *Materials Science and Engineering A* **528** 4187-4198.
[26] Inoko F, Kashihara K, Tagami M, and Okada T 2010 *Materials Transactions* **51** 597-606.
[27] Asaro R J, and Needleman A, 1985 *Acta Metallurgica* **33**, 923-953.
[28] Rossiter J, Brahme A, Simha H, Inal K, and Mishra R K 2010 *International Journal of Plasticity* **26** 1702-1725.
[29] Harren S, Lowe T.C, Asaro R J, and Needleman A 1989 *Philosophical Transactions of the Royal Society of London A* **328** 443-500.
[30] Peirce D, Asaro R J, and Needleman A 1982 *Acta Metallurgica* **30** 1087-1119.
[31] Mishra S, Pant P, Narasimhan K, Rollett A D, and Samajdar I 2009 *Scripta Materialia* **61**(3) 273-276.

IOP Conf. Series: Materials Science and Engineering **967** (2020) 012027 doi:10.1088/1757-899X/967/1/012027

The roles of yield function and plastic potential under non-associated flow rule for formability prediction with perturbation approach

Q Hu[1], J W Yoon[1,2*]

[1] Department of Mechanical Engineering, Korea Advanced Institute of Science and Technology, 291 Daehak-ro, Yuseong-gu, Daejeon 34141, Republic of Korea
[2] School of Engineering, Deakin University, Geelong Waurn Ponds, VIC 3220, Australia

*Corresponding author: j.yoon@kaist.ac.kr, j.yoon@deakin.edu.au

Abstract. In this study, the perturbation approach for predicting material's forming limit strains under non-associated flow rule (non-AFR) is proposed. The influence of yield function and plastic potential function on the forming limit curve (FLC) evaluated by the perturbation approach are discussed through analyzing the normalized growth rate of a perturbation. In the framework of non-AFR, Hill'48 and Yld2000-2d are chosen for AA5754-O. The results show that the left side of FLCs predicted with the different forms of yield function and plastic potential nearly overlap. Hence, it is concluded that the yield function and plastic potential have a negligible influence on the forming limit strain under a negative strain path. However, the FLC under a positive strain path is principally dependent on the relationship between strain ratio β and stress ratio α, which can be determined by the plastic potential. Additionally, in comparison to the FLC under AFR, an increase in the forming limits strain is observed near the plain strain region, when the derivative of the normalized yield function concerning α is positive and vice versa.

1. Introduction

To improve the reliability of the process and the product quality, the instabilities caused by necking and wrinkling should be avoided during sheet metal forming. First of all, the compressive instability like wrinkling is influenced by many factors including mechanical properties of the sheet metal, geometry of the workpiece, and contact conditions. Various analytic methods on wrinkling have been developed based on the bifurcation theory including Kim et al. [1]. Some researchers used the neural network approach and numerical analysis investigated the wrinkling behaviour of sheet metal [2,3]. Recently, there was a good progress for the compressive instability theory for hydroforming process by Chu et al. [4-6]. Second, the tensile instability in sheet metal has been a long research topic. In order to evaluate sheet metal formability, forming-limit diagram (FLD) is generally used, which is usually obtained through Marciniak [7] and Nakazima tests [8]. Leotoing et al.[9] proposed an alternative method by using an in-plane biaxial tensile test for the prediction of FLD. However, these experimental methods are costly and time-consuming. To overcome these shortcomings, several phenomenological models including diffused and localized necking theories [10,11], bifurcation analyses [12,13], imperfection approach (M-K model) [7], perturbation theories [14-16] and modified maximum force criterion (MMFC) [17] have been established for the theoretical forming limit curve (FLC).

The influences of non-linear loading paths on materials' FLCs have been studied by several researchers [15, 18-20] with different theoretical models. To improve the prediction accuracy of these models, significant improvements have been made under associated flow rule (AFR). For example, incorporating more accurate anisotropic yield criteria [21] and hardening models [19]. Stoughton [22] found that under non-associated flow rule (non-AFR), the quadratic yield criteria have a similar anisotropic characteristic predictive capability as the advanced non-quadratic associated anisotropic models. Shen [23] compared the FLCs predicted by the M-K model and MMFC with Hill'48 under AFR and non-AFR. They found that these theoretical models are more accurate under non-AFR than those under AFR. Recently, Hu et al. [15] developed an effective perturbation approach under AFR to predict the material's formability. Hence, it is worth investigating FLC predicted by the perturbation approach under non-AFR.

2. Basic theoretical fundamentals of perturbation approach under non-AFR

The perturbation theory under non-AFR is based on the work of Hu et al. [15,16]. The perturbation approach proposed that when the normalized growth rate of a perturbation $\bar{\omega}$ is negative, the plastic deformation is stable, the perturbation growth accumulation starts at $\bar{\omega}=0$. The failure criterion is proposed as Equation (1). When the perturbation growth accumulation reaches a certain value, the necking is assumed to be well developed and become observable and the strain level in the homogeneous region will be the limit strain as presented in Figure 1.

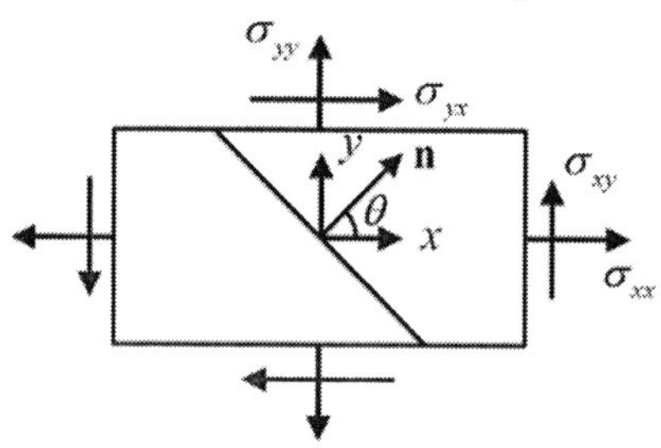

Figure 1. Schematic of strain localization in sheet metal at biaxial tension.

$$\int_{\bar{\varepsilon}_i}^{\bar{\varepsilon}_n} \bar{\omega}(\beta,\bar{\varepsilon})d\bar{\varepsilon} = W \tag{1}$$

where $\bar{\varepsilon}_n$ is the equivalent plastic strain (EPS) at necking, $\bar{\varepsilon}_i$ is the EPS at $\bar{\omega}=0$, W is perturbation growth accumulation, $\bar{\omega}$ is the function of strain path $\beta=\Delta\varepsilon_{yy}/\Delta\varepsilon_{xx}$ and $\bar{\varepsilon}$.

According to the force balance, if a perturbation is applied during deformation, the governing equation for the onset of deformation instability becomes [15,16]:

$$\begin{cases} \dfrac{\partial\tilde{\sigma}_{xx}}{\partial x} + \sigma_{xx}\dfrac{\partial\tilde{\varepsilon}_{zz}}{\partial x} = 0 \\ \dfrac{\partial\tilde{\sigma}_{yy}}{\partial y} + \sigma_{yy}\dfrac{\partial\tilde{\varepsilon}_{zz}}{\partial y} = 0 \end{cases} \tag{2}$$

where $\tilde{\sigma}_{ij}$ (i,j=x,y) are the corresponding perturbations of σ_{ij}, $\tilde{\varepsilon}_{zz}$ is the perturbation of strain along the thickness direction.

According to the [15,16], the perturbations of strain rate components are:

$$\begin{cases} \tilde{\dot{\varepsilon}}_{xx} = \delta v_1^* n_1 (iq)\exp\left(iq(\mathbf{n}\cdot\mathbf{x})+\omega T\right) \\ \tilde{\dot{\varepsilon}}_{yy} = \delta v_2^* n_2 (iq)\exp\left(iq(\mathbf{n}\cdot\mathbf{x})+\omega T\right) \\ \tilde{\dot{\varepsilon}}_{zz} = -\left(\delta v_1^* n_1 + \delta v_2^* n_2\right)(iq)\exp\left(iq(\mathbf{n}\cdot\mathbf{x})+\omega T\right) \\ \tilde{\dot{\varepsilon}}_{xy} = \dfrac{1}{2}\left(\delta v_1^* n_2 + \delta v_2^* n_1\right)(iq)\exp\left(iq(\mathbf{n}\cdot\mathbf{x})+\omega T\right) = 0 \end{cases} \tag{3}$$

The perturbations of the strains can be obtained by integrating the perturbations of strain rate:

$$\tilde{\varepsilon}_{ij} = \int \dot{\tilde{\varepsilon}}_{ij} dT = \frac{1}{\omega} \dot{\tilde{\varepsilon}}_{ij} \tag{4}$$

According to the flow theory of plasticity, the plastic strain rate $\dot{\varepsilon}_{ij}$ can be expressed as:

$$\dot{\varepsilon}_{ij} = \mathrm{d}\lambda \frac{\partial f_p}{\partial \sigma_{ij}} \tag{5}$$

where f_p is a plastic potential function.

Based upon the principle of plastic work equivalence, Equation (6) can be deduced:

$$f_y \dot{\bar{\varepsilon}} = \sigma_{xx}\dot{\varepsilon}_{xx} + \sigma_{yy}\dot{\varepsilon}_{yy} = \sigma_{xx}\left(1+\alpha\beta\right)\dot{\varepsilon}_{xx} \tag{6}$$

where f_y is yield function and $\alpha = \sigma_{xx}/\sigma_{yy}$.

The power type of constitutive law is adopted for strain rate dependent materials as in Equation (7).

$$\bar{\sigma} = \bar{\sigma}_1\left(\bar{\varepsilon}\right)\dot{\bar{\varepsilon}}^m \tag{7}$$

where m is the strain rate sensitivity parameter and $\bar{\sigma}_1(\bar{\varepsilon})$ is a function of $\bar{\varepsilon}$.

Substituting Equations (3)-(7) into Equation (2), the normalized growth rate of a perturbation $\bar{\omega}$ can be deduced after mathematical manipulation.

$$\bar{\omega} = \frac{\omega}{\dot{\bar{\varepsilon}}} = \frac{1}{m}\left[\varphi_y\left(\alpha\right)\frac{\left(\beta+1\right)}{\left(\alpha\beta+1\right)} - \frac{\partial \ln \bar{\sigma}_1\left(\bar{\varepsilon}\right)}{\partial \bar{\varepsilon}}\right], \qquad \tan^2\theta = -\beta \quad \left(\beta \le 0\right) \tag{8}$$

$$\bar{\omega} = \frac{\varphi_y\left(\alpha\right) - \dfrac{\partial \ln \bar{\sigma}_1\left(\bar{\varepsilon}\right)}{\partial \bar{\varepsilon}}\left\{1-\beta\left(\partial\beta/\partial\alpha\right)^{-1}\left(\beta - \dfrac{\varphi_y^{-1}\partial\varphi_y/\partial\alpha}{\varphi_p^{-1}\partial\varphi_p/\partial\alpha}\right)\right\}}{\left(\partial\beta/\partial\alpha\right)^{-1}\beta\dfrac{\varphi_y^{-1}\partial\varphi_y/\partial\alpha}{\varphi_p^{-1}\partial\varphi_p/\partial\alpha} + m\left[1-\beta\left(\partial\beta/\partial\alpha\right)^{-1}\left(\beta - \dfrac{\varphi_y^{-1}\partial\varphi_y/\partial\alpha}{\varphi_p^{-1}\partial\varphi_p/\partial\alpha}\right)\right]}, \qquad \theta=0 \quad \left(\beta > 0\right) \tag{9}$$

Where $\varphi_y\left(\alpha\right) = f_y/\sigma_{xx}$, $\varphi_p\left(\alpha\right) = f_p/\sigma_{xx}$ and θ is the necking angle in Figure 1.

3. Effects of f_y and f_p under non-AFR on perturbation approach

In Section 2, the expressions of $\bar{\omega}$ under non-AFR for positive and negative strain paths have been derived. To study the influence of yield function f_y and plastic potential f_p on the forming limit strains predicted by perturbation approach, the material AA5754-O [24] is adopted. The corresponding material property can be found in Table 1. By fitting the experimental forming limit strains around the plain strain with M-K model under AFR, the strain rate parameter for AA5754-O is set to 0.003. In the present study, the modified Voce hardening law is used as presented in Equation (10). As presented in Figure 2, by fitting the forming limit strains around plane strain with perturbation approach under AFR, the perturbation growth accumulation W is calculated as 2.2677.

$$\bar{\sigma}_1\left(\bar{\varepsilon}\right) = 247.53 - 149.6\left[1-\exp\left(-15.26\bar{\varepsilon}\right)\right] + 150\bar{\varepsilon} \tag{10}$$

Table 1. Material parameters of AA5754-O [24].

σ_0/σ_0	σ_{45}/σ_0	σ_{90}/σ_0	σ_b/σ_0	r_0	r_{45}	r_{90}	r_b
1	0.955	0.964	1.012	0.692	0.859	0.934	1

Here, Hill'48 and Yld2000-2d are used under non-AFR. Hill'48 parameters determined by r-values and yield stresses will be denoted as Hill'48_r and Hill'48_s, respectively. It is found that the term $\beta\left(\partial\beta/\partial\alpha\right)^{-1}\left(\beta - \dfrac{f_p}{f_y}\dfrac{\partial f_y/\partial\sigma_{yy}}{\partial f_p/\partial\sigma_{xx}}\right)$ in Equation (9) is very small, which can be ignored and modified as in Equation (11). The difference between Equation (9) and Equation (11) is very small, which also can be found in Figure 3. Henceforth, Equation (11) is used for $\bar{\omega}$ calculations for β>0.

IOP Conf. Series: Materials Science and Engineering **967** (2020) 012027 doi:10.1088/1757-899X/967/1/012027

$$\bar{\omega} \approx \frac{\varphi_y(\alpha) - \dfrac{\partial \ln \bar{\sigma}_1(\bar{\varepsilon})}{\partial \bar{\varepsilon}}}{m + \beta^2 \left(\partial\beta/\partial\alpha\right)^{-1} \dfrac{\varphi_y^{-1}\partial\varphi_y/\partial\alpha}{\varphi_p^{-1}\partial\varphi_p/\partial\alpha}}, \qquad \theta=0 \quad (\beta>0) \tag{11}$$

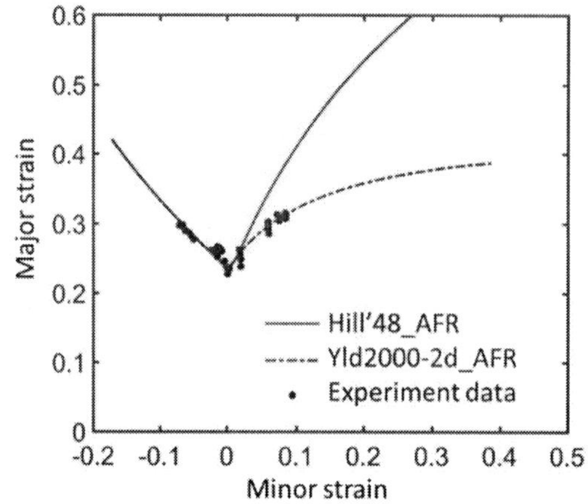

Figure 2. FLC predicted by perturbation approach with Hill'48 and Yld2000-2d under AFR.

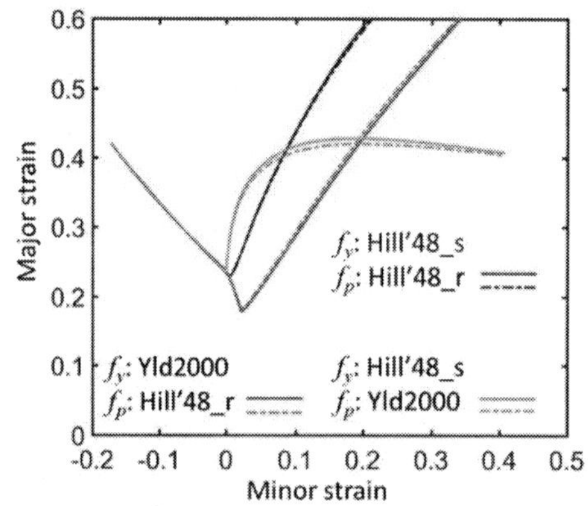

Figure 3. FLC predicted by perturbation approach under non-AFR: solid line by Eq. (9), chain-dotted line by Eq. (11).

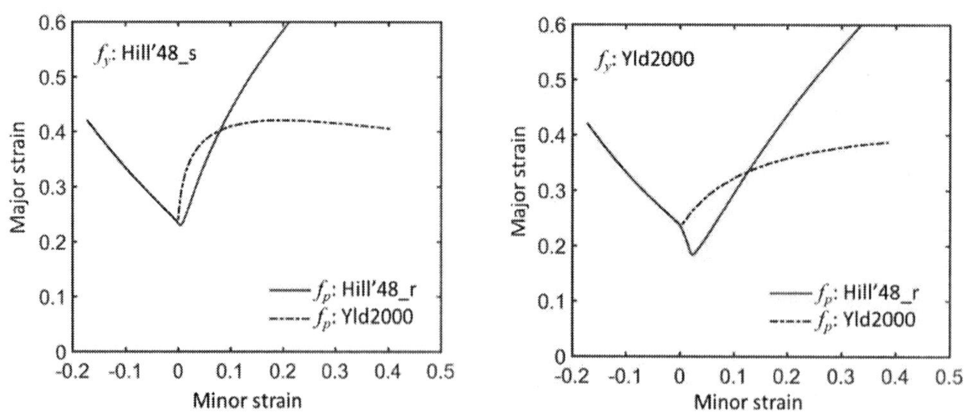

(a) (b)

Figure 4. FLCs predicted by perturbation approach with different f_p: (a) f_y is Hill'48_s, (b) f_y is Yld2000.

Comparison of FLCs predicted by the perturbation approach with different f_p is shown in Figure 4. It is found that if yield criterion is the same, the forming limit strains under negative strain path are almost the same for different plastic potential function. That means f_p does not affect the left side of FLD. For the positive strain path, regardless of the yield function, the forming limit strains near the plain strain region predicted with Hill'48_r is smaller than that of Yld2000. On the contrary, the predictions by the two potential functions near equi-biaxial tension are opposite.

Figure 5 shows the predicted FLCs with different yield function for the same plastic potential function. It's found that the yield function f_y also has no influence on the forming limit strains under negative strain path. For AA5754-O, if f_p is Hill'48_r, the predicted FLCs under a positive strain path decrease in sequence with yield functions of Hill'48_s, Hill'48_r and Yld2000 as presented Figure 5(a). If f_p is Yld2000, the forming limit strains on the right side of FLD predicted with Hill'48_s yield function will be larger than that of the Yld2000.

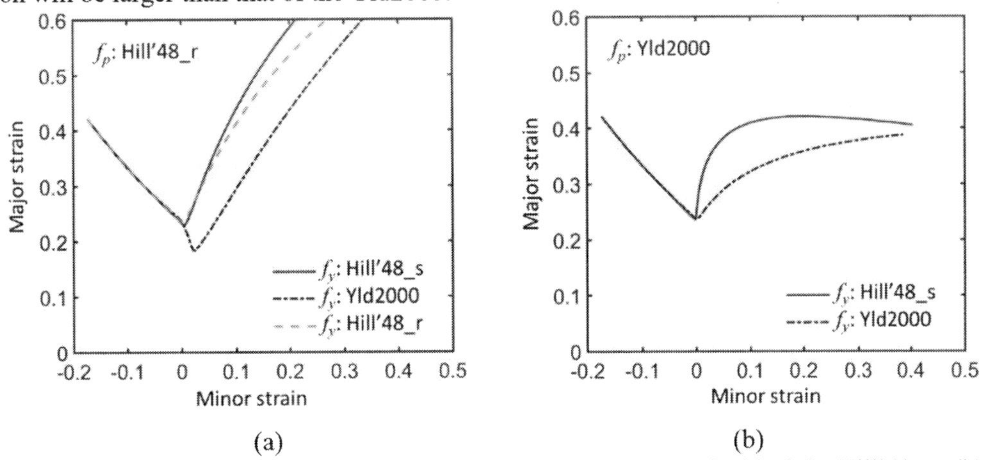

Figure 5. FLCs predicted by perturbation approach with different f_y: (a) f_p is Hill'48_r, (b) f_y is Yld2000.

To investigate the influence of f_y and f_p on perturbation approach, the effects on the normalized growth rate of a perturbation $\bar{\omega}$ are analyzed. For negative strain path, Equation (8) will be used. Under uniaxial tension along the rolling direction, the stress ratio α=0, Equation (8) will reduce to Equation (12), which has no relation with f_y. If f_p accurately predicts the Lankford coefficent along the rolling direction, the strain ratio $\beta|_{\alpha=0}$ will be the same for different f_p. For the plain strain state, β=0, Equation (8) will reduce to Equation (13), which doesn't depend on f_p. $\varphi_y(\alpha)$ is around 1 and it shows little difference for different yield criterion. Hence, for the given hardening law, $\bar{\omega}$ under uniaxial tension and plain strain states will be almost the same for different f_y and f_p. Therefore, f_y and f_p have no influences on the forming limit strains at uniaxial tension and plain strain states according to Equation (1). In addition, FLC on the left side of FLD is almost a straight line, the forming limit strain under negative stain path will be the same for different f_y and f_p as presented in Figure 4 and Figure 5.

$$\bar{\omega} = \frac{1}{m}\left[\left(\beta|_{\alpha=0} + 1\right) - \frac{\partial \ln \bar{\sigma}_1(\bar{\varepsilon})}{\partial \bar{\varepsilon}}\right] \tag{12}$$

$$\bar{\omega} = \frac{1}{m}\left[\varphi_y(\alpha) - \frac{\partial \ln \bar{\sigma}_1(\bar{\varepsilon})}{\partial \bar{\varepsilon}}\right] \tag{13}$$

For a positive strain path, Equation (11) is used. As $\varphi_y(\alpha)$ is around 1, the numerator of Equation (11) is nearly similar for the given hardening law. The term $\beta^2\left(\partial\beta/\partial\alpha\right)^{-1}\dfrac{\varphi_y^{-1}\partial\varphi_y/\partial\alpha}{\varphi_p^{-1}\partial\varphi_p/\partial\alpha}$ in Equation (11) will play an important role. In the framework of AFR, the term will reduce to $\beta^2\left(\partial\beta/\partial\alpha\right)^{-1}$. A larger value of $\partial\beta/\partial\alpha$ under positive strain path will result in larger denominator value in Equation (11).

That's the reason why FLC on the right side of FLD predicted by perturbation approach with Hill'48_AFR is higher than that of the Yld2000_AFR as presented in Figure 2. In the framework of non-AFR, as presented in Figure 5, FLC under positive strain path still principally relies on the term $\beta^2(\partial\beta/\partial\alpha)^{-1}$, i.e. plastic potential function. The term $\Omega = \dfrac{\varphi_y^{-1}\partial\varphi_y/\partial\alpha}{\varphi_p^{-1}\partial\varphi_p/\partial\alpha}$ changes the shape of FLC compared to that of AFR. As presented in Figure 6, if f_p is Hill'48_r, the numerator of Ω under non-AFR will be negative after the point A, i.e. around the plain strain region. A smaller denominator value of Equation (11) will result in larger value of $\bar{\omega}$. Then the forming limit around the plain strain region will become smaller compared that under that of AFR as presented in Figure 5(a). The lowest forming limit in Figure 5(a) depends on the range of $\alpha|_{\partial\varphi_y/\partial\alpha=0} - \alpha|_{\partial\varphi_p/\partial\alpha=0}$. If f_p is Yld2000, Ω under non-AFR will be larger than 10 after the point B. A larger denominator of Equation (11) will result in smaller value of $\bar{\omega}$, which finally leads to the higher forming limit strains around the plain strain region as present in Figure 5(b). From Figure 5, the results show that although normalized yield stress and r-values can be accurately predicted by non-AFR, the predicted FLC would be different with f_y and f_p. The normal vectors of f_y and f_p under the plane stress state play an important role in the predicted FLC.

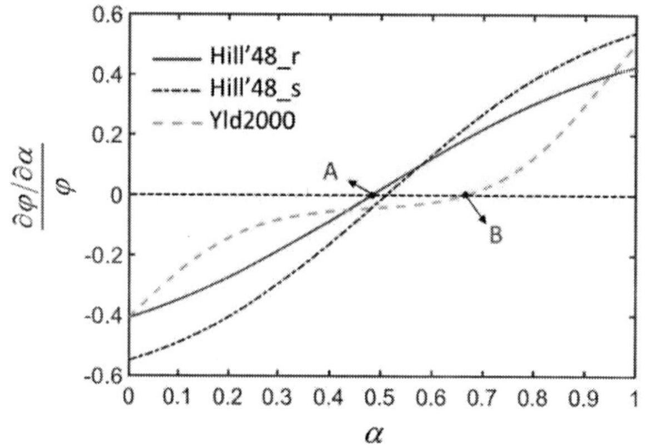

Figure 6. The relationship between $\varphi^{-1}\partial\varphi/\partial\alpha$ and α.

4. Conclusions

The perturbation approach under non-AFR has been deduced. The influences of yield function f_y and plastic potential f_p on FLC predicted by perturbation approach under non-AFR have been systematically studied. The results show that f_y and f_p almost have no influences on the left side of FLD. Although different forms of non-AFR can describe normalized yield stresses and r-values well, FLC under positive strain path still principally relies on the term $\beta^2(\partial\beta/\partial\alpha)^{-1}$. A negative $\partial\varphi_y/\partial\alpha$ around the plain strain state will decrease the forming limit strains and a positive $\partial\varphi_y/\partial\alpha$ around plain strain will improve FLC.

Acknowledgments

The research work is supported by BK21(Brain Korea21) Plus Project.

References

[1] Kim J B, Yang D Y and Yoon J W 2013 *Philos. Mag.* **93** 1914–35
[2] Wang J, Wu X, Thomson P F and Flitman A 2000 *J. Mater. Process. Tech.* **105** 215–20
[3] Youngsuk K, Youngjin, S 2000 *J. Mater. Process. Tech.* **97** 88–94

[4] Chu G N, Chen G, Lin Y L and Yuan S J 2019 *J. Mater. Process. Tech.* **265** 150-157.
[5] Chu G N, Chen G, Lin C Y, Fan Z G and Li H 2020 *J. Mater. Process. Tech.* **275** 116310.
[6] Chu G N, Chen G, Wang G D, Fan Z G and Li H 2020 *Int. J. Mech. Sci.* **165** 105216.
[7] Marciniak Z, Kuczyński K 1967 *Int. J. Mech. Sci.* **9** 609–20
[8] Nakazima K, Kikuma T and Hasuka K 1986 Yawata Tech. Report **264** 8517–30
[9] Leotoing L, Guines D 2015 *Int. J. Mech. Sci.* **99** 21–8
[10] Swift H W 1952 *J. Mech. Phys. Solids* **1** 1–18
[11] Hill R 1952 *J. Mech. Phys. Solids* **1** 19–30
[12] Stören S, Rice J 1975 *J. Mech. Phys. Solids* **23** 421–41
[13] Zhu X, Weinmann K and Chandra A 2001 *J. Eng. Mater. Tech.* **123** 329-33
[14] Li M Y, Zhu X and Chu E 2012 *Int. J. Mech. Sci.* **64** 273–9
[15] Hu Q, Zhang L, Ouyang Q, Li X, Zhu X and Chen J 2018 *Int. J. Plast.* **103** 143–67
[16] Hu Q, Li X and Chen J 2019 *Int. J. Mech. Sci.* **155** 187–96
[17] Hora P, Tong L 2008 *Proc. of the NUMISHEET 2008 Conf.* (Interlaken, Switzerland) p 205–10
[18] Stoughton T B, Zhu X 2004 *Int. J. Plast.* **20** 1463–86
[19] Yoshida K, Kuwabara T 2007 *Int. J. Plast.* **23** 1260–84
[20] Chung K, Kim H and Lee C 2014 *Int. J. Plast.* **58** 3–34
[21] Banabic D, Comsa S, Jurco P, Cosovici G, Paraianu L and Julean D 2004 *J. Mater. Process. Tech.* **157** 23–7
[22] Stoughton T B 2002 *Int. J. Plast.* 18 687–714
[23] Shen F, Münstermann S and Lian J 2019 *J. Mater. Process. Tech.* 116384
[24] Wang H, Yan Y, Han F and Wan M 2017 *Int. J. Mech. Sci.* **133** 147–66

Description of stress-strain response and evaluation of the formability for Al-Cu-Mg alloy under the impact hydroforming

Da-Yong Chen[1,2], Shi-Hong Zhang[1,*], Yong Xu[1], Yan Ma[1], Hong-Wu Song[1], Liang-Liang Xia[1,2]

1. Shi-changxu Innovation Center for Advanced Materials, Institute of Metal Research, Chinese Academy of Sciences, Shenyang, 110016, P.R. China
2. School of Materials Science and Engineering, University of Science and Technology of China, Shenyang 110016, China

* shzhang@imr.ac.cn

Abstract. This study aims at investigating on the descriptions of stress-strain response and the assessment of formability for Al-Cu-Mg alloy under impact hydroforming (IHF). IHF process effectively combines the advantages of the flexible liquid and the impact impulse. The formability of low plastic materials can be increased by using IHF process. The elongation of Al-Cu-Mg alloy can even be more than 50 % compared with the quasi-static forming means (27 %~30 %) if the strain rate was on the level of 5000/s. The stress-strain curve is characterized with S-shape. It is found that the strain hardening rate decreases linearly at the stage III, increases linearly at the stage IV and decreases non-linearly at the stage V. The deformation mechanism is the interaction of dislocation accumulation and dynamic recovery under the transmission from non-sheared to sheared precipitation when the strain rate is on the level of 3000/s~5000/s. A modified Kocks-Mecking model was established to describe the mechanical property of AA 2B06. In addition, the high strain rate formability was evaluated by the new means which revealed the relationship between the impact energies, deep drawing height ratios and the deep drawing ratios by using the finite element modelling of the solid liquid coupling.

Keywords: Al-Cu-Mg alloy, impact hydroforming, stress-strain response, modified Kocks-Mecking model, high strain rate formability

1. Introduction

Nowadays, high speed forming methods were attracting more and more attention for they contributed obviously to the increase of the formability especially for some hard-to-form materials, such as Aluminum alloy, Titanium alloy, et al [1-3]. Impact hydroforming was one classical means of high speed forming which possessed the features of flexible effect and impact wave loading [4]. This technology was an effective and potential method for manufacturing the parts used in the fields of aeronautics and astronautics, automobile which characterized with large deformation, locally small and complex structure [5,6].

It was found that the improvement of the thickness strain could be 62% for AA 5A06 by using the IHF according to the investigation of Ma, et al [4]. In addition, the novel equipment was designed and fabricated based on the investigation on the wave propagation and the fluid-solid coupling. And the aluminum part was successfully manufactured even it has complex structure and double frame by using the IHF which cannot be manufactured by the conventional dropping press and conventional hydroforming [4]. Moreover, the formability of AA 5A06 was evaluated by the means of forming limit curve (FLC) [7].

Formability evaluation was very useful especially for the sheet parts under dynamic forming. However, IHF characterized with both high strain rate and flexible contact which increased the complexity of analyzing the procedure of sheet forming. The procedure was hybrid of deep drawing and hydro-bulging which increased the difficulty of the formability evaluation only by the classical method of forming limit curve [7]. Furthermore, it had to be demonstrated that the contribution of high speed and flexible loading on the formability enhancement, respectively.

The AA 2B06 was a typical Al-Cu-Mg alloy which could be used widely on the structural parts with complicated structure in the field of aeronautics and astronautics. Its formability was very limit only using the conventional forming method such as dropping hammer, stamping and so on. So, the impact hydroforming was an effective method to increase the formability of the Al-Cu-Mg alloy.

Hence, the AA 2B06 alloy was investigated from the aspects of mechanical behavior analysis and characterization based on the modified Kocks-Mecking model. Furthermore, a novel formability evaluation was established by using the finite element modeling based on the solid liquid coupling.

2. The mechanical behavior under impact loading

AA 2B06 is employed to investigate the stress-strain response under the impact loading. It is similar with the AA 2A06 from the aspect of mechanical property and the yield stress is about 58 MPa under the condition of O state. In this section, the mechanical behavior is studied under both the quasi-static and dynamic loading conditions.

2.1. *The stress-strain curve*

The Electronic universal testing machine and Hopkinson split tensile bar are used to test the quasi-static and dynamic tensile mechanical properties. The stress-strain curves are illustrated in figure 1, which give the difference between the quasi-static and the dynamic loading. The yield and tensile stress are 58 MPa and 168 MPa, respectively. The tensile stress increased to about 175 MPa when the strain rate is 1293/s, and it continually increases when the strain rates are enhanced to 5045/s. From the figure 2, it is found that the elongation increases obviously when the strain rate increases from 3124/s to 3499/s which shows the sensitive range of plasticity enhancement. Moreover, the elongation increases almost linearly when the strain rates are in the range of 1364 /s~3124 /s. In addition, the stress-strain curve characterizes with S shape when it is under the dynamic loading.

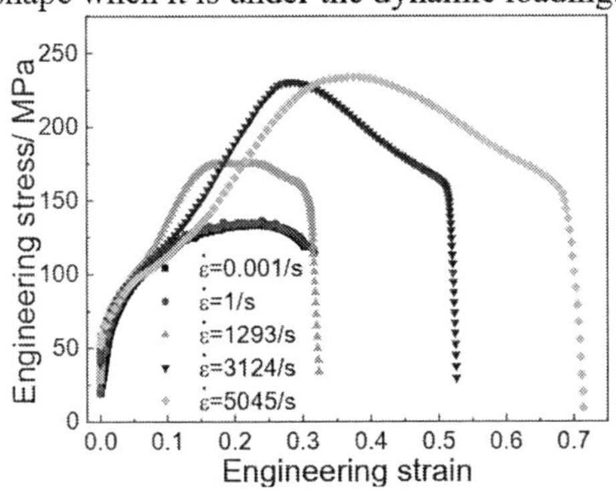

Figure 1. Stress-strain curve of AA 2B06 under both quasi-static and dynamic loading.

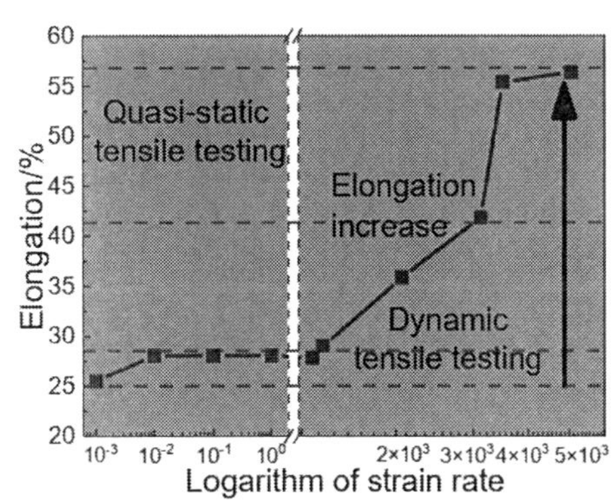

Figure 2. Elongation increase under dynamic loading with quasi-static loading.

2.2. *The necking points measurement*

The method of Sato is referenced to measure the necking points [8]. The basic principle is judging the beginning point of the obvious change of the strain and strain rate. The means of DIC (Digital image correlation) is employed to measure strain distribution under different strain rates. It is illustrated in figure 3 showing the variation of strain and strain rate with time while taking the sample of 5045/s for example. The strain of axial direction vs strain of transverse direction is illustrated in figure 3(c). As for the center area, the transverse strain increases linearly with the strain of rolling direction when the major strain is less than 0.27. However, two kinds of strain change non-linearly when the strain is larger than 0.27, which demonstrates the beginning of the necking in the local area. It is also found from the results of strain and strain rates as illustrated in figure 3 (a) and (b), the increasing rates of strain and strain rate are different for center area when the time is about 160 μs.

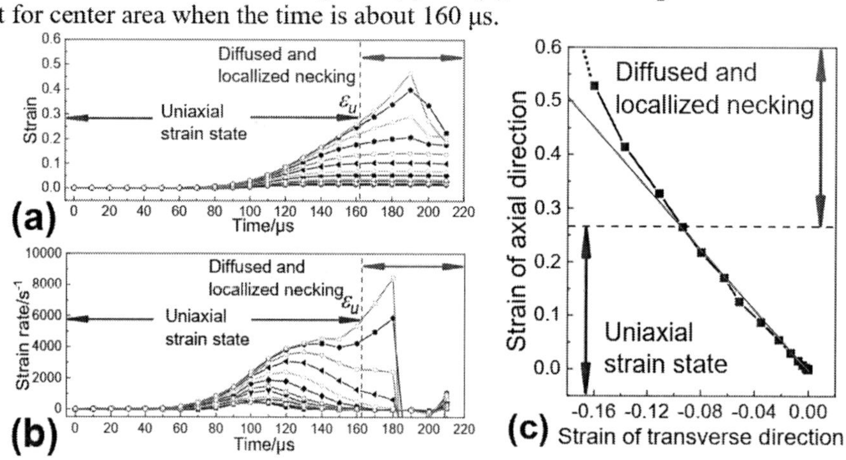

Figure 3. Determination of the necking point: (a) strain vs time, (b) strain rate vs time, (c) strain of axial direction.

The same method is used to measure the necking points at different strain rates. Then, the whole strain is divided into two part including the strains before and after the necking point. The scale of the range after the necking is calculated and listed in table 1. It is found that the strain of necking increases gradually with the strain rate especially when it is larger than 3124/s. Furthermore, it is about 28% for the scale of the strain after the necking point when the specimen is deformed in quasi-static condition. The scale increases prominently when the strain rate exceeds the level of 3000/s, which is more than 50%. It is illustrated that the material have better ability to resist the necking when the strain rates are in the range of 3000/s~5000/s.

Table 1. Calculation of the necking points.

Strain rate (/s)	The strain at the necking point	The scale of the strain after the necking point (%)
0.001	0.193	29.82
0.1	0.203	28.01
1293	0.172	42.47
3124	0.216	55.07
5045	0.271	58.31

3. The description of stress-strain curve

3.1. *The strain hardening rate curve*

As it is illustrated in figure 4, the stress-strain curves of different strain rates are extrapolated according to the method of Deng et al [9]. By this way, the hardening behavior is obtained especially when the strain is larger than the value of necking point. For the quasi-static loading, the difference is very limited compared with the original curve, however, it is obvious as for the curves under dynamic loading. This also illustrates the softening procedure under the high strain rates.

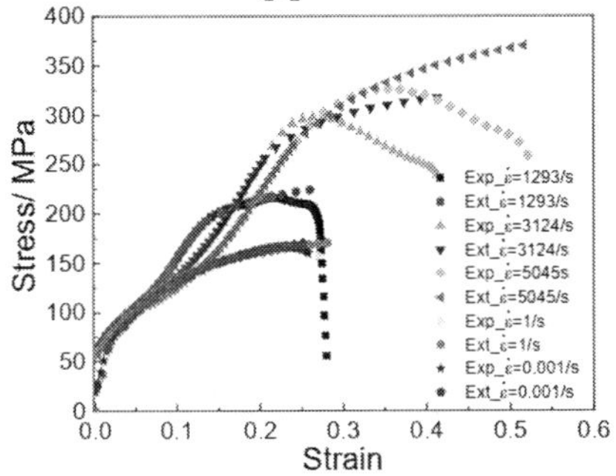

Figure 4. True stress-strain curve obtained by extrapolated method.

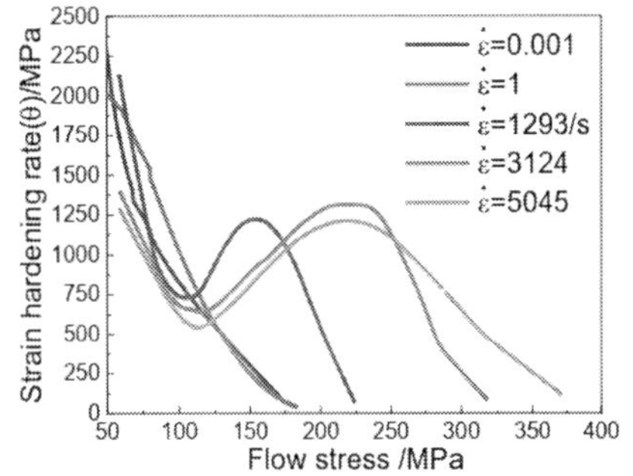

Figure 5. Strain hardening rate curves under different strain rates.

3.2. The description by modified Kocks-Mecking model

The solutes have influence on the work hardening rate according to the investigation of Bouaziz, et al [10]. The Al-Cu-Mg alloy is used in this study, the special phenomenon of strain hardening with S-shape should be studied and explained. The purpose is to establish a model that shows all hardening stages.

The strain hardening rate is defined as θ, which stands for the differentiate of stress and strain.

$$\theta = d\sigma/d\varepsilon \tag{3.1}$$

The evolution of dislocation with strain for stage III is illustrated as follows according to the classical Kocks Mecking model [11,12]:

$$\theta_{III} = B_{III} - M_{III}\sigma \tag{3.2}$$

where θ_{III} stands for the strain hardening rate, B_{III} and M_{III} are the constant.

At the stage IV, the strain hardening rate is as follows:

$$\theta_{IV} = B_{IV} - M_{IV}\sigma \tag{3.3}$$

where θ_{IV} stands for the strain hardening rate, B_{IV} and M_{IV} are the constant.

As for the stage V, the strain hardening rate decreases non-linearly. The assumption that $d\rho/d\varepsilon \sim -\rho^2$ is made to associate with climb-controlled recovery and the strain hardening rate $\theta_V = -C(\sigma - \sigma_V)^3$. Suitable switch functions H_1 and H_2 are defined to combine the model[12].

$$H_1(x) = \frac{1}{1 + e^{-2cx}} \tag{3.4}$$

and

$$H_2(x) = \frac{1}{1 + e^{-2mx}} \tag{3.5}$$

Therefore, the model can be combined as follows when it is in the $\theta - \sigma$ curve [12]:

$$\theta(\sigma) = \theta_{III}\big(1 - H_1(\sigma - \sigma_{IV})\big) + \theta_{IV}H_1(\sigma - \sigma_{IV}) + \theta_V H_2(\sigma - \sigma_V) \tag{3.6}$$

According to the experiment results, it is found that σ_V changed with strain rate which in the means of Asymptotic equation. The function can be illustrated as follows:

$$\sigma_V = a_{sr} - b_{sr} * c_{sr}^{\dot{\varepsilon}} \tag{3.7}$$

where, a_{sr}, b_{sr} and c_{sr} are corresponding constant.

The flow stress can be integrated according to the relationship of strain hardening rate and stress as follows:

$$\sigma = \int \theta d\varepsilon \tag{3.8}$$

Hence, the relationship of flow stress and strain is as follows:

$$\sigma = \int \begin{bmatrix} (B_{III} - M_{III}\sigma)\left(1 - \frac{1}{1+e^{-2cx}}\right)(\sigma - \sigma_{IV}) + (B_{IV} - M_{IV}\sigma)\frac{1}{1+e^{-2cx}}(\sigma - \sigma_{IV}) \\ -f\left(\sigma - \left(a_{sr} - b_{sr} * c_{sr}^{\dot{\varepsilon}}\right)\right)^3 \frac{1}{1+e^{-2mx}}\left(\sigma - \left(a_{sr} - b_{sr} * c_{sr}^{\dot{\varepsilon}}\right)\right) \end{bmatrix} d\varepsilon \quad (3.9)$$

The fitting is made on the strain hardening rate results under different strain rates, and the parameters are illustrated in Table 2.

Table 2 The parameters and relative values.

Parameters	B_{III}	M_{III}	c	m	B_{IV}	M_{IV}
Value	3930.2	24.4	0.03	80.5	-395.9	-8.6
Parameters	f	a_{sr}	b_{sr}	c_{sr}	m	
Value	0.0026	233.6	295.6	0.99918	0.03	

The $\theta - \sigma$ curve is illustrated in figure 5. It is found that the strain hardening rate decreases almost linearly when the strain rates are on the level of quasi-static (0.001/s and 1/s, respectively). However, the work hardening rate decreases firstly at the stage III, increases again during stage IV and decreases again to the end when the strain rates are in the range of 1000/s~5500/s.

According to the microstructure investigation by TEM, the material characterizes with precipitation of S phase (Al_2CuMg) which has high strength than the matrix. Hence, the precipitate cannot be sheared under the quasi-static loading but will be sheared under the dynamic loading which will provide enough internal stress for the hard phase. In the first stage, the precipitates become the obstacles of dislocation which will contribute as the storage by means of geometrically necessary dislocations when the strain is small. During the second stage, the interaction of dislocation accumulation and annihilation/dynamic recovery is the main procedure because of the shear of precipitate under large strain and high strain rate. The method of Cheng et al [13] is referred to determine the series of parameters. The constitutive model is established finally according to the parameters obtained.

4. The formability evaluation based on the solid-liquid coupling modeling

The impact hydroforming characterizes with flexible liquid and impact loading which is an obvious process of solid-liquid coupling. Hence, the finite element modeling is an available and useful method to be used to investigate the IHF procedure. In the first section, what is discussed is the corresponding parameters of impact hydroforming and the models that used for the simulation.

4.1. Impact hydroforming principle and modeling configuration

The principle of the impact hydroforming is illustrated as follows: The projectile is accelerated by the energy source in the manner of pressured gas or compressed liquid; and then the liquid is stimulated by the moved projectile; finally, the blank is formed by the liquid with high pressure and velocity. The principle of this process and the crucial parameters are illustrated in figure 6. The deep drawing ratio is defined to describe the intensity of the deformation as follows:

$$DR = \frac{D_i}{d_p} \quad (4.1)$$

where, DR stands for deep drawing ratio, and the limit value (LDR) represents deep drawing ratio that the blank can be formed completely without the area of flange, D_i indicates the diameter of the initial blank, and the effective diameter of liquid d_p has the relationship with the wall thickness t_b:

$$d_p = D_i - 2 * t_b \quad (4.2)$$

On the other hand, another parameter is needed to distinguish the difficulty of the deformation under the same deep drawing ratios. Hence, the parameter of deep drawing height ratio is defined to describe the degree of the height of deep drawing as follows:

$$DHR = \frac{h_f}{d_p} \quad (4.3)$$

where, DHR stands for the deep drawing height ratio, h_f represents the distance between the upper surface of flange and the lowest point of the deformed cup. The ultimate value of this parameter is defined as the limit deep drawing height ratio which is abbreviated as $LDHR$. DHR will be very useful when the forming process is conducted under the condition of uncomplete deep drawing.

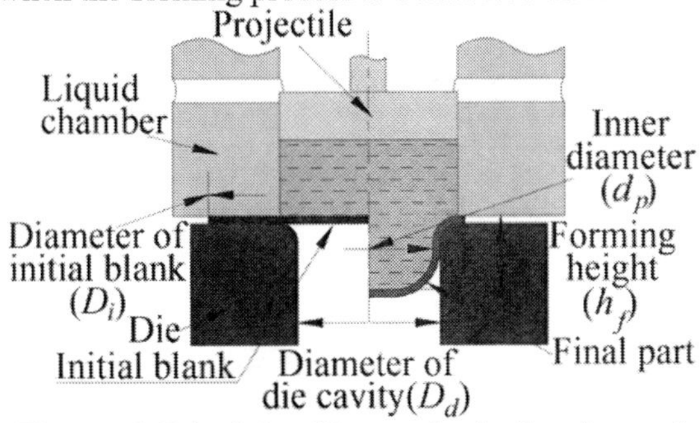

Figure 6. Principle of impact hydroforming and definition of the critical parameters.

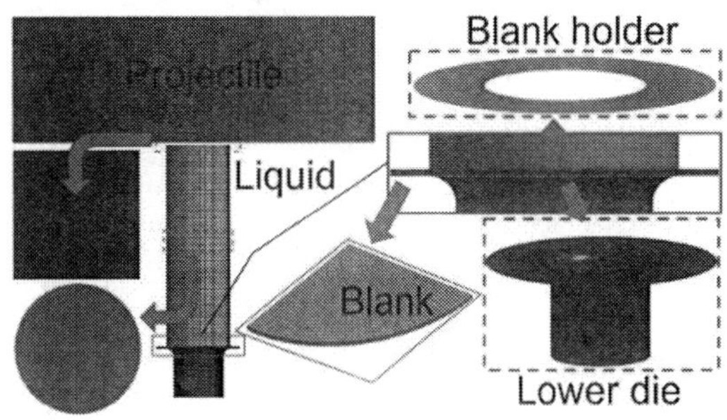

Figure 7. Model meshing for the solid-liquid coupling [5].

As it is illustrated in figure 7, the model meshing is introduced for the solid-liquid coupling finite element modeling (SLC-FEM). The projectile and liquid are treated as solid part and the same element is used with the blank from the view of vertical projection. The mass of the projectile is 60 kg, while the diameter is 250 mm. The height of the liquid is 152 mm, the effective diameter is 48.5 mm and the vertical element size is 1.2mm. As for the blank, the eight-node hexahedron solid element is employed in the Fluid-Structure Interaction (FSI) model. Furthermore, three layers are selected to investigate the thick gradient of strain and stress and the vertical size is set as 0.4mm. The blank holder and the lower die are treated as rigid parts by using the four-node quadrilateral shell element. The radius of the die is set as 5 mm.

The simplified Johnson-Cook constitutive model is employed to describe the mechanical behavior of the sheet. The keywords "MAT_SIMPLIFIED_JOHNSON_COOK_ORTHOTROPIC_DAMAGE" is used to considered the damage of the material. Directional damage begins after a defined failure strain is reached in tension and continues to evolve until a tensile rupture strain is reached in either one of the two orthogonal directions [14]. Gruneisen equation of state is utilized to express the dynamic behavior of the liquid which can correctly describe the characteristic of the shock wave [5].

4.2. *Solid-liquid coupling modeling and the results*

The formability is investigated by figuring out the relationship between the impact energy and forming ratio, meanwhile, the deep drawing height ratio and deep drawing ratio as for IHF procedure. Therefore, series of blanks with different diameters are formed under different impact energies to make them be the situation of complete, uncomplete deep drawing and crack. All the parts used to establish the formability curve are listed as in figure 8. It is found that the blank can be completely formed if the deep drawing ratio is not more than 1.77. And the blank has the possibility to be completely formed when the energy in the suitable range, it also possessed the probability to crack if the energy is too high when the forming ratio is on the level of 1.95. In addition, the blank has not any possibility to be completely formed if the deep drawing ratios are in the range of 2.13 and 2.31. The formability curve is established according to the impact energies and deep drawing height ratio which will be discussed later [5].

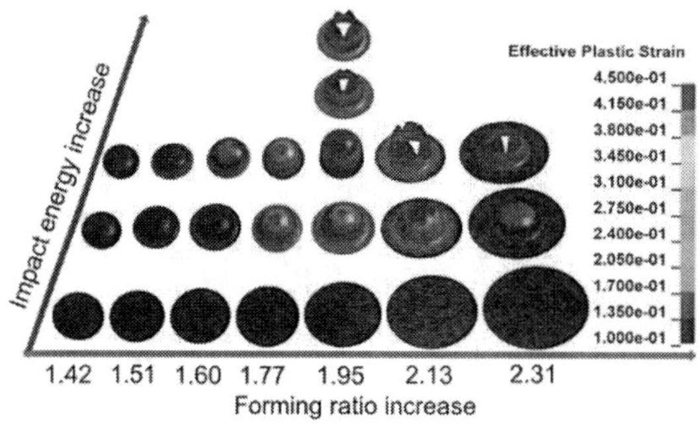

Figure 8. Parts used for establishing the curve of formability evaluation [5].

Three different deformation areas are distinguished according to the condition of the parts as shown in figure 9. They are incomplete deep drawing area, complete deep drawing area and crack area. In the same time, it is found that the limit forming ratio is 1.99 at the intersection of the two boundaries of three areas and the relative energy is 2207.2 kJ/m^2. Before the limit value, the forming energy increases with the deep drawing ratio, and it decreases with exponential trend when the forming ratio is larger than limit value [5].

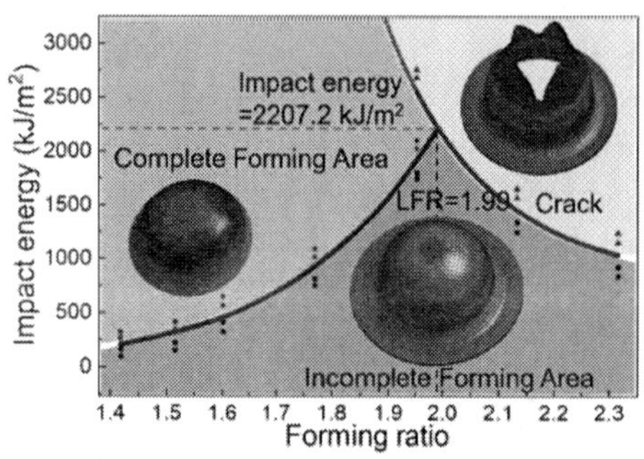

Figure 9. Formability evaluation curve describing the relationship of impact energy and deep drawing ratio.

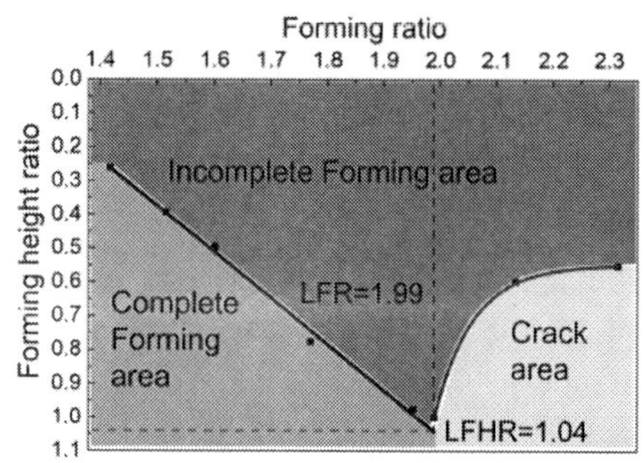

Figure 10. Formability evaluation curve describing the relationship of deep drawing height ratio and drawing ratio.

Furthermore, as illustrated in figure 10, the formability curve is established to express the relationship of deep drawing height ratio and deep drawing ratio. It is found that the limit deep drawing height ratio is 1.04 when the limit forming ratio is just 1.99. Before the value of 1.04, the DHR increases linearly with the DR, however it decreases exponentially after the limit value [5].

5. Conclusions

(1) The elongation increases obviously when the strain rates are in the level of 3000~5000/s and the material has a better ability to endure the local deformation after the necking point.

(2) The work hardening rate characterizes with decreasing linearly firstly, increasing secondly and then decreasing again with non-linear trend. A modified Kocks-Mecking model is used to describe the feature of the particular variation of flow stress.

(3) The formability can be evaluated by the means of Solid-Fluid Coupling FEM for impact hydroforming and the formability curves are established to evaluate the relationship between impact energies, the deep drawing height ratios and deep drawing ratios.

Acknowledgments

Authors wish to acknowledge the funding support from National Natural Science Foundation of China (No. 51875548), Natural Science Foundation of Liaoning Province of China (Grand No. 20180550851),

Project for transfer and transformation of scientific and technological achievements, Chinese Academy of Sciences, Henan Province (No. 2020204).

References

[1] Balanethiram V S, Hu X Y, Altynova M and Daehn G S 1994 J. Mater. Process. Technol. 45 595-600.
[2] Kosing O E and Skews B W 1998 Int. J. Impact Eng. 21 801-16.
[3] Lang L H, Li T, An D Y, Chi C L, Nielsen K B and Danckert J 2009 Mater. Sci. Eng. A 499 320-4.
[4] Ma Y, Xu Y, Zhang S H, Banabic D, Abd El-Aty A, Chen D Y, Cheng M, Song H W, Pokrovsky A I and Chen G Q 2018 CIRP Ann. Manuf. Technol. 67 281-4.
[5] Chen D Y, Xu Y, Zhang S H, Ma Y, Abd El-Aty A, Banabic D, Pokrovsky A I and Bakinovskaya A A 2020 J. Mater. Process. Technol 10. 1016/j. jmatprotec. 2019. 116553.
[6] Abd El-Aty A, Xu Y, Zhang S H, Ha S, Ma Y, Chen D Y J. Adv. Res. 18 19-37.
[7] Banabic D, Lazarescu L, Paraianu L, Ciobanu I, Nicodim I and Comsa D S 2013 CIRP Ann. Manuf. Technol. 62 255-8.
[8] Sato K, Qiang Y, Hiramoto J, Urabe T, Yoshitake A 2015 Int. J. Impact. Eng. 75: 11-26.
[9] Deng S Y, Song H W, Zheng C 2019 Int J Mater Form 12(2) 321-29.
[10] Bouaziz O, Barbier D, Embury J D and Badinier G 2013 Philos. Mag. 93 247-55.
[11] Mecking K and Kocks U F 1981 Acta. Metall. 29 1865-75.
[12] Bambach M 2013 Acta Mater 61 6222-33.
[13] Cheng L M, Poole W J, Embury J D, Lloyd D J 2003 Metall. and Mat. Trans. A. 34A 2473-81.
[14] LS-DYNA 2017 Livermore, CA: Livermore Soft-ware Technology Corporation (LSTC).

A pressure-coupled Drucker function for plasticity and fracture modelling of AA5182

Chong Zhang[1], Yue Wang[1], Yanshan Lou[1,*], Lei Fu[2], Saijun Zhang[3], and Jeong Whan Yoon[4]

[1] School of Mechanical Engineering, Xi'an Jiao Tong University, 28 Xianning West Road, Xi'an, Shaanxi 710049, China
[2] CHINALCO Materials Application Research Institute Co. Ltd, Beijing 102200, China
[3] School of Mechanical and Automotive Engineering, South China University of Technology, Guangzhou 510640, China
[4] Department of Mechanical Engineering, Korea Advanced Institute of Science and Technology (KAIST), 291 Daehak-ro, Yuseong-gu, Daejeon, 305-701, Republic of Korea

*ys.lou@xjtu.edu.cn

Abstract. In this paper, a pressure-coupled Drucker function is proposed to model the plastic deformation and fracture from shear to plane strain tension of AA5182 sheet. Experiments are conducted for AA5182 in shear, uniaxial tension and plane strain tension, and the force-stroke curves are measured during the tests. The plastic deformation is modeled by the Drucker function. The Drucker function is modified to consider the pressure effect for fracture limit stress prediction of the alloy. The pressure-coupled Drucker function for plasticity and fracture is calibrated by an inverse engineering approach. The calibrated plasticity models are then applied to numerical prediction of plastic deformation of the alloy under various loading conditions. It is observed that the pressure coupled Drucker function accurately describes the plastic deformation under large plastic deformation as well as the onset of ductile fracture under these loading conditions.

1. Introduction

Numerical simulation is now widely used in the design of tools and processes in sheet metal forming. The reliability of simulation calculations largely depends on the accuracy of material models implemented in simulations to describe the plastic behaviour of sheet metals. Accurate description of plastic yielding, strain hardening and fracture are essential to accurately model plasticity during various sheet metal forming processes. There are a class of yield functions which are formulated in forms of the stress invariants. Examples of these yield functions are referred in [1-9]. Cazacu and Barlat extended the Drucker function into an anisotropic form [2]. Lou and Yoon calibrated the Drucker function parameters for BCC and FCC metals [9]. For hardening behaviour evaluation, several hardening models including Swift, Voce, and Swift-Voce model have been widely used [10-12]. In addition, in the last decade, various fracture criteria [13-18] are also proposed for failure prediction of metals. Lou proposed the DF2012 ductile fracture criterion [13] and extended it in 2014 [14] and 2017 [16].

In this study, a yield function is proposed as an equation of the three stress invariants based on the Drucker function to take into account the pressure sensitivity. The pressure-coupled Drucker yield function is combined with the Swift-Voce hardening law to model the plasticity of AA5182 under various loading conditions of shear, uniaxial tension and plane strain tension. Fracture is also modelled by the pressure-coupled Drucker function for the alloy. The predicted force-stroke curves and the fracture strokes are compared with experimental results to check the accuracy of the pressure-coupled Drucker function for plasticity and fracture.

2. Experiments

Four kinds of specimens with different shapes for different stress states, including dog-bone specimens I, specimens with central hole II, notched specimens III and in-plane shear specimens IV, are designed to study the plasticity and fracture properties of an aluminium alloy of AA5182 with a thickness of 1.26 mm as illustrated in Figure 1. The designed specimens are tested on an electronic universal tensile testing machine of INSTRON 5182 at quasi-static condition with the testing velocities presented in [19]. Figure 1 also shows the initial gauge lengths for stroke measuring by Digital Image Correlation (DIC) of each specimens.

At least 3 experiments were conducted for dog-bone specimens along the rolling direction (RD), diagonal direction (DD), and transverse direction (TD), as shown in Figure 2(a) with good repeatability. The r-values, yield stresses and coefficients of the Swift-Voce law were measured and calibrated for dog-bone specimens along three directions as shown in Table 1. The r-values are calculated by Eq. (1). It is observed that the strength anisotropy of AA5182 in the unidirectional tensile state is not significant. Figure 2(b) shows that both Swift, Voce, and Swift-Voce hardening laws match with the experimental result very well. However, there are big differences between these hardening laws under large deformation after the necking strain.

$$r = \frac{\varepsilon_w^p}{\varepsilon_t^p} = \frac{\varepsilon_w^p}{-\varepsilon_w^p - \varepsilon_l^p} \tag{1}$$

where $\varepsilon_w^p$and ε_l^p is the transverse and longitudinal engineering strain of dog-bone specimen, respectively.

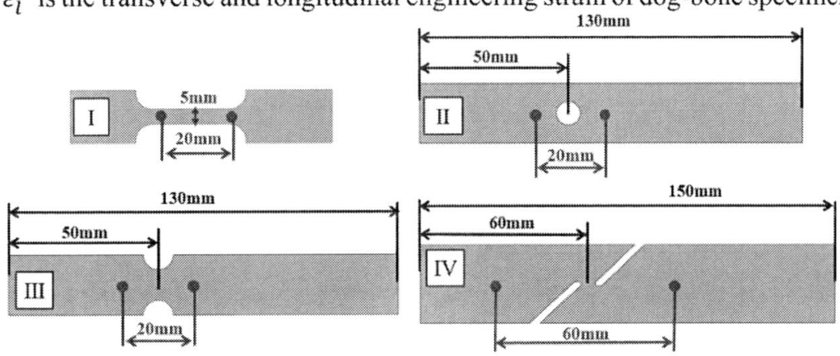

Figure 1. Four specimens for different stress state: (I) dog-bone specimens; (II) specimens with centra l hole; (III) notched specimens; and (IV) in-plane shear specimens

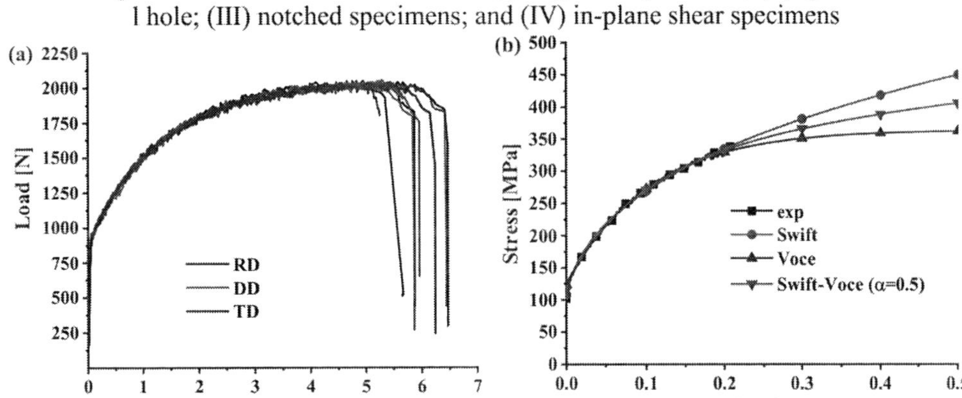

Figure 2. The experiment results of dog-bone specimens: (a) load-stroke curves along three direction; and (b) comparison of stress-strain curve with the Swift, Voce, and combined Swift-Voce laws

Table 1. The R-values, yield stresses and coefficients of the Swift-Voce law calibrated for dogbone specimens along three directions

Loading direction	r-value	Yield stress at 0.2%	Swift			Voce		
			K	e0	n	A	B	C
00	0.527 (#2)	169.8	563	0.0067	0.33	363.9	123.6	9.66
45	0.557 (#1)	166.6	561.5	0.0063	0.33	367.0	122.0	9.45
90	0.5369 (#4)	171	554.7	0.0075	0.324	363.3	129.2	9.44

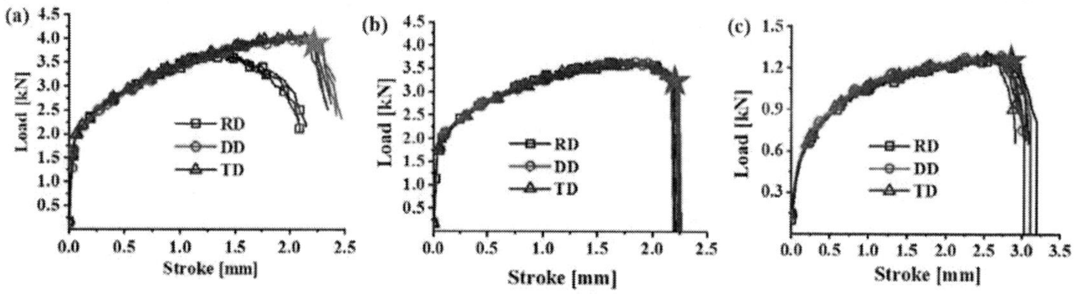

Figure. 3. The load-stroke curves along three direction: (a) specimens with central hole; (b) notched s pecimens; (c) in-plane shear specimens

The load-stroke curves of specimens with central hole, notched specimens, and in-plane shear specimens along three directions are plotted in Figure 3 and the experimental onsets of ductile fracture are indicated by solid pentagrams. For specimens with a central hole, force drops very early along RD compared with those along TD and DD. However, fracture is not observed at load drop along RD and there is no obvious difference observed during the tests for the hole specimens along RD. The mechanism is not clear currently. For other cases, the load-stroke curves show that the anisotropy in strength and fracture is not apparent for this metal. Accordingly, the metal is assumed to be isotropic.

3. Constitutive equations for plasticity and fracture

Lou et al. [20] modified the Drucker yield function to consider the pressure effect as below:

$$\bar{\sigma}(\sigma_{ij}) = a\left(bI_1 + (J_2^3 - cJ_3^2)^{1/6}\right) \tag{2}$$

where a depends on experiments to compute the stress-strain curve, b considers the effect of hydrostatic pressure on yield, and c represents the dependence of yielding on the third invariant. According to Lou and Huh [21], three principal stresses $(\sigma_1, \sigma_2, \sigma_3)$ are transformed by the stress triaxiality η, the Lode parameter L and the von Mises equivalent stress $\bar{\sigma}_{VM}$ as follows:

$$\sigma_1 = \sigma_m + s_1 = \sigma_m + \frac{(3-L)\bar{\sigma}_{VM}}{3\sqrt{L^2+3}} = \left(\eta + \frac{(3-L)}{3\sqrt{L^2+3}}\right)\bar{\sigma}_{VM} \tag{3}$$

$$\sigma_2 = \sigma_m + s_2 = \sigma_m + \frac{2L\bar{\sigma}_{VM}}{3\sqrt{L^2+3}} = \left(\eta + \frac{2L}{3\sqrt{L^2+3}}\right)\bar{\sigma}_{VM} \tag{4}$$

$$\sigma_3 = \sigma_m + s_3 = \sigma_m - \frac{(3+L)\bar{\sigma}_{VM}}{3\sqrt{L^2+3}} = \left(\eta - \frac{(3-L)}{3\sqrt{L^2+3}}\right)\bar{\sigma}_{VM} \tag{5}$$

The third stress invariant of the deviatoric tensor J_3 is derived from equation (2)-(4) as below:

$$J_3 = |s_{ij}| = s_1 s_2 s_3 = -\frac{2L(9-L^2)}{27(L^2+3)^{3/2}}\bar{\sigma}_{VM}{}^3 \tag{6}$$

According to the von Mises yield function $\bar{\sigma}_{VM} = \sqrt{3J_2}$ and the stress trixiality $\eta = \sigma_m/\bar{\sigma}_{VM} = \frac{I_1}{3\bar{\sigma}_{VM}}$, the J_2 and I_1 can be obtained as follows:

$$J_2 = \frac{\bar{\sigma}_{VM}{}^2}{3} \tag{7}$$

$$I_1 = 3\eta \cdot \bar{\sigma}_{VM} \tag{8}$$

Derived from Eqs. (5)-(7), Eq. (1) can be formulated in terms of η, L and $\bar{\sigma}_{VM}$ in a form of

$$\bar{\sigma}(\sigma_{ij}) = a\left[3b\eta + \left(\frac{1}{27} - c\frac{4L^2(9-L^2)^2}{729(L^2+3)^3}\right)^{1/6}\right]\bar{\sigma}_{VM} \tag{9}$$

If the true stress-true strain curve is measured from uniaxial tensile tests with $\eta = 1/3$ and $L = -1$, the material constant a is given by.

$$a = \frac{1}{b+\frac{1}{3}(27-4c)^{1/6}} \tag{10}$$

In this study, the pressure-coupled Drucker function is used to model the yield surface and fracture limit, while the strain hardening is described by the Swift-Voce hardening law as below:

$$\bar{\sigma} = \frac{\alpha(K(e_0+\bar{\varepsilon}^p)^n)+(1-\alpha)(A-(A-B)exp(-C\bar{\varepsilon}^p))}{2} \tag{11}$$

where the values of K, e_0, n, A, B and C are shown in the Table 1.

The pressure-coupled Drucker function is also applied to model the fracture behaviour in this study in a form of as Eq. (12).

$$\bar{\sigma}_f(\sigma_{ij}) = a\left(bI_1 + (J_2^3 - cJ_3^2)^{1/6}\right) \tag{12}$$

where $\bar{\sigma}_f$ is the equivalent stress to fracture.

4. Inverse engineering method for parameter calibration

In order to determine the appropriate parameters in constitutive equations above under large deformation, an inverse engineering method is implemented to calibrate the strain hardening properties under different loading conditions of uniaxial tension, plane strain tension and shear refer to the specimen II, III, IV as shown in Figure 1.

The experimental results of load-stroke curves of three specimens shown in Figure 1 (II)-(IV) are compared with numerical simulations, and the error of each specimens err_m is calculated by Eq. (13). The difference between experiments and prediction is normalized by the average force so that the error between difference strength tests contributes equally to the total error. The total error of three tests, denoted as $Total\_error$, is the sum of three specimens, which is minimized during parameter calibration.

$$err_m = \sqrt{\sum_{i=1}^{n}\left(\frac{F_i^{exp}-F_i^{sim}}{F_{ave}^{exp}}\right)^2} \tag{13}$$

where n is the number of samples, F_i^{sim} and F_i^{exp} is the simulated and experimental values corresponding to the i sample, F_{ave}^{exp} is the average of force values in the experimental load-stroke curves. The averaged force values F_{ave}^{exp} are taken as comparable values to evaluate the relative error between the simulation and experiments on each sample.

b and c in the pressure-coupled Drucker yield function and α value in combined Swift-Voce hardening model are taken as optimization parameters. The optimization goal is to minimize the $Total\_error$, and optimization algorithm utilizes the Downhill Simplex algorithm. The optimized b, c, and α coefficients are shown in the Table 2, as the effect of hydrostatic pressure on yielding and plastic deformation of this material is extremely small, b value is set to 0, the value of a is obtained from b and c values by Eq. (9).

Similarly, the pressure-coupled Drucker function for the fracture prediction is also optimized by comparing the predicted fracture strokes with experimental results with the inverse engineering method. The parameters are optimized as a=0.2072, b=0.0220 and c=2.6294. The details for the pressure-coupled Drucker fracture function can be referred to Lou and Yoon [9].

5. Results

The optimized coefficients are used in numerical simulation to obtain the load-stroke curves of three specimens and compared with experimental data as plotted in Figure 4. The von Mises yield function is also used in the numerical simulation for the comparison purpose. The comparison of the force-stroke curves shows that the maximum error is mostly less than 2% from the onset of yielding to the ultimate fracture. Therefore, the pressure-coupled Drucker yield function calibrated by inverse engineering approach can accurately predict the plastic behaviour up to large plastic deformation till the onset of ductile fracture. The predicted onset of fracture is also compared with experimental results and the comparison proves that the pressure-coupled Drucker function accurately models the fracture behaviour of the alloy at different loading condition of shear, uniaxial tension and plane strain tension.

Table 2. Coefficients of the pressure-coupled Drucker function and the Swift-Voce hardening law

Pressure-coupled Drucker function			Swift-Voce hardening law						
a	b	c	K	e0	n	A	B	C	α
1.8368	0	2.0045	563	0.0067	0.33	363.9	123.6	9.66	0.0032

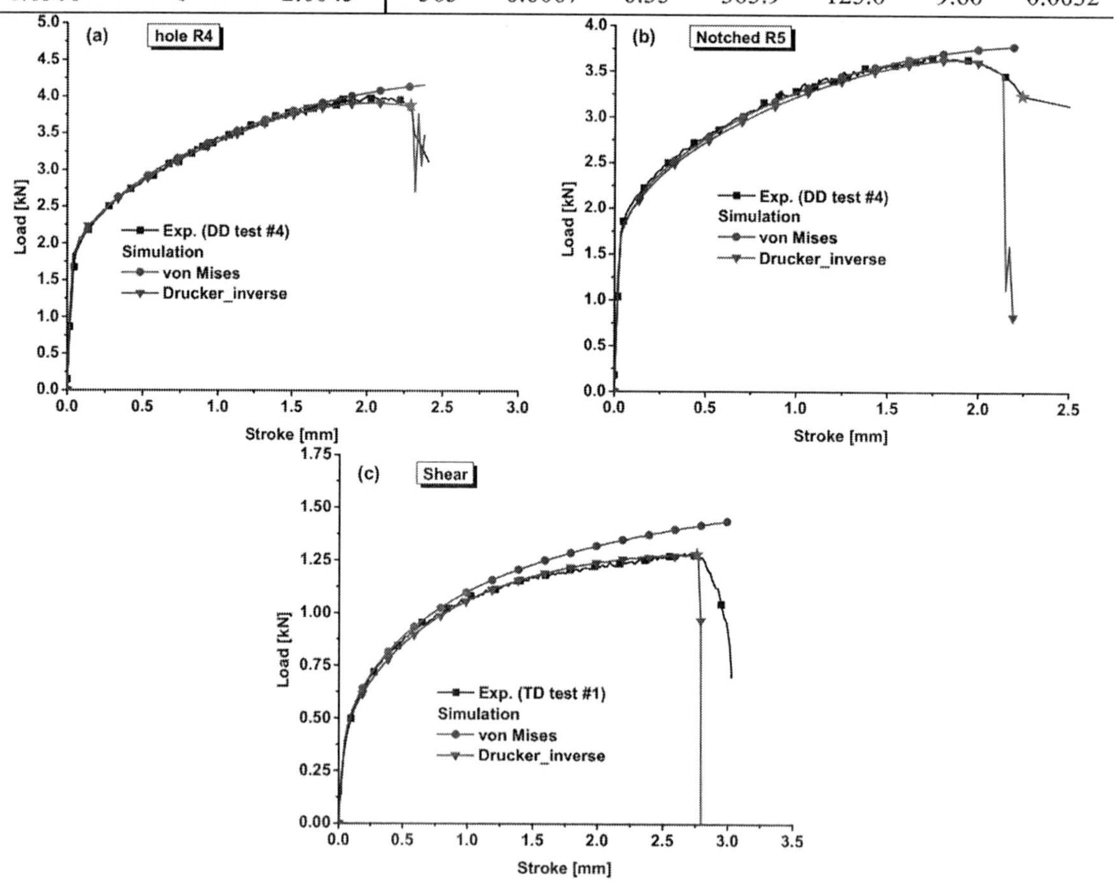

Fig. 4. Load-stroke curve comparison between experiments and prediction: (a) specimens with central hole; (b) notched specimens; (c) in-plane shear specimens

6. Conclusions

This paper modifies the Drucker function to consider the pressure effect of yielding and fracture. The pressure-coupled Drucker function is calibrated using inverse engineering approach for plasticity and fracture of AA5182. The numerical prediction results show that the pressure-coupled Drucker function reasonably predicts the plasticity and fracture stroke for the alloy under various loading conditions including uniaxial tension, plane strain tension and shear. Therefore, the pressure-coupled Drucker function is suggested to plasticity and fracture modelling at different stress states for sheet metal forming processes.

Acknowledgement

The authors acknowledge the financial support by the State Key Laboratory of Mechanical System and Vibration (Grant No. MSV202009), and the State Key Laboratory of High Performance Complex Manufacturing (Grant No. Kfkt2019-02).

References

[1] Drucker, D.C., 1949. Relations of experiments to mathematical theories of plasticity. J. Appl. Mech. 16, 349–357.
[2] Cazacu, O., Barlat, F., 2001. Generalization of Drucker's yield criterion to orthotropy. Math. Mech. Solids 6, 613–630.
[3] Cazacu, O., Barlat, F., 2004. A criterion for description of anisotropy and yield differential effects in pressure-insensitive metals. Int. J. Plast. 20, 2027–2045.
[4] Gao, X.S., Zhang, T.T., Zhou, J., Graham, S.M., Hayden, M., Roe, C., 2011. On stress-state dependent plasticity modeling: significance of the hydrostatic stress, the third invariant of stress deviator and the non-associated flow rule. Int. J. Plast. 27, 217–231.
[5] Yoshida, F., Hamasaki, H., Uemori, T., 2013. A user-friendly 3D yield function to describe anisotropy of steel sheet. Int. J. Plast. 45, 119–139.
[6] Yoon, J.W., Lou, Y., Yoon, J.H., Glazoff, M.V., 2014. Asymmetric yield function based on the stress invariants for pressure sensitive metals. Int. J. Plast. 56, 184–202.
[7] Smith, J., Liu, W.K., Cao, J., 2015. A general anisotropic yield function for pressure-dependent materials. Int. J. Plast. 75, 2–21.
[8] Cazacu, O., Revil-Baudard, B., 2017. New analytic criterion for porous solids with pressure-insensitive matrix. Int. J. Plast. 89, 66–84.
[9] Lou, Y., Yoon, J.W., 2018. Anisotropic yield function based on stress invariants for BCC and FCC metals and its extension to ductile fracture criterion. Int. J. Plast. 101, 125–155.
[10] Kim, J.H., Serpantié, A., Barlat, F., Pierron, F., Lee, M.G., 2013. Characterization of the post-necking strain hardening behavior using the virtual fields method. Int. J. Solids Struct. 50, 3829–3842.
[11] Pham, Q.T., Lee, B.H., Park, K.C., Kim, Y.S., 2018. Influence of the post-necking prediction of hardening law on the theoretical forming limit curve of aluminium sheets. Int. J. Mech. Sci. 140, 521–536.
[12] Li, X., Roth, C.C., Mohr, D., 2019. Machine-learning based temperature- and rate-dependent plasticity model: Application to analysis of fracture experiments on DP steel. Int. J. Plast. 118, 320–344.
[13] Lou, Y. Huh, H., Lim, S., Pack, K., 2012. New ductile fracture criterion for prediction of fracture forming limit diagrams of sheet metals. Int. J. Solids Struct. 49, 3605-3615.
[14] Lou, Y., Yoon, J.W., Huh, H., 2014. Modeling of shear ductile fracture considering a changeable cut-off value for the stress triaxiality. Int. J. Plast. 54, 56–80.
[15] Mohr, D., Marcadet, S.J., 2015. Micromechanically-motivated phenomenological Hosford-Coulomb model for predicting ductile fracture initiation at low stress triaxiality. Int. J. Solids Struct. 67/68, 45-55.
[16] Lou, Y., Chen, L., Clausmeyer, T., Tekkaya, A.E., Yoon, J.W., 2017. Modeling of ductile fracture from shear to balanced biaxial tension for sheet metals. Int. J. Solids Struct. 112, 169-184.

[17] Mu, L., Zang, Y., Li, X.L., Stemler, P.M.A., 2017. A micromechanically-motivated phenomenological model for predicting ductile fracture initiation. Procedia Engineering. 207, 2054-2059.
[18] Hu, Q., Li, X., Han, X., Li, H., Chen, J., 2017. A normalized stress invariant-based yield criterion: modeling and validation. Int J. Plast. 99, 248-273.
[19] Lou, Y., Huh, H., 2013. Prediction of ductile fracture for advanced high strength steel with a new criterion: experiments and simulation. J. Mater. Process. Technol. 213, 1284-1302.
[20] Lou, Y., Zhang, S.J., Yoon, J.W., 2020. Strength modeling of sheet metals from shear to plane strain tension. Int. J. Plast. doi.org/10.1016/j.ijplas.2020.102813.
[21] Lou, Y., Huh, H., 2013. Extension of a shear controlled ductile fracture model considering the stress triaxiality and the Lode parameter. Int. J. Solids Struct. 50, 447-455.

IOP Conf. Series: Materials Science and Engineering **967** (2020) 012030 doi:10.1088/1757-899X/967/1/012030

Strain hardening under large deformation for AA5182

Chong Zhang[1], Yue Wang[1], Hongchun Shang[1], Pengfei Wu[1], Lei Fu[2], Yanshan Lou[1,*], Till Clausmeyer[3], A. Erman Tekkaya[3] and Qi Zhang[1]

[1] School of Mechanical Engineering, Xi'an Jiao Tong University, 28 Xianning West Road, Xi'an, Shaanxi 710049, China
[2] CHINALCO Materials Application Research Institute Co. Ltd, Beijing 102200, China
[3] Institute for Forming Technology and Lightweight Construction, TU Dortmund University, Baroper Str. 303, Dortmund 44227, Germany

*ys.lou@xjtu.edu.cn

Abstract. In this study, an aluminium alloy of AA5182 is taken as the research object to study strain hardening under large plastic deformation. Tensile tests are done for four specimens, including dog-bone specimens, notched specimens, specimens with a central hole and in-plane shear specimens. Bulging tests are also conducted to measure strain hardening under balanced biaxial tension. In addition, an experimental method called in-plane torsion test is also used for shear loading. At least three experiments are completed for each type of specimens along the rolling direction (RD), diagonal direction (TD), and transverse direction (DD). The stroke of each tests is measured by a digital image correlation (DIC) system, and the load-stoke curves were obtained for the tests. Combined with an inverse engineering method, the strain hardening properties are calibrated for the alloy under different loading conditions of shear, uniaxial tension, plane strain tension, and balanced biaxial tension. The strain hardening under various loading conditions is compared and modelled by various yield functions to evaluate their performance. It is concluded that inverse engineering approach is a simple but powerful method to obtain the stress-strain curve up to large plastic deformation. It is also observed that it needs to develop yield functions to model yielding behaviour under complex loading conditions.

1. Introduction

Numerical simulation is now widely used in the design of sheet metal forming processes. The material models implemented in simulation need to be identified to describe the plastic behaviour of sheet metals. Accurate description of strain hardening properties is one of typical tasks for the identification of material models. The conventional method to obtain the strain hardening properties is by conducting a standard uniaxial tensile test with the standardized testing procedure and specimen geometry. However, the standard tensile test can usually only reach the equivalent plastic strains of 0.2 to 0.3 before the necking begins for good ductile metals. The large deformation conditions must be considered to meet industrial production requirements. In the last decade, the inverse method to determine the hardening properties has been widely used [1-15] to determine the hardening properties by incorporating the simulation with experimental data (e.g. global load-stroke curve, local strain fields, etc.). It is critical to apply the reliable material models in simulation.

In this study, several specimens are tested to consider the multiple loading conditions under large plastic deformation, including shear, uniaxial tension, plane strain tension, and balanced biaxial tension. In addition, an experiment called in-plane torsion test is also introduced and specimens are designed according to [16, 17]. The stress-strain curve is calculated analytically for standardized specimens while strain hardening for other specimens are obtained by inverse engineering method. Combining with an inverse engineering method, the strain hardening properties were calibrated under various loading conditions and modelled by a Drucker yield function to evaluate its performance.

2. Experiments

As illustrated in Figure 1, six kinds of specimens with different shapes, including dog-bone specimens I, specimens with central hole II, notched specimens III, in-plane shear specimens IV, bulging test specimens with a die opening diameter of 160mm and in-plane torsion shear specimens, were tested to obtain the strain hardening properties under various loading conditions. Specimens I-IV are tested on an electronic universal tensile testing machine of INSTRON 5182. In addition, Figure 1 also shows the initial gauge lengths for stroke measuring by digital image correlation (DIC) of I-IV specimens. The in-plane torsion shear specimens were tested on the in-plane torsion test device developed in Xi'an Jiaotong University.

At least three experiments were conducted for specimens I-IV along the rolling direction (RD), diagonal direction (DD), and transverse direction (TD), as shown in Figure 2 with good repeatability. Solid pentagrams for specimens II-IV indicate the experimental onset of ductile fracture. For specimens with a central hole II, force drops very early along RD compared with those along TD and DD. However, fracture is not observed at load drop along RD and the mechanism is not clear currently. To determine the material anisotropy quantitatively, the r-values and 0.2 % proof stress are computed as 0.527 and 169.8Mpa for RD, 0.557 and 166.6 Mpa for DD, 0.537 and 171.0Mpa for TD, respectively. Accordingly, the metal is assumed isotropic. For in-plane torsion shear specimens VI, the DIC measurement method is applied to calculate the torsion angle denoted as θ_{dic}.

3. Analytical calculation of stress-strain curves

For dog-bone specimens, bulging test specimens and in-plane torsion shear specimens, the hardening properties of the materials can be obtained through theoretical calculation.

By processing the load-stroke curves data in Figure 2(a), the true stress-strain curves are represented in Figure 3 with black line and it can be seen that the strain is about 0.23 before the necking occurs. The Swift, Voce and Swift-Voce models are fitted by the experimental true stress-strain curve calculated from the uniaxial tensile tests and hardening under large deformation is predicted as well in Figure 3.

Hardening under shear is computed analytically by the in-plane torsion test according to Figure 4 (a) and [16-17]. The computed stress-strain curve in shear is compared with strain hardening measured in by uniaxial tension in Figure 4 (b). It is observed that there is big difference between these two stress states. The equivalent strain value is obtained by the DIC as shown in Figure 4 (c). The strain distribution on the shear zone is not uniform, so the equivalent strain of shear zone along the transverse direction was averaged at each torsion angle. The averaged strain is combined with calculated stress to get hardening behaviour in shear from the in-plane torsion test.

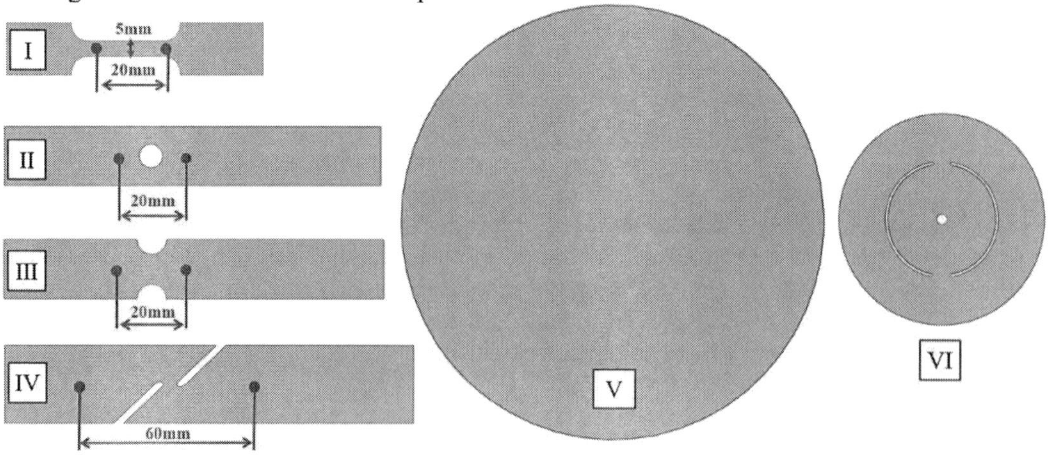

Figure 1. Six types of specimens: (I) dog-bone; (II) with central hole; (III) notched; (IV) in-plane shear; (V) bulging test; (VI) in-plane torsion

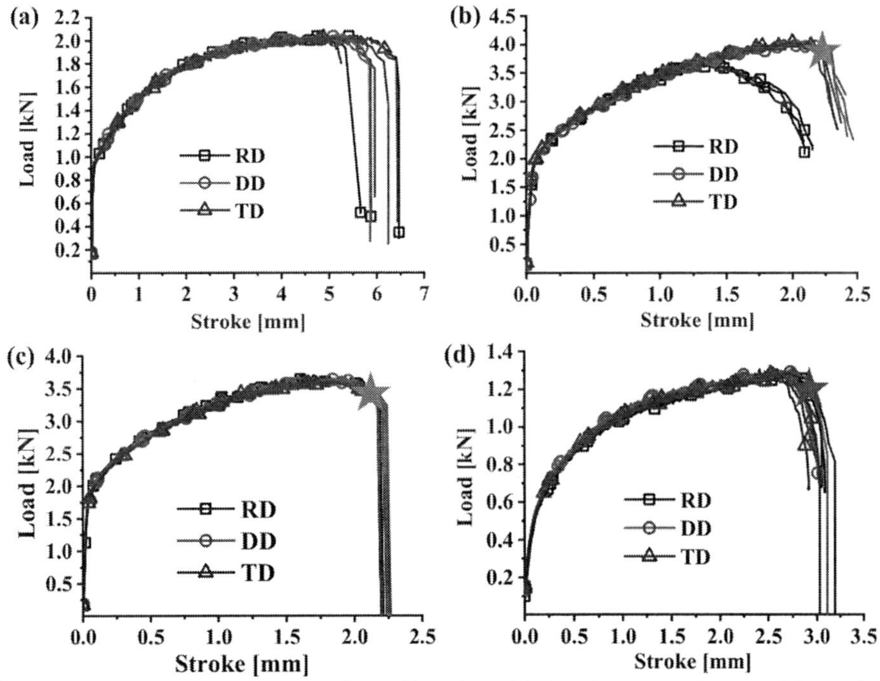

Figure 2. The load-stroke curves along three direction: (a) dog-bone specimens; (b) specimens with central hole; (c) notched specimens; (d) in-plane shear specimens

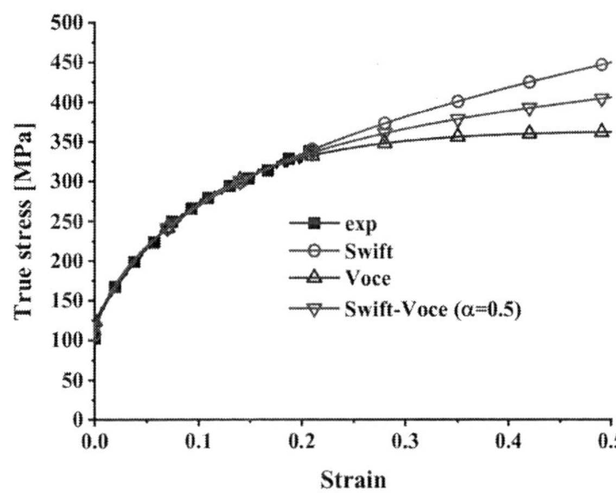

Figure 3. Stress-strain curves calibrated by Swift, Voce, and Swift-Voce laws for dog-bone specimen

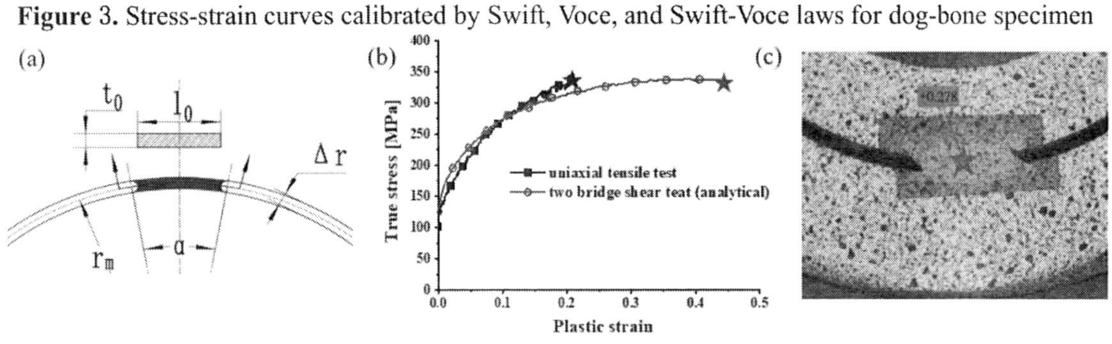

Figure 4. In-plane torsion shear test: (a) analytical computation of hardening in shear; (b) comparison of hardening between shear and uniaxial tension; (c) DIC strain distribution of shear zone

4. Inverse engineering approach for strain hardening characterization

The strain hardening properties of the material is also closely related to the choice of the yield function. Two yield functions were chosen to study the effect of yield functions on strain hardening properties. One is the commonly used von Mises yield criterion, and the other is Drucker yield function [17] which considers the effect of third invariant on yielding. The Drucker yield function is shown as below:

$$\bar{\sigma}(\sigma_{ij}) = a(J_2^3 - cJ_3^2)^{1/6} \tag{1}$$

According to Lou and Huh [18-20], Eq. (1) can be formulated in terms of Lode parameter L and von Mises equivalent stress $\bar{\sigma}_{VM}$ in a form of

$$\bar{\sigma}(\sigma_{ij}) = \boldsymbol{a}\left(\frac{1}{27} - \boldsymbol{c}\frac{4L^2(L^2-9)^2}{729(L^2+3)^3}\right)^{1/6} \bar{\sigma}_{VM} \tag{2}$$

AA5182 belongs to an aluminum alloy which is a typical FCC material. According to [18], the material constant c in the Drucker yield function is advised 2.0 for FCC materials and $a = 1.8365$.

The strain hardening in the numerical simulation is described by the Swift-Voce law as below:

$$\bar{\sigma} = \frac{K(e_0+\bar{\varepsilon}^p)^n + (A-(A-B)exp(-C\bar{\varepsilon}^p))}{2} \tag{3}$$

where K, e_0, n, A, B and C are the parameters to describe the hardening properties of AA5182.

Numerical simulations were carried out in ABAQUS/Explicit using VUMAT with solid elements (C3D8R) for tensile test specimens II-III, and in-plane torsion shear specimen VI. Load-stroke curves are extracted from numerical simulations. The numerical results are compared with experimental measurement to calibrate hardening parameters in Eq. (3) for each tests. The optimization of hardening parameters and numerical simulation are automatically conducted by a developed in-house script [15]. The "Downhill Simplex" algorithm is utilized, which is one of the search methods that can handle nonlinear problems. Initial hardening variables are obtained from the results of dog-bone specimens. The hardening parameters are optimized when the error of each specimens err_m below is minimized.

$$err_m = \sqrt{\sum_{i=1}^{n}\left(\frac{F_i^{num} - F_i^{exp}}{F_{ave}^{exp}}\right)^2} \tag{4}$$

where n is the number of samples, F_i^{num} and F_i^{exp} is the simulated and experimental values corresponding to the i sample, F_{ave}^{exp} is the average of force values in the experimental load-stroke curves.

5. Results and discussion

The optimized stress-strain curves obtained by the Drucker yield function for each specimen are shown in Figure 5(a). The maximum strain of curves is up to the strain before fracture. It can be seen that one hardening curve can describe the hardening of the loading states for three specimens under the Drucker yield function because the stress-strain curves of specimens with central hole II, notched specimens III, and in-plane torsion shear specimens VI approximately coincide with each other except for the bulging test. But there is big difference for the strain hardening curves optimized by inverse engineering approach with the von Mises yield function as shown in Figure 5(b), which indicates that the von Mises function cannot describe yielding under different stress states.

The comparison in Figure 5(a) also shows that the inverse engineering approach is very proper to get strain hardening under large plastic deformation compared with the analytical calculation for the dog-bone specimen. For dog-bone specimens, the hardening can be analytically calculated up to about 0.23 for the alloy, while with the inverse engineering approach, the strain hardening can be obtained at a strain to 0.45 for notched specimens, 0.49 for specimens with a central hole and 0.5 for the in-plane torsion test. The analytical calculation by the bulging test can also provide hardening curve up to large strain of about 0.6. It indicates that the in-plane torsion test combined with inverse engineering approach is approximately equivalent with the bulging test in term of providing strain hardening information under large strain.

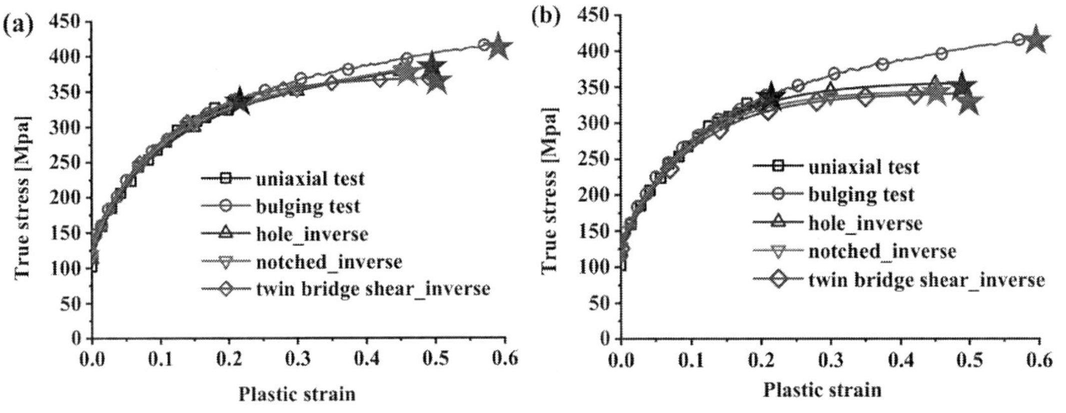

Figure 5. The optimized stress-strain curves obtained by different yield function: (a) Drucker yield function; (b) von Mises yield function

6. Conclusions

Six different tests are conducted to investigate the strain hardening behaviour under large deformation for AA5182. Analytical calculation and inverse engineering approach are used to calibrate strain hardening behaviour under large strain. The results show that the Drucker function provides much better prediction of yielding and hardening under different stress states. The inverse engineering approach is very powerful to get the strain hardening properties under large strain. Its comparison with the bulging test proves that the inverse engineering approach and the bulging test are approximately equivalent in term of largest strain up to which the hardening properties can be obtained.

Acknowledgement

The authors acknowledge the financial support by the State Key Laboratory of Mechanical System and Vibration (Grant No. MSV202009), and the State Key Laboratory of High Performance Complex Manufacturing (Grant No. Kfkt2019-02).

References

[1] Joun, M., Eom, J.G., Lee, M.C., 2008. A new method for acquiring true stress–strain curves over a large range of strains using a tensile test and finite element method. Mech. Mater. 40 (7), 586–593.

[2] Kamaya, M., Kawakubo, M., 2011. A procedure for determining the true stress–strain curve over a large range of strains using digital image correlation and finite element analysis. Mech. Mater. 43, 243–253.

[3] Kim, J.H., Serpantié, A., Barlat, F., Pierron, F., Lee, M.G., 2013. Characterization of the post-necking strain hardening behavior using the virtual fields method. Int. J. Solids Struct. 50 (24), 3829–3842.

[4] Tardif, N., Kyriakides, S., 2012. Determination of anisotropy and material hardening for aluminum sheet metal. Int. J. Solids Struct. 49 (25), 3496– 3506.

[5] Nasser, A., Yadav, A., Pathak, P., Altan, T., 2010. Determination of the flow stress of five AHSS sheet materials (DP 600, DP 780, DP 780-CR, DP 780-HY and TRIP 780) using the uniaxial tensile and the biaxial viscous pressure bulge (VPB) tests. J. Mater. Process. Technol. 210 (3), 429–436.

[6] Knysh, P., Korkolis, Y.P., 2017. Identifcation of the post-necking hardening response of rate- and temperature-dependent metals. Int. J. Solids Struct. 115–116, 149–160.

[7] Lou, Y., Huh, H., 2013. Prediction of ductile fracture for advanced high strength steel with a new criterion: experiments and simulation. J. Mater. Process. Technol. 213, 1284-1302.

[8] Pcirs, J., Verleysen, P., Van Paepegem, W., Degrieck, J., 2011. Determining the stress–strain behaviour at large strains from high strain rate tensile and shear experiments. Int. J. Impact Eng. 38, 406–415.

[9] Zhao, K., Wang, L., Chang, Y., Yan, J., 2016. Identifcation of post-necking stress–strain curve for sheet metals by inverse method. Mech. Mater. 92, 107–118.

[10] H. Zhang, S. Coppieters, C. Jiménez-Peña, D., 2019. Debruyne, Inverse identification of the post-necking work hardening behaviour of thick HSS through full-field strain measurements during diffuse necking, Mech. Mater. 129, 361-374.

[11] Rossi, M., Lattanzi, A., Barlat, F., 2018. A general linear method to evaluate the hardening behaviour of metals at large strain with full-feld measurements. Strain 54.

[12] Denys, K., Coppieters, S., Debruyne, D., 2018. On the identifcation of a high-resolution multi-linear post-necking strain hardening model. CR. Mecanique 346, 712–723.

[13] Lou, Y., Chen, L., Clausmeyer, T., Tekkaya, A.E., Yoon, J.W., 2017. Modeling of ductile fracture from shear to balanced biaxial tension for sheet metals. Int. J. Solids Struct. 112, 169-184.

[14] Brosius, A. , Yin, Q. , A. Güner, & Tekkaya, A. E., 2011. A new shear test for sheet metal characterization. Steel Research International, 82(4), 323-328.

[15] Lou, Y., Zhang, S.J., Yoon, J.W., 2020. Strength modeling of sheet metals from shear to plane strain tension. Int. J. Plast. 102813. https://doi.org/10.1016/j.ijplas.2020.102813.

[16] Yin, Q. , Soyarslan, C. , A. Güner, Brosius, A. , & Tekkaya, A. E., 2012. A cyclic twin bridge shear test for the identification of kinematic hardening parameters. International Journal of Mechanical Sciences, 59(1), 31-43.

[17] Drucker, D.C., 1949. Relations of experiments to mathematical theories of plasticity. J. Appl. Mech. 16, 349–357.

[18] Lou, Y., Yoon, J.W., 2018. Anisotropic yield function based on stress invariants for BCC and FCC metals and its extension to ductile fracture criterion. Int. J. Plast. 101, 125–155.

[19] Lou, Y., Huh, H., 2013. Extension of a shear controlled ductile fracture model considering the stress triaxiality and the Lode parameter. Int. J. Solids Struct. 50, 447-455.

[20] Lou, Y., Yoon, J.W., Huh, H., 2014. Modeling of shear ductile fracture considering a changeable cut-off value for the stress triaxiality. Int. J. Plast. 54, 56–80.

Comparison of Artificial Intelligence Methods for Prediction of Mechanical Properties

Kyungmin Lee [1], Charmgil Hong [2*], Eun-Ho Lee[3*], WooHo Yang[4]

[1] School of Mechanical and Control Engineering, Handong Global University, Pohang, Gyeongbuk, Republic of Korea
[2] School of Computer Science and Electrical Engineering, Handong Global University, Pohang, Gyeongbuk, Republic of Korea
[3] School of Mechanical Engineering, Sungkyunkwan University, Suwon, Gyeonggi-do, 16419, Republic of Korea
[4] Smart Factory Research & Development Team 2, Hyundai Motor Group, Uiwang, Gyeonggi, Republic of Korea

Email: [2*]charmgil@handong.edu, [3*]e.h.lee@skku.edu

Abstract. This paper compares artificial intelligence (AI) methods to predict mechanical properties of sheet metal in stamping processes. The deviation of the mechanical properties of each blank leads to unpredicted failures in stamping processes, such as fracture and spring back. The research team of this paper has been building a real time control system for stamping process in a smart factory. In order to facilitate that, it is necessary to predict the mechanical properties of each blank with non-destructive testing. The regression models based on the linear algebraic scheme have traditionally brought reliable results in terms of matching the measured non-destructive testing values to the mechanical properties. With a parallel to algebraic regression models, in recent studies on various domains, AI models have been adopted to improve the accuracy of the end-results and effectiveness of the models. This paper discusses the applicability of AI models for predicting the mechanical properties based on the eddy-current non-destructive testing method. For the study, 6 input features are collected through the eddy-current non-destructive testing to map eddy-current input data to mechanical properties of the blank. Yield stress and uniform elongation were predicted by using five AI methods, i.e., regularized linear regression, support vector regression, multi-layer neural network, random forest regression, and gradient boosting regression were compared. The model performance, validated with 20% of test data that are intact during the training phase, is the main discussion point of this paper. Future works to improve the predictive accuracy of AI models is also discussed.

1. Introduction

In stamping processes, sheet metals often reveal a large deviation in their mechanical properties that the deviation may cause unexpected failures, such as large spring back and fracture. The research team of this paper has been building a real time control system for stamping process in a smart factory. To facilitate that, it is necessary to predict the mechanical properties of each blank with non-destructive testing. Scientists such as García-Martín *et al* [24] describes the principle of eddy-current technique as an interaction between the magnetic field and conductive materials. This induced current is often utilized in inspecting metal cracks through flow change of the eddy-current. Many researchers such as Heingärtner *et al* [1] have attempted to predict mechanical properties based on eddy-current non-destructive testing, which presents enough anticipation that eddy-current testing with advanced regression models is able to predict mechanical properties more accurately and effectively The authors used a regression model based on the linear algebraic scheme to map the measured non-destructive testing values to mechanical properties. Their model has brought reliable fitting results in addressing the mechanical property prediction problem. Since the relationship between eddy-current and mechanical properties cannot be easily understood, this regression-based approach could be an effective solution. Several subsequent works using similar mathematical or statistical approaches also have been proposed [2,3,4,5].

Recently, Artificial Intelligence (AI) technologies are drawing much attention [6,7] as they have successfully shown their ability to find optimal mappings in many prediction problems. This work discusses the ability of AI models for mechanical property prediction. An eddy-current measurement method was employed for non-destructive testing. Data preprocessing is conducted to obtain reliable outcomes from AI methods. For the study, five different AI methods, including regularized linear regression, support vector regression, multi-layer neural network, random forest and gradient boosting, are applied to predict the yield stress and uniform elongation of sheet metals based on their eddy-current measurements. A thorough analysis on the models' predictive accuracy and applicability are demonstrated.

2. Data Gathering (Experiment)

In order to gather data, the eddy currents on samples of SS400, a widely used structural material, at 1kHz and 5kHz frequencies, are measured. A uniaxial tensile test was then conducted to check the mechanical properties of the specimen. Based on the measured eddy-current data, the real and imaginary parts of the material impedance were calculated. This forms a prediction problem, where voltages and currents at two different frequencies are fed to the models as input data (x), while the models are trained to predict the yield stress and uniform elongation (y) of metal sheet samples.

The dataset contains 190 data instances with 6 features. 80% of the data is used in training models, while 20% of data is used for testing the predictive accuracy of the models. Both train and test data are split completely and shuffled for fair comparisons. Before training and testing, the input data is normalized with the mean and standard deviation of the training data to make sure all the input values in the dataset are within a common scale, so that the training algorithms deal with the input features equally.

3. Artificial Intelligence Models

This paper aims to explore and find the most optimized mapping model for predicting mechanical properties using the AI approach. Last decades have seen a plethora of AI models developed for regression. This work studies and evaluates five widely used AI methods, including regularized linear regression, support vector regression, multi-layer neural network, random forest regression, and gradient boosting regression.

3.1 Regularized Linear Regression

Regularized linear regression (RLR) is one of the most commonly used models for regression analysis [8]. Originally, linear regression implements the least square method that minimizes the mean squared error between predictions and observations. However, insufficiency in training data often leads the model to overfit, where the resulting model cannot generalize to unseen test data. To alleviate this issue, regularization terms have been introduced to control the complexity of the model and therefore avoid overfitting. There are three types of regularizations: ridge, lasso and elastic net [8]. Below describes the objective function of RLR for regularized optimization with elastic net:

$$|y - \hat{y}_{RLR}|_2^2 + \alpha||\theta||_2^2 + \beta||\theta||_1^2 = |y - x\theta|_2^2 + \alpha||\theta||_2^2 + \beta||\theta||_1^2$$

where x and y respectively denote the observed (true) input and output value, $\hat{y}_{RLR}$ denotes the predicted output value from the regression model, θ represents the model parameter, and α and β are the regularization coefficients that control the strength of the regularization. In Section 4, RLR models predicting yield stress and uniform elongation, respectively, are trained and evaluated.

3.2. Support Vector Regression

Support vector regression (SVR) is a variant of support vector machines addressing the regression problem [9]. Training an SVR model is, in essence, determining the position of the regression hyper-plane in the feature space. The training algorithm iteratively repeats the following: 1) identify "support vectors" that lie outside the "margin" (a fixed width space along the hyper-plane) 2) find the best regression hyper-plane for the support vectors using the ε-insensitive loss function:

$$L_\varepsilon(y, \hat{y}_{SVR}) = \begin{cases} 0 & if\ |y - \hat{y}_{SVR}| \leq \varepsilon \\ |y - \hat{y}_{SVR}| - \varepsilon & otherwise \end{cases}$$

where y and $\hat{y}_{SVR}$ respectively denote the true output value and predicted output value from SVR regression model, and ε is the width of the margin (margin of tolerance). That is, the ε-insensitive loss function concerns only the errors that goes beyond the margin. After training, SVR figures out the optimal width of ε and determines the position of regression hyper-plane that has the maximum number of data points in the margin (see Figure 1).

The vanilla SVR, however, tends to perform inaccurately in modelling non-linear regression. The kernel trick is introduced to address this issue and map such non-linear relations between input and output, while not growing too much of computational overheads. A kernel broadens the original feature space to a higher-dimensional function space, where the optimal regression could be found. In Section 4, SVR with the radial basis function (RBF) kernel – $k(x_{i,}x_j) = exp\ (-\gamma \left|\left|x_{i,} - x_j\right|\right|^2)$, where γ is the parameter that determines the width of the radial basis function and x_i, x_j are input data with indices i and j – is evaluated.

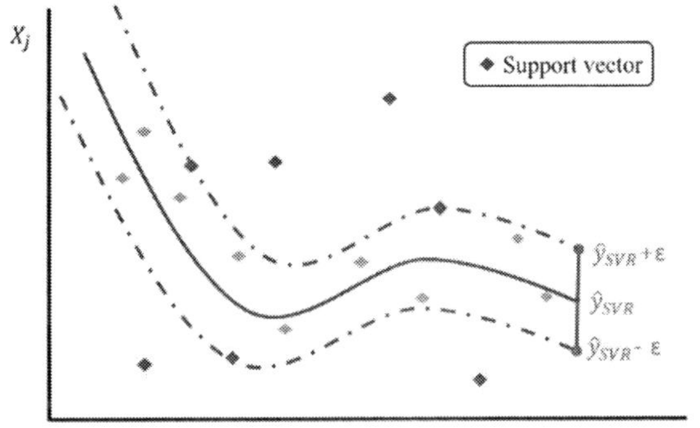

Figure 1. Support Vector Regression

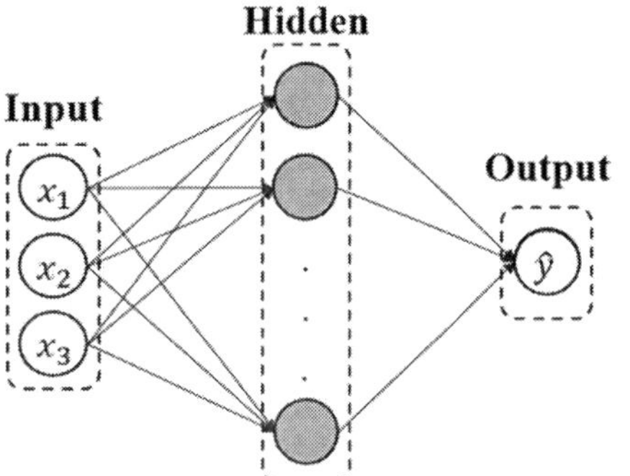

Figure 2. Multi-Layer Neural Network

3.3 Multi-Layer Neural Network

Inspired by how the human brain works, neural networks and relevant approaches, including deep learning, have achieved tremendous success in a vast number of domains. Multi-layer neural networks (NN) is one of the simplest types of predictive models in the category (Figure 2). When an input is given to the input layer, "neurons" on the hidden layers compute their values, by taking the weighted sum of the previous layer, and propagate the values to the output layer. The output layer then makes prediction by, again, taking the weighted sum of the values from the hidden layers. Between layers, non-linear mappings referred to as "activation function" are used to add non-linearity in prediction. Training a neural network consists of adjusting the weights of all its neurons so that the predictions at the output layer become close to the observed output values (the algorithm governs this process is called as "backpropagation" [10]).

Recently, the NN and deep learning-based approaches are broadly accepted in mechanical engineering industry for their reliable performance on many applications [11]. In Section 4, the performance of NN with the experiment data is evaluated and analyzed.

3.4 Random Forest Regression

Random forest regression [12] (RF) is one of the most popular off-the-shelf regression models that utilizes an ensemble method. That is, RF adopts multiple regression tree models, where individual models are trained using the classification and regression tree (CART, Figure 3) method [13]. Generally, a regression tree is prone to overfitting, which refers to a situation when a model captures too much of the training data – even noise and error – and cannot generalize to unseen data [14]. Through ensemble, RF is able to reduce the issue and, in turn, improve the overall predictive accuracy. More specifically, RF is based on a particular ensemble technique called bagging. To obtain a bagged model, multiple base regressors are trained with randomly resampled datasets (see Figure 4). This process is known to effectively reduce the variance in prediction.

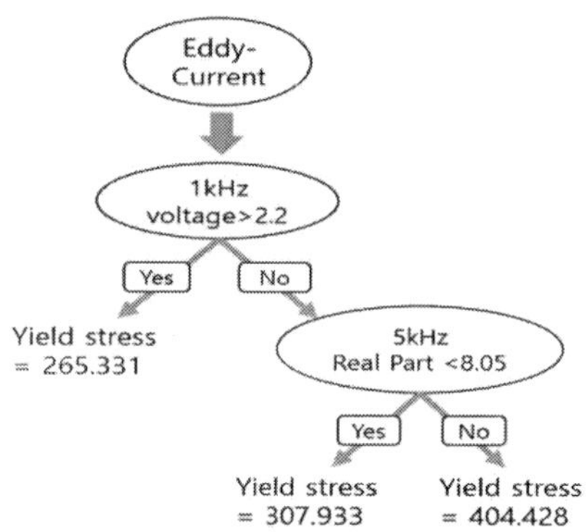

Figure 3. Example of Regression Tree

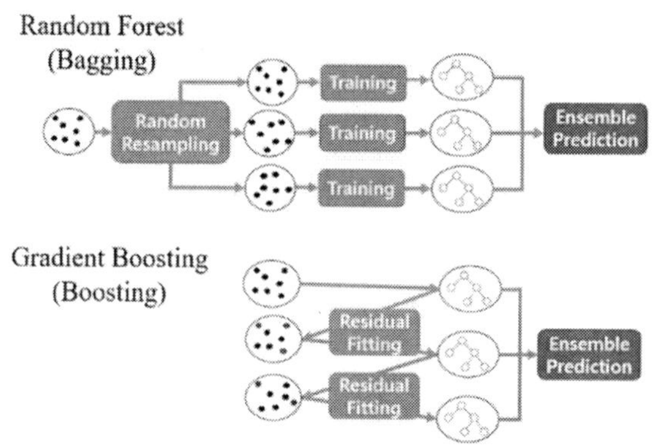

Figure 4. Data Gathering Algorithms

3.5 Gradient Boosting Regression

Gradient boosting regression (GB) [15] is another well-regarded ensemble method that combines multiple regression trees. Unlike RF, GB trains individual regression trees via so-called "residual fitting" process (see Figure 4): the first base model is trained to make prediction on all training data. Next base models are iteratively trained to fit the residuals from the previous ensemble prediction. GB generally decreases the bias of the base model, while it could produce an overfitted model. Extreme gradient boosting (XGB) [16] addresses this issue by applying regularization and subsampling to the original GB method. Moreover, XGB can ease the computational burden of GB, by allowing parallelization during training. In the next section, XGB is applied to the experiment data.

4. Results Analysis

In this section, the above described AI methods are applied to the eddy-current data to predict the mechanical properties. For evaluation, two metrics are used: root mean squared error (RMSE) and adjusted R-squared value (adj-R^2). RMSE calculates the absolute fitness of the model. It assesses absolute the accuracy of each model's prediction. Adj-R^2 measures the ratio of total errors that can be explained by the model. It evaluates relative goodness of fit of the model. Adj-R^2 utilizes the model's degree of freedom. Using both metrics, the following analyzes the results of the five AI methods in absolute and relative scale.

4.1 Prediction of Yield Stress

Table 1 summarizes the predictive accuracies of the five models on yield stress. Figure 5 shows the predicted yield stress values produced by all five models (data points shown in blue are the prediction on the training data; points shown in red are the prediction on the testing data). According to the results, most regression models, except SVR, demonstrate reliable performance in terms of both RMSE and adj-R^2. This supports the general applicability of the AI methods towards the property prediction problem. Specifically, the ensemble models (RF and XGB) and NN exhibit the best performance. In particular, RF has resulted in RMSE=4.8534. Considering the average yield stress on the test data is 328, the average error rate of RF is 1.48%. Adj-R^2=0.9921 indicates that RF explain 99.21% of the relations between the eddy-current features and yield stress properties. Although not as effective as the above three models, simple linear model (RLR) has also resulted in an acceptable performance. That is, the least square model with a proper regularization seems to find a valid mapping between eddy-current attributes to the yield stress values. On the other hand, SVR shows relatively poor performance. According to Figure 5, SVR is indeed suffering from overfitting as its prediction tends to fail more severely on the testing data.

Table 1. RMSE and Adj-R^2 on the Prediction of Yield Stress

	RLR	SVR	NN	RF	XGB
RMSE (MPa)	7.2109	23.0089	5.2933	4.8534	5.3831
Adj-R^2	0.9826	0.8226	0.9906	0.9921	0.9903

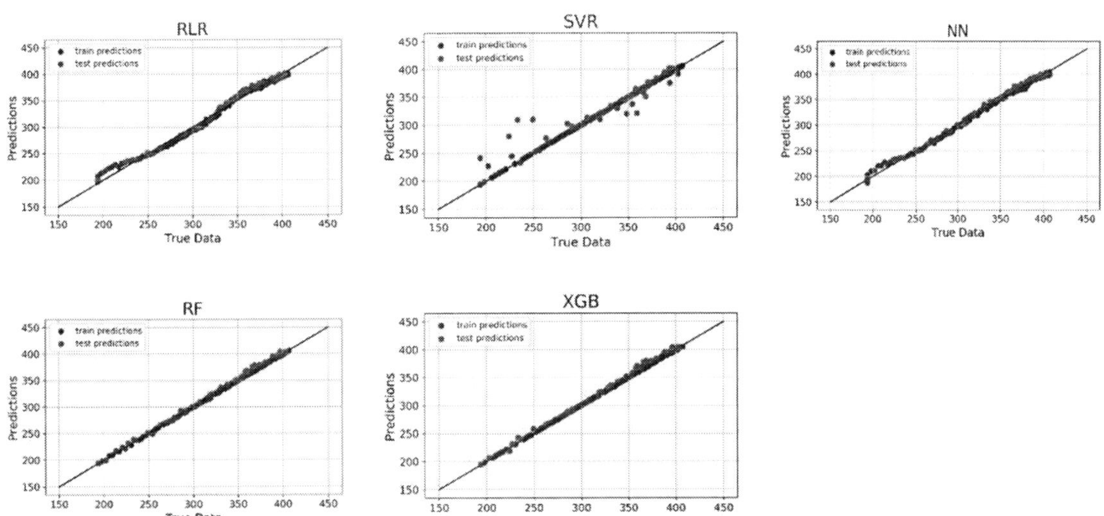

Figure 5. Prediction of Yield Stress. The points shown in blue represent the prediction on the training data; the points shown in red represent the prediction on the testing data. The diagonal lines indicate the perfect predictions compared with the true observations (ground truth values).

4.2 Prediction of Uniform Elongation

Table 2 summarizes the accuracies of the five models on uniform elongation. Figure 6 shows the predicted uniform elongation values produced by the five models. Similar to the previous results with yield stress, RF and XGB seem to successfully find the optimal mapping from input to output. Among both methods, XGB has shown slightly higher accuracy than RF as boosting (compared to bagging) generally better captures data with higher variances. Specifically, XGB has resulted in RMSE=0.1927. Given that the average uniform elongation on the test data is 5.2171, the RMSE corresponds to 3.69% of average error rate. Adj-R^2=0.9802 indicates that XGB explains 98.02% of the relation between the eddy-current data to uniform elongation property.

On the other hand, RLR and NN have demonstrated acceptable performance by resulting Adj-R^2=0.9474 and 0.9234, respectively. SVR shows lowest performance again. Its prediction on the test data is often far off from the true values (see Figure 6).

Table 2. RMSE and Adj-R^2 on the Prediction of Uniform Elongation

	RLR	SVR	NN	RF	XGB
RMSE (mm)	0.3139	0.5894	0.3788	0.2264	0.1927
Adj-R^2	0.9474	0.8145	0.9234	0.9726	0.9802

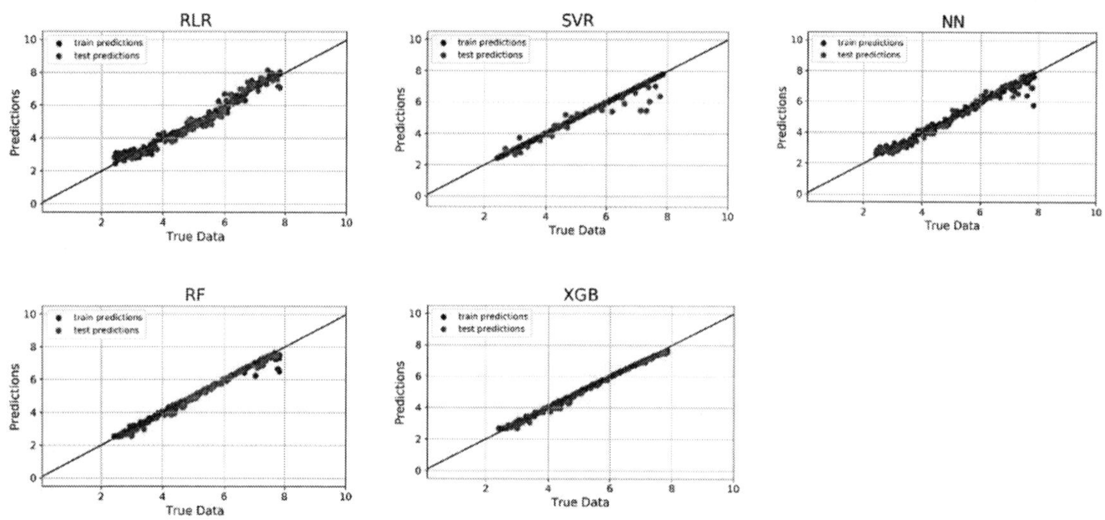

Figure 6. Predictions of Uniform Elongation. The points shown in blue represent the prediction on the training data; the points shown in red represent the prediction on the testing data. The diagonal lines indicate the perfect predictions compared with the true observations (ground truth values).

4.3 Analysis

Overall, the highest accuracy is obtained from the ensemble combinations of regression trees. By building ensembles of multiple regressors, the prediction errors on both targets are minimized. On the contrary, SVR provides poor performance due to the limits of its training algorithm. That is, the training of SVR is designed to focus on the data fall "outside" the margin, which entails that SVR does not concern how accurately individual data inside the margin is predicted. As a result, instance-wise errors smaller than the margin of tolerance (ε) do not affect the model. Using kernels even complicates the issue because it can make the model very sensitive to the training data that it focuses on (i.e., the ones fall out of the margin) [17]. All in all, the method in this study is consistently showing severe signs of overfitting.

Although it does not always produce the best performance, NN with a simple structure has shown generally reliable performance. Considering the amount of data used in the study, the results indicate huge potential and possibilities as NNs are generally capable to properly handle large amount of data with more elaborate network architecture.

5. Conclusions

This work assessed the mechanical properties prediction from the eddy-current data using five different AI methods. The results revealed that a proper use of off-the-shelf AI methods can accomplish accurate property prediction. For instance, the ensemble training methods such as bagging (random forest regression) and boosting (extreme gradient boosting) demonstrated their ability in accurate property prediction. Multi-layer neural networks also showed its effectiveness in mapping complicated relations between the electro-magnetic properties and mechanical properties of sheet metal materials. The research team plans to expand the application spectrum to predict anisotropic and asymmetric properties of other advanced materials [18-21] to use this method in real manufacturing process. The team also plans to come up with a real time control system by applying findings from this work.

In the future, more research on the interpretability of the ensemble models could be pursued to enhance the understanding towards the behavior of the prediction models. Also, further improvement of the neural networks-based approach could be realized by collecting more data, deepening and widening the network structure with an appropriate adaptation of modern learning techniques, such as batch normalization [22] and attention [23].

Acknowledgements

This work was supported by Hyundai Motors Group through Hyundai NGV.

References

[1] Heingärtner J, Neumann A, Hortig D, Rencki Y, Hora P 2013 Acquisition of material properties in production for sheet metal forming processes NUMISHEET 2014: **1567**, 671-674
[2] Sarraf IS, Green DE 2018 A numerical method to predict the rate-sensitive hardening behaviour of sheet materials using uniaxial and biaxial flow curves IOP Conf. Series: Materials Sciences and Engineering **418**
[3] Varvenne C, Curtin WA, Predicting yield strengths of noble metal high entropy alloys 2018 *Scripta Materialia* **142**, 92-95
[4] Endelt B, Part-to-Part model predictive control - using a modified Gauss-Newton scheme. 2019 IOP Conf. Series: Materials Science and Engineering **651** (2019)
[5] Sarraf IS, Green DE 2018 Prediction of DP600 and TRIP780 yield loci using Yoshida anisotropic yield function. 2018 IOP Conf. Series: Materials Sciences and Engineering **418**
[6] Chheda AM, Nazro L, Sen FG, Hegadekatte V, 2019 Prediction of forming limit diagrams using machine learning. IOP Conf Series: Materials Science and Engineering **651**
[7] Pilania G, Wang C, Jiang X, Rajasekaran S, Ramprasad R 2013 Accelerating materials property predictions using machine learning. *Scientific Report* **2810**
[8] Hastie T, Tibshirani R, Friedman J. 2009 *The Elements of Statistical Learning: data mining, inference, and prediction* (Springer)
[9] Drucker H, Burges CJC, Kaufman L, Smola AJ, Vapnik V. 2003 Support Vector Regression Machines. In Advances in Neural Information Processing Systems (NIPS'03) **9**
[10] Rumelhart DE, Hinton GE, Williams RJ 1986 Learning representations by back-propagating errors *Nature* **323** 533-536
[11] Greve L, Schneider B, Andres M, Martinez JD, Eller T, Weg Bvd, Hazrati J, Boogaard AHvd 2019 Computationally efficient necking prediction using neural networks trained on virtual test data IOP Conf. Ser. Mater. Sci. Eng. **651**
[12] Breiman L. 2001 *Machine learning* (Springer)
[13] Breiman L, Friedman JH, Olshen RA, Stone CJ 1984 *Classification and Regression Trees* (Chapman and Hall)
[14] Bishop CM 2006 *Pattern recognition and machine learning* (springer)

[15] Friedman JH. 2002 Stochastic gradient boosting *Computational Statistics & Data Analysis* **38** 367-378.
[16] Chen T, Guestrin C. 2016 XGBoost: A Scalable Tree Boosting System. Int. Conf on KDD
[17] Cawley GC, Talbot NLC 2010 Over-fitting in model selection and subsequent selection bias in performance evaluation, *Journal of Machine Learning Research* **11** 2079-2107
[18] Lee EH, Choi H, Stoughton TB, Yoon JW 2019 *Int. J. Plast.* **122** 73-88.
[19] Lee EH, Stoughton TB, Yoon JW 2018 *Int. J. Plast.* **110** 145-65.
[20] Lee EH, Stoughton TB, Yoon JW 2017 *Int. J. Plast.* **99** 120-43.
[21] Lee EH, Rubin MB 2020 *Int. J. Plast.* 102783.
[22] Ioffe S, Szegedy C 2015 Batch normalization: accelerating deep network training by reducing internal covariate shift. In Proceedings of the 32nd International Conference on International Conference on Machine Learning **37** (ICML'15). JMLR.org, 448–456.
[23] Vaswani A, Shazeer N, Parmar N, Uszkoreit J, Jones L, Gomez AN, Kaiser L, Polosukhin I 2017. Attention is all you need. In Advances in neural information processing systems (NIPS'17) 5998-6008.
[24] García-Martín J, Gómez-Gil J, Vázquez-Sánchez E 2011 Non-destructive techniques based on eddy current testing. *Sensors*, **11(3)**, 2525-2565.

IOP Conf. Series: Materials Science and Engineering **967** (2020) 012032 doi:10.1088/1757-899X/967/1/012032

A novel uncoupled ductile fracture criterion for prediction of failure in sheet metal forming

H Quach[1], J J Kim[1], J H Sung[2] and Y S Kim[3]*

[1] Graduate School of Mechanical Engineering, Kyungpook National University, 80, Daehakro, Bukgu, Daegu, 41566 Korea
[2] Korea Institute of Industrial Technology, 320, Techno sunhwan-ro, Yuga-myeon, Dalseong-gun, Daegu, 711-883 Republic of Korea
[3] School of Mechanical Engineering, Kyungpook National University, 80, Daehakro, Bukgu, Daegu, 41566 Korea
*E-mail: caekim@knu.ac.kr

Abstract. The ductile fracture behavior of anisotropic materials was modeled by the uncoupled ductile fracture criterion. The ductile fracture model was concerned with the micro mechanisms of void nucleation, void growth, and evolution of void coalescence. The advance plastic model of Hill'48 non-associated flow rule is considered in proposed ductile fracture criterion. A series of basic fracture testing covering a wide range of stress state for aluminum alloy 6016-AC200 are carried out. Strain field on the surface of specimens is captured by the non-contact measurement DIC system. The fracture locus constructed using the proposed criterion is close to the experimental data points over a wide stress triaxiality range from negative to intermediate and high stress triaxiality. Then, a square cup drawing test is conducted in order to verify efficiency of the proposed fracture criterion in predicting fracture behavior of metal sheet. The results indicate that the proposed ductile fracture criterion can be utilized for predicting initial fracture in sheet metal forming.

1. Introduction

Generally, the necking is the initial failure mode in sheet metal forming processes, which is commonly predicted by conventional Forming Limit Diagram (FLD) firstly introduced by Keeler and Backofen [1] and extended by Goodwin [2]. The FLD methods mainly describes the prediction of the failure of tension stress states in sheet metal forming. However, the shear fracture, which is observed both tension, shear and compression loading states with negligibly thickness reduction, cannot be explained by FLD.

Ductile fracture of metals is commonly induced by the micro-mechanism of the void nucleation, the void growth, and the coalescence of voids finally resulting in a macroscopic crack. Gurson [3] proposed a strain-controlled nucleation model to show that the rate of void nucleation is a function of the equivalent plastic strain. Rice and Tracey [4] suggested that the stress triaxiality is the key parameter that controls the void growth. The voids nucleation and coalescence of voids were later taken into study by Chu and Needleman [5]. Besides, fundamental work by Xue [6] confirmed that the Lode angle parameter and the stress triaxiality have essential influence on the initiation of ductile fracture.

Aluminum sheets are well known to develop considerable plastic anisotropy on the mechanical properties due to the extrusion and rolling processes. The anisotropy properties have strong effect on the fracture initiation in several manufacturing processes especially in automotive industry. The prediction of anisotropic ductile fracture of metals is critical for the design of lightweight structure. Although there are a lot of efforts in prediction of anisotropic fracture initiation in sheet metal forming, these works still remain a significant issue for automotive industry to produce car parts. Recently, Stoughton and Yoon [7] and Park and Chung [8] proved the flexibility of non-associated flow rule in modeling metal sheet forming applications In this research, the prediction initiation of anisotropic ductile fracture with Hill'48 [9] non-associated flow rule has been investigated. Hill'48 non-associated flow rule coefficients are identified by series of uniaxial tension and bugle test. The ductile fracture parameters are calculated by essential fracture tests such as in plane shear tension, central hole tension, plane strain tension for AA6016-AC200. The strain field on the surface of specimens are carried out under non-contact measurement method of digital image correlation (DIC) method. The anisotropic ductile fracture criterion is implemented into a VUMAT subroutine to simulate and predict the anisotropic initiation ductile fracture behaviors in a square cup drawing for AA6016-AC200.

2. Anisotropic ductile fracture criterion

2.1 Advanced plastic constitutive model

In the present study, the anisotropic Hill'48 function was adopted to describe the yield model and plastic flow. The Hill'48 yield function is defined as

$$\bar{\sigma}_H = \sqrt{F(\sigma_{yy}-\sigma_{zz})^2 + G(\sigma_{zz}-\sigma_{xx})^2 + H(\sigma_{xx}-\sigma_{yy})^2 + 2L{\tau_{yz}}^2 + 2M{\tau_{zx}}^2 + 2N{\tau_{xy}}^2} \quad (1)$$

Where *F, G, H, L, M, and N* are anisotropic coefficients and x, y and z are orthogonal principal anisotropic axes materially embedded along the rolling, transverse and thickness directions. Several dog-bone specimens were cut from the sheet along 0°, 45°, and 90° with respect to the rolling direction to identify the Lankford r-value and the yield stress at three directions. Besides, a bugle test were prepared to determine the biaxial yield stress. The Lankford r-values were utilized to calibrate the Hill'48 plastic potential function parameters, and the different yield stress were well known to calibrate the Hill'48 yield function parameters. It is notes that F, G, H, and N are constants and identified by above experiments, and L and M are assumed as 1.5 in this research. The Hill'48 non associated flow rule coefficients are presented in table 1.

Table1. The Hill'48 anisotropy coefficients of AA6016-AC200

AA6016-AC200	F	G	H	N	L	M
Hill'48-r	0.6593	0.5780	0.4219	1.1754	1.5	1.5
Hill'48-s	0.5421	0.4843	0.5157	1.6353	1.5	1.5

The strain hardening behavior of aluminum 6016-AC200 is determined from uniaxial tension results. The stress-strain relation is modeled by Kim-Tuan hardening equation [10] as:

$$\sigma = \sigma_0 + K(\varepsilon+\varepsilon_0)^h(1-\exp^{-t\varepsilon}) \quad (2)$$

Where σ represent the equivalent stress, σ_0 denotes the initial yield stress, K denotes a material constant to control the expansion of hardening stress, t and h denotes parameters of the hardening model, ε_0 denotes the initiation of plastic deformation, and ε_0=0.002. The flow stress of material was combined by the uniaxial tensile test and the hydraulic bugle test and fitted by the Eq. 2. Table 2 introduced the hardening parameter of AA6016-AC200

Table 2. The hardening parameters of AA6016-AC200

σ_0	ε_0	K	t	h
123.700	0.002	267.316	0.2897	28.926

2.2 Ductile fracture criterion

Ductile fracture behavior is a well-known important property of metal that leads to crack in metal parts. Microscopically, such failure due to micro mechanism of void nucleation, void growth and coalescence of void under moderate and high stress triaxiality or shear band movement under low stress triaxiality. Macroscopically, failure is associated as the progressive degradation of metals. Under larger plastic deformation the metal show a decrease in material stiffness and strength. In this investigation, a phenomenon logical uncoupled ductile fracture is proposed for predicting the onset of ductile fracture. In efforts to predict the anisotropic ductile fracture behaviors, an extend of ductile fracture criterion of Quach et al [11] work for anisotropic material is introduced as

$$\bar{\varepsilon}_f = \frac{C_1}{\left(\eta + \frac{3-\mu}{3*\sqrt{\mu^2+3}} + \frac{1}{\sqrt{\mu^2+3}}\right)^{C_2}\left(\frac{3+\sqrt{3}C_3}{\sqrt{\mu^2+3}} - C_3\right)} \tag{3}$$

Where

$$\eta = \frac{\sigma_{xx}+\sigma_{yy}+\sigma_{zz}}{3*\sqrt{F\left(\sigma_{yy}-\sigma_{zz}\right)^2 + G(\sigma_{zz}-\sigma_{xx})^2 + H\left(\sigma_{xx}-\sigma_{yy}\right)^2 + 2L{\tau_{yz}}^2 + 2M{\tau_{zx}}^2 + 2N{\tau_{xy}}^2}} \tag{4}$$

$$\mu = \frac{2\sigma_{yy}-\sigma_{xx}-\sigma_{zz}}{\sigma_{xx}-\sigma_{zz}} \qquad -1 \le \mu \le 1, with\ \sigma_{xx} \ge \sigma_{yy} \ge \sigma_{zz} \tag{5}$$

$d\bar{\varepsilon}^p = \sqrt{(2/3)d\bar{\varepsilon}^p : d\bar{\varepsilon}^p}$ is equivalent plastic strain increment, $\bar{\varepsilon}_f^p$ is equivalent plastic strain at the failure, η is stress triaxiality, μ is Lode stress parameter; C_1, C_2 *and* C_3 is material parameter which are calibrated from fracture experiments testing. Three parameters of ductile fracture criterion for anisotropic fracture model are calibrated by fracture strain from the in-plane shear, central hole and plane strain tension tests. These experimental test are selected to cover wide range of stress triaxiality and Lode parameter for sheet metal forming.

To account for non-proportional loading conditions, a linear incremental relationship is utilized between the damage indicator D and the equivalent plastic strain to fracture as

$$\mathrm{D}(\bar{\varepsilon}^{\mathrm{p}}) = \int_0^{\bar{\varepsilon}_{\mathrm{f}}^{\mathrm{p}}} \frac{\mathrm{d}\bar{\varepsilon}^{\mathrm{p}}}{\bar{\varepsilon}_f[\eta,\mu]} \tag{6}$$

When the damage indicator is reached $(D(\bar{\varepsilon}^p) = 1)$ the material element is considered to fail and delete. The average value of stress state parameters (η, μ) are calculated by

$$\eta_{av} = \frac{1}{\bar{\varepsilon}_f^p}\int_0^{\bar{\varepsilon}_f^p} \eta(\bar{\varepsilon})d\bar{\varepsilon}^p$$
$$\mu_{av} = \frac{1}{\bar{\varepsilon}_f^p}\int_0^{\bar{\varepsilon}_f^p} \mu(\bar{\varepsilon})d\bar{\varepsilon}^p \tag{7}$$

3. Calibration of the anisotropic ductile fracture model

A hybrid experimental and numerical method was applied to calibrate the fracture model. The force-displacement curves were extracted from each simulation and compared with the experimental measurements. To identify the material parameter of the proposed ductile fracture criterion, the equivalent plastic strain until fracture was measured by conducting experiments with at three different specimens, namely, in plane-shear tension, central hole tension (CH), and plane strain tension (PS). The fracture parameters obtained are summarized in table 3. Figure 1 and 2 respectively exhibit the fracture surface and fracture locus predicted by proposed ductile fracture at the rolling direction for AA6016-AC200.

Table 3. Fracture parameter of proposed fracture criterion

AA6016-AC200	C_1	C_2	C_3
	1.9328	1.8262	3.5241

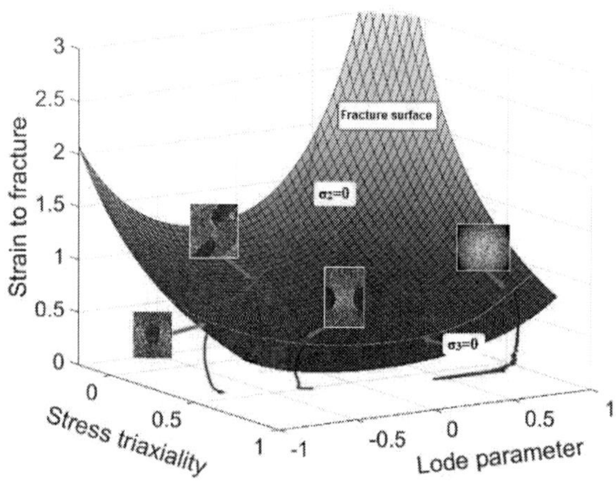

Figure 1. Fracture surface predicted by the proposed fracture criterion at rolling direction

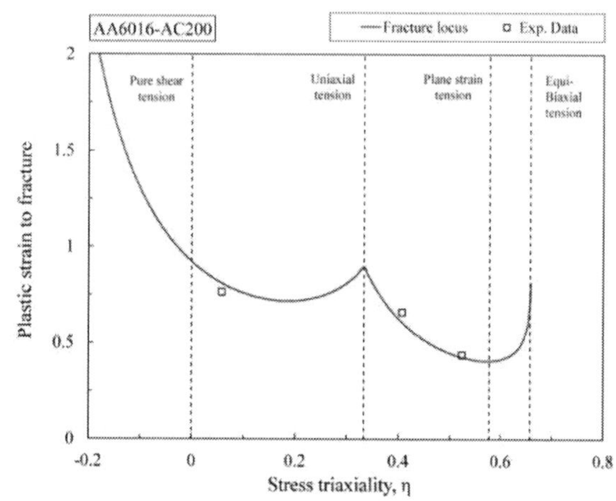

Figure 2. Fracture locus predicted by the proposed fracture criterion at rolling direction

The element contained the maximum equivalent plastic strain (PEEQ) of CH and PS are located in the central of the specimen, while its element of the in-plane shear tension stays on the surface of the specimen that can be explained by the in-plane shear tension is close to the pure shear condition where the thickness reduction behavior can be assumed to ignore. It is highlighted that the evolution of stress triaxiality and the Lode parameter moves away from the ideal-pure shear position ($\eta = 0$ and $L=0$). The reason can be seen from the shear specimen geometry. The in-plane shear tension contains both the pure shear, tension, and compression mode. Since the stress triaxiality and the Lode parameter have a tendency to move further away from the zero point. The fracture strain at the equi-biaxial mode was obtained from the bugle test by ARAMIS system.

4. Validation of anisotropic ductile fracture criterion

To verify the accuracy of proposed ductile fracture criterion, all the fracture experiment tests described above were conducted to predict the onset of fracture behavior. The proposed fracture criterion was implemented in ABAQUS/Explicit by using the VUMAT subroutine. Figures 3(a)–(c) show a comparison of the force-displacement results obtained in FEM simulations and the experimental results obtained with the in-plane shear, central hole, and plane strain tension.

IOP Conf. Series: Materials Science and Engineering **967** (2020) 012032 doi:10.1088/1757-899X/967/1/012032

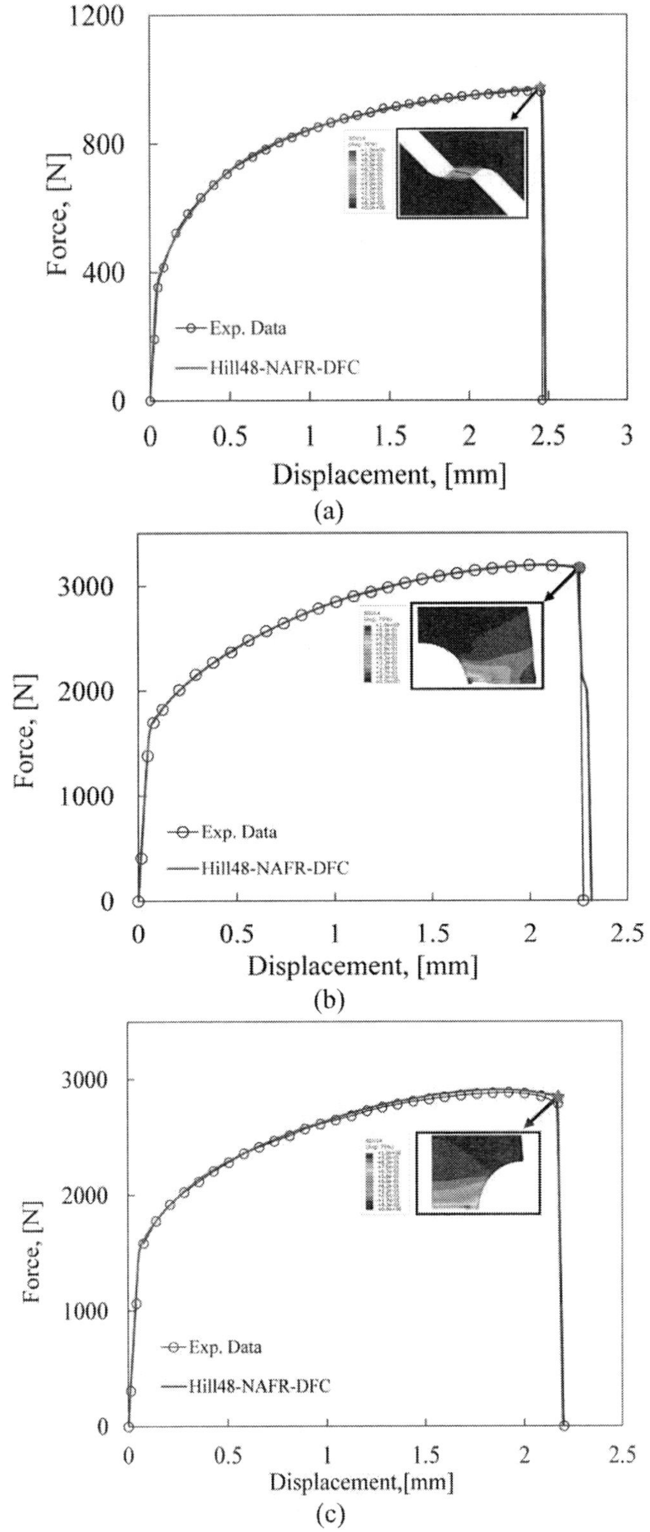

Figure 3. Comparison of load-displacement curves: (a) in plane-shear, (b) central hole, (c) plane strain.

A cup drawing test was performed to evaluate the accuracy of the proposed ductile fracture criterion. As above discussions, the FLD remains a limitation in predicting failure of metal under shear effects, while the ductile fracture criterion can capture a large range of stress state in sheet metal forming. Since, the anisotropic ductile fracture criterion was chosen to predict the fracture behavior of square cup drawing test for AA6016-AC200 in this study.

Square plates (100 mm × 100 mm) were cut from the same metal sheet of Al6016-AC200 for testing. Figure 4 and figure 5 depict schematic of the experimental setup and fracture shape of square cup drawing specimen, which consisted of a cup square punch with a width of 50 mm and an edge radius of 5 mm, and a die with a width of 54 mm and an edge radius of 5 mm. The sheet was arranged along its rolling direction in the coordinate system of the punch. During the entire experimental operation, a constant blank holding force (BHF) of 40 ton was applied on the aluminum sheet metal, while the punch moved upward and drew the blank sheet.

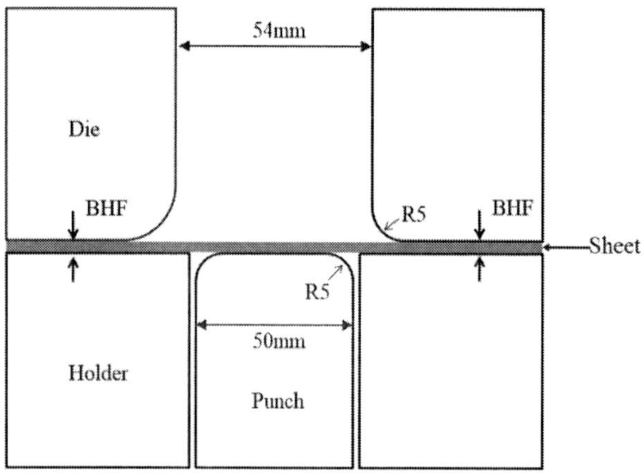

Figure 4. Schematic of the square cup drawing test

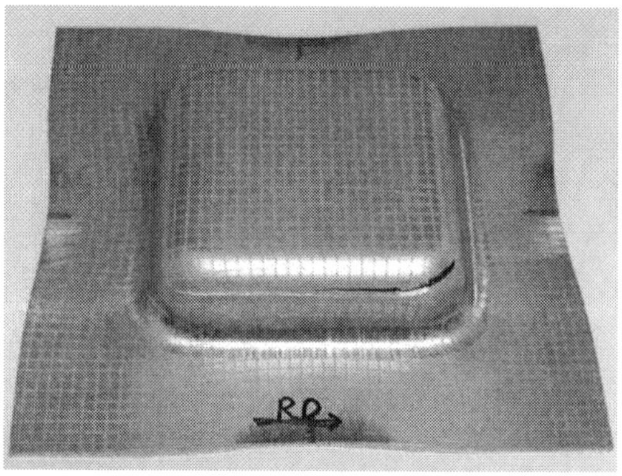

Figure 5. The square cupdrawing specimen

The FE simulation of the cup-drawing test was performed in the ABAQUS/Explicit environment. Because of the symmetric load and geometry, a quarter of the square cup geometry of the specimens was simulated to reduce simulation time. Solid C3D8R elements with reduced integration, available from the ABAQUS element library, were used in the simulation. The element size in the fracture zone was set to 0.2 mm and a friction coefficient of 0.15 is generally used in the simulations. Notablely, Hill'48 non-associated flow rule was adopted to introduce both effect of the Lankford r-values and the yield stresses. The force-displacement and fracture shape of square cup drawing simulation result are shown in figure 6 and figure 7 respectively.

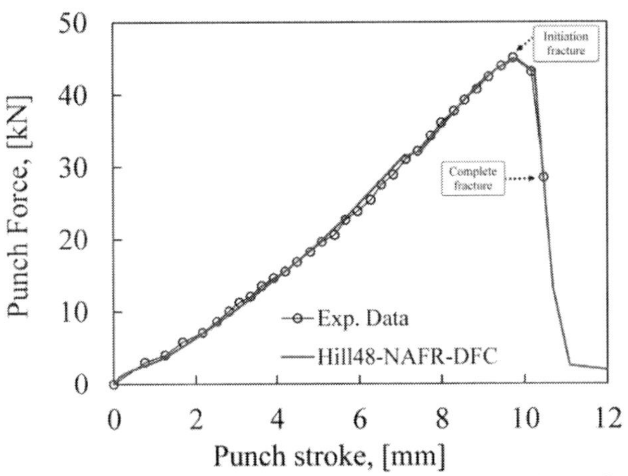

Figure 6. The comparison result of force-punch stroke for square cup drawing test

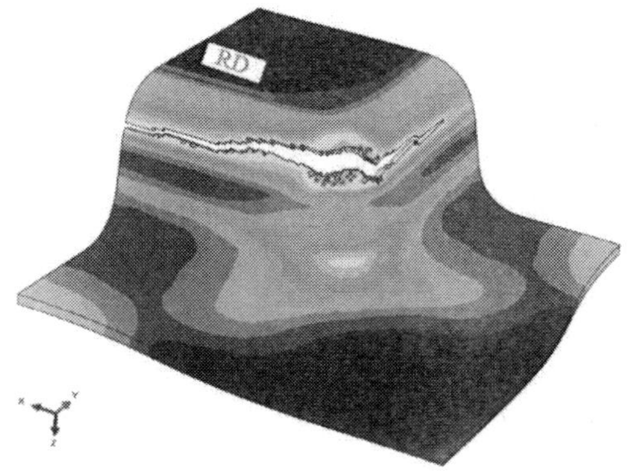

Figure 7. Fracture shape of square cup drawing simulation result

The comparison of the experimental and simulated force-strokes in the cup drawing test showed that the proposed ductile fracture criterion captured well the prediction of anisotropic fracture initiation.

5. Conclusion

An extension of new ductile fracture for anisotropic material is developed in considering effect of anisotropy property for sheet metal forming applications. The proposed ductile fracture criterion showed a good performance in predicting fracture behavior in both standard fracture test and square cup drawing test. By incorporating effect of r-value and the yield stresses in Hill'48 non-associated flow rule in proposed ductile fracture criterion, it is clearly that the proposed fracture model has capability to predict the fracture behavior of the anisotropy material in sheet metal forming.

ACKNOWLEDGEMENT

This work was supported by the Ministry of Education of the Republic of Korea and the National Research Foundation of Korea (NRF-2019R1A2C1011224).

References

[1] Keeler S P and Backofen W A 1963. *Trans. Am. Soc. Metals.* **56** 25-48
[2] Goodwin GM 1968. *SAE technical.*
[3] Gurson A L 1977. *J Eng Mater Technol.* **99**(1) 2-15
[4] Rice J R and Tracey D M 1969. *J Mech Phys Solids.* **17**(3) 201-17
[5] Chu C C and Needleman A 1980. *J. Eng. Mater. Technol.* **102** 249-56
[6] Xue L 2008. *Eng Fract Mech.* **75**(11) 3343-66
[7] Stoughton T B and Yoon J W 2004. *Int. J. Plast.* **20**(4-5) 705-31.
[8] Park T and Chung K 2012. *Int. J. Sol. Struct.* **49**(25) 3582-93
[9] Hill R 1990. *J. Mech. Phys. Solid.* **38** 405-17
[10] Pham Q T and Kim Y S 2017. *Met. Mater. Int.* **23**(2) 254-63.
[11] Quach H, Kim J J, Nguyen D T, Kim Y S 2020. *Int. J. Mech. Sci.* **169**(1)

IOP Conf. Series: Materials Science and Engineering **967** (2020) 012033 doi:10.1088/1757-899X/967/1/012033

An Infrared local heat-treatment method to improve local formability of forming and trimming processes

Ki-Young Kim[1], Soohyun Park[2], Eun-Ho Lee[1]* and Youn Hee Kang[3]

[1] School of Mechanical Engineering, Sungkyunkwan University, Suwon, Gyeonggi-do, 16419, Republic of Korea
[2] School of Mechanical and Control Engineering, Handong Global University, Pohang, Gyeongsangbuk-do, 37554, Republic of Korea
[3] Process & Engineering Research Labs. POSCO, Pohang, Gyeongsangbuk-do, 37859, Republic of Korea

* Corresponding author's email: e.h.lee@skku.edu

Abstract. This paper presents a local heat-treatment method to improve the local formability for stamping and trimming processes by using focused infrared (IR) rays. Although the demand of ultra-high strength steel (UHSS) has been increasing in order to reduce car weights, the low formability of the UHSS has been a limitation. In order to resolve the formability problem, heat treatment methods have been often studied, and the authors also have developed a local heat treatment method by using focused infrared (IR) rays for a decade. The difference of this method from the author's previous works is that the new heat treatment is completed by the supplier. The supplier of the materials completes the local heat treatment on target areas of blanks where to improve the local formability, then stamping companies can conduct cold forming and trimming. The target material of this paper is a 1.5GPa steel, CR 1480M steel supplied by POSCO. This paper discusses the local formability change caused by the presented IR heat treatment.

1. Introduction

Since the environmental problem is becoming an important issue, auto companies have tried to improve energy efficiency and reduce vehicle's weight. As part of the efforts, the automotive industry has increased the use of ultra-high strength steel (UHSS) which has a higher strength with a thinner thickness. The use of UHSS can lead to reduction of the weight of a car with maintaining the strength to improve the fuel efficiency. However, the low formability problem of the UHSS has become a limitation to expansion of the application of the UHSS. Specially, martensitic steel, consisting of martensite microstructure, has a very low formability. Consequently, roll forming process is a widely used for forming of martensitic steel since the roll forming process makes incremental forming at each stage. However, in auto industries, stamping process is popular because it can make complex shape parts under the mass production condition. For application of stamping process, martensitic steel needs to improve formability for stamping and trimming processes. Many studies have been conducted to improve formability of UHSS material by using hot press forming(HPF) and warm-forming [1-8]. The HPF is appropriate to form quenchable materials, such as boron alloyed steel sheet, and the process guarantees the hardness of the material by quenching after forming. While forming under the elevated temperature conditions drastically reduces springback, trimming of UHSS has become another issue, because the quenching process during the HPF increases the strength of the blank. The increased strength of the material results in damage of the trimming tools. A widely used solution in industry is laser trimming without trimming tools. But the laser trimming method leads to a sharp drop in productivity and the investment of the laser equipment is very high.

This paper presents a local heat-treatment process for martensitic steel in order to improve the formability of the steel for the purpose of the stamping process in the auto industry. The difference of this method from the author's previous works [9-11] is that the new heat treatment is completed by the supplier. The supplier of the materials completes the local heat treatment on target areas of blanks where to improve the local formability for stamping and trimming processes. Then, stamping companies can conduct cold forming and trimming while the previous methods conducted warm forming under the elevated temperature conditions. The target material of this paper is a 1.5GPa steel, CR 1480M steel supplied by POSCO. In this paper, V-bending and trimming tests are presented to validate the formability improvement. Microstructure change caused by the heat treatment was also observed. This study shows that this heat-treatment method has a potential to be used for stamping process of the martensitic steel.

2. Local heat treatment

This work uses the infrared (IR) local heater for the heat treatment. The heater consisted of an IR lamp and elliptical reflector. Figure 1 shows the IR local heater, and details of the IR heater is explained in the previous works [12-13]. The heater can heat a narrow heat affected zone(HAZ). The narrow HAZ effectively reduces the decrease of the whole structural stiffness and strength decrease by the heat treatment. Figure 2 shows the performance of the IR heater. The temperature profile history was collected with K- type thermocouples inserted in drilled holes of the specimen. Since the geometry of the specimen and the heater is symmetry, as shown in Figure 2(b), the thermocouples were located in 10 points on the right half of the specimen(0, 5, 10, 15, 20, 25, 30, 40, 50, and 60 mm from the symmetric axis). As observed in the Figure 2(b), temperature distribution was successfully controlled by gathering IR thermal flux at the center of the specimen. The temperature curves at 200°**C**, 400°**C**, 600°**C** and 800°**C** presents that the central temperature is the highest, and it decreases as the distance from the center increases. The reflector effectively concentrates the IR rays on the target HAZ where to improve the formability.

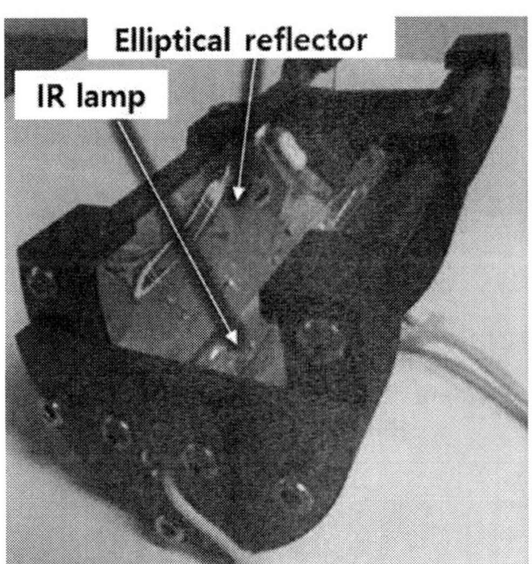

Figure 1. IR Local heater[14]

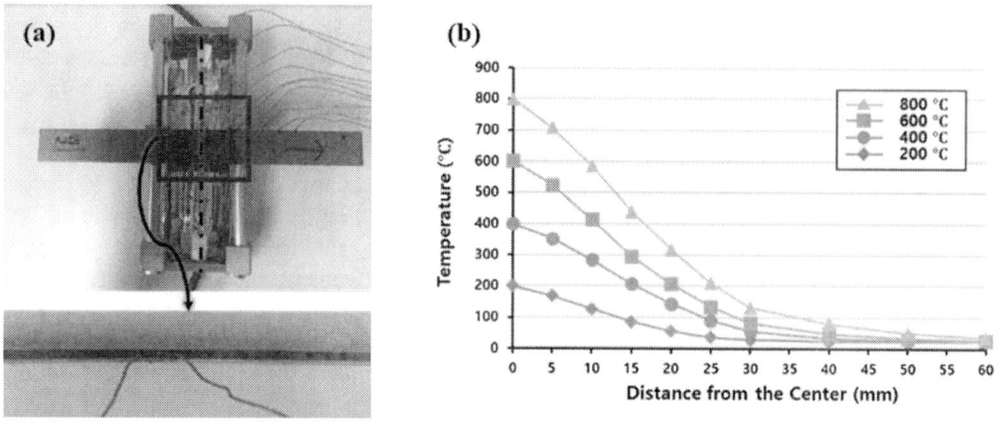

Figure 2. Temperature profile of local heating: (a) K-type thermal couple and (b) temperature profile along the longitudinal direction

The effect of the heat treatment on the microstructure was observed, as shown in figure 3. The specimen was obtained from the heated specimen after cooling to room temperature. Figures 3(a-b) shows optical micrograph images of as-received and 600°C heat-treated specimens, respectively. Figure 3(a) presents the material consists of almost of all martensite. However, the martensite was tempered after the-heat treatment, as shown in figure 3(b). This microstructural change improves the formability.

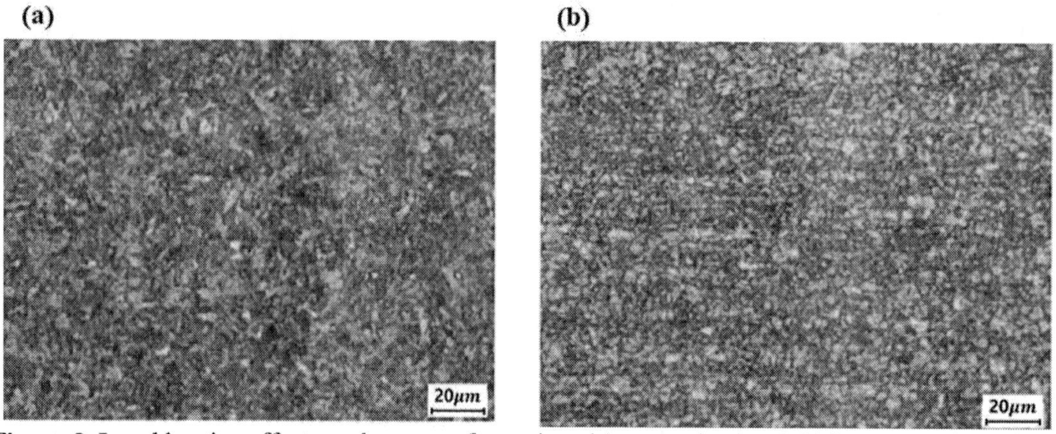

Figure 3. Local heating effect on phase transformation: (a) as-received and (b) heat-treated specimen under 600°C condition

3. Experimental Setting

The V-bending experiment was conducted and repeated three times for five heat treatment conditions (as-received, 200°C, 400°C, 600°C, and 800°C). The center of the HAZ is the reference point of the heating condition. After the central temperature reached the target temperature, each specimen was cooled in room temperature. The dimensions of the V-bending specimen were 180mm × 20mm × 1.5mm. The center line of the HAZ on the specimen was aligned along the rounded edge part of the 2mm punch with 90° angle. The punch speed was 5mm/s. With the same temperature conditions, a trimming test was conducted for three times on each condition. The clearance of the trimming blade and die was 0.12mm (8% of the thickness of the specimen) and the dimensions of the specimen were the same to the V-bending case. The trimming die was attached on 10 ton UTS, so that the maximum punch force at the trimming process could be measured. Figure 4(a-b) presents experimental settings of the V-bending and trimming, respectively.

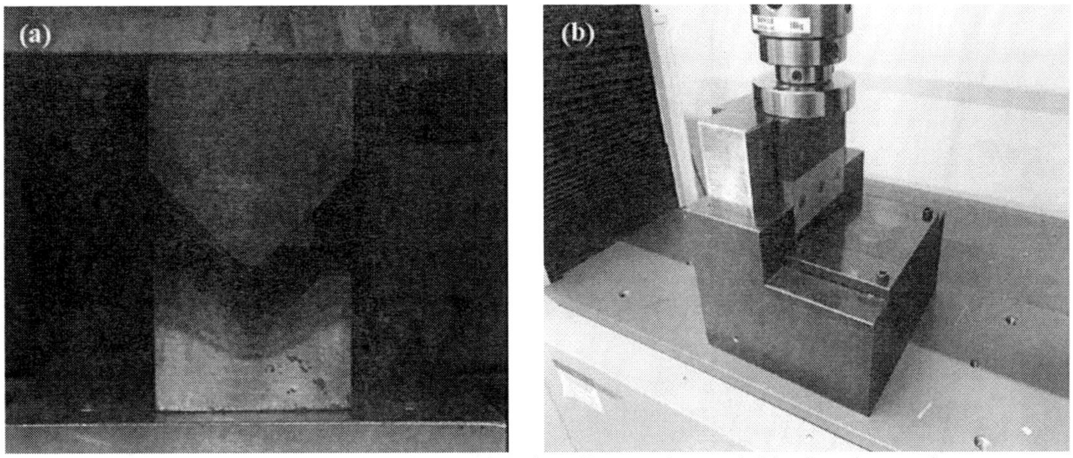

Figure 4. Experimental settings: (a) V-bending setting and (b) trimming setting.

4. Experimental results and Discussion

4.1. V-bending

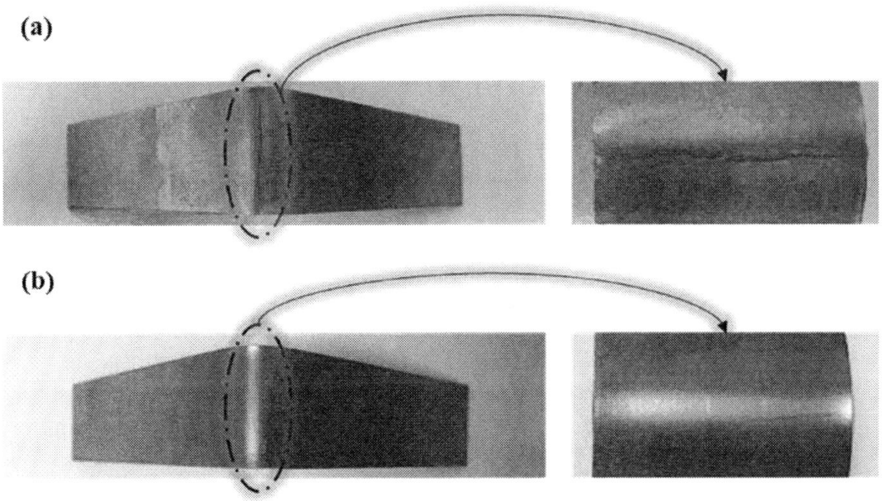

Figure 5. V-bending experiment result: (a) as-received specimen and (b) heat-treated specimen under 600°C condition

Figure 5(a-b) shows outer surfaces of the bent specimens of as-received and 600°C heat-treated, respectively. As shown in figure 5, the as-received material resulted in crack on the center of the bending zone because of the low formability. On the other hand, the heat-treated specimen had no crack in the same forming condition. This can be explained through the improved formability caused by decomposition of martensite. This comparison shows that the IR local heat treatment could improve the formability of CR1480 M. It is well reported that the 600°C condition can reduce springback and improve the formability by decreasing flow stress of advanced high-strength steels [9-10].

4.2. Trimming

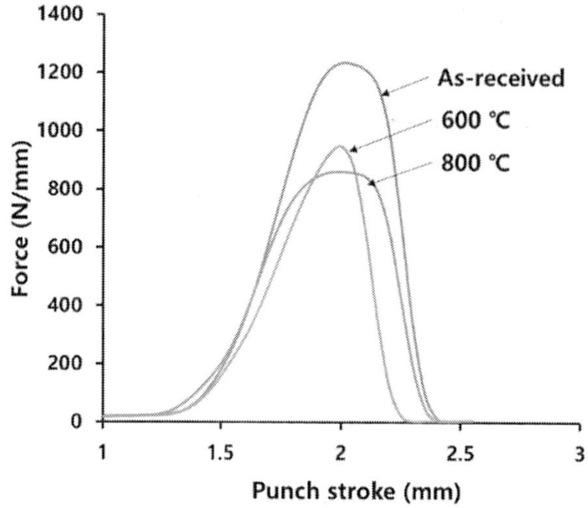

Figure 6. Trimming experiment result of representative case

During the trimming process, punch force was measured as shown in Figure 6. Figure 6 shows decrease of the maximum punch force with respect to the heat-treatment temperature condition. Note that Figure 6 shows the force per width to remove the effect of geometry. While the maximum punch force of the as-received material is 1233N/mm, those of the 600°**C** heat-treated and the 800°**C** heat-treated are 948N/mm and 861N/mm, respectively; 23% and 30% of force reduction. As discussed above, generally 600°**C** condition reduce flow stress of advanced high-strength steels [9-10], it causes the decrease of the trimming force. This can match the trimming force of CP1180 material which is used in cold trimming process in mass production processes.

5. Conclusion

This paper studies forming and trimming process of 1.5GPa steel, CR 1480M supplied by POSCO. The IR local heat treatment can improve the formability of the CR 1480M material in bending and trimming process. Details are below:

(1) The IR heat-treatment can improve the formability of the martensitic steel.
(2) This improvement of the formability can avoid the fracture in the V-bending process.
(3) Approximately, 23% of punch load decrease was observed when trimming was conducted with the 600°C heat-treated material.
(4) Trimming force reduction can match the trimming force of CP1180 material which is used in cold trimming in mass production.
(5) The bending and trimming processes were conducted with the specimens which were cooled after the heat treatment. That means the stamping companies can conduct cold stamping and trimming processes after the heat-treatment was completed by the supplier.
(6) The effect of the local heat treatment on the trimmed edge condition will be studied in a future work.

Acknowledgement

This work was supported by POSCO.

References

[1] Merklein M and Lechler J 2006 *J. Mater. Proc. Tech.* **177** 452-5
[2] Mori K, Maki S and Tanaka Y 2005 *CIRP Ann.-Manuf. Tech.* **54** 209-12
[3] Neugebauer R, Scheffler S, Poprawe R and Weisheit A 2009 *Prod. Eng.* **3** 347-51

[4] Ganapathy M, Li N, Lin J, Abspoel M, Bhattacharjee D 2019 *Int. J. Adv. Manuf. Technol.* **105** 669-682
[5] Zhang P, Zhu L, Luo S Luo J, 2019 *Int. J. Adv. Manuf. Technol.* **103** 3187-97
[6] Ma BL, Wan M, Wu XD, Cai ZY, Diao KS, Han JQ 2017 *J. Manuf. Process.* **30** 320-27
[7] Quan G, An C, Qiu H, Zhang L, Wang X 2019 *Int. J. Precis. Eng. Manuf.* **20**(7) 1169-83
[8] Kim HK, Shim HB, Cha BS, Song GH, Kim HJ 2018 *Int. J. Precis. Eng. Manuf.* **19**(5) 721-28
[9] Lee EH, Hwang JS, Lee CW, Yang DY, Yang WH 2014 *J. Mater. Proc. Tech.* **214** 784-93
[10] Lee EH, Yoon JW, Yang DY 2018 *Int. J. Mater. Form.* **11** 417-33
[11] Lee EH, Yang DY, Ko S 2017 *J. Mater. Eng. Perf.* **26** 5056–63
[12] Lee EH, Yang DY, Yang WH 2014 Int. *J. Heat. Mass. transfer.* **78** 240-50
[13] Lee EH, Yang DY, 2015 *Int. J. Heat. Mass. transfer.* **85** 860-4
[14] Lee EH, Ko S 2018 *IOP Conf. Ser. Mater. Sci. Eng. IDDRG 37th* vol 418 ed P. Hartley et al.

Additive Manufacturing of Tailored Blank for Sheet-Bulk Metal Forming Processes

R Schulte, T Papke, M Lechner and M Merklein

Institute of Manufacturing Technology, Friedrich-Alexander-Universität Erlangen-Nürnberg (FAU), Egerlandstraße 13, 91058 Erlangen, Germany

Abstract. Functional integration and lightweight construction pose increasing demands on manufacturing processes and require innovative approaches. In this context, sheet-bulk metal forming combines the advantages of conventional sheet and bulk forming processes expanding the limitations of these particular processes. However, the intended three-dimensional material flow contains major challenges regarding the material flow control. To enhance material flow control and part quality the application of process-adapted semi-finished products is an expedient approach for sheet-bulk metal forming processes. So-called Tailored Blanks have a process-adapted sheet thickness profile or combine different mechanical properties within a single blank and were applied in the 1980s for the first time. Since then tailored approaches gained in importance and find broad application in research and industry. At present, various technologies are used to manufacture Tailored Blanks. The Tailored Blanks applied in sheet-bulk metal forming are often manufactured by sheet-bulk metal forming processes themselves. The achievable gradient in thickness depends on various factors but is eventually constraint by the material volume of the initial blank. In this regard, Additive Manufacturing offers new possibilities to overcome those limitations and to expand the limits of sheet-bulk metal forming processes further. Additional material can be allocated with high geometric flexibility. Besides the allocation of additional material, this technology allows the manufacturing of discrete functional elements within the production of Tailored Blanks. In this investigation, Tailored Blanks made out of stainless steel are produced using sheet material and additively manufactured elements. These Tailored Blanks are processed in a deep drawing and a subsequent upsetting operation to manufacture a functional component with an external gearing. Conventional sheet material is also processed to compare the resulting part properties and to evaluate the different semi-finished product strategies. For this purpose, an analysis of geometrical as well as mechanical Tailored Blank properties is conducted. Furthermore, the part properties after deep drawing and upsetting are analysed. The investigation evaluates the potential for Tailored Blanks consisting of sheet metal with additively manufactured elements in sheet-bulk metal forming and conclusively points out the need for further research in this field.

1. Introduction

Nowadays, production technology is driven by increasing demands regarding lightweight construction and functional integration [1]. Conventional forming processes are often limited in terms of their flexibility as they are based on costly and time-consuming tool manufacturing. Hence, formed components can only be produced economically at high quantities. Innovative processes, such as Additive Manufacturing (AM), offer a high geometrical flexibility. However, these processes are often not profitable at high quantities. Therefore, a combination of AM and forming technology seems to be a promising approach to utilize the advantages of both processes while reducing the major disadvantages [2]. As forming technology, a combined deep drawing and upsetting process is used. This sheet-bulk metal forming (SBMF) process enables the manufacturing of externally geared functional components out of sheet metal by inducing two- and three-dimensional stress and strain states [3]. These components are manufactured with different gear geometries. To realize failure-free parts at minimum weight, Tailored Blanks are manufactured and applied [3]. These blanks are manufactured in an orbital forming or a flexible rolling process and require a deep process understanding as well as costly tools and machinery to produce various Tailored Blank geometries. In this context AM enables the flexible manufacturing of different geometries with process-adapted sheet thickness profile without costly tooling.

2. State of the Art

SBMF forming processes combine the advantages of conventional sheet and bulk forming operations. The occurring two- and three-dimensional stress and strain state enable the manufacturing of parts with functional elements having a significantly higher thickness than the initial sheet thickness. As semi-finished product, sheet material with a thickness between one and five millimetres is used, according to. [3] To extend the forming limits even further the bulk forming process is often combined with a sheet forming process, as in the process investigated. The basic shape of the component is deep drawn operation before the final contour is formed by an upsetting process within the same press stroke. As process failures for the application of conventional steel blanks with an initial diameter of $d_0 = 100$ mm and an initial sheet thickness of $t_0 = 2$ mm buckling and folding were identified in [4]. The reduction of these failures by the application of orbital formed Tailored Blanks was presented in [4] before they were avoided by adapting the Tailored Blank layout in [5].

By AM processes like powder bed fusion using a laser beam (PBF-LB) parts with high degree of geometric freedom can be produced. However, compared to conventional manufacturing processes as forming, the process time is relatively long. Therefore, it is beneficial to reduce the amount of additively manufactured part volume to a minimum. Hence, an innovative approach is the combination of additive manufacturing and forming, which is already investigated for hybrid parts consisting of a sheet metal body with an additively manufactured element [6]. Besides the application of AM material, the sheet metal is formed by bending or deep drawing, where two process chains; AM followed by forming or forming with subsequent AM are investigated [7]. In order to increase the flexibility of this process combination, incremental forming of the sheet metal body is beneficial, since geometric limitations with regard to the forming tool are avoided [8]. The approach of hybrid blanks also can be used to increase the sheet thickness locally by applying material to the sheet metal [9], which also can be used to reinforce a deep drawn cup leading to higher stiffness and load-bearing capacity of the formed part [2].
With regard to current research, the approach of hybrid parts focus on forming of the sheet metal or apply additively manufactured material on a formed part or a flat blank to increase die filling and improve part properties. A new approach is the use of additive manufacturing to accumulate material on a flat sheet metal to generate sufficient material for forming of functional elements from the AM material.

3. Process chain – Additive Manufacturing and Forming Concept

In this section the fundamentals of the processes to form and manufacture Tailored Blanks will be introduced. The conventional process chain consists of sheet bulk metal forming operations. In order to increase the material amount for forming of functional elements a material accumulation is applied to the sheet metal before the forming operation. The resulting subsequent process steps are PBF-LB for material accumulation on the sheet metal, followed by laser cutting to extract the specimens form the sheet metal body and finally forming by deep drawing and upsetting (Figure 2).

3.1. SBMF process combination

The process combination investigated combines a deep drawing and an upsetting operation within a single press stroke to enable the efficient manufacturing of externally geared functional components. The tool concept features four active components, as shown in Figure 1 a). Drawing die and upsetting punch are part of the upper tool, whereas drawing punch and upsetting punch are part of the lower tool.

In the initial position the semi-finished product is placed onto the deep drawing punch with a drawing punch radius of r = 1 mm and then clamped by the upsetting punch before the drawing die moves downwards to draw the cup. Subsequently, the upsetting punch displaces the drawing punch. As soon as the cup gets in contact with the upsetting plate, the cup wall is upset. The reduction of the cup height forces the material to flow radially into the gear cavity of the drawing die. The hydraulic press is a Lasco TZP 400/3 and the process proceeds force-controlled.

The functional component manufactured with the tool set-up described above is an externally geared cup with 80 teeth, as shown in Figure 1 b). The external diameter amounts to d_P = 82.72 mm and cup height depends on the maximum upsetting force F_U. Due to the occurrence of identified process failures according to the State of the art, Tailored Blanks are manufactured by forming and applied in the process combination. For the target geometry two different Tailored Blank layouts were used within the research project. For layout 1 the area with increased material volume begins at a radial position of r = 40 mm on a blank with an outer diameter of d = 100 mm. This enables the preventions of buckling of the cup wall in a first step. With ongoing investigations, the radial position was adjusted to r = 35.2 mm to prevent folding of the material at the drawing punch radius as well. To investigate the fundamental applicability layout 1 was chosen in the first step, because layout 2 is much more challenging for the additively manufactured structure as in this case it is drawn over the drawing punch radius causing higher stress and strain states.

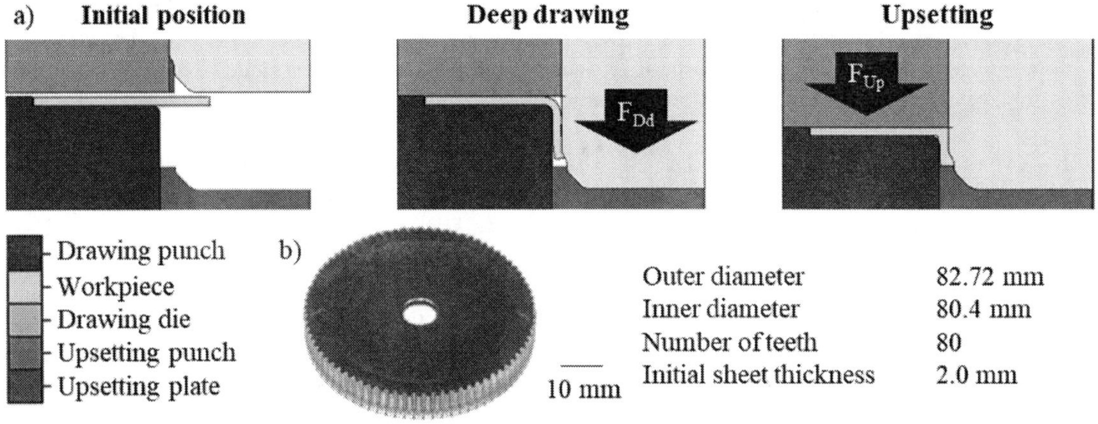

Figure 1: a) Set-up of the combined deep drawing and upsetting process and b) Functional component manufactured within the process

3.2. Additive Manufacturing

The manufacturing of Tailored Blanks is based on PBF-LB. The integration in the process chain is depicted in Figure 2. The process chain starts with laser cutting of a conventional sheet metal which is the initial plate of the PBF-LB process. The conventional sheet has the shape of the built plate of the AM machine. The second step is the manufacturing of AM structures on the sheet metal by PBF-LB. After additive manufacturing the specimens with a diameter of 100 mm are separated from the sheet metal by layer cutting. Due to the maximum size of the built plate four specimens can be manufactured on one conventional blank. These parts can be formed subsequently by deep drawing and upsetting. Regarding the final geometry of Tailored Blanks manufactured by SBMF the initial sheet thickness is t_0 = 1.5 mm with a material accumulation thickness of 0.9 mm. Since the process is a powder bed based layer by layer process starting from the surface of the conventional blank, the 316L sheet is clamped by counter sunk screws on the build plate.

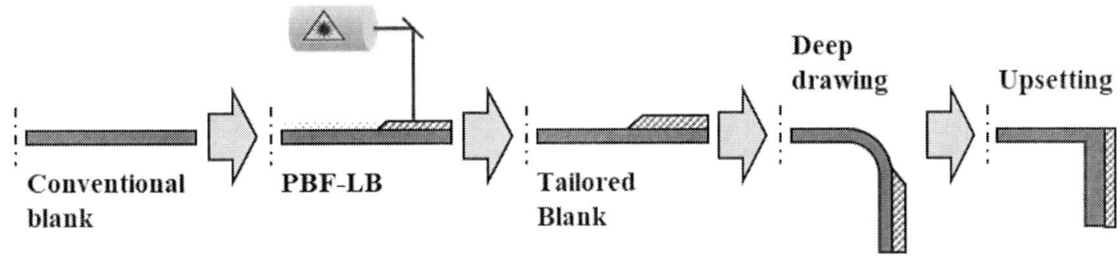

Figure 2: AM-supported process chain to manufacture functional components by SBMF

As machine for PBF-LB is used a LT 30 SLM of the 2. Generation from DMG Mori, which is equipped with a 600 W fibre laser. The beam is shaped to a Gaussian profile with a minimum diameter of approximately 70 µm. To reduce residual stresses and avoid distortion of the sheet metal the heating of the build plate is set to 200 °C. In this type of machine the build plate has a maximum size of 300 x 300 mm^2, where the sheet metal is mounted. For manufacturing the material accumulation on the sheet metal by PBF-LB a volume energy of $E_v = 51$ J/mm^3 is used. The powder size distribution ranges from 19 µm to 43 µm, which is typical for processing 316L by PBF-LB.

4. Results

The results for the combination of AM within SBMF are presented along the process chain. To evaluate the resulting part properties, the component manufactured using an additively manufactured Tailored Blank is compared to a component manufactured from a conventional stainless steel blank with an initial sheet thickness of $t_0 = 2$ mm, according to [4]. The geometrical properties of semi-finished products and components were analysed by using the optical measurement system ATOS by GoM. Furthermore, the micro hardness was measured with a Fischerscope HM2000 by Helmut Fischer to evaluate the mechanical properties.

4.1. Properties of additively manufactured Tailored Blanks

To analyse the properties of the Tailored Blank, the blanks have to be extracted from the built plate by a laser cutting process. After digitizing the Tailored Blank the geometrical properties are evaluated. The thickness profile of the Tailored Blank is shown in Figure 3. In the evaluation range the thickness corresponds to the target thickness of $t_T = 2.4$ mm. However, due to occurring distortion of the built plate during AM the sheet thickness in the outer area varies tangentially. Figure 3 presents the additively manufactured Tailored Blank, the adapted sheet thickness profile and the results for the micro hardness in the inner and outer area of the blank.

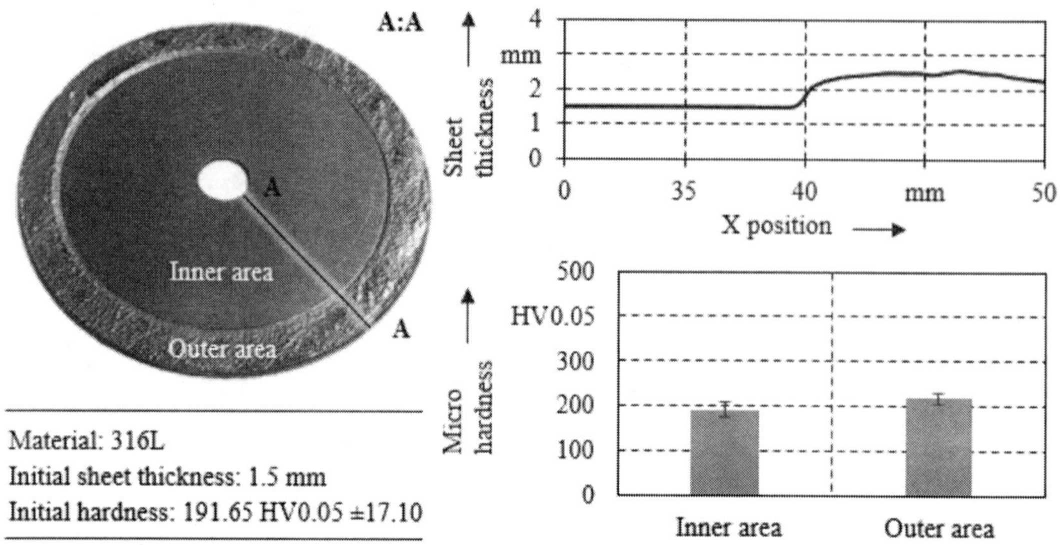

Figure 3: Additively manufactured Tailored Blanks with geometrical and mechanical semi-finished product properties

The initial hardness of the sheet material is in accordance with the requirements of conventionally manufactured 316L material [13] and is similar to the hardness in the inner area after the AM process, which is 191.35 ± 13.9 HV. This indicates a minor impact of the PBF-LB process step on the mechanical properties of the sheet outside the interaction zone. Whereas in the outer area which includes sheet metal and additive manufactured material the hardness is increased to 217.88 ± 36.4 HV. The increase in hardness is reasoned by the grain structure of the additively manufactured material. The metallographic structure of the AM part is needle-like in columnar grains throughout the weld beads [14]. This structure results from the high cooling rates during powder bed fusion.

4.2. Deep drawing of additively manufactured Tailored Blanks

In the deep drawing operation, the basic shape of the component is formed. In contrast to the application of orbital formed Tailored Blanks the deep drawing process represents the first forming operation in the AM-supported process chain. The resulting cup height after deep drawing is $h_C = 14.74$ mm and the material volume in the cup wall amounts to $V_W = 6830.35$ mm^3. The increased sheet thickness in the outer area provides an increased material volume for a higher die filling within the subsequent upsetting process, as shown in Figure 4. Nevertheless, the contour of the drawn cup shows the minor sheet thickness in area of the drawing punch radius. This minor sheet thickness leads to folding of the material according to [5]. The geometrical properties correspond to the resulting properties for Tailored Blanks manufactured by forming.

The hardness in the area of the bottom of the cup amounts to 213.98 HV0.05. Compared to the initial hardness this is an increase of almost 12%. However, the hardness in the cup wall increases by 60% during the deep drawing operation in order to higher strains in this area.

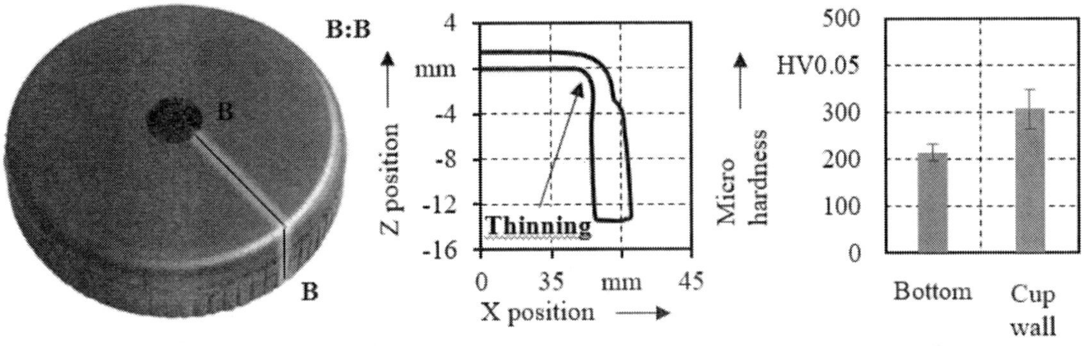

Figure 4: Geometrical and mechanical properties for the additively manufactured Tailored Blank after deep drawing

4.3. Upsetting of additively manufactured Tailored Blanks

The upsetting operation proceeds after the deep drawing operation, though within the same press stroke. As intended, the comparison of the contours in Figure 5 shows, that the increased sheet thickness prevents buckling of the cup wall. A maximum upsetting force of $F_U = 1{,}000$ kN was chosen. This results in a cup height of $h_C = 9.32$ mm for the final component. To evaluate the properties of the functional component and to analyse the applicability of AM for Tailored Blanks in SBMF processes the manufactured component is compared to a component manufactured by applying a conventional semi-finished product with an initial sheet thickness of $t_0 = 2$ mm. The conventional semi-finished product was also upset with a maximum upsetting force of $F_U = 1{,}000$ kN. Compared to the Tailored Blank, the resulting cup height is 0.43 mm higher, due to uncontrolled buckling and strain hardening of the cup wall during the forming operation. For the material volume in the gear cavity, the Tailored Blank allows an increase of up to 15%. The weight of the component can be reduced by more than 10% with improved geometrical properties for the application of the additively manufactured Tailored Blank, assuming a density of 7.8 g/cm^3. Beside geometrical properties the upsetting process also has an influence on the micro hardness in the bottom and cup wall for conventional semi-finished blank and AM Tailored Blank. Regarding the initial hardness values for the austenitic stainless steels for both types of blanks the hardness increases moderately in the bottom area but severely in the cup wall. For the conventional semi-finished product with an initial hardness of 191.65 HV0.05 the micro hardness in the bottom amounts to 302.26 HV0.05 and to 426.60 HV0.05 in the cup wall. For the AM Tailored Blank the values amount to 219.78 HV0.05 and 363.87, respectively. The high increase in hardness in the cup wall is reasoned by strain hardening of the sheet metal and additively manufactured material due to the forming operation. In [10] the influence of an upsetting operation on the work hardening behaviour of additively manufactured 316L is investigated. It is shown a significant increase in hardness in order to the forming operation. Since the stress and strains during upsetting are significantly higher compared to the deep drawn condition the work hardening and hardness are higher in this area. For the conventional sheet metal with $t_0 = 2.0$ mm the forming operation results in higher hardness than for the Tailored Blank since the buckling and folding of the sheet lead to even higher strains and increased inhomogeneity of the hardness distribution. The inhomogeneity can be evaluated over the standard deviation, shown in Figure 5. The increase of material flow control and the improved part properties represent the main advantages of the application of Tailored Blanks in SBMF processes.

IOP Conf. Series: Materials Science and Engineering **967** (2020) 012034 doi:10.1088/1757-899X/967/1/012034

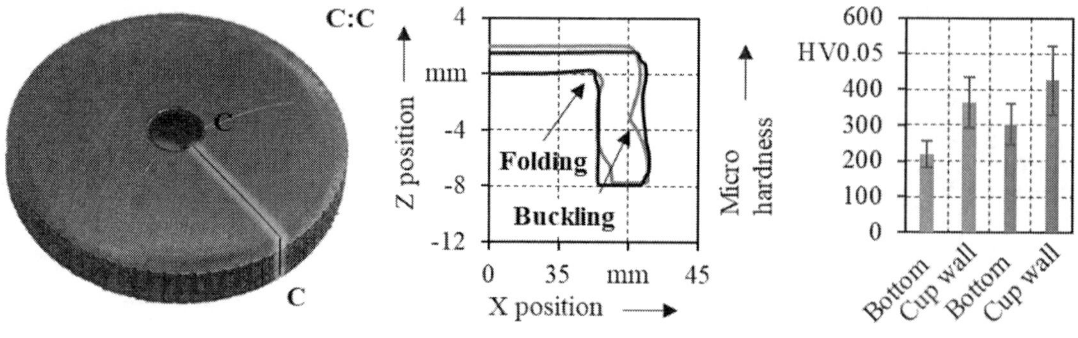

AM Tailored Blank
Material: 316L
Initial sheet thickness: 1.5 mm (built plate)
Initial hardness: 191.65 HV0.05 ±17.10

Conventional semi-finished product
Material: 1.4835
Initial sheet thickness: 2 mm
Initial hardness: 242.2 HV0.05 ±11.70

Figure 5: Resulting geometrical and mechanical properties and comparison of the contours for the application of additively manufactured Tailored Blank and conventional semi-finished product

4.4. Adaption of the Tailored Blank layout

Based on previous research, buckling was prevented by increased sheet thickness in the outer area of the blank and the fundamental applicability for additively manufactured Tailored Blanks in SBMF was verified. In a next step, the radial position of the area with increased thickness is adapted to prevent folding as well. The adapted layout was chosen according to [5] and is presented in Figure 6.

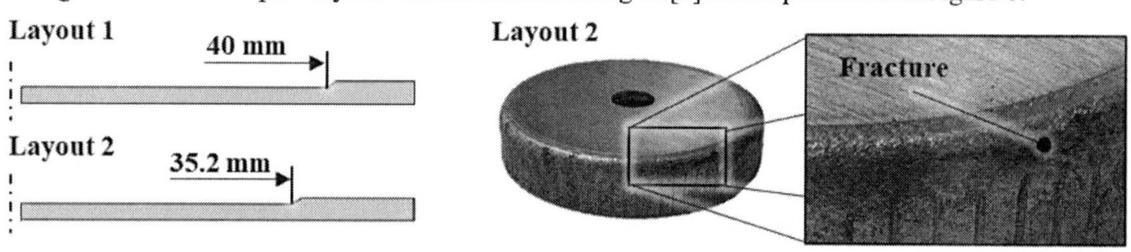

Figure 6: Adaption of the Tailored Blank layout and identification of fracture at the drawing punch radius after deep drawing

The broader radial material accumulation represented by layout 2 is beneficial for a longer tooth length and therefore results is a higher volume of functional elements along the cup wall. During deep drawing of the Tailored Blank with layout 2, fracture of the additively manufactured structure occurs on the upper side of the blank. In this area of the drawing punch bending stresses are predominant [15]. This indicates that the bending stress has an influence on the forming limits of the sheet metal with additively manufactured material. A possible explanation for the initialization of cracks is the increased stiffness in this area of the tailored blank due to the material accumulation. Regarding to the assumption of bending stresses a higher stiffness leads to higher bending resistance and with that to higher stresses on the surface. However, the fracture behaviour needs further investigation for example by analysing different forming states of the cup concerning crack initiation and fracture.

Furthermore, the high flexibility of AM enables the realization of a broader spectrum of Tailored Blank layouts. Analogous to the research on SBMF, the manufacturing of Tailored Blanks with discrete functional elements as well as a combination of additively manufactured and orbital formed Tailored Blanks is investigated. Exemplary layouts defined within the present investigations are shown in Figure 7.

Figure 7: a) Realizing discrete functional elements within the manufacturing of Tailored Blanks and b) combination of additively manufactured and orbital formed Tailored Blanks

The Tailored Blank with discrete functional elements can easily be adapted to various teeth geometries and the combination of AM and forming within the production of Tailored Blanks enables the flexible manufacturing of components with two-sided functional elements. However, the AM structures have to be manufactured first in the presented process chain and therefore, cavities for these elements have to be arranged in the tools for the subsequent forming process.

5. Conclusion and outlook

By combining additive manufacturing and sheet bulk metal forming operations as deep drawing and upsetting it is shown that the forming of tooth geometry can be improved by material accumulation compared to conventional blank without material accumulation. The presented approach offers the opportunity to manufacture different Tailored Blank geometries due to high geometric freedom of additive manufacturing, whereas the geometry is limited to the tool geometry in case of conventional forging operations. The radial material accumulation on the sheet metal surface is manufactured by PBF-LB, which results in higher hardness for AM material than for the sheet metal material. Compared to the initial state the hardness increases after deep drawing and especially after upsetting due to strain hardening. Regarding previous research two layouts of material accumulation with different radial position are investigated to evaluate the impact on the formability of the tailored blank. In case of a broader AM material accumulation the volume fraction of functional elements represented by the length of the tooth geometry is increased. This indicates that this layout is beneficial for manufacturing of functional parts made by additive manufacturing and forming. However, the bending stress in the area of the drawing punch results in fracture of the additively manufactured material characterized by cracks in the surface. Therefore, further investigations should focus on the identification of mechanisms of failure in the AM material and the sheet metal. Besides, the possibility of geometric flexibility of PBF-LB should be exploited to create a geometry which reduces the strains in the additively manufactured material. A promising approach is to process the Tailored Blanks with discrete functional elements manufactured by PBF-LB in the combined deep drawing and upsetting process. This could decrease strains in the forming zone and reduce form filling of the tooth geometry since the main part to be formed is the sheet metal. Furthermore, an upgrading of the tools for orbital forming should be taken into account for the manufacturing of two-sided functional components. In addition, the orbital forming process could also be used to calibrate and to adjust the geometrical and mechanical properties of the additively manufactured elements prior to the final processing.

[1] Cooper D R, Rossie K E, Gutowski T G 2017 The energy requirements and environmental impacts of sheet metal forming: An analysis of five forming processes *Journal of Materials Processing* **244** pp 116-135
[2] Bambach M, Sviridov A, Weisheit A, Schleifenbau J. H. 2017 Case studies on local reinforcement of sheet metal components by laser additive manufacturing *Metals* 7/4 113

[3] Merklein M, Allwood J M, Behrens B A, Brosius A, Hagenah H, Kuzmann K, Mori K, Tekkaya A E, Weckenmann A 2012 Bulk forming of sheet metal *CIRP Annals – Manufacturing Technology* **61/2** pp 725-745

[4] Schneider T, Merklein M 2013 Manufacturing of geared sheet metal components by a single-stage Sheet-bulk metal forming process *In: Dimitrov D, Schutte C (eds) Proceedings of the International Conference on Competitive Manufacturing* pp 177-182

[5] Schulte R, Hildenbrand P, Lechner M, Merklein M 2017 Designing, Manufacturing and Processing of Tailored Blanks in a Sheet-Bulk Metal Forming Process *Procedia Manufacturing* **10** pp 286-297

[6] Schaub A, Ahuja B, Karg M, Schmidt M, Merklein M 2014 Fabrication and characterization of laser beam melted Ti-6Al-4V geometries on sheet metal *Direct Digital Manufacturing Conference* pp 1–5

[7] Merklein M, Dubjella P, Schaub A, Butzhammer L, Schmidt M 2016 Interaction of additive manufacturing and forming *Proc. Int. Conf. on Additive Technologies* pp 309–316

[8] Hölker R, Khalifa N B, Tekkaya A E 2018 Verfahren und Vorrichtung zur kombinierten Herstellung von Bauteilen mittels inkrementeller Blechumformung und additiver Verfahren in einer Aufspannung *Patent* DE 10 2014 014 202 A1 2016.03.24

[9] Bambach M D, Bambach M, Sviridov A, Weiss S 2017 New process chains involving additive manufacturing and metal forming – a chance for saving energy? *Procedia Engineering* **207** pp 1176–1181

[10] Papke T, Merklein M 2019 Characterization of work hardening behavior of additively manufactured stainless steel 316L (1.4404) using bulk metal forming at elevated temperature *Lasers in Manufacturing* **153** pp 1–10

[11] Papke T, Junker D, Schmidt M, Kolb T, Merklein M 2018 Bulk metal forming of additively manufactured elements *Proc. Int. Conf. on New Forming Technologies* **190** 030021-26

[12] Bambach M, Kwiatkowski L, Meiners F, OTTO Fuchs 2019 Verfahren zum Herstellen eines Strukturbauteils aus einem hochfesten Legierungswerkstoff *Patent* PCT/EP2019/053082

[13] German Institute for Standardization 2014 DIN EN 10088-3:2014 - Technical delivery conditions for semi-finished products, bars, rods, wire, sections and bright products of corrosion resisting steels for general purposes

[14] Zitelli C, Folgarait P, Di Schino A 2019 Laser Powder Bed Fusion of Stainless Steel Grades: A Review *Metals* **9/731** pp 1-27

[15] Siegert K 2015 Blechumformung – Verfahren, Werkzeuge und Maschinen

Acknowledgments

This work was supported by the German Research Foundation (DFG) in the scope of the Transregional Collaborative Research Centre on Sheet-Bulk Metal Forming (TCRC 73) in the subproject A1 (Deep Drawing) as well as of the CRC on Additive Manufacturing (CRC 814) in the subproject B05 (Project-ID 61375930). Furthermore, the authors acknowledge the support by the DFG within the Gottfried Wilhelm Leibniz Price (DFG ME2043/38-1) for the financial support.

IOP Conf. Series: Materials Science and Engineering **967** (2020) 012035 doi:10.1088/1757-899X/967/1/012035

Development of inline closed-loop vibration control in progressive die stamping using finite element simulation

F Steinlehner[1*], M Ott[1], D Budnick[2], A Weinschenk[3], S Laumann[4], M Worswick[2] and W Volk[1]

[1] Chair of Metal Forming and Casting, Technical University of Munich, Walther-Meissner-Strasse 4, 85748 Garching, Germany
[2] University of Waterloo, 200 University Ave West, Waterloo, Ontario, Canada
[3] Hexagon Canada, 555 Michigan Drive, Oakville, Ontario, Canada
[4] Men at Work GmbH, Hertzstrasse 1, 76467 Bietigheim, Germany
[*]florian.steinlehner@utg.de

Abstract. Modern progressive dies are increasingly equipped with integrated sensors and actuators, thus enabling an improvement of the manufacturing process and an inline monitoring of the component quality. The design of the tooling is determined at an early stage and is largely based on the desired product and the specific production stages such as blanking, forming and punching. Nevertheless, malfunctions such as strip vibrations cannot always be foreseen and may occur during operation time, which may lead to reduced component quality and in the worst case to component collision or a damage of the tooling. Therefore, complex reworking or a reduction of the stroke rate and thus lower output quantity are often necessary for a reliable and stable process. Strip vibrations are often caused by the highly dynamic transportation of the strip in the tool and can have various causes. Among others, passive tooling components such as spring-loaded, hard stop limited strip lifter can be the cause of such vibrations. Strip lifter are always necessary when three-dimensional components are produced and have to be lifted out of the die for the feeding phase. The feeding phase takes place between two strokes of the stamping process. This work is aiming for a control strategy to suppress strip vibrations in various progressive die stamping processes based on closed-loop controlled active strip lifter. These strip lifter combine the spring-loaded passive standard strip lifter with an additional PID-controlled actuator. Taking the dynamics of the flexible strip during operation into account, a Finite Element Analysis (FEA) model of a progressive die tooling system is created. For the design of the control algorithm, the FEA model is connected to an environment for model-based design in a co-simulation. This approach allows modelling the influence of arbitrary control parameter settings on the movement curve of the strip, aiming for an increased stroke rate.

1. Progressive Die Stamping

Progressive die tools are used in the field of the automotive, the electronics and the household appliance industry to produce predominantly flat parts from sheet metal. A progressive die tool consists by definition of multiple individual stages, through which the strip of sheet material runs progressively. Each of these stages performs one or multiple manufacturing operations on the part. Transport of the strip is done by a carrier strip, which holds the part with stretch web connectors. The sheet metal is usually presented as coiled strips, unwound by decoiler systems and fed into the press by roll feeding machines. With each stroke, one or multiple parts are produced at the end of the tooling. In case of deep drawing operations or three-dimensional regions on the part, strip lifters are often used to lift the strip up from the die. The aim of this lifting is to avoid collisions between tooling components and the strip in order to maintain a safe part transport. Due to high acquisition costs of the tools, progressive die stamping is mainly used for high-volume production and is therefore designed to operate at a high speed. As a result, vibrations of the strip need to be reduced to enable higher stroke rates while maintaining a precise positioning of the strip in the desired position. The design of the strip-layout and the tool are the subject of current research aiming to maintain a safe operation of the tool at an early stage of the design process. [1], [2]

As shown in figure 1, a progressive die tool consists of a base die plate, a top punch plate and several single manufacturing stages which can be guided to each other by column guides. In each module, there usually is a punch plate, a die plate and a blankholder plate which is connected to the punch plate by coil or gas springs. The active parts for the forming operations of each step are mounted on the punch and the die plate.

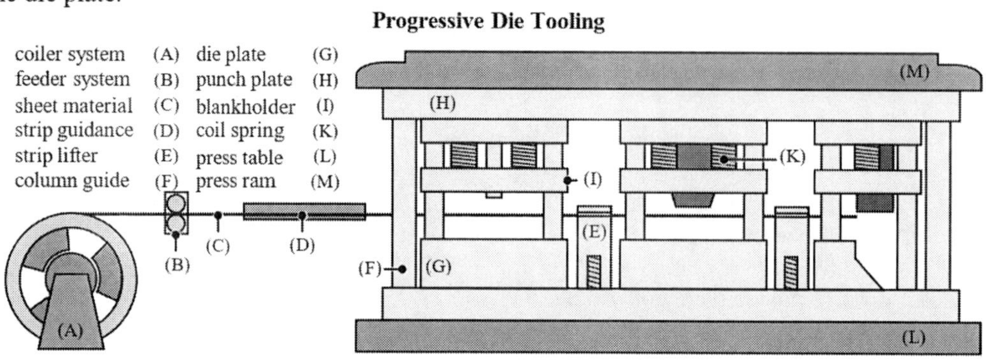

Figure 1 Components of a progressive die tool with three modules and spring loaded strip lifter.

In operation, the tool is mounted inside the press machine and a producing stroke starts at the top dead center (TDC) with the press ram going down (*phase I*). With the blankholder going down, the strip is pushed towards the die. In this phase, the strip is often positioned by pilot pins. During the downward movement of the strip, the coil loaded spring lifter guiding the strip, is compressed. After the strip is fully clamped against the die in its designated position, the actual operation process begins with the punches being pushed towards the strip. Deep drawing, cutting, embossing and other operations must be finished until the blankholder releases the strip from being clamped. The lower position of the press ram is called bottom dead center (BDC, *phase II*) and represents a half stroke of the press machine. After the forming process the tool opens and the compressed spring of the strip lifter follows the movement of the blankholder towards its upper position. The designated position of the strip for transport is maintained via stop plates, which force the strip lifter to immediately stop their movement (*phase III*). This abrupt stop induces an impact into the strip lifter and causes it to transmit vibrations on the strip. Another cause for vibrations on the strip is the inertia of the strip itself. Shortly before reaching the stop position, the strip has a velocity in the upper direction, which leads to an overshoot of the strip. During the vibrating period, the strip alternately strikes between the boundaries of the guiding gap. This introduces multiple impacts on the strip. The guiding of the strip and the operation phases of the progressive die are shown in figure 2. There are several patents for the passive design of strip lifter, whereas the active control of the movement curves is still poorly researched. [3], [4], [5]

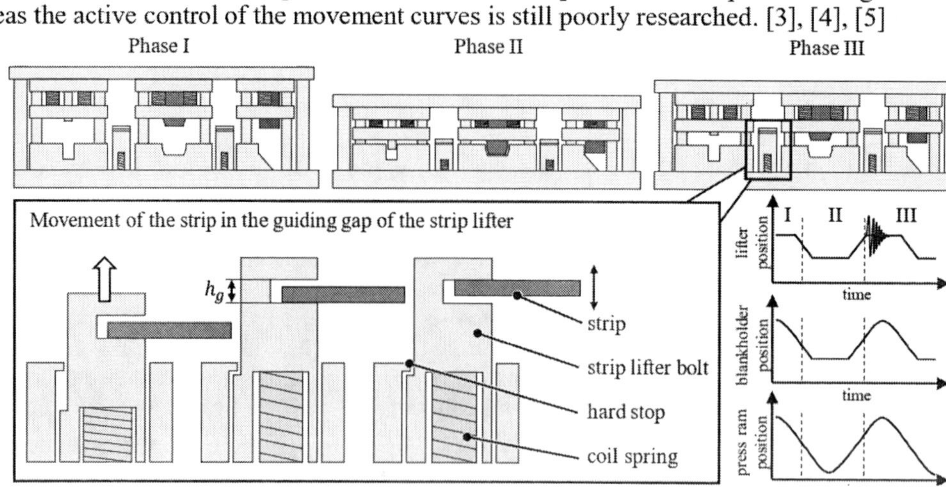

Figure 2 Phases of Progressive die operation and strip guiding gap on the strip lifter bolt.

2. Co-simulation

2.1. Overview of co-simulation structure

One aim of this work is to implement a closed-loop control using a finite element tool, a numerical computing environment and a graphical block diagramming tool.[6], [7] The possibility of coupling a finite element simulation in *Abaqus 2018* with a controller model implemented in Simulink has already been shown in [8] using subroutines. In the following, the approach of using the restart functionality in *Abaqus* and connecting it to *Matlab R2018b/Simulink 9.2* is described. The co-simulation is performed using a *Matlab* script, which is used to coordinate the different building blocks and is executed as a closed loop controller. Figure 3 represents the structure of the co-simulation as a block diagram.

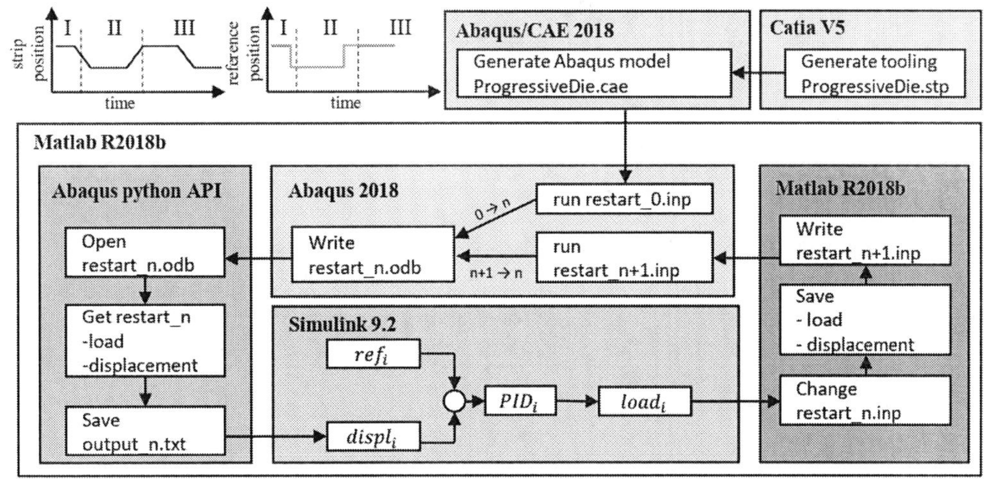

Figure 3 Block diagram of the structure of the developed co-simulation approach

Before the start of the simulation a simplified progressive die tooling was designed using *Catia V5*. The CAD design is then forwarded to the *Abaqus/CAE* environment as a step file (*ProgressiveDie.stp*). In the Abaqus model an initial input file (*restart_0.inp)* is generated. Based on this input file the simulation is initialized and the boundary conditions for modelling the behavior of the tool are set. This first input file can then be accessed from the main Matlab file (*CoSim.m*) using the command line. As an outcome of the first simulation, in Abaqus the results are written into the *restart_n.odb* file. The .odb-file is opened with a python script (*output.py*) in the Abaqus python API environment, which extracts the results information needed to control the lifted strip in the model. Specifically the load curves of the strip lifter, the displacement curves on several characteristic points of the strip and the displacement curves of the strip lifter are exported into an *output_n.txt* file. These movement curves serve as input signal for a Simulink model (*CoSim.slx*), which compares the current displacement of the strip with a reference value and determines a load with the help of a PID controller. The new load forces of the strip lifter are then forwarded to Matlab, where a new input file (*restart_n+1.inp*) for the next loop iterations is created. The load and displacement curves for this iteration are also saved. The new input file is then again processed in Abaqus and the results are saved as *restart_n.odb*, which indicates the next iteration of the co-simulation loop.

2.2. FEA Abaqus Model

The Abaqus model of the progressive die tooling used to create the input files for the co-simulation is shown in figure 4. The strip is pushed down and lifted to a height of 20 mm within each stroke. On the left side of the tool is a strip guidance, which defines the upper position of the incoming strip and two feeding rolls. They represent the feeding machine transferring the strip through the tool. The progressive die tool contains three modules consisting of a die, a punch plate and a blankholder respectively. *module A* represents a blanking station, *module B* a forming station and *module C* a cutting station, which cuts off the produced part at the end of the manufacturing process. The tooling is modelled without the active components for the actual forming and cutting processes in the modules to focus on the lifting operation during the strip movement. Each station has one strip lifter on each side of the strip with a guidance gap of 0.7 mm.

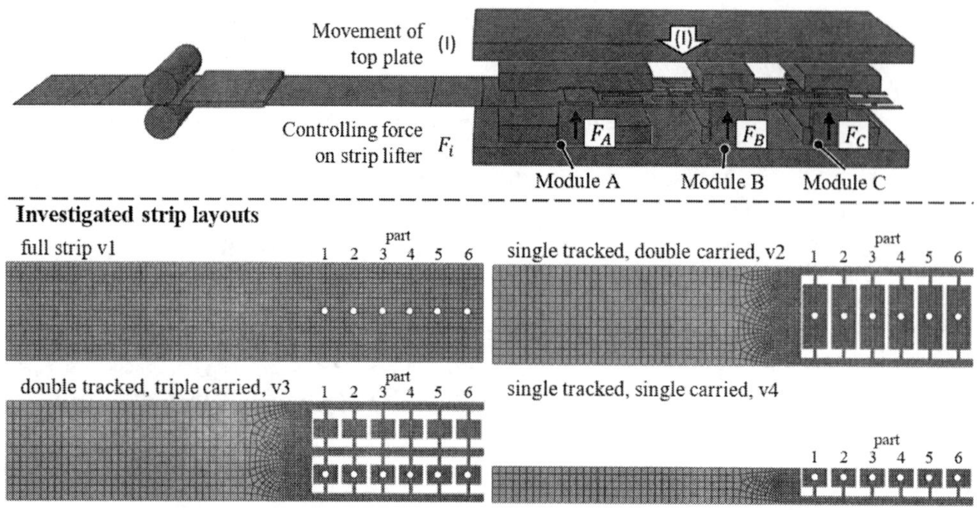

Figure 4 Assembly of the progressive die tool and layout of the investigated strips.

For the passive model configuration, the strip lifters are implemented spring-loaded with a linear spring (figure 2) with a spring rate of 5 N/mm. The force acting on the strip lifter is described in equation (1) as a function of the vertical strip lifter position z and the vertical strip lifter position z_0 at which the spring is in its designated upper position:

$$F_{SL,passive} = c * (z - z_0) \quad (1)$$

For the active strip lifter, a force $F_{control}$ acts via single point force in addition to the spring. For an actuator modelled with ideal dynamics, the total force $F_{SL,active}$ is applied to the strip lifter:

$$F_{SL,active} = c * (z - z_0) + F_{control} \quad (2)$$

The tool components are modelled as rigid shell bodies to handle simulation times and maintain a good contact behavior with the strip material. Four types of strips are modelled in order to consider different cases of progressive die stamping. Each of the strips has a thickness of 0.5 mm and is modelled using shell elements (S4R) with an AA5182 material. The meshed strips are shown in figure 4.

The *full strip, v1* is used to set up the model and to tune the PID controller implemented in the Simulink model. The strip layout *single tracked, double carried, v2* represents a configuration for a single, bigger part, which is connected to two outer carrier strips and is considerably more likely to have vibrations on the part while lifting. Whereas, the strip layout *double tracked, triple carried, v3* is supposed to have more vibrations relating the whole area of the part and stretch web connector. This configuration represents a tooling for producing two parts per stroke. A configuration which is sometimes chosen for small products is represented in the strip layout *single tracked, single carried, v4*, where the carrier is applied to just one side of the product, which leads to a high amount of vibrations on the unsupported side of the strip. Due to its geometry, this strip is only in contact with one of the two strip lifters.

2.3. Simulink Control Model

The control algorithm for reducing the vibrations in the strip was modelled in Simulink. As input signal, the Simulink model receives the vertical position at three points (part1, part4, part6 - figure 4) on the strip from the Matlab master script. This signal represents the feedback signal of the control loop. In the target curve function block, an arbitrary target trajectory can be specified as the reference variable for the vertical position of the strip. The control deviation is the difference between the reference variable and the read-in values of the controlled variable. Within the Simulink model, the force to be applied by the actuators at the three independent pairs of strip lifters is calculated from the control deviation. This actuating signal is written into the Matlab workspace and is thus available to the Matlab master script.

To calculate the actuating signal, a separate PID controller was implemented for each of the three spring lifter pairs. A PID controller is used as a universal type of control loop mechanism, which is a parallel combination of a proportional term, an integral term and a derivative term. In [9] a PI controller was used for vibration suppression of a beam structure and in [10] a PID controller was used for damping the vibrations of a smart beam. In [11] a PID controller is applied to the problem of vibration control of a cantilever beam. An ideal PID controller has the transfer function, including the three gain factors k_p, k_i and k_d, shown in equation (3):

$$K_{PID}(s) = k_p + k_i \frac{1}{s} + k_d s \tag{3}$$

The behavior of the PID controller is characterized by the value of these three gain factors:

k_p: The proportional gain factor leads to a rapid reduction of the control deviation, but cannot usually correct it completely. An increase of this factor can result in an increased overshoot of the system when a step excitation is applied.

k_i: The integral gain factor leads to a slow response of the controller. In contrast to the other two control components, however, it enables a complete correction of the system deviation for a step excitation.

k_d: The derivative gain factor cannot be used as a stand-alone controller, since it does not carry out any control intervention for any constant control deviation. A further disadvantage is that an existing measuring noise is amplified. It is advantageous that the controller provides a high control signal even in the case of a small control deviation but a high increase in control deviation.

For the real implementation of the derivative term shown in equation (4), an additional filter coefficient N is integrated. This formulation leads to a lower sensitivity to a noisy input signal. The output K_{PID} of the controller is forwarded to Abaqus as a single point force signal for each strip lifter:

$$F_{control} = K_{PID}(s) = k_p + k_i \frac{1}{s} + k_d \frac{N}{1 + N\frac{1}{s}} \tag{4}$$

Furthermore, a saturation limit of 150 N was chosen for the actuation signal. This is necessary to model the physical input limits of the actuator. To avoid integration-wind-up when the saturation limit is reached, the back-calculation anti-windup mechanism implemented in the Simulink PID controller block was chosen. This algorithm leads to a discharge of the output of the integral gain. The investigated section, the derived step excitation and the actual movement curve of the part are shown in the figure 5.

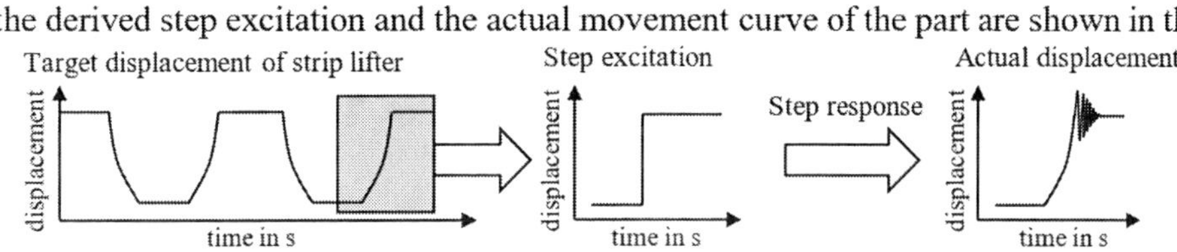

Figure 5 Investigated section of strip movement, step excitation and the actual movement of the strip

3. Results

In the following section the simulation results of the closed-loop controlled system containing active strip lifter are compared to the reference system configuration relying on passive spring strip lifter limited by a hard stop. The co-simulation model enables the automated execution of a parameter study, in which the effects of different control parameter configurations on the system behavior of the FEA model can be compared. The gain parameters of the PID controller k_p, k_i and k_d were varied within a reasonable value range for the full-strip configuration. As will be shown in the following section, the parameter configuration in table 1 allowed a reduction of the amplitude of the occurring oscillations compared to the uncontrolled model. The visualization of the simulation results is therefore carried out for all investigated strip layouts using this parameter set.

Table 1. Parameter setting of the co-simulation model.

Stroke rate	Lifting distance	Step size	k_p	k_i	k_d	Lifter spring rate	Strip thickness
120 1/min	20 mm	0.01 s	1	2	0.01	5 N/mm	0.5 mm

IOP Conf. Series: Materials Science and Engineering **967** (2020) 012035 doi:10.1088/1757-899X/967/1/012035

In the passive system, vibration of the strip is excited by reaching the hard stop at the end of the lifting phase. At a high stroke rate these oscillations can cause the strip to be released from the guide gap in the subsequent feeding phase. Furthermore, high vibration amplitudes can lead to a collision of the part with tool components. The aim of implementing closed-loop controlled active strip lifter is to increase the possible stroke rate compared to the uncontrolled configuration. This is to be achieved by lowering the maximum oscillation amplitude during the feeding phase. Therefore, the lifting and the feeding phase is examined for each of the four tested strip layouts. In figures 6 to 9, the vertical displacement of the six measurement points on the strip are shown. The lifting phase starts at t=0.32 s. The shown time steps display the moment where one of the reference points first reaches the desired 0 mm displacement.

3.1 Results full strip,v1

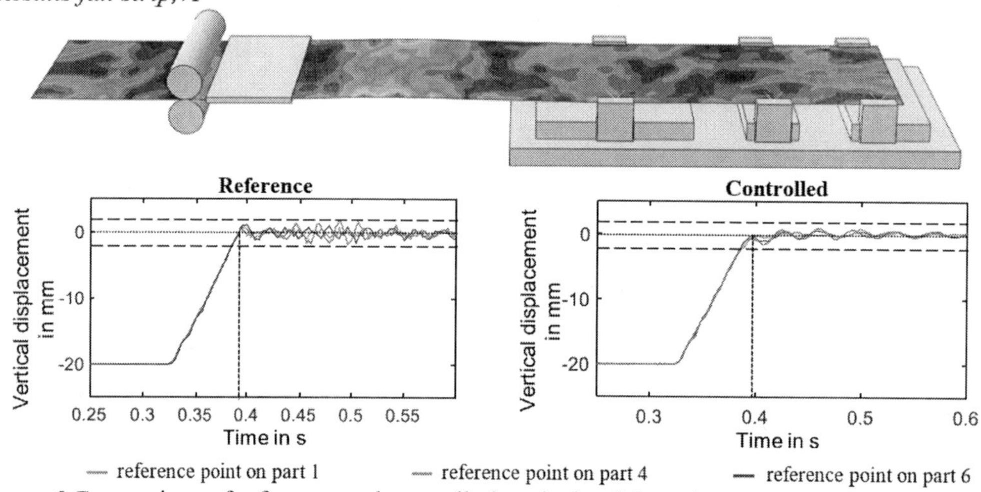

Figure 6 Comparison of reference and controlled analysis of the strip movement for strip-layout v1.

For the *full-strip* layout seen in figure 6, the uncontrolled reference system exhibits continuous strip oscillations around the reference value during the feeding phase. In the controlled system the maximum amplitude of the oscillations can be reduced by 44.01 % compared to the reference system. A comparatively high undershoot occurs immediately after the set point position is reached for the first time. Subsequently, the vibration amplitudes continue to drop. A comparison of the two configurations shows that in particular the high-frequency portion of the oscillations is compensated. The duration of the lifting phase is approximately the same for both configurations.

3.2 Results single tracked, double carries, v2

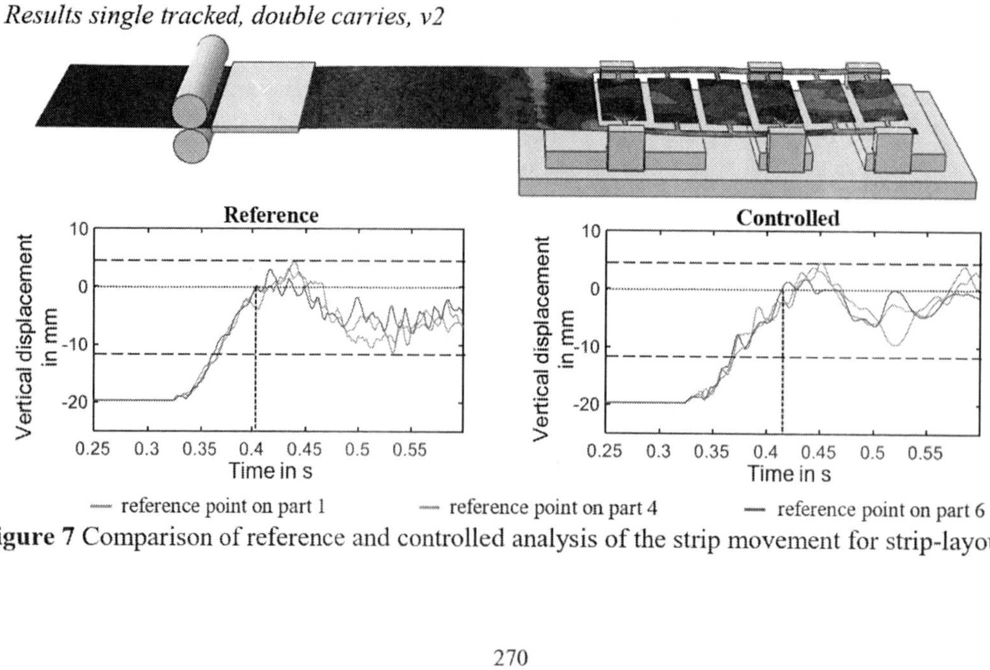

Figure 7 Comparison of reference and controlled analysis of the strip movement for strip-layout v2.

For the configuration *single tracked, double carried, v2* shown in figure 7, the vibrations occurring in the feeding phase are similarly reduced. The vibrational amplitude is decreased by 10.73 %. The duration until the feeding height is reached for the first time is slightly higher in the controlled system.

3.3 Results double tracked, triple carried, v3

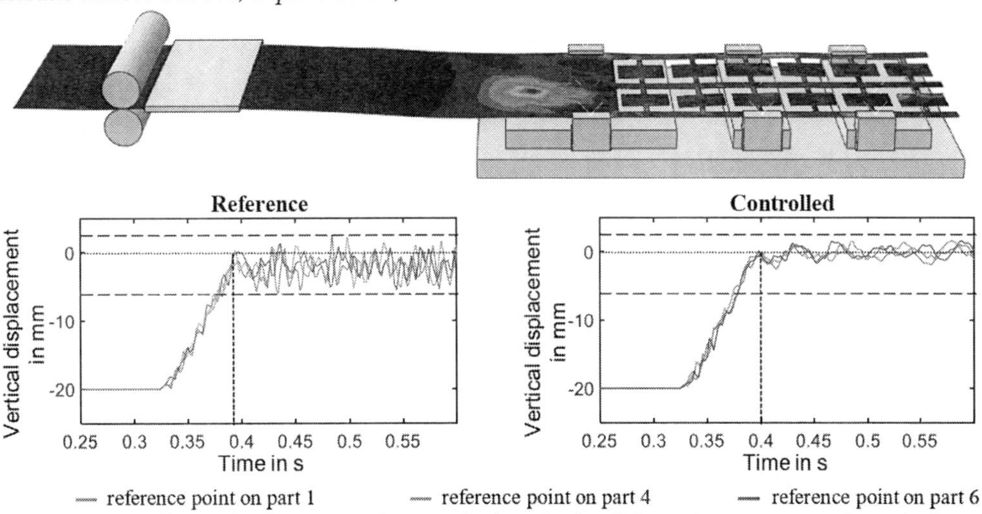

Figure 8 Comparison of reference and controlled analysis of the strip movement for strip-layout v3.

Also for the *double tracked, triple carried, v3* configuration represented in figure 8, a reduction of the maximum vibration amplitude during the feeding phase is reached by introducing the controlled strip lifter. This is a decrease of 51.32 %. However, in this case the controller introduces strip oscillations during the lifting phase. As these are less pronounced than the oscillations in the feeding phase, they do not represent a limiting factor for the maximum possible stroke rate.

3.4. Results single tracked, single carried, v4

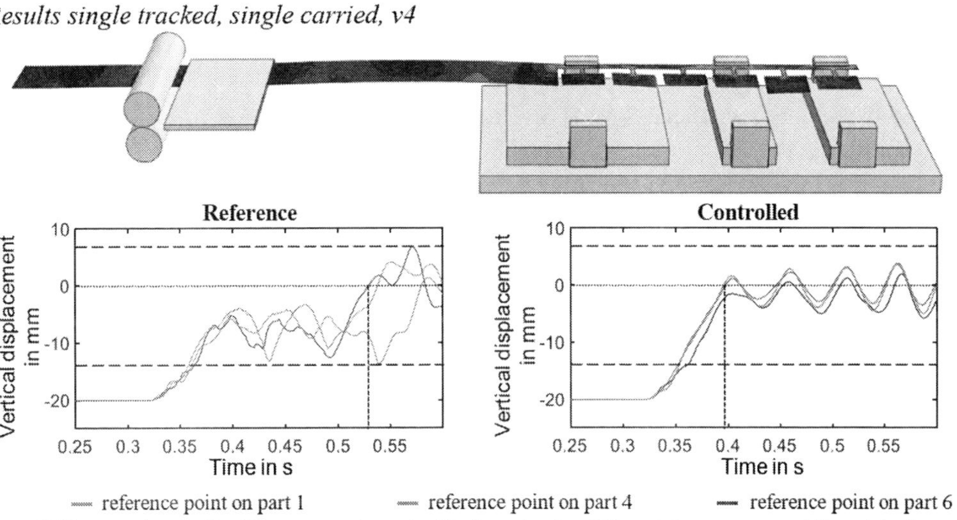

Figure 9 Comparison of reference and controlled analysis of the strip movement for strip-layout v4.

For the layout *single tracked, single carried, v4* shown in figure 9, the passive system has a comparatively high duration of the lifting phase compared to the strip geometries discussed above. By introducing the active lifter, a reduction of the rise time is achieved. Furthermore, the controller does not introduce any significant additional vibrations during the lifting phase. During the feeding phase, oscillations continue to occur in the controlled system, but these have a reduced maximum amplitude of 53.01 % compared to the passive system.

IOP Conf. Series: Materials Science and Engineering **967** (2020) 012035 doi:10.1088/1757-899X/967/1/012035

Conclusion

Closed-loop control for reducing the amplitude of strip vibrations in a progressive die tool was realised in a co-simulation model composed of a FEA model and a Simulink controller model. By using the Abaqus simulation restart functionality, the FEA simulation could be divided into single steps, whose length corresponds exactly to one time step of the controller. This approach enables to connect the FEA model with arbitrary control algorithms and to analyse the influence of various control parameter settings on the system response during the runtime of the FEA model.

Three pairs of active strip lifters combining the passive spring with a PID controlled actuator were integrated in the FEA model. Several factors, such as the alternating tapping of the strip in the guiding gap lead to a highly non-linear behavior of the system response. On four variants of progressive die strips, testing of the vibration reduction was done for various gain parameter settings of the PID controller. By this control approach, an improvement of the vibrational behavior of the investigated strip-layouts was achieved. However, depending on the strip geometry, a varying amount of residual vibrations remains. Therefore, in future work, a strip geometry specific tuning can be applied. For example, a controller design in the co-simulation model with a reinforced learning approach can be useful to deal with the complex dependencies of the system.

The results can be used in industrial applications by replacing passive, spring-loaded strip lifters with active force or displacement controlled actuators, like linear ball screw actuators or voice coil motors.

Acknowledgements

Authors acknowledge financial support of the German Aerospace Center (DLR), the German Federal Ministry of Education and Research (BMBF) and the National Research Council Canada (NRC)

References

[1] Kumar S, Singh R 2007 *Automation of strip-layout design for sheet metal work on progressive die* J. Mater Process Tec. 94–100

[2] Cheok B T, Foong K Y, Nee A Y C, Teng C H 1994, *Some aspects of a knowledge-based approach for automating progressive metal stamping die design* Comput Ind **25**, 81–96

[3] Fisk J C 1976 *Strip stock lifter*, United States Patent 4,033,168

[4] Wallis B J 1980 *Stock lifter for progressive dies*, United States Patent 4,326,402

[5] Pyper J T, Pyper J K, Breen S M, Brown C T, Pyper J S 2005 *Stock lifter for metal forming dies and method for making the same*, United States Patent 6,848,290

[6] Dassault Systèmes 2013, *Abaqus 6.13 Online Documentation*, http://dsk.ippt.pan.pl/docs/abaqus/v6.13/index.html

[7] The MathWorks Inc. *2019 Simulink Documentation*, https://de.mathworks.com/help/simulink/release-notes.html

[8] Orszulik R R and Gabbert U 2016, An interface between Abaqus and Simulink for high-fidelity simulations of smart structures, *IEEE/ASME Trans. Mechatronics*. **2** 879–87

[9] Fenik S and Starek L 2009 *Optimal PI Controller with Position Feedback for Vibration Suppression* J Vib. Control **13** 2023–34

[10] Gülbahçe E and Çelik M 2018 *Active vibration control of a smart beam by a tuner-based PID controller* J. Low Freq. Noise V A **4** 1125–33

[11] Khot S M, Yelve N, Tomar R, Desai S, Vittal S 2012 *Active vibration control of cantilever beam by using PID based output feedback controller* J Vib. Control **3** 366–72

IOP Conf. Series: Materials Science and Engineering 967 (2020) 012036 doi:10.1088/1757-899X/967/1/012036

Evolution of a forming limit curve for non-linear strain paths induced on advanced high-strength sheet steel with its proven applicability to a complex deep-drawing process

K Chongbunwatana*, S Panich and K Laokor

Department of Production Engineering, Faculty of Engineering, King Mongkut's University of Technology North Bangkok, 1518 Pracharat 1 Road, 10800 Bangkok, Thailand

*komkamol.c@eng.kmutnb.ac.th

Abstract. In this work, nonlinear forming limit curves were developed through a systematic combination of both experimental and numerical tools. FLCs derived from both intact and pre-stretched DP440 high-strength steel (HSS) sheet were involved. First of all, prepared sheet specimens were, in compliance with the Marciniak test procedure, in-plane pre-stretched in either of the following three directions, uniaxial tension, plane strain and biaxial, at varying strain levels. Afterwards, each distinguishingly pre-strained specimen was post-strained until fracture following the Nakajima test guideline. The so-called displacement function could be determined out of these experimental data. An individual nonlinear IFU-FLCs was approximated with the help of the base linear FLC, the formulated displacement function and a unique simulated strain path collectively gathered from the local fracture zone. Two yield models, Hill'48 and Yld2000-2d, were tried during strain-path calculations to observe how different yield models would alter the paths and thus developed nonlinear FLCs. From the study, notable variation was detected. In the end, validity of such nonlinear FLCs were proven through a complex-shaped industrial stamping part. The IFU-FLC noticeably-better defined forming limits of parts experiencing nonlinear strain paths than the conventional one.

1. Introduction

Nowadays, sheet metal forming processes have become more and more sophisticated. Modern part and tool geometries feature challenging manufacturing issues such as sharp radii, great draw depths and lightweight or high-strength materials to meet increasingly-demanding stiffness, eco-friendliness and design standards. These state-of-the-art issues inevitably lead to greater demands on formability prediction tools. Either trail-and-error approach or conventional FLCs are very likely prone to failure in prediction [1]. Part geometry complication, forming processes and material properties are the most decisive factors regarding formability prediction [2]. Recently, several prediction approaches have been proposed to handle nonlinear strain paths. They all better envisage failure in comparison to the classical FLC proposed by Keeler et al. [3] in the late 60s. At the moment, most of them are, however, based on a bunch of experimental data required to create and calibrate their FLCs [2, 4-6]. An empirical localised-necking based formability limit criterion was firstly introduced to particularly handle nonlinear strain path cases [2, 5]. This criterion extends conventional FLCs using successive two-stage stretching, the pre-modified Marciniak and post Nakajima stretching tests. The idea of this model is that the bottom (least major true strain) point of an FLC, usually situated in the close neighbourhood of the plane-strain loading direction for common metallic material, is secretly relatable to the equivalent strain evaluated and picked from the fracture zone of a plastically-formed part [2]. This nonlinear IFU-FLC [1, 2, 5] was of primary focus in this research. It was developed to examine formability of DP440 high-strength steel. Furthermore, influence of variation of employed anisotropic yield criteria—namely, Hill'48 and Yld2000-2d—on formability estimation performance of the formulated predictive tool was inspected. Eventually, the nonlinear FLC was validated against an industrial stamping part.

2. Anisotropic plastic deformation modelling

2.1 Hill'48 yield criterion

Having become one of the most frequently applied yield functions, a quadratic anisotropic yield model was proposed by Hill [7] in 1948. A plane stress (zero stress in the thickness direction) is assumed in the Hill'48 yield criterion. This simplification makes sense very much for common sheet metal forming applications, in which nearly no external normal force in the thickness direction (ND) is exerted on the formed sheet. The equation reads

$$2f(\sigma)=(G+H)\sigma_{xx}^2+(F+H)\sigma_{yy}^2-2H\sigma_{xx}\sigma_{yy}+2N\sigma_{xy}^2=1 \tag{1}$$

where F, G, H and N are the anisotropic coefficients, whose value is shown in TABLE 1. All four coefficients can be determined through an optimisation process based on four measured mechanical properties of the investigated material. Those properties to be measured are the uniaxial-tensile yield stress in the rolling direction (RD) and the r-value in three loading directions, 0°, 45° and 90° to the RD.

TABLE 1. Plastic deformation model parameters of the DP440 high-strength steel sheet

Model	Parameter							
	1	2	3	4	5	6	7	8
Swift	K	$\bar{\varepsilon}_0$	n	-	-	-	-	-
	741.6	0.0088	0.1866					
Hill'48	F	G	H	N	-	-	-	-
	0.4137	0.5155	0.4845	1.1073				
Yld2000-2d	α_1	α_2	α_3	α_4	α_5	α_6	α_7	α_8
	0.8974	1.0592	0.6535	0.9291	0.9349	0.6309	0.9007	1.3186

2.2 Yld2000-2d yield criterion

The Yld2000-2d yield criterion is a famous non-quadratic anisotropic yield function introduced by Barlat et al. [8]. A plane stress assumption, identical to the Hill'48 model, is also applied for this yield criterion. The equation reads

$$f(\sigma)=\left|S_1'-S_2'\right|^m+\left|2S_2''+S_1''\right|^m+\left|2S_1''+S_2''\right|^m=2\bar{\sigma}^m \tag{2}$$

where S_i' and S_i'' are the principal values of the modified deviatoric stress tensors S' and S''. More details of the Yld2000-2d yield criterion can be found in [8]. Eight material parameters pertaining to anisotropic plastic deformation such as the yield stress in four different loading directions—σ_0, σ_{45}, σ_{90} and σ_b—and the r-value in the four corresponding directions—r_0, r_{45}, r_{90} and r_b—are required for determination of the eight anisotropic coefficients, α_1 to α_8 whose value is shown in TABLE 1. The *m* exponent from equation (2) should be either 6 for the BCC crystal structure or 8 for the FCC crystal structure. In this research, 6 was applied as the DP440 steel sheet had the BCC microstructure.

2.3 Swift hardening model

The Swift hardening law [9], explaining the metal strain hardening behaviour, reads

$$\bar{\sigma} = K\left(\bar{\varepsilon}_0 + \bar{\varepsilon}_p\right)^n \tag{3}$$

where $\bar{\sigma}$ and $\bar{\varepsilon}_p$ are the equivalent flow stress and accumulative plastic strain, consecutively. The parameters *K*, *n* and ε_0 are all material constants, whose value is shown in TABLE 1. They can be derived from the measured uniaxial true stress-strain curve in the RD through a regression method.

3. Determination of material properties

3.1 Uniaxial tensile test

All uniaxial tensile tests were laboriously accomplished by a universal tensile testing machine on DP440 (SCGA440-45) high-strength steel sheet specimens. The test pieces were prepared along three sheet directions, 0°, 45° and 90°, to the RD to acquire three yield stresses and three *r*-values. A strain rate of 0.001 s^{-1} was applied and maintained all over the tests. The *r*-values were a linearly-approximated ratio between the width and thickness strains up to the overall elongation of 14% [10]. TABLE 2 reveals all those results gained from the tensile tests.

TABLE 2. Mechanical properties of the DP440 high-strength steel sheet

Loading Direction (°)	Yield Stress (MPa)	Ultimate Tensile Stress (MPa)	Elongation (%)		*r*–value (-)
			Uniform	Total	
0	331	555	16.5	30.6	0.94
45	342	564	15.5	25.4	0.69
90	326	558	16.2	30.9	1.17
Balanced biaxial	396	-	-	-	0.95[b]

[a] From the hydraulic bulge test
[b] From the disc compression test

3.2 Hydraulic bulge test

A hydraulic bulge test was carried out by a 200-ton hydraulic press equipped with a die chamber and pressure intensifier. A square DP440 steel-sheet blank with a width of 250 mm was utilised as the test workpiece. The diameter of the employed drawing die was 100 mm. The hydraulic pressure did not stay constant at a certain predetermined value but had been progressively escalated to inflate the steel blank until the working fluid leaked through initiated crack. Required by the Yld2000-2d model, the balanced biaxial yield stress of 396 MPa as shown in TABLE 2 could be obtained from this experiment. More details on the bulge test are discussed in [11].

3.3 Disc compression test

The balanced biaxial plastic-deformation anisotropy (r_b) can be attained through an execution of a disc compression test. A DP440 specimen possessed a disc geometry whose initial diameter was 10 mm. The disc was, to begin with, compressed with a medium punch force of 25.9 kN. The test was repeated ten times with an equal load increment of 1.1 kN. More details on the disc compression test procedure can be found in [8]. The balanced biaxial *r*-value (r_b) particularly needed by the Yld2000-2d model was an averaged ratio of the strain in the transverse direction (TD) over the strain in the RD among all ten samples. The balanced biaxial *r*-value of 0.95 can be gained as exhibited in TABLE 2.

4. Experimental nonlinear strain-path stretching tests

To grow a range of nonlinear (bilinear) strain paths into a number of rectangular sheet specimens, each specimen was first off pre-stretched, through the modified Marciniak in-plane stretching test, either in the uniaxial tension, plane strain or biaxial direction by a certain amount of equivalent plastic strain. All pre-stretching Marciniak tests, whose punch diameter was downscaled to 150 mm, were conducted on a hydraulic press. Their equipment setup is graphically illustrated in Figure 1 (a). The degree of pre-stretching was evaluated by an optical strain measurement system AutoGird-Vialux. Afterwards, as demonstrated in Figure 1 (b), a planar dog-bone shaped portion was taken off the centre of each pre-stretched specimen. Prepared for final subsequent stretching, those planar dog-bone shaped samples were obligated to the Nakajima specimen's geometry and downscaled size in accordance with the ISO12004 standard [12].

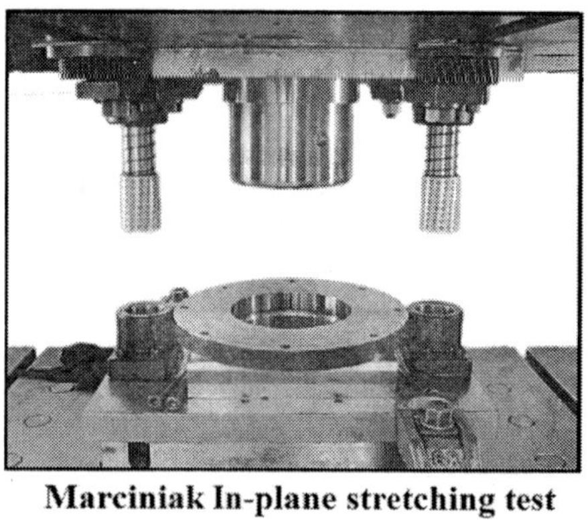

(a)

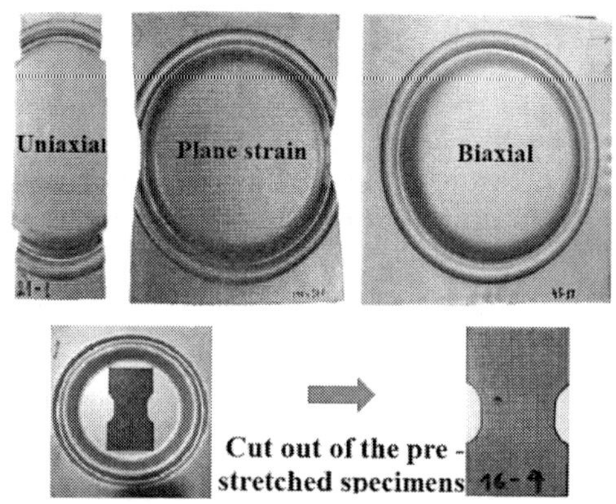

(b)

Figure 1. Demonstration of the pre-stretching Marciniak (a) test setup and (b) specimens prepared for the following Nakajima test

Further post-strain until fracture was provoked on all pre-stretched specimens by the modified Nakajima stretching test with a reduced punch diameter of 50 mm whose arrangement is shown in Figure 2 (a). The resultant strain distribution localised around the fracture area was gathered and converted into the major and minor principal strains using the AutoGird-Vialux measurement system, as illustrated in Figure 2 (b), in collaboration with the ASTM E2118-02 standard evaluation method. These critical principal strain data were collected to form FLCs (FLC50). All minimum major strain values extracted from all conventional, uniaxial-pre-stretched, plane-strain-pre-stretched and biaxial-pre-stretched FLCs were correlated with the equivalent pre-strain levels from the corresponding FLCs. The resulting correlation graph is clearly demonstrated in Figure 3.

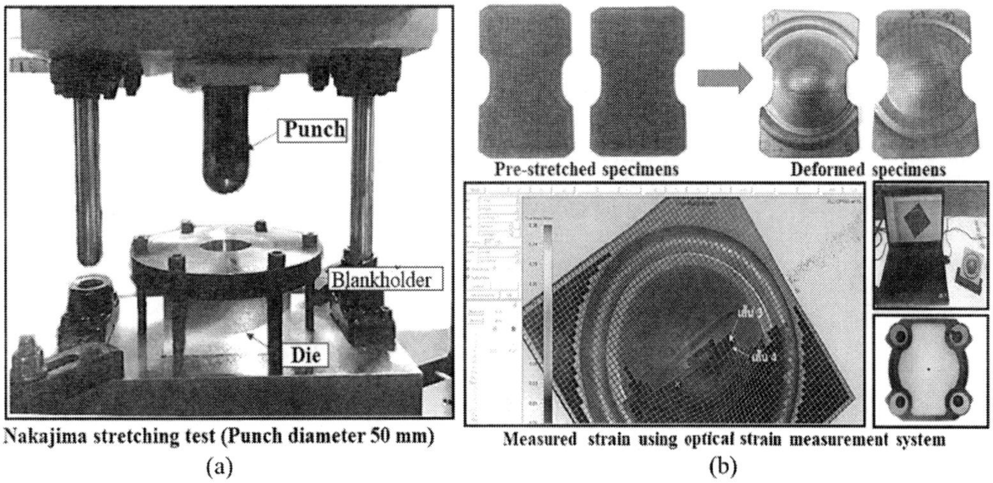

Figure 2. Nakajima (a) test setup and (b) specimens before and after deformation with strain evaluation

5. Determination of the displacement function for nonlinear FLCs and its application

The FLC of multi-step forming is usually under the conventional one. The curve shown in Figure 3 can be explained in an equation form. This equation is commonly called a displacement function, reading

$$\varphi_1 = \frac{AB}{C\bar{\varphi}^D + B} \tag{4}$$

where A is the minimum major true strain of the conventional FLC while B, C and D are all worked out to fit all data pairs. The equivalent pre-strain level ($\bar{\varphi}$) can be computed using the equation, reading

$$\bar{\varphi} = \frac{2}{\sqrt{3}}\sqrt{\varphi_1^2 + \varphi_2^2 - \varphi_1\varphi_2} \tag{5}$$

where φ_1 and φ_2 are the major and minor principal true strains, respectively.

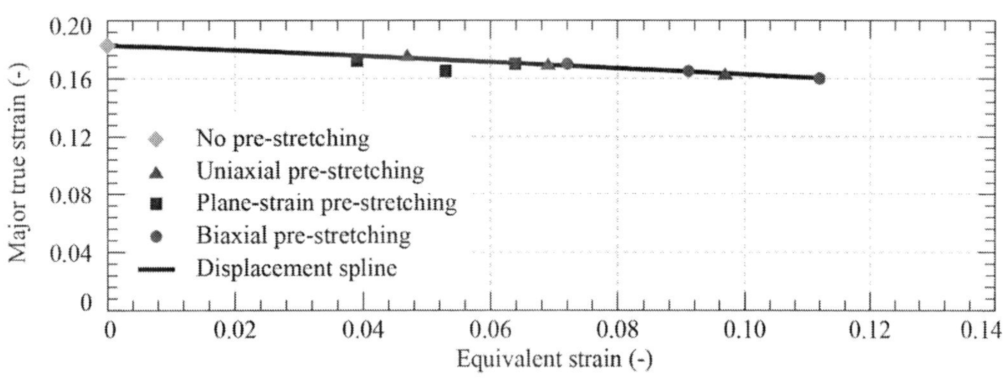

Figure 3. Correlation diagram between the minimum major true strain and equivalent pre-strain level

IOP Conf. Series: Materials Science and Engineering **967** (2020) 012036 doi:10.1088/1757-899X/967/1/012036

TABLE 3 summarises all fitted parameters *A*, *B*, *C* and *D* of the displacement function of the DP440 steel sheet.

TABLE 3. Parameters of the displacement function of the DP440 high-strength steel sheet

Displacement Function	*A*	*B*	*C*	*D*
	0.1826	998.002	1961.98	1.2028

To assess the formability of investigated material for an arbitrary nonlinear strain path using the nonlinear IFU-FLC based on an anisotropic yield criterion, the total equivalent strain of a simulated strain path from the critical area of a formed part needs to be established as presented in the procedure shown in Figure 4. To do that, a simulated nonlinear strain path must first be divided into several linear segments. Then, the equivalent strain of the linear segments up to the second last one is summed up. How the FLC should be formed and shifted is defined by the major true strain obtained through the displacement function provided such a summed equivalent strain. The modified FLC is finally used as the actual forming limit to examine the nonlinear strain path.

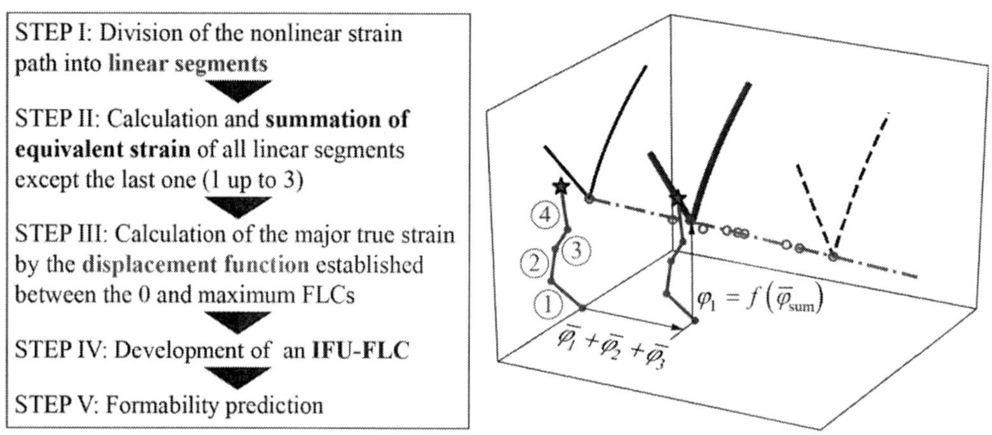

Figure 4. Development procedure for an IFU-FLC

6. FE simulation of an industrial stamping part

To obtain nonlinear strain paths, an industrial stamping part was formed on a hydraulic press until crack was initiated. In parallel, an FE simulation of the same stamping process was also conducted. The applied process parameters and boundary conditions corresponded to the experimental ones as shown in Figure 5 (a). The punch force-displacement curves from both experiment and simulation were compared to verify the FE model. An implicit formulation was applied. Considering the meshing and material properties of the DP440 sheet, the shell elements and elastoplastic material model were taken into account. Besides, the Hill'48 and Yld2000-2d anisotropic yield criteria together with the isotropic Swift hardening law were employed to describe the plastic flow behaviour of the material. The Coulomb's friction coefficient of 0.175 was applied for all contact pairs. The punch and dies were defined as an analytical rigid surface. As a result, local strains in the critical area of the formed sample were collected from the beginning until fracture. Figure 5 (b) shows the simulation result with the critical area marked as red.

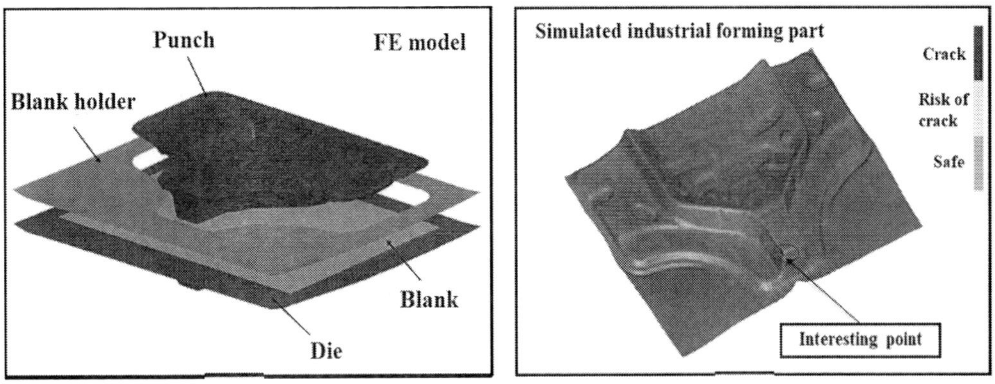

(a) (b)

Figure 5. (a) FE model of the stamping part and (b) calculation result with the red-marked critical area

According to the procedure in Figure 4, two simulated strain paths from both Hill'48-based and Yld2000-2d-based FE models were gathered. All the rest were performed exactly in the same manner as what was described earlier in Section 5 in accordance with the procedure in Figure 4. Figure 6 (a) demonstrates the experimental maximum FLC and 0-FLC overestimating the fracture point of both Hill'48 and Yld2000-2d paths. Both models obviously generate dissimilar strain trajectories.

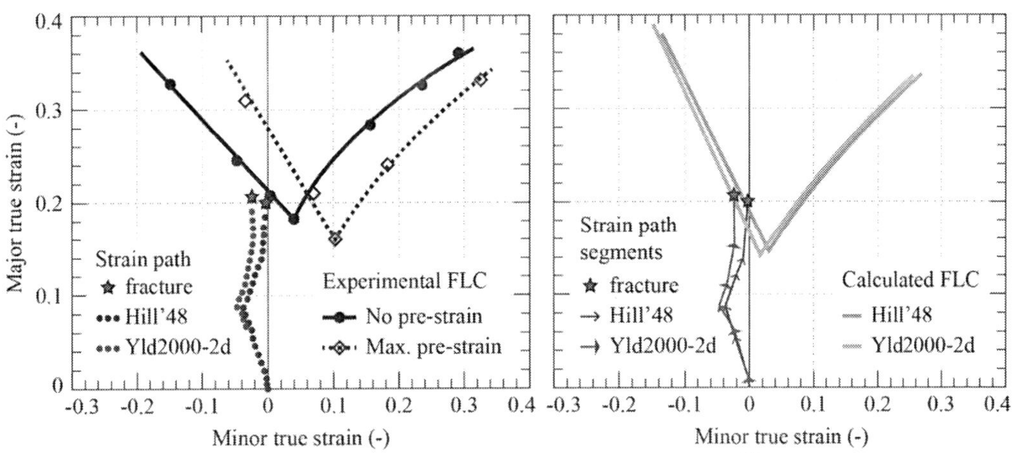

(a) (b)

Figure 6. (a) Experimental FLCs together with simulated strain paths and (b) IFU-FLCs together with exploded strain path segments

The modified FLCs based on both strain paths are demonstrated in Figure 6 (b). Both are slightly deviated in response to those two unalike routes. According to Figure 6 (b), both paths are equally broken down into 6 linear elements for the sake of prediction accuracy. The total strain summed from the first 5 pieces for the Hill'48 and Yld2000-2d models is 0.1795 and 0.2062, respectively. This consequently leads to the resulting minimum major true strains of 0.1462 and 0.1411, consecutively. Both FLCs developed with respect to the computed major true strains are clearly capable of seizing the failure moment of the stamping part as demonstrated in Figure 6 (b). The results definitely lead to the conclusion that the developed nonlinear FLCs based on both Hill'48 and Yld2000-2d yield criteria can realistically predict formability of a DP440 industrial stamping part experiencing a nonlinear strain path.

7. Summary

In this work, a bunch of forming experiments on the DP440 (SCGA 440-45) high-strength steel sheet were accomplished. A conventional FLC was firstly developed through the Nakajima stretching test with a punch diameter of 50 mm. Additionally, a number of two-step forming processes were performed to determine a displacement function. The pre-stretching process with the uniaxial, plane strain or biaxial loading direction was conducted with restriction to the Marciniak in-plane stretching procedure with a punch diameter of 150 mm in several strain levels. The subsequent Nakajima stretching test was executed on those pre-stretched specimens to obtain the minimum major strains in a wide range of equivalent pre-strain levels. The displacement function could finally be formulated by correlating those minimum major true strains with their corresponding equivalent pre-strain levels. The function was used to computationally form and locate IFU-FLCs. It can be concluded that the applied anisotropic yield criterion has a significant effect on both nonlinear FLC and simulated nonlinear strain path. The developed nonlinear FLCs were eventually verified against a complex industrial stamping part. The validation results assured that the nonlinear IFU-FLCs were undoubtedly efficient in formability estimation of the industrial part undergoing a complex nonlinear strain path.

8. Acknowledgements

This research was fully funded by Faculty of Engineering, King Monkut's University of Technology North Bangkok, under the contract number ENG-NEW-61-18. The authors wish to express their appreciation to Department of Tool and Materials Engineering, KMUTT, granting them an access to its Dynaform license. The authors wish to thank Thai Summit Automotive Co., Ltd. for all supporting. Last but not least, the authors would like to express their full gratitude towards Prof. Dr.-Ing. Mathias Liewald, M.B.A., and Dr.-Ing. Klaus Drotleff from IFU for their helpful idea on the IFU-FLC criterion.

9. References

[1] Drotleff K, Panich S, Liewald M and Uthaisangsuk V 2016 *Proc. Int. Con. on Forming Technology Forum* Sep 12 – 13 Munich Germany pp 59-64
[2] Drotleff K, Liewald M and Mosebach J.T 2015 *Proc. Int. Con. on Forming Technology Forum* Jun 29-30 Zurich Switzerland pp 69-74
[3] Keeler S P 1961 Plastic Instability and fracture in sheets stretched over rigid punches. MIT, D.Sc. thesis, May 1961 Cambridge, (Cambridge: MIT)
[4] Volk W, Hoffmann H, Suh J, Kim J 2012 Failure prediction for nonlinear strain paths in sheet metal forming *CIRP Annals - Manufacturing Technology* **61(1)** 259–262
[5] Werber A, Liewald M, Nester W, Grünbaum M, Wiegand K, Simon J, Timm J, Hotz, W 2013 Development of a new failure prediction criterion in sheet metal forming *Int. J. Mater. Form.*7 395-403
[6] Werber A, Liewald M, Nester W, Gruenbaum M,Wiegand K, Simon J, Timm J, Bassi C and Hotz W 2007 Influence of pre-streching modes on forming limit diagram of AA6014 *Key. Eng. Mater.* **344** 113-118
[7] Hill R 1948 Theory of yielding and plastic flow of anisotropic metals Proc. R. Soc. Lond. A **193** 281–297.
[8] Barlat F, Aretz H, Yoon J, Karabin M, Brem J and Dick R 2005 Linear transformation based anisotropic yield functions *Int. J. Plast.* **21** 1009–1039
[9] Swift H W 1952 Plastic instability under plane stress *J. Mech. Phys. Solids*. **1** 1–18
[10] Panich S, Liewald M and Uthaisangsuk V 2018 Stress and strain based fracture forming limit curves for advanced high strength steel sheet *Int. J. Mater. Form.* **11** 643-661.
[11] Young R F, Bird J E and Duncan J L 1981 An automated hydraulic bulge tester *J. Appl. Met. Working*. **1** 11–18
[12] Laokor K, Chongthairungruang B and Panich S 2019 Influence of Pre-Stretching Mode on the Forming Limit Strain and Stress Diagram of High strength Steel *Key. Eng. Mater.* **751** 167-172

Influence of anisotropic yield criteria on simulation accuracy of the hole-expansion test

S Panich* and K Chongbunwatana

Department of Production Engineering, Faculty of Engineering, King Mongkut's University of Technology North Bangkok, 1518 Pracharat 1 Road, 10800 Bangkok, Thailand

*sansot.p@eng.kmutnb.ac.th

Abstract. Prediction accuracy of the two well-known anisotropic yield criteria, namely, Hill'48 and Yld2004-18p, on plastic deformation of AA5052-O sheet aluminium alloy experiencing compound stress was profoundly examined and contrasted in this work. To obtain the flow behaviour and anisotropy of the test material in varying loading directions, both uniaxial tensile and balanced-biaxial tests were conducted in the corresponding directions. Furthermore, a disc-compression test was carried out to reveal the biaxial r-value. Those gained mechanical properties could therefore be used to determine the material constants required by each yield criterion. The standard conical-punch hole expansion test was chosen here to impose complex stress on the sheet specimen. Forming parameters such as the punch load, stroke paths, hole expansion ratio, sheet thickness profile around the hole circumference and flange height were all monitored and investigated. It was shown that the results from the Yld2004-18p model with an exponent of eight apparently better agreed with the experimental data than those from the Hill'48 one. A conclusion could be drawn here that the yield criterion significantly affected the accuracy of the predicted deformation behaviour of the investigated aluminium alloy grade AA5052-O nevertheless in dependence on the degree of material anisotropy.

1. Introduction

In the past, low carbon steel sheets were widely used in the automotive manufacturing industries. Recently, various lightweight materials have nevertheless been increasingly utilised and gradually replacing those conventional steels in many applications such as automotive bodies for the sake of preservation of the Earth's environment [1]. Aluminium alloy sheets are among the lightweight materials that are critically effective in reducing automotive body weight. Several drawbacks of those alloys such as inferior formability, high degree of springback and high cost have however been reported [2-4]. They are knowingly prone to fracture during common forming processes because of their low ductility. Optimal tool geometries and forming conditions, not causing fracture of the aluminium alloy sheets, are thereby definitely required [1]. In the stretch flanging process of automotive wheel parts like brake panels and hub holes, the hole expansion test has been recognised as a discriminating method for characterising the material forming behaviour. To save time and cost of the experimental procedure, a qualified finite element model, especially that able to accurately anticipate the material yield behaviour, is totally a must [5]. In this work, AA5052-O, sheet aluminium alloy, was of interest. Its anisotropic plasticity was thoroughly investigated. Three famous laboratory tests—uniaxial tension, disc-compression and hydraulic-bulge tests—were all accomplished to identify the material mechanical characteristics. Such experimentally-determined properties as flow stresses, r-values and yield loci helped formulating all anisotropic coefficients of those two yield models under scrutiny, Hill'48 and Yld2004-18p [6, 7], as well as illustrating the competence of both models in handling the aluminium-sheet anisotropy. A demonstrative hole-expansion test, acknowledged-y capable of measuring stretch-flangeability of any sheet metal, was then conducted. The test outcomes were eventually compared with those simulatively obtained from both computational models in order to study the obscured effects of both yield criteria over simulated AA5052-O stretch-flangeability.

2. Material characterisation

2.1 Uniaxial tensile test

All uniaxial tensile tests were carried out by a universal testing machine on 1-mm-thick AA5052-O specimens of the ASTM E8 standard shape. The samples were prepared along seven angles—0°, 15°, 30°, 45°, 60°, 75° and 90°—to the rolling direction (RD). Throughout the entire tests, a strain rate of 0.001 s^{-1} was kept constant. As a result, seven stress-strain curves and *r*-values were obtained. The flow curve gained from the 0° specimen is exemplarily illustrated in Figure 1.

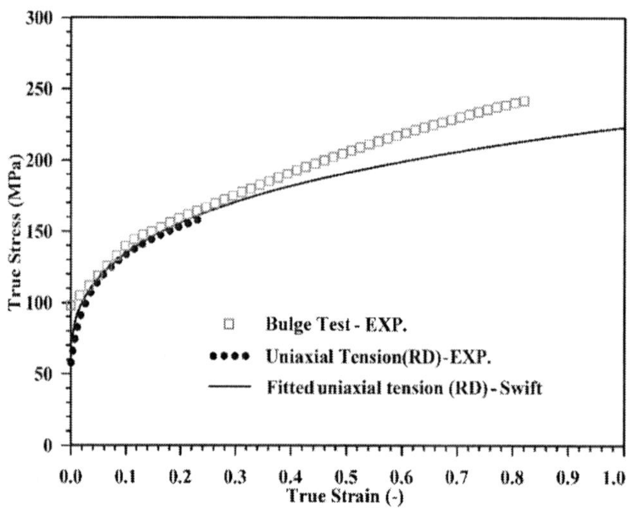

Figure 1. Experimental and numerically-Swift-fitted flow stress curves of the examined aluminium AA5052-O undergoing uniaxial tension in RD

Meanwhile, Table 1 depicts all yield stresses, extracted from the curves, normalised with respect to the yield stress at the RD and all *r*-values.

Table 1. Normalised yield stresses and *r*-values for the investigated AA5052-O sheet

Degree	0°	15°	30°	45°	60°	75°	90°	Biaxial
Normalised	1.0000	0.8719	0.8883	1.0296	0.9010	0.9182	1.0660	1.1200
r-value	0.6230	0.7070	0.7220	0.7290	0.7380	0.7160	0.6670	0.9626

2.2 Hydraulic bulge test

A hydraulic bulge test was performed on a 200-tons hydraulic press to obtain the biaxial flow behaviour of the AA5052-O sheet. The experimental procedure and tooling can be found in [11] for more details. The drawing-die diameter was 200 mm. The test specimen was a 280-mm-wide square blank. The hydraulic pressure gradually increased until crack was initiated. The pressure, dome height, membrane stress and thickness strain were all online-recorded. The resultant normalised biaxial yield stress is shown in Table 1 while the biaxial flow curve is plot in Figure 1.

2.3 Disc compression test

The through-thickness, single-layer disc compression test was executed to reveal the biaxial r-value (r_b). More details on the experimental procedure can be found in [9, 12]. An AA5052-O disc sample had a diameter of 10 mm. A compressive load was applied to the disc, starting from 5 kN up to 12 kN by a stepwise increment of 0.9 kN. The biaxial r-value could be estimated as a ratio of the strain in the transverse direction (TD) versus RD, whose consequence is shown in Table 1.

3. Anisotropic yield criteria and strain hardening model

3.1 Hill'48 yield criterion

A quadratic anisotropic yield criterion, proposed by Hill [8] in 1948, has become one of the most widely used yield functions. The Hill's 48 model assumes a plane-stress state, an absolutely-proper approximation for typical sheet metal forming. The equation

reads
$$2f(\sigma)=(G+H)\sigma_{xx}{}^2+(F+H)\sigma_{yy}{}^2-2H\sigma_{xx}\sigma_{yy}+2N\sigma_{xy}^2=1 \quad (1)$$

where F, G, H and N are the computed anisotropic coefficients. Table 2 exhibits all four coefficients of the studied material. They can be derived from a group of four experimental data, demonstrated in Table 1, such as the uniaxial yield stress measured in the RD and r-value in three varying load directions, 0°, 45° and 90° with respect to the RD.

Table 2. Anisotropic coefficients of the Hill'48 yield criterion for the AA5052-O sheet

F	G	H	N
0.5755	0.6161	0.3839	1.4645

3.2 Yld2004-18p yield criterion

The anisotropic yield function Yld2004-18p computes the scalar equivalent stress out of a full 3D stress tensor. Eighteen anisotropic coefficients are introduced as entries of the two 6×6 linear transformation matrices modifying the 3D stress tensor. More details on the definition of those parameters can be found in [9]. The Yld2004-18p yield function can be expressed as

$$\begin{aligned}\phi=4\bar{\sigma}^m=&\left|S_1'-S_1''\right|^m+\left|S_1'-S_2''\right|^m+\left|S_1'-S_3''\right|^m+\left|S_2'-S_1''\right|^m+\left|S_2'-S_2''\right|^m+\\&\left|S_2'-S_3''\right|^m+\left|S_3'-S_1''\right|^m+\left|S_3'-S_2''\right|^m+\left|S_3'-S_3''\right|^m\end{aligned} \quad (2)$$

where $\bar{\sigma}$ is the equivalent stress; m is the exponent; S_i' and S_i'' $(i=1,2 \text{ or } 3)$ are the principal values of the modified deviatoric stress tensors $S'=C'S$ and $S''=C''S$. The variable S is the deviatoric stress tensor while both C' and C'' are the modifying linear transformation matrices mentioned above. All eighteen coefficients in the matrices, describing the anisotropic deformation behaviour of the considered material, can be optimised thru an iterative regression process on Eq. (2) provided a set of measured yield stresses and r-values for several load orientations as shown in Table 1 for the AA5052-O sheet. The exponent m is set to 8 as recommended for the face-centred cubic (FCC) crystal [7, 9], which is the typical microstructure for aluminium alloys. In the meantime, Table 3 shows all those eighteen anisotropic coefficients of the examined aluminium-alloy sheet, resulting from an optimisation process.

Table 3. Anisotropic coefficients of the Yld2004-18p yield criterion for the AA5052-O sheet

α_1	α_2	α_3	α_4	α_5	α_6	α_7	α_8	α_9
1.1504	1.1504	1.3241	1.1428	0.7253	0.6311	1.0127	0.9983	0.9983
α_{10}	α_{11}	α_{12}	α_{13}	α_{14}	α_{15}	α_{16}	α_{17}	α_{17}
0.5215	0.5434	0.5387	0.4778	1.0140	0.9588	0.9789	1.0042	0.5000

3.3 Swift hardening model

The Swift hardening law [10], characterising the strain hardening behaviour, reads

$$\bar{\sigma} = K\left(\varepsilon_0 + \bar{\varepsilon}_p\right)^n \quad (3)$$

where $\bar{\sigma}$ and $\bar{\varepsilon}_p$ are the effective flow stress and plastic strain, respectively. K, n and ε_o, whose values for the AA5052-O sheet are exposed in Table 4, are all material constants. They can be figured out through curve fitting on the RD uniaxial tensile true stress-strain measurement data as illustrated in Figure 1

Table 4. The Swift hardening model's material constants for the AA5052-O sheet

K	ε_0	n
741.6476	0.0088	0.1866

3.4 Validation of anisotropic yield criteria

To prove the capability of both Hill'48 and Yld2004-18p criteria in coping with anisotropy of the AA5052-O sheet, the normalised flow stress and r-value at different loading angles numerically derived from both models will be explicitly compared with those gained from the experiments as shown in Figure 2 (a).

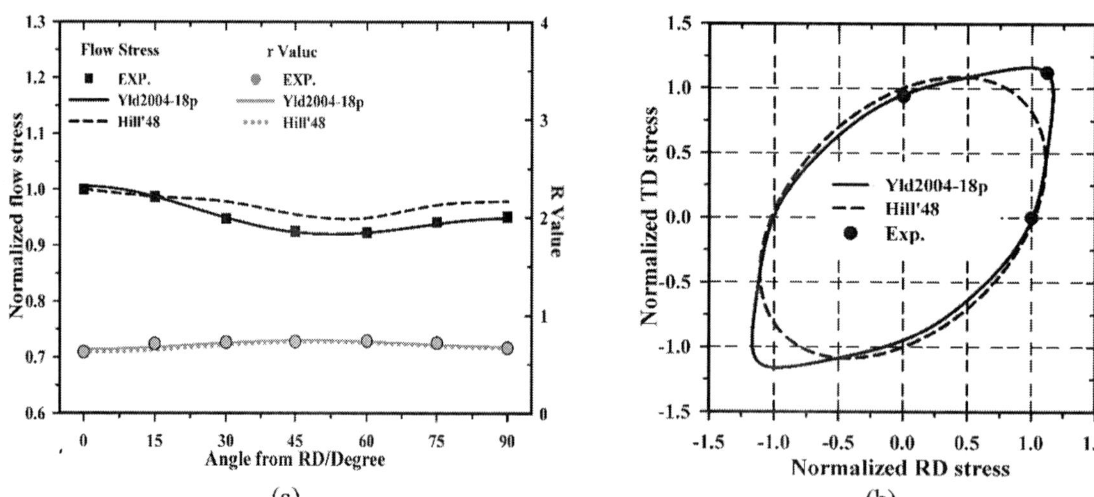

(a) (b)

Figure 2. Comparison between the experimental and model-based deformation data of the AA5052-O sheet in terms of (a) normalised flow stress and r-value and (b) yield locus

It can be obviously seen that the Yld2004-18p model provides a perfect agreement for both flow stress and r-value for all loading directions. However, the Hill'48 model overestimates the flow stress in a wide angle range from 30° up to 90° despite a faultless prediction of the r-value. This could be caused by the lack of consideration of the yield stress in all other angles except at the rolling direction during anisotropic coefficient determination. For better understanding, the comparison is also illustrated in the yield locus form on the normalised RD and TD stress space as can be seen in Figure 2 (b). Therein, the Hill'48 model slightly overvalues the uniaxial flow stress at the TD but greatly undervalues the biaxial flow stress. In contrast, the Yld2004-18p model is doing a superb flow-stress evaluation job for all loading directions.

4. Experimental hole expansion test and FE model formulation

The hole expansion test is an approved procedure for evaluating stretch-flangeability of material, in which radial compressive and tangential tensile stresses coexist simultaneously in the vicinity of the hole area [12]. In this work, a hole expansion test was conducted on an Erichsen formability tester with a conical punch. A schematic view of the hole expansion test setup is illustrated in Figure 3 (a). During the test, a square specimen, pierced with a 10-mm-diameter circular hole at the centre, was firmly clamped between a lower ring-shaped blank holder and upper cup-shaped die. Possible draw-in of the outer portion of the sample was prevented through a static blank holder force of 220 kN. The conical punch moved upwards with a constant velocity of 15 mm/min. The forming punch was suddenly stopped approximately after 17 mm once an initial crack had been detected through an instantaneous punch-force drop. On one hand, the punch force-displacement development was online monitored throughout the forming process. On the other hand, the deformed specimen's final local dimensions measured circumferentially around the flange tip such as hole diameter, flange height and sheet thickness was identified with the help of an offline 3D laser scanner.

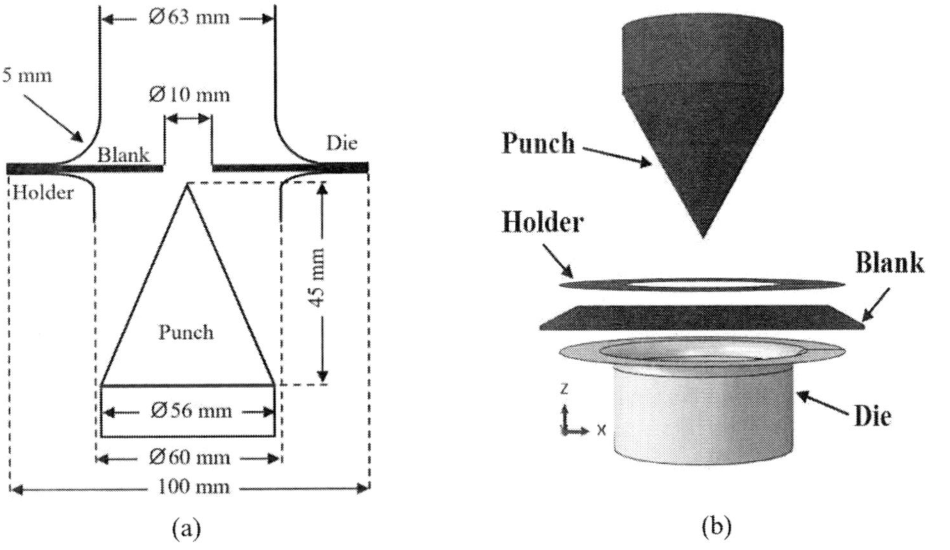

Figure 3. Hole-expansion test setup of (a) the experiment and (b) the finite element simulation model

Apart from the experiment, the hole-expansion test was correspondingly carried out on the ABAQUS2017 finite element platform to study the numerical influence of different material models. Two implicit FE models were formulated. One was based on the Hill'48 model; the other on the Yld2004-18p model. All FE simulation setups like boundary conditions, workpiece meshes and process equipment were defined precisely in accordance with the committed experiment as can be seen in Figure 3 (b). The elastoplastic material model was taken into account and the 3D hexahedral solid elements (C3D8R) were entirely applied to the AA5052-O sheet mesh. Nevertheless, the punch and dies were all modelled as an analytical rigid surface. The Coulomb's friction coefficient of 0.175 was applied to all contact pairs. During the hole expansion test, once the punch force had started to decline as a result of radial crack initiation, the punch was automatically halted. This stopping point was considered the state of fracture. The punch displacement at the point was correspondingly viewed as the simulation termination criterion. Experimental verification of the proposed FE models can therefore be realised thru comparison among the obtained force-displacement curves.

5. FE simulation results

First of all, the measured and simulated punch force-displacement developments are compared as shown in Figure 4. Both Hill'48 and Yld2004-18p models seem to perform well regarding the force-displacement aspect although the latter slightly outperforms the former, a little underestimating the punch force.

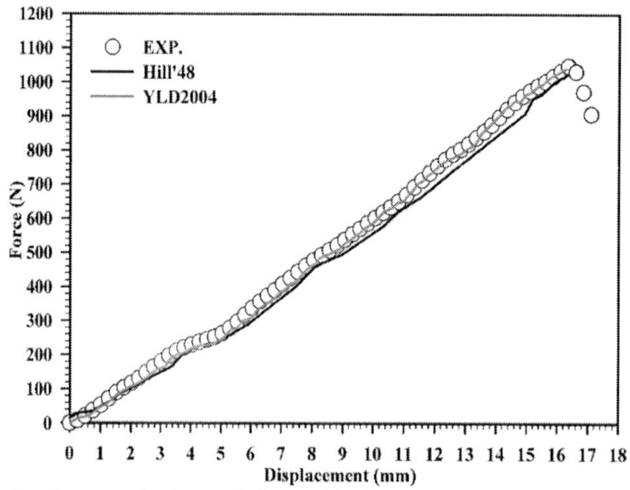

Figure 4. Experimental and numerical punch force-displacement developments

Figure 5 compares the hole expansion ratio of the initial to final hole radii along the hole fringe. The ratio usually represents formability of a blank in a hole flanging process. The greater the ratio, the greater the formability and thus expansion. Overall, the Yld2004-18p model performance surpasses that of the Hill'48 model. The Hill'48 thoroughly undercalculates the ratio. Nonetheless, the trend of its curve is still in good agreement with that of the experiment. The reason behind would be that the model considers the *r*-value at various loading orientations to account for material anisotropy. When compared at different angles, the ratio is maximum (most expanded) at 90° whilst minimum (least expanded) at 60°. This result well corresponds with the *r*-values of the AA5052-O sheet shown in Table 1. The *r*-value is maximum (0.738) at 60° while minimum (0.623 and 0.667) at 0° (RD) and 90° (TD). Therefore, the hole at 60° creeps in-plane the most and displaces inwards, while the hole at 0° and 90° is stiffer in-plane and displaces outwards. Such deformation actions lead to a bigger diameter at 0° and 90° but smaller at 60°. The next investigation point is the final altered sheet thickness along the hole circumference. According to Figure 6 (a), the predicted thicknesses from both models deviate marginally from the measured ones. However, the Yld2004-18p model once again produces more realistic results than the Hill'48 does. Considering the thickness at each angle, the trend is very much like that of the hole expansion ratio. So, the sheet tends to be thinner at the location where it in-plane extends the most on account of volume conservation.

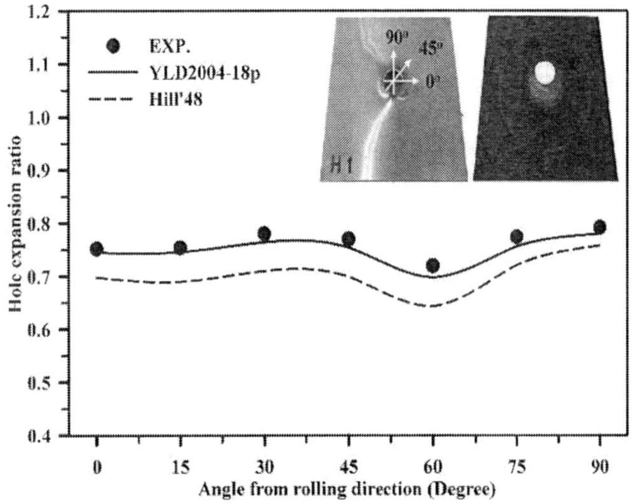

Figure 5. Experimental and numerical hole expansion ratios over the hole edge

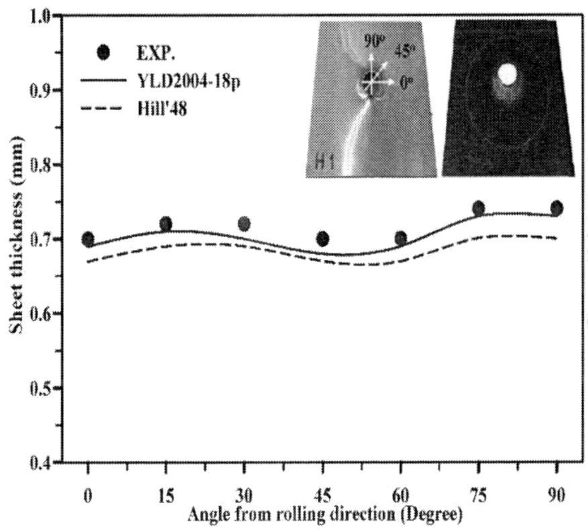

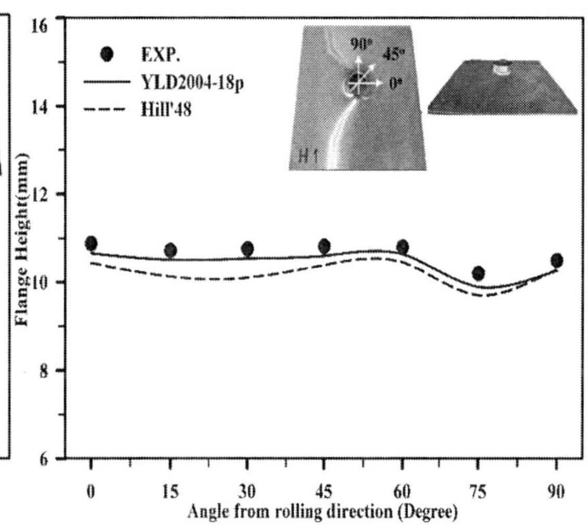

Figure 6. Experimental and numerical (a) sheet thickness over the hole edge and (b) flange height over the hole edge

The last point to be analysed is the flange height along the hole circumference. As expected, the Yld2004-18p model can generate closer consequences at every measurement point to the experimental ones than the Hill'48 can as demonstrated in Figure 6 (b). The flange height at 15°, 30°, 60° and 75° resulting from the Hill'48 model appears to be more deviated from the experimental one than that at 0°, 45° and 90°. This might probably be due to the inclusion of the r-value of only those three latter orientations when determining the anisotropic coefficients F, G, H and N.

6. Summary

In this work, a conical-punch hole expansion test was accomplished to investigate performance of the two favourite anisotropic yield criteria, namely, Hill'48 and Yld2000-18p, in modelling plasticity of AA5052-O aluminium alloy sheet. The uniaxial tension, disc compression and hydraulic bulge tests were all carried out to obtain required material properties to develop the two yield models. Normalised flow stresses, r-values and yield loci predicted by the Yld2004-18p model were in very good agreement with those from the experiments. Furthermore, FE simulations coupled with the yield criteria were carried out for the hole expansion test. The experimental and predicted hole radii in terms of the hole expansion ratio, sheet thickness and flange height all along the hole circumference were compared. It was found that the Yld2004-18p model undoubtedly provided overall simulation results with closer agreement to the measured ones than the Hill'48 model did. Nevertheless, in terms of anisotropy (curve shape), Yld2004-18p edged out Hill'48 by a very narrow margin although involving more *r*-values from eight different sheet directions. This would most likely be due to small anisotropy of the material itself as shown in TABLE 1, in which all *r*-values are a close range. A legitimate conclusion can thus be drawn that, under combined stress of radial compression and tangential tension brought about by a standard conical-punch hole expansion process, the Yld2004-18p model is preferred over the Hill'48 model due to its superior competence in characterising anisotropic plastic deformation.

7. Acknowledgements

This research was funded by the Faculty of engineering, King Mongkut's University of Technology North Bangkok (KMUTNB). This research funding is highly acknowledged. The authors also would like to acknowledge Faculty of Engineering, KMUTNB, for the financial supports on the IDDRG2020 conference participation. The authors wish to express their appreciation to KMUTNB for allowing the authors to access its commercial finite element software license ABAQUS 2017 for this research work, as well. Indispensably, the authors would like to highly express their gratitude towards Professor. Frédéric Barlat for his supports regarding the anisotropic parameter calculation code. Lastly, the authors are grateful to Professor. Jeong Whan Yoon for providing Yld2004-18p UMAT program.

8. References

[1] Kuwabara T, Mori T, Asano M, Hakoyama T and Barlat F 2016 Material modeling of 6016-O and 6016-T4 aluminium alloy sheets and application to hole expansion forming simulation *Int. J of Plasticity* **26** 1-23

[2] Mazzolani F 1994 *Aluminium Alloy Structures* (London: Taylor and Francis) p 8

[3] Kliauga A, Bolmaro R and Ferrante M 2015 The evolution of texture in an equal channel pressed aluminium AA1050 *Mater. Sci. Eng. A* **623** 22–31

[4] Smerd R, Winkler S, Salisbury C, Worswick M, Lloyd D and Finn M 2005 High strain rate tensile testing of automotive aluminium alloy sheet *Int. J. Impact Eng.* **32** 41–56

[5] Panich S, Uthaisangsuk V, Suranuntchai S and Jirathearanat S 2012 Anisotropic Plastic Behavior of TRIP 780 Steel Sheet in Hole Expansion Test *Key. Eng. Mater.* **504-506** 89-94

[6] Panich S, Uthaisangsuk V, Suranuntchai S and Jirathearanat S 2014 Investigation of anisotropic plastic deformation of advanced high strength steel *Mater. Sci. Eng. A* **592** 207–220

[7] Lee J Y, Lee K J, Lee M-G, Kuwabara T and Barlat F 2019 Numerical modeling for accurate prediction of strain localization in hole expansion of a steel sheet *Int. J. of Sol. Struc.* **156-157** 107–118

[8] Hill R 1948 Theory of yielding and plastic flow of anisotropic metals Proc. R. Soc. Lond. A **193** 281–297.

[9] Barlat F, Aretz H, Yoon J, Karabin M, Brem J and Dick R 2005 Linear transformation based anisotropic yield functions *Int. J. Plast.* **21** 1009–1039.

[10] Swift H W 1952 Plastic instability under plane stress *J. Mech. Phys. Solids.* **1** 1–18

[11] Young R F, Bird J E and Duncan J L 1981 An automated hydraulic bulge tester *J. App.l Met. Working.* **1** 11–8.

[12] Panich S, Barlat F, Uthaisangsuk V, Suranuntchai S and Jirathearanat S 2013 Experimental and theoretical formability analysis using strain and stress based forming limit diagram for advanced high strength steels *Mater. Des.* **51**, 756–766

Axial formation and cross-section distortion control for stretch bending based on batch simulations

Kaijun Lu[1], Jianxi Luo[1], Tianxia Zou[1], Dayong Li[1] and Yinghong Peng[1]

[1] State Key Laboratory of Mechanical System and Vibration, Shanghai Jiao Tong University, Shanghai 200240, China

Abstract. Stretch bending is a widely used to form long parts with a definite cross-section shape. The forming path is an effective way to eliminate the defects such as springback, distortion when the material and the die shape are determined. In this study, the stretch bending path is represented by a curve function of bending angle and stretch length. To simplify the path optimization problem, the path is represented by 4 parameters, including a newly introduced parameter of total stretch. Based on the batch simulation results, it is found that the axial springback and cross section distortion are primarily controlled by total stretch, bending angle and pre-stretch proportion. When the total stretch is determined, the different distribution schemes for pre-stretch, bending stretch and post-stretch in total stretch have a similar effect on forming quality except the schemes with extremely large proportions of pre-stretch. In most cases, with the total stretch length in growth, the axial springback decreases dramatically at first and then stays constant while the section distortion continues to rise, indicating that the optimal forming path is around the elbow point in the relationship curve of total stretch and springback angle. The optimal path is verified by experiments. It is shown that for a workpiece with determined material and die shape, the optimal path to minimize axial rebound and section distortion simultaneously does exit and the study provide an effective way to obtain it.

1. Introduction

The majority of current railroad car frames are composed of cold formed, extruded stainless steel tubes of different sections instead of unibody concepts, aiming at reducing the weight and the number of the components. The extrusions are manufactured by a series of cold forming operations, including sheet sheering, bending and stretch bending. The word "stretch" indicates that tensile force is applied to both ends of the extrusions during bending. Stretching is an effective way to avoid the cross section's compression and reduce the axial springback by adjusting the stress distribution of the section to a more uniform state. Compared with a variety of other bending process (e.g. press bending, air bending and roll bending), stretch bending provides an efficient and low-cost method in the reduction of springback. Many studies have been conducted on the possible defects in stretch bending process including torsion, crack, cross sectional distortion and especially axial springback, according to analytical, numerical or experimental method.

Analytical method is based on simplified forming process, idealized material properties and pure bending hypothesis. Ueda and Ueno[1] firstly studied the effect of stretch on axial springback in the stretch bending of rectangular channels theoretically, indicating that the axial springback can be reduced by greater tension force, and the stretch before bending is less effective than the stretch after bending. Based on pure bending assumptions, A.A.El-Domiaty et al.[2] introduced a mathematical model for stretch bending process to determine the effect of material properties on the process parameters including product geometry, residual stress and forming loads. Paulsen and Welo[3] established analytical models to predict suck-in deformation and local post-buckling of rectangular hollow sections in stretch bending by using the energy method. Corona [4] stressed the importance of appropriate tension force, indicating that higher tension reduced the axial springback while induced relatively large section distortion, and then presented an efficient formulation for cross sectional distortion prediction of thin-walled profiles. YU Cheng-long and LI Xiao-qiang[5] established a theoretical model to analyse the effect of pre-stretch force and post-stretch force on springback angle after forming in stretch bending process of L-section extrusions. Zhao et al.[6] took the profile with arbitrary cross-section as the study object and obtained the universal analytical method to solve the springback problem of stretch bending in the loading method of pre-tension and bending moment, and verified the theoretical model with experimental and finite element numerical research on rectangular section and U-section profiles. On the basis of Jun Zhao's work, Ruixue Zhai et al. [7] extended the study to the loading method of pre-bending and tension, which is far more sophisticated than the loading method of pre-tension and bending, indicating that high tension and small bending radius can prevent rebounding significantly.

Compared with the analytical method, the finite element method provides a convenient way to deal with the workpiece with complex axial curve and complex cross-section shape. Hopperstad et al. [8] studied the effect of different yield criterion on local deformations predicted in numerical simulation of stretch bending by using the software LS-DYNA3D. Hopperstad et al.[9] assigned the material parameters randomly to analyze the reliability of the stretch bending process for U-sectional aluminum extrusions by using LS-DYNA3D. J.E. Miller and S. Kyriakide [10] established 3-D finite element model to replace efficient 2-D model in which all applied loads and the cross section shape were assumed to be uniform along the axis. and studied the 3-D features problems such as springback and distortion along the axis based on 3-D model. Lasse Bjørkhaug and Torgeir Welo [11] found that the cross section distortion is over-predicted by isotropic model, and improved the accuracy of finite element model by using the anisotropic Barlet96 yield model. YU Zhong-qi and LIN Zhong-qin [12] studied the effect of processing parameters on the dimension precision of the U-shaped aluminum profile by using Dynaform, indicating that the section distortion decreases with the increase of the side pressure, whereas the axial springback increases, and both of the section distortion and axial springback derease with the increase of the tension force. Y. Okudea et al. [13] simulated the stretch bending of various channel section extrusions with laminated elastic mandrel inside by using LS-DYNA3D, proving effectiveness of the mandrel in preventing the undesirable cross section distortion. Tianjiao Liu et al. [14] simulated the displacement controlled stretch bending of the 2099-T83 and 2196-T8511 Al-Li alloys extrusions by using Abaqus, indicating that the springback decreased significantly with the increase of post-stretch or pre-stretch. Chun-guo Liu1 et al. [15] simulated the springback behavior of aluminum hollow profile under five levels of post-stretching elongations, indicating that high post- stretching elongation leads to small axial springback but large spatial twist and cross section distortion. Zhengwei GU et al. [16] simulated the stretch bending of a L-section aluminum extrusion by using Abaqus and quantified the effect of die fittingness on horizontal plate sagging distortion.

With analytical and simulation methods, the effect of various forming parameters on quality has been explored in depth, improving the accuracy of forming results prediction and the control ability of forming quality. Arild H. Clausen et al. [17] studied the effect of workpiece geometry and forming path on section distortion, indicating that the axial springback is primarily controlled by the tension force and the strain hardening exponent of the material while the cross-section distortion is primarily controlled by the workpiece geometry and tension force. By applying six forming procedures with different tension sequence into stretch bending of a car bumper numerically and experimentally, Arild H. Clausen et al. [18] studied the effect of forming path on the section distortion and axial springback. J.E. Miller et al. [19] constructed a custom bend-stretch-pressure forming facility to realize flexible control of forming path. Based on the experiments carried on the facility, the bend-stretch-pressure forming process is simplified to a 2D model, in which the cross-section shape and applied loads are assumed uniform along the axis. Based on the simplified 2D model, J.E. Miller et al. [20] studied the influence of forming process parameters on forming quality, indicating that the section distortion is primarily controlled by the tensile force and geometry while the axial springback is primarily controlled by tensile force and strain hardening exponent coefficient. Arild H. Clausen et al. [21] used factorial analysis in the planning of the experimental program and result interpretation to study the influence of the alloy, die radius and specimen wall thickness on response parameters such as die force, sagging and springback, indicating that the die force and springback is primarily controlled by alloy while the geometric parameters have the largest influence on the sagging behavior.

Previous studies have shown that the forming quality of stretch bending is primarily controlled by material properties, workpiece geometry and loading path. Compared with other means such as mold shape modification, material replacement or workpiece shape modification, path optimization is the most efficient and low-cost way to improve the forming quality.

In the present work, firstly, the finite element model is verified by the stretch bending experiment and a massive amount of data is obtained by batch simulation approach. Secondly, two new variables named distribution ratio and total stretch length are introduced to describe forming process, replacing various factors used in previous studies such as pre-stretch, bend stretch and post-stretch, and the relationship between forming path and forming quality is greatly simplified. Thirdly, the influence of forming path on forming quality is discussed. Finally, a series of experiments are carried out to confirm the conclusions drawn from analysis.

2. Stretch bending process

A typical stretch bending facility consists of four parts, as shown in Fig1: a mould is fixed on the work table, two arms for bending the workpiece into certain curvature, two tension cylinders for applying axial stretch while bending and two carriage for loading and unloading workpieces of different lengths in the appropriate positions. The forming process is usually divided into three steps: loading, forming and unloading. Firstly, the mold is installed on the work table, then two tension cylinders stretch out and two carriages are moved to appropriate positions to fix the workpiece at its ends by jaws. Secondly, the jaws moves along desired trajectories which is determined by the rotation of arms and the motion of the tension cylinders, affixing the workpiece to the mold and elongating it gradually. Thirdly, open jaws to release stress, then move the tension cylinders to the desired position for unloading.

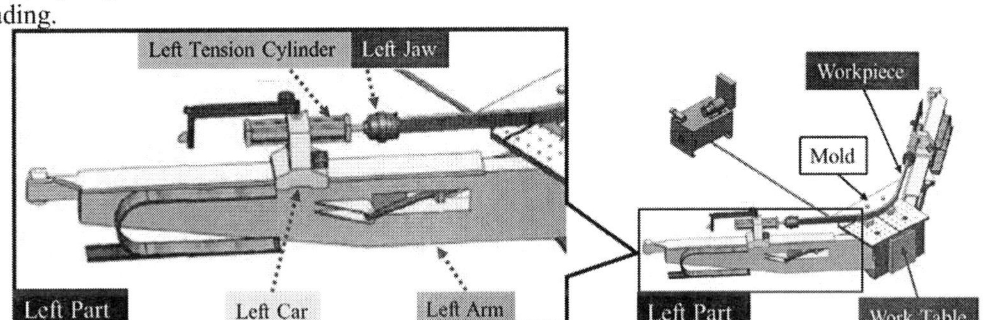

Fig 1. Schematic diagram of a typical stretch bending facility.

2.1. Motion structure of the machine and the forming path

Because the stretch bending machine is structural symmetry, analysis of forming process is performed on half of the structure. The motion structure consists of three parts: arm, carriage and tension cylinder, and their locations jointly determine the location of the jaw, which is also the location of one end of the workpiece, as shown in Fig 2. In the loading step, the tension cylinder stays still, while the arm is rotated to a certain angle to make the workpiece tangent to the die profile, and the carriage is moved to the appropriate location to load the workpiece with different lengths smoothly. In the forming step, the arm is rotated to mould the workpiece axis into the desired contour while the tension cylinder is drawn back to stretch the workpiece along its axis. In the unloading step, the jaw on the end of the tension cylinder is opened to release the product.

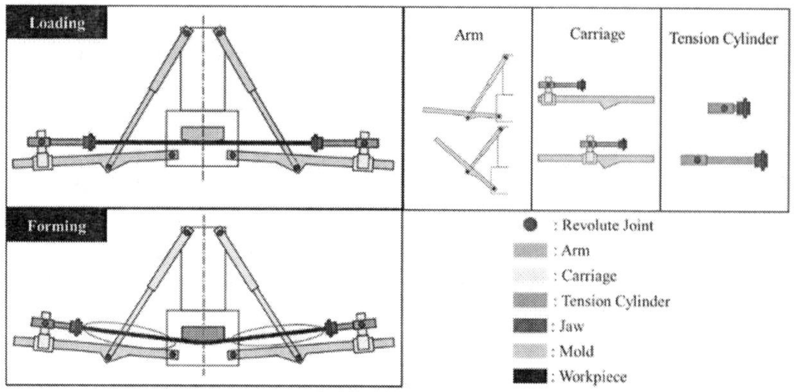

Fig 2. Motion structure of the stretch bending machine.

All axial profile outlines of the workpieces are smooth convex curves, which can be described as continuously derivable functions and can be divided into several straight lines and arcs. Fig 3 shows typical axial profile outlines of a workpiece. The target product is composed of four segments including two straight lines (S2 and S4) and two arcs (S1 and S3). Compared with the target product, the actual product is longer on the two ends. The extra linear segments of actual product are LS3 and RS4, which are used for clamping during forming process as circled in Fig 3 and will be cut off after forming.

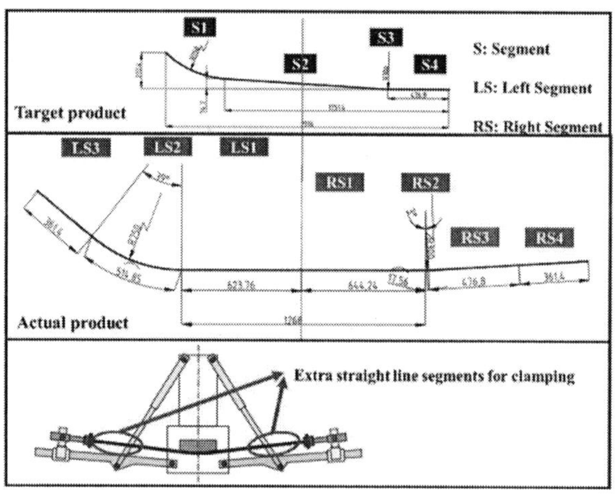

Fig 3. Axial profile outlines of target product and actual product.

The deformation of the workpiece axis during the stretch bending process can be divided into the transverse bending effect and the axial stretching effect, which can be represented by the bending angle and stretch length respectively. Fig 4 shows a new representation method of forming process. Each point in the coordinate system represents a deformation state, including the bending angle and stretch length, and it is obvious that the initial state is at the origin (0° ,0mm). The forming paths are represented by the curves connecting the initial point and the destination point, and the number of possible forming paths is obviously infinite. To simplify analysis, a variety of forming paths are standardized to polylines composed of three linear segments, as the standard forming process can be divided into three steps including pre-stretch, stretch bending and post-stretch. The workpiece can be stretched during the forming process while it is bent only in the stretch bending step. Consequently, the first and third linear segments of a standard forming path are vertical while the second segment is an oblique line, as shown in Fig 4.

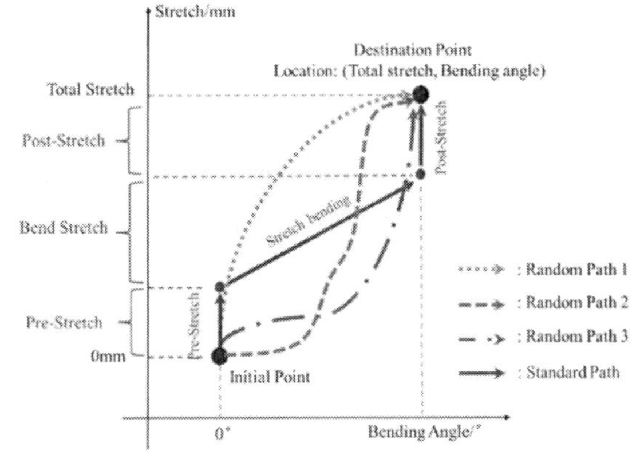

Fig 4. Schematic of possible forming paths during stretch bending process.

Therefore, the forming path problem consists of two crucial and critical parts. The first one is the final deformation state of the workpiece before rebound, which can be represented by the location of the destination point. In general, the springback angle is relatively small with a proper forming process, usually at 5% of the bending angle or lower. Accordingly, the target angle of the target product can be a suitable substitute for the bending angle when the axial rebound is ignored. The second critical part is the forming path connecting the initial point and destination point, which can be represented by three remaining variables including pre-stretch, bend stretch and post-stretch.

2.2. *Processed profile*

The product for illustration is a long rail with hat cross section which is made of stainless steel. The axis of the workpiece consists of three segments including one arc with angle 4.33° and two straight lines on two ends, as shown in Fig 5.

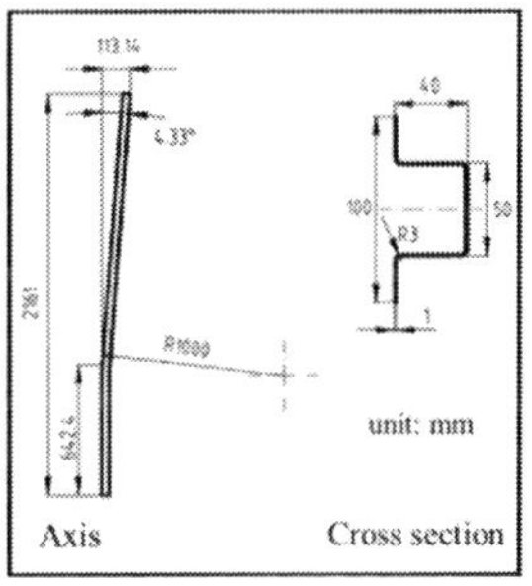

Fig 5. Sketch of the workpiece.

The purpose of the stretch bending is to form the axis of a long rod into the shape specified while keeping its cross section undeformed as far as possible. Consequently, the forming quality of stretch bending consists of the axial deformation and the cross sectional distortion, which are evaluated by θ_{as}, h and θ_e, as shown in Fig 6. θ_{as} is the axial deformation angle after springback, and the difference between bending angle (or forming angle) and θ_{as} is the springback angle. The cross section along the axis is not uniform anymore and the max deformation occurs on the arc segments of the axial curve. Meanwhile, the deformation of brim region is more server than other region in the cross section. Accordingly, the height from brim region to crown region and the angle between two hat edges are suitable to represent the distortion of the hat section.

Deformation region	axis	Section	
Symbol	θ_{as}	h	θ_e
Meaning	Axis angle after springback	the height difference	angle between two hat edges
Sampling position	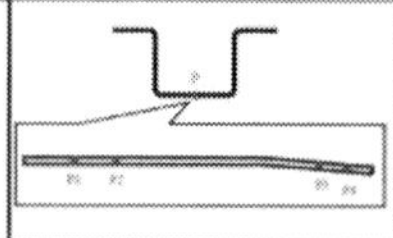	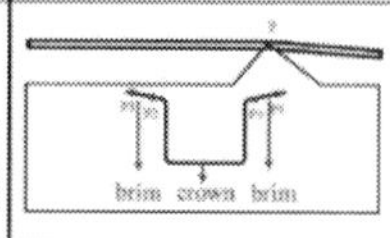	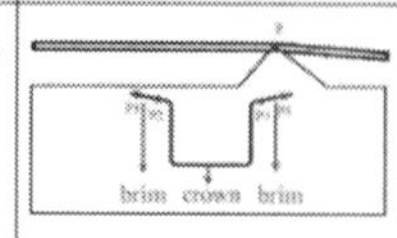
Formula	 $\theta(\vec{V1},\vec{V2})$	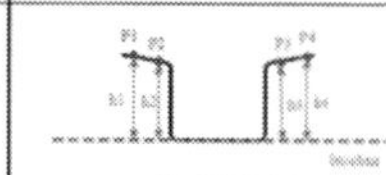 max(h1,h2,h3,h4)	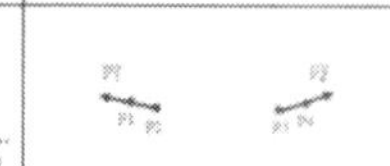 $\theta(\vec{V1},\vec{V2})$

Fig 6. Schematic of the deformation quality parameters.

3. Finite element simulation

3.1. FE modeling

A finite element model of the stretch bending process is developed by using ABAQUS/Standard software, as shown in Fig 7.

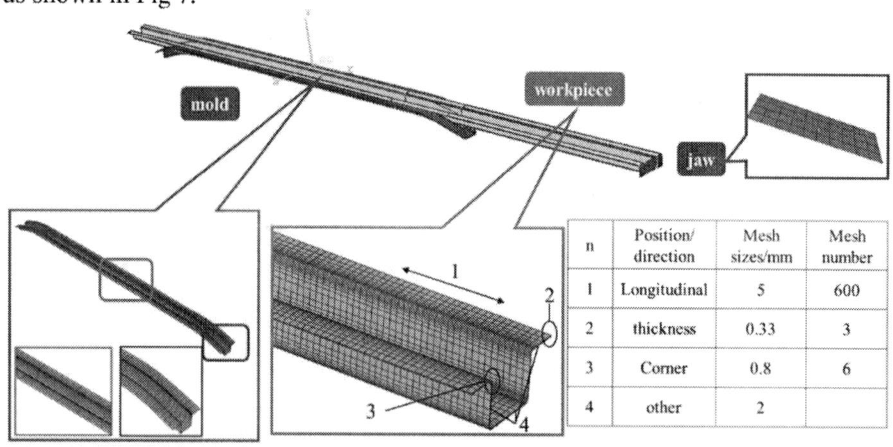

n	Position/ direction	Mesh sizes/mm	Mesh number
1	Longitudinal	5	600
2	thickness	0.33	3
3	Corner	0.8	6
4	other	2	

Fig 7. The finite element model.

The material of the present hat-shaped workpiece is SUS-301 stainless steel and its true stress-true strain curve is presented in Fig 8. The workpiece shown in Fig 7 is divided into three layers of eight-node linear brick element with reduced integration (C3D8R). After a series of mesh sensitivity check, the element sizes of 5mm, 2mm, 0.8mm for longitudinal direction, slight deformation region and corner region respectively. The blank model has 177896 nodes and 131400 elements. Mold blocks are represented as discrete rigid surfaces and discretized by using R3D4 elements. The Coulomb friction model is assumed and the friction coefficient between the workpiece and mold blocks is set as 0.1.

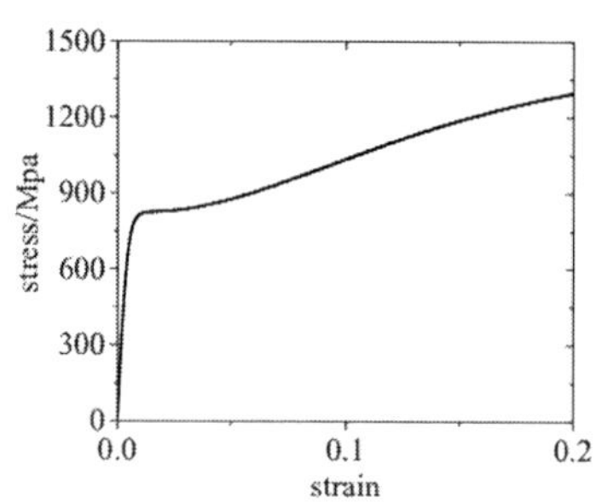

Fig 8. Stress-strain curve of stainless steel.

During the stretch bending process, the mold stays still while two jaws clamp the two ends of the profile and move in designed trajectories. Accordingly, the constraints between the ends of workpiece and two jaws are set as "tie". The trajectories are determined using the following steps, as shown in Fig 9: Firstly several dispersed featured points are selected from forming path curve to determine the total length and bending angle of the workpiece in different deform states. Secondly, the position of the jaw which is associated with the end of the workpiece is determined by the deform state of the workpiece. Thirdly, the trajectory of jaw is plotted by connecting amounts of locations of jaws in different deform states .

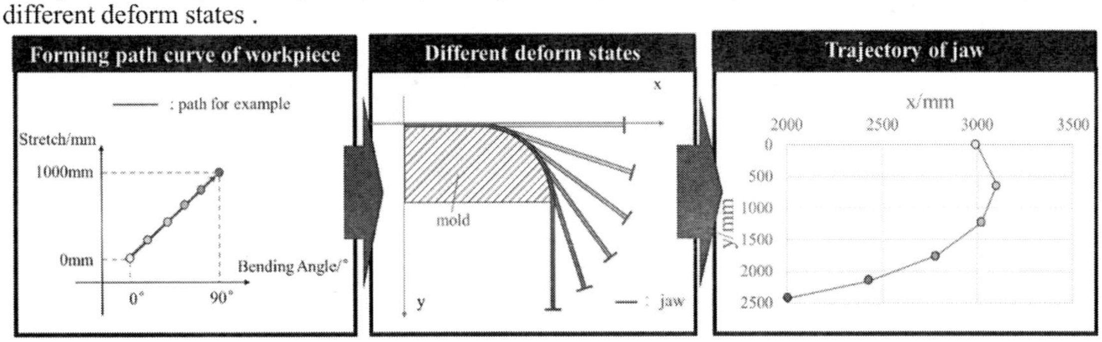

Fig 9. Three steps to obtain the trajectory of jaw.

3.2. Experimental verification

The stretch-bending experiments are carried on the equipment of Cyril bath V-50, as shown in Fig. 10. The experimental results of different forming paths agree well with the numerical results and some results are shown in Table 1.

Fig 10. The experimental machine.

Table 1. Results of FE simulations and experiments.

Number	Bend angle (°)	Pre-stretch (mm)	Bend stretch (mm)	Post-stretch (mm)	Calculated forming angle (°)	Measured forming angle (°)
1	4.4	0	5	5	2.43	2.35
2	4.4	0	6.25	6.25	3.71	3.55
3	4.4	0	7.5	7.5	4.25	4.2
4	4.4	0	10	10	4.25	4.25
5	4.4	20	0	0	0.5	0.3

4. Influence of forming path on formation quality

The forming path is a curve connecting the initial point and destination point in the coordinate of bending angle and stretch length, as shown in Fig 11.

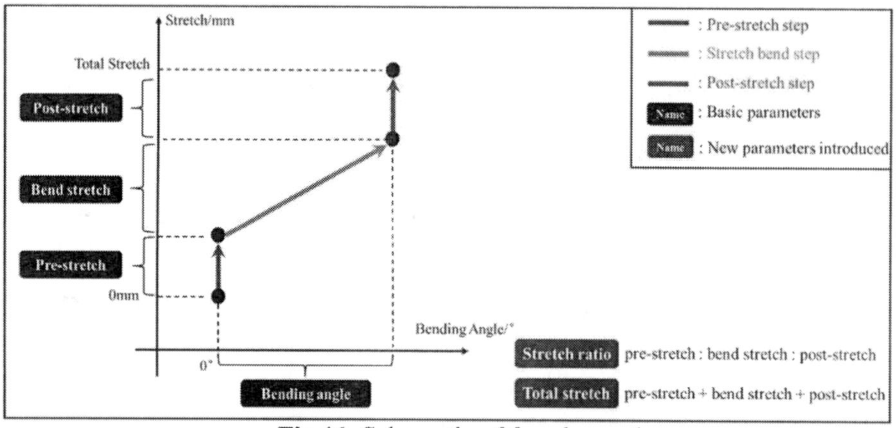

Fig 11. Schematic of forming path.

Each point in the schematic is not only the deformation state of workpiece describing its elongation and bending degree, but also represents the strain distribution of the workpiece. The curve connecting all nodes is the forming path which describes the deformation history of the workpiece. The final strain distribution of the workpiece is determined by the location of the destination point while the final stress distribution of the workpiece can be altered by changing the forming path which connects the initial point and the destination point. In the following sections, the influence of the forming parameters on the forming quality is discussed in two aspects: one is the final deformation state of the workpiece which is represented by the bending angle and total stretch, and the other one is the forming path connecting the initial and the final deformation state which is represented by the stretch ratio.

4.1. Influence of stretch distribution

As mentioned earlier, there are infinite forming path curve connecting the initial point and the destination point. To simplify the analysis, 15 forming paths of different stretch schemes are simulated as shown in Fig 12.

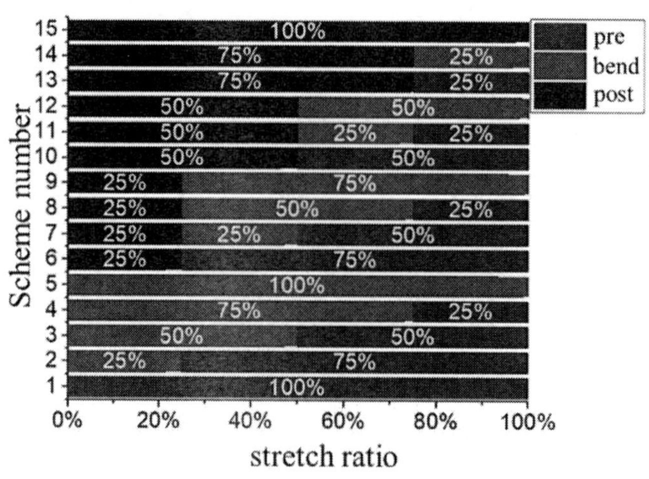

Fig 12. stretch distribution schemes.

Comparing forming results of 15 distribution schemes in Fig 12, all schemes are similar in forming quality except scheme 1 and scheme 2. Fig 13-15. shows the forming results of 15 schemes, whose bending angles are all 12° and total lengths range from 20mm to 200mm.

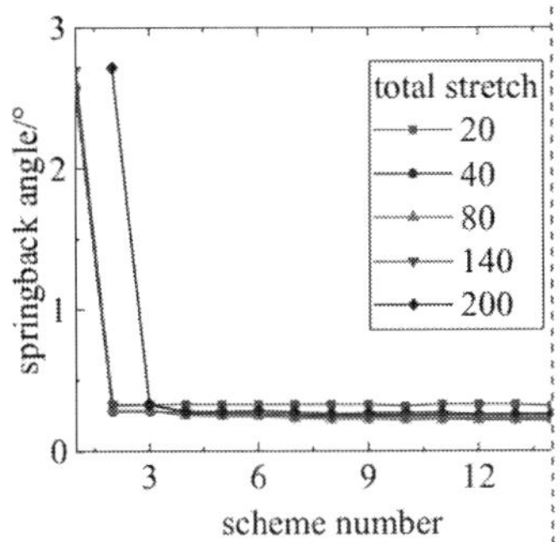

Figure 13. The influence of 15 schemes of different stretch ratio distributions on springback angle.

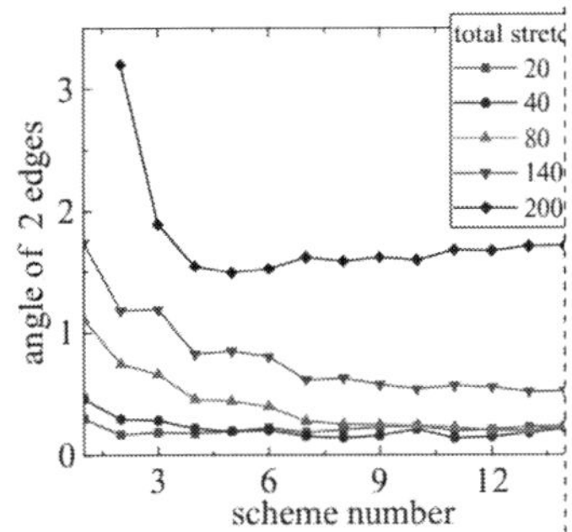

Figure 14. The influence of 15 schemes of different stretch ratio distributions on θ_e.

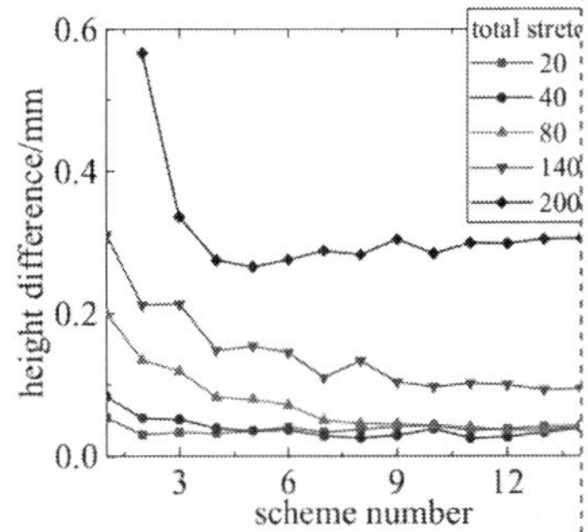

Figure 15. The influence of 15 schemes of different stretch ratio distributions on h.

The particularity of scheme 1 and 2 is mainly caused by the unevenly distributed stress in the workpiece. The bending throughout the forming process causes the reverse loading of inner layers continuously. Without continuous stretch, the inner layers will be compressed and the stress in inner layers will decrease. Compared with the other schemes, scheme 1 and 2 stretch the workpiece little in the stretch bending step and post-stretch step, causing the unbalance of the stress between the inner and outer layers of the workpiece.

4.2. *Influence of total stretch and bending angle*

As stated above, the forming results of different forming paths are similar except with small stretch bending and post-stretch. Accordingly, the result of scheme 8 can be select as an example to study the influence of total stretch and bending angle on forming qualities.

Fig. 16 shows the influence of total stretch and bending angle on forming angle, indicating that with the increase of the total stretch, the springback reduces dramatically at first and then tends to be stable. When connecting all elbow points of different bending angles, the fitting is almost linear as shown in Fig 16. According to the linear relationship, the elbow point for an arbitrary bending angle can be easily obtained. Taking the product for instance, its bending angle is about 4.3° and the total stretch of the elbow point should be about 17mm according to the proportional relationship.

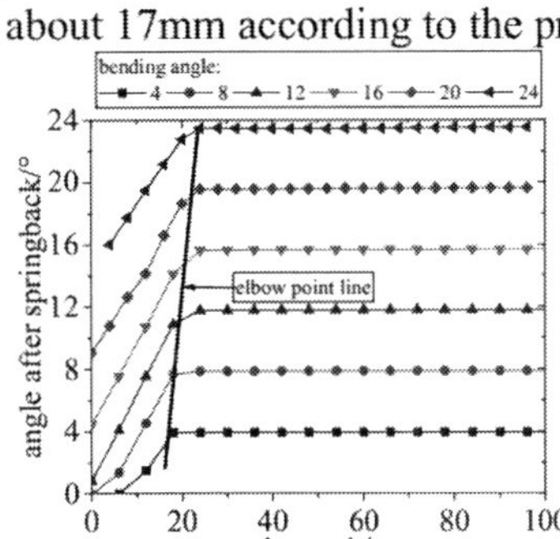

Fig 16. Influence of total stretch and bending angle on forming angle.

The springback angle, which is stable with the increase of the total stretch, is linearly proportional to the bending angle, as shown in Fig 17. According to the proportional relationship, the bending angle for an arbitrary target angle (e.g. the angle after springback) can be easily obtained. Taking the product for instance, its target angle is 4.22° and the bending angle should be 4.27° according to the proportional relationship.

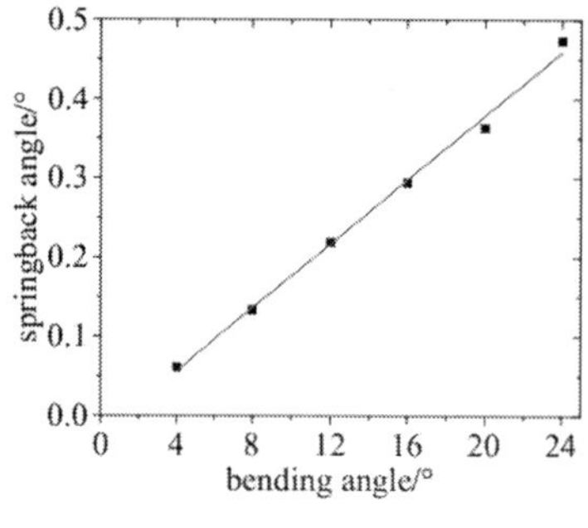

Fig 17. Relationship between bending angle and springback angle.

Fig 18 and Fig 19 shows the influence of total stretch and bending angle on section distortion, indicating that with the increase of the total stretch, the section distortion fluctuates irregularly at first and then increases continuously.

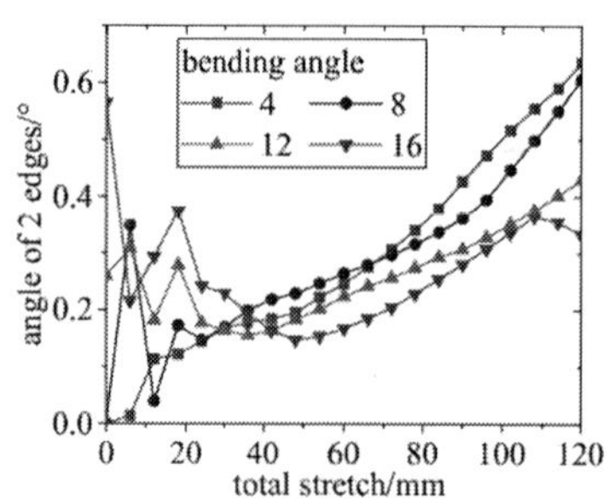

Figure 18. Influence of total stretch on section distortion parameter θ_e .

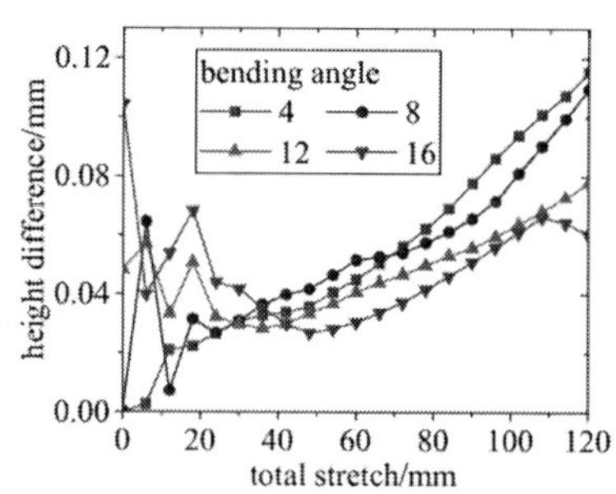

Figure 19. Influence of total stretch on section distortion parameter h.

It is obvious that the increase of total stretch in the axial direction causes more sever section distortion. The fluctuation in the initial stage of the curve is caused by the detachment between workpiece and mold. When the workpiece is bent slightly, the distance between the two ends of workpiece will be smaller than the original length, and the axis of the workpiece will be more likely to be formed into a deflection curve, rather than the target axial curve of the mold, an shown in Fig 20.

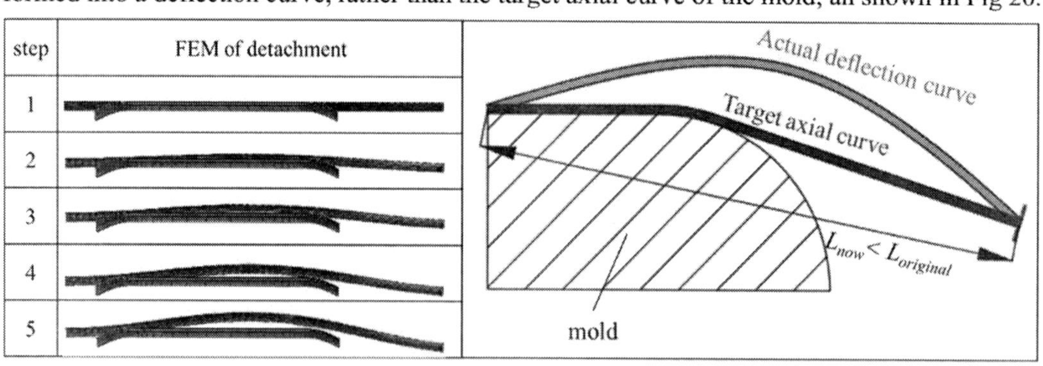

Fig 20. Detachment caused by slight bend.

4.3. *Optimum forming path*

The optimum path destination end is obviously around the elbow point. With a determined path end, the forming path, which is described by the stretch distribution, has little influence on forming quality except for some extreme cases with large ratio of pre-stretch. Accordingly, all rest paths are recommended for forming.

5. Experimental verification

A series of experiments are carried out on a Cyril bath V-50 stretch bending machine to verify the conclusions above. The forming angle is measured by inclinometer while the distortion of the section is identified by visual inspection assisted with a square ruler. The experimental results of different forming paths agree well with FEM model, as follows:

1) The results of experiments 1-4 are shown in Fig 21. With the increase of the total stretch, the forming angle increases dramatically at first and then tends to be stable while the section distortion increases continuously and obvious distortion is observed in experiment 4.

Influence of total stretch on forming angle and section distortion				
Exp num	1	2	3	4
Total stretch/mm	10	12.5	15	20
Forming angle/°	2.35	3.55	4.25	4.25

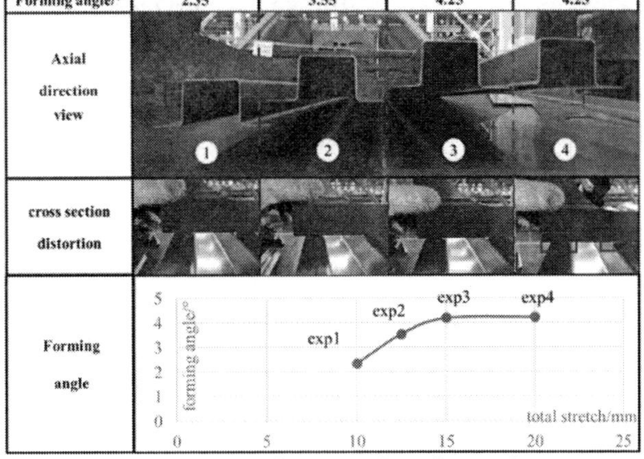

Fig 21. Influence of total stretch.

2) Fig 22 shows that the distribution of stretch does not have significant effect on forming angle except when the bend stretch and post-stretch are too small. The springback of axial angle increase dramatically when the bend stretch and post-stretch are extremely small in experiment 9.

Influence of stretch distribution on forming angle					
Exp num	9	12	11	10	4
Pre-stretch/mm	20	5	10	15	0
Bend stretch/mm	0	7.5	5	2.5	10
Post-stretch/mm	0	7.5	5	2.5	10

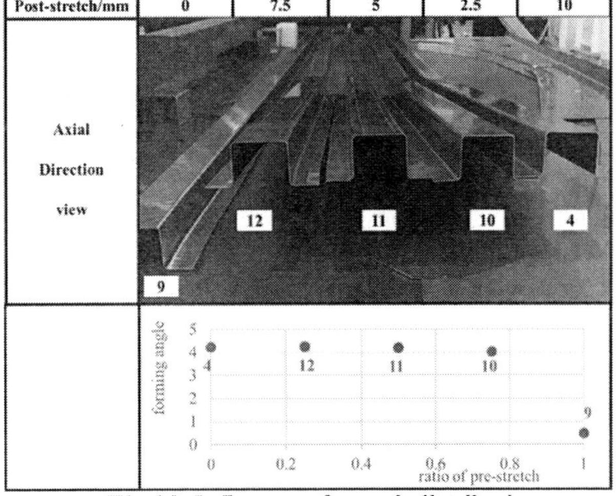

Fig 22. Influence of stretch distribution.

3) It is shown that exp4 and exp10-12 are all appropriate forming path for the illustrated product, which is small in axial springback and section distortion.

6. Conclusion

In this work, the stretch bending process of a rail made of SUS301 stainless steel with hat shape section is studied by experiment, finite element simulation, and the following conclusion can be drawn.

1) The finite element model agrees well with the experiment results.
2) With the increase of the total stretch, the springback of the axial reduces dramatically at first and then tends to be stable while the distortion of the section increases constantly.
3) The forming path (e.g. stretch distribution) hardly affect the forming quality with a few exceptions, in which the ratio of bend stretch or post stretch is large.
4) The optimal forming path can be obtained accurately and effectively by batch simulation and subsequent analysis.

[1] Ueda M, Ueno K, Kobayashi M. A study of springback in the stretch bending of channels[J]. Journal of Mechanical Working Technology, 1981, 5(3-4): 163-179.
[2] El-Domiaty A A, Shabara M A N, Al-Ansary M D. Determination of stretch-bendability of sheet-metals[J]. International Journal of Machine Tools and Manufacture, 1996, 36(5): 635-650.
[3] Paulsen F, Welo T. A design method for prediction of dimensions of rectangular hollow sections formed in stretch bending[J]. Journal of Materials Processing Technology, 2002, 128(1-3): 48-66.
[4] Corona E. A simple analysis for bend-stretch forming of aluminum extrusions[J]. International Journal of Mechanical Sciences, 2004, 46(3): 433-448.
[5] YU C, LI X. Theoretical analysis on springback of L-section extrusion in rotary stretch bending process[J]. Transactions of Nonferrous Metals Society of China, 2011, 21(12): 2705-2710.
[6] Zhao J, Zhai R, Qian Z, et al. A study on springback of profile plane stretch–bending in the loading method of pretension and moment[J]. International Journal of Mechanical Sciences, 2013, 75: 45-54.
[7] Zhai R, Ding X, Yu S, et al. Stretch bending and springback of profile in the loading method of prebending and tension[J]. International Journal of Mechanical Sciences, 2018, 144: 746-764.
[8] Hopperstad O S, Berstad T, Ilstad H, et al. Effects of the yield criterion on local deformations in numerical simulation of profile forming[J]. Journal of materials processing technology, 1998, 80: 551-555.
[9] Hopperstad O S, Leira B J, Remseth S, et al. Reliability-based analysis of a stretch-bending process for aluminium extrusions[J]. Computers & structures, 1999, 71(1): 63-75.
[10] Miller J E, Kyriakides S. Three-dimensional effects of the bend–stretch forming of aluminum tubes[J]. International journal of mechanical sciences, 2003, 45(1): 115-140.
[11] Bjørkhaug L, Welo T. Local Calibration of Aluminium Profiles in Rotary Stretch Bending—Anisotropy Effects[C]//AIP Conference Proceedings. American Institute of Physics, 2004, 712(1): 749-754.
[12] Yu Z Q, Lin Z Q. Numerical analysis of dimension precision of U-shaped aluminium profile rotary stretch bending[J]. Transactions of Nonferrous Metals Society of China, 2007, 17(3): 581-585.
[13] Okude Y, Yoshihara S, Saito A, et al. Bending deformation behavior of Aluminium extruded various sections with laminated elastic mandrel on draw bending[J]. Materials Today: Proceedings, 2015, 2(10): 4802-4811.
[14] Liu T, Wang Y, Wu J, et al. Springback analysis of Z & T-section 2196-T8511 and 2099-T83 Al–Li alloys extrusions in displacement controlled cold stretch bending[J]. Journal of Materials Processing Technology, 2015, 225: 295-309.
[15] Liu C, Zhang X, Wu X, et al. Optimization of post-stretching elongation in stretch bending of aluminum hollow profile[J]. The International Journal of Advanced Manufacturing Technology, 2016, 82(9-12): 1737-1746.
[16] Gu Z, Lv M, Li X, et al. Stretch bending defects control of L-section aluminum components with variable curvatures[J]. The International Journal of Advanced Manufacturing Technology, 2016, 85(5-8): 1053-1061.
[17] Clausen A H, Hopperstad O S, Langseth M. Stretch bending of aluminium extrusions for car bumpers[J]. Journal of Materials Processing Technology, 2000, 102(1-3): 241-248.

[18] Clausen A H, Hopperstad O S, Langseth M. Sensitivity of model parameters in stretch bending of aluminium extrusions[J]. International Journal of Mechanical Sciences, 2001, 43(2): 427-453.
[19] Miller J E, Kyriakides S, Bastard A H. On bend-stretch forming of aluminum extruded tubes—I: experiments[J]. International Journal of Mechanical Sciences, 2001, 43(5): 1283-1317.
[20] Miller J E, Kyriakides S, Corona E. On bend-stretch forming of aluminum extruded tubes—II: analysis[J]. International Journal of Mechanical Sciences, 2001, 43(5): 1319-1338.
[21] Clausen A H, Hopperstad O S, Langseth M. Stretch bending of aluminum extrusions: effect of geometry and alloy[J]. Journal of engineering mechanics, 1999, 125(4): 392-400.

Acknowledgments

The authors would like to acknowledge the support of National Science Foundation of China (Grant No.U1860110). And the authors would like to thank CRRC Changchun Railway Vehicles Co, Ltd. for providing manufacturing assistance.

The formability and application of hot-rolled dual-phase steel under nonlinear strain paths

W J Chen[1,2,3], H W Song[1*], X H Pei[3], S H Zhang[1]

[1] Institute of Metal Research, Chinese Academy of Sciences, Shenyang110016, P.R. China
[2] School of Materials Science and Engineering, University of Science and Technology of China, Shenyang110016, P.R. China
[3] Meishan Technology Center of Baosteel Central Research Institute, Nanjing 210039, P.R. China

* hwsong@imr.ac.cn

Abstract. The hot-rolled dual-phase steel 700DP was developed to reduce weight and improve safety in automobile industry. However, strain path changes inevitably occur for most of components during sheet forming process. Therefore, it's valuable to predict the cracking accurately under nonlinear strain paths for the application of 700DP steel with finite element method (FEM). In this paper, the mechanical properties of hot-rolled dual-phase steel 580DP and 700DP were studied with tensile test. Nakazima test and two-step nonlinear loading test were used to establish the modelling with the method of equivalent plastic strain-based FLC (ep-FLC). Furthermore, the ep-FLC was applied to predict the disc forming result of three-stage process, which agreed well with the experimental results for both 580DP and 700DP compared with conventional FLC.

1. Introduction

At the component design stage, numerical simulation has been used popularly to determine the forming process and die structure. Accurate description of the mechanical properties and formability is important for the process design, which indeed has great influence on the effectiveness and cost of project. Forming limit curve (FLC) is most widely used in sheet metal forming. However, it has been pointed out that conventional FLC is difficult to predict the forming failure under multi-stage forming process for the strain path changing [1-3]. Stress based FLC (FLSC) was proposed to evaluate formability during nonlinear loading process [4-6]. Nevertheless, the FLSC data cannot be determined directly from the experiments which mainly limits the application in engineering. the concept of the effective plastic strain-based FLC (ep-FLC) and Polar Equivalent Strain (PEPS) were proposed successively to estimate the failure under nonlinear forming process [7,8].

For low-cost and high property, hot-rolled dual-phase steel 580DP has been used widely for automotive part manufacturers [9,10]. Furthermore, higher strength dual-phase steel 700DP was developed to satisfy the need for further lightweight and higher property. In this page, the mechanical property and formability of 580DP and 700DP were investigated. The modelling of equivalent plastic strain-based FLC (ep-FLC) was established to describe the formability during nonlinear strain process, compared with conventional FLC.

2. Material and experimental procedure

2.1 Material and tensile test

Hot-rolled dual-phase steels usually consist of martensite and ferrite phases. The soft and ductile ferrite matrix provides dual phase steels with low yield strength and good ductility, while the hard martensite phase provides the high tensile strength mainly. Compared with cold rolled DP steel, hot-rolled DP steel has the similar property through controlled rolling and controlled cooling process, which is much more low-cost. 700DP is produced to balance the strength and plasticity, as the middle grade between 580DP and 780DP. The microstructures of 580DP steel and 700DP steel are illustrated in Fig. 1, the volume fraction of 700DP is higher than that of 580DP.

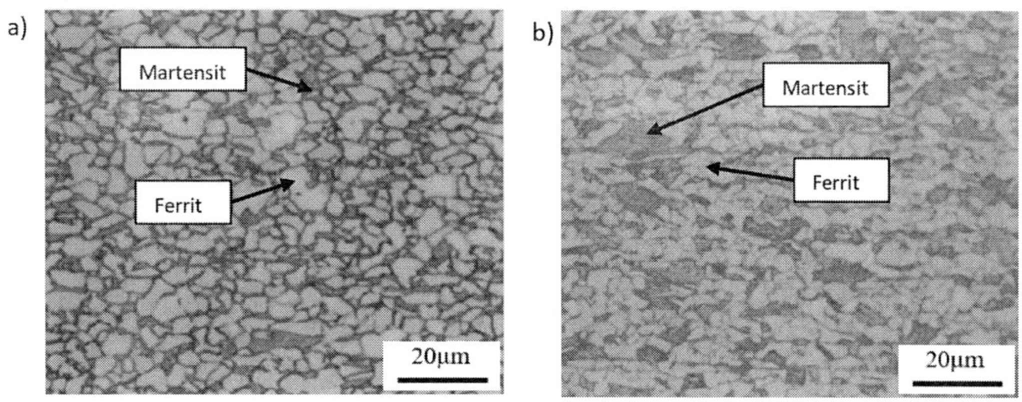

Figure 1. Microstructures of hot-rolled steels: a)580DP, b)700DP.

The sheet thicknesses of 580DP and 700DP in this investigation are 3.0mm and 2.95mm respectively. Tensile tests had been performed on INSTRON testing machine in various directions, including 0 degree, 45degree and 90 degree along rolling direction, two specimens were measured in each direction to determine the test data. As shown in Figure 2, the yield strength of 700DP is similar to that of 580DP, while the tensile strength is much higher. The material property of 580DP and 700DP are anisotropic, and the tensile strengths of two steels at 90 degree are highest, the elongations at 0 degree are greatest.

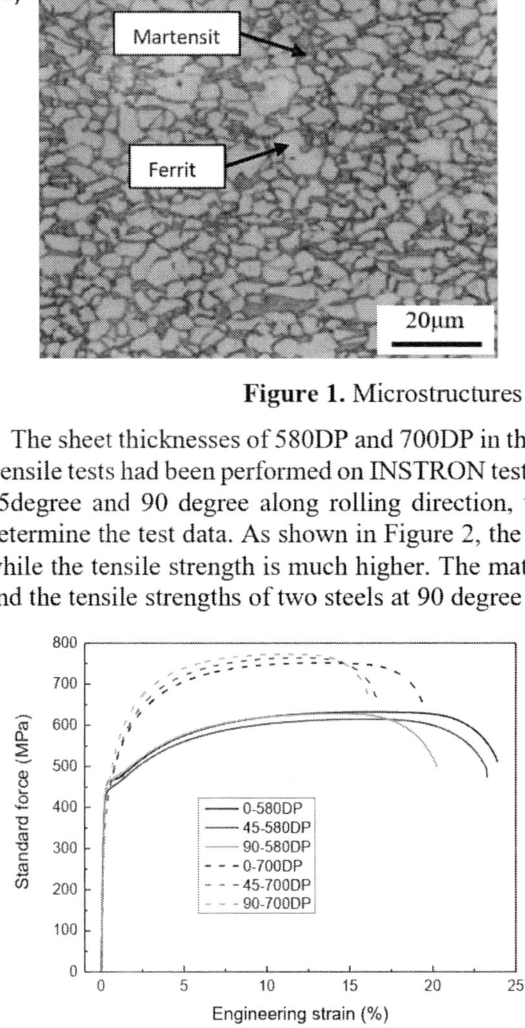

Figure 2. The engineering stress-strain curves of 580DP and 700DP in three directions.

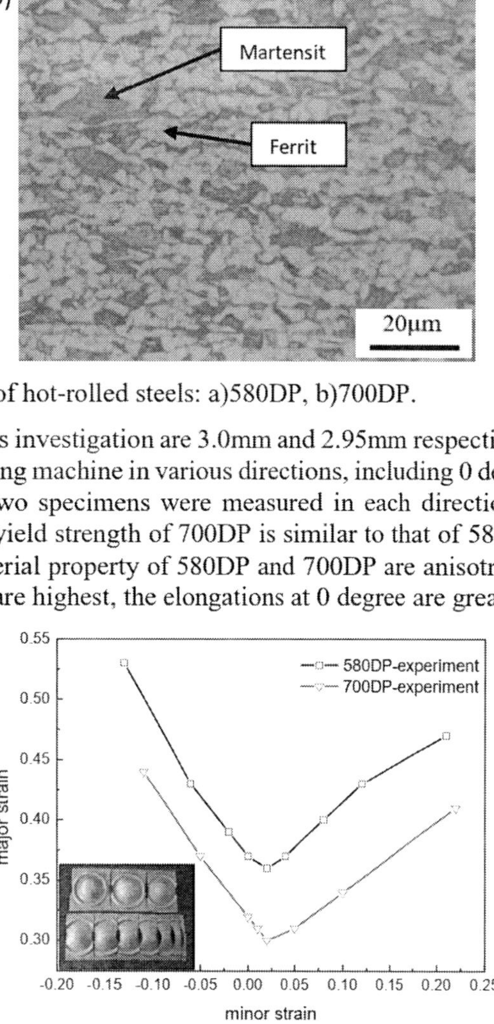

Figure 3. FLCs of 580DP and 700DP under linear strain paths.

2.2 Formability test under linear and nonlinear strain paths

According to the standard of the ISO 12004, the linear FLC was measured on Interlaken sheet metal testing machine with Nakazima test. And the whole deformation history can be recorded to identify the onset of necking directly with the VIALUX photogrammetric equipment. Friction between sample and punch was minimized by using polyurethane gasket as lubricant. The samples were deformed until fracture by a hemispherical punch with a diameter of 100 mm. In this method, ten specimens with the same length of 196 mm but different widths were selected to obtain different loading paths, varying from 20 up to 180 mm. The obtained FLCs of 580DP and 700DP under linear strain paths are showed in Figure 3.

The two-step forming method was employed to produce nonlinear strain paths, which was used to check the influence of changing strain paths of 700DP. Four different loading paths were proposed totally in this study. The path of pre-uniaxial to plane strain is defined as UT-PS, while the path of pre-uniaxial to biaxial strain is defined as UT-BS. The path of pre-biaxial strain to plane strain is called BS-PS, and the path of pre-biaxial strain to uniaxial strain is defined as BS-UT. The pre-strains under different loading conditions and limit strain results on the onset of necking were measured through optical method. The experimental results of nonlinear paths were added to the traditional FLC. As shown in Fig. 4(a), it is not unexpected that the limit principle strains of two-step path change were out of the traditional FLC. Therefore, the traditional FLC of 700DP can't predict the material formability accurately under nonlinear strain path.

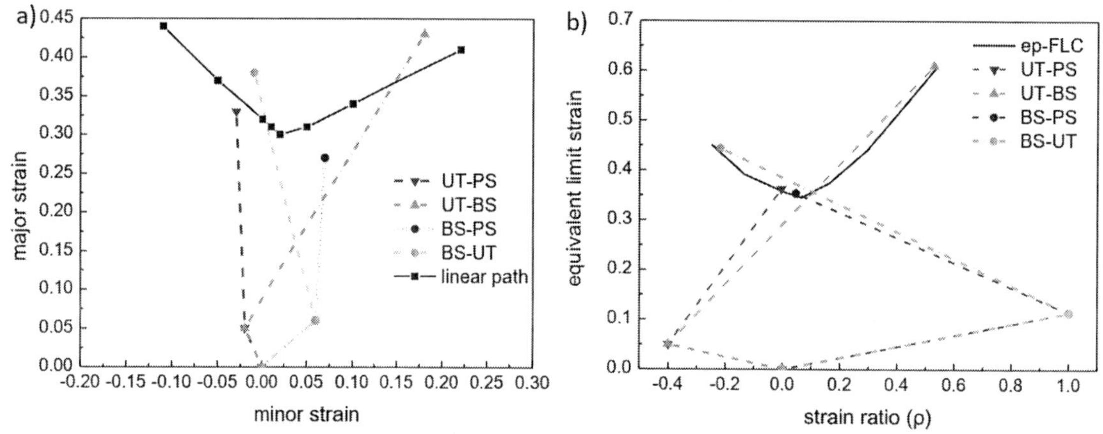

Figure 4. The two-step nonlinear strain results of 700DP: a) with linear FLC, b) with ep-FLC.

The strains under two-step loading condition could be transformed to equivalent strain form. Hill48 anisotropic yield criterion can be described as Equation (1):

$$\sigma^2 = (G+H)\sigma_1^2 + (F+H)\sigma_2^2 - 2H\sigma_1\sigma_2 + 2N\sigma_{12}^2 \tag{1}$$

Considering the associated flow, Equation (2-3) can be obtained:

$$\rho = f(\alpha) \tag{2}$$

$$\beta = g(\alpha) \tag{3}$$

where α is the ratio of principal stresses:

$$\alpha = \frac{\sigma_2}{\sigma_1} \tag{4}$$

the strain path ρ can be written as:

$$\rho = \frac{d\varepsilon_2}{d\varepsilon_1} \tag{5}$$

the strain parameter **β** can be written as:

$$\beta = \frac{d\bar{\varepsilon}}{d\varepsilon_1} \tag{6}$$

where F, G and H are the anisotropic coefficient parameters, $d\varepsilon_1$, $d\varepsilon_2$, $d\bar{\varepsilon}$, σ1 and σ2 are the principal and equivalent strain increments, and the principal stresses. If the limit principal strains and the strain path were given, the equivalent plastic strain can be calculated quickly combined of Equation (1-6). As shown in Figure 4(b), the FLC was transformed to ep-FLC, which was consisted of equivalent limit strain and strain path ρ as the vertical and horizontal coordinates. The two-step loading path change results can agree with the ep-FLC well.

3. The application of ep-FLC

3.1 The feature of wheel disc and experimental forming results

The 3D model of the wheel disc in this study is presented in Figure 5. The initial blank with the diameter of ϕ 462 mm was employed to produce the disc with max diameter of ϕ 364 mm and Max depth of 73mm. It should be difficult to form the approximate square seal seats which are arranged along the circle in the centre area of disc. The radii of the square seal seat are small, just 3mm at the bottom of seat and 5mm at the top of seat.

The wheel disc was designed to form with five stages, and the main deformation occur in the first three forming stages, as shown in Figure 6. Hot-rolled dual-phase steel 580DP and 700DP were attempted to forming the disc with that dimension. The stamping results were given in Figure 7. The wheel disc was formed perfectly with 580DP without cracking, but several cracks were observed at position A, B and C of the wheel disc formed with 700DP.

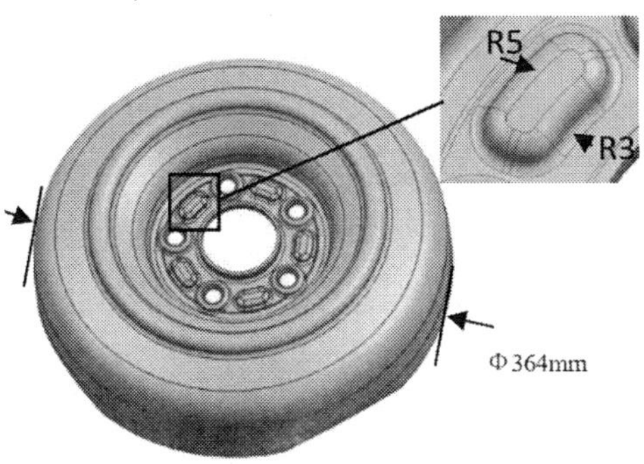

Figure 5. 3D model of the wheel disc.

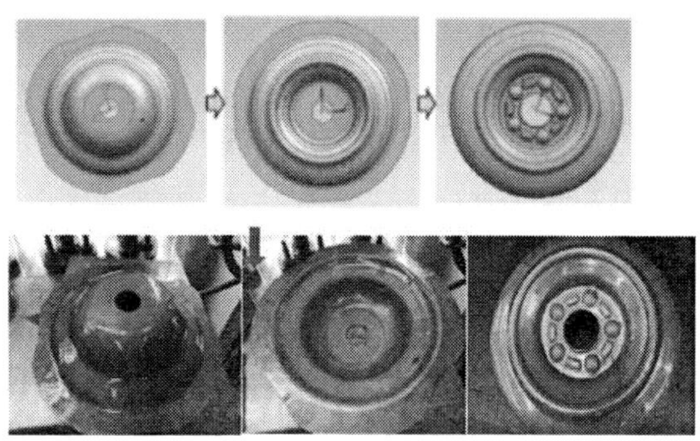

Figure 6. The three-stage forming process of disc.

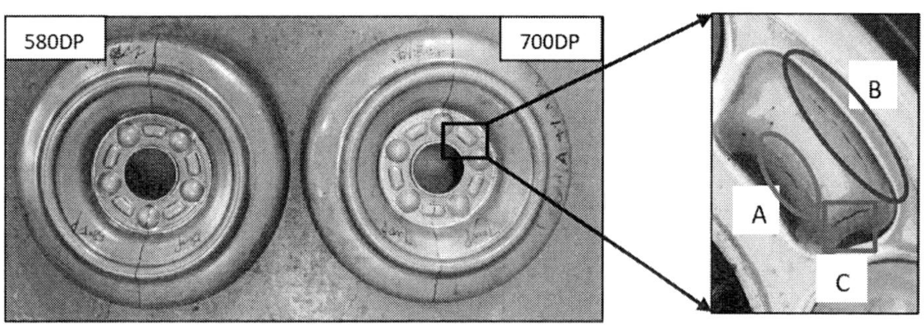

Figure 7. Forming results of wheel disc using 580DP and 700DP.

3.2The FEM results with traditional FLC and ep-FLC

FEM was used to simulate the three-stage forming process for the disc with Autoform software. The shell element was used in this simulation. The Coulomb friction coefficient between the blank and the die surfaces was set to 0.1 according to the actual lubrication condition. Hill48 anisotropic yield criterion and Swift hardening law were used to describe the yield and hardening behaviour of 580DP and 700DP.

The result of three-stage forming simulation was evaluated with traditional FLC firstly, and the strain path of disc during three-stage forming process is indeed nonlinear, as shown in Figure 8(a) and Figure 9(a). The principle strains of 700DP are almost equivalent to that of 580DP in the critical area of seal seat. The limit principle strains of 700DP still in safe area, though the maximum strain of position C is very close to the FLC. Obviously, the simulating results cannot agree with the experimental forming results using 700DP, the cracking actually occurred at position A, B and C.

The strain paths at position A, B and C of both 580DP and 700DP were respectively extracted to calculate the equivalent strains. From the Figure 8(b), the equivalent strains at position A, B and C are still under ep-FLC for 580DP, however the final equivalent strains of position A, B and C exceeded the ep-FLC for 700DP. Therefore, the high risk of cracking for 700DP at position A, B and C during three-stage forming process can be predicted accurately with ep-FLC, which was well coincident with the experimental results.

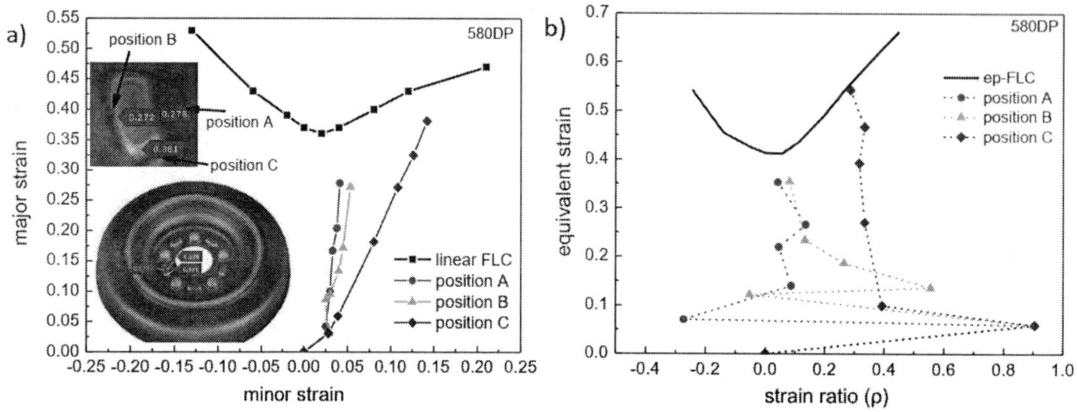

Figure 8. Simulation results of seal seat position for 580DP:
a) prediction with linear FLC, b) prediction with ep-FLC.

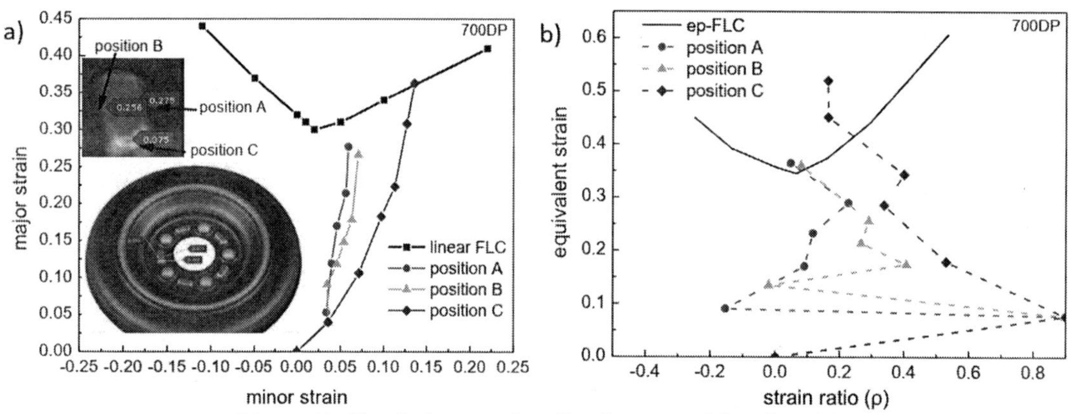

Figure 9. Simulation results of seal seat position for 700DP:
a) prediction with linear FLC, b) prediction with ep-FLC.

4. Conclusions

The mechanical property and formability under linear and nonlinear strain paths for hot-rolled steel were studied, and the main conclusions can be got:

(1) 580DP and 700DP both consist of martensite and ferrite phase. The yield strength of 700DP is similar to that of 580DP, while the tensile strength is much higher. The conventional FLC of 700DP is much lower than that of 580DP.

(2) The modelling of ep-FLC for 700DP was certified by the two-step strain path changing experimental results.

(3) The ep-FLC was applied to predict the disc forming result of three-stage process, which agreed well with the experimental results for both 580DP and 700DP compared with conventional FLC.

References

[1] Kuwabara T, Kuroda M, Tvergaard V, Nomura K (2000) Use of abrupt strain path change for determining subsequent yield surface: experimental study with metal sheets. *Acta Mater* **48** 2071–2079.

[2] Reyes A, Hopperstad OS, Berstad T, Lademo OG (2009) Forming limit diagrams with an FE-based approach for sheets under non-proportional loading. *7th European LS-DYNA conference*.

[3] Zhalehfar F, Hashemi R, Hosseinipour SJ (2015) Experimental and theoretical investigation of strain path change effect on forming limit diagram of AA5083. *Int J Adv Manuf Technol* **76** 1343–1352.

[4] Stoughton TB (2001) Stress-based forming limits in sheet metal forming. *J Eng Mater Technol* **123** 417–422.

[5] Yoshida K, Kuwabara T, Kuroda M (2007) Path-dependence of the Forming Limit Stresses in a Sheet Metal. *Int J Plasticity* **23** 361– 384.

[6] Basak S, Bandyopadhyay K, Panda S K, Saha P (2015) Prediction of Formability of Bi-axial Pre-strained Dual Phase Steel Sheets Using Stress-Based Forming Limit Diagram. *Advances in Material Forming and Joining* 167-192.

[7] Zeng D, Chappuis L, Xia ZC, Zhu X (2008) A Path Independent Forming Limit Criterion for Sheet Metal Forming Simulations. *SAE2008* **1** 1445.

[8] Dick RE, Yoon JW, Stoughton TB (2016) Path-independent forming limit models for multi-stage forming processes. *Int J Mater Form* **9** 327–337.

[9] Li CN, Li XL, Yuan G, Misra RDK, Kang J, Wang GD (2016) Precipitation behavior and mechanical properties of a hot rolled Ti-bearing dual phase steel. *Materials Science and Engineering A* **673** 213-221.

[10] Yu WH, Zhou BB, Su JJ, Li HT, Hu JP, Chen J (2012) Research on the deformation resistance of 600 MPa hot-rolled DP steel. *China Metallurgy* **22** 21-25.

IOP Conf. Series: Materials Science and Engineering **967** (2020) 012040 doi:10.1088/1757-899X/967/1/012040

Forming and trimming of 2-mm thick DP600 sheet steel in tools and dies 3D-printed in maraging steel by laser-based powder bed fusion

Thomas Skåre[1] and Nader Asnafi[2]

[1] RISE IVF, Vällaregatan 30, SE-293 38 Olofström, Sweden
[2] Örebro University, School of Science & Technology, SE-701 82, Örebro, Sweden

nader.asnafi@oru.se

Abstract. In a previous investigation, 3D-printed solid and topology optimized semi-industrial tools for forming and trimming of 2-mm thick hot-dip galvanized DP600 were certified. This certification required 50,000 strokes in U-bend forming and 100,000 strokes in trimming/cutting/blanking. The present paper focuses on the tool wear, the U-bend sheet surfaces, the shear and fracture zone lengths in trimming, and the punch forces in this certification. The 3D-printed tools behave as conventional tools do. Although small, there seems to be a difference in wear at the profile radius between the solid and topology optimized U-bending tool halves 3D-printed in maraging steel DIN1.2709.

1. Introduction

The state-of-the-art for additive manufacturing (henceforth also referred to as AM or 3D-printing) of metals is described in [1]. Additive manufacturing is subject to a technology assessment in [2]. Based on these and other relevant reviews, the research needs and challenges for the Swedish industrial use of metal additive manufacturing were identified [3]. Metal powders for tool applications, new tool design options and the shift in toolmaking are of great significance in the industrialization of additive manufacturing and for the global industrial competitiveness [1, 2, and 3].

In [4], inserts in a body panel stamping tool were 3D-printed in maraging steel DIN 1.2709. The 3D-printed inserts exhibited the same performance as the conventionally made, however, with reduced lead time and minimized internal process logistics [4]. In [5], solid and topology optimized semi-industrial tools for forming and trimming of 2-mm thick hot-dip galvanized DP600 were certified. This certification required 50,000 (50k) strokes in U-bend forming and 100,000 (100k) strokes in trimming/cutting/blanking. The present paper focuses on the tool wear, the U-bend sheet surfaces, the shear and fracture zone lengths in trimming, and the punch forces in this certification.

2. Materials

Table 1 displays the chemical composition and Table 2 shows the mechanical properties of maraging steel DIN 1.2709. To determine these properties (Table 2), 5 tensile specimens (circular cross section, ϕ5mm) per direction were 3D-printed, heat-treated, machined and tested. Table 2 displays the average values. 3D Systems ProX DMP and the AM process parameters in Table 3 were used to make these specimens (and the semi-industrial tools). The heat treatment was conducted at 490° C in 6 hours, after

Table 1. Chemical composition of maraging steel DIN 1.2709 [6].

Element	Fe	Ni	Co	Mo	Ti	Si	Mn	C
Weight %	Balance	17.0-19.0	9.0-11.0	4.0-6.0	0.9-1.0	$\leq$ 1.0	$\leq$ 1.0	$\leq$ 0.03

Table 2. Mechanical properties of maraging steel DIN 1.2709 after AM and heat treatment.

	Built vertically	Built horizontally
Yield strength, $R_{p0.2}$ (MPa)	1999	1977
Tensile strength, R_m (MPa)	2120	2167
Hardness (HRC)	56	56

Table 3. The used 3D printer and AM process parameters.

Material (tool material)	Used 3D printer	Layer thickness (μm)	Laser power (W)	Scan speed (mm/s)	Hatch distance (μm)
DIN 1.2709	3D Systems ProX DMP 300	40	185	1200	70

Table 4. Chemical composition of the workpiece (sheet) material, i.e. DP600 [7].

Element	Fe	P	S	Al	Cr	Si	Mn	C
weight %	Balance	$\leq$ 0.02	$\leq$ 0.004	$\geq$ 0.020	$\leq$ 0.50	$\leq$ 0.30	$\leq$ 1.66	$\leq$ 0.12

Table 5. Properties of the workpiece (sheet) material, i.e. 2-mm thick sheet of DP600 [7].

Sheet thickness (mm)	2.0
Yield strength, $R_{p0.2}$ (MPa)	350-480
Tensile strength, R_m (MPa)	600-700
Fracture elongation, A_{80} (%)	$\geq$ 18
Hot-dip galvanized: Layer thickness (μm)/weight (g/m^2)	10 (per side)/140

which the specimens (or the tools) were allowed to cool down in the furnace (in air).

As sheet material, 2-mm thick hot-dip galvanized DP600 was used in this certification. For the chemical composition and properties of this material, see Tables 4 & 5 respectively.

3. Experimental procedure

3.1. Tools and dies

The experimental procedure used at Volvo Cars to certify a tool concept was applied in this investigation. According to this procedure, the selected tool concept (i.e. tool material, hardening method, surface roughness and coating) is used to make [5]

- a so-called U-bend forming tool. The sheet material grade of interest is formed in a U-bend shape with a draw depth of 50 mm in this tool. The approval criterion is the surface of the stamped U-bend. Scratches on this surface cannot be accepted. On a four-level scale, only levels 0 and 1 can be accepted. The tool concept that manages 50,000 U-bends (strokes) in the selected sheet material without class 2 surface is approved. This is illustrated in Figure 1.
- a tool to trim/blank/cut the sheet material grade of interest. This sheet material is trimmed along a 150 mm long straight line. The approval criterion is the burr height on the trimmed/blanked/cut sheet. For approval, this burr height must be lower than 10% of the sheet thickness. A tool concept that manages 100,000 strokes with a burr height lower than 10% of the sheet thickness is approved. This is illustrated in Figure 2.

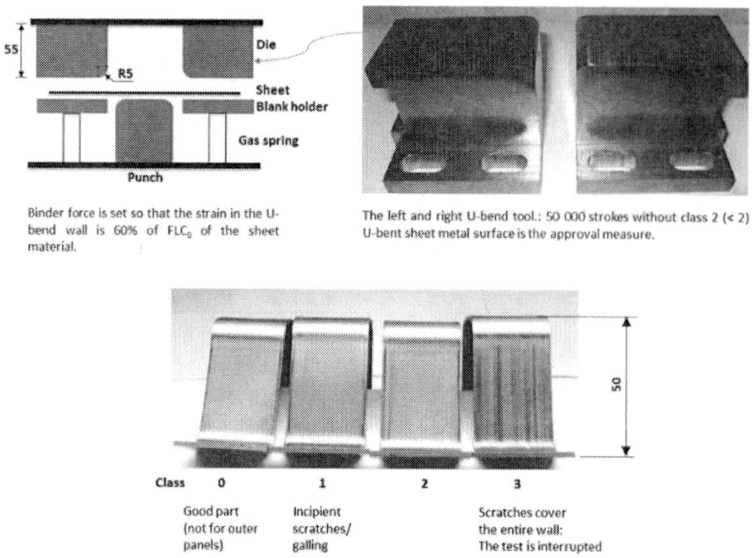

Figure 1. The experimental set-up for certification of the forming (U-bending) tool. See also [5].

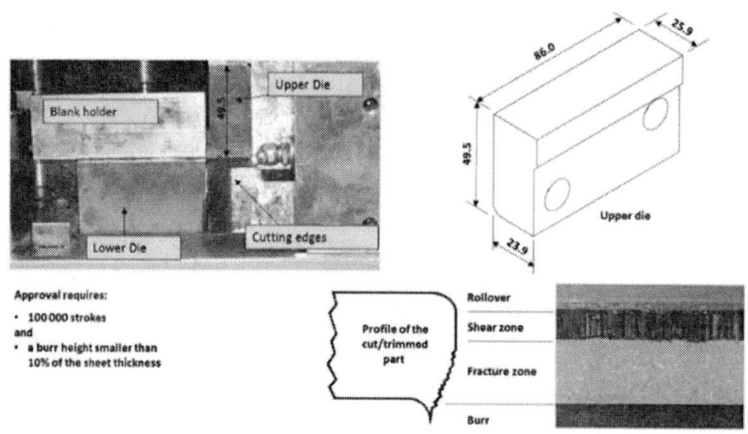

Figure 2. The experimental set-up for certification of the trimming/blanking/cutting tool. The trimming is conducted by 2 upper dies mounted along a straight line. See also [5].

For the certification in this study, the stamping tool concept comprised DIN 1.2709 (Table 2), 3D-printed both solidly and after topology optimization in 3D Systems ProX DMP 300 with the process parameters shown in Table 3, hardened to 55 HRC, and machined to a surface roughness of 0.2 μm. The hardening was conducted by heat-treatment at 490° C in 6 hours, after which the tool is allowed to cool down in the furnace (in air). None of the tools is coated. The selected sheet material is 2-mm thick hot-dip galvanized DP600 (Table 5).

To explore the industrial potential of AM in stamping tool applications, the punches shown in Figures 1 & 2 were topology optimized and 3D-printed in this investigation. The topology optimization is described in [8]. Figures 3 and 4 display the solid and topology optimized U-bending and trimming tools/dies 3D-printed in DIN 1.2709. These tools were set up in the equipment shown in Figure 5, after which the U-bending and trimming were conducted respectively in accordance to the certification procedure described above. The U-bending tool managed 50,000 (50k) strokes in accordance to the requirement in Figure 1 and was therefore approved. The trimming tool managed 100,000 (100k) strokes in accordance to the requirements in Figure 2 and was therefore approved.

Figure 3. The 3D-printed U-bending tool: the right tool half = solid piece. The left tool half = topology optimized. Both tool halves are 3D-printed in maraging steel DIN 1.2709 [9].

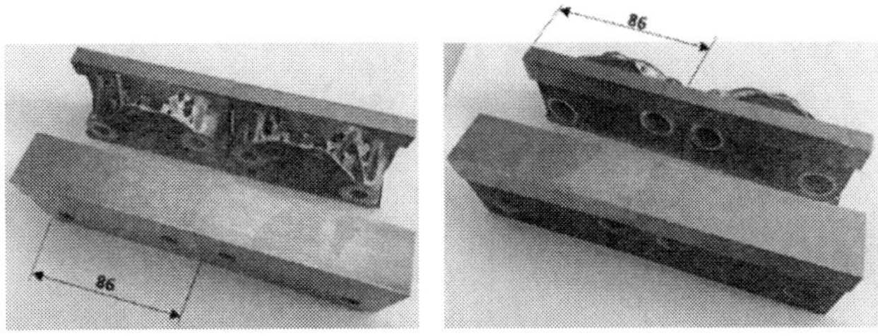

Figure 4. The 3D-printed solid and topology optimized trimming/blanking/ cutting tools. Both versions are 3D-printed in DIN 1.2709 [5].

3.2. Measurements

The wear, profile radius, and surface roughness were measured on the U-bend tool halves after 50k strokes, Figures 6-9.

The wear and edge (profile) radius on the trimming dies were measured after 100k strokes, Figures 10-11.

All of these measurements were conducted in Hommel-Etamic surfscan 120-400.

The load cell mounted in the eccentric press in Figure 5 was used to measure the U-bending (forming) and trimming forces.

Figure 5. The tools were set up in this eccentric press to form the U-bends and trim the sheet.

4. Results

Figures 12 and 13 display the U-bent sheet surface after 1k, 25k, and 50k strokes. The U-bending tool edge (profile) radius after 50k strokes is displayed in Figure 14. Figure 15 shows the U-bending maximum punch force as function of the stroke number. Figure 16 displays the roughness of the binder surfaces at different positions on the U-bending tool after 50k strokes. The wear after 50k strokes at the profile radius of the U-bending tool halves are shown in Figure 17. As displayed in this figure, there seems to be a difference in wear between the topology optimized and the solid tool halves. This difference is, however, small and difficult to measure.

Figure 18 depicts the shear and fracture zones on the trimmed sheet metal (2-mm thick hot-dip galvanized DP600) after 100k strokes. As displayed in this figure, the difference is very small between the results obtained with the topology optimized tool and those obtained with the solid tool. The sizes of the shear and fracture zones are displayed as function of the stroke number in Figure 19. Figure 20 depicts the wear after 100k strokes at the edge (profile) radius on the trim dies. The trim die edge radius was measured at different positions after 100k strokes. The obatined values are shown in Figure 21, which indicates that the topology optimized tool exhibits somewhat larger wear.

The trimming/cutting/blanking (punch) force versus the stroke number is displayed in Figure 22. The force exerted by the 3D-printed solid tool is initially approximately 10% higher and at around 65k strokes approximately 7% smaller than that exerted by the 3D-printed topology optimized tool. This difference is probably influenced by the spread in the mechanical properties of the sheet material.

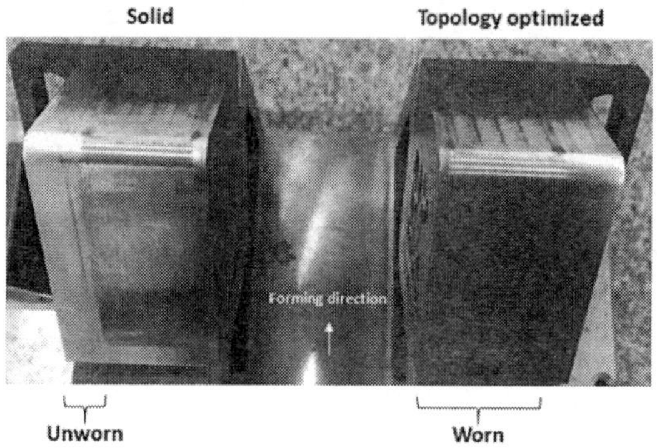

Figure 6. The profile radius on the 3D-printed solid (left) and topology optimized (right) tool halves.

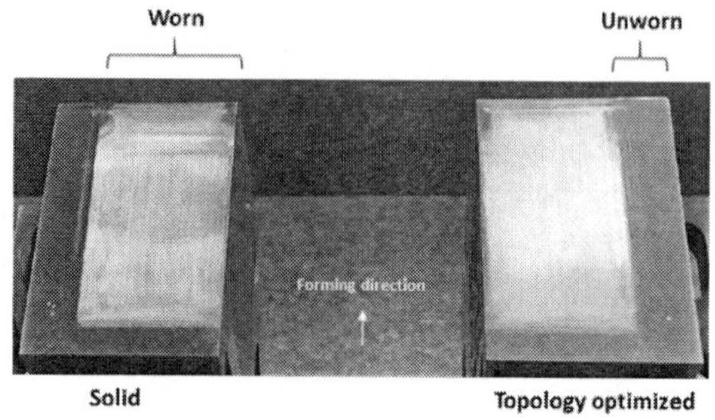

Figure 7. The top surface (the binder surface) on the 3D-printed solid (left) and topology optimized (right) tool halves.

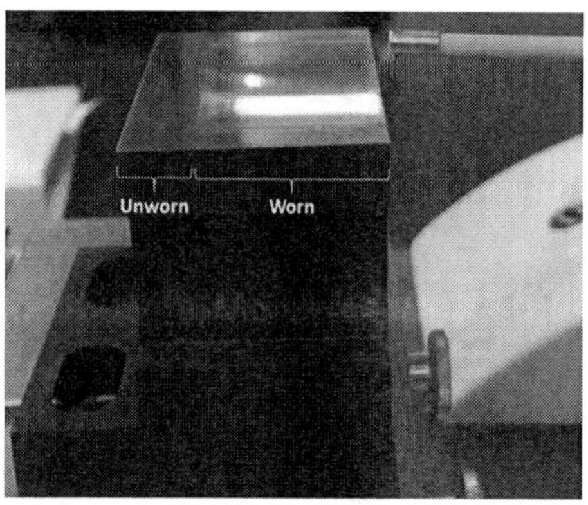

Figure 8. Surface roughness measurement across the forming (drawing) direction.

Figure 9. Profile radius & surface roughness measurement along the forming direction.

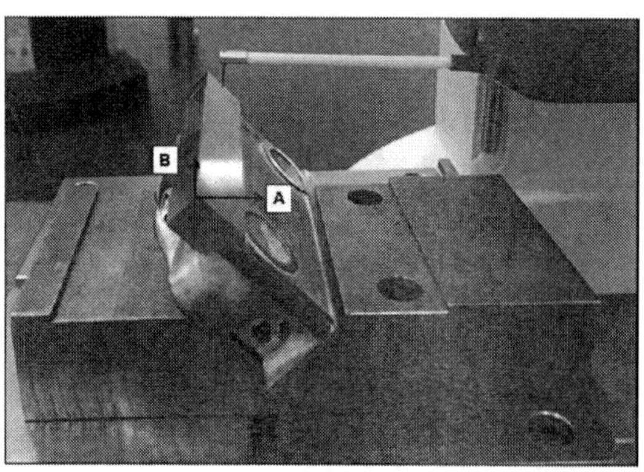

Figure 10. Measurement directions on the trimming die. Measurement tip radius = 2μm. Cone angle = 60°. The photo displays the 3D-printed topology optimized trimming die.

Figure 11. The measurement positions on the trimming die. The photo displays the 3D-printed solid trimming die.

Figure 12. The U-bent sheet surface in contact with the 3D-printed topology optimized tool half from the stroke 50k, 25k and 1k respectively.

Figure 13. The U-bent sheet surface in contact with the 3D-printed solid tool half from the stroke 1k, 25k and 50k respectively.

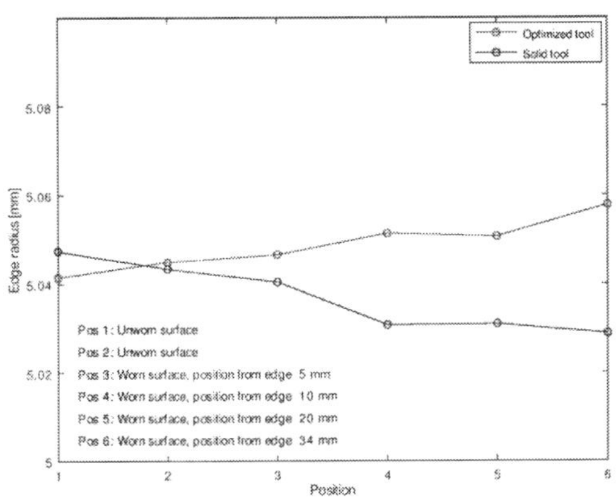

Figure 14. The U-bending tool half edge (profile) radius at different positions after 50k strokes.

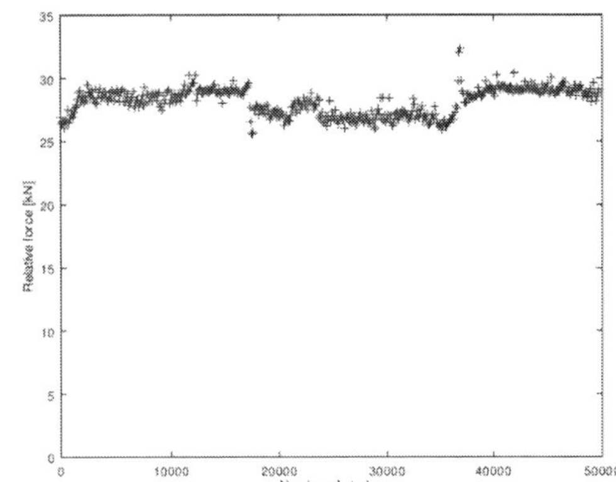

Figure 15. The U-bending maximum punch force as function of the stroke number.

For up to 2-mm thick hot-dip galvanized DP600, the results of this investigation show that topology optimized and solid tools 3D-printed in maraging steel (DIN1.2709) exhibit a similar behaviour as conventionally designed forming tools made in "conventional" Calmax and Sleipner [10] and conventionally designed trimming tools manufactured in "conventional" Fermo and Sleipner [11].

The results of this investigation were used to design and manufacture industrial tools in DIN1.2709 by additive manufacturing [12].

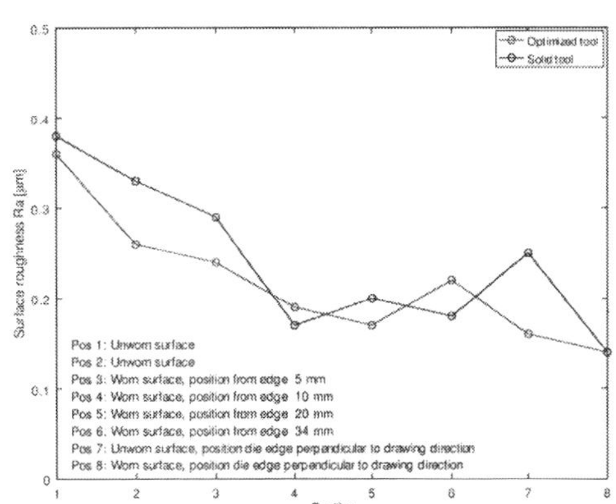

Figure 16. The binder surface roughness at different positions on the U-bend tool after 50k strokes.

5. Conclusions

3D-printed tools in maraging steel (DIN1.2709) behave as conventionally made tools do.

Although small, there is a difference in wear at the profile radius between the 3D-printed solid and topology optimized U-bending tool halves.

The trimming force exerted by the solid tool differs from that exerted by the topology optimized tool,

IOP Conf. Series: Materials Science and Engineering **967** (2020) 012040 doi:10.1088/1757-899X/967/1/012040

Figure 17. The wear at the U-bending tool profile radius after 50k strokes: Left = 3D-printed topology optimized tool half. Right = 3D-printed solid tool half. The sheet is drawn from the top to the bottom.

Figure 18. The shear and fracture zones on the trimmed sheet after 100k strokes. Left = trimmed with the 3D-printed topology optimized tool. Right = trimmed with the 3D-printed solid tool.

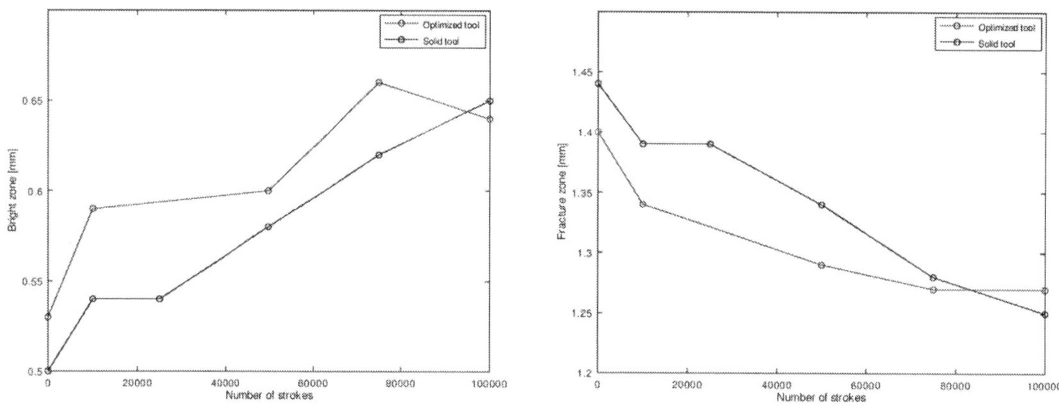

Figure 19. The size of the shear and fracture zones as function of the stroke number.

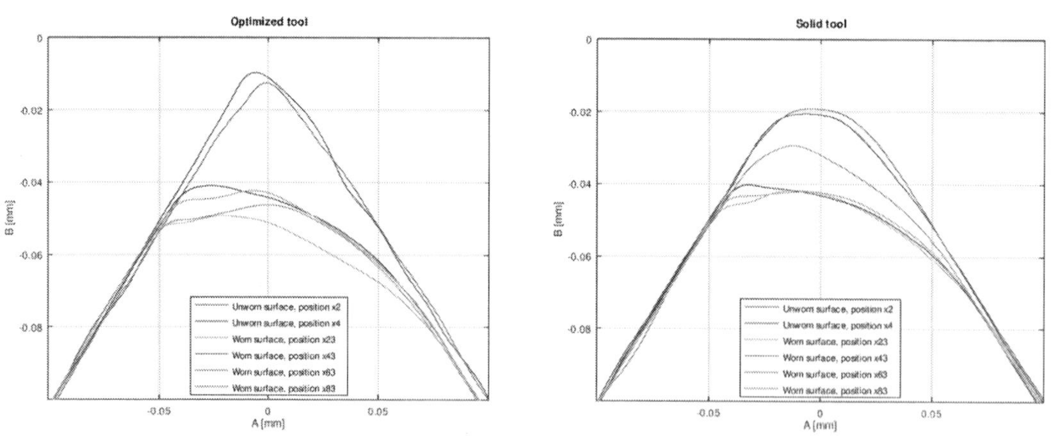

Figure 20. The wear after 100k strokes at the edge (profile) radius on the trim dies.

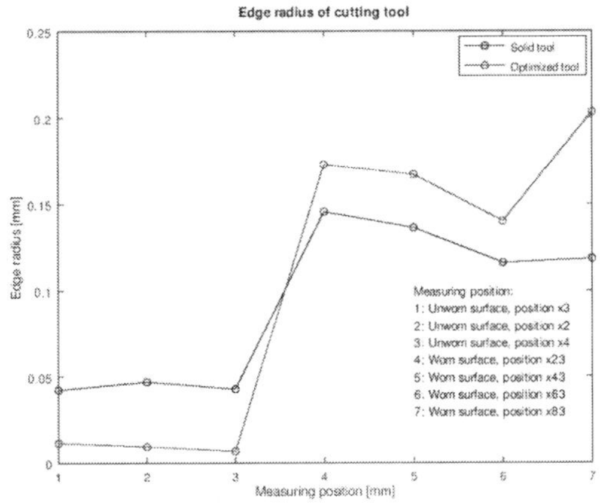

Figure 21. The trim die edge radius at different measuremet positions after 100k strokes.

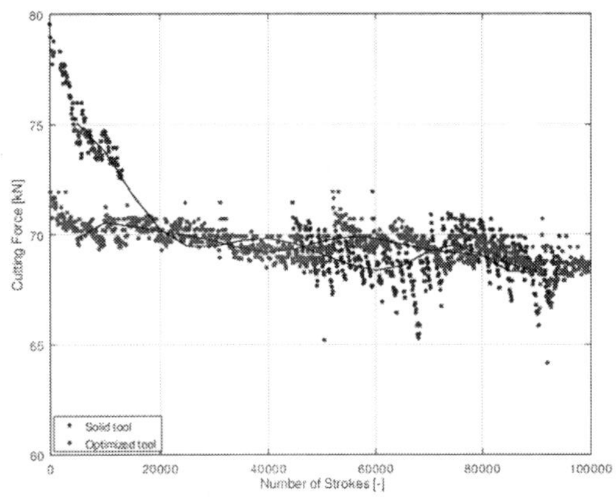

Figure 22. The cutting (punch) force versus the stroke number.

probably due to the spread in the mechanical properties of the sheet material.

Acknowledgments
The authors would like to thank Sweden's Innovation Agency Vinnova for funding this study and 3D MetPrint, Dynamore Nordic, Hydroforming Design Light, IKEA, Ionbond, Melament, Nolato Lövepac, RISE IVF, Volvo Cars, Uddeholm and Örebro University for a fruitful and efficient collaboration.

6. References

[1] Ålgårdh J, Strondl A, Karlsson S, Farre S et.al. 2017 *State-of-the-Art for Additive Manufacturing of Metals* Report 2016-03898 Swedish Arena for Additive Manufacturing of Metals
[2] Quadrennial Technology Review 2015 *Additive Manufacturing, Chapter 6 – Technology Assessment* US Department of Energy
[3] Ålgårdh J, Strondl A, Karlsson S, Farre S et.al. 2017 *Research Needs and Challenges for Swedish Industrial Use of Additive Manufacturing* RAMP-UP Report 2016-03898 Swedish Arena for Additive Manufacturing of Metals
[4] Leal R, Barreiros F M, Alves L, F. Romeiro F et.al. 2017 Additive manufacturing tooling for the automotive industry *Int. J. Adv. Manuf. Technol.* **92**:1671–1676
[5] Asnafi N, Rajalampi J and Aspenberg D 2019 Design and Validation of 3D-Printed Tools for Stamping of DP600 *IOP Conf. Ser.: Mater. Sci. Eng.* **651** 012010
[6] According to 3D System's website https://www.3dsystems.com/, July 6, 2018
[7] According to datasheet from Tibnor, July 6, 2018
[8] Aspenberg D and Asnafi N 2019 Topology optimization of a U-bend tool using LS-TaSC *Proc. of the 12th European LS-DYNA Conf.* (Koblenz, Germany May 14-16)
[9] Asnafi N, Shams T, Aspenberg D and Öberg C 2019 3D Metal Printing from an Industrial Perspective – Product Design, Production and Business Models *BHM Berg- und Hüttenmännische Monatshefte* **164** (3): 91–100
[10] Liljengren M, Kjellson K, Johansson T and Asnafi N 2006 Die materials, hardening methods and surface coatings for forming of high, extra high and ultra high strength steel sheets (HSS/EHSS/UHSS) *Proc. 2006 IDDRG conf.* (Porto, Portugal, June 19-21) pp 597-603
[11] Ullman A, Kjellsson K, Johansson T and Asnafi N 2006 On laser hardening of trimming tools and dies *Proc. 2006 IDDRG conf.* (Porto, Portugal, June 19-21) pp 613-620
[12] Asnafi N, Rajalampi J, Aspenberg D and Alveflo A 2020 Production Tools Made by Additive Manufacturing through Laser-based Powder Bed Fusion *BHM Berg- und Hüttenmännische Monatshefte* **165** (3): 125-136

IOP Conf. Series: Materials Science and Engineering **967** (2020) 012041 doi:10.1088/1757-899X/967/1/012041

New Physical Scaling Approach to compensate the part contraction in drawing operations

A Birkert[1], P Zimmermann[1], B Hartmann[2], M Scholle[1] and M Straub[2]

[1] Center for Metal Forming and Car Body Manufacturing, Faculty of Mechanics and Electronics, Heilbronn University of Applied Sciences, Max-Planck-Str. 39, 74081 Heilbronn, Germany

[2] inigence gmbh, Bernbachstr. 36, 74626 Bretzfeld, Germany

E-mail: arndt.birkert@hs-heilbronn.de

Abstract. Further increasing requirements to the quality of car body components represent a huge challenge for the process planner. Springback effects are causing dimensional deviations of the car body components from their respective target geometry. Both elastic membrane and bending stresses result in global and regional distortions and deflections. In addition to these effects in particular the membrane stresses furthermore result in a contraction of the drawshell with the effect that the drawshell is smaller than the tool after drawing. In order to compensate this contraction the drawing tool has to be increased. This process is known as the so-called scaling. An appropriate scaling should result in a contracted part which has the same surface area and unwound lengths as the target geometry. In case of deviations of the unwound lengths unwanted plastic deformations of the part in subsequent operations have to be expected and the part's assembly dimensions will differ from the target geometry which can cause significant quality issues in the car body. Today the scaling is usually done by increasing the tool with a unique global scaling factor or with different scaling factors for different directions. However, since the contraction varies along the part surface in reality the results are only suboptimal. Here a new scaling approach is being presented which uses the simulation results to scale the tool surfaces locally. By doing so the locally different contractions can be better compensated.

1. Introduction

Springback is the change of geometry which occurs in stamped parts during the opening of the dies; it is caused by the release of elastic energy. On the one hand the elastic springback results in dimensional deviations from the target part caused by bending ups, distortions or something like this. To reduce these kinds of deviations the common approach is to modify the tool geometries in the opposite direction of the dimensional deviations. This proceeding is called *springback compensation* and therefore a few of methods have been developed [1], [2].

On the other hand the elastic energy which is introduced in the forming part during the drawing operation – as a result of the increase in surface area – causes a contraction of the part during the opening of the tool. To compensate this behaviour the dimensions of the drawing tool have to be increased. This doing is the so-called *tool scaling* or only *scaling*.

In case of an inadequate *scaling* of the drawing tool there are two main problems. First the absolute part dimensions could be out of tolerance, which causes quality problems in the assembly process.

Secondly the part could even be plastically deformed by closing the pad in subsequent operations, because these operations commonly are designed with the lengths of the zero geometry [3].

The common approach to scale the drawing tool is to increase the dimensions of the tool with global factors. By doing so the selection of the global factors are based on a simple material-based approach or by practical experience. The main problem of these procedures is, that the contraction of the drawshell is not constant along its surface and so a *global scaling approach* only represents an approximation to the given conditions. In this paper a *New Physical Scaling Approach* is being presented which considers the local different contraction effects by using the results of the forming simulation and so to reduce the problems of the known scaling approaches.

2. Common scaling approaches

In [3] *Birkert et al.* it is mentioned that for steel materials in practice a scaling factor between 1.0007 to 1.0012 has proved to be useful. With these factors the tools are usually enlarged in all directions. In the simulation software *AutoForm Forming R8* an approximate approach to estimate the scaling factor is implemented which is based on *Hooke's* law. Therefore the biaxial yield stress and the Young's modulus of the respective part material is being used as in equation (1).

$$F_{scale} = 1 + \frac{\sigma_{biax}}{E} \tag{1}$$

In **Figure 1** the effect of a global scaling approach is shown at an exemplary tool section. In addition to define a scaling factor a scaling centre has to be defined. The scaled geometry is generated by increasing the distance of each point of the zero geometry according to its distance from the scaling centre by the defined factor.

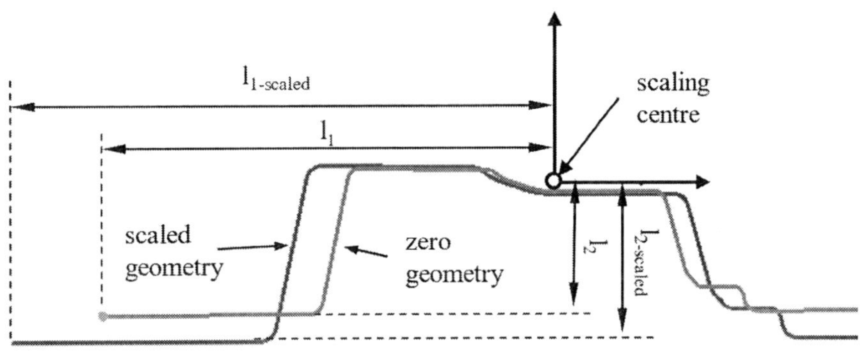

Figure 1. Effect of global scaling approach.

In addition to the approach with one common scaling factor, also different factors for different directions can be chosen. This is very useful, if the respective part is significantly more stressed in one rather than in the other direction, as hat profiles for instance. For other parts, such as roof panels or hood outer panels, tool makers quite often use their own scaling factors which are based on practical experience.

3. Deficits of today's scaling approaches

The present studies have been carried out on the example of an aluminium headboard (sheet thickness 2.5 mm). The production process of the headboard has been simulated with the FE-Software *Forming R8* from *AutoForm*. The manufacturing process is shown in **Figure 2**. It starts with a drawing operation (D20). Then the part is being separated from the addendum by a laser cut (T30). Finally two flange areas are formed in a flanging operation (F40). To evaluate the scaling quality the springback has been calculated after the laser cutting step (T30) in a virtual measurement operation (M35). Scaling quality means the difference between the springback part geometry and the zero geometry referred to the surface area and the unwound lengths.

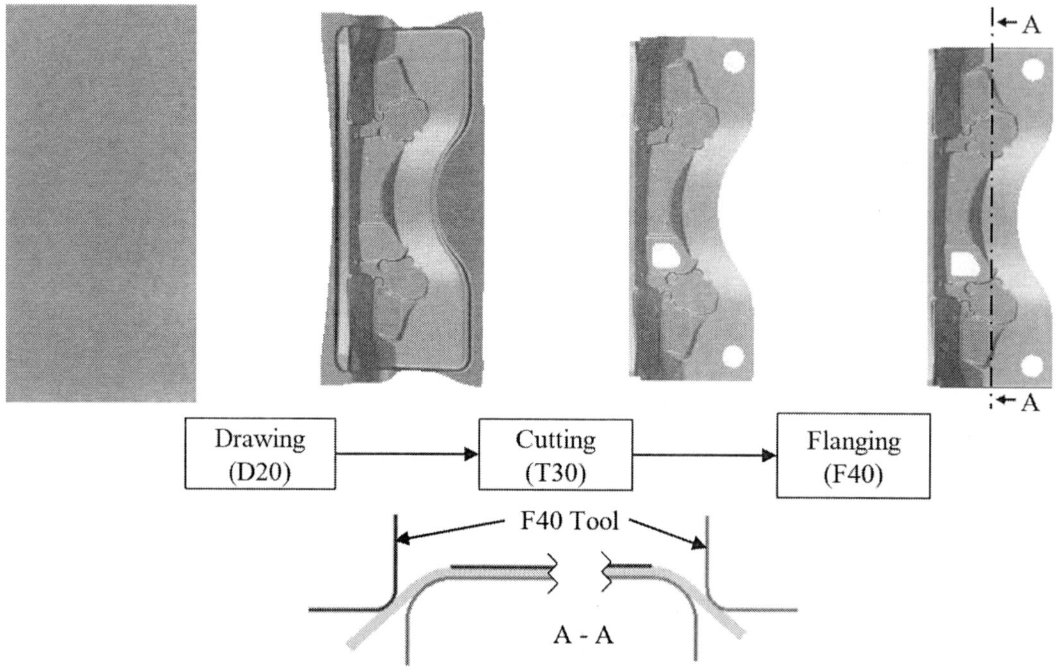

Figure 2. Manufacturing process of the headboard.

The deficit of a global scaling approach is that the tool dimensions are increased by a constant percentage in each part area. In contrast to that the part contraction is locally different. To show this part behaviour the strain change by springback of the headboard has been evaluated (**Figure 3**). In the shown section the outer part areas underlie a negative strain change which represents a contraction, whereas the centre area of the part shows a positive strain change which causes an increasing section length.

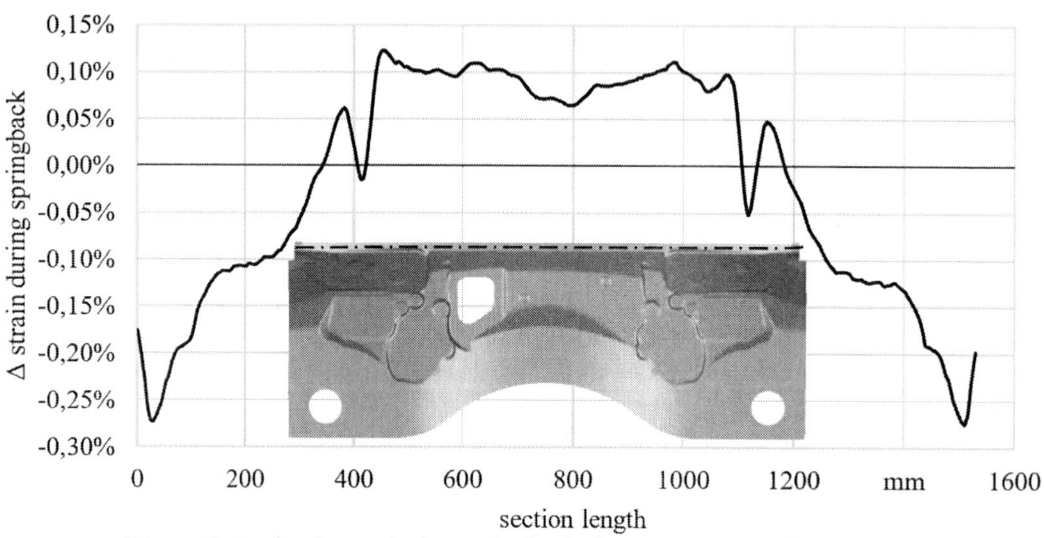

Figure 3. Strain change during springback along a representative section.

4. New Physical Scaling Approach

The *New Physical Scaling Approach* differs from the known ones insofar that it does not change the tool dimensions globally over the whole surface but takes into account the locally different surface contraction. The different contraction is caused by the locally varying stresses. To visualize the workflow of the new approach, the principle part behaviour is shown in a unidimensional stress-strain-diagram (**Figure 4**). In a first step the blank is being formed with an unscaled drawing tool (0→1). If the subsequent operations are pure cutting operations (without any forming), they should be part of this first step, but without calculating a springback step between the single operations. This is important, because the surface area of the part may be changed if the addendum is separated from the drawshell. From this state the springback can be calculated as usual (1→2) which causes a part contraction, represented by the negative strain ε_{SB} in the diagram.

According to the new approach the stress state must be adjusted (1→3) before the springback calculation step is done. To explain the following considerations the definitions of the local stresses and the local coordinate systems are done on a finite shell element. The local coordinate system (x_1, x_2, x_3) is aligned in such a way that x_1 and x_2 lie in the tangential plane of the shell element. Therefore x_3 represents the thickness direction of the shell (thickness direction of the sheet). Relating to this coordinate system the stress state can be described by the Cauchy tensor $\boldsymbol{\sigma} = (\sigma_{ij});\ i,j = 1,2,3$.

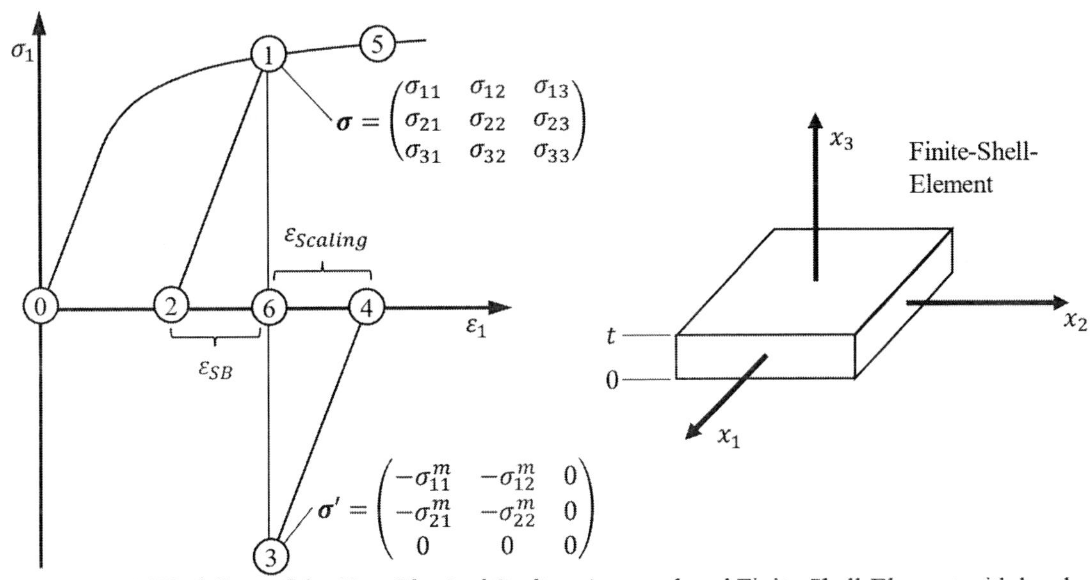

Figure 4. Workflow of the *New Physical Scaling Approach* and Finite-Shell-Element with local coordinate system.

Because only the components $\sigma_{11}, \sigma_{22}, \sigma_{12}$ and σ_{21} have an influence on the elastic surface area change, the other components are set to zero. In order to be able to map the different stresses and strains along the sheet thickness with only one element, shell elements with several integration layers are used in sheet metal forming simulations. The stresses along the sheet thickness can be split into membrane stresses on the one hand and bending stresses on the other hand. For scaling only the membrane stresses are being used. These are being handled by averaging over the shell thickness according to equation (2).

$$\sigma_{ij}^m = \frac{1}{t}\int_0^t \sigma_{ij}(x_3)\,\mathrm{d}x_3 \tag{2}$$

Subsequently the signs of the membrane stress components have to be inverted. Insofar an initially positive membrane stress component (tensile stress) turns into a negative one (compressive stress) and vice versa. With the so modified stress state ($\boldsymbol{\sigma}'$) a springback calculation step is being simulated now (3→4). By the release of the elastic energy an increased surface area is the result, which is represented by the positive strain $\varepsilon_{Scaling}$. This deformation is now used to scale the tool surface of the drawing tool (4→5). Now the manufacturing process is being simulated with the scaled tool (0→5) and after the springback calculation step (5→6) the part is expected to have the same surface area and unwound lengths as the zero geometry.

5. Comparison of the results of the new approach with those of an existent approach

In this chapter the results of the simulation setups without scaling, with *global scaling* and with the *New Physical Scaling Approach* (*NPSA*) are compared to each other. For that purpose the surface area and the unwound lengths of the tools and springback geometries are compared with the zero geometry. Furthermore the plastic deformations of the part by closing the pad in the flanging operation have been evaluated.

In the first simulation setup the process has been simulated with an unscaled drawing tool. The second setup was simulated with a globally scaled drawing tool by a scaling factor – calculated by equation (1) – of 1.001919. Finally, in the third simulation setup the *NPSA* has been used. The simulated springback results (nodal displacements) then have been used to generate springback CAD-geometries and after that the surface area and the unwound lengths of those have been compared to the zero geometry.

5.1. Evaluation of the surface areas

For the comparison of the springback surface areas out of the differently scaled tools the surface patches of the CAD-model have been numbered as shown in **Figure 5**.

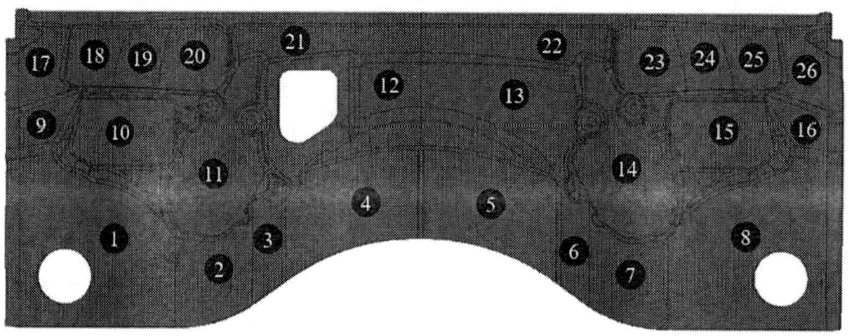

Figure 5. CAD-model and numbering of the surface patches of the headboard.

The evaluation of the scaling quality of the different scaling approaches with regard to surface area deviations is shown in **Figure 6**. The connecting lines between the single patch deviations shouldn´t describe a dependency between the single values and are only used for a better visualization. By using the unscaled tool the deviations of the respective springback geometry (*SBG-unscal*) amount to roundabout 0.3 % in average with a bandwidth of 0.2 %. By using the globally constant scaling factor the tool surfaces have been globally increased by 0.38%. The respective springback geometry (*SBG-globalscal)* shows more or less the same local contraction effect as the *SBG-unscal*, but with an offset of 0.38% to the unscaled curve. This means that the deviations after scaling move close to the zero geometry in average but still show the same bandwidth as without scaling. Insofar with this approach only a certain compromise can be achieved, as long as the contraction of the different patches is not constant, which is being expected for the most real parts.

With the *NPSA* however the tool can be increased locally differently. By doing so the maximum deviation of the *SBG* could be reduced from 0.17% *(SBG-globalscal)* to 0.07% *(SBG-physscal)*. In addition to this the bandwidth of the deviations could be reduced from 0.2% to 0.1% which corresponds to an improvement of 50%.

General findings through the evaluation of the measured curves:

1. The physically compensated curve (*Tool-physscal*) does not represent the exact mirror image of the unscaled springback curve (*SBG-unscal*). This is also expressed by the fact that the sum curve formed from both (*SBG-unscal+Tool-physscal*) does not correspond to the zero line.

2. The reason for the deviations is a non-linear reaction of the zero geometry to membrane stresses of equal magnitude and different signs.
3. Due to the fact that this sum curve (*SBG-unscal+Tool-physscal*) corresponds almost perfectly with the physically scaled springback geometry (*SBG-physscal*) shows the scaling measures is well received by the springback geometry (*SBG-physscal*).
4. Based on this findings it should be possible to adjust the for the physical scaling used stress state locally to bring the sum curve (*SBG-unscal+Tool-physscal*) – and thus also the springback geometry (*SBG-physscal*) – closer to the zero line.

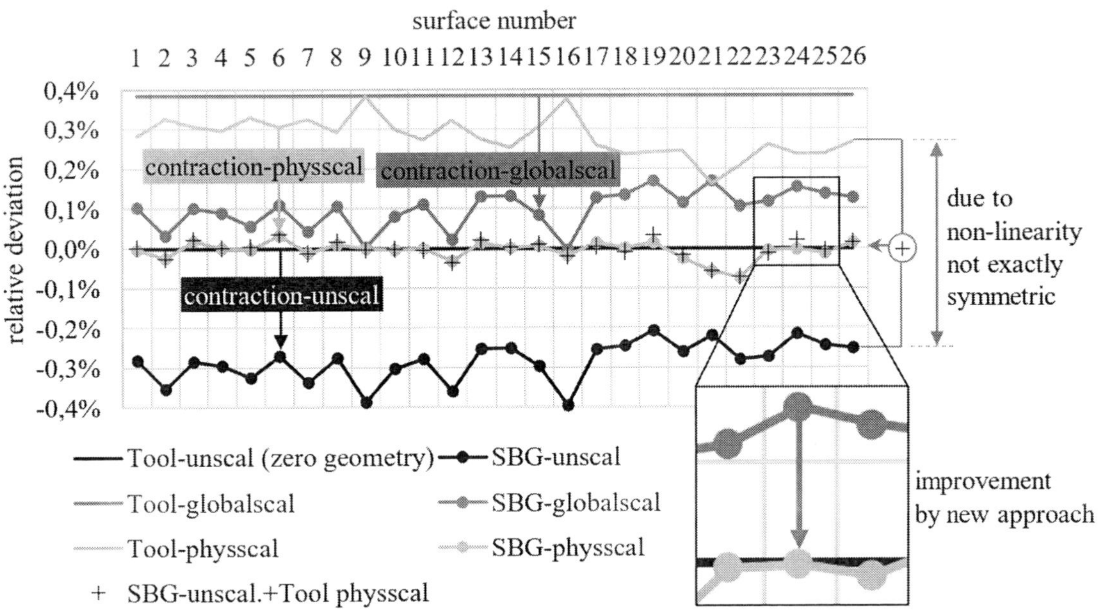

Figure 6. Evaluation of the scaling quality with regard to surface areas when using different scaling approaches.

5.2. Evaluation of the lengths of the patch boundary curves

As in chapter 5.1. the patches itself, here the boundary curves of the patches have been numbered (**Figure 7**).

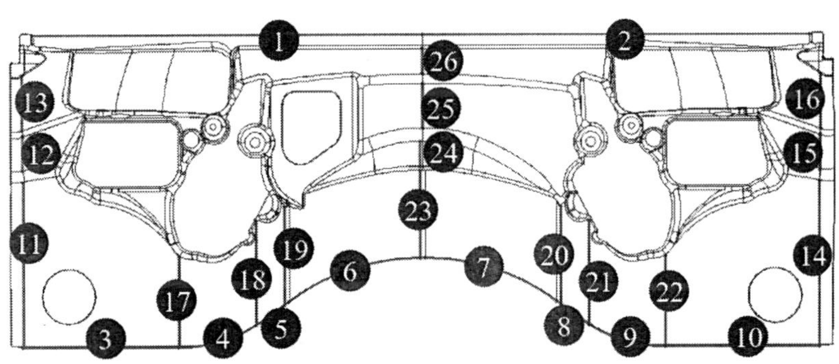

Figure 7. CAD-model and numbering of the surface boundary curves of the headboard.

The evaluation of the scaling quality of the different scaling approaches with regard to boundary curve length deviations is shown in **Figure 8**. The results are similar to the findings of the surface area deviations. By using the global scaling approach the deviations of the unscaled setup have only been shifted. This results in huge residual deviations of the *SBG-globalscal* curve lengths from zero especially in such sections where the contraction is very low (curves 1, 2, 19, 20).

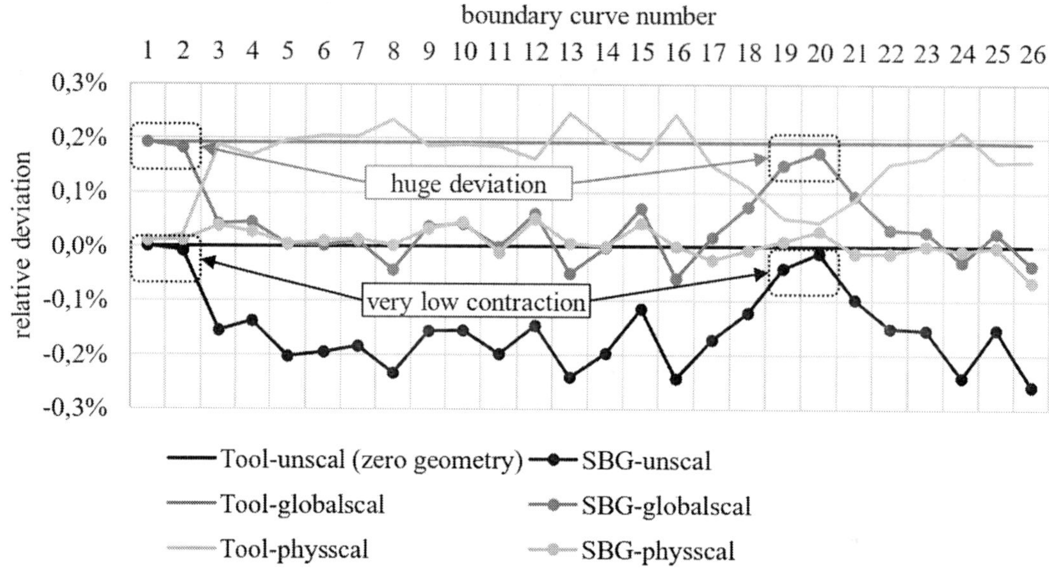

Figure 8. Evaluation of the scaling quality with regard to surface boundary curve lengths when using different scaling approaches

5.3. Evaluation of unwanted plastic deformations after pad closing

As mentioned in the previous chapters before, an inadequate scaling could result in unwanted deformations of the part when closing the pad in subsequent operations. Therefore the plastic deformations after pad closing in *F40-Flanging* have been analysed. To detect any unwanted plastic deformations the *AutoForm* postvariable *Plastic Strain Rate* in the last simulation time step of the pad closing process has been used **(Figure 9)**. With the unscaled setup there are plastic deformations of respectable size in a lot of part surface areas. By using the global scaling approach the number and the size of areas with plastic deformations could be decreased, in other areas (marked area) however plastic deformations occur where there were none before. The best result is clearly achieved by using the *NPSA*. Here in only two larger areas plastic deformations occur. A couple of smaller plastic deformations occur along some radii. These can easily be eliminated by appropriate radius clearances.

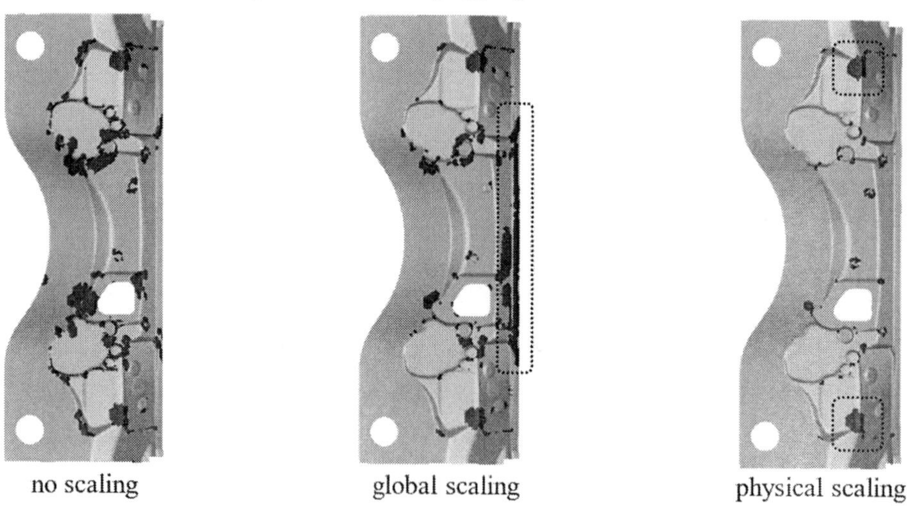

Figure 9. Unwanted plastic deformations after pad closing in the flanging operation (F40)

6. Conclusion

Only such scaling approaches which enable to increase/decrease the tool surface locally are capable to produce parts with the same surface area and unwound lengths as the *zero geometry*. If this requirement is not fulfilled, any unwanted plastic deformations in subsequent operations and problems in the assembly process may occur. In contrast to a global scaling the *New Physical Scaling Approach* does not only provide an approximation to the target surface based on practical experience or average material parameters. The new procedure uses the locally existent membrane stresses which cause the elastic part contraction to scale the drawing tool. By doing so the root cause of the problem itself is directly used to solve the problem. This approach compensates one part of the dimensional deviations caused by the elastic springback. In order to be able to transfer it to other forming processes, the respective stress states and the resulting geometrical deviations would have to be examined more closely.

The evaluation of the surface deviation after drawing shows that with the *NPSA* the maximum deviations could be reduced from 0.17% (achieved by a global scaling) to only 0.07%. Furthermore the bandwidth of the deviations along the different surface patches could be reduced from 0.2% to 0.1% which corresponds to an improvement of 50%! With regard to the boundary curve lengths the maximum deviation could be reduced from 0.19% to 0.07%. Both evaluations show, that with a global scaling only a more or less good compromise can be achieved, especially if the contraction dimensions are not homogeneous along the whole part.

In addition to the geometrical deviations the part also with regard to any unwanted plastic deformations in the flanging operation has been analysed. By using a global scaling the areas with unwanted plastic deformations could be reduced modestly whereas plastic deformations occurred in other areas where none were before. With the *NPSA* such areas have been reduced drastically.

In order to compensate the remaining dimensional deviations in a multistage manufacturing process the *NPSA* can be included in a geometrical springback compensation method. Thereby it has to be observed that the compensation method doesn't have a negative impact on the surface area and the unwound lengths because this would also result in a bad dimensional quality. The *Physical Compensation Method* [4] fulfils these requirements of equal-in-surface compensated tool geometries with high accuracy.

Current and future research work is still being done to further improve the *New Physical Scaling Approach*. Based on the findings that the local tool dimensions should be modified exactly by the amount of surface change during springback in the opposite direction, the simulated membrane stresses can be selectively modified to reach this requirement. Therefore the surface area of each element should be geometrically compared to the target geometry after springback in order to create an adjustment factor for the stresses.

Acknowledgments

These results have been developed in the course of the research project *Development of a New Method for Tool Scaling* (grant No. ZF4424802LP8). The project has been founded by the *Central Innovation Program for Small and Medium-Sized Enterprises* of the *Federal Ministry for Economic Affairs and Energy* in Germany. We would like to thank for making this project possible.

References

[1] Gan W and Wagoner RH 2004 Die design method for sheet springback International Journal of Mechanical Sciences 46(7) 1097-113

[2] Karafillis A and Boyce M 1992 Tooling design in sheet metal forming using springback calculations *Int. J.Mach. Tools. Manuf.* 34113

[3] Birkert A Straub M Haage S 2013 Umformtechnische Herstellung komplexer Karosserieteile, Springer-Verlag Berlin Heidelberg pp 674

[4] Birkert A Hartmann B and Straub M 2017 New method for springback compensation for the stamping of sheet metal components *IOP Conf. Series: Journal of Physics: Conf. Series 896 012067*

Measurement-based evaluation of interfacial polymer layer inserted in sound deadening laminated sheet

Hyeonil Park[1], Se-Jong Kim[1], Jinwoo Lee[1] and Daeyong Kim[1*]

[1]Materials Deformation Department, Korea Institute of materials Science, 797 Changwon-daero, Changwon, Gyeongnam 51508, Republic of Korea

*daeyong@kims.re.kr

Abstract. Mixed mode bending (MMB) test associated with measurement of crack tip opening displacement (CTOD) is a practical method to experimentally evaluate mixed-mode cohesive zone model (CZM) properties of polymer layers in various mixed-mode ratios. However, it has an obvious limitation that does not allow for plastic bending of adherends during the tests, making it difficult to evaluate the polymer layer inserted in the sound deadening laminated sheet (SDLS) comprising thin metal sheets which are vulnerable to the plastic bending. This paper provides an experimental technique to suppress the plastic bending of the thin metal sheets during the MMB test. Supporting tools were designed, and bonded to the both outer surfaces of the SDLS. Until delamination occurred, the evolution of fracture toughness (G) was evaluated, and the CTOD was measured in the stereo image using a digital image correlation system. Finally, the mixed-mode CZM properties was successfully established based on the differentiation of the relation between the evolution of G and CTOD.

1. Introduction

Sound deadening laminated sheet (SDLS) is a composite material with three-layered sandwich structure manufactured by bonding two metal sheets with interfacial polymer layer having excellent sound-deadening characteristics. Its applications to automotive parts that experience bending or drawing processes are still limited, because interfacial slip of the polymer layer that usually occurs upon the manufacturing processes can cause delamination defects without failure of the bonded metal sheets. In order to analysis the delamination behaviour of the polymer layer, mixed mode cohesive zone model (CZM), which can deal with mixed mode loading, is increasingly used [1]. It is defined by the traction–separation law, which represents the constitutive behavior of interfacial bonding traction (T) as function of the separation (δ) between two bonded points, with respect to loading directions (i.e. normal, shear, mixed). The δ can be obtained by the measurement of crack tip opening displacement (CTOD). The under area of T–δ law represents critical fracture toughness (G^c), and the elastic energy based calculation during the beam bending delamination tests is the well-known method to evaluate G^c. Moreover, to experimentally characterize the T–δ law, a differentiation concept has been developed between the energy based calculation and the measurement of CTOD [2].

Mixed mode bending (MMB) test is one of the beam bending delamination tests standardized in ASTM for establishing various mixed mode ratios, and Khoramishad et al. [3] firstly introduced the MMB test linked with the differentiation concept. However, the beam bending delamination tests were originally designed to target composite adherends, which allows only elastic deformation; therefore, they are not suitable for the evaluation of the ductile adhesive layer inserted in the SDLS comprising thin metal sheets that are vulnerable to plastic bending [4,5].

This paper provides an experimental technique to suppress the plastic bending of the thin metal adherends during the MMB test. The thicker supporting tools were designed, and bonded to the both outer side of the SDLS using a stiff and brittle adhesive. Then, the SDLS and supporting tools behave like a single body, and the plastic bending of the thin metal adherends is suppressed during the test. Until delamination occurred, the evolution of G was evaluated, and the CTOD was measured in the stereo image using a digital image correlation (DIC) system. The T–δ law was established based on the differentiation of the relation between the evolution of fracture toughness and CTOD. Finally, the mixed mode CZM properties were obtained from the bilinear fitted T–δ law.

2. Experimental procedure

2.1. Materials

The base material is dual phase (DP) 590 MPa grade high strength steel sheets with 0.7 mm thickness. In order to investigate the mechanical properties of the steel sheets, a uniaxial tensile test along the rolling direction was carried out on ASTM-E8 standard specimens with a universal testing machine (UTM) using the quasi-static rate of 0.001/s [6]. The SDLS was fabricated by passing through roller after inserting the acrylic-based adhesive layer between two adherend metal sheets having size of 300 mm x 300 mm.

Table 1 Characteristics of the base metal sheets

Elastic modulus (GPa)	Yield stress (MPa)	Ultimate tensile stress (MPa)	Uniform elongation (%)	Total elongation (%)	Lankford value
210.3	347.6	635.9	18.5	27.0	0.84

2.2. Design of supporting tools

To suppress the plastic bending of the thin metal sheets during the MMB test, the supporting tools were designed and made of the ion-nitrided SM45C structural carbon steel as shown in Fig. 1. They were bonded to the both outer surfaces of the SDLS using the epoxy-based strong adhesive (S/N: 3M 460NS). During the test the hinge for opening mode is fixed on the supporting tools through the bolt holes.

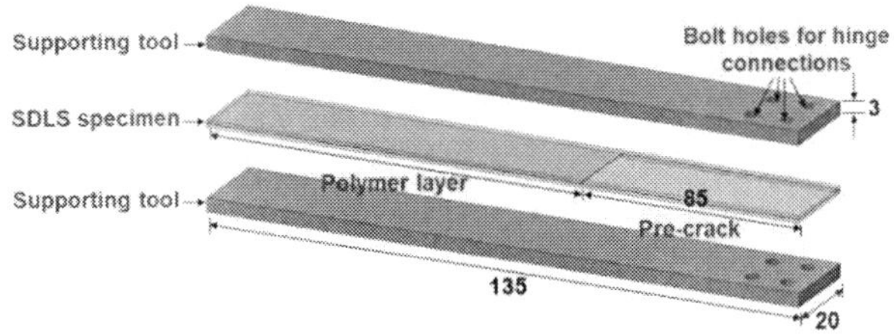

Fig. 1 Detailed geometry of the SDLS specimen with supporting tools

2.3. Mixed mode bending test

MMB test was performed using UTM to experimentally evaluate mixed mode CZM properties, and detailed apparatus is shown in Fig. 2. The arm length (c) can be changed to vary the ratio of the opening force (mode I) at the end of the adherend to the flexure force (mode II) at the middle, and the mixed mode ratio ($\eta = G^c$ of mode II / G^c of mode I+II) of 0.4 was utilized in the test. On the specimen prior to tests, a stochastic pattern was applied to the surface of side wall using white and black spray. Until delamination occurred, the applied load (P) and loading point displacement (U) were measured from the UTM, and synchronized with CTOD measured in the stereo image using the DIC system.

IOP Conf. Series: Materials Science and Engineering **967** (2020) 012042 doi:10.1088/1757-899X/967/1/012042

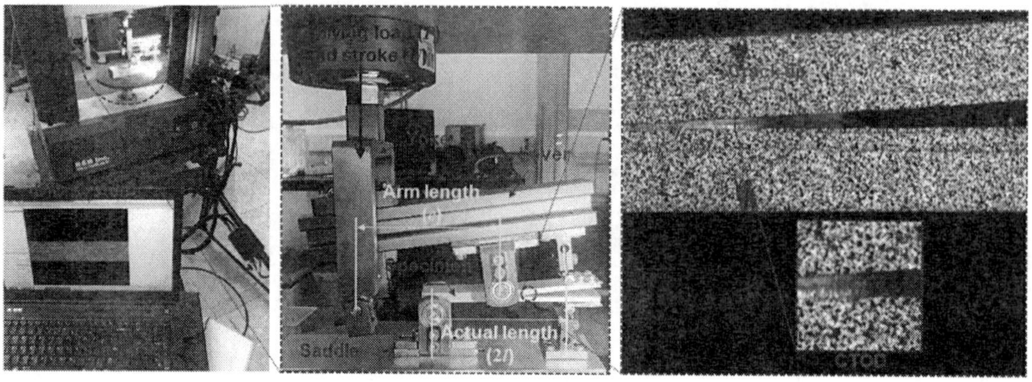

Fig. 2 Apparatus of MMB test

3. Adhesive characterization

During the MMB test, the evolution of fracture toughness (G) can be evaluated by the Irwin-Kies equation as follows [7]:

$$G = \frac{P^2}{2b}\frac{dC}{da} \tag{1}$$

where P, b, and a are the applied load, specimen width, and crack propagation length, respectively. C is the compliance, and defined by:

$$C = \frac{U}{P} \tag{2}$$

where U is the vertical displacement at the loading point. According to the Ducept et al [8], it was not easy to visually define the horizontal crack propagation length due to the unstable propagation, the equivalent crack (a_{eq}) concept based on the Timoshenko beam theory was utilized as follows:

$$a_{eq} = \left(\frac{8bh^3 E_{eq} l^2 C - 2L^3 (c+l)^2}{4(3c-l)^2 + 3(c+l)^2} \right)^{\frac{1}{3}} \tag{3}$$

where b, h, *and* l are the width, thickness, actual half length of the specimen, respectively. E_{eq} is the equivalent elastic modulus, and given by:

$$E_{eq} = \frac{(3c-l)^2 8a_0^3 + (c+l)^2 (6a_0^3 + 4l^3)}{16 l^2 b h^3 C_0} \tag{4}$$

where a_0 and C_0 are the pre-crack length and initial compliance obtained from the early linear part in Eq. (2). By introducing Eq. (3) into Eq. (1), the G can be derived as follows:

$$G = \frac{3P^2 a_{eq}^2}{16 b^2 E_{eq} h^3 l^2}\left[4(3c-l)^2 + 3(c+l)^2 \right] \tag{5}$$

Because G is determined as a function of P, the δ–P curve measured during the MMB test can be converted to the G–δ curve, then the T–δ law can be established by differentiating the G–δ curve against δ as follows:

$$T(\delta) = \frac{dG}{d\delta} \tag{6}$$

4. Results and discussion

The MMB tests were performed under the elastic deformation of the adherend sheets. Fig. 3 shows the established T–δ law of the ductile adhesive inserted in the SDLS during the MMB test. The separation (δ) was obtained from monitoring the CTOD, and the equivalent crack concept described in section 3 was applied to the test results in order to evaluate the evolution of G. The G–δ curve and its derivative, T–δ law, are shown in Fig. 3(a) and (b), respectively. The mixed mode CZM properties were extracted from the bilinear-fitted T–δ law (see the red line in Fig. 3(b)), as listed in Table 2.

Table 2 Mixed mode CZM properties (η = 0.4) of polymer layer inserted in SDLS

K (N/mm³)	T^{max} (MPa)	G^c (N/mm)
7.98	3.98	1.22

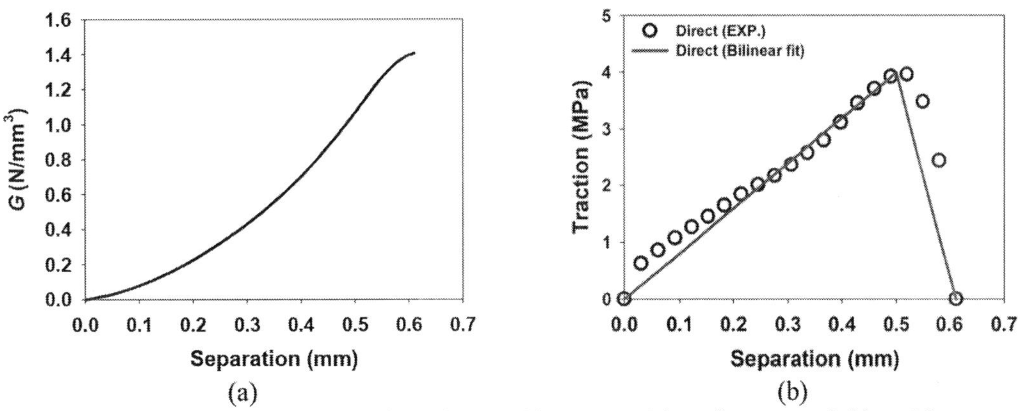

Fig. 3 Established T–δ law of polymer layer inserted in SDLS: (a) G–δ curve and (b) T–δ law

5. Summary

This study provided an experimental technique to evaluate mixed mode CZM properties of the polymer layer inserted in the SDLS. To suppress the plastic bending of the thin adherend sheets during the MMB test, the supporting tools were designed and bonded to the both outer surfaces of the SDLS. As a results, the mixed mode T–δ law was successfully established based on the differentiation of the relation between the evolution of G and CTOD. Finally, the mixed mode CZM properties (η = 0.4) were extracted from the bilinear-fitted T–δ law.

Acknowledgements

This study was financially supported by the Fundamental Research Program (PNK6000) of the Korea Institute of Materials Science funded by the Ministry of Science and ICT and the Industrial Technology Innovation program (10063579) funded by the Ministry of Trade, Industry and Energy, Republic of Korea.

References

[1] Campilho RDSG and Fernandes TAB 2015 Comparative evaluation of single-lap joints bonded with different adhesives by cohesive zone modelling *Procedia Eng.* **114** 102–9
[2] Dias GF, De Moura MFSF, Chousal JAG and Xavier J 2013 Cohesive laws of composite bonded joints under mode I loading *Compos. Struct.* **106** 646–52
[3] Khoramishad H, Hamzenejad M and Ashofteh RS 2016 Characterizing cohesive zone model using a mixed-mode direct method *Eng. Fract. Mech.* **153** 175–89
[4] Park H 2019 *Mechanical behavior of high strength laminated steel sheets with thin adhesive interlayers (Doctoral dissertation)* (Pusan National University, Republic of Korea)
[5] Park H, Kim S-J, Lee J, Kim JH and Kim D 2020 Delamination behavior analysis of steel/polymer/steel high-strength laminated sheets in a V-die bending test *Int. J. Mech. Sci.* **173** 105430
[6] Park H, Kim S-J, Lee J, Kim JH and Kim D 2019 Characterization of the mechanical properties of a high-Strength laminated vibration damping steel sheet and their application to formability prediction *Met. Mater. Int.* **25** 1326–40
[7] Kanninen MF and Popelar CH 1985 *Advanced fracture mechanics* (Oxford: University Press)
[8] Ducept F, Davies P and Gamby D 2000 Mixed mode failure criteria for a glass/epoxy composite and an adhesively bonded composite/composite joint *Int. J. Adhes. Adhes.* **20** 233–44

IOP Conf. Series: Materials Science and Engineering **967** (2020) 012043 doi:10.1088/1757-899X/967/1/012043

Enhancement of Edge Crack Predictability for Automotive High-Strength Thick Steel Sheets Considering the Pressure Effect

Gihyun Bae[1*], Namsu Park[1], Junghan Song[1], Jongsup Lee[1], Inje Jang[1] and Keunyoung Park[1]

[1] Korea Institute of Industrial Technology, 156, Gaetbeol-ro, Yeonsu-gu, Incheon 21999, Republic of Korea

Abstract. This paper deals with the prediction accuracy of edge crack of automotive high-strength thick steel sheets according to the numerical simulation method. The mechanical tests under tension, tension/compression and loading-unloading conditions were performed to evaluate anisotropy, hardening behaviour and elastic modulus change, respectively. Edge stretchability was evaluated by the HER(hole expansion ratio) test according to the hole processing method. Based on the measured data, three simulation methods were constructed with respect to the material model and the element type. The half dome test was selected for verifying edge crack predictability according to the simulation method. The simulation results show that the pressure effect along the thickness direction has significant effect on prediction accuracy of edge crack for high-strength thick steel sheets used for automotive chassis parts.

1. Introduction

The use of advanced high-strength steel sheets has been accelerating for the vehicle weight reduction and the crashworthiness improvement. Recently, this trend is spreading to automotive chassis parts considering their structural stiffness. For the weight reduction of automotive chassis parts, steel companies are developing new materials with improved strength and formability. Figure 1 shows edge crack examples of automotive chassis parts when using high-strength thick steel sheets[1-2]. This phenomenon becomes a serious problem due to the lower edge stretchability of high-strength steels. Therefore, it is very important to predict edge crack phenomenon by numerical simulation in the early manufacturing process design stage in order to reduce the development time and cost.

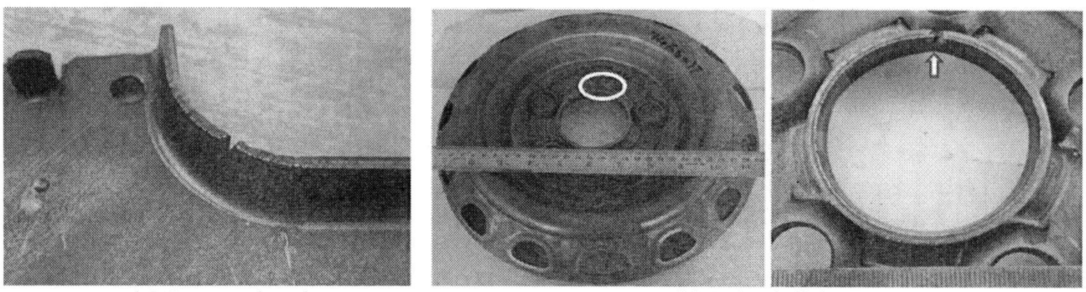

(a) Support bracket (b) Wheel disk

Figure 1. Examples of edge crack phenomenon in automotive chassis parts.

This paper suggests a more reliable simulation method for automotive thick steel sheets to increase the edge crack prediction accuracy. For the numerical simulation, various mechanical tests were performed to construct an advanced material model specialized for the sheet metal forming process. Three simulation methods were established according to the material model and the element type. The half dome test was employed for evaluating the simulation methods. The comparison result shows that the advanced material model and the pressure-dependent element are important for improving edge crack prediction accuracy.

2. Material characterization

Advanced high-strength steels have been actively adopted for the weight reduction of an auto-body. Recently, automotive companies are trying to employ 780 MPa grade thick steel sheets for the automotive chassis parts, such as sub-frame, control arm and wheel disk. Therefore, PO780CP 3.24t and PO780FB 3.30t are selected for studying the edge crack prediction accuracy by the numerical simulation. The mechanical tests were performed and material models are constructed for the numerical simulation.

2.1. Mechanical testing

In this study, three mechanical test results under tension, tension-compression, loading-unloading conditions are employed in order to use an advanced material model in sheet metal forming simulation[3]. The tensile and loading-unloading tests were conducted with MTS universal testing machine. As shown in Figure 2, DIC(Digital Image Correlation) system was utilized to measure the strain distribution and r-value during the tensile test. KS-5 standard specimen was used for tension and loading-unloaidng tests. The tension-compression test was performed with the reversal strain of 5 %. The specimen shape for the tension-compression test was specially designed for increasing the test range without buckling[4]. As shown in Figure 2(b), a laser extensometer was installed additionally in order to measure the strain at the gauge section between upper and lower anti-buckling devices. The strain rate of all mechanical tests was 0.001 /sec.

2.2. Test results and material modeling

Anisotropic tensile properties and kinematic hardening phenomenon was investigated from three mechanical tests. Figure 3(a) shows the stress-strain stress curves according to the material and loading angle. The yield point elongation was observed for PO780CP and PO780FB. Table 1 shows the anisotropic mechanical properties of two materials. The hardening curve was approximated by Swift/Hockett-Sherby model as follows:

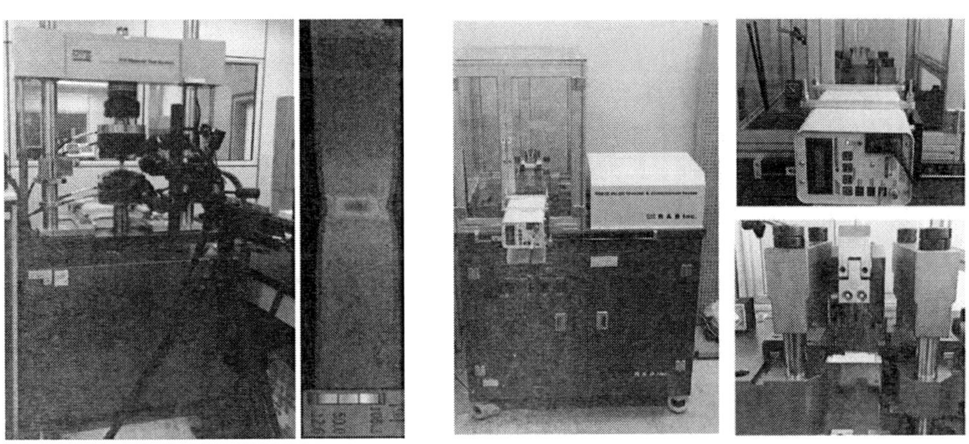

(a) MTS + DIC system (b) Tension-compression testing machine

Figure 2. Testing machines for the mechanical testing.

IOP Conf. Series: Materials Science and Engineering **967** (2020) 012043 doi:10.1088/1757-899X/967/1/012043

$$\sigma = (1-\alpha)\left\{C\left(\varepsilon_{pl}+\varepsilon_0\right)^m\right\} + \alpha\left\{\sigma_{sat} - (\sigma_{sat}-\sigma_i)e^{-a\varepsilon_{pl}^p}\right\} \tag{1}$$

Figure 3(b) shows the loading-unloading test results with a strain interval of 1 %. In order to consider the elastic modulus change according to the plastic strain, the chord modulus model was utilized as follows:

$$\mathrm{E} = E_0 - (E_0 - E_a)\{1 - e^{-\xi\varepsilon_{pl}}\} \tag{2}$$

Figure 3(c) shows the tension-compression test results with the reverse strain of 5% in order to observe the kinematic hardening characteristics. Recently, the Yoshida-Uemori model is widely-used to represent the kinematic hardening tendency[5-6]. This model is a two-surface approach using the inner surface (f) and the bounding surface(F). These surfaces can be expressed as follows

$$\mathrm{f} = \phi(\boldsymbol{\sigma} - \boldsymbol{\alpha}) - Y = 0 \tag{3}$$

$$\mathrm{F} = \phi(\boldsymbol{\sigma} - \boldsymbol{\beta}) - (B + R) = 0 \tag{4}$$

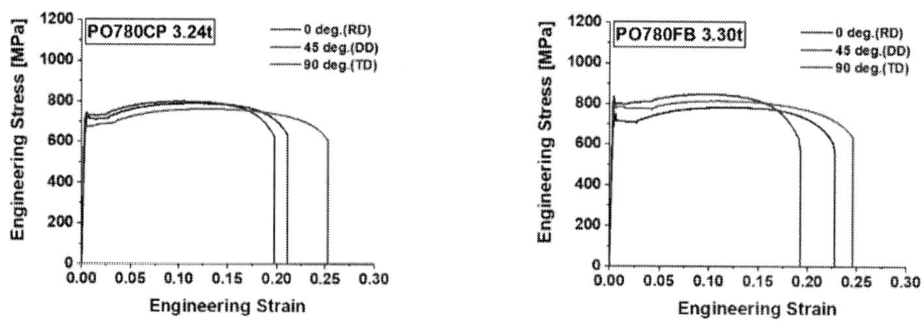

(a) Tension

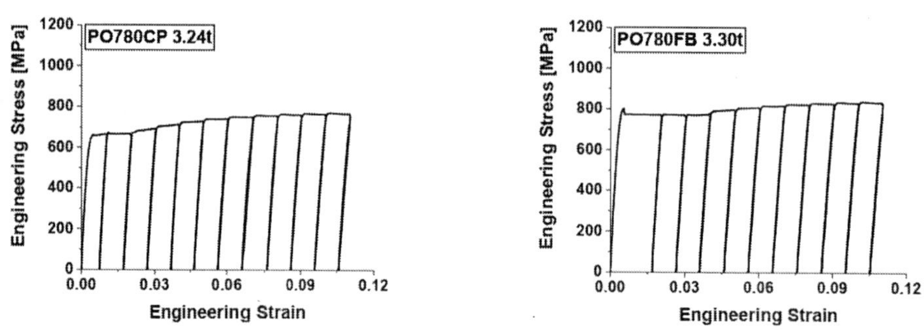

(b) Loading-unloading

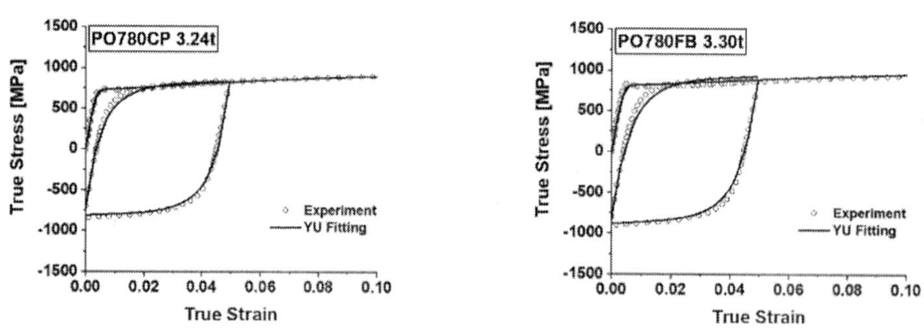

(c) Tension-compression

Figure 3. Mechanical test results of hot-rolled thick steel sheets.

Table 1. Anisotropic mechanical properties of hot-rolled thick steel sheets.

Material	Direction	Yield stress [MPa]	Tensile stress [MPa]	Uniform elongation [%]	Total elongation [%]	r-value
PO780CP 3.24t	RD	700	781	12.15	21.93	0.66
	DD	691	760	12.76	25.72	1.51
	TD	731	801	9.87	19.66	1.00
PO780FB 3.30t	RD	746	810	12.15	22.79	0.68
	DD	784	816	11.36	24.59	1.70
	TD	804	849	10.10	18.99	1.39

Table 2. Material parameters of hot-rolled thick steel sheets for simulations.

Swift / Hockett-Sherby			Yoshida-Uemori			Chord modulus		
Parameter	PO780CP	PO780FB	Parameter	PO780CP	PO780FB	Parameter	PO780CP	PO780FB
K [MPa]	1294	1313	Y [MPa]	390	355	E_0 [GPa]	210	210
e0	0.050	0.040	a_0 [MPa]	330	450	E_a [GPa]	140	155
n	0.205	0.175	b [MPa]	80	120	ξ	10	18
σ_i [MPa]	702.6	747.2	C_1	4000	4000	-	-	-
σ_{sat} [MPa]	915.6	2795.0	C_2	270	300	-	-	-
a	36.1	0.496	m	11.0	7.1	-	-	-
p	1.32	0.721	R_{sat} [MPa]	200	180	-	-	-
α	0.25	0.25	h	0.60	0.10	-	-	-

where $\boldsymbol{\alpha}$ denotes the back stress of the inner surface (f) and its size(Y). $\boldsymbol{\beta}$ corresponds to the center of the bounding surface (F), which expands isotropically at a rate defined by R from the initial size B. The relative kinematic motion $\boldsymbol{\alpha}_*$ of the yield surface and its evolution are expressed as follows

$$\boldsymbol{\alpha}_* = \boldsymbol{\alpha} - \boldsymbol{\beta} \tag{5}$$

$$\acute{\boldsymbol{\alpha}}_* = \mathrm{C}\left[\left(\frac{a}{Y}\right)(\sigma - \alpha) - \sqrt{\frac{a}{\bar{\alpha}_*}}\boldsymbol{\alpha}_*\right]\dot{\mathrm{p}} \tag{6}$$

$$\bar{\alpha}_* = \phi(\boldsymbol{\alpha}_*), a = B + R - Y \tag{7}$$

where (′) represents the objective rate, $\dot{\mathrm{p}}$ is the effective strain rate, and C is a material parameter that helps control the rate of the yield-surface translation. The increments in the size and kinematic motion of the bounding surface are defined as follows

$$\dot{\mathrm{R}} = \mathrm{m}(R_{sat} - \mathrm{R})\dot{\mathrm{p}} \tag{8}$$

$$\acute{\boldsymbol{\beta}} = \mathrm{m}\left(\frac{2}{3}b(\sigma - \alpha) - \dot{\boldsymbol{\beta}}\mathrm{p}\right) \tag{9}$$

where b is the saturation value of the bounding-surface translation, m is a material constant controlling the kinematic rate of the bounding surface, and R_{sat} is the saturation value for R. As shown in Figure 3(c), the Yoshida-Uemori model shows good approximation ability to capture the kinematic hardening of 780MPa grade hot-rolled steel sheets. Based on the test results, hardening models are constructed as shown in Table 2.

3. Edge stretching test

3.1. HER(Hole Expansion Ratio) test

The HER test is most widely-used experimental method to measure the edge stretchability of steel sheets according to the punch condition. Figure 4 shows the schematic diagram of the HER test[1]. After making the hole with the diameter of 10 mm, the conical punch expands the hole until the crack penetrates along the thickness direction. Figure 5 shows the HER test results of 780MPa grade hot-rolled steel sheets. Milling condition shows remarkable hole expanding performance comparing to the punching conditions. Between two punching conditions, the clearance increase slightly improve the HER value. The measured HER values can be used as an input for judging the edge crack in the sheet metal forming simulation.

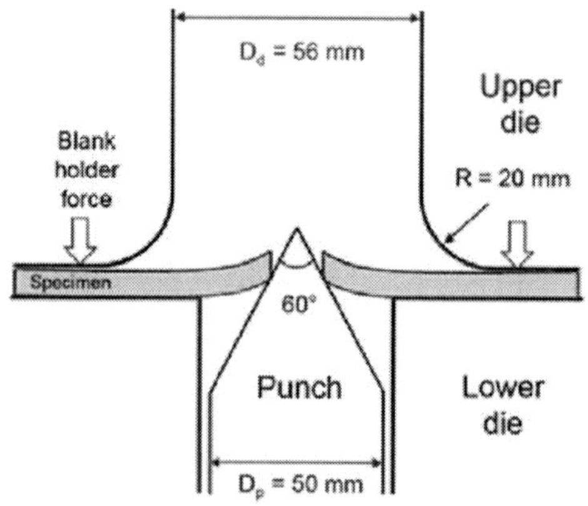

Figure 4. Schematic diagram of the HER test.

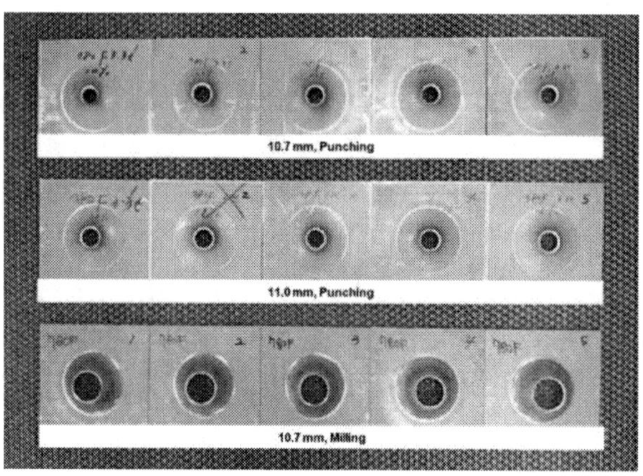

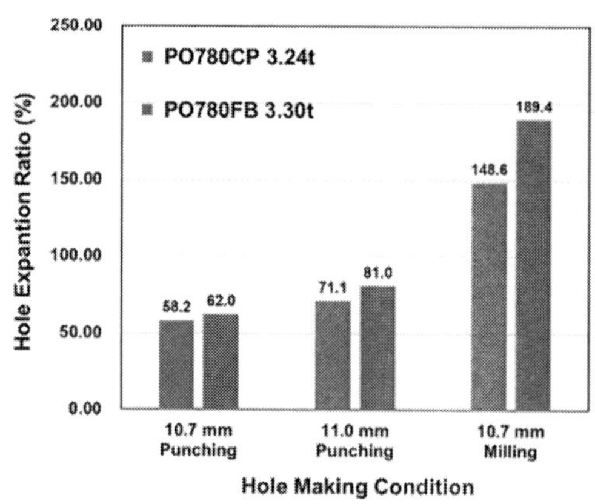

(a) Test specimens (PO780FB) (b) HER

Figure 5. HER test results according to the hole making condition.

3.2. Half dome test

The half dome test[7-8] was employed to evaluate the simulation accuracy according to the material model and finite element. As shown in Figure 6(a), 100 x 200 mm square-type specimen is deformed by the semi-spherical punch. In order to induce the early edge crack, the initial blank position was moved to 20 mm inside from the centreline. Specimens are manufactured by the milling process. From the test specimen, the edge crack was observed near the die shoulder of the upper die.

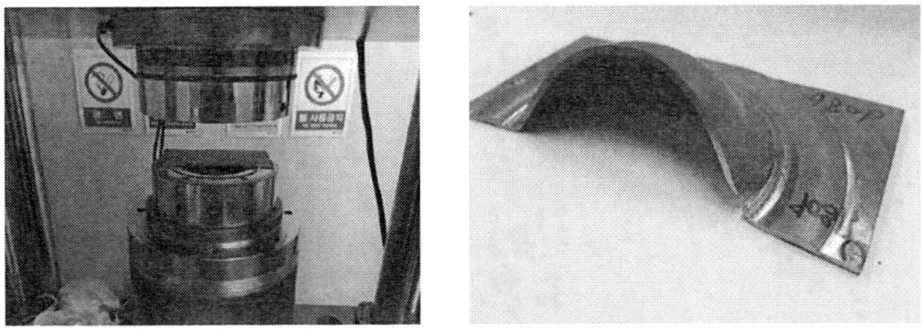

Figure 6. Half dome test setup and test specimen of PO780CP.

4. Numerical simulation for edge crack prediction

The half dome test was used to evaluate the simulation accuracy according to the material model and finite element in the edge crack prediction. The simulation software was Autoform R8. As shown in Table 3, three kinds of simulation models are established with the combination of the yield function, hardening model and finite element. In this paper, the thick shell was employed to take account of the contact pressure effect because the conventional shell element cannot calculate the thickness stress due to the plane stress assumption. Figure 7 explains the principal of the thick shell element how to calculate the thickness stress[9]. The friction coefficient was fixed as 0.15. Figure 8 shows the comparison of the limit dome height and the fracture location between experiment and simulation. The edge crack was judged with the criterion calculated from the HER value of the milling condition. All simulation methods can predict the same fracture location however the limit dome height is clearly different according to the simulation method. Figure 9 shows the force-displacement profile according to the simulation method. When the advanced material model was employed, the prediction accuracy of the edge crack was improved. And the force-displacement profile almost approaches to the experimental curve when the thick shell element is combined with the advanced material model. Table 3 presents the quantitative values of the limit dome height and the maximum punching force when the edge crack occurs. The prediction accuracy drastically enhances by using the advanced material model with the thick shell element. It means that selection of the material model and the element type is very important to obtain the accurate simulation result for the pressure-sensitive stamping problems of thick steel sheets.

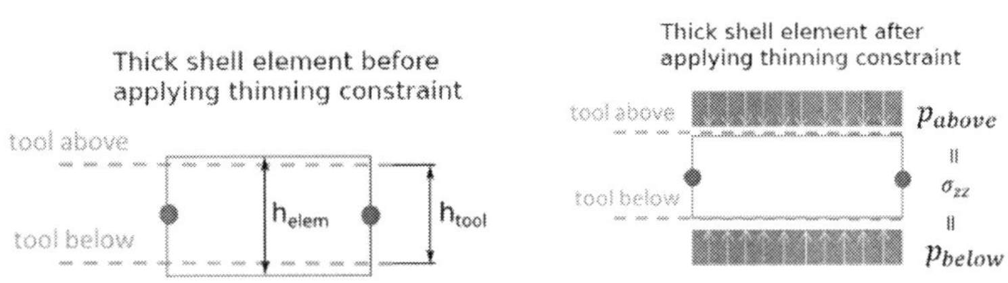

Figure 7. Principal of the thick shell element to consider the contact pressure effect[8].

(a) PO780CP 3.24t

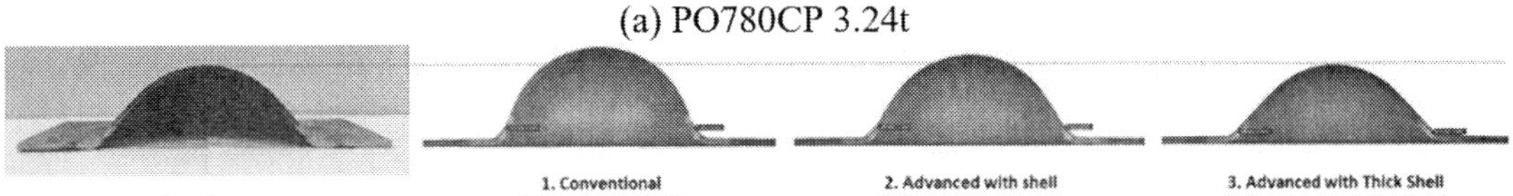

(b) PO780FB 3.30t

Figure 8. Comparison of the limit dome height according to the simulation method.

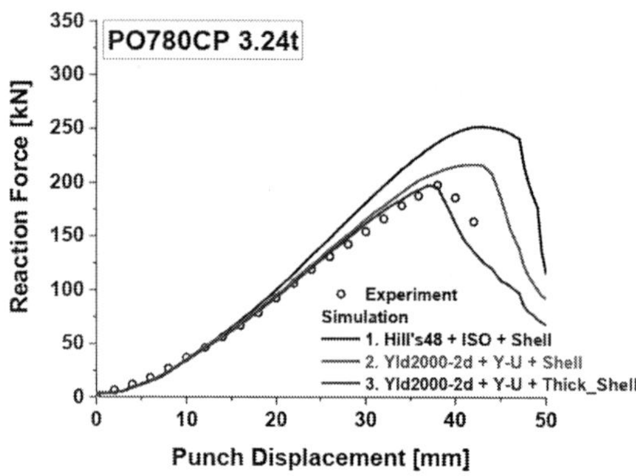

(a) PO780CP

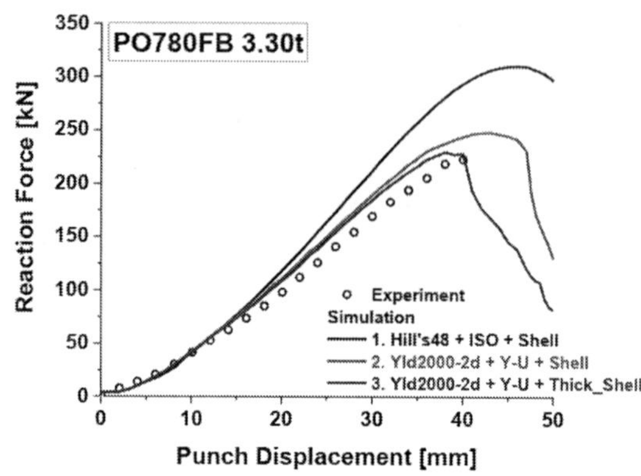

(b) PO780FB

Figure 9. Force-displacement curves of experiments and simulations.

Table 3. Comparison to experiment and simulation results.

Method		Limit dome height [mm]		Maximum force [kN]	
		PO780CP	PO780FB	PO780CP	PO780FB
Experiment		36.6	40.2	198.0	224.2
Simulation	1. Conventional (Hill's 48+ISO+Shell)	43.5 (81.1 %)	46.8 (72.6 %)	252.2 (72.6 %)	310.6 (61.5 %)
	2. Advanced with shell (Yld2000-2d+Y-U+Shell)	42.3 (84.4 %)	44.5 (90.5 %)	216.9 (90.5 %)	248.5 (89.2 %)
	3. Advanced with Thick Shell (Yld2000-2d+Y-U+Thick Shell)	37.1 (98.6 %)	38.8 (96.5 %)	197.8 (99.9 %)	230.0 (97.4 %)

5. Conclusions

In this paper, three kinds of simulation methods were evaluated for predicting the limit dome height and the maximum punching force of the half dome test. The results support that the use of the advanced material model and the thick shell element is necessary to improve the prediction accuracy of the edge crack phenomenon when the contact pressure is not negligible. The simulation scheme established can be a useful tool to reduce the development time for the sheet metal forming process of high-strength chassis parts with hot-rolled thick steel sheets in the automotive application.

References

[1] Casellas D, Antoni L, Fro´meta D, Gutie´rrez D, Molas S, Pe´rez L, Rehrl J and Suppan C 2017 *Metall. Mater. Trans.* **48A** 86
[2] Dey A, Jugade H, Jain V and Adhikary M 2019 *Engineering Failure Analysis* **105** 1273
[3] Bae G, Park N, Song J, Lee J, Jang I, Park K, Seo Y and Oh K 2020 *Proc. The 18th Int. Conf. on Metal Forming(Krakow, Poland / e-conference)* Accepted
[4] Bae G and Huh H 2011 *Proc. International Deep Drawing Research Group 2011 (Bilbao, Spain)* p213
[5] Yoshida F, Uemori T and Fujiwara K 2002 *Int. J. Plasticity* **18** 633
[6] Choi J, Lee J, Bong H J, Lee M-G and Barlat F 2018 *Int. J. Solid. Struct.* **151** 152-164
[7] Gu J C, Kim H, Dykeman J and Shih H 2018 *IOP Conf. Ser.: Mater. Sci. Eng.* **418** 012072
[8] Yin Z 2016 *Fracture of Advanced High Strength Steels (AHSS) Using the Half Dome Test* Master These of Ohio State University
[9] Carleer B, Güner A, Burchitz I and Gösling M 2018 *J. Phys.: Conf. Ser.* **1063** 012126

The mechanical properties of high strength Al alloy sheet under quasi-static-dynamic loading

L L Xia[1,2], Y Xu[1], S H Zhang[1], H W Song[1] and D Y Chen[1,2]

[1] Institute of Metal Research, Chinese Academy of Sciences, Shenyang 110016, China
[2]School of Materials Science and Engineering, University of Science and Technology of China, Shenyang 110016, China

yxu@imr.ac.cn

Abstract. To investigate the mechanical properties of high strength Al alloy, the AA2024 was selected and conducted the uniaxial tensile tests at room temperature under different loading modes, namely pure quasi-static (QSD) loading mode, pure dynamic loading mode and quasi-static-dynamic loading mode. The mechanical properties such as yield stress (YS), ultimate tensile stress (UTS), total elongation and strain rate sensitivity were compared. The results shown that the UTS under QSD loading mode is larger than that of quasi-static and pure dynamic loading mode, and dependent on prestrain levels. However, the YS at pure dynamic loading mode is less than that of QSD loading mode at prestrain of 10% but larger than that of prestrain of 6%. In addition, the QSD loading mode would further improve the total elongation of AA2024 compared to the elongation enhancement under pure dynamic loading mode. Finally, the variation of strain rate sensitivity under different loading modes was analyzed to explain the evolution of tensile properties of AA2024.

1. Introduction

As a high strength aluminium alloy belongs to Al-Cu, the AA2024 alloy possessed outstanding comprehensive properties, such as high specific strength, high fracture ductile, low density and good heat resistance compared to other commercial Al alloys. Hence, it has been widely used in aerospace field recently to manufacture the airplane parts like skins, ribs and pillar, frames and so on. However, with the development of design concept of airplane's structures, in aspect of complex thin-walled integrate sheet components especially the part with large deformation, its manufacturing characteristics has been exceeded the forming limits of AA2024 under the traditional cold forming technology.

To overcome the issues aforementioned, some technologies such as creep aging forming (CAF) [1], explosive forming (EF) [2], electromagnetic forming (EMF) [3] etc. have been developed recent years, which based on the characteristics of material's sensitivity to strain rate. In addition, Zhang [4] proposed an impact hydroforming technology to manufacture the frame parts of airplane by two forming passes that was preforming (hydroforming under quasi-static loading mode) and final forming (impact hydroforming under dynamic loading mode). Indeed, the whole forming process was similar to quasi-static-dynamic (QSD) loading mode. It has been approved that the QSD loading mode could dramatically improve the formability of some materials. Liu [5-7] studied the QSD formability of AA5052-O by uniaxial and plane-strain tension experiments. It was found that nearly two-fold increase in limit strain of plane-strain state compare to quasi-static loading mode. Li [8] pointed out that the total

IOP Conf. Series: Materials Science and Engineering **967** (2020) 012044 doi:10.1088/1757-899X/967/1/012044

tension strain under QSD loading mode of AA5182 would be the superposition between quasi-static strain and pure dynamic strain limitation. In addition, Lee [8-10] systematically investigated the mechanical behavior and the mechanism of formability improvement of 304L under QSD loading mode. However, since the difference of chemical composition, these results cannot representative the properties of 2xxx series aluminium alloys.

Accordingly, this study aims at investigating the influence of different loading modes namely quasi-static, pure dynamic and QSD loading mode on AA2024 through uniaxial tensile tests at room temperature, to compare the mechanical properties such as yield stress (YS), ultimate stress (UTS), total elongation and strain rate sensitivity. In current study, AA2024 was selected as a representative because of its widely applications in aerospace field.

2. Material and experiments

2.1. Material

The material used in current study was rolled sheet Al-Cu alloy AA2024 (O state: annealing heat treatment) with 1.2mm thick.

2.2. Experimental procedure

Although it would be perfect way to evaluate the deformation behaviours of material under a wide range of strain rate using the same testing machine, it is impractical because of the restricted range of velocity of quasi-static tensile tests machine. Thus, in current study, a 10kN Instron tension machine (Figure 1(a)) was used to perform the uniaxial tensile tests under quasi-static condition, while dynamic loading was conducted in split Hopkinson tensile bar (SHTB) (Figure 1(b)). Besides, the strain rate was affected by the dimension of specimens, to obtain the very high strain rate, the method of controlling and reducing the dimension of specimen was used. Therefore, dog-bone specimens with the gauge length and initial width of 5mm were cut from the as-received sheet with length direction parallel to the rolling direction (RD) of the sheet as depicted in Figure 1(c). In addition, The SHTB system used in current study included striker bar (projectile), incident bar and transmitter bar that made by steel with diameter of 20mm, in addition, as well as strain gauge, an oscilloscope, and data processing system. During the test, the specimens were put and fixed between the incident and transmitter bar with a special clamp apparatus. After that, the projectile was driven by the high pressure N_2 gas and impacted incident bar instantly. Thus, the specimen was deformed by the occurrence of elastic waves within very short time. The signals were captured by the strain gauges and displayed on the oscilloscope screen. Assumption the elastic waves was just transmitted along the length direction of bars, thus, the mean strain, mean stress and instant strain rate was obtained from the signals based on one dimensional wave theory. The details equations are shown as follow:

$$\varepsilon(t) = -\frac{2C}{l_0}\int_0^t \varepsilon_R(t)dt \tag{1}$$

$$\sigma(t) = E\frac{A}{A_0}\varepsilon_T(t) \tag{2}$$

$$\dot{\varepsilon}(t) = -\frac{2C}{l_0}\varepsilon_R(t) \tag{3}$$

Where $\varepsilon(t)$, $\sigma(t)$ and $\dot{\varepsilon}(t)$ are the mean strain, mean stress and instant strain rate respectively, $\varepsilon_R(t)$ and $\varepsilon_T(t)$ are the reflected strain wave and transmitted strain wave respectively, C is the velocity of elastic wave in the steel bars, l_0 is the gauge length of specimen, E is the Young's modulus of the bar's material, A and A_0 are the cross section areas of bars and the initial specimen respectively, t is the deformation time.

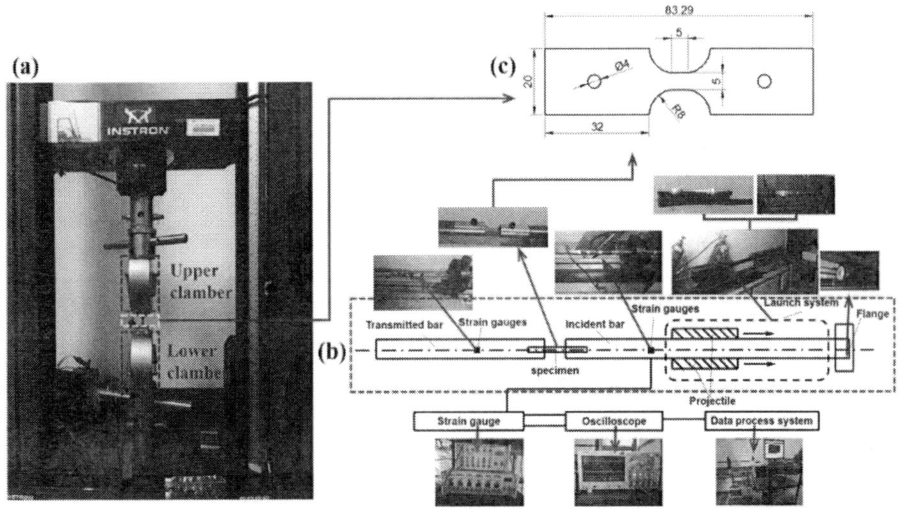

Figure 1. Testing machine and specimen used in current study; (a) 10kN tension machine for quasi-static loading mode; (b) SPHB for dynamic loading mode; (c) specimen dimension (unit in mm)

To compare the mechanical properties under different loading modes, the uniaxial tensile tests were conducted at room temperature with the strain rate of $0.001s^{-1}$ and $1000s^{-1}$~$5000s^{-1}$ corresponding to quasi-static loading mode and pure dynamic loading mode respectively. As for QSD loading mode, the prestrain levels (6% and 10%) were stretched under quasi-static loading mode with a constant strain rate of $0.001s^{-1}$, subsequently, dynamic loading was carried out with the range of strain rate of $1000s^{-1}$~$5000s^{-1}$. As summarized in table 1.

Table 1. Uniaxial tensile tests matrix; (√) means that the test was accomplished at these condition

Strain rate (s^{-1})	Loading mode			
	Quasi-static	Pure dynamic	QSD 6%	QSD 10%
0.001	√	---	√	√
1000~5000	---	√	√	√

3. Results and discussion

3.1. effect of loading modes on YS and UTS of AA2024

Figure 2 (a) shows the YS and UTS under quasi-static, pure dynamic and QSD loading mode. It is worth to notable that the stress value produced 0.2% residual deformation was specified as YS deriving from the stress strain curve. Strain rate has a dramatically influence on YS and UTS of AA2024 alloy. As depicted in Figure 1(a), while strain rate increase, the YS slightly increase till strain rate of $2500s^{-1}$ and sharply increase at strain rate beyond $2500s^{-1}$ to $4500s^{-1}$ and drop after $4500s^{-1}$. The variation of UTS is same to that of YS except the slightly decrease before 2500s-1. For QSD loading mode, the YS and UTS increase at strain rate beyond $1000s^{-1}$ to $2500s^{-1}$, and then, the YS decrease while the UTS continue increase after strain rate of $2500s^{-1}$. Furthermore, to investigate the effect of prestrain levels on the flow stress of AA2024, the variation of YS and UTS were studied at the similar strain rate level (from $2000s^{-1}$ to $3000s^{-1}$). It can conclude that the YS and UTS increase with prestrain increase.

3.2. effect of loading modes on total elongation of AA2024

Figure 2 (b) shows the total elongation under quasi-static, pure dynamic and QSD loading mode. The total elongation increases at the range of strain rate from $0.001s^{-1}$ to $4500s^{-1}$ and dramatically drop after strain rate of $4500s^{-1}$. Thus, the strain rate of $4500s^{-1}$ is the threshold for plasticity improvement of AA2024. And for QSD loading mode, the total elongation increases with strain rate increase. In addition, at the similar strain rate level (from $2000s^{-1}$ to $3000s^{-1}$), the QSD loading mode can further improve the total elongation of AA2024 compare to quasi-static and pure dynamic loading mode. Moreover, at the strain rate of $2000s^{-1}$~$2500s^{-1}$, the total elongation under the prestrain of 6% was larger than that of 10%, but the results alter to contrast while strain rate beyond $2500s^{-1}$ to $3000s^{-1}$. Hence, it is vital for AA2024 to determine the strain rate and prestrain levels when forming complex parts with two passes forming process.

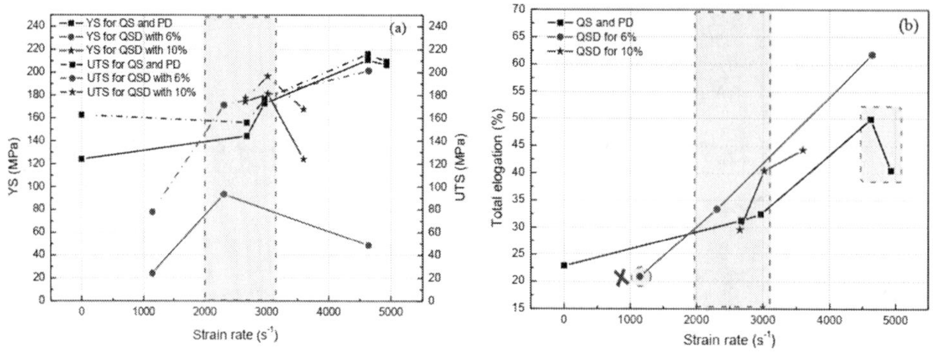

Figure 2. (a) yield stress (YS) and ultimate tensile stress (UTS) under quasi-static (QS) loading mode, pure dynamic (PD) loading mode and quasi-static-dynamic (QSD) loading mode; (b) total elongation under QS, PD and QSD loading mode

3.3. *effect of loading modes on strain rate sensitivity of AA2024*

Usually, the flow stress variation is a function of strain and strain rate at room temperature. And the strain rate sensitivity is expressed by a differential function of stress and strain rate at selected plastic strain as shown by the following equation.

$$m = \frac{\partial ln\sigma}{\partial ln\dot{\varepsilon}} \tag{4}$$

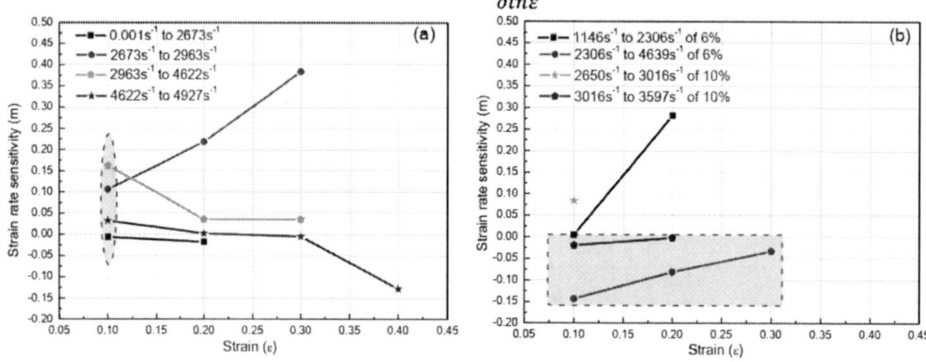

Figure 3. The variation of strain rate sensitivity at different specific plastic strain under different loading mode; (a) quasi-static and pure dynamic loading mode; (b) QSD loading mode at different prestrain levels

In current study, at the selected plastic strain ($\varepsilon = 0.1, 0.2, 0.3$ and 0.4), the strain rate sensitivity m was calculated by fitting the $ln\dot{\varepsilon}$-$ln\sigma$ curve. Figure 3 (a), (b) shows the variation of strain rate sensitivity at different specific plastic strain under different loading mode. It can be concluded that the AA2024 has a high sensitivity to strain rate no matter under quasi-static and pure dynamic loading mode or QSD loading mode. Furthermore, for quasi-static and pure dynamic loading mode, the values of m are nearly -0.128~0.384 at strain rate from $0.001s^{-1}$ to $4927s^{-1}$. With the increase of plastic strain, the strain rate sensitivity m decreases at different strain rate except for the strain rate at the range from 2673s-1 to 2963s-1. In addition, at strain rate beyond $4622s^{-1}$ to $4927s^{-1}$, the values of m change from positive to negative with the plastic rate increasing which means the flow stress decrease, this phenomenon agrees well with the variation of UTS as depicted in Figure 2(a). On the other hand, at the plastic strain $\varepsilon = 0.1$, while the strain rate increase, the values of m increase till strain rate of $4622s^{-1}$ and drop at strain rate from $4622s^{-1}$ to $4927s^{-1}$. This explain that the total elongation dramatically increases at strain rate from 3000s-1 to 4500s-1 and drop after strain rate of $4500s^{-1}$. For QSD loading mode, the values of m are nearly -0.143 to 0.282 at strain rate from $1146s^{-1}$ to $4639s^{-1}$ under different prestrain levels. And with the strain rate increase, the values of m increase and not dependent on the prestrain levels. But it is worth to notable that the values of m almost turn to negative at strain rate from

2306s^{-1} to 4639s^{-1}, which means the softening happened at this range of strain rate after pre-deformed. In addition, at the similar strain rate levels (from 2000s^{-1} to 3000s^{-1}) and $\varepsilon = 0.1$, the value of m is negative at prestrain of 6% while positive at 10%, thus, the UTS at prestrain of 10% is larger than that of 6%.

4. Conclusions

The effect of loading modes on the mechanical properties of AA2024 have been investigated and the following conclusion can be drawn:

- Under different loading modes (quasi-static, pure dynamic and QSD) at a similar strain levels (from 2000s^{-1} to 3000s^{-1}), the UTS under QSD loading mode is larger than that of quasi-static and pure dynamic loading mode. Furthermore, the UTS increases with the prestrain levels increase. However, the YS at pure dynamic loading mode is less than that of QSD loading mode at prestrain of 10% but larger than that of prestrain of 6%.
- Strain rate has a significant on the total elongation of AA2024, compare to quasi-static loading mode, the total elongation under pure dynamic loading mode would dramatically improve with strain rate increasing, but existing a threshold at strain rate of 4500s^{-1}. And the QSD loading mode would further improve the total elongation compare to pure dynamic loading mode. In addition, the prestrain levels have influence on improvement of total elongation.
- The variation of strain rate sensitivity under different loading modes agree well with the variation of UTS and total elongation. Moreover, the values of m show negative at specific condition which means softening happened.

5. References

[1] Holman M C 1989 Autoclave age forming large aluminium aircraft panels *Journal of MechanicalWorking Technology*, **20**, 477-488.
[2] Bahrani A 1986 Engineering with a bang *New scientist*, **1516**, 44-47.
[3] Kapitza P L 1924 A method of producing strong magnetic fields *Proceedings of the Royal Society of London. Series* A, **105**, 691-710.
[4] Ma Y, Xu Y, Zhang S H, Banabic D, El-Aty A A, Chen D Y and Chen G Q 2018 Investigation on formability enhancement of 5A06 aluminium sheet by impact hydroforming *CIRP Annals*, **67**, 281-284.
[5] Liu D H, Yu H P and Li C F 2009 Quasi-static-dynamic formability of AA5052-O sheet under uniaxial and plane-strain tension *Transactions of Nonferrous Metals Society of China*, **19**, 318-325.
[6] Liu D, Yu H and Li C 2011 Experimental observations of quasi-static-dynamic formability in biaxially strained AA5052-O *Journal of materials engineering and performance*, **20**, 223-230.
[7] Liu D H, Yu H P and Li C F 2012 Comparative study of the microstructure of 5052 aluminium alloy sheets under quasi-static and high-velocity tension *Materials Science and Engineering: A*, **551**, 280-287.
[8] Lee W S, Lin C F, Chen T H and Yang M C 2010 Effects of prestrain on high temperature impact properties of 304L stainless steel *Journal of Materials Research*, **25**, 754-763.
[9] Lee W S and Lin C F 2002 Effects of prestrain and strain rate on dynamic deformation characteristics of 304L stainless steel: Part 1—Mechanical behaviour *Materials science and technology*, **18**, 869-876.
[10] Lee W S and Lin C F 2002 Effects of prestrain and strain rate on dynamic deformation characteristics of 304L stainless steel: Part 2—Microstructural study *Materials science and technology*, **18**, 877-884.

Numerical modelling and formability of with and without heat-treated AA 6023-T6 alloy sheet with various necking/failure criteria

Perumalla Janaki Ramulu[1*] Habtamu Beri[1], P. Srinivasa Rao[2], Devender Kumar Sinha[1], Moera Gutu[1]

[1]Centre of Excellence and Department of Mechanical Design and Manufacturing Engineering, Adama Science and Technology University, Adama, Post Box-1888, Ethiopia
*corresponding author's E-mail: perumalla.janaki@astu.edu.et

[2]Department of Mechanical Engineering, Vardhaman College of Engineering, Kacharam, Shamshabad – 501218, Hyderabad, Telangana, India

Abstract: Sheet metal forming is a process widely used in the manufacturing industry. There are numerous sheet metal forming processes to evaluate and understand the formability. Among all formability tests, the basic formability can be formulated through tensile tests and followed with specialized tests. In the present paper, the formability of AA 6023-T6 sheet of 2mm thickness by modelling for stretching test namely limit dome height (LDH) test was performed using PAM STAMP 2G a commercial finite element software. For the simulation, input mechanical properties like yield strength (σ), material strength coefficient (K), strain hardening exponent (n), plastic strain ratio (R) etc., were considered from the existing literature. For the simulation, two different conditioned sheet such as at room temperature and annealed sheet at 400°C. For all the simulations, four strain paths 100×200mm, 125×200mm, 150×200mm, 175×200mm and 200×200 mm were taken. Results are drawn based on the three localized necking criteria namely the effective strain rate-based criterion (ESRC – R1), major strain rate-based criterion (MSRC –R2), thickness strain rate-based criterion (TSRC – R3). Form the obtained results, forming limit diagrams are developed for the both condition of sheet metal. It is observed that, formability of AA 6023-T6 sheet in-plane condition (i.e. 100×100 mm) annealed sheet at 400°C is shown better forming whereas in bi-axial condition (i.e. 200×200 mm) got reduced compared to room temperature sheet. The same phenomenon is noted in all the necking criteria too.

1. Introduction

Sheet metal forming analysis is a revolutionary for the industrial applications. The most of sheet metal applications meant for the weight reduction of automotive bodies leads to improve the fuel efficiency. Later on, industries progressed on many other applications of sheet metals emphasis ferrous sheet metal were replaced with non-ferrous metals. Among all non-ferrous metals, aluminium alloys had the great impact in majority applications. Although aluminium alloys have different series by various alloying elements such as 1xxx, 2xxx, 3xxx, 4xxx, 5xxx, 6xxx, 7xxx, 8xxx and 9xxx, in which 6xxx series aluminium alloy has better feature for applications. Knowing the localized necking limits for forming behaviour of any sheet metal is the most important before application. For evaluating the forming behaviour of sheet metal, there are many formability tests like deep drawing, stretching, incremental forming, bending, hydro forming etc., at different process conditions such as cold, warm, and hot conditions and process parameters [1]. In the literature, there were many researchers worked on the necking limits strains evaluation through different failure criteria.

A failure criterion was developed with respect to thickness gradients during stretching. This criterion was compared with the existing criteria to predict the forming limit diagram (FLD) and proved successfully [2, 3]. This criterion was used to predict the forming limit stains of welded blanks. The simulations were performed using PAM-STAMP. The un-welded blank and tailor welded blanks forming limit curves (FLCs) were predicted by the thickness gradient based necking criterion compared with the experimental FLCs obtained by limit dome height (LDH) test [4]. The applicability of necking criteria namely the effective strain rate-based criterion (ESRC), major strain rate-based criterion (MSRC), thickness strain rate-based criterion (TSRC), and thickness gradient-based criterion (TGNC) were used to quantify and predict the FLC unwelded and tailor-welded blanks at different lubricant conditions. It was identified that the FLCs predicted by ESRC, MSRC, TSRC, and TGNC were comparable with experimental FLCs. The predictions were accurate in the low-friction condition, demonstrated that the necking criteria valid under altering friction conditions. Furthermore, the overall FLC had not affected with friction conditions. In the case of laser welded blanks, modified necking criteria as $R_1>25$, $R_2>32$, and $R_3>32$, which showed a better prediction compared with the original criteria. The original and modified limit strain criteria illustrated the better correlation with results [5]. Based on the modified four necking criteria namely effective strain rate-based, major strain rate-based, thickness strain rate-based, and thickness gradient-based failure criteria, both in original and modified forms, FLCs were predicted for unwelded and tailor welded blanks (TWBs) with longitudinal weld, during stretching and drawing operations. FLC predictions were consistent with the experimental result in the drawing side, but considerable difference was observed in the stretching side in FLD of unwelded sheet. Moreover, in TWBs failures were seen near to the weld region, the strain rate-based criteria were modified as $R_1\geq 25$, $R_2\geq 32$, $R_3\geq 32$ for failure to occur. TWB FLC predicted using the modified criteria demonstrated the better precision than the original failure criteria in the stretching side of the FLD. Whereas in the drawing side, FLC predicted by original criteria was the same as that from modified criteria. From the entire work, they demonstrated that using modified failure criteria, forming limit of TWBs can be predicted with better accuracy, while the original failure criteria were sufficient for a drawing operation [6].

The validity of effective strain rate based necking criterion (ESRC) in both original and modified forms to predict the forming limit of friction stir welding (FSW) blanks made of AA6111, DP590 was analysed. The predicted FLCs were compared with thickness gradient based necking criterion and literature. It was found that the validation done with literature results were consistent and accurate from modified ESRC when compared to original ESRC. The failure pattern prediction was supportive with the literature results [7]. The influence of friction stir welding parameters such as tool rotation speed (1300 and 1400 r/min) and feed rate (90 and 100 mm/min) on the forming limit of friction stir welded Al 6061-T651 sheets investigated through TNGC necking criterion. The FLC was evaluated through limit dome height test. The forming limit of friction stir welded sheets is better than unwelded sheets. The thickness gradient after forming was severe in the cases of friction stir welded blanks made at higher feed rate and lower rotation speed. It was demonstrated that the change in the forming limit of friction stir welded sheets with respect to welding parameters due to the thickness distribution severity and strain hardening exponent of the weld region during forming. There was not much variation in the dome height among the friction stir welded sheets tested. When compared with unwelded sheets, dome height of friction stir welded sheets was higher in near-plane-strain condition, and lesser in stretching strain paths [8]. A theoretical method was proposed to predict the formability of magnesium alloy sheet at elevated temperatures by combining the Marciniak and Kuckzinsky model with the Logan–Hosford yield criterion. Forming limit tests on AZ31B magnesium alloy sheets were performed for the theoretical FLD verification at elevated temperatures of 200, 250, and 300°C and, simultaneously, the material sensitivity effect under a selective strain rate of $0.01s^{-1}$. Based on the verified FLD prediction results, numerical simulations of warm-forming AZ31B camera casing of thickness 0.8mm as an example were then carried out. The warm forming experiments for this camera casing, under the identical conditions, were also performed for verification. It was found that the effect of strain rate on the prediction of FLDs did have a significant influence with increasing temperatures.

Furthermore, the results of numerical simulations showed a good agreement with those of the warm forming experiments at different elevated temperatures. The proposed theoretical method offered an accurate prediction in warm-forming magnesium alloy sheets and should lead to a remarkable reduction of trials, at least in the sense of both time and cost benefits, before a large batch production [9]. The simulations were designed to obtain FLDs through hemispherical die stretching of low-carbon steel blanks of various thicknesses. The multiple criteria, including the second time derivatives of major strain, thickness strain and equivalent plastic strain extracted from the strain history of simulations, were used to detect the start of necking in forming limit diagrams. It was observed that necking starts when the second derivative of the thickness strain, major strain or plastic strain reaches the maximum value. Moreover, a modified Marciniak and Kuczynski method was used to predict the forming limit diagrams. The results from the proposed methods and experimental test results were compared to demonstrate the efficiency of the proposed methods [10].

A free bulge test and ductile fracture criteria were used to obtain FLD of aluminium alloy AA6063 tubes at high temperatures. Ductile fracture criteria were calibrated using the results of uniaxial tension tests at various elevated temperatures and different strain rates through adjusting the Zener-Holloman parameter. High temperature free bulge test of tubes was simulated in finite element software Abaqus, and tube bursting was predicted using ductile fracture criteria under different loading paths. FLDs were obtained from finite element simulation were compared to experimental results to select the most accurate criterion for prediction of forming limit diagram. According to the results, all studied ductile fracture criteria predict when forming condition was close to the uniaxial tension, while Ayada criterion predicts the FLD at 473 K and 573 K very well [11]. A computational approach was introduced for prediction of the forming limit diagram of Al-Cu two-layer metallic sheets. The computational approach was based on the modified Marciniak and Kuczynski theory. The FLD of Al–Cu two-layer metallic sheets were obtained through the modified Marciniak and Kuczynski theory and experimental investigations. In the modified Marciniak and Kuczynski theory, there existed four nonlinear equations which were solved simultaneously. The Quasi-Newton Method was applied for a solution to the system of equations. To verify the theoretical predictions, the experimental works were accomplished on the Al-Cu two-layer metallic sheets and a good agreement between the proposed method and experimental works was observed [12].

From the literature, it is observed that there are many failure criteria are existed to evaluate the formability of the sheet metal at various conditions like temperatures ranges, frictional conditions, strain path consideration etc. Eventhough, the present paper is induced to estimate the formability of AA 6023-T6 alloy of 2 mm thickness through modelling and simulation process analysis by considering the effective strain rate-based criterion (ESRC), major strain rate-based criterion (MSRC), and thickness strain rate-based criterion (TSRC). These all criteria are not new even-though they are the meticulousness to analyse the formability of any material.

2. Methodology

2.1 Base material chemical composition and mechanical properties

For this study, base material details were adopted from the existing literature [13]. The chemical composition of AA6023-T6 alloy sheet was shown in the Table 1.

Table 1 Chemical composition of AA6023-T6 alloy sheet

Chemical composition								
Si	Fe	Cu	Mn	Mg	Bi	Sn	Residuals	Al
0.57%	0.5%	0.19%	0.24%	0.41%	0.3%	1.2%	0.59%	96%

Considered mechanical properties were shown in the Table 2 and Table 3. Table 2 mechanical properties were at room temperature and Table 3 were mechanical properties of heat-treated alloy at 400°C.

Table 2 Mechanical property of AA6023-T6 alloy sheet at room temperature

Orientation of samples	UE (mm)	UTS (MPa)	σ_{ys}(MPa)	K (MPa)	n	R_0	R_{45}	R_{90}
0°	0.38	200 ±9	180±3	313±40	0.132	0.49	0.68	0.56
45°	0.54	208 ±4	190±18	338±20	0.134			
90°	0.52	212±12	200±12	358±20	0.133			

UE-uniform elongation; UTS-ultimate tensile strength

Table 3 Mechanical property of AA6023-T6 alloy sheet after heat treatment

Orientation of samples	UE (mm)	UTS (MPa)	σ_{ys}(MPa)	K (MPa)	n	R_0	R_{45}	R_{90}
0°	0.19	206±13	170±17	326±33	0.149	0.43	0.58	0.47
45°	0.21	197±2	168±5	323±16	0.162			
90°	0.25	193±6	166±7	315±28	0.155			

UE-uniform elongation; UTS-ultimate tensile strength

2.2 Simulation methodology

Simulations were performed using PAM STAMP 2G a finite element code. During the simulation, limit dome height test simulations were performed at a constant punch speed of 10m/s, blank holding force of 80kN and coefficient of friction 0.2 for all stretching side strain paths (100×200mm, 125×200mm, 150×200mm, 175×200mm and 200×200 mm). The base material model comprised quadrilateral shell elements of the Belytschko–Tsay formulation, with five through-thickness integration points. A constant mesh size of 2mm was used. Hollomon's power law was used to describe the strain-hardening behaviour of the base material, Hill's 1948 isotropic hardening yield criterion was used as the plasticity model.

2.3 Necking Criteria for limit stain predictions

Many failure criteria are existed to predict the local necking during sheet metal forming. In the present work, three failure criteria were used for predicting the FLC. The failure criteria were Major strain rate criterion (MSRC- R_1), Effective strain rate criterion (ESRC -R_2) and Thickness strain rate criterion (TSRC-R_3) [2,3].

The strain rate in the notch to that in the bulk, R, the ratio of major principal strain rate in the notch to that in the bulk, R1

The criterion based on major principal strain rate, was defined as follows:

$$R_1 = \frac{\text{Major strain rate in notch}}{\text{Major strain rate in bulk}} \geq 10$$

The next criterion is based on the scalar effective strain rates. The ratio of the effective strain rate in the notch to that in the bulk R2 exhibits a growth in slope at the onset of localized deformation. This criterion was mentioned as:

$$R_2 = \frac{\text{Effective strain rate in notch}}{\text{Effective strain rate in bulk}} \geq 4$$

The next criterion is based on thickness strain rates. The ratio of the thickness strain rate in notch to that in the bulk, R_3, this criterion was described as:

$$R_3 = \frac{\text{Thickness strain rate in notch}}{\text{Thickness strain rate in bulk}} \geq 10$$

3. Results and Discussion

3.1 Limit Dome height test

The limiting dome height (LDH) test is used to assess the formability of AA 6023-T6 alloy sheet of 2 mm thickness. Limiting dome height test setup has been used to carry out simulations by using PAM-STAMP (Fig. 1.). The LDH test can be used to obtain various strain conditions by changing the specimen geometry. Hemispherical punch is used to form the specimens. The sheet metal being formed is held in its position by means of a draw bead. Total of 4 strain paths are taken from 100×200 mm to 200×200mm i.e. in-plane condition to bi-axial condition. All the simulations are taken place with same conditions. Simulation phenomenon can be seen form the Fig. 2 (a to d). from the deformed strain paths, the data related to the effective stain, major stain, thickness stain are drawn after failure identification. The corresponding strain ratios from the notch to bulk are taken to evaluate the limit stains.

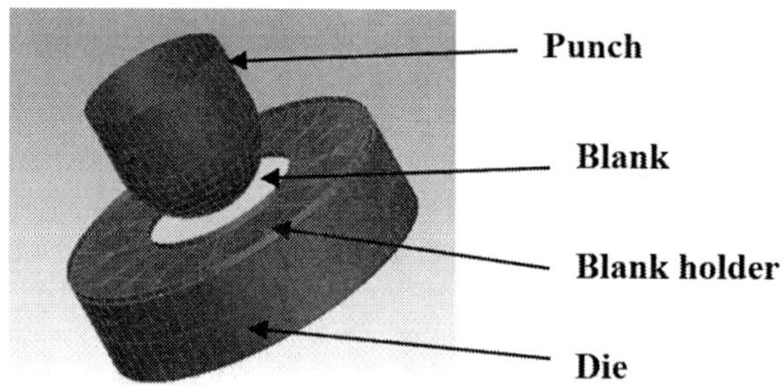

Figure 1. Limit dome height setup in PAM STAMP

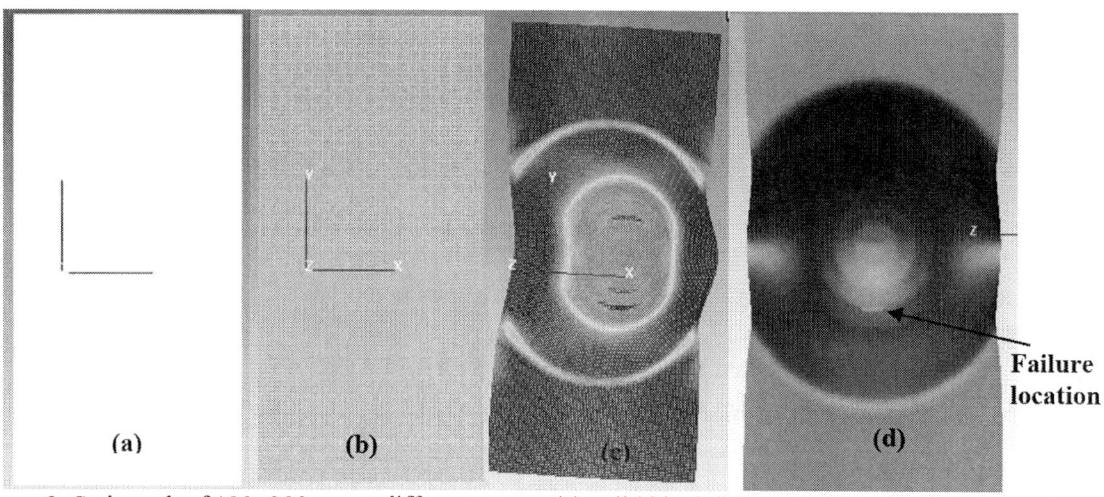

Figure 2. Stain path of 100×200mm at different stages (a) solid blank (b) meshed blank (c) Deformed blank (d) deformed with failure location blank

3.2. Limit stain Evaluations

Based on section 2.2, failure criteria the following discussion has taken place to identify the limit strains and validations of criteria through data and figures of different strain paths. Current work indicates the first criterion as MSRC i.e. ratio of major strain rate in the notch to major stain strain rate in the bulk must greater than or equal to 10 as shown below.

$$R_1 = \frac{\text{Major strain rate in notch}}{\text{Major strain rate in bulk}} \geq 10$$

From the simulation, stain path 100×200mm is observed for MRSC at ambient condition. In can be seen form the Fig. 3 that when the ratio reaches at 6.51, the data jumped to more than 10 and becomes vertical increment. From this criterion, major strains and minor strain are taken for all 100×200mm, 125×200mm, 150×200mm, 175×200mm and 200×200 mm strain paths for tested sheet at room temperature and annealed at 400°C. The corresponding data graph has been developed as shown in Fig.4. In it clear that, sheet tested at room temperature has less formability at in-plane strain path and improved at bi-axial strain path whereas the reverse phenomenon has been observed for the sheet annealed at 400°C. Similar phenomenon is seen in the Fig. 5 and Fig. 6 for the criteria ESRC and TSNC.

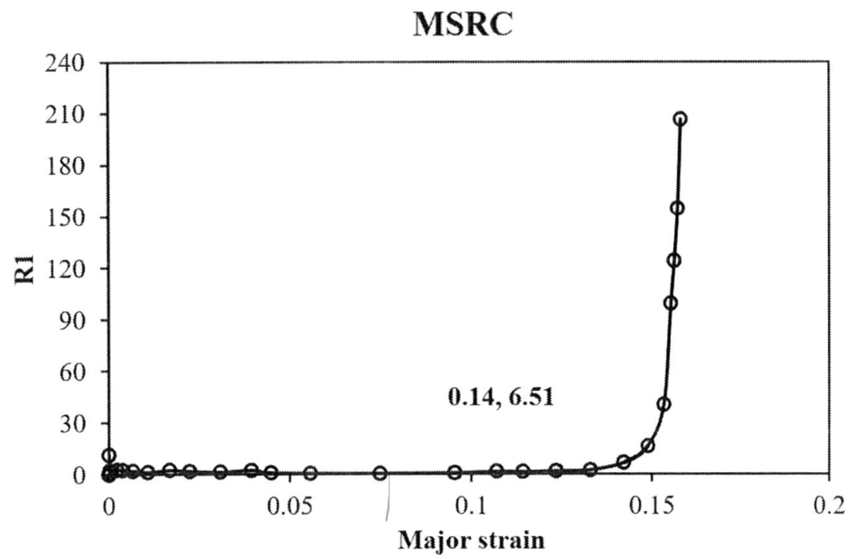

Figure 3. Major strain-corresponding R1

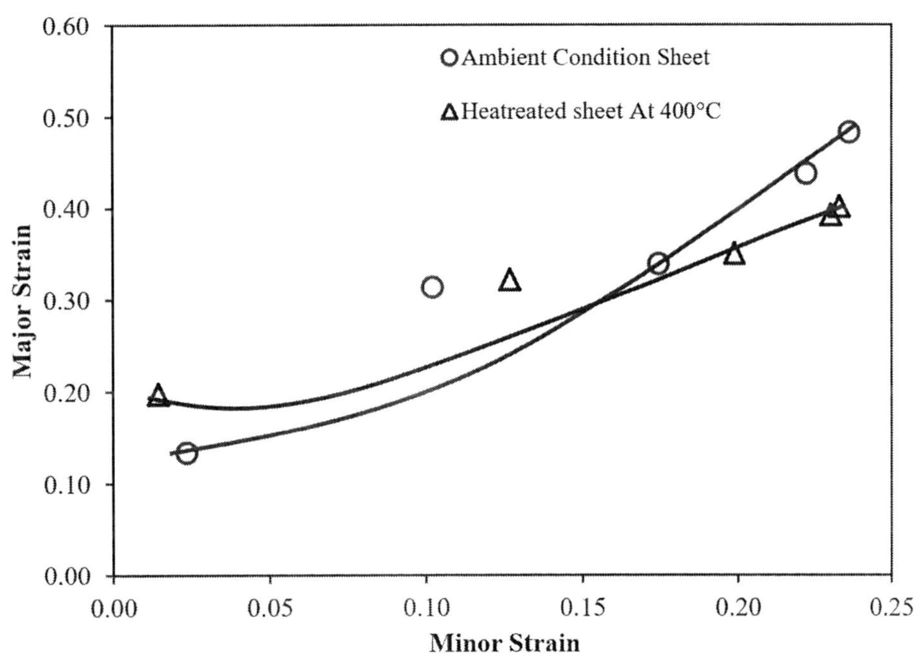

Figure 4. FLD obtained during LDH for MSRC condition

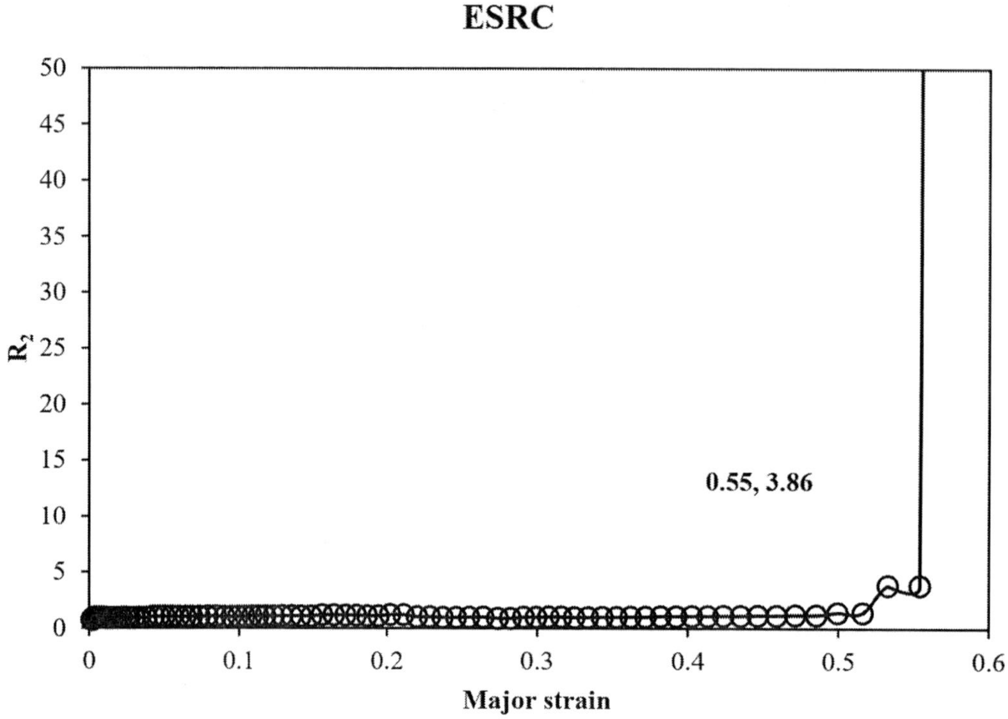

Figure 5. Effective strain rate -corresponding R_2

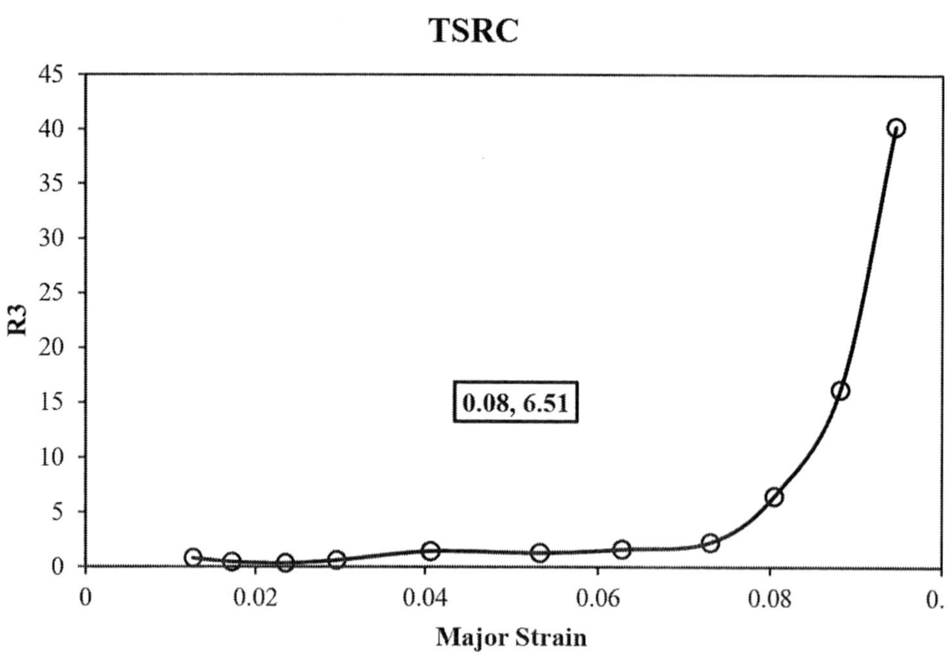

Figure 6. Thickness strain rate -corresponding R3

4. Conclusions

The present work aimed to investigate the formability of AA 6026-T6 alloys sheet of 2 mm thickness at two different temperature conditions. The formability has been predicted through various necking criteria such as MSRC, ESRC and TSRC using simulation software. Finally, conclusion is these necking criteria have the stability to evaluate the formability of any kind of sheet metal with any kind the conditions. In this work formability of AA 6029-T6 is showed better formability with the annealed sheet at in-plane stretching and reverse in the bi-axial condition. This can be extended to verify the modified criteria of MSRC, ESRC and TSRC as $R_1>25$, $R_2>32$, and $R_3>32$ along with the TGNC also can be included for the further work.

References

[1]. Perumalla Janaki Ramulu (November 27th 2019). Aluminum Alloys Behavior during Forming [Online First], IntechOpen, DOI: 10.5772/intechopen.86077. Available from: https://www.intechopen.com/online-first/aluminum-alloys-behavior-during-forming.
[2]. Narasimhan, K. and Wagoner, R.H., 1991. Finite element modeling simulation of in-plane forming limit diagrams of sheets containing finite defects. Metallurgical Transactions A, 22(11), pp.2655-2665.
[3]. Kumar, S., Date, P.P. and Narasimhan, K., 1994. A new criterion to predict necking failure under biaxial stretching. Journal of Materials Processing Technology, 45(1-4), pp.583-588.
[4]. Narayanan, R.G. and Narasimhan, K., 2008. Predicting the forming limit strains of tailor-welded blanks. The Journal of Strain Analysis for Engineering Design, 43(7), pp.551-563.
[5]. Narayanan, R.G. and Naik, B.S., 2010. Assessing the validity of original and modified failure criteria to predict the forming limit of unwelded and tailor welded blanks with longitudinal weld. Materials and Manufacturing Processes, 25(11), pp.1351-1358.
[6]. Srinivas Naik, B., Janaki Ramulu, P. and Ganesh Narayanan, R., 2010. Application of a few necking criteria in predicting the forming limit of unwelded and tailor-welded blanks. The Journal of Strain Analysis for Engineering Design, 45(2), pp.79-96.
[7]. Ramulu, P.J. and Narayanan, R.G., 2011, May. Predicting the forming limit of friction stir welded blanks. In AIP Conference Proceedings (Vol. 1353, No. 1, pp. 219-223). American Institute of Physics.
[8]. Ramulu, P.J., Kailas, S.V. and Narayanan, R.G., 2013. Influence of tool rotation speed and feed rate on the forming limit of friction stir welded AA6061-T6 sheets. Proceedings of the Institution of Mechanical Engineers, Part C: Journal of Mechanical Engineering Science, 227(3), pp.520-541.
[9]. Chan, L.C. and Lu, X.Z., 2014. Material sensitivity and formability prediction of warm-forming magnesium alloy sheets with experimental verification. The International Journal of Advanced Manufacturing Technology, 71(1-4), pp.253-262.
[10]. Hashemi, R., Mamusi, H. and Masoumi, A., 2014. A simulation-based approach to the determination of forming limit diagrams. Proceedings of the Institution of Mechanical Engineers, Part B: Journal of Engineering Manufacture, 228(12), pp.1582-1591.
[11]. Hashemi, S.J., Naeini, H.M., Liaghat, G., Tafti, R.A. and Rahmani, F., 2014. Forming limit diagram of aluminum AA6063 tubes at high temperatures by bulge tests. Journal of Mechanical Science and Technology, 28(11), pp.4745-4752.
[12]. Hashemi, R. and Karajibani, E., 2018. Forming limit diagram of Al-Cu two-layer metallic sheets considering the Marciniak and Kuczynski theory. Proceedings of the Institution of Mechanical Engineers, Part B: Journal of Engineering Manufacture, 232(5), pp.848-854.
[13]. Handiso, M.M., Ramulu, P.J. and Somasundaram, G., 2019. Stretch Formability Prediction of AA6023-T6 Alloy Sheet Under Two Different Heating Conditions. In Advances in Computational Methods in Manufacturing (pp. 37-48). Springer, Singapore.

Inelastic strain recovery of magnesium alloys and a new elastic modulus model

YU Hai Yan[1], WU Hangyu[1] and Wang Lin[2]

[1] School of Automotive Studies, Tongji University, Shanghai 201804 China
[2] SAIC Maxus Automotive Co., Ltd., Shanghai 200438, China

Abstract: Cyclic loading-unloading uniaxial tension experiments were performed for AZ31B magnesium sheets. The inelastic strain recovery behavior of the AZ31B sheets unloaded under different plastic strain was investigated. An interesting phenomenon on the elastic modulus change was presented. A new elastic modulus model was presented to characterize the inelastic strain recovery of the AZ31B sheets. Unlike the available models, this elastic modulus model takes both the instant stress and the plastic strain as the variables. The new elastic modulus model was incorporated into the ABAQUS software using the user subroutines. Validation of the subroutines and springback simulations were conducted with the presented elastic modulus models. The results showed that the inelastic strain recovery of the AZ31B sheets can reach 24% of the total strain recovery. The instant elastic modulus model can predict a perfect unloading path. The new elastic modulus model works better than the constant elastic modulus.
Keywords: inelastic strain recover; elastic modulus model; AZ31B magnesium; springback simulation

1. Introduction

During the light weighting of automobiles, adopting lightweight materials, such as high strength steels, aluminum alloys and magnesium alloys was an efficient and economical approach. Among those materials, magnesium alloys have a low density, great impact resistance, high specific strength and excellent energy absorption performance[1-2]. Magnesium alloys have been widely used in automotive products and have been proved to have serious springback problem during stamping.

Accurate springback simulation is crucial for springback control. Elastic modulus is one of the most important parameters in springback prediction. In most simulations, the elastic modulus was regarded as a constant during the plastic deformation. There are so many researches have pointed out that most of the metals have the inelastic strain recovery after plastic deformation, which directly results in the change of the elastic modulus. Cleveland and Ghosh et al.[3], stated the inelastic strain recovery mechanism of metals as early as 1980. A model was established to describe the relationship between the inelastic strain recovery and the strain. This model was inconvenient in practice because the determination of coefficients was very complicated. Morestin[4], Luo [5], Perez [6] and Sun [7] investigated the inelastic recovery behavior of steels. Li et al. [8] and Zang [9] investigated the inelastic recovery behavior of aluminum alloys. They pointed out that the variation of an elastic modulus was owing to the participation of the plastic strain in recovery. The change rate of the elastic modulus differed in different prestrain levels. If the elastic modulus was a variable, the springback simulation was proved to have better agreement with experimental results for both high strength steels [10] and aluminum alloys [11]. Yoshida [12] proposed a formula to explain the relationship between the elastic modulus and the plastic strain, which has been widely used by later researchers [13,14,15]. On this basis, the elastic plastic model [16,17] was coupled with damage.

In comparison, the study on the inelastic strain recovery of magnesium alloys has just got attention in recent years. Hama et al. [18,19] explained the inelastic behavior of AZ31B subjected to unloading from the view of crystal slip. They also compared the inelastic strain recovery behavior of magnesium alloy, aluminum alloy and steels [20]. Among which, the magnesium alloy was proved to have the largest inelastic strain recovery. Wang et al. [21] pointed out that the inelastic strain recovery of magnesium alloys attributed to the twin deformation. These researches mainly focused on the microscopic observation and theoretical interpretation. A comprehensive understanding of the inelastic strain recovery behavior of magnesium alloy is crucial to get a more accurate springback prediction.

In this study, the inelastic recovery behavior of AZ31B magnesium alloy was investigated. Cyclic loading-unloading uniaxial tension experiments were performed and the microstructures evolution was observed with SEM. The inelastic strain recovery behavior of AZ31B sheets unloaded under different plastic strain was discussed in detail. A chord elastic modulus model was presented to calculate the elastic modulus when the load was removed totally. For the case of partial unloading, an instant elastic modulus model was proposed. This model takes both the stress and the plastic strain as the variables. Then the two models were incorporated into the springback simulation of U-bending using user subroutines. U-bending experiments were conducted and compared with the simulation.

2. Experiments

2.1. Materials

The AZ31B sheets investigated in this study were produced by Beijing Yinguang Hanlin Technology Co. Ltd. Their theoretical chemical compositions were shown in Table 1. The thickness of the AZ31B sheets is 1.0mm and 1.2mm.

Table 1. Chemical compositions of AZ31B sheets in weight%

Materials	Al	Fe	Si	Mn	Cu	Zn	Mg
AZ31B	3.17	0.005	0.10	0.50	0.01	0.86	Bal.

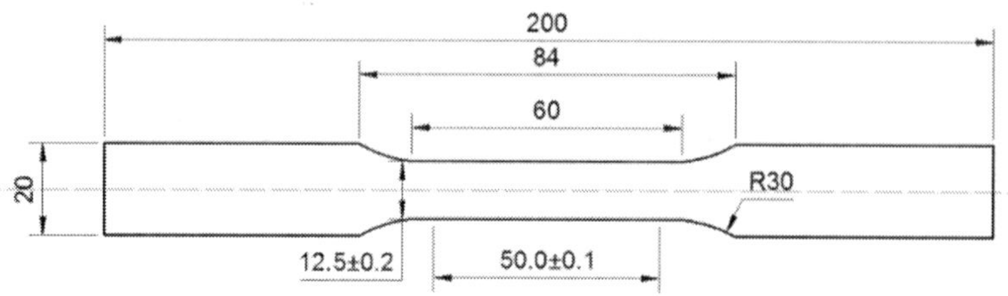

Figure 1. Dimensions of the uniaxial tension specimens

Experiments of uniaxial tension were conducted on a Zwick/Roell Z020 universal machine. As shown in Figure 1, dog-bone specimens whose gauge is 50mm were adopted according to the standard ASTM E8M-09. Each tension was repeated three times with duplicated specimens. All specimens were machined by wire cutting carefully because magnesium alloys are easily oxidized. The strain hardenability exponent (n-value) and the plastic strain ratio (r-value) of the AZ31B sheets were measured according to the ASTM E646-16 and ASTM E8M-09 standards, respectively. During the testing of r-value, the length strain and the width strain were measured using the digital image correlation (DIC) system. Table 2 showed the mechanical properties of AZ31B sheets.

Table 2. Mechanical properties of AZ31B sheets

Thickness		$\sigma_{0.2}$/MPa	σ_u/MPa	δ_u	δ_t	r-value	n-value
	0°	163.1	260.7	17.6%	28.3%	2.33	
1.0mm	45°	165.0	251.2	19.1%	30.1%	1.80	0.19
	90°	184.3	260.8	17.0%	25.5%	2.76	
	0°	182.2	269.2	17.5%	26.5%	3.19	
1.2mm	45°	191.3	270.0	17.2%	25.0%	2.29	0.18
	90°	196.4	272.8	17.0%	25.2%	3.16	

2.2. SEM observation

Scanning electronic microscopy observations of the AZ31B sheets were performed using the QUANTA FEG250 machine. The samples were eroded in a mixed acid solution composed of picric acid, acetic acid, ethanol and water after grinding and polishing[22].

2.3. Cyclic loading-unloading uniaxial tension

The cyclic loading-unloading uniaxial tension was performed on the Zwick/Roell Z020 universal machine with an engineering strain rate of 0.0003 s-1. Dimensions of the specimens are shown in Fig.1. The specimen was first loaded to a certain strain level and then it was unloaded to zero. After that, the same specimen was reloaded to a higher strain level and this specimen was unloaded to zero again. This loading-unloading process was repeated. In order to obtain enough data for further elastic modulus analysis, fourteen strain levels which ranges from 0.01 to 0.24 at 0.02 strain interval were adopted.

2.4. U-bending

In this study, the classic U-bending benchmark in Numisheet2011[23] was adopted. Dimensions of the U-bending tools were shown in Figure 2. The sheet is a rectangle plate which is 320mm long and

30mm wide. The clearance between the punch and the die is 1.4mm. The blank holder force is 2450N. The angles and curvature of the deformed parts after springback were measured using the method reported in Numisheet2011.

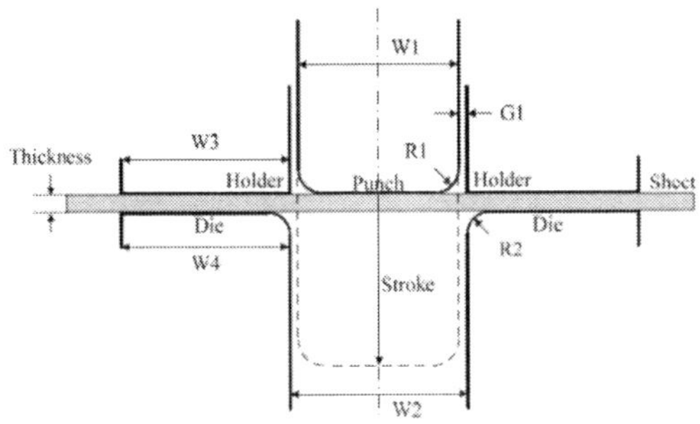

W1	W2	W3	W4
50.0	52.8	89.0	89.0
R1	R2	G1	Stroke
5.0	7.0	2.0	71.8

Figure 2. Dimensions of the U-bending tools. (Unit: mm)

3. Results and discussions

3.1. Inelastic strain recovery

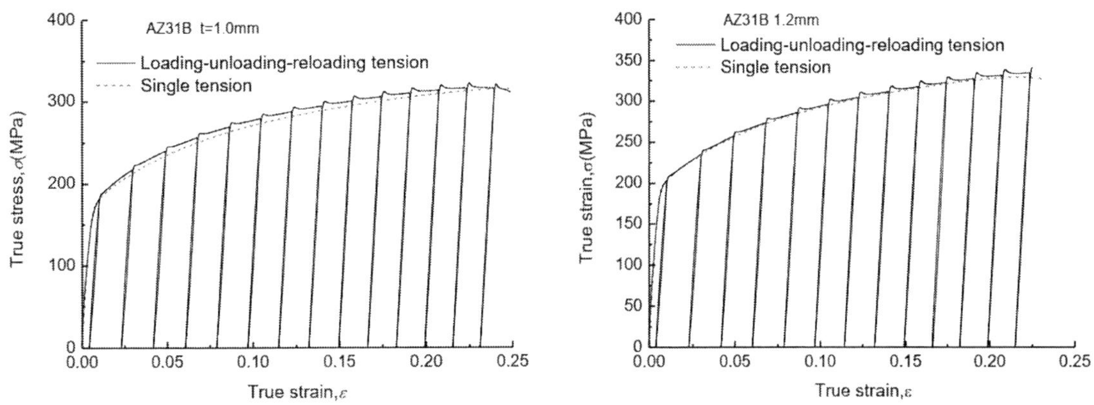

(a) t=1.0mm (b) t=1.2mm

Figure 3. Stress-strain curves of AZ31B sheets subjected to cyclic loading-unloading uniaxial tension

Figures 3(a) and 3(b) showed the true stress-strain curves of AZ31B sheets subjected to a single uniaxial tension and cyclic loading-unloading uniaxial tension along the rolling direction. From the two figures, one can find the solid curves are higher than the dashed curves. This is because of the effect of the strain hardening. From Figures 3(a) and 3(b), one can find the unloading curve doesn't coincide with the loading curve. This indicated that the slope of the unloading curve is different from that of the loading curve. In order to better understand the unloading behavior of the AZ31B sheets, the unloading stress-strain curves of the 1.0mm AZ31B at 0.02 strain were shown in Figure 4, in which the solid curve AHB is the true unloading curve and the curve BMC is the true reloading curve. The slope of the dashed line AD is equal to the initial elastic modulus. Therefore, the line segment DE represents the elastic strain recovery and the line segment BD means the inelastic component in the total strain recovery. The inelastic strain recovery has three characteristics. It is recoverable, like the elastic deformation. It dissipates work, like the plastic deformation. It is rate-independent, which is contrary to some models of anelasticity [7].

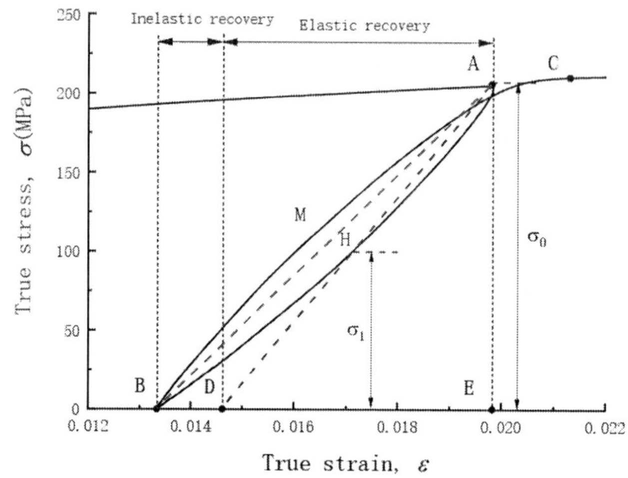

Figure 4. The unloading stress-strain curves of AZ31B at 0.02 plastic strain

3.2. Two elastic modulus models

3.2.1. The chord elastic modulus model. The slope of the line AB in Figure 4 was defined as the chord elastic modulus E_C and its value can be calculated according to the standard ASTM E 111-97. Figure 5 showed the value of E_C of two AZ31B sheets under different plastic strain. It showed that the chord elastic modulus drops dramatically first and then it increases with the plastic strain. This phenomenon is different from that of steels and aluminum alloys whose chord elastic modulus has been proved to monotonously decrease with the plastic strain [3-7].

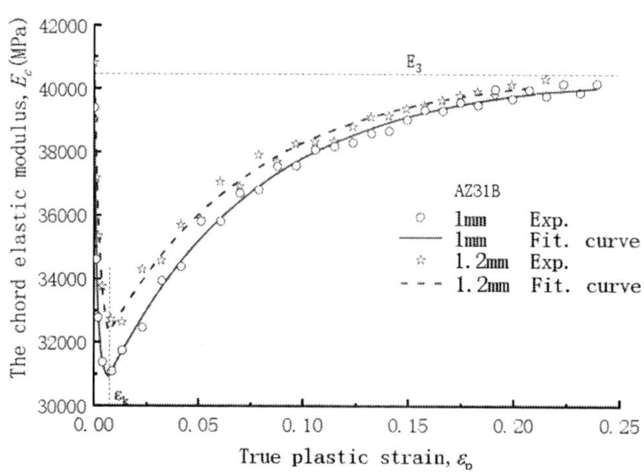

Figure 5. Variation of the chord elastic modulus with the plastic strain

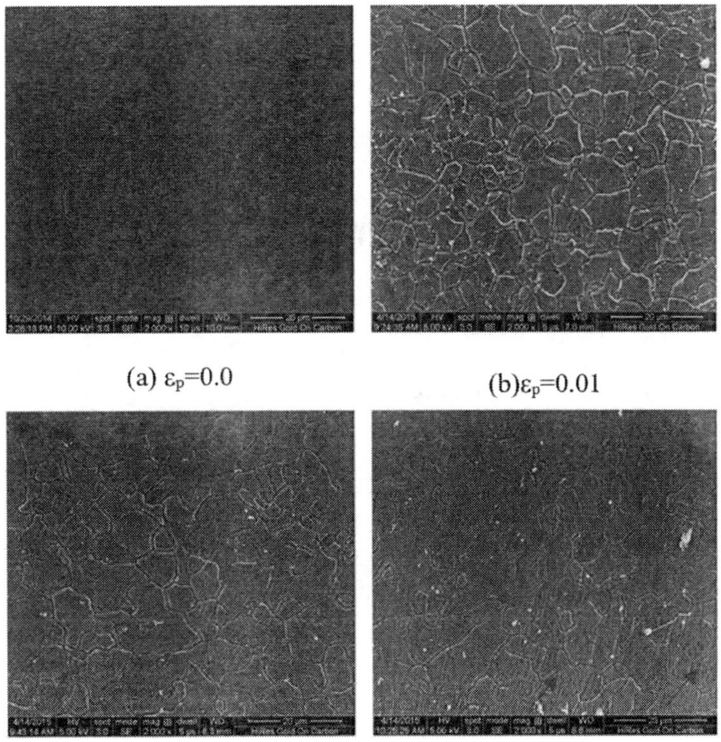

(a) ε_p=0.0 (b)ε_p=0.01

(c)ε_p=0.10 (d)ε_p=0.22

Figure 6. Microstructures evolution of AZ31B sheets subjected to different plastic strain

The reason of this phenomenon may lie in the special deformation mechanism of magnesium alloys. Magnesium (Mg) has a closed-packed hexagonal crystal structure and it has only two independent slip systems under room temperature. Disturbed by the plastic deformation, the two slip systems can't get along with each other, so the twin deformation is booted and becomes to be the dominant deformation mechanism. Furthermore, twin deformation can result in different grain orientation which will have a major influence on the formability [23,24]. In order to better understand this phenomenon, SEM observations of the AZ31B microstructures were performed. Figure 6 (a) ~ (d) showed the SEM image of the AZ31B sheets subjected to 0, 0.01, 0.10 and 0.22 plastic strain. In Figure 6 (a), the coarse equi-axial grains can be observed. The secondary $Mg_{17}Al_{12}$ particles locate among the grain boundaries, which can reduce the ductility of the material. In Figure 6 (b), some strip sub-crystal can be observed within the grains. This is just the twin crystal as shown by the arrows. The change of the deformation mechanism from slipping to twinning leads to instant stress relaxation among the grains, which may affect the atom binding force. Thus, elastic modulus decreases at this stage. Just at this moment, the twinning is booted by the plastic deformation. The twins were parallel to each other in the early stage of plastic deformation. In Figure 6(c), the amount of twins increases a lot when compared with that in Figure 6 (b). Moreover, Figure 6(d) showed a greater quantity of spindly twins in almost all grains when the strain level reached 22%. Some of them intersect, as showed in Figure 6 (d).

With the method in Yoshida's work [11], the chord elastic modulus can be expressed by Equation (1).

$$E_c = \begin{cases} E_0 - (E_0 - E_1)\left(1 - \exp(-\xi_1 \varepsilon_p)\right) & \varepsilon_p \leq \varepsilon_k \\ E_2 - (E_2 - E_3)\left(1 - \exp\left(-\xi_2(\varepsilon_p - \varepsilon_k)\right)\right) & \varepsilon_p > \varepsilon_k \end{cases} \quad (1)$$

Where, E_c is the chord elastic modulus; ε_p means the true plastic strain; E_0 represents the initial elastic modulus; E_2 and ε_k indicate the elastic modulus and the true plastic strain at the valley point P in Figure 5. E_3 is the approaching elastic modulus at the second stage. The meaning of ξ_1 and ξ_2 is same with the ones in Yoshida's equation [11]. Using the Origin software, the parameters in Equation (1) could be obtained using the exponential fitting. They were shown in Table 3.

Table 3. Parameters in Equation (1) and Equation (2)

Thickness	E_0(MPa)	E_1(MPa)	E_2(MPa)	E_3(MPa)	ε_k	ξ_1	ξ_2	A(MPa)	B
1.0mm	39391.6	30843.3	30921.4	40419.6	0.00699	671.79	13.89	10024.11	1.28
1.2mm	40803	32592.9	32306.1	40540.8	0.0074	372.12	13.97	9066.15	1.24

3.2.2. The instant elastic modulus model. The chord elastic modulus discussed above is defined as the slope of the line that connects the start point and the end point of the unloading curve. It can describe the elastic modulus of the material subjected to total unloading, i.e.,100% unloading, but it can't describe the real-time variation of the elastic modulus when the unloading is in the halfway, for example, only 80% stress is unloaded and there exists 20% residual stress. Under this case, the chord elastic modulus model can't work well because it is presented under the condition of total unloading. Another item was added to the chord modulus and an instant elastic modulus E_i was proposed. It can be expressed as,

$$E_i = \begin{cases} E_0 - (E_0 - E_1)\left(1 - \exp(-\xi_1 \varepsilon_p)\right) + A\left(^{\sigma_1}/_{\sigma_0}\right)/\left(B - {^{\sigma_1}/_{\sigma_0}}\right) & \varepsilon_p \leq \varepsilon_k \\ E_2 - (E_2 - E_3)\left(1 - \exp\left(-\xi_2(\varepsilon_p - \varepsilon_k)\right)\right) + A\left(^{\sigma_1}/_{\sigma_0}\right)/\left(B - {^{\sigma_1}/_{\sigma_0}}\right) & \varepsilon_p > \varepsilon_k \end{cases} \quad (2)$$

Where the parameters E_0, E_1, E_2 and E_3, ε_p, ε_k, ξ_1 and ξ_2 are same with the ones in Equation (1). In addition, σ_0 and σ_1 represent the equivalent stress of the start point and the end point of the unloading curve, respectively. Compared with Eq.(1), another item was added in Eq.(2). According to Fig.4, the stress σ_0 at the beginning of unloading and instant stress σ_1 during unloading were adopted to characterize the role of unloading stress. The supplemented item could characterize the role of stress level during unloading. The ratio of σ_1/σ_0 means the percentage of unloading and its value can vary from 0 to 1. If this ratio is equal to 0, then Equation (2) is same with Equation (1). Therefore, under the case of 100% unloading, Equation (2) is equal to Equation (1). If the value of σ_1/σ_0 is 1, then it hints no unloading. The last item in Equation (2) is equal to A (B-1) and it can be regarded as an extra constant which represents the elastic modulus enhancement caused by the strain hardening. Parameters A and B are material constants. They can be fitted using the data in Figure 5. For the investigated AZ31B sheets, all parameters in Equation (1) and Equation (2) were shown in Table 3.

Compared with Equation (1), Equation (2) can calculate the true elastic modulus when the stress is not unloaded totally. Partial unloading can lead to the existence of residual stress which is a common phenomenon in stamping because of the incompatible deformation caused by geometry. From this view, the instant elastic modulus model expressed by Equation (2) is more suitable for practical use in springback analysis.

4. Springback simulation

4.1. Elastic modulus subroutines

In order to get a better springback simulation, the proposed elastic modulus models Equation (1) and Equation (2) were incorporated into the ABAQUS software by combing use the field variables and user defined subroutines. The variable elastic modulus is defined as a field invariant. A user subroutine USDFLD for explicit forming simulation and a user subroutine VUSDFLD for implicit springback simulation were coded. The flow chart of the instant elastic modulus calculating is shown in Figure 7. Where, all stress and strain are defaulted as the equivalent stress and the equivalent strain. The difference between calculating E_c and calculating E_i lies in the assessment condition. When calculating E_c, the assessment of $\sigma<\sigma_0$ is changed to the evaluation of $\varepsilon<\varepsilon_p$.

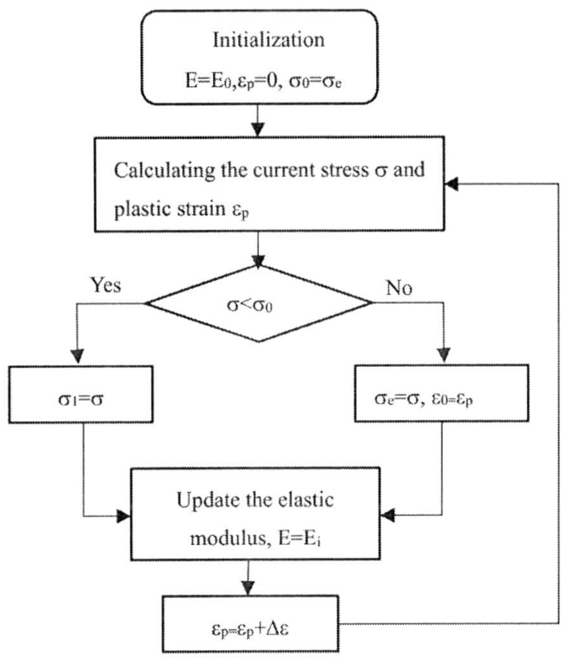

Figure 7. Flow chart of the user subroutine of the instant elastic modulus model

4.2. Hardening model

Lee and Kim [25] concluded that the Chaboche model can predict the hardening performance of AZ31B magnesium alloy very well. And the material parameters of the Chaboche model can be assumed constants since the AZ31B alloy has no fast-transient behavior. So the exponential back stress evolution can be analytically fitted. According to the above assumption, the isotropic hardening stress(X) and the kinematic hardening stress (α) can be written as [25],

$$X = 148 + 88.4(1 - e^{-13.3\varepsilon})\text{MPa} \tag{3}$$

$$\alpha = 72.2(1 - e^{-113.7\varepsilon})\text{MPa} \tag{4}$$

4.3. Validation of the subroutine

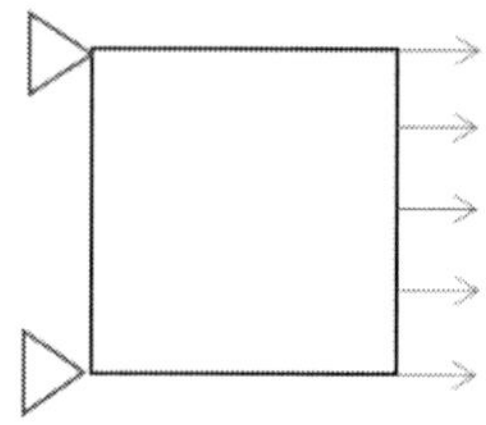

Figure 8. Finite element model adopted in validating the subroutine

In order to validate the subroutines presented in section 4.1, a one-element model subjected to cyclic loading-unloading uniaxial tension was simulated using the presented subroutines, as shown in Figure 8. The left end was fixed and the right end was loaded by enforced displacement. The element size was 1mm × 1mm and it was meshed with the S4R shell elements having 5 integration points along the thickness. Three types of elastic modulus were adopted, namely, a constant elastic modulus E_0, the chord elastic modulus, E_c, expressed by Equation (1) and the instant elastic modulus, E_i, expressed by Equation (2). The last two simulations should use the user-defined subroutines which have been discussed in section 4.1.

When the one-element model was loaded to 0.02 true strain, unloading was triggered and the stress-strain curve was obtained. The stress-strain curves simulated using the three elastic modulus models were shown in Figure 9. From which, one can find the loading curves simulated with the three elastic modulus models almost coincide with the experimental curve. This demonstrated that the presented subroutine can work well and predict the stress-strain curve correctly.

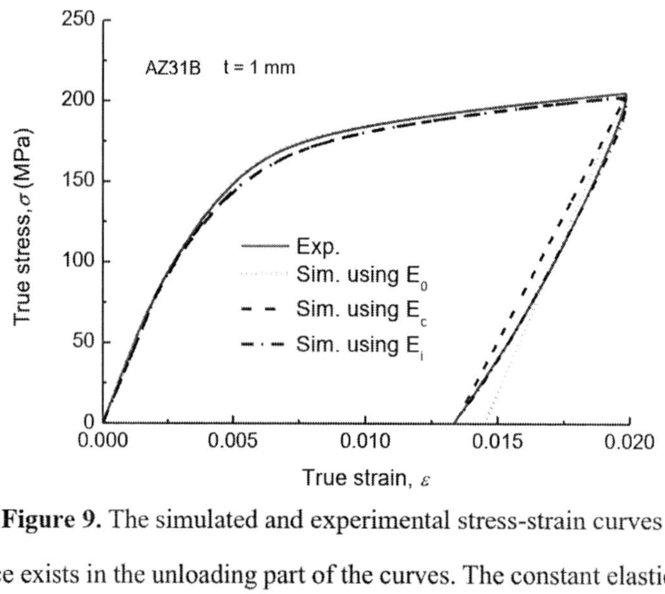

Figure 9. The simulated and experimental stress-strain curves

Major difference exists in the unloading part of the curves. The constant elastic modulus predicts a linear unloading stress-strain relation and underestimates the total strain recovery when compared with the experimental value. The presented chord elastic modulus and the instant elastic modulus model can predict the nonlinear unloading stress-strain curves correctly. The total strain recovery predicted by the two elastic modulus models is equal to the experimental data. The reason why these two models predict the same amount of strain recovery is that the simulation is performed under 100% unloading. Therefore, the chord elastic modulus and the instant elastic modulus have the same precision in the case of total unloading. However, the unloading path predicted by the instant elastic modulus model almost coincides with the experimental path, while the paths simulated with the chord elastic modulus and the constant elastic modulus departed remarkably from the experimental one. Because the precision of springback simulation is strongly dependent on the precision of the stress unloading, the instant elastic modulus is more accurate for springback simulation.

4.4. U-bending springback simulation

The subroutines were adopted to simulate the 2D-bending springback which has been used as a benchmark in Numisheet 2011. The 2D-bending finite element model was created according to the tool dimensions described in Figure 2. Seven integration points were adopted. The friction coefficient was assumed to be 0.125.

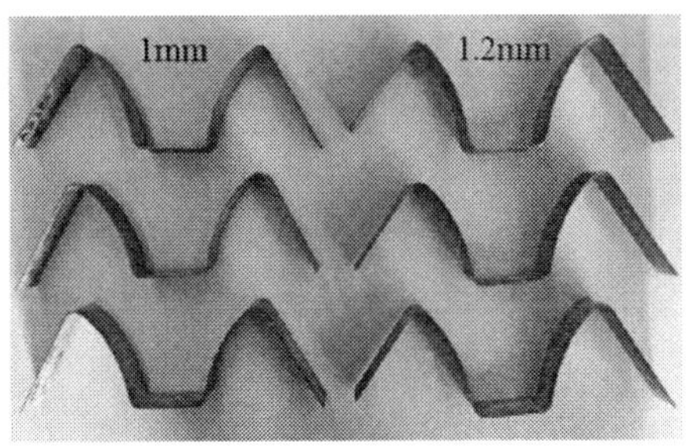

Figure 10. The experimental U-bending AZ31B alloy parts after springback

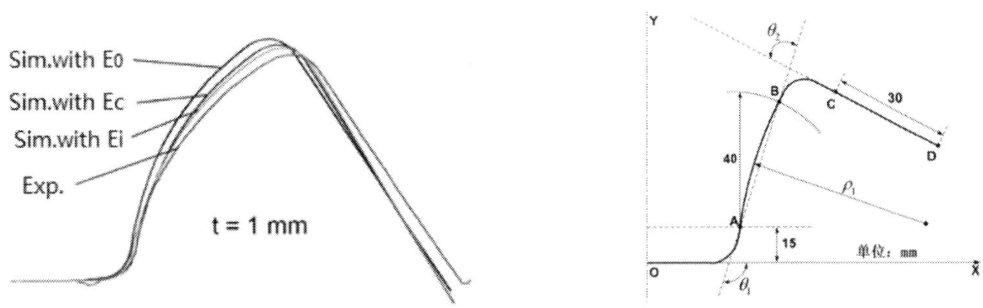

(a) Profiles of the U-shaped parts (b) Illustration of springback measurement

Figure 11. The simulated and experimental profiles of the U-shaped parts after springback

Figure 11(a) showed the experimental U-bending parts after springback. Their section profiles were obtained and compared with the simulated outline. From which, one can find that the simulation using the instant elastic modulus E_i has the highest precision while the simulation using the constant elastic modulus E_0 has the lowest precision when they were compared with the experimental profile. The simulations using E_c and the E_i almost predicted similar profiles. This demonstrated that the variable elastic modulus models, Equation (1) and Equation (2), can predict a more accurate springback than the constant elastic modulus can do. The discrepancy between experimental and simulated outlines was mainly due to the following reasons. First, the experimental outline is difficult to be measured. As shown in Fig.11(b), points A and B was identified on the basis of the bottom line. As a matter of fact, experimental U-bending parts often had a curved surface instead of a flat bottom. So point A is difficult to be distinguished. This was a main error source from the view of experiments. Second, suitable correct constitutive model and correct material parameters are the two main factors that decided the precision of springback simulation. Among the material parameters, elastic modulus, yield strength, strain hardening coefficient and anisotropic factor were important parameters that decided the precision of the stress strain level in the stamped part. Although these four parameters were measured by experiments in this work, there were still errors during the measurement.

The U-bending springback was assessed by the angles and curvature shown in Figure 11(b) [22]. Table 4 showed the simulated and the experimental results. In Table 4, $\Delta\theta_1$ and $\Delta\theta_2$ are equal to θ_1-90 and 90-θ_2, respectively. Both of them mean the true springback amount of the two angles. From Table 4, one can find that the simulated springback amount of θ_1, θ_2 and ρ using E_c and E_i are closer to the experimental data than that simulated using E_0.This demonstrated that the simulations with the chord elastic modulus and the instant elastic modulus models can predict a more accurate springback amount.

Table 4. Simulated and experimental springback amount

	θ_1/ °	θ_2/ °	ρ_1/mm	$\Delta\theta_1$/ °	$\Delta\theta_2$/ °
Experimental data	127.2	75.9	73.2	37.2	14.1
Simulated using constant elastic modulus E_0	121.4	65.1	62.9	31.4	24. 9
Simulated using the chord elastic modulus E_c	124.0	71.3	72.9	34.0	18.7
Simulated using the instant elastic modulus E_i	124.1	72.0	73.4	34.1	18.0

5. Conclusions

(1) The inelastic strain recovery of the investigated AZ31B sheets can reach 24% in the total strain recovery. Different from the published results, the elastic modulus of the AZ31B drops sharply first and then it increases with the increase of plastic strain instead of keeping decrease. This uncommon phenomenon may be related to the twin deformation mechanism of magnesium alloy. This explanation needed to be verified in the further investigation.

(2) A new elastic modulus model was presented to characterize the inelastic strain recovery of the AZ31B sheets. Unlike the existing models, this elastic modulus model takes both the instant stress and the plastic strain as the variables. This instant elastic modulus model can predict a perfect unloading path which almost overlaps with the experimental one.

(3) Springback simulations with the new elastic modulus model is proved to be more accurate than the simulation with a constant elastic modulus is.

Acknowledgments

The authors acknowledge the financial support from the Natural Science Foundation of Shanghai, China (19ZR1460100).

References

[1] Wang W R, Huang L, Tao K H, et al. Formability and numerical simulation of AZ31B magnesium alloy sheet in warm stamping process. *Materials & Design*, 2015, 87(15): 835-844.

[2] Abdullah M F, Abdullah S, Omar M Z, et al., Fatigue life of the magnesium alloy AZ31B under specific spectrum loading. *Materials Testing*, 2016, 58(3):200-205.

[3] Cleveland R M, Ghosh A K, Inelastic effects on springback in metals. *International Journal of Plasticity*, 2002, 18(6):769-785.

[4] Morestin F, Boivin M, On the necessity of taking into account the variation in the Young modulus with plastic strain in elastic–plastic software. *Nuclear Engineering and Design*, 1996, 162:107-116.

[5] Luo L, Ghosh A K, Elastic and inelastic recovery after plastic deformation of DQSK steel sheet. *Journal of Engineering Materials and Technology*. 2003, 125(3), 237-246.

[6] Perez R, Benito JA, Prado JM. Study of the inelastic response of TRIP steels after plastic deformation. *ISIJ International*, 2005, 45(12):1925–33.

[7] Sun L, Wagoner R H, Complex unloading behavior: Nature of the deformation and its consistent constitutive representation. *International Journal of Plasticity*, 2011, 27(7):1126-1144.

[8] Li X C, Yang Y Y, Wang Y Z, et al. Effect of the material hardening mode on the springback simulation accuracy of V-free bending. *Journal of Materials Processing Technology*, 2002, 123: 209-211.

[9] Zang S L, Guo C, Wei G-J, et al. A new model to describe effect of plastic deformation on elastic modulus of aluminum alloy. *Transactions of Nonferrous Metals Society of China*, 2006, 16s:1314-1318.

[10] Thibaud S, Boudeau N, Gelin J C, TRIP steel: Plastic behavior modelling and influence on functional behavior, *Journal of Materials Processing Technology*. 2006, 177(1), 433-438.

[11] Yoshida F, Uemori T, Fujiwara K. Elastic-plastic behavior of steel sheets under in-plane cyclic tension-compression at large strain. *International Journal of Plasticity 2002*, 18:633-659.

[12] Chongthairungruang B , Uthaisangsuk V , Suranuntchai S , et al. Springback prediction in sheet metal forming of high strength steels. *Materials & Design*, 2013, 50:253-266.

[13] Lee J, Lee J Y, Barlat F, et al., Extension of quasi-plastic-elastic approach to incorporate complex plastic flow behavior-application to springback of advanced high-strength steels. *International Journal of Plasticity*, 2013, 45, 140-159.

[14] Zhan M, Huang T, Zhang P P, Yang H, Variation of Young's modulus of high-strength TA18 tubes and its effects on forming quality of tubes by numerical control bending, *Materials & Design*, 2014, 53, 809-815.

[15] Chatti S, Modeling of the elastic modulus evolution in unloading-reloading stages, *International Journal of Material Forming*, 2013, 6(1), 93-101.

[16] Chatti S, Fathallah R, A study of the variations in elastic modulus and its effect on springback prediction, *International Journal of Material Forming*, 2014, 7(1), 19-29.

[17] Hama T, Kariyazaki Y, Ochi K, et al., Springback characteristics of magnesium alloy sheet AZ31B in draw-bending, *Materials Transaction*, 2010, 51(4), 685-693.

[18] Hama T, Takuda H, Crystal plasticity finite element analysis of inelastic behavior during unloading in a magnesium alloy sheet, *International Journal of Plasticity*. 2011, 27(7), 1072-1092.

[19] Hama T, Matsudai R, Kuchinomachi Y, et al., Non-linear deformation behavior during unloading in various metal sheets. *ISIJ International*, 2015, 55(5): 1067-1075.

[20] Wang H, Wu P D, Wang J, Modeling inelastic behavior of magnesium alloys during cyclic loading-unloading, *International Journal of Plasticity 2013*, 47, 49-64.

[21] Li S M, Wang M P, Zhang Z, et al., Compression mechanism of AZ31B Mg alloy at low

temperature. *Chinese Journal of Nonferrous Metals*, 2010, 20(7):1254-1259.

[22] Eggertsen P A, Mattiasson K, Myoung-Gyu Lee. A comprehensive analysis of benchmark 4: Pre-strain effect on springback of 2D draw bending. *AIP Conference Proceedings*, 2011, 1383:1064-1071.

[23] Hama T, Kitamura N,Takuda H, Effect of twinning and detwinning on inelastic behavior during unloading in a magnesium alloy sheet. *Materials Science and Engineering: A*, 2013, 583:232-241.

[24] Lee S Y, Wang H M, Gharghouri M A, Twinning-detwinning behavior during cyclic deformation of magnesium alloy. *Metals*, 2015, 5(2):881-890.

[25] Lee M G, Kim S J, Wagoner R H, et al., Constitutive modeling for anisotropic/asymmetric hardening behavior of magnesium alloy sheets: Application to sheet springback. *International Journal of Plasticity*, 2009, 25:70-104.

IOP Conf. Series: Materials Science and Engineering **967** (2020) 012047 doi:10.1088/1757-899X/967/1/012047

Investigation on ductile fracture of an aluminium alloy using a mean-field crystal plasticity framework

J Lee[1], H Park[2], M Joo[1] and Y Jeong[1,*]

[1]20 Changwondaehak-ro, Changwon, Gyeongnam 51140, Republic of Korea
School of Materials Science and Engineering, Changwon National University
[2]797 Changwon-daero, Changwon, Gyeongnam 51508, Republic of Korea
Materials Deformation Department, Korea Institute of Materials Science
*Corresponding author email: yjeong@changwon.ac.kr

Abstract. The onset of ductile fracture can be described by a simple damage parameter, which accounts for the accumulation of damage under arbitrary deformation. The damage parameter is usually given as integral form of strain over loading history which also has been described as a function of stress state. While the evolution of strain field can be experimentally determined using digital image correlation, experimental measurement of the multiaxial stress state is still a challenging task. Therefore, various uncoupled fracture criteria rely on stress states calculated from phenomenological plasticity model. As a result, the number of mechanical tests required to calibrate the fracture criterion may significantly increase when an anisotropic constitutive model is used. We propose an alternative approach on the basis of a mean field crystal plasticity (VPSC) model, which accounts for the microstructural features such as slip system, crystallographic texture and its evolution. While stress fields can be obtained from the use of full-field crystal plasticity framework, the proposed method utilizes the mean field crystal plasticity framework and repeat the stress estimation on various spatially resolved locations to which DIC technique provides strain history. The repeated VPSC calculations at various locations efficiently provide the map of stress evolution. The resulting map of spatially resolved stress response is further validated by comparing with the bulge stress strain curves.

1. Introduction

Uncoupled fracture criteria introduce a damage indicator which indicates the onset of fracture by a scalar function of stress and strain history [1]. Doing so, it requires the constitutive behaviour of material that is calculated by phenomenological plasticity model [2]. The material parameters associated with strain evolution can be easily obtained by the digital image correlation (DIC). However, it is still difficult to accurately measure multiaxial flow stresses. Therefore, the stress and other stress-associated quantities such as triaxiality and Lode angle usually rely on the calculation by a chosen phenomenological plasticity model. In conventional phenomenological plasticity theory, the material anisotropy is regarded constant during the plastic deformation, and this assumption may not be appropriate because a large amount of plastic strain occurs for ductile materials until fracture. In this study, a preliminary study is conducted to investigate the possibility to use a mean-field self-consistent crystal plasticity model instead of the phenomenological plasticity model to estimate stress states. A simple and direct method to utilize the field strain data from DIC for a mean-field crystal plasticity is introduced.

2. Method

2.1. Models

The mean-field crystal plasticity framework of Visco-Plastic Self-Consistent (VPSC) model [3] was utilized in this study. In it, a power-law relationship between the local strain rate $\dot{\boldsymbol{\varepsilon}}$ and local stress $\boldsymbol{\sigma}$ is assumed:

$$\dot{\varepsilon} = \dot{\gamma}_0 \sum_s \boldsymbol{m}^s \left(\frac{\boldsymbol{m}^s : \boldsymbol{\sigma}}{\tau_c^s}\right)^n \qquad \text{Eq. 1}$$

where $\dot{\gamma}_0$, $\boldsymbol{m}^s$ and τ_c^s denote the reference shear strain rate, Schmid tensor and the critical resolved shear stress (CRSS), respectively. The exponent n is related to strain rate sensitivity and fixed as 20 in this study. The Schmid tensor $\boldsymbol{m}^s$ were constructed by assuming that {111} <110> slip systems are available for a chosen aluminium alloy AA6061-T6 sample. The CRSS τ_c^s may evolve using a Voce-hardening law:

$$\tau_c^s = \tau_0^s + (\tau_1^s + \theta_1^s \Gamma)\left(1 - \exp\left(-\Gamma \left|\frac{\theta_0^s}{\tau_1^s}\right|\right)\right) \qquad \text{Eq. 2}$$

in which the symbol Γ denotes the accumulated shear strain; The symbols $\tau_0^s, \tau_1^s, \theta_0^s$ and θ_1^s are determine the strain hardening behaviour. The equivalent inclusion method [4,5] for visco-plastic inclusion embedded in the visco-plastic medium [3,6] leads to the visco-plastic macroscopic constitutive description:

$$\bar{\varepsilon}_{ij}^{vp} = \bar{\mathbb{M}}_{ijkl}^{vp} \bar{\sigma}_{kl}' + \bar{\varepsilon}_{ij}^{0,vp} \qquad \text{Eq. 3}$$

where $\bar{\boldsymbol{\sigma}}'$ and $\bar{\boldsymbol{\varepsilon}}^{vp}$ are the macroscopic stress and strain tensor, respectively; and $\bar{\mathbb{M}}^{vp}$ and $\bar{\boldsymbol{\varepsilon}}^{0,vp}$ are the macroscopic visco-plastic compliance and back-extrapolated term, respectively. The macroscopic quantities (i.e., $\bar{\mathbb{M}}^{vp}$ and $\bar{\boldsymbol{\varepsilon}}^{0,vp}$) are functions of micromechanical properties, which are priori unknowns and are iteratively determined.

2.2. Experiments

Experimental data obtained from X-ray diffraction, bulge tests, uniaxial tension tests, and Nakajima tests were conducted. X-ray diffraction was performed using a Bruker D8 Discovery machine with a Co Kα target. The crystallographic orientation distribution was obtained from experimental pole figures using MTEX, from which a population of 6000 discrete orientations was extracted. The reconstructed pole figures from the orientation population are shown in Figure 1.

Bulge test was performed using an Erichsen bulge tester model 161 on specimens with a dimension of 300×300 mm^2. Hydraulic pressure and dome height together with the images obtained from 3D stereo camera system were used to obtain the flow stress-strain curve. Uniaxial tension tests along rolling, diagonal, and transverse directions were performed according to the ASTM E8 standard. Strain rate was about $4.3 \times 10^{-4}\ s^{-1}$.

A Nakajima test was conducted following the ISO standard 12004-2. The diameter of punch was 100 mm, and the stroke speed was fixed as 0.5 mm/s. Square specimens were tested to induce a near balanced biaxial state. DIC analysis was applied to obtain strain fields that develop on the top surface of specimens subjected to the Nakajima test. The evolution of deformation gradient tensor $\boldsymbol{F}$ with respect to test time was collected at a set of spatial coordinates of interest near fracture site.

2.3. VPSC simulation condition

The polycrystal aggregate used in VPSC calculation was characterized by the discrete orientations and the Voce hardening parameters calibrated by fitting with the flow stress-strain curve resulting from the bulge test. The tangent linearization method was assumed in the current study. The goodness of the VPSC fit is well illustrated in Figure 2. The calibrated parameters are given in Table 1.

The VPSC simulation was conducted to estimate the stress fields corresponding to a set of spatial locations at which the digital images were collected during the Nakajima tests. As the deformation obtained from DIC includes both strain and spin, the deformation gradient tensor is required to fully reflect the behaviour of each spatial point. However, as the DIC was applied to the top surface of specimen, only F_{11}, F_{12}, F_{21} and F_{22} components can be experimentally obtained. The other components associated with the thickness direction (namely $F_{13}, F_{23}, F_{31}, F_{23}$ and F_{33}) cannot be experimentally determined. To that end, the assumption of incompressibility and zero thickness-related shear components were applied. The determinant of $\boldsymbol{F}$ is related to the volume change:

$$\frac{dv}{dV} = \det(\boldsymbol{F}) \qquad \text{Eq. 4}$$

where dv and dV denote the volume elements before and after the deformation, respectively. The incompressibility condition (i.e., $dv = dV$) leads to $\det(\boldsymbol{F}) = 1$. Additionally, by assuming $F_{13} = F_{31} = F_{23} = F_{32} = 0$, the determinant is expressed as:

$$\det(\boldsymbol{F}) = F_{11}F_{22}F_{33} - F_{12}F_{21}F_{33} = F_{33}(F_{11}F_{22} - F_{12}F_{21}). \qquad \text{Eq. 5}$$

As a result of this assumption, the thickness component of deformation gradient can be estimated as $F_{33} = 1/(F_{11}F_{22} - F_{12}F_{21})$.
The deformation gradient tensor may contain a significant amount of experimental noise. Savitzky-Golay filter [7] as implemented in SciPy package [8] was applied in order to reduce the noise in the experimental data. A series of discrete F_{ij} values corresponding to a batch of sequential images are fitted to a polynomial function of a chosen order. The smoothed $\boldsymbol{F}$ tensor is used to determine the boundary condition, to which the VPSC polycrystal aggregate is subjected. In a recent study, it was shown that although the degree of smoothing is affected by the choice of the order of polynominal and the size of batch used in the Savitzky-Golay filter, the VPSC-calculated stress is not significantly affected [9]. The smoothed deformation gradient is multiplicatively decomposed to stretch $\boldsymbol{U}$ and rotation $\boldsymbol{R}$ tensors such that:

$$\boldsymbol{F} = \boldsymbol{R} \cdot \boldsymbol{U} \qquad \text{Eq. 6}$$

From the stretch tensor, the engineering strain $\boldsymbol{\epsilon}$ is obtained as:

$$\epsilon_{ij} = U_{ij} - \delta_{ij}\,, \qquad \text{Eq. 7}$$

where δ_{ij} is the Kronecker delta. The logarithmic strain $\boldsymbol{\varepsilon}$ is:

$$\varepsilon_{ij} = \delta_{ij}\ln(1 + \epsilon_{ij}) + (1 - \delta_{ij})\epsilon_{ij}\,.$$

The rate of rotation $\dot{\boldsymbol{R}}$ can be estimated using the Euler method:

$$\dot{R}_{ij}^{(n+1)} = \frac{R_{ij}^{(n+1)} - R_{ij}^{(n)}}{t^{(n+1)} - t^{(n)}} \qquad \text{Eq. 8}$$

in which the superscripts (n + 1) and (n) denote the time steps at which the relevant digital images were acquired. The same was applied to obtain the logarithmic strain rate $\dot{\boldsymbol{\varepsilon}}$. Both rates of rotation and strain were directly employed to VPSC at the various spatial locations selected on the top surface of specimen deformed by the Nakajima test.

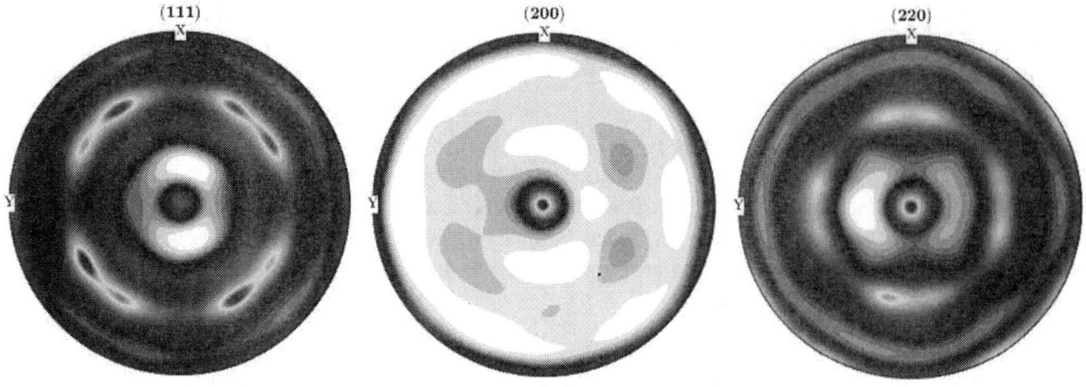

Figure 1 (111), (200), and (220) pole figures reconstructed from the population of 6000 discrete orientations

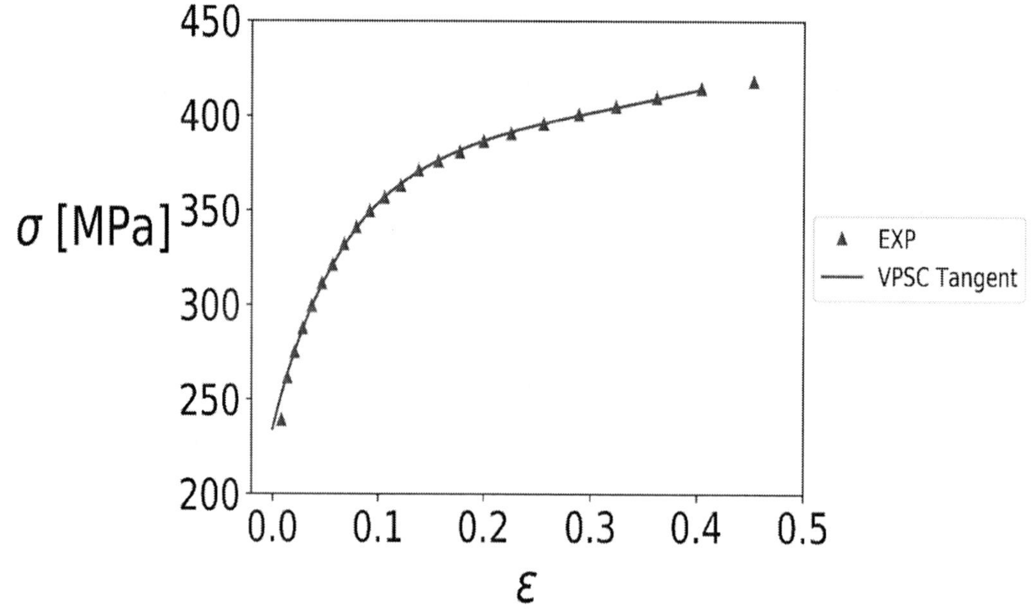

Figure 2 Experimental flow stress-strain curves from bulge test and the VPSC model fit

τ_0	τ_1	θ_0	θ_1
92.72	58.85	391.6	11.26

Table 1 Voce hardening parameters

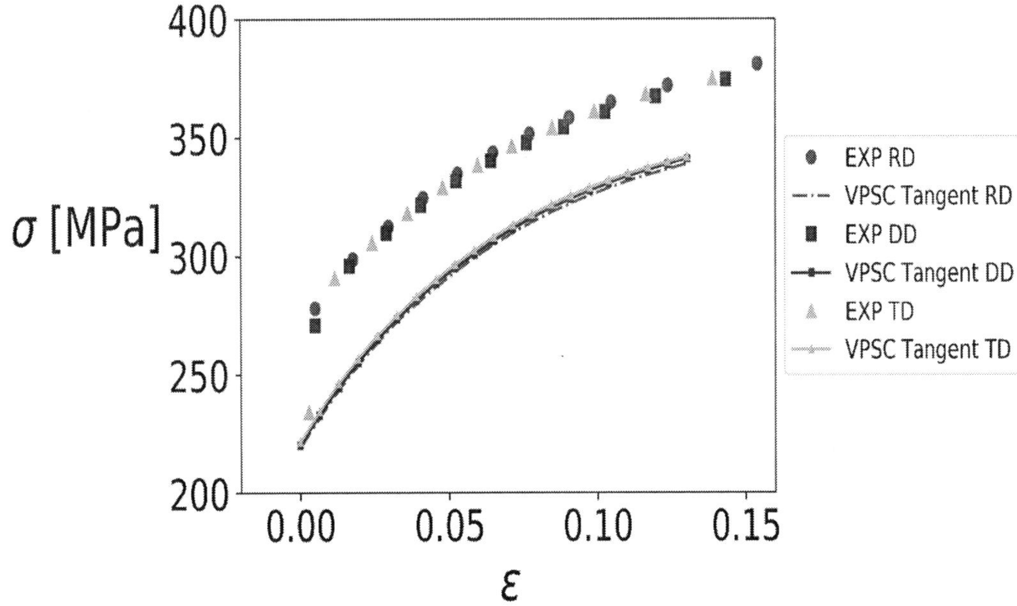

Figure 3 Experimental true stress-strain curves and model predictions

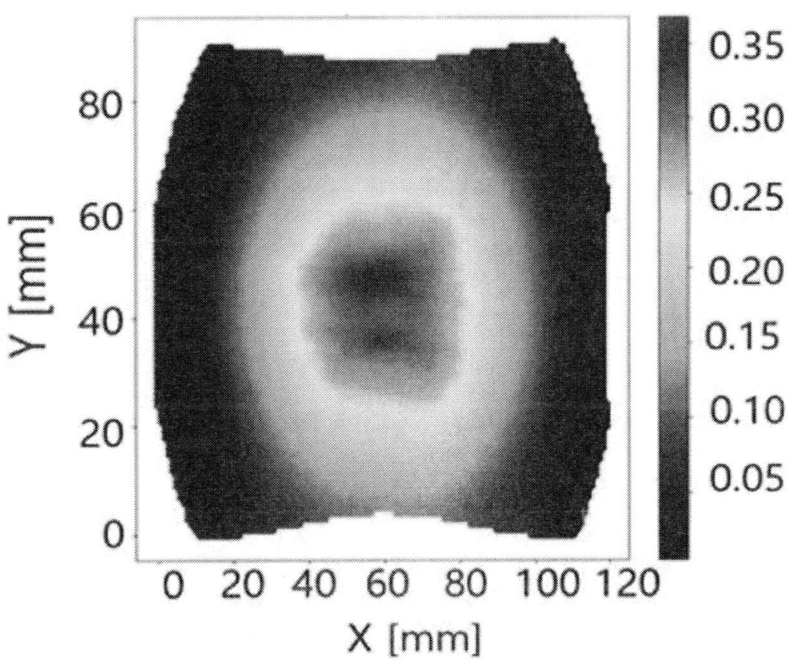

Figure 4 Strain map obtained from the DIC at the final step prior to the onset of fracture

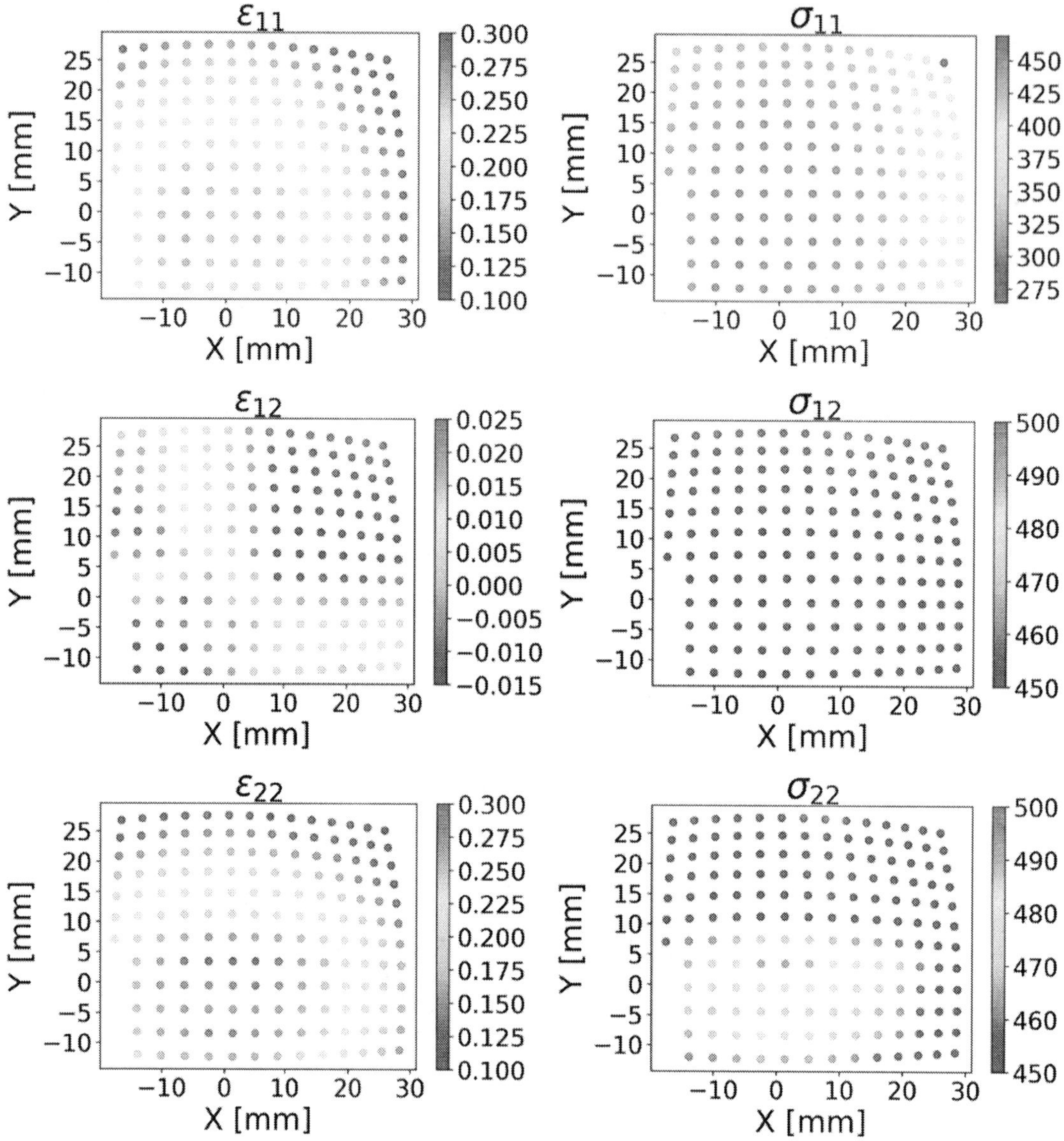

Figure 5 Nominal components of strain and stress corresponding to the digital image just prior to fracture.

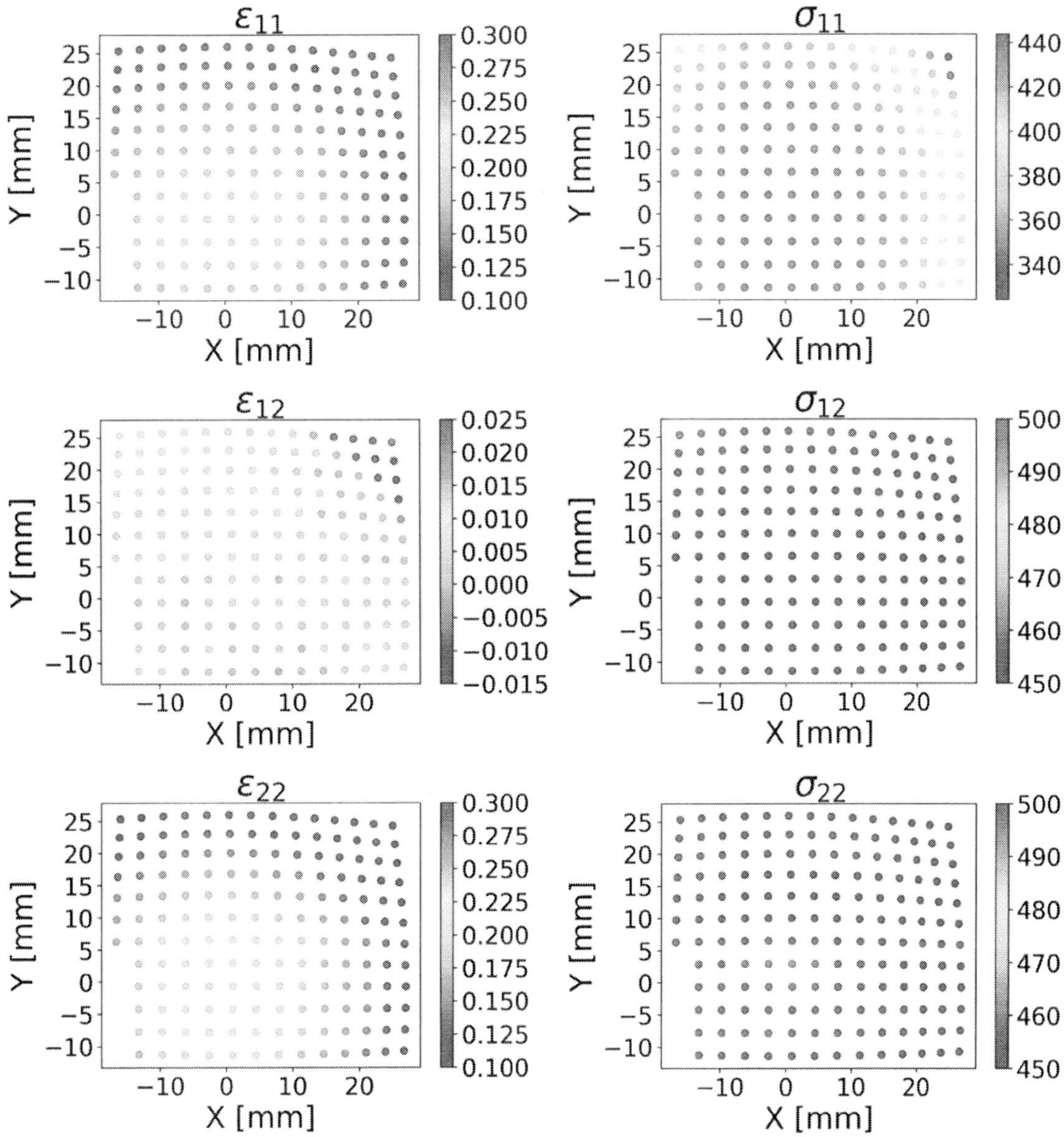

Figure 6 Nominal components of strain and stress corresponding to the equivalent strain of 0.4

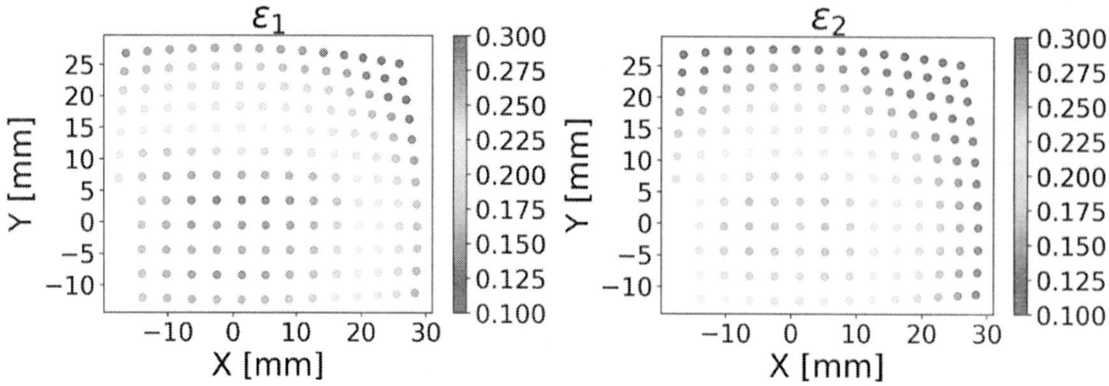

Figure 7 Major and minor principal strain obtained by DIC on spatially resolved points, from which VPSC inputs were obtained

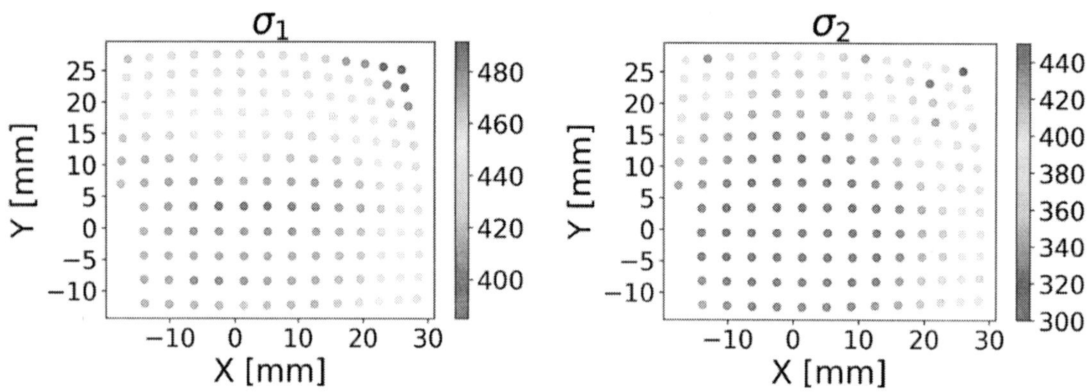

Figure 8 Major and minor principal stresses calculated by VPSC

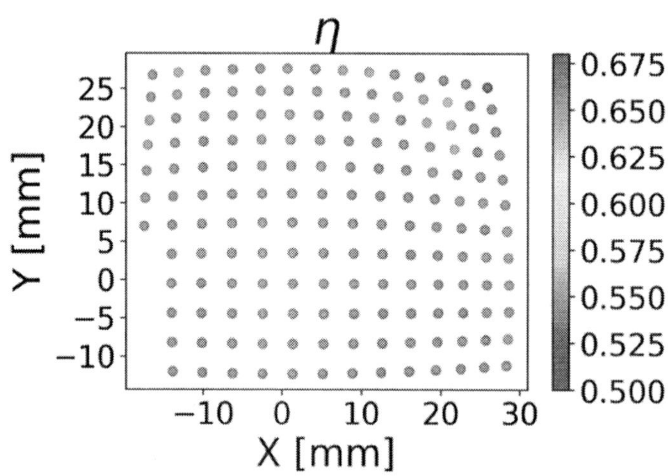

Figure 9 Triaxiality calculated by VPSC

3. Results and discussion

The VPSC model with the parameters calibrated by the bulge test systematically underestimated the uniaxial tension flow stress responses as shown in Figure 3. Nevertheless, the isotropic response regardless of direction was captured by the VPSC model. Figure 4 shows the entire strain map resulting from the DIC analysis using the images obtained during the Nakajima test. The VPSC simulation was conducted for a more confined region near the fracture. To that end, a grid consisting of 163 locations that are equally distanced within a 46 mm × 40 mm area was selected near the fracture site. The strain components obtained from DIC analysis corresponding to this selected area are shown in terms of $\varepsilon_{11}, \varepsilon_{12}$, and ε_{22} in the left column of Figure 5. It should be noted that the coordinate origin ($X = 0, Y = 0$) in Figure 5 corresponds to the coordinate ($X = 64, Y = 52$) in Figure 4. The same applies to X, Y coordinates in Figures 5-9 as well. The stress resulting from VPSC calculation for the chosen grid is shown on the right column of Figure 5-6. The major and minor components of principal strain and stress are shown in Figures 7 and 8, respectively.

Figure 5 shows that the normal stress components σ_{11} and σ_{22} near the fracture amounts to 400~500 MPa. The corresponding normal strain components ε_{11} and ε_{22} shown in Figure 5 are around 0.25~0.30, which correspond to an equivalent strain of 0.5~0.6 for the case of bulge test. The maximum strain available obtained from the bulge test was limited to 0.4 as shown in Figure 2, so that a direct comparison between the bulge and the VPSC-calculated result at fracture is not plausible. Nevertheless, Figure 6 indicates that the VPSC-calculated stress at the equivalent strain of 0.4 amounts to 420~460 MPa, which is marginally different from the corresponding stress from bulge test (i.e., 410 MPa). This suggests that the stress estimated by VPSC at fracture should not be significantly different from the actual stress state. Figures 7-8 show the map of principal strains and the corresponding VPSC-calculated principal stresses. Although the VPSC-predicted principal stresses are a bit noisy, the region of the maximum stress well corresponds to the one with maximum principal strain shown in Figure 7. Figure 9 shows that the stress triaxiality predicted by VPSC amounts to around 0.66, which is reasonable considering that the circular specimen was used in the Nakajima test to induce balanced biaxial stress state.

The stress states and the triaxiality calculated by VPSC model through the direct employment of DIC raw data after smoothing by Golay-Savitzky method seem quite reasonable. Therefore, it is expected that an alternative uncoupled ductile fracture criterion can be suggested by repeatedly conducting the mean-field VPSC calculations for various locations near fracture. An advantage of using the mean-field VPSC model over the use of full-field crystal plasticity is the efficiency of calculation: the total running time of VPSC was about 4 minutes for the entire spatial locations shown in Figure 5-9 using a work station with 48 CPU core units.

4. Conclusions

In this study, a mean-field self-consistent crystal plasticity model has been applied to investigate its applicability for ductile fracture criterion, which requires the evolution of stress states corresponding to arbitrary strain history. A Visco-Plastic Self-Consistent polycrystalline aggregate was calibrated for an aluminium AA6061-T6 sample by bulge test and X-ray diffraction, which shows a reasonable predictive accuracy on uniaxial tension flow stress-strain curves. The stresses calculated by VPSC at various locations near fracture seem reasonable in comparison with the bulge test result. Also, the triaxiality map estimated by the VPSC-predicted stress seems reasonable for a near balanced biaxial state realized through a Nakajima test. Therefore, it is expected that the method introduced in this study can provide an efficient alternative plasticity model for an uncoupled ductile fracture criterion, in which both strain and stress states are required. As a follow-up study, further investigation to develop a fracture criterion applicable for a wider range of stress and strain states is required.

5. References

[1] F.D. Fischer, O. Kolednik, G.X. Shan, F.G. Rammerstorfer, A note on calibration of ductile failure damage indicators, Int. J. Fract. 73 (1995) 345–357. https://doi.org/10.1007/BF00027274.

[2] Y. Bai, T. Wierzbicki, Application of extended Mohr–Coulomb criterion to ductile fracture,

Int. J. Fract. 161 (2010) 1–20. https://doi.org/10.1007/s10704-009-9422-8.
[3] R.A. Lebensohn, C.N. Tomé, A self-consistent anisotropic approach for the simulation of plastic deformation and texture development of polycrystals: Application to zirconium alloys, Acta Metall. Mater. 41 (1993) 2611–2624. https://doi.org/10.1016/0956-7151(93)90130-K.
[4] J.D. Eshelby, The determination of the elastic field of an ellipsoidal inclusion, and related problems, Proc. R. Soc. London. Ser. A. Math. Phys. Sci. 241 (1957) 376–396.
[5] T. Mura, Micromechanics of defects in solids, Springer Science & Business Media, 2013.
[6] A. Molinari, G.R. Canova, S. Ahzi, A self consistent approach of the large deformation polycrystal viscoplasticity, Acta Metall. 35 (1987) 2983–2994. https://doi.org/10.1016/0001-6160(87)90297-5.
[7] A. Savitzky, M.J.E. Golay, Smoothing and Differentiation of Data by Simplified Least Squares Procedures., Anal. Chem. 36 (1964) 1627–1639. https://doi.org/10.1021/ac60214a047.
[8] E. Jones, T. Oliphant, P. Peterson, SciPy: Open source scientific tools for Python, (2001).
[9] Y. Jeong, D. Steglich, Modelling-assisted description of anisotropic edge failure in magnesium sheet alloy under mixed-mode loading (submitted).

6. Acknowledgments

This research was supported by Changwon National University in 2019~2020

IOP Conf. Series: Materials Science and Engineering **967** (2020) 012048 doi:10.1088/1757-899X/967/1/012048

Springback and end flare compensation in flexible roll forming

B Abeyrathna*[1], S Ghanei[1], B Rolfe[2], R Taube[3] and M Weiss[1]

[1] Institute for Frontier Materials, Deakin University, Waurn Ponds, Pigdons Rd. VIC. 3216, Australia
[2] School of Engineering, Deakin University, Waurn Ponds, Pigdons Rd. VIC. 3216, Australia
[3] Ford Motor Company of Australia Ltd, 1735 Sydney Rd, Campbellfield VIC. 3061

*Corresponding author e-mail: buddhika.abeyrathna@deakin.edu.au

Abstract. Flexible roll forming enables the forming of complex automotive components from Advanced High Strength Steels (AHSS) and Ultra High Strength Steel (UHSS) but common shape defects such as end flare and springback occur and need to be compensated. In flexible roll forming a pre-cut blank is fed through a set of rolls which are computer numerically controlled (CNC) to follow the 3D contour of the part. The company, dataM Sheet Metal Solutions, have established a new 3D flexible roll forming facility at Deakin University which allows the manufacture of flexible roll formed prototypes.
In this study, Deakin's flexible roll forming facility is employed to implement an approach to overcome springback and end flare defects in a flexible roll formed automotive component of variable width. To do this, the flexible roll forming process is simulated using the commercial software package Copra FEA. Non-uniform springback due to end flare is observed and a variable overbending approach developed and implemented to overcome this defect. Experimental validation is performed on Deakin`s flexible roll forming facility and the results suggest that the flexible overbending approach can be used to control final shape in flexibly roll formed components.

1. Introduction

Advanced High Strength Steels (AHSS) and Ultra High Strength Steel (UHSS) are increasingly used in the automotive industry. Their higher strength enables the manufacture of lighter vehicles using lower thickness sheet while maintaining the same part performance and strength. Due to their higher strength and limited formability, the forming of AHSS and UHSS with conventional cold stamping methods is difficult [1].

Flexible Roll Forming (FRF) has been developed as an alternative and flexible method to form AHSS and UHSS. In conventional roll forming, a long strip of sheet is formed into a long component with a simple cross section by a set of fixed consecutive profiled rollers. In flexible roll forming, those forming rolls can move to follow the contour of the component to form a more complex shape where the cross-section changes over the length of the part. The forming rolls are CNC controlled and this allows the forming of multiple component shape variations on one roll forming line with minimum change in tooling required. This promises higher process flexibility and reduces tooling costs.

Abee et al. [2] have demonstrated the FRF of several prototype profiles with variable cross-sections while in [3] various automotive components such as crash barriers, frame and panel members, window components, seat adjusting parts and bumpers were suggested as potential candidates for the technology. A bus frame was successfully formed using FRF in [4].

Recently, Deakin University has developed a prototyping FRF facility that enables the forming of variable depth and width profiles from AHSS and UHSS. This prototyping facility simulates the forming conditions present in FRF and can be used to develop new forming approaches and to manufacture prototype components to prove process feasibility [5].

Web warping which is the deviation of height in the web area of the profile as well as wrinkling in the flange, are the two most critical problems in the FRF field. Both increase with material strength [6]. There are numerous studies regarding web warping compensation which include the heating of critical flange regions [7], the use of a blank holder [2, 8] and over-bending [9]. Wrinkling compensation using special rolls [6, 10] and a blank holder has also been studied [5].

In this study Deakin's flexible roll forming facility has been used to manufacture a simplified automotive component of variable width. While web-warping and wrinkling issues were overcome, significant shape issues in the form of end flare and springback were observed. A new overbending approach is introduced and successfully applied to compensate for springback and end flare.

2. Material and part geometry

A 1mm CR2 cold-rolled mild steel (per Ford Material Specification WSS-M1A365-A12) was used in this study. Standard tensile tests were performed in an Instron 5967 machine with a 30kN load cell according to ASTM E8/E8M [11] with a cross head speed of $0.025 mms^{-1}$. The resulting averaged true stress strain curve along the rolling direction is shown in Figure 1(a). The original automotive component considered in this work and the simplified version that was flexibly roll formed are shown in Figure 1(b) and (c) respectively. The dimensions of the front cross section and the front half of the unfolded pre-cut blank are shown in Figure 2(a) and (b) respectively.

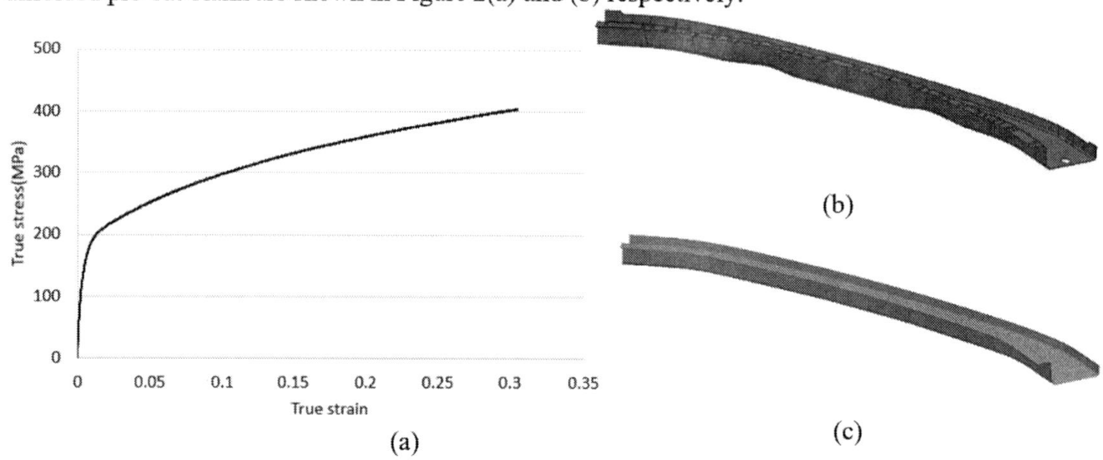

Figure 1: (a)Averaged true stress strain curve along the rolling direction for Ford WSS-M1A365-A12 (a) geometry of the original part to be formed (c) geometry of the simplified part for flexible roll forming

IOP Conf. Series: Materials Science and Engineering **967** (2020) 012048 doi:10.1088/1757-899X/967/1/012048

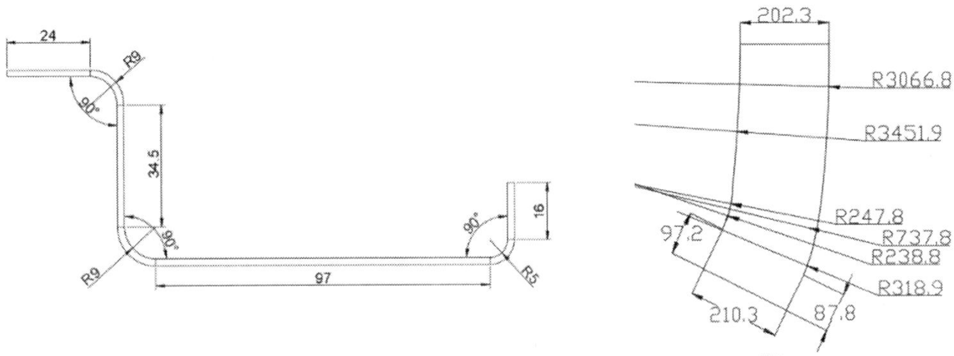

(a) (b)

Figure 2: Dimensions of the (a) front cross section (b) front half of the pre-cut blank

3. Flexible roll forming experiments

The flexible roll forming prototyping facility developed by dataM Sheet Metal Solutions was installed at Deakin University in 2017 [12]. The facility can be used for "proof of concept" studies in both conventional and flexible roll forming. A schematic of the facility is shown in Figure 3(a). The pre-cut blank is held between the top and bottom clamps. The clamps and the blank are then driven back and forth by a servo motor driven leadscrew. The two forming rolls are located on either side of the carriage on two robotic aims. A forming roll has 6 degrees of freedom which allow it to change the orientation as directed by the control programme and to form the part incrementally in several passes. The control programme uses a contour to define the tool orientation and movement. This "tool path" is defined by the shape of the component in the longitudinal direction (Figure 3(a)).

The forming sequence and the corresponding symmetric halves of the tooling are shown in Figure 3(b) and Figure 4 respectively. According to the forming sequence, the left flange is formed up to 43° in the first 4 passes with the tools shown in Figure 4(a) and then the top hat is formed completely in 5 passes. To accomplish this, the tools shown in Figure 4(b) and (c) were used. After that the left flange with the top hat is fully formed up to 90° in the last 4 passes using the tooling shown in Figure 4(d). The right flange is gradually formed in the first four passes and the last four passes with the same angle increments as the left flange. For the single bottom roll shown in Figure 4(a) was used.

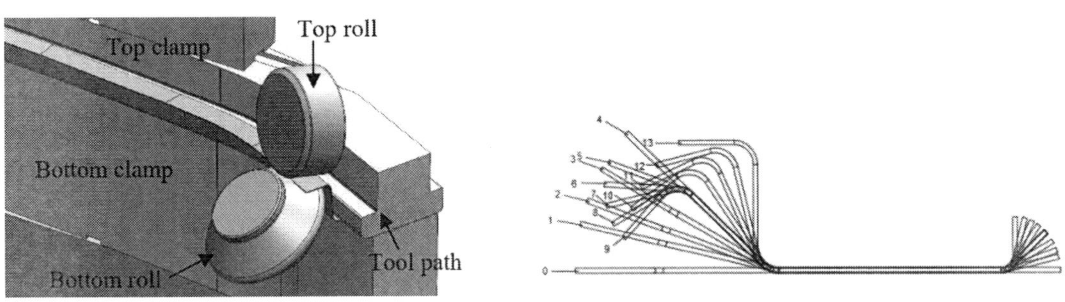

(a) (b)

Figure 3: (a) Schematic of the flexible roll forming prototyping equipment at Deakin University (b) forming sequence

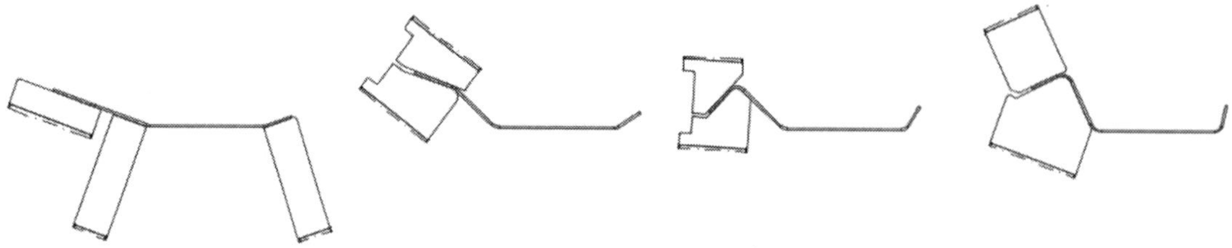

(a) (b) (c) (d)

Figure 4: Tooling used for forming passes (a) 1-4 (b) 5-7 (c) 8-10 (d) 11-13

After forming, the outer surface of the flexibly roll formed parts was scanned with a HandySCAN 3D [13] and in three different locations (Figure 5(a)) the critical angles measured (Figure 5(b)).

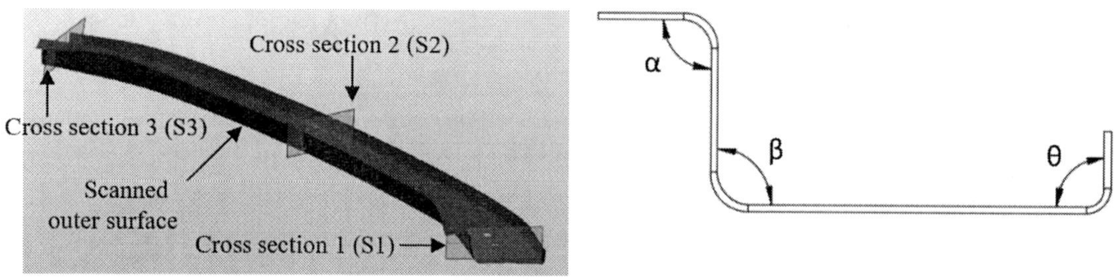

(a) (b)

Figure 5: (a) Different cross sections considered for the evaluation of (b) the three critical forming angles

4. Finite element analysis (FEA)

The finite element simulation of the flexible roll forming process was performed with Copra RF/FEA [14] which uses an implicit solver. In the actual process, the rolls are stationary and the carriage moves back and forth. However in the FEA, the pre-cut sheet was kept stationary and the rolls move along the specified path to incrementally form the flanges (see Figure 6(a)). The assigned tool paths for the left and right tools are shown in Figure 6(b). The rolls were modelled as rigid bodies and the pre-cut sheet was discretised as full integration, hexahedral, type 7 arbitrarily distorted brick elements [15]. Two elements through the thickness were used and a longitudinal half of the blank was modelled due to symmetry. The effect of friction was neglected. The Young's modulus and the Poisson's ratio were taken as 200GPa and 0.3 respectively while the von Mises criterion was used to define plastic material behaviour using the plastic part of the true stress strain curve shown in Figure 1.

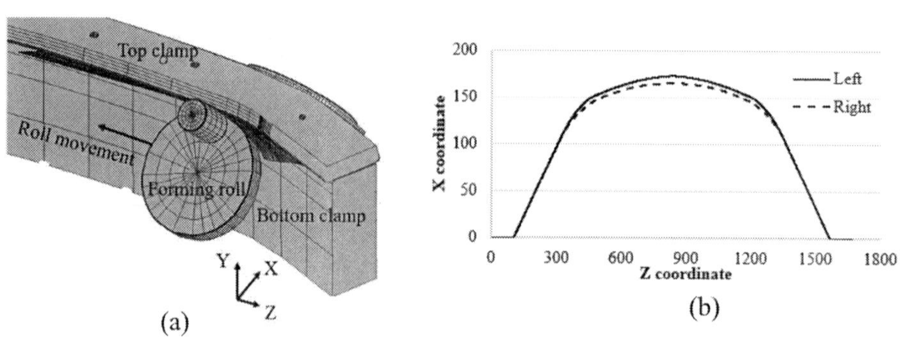

(a) (b)

Figure 6: (a) FEA model (b) tool path used for left and right tool movement

5. Results and discussion

The final formed part is shown in Figure 7(a) and the modelled symmetric half is shown in Figure 7(b). Figure 8(a) shows the experimental results for the final angles measured at 3 locations. By design, all angles α, β and θ should be 90° (see Figure 5(b)). However a significant deviation of angle β in the front and the back cross section (S1 and S3) can be observed. Angles α and θ also show some deviation from the ideal shape but at a lower magnitude compare to angle β. In the centre blank location, S2, angle β after springback is very close to the required angle of 90°. The situation for angle β with larger springback on both component ends is very similar to the end flare defect observed in conventional roll forming [16]. Here flaring out of two ends is generally a result of the bending moment created by the longitudinal residual stresses [17]. Figure 8(b) compares the measured angles between the FEA and the experiments for locations S1 and S2 (results for location S3 are not shown in the FEA due to the symmetry condition that was used). While there is a good agreement for location S1, some deviation can be observed, especially for angles β and θ for the blank centre position S2.

(a)

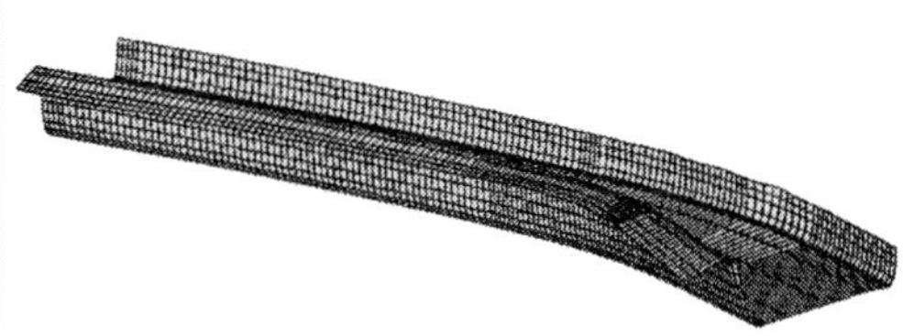

(b)

Figure 7: Final formed part in (a) experiments (b) FEA result

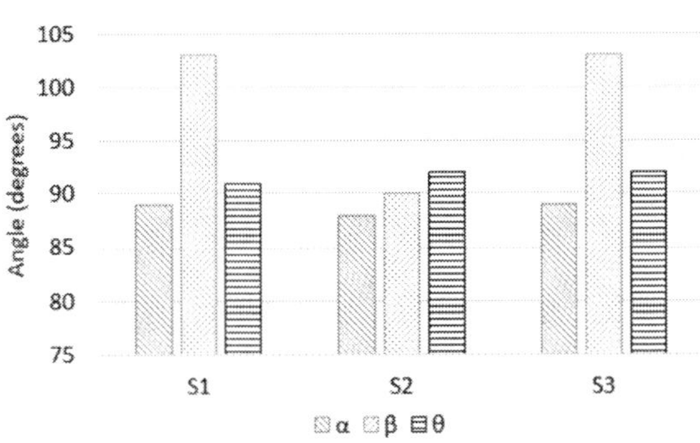

(a)

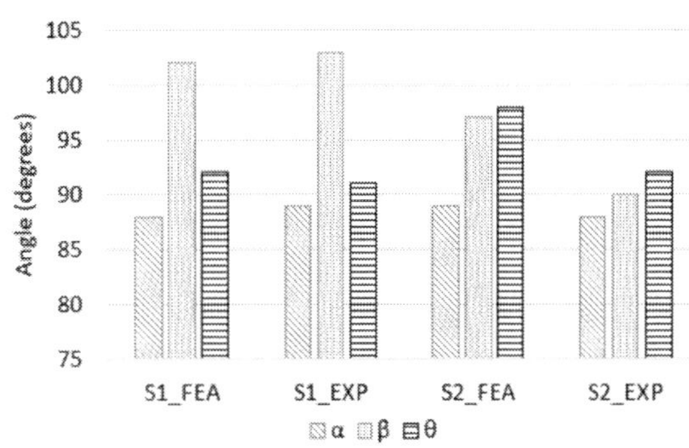

(b)

Figure 8: Angle measurements at different locations in (a) experiment (b) experiments and FEA result

Overbending is an approach which is commonly used in conventional roll forming to compensate for springback. However, if springback varies from one cross section to another due to end flare, a flexible overbending approach may be required; such a technique can be implemented in Deakin's flexible roll forming facility where the tool movement is numerically controlled. The tool orientation for the regular overbending is shown in Figure 9(a). If the tools follow the regular tool path, then it will produce the same overbending angle along the whole length of the part. To have more overbending at the ends and less at the centre, the tool path needs to deviate from the part contour towards the component centre as it is shown in Figure 9 (b). For this approach the top roll shown in Figure 9 (a) needed to be removed since otherwise the change in tool path would have affected the bend line of the top hat. The simplified roll tooling and the changed tool path are illustrated in Figure 9 (b).

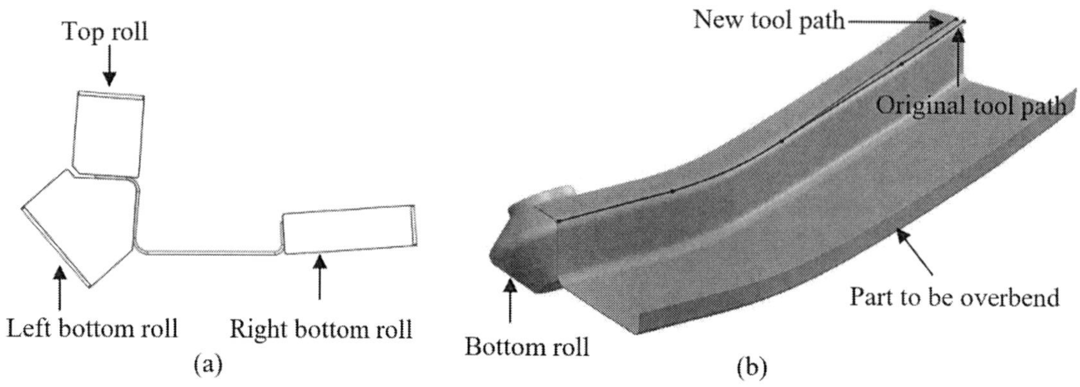

Figure 9: (a) Tool orientation for regular overbending (b) schematic of flexible overbending approach

The measured angles after the overbending pass are shown in Figure 10(a). It can be seen that the angle β has been significantly improved in both component end locations S1 and S3. However comparing Figure 8(a) and 10 (a), it can be seen that the deviation of angle α from the ideal 90° has now increased. This is likely due to angle α being unsupported during overbending given that the top roll could not be used. To compensate for this, forming pass 13 (see Figure 3(b)) was repeated after the variable overbending pass with the top roll included. The final result is illustrated in Figure 10(b). It can be observed that angle α has been improved due to the additional support provided by the top roll and all angles are within ±1° to the ideal 90° (see Figure 10(b)).

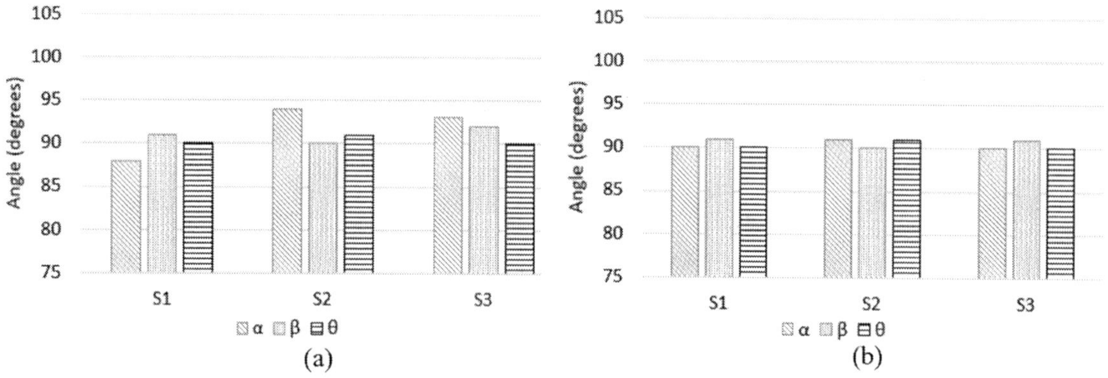

Figure 10: Angle measurement (a) after overbending and (b) repeated pass 13

6. Conclusion

- A simplified automotive cross member was successfully formed by flexible roll forming from 1mm Ford WSS-M1A365-A12.
- Springback and end flare were observed to be the main shape defects.
- A flexible shape compensation approach has been proposed and experimentally validated.
- This approach can be applied to complex parts formed from AHSS and UHSS where the end flare and the springback are more prominent.

7. Acknowledgment
The authors would like to thank ARC Training Centre in Lightweight Automotive Structure (project number IC160100032) and Ford Motor Company for the financial support. The authors further acknowledge the technical and software support provided by data M Sheet Metal Solutions GmbH.

8. References
[1] Groche P and Christiany M, 2013, "Evaluation of the potential of tool materials for the cold forming of advanced high strength steels," *Wear,* vol. **302**, pp. 1279-1285, 2013/04/01/.
[2] Abee A, Berner S, and Sedlmaier A, 2008, "Accuracy improvement of roll formed profiles with variable cross sections," presented at the ICTP 2008 (The 9th International Conference on Technology of Plasticity).
[3] Wenting S, Qiang L, and Bo Q, 2011 "Control Strategy for the Flexible Roll Forming Machine with the Structure of Double Rack and Single Gear," presented at the International Conference on Transportation, Mechanical, and Electrical Engineering (TMEE), Changchun, China.
[4] Kim D, Cha M, and Kang Y S, 2017, "Development of the Bus Frame by Flexible Roll Forming," *Procedia Engineering,* vol. **183**, pp. 11-16, 2017/01/01/.
[5] Ghanei S, Abeyrathna B, Rolf B, and Weiss M, 2019, "Analysis of material behaviour and shape defect compensation in the flexible roll forming of advanced high strength steel," *IOP Conference Series: Materials Science and Engineering,* vol. **651**,
[6] Abeyrathna B, Rolfe B, Pan L, Ge R, and Weiss M, 2016, "Flexible roll forming of an automotive component with variable depth," *Advances in Materials and Processing Technologies,* vol. **2**, pp. 527-538, 2016/10/01.
[7] Larrañaga J, Berner S, Galdos L, and Groche P, 2011, "Geometrical accuracy improvement in flexible roll forming lines," *AIP Conference Proceedings,* vol. **1315**, pp. 557-562,
[8] Park J-C, Yang D-Y, Cha M, Kim D, and Nam J-B, 2014, "Investigation of a new incremental counter forming in flexible roll forming to manufacture accurate profiles with variable cross-sections," *International Journal of Machine Tools and Manufacture,* vol. **86**, pp. 68-80,
[9] Berner S, Storbeck M, and Groche P, 2011, "A Study on Flexible Roll Formed Products Accuracy by Means of FEA and Experimental Tests," *AIP Conference Proceedings,* vol. **1353**, pp. 345-350,
[10] Abeyrathna B, Abvabi A, Rolfe B, Taube R, and Weiss M, 2016, "Numerical analysis of the flexible roll forming of an automotive component from high strength steel," *IOP Conference Series: Materials Science and Engineering,* vol. **159**, p. 012005,
[11] ASTM, "Standard test methods for tension testing of metallic materials," in *https://www.astm.org*, ed: ASTM International.
[12] Sedlmaier A, Dietl T, and Ferreira P, 2017, "Digitalization in roll forming manufacturing," in *Journal of Physics: Conference Series.*
[13] (2020, 13/01/2020). *HandyScan3D.* Available: https://www.creaform3d.com/en
[14] Solutions d S M, "COPRA® RF - Software for Roll Forming," 2013 ed. Germany.
[15] "Marc® User Manual," ed: MSC Software Corporation.
[16] Moneke M and Groche P, 2017, "Counter measures to effectively reduce end flare," in *AIP Conference Proceedings*, p. 020006.
[17] Nagamachi T and Ona H, 2015, "Mechanism of end deformation after cutting of light gauge channel steel formed by roll forming," *Materials transactions,* vol. **56**, pp. 187-192,

IOP Conf. Series: Materials Science and Engineering **967** (2020) 012049 doi:10.1088/1757-899X/967/1/012049

Finite element analysis of self-piercing riveting of aluminum alloy and carbon fiber reinforced polymer composites

Y C Hur[1], D Kim[2], K S Lee[2], M G Bae[3], S E Park[3] and J H Kim[1]*

[1] Pusan National University, 2 Busandaehak-ro 63beon-gil, Geumjeong-gu, Busan, 46241 Republic of Korea
[2] Korea Institute of Material Science, 797 Changwon-daero, Seongsan-gu, Changwon-si, Gyeongsangnam-do, Republic of Korea
[3] SUNGWOO HITECH, 18-2, Sojugongdan 6 street, Yangsan, Gyeongsang, Republic of Korea

Abstract. Self-piercing riveting (SPR) is a high-speed fastening process that can join similar and dissimilar sheet materials without the need for pre-processing such as drilling or punching. During SPR processes, two overlapping sheets are joined by a rivet. The upper sheet is punched first by the rivet and then the lower sheet is deformed between the rivet and the die, creating a mechanical interlock. In this study, self-piercing riveting of aluminum alloy and carbon fiber reinforced polymer composites (CFRP) sheets was analysed using finite element simulations. For the finite element simulation of SPR processes, the orthogonal elasticity, the fracture model, and the cohesive zone model were used for describing the behaviour of CFRP. For validation of the composite material model, the punching process of CFRP was performed and the results were compared with FE predictions. The SPR process of the aluminum alloy and CFRP was simulated numerically and the performance of the joint was evaluated.

1. Introduction

Currently, many researches are being conducted on utilizing various materials such as aluminum, magnesium, high strength steel, and composite materials for achieving light weight and high strength. n particular, for automotoive body parts such as A-pillar, B-pillar, and side seal, carbon fiber reinforced polymer (CFRP), aluminum alloy, and advanced high strength steel sheets are attempted to be used to reduce weight. In order to use various materials together, not only forming technology but also technology for joining different materials must be developed.

Conventional joining techniques have been used to join similar materials mainly using heat sources. Spot welding, arc welding, laser welding, and friction stir welding are the typical welding methods using these heat sources. However, they are not suitable for joining materials with different thermal and mechanical properties. In addition, although there is a method of using an adhesive, this method generally does not satisfy the bonding strength.

One method of joining different types of materials is self-piercing riveting (SPR). Because of the mechanical bonding, it can be used with adhesive bonding, and materials with different thermal and mechanical properties can be easily bonded. SPR is widely used because the materials can be joined in a very short time so that it can easily be added to a part production line.

Various studies related to SPR have been conducted. In addition to the study of SPR joining similar or different materials, the effect of plate thicknesses and combinations in the SPR process, the joining ability of rivet material to aluminum, and the effect of friction were studied [1-4]. Studies on the strength of SPR junctions and numerical analysis of SPR joint were conducted [5-12].

In this study, the SPR joint of CFRP plate and Al7XXX alloy plate is analysed numerically. There are many experimental papers on this topic, but not many numerical studies were conducted. Since the physical properties of the composite material and the metal sheet are very different, it is challenging to analyse the joining process numerically.

2. Process of Self-Piercing Riveting

SPR consists of four processes. The first process is the clamping process, where the top and bottom materials are held in place by the holder to prevent movement between the rivet and the die, and the rivet is placed perpendicular to the top material. The next step is piercing, through which the rivet penetrates the top material by the force that goes into the punch. When the rivet that completely penetrates the top material meets the bottom material, the rivet tail opens and the rivet and the plates undergo deformation together. This is called the flaring process. Finally, a compression process ends where the rivet is compressed to the end with the specified stroke or force. The shape of the rivet used in this study was a semi-tube and a flat die was chosen for the Al7XXX alloy sheet to avoid the fracture.

3. Material Properties

3.1. Sheet material

The plates used in this study are CFRP composite plates and Al7XXX alloy plates. CFRP is a brittle material with no plastic deformation. The brittle material of the SPR process cannot be used for the bottom material. The CFRP composite plate used in this study is a uni-direction plate and has anisotropic mechanical properties. The physical properties of CFRP composites were assumed to be orthotropic and the strength and stiffness values for each direction were used (Figure 1). The direction of piercing is the direction of thickness, and the physical property in the thickness direction is difficult to measure or estimate. The stiffness values are calculated as follows.

$$\nu_{21} = \nu_{12}\left(\frac{E_{22}}{E_{11}}\right) \tag{1.1}$$

$$\nu_{23} = \nu_{32} = \nu_{12}\frac{(1-\nu_{21})}{(1-\nu_{12})} \tag{1.2}$$

$$G_{23} = \frac{E_{22}}{2(1+\nu_{23})} \tag{1.3}$$

where the indices 1, 2, and 3 represent the fiber, transverse and thickness directions, respectively. E_{ii} is the stiffness in the i direction, ν_{ii} is the Poisson ratio in the i direction, and G_{ii} is the shear stiffness in the i direction. The criterion for determining fracture is calculated by the strength and shear stiffness in each direction using the Hashin's damage model.

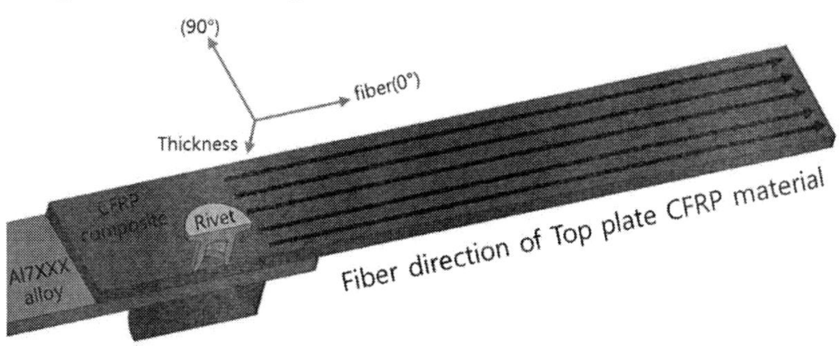

Figure 1. Schematic description of CFRP composite and Al7XXX SPR setting.

Aluminum 7XXX alloy sheet was evaluated for mechanical properties by uniaxial tension test. The strain was measured using a digital image correlation system (Figure 2).

3.2. Rivet material

IOP Conf. Series: Materials Science and Engineering **967** (2020) 012049 doi:10.1088/1757-899X/967/1/012049

Rivets should be stronger than the base sheet used in the process because they must not break during the SPR process. In this study, high-strength steel with a strength higher than Al7XXX alloy plate or CFRP composite plate was used. The mechanical properties were evaluated by the uniaxial tension test (Figure 3).

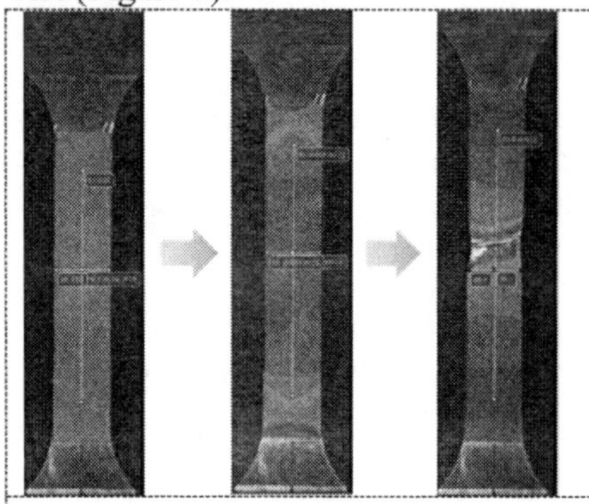

Figure 2. Image of Al7XXX alloy sheet uniaxial tension test digital image correlation system.

Figure 3. Uniaxial tension test of cylindrical specimen made by rivet steel material.

4. Numerical Analysis

Numerical analysis was done using a commercial FE program, LS-DYNA explicit. The 3D model was used to implement the CFRP composite properties with significant deformation in the thickness direction. The die was assumed to be rigid, and a uniform force was applied to the rivet head (Figure 4). The measured properties of rivets and plates were used. The half model was used to reduce the computation time.

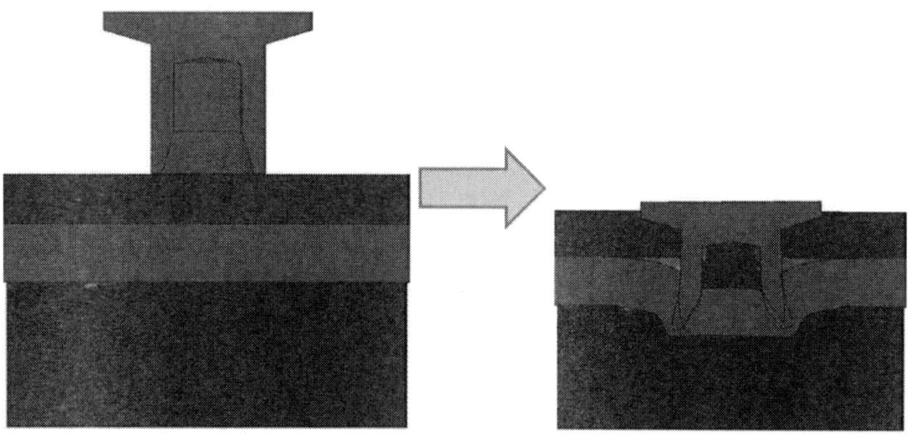

Figure 4. Schematic description of CFRP composite and Al7XXX SPR setting.

5. Conclusion

In this study the self-piercing riveting process of CFRP composite plate and Al7XXX alloy plate were simulated numerically. It can be seen that the composites on the top are easily punched and have little effect on the deformation of the semi-tube rivet. After that, when the rivet meets the bottom sheet, the rivet tail opens, The amount of opening was smaller than that of the metal-metal joint, interlock distance [13]. Figure 5 shows an experiment in which the SPR process was performed by changing only the top plate to a DP780 steel sheet with a thickness of 1.6 mm and a model that analyzed it numerically. When the top plate was CFRP composite, the interlock distance was 0.35 mm on average, whereas when the top plate was DP780, the experimental and numerical values were larger, 0.74 and

0.63 mm, respectively, despite the smaller thickness. It seems that the deformation of the top plate and rivet when the rivet penetrates the top plate affects the interlock distance.

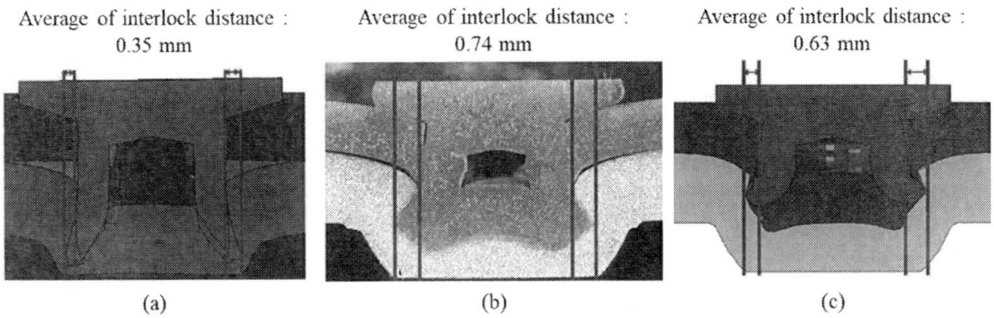

Figure 5. Average interlock distance of corss-section (a) numerical model of CFRP composite 2.3 mm at top plate, (b) experiment of DP780 1.6 mm at top plate, (c) numerical model of DP780 1.6 mm at top plate SPR joint.

6. References

[1] Porcaro R, Hanssen AG, Aalberg A, Langseth M 2004 Int. J. Crashworthines. 9:2 141-154
[2] Xin S, Mohammad AK 2005 J. Manuf. Process. 7(1) 83-93
[3] Haque R, Duradet Y 2017 J. Manuf. Process. 30 148-160
[4] Cacko R 2008 Arch. Civ. Mech. Eng. 8(2) 21-30
[5] Settineri L, Atzeni E, Ippolito R 2010 Int. J. Mater. Form. 3 995-998
[6] Iyer K, Hu SJ, Brittman FL, Wang PC, Hayden DB, Marin SP 2005 Fatigue. Fract. Engng. Mater. Struct. 28 997-1007
[7] Porcaro R, Langseth M, Hanssen AG, Zhao H, Weyer S, Hooputra H 2008 Int. J. Impact. Eng. 35 1251-1266
[8] Bouchard PO, Laurent T, Tollier L 2008 J. Mater. Process. Tech. 202 290-300
[9] Casalino G, Rotondo A, Ludovico A 2008 Adv. Eng. Softw. 39 787-795
[10] Atzeni E, Ippolito R, Settineri L 2009 CIRP Ann-Manuf. Techn. 58 17-20
[11] Kang SH, Kim HK 2015 Int. J. Fatigue. 80 58-68
[12] Hirsch F, Müller S, Michael M, Staschko R, Fuchs N, Kästner M 2017 J. Mater. Process. Tech. 8 21-30
[13] Haque R 2018 Arch. Civ. Mech. Eng. 18 83-93

Evaluation of the stress-strain relationship in the high strain region of high strength materials by using the shear stretch test

Y Taguchi[1], Y Maeda[1], Y Maeda[2]

[1]Forming Research Station, Application Technology Center,
Kobe Steel, Ltd., 1-5-5 Takatsukadai, Nishi-ku, Kobe, Hyogo 651-2271 Japan

[2]Materials Research Laboratory, Technical Development Group,
Kobe Steel, Ltd., 1-5-5 Takatsukadai, Nishi-ku, Kobe, Hyogo 651-2271 Japan

taguchi.yuki@kobelco.com

Abstract. Stress-strain relationships are used in FEM analysis. After uniform deformation, they are simply extended values expected from ones measured without experiment. The purpose of this research was to understand the stress-strain relationship in the high strain region after uniform deformation. We investigated a method for understanding this relationship in high strength materials using a simple shear test. The relationship in this region was calculated by determining a constant parameter κ as the coefficient between equivalent stresses and shear stresses. Using this constant, the relationship after uniform deformation was extended from the result of a stretching test. This method is called the κ method. In the case of high-strength materials, it is difficult to avoid rotating the specimen under a high testing load. Because of this problem, we designed the shape of the specimen so that it had two symmetrical deformation areas. The notch shapes for the deformation areas on the specimen were designed to reduce the load and provide a strong gripping force. In terms of shortening the deformation region, it is difficult to ignore a non-uniform deformation. To reduce the influence of non-uniform deformations, the effect of the R shape on the notch end was evaluated. The uniformity of two specimens with R1.5 mm and R7.0 mm on the notch end was examined by FEM. The values of κ, which were determined from the shear stress and shear strains based on the results of FEM, depended on the equivalent strain. In the case of R7.0 mm, the κ results were almost constant, and this was a suitable result for applying the κ method. The change in κ was affected by the non-proportional stress ratio, the stress error of the observed shear stress, and the strain error of observed shear strain. In the case of R7.0 mm, it was found that there was an advantage for the constant stress error in the higher strain region. Using R7.0 mm and R1.5 mm specimen shapes, we applied a shear test with two kinds of high-tensile steels and aluminum materials. From experimental results, the values for κ on R7.0 mm were also more constant than on R1.5 mm in the higher strain region. A high stress-strain relationship after uniform deformation was obtained in high-tensile steels and aluminum materials. This research could therefore be used to understand the stress-strain relationship in the high strain region of high strength materials.

1. Introduction

In order to reduce the cost of the dies with trial and error, a more accurate prediction for sheet metal forming analysis is required. Constitutive equations for material modeling are important for accurately predicting defects such as cracks and wrinkles during the press forming process. One of the most important constitutive equations is the work hardening law. In general, this law is determined by using the results of uniaxial tensile tests. However, it is difficult to determine the work hardening law in the high strain region, because of diffusion necking after uniform deformation. During usual sheet metal formation, strain over uniform elongation sometimes occurs. The stress-strain relationship after uniform elongation has to be decided on for metal forming analysis, but it is impossible to use a tensile test alone.

Therefore, instead of a tensile test, a shear test [1, 2] has been proposed as a κ method, where the stress-strain relationship in the high strain region can be measured without diffusion necking. Shirakami et al.[1] introduced a conversion method called the κ method, which decided the proportional constant between equivalent stress and shear stress by comparing results of a simple shear test with that of a tensile test in the lower strain region.

In this paper, to obtain the stress-strain relation of higher strength materials following their method, an optimum specimen shape with a lower test force and accurate determination of κ was investigated both experimentally and numerically.

2. Simple shear test

2.1. Derivation of stress-strain relationship using the κ method

In a simple shear test (Fig. 1), shear deformation occurs in the specimen due to the jig moving up and down. Shear strain γ is measured from the deviation (θ) of the horizontal straight line at the center of the deformation area. Shear stress τ is measured from the load F on the longitudinal deformation area A_0 [3].

$$\tau = \frac{F}{A_0} \quad \gamma = \tan\theta \tag{1}$$

The obtained shear stress-strain relationship is converted into an equivalent stress-strain relationship by the κ method. (Lee et al. [4]) The κ method decides the proportional constant κ between equivalent stress-strain and shear stress-strain by the assumption of plastic work conjugation and isotropic hardening.

$$\bar{\sigma} = \kappa_\sigma \tau \quad d\bar{\varepsilon}^p = \kappa_\varepsilon d\gamma \quad \vdots\, \kappa_\sigma = \frac{1}{\kappa_\varepsilon} = \kappa \tag{2}$$

$\bar{\sigma}$, $d\bar{\varepsilon}^p$ are the equivalent stress and the equivalent plastic strain increments. The constant κ agrees with κ_σ and $1/\kappa_\varepsilon$ from assuming plastic work conjugation (Equation (2). The constant κ is decided using the least-squares method so that shear and tensile deformation agree within the uniform elongation.

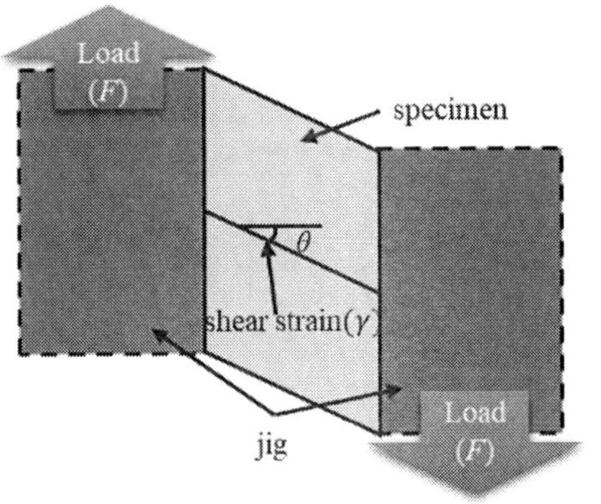

Fig.1 Schematic illustration of simple shear test

2.2. Experimental equipment and method

In this paper, the specimen and test jig shown in Fig. 2 were used for the simple shear test. The specimen was made symmetrical with two shear deformation parts on the left and right to prevent rotation of the specimen during the experiment. The test jig has a mechanism that can apply sufficient surface pressure (84 MPa) with bolts and can constrain the specimen in a vertical orientation, thereby suppressing slippage of the test specimen.

The shear strain was measured by digital image correlation (DIC) from the deviation of the horizontal straight line in the centre of the deformation area. Shear stress was measured from the test load obtained from the load cell.

(a) Appearance of specimen (b) Appearance of test jig

Fig. 2 Simple shear test.

2.3. Problems with the shear test method

The deformation area of the shear test becomes inhomogeneous due to the edge face effect. The effect of non-uniform deformation is reduced by increasing the deformation area, but the capability of the test equipment reaches its limit when the material is a high-strength material. Therefore, the shape of a specimen that could reduce the test load and non-uniform deformation was examined by numerical analysis.

3. Examination of shear specimen shapes by FEM analysis

3.1. FEM analysis conditions

Notches are provided along both edges of the deformation area to reduce the test load. The notched R shape was examined by FEM analysis (LS-DYNA9.3.0: solid element). The effects of the notch shape were valuated from two types of notch shapes (R1.5, R7.0 mm) as shown in Fig. 3. As a material model, the stress-strain relationship was extrapolated from a uniaxial tensile test, and Mises and isotropic hardening were assumed for the yield function.

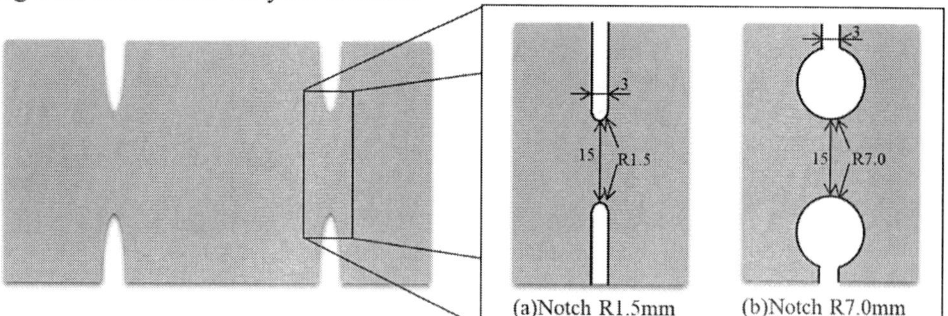

(a)Notch R1.5mm (b)Notch R7.0mm

Fig. 3 Two specimens with R1.5 mm and R7.0 mm on the notch end.

3.2. *Results of the supposition experiment using FEM analysis*

The shear stress-strain was measured from FEM analysis, with the same definition as the experiment, and the constants κ and κ_σ, ${}^{1}/_{\kappa_\varepsilon}$ were determined (Fig. 4).

κ_σ and ${}^{1}/_{\kappa_\varepsilon}$ for both R1.5 mm and R7.0 mm indicated a constant value until uniform elongation ($\bar{\varepsilon}^p < 0.1$), and R1.5 mm was closer to the ideal $\kappa = 1.73$. However, as deformation progressed, the κ_σ and ${}^{1}/_{\kappa_\varepsilon}$ of R1.5 mm were not constant, and the strain dependence of κ was confirmed at R1.5 mm. On the other hand, κ_σ and ${}^{1}/_{\kappa_\varepsilon}$ of R7.0 mm indicated a constant value in the high strain region ($\bar{\varepsilon}^p = 0.45$), indicating that the strain dependence of κ was small. Since the κ value in the high strain region cannot be measured experimentally, it must be estimated from the low strain region until uniform elongation is achieved. Therefore, the strain dependence of κ in the high strain area is important, and it was found that R7.0 mm, which had a small strain dependence of κ, was superior.

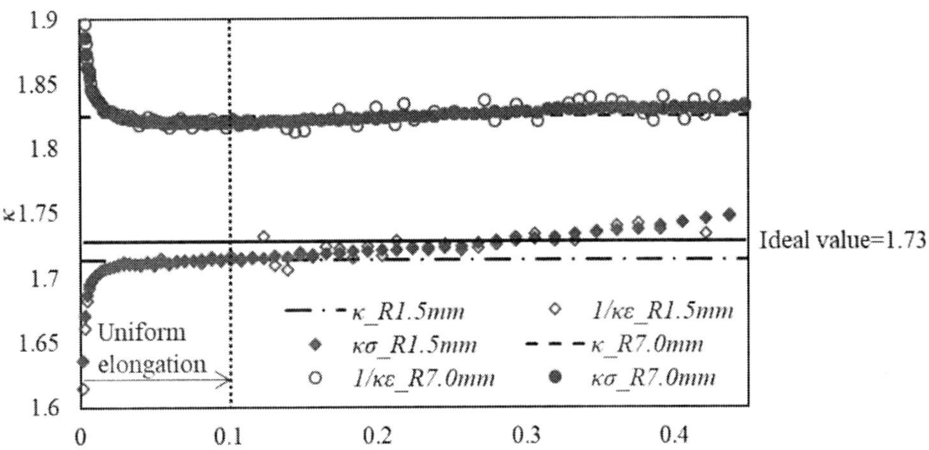

Fig. 4 Changes in κ, κ_σ, ${}^{1}/_{\kappa_\varepsilon}$

4. Investigation of the effect of notched R shape on the κ value

From the results of FEM analysis with a different Notch R, it was found that there is a difference in the strain dependence of κ depending on the R notch shape. Therefore, the difference in the strain dependence of κ due to R notch shape can be considered as follows.

4.1. *Effects of non-proportional loading*

In this paper, plastic work conjugation assumes two things: uniaxial tensile test and pure shear test. However, pure shear tests are difficult, we used a simple shear test in which shear deformation and tensile deformation are mixed. We therefore discuss the effect of mixing between shear and tensile deformation. Even in simple shear deformation, if the shear stress and tensile stress and the shear strain and tensile strain pass through the proportional loading path, the value of κ can be considered constant, though different from the ideal value. Therefore, the κ value was evaluated when all stress and strain states could be measured with this experimental method.

The equivalent stress $\bar{\sigma}$ and the equivalent plastic strain increment $d\bar{\varepsilon}^p$ at the center of the deformation were obtained from the FEM analysis results, and κ_σ , ${}^{1}/_{\kappa_\varepsilon}$ were determined (Fig. 5). To calculate the equivalent stress $\bar{\sigma}$ and the equivalent plastic strain increment $d\bar{\varepsilon}^p$, the shear stress and strain (τ_0, γ_0), the stress and strain in the X direction (σ_x, ε_x), and the stress and strain in the Y direction (σ_y, ε_y) were used.

$$\bar{\sigma} = \frac{1}{\sqrt{2}}\sqrt{2{\sigma_x}^2 + 2{\sigma_y}^2 - 2\sigma_x\sigma_y + 6{\tau_0}^2}, d\bar{\varepsilon}^p = \frac{\sqrt{2}}{3}\sqrt{2{d\varepsilon_x}^2 + 2{d\varepsilon_y}^2 - 2d\varepsilon_x \times d\varepsilon_y + 6{d\gamma_0}^2} \quad (3)$$

The κ_σ and $^1/_{\kappa_\varepsilon}$ for R1.5 mm were constant in the high strain region, and the strain dependence of the κ was small. On the other hand, κ_σ and $^1/_{\kappa_\varepsilon}$ of R7.0 mm were not constant due to deformation, indicating that the strain dependence of the κ value was larger. Therefore, when the stress and strain can be measured at the center of the deformation area, the strain dependence of the κ value is smaller for R1.5 mm, and it was found that R1.5mm formed a proportional load path showing a constant κ value in the high strain area.

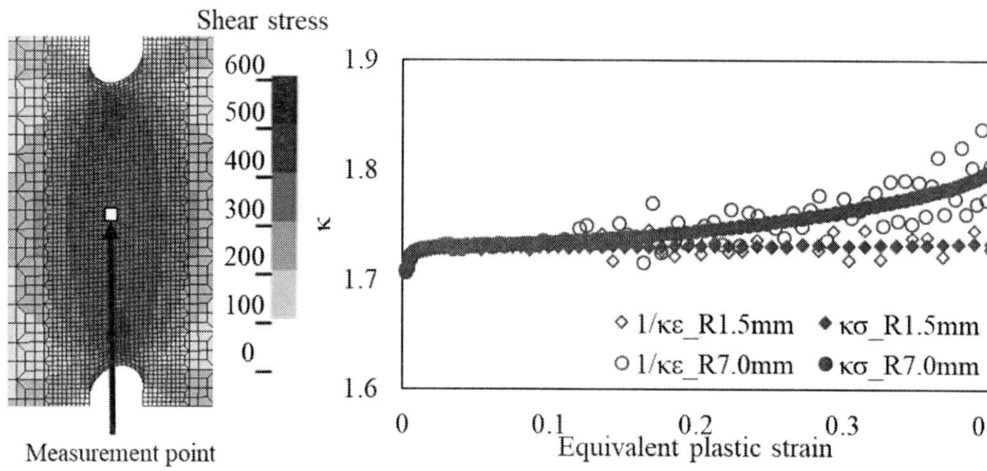

Fig. 5 Changes in the κ value calculated from stress-strain at the center of the deformation area.

4.2. Effects of stress and strain measurement errors

The stress in the experiment was measured by the force acting on the jig in the deformation area (Fig. 6). However, during the shear deformation, non-uniform stress-strain distribution occurs in the deformation area, so that the measured stress in the experiment is the average value in the deformation area (Fig. 6). Therefore, it is considered that this measurement error affects the conversion accuracy of the κ value.

The shear stress τ_0 at the center of the deformation area and the shear stress $\bar{\tau}$ which was calculated from the load were obtained by the results of the FEM analysis. The stress error rate of the shear stress was calculated from equation (4) (Fig. 7 (a)). For the shear strain, shear strain $\bar{\gamma}$ was calculated from the deviation of the horizontal straight line at the center of the deformation area, and the shear strain γ_0 at the center of the deformation area was obtained by the results of the FEM analysis. The strain error rate was calculated from equation (5) (Fig. 7 (b)). When both the stress error rate α and the strain error rate β are close to 0.0, the measured stress and strain are close to the exact values on the center of the deformation area.

$$\text{Stress error } \alpha(\%) = \left(\frac{\tau_0 - \bar{\tau}}{\tau_0}\right) \times 100 \quad (4)$$

$$\text{Strain error } \beta(\%) = \left(\frac{\gamma_0 - \bar{\gamma}}{\gamma_0}\right) \times 100 \quad (5)$$

The stress error rate is closer to 0.0 for R1.5 mm, but α changed with deformation. On the other hand, at R7.0 mm, the stress error rate is larger, but α didn't change during the deformation. It was found that the shear strain error rate changed during the deformation at both notches R. Also, the difference between the two notches was small.

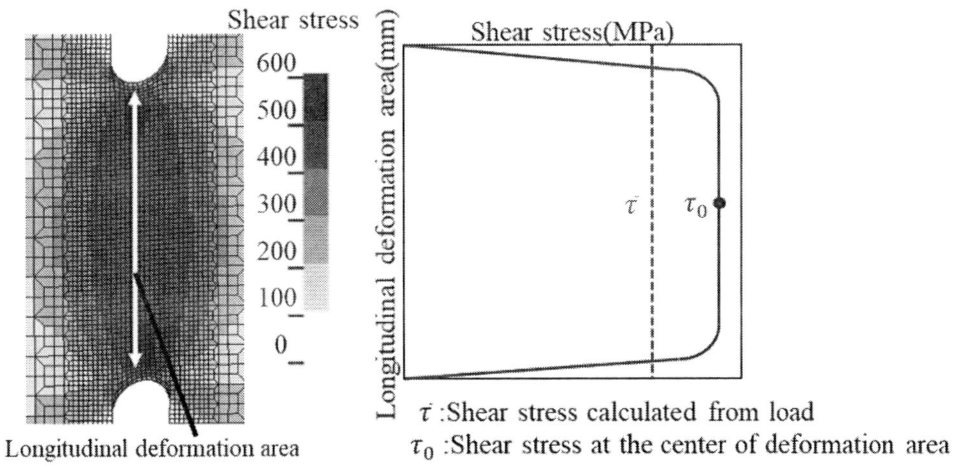

Fig.6 Shear stress distribution in deformation area.

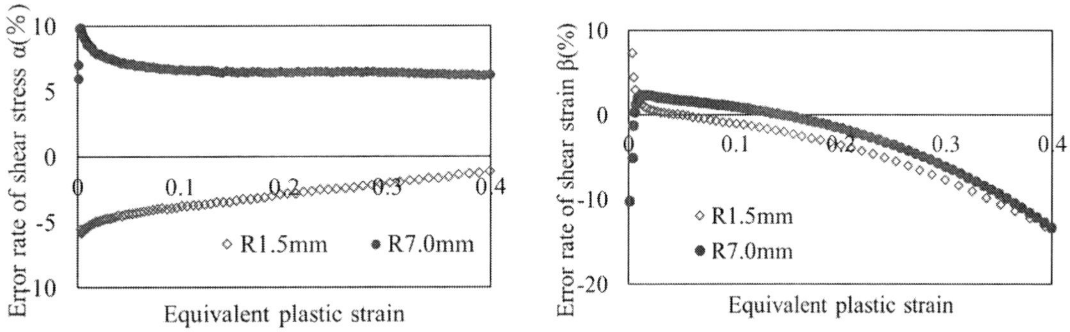

(a) Error rate of shear stress (b) Error rate of shear strain

Fig.7 Error rate of shear stress and shear strain during the deformation with different notch shape.

4.3. Influence of factors on the κ value calculated by experimental method

It was found that the notch shape changed the strain dependence of κ and the error rate during measurement. Therefore, it is necessary to select an R notch shape that is less affected by the strain dependence of κ and measurement errors. In this section, we investigated the relationship between the change in the κ value when calculating the shear stress-strain by the same method as in the experiment and the strain dependence of κ, the measurement error. A regression equation was used for evaluation. (Equation 6)

$$\Delta\kappa = 1.0 \times \Delta\kappa_c + 1.726 \times \Delta\alpha - 0.1709 \times \Delta\beta \tag{6}$$

$\Delta\kappa$ is the change in the κ value calculated by the same method as in the experiment, and $\Delta\kappa_c$ is the change in the κ value calculated at the centre of the deformation area. The amount of change of both κ values was calculated from the reference value of equivalent strain $\bar{\varepsilon}^p = 0.1$.

The regression equation accurately reproduced the change in the κ value calculated by the experimental method (Fig. 8). The stress error rate was the largest factor among these coefficients

Therefore, the stress error rate is important with respect to the change in the κ value due to the shear stress-strain calculated in the same way as the experiment, and R7.0 was a better notch shape.

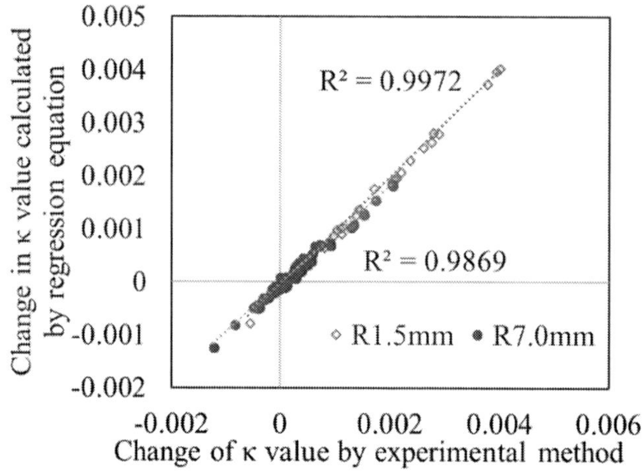

Fig.8 Accuracy of κ value calculated by regression equation.

5. Experimental results

Based on the results from the previous section, two different notches were used to test different materials (Table 1). Fig. 9 shows the constants κ, κ_σ, and $^1/_{\kappa_\varepsilon}$ for each material. In order to examine the strain dependence of κ, the increment of κ_σ for the equivalent plastic strain was calculated from equation (7) as an average slope of 15 points (Fig. 10).

$$Incremental\ of\ \kappa_\sigma = \frac{\sum(\bar{\varepsilon}^p - \bar{\varepsilon}^p{}_{average})\sum(\kappa - \kappa_{average})}{\sum(\bar{\varepsilon}^p - \bar{\varepsilon}^p{}_{average})^2} \quad (7)$$

In all materials, the increase of κ_σ in the high strain region was smaller at R7.0 mm. Obtaining similar results of the FEM analysis, it can be said that the strain dependence of κ is small and κ can be estimated stably at the notch tip R7.0 mm.

Fig. 11 shows the stress-strain relationship for each material. The points up to uniform elongation in the tensile test are also shown for reference. For all materials, stress-strain curves beyond the uniform elongation were obtained.

Table 1 Material properties of test materials.

Materials	t_0(mm)	YS(MPa)	TS(MPa)	U-EL(%)
6000-series Al	1.2	134	250	20.2
590 grade steel	1.4	361	608	15.7
980 grade steel	1.4	598	971	8.84

IOP Conf. Series: Materials Science and Engineering **967** (2020) 012050   doi:10.1088/1757-899X/967/1/012050

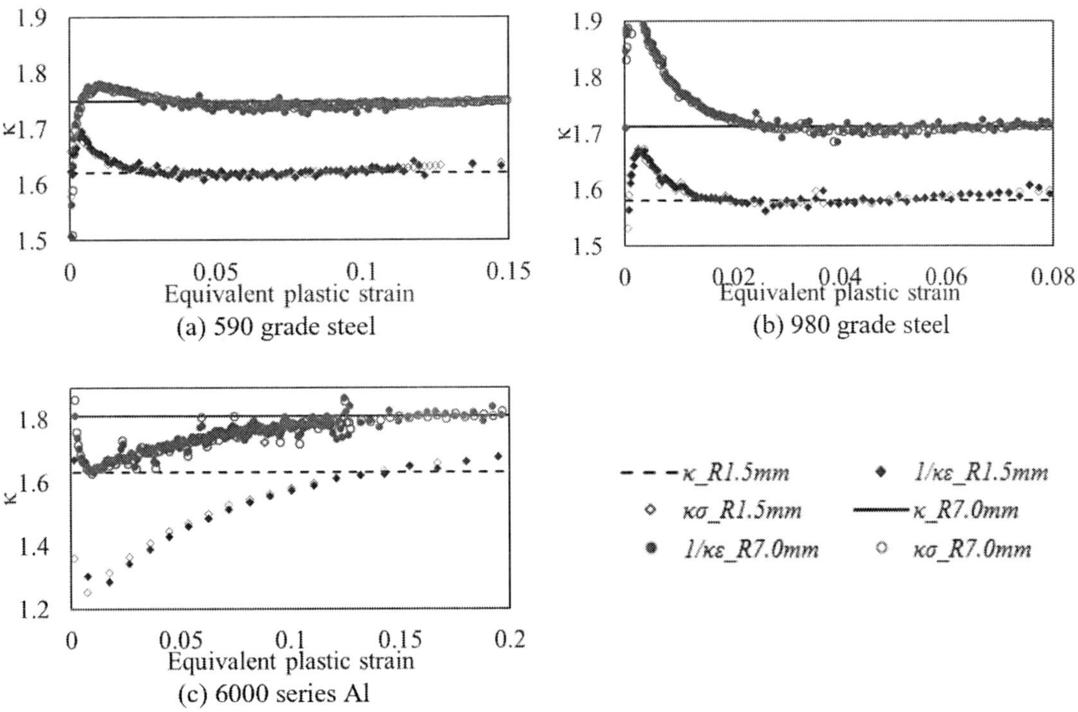

Fig. 9 Changes in κ, κ_σ, $1/\kappa_\varepsilon$ for each material with different R shapes.

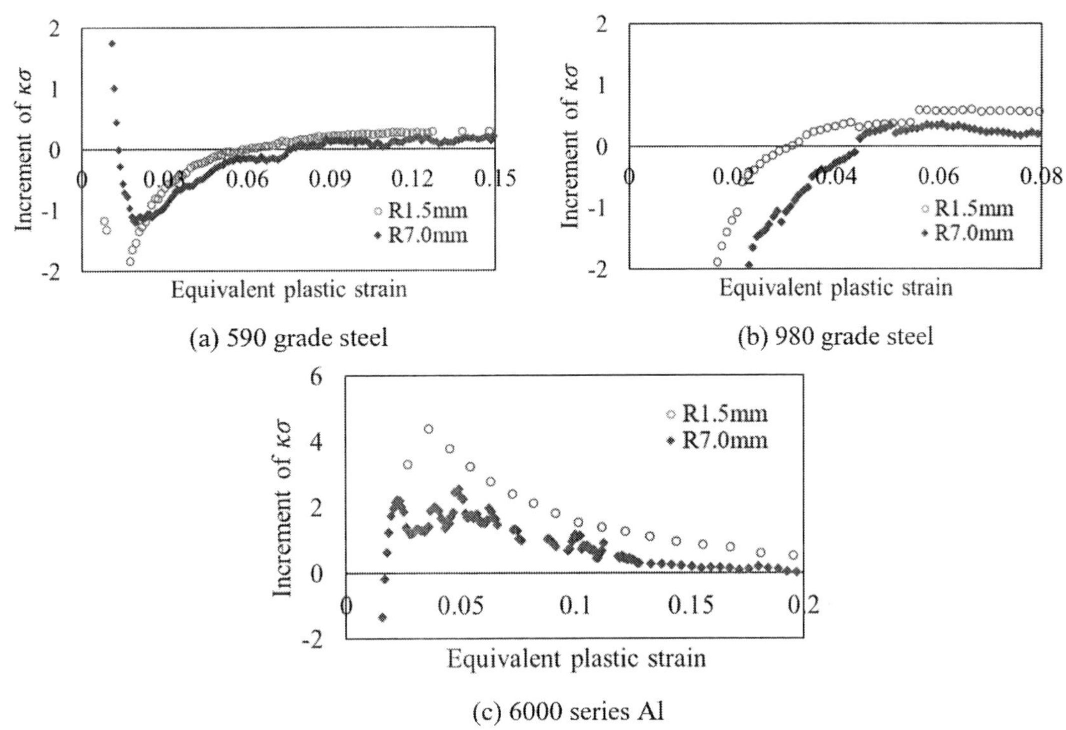

Fig.10 Changes in increment of κ_σ for each material with different R shapes.

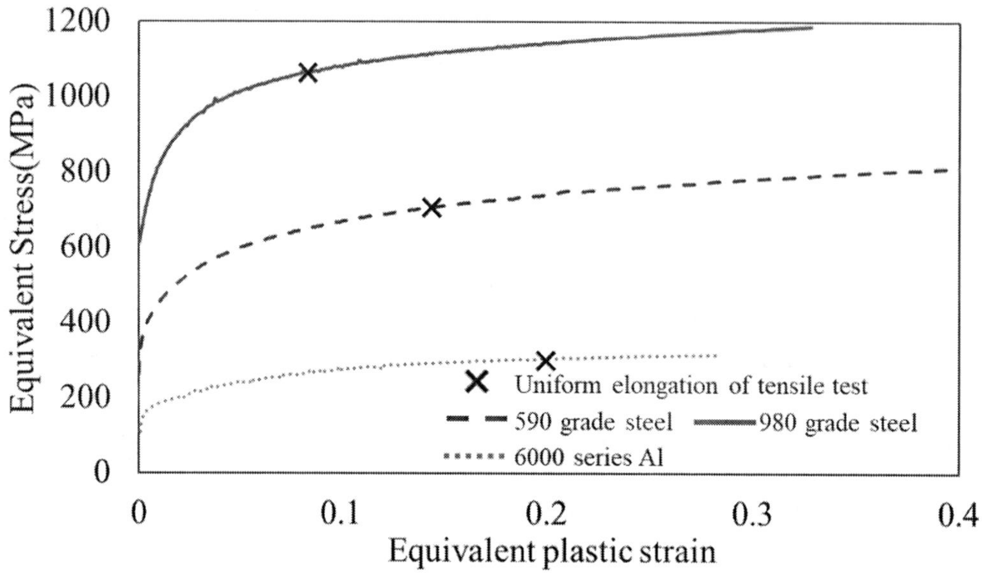

Fig.11 Stress-strain curves up to high strain region for each material.

6. Conclusion

1) The appropriate specimen shape for obtaining the stress-strain curve of a high-strength material from a shear test was investigated by using the κ method. These specimens were used to obtain stress-strain curves for various materials in the high-strain region.
2) The behavior of κ in the κ method changes depending on the notch R shape. As a result of FEM analysis, it was found that the influence factors were non-proportional loading and the accuracy of stress and strain measurements.
3) Due to these effects, κ changes with deformation depending on the material, but by adopting an appropriate R shape, it was found that κ is almost constant in the high strain region.

References

[1] Shirakami et al.: Iron and Steel, 103-10(2017), 589-916.
[2] G'Shell et al.: J. Mater. Sci., 18(1983), 903-918.
[3] Bouvier et al.: J. Mater. Process. Technol., 172(2006), 96-103.
[4] SigvantShek et al.: Int. J. Mater. Form., 2(2009), 235-242.

A fully coupled damage model with stress triaxiality and Lode dependence

K Zhang[1], H Badreddine[2], N Hfaiedh[3], K Saanouni[2] and J Liu[1]

[1]Department of Engineering Mechanics, College of Pipeline and Civil Engineering, China University of Petroleum (East China), Qingdao, China

[2]LASMIS, University of Technology of Troyes, Troyes Cedex, France

[3]Léonard de Vinci Pôle Universitaire, Research Center, Paris, France

E-mail: kai.zhang@upc.edu.cn

Abstract. In the present work, an advanced CDM model considering stress triaxiality and Lode angle effect is proposed. The framework of thermodynamics of irreversible processes with state variables is used to build the constitutive equations accounting for strong and full coupling between all the dissipative phenomena. The proposed model is implemented into Finite Element (FE) code ABAQUS/Explicit via a user material subroutine (VUMAT). A detailed parametric study with various values of the new material parameters is conducted in order to show the predictive capability of the proposed model. Applications to sheet metal forming simulation have been performed to validate the damage prediction capability of the proposed model, and the numerical simulation results are analysed and discussed.

1. Introduction

Modeling and simulation of damage and fracture of metallic materials are important issues in sheet metal forming industry[1]. In complex forming processes, reliable and accurate predictions of ductile damage is still a challenging task, and advanced constitutive models are still required. Up to date, two kinds of methodologies based on uncoupled and coupled models are used by many researchers and engineers. Among these models, the CDM (Continuum Damage Mechanics) framework is increasingly used in damage prediction for sheet metal forming, since it is convenient in coupling with different physical phenomena[2].

However, the classical CDM model shows limitations in damage prediction under shear dominated and low stress triaxiality stress sates[3]. As observed by experiments, the equivalent plastic strain at fracture is highly dependent on the Lode angle and stress triaxiality for many sheet metals[4]. Therefore, more advanced fracture models have been proposed to consider the stress state, i.e. Wierzbicki-type models[4], Lou-Huh model[5], modified GTN model[6], enhanced CDM models[7, 8]. Based on these models, the improvement of the damage prediction accuracy were evident in sheet metal forming processes[9].

The objective of the present study is to enhance the fully coupled damage model by accounting for stress state dependence. The formulation of the model is performed in the framework of thermodynamics of irreversible processes with state variables. The proposed fully coupled damage model is then implemented into Abaqus/Explicit finite element code. After a relatively exhaustive parametric study, the material parameters are identified using experimental results taken from literature for application purpose. A square cup deep drawing simulation is made, which shows good agreement with the experimental results.

2. Fully coupled damage model

The Helmholtz free energy, a convex and closed function of strain-like state variables in the effective strain space, is taken as a state potential. The elastic strain and Cauchy stress tensors are decomposed into hydrostatic part ($\underline{\varepsilon}^{eH}, \underline{\sigma}^{H}$) and deviatoric part ($\underline{\varepsilon}^{eD}, \underline{S}$); ($r, R$) denotes isotropic hardening; ($\underline{\alpha}, \underline{X}$) represents kinematic hardening and finally (d, Y) represents isotropic ductile damage. The damage scalar variable d records the isotropic damage state on a material point and evolves from initial undamaged state (d = 0) to fully damaged state (d =1) indicating the final fracture of that material point. To the above presented state variables, appropriate effective state variables can be defined based on the total energy equivalent assumption in the following form:

$$\underline{\tilde{\varepsilon}}^{eH} = \sqrt{1-h(\eta)d^{\gamma_e}}\,\underline{\varepsilon}^{eH} \text{ and } \underline{\tilde{\sigma}}^{H} = \frac{\underline{\sigma}^{H}}{\sqrt{1-h(\eta)d^{\gamma_e}}} \tag{1}$$

$$\underline{\tilde{\varepsilon}}^{eD} = \sqrt{1-h(\eta)d}\,\underline{\varepsilon}^{e} \text{ and } \underline{\tilde{S}} = \frac{\underline{S}}{\sqrt{1-h(\eta)d}} \tag{2}$$

$$\underline{\tilde{\alpha}} = \sqrt{1-h(\eta)d}\,\underline{\alpha} \text{ and } \underline{\tilde{X}} = \frac{\underline{X}}{\sqrt{1-h(\eta)d}} \tag{3}$$

$$\tilde{r} = \sqrt{1-h(\eta)d^{\gamma_r}}\,r \text{ and } \tilde{R} = \frac{R}{\sqrt{1-h(\eta)d^{\gamma_r}}} \tag{4}$$

The parameters γ_e and γ_r are used to differentiate between the effect of isotropic damage on elastic hydrostatic stress and isotropic hardening respectively. The microcracks closure parameter $h(\eta)$ is dependent on stress triaxiality $\eta = \dfrac{tr(\underline{\sigma})}{J_2(\underline{\sigma})}$ as follows:

$$h(\eta) = \frac{1+h_c}{2} + \frac{1-h_c}{2}\tanh(\xi_h \eta) \tag{5}$$

where h_c is the value of h at negative stress triaxiality and the parameters ξ_h control the evolution of h from positive to negative stress triaxiality.

The specific Helmholtz free energy $\Psi(\underline{\varepsilon}^{e}, \underline{\alpha}, r, T, d) = \Psi(\underline{\tilde{\varepsilon}}^{e}, \underline{\tilde{\alpha}}, \tilde{r}, T)$, defined in the fictive undamaged configuration, is taken as a state potential. In which, Ψ^{e} is the elastic part and Ψ^{p} plastic part:

$$\begin{aligned}\rho\Psi &= \rho\Psi^{e}(\underline{\tilde{\varepsilon}}^{eH}, \underline{\tilde{\varepsilon}}^{eD}) + \rho\Psi^{p}(\underline{\tilde{\alpha}}, \tilde{r}) = \rho\Psi^{e}(\underline{\varepsilon}^{eH}, \underline{\varepsilon}^{eD}, d) + \rho\Psi^{p}(\underline{\alpha}, r, d) \\ &= \frac{1}{2}\left(1-h(\eta)d^{\gamma_e}\right)\kappa_e(\underline{\varepsilon}^{eH} : \underline{\varepsilon}^{eH}) + \left(1-h(\eta)d\right)\mu_e(\underline{\varepsilon}^{eD} : \underline{\varepsilon}^{eD}) + \frac{1}{3}\left(1-h(\eta)d\right)C\underline{\alpha} : \underline{\alpha} + \frac{1}{2}\left(1-h(\eta)d^{\gamma_r}\right)Qr^2\end{aligned} \tag{6}$$

Based on the state potential defined above, the stress-like variables are derived and summarized in Table 1. In these equations κ_e is the compressibility modulus defined by the following equation $\kappa_e = (3\lambda_e + 2\mu_e)/3 = E/(3(1-2\nu))$, μ_e and λ_e are the Lame's constants ($\lambda_e = \nu E/\left((1+\nu)(1-2\nu)\right)$ and $\mu_e = E/\left(2(1+\nu)\right)$), E and ν are the Young's modulus and the Poisson's ratio, C and Q are the kinematic and isotropic hardening moduli, ρ is the material density.

Since this paper focuses on developing damage model with stress state dependence, the classical isotropic von Mises quadratic criterion is used. Remarkably, more advanced quadratic and non-quadratic anisotropic yield function can be easily adopted. Assuming the non-associative plasticity theory, the quadratic von Mises yield criterion $f(\underline{\sigma},\underline{X},R,d)$ and plastic dissipation potential $F(\underline{\sigma},\underline{X},R,Y,d)$ are defined as follows:

$$f(\underline{\sigma},\underline{X},R,d)=\frac{\|\underline{\sigma}-\underline{X}\|_M}{\sqrt{1-h(\eta)d}}-\frac{R}{\sqrt{1-h(\eta)d^{\gamma_r}}}-\sigma_y\leq 0 \tag{7}$$

$$F(\underline{\sigma},\underline{X},R,Y,d)=f+\frac{3a(\underline{X}:\underline{X})}{4C\left(1-h(\eta)d\right)}+\frac{bR^2}{2Q\left(1-h(\eta)d^{\gamma_r}\right)}+\frac{S(\bar{\theta})}{(s+1)\left(1-h(\eta)d\right)^{\beta}}\left\langle\frac{Y-Y_0}{S(\bar{\theta})}\right\rangle^{(s+1)} \tag{8}$$

where σ_y is the yield stress, a and b are material coefficients defining the nonlinear evolution of isotropic and kinematic hardening, and S, s, Y_0, β are damage parameters. From Eq. (8), all the fluxes variables governing the evolution of the dissipative phenomena can be derived, as given in Table 1.To enhance the shear fracture prediction of CDM model, the damage parameter S is defined as appropriate function of normalized Lode angle $\bar{\theta}$ to differentiate between the damage evolution under tension and shear loading paths:

$$S(\bar{\theta})=S_{sh}+\left(S_{ten}-S_{sh}\right)\tanh(|\bar{\theta}|^{\xi_s}) \tag{9}$$

where the normalized Lode angle $\bar{\theta}$ ($-1\leq\bar{\theta}\leq 1$) is defined as follows:

$$\bar{\theta}=1-\frac{6\theta}{\pi}=1-\frac{2}{\pi}\arccos\left(\frac{3\sqrt{3}J_3}{2J_2^{3/2}}\right) \tag{10}$$

The parameters S_{sh} and S_{ten} represent ductility of material in shear (i.e. $S(\bar{\theta}=0)=S_{sh}$) and in tension (i.e. $S(\bar{\theta}=\pm 1)=S_{ten}$) respectively, while ξ_s adjust the effect of $\bar{\theta}$.

Table 1. Stress-like variables and evolution equations.

	Stress-like variables	Evolution equations
Cauchy stress	$\underline{\sigma}^H=\rho\frac{\partial\Psi}{\partial\underline{\varepsilon}^{eH}}=\left(1-h(\eta)d^{\gamma_e}\right)\kappa_e\underline{\varepsilon}^{eH}$ $\underline{S}=\rho\frac{\partial\Psi}{\partial\underline{\varepsilon}^{eD}}=2\left(1-h(\eta)d\right)\mu_e\underline{\varepsilon}^{eD}$	$\underline{D}^p=\dot{\lambda}\frac{\partial F}{\partial\underline{\sigma}}=\dot{\lambda}\underline{n}=\frac{\dot{\lambda}}{\sqrt{1-h(\eta)d}}\underline{\tilde{n}}$ $\underline{\tilde{n}}=\frac{\underline{\underline{H}}(\underline{\sigma}-\underline{X})}{\|\underline{\sigma}-\underline{X}\|_M}$
Kinematic hardening	$\underline{X}=\rho\frac{\partial\Psi}{\partial\underline{\alpha}}=\frac{2}{3}(1-h(\eta)d)C\underline{\alpha}$	$\underline{\dot{\alpha}}=-\dot{\lambda}\frac{\partial F}{\partial\underline{X}}=\dot{\lambda}\left(\frac{\underline{\tilde{n}}}{\sqrt{1-h(\eta)d}}-a\underline{\alpha}\right)$
Isotropic hardening	$R=\rho\frac{\partial\Psi}{\partial r}=(1-h(\eta)d^{\gamma_r})Qr$	$\dot{r}=-\dot{\lambda}\frac{\partial F}{\partial R}=\dot{\lambda}\left(\frac{1}{\sqrt{1-h(\eta)d^{\gamma_r}}}-br\right)$
Damage	$Y=-\rho\frac{\partial\Psi}{\partial d}=\frac{1}{2}h(\eta)\gamma_e d^{\gamma_e-1}\kappa_e(\underline{\varepsilon}^{eH}:\underline{\varepsilon}^{eH})+\frac{1}{3}h(\eta)C\underline{\alpha}:\underline{\alpha}$ $+h(\eta)\mu_e\left(\underline{\varepsilon}^{eD}:\underline{\varepsilon}^{eD}\right)+\frac{1}{2}h(\eta)\gamma_r d^{\gamma_r-1}Qr^2$	$\dot{d}=\dot{\lambda}\frac{\partial F}{\partial Y}=\frac{\dot{\lambda}}{(1-h(\eta)d)^{\beta}}\left(\frac{\langle Y-Y_0\rangle}{S(\bar{\theta})}\right)^s$

3. Parametric study of the proposed model

In this work, the stress triaxiality effect is embedded in the microcracks closure effect with the function h depending on stress triaxiality as shown in Eq. (6). From this dependent function, the desired values of h in tension, shear and compression can be achieved by adjusting the two parameters h_c and ξ_h. Figure 1a displays the effect of parameter h_c on the evolution of accumulated plastic strain (computed by time integration of $\dot{p}=\sqrt{(2/3)\underline{D}^p:\underline{D}^p}$) at fracture according to stress triaxiality. Note that the case h_c=1.0 corresponds to h independent from stress triaxiality (h=1.0). While with the decrease of h_c, the fracture strain at low triaxiality is increased. As h_c decreases from h_c=1.0 to h_c=0.1, the final fracture plastic strain (i.e. ductility) increases significantly for lower stress triaxiality. Since the negative value and positive value of triaxiality represent the compressive and tensile loading respectively, the parameter h_c controls the different damage evolutions from tension to compression.

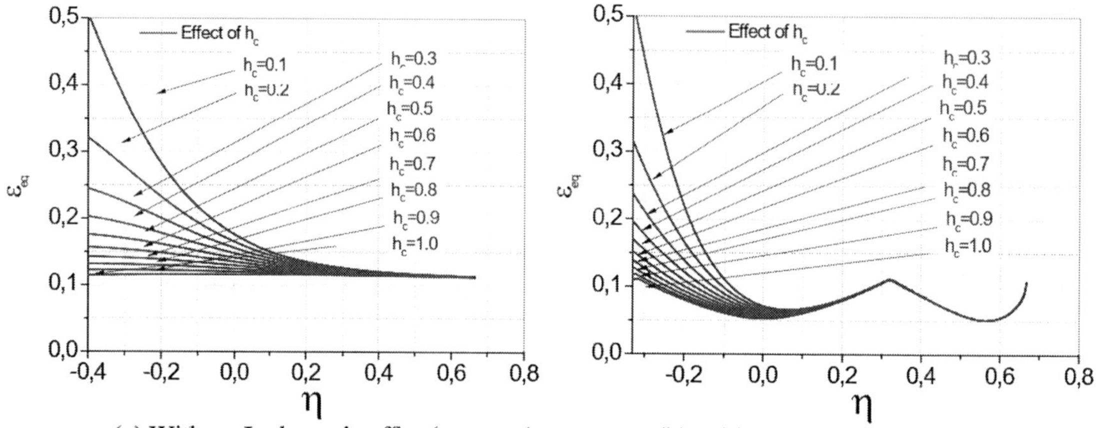

(a) Without Lode angle effect(S_{sh}=S_{ten}) (b) With Lode angle effect($S_{sh} \neq S_{ten}$)

Figure 1. Effect of parameter h_c on the curves accumulated plastic strain vs. stress triaxiality.

The Lode angle effect is taken into account in ductility parameter S which becomes function of the normalized Lode angle as given in Eq. (9). The parameter S_{sh}, S_{ten} and ξ_s are used to adjust the value of the damage parameters S according to different stress states. When the case S_{ten}=S_{sh} is considered the Lode angle will not have any effect on the damage evolution. When the Lode angle effect is taken into account (with S_{ten}=0.5 and S_{sh}=0.1), the equivalent plastic strain at fracture is no longer a monotonic evolution with stress triaxiality, as shown in Figure 1b. It increases with the increase of stress triaxiality within the range ($0 \le \eta \le 0.33$) and ($0.566 \le \eta \le 0.66$), meanwhile, it decreases with the increase of stress triaxiality within the remaining ranges ($-0.33 \le \eta \le 0$) and ($0.33 \le \eta \le 0.566$). The parameter h_c does not affect the value of fracture strain when $\eta \ge 0.33$, a smaller value of h_c leads to a high equivalent plastic strain at fracture for the lower stress triaxiality range. The parameters S_{ten} and S_{sh} in the CDM damage model make it possible to describe the Lode angle dependence of fracture strains.

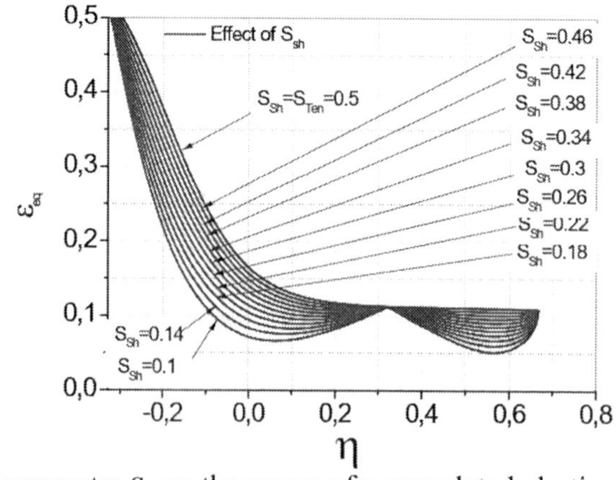

Figure 2. Effect of parameter S_{sh} on the curves of accumulated plastic strain at fracture vs. stress triaxiality considering S_{ten}=0.5 and h_c=0.1.

Figure 2 shows the effect of the parameter S_{sh} (under the condition $S_{sh} < S_{ten}$) on the evolution of accumulated plastic strain at fracture versus stress triaxiality. This figure shows a local maximum for positive triaxiality corresponding to both simple tension ($\eta = 1/3$) and equi-biaxial tension ($\eta = 2/3$) loading paths. The difference between the fracture plastic strain increases when the difference between S_{ten} and S_{sh} is significant. When $S_{ten}=S_{sh}$, Lode angle effect disappears from the accumulated plastic strain at fracture. While with the decrease of the S_{sh}, the Lode angle effect is more obvious.

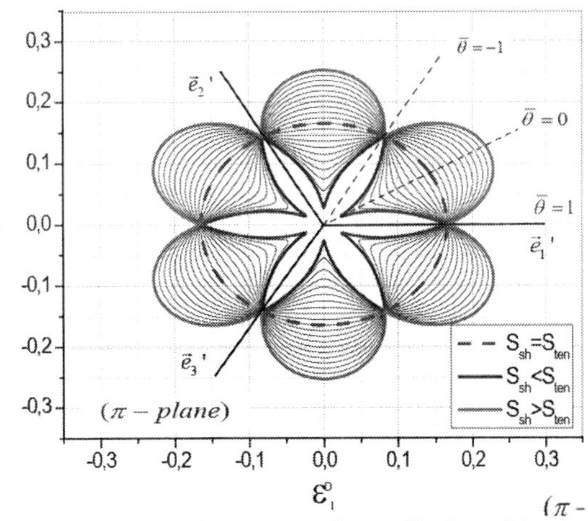

Figure 3. Effect of Lode angle on the damage surfaces displayed in the deviatoric strain plane[8].

The fracture (or damage) loci including the Lode angle effect are plotted in Figure 3. When $S_{ten}=S_{sh}$, there is no Lode angle effect, the fracture surface is a circular surface (dashed line). When $S_{sh} < S_{ten}$, the shape of fracture locus at constant triaxiality varies from a right hexagon to a six-point star. When $S_{sh} > S_{ten}$, the fracture locus at constant stress triaxiality is transferred to a shape of flower with six petals. Note that for this case of initially isotropic von Mises plasticity, the iso-damage surfaces has a $\pi/3$ symmetry so that fracture locus at six vertices with $\bar{\theta} = \pm 1$ are the same. It is observed that the ductile fracture prone to happen at the region between the two vertices (shear or plane strain mode), which was observed experimentally[5].

Through the parametric study, the combined effect of stress triaxiality and Lode angle are discussed above. The Lode angle and triaxiality dependence of the damage surfaces is included in our proposed fully coupled CDM model, which could make accurate predictions of damage in sheet metal forming simulations.

4. Application

The proposed isotropic and isothermal model is applied to a square cup drawing process for Al6014-T4. The material parameters are identified using the available experimental results including various tensile tests from literature[10], as given in Table 2.

Table 2. Calibrated material parameters.

E (GPa)	υ	σ_y (MPa)	Q (MPa)	b	C (MPa)	a	S_{sh}	S_{ten}
69.0	0.3	122.37	263.5	1.2	1725.4	16.2	7.2	8.9
h_c	s	γ_e	γ_r	β	Y_0 (MPa)	ξ_s	ξ_h	
0.2	1.0	1.0	4.0	2.0	0.0	1.0	8.0	

As shown in Figure 4, the force-displacement curve from simulations has good accordance with the experimentally observed one. The distribution of equivalent plastic strain (SDV15) at final rapture is also presented, strain localizations were observed at the four corners of the punch. When the displacement reached at 15 mm, the predicted crack appears at the punch corner of the sheet and the predicted crack path matches well with the experimental observations. From all the comparison of simulation results and experiments, the proposed fully coupled damage model with stress state dependence could correctly predict the fracture of deep drawing of Al6014-T4 sheets.

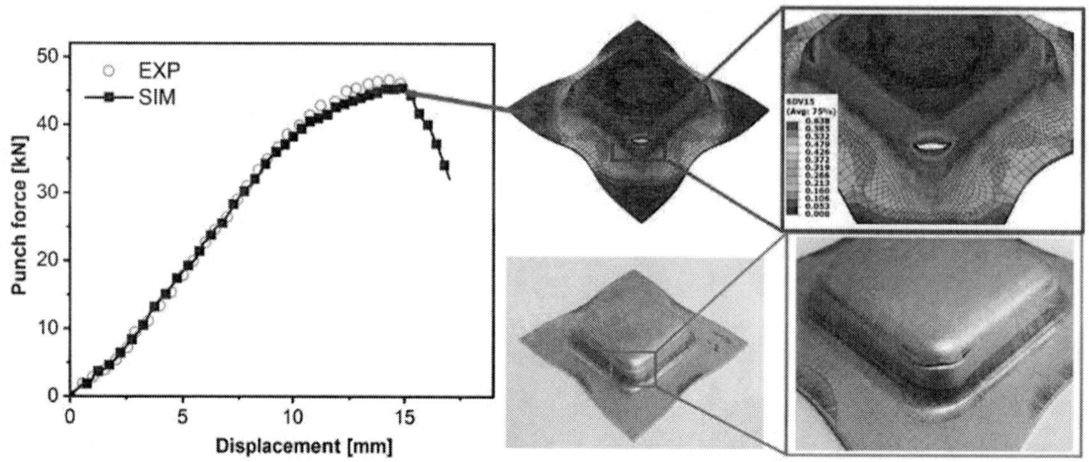

Figure 4. Comparison of numerically predicted and experimentally observed results[10].

Acknowledgments
Financial support by the Qingdao Postdoctoral Applied Research Program (QD20190014), Fundamental Research Funds for the Central Universities (20CX06023A), the European Funds and the Alsace-Champagne-Ardenne-Loraine Region (ESSAIMAGE D201600778) is gratefully acknowledged.

References

[1] Lee J Y, Steglich D and Lee M G 2018 *Int. J. Plast.* **105** 1-23
[2] Saanouni K 2012 *Damage mechanics in metal forming: Advanced modeling and numerical simulation* ed P Devalan (London: John Wiley & Sons)
[3] Yue Z, Cao K, Badreddine H, Saanouni K and Gao J 2019 *Int. J. Mech. Sci.* **153-154** 1-9
[4] Bai Y and Wierzbicki T 2008 *Int. J. Plast.* **24** 1071-96
[5] Lou Y and Huh H 2013 *J. Mater. Process. Technol.* **213** 1284-302
[6] Nahshon K and Hutchinson J 2008 *Eur. J. Mech. A-Solid.* **27** 1-17
[7] Cao T S, Gachet J M, Montmitonnet P and Bouchard P O 2014 *Eng. Fract. Mech.* **124-125** 80-96
[8] Zhang K, Badreddine H and Saanouni K 2020 *J. Mater. Process. Technol.* **277** 116462
[9] Gu B, He J, Li S and Lin Z 2020 *Int. J. Solids Struct.* **182-183** 112-40
[10] Quach H, Kim J J, Nguyen D T and Kim Y S 2020 *Int. J. Mech. Sci.* **169** 105297

IOP Conf. Series: Materials Science and Engineering **967** (2020) 012052 doi:10.1088/1757-899X/967/1/012052

Distortional Hardening Behavior and Strength Different Effect of Pure Titanium Grade 1 Sheets: Experimental Observation and Constitutive Modeling

Quoc Tuan Pham[1,2], Myoung Gyu Lee[3], Young Suk Kim[4]

[1] Division of Computational Mathematics and Engineering, Institute for Computational Science, Ton Duc Thang University, Ho Chi Minh City, Vietnam

[2] Faculty of Civil Engineering, Ton Duc Thang University, Ho Chi Minh City, Vietnam

[3] Seoul National University, Gwanak-ro, Gwanak-Gu, Seoul 08826, Korea

[4] Kyungpook National University, 80 Daehak-ro, Bukgu, Daegu 41566, Korea

E-mail: phamquoctuan@tdtu.edu.vn, myounglee@snu.ac.kr, caekim@knu.ac.kr

Abstract. This study presents a systematic methodology that used to identify the subsequent yield surfaces of a pure titanium grade 1 sheet at different levels of equivalent plastic work. Several experimental tests including uniaxial tensile tests (UT), hydraulic bulge test (BT), simple shear tests (SS), and uniaxial compressive tests (UC) have been conducted for samples prepared in different orientations to achieve a comprehensive experimental data of yielding behaviors observed in different forming modes. Under the condition of equivalent plastic work, the yielding behaviors are characterized and normalized to clarify the distortional hardening behavior and strength different effect for the tested material. It is seen that the yielding surface of the tested material distorts largely during plastic deformations and approaches to its final shape at an equivalent plastic work value of 40 MPa. Under the plane-stress assumption, experimental data obtained from on-axis tests (UT, BT, UC) are used to calibrate three constitutive models: Yld2k, CPB06, and CB04 . Finite element analyses for a simple shear test have been performed in Abaqus/explicit software to validate the developed material models. In conclusion, the CPB06 model provides the best prediction for plastic yielding behaviors of the tested material.

1. Introduction

Pure titanium Grade 1 sheet material is more and more utilized in industrial to make plate type heat-exchanger owing to its high heat-transfer rate, high formability during press forming, and excellent corrosion resistance. Recent efforts have been made to provide better understanding on material behaviors when they are subjected to press forming processes, which are believed to reduce the experimental cost for designing and manufacturing sheet metal products as well as plate heat exchangers [1].

Due to their hexagonal close packed structure, yielding responses of commercially pure Titanium (CP Ti) sheets subjected to tensile and compressive loads are different. The difference between yield stresses in tension and compression of sheet materials is conventionally called strength different (SD) effects. Additionally, the yielding surface of the tested material is found to be distorted under external loads. As consequences, the SD effects of CP Ti sheet are gradually changed during plastic deformation [2].

This study investigates the relation between the distortional hardening behavior and the SD effect of CP Ti sheets subjected plastic deformation. To this goal, a series of standard testing methods have been conducted to observe material responses in different forming modes. Based on these observations, various symmetric and asymmetric yield functions including Yld2k [3], CP04 [4], and CPB06 [5] have been applied to describe the subsequent yielding behaviors of the examined material. These models are then adopted in Abaqus/Explicit package to simulate several forming processes as validations.

2. Experimental procedures

The tested material in this study is pure titanium Grade 1 sheets with a thickness of 0.5 mm which were produced at POSCO, Korea with 600 mm in width and length. To characterize the in-plane anisotropy of the tested material, a series of uniaxial tensile (UT) tests was conducted in the rolling direction (RD), diagonal direction (DD), and transverse direction (TD). In addition, the SD effect was characterized by performing uniaxial compressive (UC) tests along the rolling and transverse directions. Fig. 1 shows the geometry of UT and UC specimens used in this study. All these tests were conducted at a deformation speed of 1 mm/min to achieve material's responses under an assumption of quasi-static deformation.

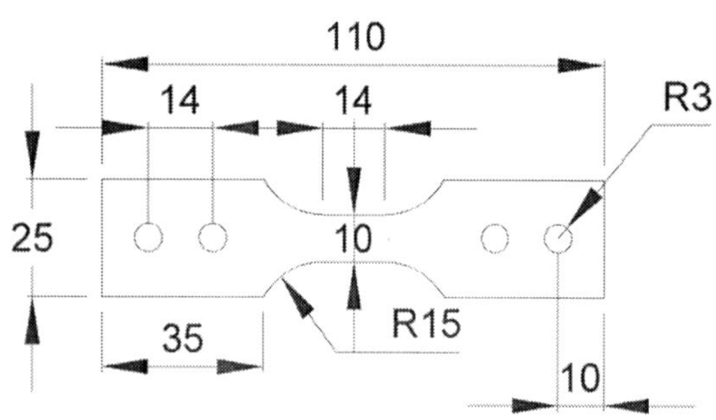

Figure 1. Geometry of the specimens used in uniaxial tensile and uniaxial compressive tests. Unit in mm.

Material characterization under balanced-biaxial tension mode was estimated through hydraulic bulge tests. These tests also provide information of stress-strain relationship over large strain ranges. Fig. 2 shows experimentally measured stress-strain curves obtained from these tests.

Hence, these curves are converted into an effective stress-strain curve under the theory of equivalent plastic work. To this goal, each data point is converted into a corresponding effective data point [6] as follows

$$\sigma_e = \sigma_i / k_i \tag{1a}$$

$$\varepsilon_e = \varepsilon_i . k_i \tag{1b}$$

In these equations, subscript i denotes the considering forming mode, i.e. RD-UT, TD-UT,RD-UC, TD-UC, etc; k_i denotes the corresponding convert factor. It is noted that the stress-strain curve obtained from the UT-RD test was used as the reference stress-strain curve to reproduce the hardening behavior of the tested material. Therefore, its corresponding convert factor always is the unit, as shown in Fig. 3.

It is seen from this figure that the variations of these convert factors according to the increment of the equivalent plastic work are significant and un-ignorable. That further result in the distortion of the yield surface, as reported in our previous work [7]. According to Fig. 3, these convert factors seem to be gradually decreased when plastic deformations are generated and finally approach to their saturated values in large-deformation stages.

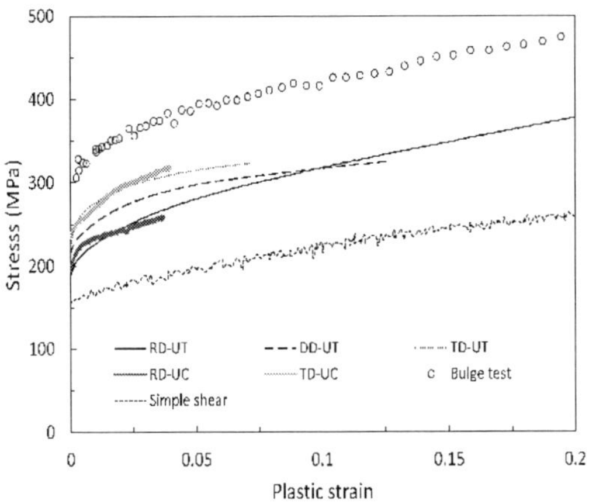

Figure 2. Experimental stress-strain curves obtained from several standard testing methods.

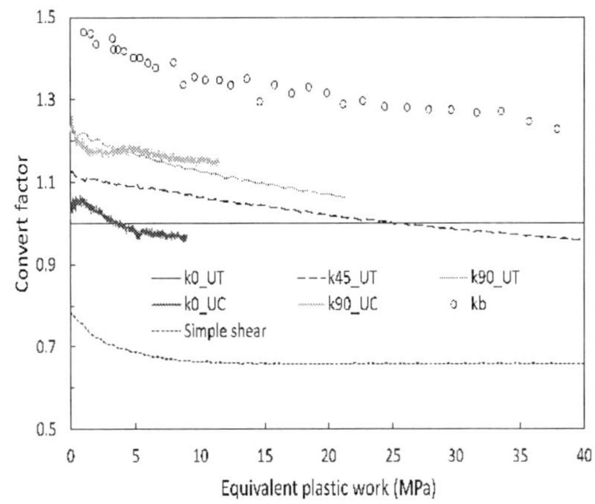

Figure 3. Evolution of convert factors according to increment of equivalent plastic work.

3. Constitutive modeling

The hardening law is one of the essentials for simulating sheet metal forming processes with finite element method (FEM). To identify the hardening law of the tested material, the Kim-Tuan hardening model is adopted to fit with experimental stress-strain data obtained from UT in RD. This hardening model is able to provide good estimation for stress-strain relationship in large strain ranges [8-9].

In order to describe the elastoplastic behavior of the tested material, three constitutive material models are developed under an assumption of associated flow rule. These models include Yld2000-2d symmetric function (Yld2k) [3], Cazacu-Barlat 2004 asymmetric function (CB04) [4], and Cazacu-Plunkett-Barlat 2006 function (CPB06) [5]. Details on each model can be found in references. The formulation of Yld2000-2d model can be expressed in following equation:

$$\text{Yld2k: } \bar{\sigma} = \left\{ \frac{1}{2} \left[|X_1' - X_2'|^m + |2X_1'' - X_2''|^m + |2X_2'' - X_1''|^m \right] \right\}^{1/m}, \tag{2}$$

In this equation, $X_{1,2}^{','' }$ are principals of two tensor transformations: $\boldsymbol{X}' = \boldsymbol{L}'.\boldsymbol{\sigma}$ and $\boldsymbol{X}'' = \boldsymbol{L}''.\boldsymbol{\sigma}$; L' and L'' are two transform operators, which include eight parameters i.e. $\alpha_1 \sim \alpha_8$. Using a tensor transformation approach, Cazacut et al. [5] estimated the equivalent stress in the following formula:

$$\text{CPB06: } \bar{\sigma} = A\left[(|S_1| - kS_1)^a + (|S_2| - kS_2)^a + (|S_3| - kS_1)^a\right]^{1/a}, \tag{3}$$

where A is a constant used to match the calculated stress-strain curve in RD_UT with the referenced data; $S_{1,2,3}$ are principals of the transformed tensor $\boldsymbol{S} = \boldsymbol{L}.\boldsymbol{\sigma}$; $\boldsymbol{L}$ is the transform operator including seven parameters i.e. $c_1 \sim c_7$; k is the parameter describing the SD effect.

Using different approach, the equivalent stress in the CB04 model can be calculated as follows

$$\text{CB04: } \bar{\sigma} = \tilde{J}_2^{3/2} - c\tilde{J}_3 \tag{4}$$

where $\tilde{J}_{2,3}$ are anisotropic forms of stress invariants $J_{2,3}$ which include ten parameters i.e. $a_1 \sim a_{10}$; c is a parameters for describing the SD effect.

A Matlab code is developed to identify values of parameters of each model following a strategy presented in the work of Pham et al. [7]. In this approach, value of parameters in Eqs. (2-4) should be determined at several discrete levels of equivalent plastic work (W^p). Hence, their variations according to increment of W^p reproduce the distortion of the calculated yield surface during plastic deformations.

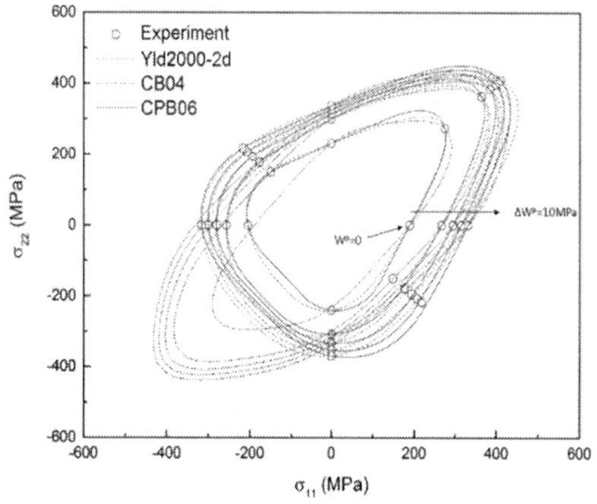

Figure 4. Evolution of yield loci predicted by different models

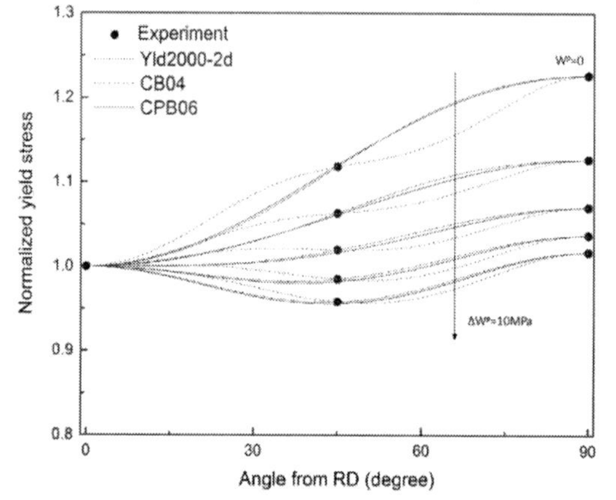

Figure 5. Evolution of normalized yield stress.

Fig. 5 shows predictions of yield loci at several levels of W^p based on different models. All of three models present good prediction for experimental data points. Due to its formula, Yld2k model provides symmetric yield loci, which are unable to capture the SD effect of the examined material. Unlike the Yld2k model, CPB06 and CB04 models predict plat loci in the biaxial-compressive regions. In addition, Fig. 5 compares the predictions of these models for the normalized yield stresses obtained from UT tests with experimental data. All of these models capture well the yield stresses in RD, DD, and TD. The predictions of Yld2k and CPB06 models are almost similar. However, the predictions of CB04 model are largely different from the others. Their predictions of yield loci and normalized yield stresses may affect on the simulated results with finite element method.

4. Numerical verification

This section provides numerical verification for the developed constitutive models. To this goal, simulations of a simple shear test in RD have been performed in Abaqus/Explicit package through VUMAT user's subroutines. Fig. 6 shows an image of un-deformed mini-shear specimen [10] and full mesh on the numerical specimen. In this study, shell element (S4R) with 9 integration points through thickness was adopted to model the specimen.

During these simulations, loading forces according to the displacement of an initial gauge length ($L_0 = 8$ mm) were summarized and reported in Fig. 7. As seen in this figure, Yld2k and CB04 models seem to underestimate the measured loading forces. The observation is relevant to their predictions of yield loci shown in Fig. 4 since these models estimated lower yield stresses than the measured data of TD-UC and SS specimens.

Prediction of the CPB06 model shows a good agreement with the measured loading forces in the test. In this manner, the CPB06 model provides the best proper constitutive modeling for the tested material. However, simulations for other testing methods, for example, in-plane plane-strain tension or punch-stretching tests should be performed to validate its accuracy.

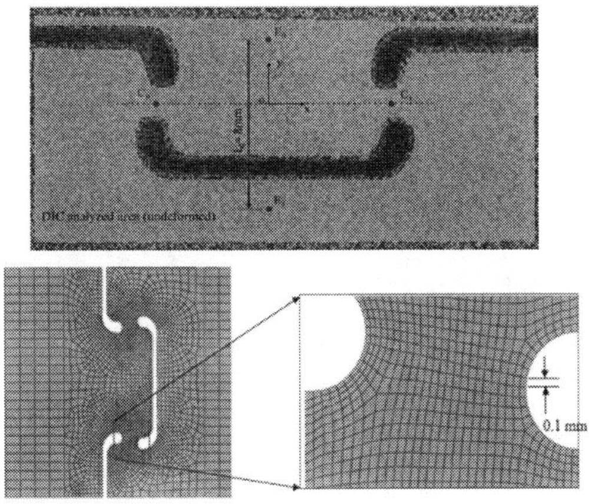

Figure 6. Full mesh on the simple shear specimen

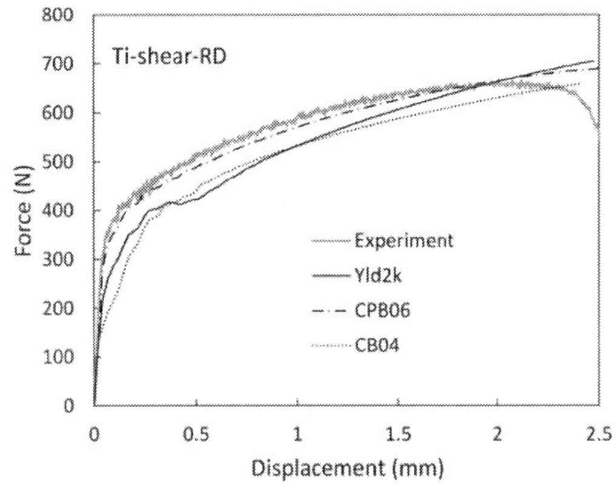

Figure 7. Loading force predictions of different material models and experimentally measured data

5. Conclusions

This study presents experimental observations of distortional hardening behavior and strength different effect in a pure titanium grade 1 sheets by conducting a series of standard testing methods such as uniaxial tensile tests, uniaxial compressive tests, bulge tests, simple shear tests. Based on the observations, three different constitutive models have been developed to describe material's responses during the tests. The developed material models are implemented in Abaqus/Explicit to simulate a simple shear test to verify their accuracy. It is found that the CPB06 model provides the best prediction for loading force during the test. Further studies are needed to validate the applications of the developed models in simulating real forming processes for the tested material.

Acknowledgments

This work was supported by the Ministry of Education of the Republic of Korea and the National Research Foundation of Korea (NRF-2019R1A2C1011224) and Brain Korea 21 (BK21) project at Kyungpook National University.

References

[1] Hama T, Kobuki A and Takuda H 2017 *Int. J. Plast.* **91** 77-108
[2] Okulov I, Kuhn U, Romberg J, Soldatov I, Freudenberger J, Schultz L, Eschke A, Oertel C G, Skrotzki W, and Eckert J 2014 *Mat. Design* **62** 14–20
[3] Barlat F, Brem J C, Yoon J W, Chung K, Dick R E, Lege D J,Pourboghrat F, Choi S H and Chu E 2003 *Int. J. Plast.* **19(9)**1297– 1319
[4] Cazacu O and Barlat F 2004 *Int. J. Plast.* **20(11)** 2017–45
[5] Cazacu O, Plunkett B and Barlat F 2006 *Int. J. Plast.* **22(7)** 1171–94
[6] Kuwabara T, Mori T, Asano M, Hakoyama T and Barlat F 2017 *Int. J. Plast.* **97** 164-186
[7] Pham Q T, Lee M G and Kim Y S 2019 *Int. J. Mech. Sci.* **10** 90–102
[8] Pham Q T and Kim Y S 2016 *Key Eng. Mat.* **716** 87–98
[9] Pham Q T, Lee M G and Kim Y S 2018 *IOP Conf. Series: Journal of Physics: Conf. Series* **1063** 012024
[10] Abedini A, Butcher C and Worswick M J 2017 *Experimental Mechanics* **57** 75-88

IOP Conf. Series: Materials Science and Engineering **967** (2020) 012053 doi:10.1088/1757-899X/967/1/012053

Dynamic hardening properties identification utilizing acceleration data by the Virtual Fields Method

Ji-Min Kim[a*], Jin-Hwan Kim[a] and Frédéric Barlat[a]

[a]GIFT, Pohang University of Science and Technology, San 31 Hyoja-dong, Nam-gu, Pohang, Gyeongbuk 790-784, Republic of Korea

* Corresponding author: skycrossing@postech.ac.kr

Abstract. Finite element (FE) simulations are very important in the field of automotive, aerospace and defense to evaluate crashworthiness of high strength reinforcing parts. Reliable material properties at high strain rates should be used for the FE simulation to acquire accurate simulation results. However, at high strain rates, it is difficult to obtain accurate stress-strain data because reasonable load data is not easily achieved from the experiments due to ringing problem caused by inertia effect. In this study, the virtual fields method (VFM) which is one of the inverse methods suggests another possibility of identifying hardening properties by utilizing acceleration data from the experiments without using load data. In the current study, a methodology is introduced for this purpose. The minimum magnitude of acceleration necessary to retrieve the hardening properties at high strain rate testing is investigated numerically and experimentally. In addition, a new type of high strain rate testing equipment, impact frame high speed tester (IFHS) is described. Various aspects allowing an increase of the acceleration magnitude in the IFHS are discussed for an optimum application of the methodology. Lastly derived hardening properties with the acceleration by the VFM are discussed.

1. Introduction

Steel is one of the mostly used resources around the globe. Many industries for example, automotive, electronics and defense industries are taking advantage of steel. To guarantee the safety of products, demand on accurate dynamic characteristics of steel became significant.

In evaluating the deformation rate of the steel, strain rate is an important criterion. When the deformation speed changes, flow stress is affected which means that flow stress of the steel depends on the strain rate. Thus, many attempts have been made to understand the high strain rate properties. Charpy [1], Izod [2] test or other types using hydraulic machines are well known methods. However, those methods have several disadvantages like spatial limit and difficulty of obtaining accurate load data. Therefore, it is difficult to identify precise dynamic properties with aforementioned methods.

In order to solve this problem, the virtual fields method (VFM), a material property evaluation method, was used in this study. Especially for high strain rate case, a strong advantage of the VFM is that it does not need load data for evaluation of dynamic properties. When sufficient acceleration data is given, the VFM fully utilizes this data, so that dynamic properties can be identified from only acceleration.

Before real experiments, simulations using ABAQUS were carried out. Many kinds of mechanical tests can be done virtually from FE simulations. In this study, the minimum magnitude of acceleration to identify dynamic property was figured out through the FE simulations.

To obtain the sufficient acceleration data from the real experiment, the impact frame high speed tester (IFHS) [3] was used for carrying out dynamic tensile experiments. Attempts to evaluate the dynamic properties of materials using the VFM under high strain conditions using the IFHS have already been made in several previous researches [4, 5]. The IFHS uses the stored elastic strain energy allowing dynamic deformation which is different from conventional ones.

A high speed camera and the digital image correlation technique (DIC) were adopted to capture the deformation and to obtain displacement fields. With a high speed camera, displacement information during whole dynamic deformation was acquired and, acceleration data was captured using the displacements.

An advanced high strength steel (AHSS), DP980 was used for this study.

2. Methodology

2.1. Logarithmic strain

Dense data on the area of interest (AOI) was achieved by using the DIC technique. The whole AOI was covered with triangle mesh. Plane stress state and plastic incompressibility were assumed. As whole deformation process was captured, deformation gradient F for each triangle element was calculated based on the finite deformation theory [6]. The logarithmic strain tensor ε_L was calculated from the deformation gradient F as in equation (1).

$$\varepsilon_L = \sum_{i=1}^{3} \boldsymbol{ln(\lambda_i) r_i \otimes r_i} \tag{1}$$

where λ_i and r_i indicates eigenvalues and eigenvectors of the left stretch tensor V respectively [6].

2.2. Constitutive model

In this study, the Johnson-Cook model with Swift hardening was used to describe the dynamic hardening behavior appropriately.

$$\sigma = (X_1(\varepsilon_P + X_2)^{X_3}) * (1 + Cln\frac{\dot{\varepsilon}}{\dot{\varepsilon}_0}) \tag{2}$$

where $X_1 \sim X_3$ are from the Swift hardening law, ε_P the equivalent plastic strain , $\dot{\varepsilon}$ target strain rate, and $\dot{\varepsilon}_0$ is the reference strain rate. The Johnson-Cook parameter C was identified through the VFM which will be introduced below.

2.3. The virtual fields method

In this study, the virtual fields method (VFM) was used to identify constitutive parameters inversely. The principle of virtual work (PVW) makes possible to link extracting the parameters of the constitutive equations from the experimental data such as displacement or strain obtained through the full-field measurement techniques. In case of high strain rate plasticity case, assuming no body force, the PVW can be written as follows:

$$-\int_V \left[\int_0^t \dot{\sigma}_{ij} dt\right] \varepsilon_{ij}^* dV + \int_{S_f} T_i \cdot u_i^* dS = \int_V \rho a_i u_i^* dV, \quad \forall\, u^*\ KA \tag{3}$$

where $\dot{\sigma}_{ij}$ is the stress rate function of actual strain rate, actual stress and unknown constitutive parameters. $\varepsilon^*$is the virtual strain field calculated from the virtual displacement u^*. V is measured volume, ρ, a means the density and acceleration respectively. The PVW is consist of three terms internal virtual work (IVW), external virtual work (EVW) and acceleration virtual work (AVW). In Equation (3), the external virtual work term can be eliminated with using proper virtual displacement fields. So only the IVW including stress information and AVW including acceleration information are left in the PVW. Thus the objective function only includes the IVW and AVW. The chosen u^* was as follows:

$$u_x^* = 0 \text{ and } u_y^* = (y - y_{min})(y - y_{max}) \tag{4}$$

where x and y means the horizontal and vertical coordinates in deformed configuration respectively.

2.4. Impact frame high speed tester (IFHS)
Several types of experiments can be conducted to study the dynamic behavior of materials. In this study, a new type of high speed tensile tester, impact frame high speed (IFHS) tester was used for the experiments [3]. Unlike ordinary high strain rate tensile tester, the IFHS utilizes the elastic strain energy. The original IFHS tester was invented for cement fiber specimens and in this study it was modified to use the steel sheet specimens.

When the hydraulic pump pulls the coupler, elastic strain energy is stored in the two energy frame bars. When coupler breaks, the stored energy is converted into kinetic energy to pull the specimen. The number 1 to 7 in Figure 1 means a high speed camera, an elastic energy frame bar, a transmitter bar, a hydraulic motor, a coupler, a lighting system and a specimen respectively.

Figure 1. IFHS tester set-up for high strain rate tensile test.

3. Simulation
ABAQUS dynamic explicit analysis mode was used for simulation. The objective of simulation was to determine minimum acceleration magnitude for obtaining constitutive parameter with the acceleration information by the VFM. Then, the Johnson-Cook parameter was identified by the VFM based on the simulation results. Finally, two stress-strain curves from the input one and identified one by the VFM were compared.

In order to reproduce real experimental features, the velocity was extracted from the real experiments then this data was applied as a velocity boundary condition for simulations. Figure 2 describes the velocity boundary condition used for the simulation.

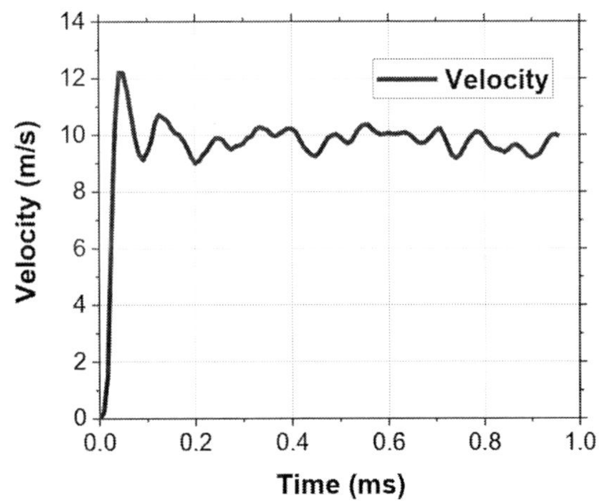

Figure 2. Experimental result based velocity boundary condition.

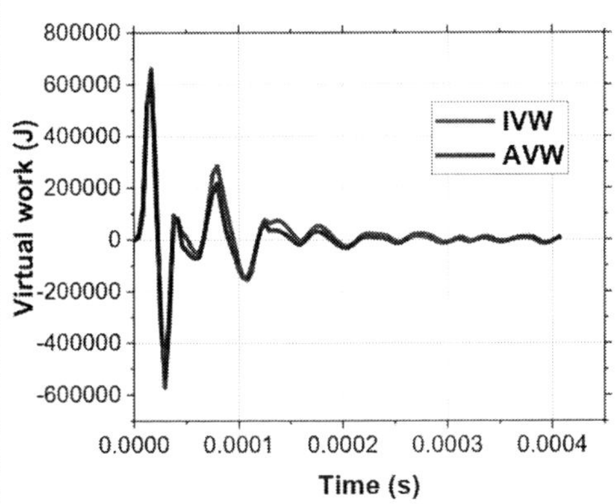

Figure 3. The IVW and AVW convergence result.

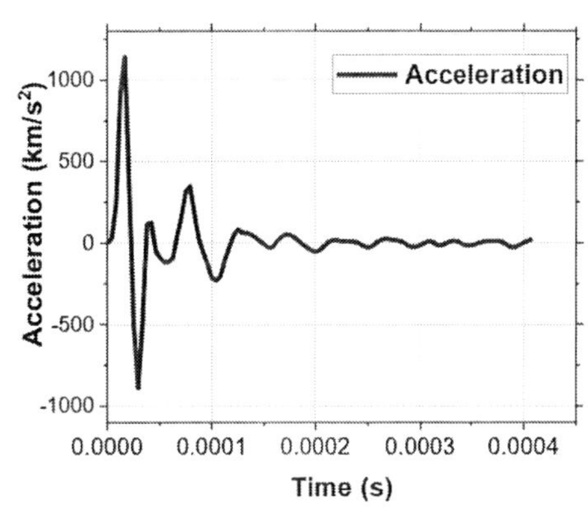

Figure 4. The acceleration data from the simulation.

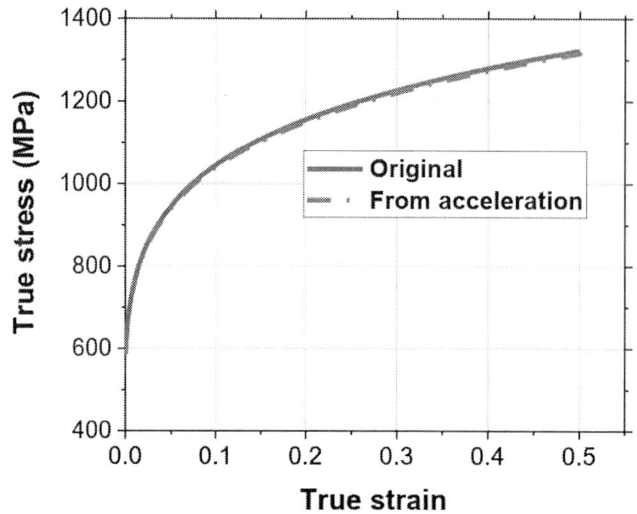

Figure 5. The stress strain curve comparison. Red one (Original) is plotted with the input parameter in simulation and green dotted one is plotted with the identified parameter.

Simulation result proved that constitutive parameter can be identified by the VFM, when sufficient magnitude of acceleration is provided. The objective function which is consist of the IVW and AVW showed good convergence. Figures 3 and 4 show the convergence of the objective function and the amount of acceleration captured from the simulation. In Figure 5, red curve means the curve from the input parameter and green dotted curve is the curve with the parameter identified by the VFM. Finally, Figure 5 verifies that the parameter identified by the VFM using the acceleration is reasonable.

4. Methods to increase the acceleration

The IFHS tester was used for high strain rate tensile test. But, captured magnitude of acceleration was not sufficient for successful identification. Thus, methods to increase acceleration were needed to be applied. In this study, two methods were considered.

4.1. Increase capacity of coupler

When the capacity of coupler increases, the amount of stored elastic strain energy increases naturally. However, it might be lead to unexpected failure of other parts of the IFHS tester. Thus it is not a recommended method.

4.2. Decrease the diameter of an energy frame bar

Two energy frame bars store elastic strain energy when they are pulled by the hydraulic pump. It means that if the stored energy can be increased, the higher impact velocity can be obtained. What is important is that the decreased diameter should be strong enough to bear high tensile load from the motor. The optimized diameter should be carefully selected based on precise calculations.

5. Conclusion

In this study, a method to identify constitutive parameter inversely by the VFM was introduced. The advantage of the VFM in high strain rate case is that the load data is not necessary when sufficient amount of acceleration information can be applied. Simulation result from the ABAQUS verified numerically.

IOP Conf. Series: Materials Science and Engineering **967** (2020) 012053 doi:10.1088/1757-899X/967/1/012053

To verify experimentally, the IFHS tester was used. However, with the current setup of the IFHS tester, the magnitude of acceleration was not enough. Therefore, to increase the acceleration magnitude, two possible methods were suggested.

In the presentation, the experimental results will be presented and some aspects for the importance of sufficient acceleration magnitude will be delivered.

References

[1] G. Charpy, 1901 *Mémoire compte-rendus la Société des ingénieurs Civ.* Fr. 848–77
[2] E. G. Izod, 1903 *Engineering.* **76** 431-2
[3] Kim D J, K. Wille, S. El-Tawil, and A. E. Naaman, 2011 *J. Eng. Mech.* **137** 268–75
[4] Leem D H, 2016 M.S. Thesis, (GIFT, POSTECH, Pohang)
[5] Park J S, 2018 M.S. Thesis, (GIFT, POSTECH, Pohang)
[6] Dunne F and Petrinic N 2005 *Introduction to Computational Plasticity* (Oxford University Press on Demand)
[7] Kim J H, A. Serpantié, F. Barlat, F. Pierron, and Lee M G 2013 *Int. J. Solids Struct.* **50** 3829–42

Study of Khan-Huang-Liang (KHL) Anisotropic Deformation Model for Deep Drawing Behaviour of Inconel 718 Alloy

Gauri Mahalle[1], Omkar Salunke[1], Nitin Kotkunde[1*], Amit Kumar Gupta[1] and Swadesh Kumar Singh[2]

[1] Department of Mechanical Engineering, BITS Pilani, Hyderabad Campus, Hyderabad-500078, Telangana, India
[2] Department of Mechanical Engineering, GRIET, Hyderabad - 500072, Telangana, India

E-mail: p20160415@hyderabad.bits-pilani.ac.in

Abstract. Study of anisotropic deformation behavior of a material plays a crucial role in optimizing hot working process parameters and trustworthy Finite Element (FE) analysis in sheet metal forming processes. In this work, Khan–Huang–Liang (KHL) phenomenological based constitutive model and anisotropic yield criteria has been formulated for Inconel 718 alloy. Firstly, uniaxial tensile tests have been conducted at different temperatures (room temperature - 700°C) and slow strain rates (0.0001-$0.1 s^{-1}$) conditions. KHL constitutive model has been formulated and validated with experimental flow stress data. The prediction capability of the model is evaluated based on correlation coefficient (R), average absolute error, AAE (Δ) and its standard deviation (s). Subsequently, anisotropic yielding behavior of Inconel 718 alloy is predicted based on KHL yield criterion. Anisotropic coefficient (Lankford parameters) and tension compression asymmetry parameters have been calculated experimentally. The prediction capability of KHL yield criterion is analyzed based on yield locus, yield stress variation and anisotropic coefficient variation. The quality index of performance, namely global accuracy index (β) is evaluated. Further, Finite Element (FE) analysis has been carried out for deep drawing of Inconel 718 alloy using commercially available ABAQUS software. The developed KHL constitutive model and anisotropic yield criterion has been incorporated in FE simulation using UMAT/VUMAT code. The FE results are validated with experimental deep drawn cups at different process conditions.

Keyword: Inconel 718 alloy, Deformation Behavior, Constitutive Model, Anisotropy, Deep Drawing.

1. Introduction

The sheet metal forming process is mostly used in manufacturing industries, mainly to switch conventional welding process. Critical components (with definite dimension, mechanics, and appearance requirements), are certainly manufactured with the help of various advanced forming procedures. These processes are mostly limited by the instability occurrence or flow localization. It is strongly depends on intrinsic mechanical properties of material and extrinsic factors such as operation/process conditions, selection of different process parameters and tooling material, etc. The higher performance necessities, safety and environmental concerns have been forcing industry to choose correct parameters to satisfy ever-increasing market needs [1]. But forming of high strength and limited ductility material is sometimes an unfeasible task. In literature, warm/elevated temperature forming is proposed as one of the proven technique to produce complex shapes of these materials. The elevated temperature forming facilitate easy material flow in the die cavity and substantially decrease the amount of elastic recovery (springback) during the plastic deformation. But, the major issues in industrial utilities are high tooling and material cost, complex die setup, additional temperature and cooling control arrangement [2–4].

In order to explain complicated deformation behavior of different matels/alloys by the process variables such as strain, strain rate & temperature, extensive investigations have been conducted to improve constitutive relations. Lin and Chen presented a study of different constituent models, further split down into 3 groups, namely phenomenological, physical and Artificial Neural Network (ANN) based models [1]. Phenomenological based model gives a classical approach for modeling material behavior by viscoplastic theory. This theory aims at simulating the processing parameters with other variables by considering the hardening and softening [1]. Popular models were Johnson–Cook, Khan Huang Liang (KHL) and Arrhenius [1]. Physical-based models are focused on the physical aspects wherever dislocation movement activated thermally, thermodynamics theory, and kinetics of slips were involved. Presently, artificial intelligence technology, primarily neural networks based on genetic algorithms and back propagation algorithms, provides promising results in the fields of materials science, material processing and shape [2]. These have been well-defined algorithms which reduce complex calculations and require many experiments to find material constants. Limited studies on Inconel 718 alloy have been reported to define relation between deformation parameters and uniaxial flow stress at high temperature. A new constitutive model is developed by Lin et al. [3] to define deformation behavior at slow strain rates of aged Inconel 718. This model is combination of two models namely viscoplastic model (to define work hardening stage and DRX) & phenomenological model (to define dynamic softening stage). But, this new model unable to envisage flow behavior for high temperatures and different strain rates.

Quality of formed product is significantly influenced by anisotropy behaviour.Many researchers have made several efforts on development of the anisotropic yield criterion. Hill and Barlat criteria were popularly in sheet metal productions due to effortlessness to define material constants and easy to apply in FE analysis[4,5]. However, these yield criteria unable consider the biaxial deformation (prominent in the forming processes), deviation in uniaxial tensile strength and anisotropy coefficients is not captured [6]. A new yield criterion proposed by Khan and Yu based on the KHL constitutive model. This model not only describe anisotropic yield behaviour but also compression-tension asymmetricity of alloy. This proposed KHL model is used to predicted yielding behaviour at diverse strains, temperatures and strain rate with excellent relationship with experimental data. Model is generalized to introduce shape independent term for predicting yielding behaviour of different metals [7]. It is noticed that, recently few attempts were made to develop a constitutive models and yield criterion for Inconel alloys [8]. However, anisotropic yield criterion depends mainly on strain rates and temperatures variation dependent of Inconel 718 is not reported yet. Thus, present work mainly focuses on the thorough investigation of anisotropic yield criterion based on KHL constitutive model at different strain levels, temperatures & strain rates.

2. Materials and Experimental Methods

Inconel 718 sheet (commercially available) of 1 mm thick was used in the current study. Major alloying elements of Inconel by % mass are 51.5%Ni, 20%Fe,18.4%Cr, 5%Nb 3%Mo along with balancing elements (≤ 0.17%).Uniaxial tensile tests were conducted at different test temperatures (room temperature(28°C) -700°C) and slow strain rates (0.0001-0.1s^{-1}) conditions. Authors in their previous work discussed detail experimental procedure[8–10].Tensile strength were evaluated by sequence of experiments as listed in Table 1. Lankford Coefficients (r values) (r_0, r_{45}, r_{90}), which defines resistance to thinning and further promotes the material flow into the die cavity, were also evaluated at different tension axis/sheet orientations of RD(0°), DD(45°) & TD(90°) to RD. Due to unavailability of biaxial test facility, GN738 superalloy (similar chemical composition and mechanical properties) , biaxial (r_b and σ_b) properties have been used in present work [11].

Table 1: Representative mechanical properties of Inconel 718 alloy

Temp.	σ_0(GPa)	(r_0)	σ_{45}(GPa)	(r_{45})	σ_{90}(GPa)	(r_{90})	σ_b(GPa) [11]	% elong
RT(28 °C)	0.545	0.781	0.5006	0.941	0.504	1.043	0.502	42
300 °C	0.424	0.997	0.4109	1.095	0.396	0.969	0.404	48
500 °C	0.371	1.142	0.3478	0.969	0.328	0.917	0.338	52
700 °C	0.324	1.123	0.3100	1.336	0.295	1.132	0.281	55

To perform warm deep drawing, laboratory-scale tool setup (consisting flat bottom punch, blank holder and die) is designed and fabricated as per schematic shown in Fig 1a. The warm deep drawing set up is designed with independent heating of die with induction heating elements with K-type thermocouples at three different locations. The warm deep drawing experiments are conducted using a 40-Ton double acting hydraulic press. All experiments are performed with applying graphene-based moly-spray as lubricant. Mainly deep drawing experiments performed under different die-punch temperature combination at 28-28°C, 300-28°C and 500-28 °C with fixed blank holding pressure (BHP = 2.5 MPa). Inconel 718 blanks of different diameter namely 58, 60 and 62 mm are used for drawing (Fig.1b). Specimens are wire cut out from parent sheet and centering to Inconel 718 blanks is done so as to reduce deep drawing defect. For each setting, 3 test were performed and avg. value was stated for consistent results. Fig 1c shows the drawn Inconel cup for 60 mm blank diameter at room temperature.

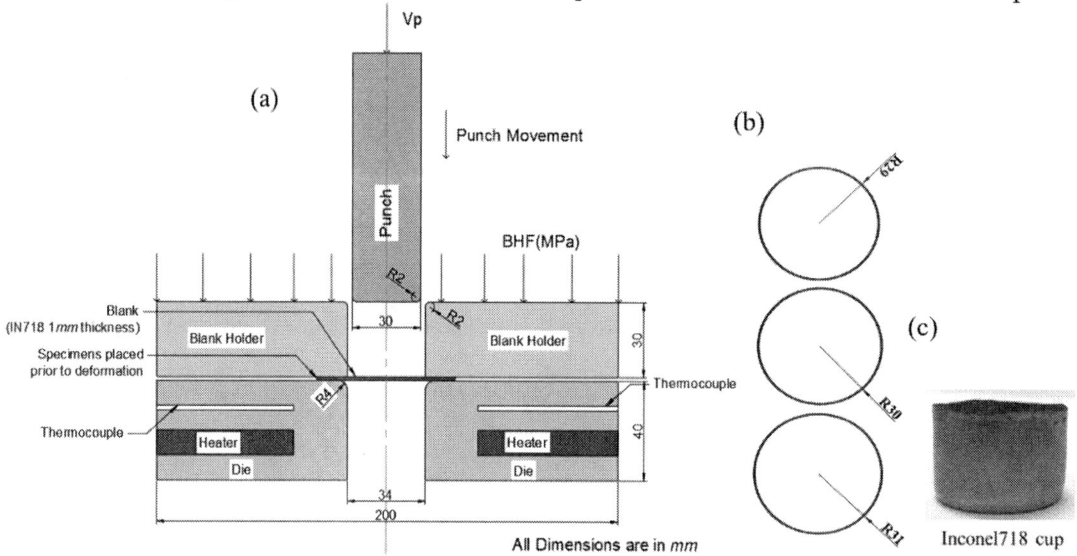

Fig. 1: Schematic of (a) warm deep drawing set up (b) Inconel blank dimensions (c) Inconel 718 drawn cup (D = 60 mm)

3. Material Modeling

3.1. Khan–Huang–Liang (KHL) Phenomenological Constitutive Model

The constitutive model expresses the flow behaviour as a function of processing variables, like strain, strain rate, and test temperature. Predicted flow stress by KHL model [9] is stated as,

$$\sigma = \left[A + B\left(1 - \frac{\ln \dot{\varepsilon}}{D_p{}^0}\right)^{n_1} \varepsilon_p^{n_0}\right]\left(\frac{\dot{\varepsilon}}{\dot{\varepsilon}^*}\right)^C \left(\frac{T_m - T}{T_m - T_{ref}}\right)^m \tag{1}$$

where, ε_p & σ are true plastic strain & true stress. T, T_{ref} & T_m are the current, reference (28°C) and melting (1337°C) temperatures of the Inconel 718 alloy respectively. $D_p^0 = 10^{-6}$ s^{-1}, called as the deformation rate (subjectively selected strain rate upper bound) and $\dot{\varepsilon}^*$(reference strain rate) = 0.01 s^{-1}, at T_{ref}, where A, B and n_0 constants were evaluated). Additional constants are n_0, n_1, C and m. Material constant are evaluated by detailed procedure followed by Khan et al.[12] and listed in the Table 2.

Table 2: Calculated constants in KHL constitutive model

C	B (Pa)	A (Pa)	m	n_0	n_1
1.3417	580.06 $\times 10^6$	828.9 $\times 10^6$	1.385	0.62	0.0135

The equation for KHL for Inconel alloy is given as

$$\sigma = \left[828.9 \times 10^6 + 580.06 \times 10^6 \left(1 - \frac{\ln \dot{\varepsilon}}{\ln D_p^0}\right)^{0.0135} \varepsilon_p^{0.62}\right] \left(\frac{\dot{\varepsilon}}{\dot{\varepsilon}^*}\right)^{1.3417} \left(\frac{T_m - T}{T_m - T_{ref}}\right)^{1.385} \quad (2)$$

3.2. Khan–Huang–Liang (KHL) Anisotropic Yield Criterion

A novel anisotropic yield criterion is suggested by Khan et. al. [7] disassociating not only anisotropic yielding behaviour but also compression-tension asymmetricity. Generalized form of yield function is given by

$$\left[\frac{2}{3}\sin\left(\theta + \frac{\pi}{3}\right)\right]^k \left[e^{-C(\xi+1)}(F\sigma_1^2 + G\sigma_1\sigma_2 + H\sigma_2^2 + P\sigma_{12}^2)\right] = 1 \quad (3)$$

where θ - Lode angle, k-material constant (calculated by fitting the plane strain experimental data, k = 0.5 considered in the present study). Lode angle in the deviatoric stress plane, expressed as $\cos\theta = \frac{\sqrt{3}}{2}\frac{S_1}{\sqrt{J_2}}$ [13], where S_1 & J_2 - first and second deviatoric stress invariant. First term in yield function$\left[\frac{2}{3}\sin\left(\theta + \frac{\pi}{3}\right)\right]^k$, is called as shape-dependent term. Further, Eq. (3) can modify as function temperature, strain rate or strain to consider subsequent shape revolution in the work hardening. Ratio $e^{-C(\xi+1)}$ describe tension compression asymmetry, where, ξ = 1. Material constants are given as,

$$F = \frac{1}{X^2};\ H = \frac{1}{Y^2};\ P = \frac{1}{\tau^2};$$
$$G = \frac{1}{Z^2} - \frac{1}{X^2} - \frac{1}{Y^2};\ C = -\frac{1}{2}\ln\left(\frac{X^2}{\sigma_{1T}^2}\right) = -\ln\left(\frac{X}{\sigma_{1T}}\right) \quad (4)$$

where, X, Y and Z - 0.2% of yield strengths (compressive) along RD(0°), DD (45°) & TD (90°) to loading direction respectively, τ - initial yield strength (shear). σ_{1T} - 0.2% of yield strengths (tensile) along RD. Initial value of C is -0.074 (average of evaluated yield results along RD and TD). Values of G, H, F and P are listed in Table 3.

Table 3: Anisotropy coefficients for KHL yield model

Temp.	G	H	F	P
RT(28°C)	-0.5171	1.185	1	1.1769
300°C	-0.3085	1.759	1.6506	1.8229
500°C	-0.2251	2.455	2.1599	2.5947
700°C	-0.1204	2.754	2.3571	2.9575

4. Result and Discussion

Fig.2a shows the plots of predicted flow stress and experimental stresses at different temperatures and strain rates. KHL constitutive model displays comparatively superior agreement for the predicted flow stress. Fig.2b shows plots between yield loci with experimentally measured stresses at different test strain rates & temperatures. It is clear from plots that yield loci of KHL model passes well through uniaxial and biaxial tensile points. This confirms suitability of KHL yield model for Inconel 718 alloy. A statistical parameters namely correlation coefficient (R), is generally used for the suitability. Fig. 3(a) shows evaluated correlation coefficient among predicted and measured (experimental) flow stress of

KHL model at tested strain rates. As, correlation coefficient was biased near higher or lower values. Therefore, it is essential to calculate other statistical parameters namely average absolute error (AAE) (Δ) and its standard deviation (s) [14]. Here, KHL constitutive model show moderate values for lower error (Δ), and its standard deviation. It is also possible to test the capability of any yield parameters by comparing the estimated values of r and σ values with experimental values. Fig. 3(b-c) show deviation of normalized σ and r-value with the loading axis. It has been found that yield function has good agreement for stress and significant inconsistency for r-variation. Therefore, it is helpful to estimate numerically the yield criteria performance before finalizing their suitability for the yielding behaviour prediction for Inconel 718 alloy. To evaluate performance of the yield criteria, global accuracy index (ω) is used to analysis the yield locus shape, and to visualize the planer distribution of uniaxial strength and the anisotropic coefficients [15]. Mathematically, it is given as,

$$\omega = \varphi + \delta + \gamma \ [\%] \qquad (5)$$

$$\text{where, } \varphi = \frac{\sqrt{\sum_{i=0}^{j} d^2(P_i, Q_i)}}{Y} \times 100\ \% \qquad (6)$$

$$\delta = \sqrt{\sum_{i=0}^{j} \left(\frac{\sigma_{\theta_i}^{exp} - \sigma_{\theta_i}^{pre}}{\sigma_{\theta_i}^{exp}} \right)} \times 100\ \% \qquad (7)$$

$$\gamma = \sqrt{\sum_{i=0}^{j} \left(\frac{r_{\theta_i}^{exp} - r_{\theta_i}^{pre}}{r_{\theta_i}^{exp}} \right)} \times 100\ \% \qquad (8)$$

Where, $d^2(P_i, Q_i)$ - Square of distance from the P_i (experimental to Q_i (its projection) in the predicted yield loci, j is total number of considered experimental points. θ_i is angle measured from loading axis. $\sigma_{\theta_i}^{exp}$, $r_{\theta_i}^{exp}$, $\sigma_{\theta_i}^{pre}$ & $r_{\theta_i}^{pre}$ are experimental and predicted yield strength and anisotropy coefficients along $\theta_i$respectively. Fig. 3d shows error function comparted at different temperature and strain state. It specifies suitability of the yield criterion for Inconel 718 at different temperatures with error function $15 < \omega < 29\%$

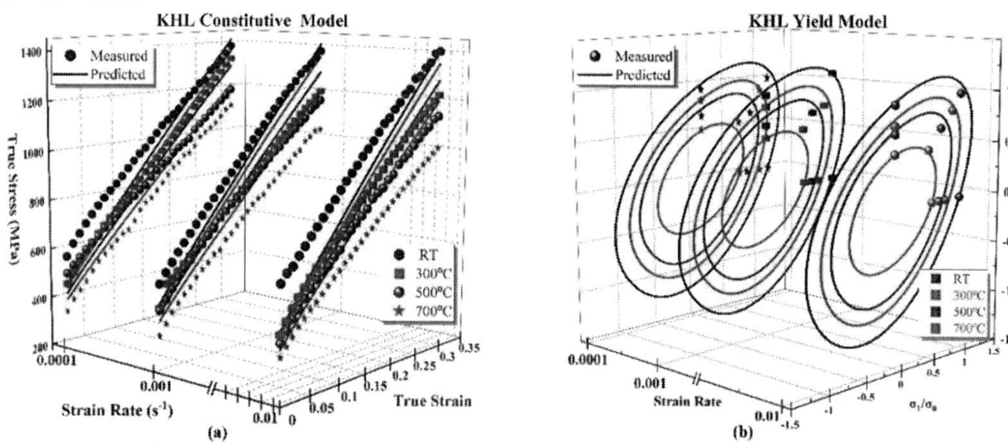

Fig. 2: Comparison of predicted and measured (experimentally) (a) flow stresses and (b) Yield loci at different strain rates

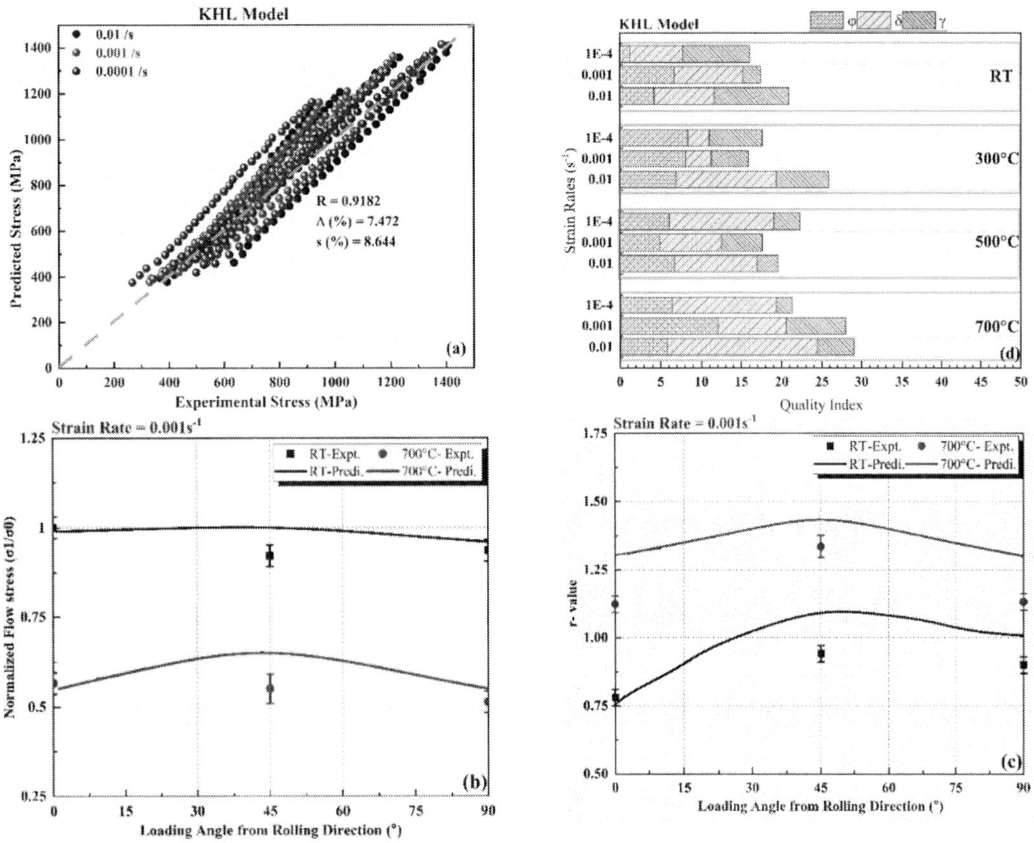

Fig. 3: (a) Correlation coefficient, deviation of (b) Normalized flow stress (c) r-value directionality and (d) Quality index at different test temperature

A quarter input FE model was developed in Abaqus-Explicit-SIMULIA™ version 6.14 due to symmetric geometry and to reduce total computational time. Fig.4a gives FE simulation model of deep drawing set up with blank, die, blank holder and flat bottom punch where except Inconel blank, and all tools are assigned as rigid bodies. In FE model, punch is allow to move downward (y-direction) with a velocity profile (trapezoidal) with fixed die. In order to restrict material flow in flange portion, blank holding force of 2.5MPa was considered in y-direction. Inconel blank is considered as deformable body with planar and normal anisotropy properties and material properties from Table 1 are considered as input. The developed KHL constitutive model and anisotropic yield criterion has been incorporated in in user defined material subroutine (UMAT) for FE simulation. The FE results are validated with experimental deep drawn cups at different process conditions in terms of limit dome height (LDH) and average thickness (Fig 4b). LDH is measured as drawn height of formed cup just before occurring of fracture. Comparison of experimental and numerical results for punch displacement in terms of limit dome height at different test temperature is shown in Fig 5a. It is noticed that the FE predicted results are well comparable with experimental values. Average thickness distribution for specimen with diameter 60 mm with distance from the centre of cup is shown in Fig 5b. It is observed that thickness of drawn specimen remains constant initially and then it underway declining. Minimum thickness is perceived at a bending location wherever friction between punch and blank was maximum (as shown in insert in Fig.4b). FE predicted results are well comparable. Uniform thickness distribution is main concern in deep drawing, thus relative error, between experimental and FE predicted results, was evaluated. The avg. error (in %) for limit dome height and average thickness distribution were 2.93% and 4.64 % respectively. As error is within acceptable range, hence KHL constitutive model coupled with KHL yield criteria are best suited for execution of deep drawing of Inconel 718 alloy by FE analysis.

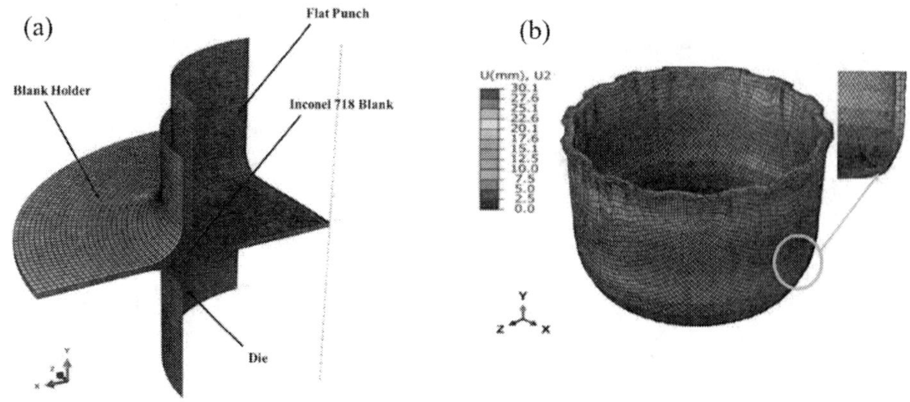

Fig 4: (a) FE model of deep drawing setup (b) Deep drawn cup at 28°C (D= 60 mm)

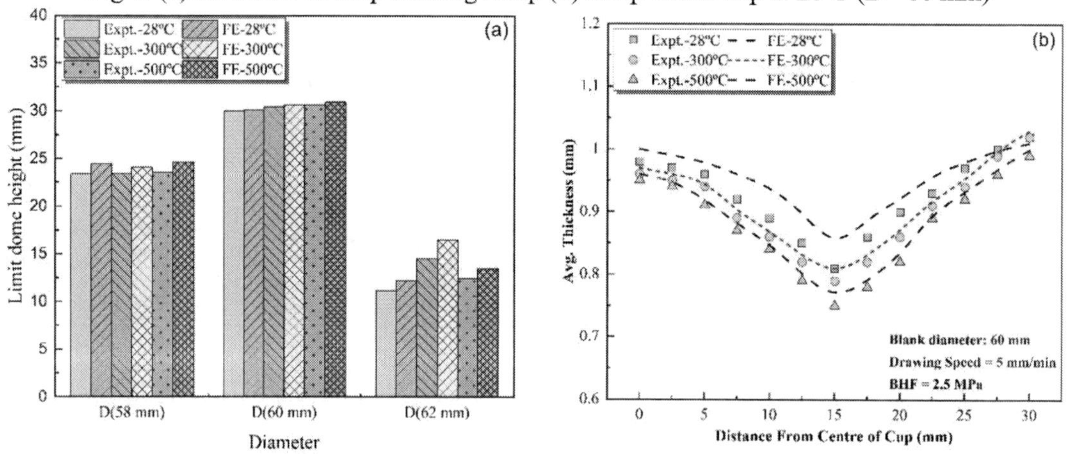

Fig 5: Experimental and FE predicted (a) Limit dome height (b) Average thickness at different test temperature.

5. Conclusion

Khan–Huang–Liang (KHL) constitutive and yield model has been formulated for Inconel 718 alloy. Some of the important conclusion:

1. Uniaxial tensile flow stress of Inconel 718 have been considerably affected by test temperatures and strain rates. KHL constitutive model display good agreement with higher correlation coefficient (R > 0.9182), lower Average absolute error (AAE) and its standard deviation with experimental flow stress.
2. Anisotropic yield behaviour for Inconel 718 have been calculated by KHL yield criteria. The criteria were assessed based on the experimental yield strength values, yield stress and anisotropic coefficient variation. It was noticed that yield function a good agreement for σ-variation and significant inconsistency has been perceived for *r*-variation.
3. The developed KHL constitutive model and anisotropic yield criterion has been incorporated in FE simulation using user defined material subroutine UMAT code. The avg. error (in %) for limit dome height and average thickness distribution were 2.93% and 4.64 % respectively.

References

[1] Lin Y C and Chen X M 2011 A critical review of experimental results and constitutive descriptions for metals and alloys in hot working *Mater. Des.* **32** 1733–59
[2] Lin Y C, Zhang J and Zhong J 2008 Application of neural networks to predict the elevated temperature

flow behavior of a low alloy steel *Comput. Mater. Sci.* **43** 752–8

[3] Lin Y C, Li K-K, Li H-B, Chen J, Chen X-M and Wen D-X 2015 New constitutive model for high-temperature deformation behavior of inconel 718 superalloy *Mater. Des.* **74** 108–18

[4] Mahalle G, Salunke O, Kotkunde N, Gupta A K and Singh S K 2019 Anisotropic yielding behaviour of Inconel 718 alloy at elevated temperatures *ASME International Mechanical Engineering Congress and Exposition, Proceedings (IMECE2019)* vol 2A (Salt Lake City, Utah, USA) pp 1–5

[5] Mahalle G, Kotkunde N, Gupta A K and Singh S K 2020 Efficacy of semi-empirical models for prediction of forming limit curve of IN718 alloy at elevated temperatures *Adv. Mater. Process. Technol.* **00** 1–13

[6] Plunkett B, Cazacu O and Barlat F 2008 Orthotropic yield criteria for description of the anisotropy in tension and compression of sheet metals *Int. J. Plast.* **24** 847–66

[7] Khan A S, Yu S and Liu H 2012 Deformation induced anisotropic responses of Ti-6Al-4V alloy Part II: A strain rate and temperature dependent anisotropic yield criterion *Int. J. Plast.* **38** 14–26

[8] Mahalle G, Kotkunde N, Gupta A K, Sujith R, Singh S K and Lin Y C 2019 Microstructure Characteristics and Comparative Analysis of Constitutive Models for Flow Stress Prediction of Inconel 718 Alloy *J. Mater. Eng. Perform.* **28** 3321–3321

[9] Mahalle G, Kotkunde N, Gupta A K and Singh S K 2020 Prediction of flow stress behaviour by materials modelling technique for Inconel 718 alloy at elevated temperature *Adv. Mater. Process. Technol.* **00** 1–8

[10] Mahalle G, Kotkunde N, Shah R, Gupta A K and Singh S K 2018 Analysis of Flow Stress Behaviour of Inconel Alloys at Elevated Temperatures Using Constitutive Model *Journal of Physics: Conference Series* vol 1063

[11] Xiao R, Li X X, Lang L H, Chen Y K and Yang Y F 2016 Biaxial tensile testing of cruciform slim superalloy at elevated temperatures *Mater. Des.* **94** 286–94

[12] Khan A S and Liang R 1999 Behaviors of three BCC metal over a wide range of strain rates and temperatures: Experiments and modeling *Int. J. Plast.* **15** 1089–109

[13] Khan A S and Huang S 1995 Continuum Theory of Plasticity *John Wiley Sons, INC, New York*

[14] Mahalle G, Salunke O, Kotkunde N, Gupta A K and Singh S K 2019 Neural network modeling for anisotropicmechanical properties and work hardeningbehavior of Inconel 718 alloy at elevatedtemperatures *J. Mater. Res. Technol.* **8** 2130–40

[15] Mahalle G, Morchhale A, Kotkunde N, Amit Kumar Gupta S K S and Lin Y C 2020 Forming and Fracture Limits of IN718 Alloy at Elevated Temperatures: Experimental and Theoretical Investigation *J. Manuf. Process.* **56** 482–99

Deformation characteristics in 3D free-bending forming of complex spiral tubes

A Abd El-Aty[1, 2, 4*], W Wei[1, 3], X Guo[1, 3*], H Wang[2, 3], J Tao[1, 3], H Chen[1, 3], Y Qin[1, 3]
[1]College of Material Science and Technology, Nanjing University of Aeronautics and Astronautics, Nanjing 211100, P.R. China.
[2]College of Mechanical and Electrical Engineering, Nanjing University of Aeronautics and Astronautics, Nanjing 210016, P.R. China.
[3]Jiangsu Key Laboratory of Nuclear Energy Equipment Materials Engineering, Nanjing University of Aeronautics and Astronautics, Nanjing 211100, P.R. China.
[4]Mechanical Engineering Department, Faculty of Engineering-Helwan, Helwan University, Cairo, Egypt.

* E-mail: guoxunzhong@nuaa.edu.cn
aabdelaty@nuaa.edu.cn

Abstract. Free-bending is a novel bending technology suitable for tube and profile bending. It can achieve accurate bending of tubes or profiles with different bending radii without die. In current study, a new theoretical analysis of a spiral tube formed by 3D free-bending is proposed to reveal the relationship between the spiral tube and the parameters of free-bending forming technology. The results obtained from the proposed theoretical analysis were compared with those achieved from FE modelling and experimentation. The maximum error of forming was less than 7%, which verified the accuracy and the reliability of the proposed theoretical model. The forming limits of the spiral tubes based on 3D free-bending were explored. It was noticed that the spiral diameter was mainly influenced by the eccentricity and the distance between the center of the bending die and the front end of the guide mechanism in the Z–direction (distance A). Furthermore, the spiral diameter is inversely proportional to the eccentricity, and directly proportional to the distance A. In addition, the screw pitch is mainly influenced, and directly proportional to the spiral diameter and the single spiral circle swing angle of the spherical bearing.

Keywords: Free-bending; Trajectory of bending die; Spiral tubes; Theoretical analysis; forming limit

1. Introduction

Spiral tubes are 3D spiral structures mainly used to transfer oil and gas, and play an important role in petroleum, chemical, and power industries [1-3]. Spiral tubes are usually used as buffer tubes in the fluid tube and are generally placed in the front section of a pressure gauge to reduce the impact force on a pipeline conveying fluid and facilitate the calculation of normal pipeline pressure [4]. Therefore, it is of great significance to study the forming of spiral tubes. The forming of a spiral tube involves space-bending, which is difficult to perform. At present, spiral tubes are mainly formed through traditional bending methods, including roll bending [5], and numerical control bending [6]. Wang [7] studied and

analyzed the roll bending forming process of spiral tubes, established a theoretical model of space bending forming, and put forward the influence law of roll bending on the forming quality of spiral tubes. Jing et al. [8] investigated the control of winding process of thin wall spiral tubes using an elastic-plastic finite element method, and analyzed the location and magnitude of the maximum tensile stress and maximum compressive stress, thereby providing a theoretical base for the control of working forming course of thin wall spiral tubes. Cui et al. [9] designed and investigated a spiral tube structure working under high temperature and high-pressure loads. A CAE method was used to calculate and analyze the performance of this spiral structure under working conditions. In addition, a number of studies have explored the forming limit of tube bending. Daxin et al. [10] established a relationship between the material hardening index and the minimum relative bending radius. The calculation formula of the minimum relative bending radius of the tube was derived according to the theory of concentrated instability and dispersion instability. Hu et al. [11] developed a computer simulation system based on stress state and deformation analysis of tube bending, using local induction heating with small bending radius. Langenbrunner et al. [12] investigated a method which tube is filled by sand before bending to avoid the wrinkling in thin-walled tubes in bending process. This method can effectively increase the forming limit of tube bending, but it cannot be used in the forming process of complex tube fittings. Hasanpour et al. [13] studied the influence of anisotropy on the ellipsis, thickness and wrinkle of tube by finite element method, and explored the forming limit of tube. The newly developed free bending process has proved to have many advantages. Free-bending is a continuous bending and die-less forming method [14-17]. Compared to traditional bending, it has the advantage to be a low-cost continuous forming method [18, 19]. In general, it has been demonstrated that the theory of bend forming of spiral tubes is not perfect, and especially bend forming based on the 3D free-bending technology has not been investigated. Moreover, the forming limit of spiral tubes has been seldom studied.
In this study, the 3D free-bending technology was used to form spiral tubes. However, the free-bending technology is still in the research stage and its theory is far from perfect. Therefore, in the present study, a theoretical model of spiral bending based on the free-bending technology is presented, and the relationship between pitch, diameter, and free-bending parameters is revealed. At the same time, the accuracy of the theoretical model was validated by simulations and experiments, and the forming limit was explored based on the theoretical model

2. Finite element modelling
Firstly, a three-dimensional model of the free-bending system was designed, and then it was imported into the finite element software ABAQUS (Fig. 1).

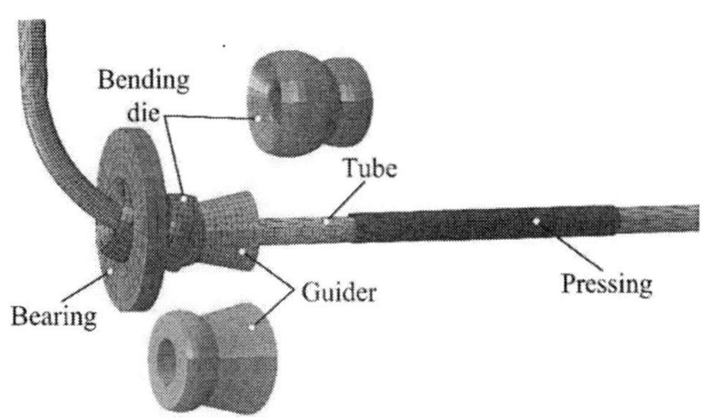

Figure 1. Finite element model

The analysis step was set as dynamic explicit, the interaction was defined as universal contact, and the global friction coefficient between the tube and die was set to 0.2. Moreover, the diameter of the tube was 15 mm, and the thickness of the wall was 1 mm. 6061-T6 aluminium was used as the material model of the tube, and YG8 cemented carbide was used as the bending die material. The specific parameters are presented in Table 1.

Table 1 Materials' mechanical properties at room temperature

Materials	Poisson's Ratio	Young's Modulus Y (GPa)	Tensile Strength R_m (MPa)	Density ρ (g/cm^3)
6061-T6 Aluminum	0.3	80.7	260	2.7
YG8	0.3	71	*	14.5

As it can be seen in Fig. 2, the simulation results are consistent with the theoretical analysis. More specifically, in Fig. 2(a), it can be observed that there is stress concentration in the position where the bending die swinging. On the lateral bend side of the spiral tube, there is tensile stress concentration at the swinging position. On the contrary, at the inner bend side of the spiral tube, compressive stress concentration is developed.

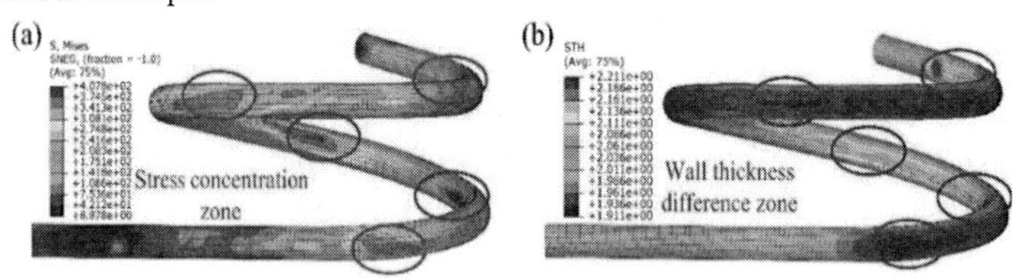

Figure 2. Spiral free bending forming simulation: (a) Stress cloud; (b) Wall thickness cloud

Under the action of stress, the wall thickness distribution in the spiral tube also has a certain influence at the position of bending die swinging. As it can be seen from Fig. 2(b), the thinning rate of the outer bend side is greater and the thickening rate of the inner bend side is greater in the swinging position of the bending die. The results of the simulation were basically consistent with the theoretical analysis.

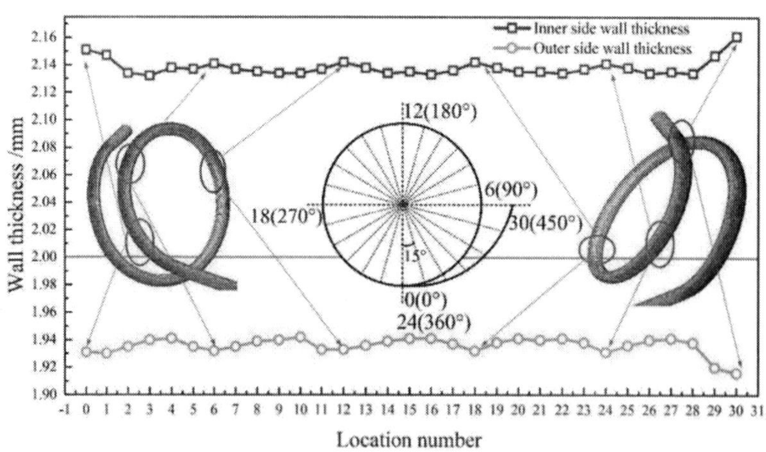

Figure 3. Simulated wall thickness distribution of spiral tubes

The stress state has a significant effect on the wall thickness distribution. As it can be seen from Fig. 3, there is an uneven wall thickness distribution in the spiral tubes. At the position where the bending die swinging, the maximum thickening of the inner side wall was 1.4% higher than that of the other parts, and the maximum thinning of the lateral side wall was 1.2% higher than that of the other parts. However, in general, the thickening and thinning of the wall at the swinging position of the bending die appeared to have little effect on the overall forming quality of the spiral tubes.

3. Experiments of free-bending forming

In this study, based on the forming process of spiral tubes, free-bending forming experiments were carried out (Fig. 4). In these experiments, a three-axis free-bending forming equipment was used, and the required spiral tube parts were formed. According to the theoretical analysis and simulation results, the swinging angles of the bending die will be produced by a large force, which has a certain influence on the wall thickness distribution of the tube. As it can be seen in Fig. 5, the wall thickness distribution resulted from the spiral bending experiments was basically in line with the theoretical and simulation results. At the swinging position of the bending die, the maximum thickening of the inner wall was 1.5% higher than that of the other parts, and the maximum thinning of the outer wall was 0.8% higher than that of the other parts. However, these changes have no significant effect on the forming quality of the final spiral tubes.

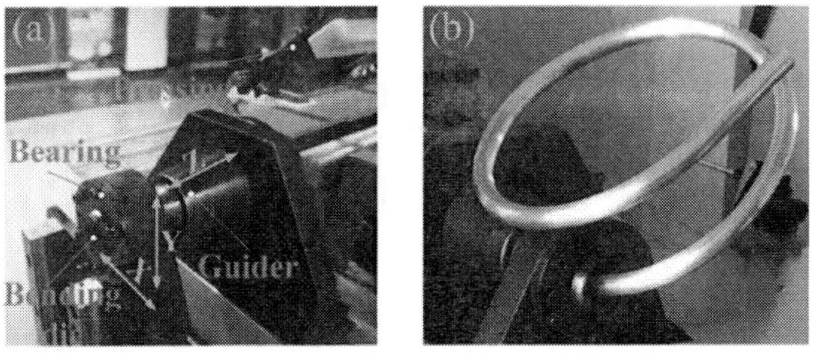

Figure 4. Bending forming experiments: (a) Forming equipment; (b) Free bending forming of spiral tubes.

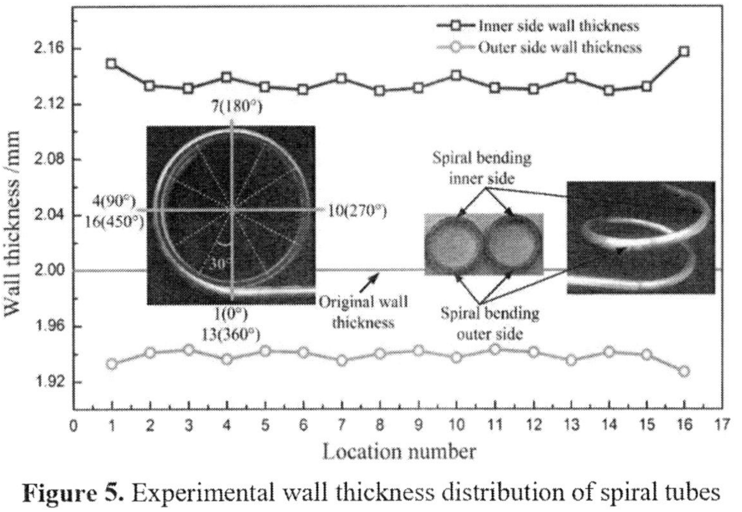

Figure 5. Experimental wall thickness distribution of spiral tubes

IOP Conf. Series: Materials Science and Engineering **967** (2020) 012055 doi:10.1088/1757-899X/967/1/012055

After comparing the spiral tube geometric model, the simulation results, and the experimental results, it was found that the overall forming effect was good with no apparent forming defects (Fig. 6). In Table 2, it can be seen that the maximum error of the spiral diameter of the simulation and experiment does not exceed 6.5%, and the maximum error of the spiral pitch does not exceed 7% compared to the three-dimensional model. As it can be observed, the reliability and accuracy of the theoretical process was strongly validated by the computational and experimental results.

Figure 6. Spiral tubes comparison: (a) 3D model; (b) Finite element simulation; (c) Actual bending forming

Table 2 Comparison of parameters of free bending forming of spiral tubes

Parameters	Spiral tube model	Simulation result	Experimental result
Spiral diameter D	130	138	136
Screw pitch S	60	61	64

In order to further verify the reliability of the theoretical process, a more complex multi-loop spiral free-bending forming was performed (Fig. 7). Through measurements and comparisons, it was found that the forming effect was adequate and the forming error was small, conforming to the expected forming result. It has been demonstrated that the theory of spiral free-bending forming, proposed in this paper, has strong practicability and reliability.

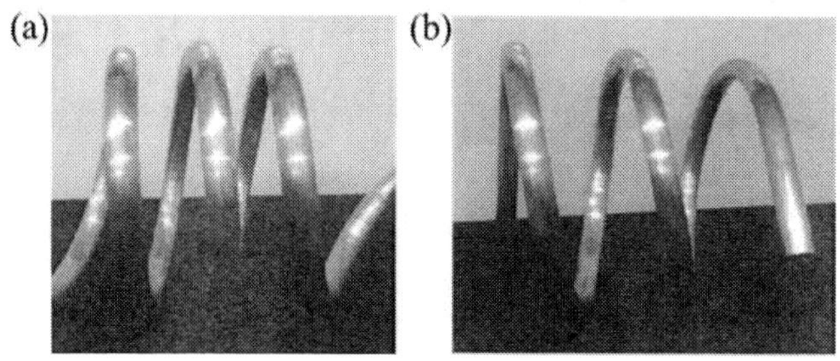

Figure 7. Complex spiral tubes: (a) D=110mm, S=45mm; (b) D=110mm, S=70mm

The diameter D of a spiral column is an important parameter of spiral tubes. For precision and small parts, the spiral diameter D of the mounted spiral tube is required to be small. The simulation results indicated that the values of eccentricity U and distance A are the main factors affecting the spiral diameter, as shown in Fig. 8. The spiral diameter D decreases with the increasing eccentricity U. Moreover, the decrease in spiral diameter D is faster when the eccentricity is small, while it slows down when the eccentricity is larger. That is, the eccentricity U is inversely proportional to the spiral diameter D. On the contrary, the spiral diameter D increases with the increase of distance A. That is, the eccentricity U is directly proportional to the distance A. In the simulation and the forming process, it was found that the eccentricity U cannot increase indefinitely. There is a maximum eccentricity U_{max}, and its size is closely related to the structure of the die (Fig. 9).

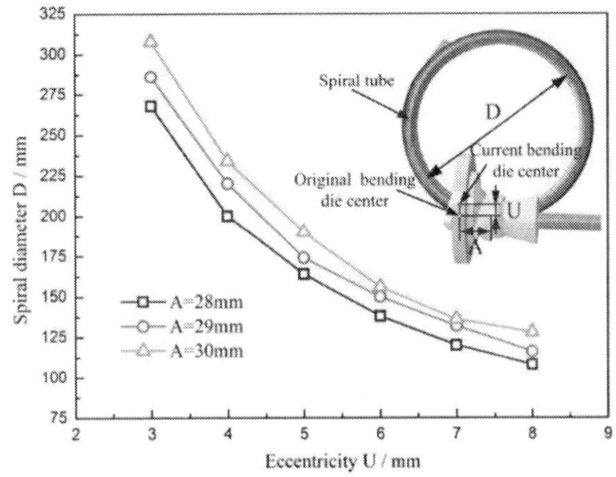

Figure 8. Influence of eccentricity U and A on spiral diameter

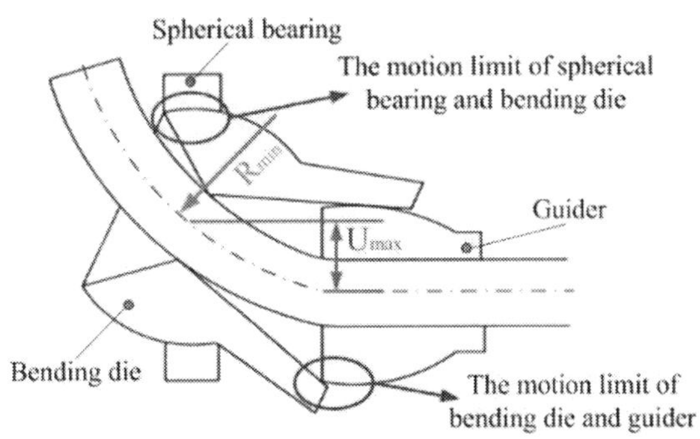

Figure 9. The motion limit diagram of free bending device

As it can be seen in Fig. 9, there are two positions where a limit of motion exists. One of them is the motion limit of the spherical bearing and the bending die, and the other is the motion limit of the bending die and the guider. These two limits constrain the upward movement of the spherical bearing, thus the eccentricity cannot further increase. When the distance A is fixed, there is a maximum eccentricity U_{max}, which defines the minimum bending radius R_{min}. Different free-bending devices have different U_{max}.

According to the forming procedure of the existing free-bending equipment, the minimum spiral diameter D is:

$$D \geq 6d \quad (1)$$

where d is the tube diameter.

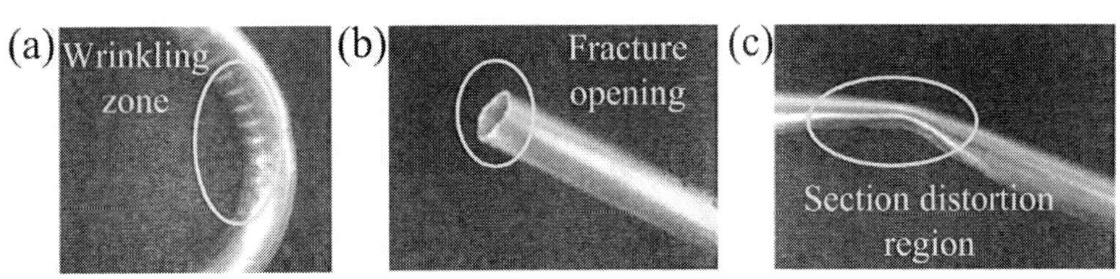

Figure 10 Forming defects caused by increasing eccentricity

In addition, the spiral diameter is affected by the mechanical properties of the material. Materials with good bending properties provide spiral tubes with smaller spiral diameters. On the other hand, materials with poor bending performance, generally increase the eccentric distance U and decrease the distance A, leading to a series of quality defects in the formed workpieces. As it can be seen in Fig. 10,theforming process can cause wrinkling at the inside of the bend, cause large cross-section distortion, and even break the tube.

4. Conclusion

In this study, the process analysis of spiral tubes based on the 3D free-bending technology was presented. Moreover, the relationship between the free-bending parameters and the spiral bending parameters was investigated. According to the simulation results, the main factors that influence the spiral diameter include eccentricity U and distance A. More specifically, the diameter of the spiral decreases continuously with the increase of the eccentricity, showing an inverse correlation. On the contrary, the diameter of the spiral increases with the increase of the distance A, showing a direct correlation. Based on the existing 3D free-bending equipment, the minimum spiral diameter D was 6d.

Acknowledgements

The authors greatly acknowledge the financial support from the National Natural Science Foundation International (regional) cooperation and exchange project (Grant No.51711540301), the National Natural Science Foundation of China (Grand No. 51875548), "the Fundamental Research Funds for the Central Universities No. kfjj20180612), the Open Research Funds for the Jiangsu Key Laboratory of Nuclear Energy Equipment Materials Engineering (No. NJ20170012), and the China Aviation Science Foundation (No. 2016ZE52047).

References

1. Z Cui, Y Zhao, Z Zhang, X He, S Guo. Structure design and research on a spiral pipe type reactor, Advanced Materials Research, 476-478 (2012):5.
2. M Lee, T Kang, Y Kim. Air-side heat transfer characteristics of spiral-type circular fin-tube heat exchangers, International Journal of Refrigeration, 33.2 (2010):313-320.
3. S Li, W Cai, J Chen, H Zhang, Y Jiang Numerical study on condensation heat transfer and pressure drop characteristics of ethane/propane mixture upward flow in a spiral pipe, International Journal of Heat & Mass Transfer, 121 (2018):170-186.
4. P Yanuar, B Gunawan. Flow and convective heat transfer characteristics of spiral pipe for nanofluids, Int J Res Rev Appl Sci, 7(2011): 236-48.
5. Y Dong, Y Yang, L Zhao. Research on the relationship between the number of rotary draw bendings and the shape precision of a bent hat-section profile, Journal of Materials Processing Technology, 151(2004):307-311.
6. H Li, G Ren, Z Li, Z Li, L Feng, H Yang. Forming mechanism and characteristics of a process for equal-thickness in-plane ring roll-bending of a metal strip by twin conical rolls, Journal of Materials Processing Technology, 227(2016):288-307.
7. Wang Lin. Technological Research and Numerical Simulation of Spiral Tube, Yanshan University, 2006.
8. J Zheng, M Guang, Y Wang, Y Li, Z Jia, J Li. Study on Numerical Simulation for Control of Winding Process of Thin Wall Spiral Tube, Precious Metals, 33(2012):227-232.
9. Z Cui, Y Zhao, Z Zhang, X He, S Guo. Structure Design and Research on a Spiral Pipe Type Reactor, Advanced Materials Research, 476-478(2012):1197-1201.
10. D E. Minimum relative bending-radius of tube-bending without mandrel, Journal of Mechanical Engineering, 43(2007):219-222.

11. Z Hu, J Li. Computer simulation of pipe-bending processes with small bending radius using local induction heating, Journal of Materials Processing Technology, 91(1999):75-79.
12. A Langenbrunner, M Trautwein. Modeling and Experimental Validation of the Effect of Sand Filling on Avoiding Wrinkling Phenomenon in Thin-Walled Tube Bending Process // Asme Biennial Conference on Engineering Systems Design & Analysis, 2010.
13. K Hasanpour, B Amini, M Poursina, M Barati. The Effect of Anisotropy on Thin-Walled Tube Bending[J], 2011.
14. P Li, L Wang, M Li. Flexible-bending of Profiles and Tubes of Continuous Varying Radii, International Journal of Advanced Manufacturing Technology, 88(2016): 1-7.
15. S Kawasumi, Y Takeda, D Matsuura. Precise pipe-bending by 3-RPSR parallel mechanism considering springback and clearances at dies, Transactions of the Japan Society of Mechanical Engineers, 80(2014):TRANS0343-TRANS0343.
16. H Goto, K Ichiryu, H Saito, Y Ishikura, Y Tanaka. Applications with a new 6-DOF bending machine in tube forming processes, // Jfps International Symposium on Fluid Power. The Japan Fluid Power System Society, 2008(1-7):183-188.
17. Y Ma, H Xiong, H Wang, X Luo, K Jin, Y Xu, J Tao, X Guo. Simulation and Experimental Study on Three Dimensional Free Bending of Complex Space Elbow, Journal of Netshape Forming Engineering, 2017.
18. M Murata. Analysis of Circular Tube Bending by MOS Bending Method, Proc Ictp, 1996:505-508.
19. X Guo, W Wei, Y Xu, A El-Aty, H Liu, H Wang, X Luo, J Tao. Wall thickness distribution of Cu–Al bimetallic tube based on free bending process. International Journal of Mechanical Sciences, 150(2019):12-19.

IOP Conf. Series: Materials Science and Engineering **967** (2020) 012056 doi:10.1088/1757-899X/967/1/012056

The determination of friction coefficient in Nakazima test and its application in predicting FLCs with modified M-K model

Yubao Wang[1] **and Cunsheng Zhang**[1, 2]

[1] Key Laboratory for Liquid-Solid Structural Evolution & Processing of Materials (Ministry of Education), Shandong University, Jinan, Shandong 250061, P.R.China
[2] Author to whom any correspondence should be addressed.

zhangcs@sdu.edu.cn

Abstract. In this work, the friction coefficient between the punch and the specimen during the Nakazima tests is determined by comparing the experimental and the numerical strain paths. Afterwards, the FLCs of the Al-Mg-Li alloy sheet are predicted with the modified M-K model, which considers the through-thickness normal stress and the friction through introducing the expressions of the normal stress and the friction stress into M-K model. Finally, the predicted FLCs of the sheet are compared with the forming limits determined by Nakazima tests.

1. Introduction

The Al-Mg-Li alloy sheet has obtained wide application in many fields such as aircraft for its excellent properties of low density, high specific stiffness and high specific strength [1-3]. Forming limit curves (FLCs) are important methods to judge the formability of sheet metals and it is specified in ISO 12004-2 standard that the FLCs of sheet metal can be determined by the Nakazima test. However, the experimental procedure to obtain FLCs is complicated and time-consuming, because large quantities of specimens with a wide range of sizes are needed to take various strain paths into account [4]. Therefore, several theoretical models such as the MMFC model [5] and the M-K model [6] have been proposed based on the deformation theory to calculate the forming limits of sheet metal. Therein, the M-K model which is established based on the groove theory has been widely applied to predict the forming limits of different sheet metals such as AA5086 [7], Ti6Al4V alloy [8] and Qste600TM steel [9] for its simplicity in calculation process and high prediction accuracy.

Although the M-K model has been used in many literatures to predict the forming limits of different sheet metals, there are still differences between the predicted FLCs and the experimental forming limits. As a result, scholars have made many resourceful improvements on the original M-K model to improve the prediction accuracy of the FLCs. In the original M-K model, the stress state of the specimen is supposed to be planar, but the sheet during the Nakazima test is under through-thickness normal stress, which has been proved to affect the FLCs of sheet metal [10]. Given the ratio of the normal stress to the major stress, or simplifying the normal stress in the deformation process to be constant is one method to consider the through-thickness normal stress in the M-K model [11, 12]. Besides, Wang et al. [13] established the expression of normal stress through analyzing the stress state in the Nakazima test and introduced it into the M-K model. During the Nakazima test, the friction between the punch and the specimen is inevitable and it has been identified that the friction has influence on the strain paths and the FLCs of sheet metal [14]. Thus, it should be considered in the M-K model to improve the accuracy of FLCs predicted by M-K model. Based on the experimental and

IOP Conf. Series: Materials Science and Engineering **967** (2020) 012056 doi:10.1088/1757-899X/967/1/012056

numerical investigation for the friction influence on the FLCs [15, 16], Wang et al. [13] established the friction stress expression and introduced it into the M-K model. Results showed that this modified M-K model where the expressions of the normal stress and the friction stress were introduced could improve the prediction accuracy of FLCs for the IF and DP600 steels. However, the friction coefficient in the friction stress expression which denotes the lubrication condition between the specimen and the tooling during Nakazima tests is difficult to directly determine through experiments.

In this work, the friction coefficient is determined by comparing the strain paths of specimens which are respectively obtained by the finite element simulation method and the Nakazima tests. Furthermore, the FLCs of the Al-Mg-Li alloy sheet are predicted by the modified M-K model and compared with the experimental forming limits.

2. The through-thickness normal stress and friction stress expressions

During Nakazima test process, the fracture of specimen does not always occur at the apex of the dome because of the friction between the specimen and the punch. Thus, as shown in figure 1, one small force body near the apex of the dome is chosen, where the fracture is assumed to occur.

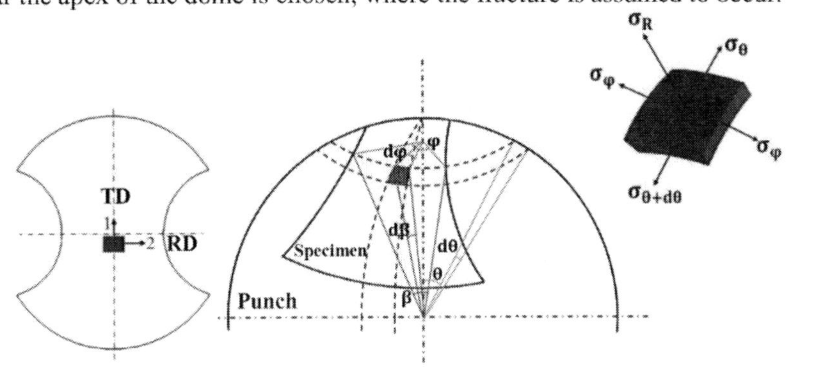

Figure 1. The location and the stress state of the force body in the Nakazima test [13].

The force body is assumed to be small enough that the increments of radians $d\theta$ and $d\varphi$ are infinitesimal. Based on this assumption and considering the stress gradient along the thickness direction, the force equilibrium equation is established as shown in equation (1) and the through-thickness normal stress expression is established in the form of equation (2).

$$\frac{\int_0^t \frac{\sigma_R}{t} x dx}{t} S_R = 2\sigma_\theta S_\theta \sin\frac{d\theta}{2} + 2\sigma_\varphi S_\varphi \sin\frac{d\beta}{2}, \quad (1)$$

$$\sigma_3 = -\frac{2t}{R}(\sigma_1 + P\sigma_2), \quad (2)$$

where the parameter P represents the position of sheet failure in the deformation process and it is associated with the geometry around the apex. When the fracture occurs within a distance less than 15% of the punch diameter away from the apex of the dome, the φ value is close to π, so the value of P could be set to 1 in the computational process [13].

Based on the analysis of the contact condition between the punch and the specimen during Nakazima tests, the friction between the sheet and the punch is usually considered as the Coulomb friction and the friction force can be defined by

$$f = \mu F, \quad (3)$$

where f is the friction force, μ is the friction coefficient and F is the compression force on the contact surface, respectively. The friction force on the contact surface of the force body is shear force, but it is difficult to be directly considered in the M-K model. In order to take into account the effect of friction on the forming limits in the M-K model, the symbol σ_f is defined as the friction stress along the principal stress σ_1 direction by means of equaling the friction force f on the internal surface to the increment along the principal stress [13]. The defined friction stress σ_f is determined by

$$\sigma_f = 2\mu d\theta(\sigma_1 + P\sigma_2). \quad (4)$$

3. The determination of friction coefficient between the punch and the specimen

In this section, the friction coefficient between the punch and the tooling in the Nakazima test is determined by comparing the strain paths of samples, which are obtained by the experiment and the finite element simulation method, respectively.

Based on the EC600 sheet test platform and the ARAMIS non-contact strain measurement system, the Nakazima tests are performed with the samples of two sizes which are named R60 and R0 as shown in figure 2. According to our previous study, the R60 and R0 specimens are under tensile-compressed and tensile-tensile stress states during the Nakazima test, which indicates that the strain paths on the left and the right sides of forming limit diagram (FLD) are described, respectively. In the Nakazima test, a composite lubrication of grease and 0.05mm thick polytetrafluoroethylene (PTFE) film is used. The ARAMIS non-contact strain measurement system is used to continuously record the deformation state of the samples during experiment process.

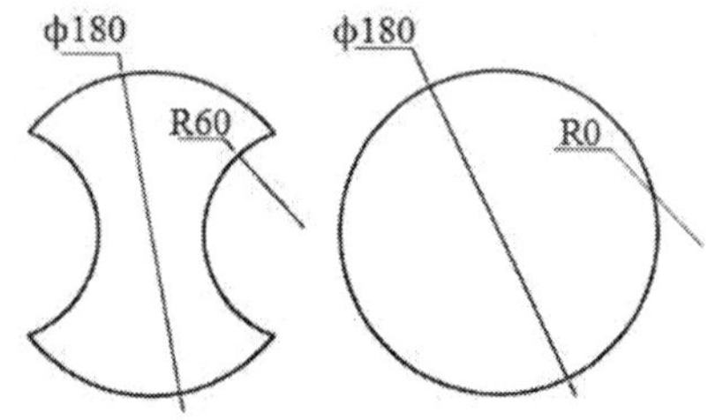

Figure 2. Dimensions of the specimens for Nakazima tests.

Afterwards, with the ABAQUS finite element software, the numerical simulation of the Nakazima test process is performed, which includes a spherical punch, a die and a specimen as shown in figure 3. In order to shorten the calculation time, a 1/4 model is used for the simulation considering the axial symmetry of the sample in the Nakazima test, and the blank holding force on the specimen is applied by dividing the specimen surface instead of defining the blank holder parts. In the finite element simulation, the boundary conditions and the load settings are consistent with the experiment to obtain more accurate simulation results. In the Nakazima test, there are two kinds of friction contacts between the sheet and the tooling, one is between the punch and the sheet, the other is between the sheet and the die. Here, for simplification, the same value is defined for these two kinds of friction coefficients. In the simulation, three different friction coefficients of 0, 0.1 and 0.2 are used, respectively.

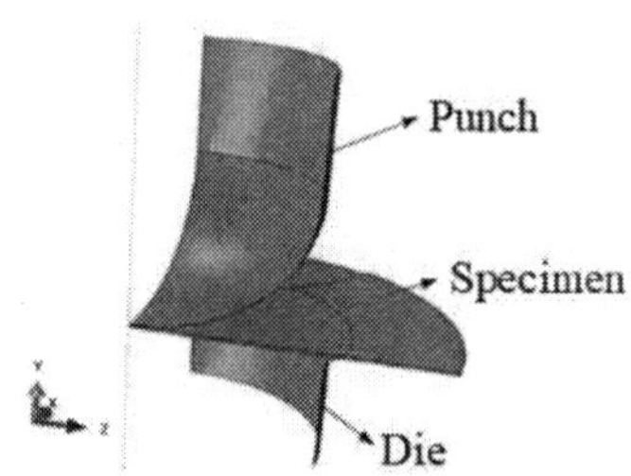

Figure 3. The finite element simulation model of Nakazima test.

The strain paths at the fracture locations of samples are shown in figure 4, which are respectively determined by the experiment and the finite element simulation with different friction coefficients of 0, 0.1 and 0.2. It can be found from the comparison between simulated and experimental strain paths that the experimental strain paths of the R0 and R60 samples are between the simulated strain paths obtained at the friction coefficient of 0 and 0.1. As a result, it can be concluded that the friction coefficient on the contact surface of the specimen and the tooling is between 0 and 0.1 when the composite lubrication of PTFE film and grease is applied in the Nakazima tests. Furthermore, the friction coefficient in the friction stress expression should be set according to the comparison results when the modified M-K model is applied to improve the prediction accuracy of FLCs.

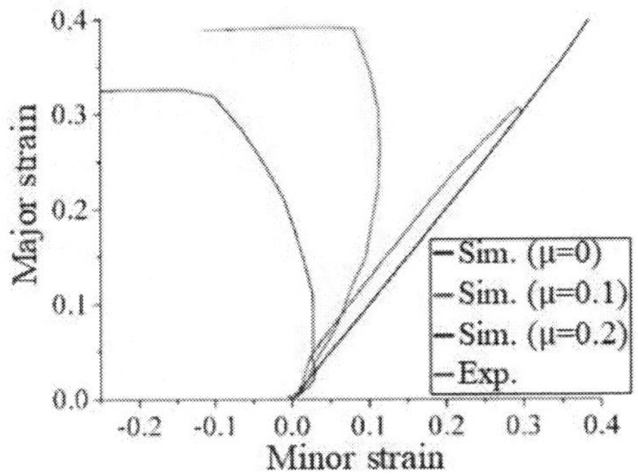

(a) R0

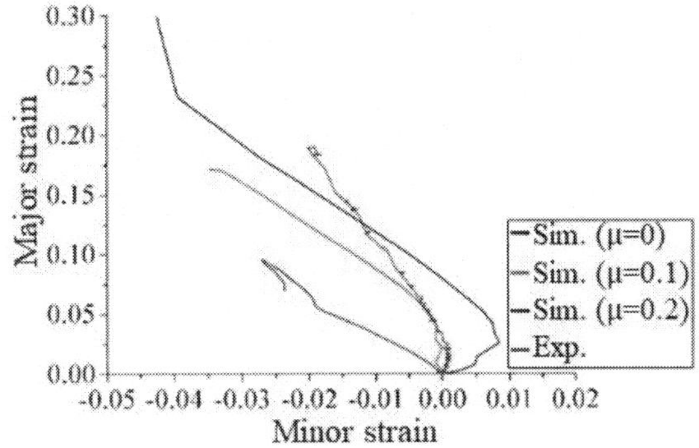

(b) R60

Figure 4. Comparison between the strain paths of simulation and Nakazima test (a) R0 (b) R60.

4. The predicted FLCs of Al-Mg-Li alloy sheet

With the modified M-K model and the determined friction coefficient, the FLCs of the Al-Mg-Li alloy sheet are predicted and compared with the experimental results of Nakazima tests in this section.

The modified M-K model is established by introducing the expressions of normal stress and friction stress into the original M-K model as shown in figure 5. The Al-Mg-Li alloy sheet has the thickness of 1.8 mm and table 1 shows its composition. Moreover, the Swift constitutive model as shown in equation (5) and the Hill's 48 yield criterion as shown in equation (6) are applied to respectively describe the stress-strain relationship and the anisotropy of the sheet, where the material parameters have been obtained in our previous work as listed in table 2.

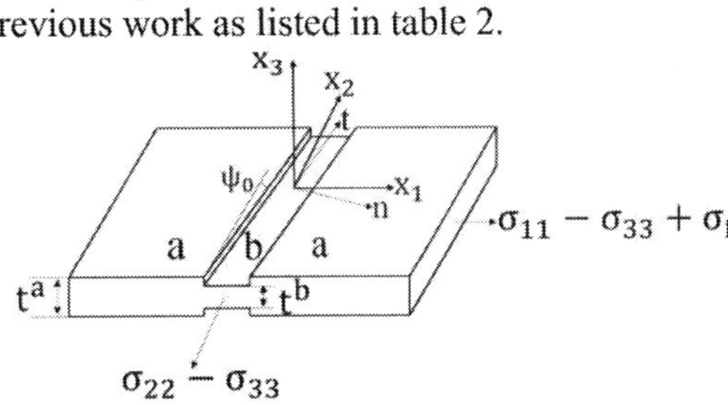

Figure 5. The schematic diagram of the modified M-K model [13].

Table 1. The constituent elements and corresponding mass fraction of the Al-Mg-Li alloy sheet.

Element	Mg	Li	Zr	Fe	Si	Cu	Ti	Al
wt %	5.22	2.19	0.10	0.04	0.02	0.01	0.01	Bal

$$\bar{\sigma} = K(\bar{\varepsilon}_0 + \bar{\varepsilon})^n, \tag{5}$$

$$\bar{\sigma} = \sqrt{F\sigma_2^2 + G\sigma_1^2 + H(\sigma_1 - \sigma_2)^2 + 2N\sigma_{12}^2}, \tag{6}$$

where K is hardening coefficient, n is hardening index, F, G, H and N are anisotropy coefficients.

Table 2. The material parameters of Al-Mg-Li alloy sheet in the Swift constitutive model and the Hill's 48 yield criterion.

Swift constitutive model			Hill's 48 yield criterion			
K	n	$\bar{\varepsilon}_0$	F	G	H	N
790	0.19	0.009	0.36	0.75	0.25	2.32

Finally, the comparison between the predicted FLCs of modified M-K model and the experimental results are shown in figure 6. It can be found that the modified M-K model can accurately predicted the FLCs of the Al-Mg-Li alloy sheet at the left side and under small strain ratios. When the strain path increases, the difference between the predicted FLCs and the experimental forming limits becomes significant. The reason that causes this problem may be the material parameters in the constitutive model, which will be analyzed in the next work.

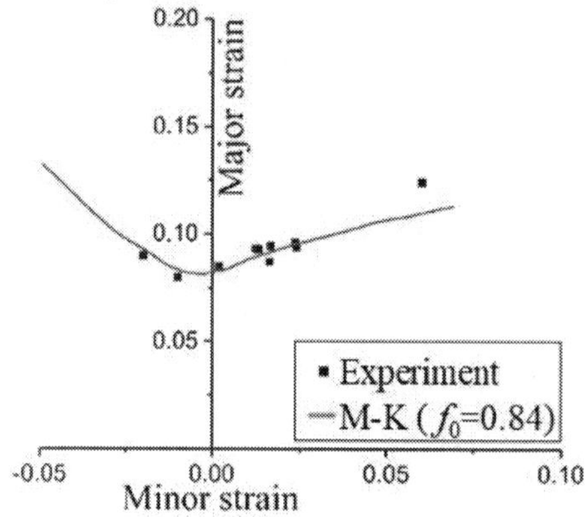

Figure 6. The predicted FLCs and the experimental forming limits of the Al-Mg-Li alloy sheet.

5. Conclusions

In this work, the friction coefficient in the friction stress expression is determined between 0 and 0.1 when the composite lubrication of PTFE film and grease is applied in the Nakazima tests. Furthermore, the forming limits of the Al-Mg-Li alloy sheet are predicted by the modified M-K model and compared with the experimental results of the Nakazima tests. The comparison shows that the FLCs of the Al-Mg-Li alloy sheet are basically accurately predicted.

References

[1] S.Y. Betsofen, V.V. Antipov, M.I. Knyazev, Al-Cu-Li and Al-Mg-Li alloys: Phase composition, texture,and anisotropy of mechanical properties (Review), Russ. Metall. (Metally) 4 (2016) 326-341.

[2] A.A. El-Aty, Y. Xu, S. Zhang, Y. Ma, D. Chen, Experimental investigation of tensile properties and anisotropy of 1420, 8090 and 2060 Al-Li alloys sheet undergoing different strain rates and fibre orientation: a comparative study, Procedia Eng. 207 (2017) 13-18.

[3] A. A. El-Aty, Y. Xu, X. Guo, S.H. Zhang, Y. Ma, D. Chen, Strengthening mechanisms, deformation behavior, and anisotropic mechanical properties of Al-Li alloys: A review, J. Adv. Res. 10 (2018) 49-67.

[4] Barata da Rocha, A., Santos, A.D., Teixeira, P., Butuc, M.C., 2009. Analysis of plastic flow localization under strain paths changes and its coupling with finite element simulation in sheet metal forming. J. Mater. Process. Technol. 209, 5097-5109.

[5] Manopulo, N., Hora, P., Peters, P., Gorji, M., Barlat, F., 2015. An extended Modified Maximum Force Criterion for the prediction of localized necking under non-proportional loading. Int. J. Plast. 75, 189-203.

[6] Marciniak, Z., Kuczyński, K., 1967. Limit strains in the processes of stretch-forming sheet metal. Int. J. Mech. Sci. 9, 609-620.

[7] Ding, J., Zhang, C., Chu, X., Zhao, G., Leotoing, L., Guines, D., 2015. Investigation of the influence of the initial groove angle in the M–K model on limit strains and forming limit curves. Int. J. Mech. Sci. 98, 59-69.

[8] X. Li, G. Guo, J. Xiao, N. Song, D. Li, Constitutive modeling and the effects of strain-rate and temperature on the formability of Ti-6Al-4V alloy sheet, Mater. Des. 55 (2014) 325-334.

[9] Ma, B., Diao, K., Wu, X., Li, X., Wan, M., Cai, Z., 2016. The effect of the through-thickness normal stress on sheet formability. J. Manuf. Process. 21, 134-140.

[10] Wu, P.D., Embury, J.D., Lloyd, D.J., Huang, Y., Neale, K.W., 2009. Effects of superimposed hydrostatic pressure on sheet metal formability. Int. J. Plast. 25, 1711-1725.

[11] Zhang, F., Chen, J., Chen, J., 2014. Effect of through-thickness normal stress on forming limits

under Yld2003 yield criterion and M-K model. Int. J. Mech. Sci. 89, 92-100.
[12] Zhang, F., Chen, J., Chen, J., Zhu, X.. Forming limit model evaluation for anisotropic sheet metals under through-thickness normal stress. Int. J. Mech. Sci. 89, 40-46.
[13] Y.B. Wang, C.S. Zhang, Y. Yang, S.M. Fan, G.C. Wang, G.Q. Zhao, L. Chen, The integration of through-thickness normal stress and friction stress in the M-K model to improve the accuracy of predicted FLCs, I. J. Plast. 120 (2019) 147-163.
[14] Zhang, L., Min, J., Carsley, J.E., Stoughton, T.B., Lin, J., 2017. Experimental and theoretical investigation on the role of friction in Nakazima testing. Int. J. Mech. Sci. 133, 217-226.
[15] Ma, X., Li, F., Li, J., Wang, Q., Yuan, Z., Fang, Y., 2015. Analysis of forming limits based on a new ductile damage criterion in St14 steel sheets. Mater. Des. 68, 134-145.
[16] Ozturk, F., Lee, D., 2005. Experimental and numerical analysis of out-of-plane formability test. J. Mater. Process. Technol. 170, 247-253.

IOP Conf. Series: Materials Science and Engineering **967** (2020) 012057 doi:10.1088/1757-899X/967/1/012057

Material testing of magnesium alloy AZ31B using a finite element polycrystal method based on a rate independent crystal plasticity model

Giorgio Vago[1 2,a] and Tetsuo Oya[1,b]

[1]Keio University,3-14-1 Hiyoshi, Kohoku-Ku, Yokohama 223-8522,Japan
[2]Politecnico di Milano, Via Privata Giuseppe La Masa, 20156 Milano MI, Italia
[a]giorgio.vago@mail.polimi.it,[b]oya@sd.keio.ac.jp

Keywords: Multi-scale, Finite element polycrystal method, twinning, magnesium,material modelling, hcp materials, numerical investigations.

Abstract

Magnesium and its alloys are attractive structural light materials because they have a high specific strength and high specific stiffness. That is why their usage in high performance automotive and aerospace applications has been increased. However, their high anisotropy and low ductility could bring to forming failure, to avoid these a better accuracy for metal forming simulation is required. The identification of macroscopic parameters is expensive due to the experiments needed to find them, the use of multi-scale approach allows to find parameters for the model through easier experiments. A finite element polycrystal method, based on a rate independent polycrystal plasticity model, is implemented in order to perform the material testing of a cast magnesium alloy AZ31B . The parameters of the model are adjusted to better fit the experimental data through a trial and error process.

1 Introduction and Deformation texture model

Tension-compression asymmetry of mechanical properties of textured magnesium alloys is one specific phenomenon, which should be taken into account on designing the machine components. In this paper, the study of the behaviour of a cast alloy of the magnesium alloy AZ31B when undergoes simple tension and compression tests is conducted. The importance of texture in magnesium and its alloys has been proved by many researchers, the study of constitutive model that could predict the evolution of magnesium with a random texture is important in order to perform a numerical rolling to numerically calculate the evolution of the texture without conducting expensive experiments on rolled sheet to evaluate it. The active deformation systems considered are the basal, prismatic and pyramidal II slip systems and the tensile twinning. To identify the rate-independent slip/twin systems and to determine the plastic strain, a numerical scheme called successive integration method has been adopted. To evaluate the accuracy of this constitutive model in predicting the stress response in a magnesium alloy at room temperature, the numerical predictions and the literature data are compared. The influences of microstructural parameters on macroscopic mechanical properties are investigated. In particular, expression of latent hardening, work-hardening, CRSS values are investigated, to better fit our predictions with the experimental data.

2 Finite element polycrystal model

2.1 Successive integration scheme

To simulate the deformation of the texture a rate independent crystal plasticity model (RICP) has been applied in this study, thus the stress-strain relation is described as:

$$\dot{\sigma}_{ij} = C^{e}_{ijkl}\dot{\varepsilon}^{p}_{ij}. \tag{1}$$

Where C^{e}_{ijkl} is the elastic constitutive tensor, $\dot{\varepsilon}^{p}_{ij}$ is the elastic strain increment that is calculated in this way:

$$\dot{\varepsilon}^{e}_{ij} = \dot{\varepsilon}_{ij} - \dot{\varepsilon}^{p}_{ij}, \tag{2}$$

where $\dot{\varepsilon}_{ij}$ can be transformed from the macroscopic strain increment, and plastic part $\dot{\varepsilon}^{p}_{ij}$ is calculated using the successive integration method[1], which can be expressed as following: We call τ^{s} the resolved shear stress on the sth slip/twin system:

$$\tau^{s} = L^{s}_{ij}\sigma_{ij}, \tag{3}$$

where:

$$L_{ij} = (a_i b_j + a_j b_i)/2 \tag{4}$$

and a and b are respectively the unit vector normal on a slip/twin system and the unit vector parallel to the slip/twinning both in a deformed configuration. The conditions for judging the activity of a given slip/twinning system are as follow. When we have sign$(\tau^{s})\dot{\gamma}^{s} > 0$ the slip system is active,when we have sign$(\tau^{s})\tau^{s} < k^{s}$ it means that the slip system is inactive. Where $\dot{\gamma}^{s}$ is the shear strain increment on the sth deformation system, k^{s} is the critical yielding stress of that slip/twinning system.
Due to the directionality of the twinning, to consider whether the twin occurs or not the condition that the shear stress along it must be positive($(\tau^{s})> 0$) needs to be applied. For all the systems the following differential equations are used:

$$2G(dX^{s}/d\rho) = Y^{s}, \tag{5}$$

where $\rho(\geq 0)$ is a monotonously increasing parameter such as time, G is the shear modulus and

$$X^{s} = \text{sign}(\tau^{s})\dot{\gamma}^{s}, \tag{6}$$

$$Y^{s} = \text{sign}(\tau^{s})\tau^{s} - k^{s}. \tag{7}$$

The above equations are successively integrated under the following conditions:$X_r = 0$ at $\rho = 0$ and when the value of X_r becomes negative during the integration, set $X_r = 0$.
Continue the above integration until all X_r remain unchanged. After obtaining dX_r we get $\dot{\gamma}^{r}$

$$\dot{\gamma}^{r}_{\rho+d\rho} = \dot{\gamma}^{r}_{\rho} + d\dot{\gamma}^{r}. \tag{8}$$

Then the plastic strain in a crystal is calculated as follows:

$$\dot{\varepsilon}_{ij} = (1 - \sum_{\beta=1}^{N^{tw}} g^{n,\beta}) \sum_{\alpha=1}^{N^{s}} \dot{\gamma}^{\alpha} L^{\alpha}_{sl} + \sum_{\beta=1}^{N^{tw}} g^{n,\beta}\dot{\gamma}^{\beta}_{tw} L^{\beta}_{tw}, \tag{9}$$

where $\sum_{\beta=1}^{N^{tw}} g^{n,\beta}$ is the total volume fraction of deformation twinned region in the grain, $\dot{\gamma}^{\alpha}$ and $\dot{\gamma}^{\alpha}_{tw}$ denote the shear strain increment on slip system α and twinning system β, respectively, N^{s} and N^{tw} represent the number of potentially active slip and twinning systems and L defines the tensorial direction of the shear caused by an active deformation system.
This algorithm has been implemented where a single cubic element composed by 5 tetrahedral elements is used to simulate the single crystal of the magnesium alloy AZ31B.

2.2 Hardening model

The strain hardening can be described by the sum of the self hardening and latent hardening that is caused by the interaction among different slip/twin systems. The change of the critical resolved stress on the sth slip/system is described by:

$$\dot{\tau}^s = \sum_t h^{st}|\dot{\gamma}^t|, \tag{10}$$

with

$$h^{st} = h(\bar{\gamma}^t)q^{st}, \tag{11}$$

where h^{st} is the hardening coefficient and q^{st} is the interaction matrix describing the self and latent hardening.

To reproduce the complex mechanical response of the single crystal magnesium, the following hardening laws are used:
-Linear hardening:

$$h(\bar{\gamma}) = h_0, \quad \bar{\gamma} = \sum_\alpha \int_0^t |\dot{\gamma}^\alpha| d\tau, \tag{12}$$

where h_0 is a constant and $\bar{\gamma}$ is the equivalent shear strain, which is the sum of shear strains in all the slip/twin systems.The linear hardening has been used just to describe the twin hardening paired with the twin scheme.
-Voce hardening ([2],[3]):

$$h(\bar{\gamma}) = h_0(1 - \frac{\tau_0}{\tau_\infty})\exp(-\frac{h_0\bar{\gamma}}{\tau_\infty}), \tag{13}$$

where τ_0 is the initial CRSS and τ_∞ is the saturation CRSS

2.3 Twinning formulation

Since magnesium does not have enough independent slip systems to accumulate arbitrary deformation, twinning is an important deformation mechanism, especially at room temperature, and it should be considered in the simulations. In this study, the contribution of twinning to the grain reorientation is treated according to a modified version of the PTR scheme [4]. The procedure is summarized as follows.

At each deformation step the increment of the twinned volume fraction in the grain n associated with each twinning system β is:

$$\Delta g^{n,\beta} = \frac{\Delta\gamma^{n,\beta}}{\gamma^\beta}. \tag{14}$$

Then the twinned fraction is accumulated after each deformation step

$$g^{n,\beta} = \sum_{steps} \Delta g^{n,\beta}, \tag{15}$$

here $\Delta\gamma^{n,\beta}$ is shear strain increment contributed by the twinning system β in the grain n, and γ^β is the characteristic shear strain associated with twinning, it is a constant and its value is $\gamma^\beta = 0.13$. A summation over all twinning systems gives the twinned volume fraction in the grain f^n

$$f^n = \sum_\beta^{N_{Tw}} g^{n,\beta}, \tag{16}$$

where f^n is the volume fraction of grain n. A threshold value for reorientation is empirically set as

$$F_T = 0.9 \tag{17}$$

At every deformation step the twinned fraction of each grain f^n is compared with the threshold value F_T. If f^n is greater than F_T, the grain is completely reoriented to the orientation of the twin and the twin system will stop its growth in the next steps.

3 Model Parameters

The simulation has been conducted on a cube sample composed of a 1000 grains (10x10x10) with random texture to simulate the cast alloy, the pole figures are shown below. The boundary condition

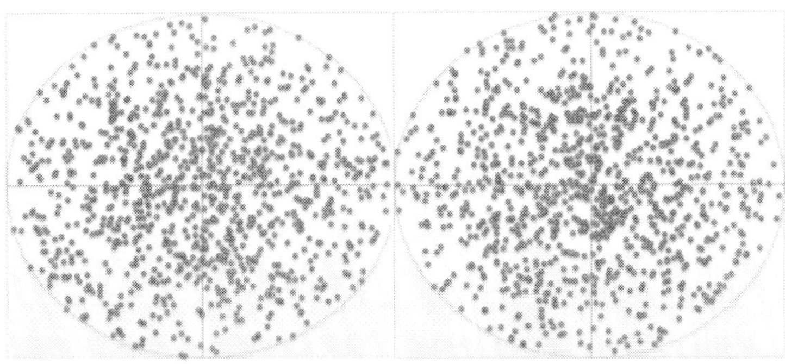

{0001} {11-20}

Fig. 1 – Pole Figures of the sample used for the simulations

for the uniaxial or compression in the z direction all components of the displacements are prescribed at the block surfaces by the following.

$$\begin{bmatrix} \dot{u} \\ \dot{v} \\ \dot{z} \end{bmatrix} = \begin{bmatrix} \dot{E}_x & \overline{\dot{\gamma}}_{xy}/2 & 0 \\ \overline{\dot{\gamma}}_{xy}/2 & \dot{E}_y & 0 \\ 0 & 0 & \dot{E}_z \end{bmatrix} \begin{bmatrix} x \\ y \\ z \end{bmatrix}$$

$\dot{E}_z = \dot{e}$ is prescribed as time step and we set $\dot{e} = 0.05\%$. All the components of the macroscopic stresses except z component are forced to vanish by the following.

$$\dot{E}_x = -\nu(\dot{e} - \dot{E}_z^p) + \dot{E}_x^p, \qquad \dot{E}_y = -\nu(\dot{e} - \dot{E}_z^p) + \dot{E}_y^p, \tag{18}$$

where $\nu = 0.35$ is the poisson ratio. For evaluating the accuracy of the constitutive model in predicting the stress response in a polycrystal of magnesium AZ31B at room temperature, the numerical predictions and experimental results of simple tensile and compression tests taken from [5] are compared. The parameters used in the this model have been founded through a trial and error model to achieve a better fit of the experimental data. Here below the parameter found in this research are shown, where Table 1 contains the microscopic parameters, while Table 2 represents the self and latent hardening matrix.

IOP Conf. Series: Materials Science and Engineering **967** (2020) 012057 doi:10.1088/1757-899X/967/1/012057

	τ_0	τ_∞	h_0
basal	10.52	36.85	105
prismatic	25	57.8	195.8
pyramidal II	48.7	87	334
twin	27.43	76.3	131.3

Table 1 – Microscopical parameters

$q(\alpha \backslash \beta)$	slip	prismatic	pyramidal II	twin
slip	1	0.2	0.2	0.5
prismatic	0.2	1	0.25	0.2
pyramidal II	0.2	0.25	1	0.2
twin	0.5	0.4	0.4	1

Table 2 – Self and latent hardening matrix

4 Verification of the constitutive model

The simulation has been conducted performing both the compression and the tension test on all the three axes, but because of the isotropic nature of the sample due to the random texture, the results are almost the same on all the three axes.

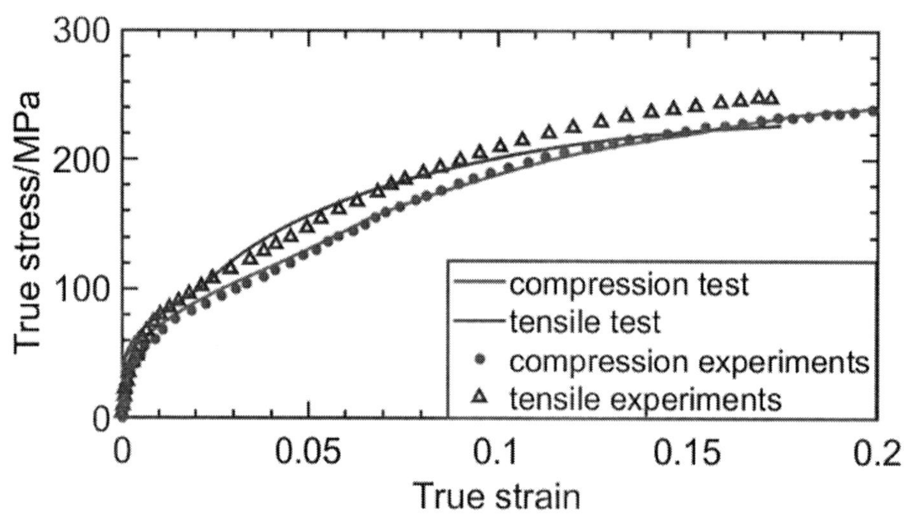

Fig. 2 – Evaluation of the fitted curves.

As it can be noticed the compression curve is better approximated that the tension one, but we can state that our model can represent the trend of the cast AZ31B under tension and compression. In Fig. 3 and Fig.4 the relative activities of each slip system during the deformation are shown, as it can be forseen the activities are very similar, this because having the random distribution of the texture gives the sample an almost isotropic nature, the basal slip and the twinning occur first due to the lower CRSSs, as expected, once the twinning reach the saturation and they stop the other two slips take over the deformation, specially the pyramidal II,that it is indeed only left deformation that occurs along c-axes of the grains.

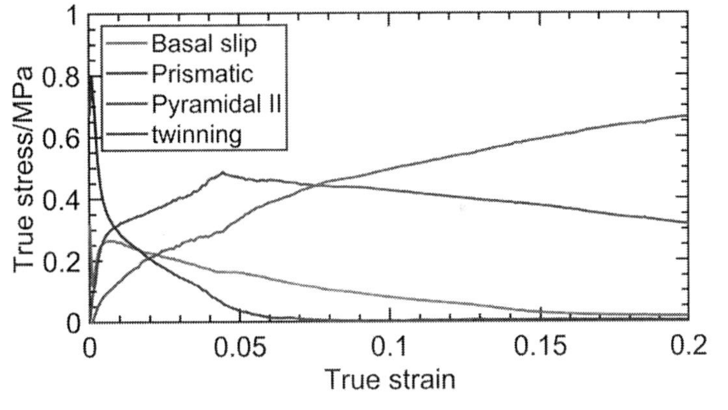

Fig. 3 – Relative activities during compression

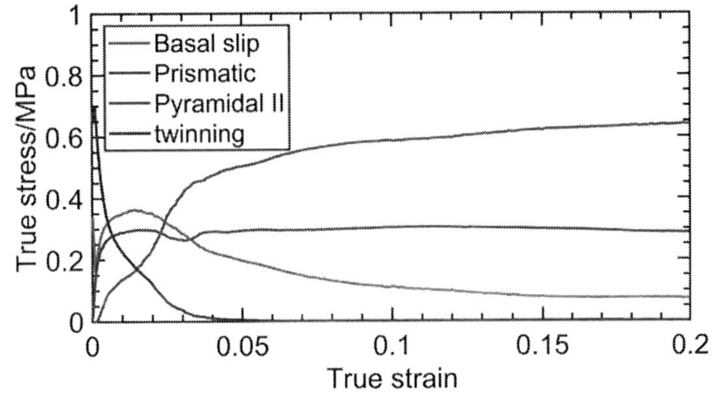

Fig. 4 – Relative activities during tension

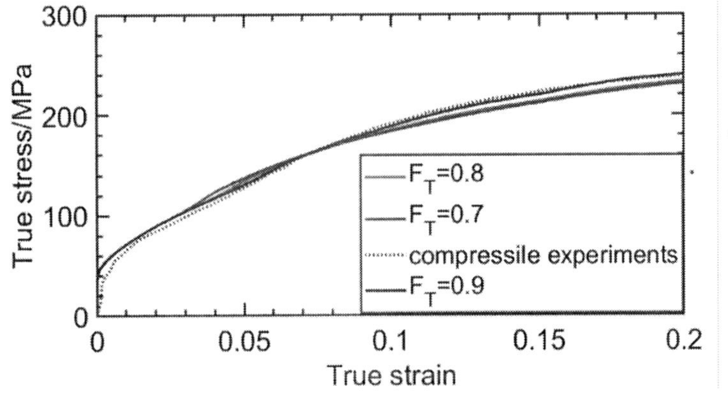

Fig. 5 – Comparisons of F_T 's values

The value of the threshold $F_T = 0.9$ has been set empirically to have the best fit with the experimental data. Here above the Fig.5 shows the effects of the twin's parameter, as it could be expected, the higher it is the later the twin reaches its saturation and at this point corresponds the increase of steepness of the curves.

Increasing the threshold value means that the twins will stop their growth later on during the deformation, when this happens we can notice that the stress required to keep the plastic deformation occurring is higher, this because normally the twin would occur because of its lower CRSS, but blocking it implies that the system has less deformation systems available and this will make the material stiffer. Before the strain reaches 0.75% the lower the F_T curve the higher the stress needed, after that point the trend is the opposite, this happens because the effect of the latent hardening of the twin has an important role and a larger deformation along twin strain would result in a stiffer matrix.

5 Conclusion

As written before, this model can represent the deformation of the cast AZ31b with good precision, a better fitting could be reached, but it would not be worth it to be found in the case of the cast sample, because of two reasons: the first one is that is difficult to isolate the contribution of each parameter to the curve if every slip system is active at the same time with almost the same magnitude, as illustrated in Fig.3 and Fig.4 ,in the case of an extruded tube or a rolled sheet the anistropy in various direction would make easier to evaluate the parameters giving that for each angles there are different slip systems activated. The second one is that the magnesium and its alloys are usually used when in the form of extruded tubes or rolled sheet, so a deeper investigation of those it would be preferred. In the future this constitutive model will be applied to the numerical testing of a rolled sheet of AZ31B.

References

[1] H. Takahashi, H. Motohashi, M. Tokuda and T. Abe. (1994) Elastic-Plastic Finite Element Polycrystal Model International Journal of Plasticity 10 63.

[2] S.B. Yi, C.H.J. Davies, H.-G. Brokmeier, R.E. Bolmaro, K.-U. Kainerand J. Homeyer.(2006) Deformation and texture evolution in AZ31 magnesium alloy during uniaxial loading. Acta Materialia 54, 549–562.

[3] S.R. Agnew, M.H. Yoo and C.N Tome'. (2001) Application of texture simulation to understanding mechanical behavior of Mg and solid solution alloys containing Li or Y. Acta Materialia 49, 4277–4289.

[4] Jalili, M., Soltani, B., (2019) Investigation the micromechanisms of strain localization formation in AZ31 Mg alloy: A mesoscale 3D full-field crystal plasticity computational homogenization study, European Journal of Mechanics / A Solids , doi: https://doi.org/10.1016/ j.euromechsol.2019.103903.

[5] Shimizu I. (2018), Biaxial Compressive Behavior and Tension-Compression Asymmetry on Plastic Deformation of Cast and Extruded AZ31 magnesium Alloys Advanced Experimental Mechanics, Vol.3 , 141-146

[6] Paramatmuni, C., Kanjarla, A.K, A crystal plasticity FFT based study of deformation twinning, anisotropy and micromechanics in HCP materials: Application to AZ31 alloy, International Journal of Plasticity, https://doi.org/10.1016/j.ijplas.2018.10.007

IOP Conf. Series: Materials Science and Engineering **967** (2020) 012058 doi:10.1088/1757-899X/967/1/012058

A simple device to measure bend limit of sheet metals

A Deole[*1], **N Etxebarria**[*2,] **J Mendiguren**[2], **A Ilinich**[3], **M Weiss**[1]

[1] Deakin University, Geelong, Australia, Institute for Frontier Materials, Waurn Ponds, Pigdons Rd., VIC. 3216
[2] Mondragon Unibertsitatea, 4, Loramendi kalea, 20500 Mondragon, Spain
[3] Research and Innovation Center, Ford Motor Company, Dearborn, 2101 Village Rd., MI 48121, USA.

*Corresponding author e-mail: adityad@deakin.edu.au, nagore.etxebarria@alumni.mondragon.edu

Abstract. The analysis of bend fracture limits of sheet metals is important for the automotive industry. For high and mid-ductility materials, current methods are often unreliable due to the difficultly of detecting the initiation of a crack on the outer surface of the bend. Sophisticated techniques that rely on the strain evolution in the material are capable to detect the cracks and establish the bend limits using digital image correlation (DIC) strain measurement systems. However, these techniques are complex and sometimes impractical for industrial use. This study uses a simplified bend test technique that uses optical image processing to detect crack initiation to then determine the bend limits of aluminum alloys using a modified angled line method (MALM). The findings of this study indicate that the optical method is inconsistent in detecting the crack initiation and likely to result in a large deviation in detected the bend limits.

1. Introduction

The standard vee bend test [1,2] is commonly used to determine the minimum radius, ρ, in bending before the sheet fractures. The ratio of this radius, ρ, and the thickness, t, gives the min bend limit, $\frac{\rho}{t}$. A limitation of the standard bend test is that the measurement of the bend limit often relies on the punch radius while the radius of the actual sheet is not considered [3]. This leads to inaccurate estimates of the minimum radius in bending since the sheet is likely to deform to a smaller radius than that of the punch [3,4]. The VDA238-100 standard [2] has been developed to identify fracture in tight radius bending of sheet materials. It is a 3-point bend test with a sharp punch radius and two rolls which act as the bend supports. Fracture is determined using a predefined force threshold. Newly developed setups combine the 3-point bend test with digital image correlation (DIC) systems [2, 3]; this enables analyzing bending radius and strain during testing.

Mattei et al., showed that the fracture in bending is mainly controlled by the fracture limit of the surface grain. The damages in bending first initiated by formation of the shear bands. This then leads to surface undulations resulting in necking on the surface and then the final fracture[5]. This means the surface roughness in the bend corners grows into the formation of a crack. As bending progresses, the distributed micro cracks propagate and merge into wider, visible macro-cracks. This transition of the surface roughness to the onset of macro cracking is gradual and this makes the detection of the crack difficult [6,7]. The bend limit is directly affected by the accuracy of the crack detection and there is a high possibility for overestimating the bend limit. The forming processes when designed based on such overestimated bend limit may then result in poor quality of the components. This issue is likely to occur when the detection of the crack is done by visual inspection which subjected to the operator's perception of a crack. Kaupper et al., addressed this issue and proposed to identify the crack initiation by identifying the stagnation of the strain evolution curve in the material neighboring to a visible crack[7]. A practical limitation of this method is that the method involves complex DIC strain measurement.

Swillo et al. [6] investigated fracture initiation in aluminum with an optical image processing algorithm The algorithm converts the grayscale image into a binary image using a histogram distribution of the original image. Due to the inherent color of the aluminum alloys, the cracks can be distinguished from the undamaged regions in the image processing routine. Another problem addressed in their study [6][8] was the in-situ measurement of strain. They used a modified angled line method (MALM) [6] and validated the approach in bending tests performed on two aluminum alloys. Above methods for detecting the crack and to measure forming strains in the bend test can be integrated into a simple bend test. This may provide a promising solution for estimating bend fracture limits in industrial production.
This paper presents a simple mechanical device combined with a camera that detects the crack initiation and measures strain at the onset of the crack. An optical image processing algorithm is developed in MatLab2019 that enables both crack detection and the strain measurement with the MALM method. The proposed device is tested for its robustness and repeatability using specimens of high and low ductile aluminum.

2. Materials

Two different aluminum alloys A1 with 2mm thickness and A2 with 1.5mm thickness were used. The tensile tests were performed according to ASTM E8/E8M-11 standard and with the specimen shape given in **Figure 1**. The crosshead speed was 3mm/min and the engineering strain was measured with a video extensometer. The tensile parameters are given in Table 1.

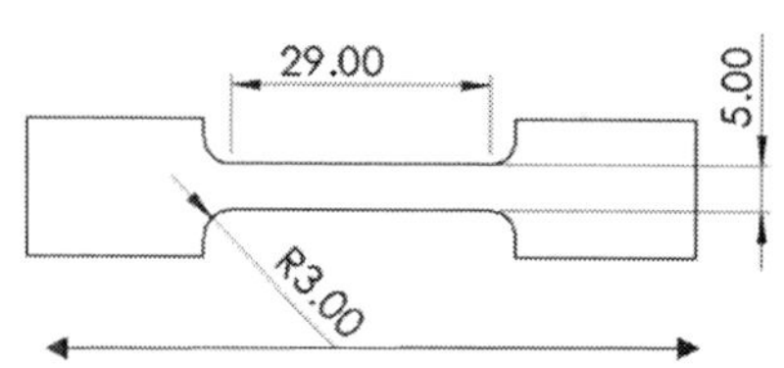

Figure 1. Tensile test specimen.

Table 1. mechanical properties of the aluminum alloys.

Material	$YP_{0.2\%}$ (MPa)	Ultimate tensile stress (MPa)	Uniform elongation (%)	Total elongation (%)
A1	224	339	19.8	27.1
A2	498	556	11.5	15.8

3. Bending experiments

Rectangular strips of 90 mm length and 20 mm width (Figure 2b) were tested. The samples were cut using water jet and the rough edges cleaned using polishing paper. Two black lines with an angle inclination of 45^0 were marked on the sample surface. These lines were used to determine the bend strain using a modified angled line method explained in the section 5.

The bending device used in this study is shown in Figure 2a. The tooling is assembled in an Instron testing machine (see Figure 2a) which is equipped with a 30 kN load cell. It consists of a base plate with two vertical guide rods and a punch positioned in the centre having a 0.2 mm radius that provides the support point to the sample. The rollers are connected to the movable crosshead of the device with roller bearings to reduce friction. The diameter of the roller and the centre-to-centre distance are 25 mm and 33 mm, respectively. The specimen is inserted and initially held in place between the rollers and the punch by applying a small force. A crosshead movement of 5 mm/min is applied. A camera facing perpendicular to the sample surface is positioned at a distance of 40mm (Figure 2a). The 2D images of the sample surface having a resolution of 1600 pixel x 1200 pixel were recorded with rate of 1picture/second. The punch remains stationary while the rollers move downwards (see Figure 2a). In this way the area of the sample in the punch contact region remains stationary and at a constant distance to the camera. This allows deformation being recorded without the risk of the viewing area moving out of focus. The tests were carried out with a speed of 5mm/min until fracture occurred. 3 to 4 tests were performed for each material.

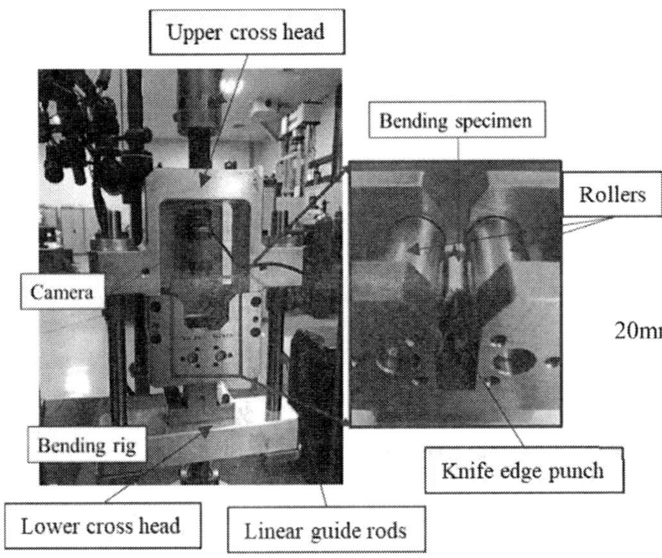

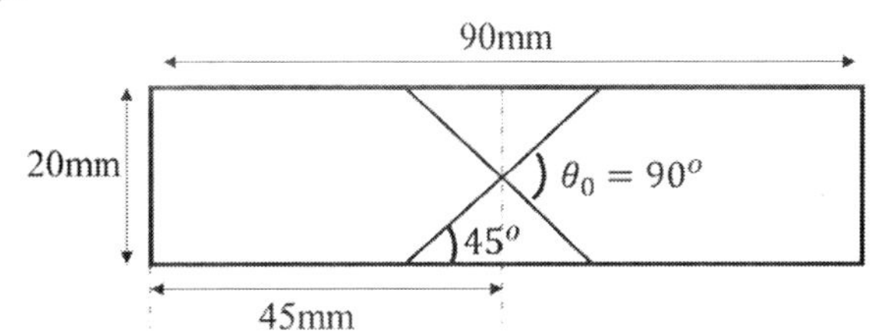

Figure 2. (a) Bend test rig with camera attached on the top (b) bend test specimen.

4. Automatic Crack Detection (ACD) software

A series of images captured during the bend test is processed in the automatic crack detection (ACD) software developed based on the pattern recognition algorithm presented in [6,8]. The algorithm differentiates the black and white pixels in the image and considers white pixels to be the crack (Figure 3d). The cracks developed are counted in each picture (Figure 3a and b). In the beginning the crack quantity increases due to number of microcracks emerging on the surface. The micro cracks then join to form a macro crack and the crack quantity decreases [6,8]. The crack is assumed to be initiated at an instance when the quantity of the cracks reaches a plateau and decreases. The software allows to change the threshold in order to change the curve shown in Figure 3c to show a distinct peak. This threshold is the minimum pixel size that will be considered as a crack.

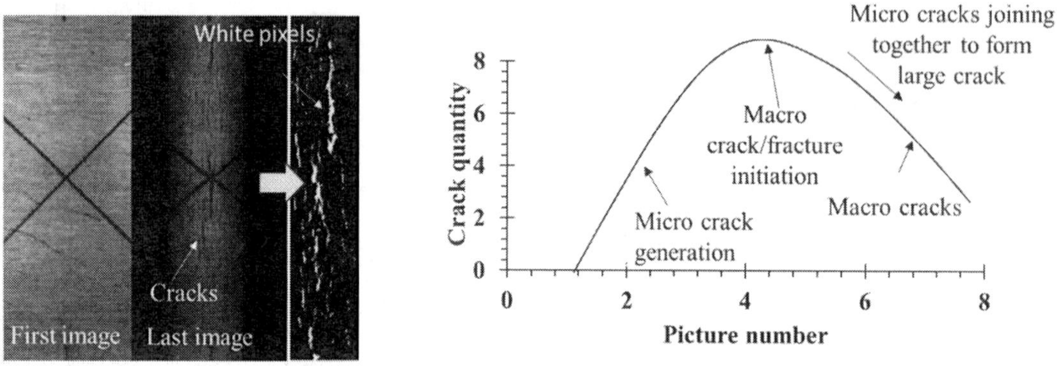

(a) (b) (c) (d)

Figure 3. The sample surface (a) before crack initiation (b) after crack (c) greyscale converted image (d) the evolution of the crack quantity.

5. Modified angled line method

The surface strain of the sample was measured using the modified angled line measurement method presented in [6,8]. For this, two inclided lines with an initial angle θ_o of 90^0 were drawn on the sample suface. The ACD software then detects these lines and calculates the angle, θ, between the two lines at every stage of bending. The angle, θ, is then used to determine the maximum strain using Eq. 1. The change in angle during deformation is visually shown in Figure 4a,b and c.

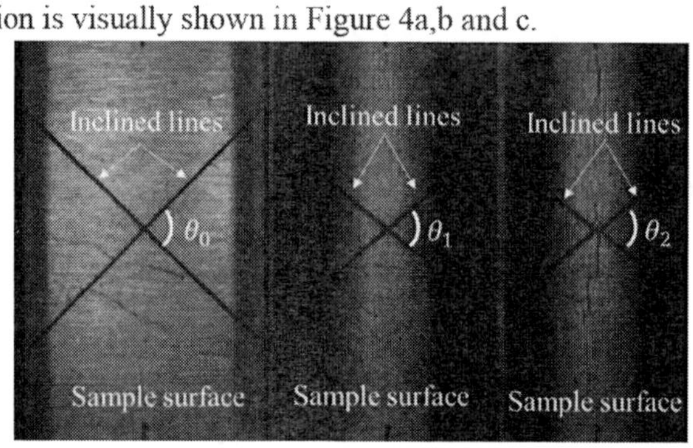

(a) (b) (c)

Figure 4. The displacement of the inclined lines (a) before bending (b) during bending (c) after fracture initiation.

$$\varepsilon_1 = \ln\frac{\cot\theta}{\cot\theta_o} \tag{1}$$

6. Result and Discussion

6.1 Crack detection with the conventional load drop method

Figure 5a shows that load displacement curve for material A1. Using the recommended load drop of 30 N indicates fracture to occur at a punch displacement of approximately 10 mm, point A. However, no visual crack can be observed at this point (**Figure** 5b). Fracture initiation is visually observed at a later punch displacement of 12 mm (point B) representing a load drop of 230N.

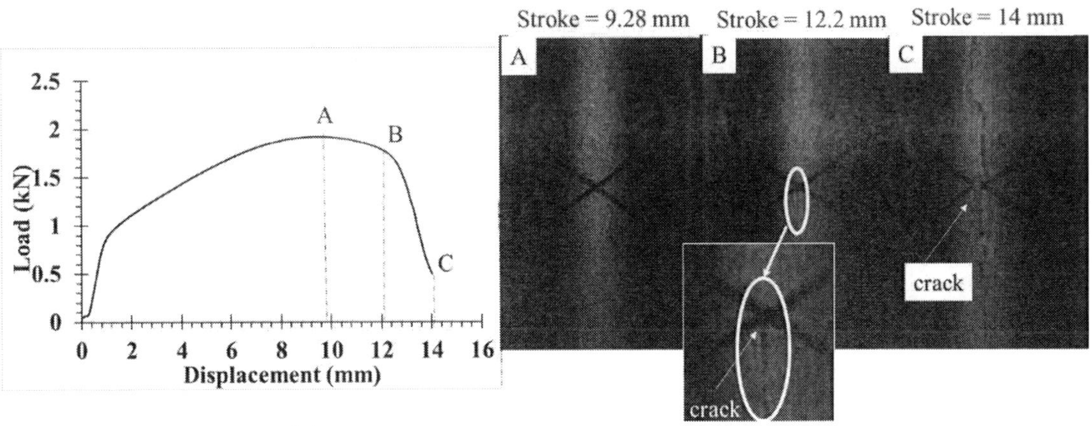

(a) (b)

Figure 5. (a) Load-displacement curve for material A1 (b) crack initiation and growth on the outer sample surface.

In case of material A2, the onset of cracking was clearly identified at the peak load ata punch stroke of approximately 7 mm, point A (**Figure** *6*b)

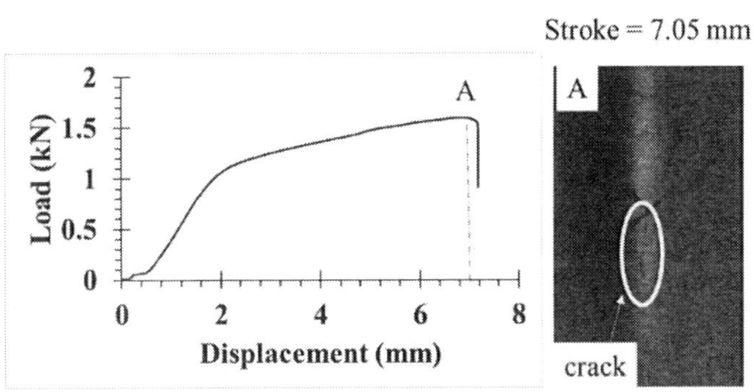

(a) (b)

Figure 6. (a) Load-displacement curve for material A2 (b) crack initiation and growth on the outer sample surface.

IOP Conf. Series: Materials Science and Engineering **967** (2020) 012058 doi:10.1088/1757-899X/967/1/012058

6.2 Crack quantity

Figure 7a shows an example of the evolution of the number of cracks determined with the ACD method for the ductile aluminum alloy, A1 for different thresholds. It can be seen that the number of cracks does not show a distinct maximum when the crack occurs (Figure a and b).

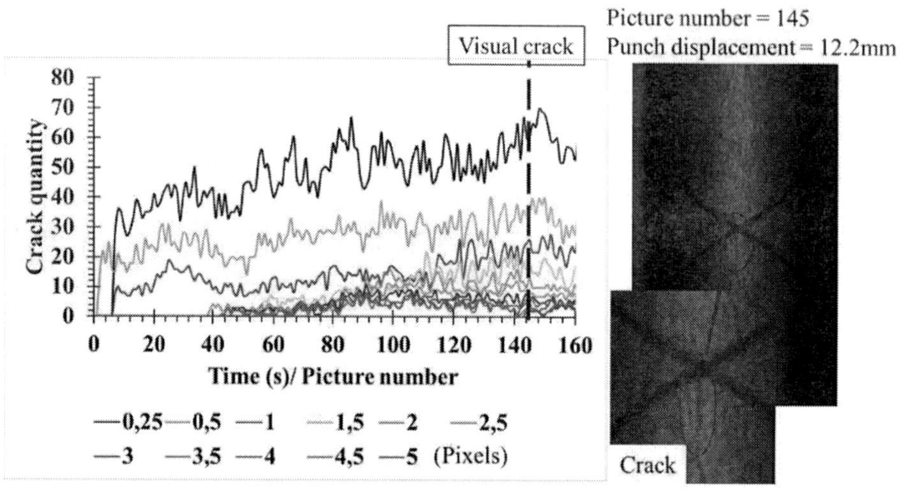

(a) (b)

Figure 7. (a)Example of the evolution of the number of cracks for A1 with no black paint applied (b) crack initiation on the sample surface.

To facilitate the distinction between white and dark sample areas in the tests were repeated with the brittle A2 material and the outer surface painted with black ink. Figure 7a shows an example of the crack quantity evolution measured using ACD software for this condition for different threshold values. The results show that that the crack quantity increases initially and then drops after showing a peak. The crack quantity is higher for a lower value of threshold and vice-versa. Based on the theory the maximum in crack quantity represents the point where macro cracking initiates. The punch displacement at crack initiation measured with the ACD method is compared to that measure with the conventional force-drop approach in Figure 7b. A significant scatter in results can be observed for the ACD method. In addition the ACD method appears to significantly overestimate the punch displacement at crack initiation which was accurately estimated with the conventional load-drop method for the brittle A2 alloy. Further calibration of the ACD method may be possible to identify the optimum threshold. However, given that for all thresholds the punch displacement at cracking initiation has been overestimated the ACD method may not represent a valuable solution.

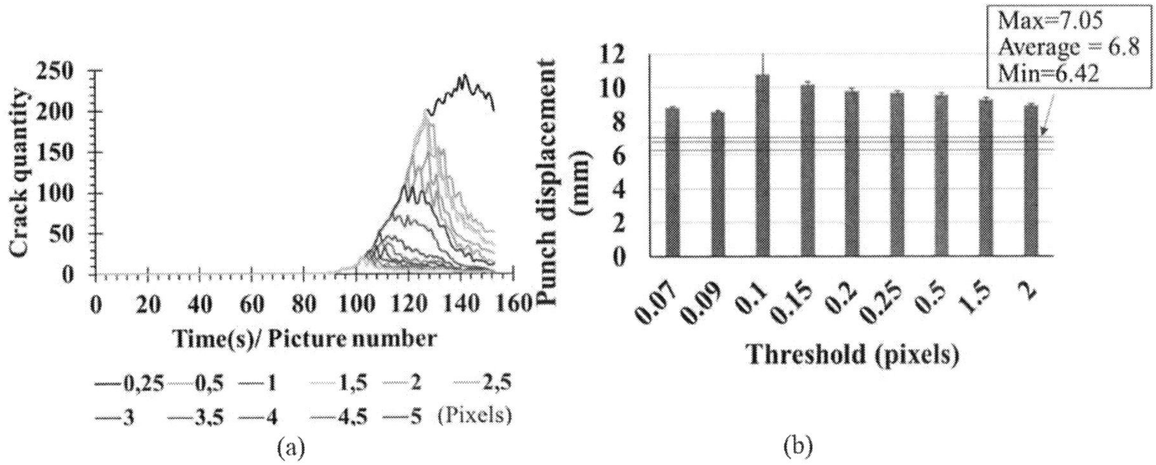

Figure 8. (a)The crack quantity measured for material A2 (b) the punch displacement averaged for four alloys for different threshold with red lines indicating the displacement at which the fracture detected with the conventional force-drop method.

6.3 Strain measurement using MALM

The surface strain measured with the MALM method in the bending tests performed on the ductile A1 material are given in Table 2. There is some deviation between the strain measurements which may be due to crack initiation being identified visually. The min bend ratio , $\frac{R_{in}}{t}$, that can be formed before fracture can be estimated based on the strain measured on the outer sample surface, ε_{major} using the relation [3]

$$\frac{R_{in}}{t} = \frac{1}{2\varepsilon_{major}} - \frac{1}{2} \quad (2)$$

with t being the material thickness. This gives a minimum bend ratio between 0.07 and 0.45 which is common for an aluminum alloy with high tensile ductility such as A1 (Table 1). Nevertheless future work needs to be aimed at validate the bending strain values determined with the MALM method using in-situ DIC strain measurement. To enable MALM strain measurement in combination with ACD crack detection will require future tests using white lines for strain measurement in combination with a black sample surface to facilitate micro crack identification.

Table 2. The strain measured using ACD software when the crack was visually detected for material A1 with no black paint.

	Sample 1	Sample 2	Sample 3	Deviation
ε_{major} **(mm/mm)**	0.71	0.88	0.53	0.35
Min bend radio	0.2	0.07	0.45	

7. Conclusions

Bend tests were performed on ductile and brittle Aluminium to validate an automatic crack detection software that identifies the initiation of fracture and measures the outer fibre strain with an optical approach. The main conclusions from this study are

1. The bend test device presented in this work maintains a fixed distance between the camera and the sample surface during the test. This leads to high quality imaging that allows optical crack detection and strain analysis.
2. The proposed method and software can be used to identify the crack initiation point but needs further calibration to achieve sufficient accuracy.
3. The results suggest that painting the outer sample surface with black ink is required to enable accurate identification of micro cracks. This requires the use of white lines for optical strain measurement which needs to be validated with additional trials.

Acknowledgements: The authors would like to acknowledge the University Research Project (URP) scheme of the Ford Motor Company for funding this work.

References

[1] ASTM S T M 2014 *Bend Testing of Material for Ductility- E290 – 14*
[2] Anon VDA 238-100 Plate bending test for metallic materials - VDA
[3] Deole A D, Barnett M and Weiss M 2018 Analysis of fracture in sheet bending and roll forming *AIP Conference Proceedings* vol 1960
[4] Yu T X and Johnson W 1981 The press-brake bending of rigid/linear work-hardening plates **23** 307–18
[5] Mattei L, Daniel D, Guiglionda G, Klöcker H and Driver J 2013 Strain localization and damage mechanisms during bending of AA6016 sheet *Mater. Sci. Eng. A* **559** 812–21
[6] Swillo S J, Iyer K and Jack Hu S 2006 Angled line method for measuring continuously distributed strain in sheet bending *J. Manuf. Sci. Eng. Trans. ASME* **128** 651–8
[7] Kaupper M and Merklein M 2013 Bendability of advanced high strength steels - A new evaluation procedure *CIRP Ann. - Manuf. Technol.* **62** 247–50
[8] Lin G, Jack Hu S and Cai W 2009 Evaluation of formability in bending/hemming of aluminum alloys using plane-strain tensile tests *J. Manuf. Sci. Eng. Trans. ASME* **131** 0510091–9

Effect of cooling system in hot stamping process

P Srimon Narayanasamy, Amarjeet Kumar Singh, K Narasimhan

Department of Metallurgical Engineering & Materials Science

IIT Bombay, Mumbai, Maharashtra – 400076

nara@iitb.ac.in

Abstract: Hot stamping process is used to produce automotive components that give a high strength-to-weight ratio, therefore it reduces fuel consumption and improves crashworthiness. Hot stamping is a non-isothermal process where forming and quenching takes place simultaneously. One of the major factors affecting the mechanical properties of components is the cooling rate which is controlled by the cooling systems. Direct cooling and indirect cooling (cooling channel) are the two cooling methods used in this process. In direct cooling, water is used as the cooling medium that gives higher cooling rate, however it induces large residual stress in the final component. In indirect cooling, cooling channels are used in die and punch for cooling purpose. In indirect cooling system residual stresses are not developed, but thinning occurs in wall region of the blank due to inhomogeneous contact between blank and die. Therefore, in this work, a new method has been proposed with the combination of direct and indirect cooling. In this combined cooling method, different coolant like spray water, air mist and compressed air was used for direct cooling. Simulation of the proposed approach was done by using PAMSTAMP software. The effect of the combined cooling system with various cooling medium was investigated and it was found that the air mist cooling with indirect cooling system resulted in 7% improvement in thickness distribution and 20% improvement in temperature distribution compared to the conventional method.

Keywords: - Combined cooling, Temperature gradient, Thickness distribution, Hot stamping

1. Introduction

Lightweight automotive vehicles release less CO_2 due to low fuel consumption but problem associated with lightweight component is less strength. Therefore to achieve high strength and crashworthiness, hot stamping process was developed [1]. Hot stamping is a non-isothermal process in which forming and quenching takes place simultaneously. 22MnB5 steel is used in the hot stamping process because a small quantity of boron addition can increase the hardenability of the material. This process is commonly used to produce components like B-Pillar, A-Pillar, Front bumper, Side rails, etc. Final mechanical properties and microstructure of the component in hot stamping process depends on various factors like deformation temperature, rate of deformation, cooling etc. Our main concern in this work is cooling system. Hot stamping process consist of two types of cooling system namely direct cooling and indirect cooling. In direct cooling, the cooling medium directly contacts with blank surface which increases the heat transfer and production rate. However in this method, due to high cooling rate, residual stress is induced in the blank. Hence, indirect cooling method is used conventionally, where inbuilt cooling channels in die and punch are used for cooling. Due to lower cooling rate in indirect cooling, inhomogeneous temperature distribution develops in the blank. Due to temperature gradient in the blank, excessive thinning occurred in the wall region. André et al [2] studied the gap behaviour between die and blank and found that the cooling rate reduced when gap thickness increased. Nakagawa et al [3] performed experiment in water submerged die to avoid local thinning but static water in the die cavity created a thermal shock as well as scale formation in the component. Ganapathy et al [4] rapidly pre-cooled the austenized blank in the transfer stage to acquire the minimum initial temperature for higher formability. However temperature control should be accurate in pre-cooling to avoid ferrite and pearlite formation. Maeno et al [5] used both indirect quenching and static water in a die cavity to achieve better formability and observed that the thickness distribution completely depends on the height of water in the die cavity. Eiichi et al [6], in their work air was impinged in the wall region of the blank to increase the flow stress. But air cooling did not give a significant improvement in thickness distribution. To reduce the temperature drop, Maeno et al [7] used a high speed mechanical servo press, but investment cost for servo press is too high. Suzuki et al [8] used specially designed transfer devices to clamp the blank in the wall region to reduce its temperature. However different arm holder needed for different shaped blank in this approach.

To overcome the above problem, combination of direct and indirect cooling (i.e. Combined cooling) was established. The principal objective of this proposed approach was to reduce the temperature gradient between the flange and wall region (Figure 1). Indirect cooling was implemented by cooling channel whereas spray nozzles were fixed in the desired position of the die wall for direct cooling. During stamping, direct cooling was done for few seconds to reduce the temperature in wall region of the blank. Different coolant like compressed air, air-mist and water spray were used in direct cooling. The simulation of the proposed method was carried out using PAMSTAMP software. The outcome of combined cooling with different cooling medium were investigated by thickness, hardness, and temperature distribution along the blank.

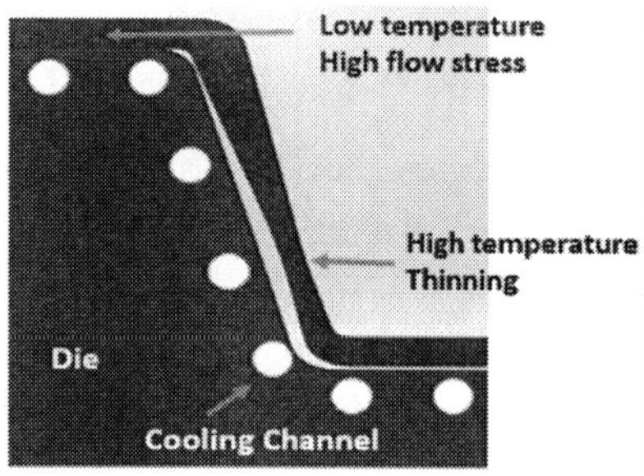

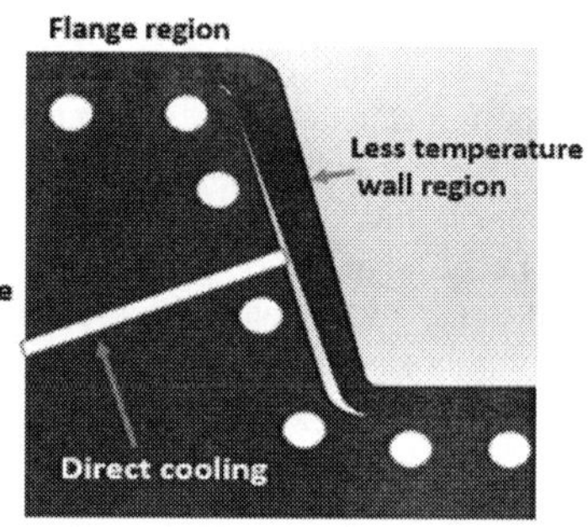

Figure 1. (a) Conventional cooling (b) Combination of direct and indirect cooling

2. Methodology

2.1 FE Simulation

The combined cooling method was simulated by FE simulation to study the temperature distribution and thinning behavior. The commercial FE code PAMSTAMP was used for thermo-mechanical simulation, which comprises different stages(Process cycle) like Transport, Holding, Stamping, Quenching and Cooling on air. Tools for the hat shaped part was designed in SolidWorks and imported to PAMSTAMP. Punch, die and blankholder were assigned as a rigid body. 10 mm^2 mesh size was used for the surface blank. To capture the effect of strain rate and tempeature, various experiments were performed in Gleeble to find the flow behaviour of boron alloyed steel (22MnB5). This flow behaviour of material is used for simulation. Blank holding force, coefficient of friction and other parameter values are given in table 1. Heat transfer coefficient is the most important parameter in the simulation for analysis of thermal behaviour. Except heat transfer coefficient to each combined cooling method, all other simulation parameters were kept identical in all simulation. For all simulations, direct cooling was assumed for one second.

Table 1. Properties used for FE simulation

Properties	Values
Dimension of blank/ specimen	300*140*2.4 mm
Blank-holding force	150kN
Tool temperature	70 ℃
Coefficient of friction	0.4
Yield criteria	Hill 48
Punch velocity	10mm/s
Time for austenization	3min
Transfer time	5s
Stamping temperature	800 ℃
Quenching time	15s
Cooling on air	120s

2.2 Heat Transfer Coefficient

The cooling behaviour of the stamped component is determined by the heat transfer coefficient. Heat transfer coefficient decides the temperature distribution that impacts the microstructure and mechanical property of the end product. Selection of proper heat transfer coefficient is essential for the simulation of heat transfer problems to get nearly precise results. Therefore, heat transfer coefficient calculation method for different methods is discussed below.

Indirect cooling heat transfer coefficient was calculated based on the energy balance method. The convection heat transfer coefficient was calculated by using Dittus-Bolter equation for the internal flow of water[9]. Sudhanshu et al [10] suggested that cooling channel diameter and the water flow rate were 12 mm and 30 l/min respectively for better properties of end product therefore in this work dimension of cooling system were taken according to this analysis.

$$\mathrm{Nu_D} = \frac{h_c * D}{k_f} = 0.023 * \mathrm{Re_D^{0.8}} * \mathrm{Pr^{0.3}} \qquad \ldots(1)$$

An array of round shaped nozzles were placed in the wall region to impinge the air. The heat transfer coefficient expression is based on the flow rate, diameter, and distance between the nozzles which were selected as 100m/s, 1.5mm, 3mm respectively[11]. K and G are the dimensionless coefficients.

$$\frac{\mathrm{Nu}}{\mathrm{Pr^{0.42}}} = 0.5K * G * \mathrm{Re}^{2/3} \qquad \ldots(2)$$

Breitenbach et al [12] developed a theoretical model for water drop impact on film boiling regime and expression for heat transfer coefficient. Water flux, drop diameter, velocity and saturation temperature were selected as 3 l/m^2.s, 350μm, 15m/s and 150℃ respectively.

$$\propto_{ht} = 8.85C \frac{mGe_w(T_0 - T_{sat})}{\rho_f \Delta T(K + 2G) d_0^{0.5} u_0^{0.5}} \eta_{wet} \qquad \ldots(3)$$

Constantin et al [13] did the heat transfer experiment in a platinum strip from 1200℃ with air mist cooling and obtained an expression. It mainly depends on water flux and velocity. The water flux, drop mean diameter and drop velocity were selected as 3 l/m^2.s, 30μm and 10m/s respectively.

$$h = 379.93 w^{0.318} u_{z,v}^{0.330} T_w^{-0.895} d_{30}^{-0.024} \qquad \ldots(4)$$

Heat transfer coefficient of indirect cooling, air cooling, water spray cooling and air mist cooling were calculated by equation 1,2,3 and 4 respectively. Based on heat transfer coefficient(HTC) value, cooling rate was calculated for direct cooling. Reduction in initial temperature of the blank in wall region was calculated by using cooling rate and time.

3. Results & Discussion

3.1 Temperature distribution:

The simulation was done for each set of combined cooling and conventional cooling method. Half of the hat shaped component which is highlighted in figure 2a was selected for analysis due to symmetric nature. The initial temperature at the wall region of blank was decreased based on the heat transfer coefficient value of direct cooling. Temperature distribution of the stamped component after quenching for 2.5s is shown in figure 2b. High temperature gradient was observed in indirect cooling and it was compared with combined cooling method. Figure 2b clearly shows that the temperature gradient in air cooling is not significant due to less heat transfer. The water spray method resulted in 6% improvement when compared to air cooling. But overall air mist cooling provided 20% improvement in temperature distribution from the conventional method. Because, air disintegrates the water into small sized droplets which improves the heat flux by increasing the surface area.

IOP Conf. Series: Materials Science and Engineering **967** (2020) 012059 doi:10.1088/1757-899X/967/1/012059

(a)

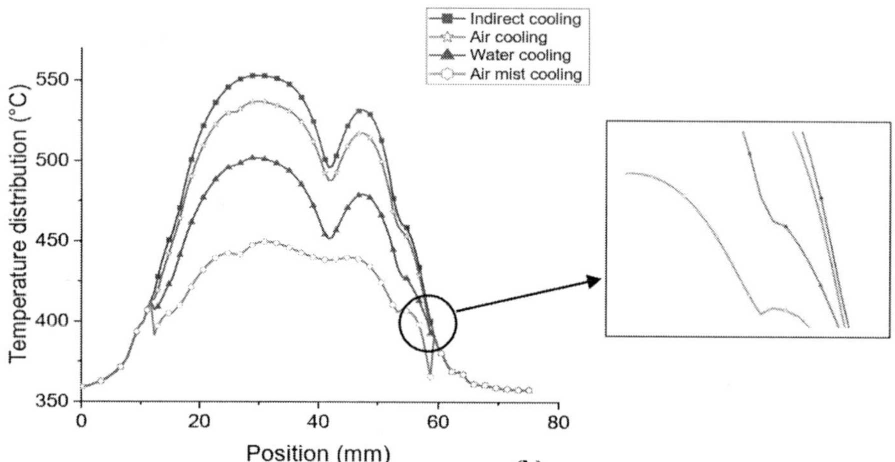

(b)

Figure 2. (a) Formed component (b) Comparision of temperature distribution in quenching for 2.5 sec from flange to center of the blank for indirect cooling with air cooling, water spray cooling and air mist cooling

3.2 Thickness distribution

Thinning was observed near the punch corner area in the conventional method. It was shown in figure 3 that air cooling with indirect cooling resulted in negligible amount of improvement in thickness distribution. But in case of air mist cooling with indirect cooling as shown in figure 3 that 7% improvement in thickness distribution that reduces the thinning problem and also indirectly increases the forming depth.

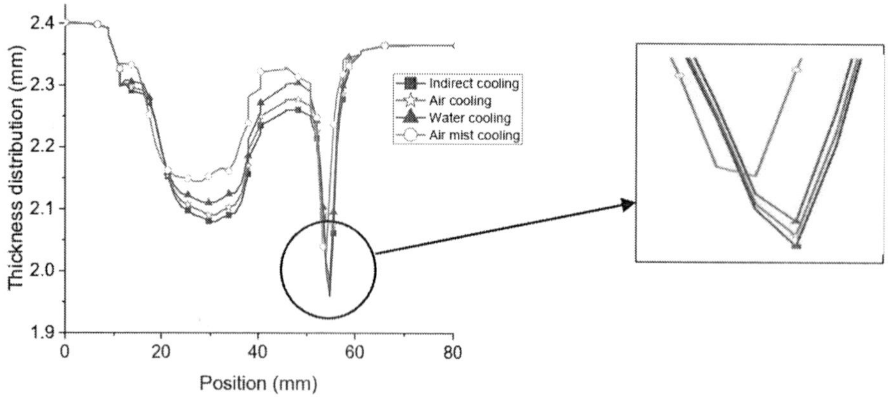

Figure 3. Comparision of thickness distribution from flange to center of the blank for indirect cooling with air cooling, water spray and air-mist cooling.

Temperature and thickness distribution values obtained in table 2 are calculated using

$$\frac{Combined\ cooling - conventional\ cooling}{Conventional\ cooling}\ X\ 100$$

which shows the improvement % in temperature and thickness of combined cooling.

Table 2. Improved property percentage of combined cooling compared with conventional method

Types of cooling system	Temperature distribution	Thickness distribution
Air cooling with indirect cooling	3.25	0.55
Water spray cooling with indirct cooling	9.37	1.44
Air mist cooling with indirect cooling	19.54	7.25

3.3 Hardness distribution

No significant variation in hardness distribution was observed for different types of cooling system as shown in figure 4. This clearly indicates the uniform martensite formation at the end of varing cooling cycle.

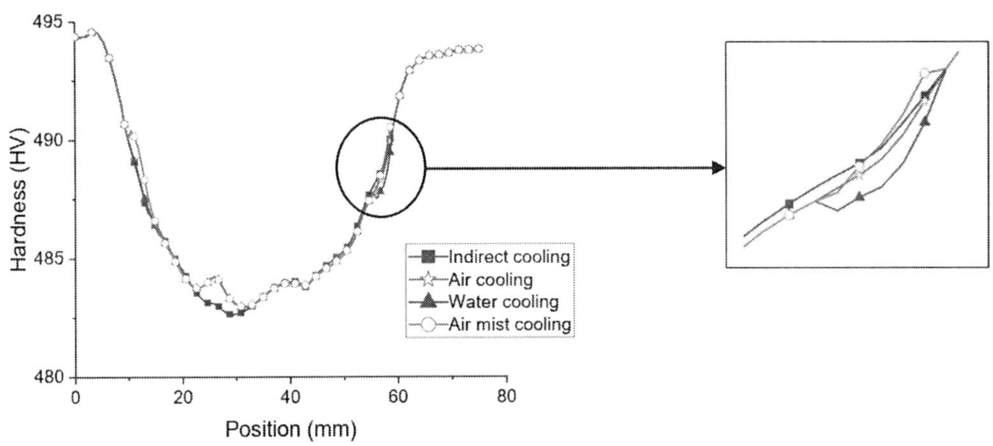

Figure 4. Hardness distribution from flange to center of the component for different cooling systems

4. Conclusion

In this work, different types of cooling system were simulated under different cooling medium to reduce non uniform thickness and temperature distribution. In combined cooling, three different cooling medium were compared to find out the medium that gives the best thickness and temperature distribution. Some of them are new approach to reduce the temperature gradient throughout the blank as seen in simulation. If it used in experiment, it will give better results. Conclusions from the above work is as follows:

- The temperature gradient in the blank was less in case of combined cooling as compared to the conventional cooling method.
- Air-mist cooling provides a 7% improvement in thickness distribution and 20% improvement in temperature distribution compared to conventional methods which is also higher than the other cooling medium.
- No significant changes were observed in hardness distribution for different cooling system.

Nomenculture

Nu	Nusselt number	Pr	Prandtl number
D	Diameter of the nozzle	H	Distance between the nozzle and blank
A_r	Area covered by air	d_0	Drop diameter
K,G	dimensionless coefficient	u_0	Drop velocity
α_{ht}, h	Heat transfer coefficient	$\acute{\eta}_{wet}$	effective wetted ratio
C,G	dimensionless coefficient	w	local water impact density (or flux))
m	mass flux density	T_w	Surface temperature
e_w	thermal effusivity of wall	d_{30}	Volume mean diameter
T_0	Initial wall temperature	$u_{z,v}$	volume-weighted mean of velocity magnitude of z-velocity component
T_{sat}	Saturation temperature	Re	Reynolds number
ρ_f	Density of liquid		
k_f	Thermal conductivity		

References

[1] H. Karbasian and A. E. Tekkaya, "A review on hot stamping", *J. Mater. Process. Technol.*, **210**, no. 15, pp. 2103–2118, 2010.

[2] O. Neto, G. O. Verran, G. C. Pissolatto, and R. R. Dutra, "Effect of necking behavior and of gap size on the cooling rate in hot stamping", *Mater. Res.*, **19**, no. 5, pp. 1138–1143, 2016.

[3] Y. Nakagawa, K. ichiro Mori, T. Maeno, and Y. Nakao, "Reduction in holding time at bottom dead centre in hot stamping by water and die quenching", *Procedia Manuf.*, **15**, pp. 1111–1118, 2018.

[4] B. J. Ganapathy M, Li N, Lin J, "Investigation of a new hot stamping process with improved formability and productivity", *International conference on the technology of plasticity*, **207**, pp 771-776, 2017.

[5] T. Maeno, K. I. Mori, and M. Fujimoto, "Improvements in productivity and formability by water and die quenching in hot stamping of ultra-high strength steel parts", *CIRP Ann. - Manuf. Technol.*,**64**, no. 1, pp. 281–284, 2015.

[6] E. Ota, Y. Yogo, T. Iwata, N. Iwata, K. Ishida, and K. Takeda, "Formability improvement technique for heated sheet metal forming by partial cooling", *Key Eng. Mater.*, **622–623**, pp. 279–283, 2014.

[7] T. Maeno, K. I. Mori, and T. Nagai, "Improvement in formability by control of temperature in hot stamping of ultra-high strength steel parts", *CIRP Ann. - Manuf. Technol.*, **63**, no. 1, pp. 301–304, 2014.

[8] Y. Suzuki, K. I. Mori, T. Maeno, K. Sakakibara, and Y. Abe, "Improvement of formability using partial cooling during transfer in hot stamping of ultra-high strength steel parts", *Procedia Manuf.*,**15**, pp. 1119–1126, 2018.

[9] P. Hayes, 2014, *Transport Phenomena and Kinetics in Process Metallurgy Rate Phenomena in Process Metallurgy,* Vol 1, Chapter 4 (Brisbane / Elsewiser) p 658.

[10] S. Narkade, "Component Level Simulation and Forming Limit Prediction in Hot Stamping of 22MnB5 Steel", Master's Thesis, Indian Institute of Technology, Bombay, 2017.

[11] D. P. D. Theodore L. Bergman, Adrienne S. Lavine, Frank P. Incropera, 2017 *Fundamentals of Heat and Mass Transfer*, 7th ed.(Hoboken/John Wiley & Sons. Inc) p 477.

[12] J. Breitenbach, I. V. Roisman, and C. Tropea, "Heat transfer in the film boiling regime: Single drop impact and spray cooling", *Int. J. Heat Mass Transf.*, **110**, pp. 34–42, 2017.

[13] C. A. Hernández-Bocanegra, J. I. Minchaca-Mojica, A. H. Castillejos E., F. A. Acosta-González, X. Zhou, and B. G. Thomas, "Measurement of heat flux in dense air-mist cooling: Part II - The influence of mist characteristics on steady-state heat transfer", *Exp. Therm. Fluid Sci.*, **44**, pp. 161–173, 2013.

Identification of the anisotropic plasticity using the virtual fields method for 2024 aluminium alloy

J W Fu*, W W Xie, Z Ma and L H Qi*

School of Mechanical Engineering, Northwestern Polytechnical University, Xi'an, 710072, PR China

* Corresponding author. *E-mail address:* jiawei.fu@nwpu.edu.cn, qilehua@nwpu.edu.cn

Abstract. In this work, the virtual fields method was applied to identify the anisotropic plastic constitutive parameters for the wrought 2024 aluminium alloy. First, low cyclic tension-compression tests were conducted for the specifically designed aluminium alloy specimens on a hydraulic universal testing machine and the full-field deformations measured by digital image correlation. Then, the measured experimental data were used to simultaneously identify the constitutive parameters of the Hill 1948 yield criterion and the nonlinear kinematic hardening model based on the virtual fields method. Reasonable identified parameters were obtained. The fitting curves of the virtual work during the minimization process verify the reliability of the identification results.

1. Introduction

2024 aluminium alloys have been widely used as important structural materials in industries due to their excellent mechanical properties and lightweight. It is frequently used in the sheet form such as automotive shell parts, aircraft skin etc. Aluminium alloy sheets usually present high plasticity anisotropy such as the different yield strengths along different material orientations as well as the early yielding (Bauschinger effect) during a load reversal [1-2]. To improve the forming precision, an effective way is to perform finite element simulations of the forming process, which, however require proper material constitutive models and the corresponding constitutive parameters. Using conventional methods to characterize material anisotropic plasticity usually involve multiple experiments and sophisticated machines, which are time consuming but also expensive.

As an effective inverse identification strategy, the Virtual Fields Method (VFM) is increasingly applied to characterize the mechanical properties for various materials and constitutive models from experimentally measured strain fields [3], such as the calibration of the anisotropic yield criteria from notched specimen uniaxial tension tests and the anisotropic hardening models from cyclic forward-reverse simple shear tests [4-6]. High computational efficiency and less sensitivity to loading boundary conditions make it more and more popular in the material science and engineering community along with other inverse identification strategies such as the Finite Element Model Updating and the Constitutive Equation Gap Method, *etc.* [7].

In this paper, the VFM is utilized to characterize the anisotropic plasticity of the 2024 aluminium alloy. A specific tension-compression test configuration is adopted to simultaneously identify the constitutive parameters of the Hill 1948 yield criterion and the nonlinear kinematic hardening model for the rolled 2024 aluminium alloy sheet specimen, which effectively simplifies the material constitutive characterization process.

2. Methodology

2.1. *The virtual fields method*

The basic idea of the VFM is to establish an equilibrium equation between the internal virtual work (IVW) and the external virtual work (EVW). In quasi-static state, the principle of the VFM can be expressed as [3]:

$$-\int_V \sigma : \varepsilon^* dV + \int_{\partial V} \boldsymbol{T}.\boldsymbol{u}^* dS = 0 \tag{1}$$

For the nonlinear case of plasticity, a cost function between the IVW and EVW can be defined as:

$$C(\boldsymbol{X}) = \sum_{i=1}^{N} \left(-t \int_A \sigma : \varepsilon^* dS + \int_{\partial V} \boldsymbol{T}.\boldsymbol{u}^* dS \right)^2 \tag{2}$$

where N represents the number of loading steps, t the specimen thickness, A the area of interest, $\boldsymbol{X}$ the unknown parameters, $\boldsymbol{T}$ the distributed load on the loading boundary ∂V, $\boldsymbol{u}^*$ the defined virtual displacement vector, $\boldsymbol{\varepsilon}^*$ the corresponding virtual strain tensor, $\boldsymbol{\sigma}$ the stress tensor calculated from real strain tensor $\boldsymbol{\varepsilon}$ through the constitutive relation. The unknown parameters $\boldsymbol{X}$ can be obtained by minimizing $C(\boldsymbol{X})$ using appropriate optimization algorithms.

2.2. *Constitutive models*

Hill1948 yield criterion was chosen to describe the anisotropic yield behaviours of the 2024 aluminium alloy specimens, which writes:

$$f\left(\sigma_{ij}, \varepsilon_p\right) = \sqrt{(G+H)\sigma_{xx}^2 + (F+H)\sigma_{yy}^2 - 2H\sigma_{xx}\sigma_{yy} + 2N\sigma_{xy}^2} - \bar{\sigma}_s\left(\varepsilon_p\right) = 0 \tag{3}$$

where the Swift hardening law is adopted for the current flow stress, namely

$$\bar{\sigma}_s\left(\varepsilon_p\right) = K_0\left(\varepsilon_0 + \varepsilon_p\right)^n \tag{4}$$

The nonlinear kinematic hardening law was adopted to capture the anisotropic hardening behaviour of the material. Thus,

$$f\left(\boldsymbol{\sigma}, \boldsymbol{\alpha}, \varepsilon_p\right) = \sqrt{\frac{3}{2}\left(\boldsymbol{\sigma}^{dev} - \boldsymbol{\alpha}^{dev}\right) : \left(\boldsymbol{\sigma}^{dev} - \boldsymbol{\alpha}^{dev}\right)} - \bar{\sigma}_V\left(\varepsilon_p\right) = 0 \tag{5}$$

and the Voce hardening law is used,

$$\bar{\sigma}_v\left(\varepsilon_p\right) = Y_0 + R\left(1 - e^{-m\varepsilon_p}\right) \tag{6}$$

In equation (5), the evolution of the backstress tensor $\boldsymbol{\alpha}$ can be expressed as:

$$d\boldsymbol{\alpha} = \frac{C}{\sigma_v}\left(\boldsymbol{\sigma} - \boldsymbol{\alpha}\right) d\varepsilon_p - \gamma \boldsymbol{\alpha} d\varepsilon_p \tag{7}$$

More details regarding the above constitutive models can be found in [8, 9].

3. Experimental results

In the present study, a low cycle tension-compression test configuration was used to characterize the anisotropic plasticity of the 2024 aluminium alloy, based on which the constitutive parameters of the Hill1948 and the nonlinear kinematic hardening models were identified simultaneously from the same tests. To realize this, a specimen configuration was designed that can generate a balanced heterogeneous tension/shear stress state under uniaxial loading and is not liable to buckle when compressed, as shown in Figure 1. More details regarding the design of the test configuration have been provided elsewhere in a submitted work.

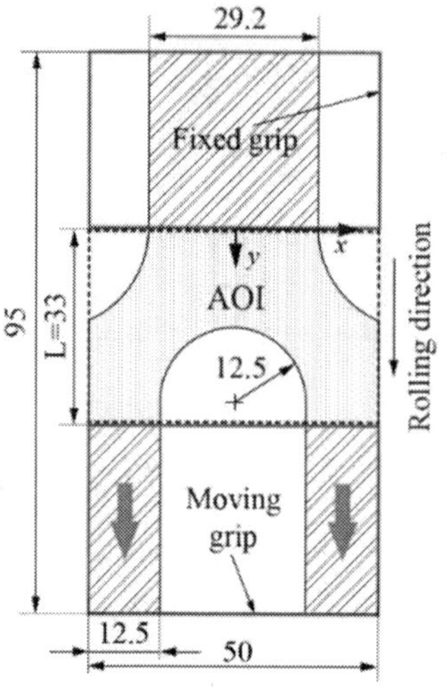

Figure 1. Schematic of the current tension-compression test configuration

Several groups (for checking the repeatability of the test results) of rolled 2024 aluminium alloy specimens were machined by wire cutting. Each group includes one specimen in the rolling direction and one in the transverse. Cyclic tension-compression tests were carried out for each group on the MTS universal test machine. Full-field deformations were measured using digital image correlation (VIC-3D, Correlated Solution). The load-displacement curves of one tension-compression test is shown in Figure 2. The measured full-field strain and the load data were then input into the VFM program to extract the anisotropic yield and hardening constitutive parameters.

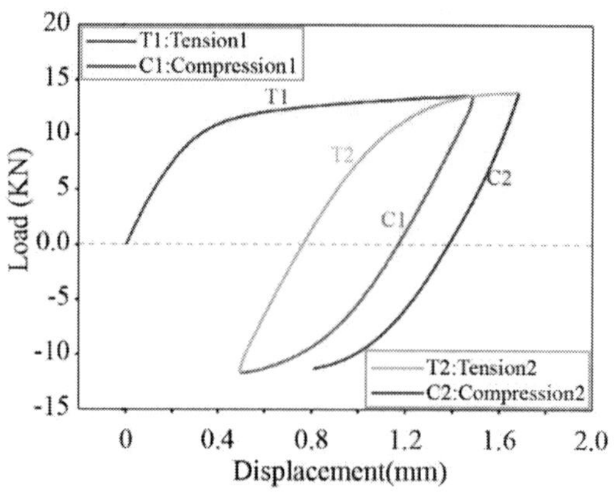

Figure 2. Load-displacement curves of the tension-compression in the rolling direction

Table 1 lists the identified parameters of the Hill 1948 yield criterion and the Swift hardening law, which were obtained by combining the experimental data of the first tensile stage of the two specimens in the rolling and the transverse directions. The virtual fields used here were manually defined, which writes:

$$u_x^* = (y*(y-l)*x)/200,\ u_y^* = (l*y^2/2 - y^3/3)/200 \tag{8}$$

where l represents the length of the AOI along the loading direction.

To check the robustness of the results, three different initial estimates were randomly chosen. In Table 1, one can see that the identified results are stable. The fitting curves of the IVW and the EVW when the minimization process reached its convergence is illustrated in Figure 3. One can see that the IVW and the EVW fit well with each other, indicating the reliability of the identified results.

Table 1. Identified parameters of the Hill 1948 yield criterion

direction		K	ε_0	n	H	F	N
0°+90°	I.E.1	600	0.002	0.25	0.5	0.5	2.5
	I.R.1	639	0.006	0.15	0.15	0.87	0.62
	I.E.2	800	0.001	0.3	0.8	0.8	5
	I.R.2	638	0.006	0.15	0.14	0.88	0.61
	I.E.3	1000	0.001	0.4	0.3	0.3	7.5
	I.R.3	653	0.008	0.16	0.12	0.90	0.60

(Notes: I.E. — Initial Estimates, I.R. — Identified Results)

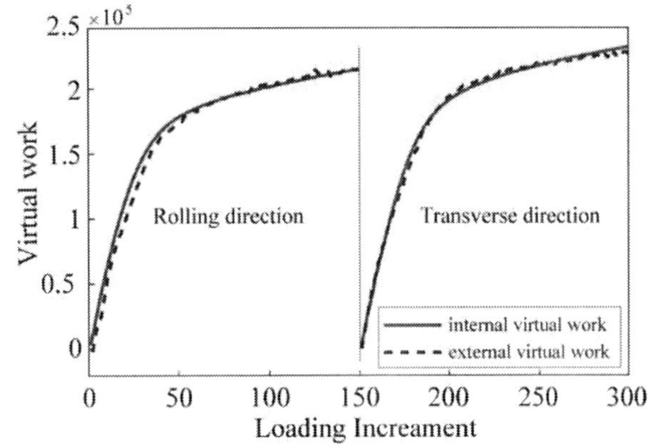

Figure 3. Fitting curves of the IVW and EVW for the Hill 1948 yield criterion when the minimization converges

Table 2 lists the identification results of the nonlinear kinematic hardening model. These were obtained using the whole experimental data of the cyclic tension-compression in the rolling direction. As can be seen, very stable results were obtained from different initial estimates. The good match between the IVW and EVW curves in Figure 4 also verifies the reliability of the results. It should be mentioned that the current identification algorithm is highly efficient. Using the trust-region-reflective algorithm and a regular desktop, the minimization can reach its convergence within approximately 3 minutes.

Table 2. Identified results of the nonlinear kinematic hardening model

direction		C	γ	Y_0	R	m
0°	I.E.1	10000	150	180	50	30
	I.R.1	8117	87.8	304	10^{-5}	10^{-3}
	I.E.2	15000	200	120	20	20
	I.R.2	8111	87.7	304	10^{-5}	10^{-5}
	I.E.3	5000	100	100	100	100
	I.R.3	8117	87.8	342	10^{-5}	10^{-5}

(Notes: I.E. — Initial Estimates, I.R. — Identified Results)

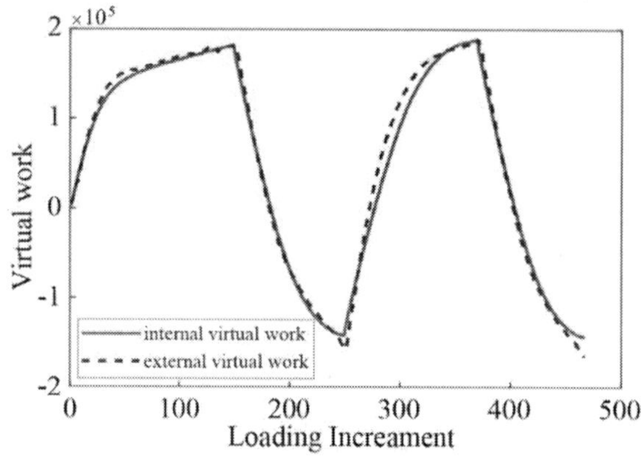

Figure 4. Fitting curves of the IVW and EVW for the nonlinear kinematic hardening law when the minimization converges

4. Conclusions

In this paper, a method that combines the low cycle tension-compression test and the VFM inverse identification strategy was applied to characterize the anisotropic plasticity of the 2024 aluminium alloy sheet specimens. A specific test configuration was adopted so that the constitutive parameters of the Hill 1948 yield criterion and the nonlinear kinematic hardening law can be identified simultaneously. It was found that using the current method, reliable and robust identification results were obtained. This can simplify the conventional complex identification procedures of the material anisotropic plasticity.

5. Acknowledgments

The financial supports from the Fundamental Research Funds for the Central Universities (No.31020190503006) and the Natural Science Foundation of China (No.51805439) are gratefully acknowledged.

References

[1] Bron F and Besson J 2004 A yield function for anisotropic materials Application to aluminum alloys *Int. J. Plasticity* **20** 937-963.

[2] Lee M G, Kim D, Kim C, Wenner M L, Wagoner R H and Chung K 2005 Spring-back evaluation of automotive sheets based on isotropic-kinematic hardening laws and non-quadratic anisotropic yield functions: part II: characterization of material properties *Int. J. Plasticity* **21** 883-914.

[3] Pierron F and Grediac M 2012 *Virtual Fields Methods: Extracting Constitutive Mechanical Parameters From Full-Field Deformation Measurements* (New York: Springer) p 1-517.

[4] Rossi M, Pierron F and Štamborská M 2016 Application of the virtual fields method to large strain anisotropic plasticity *Int. J. Solids. Struct.* **97-98** 322-335.

[5] Fu J, Barlat F, Kim J and Pierron F 2016 Identification of nonlinear kinematic hardening constitutive model parameters using the virtual fields method for advanced high strength steels *Int. J. Solids. Struct.* **102-103** 30-43.

[6] Fu J, Barlat F and Kim J, Pierron F 2017 Application of the virtual fields method to the identification of the homogeneous anisotropic hardening parameters for advanced high strength steels *Int. J. Plasticity* **93** 229-250.

[7] Martins J.M.P, Andrade-Campos A and Thuillier S 2018 Comparison of inverse identification strategies for constitutive mechanical models using full-field measurements *Int. J. Mech. Sci.* **145** 330-345.

[8] Hill R 1948 A theory of the yielding and plastic flow of anisotropic metals *Proceedings A* **193** 281-297.

[9] Chaboche J L 1986 Time-independent constitutive theories for cyclic plasticity *Int. J. Plasticity* **2** 149-188.

IOP Conf. Series: Materials Science and Engineering **967** (2020) 012061 doi:10.1088/1757-899X/967/1/012061

Evaluation of Sliding Wear Behaviour for STD11 Coated with TiCN Using the Pin-on-Disk Test

J H Bang[1,2], J H Song[2], G H Bae[2], N S Park[2], S O Choi[2], M G Lee[1] and H G Kim[3]

[1] Department of Materials Science and Engineering, Seoul National University, 1, Gwanak-ro, Gwanak-Gu, Seoul 08826, South Korea

[2] Korea Institute of Industrial Technology, 156, Gaetbeol-ro, Yeonsu-gu, Incheon, 21999, South Korea

[3] POSCO Global R&D Center, 100, Songdogwahak-ro, Yeonsu-gu, Incheon, 21985, South Korea

Abstract. The forming process using ultra high strength steel (UHSS) may cause early damage than the expected lifetime of the tool surface due to the high reaction force of the material compared to the conventional steel sheet. This leads to the increase in the cost of maintenance of the tool and the decrease in the product productivity and quality. In this study, the sliding wear behavior of the titanium carbonitride (TiCN) physical vapor deposition (PVD) coated pin against UHSS was quantitatively evaluated from the pin-on-disk test based on the Taguchi method. S/N ratio and ANOVA are used to evaluate the sensitivity of process parameters affecting on the pin wear depth. Finally, a wear prediction model for the TiCN coated pin against UHSS was constructed by the power law equation.

1. Introduction

The demand for the weight reduction has been continuously increasing in the automotive industry over recent years. As a result, the proportion of UHSS(Ultra High Strength Steel) in an auto-body tends to increase, which also could satisfy the strengthened regulations regarding carbon emission and passenger safety [1-2].

UHSS generates tremendous stamping pressure on the tool surface owing to its higher strength compared to conventional steel sheets. Therefore, tool damage such as galling, chipping, and cracking occurs due to the frictional resistance between the sheet metal and tool surface during the cold stamping process [3-5]. These defects lead to numerous problems in the surface quality, geometric accuracy, maintenance cost, and productivity. To overcome these problems, special surface coatings should be applied to the tool surface to improve wear resistance. However, this may increase the overall manufacturing cost of the product.

To improve productivity and tool maintenance, the process management based on the accurate prediction of tool wear is required. To achieve the accurate tool wear prediction, it is necessary to obtain the experimental wear database and analyze the wear behavior for tools specially coated for the wear resistance enhancement. The aim of this study is to analyze the wear behavior quantitatively that occurs between the UHSS sheet and the tool for the stamping die using the Pin-on-disk test. The Pin is manufactured by STD11 tool steel coated with TiCN (Titanium Carbonitride). The disk material is selected as DP980 1.4t which is the most widely-used UHSS sheet in auto-body components. The quantitative wear behavior of the pin was confirmed by measuring the wear depth after the pin-on-disk test. Wear characteristics and behaviors for TiCN coated STD11 were evaluated by analyzing the experimental result for the test set established by the Taguchi method and by performing ANOVA for the sensitivity analysis of process parameters. Finally, the wear prediction model for TiCN coated STD11 tool steel was constructed by power law type equation.

2. Experimental setup for the Pin-on-disk test

2.1. Materials

The material of the pin is STD11 tool steel with the elastic modulus of 210 GPa which is the general property of the steel. To improve the wear resistance and hardness, heat treatment is applied to STD11 tool steel. First, quenching was applied to introduce martensite for hardening the STD11 steel, which were heated up from 600°C, 850°C to 1030°C for 1hr 30min at each temperature. After that, tempering was performed twice for 2hr to reduce brittleness occurring from quenching process. The hardness of the heat-treated STD11 tool steel was measured to be 63.5 HRC, which is higher than the hardness value of 58 HRC specified in KS D 3753 [6].

Pin surfaces were coated with TiCN using the PVD process. The hardness of the TiCN coating is 3000 Hv [7]. DP980 1.4t was selected as the disk material. The mechanical properties of DP980 is presented in Table 1.

Table 1. Mechanical properties of DP980 1.4t.

			Swift hardening law $\sigma = k(\varepsilon_0 + \varepsilon)^n$			R-value		
Thickness (mm)	YS (MPa)	UTS (MPa)	k (MPa)	ε_0	ε_0	0°	45°	90°
1.4	721.60	1042.0	1499.11	0.00108	0.107	0.956	0.996	1.098

2.2. Test method and configuration

The pin-on-disk test is a representative experimental method to analyze, quantify and compare the wear characteristics for various environmental conditions and process parameters. In this test, the tool wear behaviour was investigated by using the pin-on-disk test. Figure 1 shows the pin-on-disk tester consists of pin, disk and weights. Pin contacts with the metallic disk under the normal force loaded by weights. Wear occurs at the contact surface between the pin and the disk due to the relative rotational movement of the disk. Test instrument is TRIBOMETER of CSM Instrument SA.

Figure 1. Experimental setup for Pin-on-disk test.

2.3. Test conditions

The most significant factors on the tool wear are the normal force and the sliding distance [8-12]. Therefore, these two factors were selected as the control parameter for the Pin-on-disk test. The test range of the normal force was set up to 10 N. The sliding distance was performed up to 1000 revolutions (94.2 m) in order to cause the sufficient wear occurrence of pin. Because the effect of the rotational velocity on the tool wear is relatively smaller than the normal force and the sliding distance, it was fixed to 100 mm/s for all experimental conditions. Table 2 shows the value of control parameters with three levels.

Table 2. Control parameters and their levels.

Control parameter	Level 1	Level 2	Level 3
Normal force (F)	2 N	6 N	10 N
Sliding distance (L)	200 rev. (18.84 m)	600 rev. (56.52 m)	1000 rev. (94.20 m)

2.4. Measurement of pin wear depth

In order to measure the wear depth of pin coated with TiCN, the wear area of pin has to be determined. It was measured by using the optical electron microscope (OEM), as shown in Figure 2. The pin wear depth was derived from the measured wear area using the following equations. The wear scar diameter d was calculated by assuming that the pin wear area measured from OEM is a circle.

$$A_p = \frac{\pi d^2}{4} \tag{1}$$

The pin wear depth h is a function of the wear scar diameter d

$$h = r - \left(r^2 - \left(\frac{d}{2} \right)^2 \right)^{\frac{1}{2}} \tag{2}$$

where r is pin end radius [13].

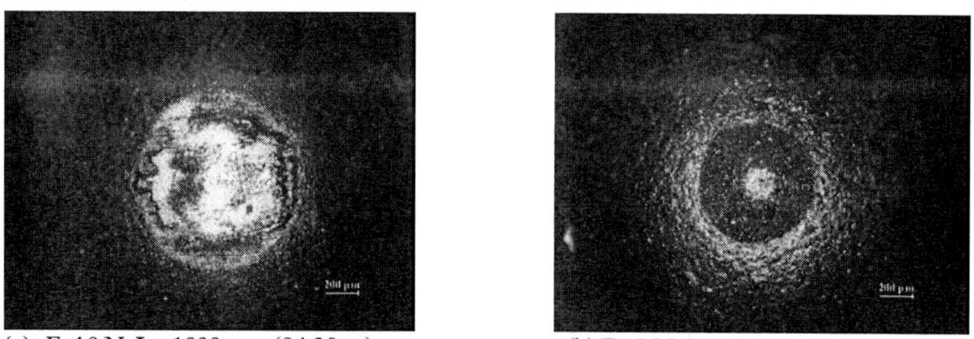

(a) F: 10 N, L : 1000 rev. (94.20 m) (b) F : 2 N, L : 600 rev. (56.52 m)

Figure 2. Pin wear area measured by OEM.

3. Experimental results

The Taguchi method developed by Taguchi and Konishi [14] is used to identify the characteristics of process parameters and the sensitivity effect. It is a powerful method that has been widely used in engineering problems because an efficient and systematic approach is possible to evaluate the control parameters. In this study, the experimental design set for the Pin-on-disk test was established based on the Taguchi method. The orthogonal array L_9 (3^2) was employed by using three factor levels selected in previous section. Table 3 shows the experimental designs based on the orthogonal array L_9 and the measured pin wear depth obtained from the pin-on-disk test.

Table 3. Orthogonal array L_9 (3^2) and results of experimental designs.

Run	Normal force (F)	Sliding distance (L)	Pin wear depth (h)
1	10 N	1000 rev. (94.20 m)	45.678 μm
2	10 N	600 rev. (56.52 m)	37.864 μm

3	10 N	200 rev. (18.84 m)	36.000 μm
4	6 N	1000 rev. (94.20 m)	40.692 μm
5	6 N	600 rev. (56.52 m)	35.345 μm
6	6 N	200 rev. (18.84 m)	31.526 μm
7	2 N	1000 rev. (94.20 m)	31.440 μm
8	2 N	600 rev. (56.52 m)	26.646 μm
9	2 N	200 rev. (18.84 m)	23.763 μm

The signal-to-noise ratio (S/N ratio) is defined as the ratio of signal (meaningful input) to noise (meaningless input), which checks how much the response of the pin wear depth varies relative to change of each control parameter. The purpose of this study is to identify the response behavior of the pin wear depth as the each control parameter value increase. Therefore, larger-the-better characteristics were chosen as shown by

$$\eta = -10\log\left(\frac{1}{n}\sum_{i=1}^{n}\frac{1}{y_i^2}\right) \tag{3}$$

where n is the number of experiments and y_i is the experimental results in ith experiment.
The results of S/N ratio is shown in Figure 3. As the normal force and the sliding distance increase, the S/N ratio increases. The mean of SN ratios for two control parameters is large enough to change the pin wear depth sensitively. Therefore, it was confirmed that the two control parameters are significant process parameters for the tool wear in the stamping process of UHSS sheets.

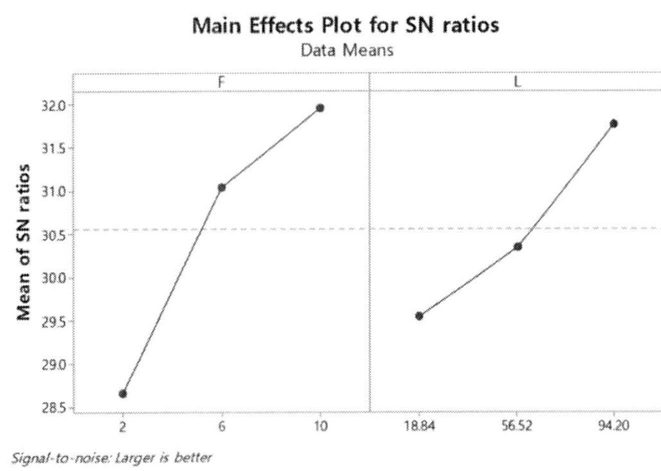

Figure 3. Results of S/N ratio on pin wear depth.

ANOVA is a method for evaluating the statistical significance of control parameters. The significance of the normal force and the sliding distance on the pin wear depth was confirmed by F-value and P-value. F-value evaluates how much the control parameter affects the response. P-value is to check whether the control parameter is significant in response. If the P-value is less than 0.05, it means that the control parameter is significant in the response at the 95% confidence level. The ANOVA results for the wear depth of pin are given in Table 4. F-value of the normal force and the sliding distance is 160.49 and 79.26, respectively. F-value of the normal force is larger than that of the sliding distance, which means that the normal force has a more significant effect on the pin wear depth than the sliding distance in the test range.

Table 4. Results of ANOVA for the wear depth of pin.

Variance source	Degree of freedom (DOF)	Adjusted sum of squares (Adj. SS)	Adjusted mean squares (Adj. MS)	F-value	P-value
Normal force (F)	2	247.296	123.648	160.49	0.000
Sliding distance (L)	2	122.125	61.062	79.26	0.001
Error	4	3.082	0.770	-	-
Total	8	372.502	-	-	-

4. Wear prediction model

The wear prediction model of TiCN coated STD11 tool steel was constructed by using the pin-on-disk test results for the orthogonal array established by the Taguchi method. The wear depth is a function of the normal force and the sliding distance, and it is formulated by the power law type equation as follows:

$$h = 12.94F^{0.2342}L^{0.1523}\,\mu m \tag{4}$$

where h is pin wear depth, and F and L are the normal force and sliding distance, respectively. Figure 4 shows the regression result of the formulated wear prediction model. The R^2 and $Adj.R^2$ values of the wear prediction model are found to be 94.58% and 92.77%, respectively. These values supports that the power law type regression equation can predict the reliable wear amount of the TiCN coated STD11 tool steel for the stamping process of UHSS sheets. The wear prediction model can be used when designing the tools of the drawing process of actual auto-body parts.

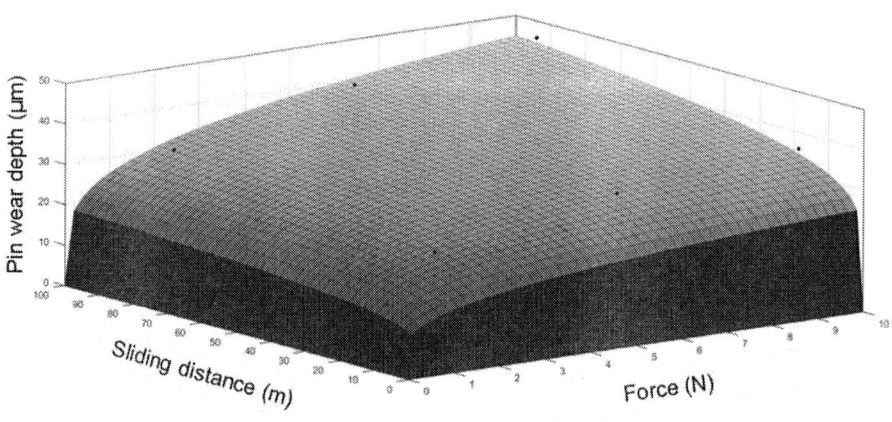

Figure 4. Regression curve fitted by experimental results.

5. Conclusion

This paper identifies the correlation between pin wear depth and control parameters (normal force and sliding distance) by the sensitivity analysis and proposes the wear prediction model for TiCN coated STD11 tool steel. The pin-on-disk test was performed based on the Taguchi method in order to evaluate and quantify the wear behavior according to the normal force and the sliding distance. The statistical analysis by the S/N ratio and ANOVA was found that the normal force and the sliding distance have a significant influence on the wear behavior of the TiCN coated STD11 tool steel. The power law type regression model was constructed to predict the pin wear by using experimental results of pin-on-disk test. It can be effectively utilized to predict the wear amount of TiCN coated STD11 tools in the stamping process of UHSS sheets.

6. References

[1] Jambor A and Beyer M 1997 *Materials & Design* **18** 203
[2] Li Y, Lin Z, Jiang A and Chen G 2003 *Materials and Design* **24** 177

IOP Conf. Series: Materials Science and Engineering **967** (2020) 012061 doi:10.1088/1757-899X/967/1/012061

[3] Kim S G, Choi H S, Seo P K, Kim B M and Ko D C 2012 *KSTP Autumn Conf.* 436
[4] Bayer R G 2004 *Mechanical Wear Fundamentals and Testing* (New York: CRC Press)
[5] Williams J A 1999 *Wear* **225–229** 1
[6] KS D 3753:2008, *Alloys tool steels*
[7] www.matweb.com/search/DataSheet.aspx?MatGUID=5ffb257a8cc74b69a140046a34b5d61c
[8] Pereira M P, Yan W and Rolfe B F 2008 *Wear* **265** 1687
[9] Pereira M P, Yan W and Rolfe B F 2010 *Wear* **268** 1275
[10] Rhee S K 1970 *Wear* **16** 431
[11] Bayer R G 1993 *Wear* **162–164** 913
[12] Archard J F 1953 *Journal of Applied Physics* **24** 981
[13] ASTM G99, *Standard Test Method for Wear Testing with a Pin-on-Disk Apparatus*
[14] Taguchi G and Konishi S 1987 *Taguchi Methods, orthogonal arrays and linear graphs, tools for quality* (American Supplier Institute) pp 8–35

IOP Conf. Series: Materials Science and Engineering **967** (2020) 012062 doi:10.1088/1757-899X/967/1/012062

Work-hardening behavior prediction model of arbitrary reloading process based on material crystallographic structure

T Oya[1], J Yanagimoto[2], K Ito[3], G Uemura[3] and N Mori[3]

[1] Keio University, 3-14-1 Hiyoshi, Kohoku-ku, Yokohama, Japan
[2] The University of Tokyo, 7-3-1 Hongo, Bunkyo-ku, Tokyo, Japan
[3] M&M Research Inc., 4-15-3 Jingumae, Shibuya-ku, Tokyo, Japan

E-mail: oya@sd.keio.ac.jp

Abstract. In the case of complex forming process that receives a secondary loading after the primary loading, it is necessary to appropriately represent the work-hardening state according to the forming history. Namely, it is necessary to accurately model complex forming process including reversal loading (Bauschinger effect) and orthogonal loading (cross-hardening). In this study, a composite anisotropic hardening expression based on the crystallographic structure using the finite element polycrystal model is proposed. Numerical investigations are conducted to consider the effectiveness of the proposed model. In which, an alternation rate is introduced to capture the activity of the slip systems to support the idea behind the proposed model. In addition, flow curves of the secondary loading are produced by the proposed model, which is compared with the conventional model.

1. Introduction

The importance of material models for improving the accuracy of forming simulation is increasing. However, since multi-path forming has increased to realize complex shape, achieving accurate simulation has become difficult. Therefore, it is necessary to construct a model that can handle not only monotonic loading but also complex loading histories including reverse and cross-loadings. In such a case, a kinematic hardening model is often used[1][2][3], but there are problems such as insufficient consideration of the microstructure of the material. In these models, many parameters are used to achieve better fit to the experimental data; however, some of them do not have explicit physical meaning. Therefore, if these parameters are determined by multivariable optimization technique, it is difficult to confirm that the obtained solution is the globally optimized solution. The other problem is that a back stress is frequently used to express the reduction of the yield stress at the secondary loading phase know as Bauschinger effect; however, this type of model cannot be applied to arbitrary reloading process because another hardening mechanism could be activated when the reloading path is deviated from pure reverse.

The authors have studied for the purpose of establishing a composite anisotropic hardening expression based on the crystallographic structure of polycrystalline materials. In this paper, a verification of the proposed work-hardening model of the arbitrary load path is conducted. This paper also describes the results of numerical analysis using the finite element polycrystal model.

2. Representation of a general reloading process

In this study, two-stage loading paths are considered; namely, there are secondary loading paths subsequently continued after the primal monotonic loading. The following quantity is used as an indicator of different strain paths[4]:

$$\beta = \frac{\boldsymbol{D}_1 : \boldsymbol{D}_2}{||\boldsymbol{D}_1||||\boldsymbol{D}_2||}, \tag{1}$$

where $\boldsymbol{D}_1$ and $\boldsymbol{D}_2$ are the strain rate tensors under the primary loading and the secondary loading, respectively. In this way, switching of an arbitrary load path can be expressed. Any reloading processes ($-1 < \beta < 0$) is called a general reloading process. It is important to define a representation of the work-hardening curve in this general reloading process in the sheet forming process. The general reloading process is characterized by the strain change parameter β. It can be considered that Bauschinger effect appears most remarkably at the reversal loading ($\beta = -1$), and the cross-hardening is caused most apparently at the orthogonal loading ($\beta = 0$). When β changes from -1 to 0, the work-hardening curve would gradually change. Therefore, the general work-hardening curve can be represented by a combined curve of the Bausinger curve and the cross-hardening curve with the weighting factor according to the value of β.

The reload path for the preload path is represented by the angle α between the preload deviatoric stress tensor $\boldsymbol{\sigma}'_P$ and the reload deviatoric stress tensor $\boldsymbol{\sigma}'_R$, which is expressed as

$$\cos\alpha = \frac{\boldsymbol{\sigma}'_P : \boldsymbol{\sigma}'_R}{||\boldsymbol{\sigma}'_P||||\boldsymbol{\sigma}'_R||}, \tag{2}$$

where $\alpha = 0, \pi/2, \pi$ represent the continued preloading, orthogonal reloading and reverse reloading, respectively. Thus, the general work-hardening curve $H_R(\int d\varepsilon^P)$ is expressed as

$$H_R(\int d\varepsilon^P;\alpha) = H_C(\int d\varepsilon^P)\sin^2\alpha + H_B(\int d\varepsilon^P)\cos^2\alpha, \quad \left(\frac{\pi}{2} < \alpha < \pi\right), \tag{3}$$

where H_C represents a cross-hardening curve and H_B represents a Bauschinger curve, respectively. If an expression such as Eq. (3) can be established, once a Bauschinger and cross-hardening curves are obtained, the work-hardening curve for the general reloading paths can be predicted as an arbitrary curve between them by designating α.

The Bauschinger effect is considered to be caused by the reversal of the slip direction while the active slip system remains unchanged, on the other hand, the cross-hardening is considered to be caused by the alternation of the active slip system. If the correlation between the change of the alternation ratio of slip systems and $\sin^2\alpha$ is shown, the proposed model is physically valid. Here, an alternation ratio, which is explained in section 4, means a ratio of slip systems whose activeness is switched at the load-path change.

3. Description of models that express reverse and orthogonal loadings

The crystal plasticity code used in this study is based on a successive integration scheme with implicit finite element method, which is called FEPM (finite element polycrystal model [5]). To express Bauschinger effect and cross-hardening effect, two extra models were introduced by Takahashi. In this study, these models were implemented to compute the proposed model. In this section, these models are briefly explained.

3.1. Backlash model

When a reverse load such as compression is applied after tension, the Bauschinger effect occurs in which the flow stress of the reverse load decreases. This effect can be expressed by FEPM using the backlash model [6]. The Bauschinger curve has a stagnation in flow stress, which seems to correspond to the process of collapse and reconstruction of the cell structure of dislocation. In the larger strain region, it can be considered that the primal loading curve is parallel to the Bauschinger curve with a same gradient. The amount of difference of these two curves can be considered as a preparation period for resuming work-hardening in the reverse load direction due to collapse and reconstruction of the cell structure after stress reversal. Based on the analogy to the backlash of mechanical gears, this amount is called the backlash quantity. The Bauschinger effect can be expressed by introducing such a concept of backlash into a crystallographic work-hardening mechanism.

First, a strain γ_a which is effective to work-hardening in each slip system is considered. As shown in Fig.1, γ_a increases with shear strain γ at the beginning of loading, but when the stress is reversed at point A, a non-hardening region, i.e. backlash region, appears. The backlash region is represented by γ_b. Let $\gamma_a|_{\max}$ be the largest effective strain γ_a that a slip system has ever experienced till then, and γ_b is defined as a function B of this value, which is given by the following equation.

$$\gamma_b = B(\gamma_a|_{\max}). \tag{4}$$

If γ_a is constant in the backlash region, the stress drop due to the Bauschinger effect cannot be expressed. Then, as shown in Fig.1, γ_a decreases slightly. An arbitrary position in the backlash region is presented by a parameter ξ.

$$\xi = 2\frac{\gamma - \gamma_c}{\gamma_b}, \tag{5}$$

where ξ takes a value between -1 and $+1$ in the backlash region, γ_c is the midpoint of the backlash region. Then, the work-hardening rule is formulated as follows.

(1) $|\xi| = 1$ (work-hardened state)

$$\dot{\gamma}_a = |\dot{\gamma}|. \tag{6}$$

(2) $|\xi| < 1$ (backlash region)

$$\dot{\gamma}_a = |\dot{\gamma}|\frac{\xi \mathrm{sign}(\tau) + \lambda}{1 + \lambda}, \tag{7}$$

where $\lambda = 1$ $(\gamma_a|_{\max} < 0.04)$, $\lambda = 0$ $(\gamma_a|_{\max} > 0.04)$. This equation is defined to express softening of the flow curve during backlash region.

By incorporating this model, it is possible to express the Bauschinger effect using a finite element polycrystal model.

3.2. Maximum forest dislocation model

In the case of an orthogonal loading, it is observed that the initial flow stress temporarily rises at the secondary yielding. This phenomenon is called cross-hardening, which is considered to be caused by an abrupt change of active slip systems. In order to express this cross-hardening, a maximum forest dislocation model [7] was introduced.

First, the slip planes are designated by $p = 1 \sim 4$, the slip directions are represented by $q = 1 \sim 3$, and the slip systems are represented by $r = 3(p-1)+q = 1 \sim 12$. The strain of each slip system is represented by $\gamma^{(p,q)}$ or $\gamma^{(r)}$. The dislocation accumulated around the obstacle on the slip plane (p) is expressed by the following equation.

$$\overline{\gamma_a^{(p)}} = \sum_q \gamma_a^{(p,q)}. \tag{8}$$

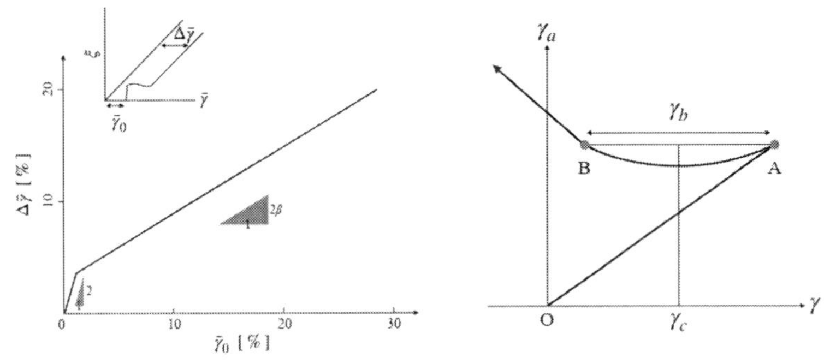

Figure 1. Work-hardening diagram in slip sysems. (a) Backlash strain, (b) hardening process.

For example, consider the slip surface $p = 1$. $\overline{\gamma_a^{(2)}}$, $\overline{\gamma_a^{(3)}}$, $\overline{\gamma_a^{(4)}}$ can be interpreted as a forest dislocation for slip plane $p = 1$. Thus, it is assumed that the flow stress $k^{(1)}$ depends on the sum of them. As an improved model, it is assumed that the largest value of these forest dislocations controls the flow stress. Therefore, the critical yielding shear stress $k^{(p)}$ is represented by

$$k^{(p)} = H(\Gamma^{(p)}) \quad \text{where} \quad \Gamma^{(p)} = \max\overline{\gamma_a^{(p')}} \ \ (p' \neq p). \tag{9}$$

4. Introducing an alternation rate

The alternation rate 1: $\rho_1(\alpha)$ and the alternation rate 2: $\rho_2(\alpha)$, which represent the switching ratio of the active slip systems during primary and secondary loadings, are defined as follows:

$$\rho_1(\alpha) = \frac{N_{10}}{N_{10} + N_{11}}, \quad \rho_2(\alpha) = \frac{N_{01}}{N_{01} + N_{11}}. \tag{10}$$

Here, N represents the number of slip systems (or slip planes), subscripts 0 and 1 mean active and inactive, respectively. For example, N_{10} means the number of slip systems (planes) that are active slip systems (planes) under the primary load and subsequently become inactive under the secondary load. In other words, the alternation rate 1 indicates the ratio of the slip system (plane) that was active under the primary load and became inactive under the secondary load. In this study, two definitions are used to investigate which expression captures the phenomenon better.

5. Numerical study of the proposed model

5.1. Verification using the alternation rate

After applying a pre-strain of 0.1, reverse loading ($\alpha = \pi$), 45-degree shear loading ($\alpha = 5\pi/6$), and orthogonal loading ($\alpha = \pi/2$) were applied. As an example, Fig. 2 shows the change of the alternation rate under the reverse loading case. Alternation rates exhibit rapid reduction and converge to the value around 0.2. This means that switching of active slip system occurred at a relatively lower degree. This calculation was also conducted for the other cases, as shown in Fig.3, and the converged alternation rates are collected, resulted in the relationship as follows.

$$\rho(\pi) < \rho\left(\frac{5\pi}{6}\right) < \rho\left(\frac{\pi}{2}\right). \tag{11}$$

From this magnitude relation, as described above, the change of the active slip system shows the transition according to the change of the load path. Namely, this result state that the alternation rate increases from the minimum value that is observed in reverse loading as the angle α increases toward the maximum value that is observed in orthogonal loading. Therefore, at least from the viewpoint of the switching of the active slip system, the physical validity of the proposed model is shown.

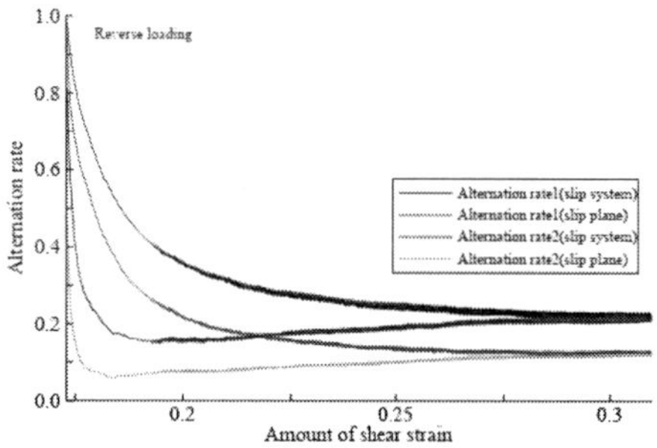

Figure 2. Alternation rate of reverse loading.

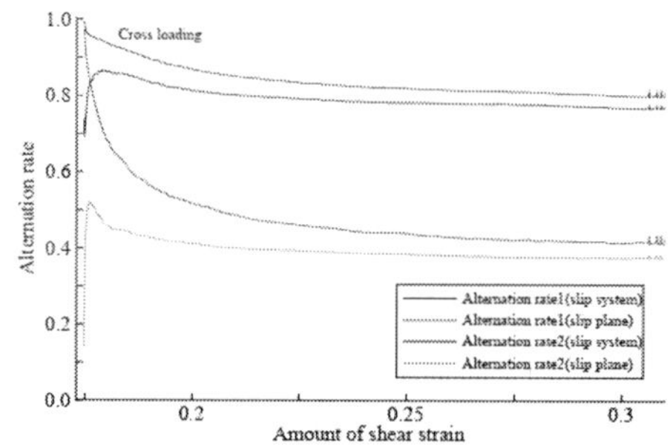

Figure 3. Alternation rate of orthogonal loading.

5.2. Generation of secondary loading stress-strain curves

If a model that can express the general reloading process can be established, it is possible to predict any work-hardening curve between orthogonal and reverse reloading. The necessary data are flow curves of reverse and orthogonal loading. To verify the proposed model, a material data of DC06 is used and the results are compared with those of the Teodosiu-Hu (TH) model that is recognized as a reference curve in this analyses [3]. Note that the implemented FEPM model is based on the material structure of FCC (face-centered crystal) but DC06 is a BCC (body-centered crystal) material. Therefore, the following simulations only serves as qualitative investigation; however, the proposed concept is assumed to be effective for any crystalline metals.

Figures 4, 5, 6, and 7 show the verification results of the model for each load path. Prestrain of 0.1 and 0.2 are applied and represented by solid and dashed curves, respectively. As shown in Figs.4 and 5, typical drop and overshoot of flow stress are observed in both TH model and the proposed model. In each case, the qualitative trend represents the material response well, and the results are quantitatively similar. Figs.6 and 7 show examples of work-hardening curve prediction based the proposed model, which represent similar ones to the reference curve. Although further investigation and parameter adjustment are necessary, it is considered that the effectiveness of the proposed model could be shown.

6. Conclusion

A novel work-hardening model that can express arbitrary loading paths has been proposed. This model requires only Bauschinger and cross-hardening curves for the initial input, and an arbitrary reloading process can be predicted by designating an angle between the primary and secondary loading paths. This concept was confirmed by using the alternation rate that can represent how much slip systems change its activity. In this paper, numerical analyses of the proposed model for predicting the work-hardening behavior in the arbitrary secondary reloading process was performed using a finite element polycrystal model, and valid results were obtained.

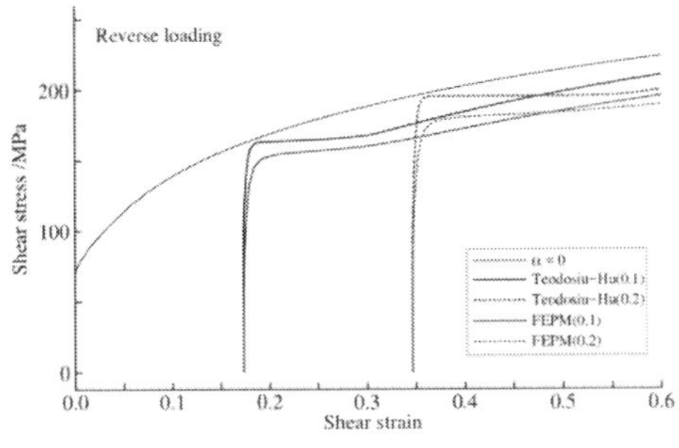

Figure 4. Comparison of shear stress-strain curve of FEPM and Teodosiu-Hu for reverse loading.

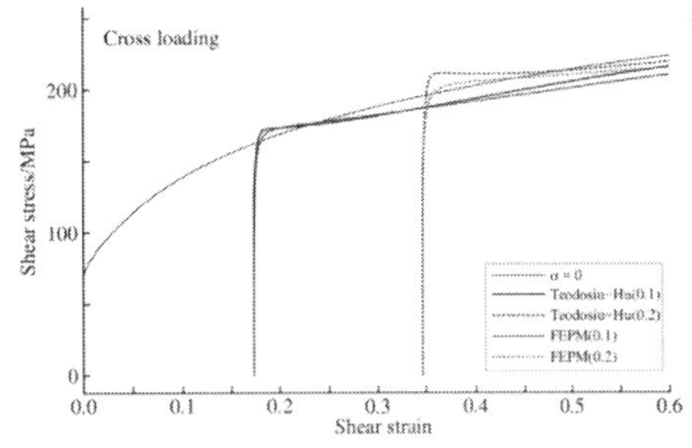

Figure 5. Comparison of shear stress-strain curve of FEPM and Teodosiu-Hu for cross loading.

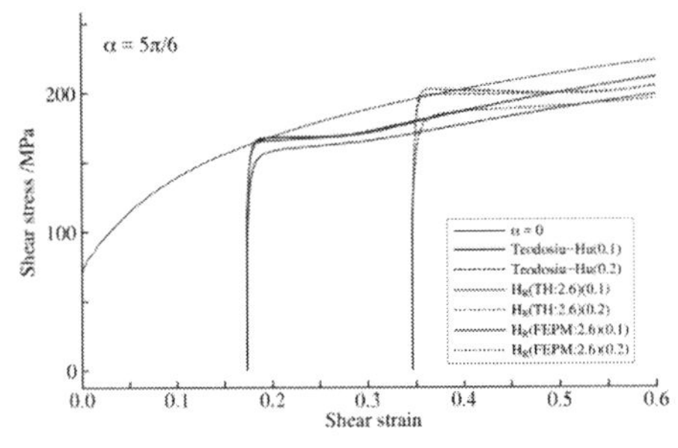

Figure 6. Comparison of shear stress-strain curve of FEPM, Teodosiu-Hu and proposed model for $\alpha = 5\pi/6$ loading.

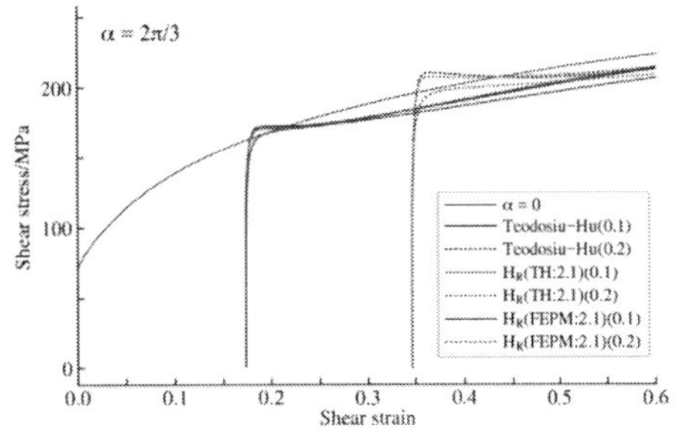

Figure 7. Comparison of shear stress-strain curve of FEPM, Teodosiu-Hu and proposed model for $\alpha = 2\pi/3$ loading.

References

[1] Yoshida, F., Uemori, T. 2002 A model of large-strain cyclic plasticity describing the Bauschinger effect and workhardening stagnation *International Journal of Plasticity*, 18, 661–686.

[2] Chaboche, J. L. 1991 On some modification of kinematic hardening to improve the description of ratcheting effects *International Journal of Plasticity*, 7, 661–678.

[3] Haddadi, H., Bouvier, S., Banu, M., Maier, C., and Teodosiu, C. 2006 Towards an accurate description of the anisotropic behavior of sheet metals under large plastic deformations: Modelling, numerical analysis and identification *International Journal of Plasticity*, 22, 2226–2271.

[4] Schmitt, J. H., Aernoudt, E., and Baudelet, B., 1985 Yield loci for polycrystalline metals without texture *Materials Science Engineering*, 75, 13–20.

[5] Takahashi, H., Motohashi, H., Tokuda, M., and Abe, T. 1994 Elastic-plastic finite element polycrystal model *International Journal of Plasticity*, 10, 63–80.

[6] Takahashi, H. and Shiono, I. 1991 Backlash model for large deformation behavior of aluminum under torsional cyclic loading *International Journal of Plasticity*, 7, 199–217.

[7] Takahashi, H., Fujiwara, K., and Nakagawa, T. 1998 Multiple-slip work-hardening model in crystals with application to torsion-tension behaviors of aluminum tubes *International Journal of Plasticity*, 14(6), 489–509.

IOP Conf. Series: Materials Science and Engineering **967** (2020) 012063 doi:10.1088/1757-899X/967/1/012063

A novel testing method to evaluate edge formability non-destructively with eddy current

J Gu[1], N Pathak[1], S Freed[1], E Todorov[1] and H Kim[1]
[1] EWI - Forming Center (EWI-FC), Columbus, Ohio, United States

Abstract. Eddy current (EC) testing is one of the most extensively used NDE techniques in the automotive industry for automatic in-line inspection of ferrous materials such as advanced high strength steels (AHSS). In addition, shearing is a very common forming operation in the automotive industry. With the increase of shearing clearance, the sheared-edge experiences significant work-hardening that normally decreases the formability of the sheared edge. This paper introduces a novel and real-time monitoring NDE method based on the EC sensor to characterize variations in shear edge quality for two different grade advanced high strength steels, DP780 and 980GEN3. The developed NDE method was applied to scan the edges sheared at various clearances between 5 and 25% of the material thickness. The NDE signal received was correlated with pre-straining introduced during the shearing process at various clearances. Microhardness measurements were taken to compare the trends obtained from the NDE tool with the hardness values. To evaluate the edge formability, half-specimen dome testing (HSDT) was conducted for the edges sheared at various clearances. A digital image correlation (DIC) system was used to measure failure strain during the HSDT. The failure strain of sheared edges was correlated with the NDE measurements for each clearance to assess the application of an NDE measurement in determining edge quality. The NDE method can potentially be used to inspect the sheared edge quality for sheet metal stamping process.

1. Introduction

The eddy current (EC) technology has been widely used for detecting surface and internal flaws of metal structures nondestructively. Eddy current is the induced current with a closed-loop contour pattern generated by the alternating magnetic field, according to Faraday's law [1]. The EC signal contour is affected by the electric property of the material, stand-off distance and sample geometries. Discontinuity in the microstructure such as micro-cracks can significantly distort the contour of the EC signal, shown in figure 1 [2]. Therefore, the work hardening induced by the shearing can be evaluated by EC based nondestructive evaluation (NDE).

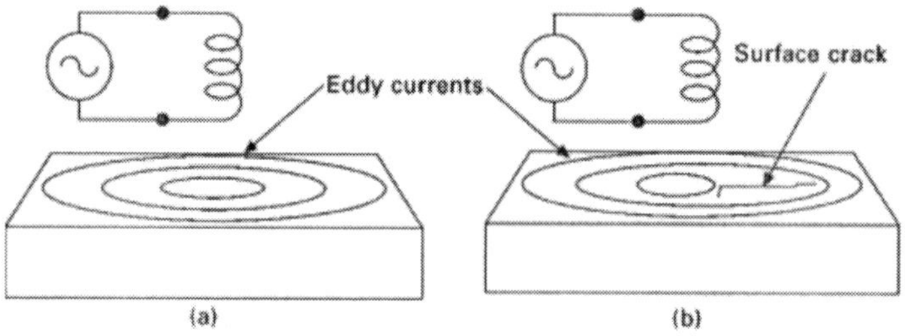

Figure 1. Schematics of the eddy current contour distorted by the surface crack.

Edge cracking has been widely observed during the stamping process of advance high strength steel (AHSS). Local edge formability after shearing is observed to have a significantly lower forming limit than the base metal. The sheared edges experience out of plane local pre-straining due to the shearing process, which is subsequently stretched in stamping. Therefore, the process is challenging to simulate accurately. Various research has been performed to evaluate and predict edge formability. The critical parameters that affect local edge formability were investigated by Pathak, et.al [3]. The study concludes that the edge formability was mainly affected by the work-hardening introduced by the shearing process. Various lab-scale testing methods were introduced to evaluate work hardening and edge formability, such as hardness test, grain rotation and half specimen dome test [4, 5, 6]. However, most testing methods are difficult to use for inspecting the sheared edge quality in the industrial production environment.

The test was conducted based on the previous work published by EWI [7]. The study contributed further developments in correlation of EC based NDE on the sheared edge. However, this setup requires tight tolerance of the sensor alignments, and is limited by the sheared edge geometry. In this study, NDE setup was upgraded to scan the sheared edge directly across the fracture surface. In the upgraded setup, multiple stacked samples can be measured together using a high-resolution sensor and computer numerical control (CNC) fixture as shown in figure 2.

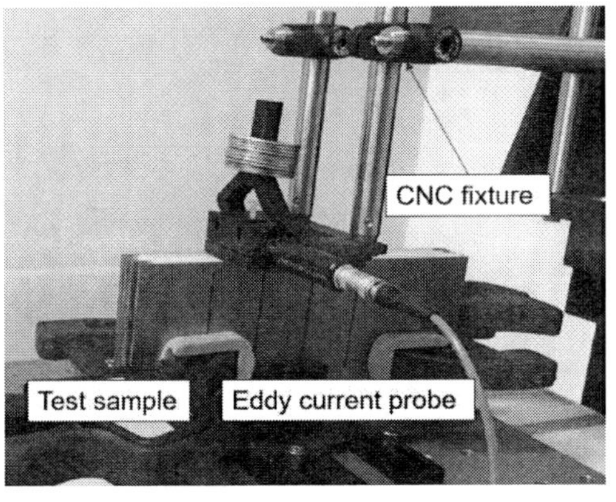

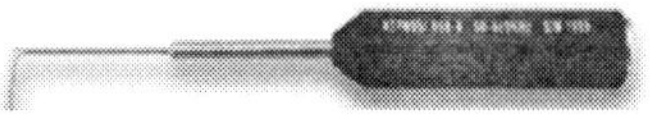

Figure 2. Eddy current probe setup and the eddy current probe.

2. Shearing test

In this study, a DP780 material with 1.2-mm thickness was selected. The following edge conditions were considered to create the different sheared edge conditions in table 1.

Table 1: Edge condition variations of the sample for NDE.

Sample Label	Edge condition
5%	Die shearing with 5% shearing clearance
10%	Die shearing with 10% shearing clearance
15%	Die shearing with 15% shearing clearance
20%	Die shearing with 20% shearing clearance
25%	Die shearing with 25% shearing clearance
Machined	Milled edge
Sheared	Sheared using a conventional lab shear machine with unknown shear clearance

The shear clearance was calculated with percentage respect to the material thickness (1.2 mm). US Steel supplied EWI most die-sheared samples. EWI prepared the machined and conventionally sheared samples. Most samples were prepared to have the sheared edges perpendicular to the rolling direction.

3. Eddy current based NDE

3.1. Test setup

An extra-small sized EC probe was used to collect the NDE signal at the edge, as shown in figure 2. The test samples were stacked with interested sheared edge aligned. Each sample was separated with non-conductive spacer layer by layer to eliminate the interference of signal across the neighboring samples. The CNC fixture allows the probe to move in x and y direction horizontally, as shown in figure 3. The probe was set to travel 25-mm horizontally, limited by the CNC fixture. Each horizontal measurement path in x-direction collected approximately 16,000 data points. 1000 data points were collected every second. The horizontal san paths had 0.2-mm index distance with each other in x-direction. A total of 200 scan paths were measured for five stacked samples.

The test setup uses the highest frequency setting of the EC coil, which is 500 kHz, to obtain the high-resolution NDE data. The EC signal that was sensitive to the edge hardening was calibrated as the representative EC parameter indicating edge hardening level. This EC signal component was referred to as the vertical component.

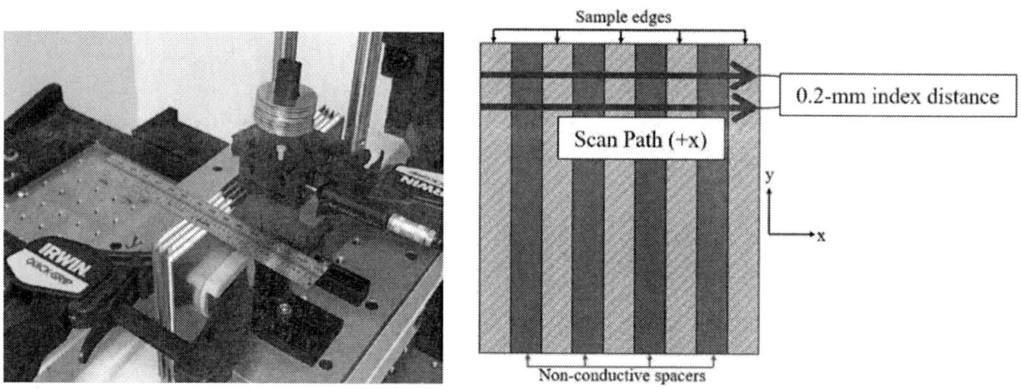

Figure 3. EC measurement setup and schematic of sample line-up.

3.2. Test result

Figure 4(a) shows a colormap of the NDE data collected with five samples stacked with five different edge conditions. Red color indicates higher EC parameter, while blue color indicates lower EC measurement parameter. The data were sliced and plotted in figure 4(b). The five peaks correspond to the five samples stacked and measured together as shown in figure 3. The peak NDE value occurs around the middle point of the material thickness. The EC peak values for each edge condition are summarized in figure 5.

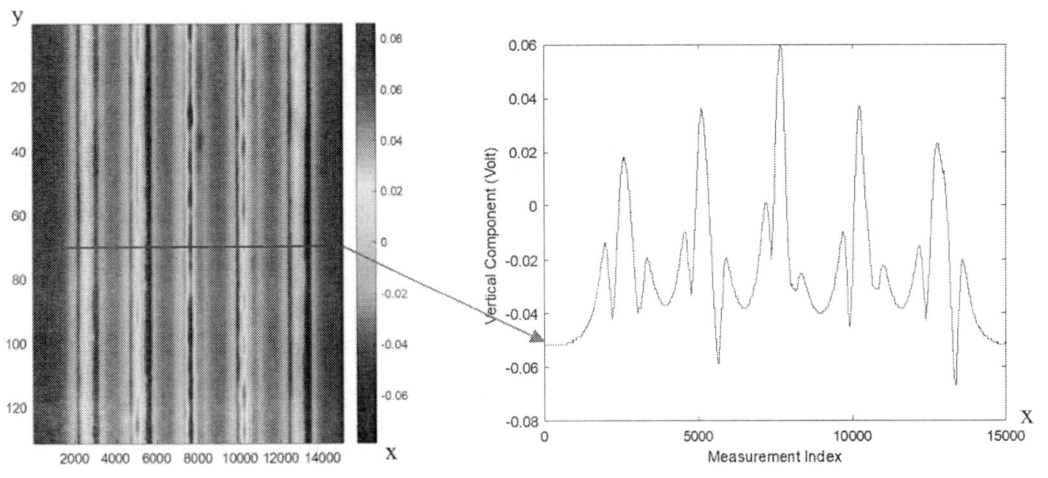

(a) **(b)**

Figure 4. (a) Colormap of EC scanning and schematic of selection of data slices from EC scanning of machined edge, 20%, 25%, 5% and 10% sheared clearance (L-R). (b) Example of peak data across the thickness in the same order.

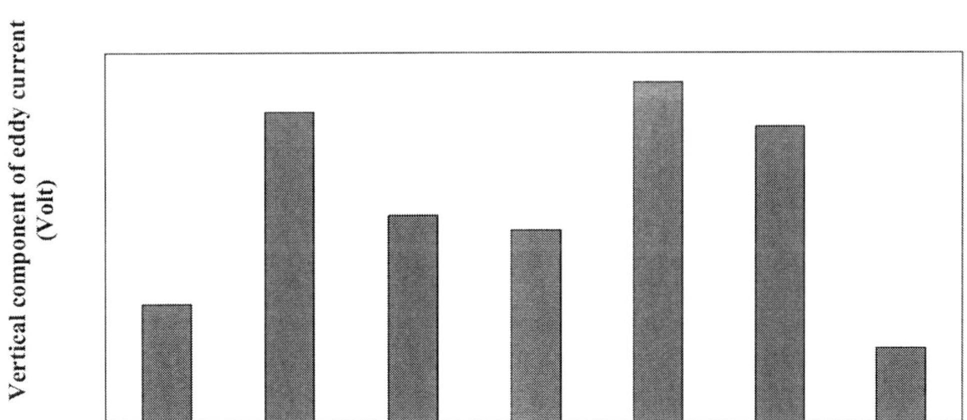

	5%	10%	15%	20%	25%	Sheared	Machined
EC peak data	0.008	0.021	0.014	0.013	0.023	0.02	0.005

Figure 5. The EC peak values of different edge conditions

4. HSDT and hardness test

4.1. HSDT

HSDT was used to evaluate the edge formability by stretching the straight sheared edge with a 100-mm diameter dome punch as shown in figure 6. The local strain distribution was captured using DIC cameras as shown in figure 6. The maximum equivalent strain before cracking initiation was used as the important indicator to compare the local formability. HSDT was conducted with three repetitions for each edge condition.

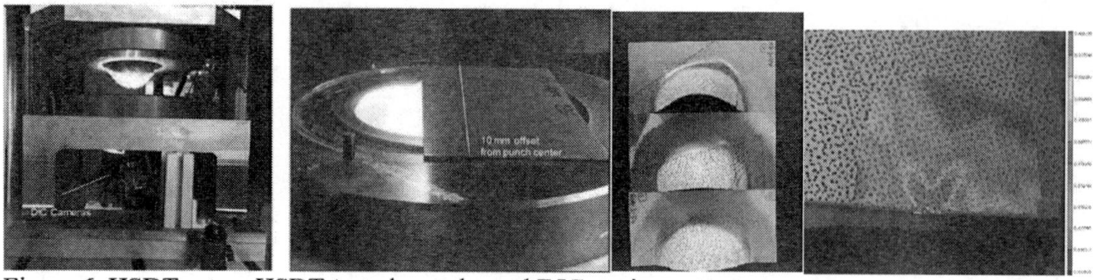

Figure 6. HSDT setup, HSDT tested sample, and DIC strain contour.

4.2. Hardness test

Vickers microhardness test was also conducted on the sample cross section for each edge condition to evaluate the work hardening level. Ten indentations were made in every 0.15-mm interval for each sample from the sheared or milled edge to the base metal, as shown in Figure 7. Each measurement pattern started at the transition of burnished zone to fracture zone.

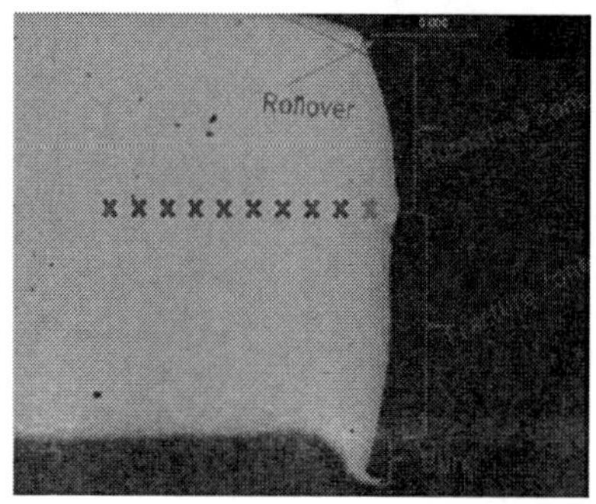

Figure 7. Locations for microhardness indentation testing.

4.3. Results of HSDT and microhardness test

The failure strain and hardness measurement from the HSDT are summarized in figure 8. The normalized hardness was calculated using Equation (1) to compare the increased hardness on the sheared edge.

$$Normalized\ Hardness = {HVmax}/{HV_{base}} \tag{1}$$

The failure strain from the HSDT and hardness test result have a negative correlation with each other. Lower HSDT equivalent failure strain and higher normalized hardness indicates poor edge quality, which has lower edge local formability and higher work hardening level.

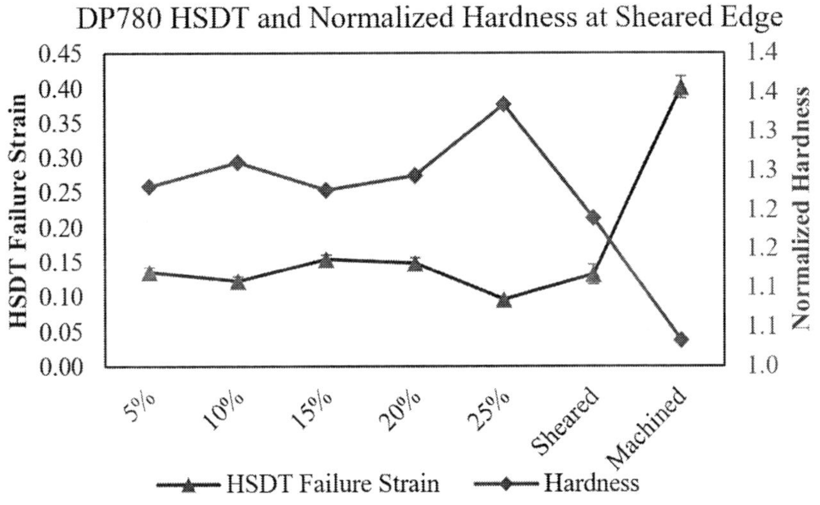

Figure 8. HSDT result and normalized hardness of the sheared edge.

4.4. Correlation of NDE result with HSDT and hardness test

The NDE data were compared with the failure strain from HSDT at the onset of cracking and normalized edge hardness from the microhardness test. Figure 9 compares the parameters relating to the work hardening level from NDE and the two destructive tests. EC peak parameter shows the consistent negative correlation with the HSDT failure strains and positive correlation with hardness results. The samples showing the higher equivalent failure strain indicate better edge formability and gave the lower NDE peak values.

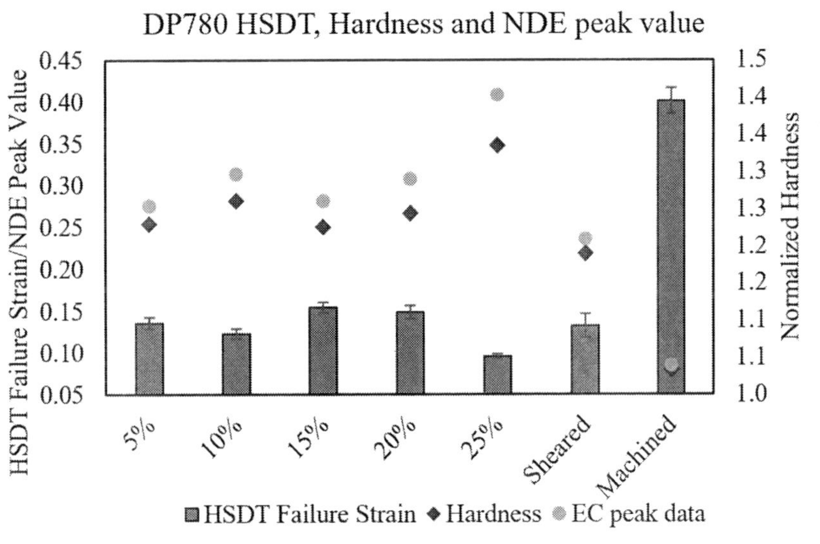

Figure 9. Comparison of NDE result and destructive evaluations of edge formability.

The relationship among EC measuremnt parameter, the HSDT failure strain, and hardness value was further analyzed using a power regression model as shown in figure 10. The r-squared value of the regression model of NDE results is 80.2% with HSDT failure strains and 94.7% with normalzed hardness values. This method can be used for predicting edge local formability and evaluating work hardening level nondestructively. However, to validate this method, more experiments should be considered as future work.

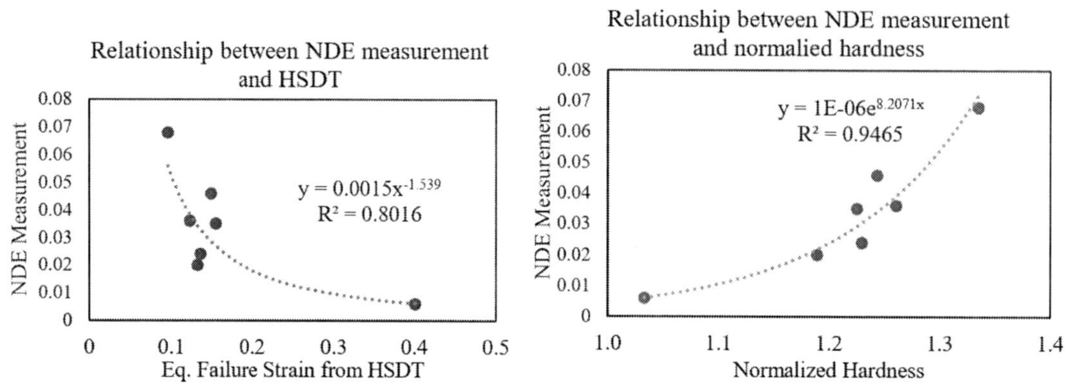

Figure 10. Correlation of NDE measurement with HSDT failure strains and hardness values.

5. Conclusions

The following conclusions can be drawn from the study:

- The NDE measurement parameter has strong correlation with the edge formability from HSDT and the microhardness test.
- The NDE value increases as the HSDT strain decreases and edge hardness increases, which indicates higher edge work hardening level.
- Eddy current based NDE can be used to evaluate the edge work hardening level on the sheared edges.
- The NDE measurement parameter might predict the edge failure by correlating with the edge local formability result. However, it needs more experimentation and validation.

6. References

[1] Innerspec Technologies., Inc. 2017 EMAT technology Retrieved August 30, 2017 http://www.innerspec.com/knowledge/emat-technology/

[2] Mouritz, A P 2012 *Introduction to Aerospace Materials* (Elsevier) pp 534–557

[3] Pathak, N, Butcher, C and Worswick, M 2016 Assessment of the critical parameters influencing the edge stretchability of advanced high strength steel *Journal of Materials Engineering and Performance* **25** 11

[4] Wu, X, Bahmanpour, H and Schmid, K 2012 Characterization of mechanically sheared edges of dual phase steels *Journal of Materials Processing Technology* **212** 6 pp 1209–1224

[5] Levy, B and Van Tyne, C 2012 Review of the shearing process for sheet steels and its effect on sheared-edge stretching *Journal of Materials Engineering and Performance* **21** pp 1205 –1213

[6] Gu, J, Kim, H, Shih, H C and Dykeman, J 2018 Effects of Blanking Conditions to Edge Cracking in Stamping of Advanced-High Strength Steels (AHSS) (SAE Technical Paper 2018-01-0626)

[7] Pathak, N, Gu, J and Kim, H 2019. A Novel Method to Nondestructively Measure the Shear Edge Properties for Edge Cracking Evaluation with Advanced High Strength Steels (SAE Technical Paper 2019-01-1090)

Acknowledgments
The authors acknowledge great support from the Ohio Development Service Agency on Edison Advanced Manufacturing Program (Grant # AMP 17-103-2017) and wonderful collaborations with the industry partners: General Motors Company, Fiat Chrysler Automobiles (FCA), Cosma International, POSCO, Novelis, ESI, AutoForm, Diversified Tooling Group, AMEPA, Fuchs Lubricants Co., Selmatec, and GBTP.

IOP Conf. Series: Materials Science and Engineering **967** (2020) 012064 doi:10.1088/1757-899X/967/1/012064

A Study on Effects of the Press Speed on Sheared Edge Formability

J Gu*[1], F Alamos[1], D Schoch[2], J Bornhorst[2], H Kim[1]

[1] EWI - Forming Center (EWI-FC), Columbus, Ohio, United States 43221
[2] AIRAM, Covington, Ohio, United States 45318

Abstract. In this study, the effect of shearing speed on edge formability was investigated with two different grades of advanced high strength steel (AHSS), DP780 and GEN3-980. A straight blanking tool with adjustable shear clearance was used to shear the selected materials in two different type presses, a high-speed pneumatic press, and a mechanical servo-press. The four different shearing speeds from 0.01 meter per second up to 1.5 meters per second were realized and measured. The different shearing speed conditions result in different edge quality and work hardening levels. The half specimen dome test and micro-hardness analysis were conducted to evaluate the edge formability. The samples sheared with a pneumatic press at a higher speed showed significantly better edge quality, less hardness increase, and smoother sheared edge than the samples sheared with the servo press. The developed high-speed shearing condition has a very good potential to shear AHSS without creating significant damage and work hardening on the sheared edge, providing further advantages for the post stamping operations.

1. Introduction

The edge cracking issue has been observed in the sheet metal stamping process, especially in the advance high strength steels (AHSS). The edge cracking does not typically follow the forming limit diagram or other failure criteria during forming. As more AHSS has been considered for the lightweighting of motor vehicles, the edge cracking issue needs to be addressed. Various studies have been done to evaluate and improve the sheared edge quality of AHSS.

To evaluate the factors that influences the edge local formability, the effect of die clearances has been reported by Henderson et al. [1] and Lee et al. [2]. The critical parameters that influence the local edge formability was investigated by Pathak et al. [3] The study concluded that the sheared edge formability was influenced by the edge work hardening.

Interestingly, a recent finding by Fraunhofer IWU has shown that through high-speed impact shearing, the pierced hole has a much smoother surface than the conventional hole piercing process [5]. The mechanism behind this could be the adiabatic softening effect from the high-speed impact. However, little research has been done on the effect of formability in various shearing speed. Therefore, a study on the influence of different shearing speed on the edge local formability was conducted at EWI.

2. Sample preparation

2.1. Test matrix

This study was conducted with two AHSS materials, DP780 and GEN3-980. The material properties are given in table 1.

Table 1: Material properties of the selected sheet metal

Material	Thickness (mm)	Yield Strength (MPa)	Tensile Strength (MPa)	Total Elongation
DP780	0.9	499	855	18.0 %
GEN3-980	1.2	609	1022	22.1 %

Each material was sheared using the same test tooling with four different die clearances and four shear speeds. The shear die breakages are adjusted by the percentage of the material thickness. The shear die can be adjusted with different shim combinations to accommodate different material thickness and different shear clearances.

The edge local formability of the sheared sample was be evaluated using HSDT, microhardness test and optical microscopy. The test matrix is shown in table 2.

Table 2: Test matrix of shearing test and edge evaluation

DP780	**GEN3-980**	**Shearing speed**	**Repetition**		
Shear clearance %			Shearing	HSDT	Hardness test and Optical Microscopy
9%[a]	9%	4	Minimum 5	Minimum 3	1
12%	11%				
15%[b]	15%				
18%	18%				

a. Shear clearances are calculated by $\boldsymbol{Shear\ clearance}\% = \frac{Die\ clearance}{Material\ thickness} * \mathbf{100}\%$

b. This condition was sheared

2.2. Shear test

Each material was sheared with four different press speeds using AIDA 300-ton servo press and AIRAM 12-ton pneumatic press using the same set of tooling shown in figure 1. Two press motions, silent deceleration and crank motion, were used in AIDA servo press, shown in figure 2. Crank motion imitates the production of conventional mechanical blanking press using 20-spm press setting, while silent deceleration decelerates the ram to 10% initial speed of the crank motion before initial contact of the blank. Two pressure settings of the pneumatic press, 30 psi and 50 psi, were used to provide the shearing speed up to 1.8 m/s. The speed of the pneumatic press can be adjusted by changing air pressure setting.

Figure 1. Servo-press (left), pneumatic press (middle) and shear test tooling (right)

IOP Conf. Series: Materials Science and Engineering **967** (2020) 012064 doi:10.1088/1757-899X/967/1/012064

The instantaneous shearing speed of the servo press was provided by the servo control module. The displacement and speed vs time of the two different press motions are shown in figure 2. For the pneumatic press, the speed was measured by a laser sensor with 1000-Hz measuring frequency. An example of the press motion and speed measurement is shown in figure 3. The shearing speeds of all press options were measured when the shear die makes initial contact with the blank.

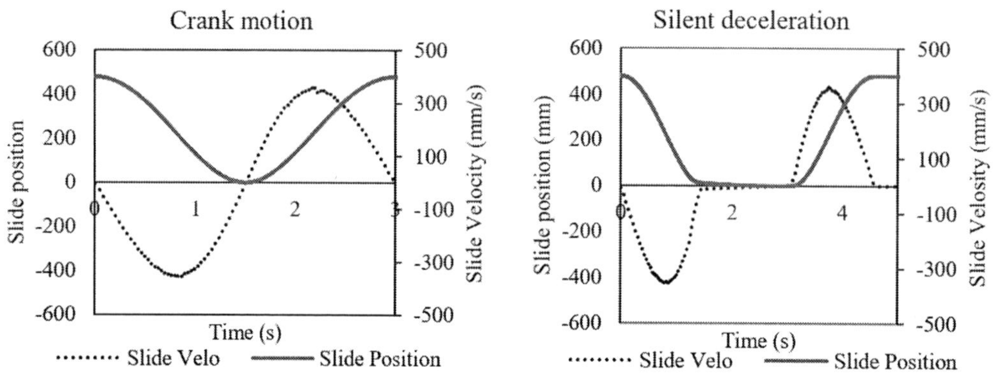

Figure 2. AIDA servo-press speed and position vs time of crank motion and silent declaration

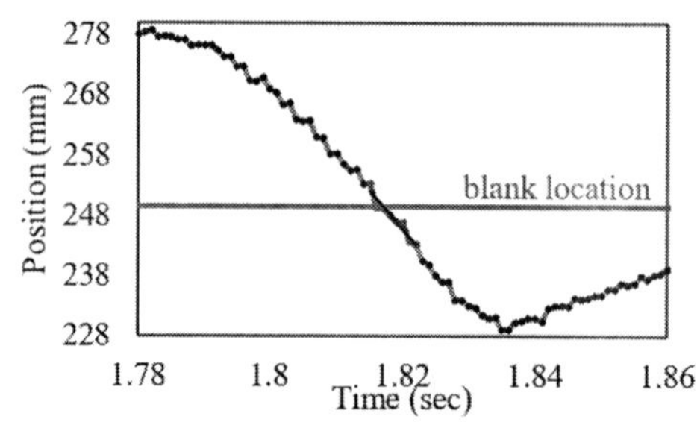

Figure 3. AIRAM pneumatic press position vs time and speed measurement

The measurement of average press speed and standard deviation are shown in table 3. The silent deceleration press motion carries less than 10% the speed of crank motion, while the pneumatic press motion has about 10 times the speed of the crank motion.

Table 3: Shearing speed measurement at tooling contact for each press motion

	Press motion	*Speed (mm/s)*	*St. dev*
Servo-press	Silent deceleration	12.7	0.3
	Crank motion	144.7	1.6
Pneumatic press	30psi-AIRAM	1283.1	142.2
	50psi-AIRAM	1810.7	245.8

3. Evaluation of the sheared edge quality

3.1. Half Specimen Dome Test (HSDT)

An alternative method, HSDT, was conducted to evaluate the edge local formability. HSDT was extensively developed and evaluated by Kim et al. [6] and Gu et al. [7]. Different from the ISO standard hole expansion test, HSDT can evaluate the stretchability of a straight sheared edge. The test sample was sheared into a 100-mm by 200-mm rectangular shape with the interested sheared edge on the length of the rectangle. The 100-mm diameter Nakajima punch was used in HSDT to expand the sheared edge.

3D-DIC was used to capture the strain fields of the sample. The tooling setup and example of strain field post processing are shown in figure 4.

The dome punch expands the edge and creates stress concentration at the sheared edge between the apex of the dome and the die, without contact to the tooling. The DIC captured equivalent strain at the onset of edge cracking was used as main criteria to evaluate the sheared edge formability.

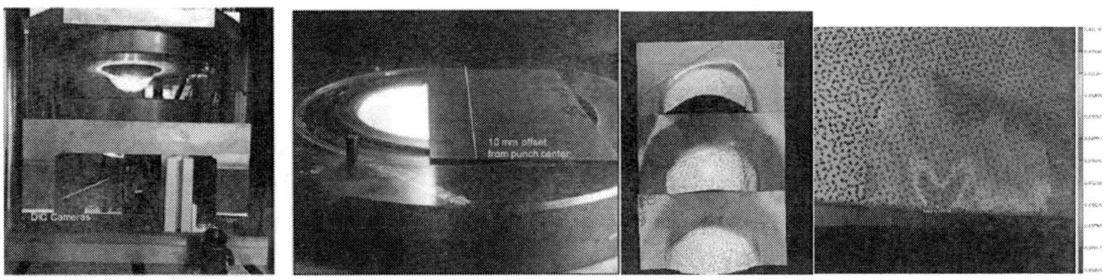

Figure 4. HSDT setup, HSDT tested sample, and DIC strain contour

3.2. Microhardness and Optical Microscopy
The hardness of the shear affected zone can be measured using a microhardness test. The edge hardness was compared to the base metal hardness. The sample was cut in the cross-section direction, mounted and polished. A microscopic image of the edge topography was also captured during the hardness test.

Ten indentations were made from 0.10 mm from the edge and has index distance of 0.15 mm from each other to affect the interference of the shear field from the indentation. The starting point of the hardness measurement is indicated in figure 5. The hardness measurement has the first indentation at the transition of burnished zone and fracture zone, where the highest hardness value is usually observed along the thickness direction.

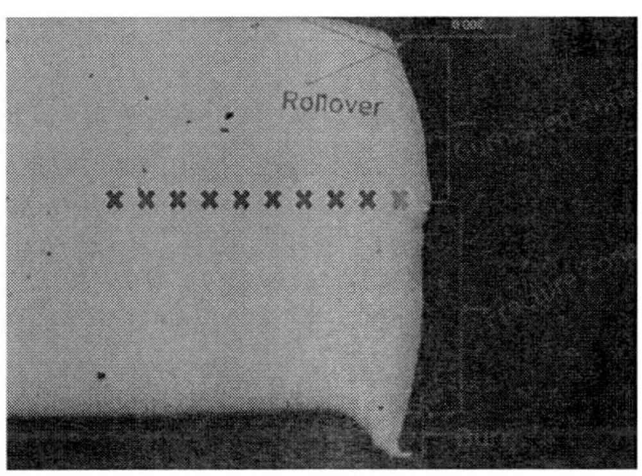

Figure 5. Starting point selection of hardness test

4. Results and Discussions
4.1. Half specimen dome test result
Figure 6 shows the summary of equivalent strain at the onset of cracking for the sample sheared with different clearances and in different speed. Higher equivalent strain indicates better edge formability. In each shear clearances for both materials, the higher two shearing speeds, 1283 mm/s and 1811 mm/s, show higher equivalent failure strain compared to the lower-speed cases. In the lower speed cases, silent deceleration (12.7 mm/s) shows slightly worse edge stretchability. Overall, the high-speed cases show improved edge failure strain compared to the conventional and low speed cases, but the result cannot differentiate a preferred case in the two high-speed settings.

IOP Conf. Series: Materials Science and Engineering **967** (2020) 012064 doi:10.1088/1757-899X/967/1/012064

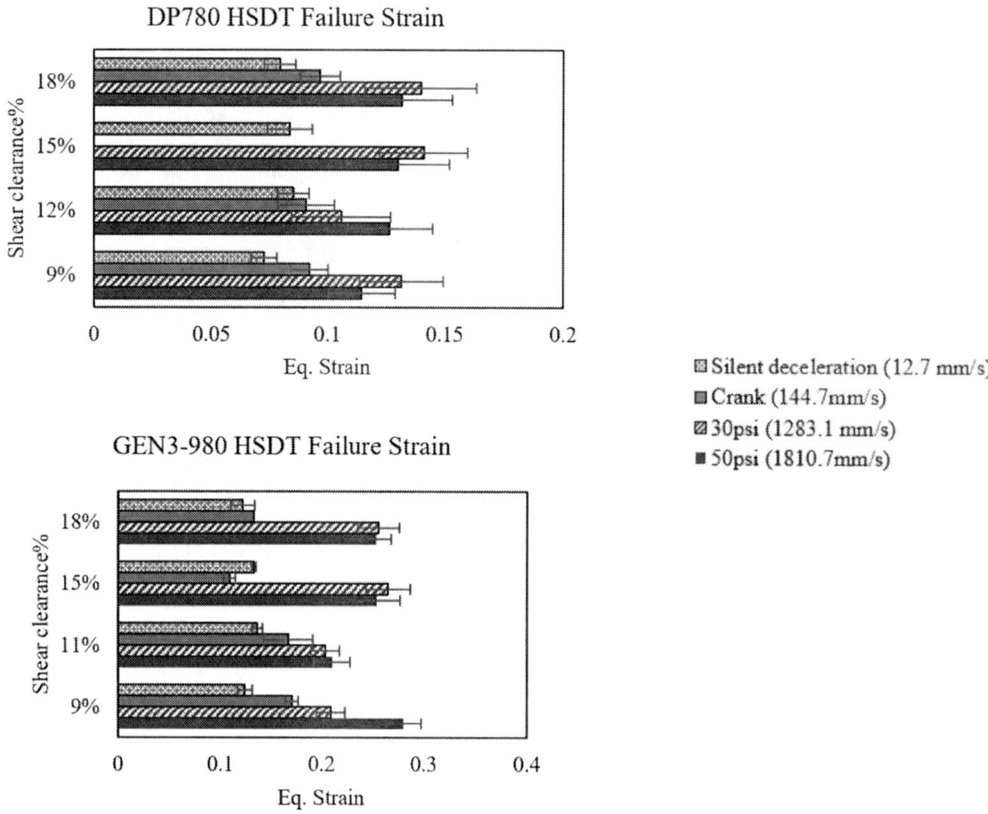

Figure 6: HSDT equivalent strain in each shear clearance and speed combination.

4.2. Microhardness measurements

An example of the microhardness of DP780 and GEN3-980 is shown in figure 7. The first hardness at the edge shows the highest value and the hardness decreases and approaches base metal hardness. The hardness measurement in other clearance and shearing speed combinations follow a similar trend. The measurement from the first indentation is identified as the critical edge hardness that represents the work hardening level at the sheared edge. The normalized critical hardness was calculated by equation 1. Higher Normalized hardness indicates higher work hardening induced from shearing process.

$$Normalized\ Critical\ Hardness = \frac{Critical\ hardness}{Base\ metal\ hardness} \quad (1)$$

IOP Conf. Series: Materials Science and Engineering **967** (2020) 012064 doi:10.1088/1757-899X/967/1/012064

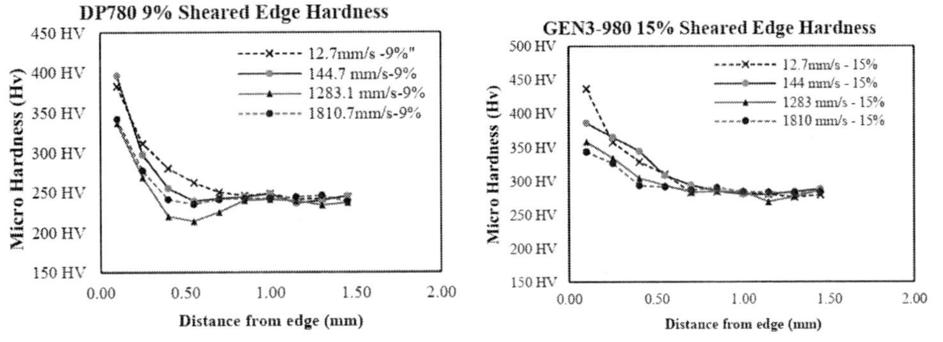

Figure 7. Microhardness measurement of DP780 with 9% shear clearance and GEN3-980 with 15% shear clearance

All combinations of normalized critical hardness with different shear clearances and speeds are summarized in figure 8. Although either within the two higher speed settings or the lower tow speed settings, the preferred case varies depending on the shear clearances. It is clear that the sample sheared with higher speed from pneumatic press shows lower normalized hardness. For both materials, the sample produced by pneumatic press with high-speed shearing shows lower work hardening, which indicates preferred edge condition.

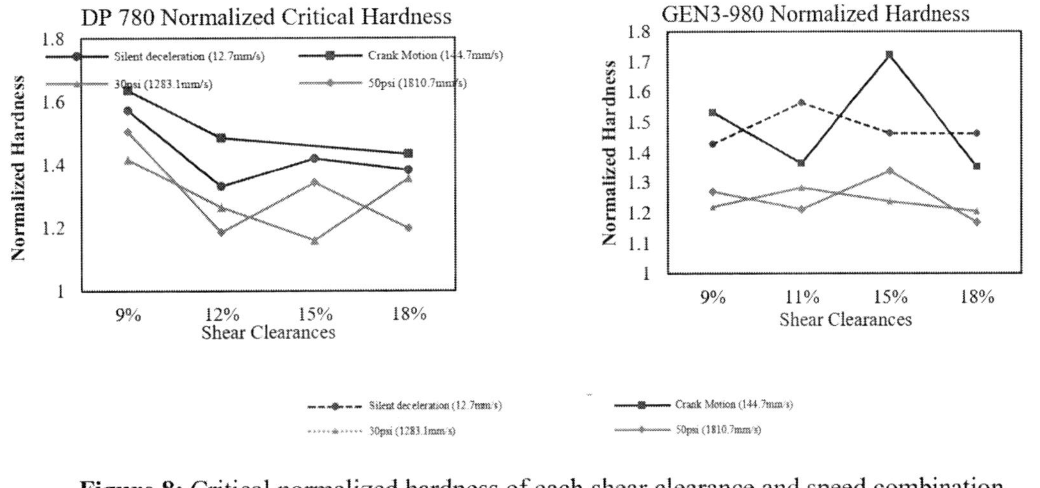

Figure 8: Critical normalized hardness of each shear clearance and speed combination

4.3. Optical microscopy

The optical microscopy images were taken for each shearing conditions to qualitatively evaluate the edge quality. Figure 9 shows an example of the DP780 sheared in different speeds with 11% die clearance. Both materials show similar edge profile. The sample sheared with the two press settings from pneumatic press have a smooth transition between burnished zone and fracture zone, as well as minimal rollover zone and burr.

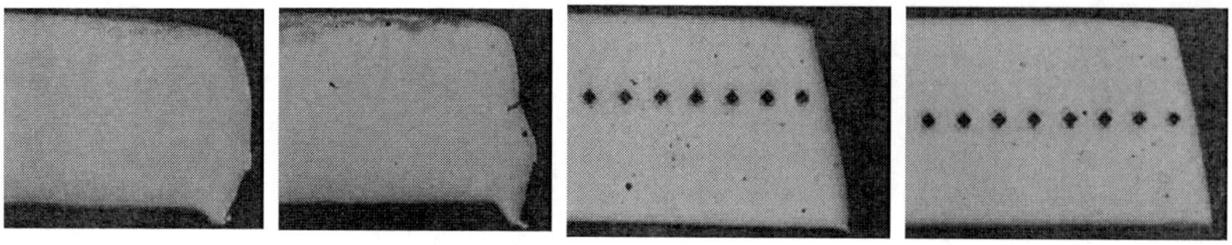

Servo silent deceleration	Servo crank motion	Pneumatic press – 30 psi	Pneumatic press – 50 psi
12.7 mm/s	144.7 mm/s	1283.1 mm/s	1810.7 mm/s

Figure 9: Optical microscopy images of DP780 material sheared with 11% die clearance in different speeds

5. Conclusions

The following conclusions can be drawn from this study:

- Sheared edge quality can be significantly influenced by shearing speed. With all clearances, the sample sheared with a high-speed process using pneumatic press shows improved sheared edge local formability.
- From optical microscopy analysis, the sample sheared by both high-speed settings of the pneumatic press shows minimal rollover height, smoother sheared surface and negligible burr. The topography of the edges sheared in high speed is different from lower shearing speed using servo press.
- The preferred shear clearances vary with different speed settings. The preferred shear condition for DP780 and GEN3-980 are:

  - DP780: 1283 mm/s shearing speed with 15% die clearance (92% higher HSDT failure strain than that of crank motion average)
  - GEN3-980: 1810 mm/s shearing speed with 9% clearance (50% higher HSDT failure strain than that of crank motion average)

6. References

[1] Konieczny, A and Henderson 2007 Formability Limitations in Stamping Involving Sheared Edge Stretching, *SAE Technical Paper*

[2] Lee, J S, Ko, Y K, Huh, H, Kim, H K, and Park, S H 2007 Evaluation of Hole Flangeability of Steel Sheet with Respect to the Hole Processing Condition *Key Engineering Materials*, 340: 665-670

[3] Pathak, N, Butcher, C, and Worswick, M 2016 Assessment of the Critical Parameters Influencing the Edge Stretchability of Advanced High Strength Steel *Journal of Materials Engineering and Performance*, 25(11)

[4] Shih, H, C, and Shi, Ming F 2012 Robust Shearing Process for Improving AHSS Sheared Edge Stretchability *Proc. of the ASME 2012 International Manufacturing Science and Engineering Conf. collocated with the 40th North American Manufacturing Research Conf. and in participation with the International Conf. on Tribology Materials and Processing (Notre Dame, Indiana)* June 4–8, 2012. pp. 363-369. ASME.

[5] Muller, R High Speed Impact Cutting of Sheets and Tubes. Fraunhofer IWU, www.iwu.fraunhofer.de/content/dam/iwu/en/documents/Infosheet/Infosheet-High-Speed-Impact-Cutting-Of-Sheets-And-Tubes.pdf

[6] Kim, H, Shang, J, Dykeman, J, Samant, A, and Hoschouer, C 2017 Practical Evaluation and Prediction of Edge Cracking in Forming Advanced High Strength Steels (AHSS), *SAE Technical Paper* 2017-01-0308

[7] Gu, J C, Kim, H, Dykeman, J, Shih, H, 2018 A Practical Methodology to Evaluate and Predict Edge Cracking for Advanced High-Strength Steel *IOP Conference Series: Materials Science and Engineering,* vol. 418, p. 012072., doi:10.1088/1757-899x/418/1/012072

Development of springback tester and the effect of forming speed on the springback of four sheet metal bending processes

DOU Luyi, LI Xiaoqiang*, DONG Hongrui, LI Dongsheng and PENG Xingyi
School of Mechanical Engineering and Automation, Beihang University, Beijing, CN 100191, China

E-mail: lixiaoqiang@buaa.edu.cn

Abstract. With the current increasing emphasis on issues of energy saving, emission reduction, driver and passenger safety in automotive industry, lightweight bodywork is an effective measure to reduce vehicle energy consumption. And also, high-strength bodywork structure can improve automobile safety performance because of its high impact absorption energy, and high fatigue resistance. In sheet metal forming, springback is a main defect with using high strength steel. Aiming to investigate sheet metal springback phenomenon for high-strength steel, a multi-funtional springback test machine is built. The testing machine can perform four types of springback tests: V-bend, U-bend, stretchbend and drawbend, and all these tests have been carried out to investgate the effect of springback factor in sheet metal forming processes for example forming speed. The research found that the springback angle of U-bend and drawbend increase with enlarging speed, however, V-bend and stretchbend show the opposite trend.

1. Introduction

At present, one of the main ways to solve the problem of energy saving and emission reduction of automobiles is to use high-strength steel instead of traditional steel plates. The application of high-strength steel can not only improve the lightweight of the car, but also improve the safety performance. However, springback is one of the most serious problems in the current forming of high-strength steel sheets for automobiles, and it has become a bottleneck restricting application of high-strength steels [1]. Therefore, it is of great significance to study the springback of high-strength steels in order to improve the forming accuracy of high-strength steel parts and to promote their applications.

Drawbend was proposed by R.H.Wagoner et al [2]. It can describe the complex deformation of the material when it flows into the inner wall through the fillet radius of the die, and can control the axial tensile stress individually, accurately and quantitatively. The test principle is shown in figure 1 (a). T. Kuwabara et al [3] proposed the stretchbend springback. The test principle is shown in figure 1 (b). At present, this test method has become Japan's national industrial standard JIS H 7702 and international standard ISO_24213_2008 (E). V-bend and U-bend are also commonly used test methods to measure the springback. They are shown in figure 1 (c) and figure 1 (d) respectively. It's known that different springback tests need different testers. So it's necessary to develop one tester which can do four bending tests such as drawbend, stretchbend, V-bend and U-bend.

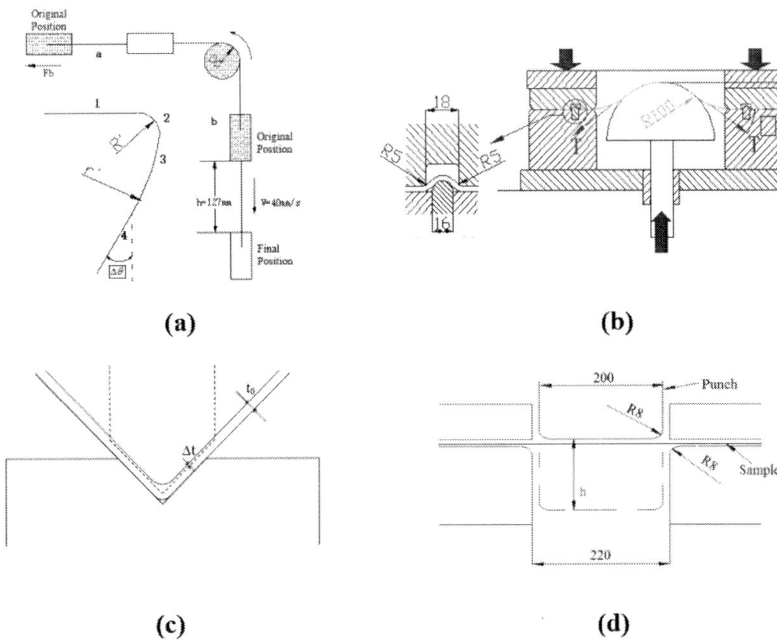

Figure 1. Schematic of four tests: (a) V-bend. (b) U-bend. (c) Stretchbend. (d) Drawbend. (unit: mm)

Scholars from various countries have conducted different bending test to analyze the influence of die parameters (male die fillet radius, female die fillet radius, female die's angle or curvature), process parameters (blank holder force, punch displacement), and sheet parameters (sheet thickness, mechanical property) on the springback amount. However, there is not much research on the influence of forming speed on springback. Mori K et al [4] studied the effects of three high-strength steel sheets (SPFC440, 780Y, 980Y) properties, excessive displacement of punches, forming rate and holding time on the V-bend springback using a CNC servo press. It's found that the effect of forming rate on springback is not obvious. Min Kuk Choi et al [5] carried out U-bend with different punch speed (0.07, 0.7, 7, 70mm/s) of 1mm thick DP780 steel sheets. The springback angle of the side wall increased from 117 ° to 118.57 °, while the angle between the side wall and the blank holder decreased from 68.39 ° to 66.72 °. O Çavuşoğlu et al [6] conducted a 60-degree V-bend test at different strain rates ($0.0083s^{-1}$,$0.16s^{-1}$) for 1mm thick DP1000 steel sheets, and found that the springback increased from 9 ° to 13 ° as the strain rate increased.

Most of researches about effect of forming speed on springback are carried out by using V-bend or U-bend. With the application of servo presses in the field of high-strength steel sheet stamping, it is possible to flexibly adjust the stamping speed. So it's meaningful to investigate the effect of forming speed on springback by using different bending tests including drawbend and stretchbend.

The article consists of two parts. The design principle and implementation process of the test platform are introduced in first part. The effect of forming speed on springback by using four kinds of bending tests are analyzed.

2. Design and implementation of test platform

2.1. Overall plan

The developed testing machine can be divided into two sets of test modes: stretchbend and drawbend. The tensioning and blank holder cylinders are controlled by force, and the drawing and punching cylinders are controlled by speed. Stretchbend, U-bend, and V-bend adopt a set of test modes, and drawbend has a separate set of test modes, and the die can be easily changed when performing different tests. Mechanical system adopts three beams and four columns structure. The device is used to automatically or manually control the movements of the various components of the hydraulic system by means of "computer/manual control panel-PLC-electro-hydraulic proportional control". The overall schematic is shown in figure 2 and each die set is shown in figure 3.

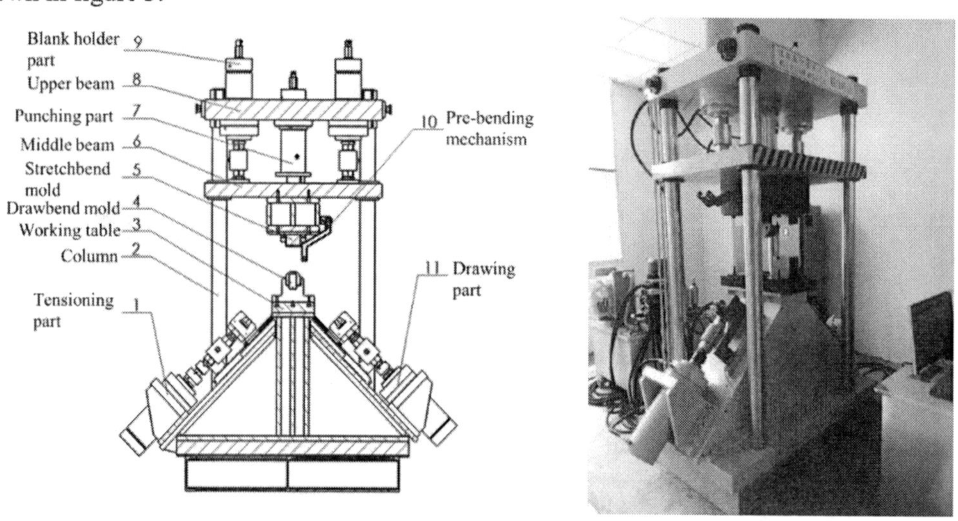

(a) (b)

Figure 2. Test machine overall: (a) Schematic. (b) Physical.

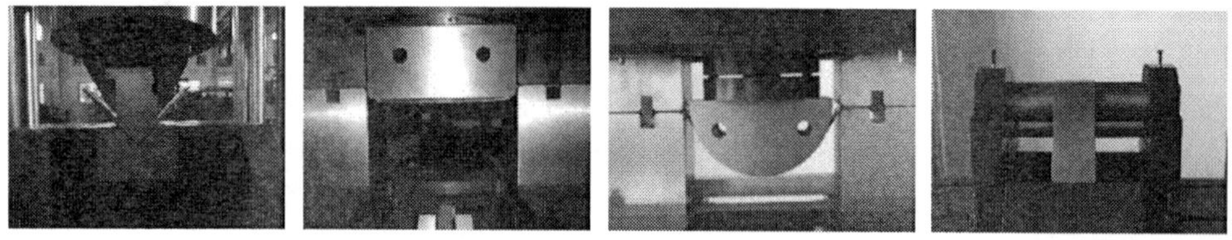

(a) (b) (c) (d)

Figure 3. Various dies: (a) V-bend. (b) U-bend. (c) Stretchbend. (d) Drawbend.

2.2. Design and implementation

2.2.1. Mechanical system. (1) Stretchbend module. It includes two blank holder and one stamping cylinders, upper beam and movable beam, and die assembly. Magnetostrictive displacement sensors are installed at the rear of the blank holder and the punching cylinder, and the force sensors are installed in the middle. The upper beam and the movable cross beam are connected by bolts, and an upper die is installed under the movable beam. The die part includes three types: U-bend, V-bend and stretchbend. The male die is installed under the punching cylinder's force sensor, and the female die is installed on the work platform. Different dies need to be replaced during different tests. (2) Drawbend module. Tensioning cylinder, drawing cylinder, work platform and die assembly are includws. The two inclined planes of the working platform are at 90 °, and the tensioning cylinder and the drawing cylinder are respectively installed. The tensioning cylinder adopts a compound structure of large and small cylinders to meet the wide range of controllable tension force. A clamping cylinder assembly is installed at the front of the force sensor, and the scale is connected to the moving part through an adapter card to control the displacement of the hydraulic cylinder. The die part includes five sets of round rolls with different radii, a series of pads of different heights to accommodate different thickness samples, and it is ensured that the initial angle of the test sheet are accurate 90 °.

2.2.2. Hydraulic system. The hydraulic system uses dual pumps to supply oil to provide the required power for the test machine to achieve various actions. The movement of the hydraulic cylinder is controlled by the action of the electromagnetic reversing valve. The German Rexroth proportional valve is selected as the main control valve, the tension and blank holding force is adjusted by the proportional relief valve, and the speed of the drawing cylinder and the punching cylinder is adjusted by the proportional speed regulating valve. The hydraulic system consists of a pumping station, a cooler, an oil pipeline, a hydraulic valve, a valve plate and related accessories.

2.2.3. Control System. The control system is composed of Delphi software, OMRON PLC, Advantech data acquisition control card, etc., which can realize closed-loop control of tension and blank holder force, and close-loop control of punching displacement and drawing displacement. The testing machine can send the signal to PLC and proportional valve through the control software to realize automatic operation, and also can realize the manual operation directly through the button on the manual console. The software system is equipped with force sensor out-of-tolerance alarm and break alarm to provide safety guarantee.

3. The springback tests by using different bending methods

In order to verify the reliability of the testing machine, V-bend, U-bend, stretchbend and drawbend were tested in multiple groups to study the effects of conditions such as force, distance, speed, and lubrication on different springback tests, as shown in table 1. The samples after test are shown in figure 4. It shows that the testing machine can complete the four types of tests. Springback index of V-bend, U-bend and stretchbend is shown in figure 5 and one of drawbend is shown in figure 1 (a).

Table 1. Experiment matrix to verify the reliability of the test machine.

	Material	Sample size (mm)	Test conditions
V-bend	DP980/ 780/500	150 × 50	Die angle: 60 °, 90 °, and 120 °
U-bend	DP590	560 × 50×1.6	Punch depth: 100mm, speed: 40mm/s, blank holding force: 30, 100, and 200kN from left to right.
Stretchbend	DP590	560 × 50×1.6	Speed: 20mm/s, blank holding force: 30kN, punch depth: 50mm, 77mm, and 100mm from left to right.
Drawbend	DP980	660 × 50×1.2	Tension: 11.5kN, speed: 40mm/s, radius of the round roller: 25.4mm, drawing distance: 25.4mm, 76.2mm, and 127mm from top to bottom

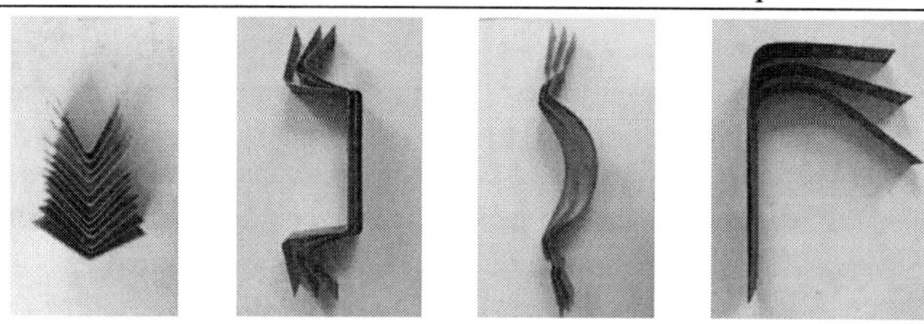

(a) **(b)** **(c)** **(d)**

Figure 4. Various samples: (a) V-bend. (b) U-bend. (c) Stretchbend. (d) Drawbend.

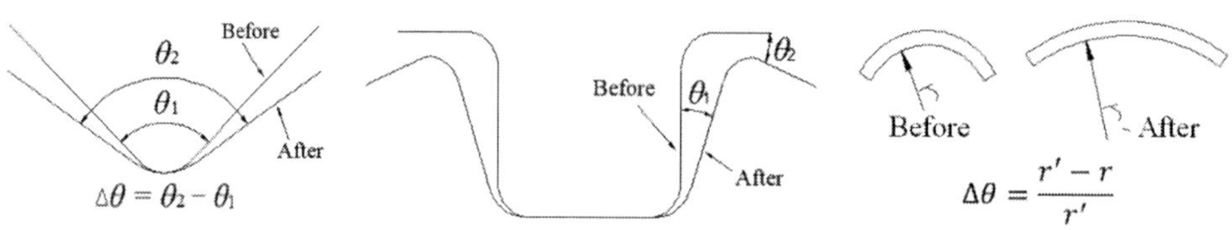

(a) **(b)** **(c)**

Figure 5. Springback index: (a) V-bend. (b) U-bend. (c) Stretchbend.

This paper studies the effect of forming speed on the springback of each test. The size of the V-bend springback specimen is 150 × 50×1.2mm, the punch angle is 90 °, and the test termination condition is that the punch pressure reaches 6kN, forming speeds are 2, 30, and 60mm/min. The size of the U-bend sample is 150 × 50×1.5mm, the blank holding force is 50kN, and the punching depth is 80mm, forming speeds are 10, 20, and 40mm/s. The size of stretchbend sample is the same as that of the U-bend, blank holder force is 30kN, and the punching depth is 77mm, forming speeds are 2, 20, and 40mm/s. The drawbend sample size is 660 × 50×0.8mm, tension force is 6.24kN, drawing distance 127mm, round roller radius is 25.4mm, and speeds are 5, 20, and 40mm/s. The above test materials are DP780, and the test is performed once. The results of the test are shown in figure 6.

It can be seen from the test results that V-bend and stretchbend decrease the springback amount with increasing speed, while U-bend and drawbend are opposite. The contrary is interesting because it is also found in testing of DP590 although the test of DP780 here are not repeated. The cause will be discovered by using FE modelling and experiment in the future.

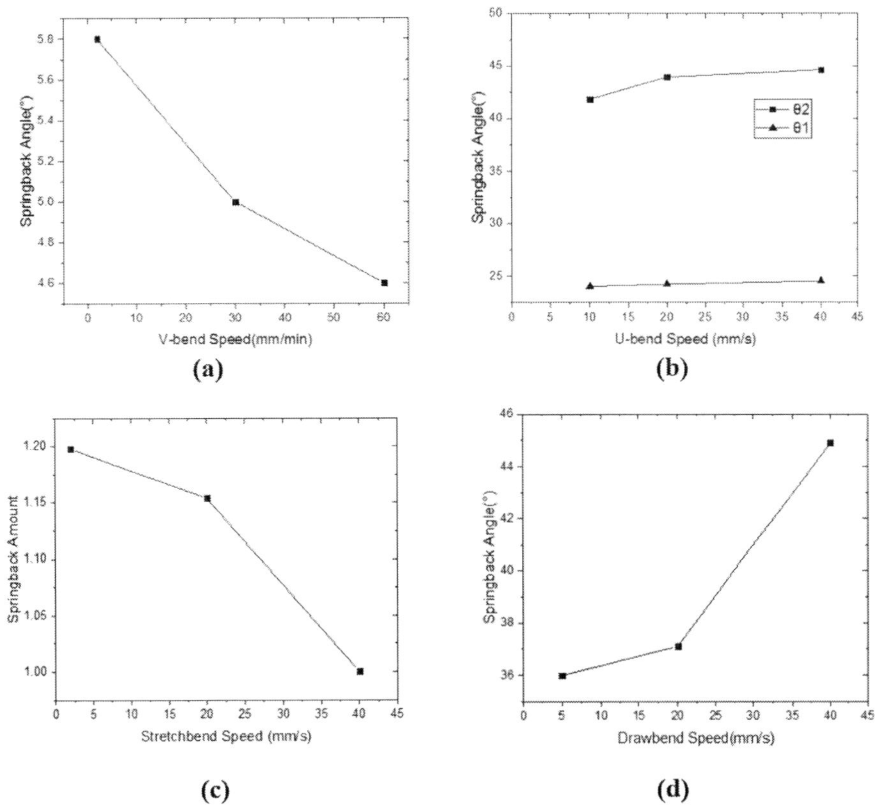

Figure 6. Springback amount of each test: (a) V-bend. (b) U-bend. (c) Stretchbend. (d) Drawbend.

4. Summary

- The multifunctional springback tester is developed which can do V-bend, U-bend, stretchbend and drawbend tests in the same machine.
- The tester's reliability are validated by different springback test of high strength steel. Especially, effect of forming speed on the springback amount are investigated using different method. It's found that the springback amount of the V-bend and stretchbend decreases as the forming speed increases and the U-bend and the drawbend are opposite.

References

[1] Bingkun X and Fazhong S 2001 Numerical Simulation of Springback in Sheet Metal Forming *J B UNIV AERONAUT ASTRONAUT* **27(2)** 194-97

[2] L.M.Geng 2000 Application of plastic anisotropy and non-isotropic hardening to springback prediction The Ohio State University

[3] Kuwabara T, Ikeda S and Asano Y 2004 Effect of Anisotropic Yield Function on The Accuracy of Springback Simulation *8th Int.Conf.on Numerical Methods in Industrial Forming Processes* (Columbus Ohio) (6) p 887-92

[4] Mori K, Akita K and Abe Y 2007 Springback behavior of ultra-high-strength steel sheets using CNC servo press *INT J MACH TOOL MANU* **47** 321-25

[5] Min Kuk Choi and Hoon Huh 2014 Effect of punch speed on amount of springback in U-bending process of auto-body steel sheets *Procedia Eng* **81** 963-68

[6] Çavuşoğlu O, Toros S and Gürün H 2019 The effects of temperature and strain rate on

yielding and springback behaviour of DP1000 dual phase steel *Mater. Res. Express* **6** 026509

A new workflow for the effective distinction between necking induced spilts and direct fracture phenomena

N Manopulo[1*] and B Carleer[2]

[1]AutoForm Development GmbH, Technoparkstrasse 1, 8005, Zurich, Switzerland
[2]AutoForm Engineering Deutschland GmbH, Joseph-von-Fraunhofer-Straße 13a, 44227, Dortmund, Germany

Abstract. The predominant mode of failure in stamping applications is still splitting due to localized necking. This kind of failure is very effectively captured by the forming limit curve, especially if nonlinear deformation effects are modelled. With the sharp increase in low-ductility (weight efficient) materials, direct fracture phenomena (surface cracks, edge cracks, shear cracks, etc.) increasingly enter the daily industrial stamping practice, especially in follow operations. Contrary to the classical FLC analysis, there exists nowadays no industry-wide standardization for characterizing these phenomena. This drives up modelling costs and reduces the expected improvement in the overall modelling uncertainty. It therefore becomes crucial to understand and detect the circumstances in which an advanced fracture characterization is necessary, such that limited resources can be effectively allocated. The present contribution aims to propose such a workflow. Firstly a reasonable lower bound for the fracture limits is estimated based on theoretical considerations and evidence from scientific literature. This estimation is used to conduct a first analysis step which is used to categorize the process based on whether a standard modelling approach is sufficient to deliver the required accuracy. A second step of accurate fracture identification is therefore only recommended for the subset of geometries with particularly high risk. The approach is tested on data for AA6016 and DP980 from literature and validated based on published experimental results.

1. Introduction

As it is the case with all engineering activities, the primary aim of numerical simulations is the minimization of uncertainties during the design of a physical system, such that it can be realized effectively and without failure. Simulation tools provide nowadays excellent possibilities for accurately modelling very complex engineering systems and deliver an unprecedented level of accuracy. On the other hand this increased effectiveness is most of the time consumed to design increasingly complex systems in shorter times, such that the tolerance for inaccurate results is also steadily decreasing. Understanding, quantifying and managing uncertainties therefore continue to be in the primary focus of the engineering endeavour. Failure of the stamping process due to local necks or spilts, can incur significant costs during tryout and/or production and can cause disruptions in the product schedules. An accurate modeling of these phenomena, therefore receives a lot of industrial attention. As a consequence the recent scientific literature has been very prolific in this direction. This is especially the case for direct fracture phenomena such as surface cracks, edge cracks and shear

* Corresponding author: Dr. Niko Manopulo, Technoparkstrasse 1, 8005, Zürich, Switzerland, +41 43 444 61 90
niko.manopulo@autoform.ch

IOP Conf. Series: Materials Science and Engineering **967** (2020) 012066 doi:10.1088/1757-899X/967/1/012066

cracks. Surprisingly, only few of the proposed approaches have made it nowadays into the industrial practice. This is primarily due to the following two reasons:

1. The predominant failure mechanism in the first drawing stage is still "failure due to localized necking" which is very well captured using a (nonlinear) FLD analysis. Direct fracture phenomena occur mostly in follow operations, which are relatively easier to optimize during tryout.

2. Experimental methods for characterizing direct fracture phenomena are not standardized; they are therefore costly and nevertheless feature relatively high uncertainty.

Both of these points limit the incentive of engineers to use direct fracture models, as the overhead required many times does not counterbalance the expected reduction in the overall uncertainty levels. An effective way for resolving this dilemma is the reliable distinction of cases which are not prone to direct fracture from those which do exhibit this risk. As the majority of stamping operations fail under localized necking, this would reduce the number of cases requiring additional modelling drastically, making the cost of the latter feasible as a whole.

2. A two-step strategy for streamlining fracture prediction

The proposed implementation of the approach introduced in the previous section is depicted in Figure 1. The first step in the analysis is the carrying out of an FEM analysis based on standard material / failure data, which have been enhanced using an estimation of the fracture limits. A preliminary analysis is thus carried out on the part, which leads to two possible outcomes. Either, fracture phenomena are irrelevant for this particular case and the classical FLD analysis is sufficient to capture failure or additional fracture modelling is necessary to improve prediction.

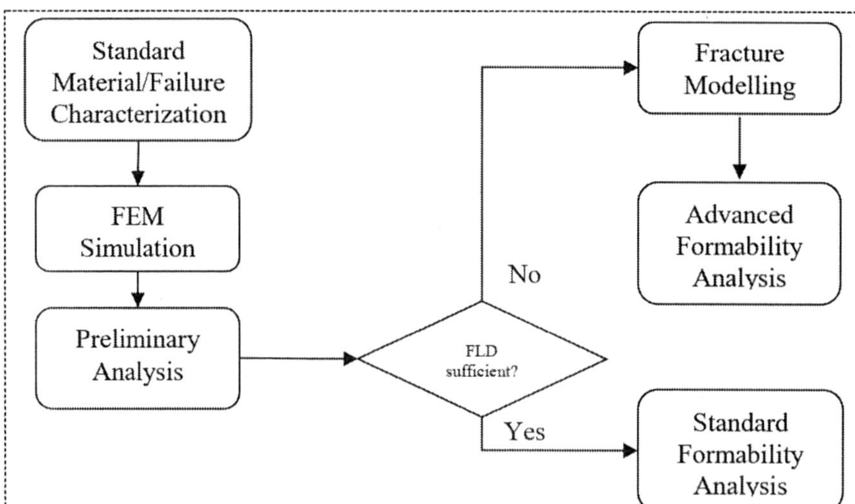

Figure 1. Advanced Formability Analysis Workflow

Clearly, the decisive component of this modelling approach is the preliminary analysis used to assess the necessity for additional modelling. A reasonable *lower bound fracture* (LBF) limit curve (see Figure 2) therefore needs to be estimated which effectively detects *the largest majority* of direct fracture cases even at the cost of occasional false positives. This way if the analysis indicates no direct fracture the user can reliably use the classical FLD for predicting failure.

2.1. FLC based estimation of fracture curves

The estimation of fracture limits is no trivial task. Direct fracture phenomena are strongly local and thus strongly dependent on the materials microstructure. There are nevertheless theoretical and heuristic considerations, which can be used to estimate a reasonable low bound fracture (LBF) curve based on four cardinal directions in the FLD. These points are equibiaxial tension (EB), plane strain (PS), uniaxial tension (UT) and pure shear (SH) as depicted in figure 2.

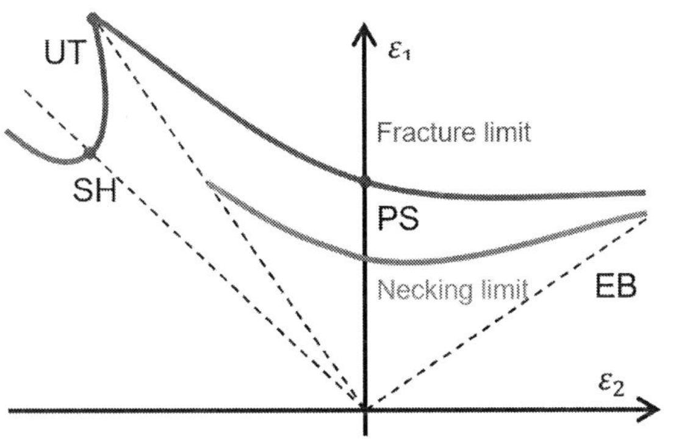

Figure 2. Limiting loading states in sheet stamping.

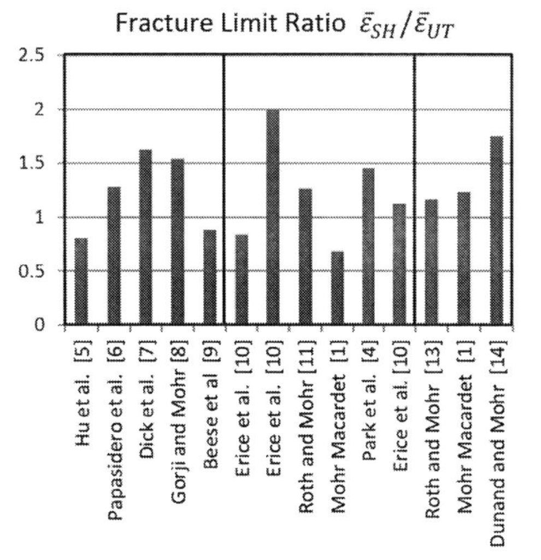

Figure 3. The ratio of eq. plastic strain at failure in shear to the same measure in uniaxial tension.

Fracture limit in equi-biaxial loading (EB). The starting point of the estimation is the EB loading direction. It is well-known from Nakajima experiments that fracture under this mode occurs almost immediately after necking in terms of major strain. It is therefore reasonable to assume that the localized necking major strain of the FLC will provide an acceptable lower bound for the fracture major strain as well. Note that the EB conditions as well as the UT condition are purely theoretical. Devising experiments that remain purely uniaxial or purely equi-biaxial is very challenging. In most tests carried out with sheet materials there will be an observable drift towards plane strain tension loading before fracture occurs.

2.1.1. Fracture limit in uniaxial tension loading (UT). The fracture limit for uniaxial tensile loading is assumed to have the same *equivalent plastic strain* as the EB loading. This corresponds to the assumption that fracture phenomena are essentially isotropic. Many of the popular fracture models such as Mohr-Coulomb and Hosford-Coulomb [1] are based on this assumption and will predict identical equivalent strains for UT and EB. This point will be elaborated based on the HC model which reads as follows in its simplest parametrization

$$\bar{\sigma}_{HF} + c(\sigma_1 + \sigma_3) = b \tag{1}$$

where $\bar{\sigma}_{HF}$ is the Hosford equivalent stress defined as

$$\bar{\sigma}_{HF} = \left[\frac{1}{2}((\sigma_1 - \sigma_2)^a + (\sigma_2 - \sigma_3)^a + (\sigma_1 - \sigma_3)^a)\right]^{\frac{1}{a}} \tag{2}$$

And a is a,b and c are material parameters. Introducing the uniaxiality condition ($\sigma_2 = \sigma_3 = 0$) in (2) and then in (1) we obtain:

$$\sigma_1 = \frac{b}{1+c} \tag{3}$$

The same result is obtained by enforcing the equibiaxiality condition ($\sigma_1 = \sigma_2$) together with the plane stress ($\sigma_3 = 0$) assumption.

2.1.2. Fracture limit in plane strain tension (PS). To estimate the plane strain tension point we will rely on the linearity of the fracture curve in the FLD between UT and EB. A linear behaviour in this region has been reported by many authors (see e.g. [2] and [3]). The limit strain is therefore:

$$\varepsilon_{1,PS} = \varepsilon_{1,EB}\left(1 - \frac{\varepsilon_{1,UT} - \varepsilon_{1,EB}}{\varepsilon_{2,UT} - \varepsilon_{2,EB}}\right) \tag{4}$$

$$\varepsilon_{2,PS} = 0 \tag{5}$$

2.1.3. Fracture limit in in-plane shear (SH). The estimation of the SH point is particularly uncertain. In fact it is strongly dependent on the material. Figure 3 depicts the shear limit strains reported by various authors in the literature. It is seen that there exist strong scatter even for similar or equivalent materials. In absence of better information the only conclusion that can be drawn from Figure 3 is that the fracture limit in in-plane shear is higher than in uniaxial tension for most of commonly used materials. This means that the strain limit in uniaxial tension also provides a reasonable lower bound for the in-plane shear case.

2.1.4. Edge fracture limits (EF). Edge fracture is a phenomenon that is frequently encountered especially with high strength steels and some aluminum alloys. The limiting strains for the edge formability are primarily dependent on the blanking process and on the resulting edge condition, which is mostly unknown at the simulation stage. A viable approach is to analyze edge crack conditions for various different edge conditions such as "laser cut", "sharp edge" and "worn edge". Heuristics can be applied for these limits relating them to the uniaxial fracture limits. A sound experimental base is in any case recommendable.

3. Results and Discussion

The proposed approach for generating a lower bound fracture curve estimate has been applied to the results published by Pack et al. [4].

3.1. Application to AA6016

Figure 4 illustrates the LBF curve in comparison to the Forming Limit Curve, whereas figure 5 compares the LBF with the fracture envelope as published in [4], plotted in the eq. fracture strain vs. stress triaxiality diagram. It is noted that the LBF estimation always remains, as by design, below the measured fracture curve. It is furthermore seen that the plane strain point is quite accurately predicted but the remaining points feature a larger safety margin up to about 33% as for the SH point.

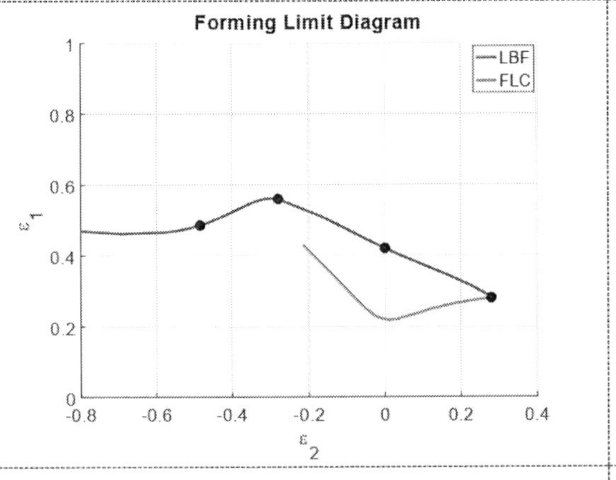

Figure 4. Representation of LBF in principal strain space for AA6016. The black circles represent the estimated limit strains.

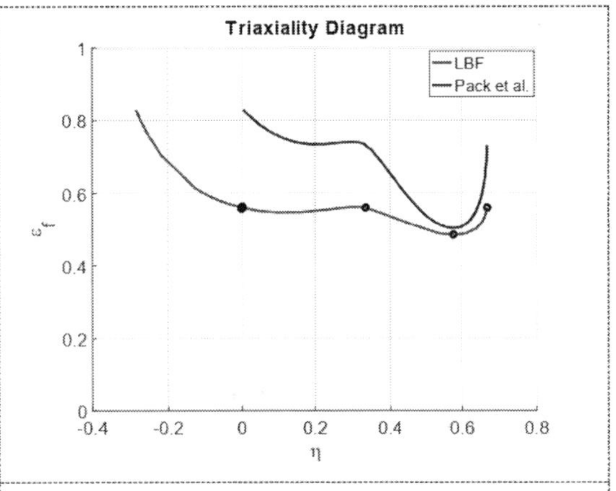

Figure 5. Representation of LBF in eq. fracture strain vs. triaxiality for AA6016 in comparison to the measured fracture envelope in [4].

This indicates that both the assumption of congruency of the necking and fracture points in EB and the linearity between EB and UT are quite strong. However, they err on the safe side and as such remain suitable for the type of analysis envisaged. If the EB limit were to be heuristically lifted an additional parameter would have to be introduced in the approach and the lower bound requirement would not be strictly guaranteed.

3.2. Application to DP980

A similar behavior can be seen in figures 6 and 7. Analogously to the AA6016 material the LBF curve for DP980 also consistently remains underneath the fracture envelope reported in [4].

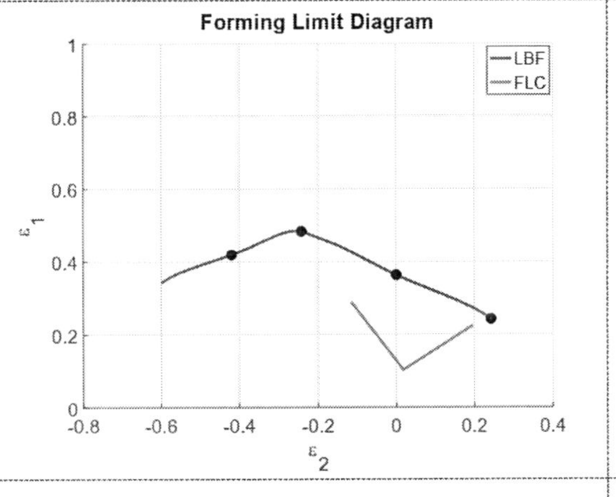

Figure 6. Representation of LBF in principal strain space for DP980. The black circles represent the estimated limit strains.

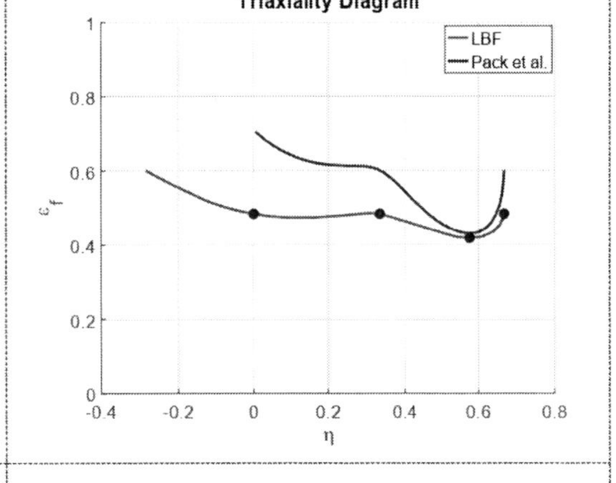

Figure 7. Representation of LBF in eq. fracture strain vs. triaxiality for DP980 in comparison to the measured fracture envelope in [4].

3.3. Validation with AA6016

Gorji [2] proposed the use of a triangle die to test for direct fracture phenomena not captured by the FLC. He reported that fracture at the die radius was observed for a die radius of 3mm, whereas the part could be drawn flawlessly until 45mm depth with a die radius of 5mm. The procedure has been replicated using a development version of AutoForm R8 in which the proposed LBF curve has been implemented. A damage accumulation approach of the form

$$D = \int \frac{d\bar{\varepsilon}}{\bar{\varepsilon}_f(\eta)} \tag{6}$$

has been used to account for path nonlinearities. For the material data and process conditions the reader is referred to the original publication.

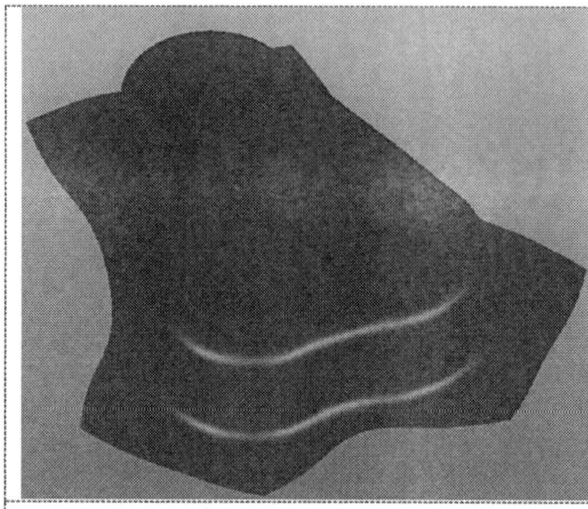

Figure 8. Triangle test die [2] with 5mm die radius evaluated with the LBF curve. No excessive damage observed.

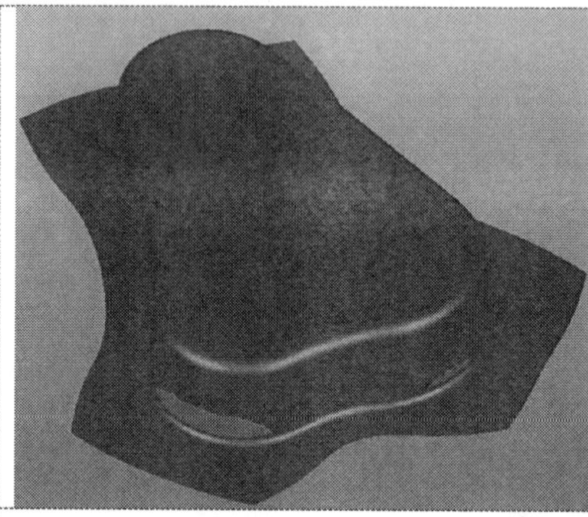

Figure 9. Triangle test die [2] with 3mm die radius evaluated with the LBF curve. Orange regions depict excessive damage.

Figures 8 and 9 respectively illustrate the results obtained for a 5mm die edge radius and 3mm die edge radius. In case of the edge radius of 5mm the LBF delivered a fracture free result. Referring to figure 1 we can than stop the analysis and avoid a deeper look into fracture phenomena. In case of figure 9, however, the results indicate that the damage accumulated towards the LBF will be critical. The simulation engineer should in this case invest additional characterization/ modelling efforts and evaluate the example with a more precise fracture curve in order to avoid any fracture occurrences.

4. Conclusions

Direct fracture phenomena such as surface cracks, edge cracks and shear cracks, are increasingly entering the daily industrial stamping practice, especially with low ductility materials such as high strength steels and aluminium. The experimental characterization of the corresponding limit strains is however nowadays not completely standardized. There is a large variety of different approaches and testing equipment, which lead to significant uncertainties in the measured fracture limits published din the literature. Put differently measuring fracture limits is expensive and the results obtained show high uncertainty. This creates a strong incentive to distinguish industrial cases which fail under standard conditions (splits due to localized necking) from those which are prone to fail under direct fracture. The proposed LBF approach aims taking a step in this direction. The method is effective in separating standard cases from cases requiring complex modelling. Considering that localized necking is still the most widespread mode of failure in stamping applications, this effectively reduces the characterization efforts to a small subset of stamping applications.

5. References

[1] Mohr, D., Marcadet, S.J., 2015. *Micromechanically-motivated phenomenological Hosford–Coulomb model for predicting ductile fracture initiation at low stress triaxialities*. International Journal of Solids and Structures 67, 40–55.

[2] Gorji, M., B.: *Instability and Fracture Models to Optimize the Metal Forming and Bending Crack Behavior of Al-Alloy Composites*, Diss. ETH Zurich, Nr. 23066, 2015.

[3] Isik, K., Silva, M.B., Tekkaya, A.E., Martins, P.A.F., 2014. Formability limits by fracture in sheet metal forming. Journal of Materials Processing Technology 214, 1557–1565.

[4] Pack, K., Tancogne-Dejean, T., Gorji, M.B., Mohr, D., 2018. *Hosford-Coulomb ductile failure model for shell elements: Experimental identification and validation for DP980 steel and aluminum 6016-T4*. International Journal of Solids and Structures, Special Issue with a Selection of Papers from the NUMISHEET 2016 Conference 151, 214–232.

[5] Hu, Q., Li, X., Han, X., Chen, J., 2017. *A new shear and tension based ductile fracture criterion: Modeling and validation*. European Journal of Mechanics - A/Solids 66, 370–386.

[6] Papasidero, J., Doquet, V., Mohr, D., 2015. *Ductile fracture of aluminum 2024-T351 under proportional and non-proportional multi-axial loading: Bao–Wierzbicki results revisited*. International Journal of Solids and Structures 69, 459–474.

[7] Dick, R.E., Yoon, J.W., 2018. *Plastic anisotropy and failure in thin metal: Material characterization and fracture prediction with an advanced constitutive model and polar EPS (effective plastic strain) fracture diagram for AA 3014-H19*. International Journal of Solids and Structures.

[8] Gorji, M.B., Mohr, D., 2018. *Predicting shear fracture of aluminum 6016-T4 during deep drawing: Combining Yld-2000 plasticity with Hosford–Coulomb fracture model*. International Journal of Mechanical Sciences 137, 105–120.

[9] Beese, A.M., Luo, M., Li, Y., Bai, Y., Wierzbicki, T., 2010. *Partially coupled anisotropic fracture model for aluminum sheets*. Engineering Fracture Mechanics 77, 1128–1152.

[10] Erice, B., Roth, C.C., Mohr, D., 2018. *Stress-state and strain-rate dependent ductile fracture of dual and complex phase steel*. Mechanics of Materials, IUTAM Symposium on Dynamic Instabilities in Solids 116, 11–32.

[11] Roth, C.C., Mohr, D., 2016. *Ductile fracture experiments with locally proportional loading histories*. International Journal of Plasticity 79, 328–354.

[12] Park, N., Huh, H., Lim, S.J., Lou, Y., Kang, Y.S., Seo, M.H., 2017. *Fracture-based forming limit criteria for anisotropic materials in sheet metal forming*. International Journal of Plasticity 96, 1–35.

[13] Roth, C.C., Mohr, D., 2014. *Effect of strain rate on ductile fracture initiation in advanced high strength steel sheets: Experiments and modeling*. International Journal of Plasticity 56, 19–44.

[14] Dunand, M., Mohr, D., 2011. *On the predictive capabilities of the shear modified Gurson and the modified Mohr–Coulomb fracture models over a wide range of stress triaxialities and Lode angles*. Journal of the Mechanics and Physics of Solids 59, 1374–1394.

IOP Conf. Series: Materials Science and Engineering **967** (2020) 012067 doi:10.1088/1757-899X/967/1/012067

Crystal plasticity as complementary modelling technique for improved simulations results of anisotropic sheet metal behaviour in forming processes

Sebastian Hirsiger[1], Bekim Berisha[1,2], Holger Hippke[1], Pavel Hora[1]
[1]ETH Zurich, Institute for Virtual Manufacturing, Tannenstrasse 3, 8092 Zurich, Switzerland
[2]inspire AG, inspire-ivp, Tannenstrasse 3, 8092 Zurich, Switzerland

E-mail: hirsiger@ivp.mavt.ethz.ch

Abstract. The accuracy of the simulation results in terms of metal sheet forming strongly depends on the capability of modelling the anisotropic material behaviour. In addition, predictive capabilities of the models are strongly influenced by the way how the constitutive model parameters are calibrated. Macroscopic models lean towards to become more complex in order to map the material behaviour more precisely. As consequence the amount and complexity of the experiments is increasing as well. In addition, it is well known that, some of the experiments, for example the equibiaxial compression test, are difficult to perform and therefore, a reasonable coupling of crystal plasticity (CP) modelling and macroscopic models is proposed. It is worth to mention that, in the domain of CP, arbitrary load cases are possible and therefore, any stress ratio of the yield criterion can be used for calibration. Prediction of anisotropic material behaviour of AA6016-T4 and DC05 sheets based on CP simulations were previously presented and compared with the macroscopic Yld2000-2d model. Their data set is now used for the calibration of the parameters of the macroscopic model, where in contrast to the classical procedure, the exponent of the yield locus is defined as a fitting parameter. The strain distributions predicted by the models have been compared with DIC-measurements of Nakajima samples. The predictive capabilities of the CP-based fitting procedure, compared to the classical fitting, are highlighted. Additionally, a comparison of the strain distribution prediction between all model variants is performed on a cruciform shaped deep drawing part. It underlines the importance of the correct prediction of the yield normal, as it is given by the crystal plasticity computation.

1. Introduction

In the era of climate change and increased sensibility for material efficiency the demand and requirements for more safe and light structures has led to more sophisticated material models and simulation tools. Nowadays it is state of art for most metal forming process like e.g. deep drawing, to use numerical simulations based on finite element software. In this context the complexity of the available constitutive models has increased a lot but only a small number of them are actually established in industry. In addition, the effort of the parameter determination has increased too, making them not appropriate for the daily use in industry, because of the required complex experiments and their evaluation. In this context, the focus of the research is the development of a new methodology of incorporating texture data into the set of available experimental data. It was already shown [1, 2] that proper calibrated CP models can be used to predict the metal sheet anisotropy. In this work we use the virtually created simulation results by CP to improve the model parameter estimation. This methodology is exemplary shown for two different commonly used metal sheets in industry. To be precise, the method is applied on the aluminium alloy AA6016-T4 with dominant Cube texture $\{001\}\langle 100\rangle$. The second material is a deep drawing steel, labeled with DC05 and has a strong γ-fiber $\{111\}//$ND texture. The novel methodology to calibrate the constitutive model parameters is tested in the prediction of strain distribution for selected NAKAJIMA samples. DC05 is additionally validated by predicting strain patterns on a deep drawn part.

2. Numerical methods

For the CP calculations the physically motivated phenomenological model from DAMASK [3] is used. For this work, dislocation glide is considered as exclusive source of plasticity.

2.1. Phenomenological crystal plasticity

The plastic velocity gradient $\mathbf{L}_\mathrm{p}$ is expressed in the sense of [4] as a product of the plastic shear rate $\dot{\gamma}$ with the slip system projection tensor by

$$\mathbf{L}_\mathrm{p} = \sum_{\alpha \in \mathcal{A}}^{N_\mathrm{s}} \dot{\gamma}^\alpha \mathbf{m}^\alpha \otimes \mathbf{n}^\alpha. \tag{1}$$

In equation (1), $\mathcal{A}$ denotes all active slip systems α's with $|\dot{\gamma}^\alpha| > 0$ and N_s the amount of theoretically possible slip systems for the specific unit cell type. The vector $\mathbf{m}^\alpha$ represents the direction of slip and $\mathbf{n}^\alpha$ the plane normal of the slip plane respectively. The evolution of the shear rate is modelled with a power-law after [5] with

$$\dot{\gamma}^\alpha = \dot{\gamma}_0 \left| \frac{\tau^\alpha}{\xi^\alpha} \right|^{m_\mathrm{slip}} \cdot \mathrm{sgn}(\tau^\alpha). \tag{2}$$

In equation (2) the shear stress τ acting on the slip system α acts as driving force for the phenomenological description of dislocation glide. The exponent m_slip describes the strain rate sensitivity (SRS) behaviour of the slip systems and the value of ξ^α denotes the strength or necessary resolved shear stress to start a dislocation. In the plastic domain, metals usually show a nonlinear work hardening behaviour. This process is phenomenologically described in the sense of CP with

$$\dot{\xi}^\alpha = h_{0,\text{slip-slip}} \sum_{\alpha'=1}^{N_\mathrm{s}} \left|\dot{\gamma}^{\alpha'}\right| \left|1 - \frac{\xi^{\alpha'}}{\xi^{\alpha'}_\infty}\right|^\mathrm{a} \mathrm{sgn}\left(1 - \frac{\xi^{\alpha'}}{\xi^{\alpha'}_\infty}\right) h_{\alpha\alpha'} \tag{3}$$

Due to the nature of equation (3), the value of ξ^α asymptotically increases until it reaches the saturation value of slip resistance $\xi^{\alpha'}_\infty$. The slope of this increase is controlled by several coefficients. $h_{0,\text{slip-slip}}$ is the reference slip-slip hardening parameter, it denotes the minimum stress increase of the interaction of two slip systems. $h_{\alpha\alpha'}$ is dimensionless latent hardening factor, which denotes the specific stress increase factor by the interaction of two distinct slip systems. The value of this interaction is usually taken with $h_{\alpha\alpha'} = 1.0$ for a coplanar slip system interactions and $h_{\alpha\alpha'} = 1.4$ for non-coplanar interactions [6].

2.2. Mechanical boundary problem solver

To numerically solve the conservation of linear momentum for a representative volume element (RVE), the highly efficient spectral solver of DAMASK is used . This solver is based on the fast Fourier transformation (FFT) and is considerably faster than the finite element method (FEM) [7]. FFT requires that the applied boundary conditions on the RVE are of periodic nature, as the FFT only works for periodic functions.

3. Calibration of macroscopic yield criteria by virtual experiments

Based on calibrated CP models for AA6016-T4 and DC05, virtual experiments with varying boundary conditions are performed. Details about the calibration of the phenomenological CP parameters are given in [8]. In the plane stress yield domain 72 biaxial simulations are performed. In addition 18 RVE simulations loaded by uniaxial tension and with rotated texture between rolling and transverse direction are performed. For all simulations the stress state $\boldsymbol{\sigma}$ and the plastic strain rate tensor $\mathbf{D}_p$ are extracted at the same level of plastic strain energy density. The obtained simulation data is then used in a least squares method to calibrate the Yld2000-2d [9] model parameters. The classical calibration of Yld2000-2d is done by only using the experimental data of three uniaxial tensile tests (σ_0, σ_{45}, σ_{90}) and an equibiaxial experiment (σ_b), in combination with their measured plastic strain anisotropy values (r_0, r_{45}, r_{90}, r_b). These values are used to estimate the values of $\alpha_1 - \alpha_8$, where the exponent a is fixed with respect to crystalline structure (fcc: $a = 8$, bcc: $a = 6$). To calibrate all constitutive model parameters simultaneously, including the exponent a, the equation (4) is introduced.

$$\begin{aligned} e_{\text{Calibration}}(\alpha_i, W_{\text{pl}}) = {} & w_\sigma \left\| \frac{\overline{\sigma}\,(\alpha, W_{\text{pl}}) - \langle \overline{\sigma} \rangle\,(W_{\text{pl}})}{\sigma_0} \right\|_2 \\ & + w_r \left\| \overline{r}\,(\alpha, W_{\text{pl}}) - \langle \overline{r} \rangle\,(W_{\text{pl}}) \right\|_2 \\ & + w_\theta \left\| \frac{\sigma_\theta\,(\alpha, W_{\text{pl}}) - \langle \sigma_\theta \rangle (W_{\text{pl}})}{\sigma_0} \right\|_2 \\ & + w_{\text{dir}} \left\| \frac{d\varepsilon_2}{d\varepsilon_1}\,(\alpha, W_{\text{pl}}) - \left\langle \frac{d\varepsilon_2}{d\varepsilon_1} \right\rangle (W_{\text{pl}}) \right\|_2 \end{aligned} \tag{4}$$

Please note, that $\langle \bullet \rangle$, denotes the volume average value of the RVE quantity $\bullet$. Equation (4) contains four terms, where the first one describes the error of the biaxial stress results. The second term takes the plastic strain ratios into account. The third one quantifies the error of the observed uniaxial flow stresses in different sheet directions. The novel fourth term, takes the plastic flow direction on the yield surface into account. All terms can be individually weighted (w_i), however for the sake of simplicity all weights were chosen to be identical. The results of the CP simulations and the novel calibrated constitutive model for AA6016-T4 at a level of $\varepsilon_{pl} = 0.0219$ are illustrated in Figure 1, where the results are compared to the classical macroscopically model variant of [10]. The calculated stress points (Fig. 1a) from the CP simulations show small deviations compared to the macroscopically calibrated constitutive model, whereas the experimental data is properly matched. Most pronounced are the differences in the shape of the yield locus itself, which can be observed in the diagram for the plastic flow direction (Fig. 1c). This also applies to the prediction of the plastic strain anisotropy (Fig. 1b), where the predicted shape is similar to the macroscopic one but at slightly higher values. In terms of model parameters (Tab. 1), it can be observed that the biggest difference is given in the value of the exponent a. The same methodology has been applied for DC05. The results at a level of $\varepsilon_{pl} = 0.0274$ are illustrated with Figure 2. It can be observed that the error for the plastic strain anisotropy (Fig. 2b) is slightly bigger than previously seen in AA6016-T4. The yield locus and the plastic flow direction show almost no difference. In terms of anisotropy

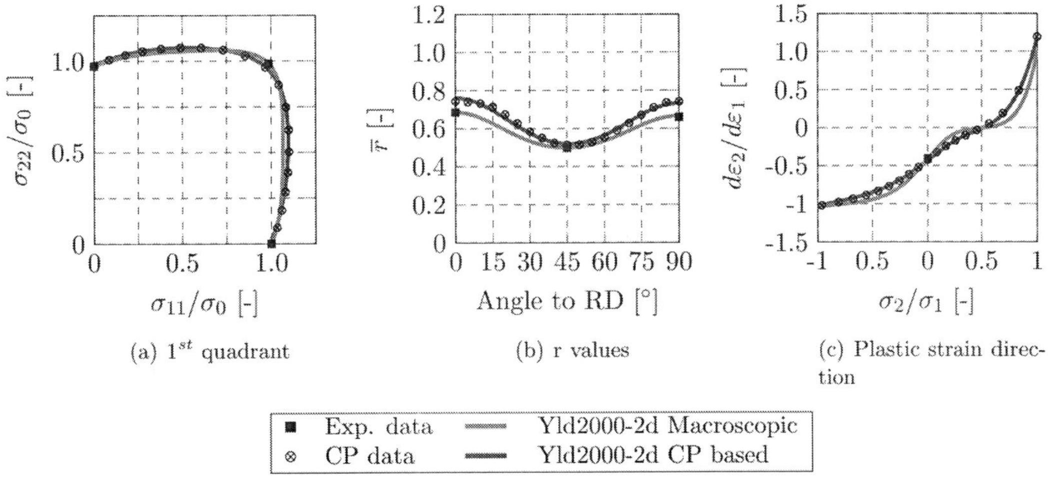

Figure 1. Response of calibrated Yld2000-2d model for AA6016-T4.

Type	a	α_1	α_2	α_3	α_4	α_5	α_6	α_7	α_8
Macroscopic	8	0.947	1.107	0.961	1.032	1.021	1.013	0.967	1.153
CP based	5.697	0.938	1.030	1.063	1.046	1.022	0.969	0.928	1.083

Table 1. Yld2000-2d parameters for AA6016-T4.

coefficients Table 2 shows that no big differences between the macroscopic and CP based variant exist. The assumption of an exponent value of $a = 6$ seems to be sufficiently good. For DC05 also the Yld2000-2d model variants with strain dependent coefficients as introduced in [11, 12], were calibrated. For the sake of brevity their results in terms of yield locus prediction are not

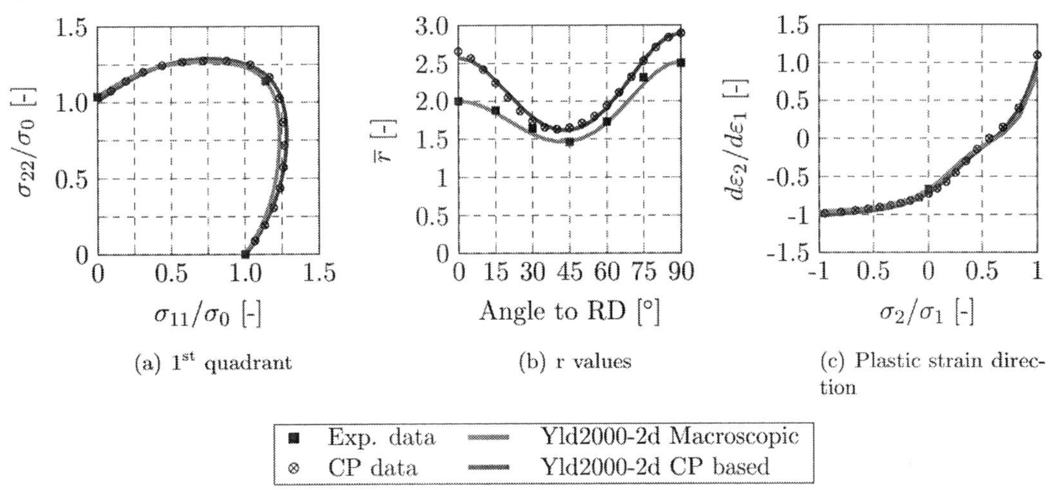

Figure 2. Response of calibrated Yld2000-2d model for DC05.

Type	a	α_1	α_2	α_3	α_4	α_5	α_6	α_7	α_8
Macroscopic	6	1.084	0.982	0.846	0.881	0.908	0.836	0.972	0.975
CP based	5.454	1.030	1.095	0.875	0.871	0.880	0.781	0.983	0.979

Table 2. Yld2000-2d parameters for DC05.

presented in this work, however they can be accessed in [8]. The performance of the strain-dependent Yld2000-2d variants can be summarized as being able to represent the deformation of the yield locus more accurately at higher plastic strains, since the anisotropic strain hardening effect in DC05 is significant.

4. Validation

The calibrated macroscopic yield criteria are validated in terms of strain distribution prediction in finite element (FE) simulations. FE simulations of a standardized ductility test for metal sheets [13] with so-called NAKAJIMA specimens are used. Experimental strain distribution was gathered for AA6016-T4 by [10] and for DC05 by [11]. Their results and the performance of the classically derived model parameters will be compared to the ones obtained by the virtual experiments. The FE simulation is set-up in LS-DYNA[1] using standard fully integrated shell element for the blank discretization, while having seven integration points over the sheet thickness. For both materials, the model performance is additionally validated using a deep drawing process with a cruciform shaped die geometry. The setup of the deep drawing process is illustrated in Figure 3.

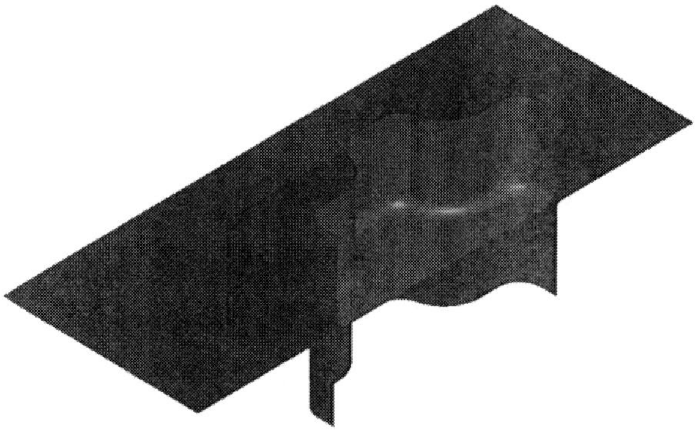

Figure 3. Model illustration of deep drawing processes with a cruciform geometry, with faded out blankholder.

4.1. Major strain distribution on NAKAJIMA *specimens*

The strain distribution is exclusively analysed on the sample types of B20, B100 and B200. For the sake of brevity only the simulation results of major strain (ε_1) are presented. The strain distributions are evaluated at several punch heights, along the center line of the sample itself. The results for AA6016-T4 (Fig. 4) clearly show the improved strain prediction of the

[1] www.lstc.com

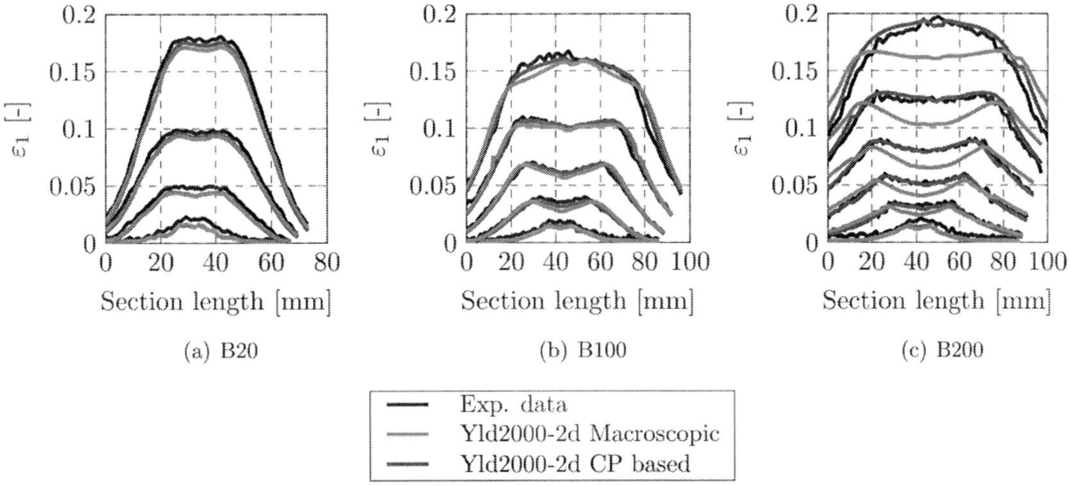

Figure 4. Major strain ε_1 distribution evaluated at several punch levels for AA6016-T4 [14].

CP based model coefficients. In particular, the results for the B200 sample (Fig. 4c) show the substantially better performance. In the other two sample types the differences between the model variants are smaller. The results for DC05 (Fig. 5) show a more complex behaviour. For B20 no differences are observable, for the B100 it can be seen that at the highest punch level the model with strain dependent coefficients works best. Similar results can be seen in the B200 sample. However the experimental data is suffering from strain localization outside the specimen center.

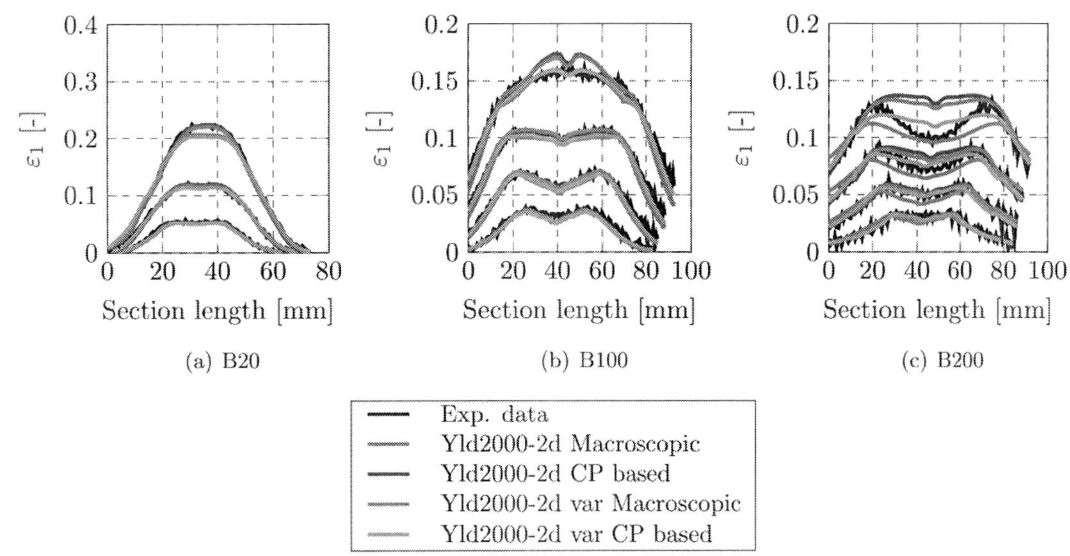

Figure 5. Major strain ε_1 distribution evaluated at several punch levels for DC05 [14].

IOP Conf. Series: Materials Science and Engineering **967** (2020) 012067 doi:10.1088/1757-899X/967/1/012067

4.2. Strain distribution of the drawn cruciform profile along a section

Simulations results are compared with experimental data at a parallel cut to the base of the drawn part, at a height of 5mm. Unfortunately for AA6016-T4 no experimental data is available, therefore only a qualitative comparison is made. It is seen for both materials, that the biggest differences occur when the minor strain component is positive. This coincidences with the range, where the predicted plastic flow direction on the yield locus has the biggest difference between both model variants.

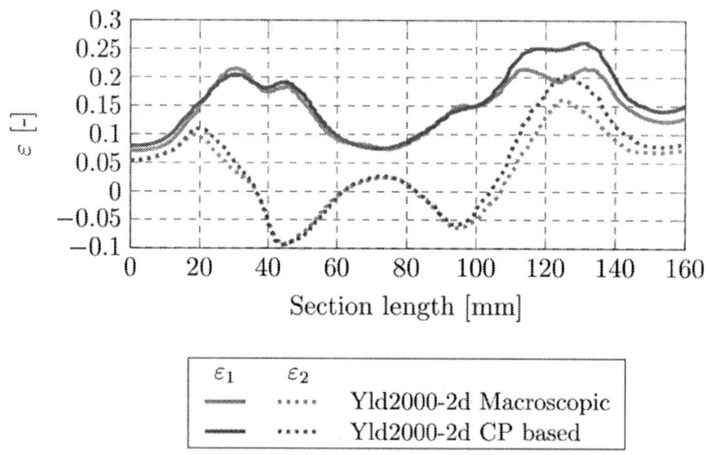

Figure 6. Simulation results for AA6016-T4 from a section parallel to punch head surface, 5.0mm above the base.

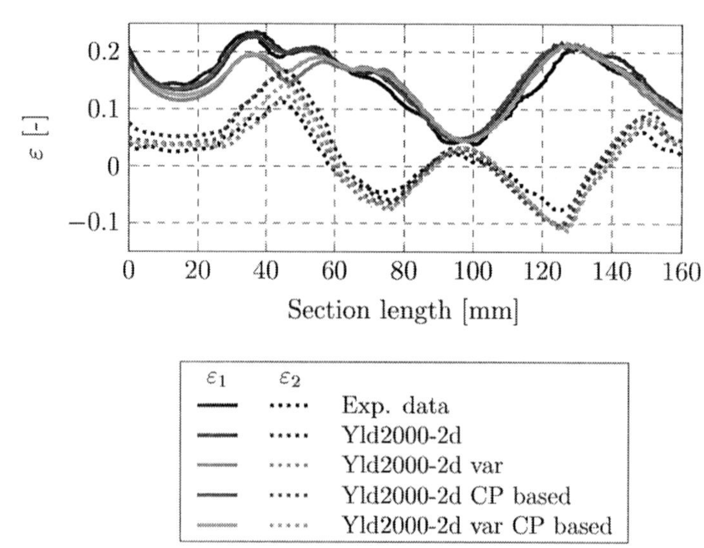

Figure 7. Simulation results for DC05 from a section parallel to punch head surface, 5.0mm above the base.

IOP Conf. Series: Materials Science and Engineering **967** (2020) 012067 doi:10.1088/1757-899X/967/1/012067

5. Conclusion

The presented work illustrates the importance of an accurate description of the plastic flow direction in macroscopic constitutive models. CP simulations based on EBSD data allow us to consider the influence of grain orientations on plastic flow. Therefore, a more accurate calculation of the plastic flow direction is given if standard, not highly flexible flow criteria such as the Yld2000-2d criterion are still used. A fitting methodology was presented to determine the anisotropic parameters as well as the yield locus exponent while additionally using the information of plastic flow direction, predicted by CP calculations. Based on this methodology, it was shown that it is not recommended to use an exponent of 8 for aluminum alloys, whereas the suggestions for steel with $a = 6$ is sufficiently appropriate.

References

[1] Hirsiger S, Berisha B, Raemy C and Hora P 2018 *Journal of Physics: Conference Series* **1063** 012056 ISSN 1742-6588
[2] Zhang H, Diehl M, Roters F and Raabe D 2016 *International Journal of Plasticity* **80** ISSN 07496419
[3] Roters F and et al 2019 *Computational Materials Science* **158** 420–478 ISSN 09270256
[4] Rice J 1971 *Journal of the Mechanics and Physics of Solids* **19** 433–455 ISSN 00225096
[5] Hutchinson J W 1976 *Proceedings of the Royal Society A: Mathematical, Physical and Engineering Sciences* **348** 101–127 ISSN 1364-5021
[6] Peirce D, Asaro R and Needleman A 1982 *Acta Metallurgica* **30** 1087–1119 ISSN 00016160
[7] Eisenlohr P, Diehl M, Lebensohn R and Roters F 2013 *International Journal of Plasticity* **46** ISSN 07496419
[8] Hirsiger S 2020 *A virtual testing method based on crystal plasticity for macroscopic constitutive models* (ETH Zürich) ISBN 978-3-907234-07-5
[9] Barlat F and et al 2003 *International Journal of Plasticity* **19** 1297–1319 ISSN 07496419
[10] Gorji M 2015 *Instability and Fracture Models to Optimize the Metal Forming and Bending Crack Behavior of Al-Alloy Composites* (ETH Zürich) ISBN 978-3-906327-20-4
[11] Peters P R 2015 *Yield functions taking into account anisotropic hardening effects for an improved virtual representation of deep drawing processes* (ETH Zürich)
[12] Peters P R, Manopulo N and Hora P 2012 *Proceedings of IDDRG 2012* pp 111 – 120
[13] DIN EN ISO 12004-2:2019-06, Metallic materials - Sheet and strip - Determination of forming-limit curves - Part 2: Determination of forming-limit curves in the laboratory
[14] Hirsiger S, Berisha B, Hippke H and Hora P 2019 *Proceedings of COMPLAS 2019*

IOP Conf. Series: Materials Science and Engineering **967** (2020) 012068 doi:10.1088/1757-899X/967/1/012068

Potentials of an adaptive blank positioning to control material and process fluctuations in deep drawing

David Briesenick[a*], Mathias Liewald[a] and Patrick Cyron[a]

Institute for Metal Forming Technology, University of Stuttgart
Holzgartenstr. 17, 70174 Stuttgart, Germany

[*]david.briesenick@ifu.uni-stuttgart.de

Abstract. Series production of sheet metal components in press shops is permanently subjected to an increasing pressure in time, cost and quality. Varying sheet metal properties due to batch fluctuations and changing process conditions lead to an increase of try out time phase and scrap rate, resulting in a demand for adaptive control strategies for deep drawing. Today, these challenges are met by integrating sensors and actuators into tool structure which measure and control part quality, mainly through blank draw-in. However, such tool-based control systems require a complex and cost-intensive modification of the existing die or press technology. Against this background, the active adjustment of blank position prior the deep drawing process realized with an intelligent transfer and positioning system as a promising approach towards economic and technical aspects. This paper deals with blank positioning and its sensitivity to quality-related failures of deep drawn sheet metal components like splits and wrinkles. Therefore, a numerical study was conducted on a deep drawing process of an exemplary structural part geometry, wherein a typical fluctuation of material parameters and variation of local friction conditions was simulated. Subsequently, correlations between process disturbance and blank draw-in were elaborated, thus using local draw-in values as controlled variables for a closed loop system. Simulation results showed that the manipulating parameter, i.e. the blank position prior deep drawing, reveals a significant influence on the draw-in and therefore on the component´s quality. An essential finding of this study is the numerical proof of concept for this new deep drawing control strategy, demonstrated for different process conditions. Finally, disturbances in the deep drawing process considered could be successfully controlled by adapting the blank position.

1. Introduction

In large-scale production, quality of deep drawn sheet metal components is substantially affected by material and process fluctuations. Controlling such production fluctuations therefore poses a major challenge for the robust and economical design of respective series forming processes [1]. In this respect, process monitoring strategies and active control systems for forming processes have been developed in the course of various research activities and constantly advanced during the recent decades.

Implementation of closed loop control or plain monitoring requires measurement of quality related control variables, allowing direct or indirect identification of commonly used failure criteria such as splits and wrinkles. Here, indirect identification of splits and wrinkles is enabled, for example, by measuring process forces or blank draw-in. Latter is frequently used in practice and can be detected by using laser-optical [2], tactile [3], inductive [4] or piezo-resistive [5] measuring methods. Reliable detection of component defects via force measuring depends heavily on the positioning of the sensors and is limited for irregularly shaped components [6]. On contrary, measuring the part wall stress by using piezo force transmitters integrated beneath tool surface allows reliable assessment of emerging splits and wrinkles [7], [8]. Split detection during deep drawing can additionally performed by using highly sensitive sensors for noise emissions [9], [10]. In-situ measurement of wrinkle formation in part wall and flange areas, however, requires sophisticated sensor concepts applying for example eddy current [11] or strain gauge sensors [12], [13] and is therefore today only used in research applications.

Tool based measuring devices, such as those mentioned above, are usually subjected to high mechanical and thermal loads and can only detect small gauge regions. Thus, additional optical measuring technology is regularly integrated prior to or directly in subsequent operations in order to detect blank draw-in with wide resolution [14], [15]. An immediate classification of part quality can also be based on visual and contactless sensors such as thermal imaging cameras [16] or monochrome cameras [17]. Common practice in press shops is nevertheless a visual inspection by workers at the end of the line or off-line sampling [15].

Active control interventions for deep drawing mainly address adjustment of the sheet metal blank's draw-in. Therefore, different actuator concepts have been developed in the past years which are used to influence restraining forces in flange areas. In this respect, segment elastic blankholders and correspondingly placed hydraulic pistons enable a customized pressure distribution [18]. Height adjustable draw beads also provide a sensitive but technologically complex controllability of restraining forces [19]. Active spacers, which are mounted on the binder along the blank boundary, allow a controlled bypass of blankholder forces and thus a local controlling of material flow [20]. A comprehensive summary of sensor and actuator concepts for closed loop process control in press shops is given by [1], [6] and [21].

Variation of blank position prior to deep drawing is usually classified as process uncertainty and therefore prevented in most tool structures by using gages. However, the study presented in this paper aims to identify existing correlations between an adaptive blank position and part quality related failure criteria such as splits and wrinkles in order to use them for active process control.

2. Modelling and numerical study

The numerical study presented here included a sensitivity analysis and an exemplary compensation of process disturbances for the deep drawing process of the complex sheet metal component "Mini-Tunnel". For modelling, simulating and evaluating this deep drawing process considered, commercial FEM code was used in combination with the AutoForm R8 user interface. The tooling setup consisted of a multiple curved blankholder, a tiered die face with different drawing depths and an octagonal trimmed blank (see Figure 1a)). Process fluctuations such as locally increased tool wear or inhomogeneous lubrication were modelled by defining two exemplary local friction areas on the punch surface. Here, the local friction coefficients $\mu_{concave}$ and μ_{convex}, which were assigned to the punch radii as shown in Figure 1a), infact diverged from the standard value $\mu_{global} = 0.15$. Figure 1 b) further shows the geometric parameter space within which the blank position before deep drawing was varied during the numerical investigations. While the shape of the blank maintained constant, center of inertia was varied between + 6 mm and -6 mm in x-direction, respectively in y-direction (see Table 2).

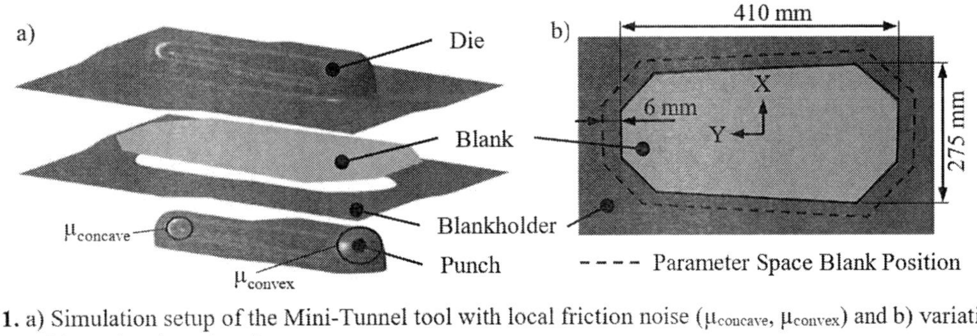

Figure 1. a) Simulation setup of the Mini-Tunnel tool with local friction noise ($\mu_{concave}$, μ_{convex}) and b) variation of blank position prior deep drawing.

For the numerical investigations, cold rolled mild steel CR4 having a sheet thickness of t = 1 mm was selected as blank material. Properties of this sheet metal material are summarized in Table 1 and available in current standard database of AutoForm R8. To simulate process related batch variations of the material, fluctuations were estimated for the hardening exponent n (15 %) and the anisotropy coefficient r_m (40 %) according to [22]. The other material parameters maintained constant in order to limit non-identifiable noises and cross correlations and thus to directly assign calculated process influences to the simulation parameters under consideration.

Table 1. Material properties of CR4 and variation of hardening exponent n and anisotropy r_m according to [22].

Material	t (mm)	E (GPa)	Yield Strength (MPa)	UTS (MPa)	U.E. (%)	n	r_m
CR4	1	210	146	286	25.4	0.22 (15 %)	1.87 (40 %)

Gained simulation results discussed in the following chapter were obtained by a numerical study in which both the local friction conditions and the described material parameters (n and r_m) were varied within mentioned limits in Table 2. The parameter setup was based on a Latin Hypercube Sampling (LHS) with over 100 runs and implemented in AutoForm-Sigma. For evaluation, result variables such as draw-in, splits and wrinkles were analyzed with regard to their correlation to design (blank position) and noise variables (r_m, n, $\mu_{concave}$, μ_{convex}). As reference, a baseline simulation was run with nominal process parameters (see Table 2).

Table 2. Process parameters and corresponding variations for deep drawing of a Mini-Tunnel.

Parameter	Nominal	Variation
Local friction $\mu_{concave}$, μ_{convex} (-)	0.15	0.05 ... 0.25
Blankholder force F_{BH} (kN)	530	constant
Blank position X-direction (mm)	0	-6 ... +6
Blank position Y-direction (mm)	0	-6 ... +6

3. Results and discussion

3.1. Sensitivity of process variations on blank draw-in

The above mentioned setup of the numerical study was initially evaluated in terms of correlations between process or material variations and the blank draw-in after deep drawing. Therefore, friction and material parameters were varied individually and independently within the boundaries defined in Table 1 and 2. Subsequently, the simulation results obtained via this parameter setup were assessed regarding calculated blank draw-in and measurement distance between blank edge and punch opening line (POL). Figure 2 reveals the influence of local changes in friction and material batch variations on the blank draw-in at four designated positions. This was done by varying only one parameter at a time while the others remain constant. In particular, it can be seen that the friction conditions at the large punch radii (μ_{convex}) lead to a wide scatter of draw-in in the straight part areas (see Position 2 and 4). In worst case, the asymmetric part geometry combined with wide variation of friction conditions can cause edge movements beyond the punch opening line, thus resulting in scrap. Another remarkable finding is the impact of the fluctuating anisotropy (r_m) and material hardening behaviour (n-value), which can also lead to critical draw-in in straight part areas (see Position 2 and 4). Summarizing, the simulation results presented here reveal the sensitivity of local friction conditions and material fluctuations on the blank draw-in during deep drawing, which may lead to component failure. For this reason, an adaptive control of the blank positioning is recommended to compensate the mentioned process fluctuations.

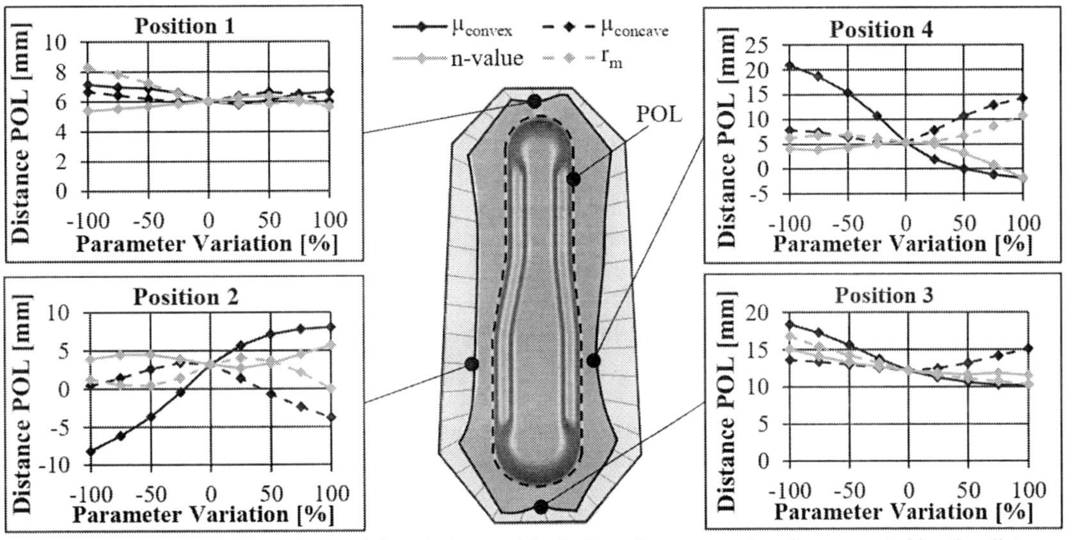

Figure 2. Influence of process and material variation on blank draw-in measured and represented by the distance between blank edge and Punch Opening Line (POL).

3.2. Sensitivity of blank position on part quality

Further, on in the numerical study, the impact of blank positioning prior to deep drawing on the part quality was investigated. To identify existing correlations between blank position and part quality, high-stress part areas were first identified (see Figure 3 a), b)). Subsequently, the simulation results obtained were evaluated in terms of split and wrinkle formation, especially in high-stress areas. For all simulation runs (m) and each parameter value (x_i), maximum failure values (y_i) were extracted and accumulated to expected values $\overline{x}, \overline{y}$. Afterwards, the Pearson's linear correlation coefficient r was calculated according to formula (1) in order to determine strong positive (+1) or negative (-1) correlations between input and output variables.

$$r = \frac{\sum_{i=1}^{m}(x_i-\overline{x})(y_i-\overline{y})}{\sqrt{\sum_{i=1}^{m}(x_i-\overline{x})^2 \sum_{i=1}^{m}(y_i-\overline{y})^2}} \quad (1)$$

Regarding splits, Figure 3 c) reveals a pronounced sensitivity to the blank positioning in X-direction in the blank area S-2 and to positioning in Y-direction in S-1. For example, a displacement of the blank in positive X- and Y-direction leads to a risk of splits in S-2 but increases safety margin in area S-1. On the opposite, wrinkling occurs reciprocally in W-1 and W-2 when moving the blank in X-direction. This behaviour corresponds to the state of knowledge and proves the validity of simulation but raises questions whether an optimal blank position exists for different process conditions.

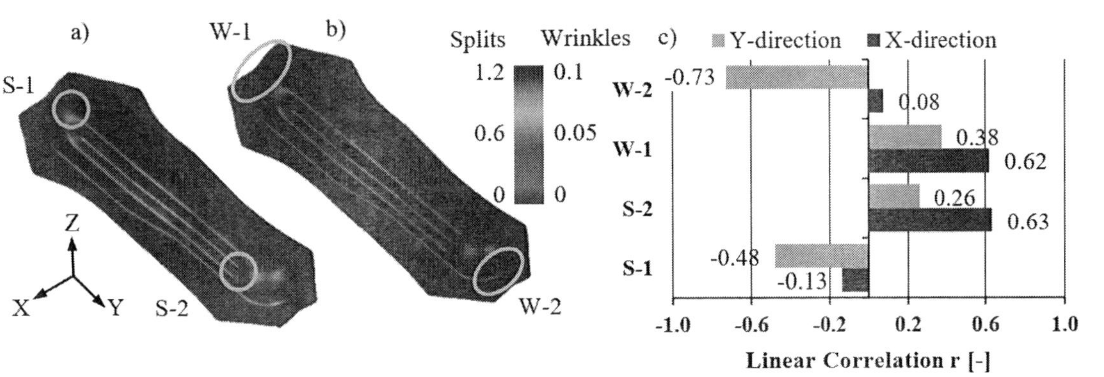

Figure 3. Impact on a) splits and b) wrinkles of blank position prior deep drawing and c) corresponding linear correlation coefficients.

3.3. Exemplary disturbance and process control

For assessing the potential of adaptive blank positioning with regard to forming process control, a metamodel-based optimization study was performed using exemplary practical use cases. The parameter variations assumed in these cases are described in the following and refer to nominal settings already listed in section 2. Firstly, it was assumed that due to tool wear and lack of lubrication, friction ($\mu_{concave}$) is locally increased up to 60 % in the tiered punch area. Secondly, batch change was considered to cause fluctuations in the anisotropy r_m of up to 6.1 % and in the hardening exponent n of up to 1 %. Compared to the nominal process design (Figure 4 a)), such process and material fluctuations assumed do result in part defects like splits and excessive thinning localized on the tiered part geometry, as shown in Figure 4 b). Afterwards, a new blank position was searched to compensate the above mentioned defects based on a calculated metamodel and the sensitivities received before. Figure 4 c) shows the simulated forming results for this adapted blank position, whereby the blank is moved 3.5 mm in negative X-direction and 1.5 mm in positive Y-direction. This displacement leads to a change of restraining forces in the flange area and entails different strain paths. Consequently, the risk of splits and excessive thinning is strongly reduced and the process is stabilized again. Furthermore, part quality relevant shape deviations due to springback are kept constant at a maximum of 0.9 mm at the left straight part area. In addition, the edge of the blank still provides a safety margin towards POL and part quality is ensured. In the end, the flange area reveals a distribution of thickening and compression comparable to the nominal setup.

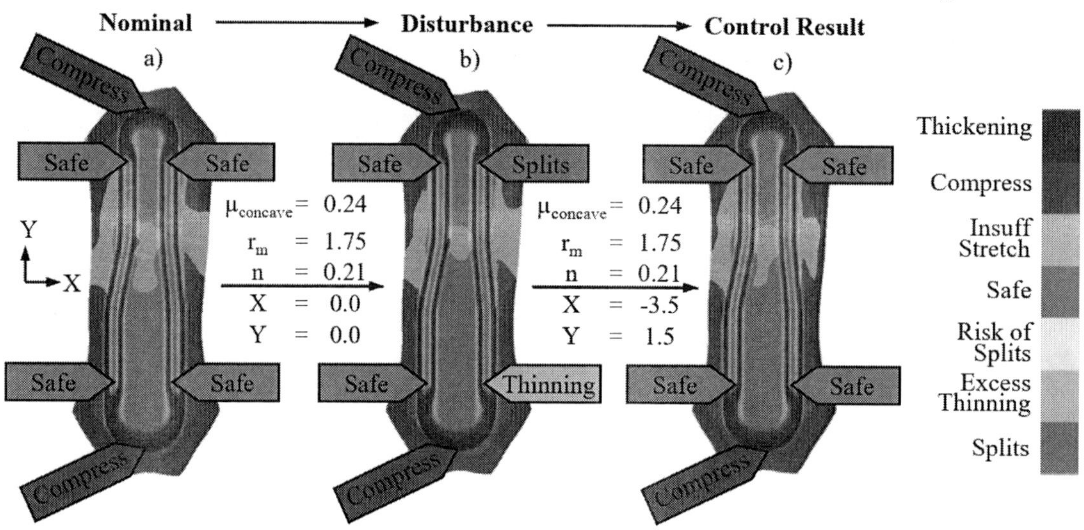

Figure 4. Nominal process design a) and the impact of process variations b) and blank positioning c) on the forming result.

4. Conclusions and outlook

The numerical study presented in this paper demonstrated a new approach for controlling the deep drawing processes of complex sheet metal parts using adaptive blank positioning. To identify correlations between part defects (split, wrinkles) and blank position prior deep drawing, a sensitivity analysis based on numerical studies was carried out, taking into account commonly appearing material and process fluctuations. Overall, a dominant and sensitive correlation was observed between blank position and failure criteria in critical part areas of the given tool geometry. Based on these findings and as a proof of concept, exemplary process disturbances were numerically induced and compensated by a metamodel based optimization of blank positioning. Further, the investigations revealed and confirmed a detectable influence of commonly process disturbances on the blank draw-in. Thus, the manipulating parameter blank position and a simple and state of the art measurement of controlled variable blank draw-in provide a solid basis for such applicable closed loop system in future.

Figure 5 shows concepts of practical implementations of such a new control approach. Today, many press lines are already equipped with flexible transfer systems such as multi-axis robots which could be used to easily adapt blank position prior to the deep drawing process (Figure 5 a)). In general, robust and high productivity is ensured by fixed gages installed along blank circumference. According to Figure 5 b), the implementation of the new control approach for the given blank guide system could be achieved by using uniaxial movable and cooperating gages. The proposed control system is able to adapt blank position prior to deep drawing and may even react automatically to changes of blank size, by using control concepts as analytical models or artificial neural networks. In comparison to conventional tool based control systems, an adaptive blank positioning system promises a cost-effective, insignificant stressed and tool independent applicable solution. In future work, extended numerical studies and experiments for different part geometries need to demonstrate the viability of the proposed control strategy. Development of a blank positioning system like the below proposed adaptive gages enables endurance testing to evaluate the potentials of scrap reduction.

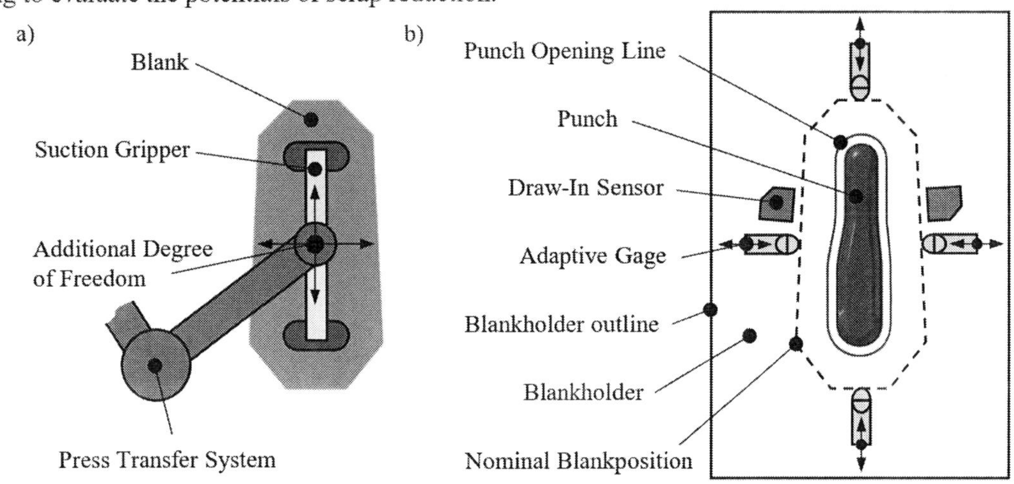

Figure 5. Concepts of the implementation of adaptive blank positioning with a) press transfer system and in case of gages b) a linear movable and controlled blank guiding system.

References

[1] Allwood, J.M. *et al.*, 2016, Closed-loop control of product properties in metal forming, CIRP Annals - Manufacturing Technology, vol. 65, no. 2, pp. 573–596.
[2] Munser, R., Jacubasch, A., Wagner, U., 2004, Messen beim Pressen, wt Werkstattstechnik onlline, vol. 94, pp. 544–545.
[3] Straube, O., 1994, Untersuchungen zum Aufbau einer Prozeßregelung für das Ziehen von Karosserieteilen, Dissertation, Technische Universität Berlin.
[4] Forstmann, U., 2000, Induktive Wegsensoren zur Überwachung und Regelung des Blecheinzugs beim Tiefziehen, Dissertation, Technische Universität Berlin.
[5] Biehl, S., Staufenbiel, S., Hauschild, F., Albert, A., 2010, Novel measurement and monitoring system for forming processes based on piezoresistive thin film systems, Microsystem Technologies, vol. 16, pp. 879–883.
[6] Beck, S., 2004, Optimierung der Zargenspannung beim Ziehen unregelmäßiger Blechformteile, Dissertation, Universität Stuttgart.
[7] Blaich, C., 2012, Robuster Tiefziehprozess durch Erfassung und Optimierung der örtlichen Bauteilqualität, Dissertation, Universität Stuttgart.
[8] Barthau, M., Liewald, M., Held, C., 2017, Improved process robustness by using closed loop control in deep drawing applications, Journal of Physics: Conference Series, vol. 896, no. 1.

[9] Haupt, H., 2003, Ein auf der Schallemissionsanalyse basierendes Verfahren zur Risserkennung in Umformprozessen, Dissertation, Universität Paderborn.
[10] Behrens, B.A., Hübner, S., Wölki, K., 2017, Acoustic emission—A promising and challenging technique for process monitoring in sheet metal forming, Journal of Manufacturing Processes, vol. 29, pp. 281–288.
[11] Hengelhaupt, J., Vulcan, M., Darm, F., Ganz, P., Schweizer, R., 2006, Robust Deep Drawing Process of Extensive Car Body Panels, in *Neuere Entwicklungen in der Blechumformung*, pp. 277–304.
[12] Han, F., Radonjic, R., 2015, New approach for wrinkle prediction in deep drawing process, Key Engineering Materials, vol. 639, pp. 459–466.
[13] Liewald, M., Han, F., Radonjic, R., 2015, New criterion for prediction of the wrinkle formation in deep drawing process, Key Engineering Materials, vol. 651–653, pp. 71–76.
[14] Kraft, M., Bürgel, U., 2017, Novel concept for measurement of global blank draw-in when deep drawing outer skin automotive components, Journal of Physics: Conference Series, vol. 896, no. 1.
[15] Maier, S.J., 2018, Inline-Qualitätsprüfung im Presswerk durch intelligente Nachfolgewerkzeuge, Dissertation, Technische Universität München.
[16] Thamm, U., 1998, Bewertung von Tiefziehprozessen durch Infrarot-Thermografie, Dissertation, Technische Universität Chemnitz.
[17] Gayubo, F., González, J.L., De La Fuente, E., Miguel, F., Perán, J.R., 2006, On-line machine vision system for detect split defects in sheet-metal forming processes, in *International Conference on Pattern Recognition*, pp. 723–726.
[18] Siegert, K., Ziegler, M., Wagner, S., 1997, Closed loop control of the friction force. Deep drawing process, Journal of Materials Processing Technology, vol. 71, pp. 126–133.
[19] Siegert, K., Liewald, M., Blaich, C., Kauschinger, B., 2007, Robuster Tiefziehprozess durch Ziehsickenstabhöhenregelung, vol. 97, pp. 781–791.
[20] Faaß, I., 2009, Prozessregelung für die Fertigung von Karosseriebauteilen in Presswerken, Dissertation, Technische Universität München.
[21] Liewald, M., Barthau, M., 2018, Adaptive control strategies for deep drawing of high performance sheet metal materials, in *presented at the IDDRG 2018, Waterloo, Canada.*
[22] Sturm, V., 2013, Einfluss von Chargenschwankungen auf die Verarbeitungsgrenzen von Stahlwerkstoffen, Dissertation, FAU Erlangen-Nürnberg.

IOP Conf. Series: Materials Science and Engineering **967** (2020) 012069 doi:10.1088/1757-899X/967/1/012069

Prediction of limit strains during non – proportional load paths with a change in loading direction

W. Volk[1*], M. Gruber[1], R. Norz[1]

[1]Chair of Metal Forming and Casting, Technical University Munich, Walther-Meissner Strasse 4, 85748 Garching, Germany

*wolfram.volk@utg.de

Abstract. Many different models have been published to predict failure after non-proportional load paths. Most of those models are phenomenological and heuristical models. They require a profound knowledge about the material. Examples are the enhanced Modified Maximum Force Criterion (eMMFC), the Polar Effective Plastic Strain-model (PEPS) or the Generalized Forming Limit Concept (GFLC). In addition to the load path, the loading direction has a significant influence on the formability of sheet metals. The mentioned models currently neglect this influence. By extending the GFLC-model by the parameter of loading direction, this influence is taken into account. By analyzing an acceptable number of bi-linear experiments, it is possible to calibrate the proposed model for a micro-alloyed steel HC340LA. Therewith an arbitrary load path with a change in loading direction can be evaluated. The results of this contribution show the effectiveness of this approach by different experiments.

1. Introduction

The prediction of localized necking is topic of many different research projects. For linear load paths the Forming Limit Curve (FLC) is widely used. Graf and Hosford [1] have found, that the loading history has a significant influence on the formability of AL2008-T4 sheet metal. A biaxial pre-forming state generally lowers the FLC, while a uniaxial pre-forming increases the FLC. They confirmed the results of Bergström and Ölund [2] who investigated an aluminium alloy and an AK – steel. Both investigated materials showed a significant influence of pre-forming on the remaining formability. The prediction of necking after a non – proportional load history has been in the focus of research for several years. Different models have been published based on different approaches. Stoughton and Yoon [3] used the Polar Effective Plastic Strain – model (PEPS) to predict non-linear effects. They found, that a big merit of the PEPS-diagram is the strain path independence. Nevertheless, the constitutive law has a strong influence on the effective strain. Hora et al. [4] presented the Modified Maximum Force Criterion (MMFC-model) which was extended to the enhanced Modified Maximum Force Criterion (eMMFC-model) [5]. The presented approaches are theoretical models and therefore require a very accurate description of the material behaviour [6]. A different model was published by Volk et al. [7]. The Generalized Forming Limit Concept (GFLC-model) is a phenomenological model to predict bi-linear strain paths. It has already been extended to arbitrary strain paths [8]. In this model, a database of multiple linear and non-proportional load paths are used to predict necking of arbitrary load paths accurately.

All the presented models are limited to strain-paths with no change in loading direction. However, industrial applications show that especially these changes significantly influence the remaining formability. Therefore, a novel approach is presented in this contribution. The GFLC-model is extended by the parameter of loading direction. By creating a database with multiple experiments, an accurate prediction of failure is possible.

2. Experimental setup

In this investigation, a micro-alloyed steel HC340LAD with an initial thickness of 1.0 mm was used. The presented database for the linear GFLC, as in [8], is extended by conducting experiments with a new specimen geometry. Within this new geometry pre-strain ratios $\beta = 0.5$ ($\beta = \varphi_2/\varphi_1$) can be created. This allows a more precise prediction of strain paths in between the plane strain and biaxial strain state. The used specimen geometries can be found in Figure 1. These geometries have to be adapted for each material individually. Using the same geometry for different materials, will lead to deviations in the resulting strains. To pre-form the specimens a modified Marciniak tool, proposed by Weinschenk and Volk [9],was used. The punch speed was set to 15 mm/s. To reduce the friction between the punch and the specimen, deep drawing oil was applied on the punch and the specimen.

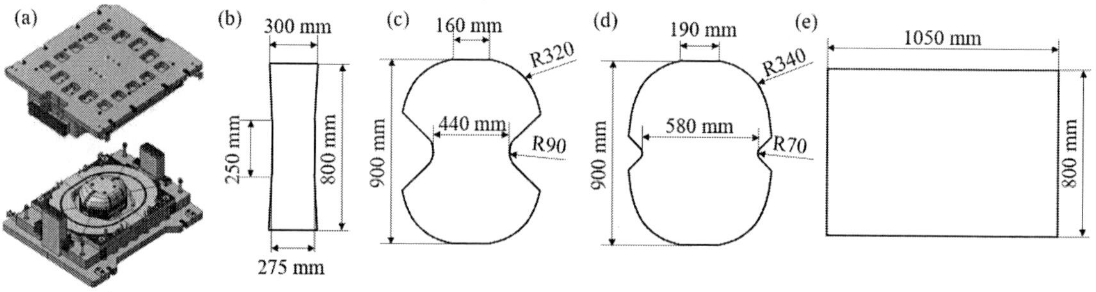

Figure 1. (a) Tool used for the pre-forming, (b) uniaxial specimen geometry, (c) plane strain geometry, (d) $\beta = 0.5$ geometry and the biaxial specimen geometry (e).

The initial specimens for the pre-forming step were manufactured by laser-cutting under three different directions (0°, 45° and 90°) with respect to the rolling direction. A total of 10 different initial specimens are manufactured. These specimens are pre-formed to two different strain heights. The height of the strain is approximately 25 % and 50 % of the linear FLC. After the pre-forming step, four different Nakajima specimen geometries were extracted by laser-cutting from the formed specimens. This allows the identification of the pre-forming influence on a complete FLC. The Nakajima specimens were rotated with regard to the initial forming direction, see Figure 2.

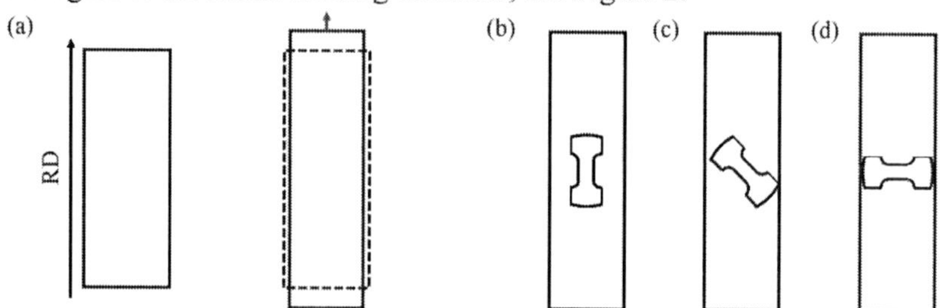

Figure 2. Schematic pre-forming process for a uniaxial specimen (a) and the specimen extraction under different angels with respect to the initial forming direction (b) 0°, (c) 45° and (d) 90°.

The chosen specimen geometries have a specimen width of 30 mm, 100 mm, 130 mm and a circle with a diameter of 235 mm for the biaxial specimen. The Nakajima tests were conducted on a BUP1000 from ZwickRoell™. For the strain measurement an ARAMIS™ 4M system with a measuring frequency of 10 frames per second was used. To reduce the friction between the Nakajima punch and the specimens, lubrication paste in combination with a PVC-pad was applied. The punch speed was set to a constant value of 1.0 mm/s. To determine the onset of localization, the Time Dependent Evaluation Method was used [10]. A minimum of three valid experiments for each specimen combination was tested.

In a first validation step Nakajima experiments with geometries not used for the meta-model were performed. To validate the proposed model, also experiments with a tool proposed by Eder et al. [11] were conducted. This tool allows the creation of a wide spread strain distribution by forming sheet metal strips. The tool geometry can be found in Figure 3. As this tool is also used on the BUP1000, the same experimental setup with a measuring frequency of 10 frames per second and a constant punch velocity of 1.0 mm/s as for the Nakajima experiments was used. To reduce the friction deep drawing foil and lubrication paste was applied on the punch.

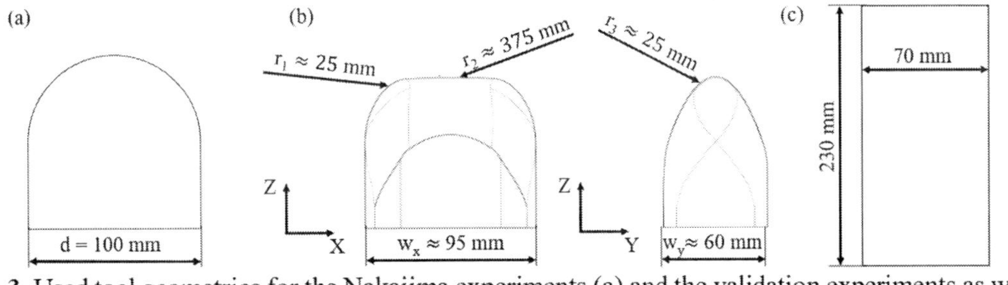

Figure 3. Used tool geometries for the Nakajima experiments (a) and the validation experiments as well as the specimen geometry (b)

3. Results

3.1. Pre-forming

The specimens are pre-formed to 25 % and 50 % of the linear FLC. To pre-form all directions of the specimens equally, the linear FLC is determined for the three different directions. The resulting pre-forming states can be seen in Figure 4. A total of 20 pre-forming states is needed for a sufficient database. The nomenclature of the pre-forming states is "P" for pre-forming, followed by the pre-forming state (U for uniaxial, PS for plane strain, BZF for the $\beta = 0.5$ strain state and B for biaxial), the angle with respect to the initial rolling direction and the height of the strain (25 % as low and 50 % as high). Due to the different punch geometry in the pre- and post-forming experiments, the biaxial pre-strains of the linear FLCs are corrected by the GFLC according to [12].

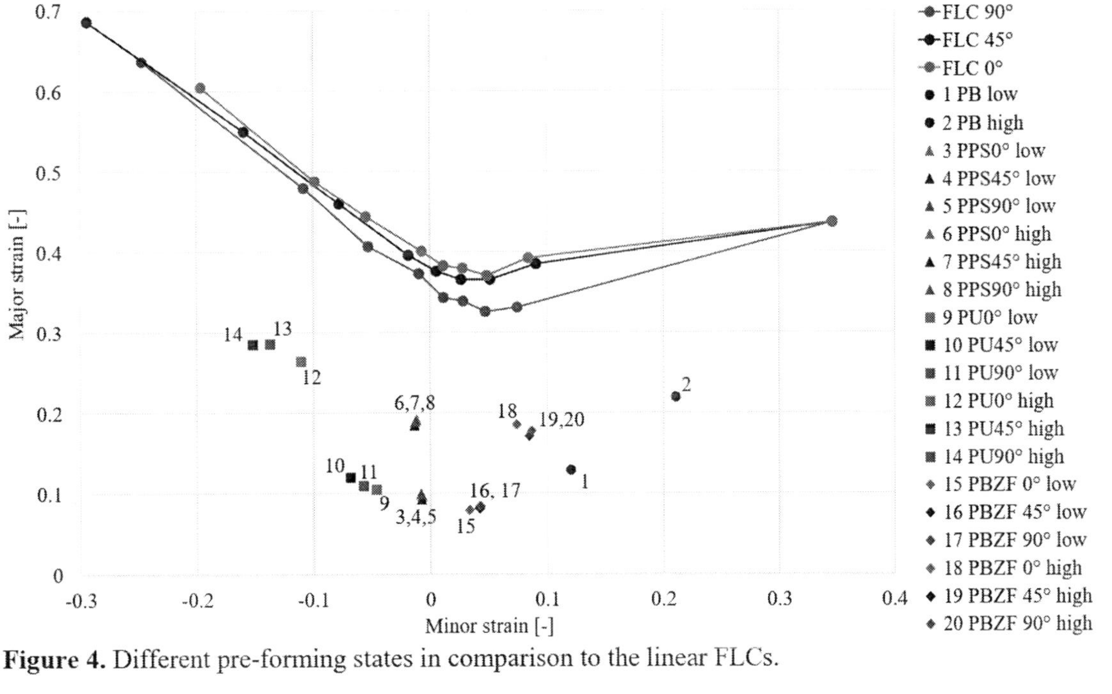

Figure 4. Different pre-forming states in comparison to the linear FLCs.

3.2. *Post-forming Nakajima experiments*

The Nakajima test results show a significant influence of the pre-strain height, the pre-form state and the change in loading direction on the formability of the investigated material. A uniaxial pre-forming leads to failure beyond the linear FLC while a biaxial pre-forming leads to failure below the linear FLC, shown in Figure 5. The post-forming direction is denoted as PF0°, PF45° and PF90° with respect to the initial rolling direction.

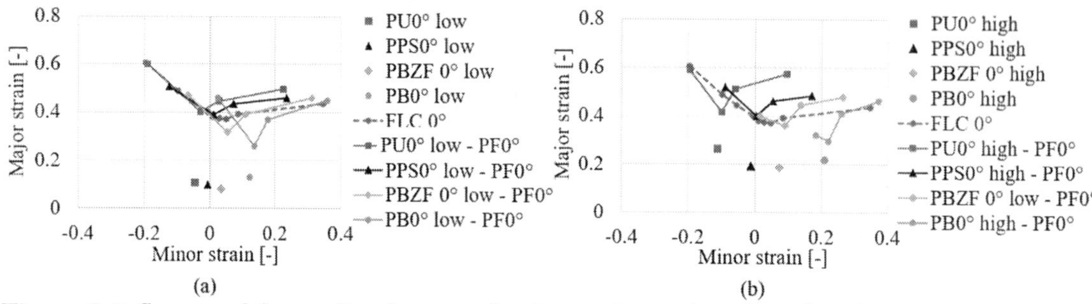

Figure 5. Influence of the pre-forming state for the specimens in rolling direction on the FLC for the 25 % pre-formed specimen (a) and the 50 % pre-formed specimen (b) for specimens with no change in loading direction.

The onset of necking for specimens with no change in loading direction and the same strain state are in good agreement with the linear FLC, as expected. The higher the pre-strain the more significant the influence of a change in loading direction is, see Figure 6. It is also noticeable, that the left hand side of the FLD is stronger affected than the right hand side.

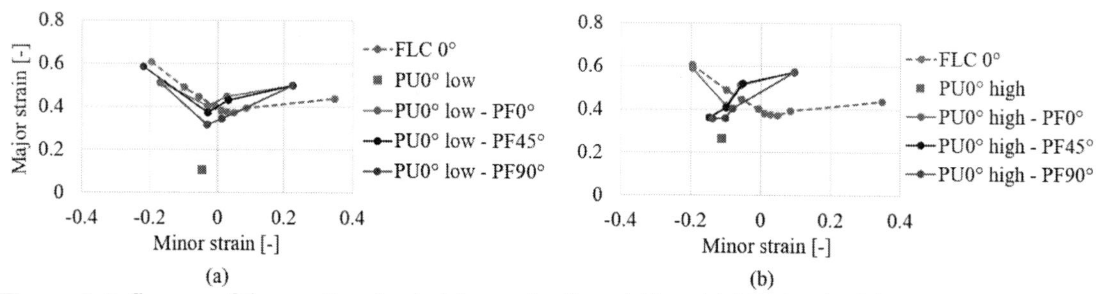

Figure 6. Influence of the pre-forming height on the formability, (a) 25 % uniaxial pre-formed, (b) 50 % uniaxial pre-formed specimens in rolling direction.

As the biaxial specimen has a circular shape, the biaxial necking point is the same for all tested loading directions. The conducted Nakajima experiments are used as the database for the proposed extension of the GFLC-model.

4. Prediction of necking by the modified GFLC-model

To use the database in the extended GFLC-model, the strain paths have to be parameterized. Hence, each strain ratio β has a strain path length l_{FLC}. The strain path length ratio is calculated with a corresponding strain ratio and a corresponding loading direction. The total strain path ratio λ is defined as:

$$\lambda = f\left(l_{pre},\beta_{pre},\varphi_{pre},l_{post},\beta_{post},\varphi_{post}\right) = \lambda_{pre} + \lambda_{post} = \frac{l_{pre}(\beta_{pre},\varphi_{pre})}{l_{FLC}(\beta_{pre},\varphi_{pre})} + \frac{l_{post}(\beta_{post},\varphi_{post})}{l_{FLC}(\beta_{post},\varphi_{post})} \quad (1)$$

Each strain increment can be described with the parameter β, λ and φ. The following meta-model is shown in Figure 7. Compared to the standard GFLC-model, the 3D-GFLC is extended by the parameter of loading direction. The shown diagram consists of 48 data points. 24 of them are from the linear FLC while the other 24 are determined by the pre-forming. The database is therefore significantly bigger then the database for the linear GFLC. The strain path length of a bi-linear strain path is calculated by linear

Lagrange interpolation of the surrounding data points of the element. The resulting post strain path length ratio λ_{post} is approximated by the scalar product of the calculated path length ratio and the interpolation function of the used data points. The onset of necking is calculated inversely

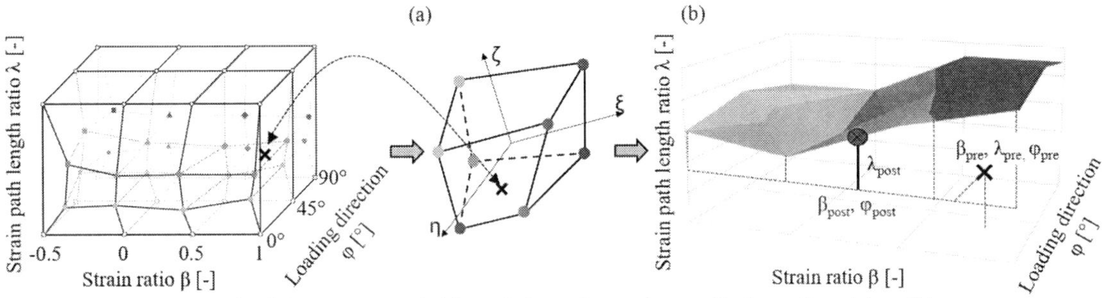

Figure 7. Meta-model of the 3D-GFLC (a) and the schematic prediction of necking (b).

For a given pre-forming state with a certain strain path length ratio λ, strain ratio β and loading direction φ, the remaining formability for an arbitrary strain path can be calculated, see Figure 8. The reduced formability for a change in loading direction is clearly visible, where a biaxial post-forming leads to an increased formability compared to the linear FLC. A strain path length ratio of 1 indicates the formability of the linear FLC. Therefore, the value can be bigger as 1.

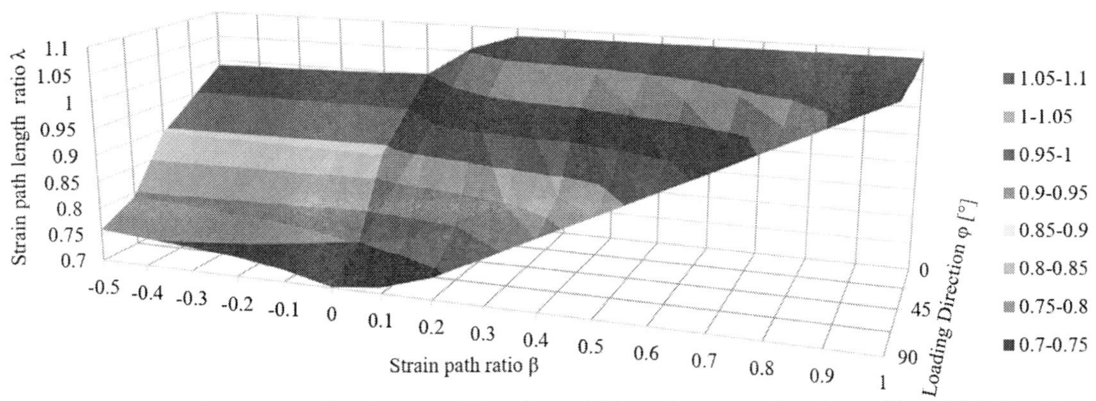

Figure 8. 3D-Surface to predict the remaining formability after a pre-forming of $\lambda = 0.35$, $\beta = 0$ and $\varphi = 0°$

5. Validation of the proposed 3D-GFLC model

To validate the proposed 3D-GFLC model, different validation experiments are conducted. For a strain path with no change in loading direction, the onset of necking can accurately be predicted, see Figure 9. A biaxial post-strain is applied to a plane strain pre-formed specimen. The plane strain pre-forming has used 38 % of the formability. The remaining formability until necking is therefore 62 % of the linear FLC.

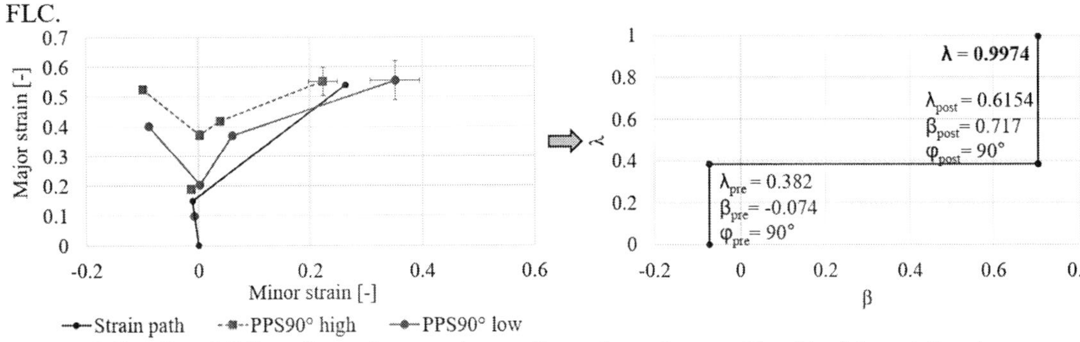

Figure 9. Predicted failure for a plane strain pre-formed specimen with a biaxial post-forming

The validation for the 3D-GFLC is done by experiments which are not used for the meta-modelling. Pre-forming height and the loading direction for the post forming step are therefore changed. A uniaxial post-strain is applied to a biaxial pre-formed specimen. The loading direction is changed from 90° to 22.5°. The small deviation of the strain path length ratio shows, that an accurate prediction of the necking is possible as shown in Figure 10.

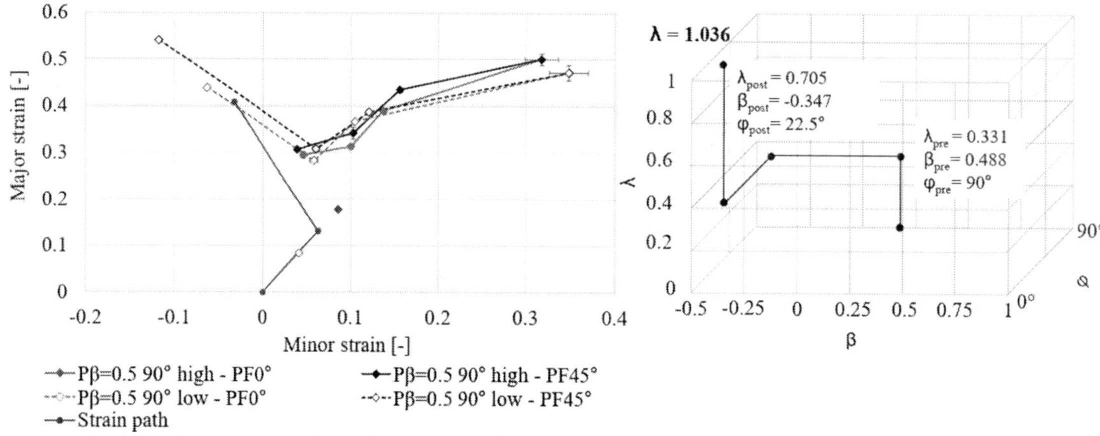

Figure 10. Predicted failure for a $\beta = 0.5$ pre-formed specimen under 90° with a uniaxial post forming under 22.5°

To proof, that also more complex forming operations are accurately predicted, a tool proposed by Eder et al. [11] is used. Specimens are uniaxial pre-formed under 0° to the rolling direction. The specimens are formed until fracture and the strain path of the necking zone is investigated. The influence of a change in loading direction on the drawing depth is clearly visible in Figure 11.

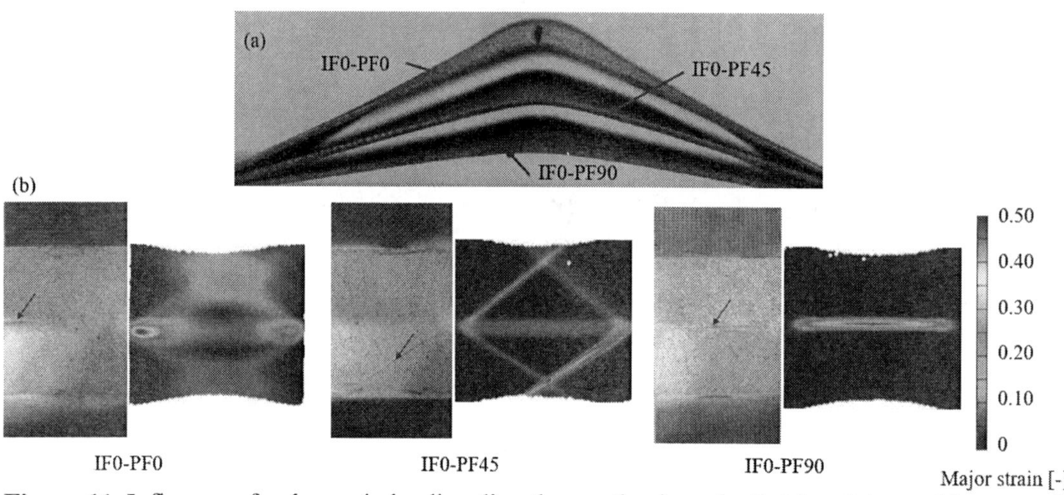

Figure 11. Influence of a change in loading direction on the draw depth (a) and the necking zone (b)

The higher the change in loading direction, the smaller is the remaining formability. Also the necking zone is affected by the pre-forming and the change in loading direction. Thus all three investigated loading directions show a different necking zone. To predict this behaviour accurately the usage of the standard GFLC-model is not possible. The formability for strain paths with a change in loading direction would be overestimated. By using the proposed 3D-GFLC method the different failure points are predicted accurately.

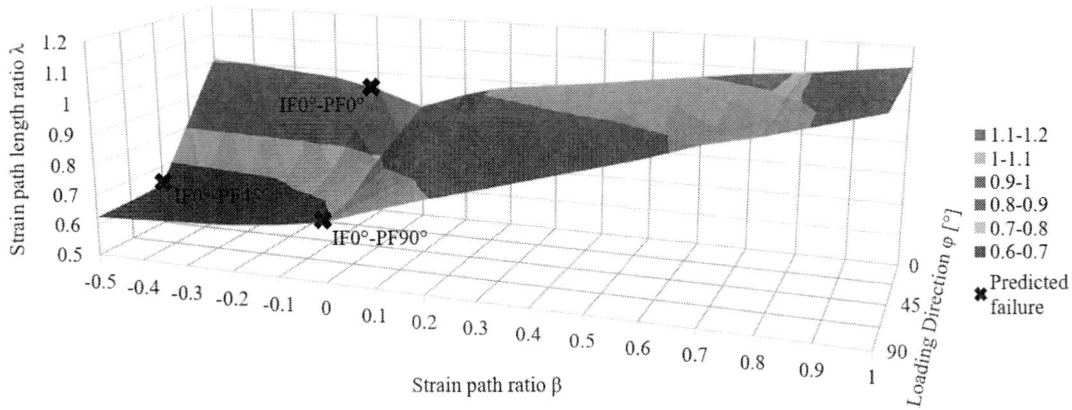

Figure 12. Predicted failure for the validation experiments with the tool proposed be Eder et al. [11]

6. Conclusion and outlook

The influence of the strain path history on the remaining formability of sheet metals is well investigated nowadays. In this contribution the significant influence of a change in loading direction on the remaining formability of a micro-alloyed steel HC340LA has been investigated. It is found, that the bigger the pre-forming, the bigger the influence of a change in loading direction is on the remaining formability. A change of the loading direction of 45° or 90° leads to a significant reduction in remaining formability. To predict this material behaviour, the GFLC-model is extended by the parameter of loading direction. As the GFLC-model is a phenomenological model, the extension is possible with only little effort. To proof the accuracy of the proposed 3D-GFLC, validation experiments have been conducted. The different strain paths are predicted accurately.
In future investigations the influence of a change in loading direction has to be shown for more materials. Also the reason for the significant loss of formability has to be further investigated. Nevertheless, as the 3D-GFLC model is a phenomenological model, only a solid database without any further knowledge about the material is needed to predict necking accurately. The significant influence has to be taken into account for industrial forming operations.

Acknowledgement

The authors would like to thank the German Research Foundation (DFG) for the financial support under the grant number VO1487/34-1.

References

[1] Graf A. and Hosford W. 1993 Effect of changing strain paths on Forming Limit Diagrams of Al 2008-T4. *Metallurgical Transactions A* **24:**671
[2] Bergström Y. and Ölund S. 1982 The forming limit diagram of sheet metals and effects of strain path changes on formability: a dislocation treatment. *Materials Science and Engineering* **56:**47–61
[3] Stoughton T.B. and Yoon J.W. 2012 Path independent forming limits in strain and stress spaces. *International Journal of Solids and Structures* **49:**3616–3625
[4] Hora, P.,Tong, L., Reissner, J. 1994 *Prediction Methods for Ductile Sheet Metal Failure using FE-Simulation.* 18th IDDRG Congress, Lisbon
[5] Hora P 2008 *Theoretical prediction of the influence of curvature and thickness on the FLC by the enhanced modified maximum force criterion* (Interlaken)
[6] Volk W., Groche P., Brosius A., Ghiotti A., Kinsey B.L., Liewald M., Madej L., Min J. and Yanagimoto J. 2019 Models and modelling for process limits in metal forming. *CIRP Annals* **68:**775–798

[7] Volk W., Hoffmann H., Suh J. and Kim J. 2012 Failure prediction for nonlinear strain paths in sheet metal forming. *CIRP Annals* **61**:259–262

[8] Volk W. and Suh J. 2013 Prediction of formability for non-linear deformation history using generalized forming limit concept (GFLC). *AIP Conference Proceedings* **1567**:556–561

[9] Weinschenk A. and Volk W. 2017 FEA-based development of a new tool for systematic experimental validation of nonlinear strain paths and design of test specimens. *AIP Conference Proceedings* **1896**

[10] Volk W. and Hora P. 2011 New algorithm for a robust user-independent evaluation of beginning instability for the experimental FLC determination. *Int J Mater Form* **4**:339–346

[11] Eder M., Gruber M. and Volk W. 2019 Innovative Tool for Material Model Assessment and Improvement. *Proceedings of the Forming Technology Forum 2019*

[12] Volk W. and Gaber C. 2017 Investigation and Compensation of Biaxial Pre-strain During the Standard Nakajima- and Marciniak-test Using Generalized Forming Limit Concept. *Procedia Engineering* **207**:568–573

IOP Conf. Series: Materials Science and Engineering **967** (2020) 012070 doi:10.1088/1757-899X/967/1/012070

Development of rapid heat solid solution process of Al 7075-T6 alloy applying near-infrared ray heater

Seon-Ho Jung [1, 2], Yongbae Kim[1], Seogou Choi[1], and Jongsup Lee[1*]
[1]Korea Institute of Industrial Technology 156, Gaetbeol-ro, Yeonsu-gu, Incheon 21999, Republic of Korea
[2]Department of Mechanical Engineering, Inha University, 100, Inha-ro, Nam-gu, Incheon, 22212, Republic of Korea

[*]jongsup@kitech.re.kr

Abstract. To improve the formability and productivity in cold forming of aluminum alloy, w-temper forming process was developed. In this process, the T6-temper blank is subjected to the solution heat treatment (SHT) and then quenched to achieve the w-temper. At present, the main problem of this process is that too long SHT time cannot meet the manufacturing cost. For example, the quenching time is 10 sec, while the SHT time is 30 min. In this paper, rapid solid solution of Al 7075-T6 alloy sheet was implemented in near-infrared ray(IR) heater. Applying IR, contact heater was developed instead of the traditional heater using cartridge heater. The results have shown that it is possible to accurately control the sheet temperature for solid solution temperature in short time. Also, compared to SHT using traditional heater, the rapidly heating using IR heater could make the mechanical properties of w-temper.

1. Introduction

Global warming is a serious concern where it is driven in part by the exhaust gases from automotive vehicles. For both fuel economy and emissions control, reducing the weight of car bodies has been a focused issue in the automotive industry [1-3]. For weight reduction, Al7xxx alloys have attracted much attention from the automotive industry due to its superior specific strength compared to ultrahigh-strength steel (UHSS). But, a disadvantage with high-strength these alloys compared with UHSS is that the formability is low. In order to improve the formability and productivity in the cold forming of these alloys, a limited study has suggested the application of W-temper forming (WF).

In WF process, the T6-temper blank is subjected to the SHT and then quenched to achieve the w-temper. After that, the blank is transferred into the die and deformed at RT before the precipitation starts. At present, the main problem of this process is that too long solid solution time cannot meet the manufacturing cost. Several studies have demonstrated that the rapid heating of aluminum alloys is significantly influenced by short heating time. In practice, Xu, Xiaofeng et al. examined that heating speed by contact solid solution treatment show a critical influence on microstructure and mechanical properties of Al 7075-T6 alloy [4]. Maeno, Tomoyoshi et al. demonstrated that quick heating provides similar mechanical propertied and formability [5].

In this paper, the innovative heating methods in a contact IR heating was utilized to reduce the solid solution time of Al 7075 alloy sheets. First, a contact IR heater was designed and manufactured. And then the research on the rapid SHT of Al 7075 alloy by contact heating method was conducted.

2. Experiment

A commercial Al7075 alloy in T6 condition with a thickness of 2.0 mm has been used in this research, having the following chemical composition given in table 1. In order to simulate the WF process, Al 7075 alloy samples were solution heat treated in a contact IR heater for 150 sec at 475°C. Then, all the samples were quenched by a cold flat die. The WF process is schematically shown in Figure 1. Hardness test was conducted to observe the change in mechanical property in the alloy with different solid solution time. The measurements were operated using a HM-100 Vickers microhardness tester with a load of 100 gf and a dwell time of 12 sec.

Thermal analysis was carried out using the differential scanning calorimetry (DSC) system (TA-100, TA Instruments). High purity Al 99.9% was used as a reference material. The specimens were heated from 60 to 300 °C at a constant heating rate of 10 °C min^{-1}. The selected conditions of samples were prepared for the DSC testing to have thicknesses of ~0.6 mm with weights of ~20 mg.

Table 1. Chemical compositions of Al 7075 alloy (wt.%).

Mg	Zn	Cu	Si	Fe	Mn	Ti	Zr	Cr	V	Etc.
2.39	5.61	1.36	0.095	0.18	0.042	0.032	0.016	0.2	0.011	0.064

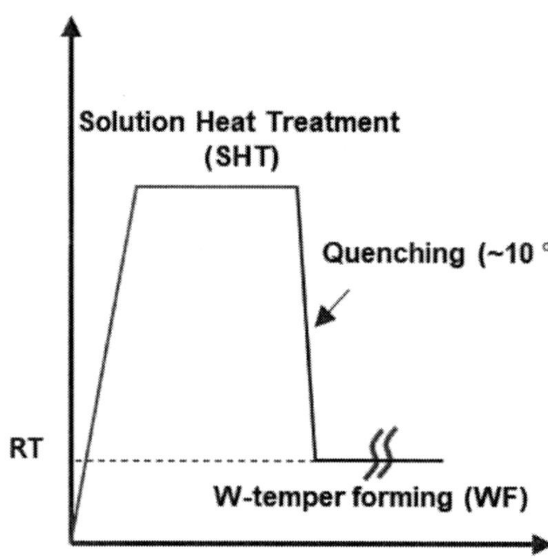

Figure 1. The schematic view about WF process

3. Results and discussion

3.1. Contact IR heater setup

IR heating has more advantages than traditional heating. However, due to the temperature accuracy of sheets during heating, there are some requirements for the design of contact IR heater such as excellent temperature uniformity, minimize overshoot with temperature controllers and heating time. On the basis of important points, a contact IR heater was designed, as shown in Figure 2(a). It consists of the upper IR heater and lower cartridge heaters. The upper was designed to rapidly heat on the sheets by the output power of the IR heater controlled by the PID control system (Figure 2(b)). The lower die was selected for copper because of its high thermal conductivity and designed to maintain the temperature uniformity of the sheets during the control IR heater by cartridge heaters. Spring loaded K-type thermocouple were installed inside the lower die between each of cartridge heaters feedback adjustment of temperature uniformity and minimizing overshoot of the sheets. Guide pins were arranged in the lower die to assist the placement of the sheets (Figure 2(c)). Finally, the contact IR heater did not need to be installed in the press and equipment was driven to open and close by the movement of the pneumatic cylinder. The sheets were placed on the lower die, and the heating is completed after the equipment is closed, as shown in Figure 2(d).

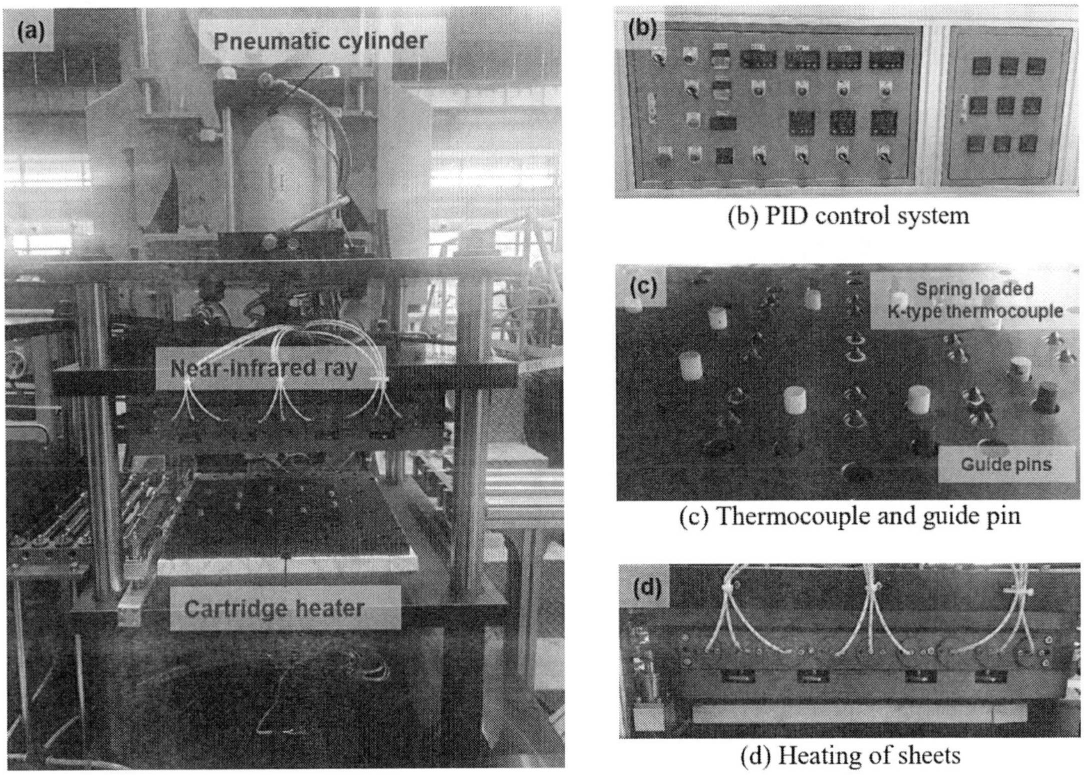

(b) PID control system

(c) Thermocouple and guide pin

(d) Heating of sheets

Figure 2. Contact IR heater

3.2. Measurement of temperature

The temperature evolution of the sample at 475 °C with a holding time of 150 sec is shown in Figure 3. The temperature-time curve can be divided into two stage. In the first stage, when the sample temperature increased to 475 °C, the heating rate was very fast, and the sample reached 475 °C within 90 sec. In the second stage, when the samples reach target temperature, under the controlling of PID control system, the temperature fluctuates up and down at 475°C and gradually stabilizes. The temperature of the sample is relatively uniform and the temperature difference is about 2 °C.

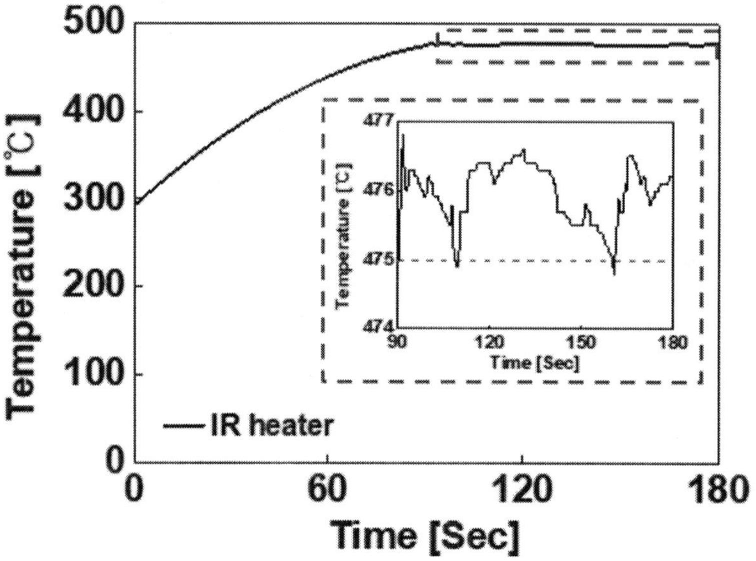

Figure 3. The schematic view about WF process

3.3. Effect of solid solution time

3.3.1. Result of Hardness

Before forming at room temperature, the samples require sufficient solid solution time. In order to study the effect of solid solution time, the hardness of the w-temper with different solid solution time was measured. Figure 4 shows the measurement of hardness for four different times, 90, 105, 120 and 150 sec. It can be seen that hardness has decreased compared with the as-received condition (T6 ≈ 178Hv). Also, all samples show the hardness of Hv ≈ 88 immediately after quenching. The decreased hardness with different solid solution times is attributed to microstructure immediately after quenching. It is implied that the dissolution of $MgZn_2$ precipitation occurred.

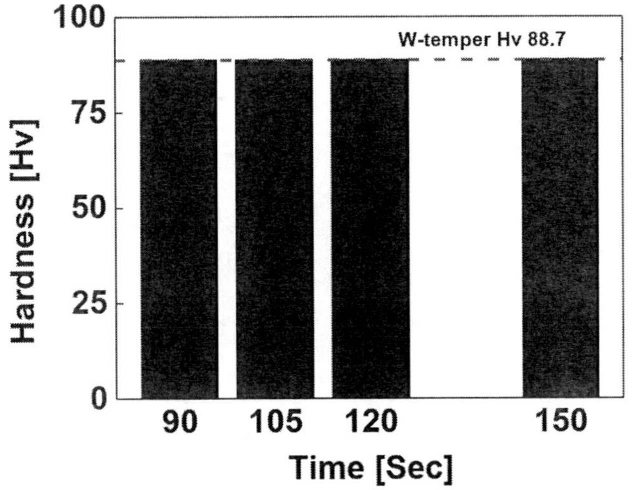

Figure 4. Hardness with different solid solution time

3.3.2. Result of DSC

Significance of solid solution time on the precipitation behavior was examined through the DSC analysis for WF capability on the Al 7075 alloy. It is well described that the precipitation sequence of Al 7xxx alloys is expressed as follows [6,7]: Super saturated Solid Solution → G.P. Zones (GPZs) → metastable η' phase → stable η phase ($MgZn_2$). The DSC curves of the WF samples measured at four different times, 90, 105, 120 and 150 sec after SHT are given in Figure 5. It was reported earlier through the DSC examinations that the first peak reaction (stage I) at ~200 °C corresponds to the formation of the GPZs, the following double exothermic reactions (stages II and III) at ~210 and ~230 °C are attributed to the formation of the η' phase and the transformation of η' into the η phase. All samples demonstrated that first peak reaction appears at approximately 95 °C, most likely due to the formation of GPZs. The GPZs peak could be detected at DSC curve, which means the soluble Al matrix were completely dissolved.

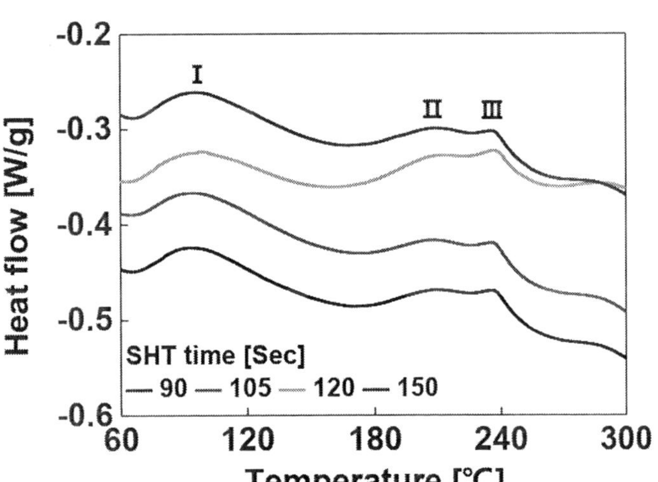

Figure 5. The DSC curves with different solid solution time

4. Conclusion

The effect of solid solution time on the hardness using of Al 7075 alloy using contact IR heater was investigated. The following conclusions can be drawn.

1. The results showed that contact IR heater could reach the target temperature in a short time and greatly shorten the solid solution time.
2. The mechanical property resulted in reduced hardness compared to as-received condition (T6) and it was similar in hardness to WF using traditional heater.
3. The measured solid solution time will be useful database for the amount of pre-strain and aging condition for WF.

5. References

[1] Zheng, Kaijun, et al., 2018, *International Journal of Lightweight Materials and Manufacture* 1.2 55-80
[2] M. Goede, M. Stehlin, L. Rafflenbeul, G. Kopp, E. Beeh, 2009, *European Transport Research Review* 1, 5-10
[3] Harrison, Nia R., and S. George Luckey., 2014, *SAE International Journal of Materials and Manufacturing* 01-0981 567-573
[4] Xu, Xiaofeng, et al., 2016, *Materials Science and Engineering: A* 654 278-281
[5] Maeno, Tomoyoshi, Ken-ichiro Mori, and Ryosuke Yachi., 2017, *CIRP Annals* 66.1 269-272
[6] G. Sha, A. Cerezo, 2004, *Acta Materialia* 52 4503-4516
[7] N.Q. Chinh, J. Lendvai, D.H. Ping, K. Hono, 2004, *Journal of Alloys and Compounds* 378 52-60

Reduced texture approach for crystal plasticity finite element method toward macroscopic engineering applications

D Noh[1] and J W Yoon[1,2*]

[1] Department of Mechanical Engineering, Korea Advanced Institute of Science and Technology, 291 Daehak-ro, Yuseong-gu, Daejeon 34141, Republic of Korea
[2] School of Engineering, Deakin University, Geelong Waurn Ponds, VIC 3220, Australia

E-mail: j.yoon@kaist.ac.kr, j.yoon@deakin.edu.au

Abstract. In Crystal Plasticity Finite Element Method (CPFEM), normally over thousands Euler angles are used. It leads to high computational cost. To efficiently solve this problem, a reduced texture approach was implemented through User MATerial Interface (UMAT). Specific material parameters including the texture information were calibrated to characterize anisotropic behavior. For the calibration, it is used the stress-strain curves and r-values along the rolling, diagonal, and transverse directions. In this study, AA 2090-T3 was modelled with the reduced texture approach by characterizing 12 parameters. Single element simulation result from the reduced texture approach shows a good agreement with the experimental data. In addition, a deep drawing simulation for AA 2090-T3 was performed. The simulation results from the reduced texture approach were compared with those from the advanced constitutive models such as Yld2000-2d and Yld2004-18p in terms of accuracy and time efficiency. It shows a great potential that the reduced texture approach based on the crystal plasticity theory could be applied to macroscopic engineering problems as an alternative solution for continuum level advanced constitutive models.

1. Introduction

In large scale engineering applications, many grains are employed for CPFEM leading to texture mapping problems. The major issue is how to map the intensive information of crystallographic orientations to integration points. In the view of computational cost, it is inefficient to use the huge amount of grains in CPFEM. Therefore, an effective way of mapping of crystallographic orientations to the integration points is recommended for the large scale CPFEM.

Many research has been devoted to achieve reasonable computational cost with the reduction of crystallographic orientations. Among them, a reduced texture methodology was proposed by Rousselier et al. [1, 2]. In the work, a strongly anisotropic and an initially isotropic material were firstly attempted with the self-consistent model. Moreover, the ductile fracture of sheet metals and the multi-axial deformation with the strength-differential effect were predicted with the reduced texture methodology [3, 4]. However, many parameters were required to implement the reduced texture methodology.

In this study, a reduced texture approach was implemented with simple crystal plasticity constitutive model instead of the self-consistent model. Therefore, the number of parameters was reduced. To describe anisotropic behavior, total 12 parameters were calibrated using a single element simulation. In addition, a circular cup drawing simulation was performed and simulation results were evaluated in terms of accuracy and time efficiency.

2. Crystal plasticity model

In this study, the crystal plasticity model described by Yoon et al. [5] was used. Total deformation gradient ($\mathbf{F}$) is decomposed into the elastic ($\mathbf{F}^e$) and plastic ($\mathbf{F}^p$) parts as shown in Equation (1).

$$\mathbf{F} = \mathbf{F}^e \cdot \mathbf{F}^p \tag{1}$$

The velocity gradient ($\mathbf{L}$) is also decomposed into the elastic ($\mathbf{L}^e$) and plastic ($\mathbf{L}^p$) parts, i.e.,

$$\mathbf{L} = \dot{\mathbf{F}} \cdot \mathbf{F}^{-1} = \dot{\mathbf{F}}^e \cdot \left(\mathbf{F}^e\right)^{-1} + \mathbf{F}^e \cdot \dot{\mathbf{F}}^p \cdot \left(\mathbf{F}^p\right)^{-1} \cdot \left(\mathbf{F}^e\right)^{-1} = \mathbf{L}^e + \mathbf{L}^p \tag{2}$$

Because the plastic velocity gradient is the result of the shear strain contributions over all the slip systems, it can be expressed as:

$$\mathbf{L}^p = \sum_{\alpha=1}^{NSYS} \dot{\gamma}^{(\alpha)} \mathbf{s}^{(\alpha)} \otimes \mathbf{m}^{(\alpha)} \tag{3}$$

where $NSYS$ is the number of the slip systems, $\dot{\gamma}^{(\alpha)}$ is the slip rate, and $\mathbf{s}^{(\alpha)}$ and $\mathbf{m}^{(\alpha)}$ are the vectors along the slip direction and normal direction of the slip plane. When the constitutive relations are formulated, the objectivity should be satisfied. The Jaumann rate is generally used as the objective stress rate and the Jaumann rate of the Kirchhoff stress was well derived by Kim and Yoon [6]. After calculating the Jaumann rate of the Kirchhoff stress, the stress state at the grain level is updated, i.e.,

$$\boldsymbol{\tau}_{n+1} = \mathbf{R} \cdot \boldsymbol{\tau}_n \cdot \mathbf{R}^T + \Delta t \breve{\boldsymbol{\tau}} \tag{4}$$

where $\breve{\boldsymbol{\tau}}$ is the Jaumann rate of the Kirchhoff stress and $\mathbf{R}$ is the rotation tensor which is obtained from the polar decomposition .

In this study, rate dependent hardening behavior was used to compute the slip rate and the phenomenological models were employed as follows:

$$\dot{\gamma}^{(\alpha)} = \dot{\gamma}_0 \cdot \left[\frac{\left|\tau_{rss}^{(\alpha)}\right|}{g^{(\alpha)}}\right]^{\frac{1}{m}} \cdot sign\left(\tau_{rss}^{(\alpha)}\right) \tag{5}$$

where $\dot{\gamma}_0$ is the reference slip rate, $\tau_{rss}^{(\alpha)}$ is the resolved shear stress, m is the rate sensitivity, and $g^{(\alpha)}$ is the slip resistance. Also, the slip resistances are evolved as follows:

$$\dot{g}^{(\alpha)} = \sum_{\beta=1}^{NSYS} h^{\alpha\beta} \left|\dot{\gamma}^{\beta}\right| \tag{6}$$

$$h^{\alpha\beta} = qh + (1-q)h\delta_{\alpha\beta} \tag{7}$$

$$h = h_\infty + (h_0 - h_\infty)\operatorname{sech}^2\left[\left(\frac{h_0 - h_\infty}{g_\infty - g_0}\right) \cdot \gamma\right] \tag{8}$$

where h_0, h_∞ are the initial and saturated hardening rates, g_0, g_∞ are the initial and saturated value of the slip resistances, q describes the latent hardening, and γ is the summation of the slips over all the slip systems. In addition, the stress state at integration point level is volume averaged:

$$\boldsymbol{\sigma} = \sum_{g=1}^{N} f_g \boldsymbol{\sigma}_g \tag{9}$$

where N indicates the number of grains in each integration point, g means each grain, and f_g is the volume fraction of g.

3. Reduced texture approach

3.1. Experimental data

In this study, AA 2090-T3 was modelled. The stress ratios $\left(s_\theta = \sigma_\theta / \sigma_0\right)$ where σ_θ indicates the stress anisotropy in θ degree from the rolling direction and r-values were measured from the tensile tests conducted along the rolling, diagonal, and transverse directions. The results are shown in Table 1 [2]. However, the complete experimental data does not exist. The experimental data were built by using the yield stress ratios, r-values, and the stress-strain hardening curve along the rolling direction. Therefore, the stress-strain curves in θ degree from the rolling direction were defined as:

$$\overline{\sigma}_\theta = 646\left(0.025 + \overline{\varepsilon}^p\right)^{0.227} \cdot s_\theta \tag{10}$$

In addition, the transverse strain curves are shown in Figure 1(b) assuming that r-values are independent of the strain value. All of the data were built up to $\overline{\varepsilon}^p = 0.28$ to implement large strain results into deep drawing simulations.

Table 1. Experimental results of AA 2090-T3.

	Rolling	Diagonal	Transverse
Stress ratio	1.0000	0.8148	0.9115
r-value	0.20	1.57	0.70

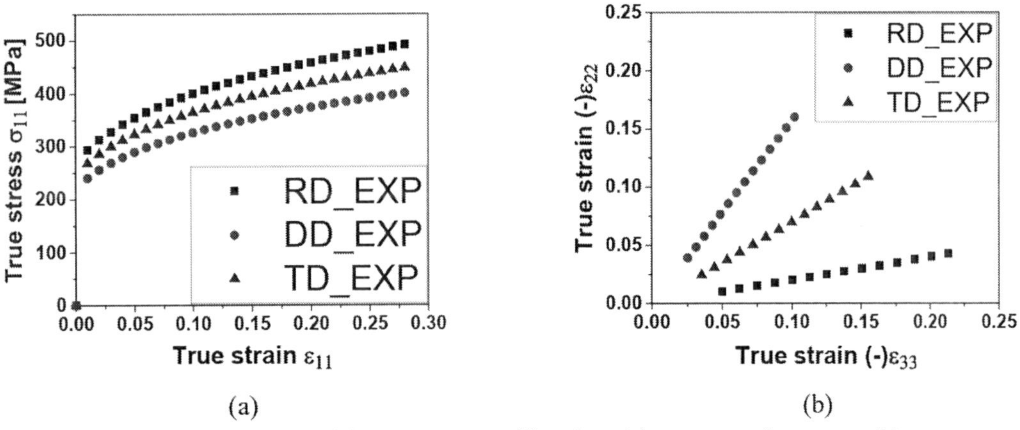

Figure 1. Experimental data used for parameter calibration: (a) stress-strain curves; (b) transverse strain curves.

3.2. Calibration

Parameters were calibrated with Abaqus2016/Standard and Isight2016. The experimental data used for parameter calibration were as follows:

IOP Conf. Series: Materials Science and Engineering **967** (2020) 012071 doi:10.1088/1757-899X/967/1/012071

- Stress-strain curves along the rolling, diagonal, and transverse directions
- Transverse strain curves along the rolling, diagonal, and transverse directions

Total six curves were used for material characterization. For material constants, the rate sensitivity was set to $m = 0.04$ because the aluminium alloy is almost rate independent at the room temperature. For the reference slip rate, $\dot{\gamma}_0 = 0.001$ [/s] was employed. Elasticity tensor constants for aluminium alloy were set to $C_{1111} = 108000, C_{1122} = 62000, C_{1212} = 28300$ [MPa]. In the framework of the reduced texture approach, there are some parameters related to crystallographic orientations. As shown in the result of Rousselier et al. [1], two texture representatives were employed. For orthotropic symmetry, each texture representative consists of four grains which are expressed as $(\varphi_1 \quad \phi \quad \varphi_2)$, $(-\varphi_1 \quad \phi \quad -\varphi_2)$, $(\varphi_1 \quad -\phi \quad \varphi_2)$, and $(-\varphi_1 \quad -\phi \quad -\varphi_2)$. Therefore, total eight grains were included in each integration point. In conclusion, the parameters to be calibrated were as follows:

- Slip resistance ($h_0, h_\infty, g_0, g_\infty$, four parameters)
- Grain information ($\{\varphi_1(1), \phi(1), \varphi_2(1)\}, \{\varphi_1(2), \varphi_1(2), \varphi_1(2)\}, f(1)$, seven parameters)
- Latent hardening (q, one parameter)

Total twelve parameters were calibrated for the material characterization. In addition, q was bounded between 1 and 1.4 [7]. Although it is hard to find the exact latent hardening parameter with only tensile test, it was used as a fitting scheme. Then, Single element tensile simulation was conducted for the parameters' calibration.

The calibrated parameters are shown in Table 2. Figure 2 shows the stress-strain and transverse strain curves from the experimental data and simulation results. The reduced texture approach (CP) shows a good agreement with the experimental data.

Table 2. Results of the calibrated parameters.

g_0 [MPa]	g_∞ [MPa]	h_0 [MPa]	h_∞ [MPa]	q	$f(1)$
99.69	130.21	199.32	37.23	1.0470	0.0882
$\varphi_1(1)$ [°]	$\phi(1)$ [°]	$\varphi_2(1)$ [°]	$\varphi_1(2)$ [°]	$\phi(2)$ [°]	$\varphi_2(2)$ [°]
62.66	13.59	51.02	51.08	32.07	4.58

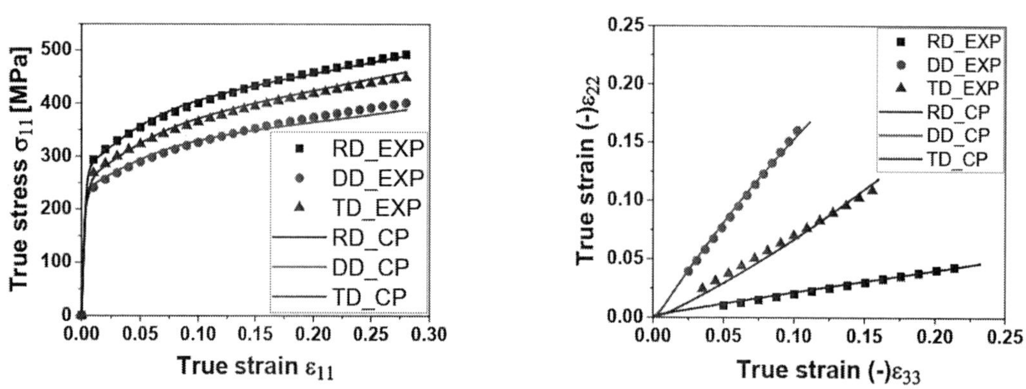

(a) (b)

Figure 2. Experimental data and simulation results: (a) stress-strain curves; (b) transverse strain curves.

4. Deep drawing simulation

A circular cup drawing simulation was conducted to evaluate the applicability of the reduced texture approach for the large scale engineering problems. The schematic view of the circular cup drawing simulation is illustrated in Figure 3. The dimensions and process variables are specified in Table 3 [8].

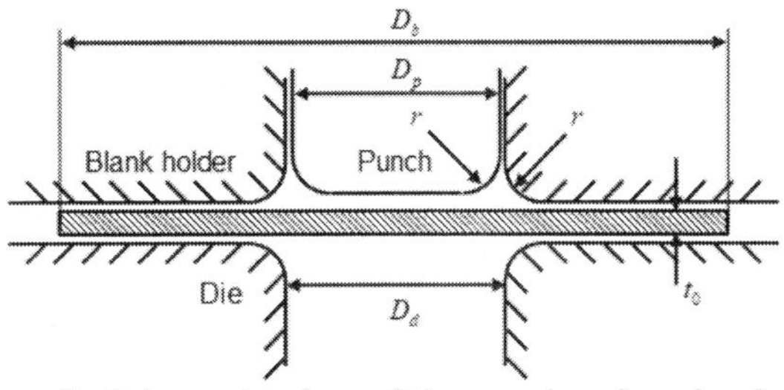

Figure 3. Schematic view of the cup drawing simulation.

Table 3. Dimensions and process variables for the cup drawing simulation.

Dimension [mm]	
Punch diameter (D_p)	97.46
Die opening diameter (D_d)	101.48
Blank diameter (D_b)	158.76
Fillet radius (r)	12.70
Initial blank thickness (t_0)	1.60
Process variables	
Coulomb coefficient of friction	0.1
Blank holding force [kN]	22.2

Because the material was assumed to be orthotropic symmetry, only a quarter section of the blank was analyzed. Total 2400 and 100 elements were employed for C3D8R and C3D6, respectively.

Two comparisons are following; First, the earing profile from the reduced texture approach was compared with that from the Yld2000-2d [9] because both of them used the stress ratios and r-values along the rolling, diagonal, and transverse directions. Second, the computational time from the reduced texture approach was compared with that from the Yld2004-18p since they are based on 3D solid element. The results are shown in Figure 4 and Table 4, respectively. Although the experimental r-value in 75 degree was not used in the reduced texture approach, the current model can predict six earing including a small ear along 15 degree which comes from the r-value in 75 degree from the rolling direction. In addition, the predicted earing profile shows a good agreement with the experimental data. Moreover, the computational time from the current model was less than twice that from the Yld2004-18p.

5. Conclusion

In this study, the reduced texture approach was employed to apply the crystal plasticity theory to large scale engineering problems with reduced computational cost. Specific parameters including texture representatives were calibrated and anisotropic behavior of AA2090-T3 was well described. It is concluded that the reduced texture approach could be a good alternative way to advanced continuum constitutive models such as Yld2000-2d or Yld2004-18p in terms of time efficiency and accuracy.

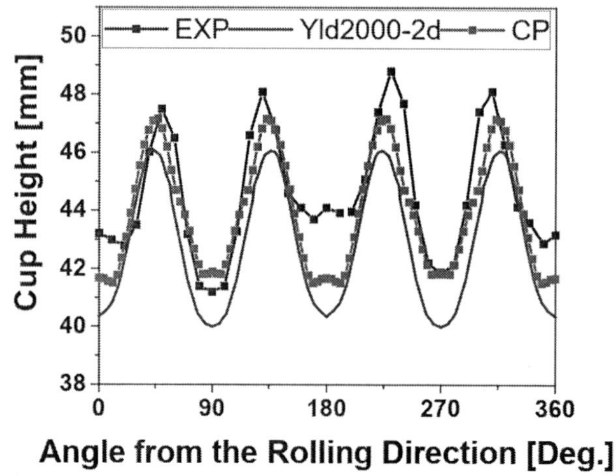

Figure 4. The experimental and predicted cup height.

Table 4. Normalized simulation time (wallclock time) between the models.

	Yld2004-18p	Reduced texture approach
Normalized simulation time	1.000	1.817

References

[1] Rousselier G, Barlat F and Yoon J W 2009 A novel approach for anisotropic hardening modeling. Part I: Theory and its application to finite element analysis of deep drawing *International Journal of Plasticity* **25** 2383-409

[2] Rousselier G, Barlat F and Yoon J W 2010 A novel approach for anisotropic hardening modeling. Part II: Anisotropic hardening in proportional and non-proportional loadings, application to initially isotropic material *International Journal of Plasticity* **26** 1029-49

[3] Rousselier G and Luo M 2014 A fully coupled void damage and Mohr–Coulomb based ductile fracture model in the framework of a Reduced Texture Methodology *International Journal of Plasticity* **55** 1-24

[4] Luo M and Rousselier G 2014 Modeling of large strain multi-axial deformation of anisotropic metal sheets with strength-differential effect using a Reduced Texture Methodology *International Journal of Plasticity* **53** 66-89

[5] Yoon J W, Barlat F, Gracio J J and Rauch E 2005 Anisotropic strain hardening behavior in simple shear for cube textured aluminum alloy sheets *International Journal of Plasticity* **21** 2426-47

[6] Kim J-B and Yoon J W 2015 Necking behavior of AA 6022-T4 based on the crystal plasticity and damage models *International Journal of Plasticity* **73** 3-23

[7] Asaro R J 1983 Crystal plasticity *Journal of applied mechanics* **50** 921-34

[8] Yoon J W, Barlat F, Dick R and Karabin M 2006 Prediction of six or eight ears in a drawn cup based on a new anisotropic yield function *International Journal of Plasticity* **22** 174-93

[9] Choi H and Yoon J W 2019 Stress integration-based on finite difference method and its application for anisotropic plasticity and distortional hardening under associated and non-associated flow rules *Computer Methods in Applied Mechanics and Engineering* **345** 123-60

Lubricant Selection and Post Forming Material Characterization in Incremental Sheet Forming

Amrut Mulay[1], Satish Ben[2] and Syed Ismail[2]
[1]Mechanical Engineering Department, Sardar Vallabhbhai National Institute of Technology, Surat 395007, Gujarat, India

[2]Mechanical Engineering Department, National Institute of Technology, Warangal 506004, Telangana, India

Abstract. Single Point Incremental Forming (SPIF) is a flexible, novel and rapid sheet forming process to deform extra thin to the thick sheet with the help of computer numerical control unit, hemispherical tool and fixture. A tool follows a series of small incremental steps which deforms the sheet into the desired shape. Despite of recent development occurred in past, the little literature was reported which investigate the effect of various types of lubricants in SPIF process. This study aims to address the distinct type of lubricant to accomplish better surface roughness and formability of final parts while incremental forming of aluminium 5052 H32 sheets. After the forming process, the fractographies of the specimen were studied at optimal condition. The study is also extended to determine residual stresses generated due to incremental deformation and its effect on corrosion phenomenon with the environment. It has been found that the fracture zone material is likely to corrode more rapidly due to increased plastic deformation than the material at initial and middle forming zones.

1. Introduction

Incremental Sheet Forming (ISF) is an unconventional metal forming process which is extensively used to manufacture products with free form shapes without using any specialized dies. A computer numerical controlled (CNC) tool traces a series of contour, after completing each contour, the tool takes a small depth and deforms the sheet into the product. Single Point Incremental Forming (SPIF) is a highly flexible sheet metal forming process which uses rapid prototyping techniques to produce parts in small batches economically. The working principle of the SPIF process has been explained in the published article [1]. The basic components of the process are sheet metal blank, blank holder, backing plate and hemispherical tool. The blank holder is generally used in holding and clamping of the blank metal sheet during the process. The backing plate acts as a support to sheet metal to avoid initial sheet bending and also provides the open area during forming. The single point tool rotates over the sheet and its incremental depth forms the sheet. The incremental depth and movement of the tool are controlled by CNC machine control unit. During the forming process, die sets are not used, hence, it is a cheaper process. The process is still under its development due to some limitations such as slow operation, geometrical inaccuracy, no uniform thinning of formed sheet and wavy surface texture.

Some researchers have paid attention to investigate the effect of SPIF process parameters on the response of sheet metal by incorporating various strategies. Cerro et al. [2] investigated the SPIF process by the numerical and experimental techniques to predict the surface roughness and stress distribution in the sheet during forming. Hussain et al. [3] have found that the good surface quality of deformed pure titanium sheet is obtained by using surface hardened high-speed steel tool and molybdenum disulphide paste with petroleum jelly in a specific proportion. Shanmugananatan and Senthil Kumar [4] carried out an experimental and numerical simulation to investigate maximum wall angle, surface roughness and sheet thinning of Al 3003(O). Zhaobing et al. [5] studied the impact of input parameters on surface roughness using the design of experiments (DOE) along with multi-objective function. Thickness distribution and failure limited diagram are studied by Malwad and Nandedkar [6] to understand the deformation mechanism behaviour of commercial AA8011. Desai et al. [7] presented one factor at a time approach for dieless rapid prototyping process to study the effect of process parameters on forming characteristics. Yanle et al. [8] performed some tests using RSM with Box-Behnken design to study the effects of input parameters on both deformation energy and geometrical accuracy in ISF process. Golabi and Hossain [9] statistically analyzed the incremental forming of SS304. They studied the effect of some parameters on forming the height of the component. Gulati. et al. [10] concluded the effect of six different input process parameters like tool radius, sheet thickness, tool rotational speed, feed rate and lubrication on performance measures using Taguchi's method. Aerens et al. [10] devised a novel technique to predict the developed forces in SPIF process based on the results obtained from experiments and finite element analysis. The ultimate goal of the work was to obtain an equation for different materials, which could help in the prediction of forces generated during forming. Five different materials and sheet blanks having different thickness were used to form a truncated cone. Equations were developed with the partial involvement of regression modelling techniques, results of finite element analysis and physics of the ISF process. Zhang et al. [12] investigated the more suitable lubricants and lubricating methods to be employed in case of magnesium alloy AZ31 sheet with warm negative incremental forming. The combined effect of tool size and lubricants was also discussed in the literature. The effect of lubrication, punch material and spindle velocity on the surface finish with CP titanium Gr.2 sheet of 1.0 mm thickness for SPIF of biomedical devices was studied by Fiorentino et al. [13]. Micari et al. [14] also studied the shape and dimensional accuracy of formed geometry by SPIF process. They defined the geometrical error as the distance between the ideal and actual profile. Bending of the sheet, spring back and the pillow effect at the base of formed geometry are some of the errors associated with SPIF process. They also addressed some strategies for error minimization in SPIF process. Mugendiran and Gnanavelbabu [15] compared the measurement accuracy of the developed plastic strain in SPIF using conventional, least-square and digital-based processing method. The result shows that the digital-based processing method is found to be more accurate than the other two. They claim that the strain measurement accuracy can be improved using digital-based processing method. Allwood et al. [16] proposed a novel blank shape, termed as partially cutout blanks to improve the accuracy of the final part. They observed that partially cutout blanks did not provide many benefits in the improvement of dimensional accuracy of the formed components. However, the use of stiff backup plate was the key solution in enhancing the part accuracy than the partial cut-outs.

The metal forming industries are moving towards customized production of sheet metal parts. In recent years, sheet metal industries demand flexibility, a novelty in the products to satisfy diversified customers worldwide. A lot of studies had reported to investigate dimensional accuracy, execution time, part formability and surface integrity of the deformed part. The attention was also sought for process optimization, FE simulation, material characterization, high speed, multi-sheet- and, multi-step- SPIF process. However, not much work had reported on the prerequisite lubrication condition and post investigations of parts resulting out of SPIF process. The use of suitable lubricant is the vital step to improve forming performance. The several experiments were performed and authors have identified set of process parameters to achieve better process performance in terms of minimum surface roughness value and maximum formability. The corrosion and residual stress behaviour study were carried out at optimized conditions of process parameters.

2. Selection of proper lubricant

The ASTM D4172-94 standard was used to select proper lubricant for the SPIF process to evaluate antiwar characteristics of several lubricants using four-ball tester (Model: TR-30L-IAS, Make: DUCOM) as shown in Figure 1. In the tests, three 12.7 mm diameter steel balls made of AISI standard steel No. E-52100 with Grade 25 EP (Extra Polish) are clamped together and covered with the lubricant to be evaluated. A top steel ball of the same diameter will be the fourth ball which presses into the cavity made by the three clamped balls for three-point contact. The test load, duration, temperature and rotational speed are defined in accordance with set standards. As shown in Figure 2, the real-time data of frictional torque vs. time data was recorded. The average scar diameter on the bottom three balls is obtained using an image acquisition system. An optional scar measuring image acquisition system with an image sensor allows acquiring scar images on the computer. The ability of the lubricant to avoid wear can be indicated by the area of the scar. A larger diameter of wear scar indicates poor wear resistance of the lubricant and vice versa.

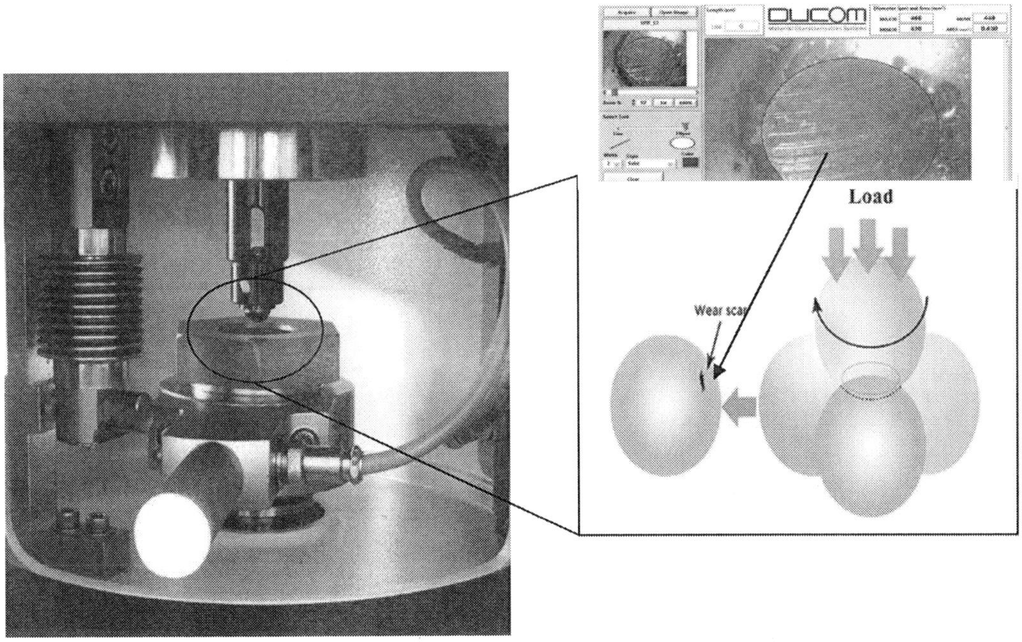

Figure 1. Four-Ball tester equipment

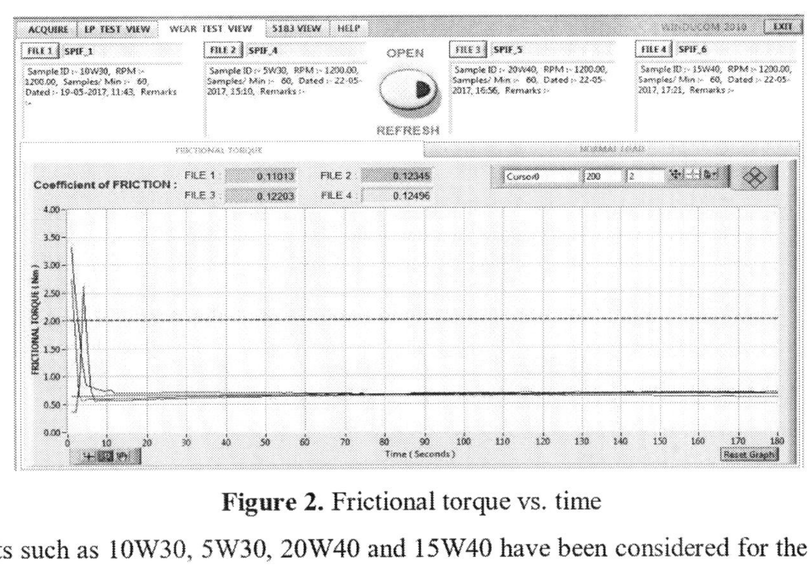

Figure 2. Frictional torque vs. time

The lubricants such as 10W30, 5W30, 20W40 and 15W40 have been considered for the investigation. The lubricants are compared by using the average size of the scar diameters worn on the three lower clamped balls. From Figure 3 and Figure 4, it has been found that the coefficient of friction value obtained with 10W30 lubricant is relatively less as compared to other synthetic lubricants. For 5052 H32 aluminium, tests reveal that the 10W30 oil gives the best surface finish than other oils. The coefficient of friction of 0.11013 was obtained with 10W30 oil which has lesser in magnitude. The efforts have also extended by comparing synthetic oil with grease and dry lubricating techniques. The synthetic oil lubricant is found to be a good choice for SPIF process to get a finished surface and formability of deformed specimen as compared to dry, grease and paste of graphite and grease.

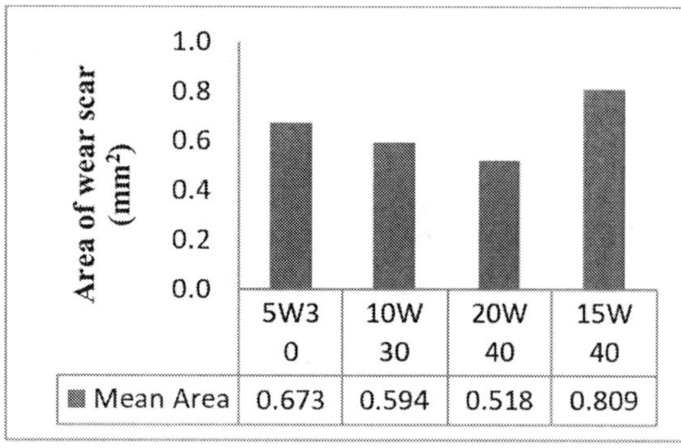

Figure 3. Results of wear test conducted as per ASTM D 4172

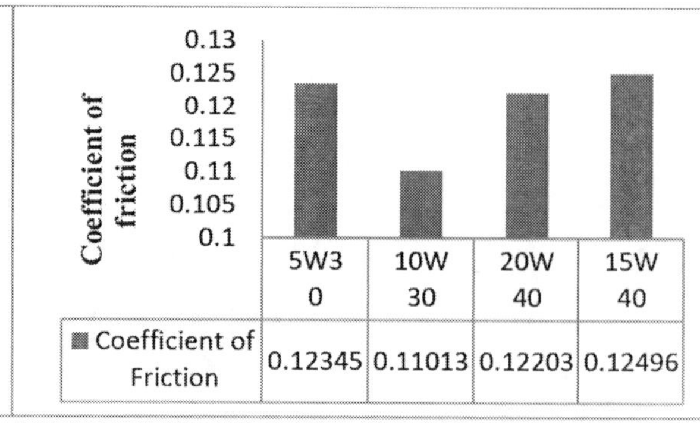

Figure 4. Results of the coefficient of friction conducted as per ASTM D 4172

At the same forming condition, the process shows different behavior with a change in the lubricant as illustrated in Figure 5. The formability has been indicated by the maximum angle of forming without getting a fracture in SPIF process with varying wall angle conical frustum. According to the results obtained, it was possible to note that 10W30 provides better results during forming operation. The 10W30 oil has a viscosity of 71.5mm^2/s at 40^{0}C, the relative density of 0.856 g/ml at 15^{0}C. Servo 10W30 oil lubricant gives good results as compared to other oil lubricants. It is viscous oil generally used in automobile engine to attain lubrication in the service conditions. Hence, the synthetic oil 10W30 was used in the experimental investigation.

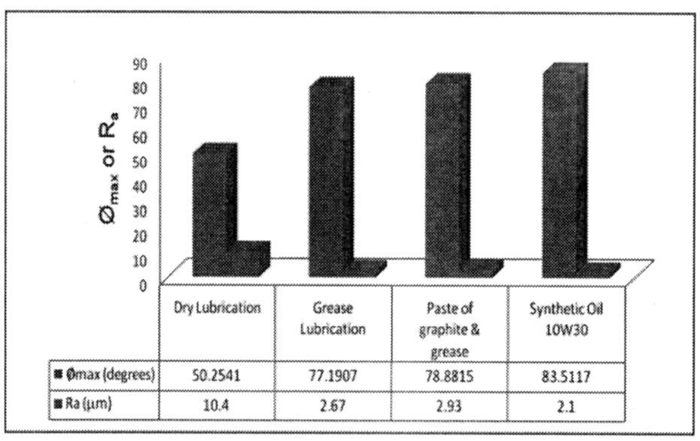

Figure 5. Selection of type of lubricant to SPIF process

3. Experimental investigations

The experimental setup of SPIF forming fixture mounted on CNC machine bed (Model: Surya VF 30 CNC VS, Make: Bharat Fritz Werner LTD, Bangalore, India). The experiments were carried out to evaluate the effect of input process parameters such as spindle speed (N), table feed (f), step depth (p), tool diameter (d) and sheet thickness (t_o) on surface finish and formability. The process parameters and their levels have been shown in Table 1. The SPIF experiments were performed on AA5052 H32 material as per Taguchi L_{27} design of experiments. The experimental setup, specimen geometry, fixture and statistical analysis can be found in the article [1]. According to the investigation carried out by the author and his team, the table feed and step depth have a major contribution in affecting the surface finish of 71.11% and 18.24% among the total value. The spindle speed has played very little role in affecting the surface finish and sheet formability. Table 2 depicts the percentage error corresponding to average surface roughness and the maximum forming angle at their optimum conditions. The optimized conditions for average surface roughness and formability are $N_3f_1p_1d_2(t_o)_1$

and $N_3f_2p_1d_1(t_o)3$, respectively. In notation N_3, 'N' indicates spindle speed whereas '3' indicates level three. The prediction model was developed within 95% of confidence level. It has been found that the experimental response values obtained within a specified range of input process parameters were closely matched with the predicted outcome. It is also understood that the optimum combinations suggested are reasonably acceptable. Analysis of S/N ratio concludes that the optimum forming parameters are spindle speed at 3500 rpm, table feed at 600 mm/min, step depth at 0.2 mm, tool diameter 10 mm and sheet thickness at 0.8 mm for optimum value of average surface roughness (1.42 μm) while the optimum value of the maximum forming angle was found to be 85.40 degrees at spindle speed of 3500 rpm, table feed of 1400 mm/min, step depth of 0.2 mm, tool diameter of 8 mm and sheet thickness of 1.2 mm. The parts obtained from the confirmation experiments have shown in Figure 6.

Table 1. SPIF input parameters and their levels

Sr. no.	Factors	Units	Symbol	Low	Medium	High
1	Spindle speed	rpm	N	1500	2500	3500
2	Table feed	mm/min	f	600	1400	2200
3	Step depth	mm	p	0.2	0.4	0.6
4	Tool diameter	mm	d	8	10	12
5	Sheet thickness	mm	t_o	0.8	1	1.2

Table 2. Results of confirmation experiments

Responses	Optimum conditions	Predicted Values	Experimental Values	% of error
Surface Roughness	$N_3f_1p_1d_2(t_o)_1$	1.35	1.42	4.24
Maximum forming angle	$N_3f_2p_1d_1(t_o)_3$	85.40	84.26	1.35

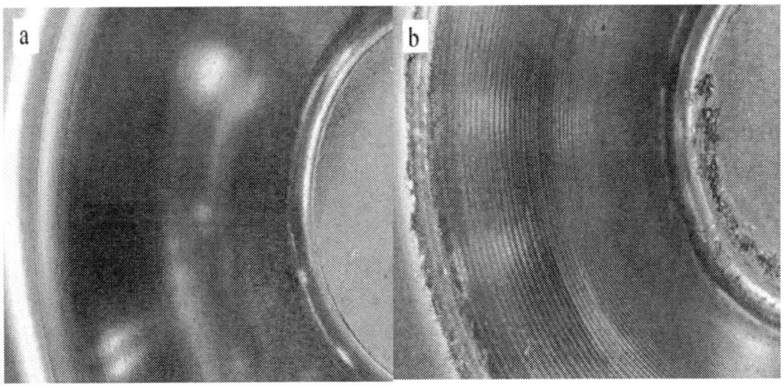

Figure 6. Confirmation experiments (a) average surface roughness (b) maximum forming angle

4. Fractography analysis

SEM micrographs of the cross-sectional view of SPIF fractured surface have been taken with the help of VEGA3TESCAN, Scanning Electron Microscope (SEM), TESCAN, CZECH REPUBLIC. After SPIF forming, the fractured surfaces at optimum conditions were cut precisely at fracture using wire Cut EDM to analyze fracture surface. The fracture surface of the incrementally formed part has analyzed using SEM photographs. The SEM images of the fractured surface at a magnification of ×500 at the optimum condition of surface roughness and formability are shown in Figure 7 (a) and 7 (b), respectively. The scanning electron microscope (SEM) photographs of the fractured surface of surface roughness and formability at ×2500 magnification are shown in Figure 8 (a) and 8 (b), respectively. The cup-like impressions (dimples) can be observed from the microstructure and they are nearly circular holes on the ruptured surface. Due to the presence of live local stresses and state of strain, the dimples of various sizes appeared on the yielding fracture surface. It indicates that the fracture is predominantly ductile in nature. In ductile fracture, damage accumulates due to nucleation, growth and coalescence of voids [17]. The fracture has occurred during the process after a considerable degree of deformation called severe plastic deformation. The results depict that the material deformed plastically before the fracture has happened.

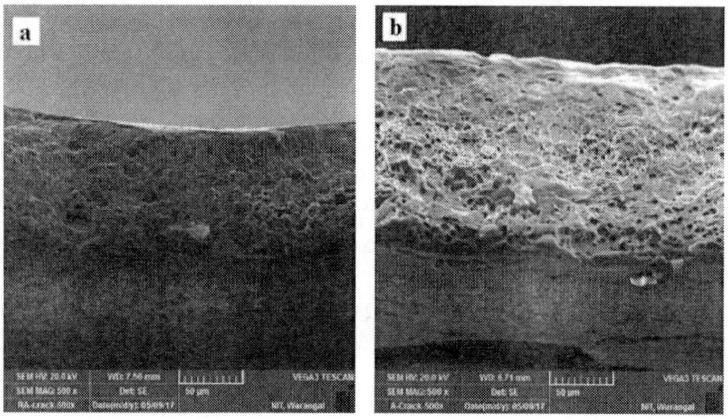

Figure 7. SEM photographs of fractured surface at ×500 magnification (a) average surface roughness (b) maximum forming angle

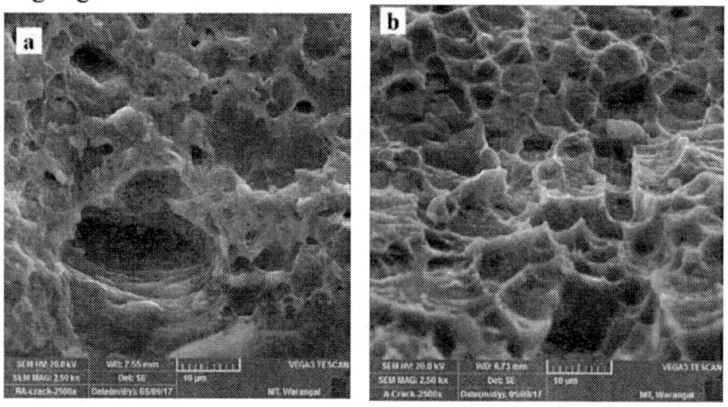

Figure 8. SEM photographs of fractured surface at ×2500 magnification (a) average surface roughness (b) maximum forming angle

5. Corrosion resistance and residual stress

The corrosion leads to material degradation, wastage of valuable resources, and loss in material efficiency. Polarization methods are faster experimental techniques as compared to classical weight loss estimation. All electrochemical measurements have been performed on the computerized electrochemical analyzer (supplied by Autolab Instruments, Netherlands). AA5052 H32 deformed sheets were used as the working electrode with a platinum wire as the counter and a saturated calomel electrode as the reference electrode. The AA5052 was cleaned initially and about 2 cm^2 of surface area was used for sample testing as the working electrode. Polarization measurements were taken after exposure to the 3.5% NaCl solution. The potential was scanned between −1 V to +1 V at a rate of 10 mV/s. The electrochemical measurements were carried out at room temperature (25^{0}C). For an electrochemical reaction under activation control, polarization curves exhibit linear behaviour in the E vs. log (i) plots called Tafel behaviour. The results obtained for the deformed SPIF parts at optimum conditions are shown in Figure 9(a) and 9(b).

The deformation of the AA5052 H32 sheet can significantly affect the corrosion resistance. As like other cold forming techniques, SPIF can also introduce residual stresses. The magnitude and significance of such stresses are always underestimated. The combination of induced stress and the corrosive atmosphere has a synergistic effect and the material corrodes faster than if unstressed. The combined effect is generally called as stress-corrosion or environmentally assisted cracking. From Figure 9, these results indicate that the degradation of the localized corrosion resistance can be caused by increases in the level of residual stress due to the incremental deformation. The residual stresses are always caused by inhomogeneous elastic and/or elastic-plastic changes of grain shape.

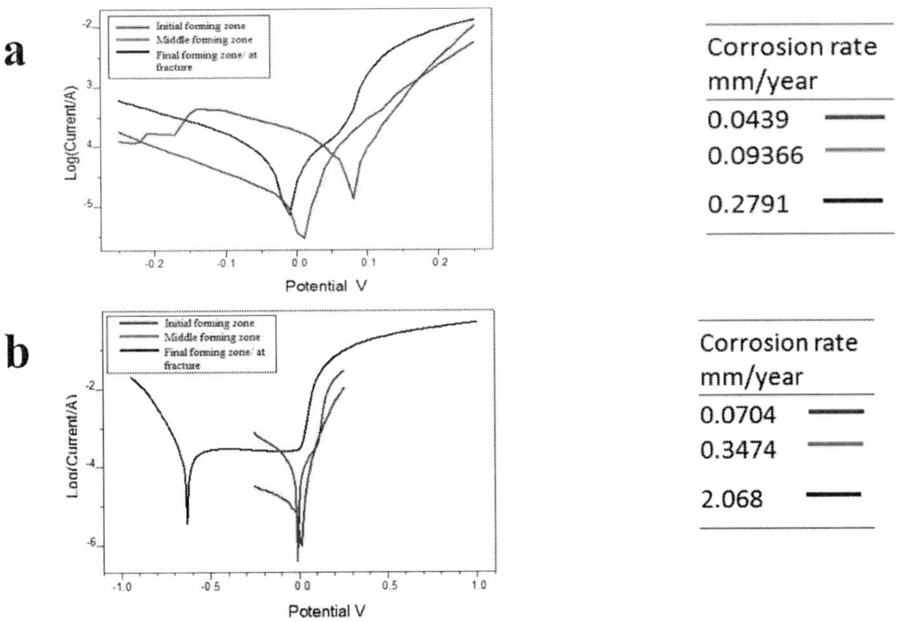

Figure 9. Tafel plots for the AA5052 optimum samples in 3.5% NACL solution (a) average surface roughness (b) maximum forming angle

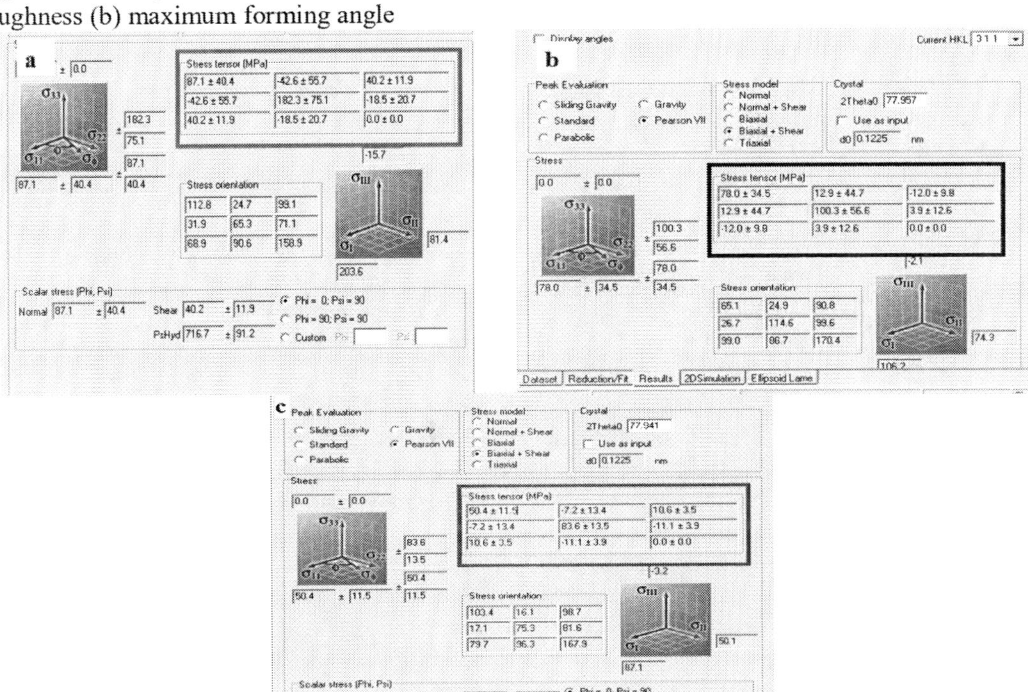

Figure 10. Residual Stress (a) average surface roughness optimum condition (b) formability optimum condition (c) as-received sheet.

The non-destructive x-ray diffraction method is used to find induced residual stress due to the forming operation. It has been found that the corrosion rate of the fully deformed sheet is higher than that of the initial deformed sheet in case of SPIF. It is well known that stress is a major factor in the rate of corrosion. Presence of any stress in the material will accelerate corrosion phenomena with the environment. Therefore the study is further extended to determine residual stresses present near to fracture zone. It was observed that residual stresses are developed in the material due to forming operation which eventually leads to lower corrosion resistance than as-received material (See Figure 10).

6. Conclusions

Even if the geometrical, forming and material parameters are the most significant parameters, the use of a proper lubricant may significantly improve the SPIF process performance. The preliminary study presented here provides a base for research engineer on lubricant should they select in the SPIF process. The research work contributes to understand the lubricant selection, residual stress and corrosion behavior. The following conclusions are drawn from the research results presented in this study.

1. It has been found that the performance 10W30 lubricant is relatively better to get a smooth surface and formability of deformed specimen as compared to dry, grease and paste of graphite and grease. The synthetic oil lubricant is found to be a good choice for incremental forming of AA5052 H32 alloy.
2. The fracture surfaces with respect to optimum conditions contained large and wide voids and dimples indicating fracture occurred at larger plastic strains.
3. Presence of any stress in the material will accelerate corrosion phenomena. It was observed that residual stresses are developed in the material due to forming operation which eventually leads to lower corrosion resistance than as-received material.
4. The accumulation of residual stress near fracture leads to generate rapid corrosion as compared to middle and initial zone specified in the study.

Acknowledgement

The Authors are thankful to the Department of Mechanical Engineering, National Institute of Technology, Warangal and Ministry of Human Resource Development, Government of India for support.

References

[1] Mulay A, Ben B S, Ismail S and Kocanda A 2017 *Arab. J. Sci. Eng.* **42**
[2] Cerro I, Maidagan E, Arana J, Rivero a. and Rodríguez P P 2006 *J. Mater. Process. Technol.* **177** 404–8
[3] Hussain G, Gao L, Hayat N, Cui Z, Pang Y C and Dar N U 2008 *J. Mater. Process. Technol.* **203** 193–201
[4] Shanmuganatan S P and Senthil Kumar V S 2014 *Procedia Eng.* **97** 346–56
[5] Liu Z, Li Y and Meehan P A 2013 *Int. J. Precis. Eng. Manuf.* **14** 1891–9
[6] Malwad D S and Nandedkar V M 2014 *Procedia Mater. Sci.* **6** 1505–10
[7] Desai B V., Desai K P and Raval H K 2014 *Procedia Mater. Sci.* **6** 666–73
[8] Li Y, Lu H, Daniel W J T and Meehan P A 2015 *Int. J. Adv. Manuf. Technol.* 2041–55
[9] Golabi S and Khazaali H 2014 *J. Mech. Sci. Technol.* **28** 3273–8
[10] Gulati V, Aryal A, Katyal P and Goswami A 2015 *J. Inst. Eng. Ser. C* **97** 221–9
[11] Aerens R, Eyckens P, Van Bael A and Duflou J R 2010 *Int. J. Adv. Manuf. Technol.* **46** 969–82
[12] Zhang Q, Xiao F, Guo H, Li C, Gao L, Guo X, Han W and Bondarev A B 2010 *J. Mater. Process. Technol.* **210** 323–9
[13] Fiorentino A, Marzi R and Ceretti E 2012 *Int. J. Mechatronics and Manufacturing Systems* **5** 36–45
[14] Micari F, Ambrogio G and Filice L 2007 *J. Mater. Process. Tech.* **191** 390–5
[15] Mugendiran V and Gnanavelbabu A 2017 *J. Mech. Sci. Technol. 31* **31** 2943–9
[16] Allwood J M, Braun D and Music O 2010 *J. Mater. Process. Technol.* **210** 1501–10
[17] Kurra S and Regalla S P 2014 *Integr. Med. Res.* **3** 158–71

Simulation of springback in cyclic wipe-bending

H H Bok*, K S Oh and Y S Kang

POSCO Global R&D Center, 100, Songdogwahak-ro, Yeonsu-gu, Incheon 21985, Republic of Korea

Email: bokhh@posco.com

Abstract. This paper proposes a bending method to measure the robustness of advanced material constitutive models, which are developed aiming to precisely express material responses including the Bauschinger effect, permanent softening, etc. Such constitutive models, in general, require multiple sophisticated material tests, e.g. cyclic uniaxial tension-compression, followed by appropriate numerical optimizations to identify corresponding material model parameters. These tests, for example, intrinsically require anti-buckling measures for specimens that can lead to redundant frictional forces and biaxial effects. Besides, the strategy for the optimization of material parameters or the selection of elastic modulus model is also an important point to influence the final performances. In an effort to provide more robust reliability for springback simulation results, a wipe-bending tool enabling cyclic load reversals has been developed as an intermediate validator positioned between the role of the cyclic uniaxial tension-compression and that of the U-drawbending. Predicted springback results employing the Yoshida-Uemori model have shown that this approach can be considered as an effective way to confirm both the reliability of material model parameters and the capability of a selected constitutive model in a part development or tooling stage.

1. Introduction

High strength steels for automobile structural parts often lead to unwanted deterioration of shape fixability due mainly to springback in cold forming. Concerning this, FE analyses are being widely used, employing advanced constitutive descriptions such as the Yoshida-Uemori model [1,8], Barlat's homogeneous anisotropic hardening model [2], etc. as well as the works of Chaboche [3,4] to precisely express complicated hardening behavior of the steels.

These constitutive models, providing accurate springback predictions, require multiple elaborate material tests to identify their material model parameters, which are closely related to load reversals or the change of loading in various directions [2]. While carrying out such tests, and doing subsequent data treatments, unintended uncertainties or redundant data can be produced by certain sources such as uneven distribution of material properties, the complexity of reference data acquiring (e.g., the correction of the friction and biaxial effects from anti-buckling plates [1,10] attached to tension-compression specimens), the following numerical optimizations for material model parameters, etc.

Regarding this, a cyclic wipe-bending tool has been introduced in order to help confirming not only the reliability of acquired material model parameters but also the performance of selected constitutive descriptions. This unique bending tool, specifically, aims to generate an elastic recovery zone only by bending moment after a cyclic loading, unlike the U-drawbending [5] by which the superposition of tensile stress is also allowed, thus diminishing the bending moment. Then an application with a DH780 steel sheet is shown together with corresponding FE simulations as described in the following sections.

2. Springback in cyclic wipe-bending

Wipe-bending (also, L-bending) is widely used in the automobile industry as a process to shape straight flanges in parts. Tools are basically consisted of wiping die, a punch and a pressing pad. This tool is useful to study the effect of tool conditions: as it is seen in other cases, the springback in wipe-bending is also largely influenced by the radius of die, tool clearance, pad pressure, punch radius, etc. [13]. In this paper, an improvement on the conventional tool concept has been made to add a cyclic loading function as follows.

2.1. Cyclic wipe-bending device

This new tool consists of a pair of rolling pins (rolls) and another pair of bending dies to perform a serial bending including load reversals, for which no unclamping and reclamping of the specimen are required.

20-mm-diameter rolls are mounted through four roller bearings inside the surrounding rigid steel frame with a fixed distance (100 mm) between the axes of the two rolls as shown in figure 1. These synchronized rolls travel vertically by a connected actuator according to programmed forward-reverse bending cycles. The upper and lower bending dies, corner radii (R_d) of 10 mm, tightly hold the rectangular specimen (40 mm × 140 mm) with two fastener bolts. The gap between the outer surfaces of the rolls and the die wall is fixed as 3 mm. Figure 2 schematically explains the way the specimen is bent by the lower and upper rolls for bending up and down, respectively. Here, the specimen undergoes springback in a gradual manner because either the upper or the lower roll maintains its contact with the specimen while the rolls are moving along the opposite direction. This gradual springback has to form the required opening angle in order for the approaching roll to successfully intrude into the narrow open space between the specimen and the die walls in the initial stage of the bending reversal.

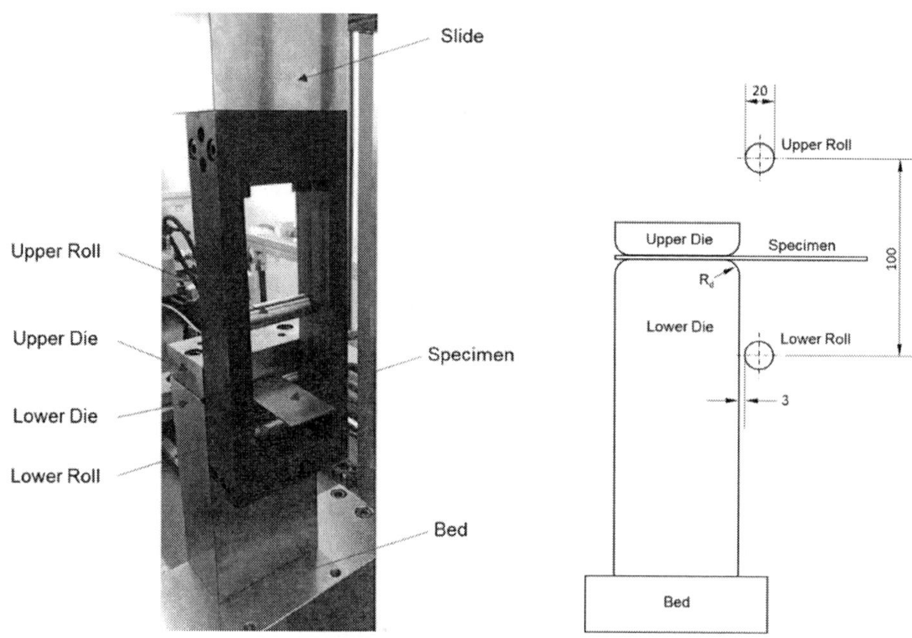

Figure 1. Cyclic wipe-bending device (left) and its side view with dimensions (right)

Through this way, the system can repeat a given number of bending reversals unless the specimen concerned shows enough amount of springback; in other words, in case the specimen is relatively thick or its strength very low (especially, mild steel cases).

A hot dip galvanized 780 MPa (minimum tensile strength) steel, DH780, sheet was used as the specimen material. The thickness of the specimen is 1.5 mm, of which the ratio between the specimen thickness and the tool gap (3 mm) leads to 0.5. The rolling direction of the specimen is perpendicular to the bending axis. Here, DH steel is a new concept of DP steel that improves formability by introducing a small amount of retained austenite into the microstructure, and thus can be applied to various cold stamped automobile body parts with complicated geometries.

Figure 3 presents the specimens produced by Forward (bending down), Forward (bending down)-Reverse (bending up) and Forward-Reverse-Forward bending processes involving zero, one and two load reversals, respectively. Three specimens were used for each bending condition to consider the deviation of the test data. Note that a negative sign of Z-coordinate is given only to the Forward-Reverse case in figure 3 for convenience because the final tool movement is the opposite of the others, leading to an upward flange.

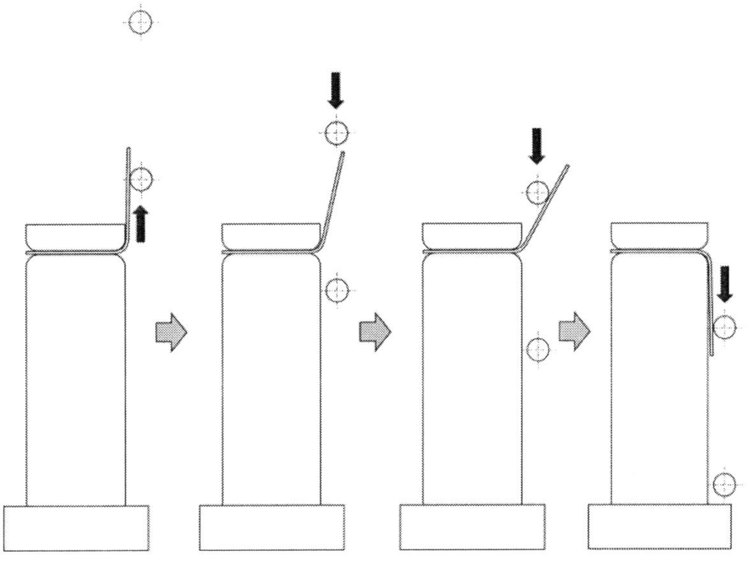

Figure 2. Schematic explanation of cyclic wipe-bending process

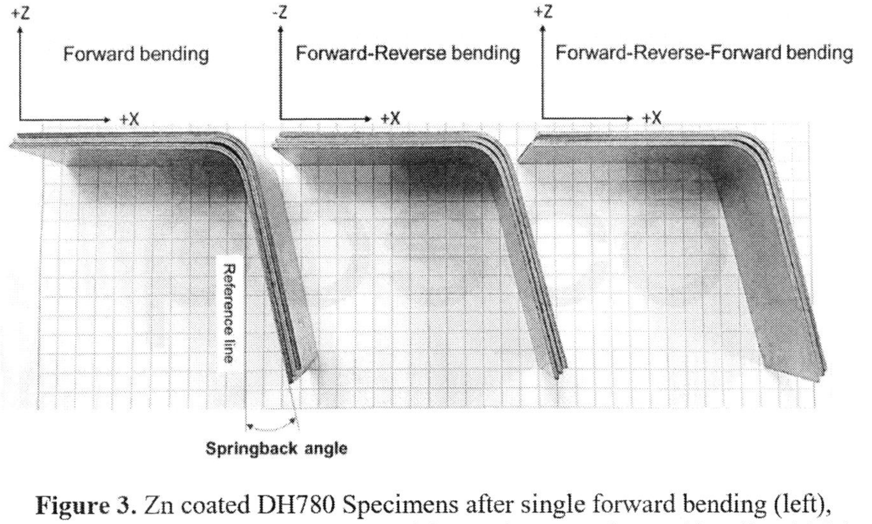

Figure 3. Zn coated DH780 Specimens after single forward bending (left), forward-reverse bending (center), and forward-reverse-forward bending (right)

In this study, the springback is simplified as the angle between the vertical reference line along Z coordinate and the line representing the straight portion of the flange region as explained in figure 3. This is not the true springback but an apparent measure because the flange at the end of bending is not exactly aligned with the reference line due to the tool clearance, which is larger than the specimen thickness. Consequently, almost the same springback angles were successfully reproduced in each specimen group subjected to their given bending conditions. These experimental results are summarized in figure 11 in section 3.

2.2. FE simulation

The developed wipe-bending process introduced in the previous section was implemented through a commercial FE software, Pam-Stamp2G^{TM} [6], as depicted in figure 4, employing the Belytschko-Tsay shell [7] for the element type of the specimen. The initial size of the element, the number of integration points through thickness and the vertical velocity of the rolls were optimized as 1 mm × 1mm, seven and 2 m/sec, respectively, through preceded springback sensitivity analyses. A negligibly low friction coefficient (0.001) was applied to the contact of the roll and the specimen surfaces to simplify the lubricating effect by the bearings.

The Yoshida-Uemori model (Y-U model) [1,8], a two surface model which is associated with the motions of both yield and bounding surfaces, by which cyclic plasticity relevant hardening behaviours can be precisely expressed, was considered. The two yield surfaces in the Y-U model are described by $f = \Phi(\boldsymbol{\sigma} - \boldsymbol{\alpha}) - Y = 0$ and $F = \Phi(\boldsymbol{\alpha} - \boldsymbol{\beta}) - (B + R) = 0$ for the yield and the bounding surfaces, respectively. In this, f moves only kinematically while F facilitates mixed isotropic-kinematic hardening. Here, $\boldsymbol{\alpha}$ and $\boldsymbol{\beta}$ are the back stresses of the yield and bounding surfaces, giving the relative kinematic motion of f and F as $\boldsymbol{\alpha}^*$ $(= \boldsymbol{\alpha} - \boldsymbol{\beta})$. Y represents the radius of f, and B and R are the initial size and the isotropic hardening (IH) component of F, respectively.

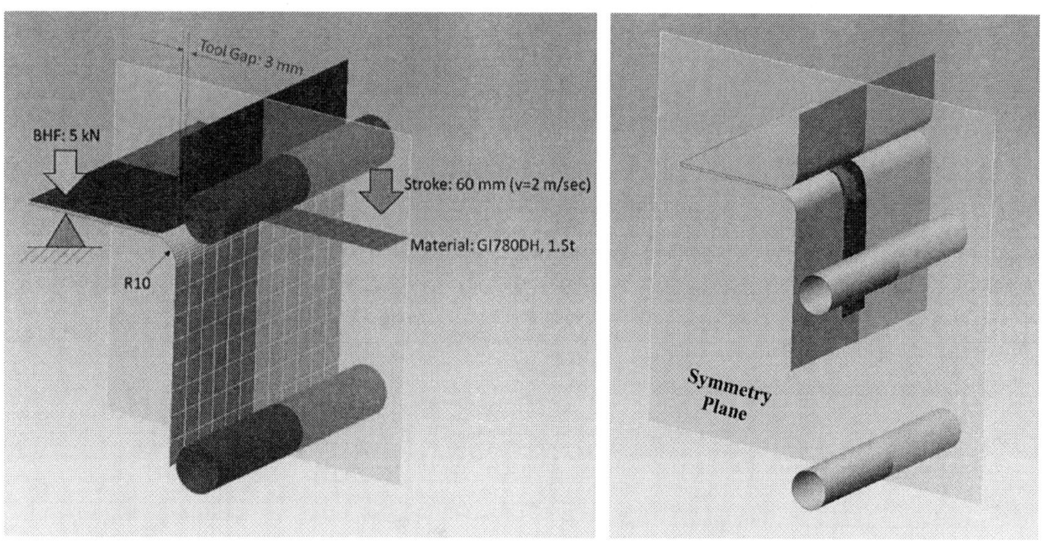

Figure 4. FE model (left) and its deformed configuration after forward bending (right)

Barlat's anisotropic yield surface model, the YLD2000-2d, was incorporated into the Y-U model to describe the plastic anisotropy (r_0, r_{45} and r_{90} of 0.789, 1.001 and 1.072, respectively). More detailed information to the Y-U and the YLD2000-2d can be found in the relevant works [1,8,9]. Then all material model parameters were numerically identified as listed in tables 1 and 2, associated with required material tests: namely, uniaxial tension-compression-tension (T-C-T) tests, tensile tests for r-values, etc.

Meanwhile, the parameters for the elastic modulus model (appeared in the bottom of table 1) as a function of plastic strain (prestrain) proposed by Yoshida et al. [8], $= E_0 - (E_0 - E_a)[1 - \exp(-\xi\varepsilon_p)]$ were identified by a direct fitting to the experimental unloading elastic moduli data measured along the rolling direction. In this, the stress points used to extract every average elastic modulus (chord modulus) required to constitute $E(\varepsilon_p)$ were taken at each load reversal from tension or compression state produced by another cyclic T-C-T test with a load reversing increment of 0.02. In other words, a chord modulus in this paper is the slope of a line bisecting the narrow loop (due to the nonlinear nature of elastic modulus) of stress-strain curve, formed while reversing load. This loop is also introduced as the so-called Quasi-Plastic-Elastic (QPE) strain [10,11]. This choice leads to the largest elastic recovery in a given stress drop in a course of unloading.

Here, the specimens for all the experimental T-C-T tests were specially designed to effectively adapt to their anti-buckling plates that covered the entire surfaces of the parallel regions of the specimens (gage length: 20 mm, width: 15 mm).

Figures 5 and 6 present the performances of these optimized material model parameters incorporated into the Y-U model in the case when they are applied to a single element FE model, of which its size is 1 mm× 1 mm, for the T-C-T test and the tensile tests regarding r-values along three different test directions. The Bauschinger effect followed by the non-linear kinematic hardening during the compression was appropriately captured by this simulation in figure 5.

Table 1. Material model parameters for Yoshida-Uemori model regarding DH780 steel

Description	Parameter	Typical Range	Value	Unit
Yield surface (f)	Y_0		0.4642	GPa
	C_1	200~2000	104.0	
	C_2	100~800	131.8	
	$\varepsilon_{p\ ref}$	0.005	0.02	
Bound surface (F)	B_0			
	$B_0 - Y$	0.1Y~0.5Y	0.0002	GPa
	R_0		0.0001	GPa
	b_0	0.1Y~0.5Y	0.07557	GPa
	M	2~30	6.224	
Hardening law (Swift type)	K		2.784	GPa
	e_0		0.0008819	
	N		0.03641	
Swift/Voce Weight	mu		1	
Non-IH	h_2	0~1	0	
	r_0	0.01	0.00001	GPa
Elastic modulus	E_0		207	GPa
	E_a		136.7	GPa
	ξ		23.1	

Table 2. Optimized parameters for YLD2000-2d

m	α_1	α_2	α_3	α_4	α_5	α_6	α_7	α_8
6	0.9094	1.0644	1.0565	0.9909	1.0106	0.9445	1.0107	1.0299

Overall, the one element shows good agreements with the experiments in this basic validation stage. However, a visible deviation from the experiment still remains: especially, a slightly higher hardening rate of the predicted curve is clearly observed in the compression stage (figure 5).

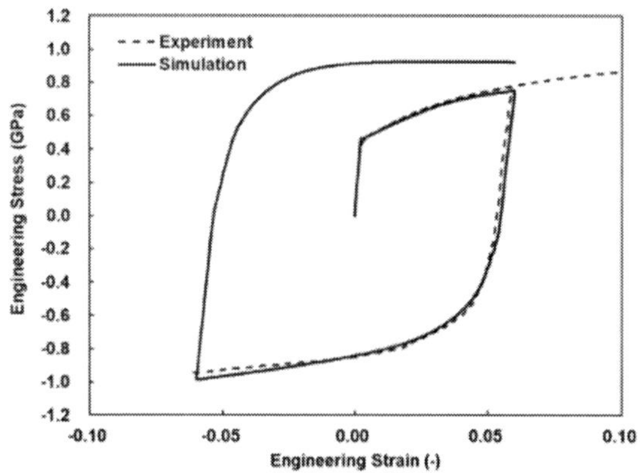

Figure 5. Simulated T-C-T curve

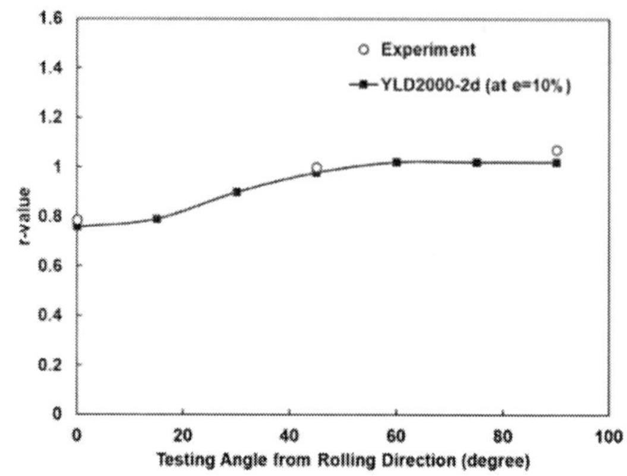

Figure 6. Simulated r-values of DH780

3. Results and Discussions

Figure 7 shows that the range of the strain change on the surfaces of the element at position P subjected to bending is similar to that of the T-C-T curve in figure 5, which are ± 0.068 and ± 0.060 for the former and the latter, respectively. In this, the coordinate X defined in the local (element) frame, which is initially aligned to the fixed global frame, is parallel to the longitudinal direction of the blank. Note that these stress and strain components are manually converted from those projected to the fixed global frame, based on the coordinate transformation rule. Besides, the shear components are ignored in this operation due to lack of related information from the software. This assumes that the monitoring element located at P undergoes negligible amount of shear.

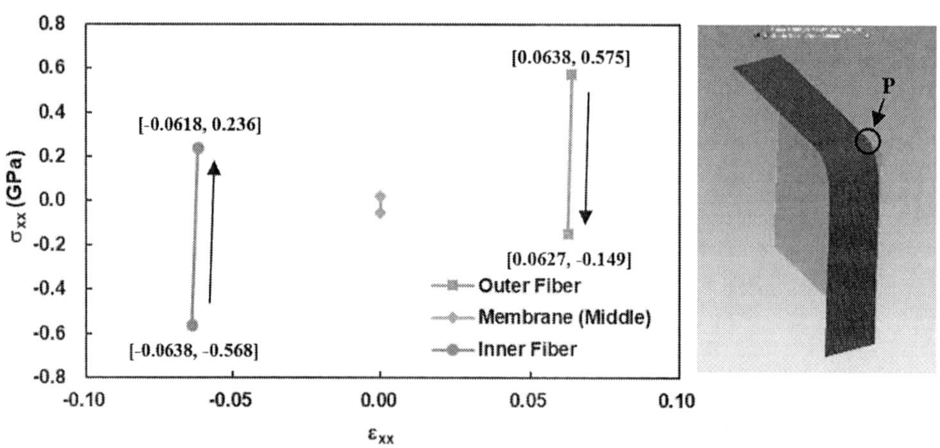

Figure 7. Change in stress-strain in material coordinate X at P during springback after Forward bending

The corresponding axial stress along X in the local frame is relieved from 0.575 to -0.149 GPa on the outer fiber (outer surface), and is from -0.568 to 0.236 GPa on the inner fiber (inner surface) of the monitoring position, P. The stress in the membrane (the middle fiber), meanwhile, remains almost unchanged compared to those on the surfaces.

The resultant springback angle corresponding to this stress change (figure 7) after Forward bending is depicted in figure 8, and those after Forward-Reverse and Forward-Reverse-Forward bending processes are shown in figures 9 and 10, respectively.

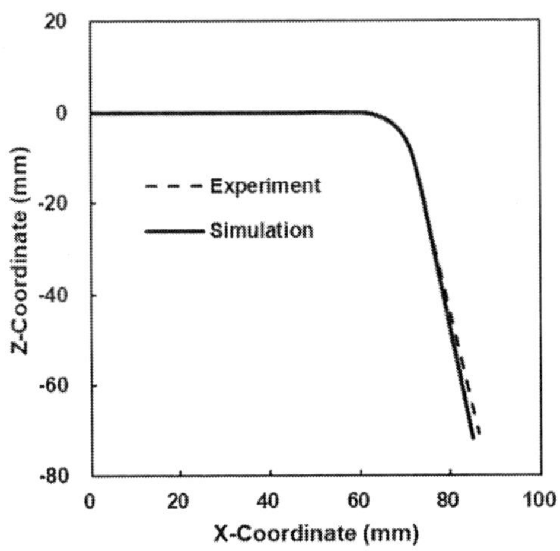

Figure 8. Springback after forward bending

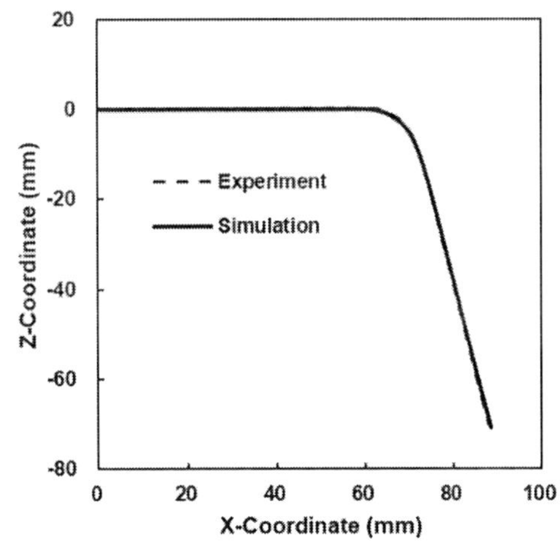

Figure 9. Springback after forward-reverse bending

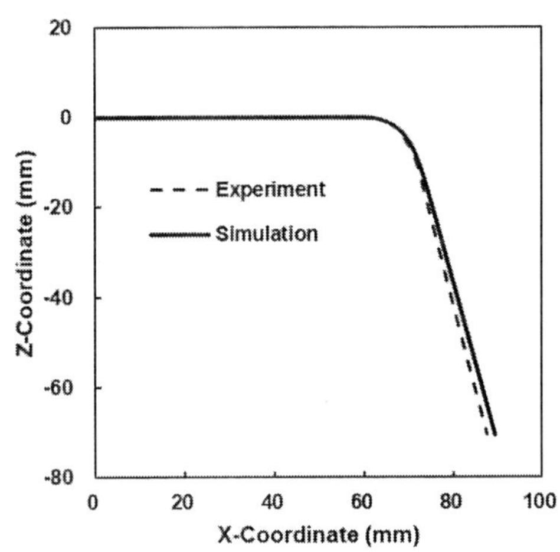

Figure 10. Springback after forward-reverse-forward bending

The springback trend, the change in the springback angle increases as the number of bending increases, is reasonably captured by the simulation as summarized in figure 11 in accordance with the sluggish material hardening in a large deformation stage. The range of the discrepancy from the experimental springback angles is -1.0 to 1.5 degree. The simulation leads to a larger increasing trend in springback angle as the number of bending increases. The similar discrepancy observed in the one element validation (section 2.2)—namely, the slightly higher hardening rate—can partly account for this difference. Predicted springback can also be influenced by not only the degree of fitness for an experimental cyclic flow curve depending on its parameter optimization strategy, but also by the expression of unloading elastic modulus. Especially, the way to define elastic modulus is a crucial factor to elastic recovery due to the nonlinear nature of elastic modulus, i.e., the QPE [10,11]. The chord modulus used in this work is, however, linear, and its slope can be chosen by the selection of the stress points in a stress-strain loop of unloading/reloading. The choice of this paper expects the largest elastic recovery as explained in section 2.2.

In addition, a recent study [12] showed that the prestrain dependent elastic modulus can also be modeled as function of testing direction. These aspects might be considered as a topic of further study with this bending tool.

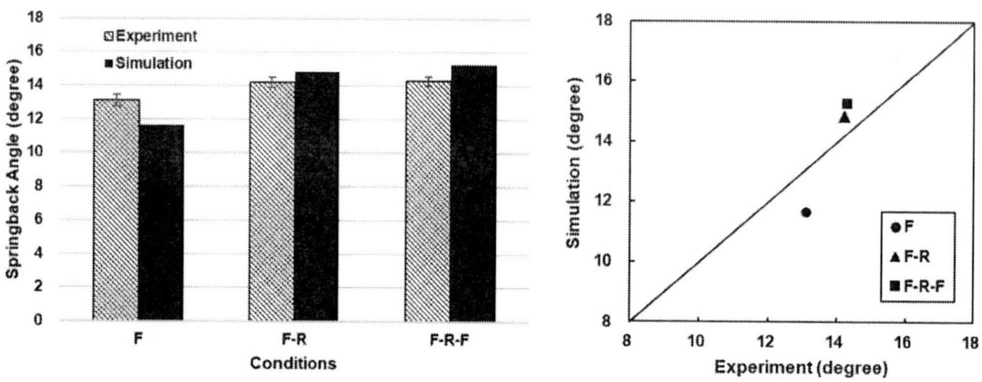

Figure 11. Comparison of springback angles (F: Forward, F-R: Forward-Reverse, F-R-F: Forward-Reverse-Forward)

4. Conclusions

A cyclic wipe-bending tool that can impose multiple bending reversals on sheet specimens is proposed for advanced material models to be applied more reliably. As the initial application of this device, the evaluations of springback angles after Forward, Forward-Reverse and Forward-Reverse-Forward bending processes were successfully carried out experimentally. Then, corresponding FE simulations were also implemented adopting the Yoshida-Uemori model [1,8] combined with the YLD2000-2d [9] employing numerically identified material model parameters, referring to the uniaxial tension-compression-tension response of a DH780 steel sheet. As a result, the predicted springback angles showed a good agreement with the experiments, and the sluggish trend in the springback angle development, which increases as the number of the bending reversal increases, was also properly captured by the simulation, though a small amount of discrepancy still remained due to various aspects, e.g., the strategies for the hardening parameter optimization, elastic modulus modelling, etc.

Nevertheless, it is worth using the developed procedure to check the performance of advanced material models specialized for springback evaluations such as the Chaboche model [3, 4], the Yoshida-Uemori model, Barlat's homogeneous anisotropic hardening model [2], etc., especially in need of an additional validation before applying them to full-scale practices.

References

[1] Yoshida F and Uemori T 2002 *Int. J. Plasticity* **18** p 661
[2] Barlat F, Gracio JJ, Lee MG, Rauch EF and Vincze G 2011 *Int. J. Plasticity* **27** p 1309
[3] Chaboche JL 1986 *Int. J. Plasticity* **2** p 149
[4] Chaboche JL 1991 *Int. J. Plasticity* **7** p 661
[5] Taylor L, Cao J, Karafillis AP and Boyce MC 1995 *J. Mater. Proc. Tech.* **50** p 168
[6] Pam-Stamp^{TM} User's Guide 2018 ESI Group
[7] Belytschko T and Tsay CS 1983 *Int. J. Num. Methods. Eng.* **19** p 405
[8] Yoshida F, Uemori T and Fujiwara K 2002 *Int. J. Plasticity* **18** p 633

[9] Barlat F, Brem JC, Yoon JW, Chung K, Dick RE, Choi SH, Pourboghrat F, Chu E and Lege DJ 2003 *Int. J. Plasticity* **19** p 1297
[10] Sun L and Wagoner RH 2011 *Int. J. Plasticity* **27** p 1126
[11] Wagoner RH, Lim H, Lee MG 2013 *Int. J. Plasticity* **45** p 3
[12] Sumikawa S, Ishiwatari A, Hiramoto J and Urabe T 2016 *J. Mater. Proc. Tech.* **230** p 1
[13] Livatyali H and Altan T 2001 *J. Mater. Proc. Tech.* **117** p 262

Compensation of elastic die and press deformations during sheet metal forming by optimizing blank holder design

M Burkart[1]*, P Essig[1], M Liewald[2], M Beck[1] and M Mueller[1]

[1]Mercedes-Benz AG, HPC C133, 71059 Sindelfingen, Germany
[2]Institute for Metal Forming Technology, University of Stuttgart, Holzgartenstr. 17, 70174 Stuttgart, Germany

E-Mail: maximilian.burkart@daimler.com

Abstract. Shortened product development processes of car body components made of sheet metal combined with the upcoming lack of experts and craftsmen's in that field do challenge car manufacturers fundamentally. Required tryout time phases and manufacturing costs of complete die sets for car body panels are significantly influenced by their proper function in production and therefore by their highly developed manufacturing process. Looking in more detail, elastic deformation of costed two components and forming press behaviour under load impacts the contact areas between the sheet metal and the load specific spotted areas of the tool. Especially in deep drawing such contact areas between matrice and blank holder are strongly influenced by the sheet metal properties and the part flange behaviour in terms of thickening and thinning during the process. Modern advanced forming simulation models indeed increase precision of sheet metal forming simulation remarkably. For example, new friction models and structural modelling of forming die and forming press components do improve simulation accuracy, while still using rigid tool components. Considering tool stiffness and press deformation in the die design by generating specific blank or geometries via stiffness optimisation, elastic deformations and pressure distribution presumably can be homogenised. In this respect, this paper presents a numerical approach for minimising elastic deformations in drawing operations and thus reducing trial time in the early engineering face. Here a stiffness optimised blank holder is designed based on topology optimisation, taking into account press deformation characteristics such as used drawing pins underneath the blank holder and the drawing cushion box integrated in the press bolster. Afterwards, the new designs compared to conventionally designed die based on structural simulations. The main finding of this contribution is that new approach presented on the one hand leads to improved tool stiffness and reduced tryout effort and more robust production conditions on the other hand.

1. Introduction

Facing major challenges such as the lack of experts and craftsmen's or the shortening of product development processes of car body components, manufacturing companies are forced to continuously improve their development processes, especially in the automotive industry. This particularly applies to the process chain in die manufacturing, which is accompanied by design and part development departments. Here, after the generation of sheet metal part CAD models, die engineering defines the process plan, consisting of numerous sheet metal operations such as forming, cutting, and flanging in order to produce a workpiece according to specification. The process plan is therefore crucial for robust and feasible part production. When transferring digital results into physical manufacturing, die tryout depicts an essential stage due to cost and time consuming manual spotting efforts [1]. These efforts are based on the differences between the forming simulation results and the manufactured geometries. Elastic deformations due to process forces induced by press ram action and die during sheet metal part forming are unknown. For this reason, surface pressures within the die are homogenized by iterative spotting steps during die tryout. Homogeneous surface pressures in contact areas reduce wear in series production to a minimum and ensure required part quality and die performance [2]. However, the initial contact areas at the beginning of the drawing process change due to straining effects of the sheet during the forming process. Furthermore, draw-in is influenced by part-dependent bearing areas and draw beads in the blank holder region. In turn, inhomogeneous pressure distribution in bearing areas leads to irregular draw-in of the blank and undesired surface defects of press part due to local pressure areas. Therefore, it is crucial to assure a homogeneous pressure distribution in bearing areas between matrice and blank holder to control tribological conditions and flange draw-in during the forming operation.

2. State of the art

To reduce tryout effort, present research work focuses on compensating specific press characteristics or unknown tool stiffness. Here, the main objective in published state of the art is to extend simulation models by considering press components in order to efficiently perform manual spotting. Pilthammar added components like the blank holder, the drawing cushion box, and the cushion pins to the forming simulation model in order to predict a virtual spotting image. However, the model is restricted due to the lack of information about component characteristics such as the exact design of the press components [3, 4].

Zgoll developed a coupled simulation model, which combines a forming simulation with a static FEM-simulation improving the engineering process. A so-called *fingerprint* tool is used to measure elastic deformations of the upper and lower press components, such as the press ram and the press table, enhancing his model. Simulation results led to new forming surfaces compensating elastic deformations in die design. The models' computing time is approximately 50 % higher than that of standard sheet metal forming simulation. So far, Zgoll could not approve positive effects on cost or time by use of his novel simulation approach [5].

In the context of component stiffness optimization, different design concepts such as bionic and topology optimization approaches can be distinguished [6, 7]. Topology optimization is represented by numerical calculation methods determining an optimized component design, based on specified optimization objectives. Possible design objectives are the maximization of stiffness or the reduction of mass or volume. Topology optimizations algorithms describe structures by relocating and rearranging originating design elements [8]. Thereby, resulting component designs are restricted by predefined boundary conditions, load cases, and of course available design space. The structural topology optimization calculation considers the components' internal force/stress distribution, regarding its stiffness or mass [9].

Figure 1 represents a characteristic process for design optimization, using the core element topology optimization. The principle of this topology optimization is based on the numerical Finite Element Method (FEM), which mainly applies the Solid Isotropic Material with Penalization (SIMP) method. Here, the elements' densities are adjusted according to the internal force distribution of the optimized structure [10]. After smoothing the component surface, the SIMP-method generates an output STL-mesh representing the optimized design. Subsequently, the proposed surface model has to be transferred to a solid body that specifies the final component design using a Computer-Aided Design (CAD) software. Therefore, it is crucial to consider the manufacturability as well as appropriate machining processes when re-designing the resulting geometry. Finally, a structural simulation and strength calculation validate the design of the optimized component [9].

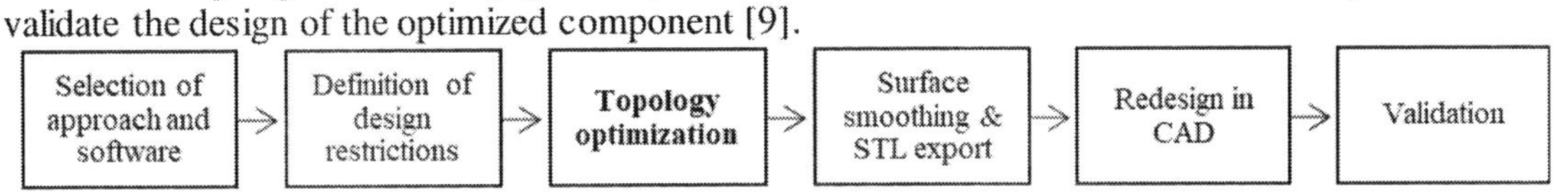

Figure 1. Structural design process using topology optimization [6, 8, 9].

In recent years, research in the field of topology optimization of tool structures associated with sheet metal forming disclose a rapidly increasing number of publications. Thereby, research work particularly focuses on the design of forming tools as well as press system components. In the context of the optimization of sheet metal forming tools by topology optimization, the main focus is on mass reduction. On the one hand, this leads to cost savings through reduced material use and energy consumption. On the other hand, the systematic reduction of accelerated masses can influence the dynamic tool behavior and does minimize reaction forces, such as part holder forces [11]. The research work of Dongkai presents a plane tool for deep drawing. The blank holder structure of this rectangular cup tool was optimized to decrease its mass. Within the optimization approach, loads and contact areas resulting from the sheet metal forming simulation were specified as design restrictions for complying with required forming conditions. The weight of the blank holder could be reduced by 28.1 % [12]. Further publications present mass reductions of a lower die by 20 % [13]. In general, topology optimization offers excellent potential for mass reduction while maintaining the stiffness of tool components [14].

Another topology optimization approach is improving press component stiffness. By optimizing the forming press, Ráž and Čechura improved the stiffness of a press crossbeam by up to 50 %. As a result, the required process precision during part production could be achieved [15]. Further research work presented similar results, for example, by reducing the weight of the press frame by 13,66 % and simultaneously increasing its stiffness [16]. However, an adaption of the press components is not sufficient for series part production, because of part-specific elastic deformation characteristics and long term invests in press machine standards.

Nevertheless, Pilthammer's research presents how the elastic deformation behavior of the press table can be considered in die manufacturing process [17]. By simulating real table deformations with a topology-optimized simplified model, elastic table behavior could be characterized and reproduced. Generating such an inverse substitute model enabled the prediction of elastic tool deformations via virtual die-tryout for an enhanced rework of forming contact areas.

In conclusion, contact areas between blank, blank holder and matrice during sheet metal forming are not only influenced by material thickening effects, but also by elastic press and tool deformations. These deformations are mainly resulting from the superposition of blank holder and drawing cushion box deflections, due to their stiffness characteristics and process loads. Therefore, topology optimization methods can lead to improved process characteristics by adapting die design.

Against this background, research work reported about in this paper has concentrated on the topology optimization of a blank holder in order to increase its stiffness regarding the drawing cushion box and cushion pins behavior. The main objective was to ensure simulated blank holder contact areas by compensating inhomogeneous drawing cushion box pressure distribution. The presented holistic optimization process for a new blank holder design was evaluated by a nominal forming simulation and a structural FEM simulation of the nominal blank holder, regarding its elastic deformation. As shown in figure 2, the total tryout effort could be reduced by improved sheet metal forming simulations or die designs. By implementing both simulation approaches, benefits could be combined in a holistic process chain improvement without increasing engineering effort significantly.

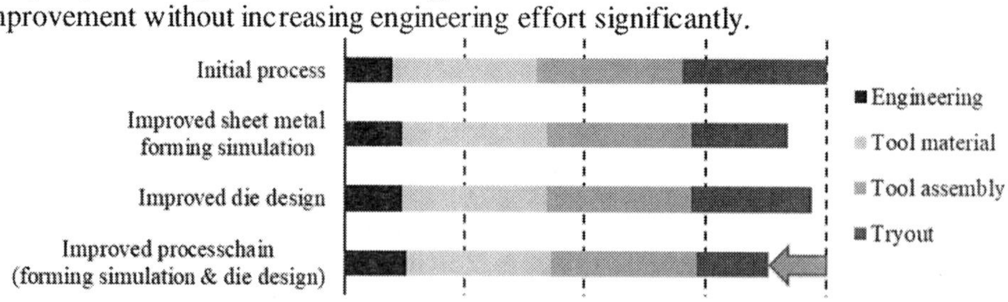

Figure 2. Improvement in die manufacturing process by decreasing tryout effort by the implementation of new methodologies considering sheet metal forming simulation and die design.

In the following section, the topology optimization process of an outer hood panel blank holder is presented. The required boundary conditions for topology optimization are specified regarding the blank holders' geometry and optimization process. Furthermore, different load cases are defined considering two stages of contact areas during forming process. Different contact areas result from homogeneous pressure distribution and from the consideration of straining effects The resulting optimized blank holder geometries are finally compared to the nominal blank holder geometry.

3. A general blank holder design optimization approach

The following section illustrates the proposed optimization methodology for blank holder structure. The method combines sheet metal forming simulation results with topology optimization using an approach including press characteristics, such as the drawing cushion box and cushion pins. The blank holder investigated and improved in this work is designed for drawing an outer hood panel. The nominal blank holder consists of EN- JS 2070 material and was already manufactured for series production. Due to intensive manual spotting work required, it became apparent that the stiffness of the blank holder should be improved. Results gained by structural FEM simulation in figure 3 shows nominal elastic deformations resulting from the forming load for the hood panel. Inhomogeneous deformation is caused by the critical rear area of hood due to the extremely reduced blank holder stiffness. Therefore, topology optimization offers potential for a re-designed blank holder to generate a homogeneous pressure distribution, which leads to homogenous deformations. Thereby, *Siemens NX 12.0* software system was used for the blank holder design and *Altair Inspire 2018.3* software for topology optimization targeting performed structural stiffness in this rear area.

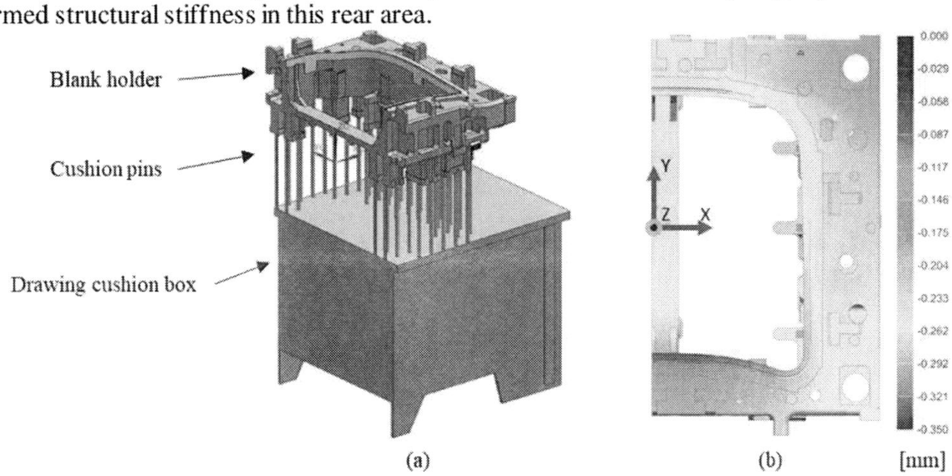

Figure 3. Structural FEM analysis of (a) CAD model of nominal blank holder and drawing cushion box components under load, resulting in (b) elastic deformation in z-direction.

The flow chart in figure 4 describes the method of the blank holder stiffness optimization performed. The main stages are listed on the left-hand side of the figure. The right-hand side illustrates the iteration process using the SIMP-based topology optimization method considering multiple load cases:

- Step 1: As a reference for comparison and validation, a nominal geometry is defined.
- Step 2: Contact areas with corresponding pressure values are derived for different load cases, based on a nominal forming simulation.
- Step 3: SIMP-based topology optimization is used to generate an optimized component design.
- Step 4: Design restrictions and design space are determined due to the die specification standards and given boundary conditions as well as press specifications. These specifications ensure a final design that meets series production conditions.
- Step 5: After smoothing the component surface, the SIMP-method generates an output STL-mesh of coordinate surface representing the optimized design.
- Step 6: Using CAD-software, the proposed surface model is transferred to a solid body that specifies the final component design. Considering the manufacturability while re-designing the geometry is crucial.

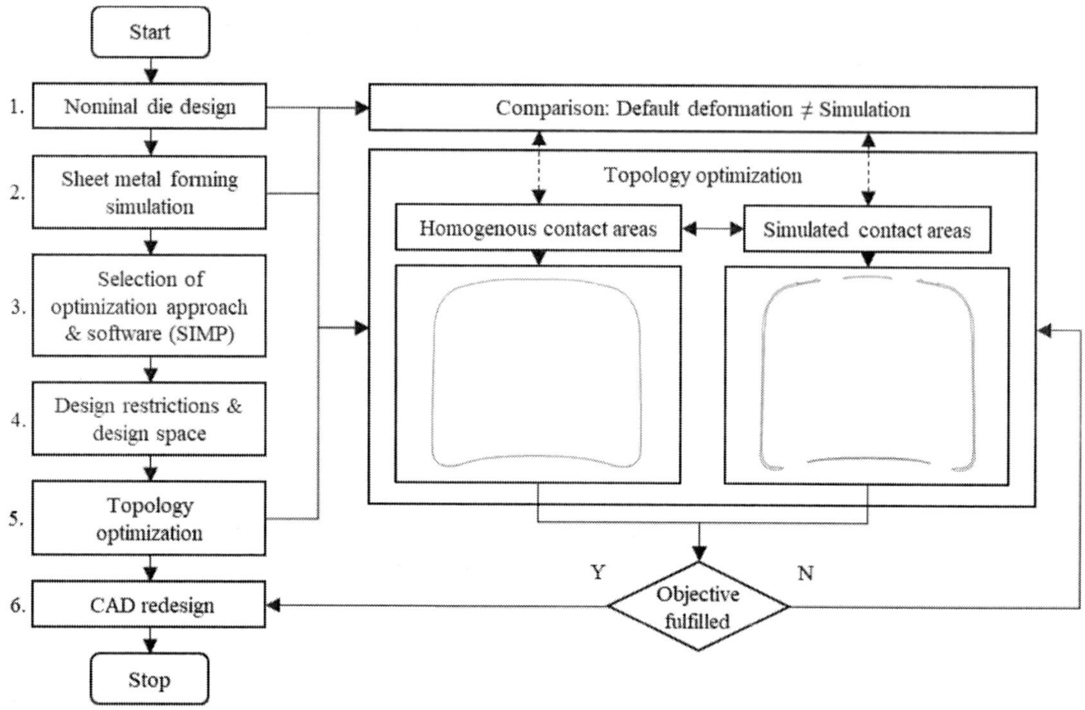

Figure 4. Die design optimization approach applied on blank holder structure using topology optimization and variation of pressure distribution.

4. Blank holder topology optimization regarding stiffness

Following sections describe the implementation of the topology optimization of the blank holder based on given boundary conditions. Furthermore, drawing cushion forces and two different load cases were considered in the optimization process. Finally, the proposed optimization results are described and compared, regarding their masses. The final stiffness evaluation will be carried out in section 5.

4.1. Boundary conditions regarding stiffness optimization (according to step 1, 3 & 4)

The optimization of the blank holder geometry was based on numerous design restrictions. Implemented boundary conditions and specifications for optimizing the blank holder originated from the process planning and the die design department. The process planning initially defined the active tool surfaces as well as the blank holder force required for ensuring a robust series part production. The design department engineer provided spatial boundary conditions, for example, the maximum tool space or guiding positions within the tool depending on the target press. The resulting design space for optimization was transferred to *Inspire* software, and functional areas were defined. These areas are represented by contact areas or cushion pin positions.

Based on given explicit design objectives, the design space subsequently was adapted via topology optimization. The first objective was to increase blank holder stiffness, which is significant to reduce elastic deformation during part production. Furthermore, this ensures homogeneous pressure distribution in the active contact areas during forming. A further objective of the optimization process was a mass reduction of 30 %, taking into account the restriction of minimum wall thickness required for the casting process.

4.2. Elastic characteristics of cushion box and cushion pins (according to step 2 & 3)

Since the stiffness behavior of a press generally is unknown, it is necessary to consider the elastic deformations of the press components, especially its lower components such as drawing cushion box and cushion pins. On the one hand, the approach presented in this paper assumes that the deflections of

IOP Conf. Series: Materials Science and Engineering **967** (2020) 012074 doi:10.1088/1757-899X/967/1/012074

the die emerge as specific to the sheet metal part. On the other hand, cushion pins and the drawing cushion box are also considered, as these have a significant influence on the size and location of merging contact areas. For the consideration of the elastic deformation, static FEM simulations of the drawing cushion box, including the cushion pins, were carried out based on the cushion pin pattern used for this hood. The force applied onto the structure in following topology optimization was implemented considering the drawing cushion box deformation, which results in individual forces for each cushion pin. Figure 5 shows the resulting elastic deformations based on a static structural simulation. The local deformation reveals how press structure is influencing pressure distribution during forming.

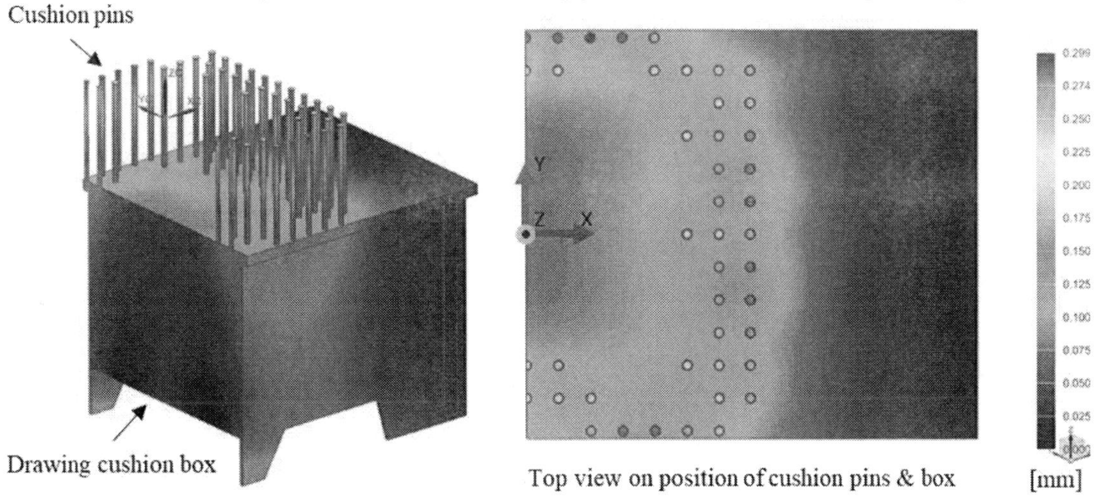

Figure 5. Structural FEM for drawing cushion box and cushion pins resulting in local elastic deformation in z-direction regarding part specific drawing cushion load.

4.3. Simulated load cases (according to step 2)

For the topology optimization of the blank holder geometry, two different load cases, represented by different contact areas, were investigated and implemented in the optimization software in order to generate a sufficient tool structure. During the deep drawing of sheet metal parts, the active contact areas are mainly influenced by the elastic deformations of tool and the press components as well as the thickening or thinning of the sheet metal due to the straining effect by deep drawing [18]. Based on these input parameters, resulting local surface load conditions onto the blank holder structure are calculated and corresponding contact surfaces are derived as boundary constraints via sheet metal forming simulation. According to these specific constraints, die structure analysis determines the adaption of design space for die structure topology optimization. The load cases investigated differ in pressure values and areas, leading to an extended investigation of the optimization process, represented in figure 6.

- **Load case 1.** Assuming a homogenous pressure distribution calculated by conventional sheet metal forming simulation specifies the first load case. In this case, a numerical factor in the simulation model, called *tool stiffness*, compensates emerging straining effects, resulting in the homogenous pressure distribution shown in figure 6 (a).
- **Load case 2**. In contrast to *load case 1 load case 2* ,as shown in figure 6 (b), differs mainly in the compensation of the straining effects. By changing the *tool stiffness* value in simulation, pressure distribution is shifting locally. Areas with higher thinning rates do not remain in contact, and therefore, pressure values decrease. As a result, contact pressures of remaining areas increase.

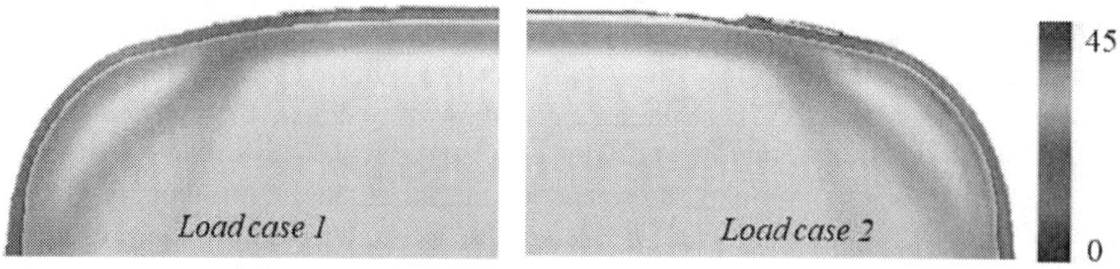

(a) (b)

Figure 6. Simulated pressure distribution for the two defined load cases represented by (a) *load case 1* homogenized spotting and (b) *load case 2* by variation of *tool stiffness* value based on thinning/thickening effects.

The load cases shown in figure 7 were implemented in *Inspire* software, using the defined boundary conditions. From these load cases the design space required for topology optimization was derived. Figure 7 (a) shows the nominal model used for evaluation. Figure 7 (b) represents the simulation model for *load case 1* and 2, whereby only the constraint surfaces were varied. The applied forces were obtained from static FEM simulation performed in section 4.2.

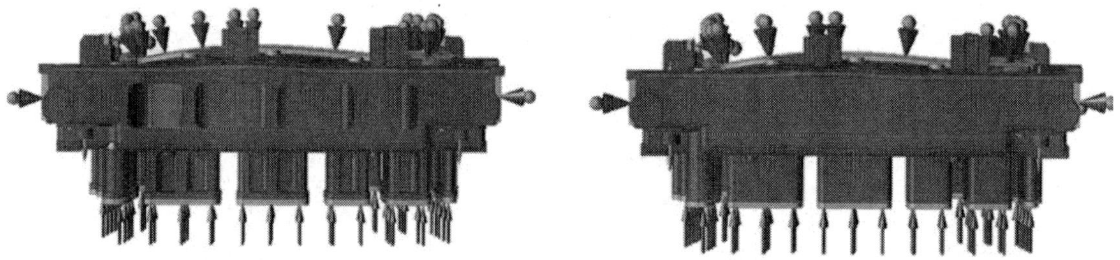
(a) Nominal design space (b) Extended design space

Figure 7. Implemented geometries used for *Inspire* with (a) nominal geometry as a structural comparison and (b) extended design space for *load case 1* and *2* for topology optimization.

4.4. Optimization results (according to step 5 & 6)

Based on the load cases, the optimized solutions were calculated by the SIMP optimization method as shown in figure 4. Compared to the nominal blank holder, the results shows significant differences in design. The calculated reaction force distribution is dependent on assumed load cases and the optimization objectives and is represented by the new structural design. Figure 8 shows the optimization result for *load case 1* and 2 in different views. Regarding the optimization results depicted in figure 8 (a) and (b), a minor difference between the optimizations becomes apparent. The optimization results are basically characterized by a material accumulation between the cushion pins and the contact areas, as well as, the remaining design features of the blank holder.

Due to the decreased contact area of *load case 2*, the local bearing forces are higher, which leads to additional structural support. According to the optimization method, the connection directed to the guiding column in the corner areas is reduced to a minimum and therefore is only relevant for blank holder positioning.

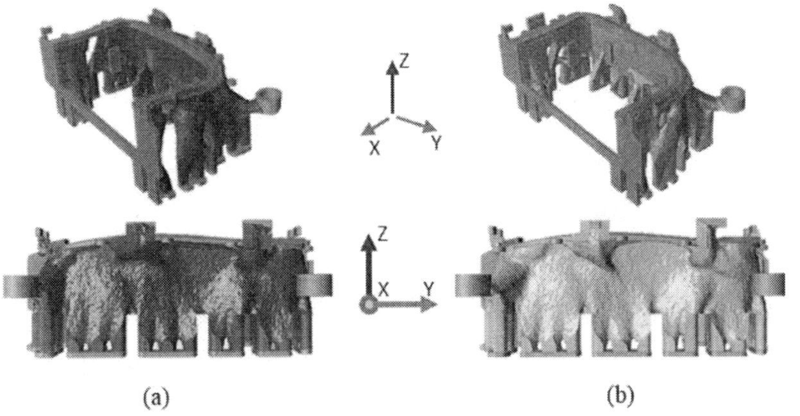

Figure 8. Proposed topology optimization results for (a) *load case 1* and (b) *load case 2* for blank holder shown in figure 3.

The two resulting design proposals show a changed structural geometry with different masses compared to the nominal geometry. In figure 9 the different masses are compared. The new design based on *load case 1* shows a reduction of 44 % in mass. *Load case 2* led to a mass reduction of 43 %, caused by the additional stiffness support. Regarding manufacturability and die design standardization requirements an empirical redesign factor of 1.3 was used, adapting the optimized geometries mass. The mass could be significantly decreased compared to the mass of the nominal blank holder. In addition to stiffness optimization, this leads to material and energy efficiency in series part production, due to reduced mass of blank holder. Especially concerning servo press technology, mass reduction enables process potentials. The resulting stiffness of the optimized blank holder geometries compared to the nominal design is evaluated in the following section.

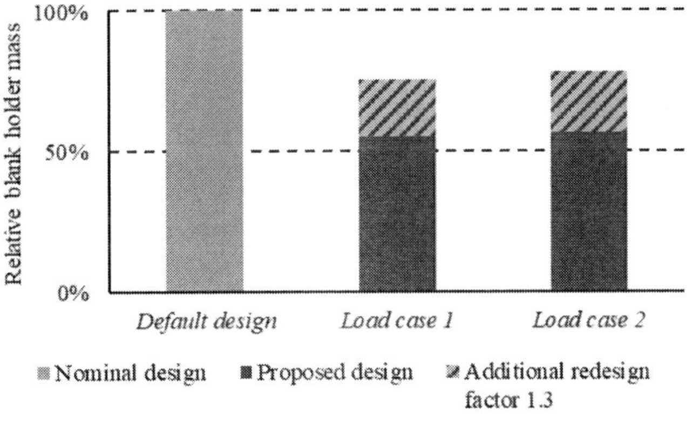

Figure 9. Mass comparison of proposed design for both load cases including the redesigning factor, in relation to the nominal design mass.

5. Stiffness evaluation

The main objectives of the improved blank holder structure are to ensure an increased stiffness and a homogenous pressure distribution during the forming process. For evaluating these objectives the elastic deformations resulting from implemented load cases were used. Based on boundary conditions, load cases and applied forces, the *Inspire* software calculates displacement via FEM. Figure 10 shows the nominal displacement compared to the optimized designs regarding their specific load behavior.

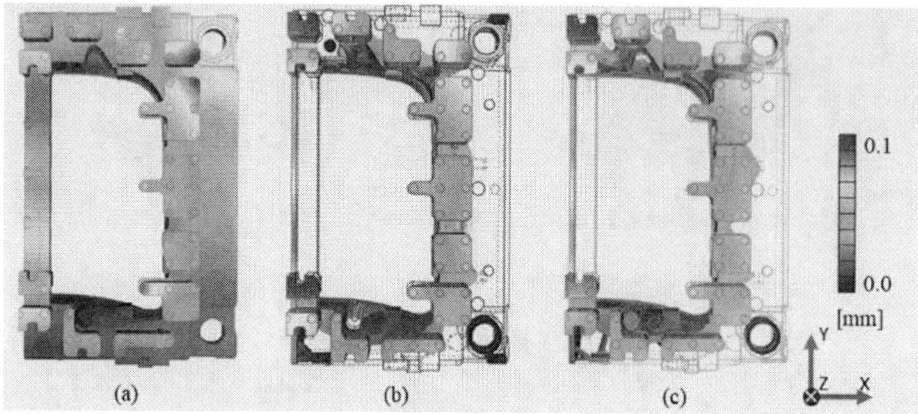

Figure 10. Calculated displacement in z-direction for (a) nominal design, (b) *load case 1* and (c) *load case 2* by *Inspire* software FEM simulation.

Due to the inhomogeneous force application resulting from the press characteristics, an inhomogeneous displacement of the blank holder results for each individual laod case. The deformation behavior of the nominal geometry does not correspond to the force distribution of the cushion pins as shown in figure 5. The stiffness optimized geometries compensate for the cushion pin force differences by means of specific material accumulation. Thus it becomes apparent that in both *load case 1* and 2 the elastic blank holder deformation in relevant regions corresponds to the force distribution resulting from the drawing cushion box and cushion pins deformation. Nevertheless, the presented deformations in both load cases differ according to the local structure.

The middle areas of the blank holder are characterized by increased deformations, as indicated by the red areas in figure 10 (b) and (c). In order to verify whether the deflection of these areas has a significant effect on the required stiffness, an additional evaluation criterion was applied. The deformation behavior of the bearing area can be evaluated by a load simulation of blank holder structure simulation by *Siemens NX* software, in which the drawing cushion box and cushion pins, as shown in figure 3, are positioned beneath the blank holder structure. The forming force is introduced into the tool via the simulated contact areas. Therefore the contact areas deformation in negative z-direction. The simulation result, shown in figure 11, indicates that the critical rear area displacement is improved by the optimization. Due to this, the global blank holder deformation is homogenized, enabling much more homogeneous contact areas and pressure distribution.

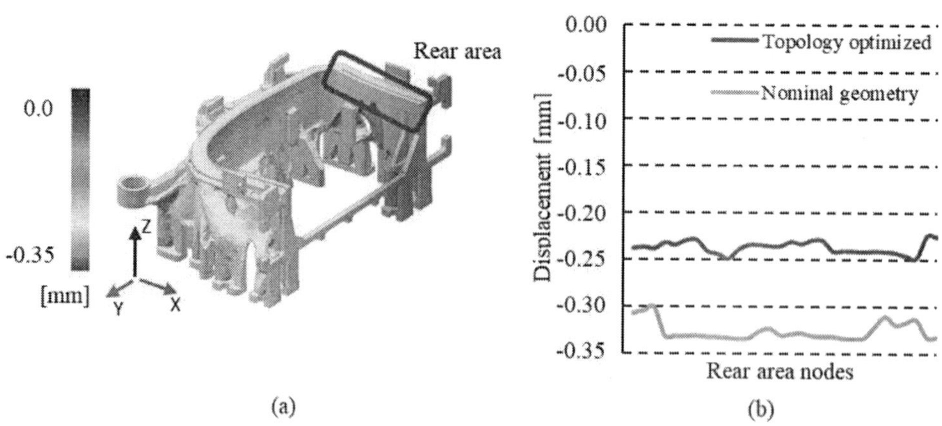

Figure 11. Simulation result as (a) elastic deformation of optimized blank holder design simulated with drawing cushion box components and (b) comparison of bearing area displacement in z-direction between nominal and optimized geometry regarding the rear part of the blank holder.

IOP Conf. Series: Materials Science and Engineering **967** (2020) 012074 doi:10.1088/1757-899X/967/1/012074

6. Conclusion

This paper shows that the stiffness of blank holder structure can be topology optimized regarding applied process characteristics, not only reducing mass/volume. Since the deformation of the blank holder during the sheet metal forming process is significantly influenced by its structure and the drawing cushion box as well als cushion pins. Drawing cushion box and pins deflections were simulated and considered by means of static FEM. Presented investigations indicate that, on the one hand, resulting reaction forces and, on the other hand, local displacement of the drawing cushion box must be taken into account for the part specific die design due to their influences on emerging blank holder contact areas. By topology optimization of the blank holder, its stiffness is adjusted to the inhomogeneous force application and the desired contact area. Homogeneous or straining of sheet metal part flange based contact areas can be used as target values. In addition, the mass of the blank holder is reduced by the optimization approach, as well as relevant guiding areas are highlighted and can be prioritized during the design process. By redesigning the optimization proposal and manufacturing by sand casting, a force distribution optimized blank holder structure could be enabled. This is especially crucial for the reduction of time and costs during die tryout. Further research will focus on validating and evaluating the presented optimization and simulation approach. Drawing cushion box deformation measurement as well as the digitalization of spotting images can be used to improve the boundary conditions for blank holder topology optimization. Finally, a holistic implementation by the manufacturing and analysis of an optimized series die can validate more the thoroughly presented design process.

7. References

[1] Birkert A R, Haage S and Straub M 2013 Umformtechnische Herstellung komplexer Karosserieteile: Auslegung von Ziehanlagen (Berlin, Heidelberg: Springer Berlin Heidelberg)

[2] Lange K (ed) 1990 Blechbearbeitung (Umformtechnik Handbuch für Industrie und Wissenschaft / hrsg. von Kurt Lange ; Bd. 3) 2nd edn (Berlin: Springer)

[3] Pilthammar J, Sigvant M and Kao-Walter S 2016 J. Phys.: Conf. Ser. **734** 32036

[4] Pilthammar J, Sigvant M and Kao-Walter S 2018 Int J Solids Struct **151** 76–90

[5] Zgoll F, Kuruva S, Götze T and Volk W 2019 IOP Conf. Ser.: Mater. Sci. Eng. **651** 012021

[6] Lippert R B 2016 Gestaltung von Additive Manufacturing Bauteilen 3D-Druck beleuchtet ed R Lachmayer et al (Berlin, Heidelberg: Springer Berlin Heidelberg) pp 45–55

[7] Mattheck C 2006 Verborgene Gestaltgesetze der Natur: Optimalformen ohne Computer 1st edn (Karlsruhe: Forschungszentrum Karlsruhe)

[8] Schumacher A 2013 Optimierung mechanischer Strukturen: Grundlagen und industrielle Anwendungen 2nd edn (Berlin, Heidelberg: Springer)

[9] Klahn C, Meboldt M, Fontana F, Leutenecker-Twelsiek B and Jansen J (eds) 2018 Entwicklung und Konstruktion für die Additive Fertigung: Grundlagen und Methoden für den Einsatz in industriellen Endkundenprodukten 1st edn (Würzburg: Vogel Business Media)

[10] Goelke M 2018 Practical Aspects of Structural Optimization: A Study Guide 3rd edn (Academic Program Altair University)

[11] Burkart M, Liewald M, Wied J, Todzy T, Hartmann M and Müller M 2020 Procedia Manufacturing 47 861–6

[12] Xu D, Chen J, Tang Y and Cao J 2012 International Journal of Mechanical Sciences **59** 73–82

[13] Kruschke G 2016 Enorme Kräfte von bis zu 2.500 Tonnen: Schneller und präziser fertigen https://www.produktion.de/wirtschaft/audi-entwickelt-neue-werkzeuggeneration-311.html?page=2 (accessed 7 Nov 2019)

[14] Eracheri J, Venkatesan R and Kumar N 2018 15th International LS-DYNA® Users Conference 1–6

[15] Ráž K and Čechura M 2019 IOP Conf. Ser.: Mater. Sci. Eng. **651** 012015

[16] Zhao X, Liu Y, Hua L and Mao H 2016 Struct Multidisc Optim **54** 375–89

[17] Pilthammar J, Sigvant M, Hansson M, Pálsson E and Rutgersson W 2017 J. Phys.: Conf. Ser. **896** 12068

[18] Fann K-J, Liewald M, Riedmüller K, Randonjic R and Karadogan C 2019 Proc. of the 2nd APSTP

Inferring the surface roughness of Al-Si coated 22MnB5 steel using an *in situ* laser speckle characterization technique

C M Klassen[1,3], J Emmert[2] and K J Daun[1]

[1]Department of Mechanical and Mechatronics Engineering, University of Waterloo, 200 University Ave. West, Waterloo, ON, N2L 3G1, Canada
[2]Department of Mechanical Engineering - Reactive Flows and Diagnostics, Technical University of Darmstadt, Otto-Berndt-Str. 3, Darmstadt, HE, 64287, Germany
[3]Corresponding author: cameron.klassen@uwaterloo.ca

Abstract. Hot stamping of aluminium-silicon (Al-Si) coated 22MnB5 steel blanks is widely used in the automotive industry to produce light yet crashworthy parts. However, the coating melts at ~577°C and transforms into a rough intermetallic layer as iron from the base steel diffuses towards the surface. The blank surface roughness impacts the radiative properties during heating as well as weldability, paint adhesiveness, and cooling rate during forming and quenching. This study pioneers the use of laser speckle patterns, caused by the constructive and destructive interference of collimated light reflected off the blanks, to infer the evolving surface roughness of Al-Si coated steel coupons *in situ*. The results reveal a significant increase in surface roughness once intermetallic compounds reach the surface and that higher furnace set-points produce rougher parts.

1. Introduction

To combat greenhouse gas emissions and improve fuel economy, many automotive manufacturers use thinner structural components made of stronger steels to lighten vehicles [1]. In particular, ultra-high strength steel (UHSS) parts made from aluminium-silicon (Al-Si) coated 22MnB5 steel are ubiquitous for reducing vehicle weight without sacrificing crash protection [2]. These UHSS parts are made through hot stamping, a process in which blanks are heated to ~950°C in a long roller furnace to austenitize the steel and then transferred to a press where they are simultaneously formed and quenched [2]. The protective Al-Si layer is used to prevent the blanks from oxidizing and decarburizing in the furnace [3] while additionally providing moderate long-term corrosion resistance [4].

The Al-Si coating has a melting point of ~577°C, which will cause a brief liquid phase during heating [5]. The coating then re-solidifies as it reacts with iron that diffuses from the substrate steel, forming a variety of intermediate intermetallic phases during the heating cycle [6]. After the steel is austenitized, the surfaces of the coated samples transform into a complex microstructure of various Al-Fe-Si and Al-Fe phases [7]. While the liquid phase of the coating can have detrimental effects on the furnace [8], its transformation also drastically alters the surface roughness of the steel.

Surface roughness is often quantified by the arithmetic average roughness,

$$R_a = \frac{1}{L}\int_0^L |z(x)|\mathrm{d}x \tag{1}$$

where $z(x)$ is the difference between the profile height at location x and its overall mean height and L is the sampling length. Iron diffusion into the Al-Si coating significantly increases the surface roughness of the samples [9, 10]. Borsetto et al. showed that a longer holding time resulted in increased surface roughness and a higher holding temperature yielded samples with larger coefficients of friction [11]. Liang et al. found surface roughness to increase with higher dwell temperatures and heating rates, until reaching a maximum value at approximately 700°C [12]. Jenner et al. showed a distinct increase in the surface roughness of the samples shortly after the melting point of the coating as well as enhanced void formation within the coating at higher hold temperatures [13]. Conversely, they showed that a lower heating rate produced a rougher surface. Ghiotti et al. also found that lower soaking temperatures resulted in surfaces with higher friction coefficients once they had been quenched [14]. Barreau et al. investigated the surface roughness of hot stamping blanks with different heating rates and noticed a general increasing proportionality but could not identify a clear trend [15]. Finally, Podor et al. were able to estimate the topography of Al-Si samples using a high temperature environment scanning electron microscope and 3D image reconstruction based upon the 2D images to acquire height maps [16]. They discovered that the surface roughness of the samples remained relatively constant until the 675°C mark, where it increased greatly until 715°C. From 715°C to the holding temperature of 900°C, the roughness of the specimens decreased slightly.

The surface roughness of the blanks significantly affect various industrial processes that occur during and after heating, including paint adhesion [17], the interfacial heat transfer coefficient between the quenching die and the hot stamping blank [18], and weldability [19]. Furthermore, the evolving surface roughness of steel blanks greatly affects their radiative properties [9, 10], which impacts the temperature profiles of the blanks in the furnace.

Traditionally, surface roughness is evaluated using contact profilometry via a stylus instrument or with an optical profilometer, but these techniques are limited to *ex situ* characterization. While useful, *ex situ* measurements may be inaccurate due to phase transformations undergone by the surface as the samples are extracted from the furnace. Various optical procedures have been developed to produce non-contact, in-process methods of inferring roughness. One such procedure is the analysis of the speckle pattern produced when collimated laser light is reflected off a rough surface. The pattern consists of a series of bright and dark spots, as seen in figure 1, which is caused by the constructive and destructive interference of the scattered electromagnetic waves [20].

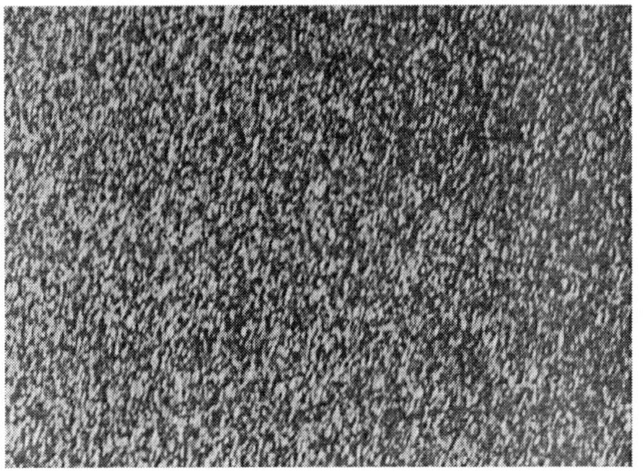

Figure 1. Example of a speckle pattern.

While several statistical processes exist to quantify the relationship between speckle patterns and surface topography, many researchers have related the contrast of the speckle pattern, i.e. the standard deviation of the pixel intensities within the image, directly to the surface roughness [21, 22, 23]. Rougher surfaces will scatter collimated light to a greater degree, yielding speckle patterns with more frequent deviations from the mean intensity value and higher contrast parameters. Surface roughness is proportional to the contrast of the speckle pattern provided that

$$R_a \leq \frac{\lambda}{4\cos\theta} \tag{2}$$

where λ is the wavelength of the laser and θ is the angle of incidence [21]. The correlation is invalid for rougher surfaces due to the enhanced scattering and resultant diffuse component of the laser illumination [24]. To evaluate rougher surfaces, Persson implemented a correlation method that compares two speckle patterns from the same surface at different angles of incidence and laser wavelengths [20].

Other researchers have employed the "bright-dark ratio method", which has a larger measurement range. This procedure converts the speckle pattern into a binary image of bright, B, and dark, D, pixels based on one or more cut-off values. If the intensity of an individual pixel is greater than the specified threshold, it is counted as bright and vice versa. Kayahan et al. compared the ratio of B and D to surface roughness and found an exponentially decreasing relationship to roughness values of 3λ [24]. Meireles et al. found an exponential proportionality between B/D and surface roughness values up to 12.75 μm for various surface types, but showed that the B/D ratio increased with increasing surface roughness [25]. They attributed this to their measurement setup as the image plane was aligned normal to the specimen and not in the specular direction of the reflected laser light. Using a similar setup, Hamed et al. related the surface roughness of aluminum samples to the proportion of bright values within the speckle images and also discovered a proportional relationship between the B/D ratio and surface roughness [26]. Xu et al. used a modification of this technique to yield linear relationships between the surface roughness of metals and the B/D ratio [27]. Finally, Wong and Li used this technique to analyze the changing surface roughness of parts moving through a machining process [28].

This paper documents the first application of laser speckle analysis to investigate the evolving surface roughness of the Al-Si coated 22MnB5 steel within a furnace. A laser was shone onto the surface of coated blanks heated within a laboratory muffle furnace via its exhaust port and the reflected light was imaged onto a camera sensor. A measurement model is derived by comparing B/D values of images against *ex situ* contact profilometry measurements. The surfaces of the extracted samples were further characterized using scanning electron microscopy (SEM) and energy-dispersive X-ray spectroscopy (EDS) to compliment the measurements. The speckle measurement model is then used to provide an *in situ* estimate of the surface roughness of the blanks during heating. These results provide important new insights into the evolving surface state of the steel and may improve the reliability of furnace modeling techniques and proceeding industrial processes.

2. Material and Methods

The experimental apparatus is shown in figure 2, which is built around a laboratory scale muffle furnace. An optical breadboard is custom fit around the exhaust port on the furnace roof, which provides optical access to the coupon. Light from a green laser diode (Thorlabs® CPS520) with a wavelength of 520 nm is imaged through the exhaust port onto the sample using a flat mirror. A portion of the reflected light from the sample is then projected onto the imaging chip of a Nikon® D7200 camera and the resultant speckle image was analysed using MATLAB® software. An image of the evolving surface state was taken approximately every two seconds.

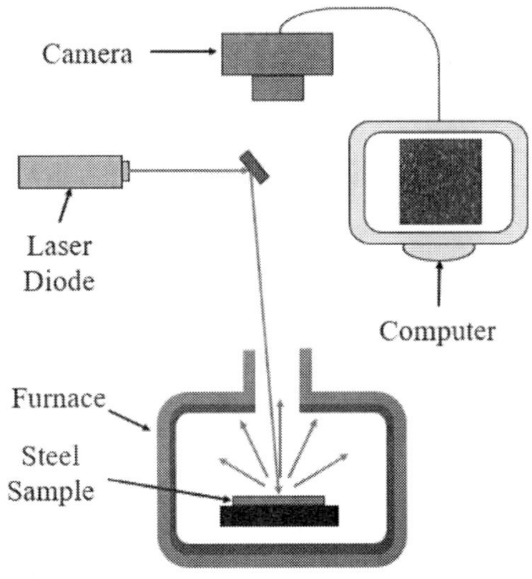

Figure 2. Speckle characterization experimental setup schematic.

Coupons consisted of 38 mm x 38 mm x 2.0 mm Usibor® 1500-AS150 blanks. The coupons had a nominal coating thickness of approximately 25 μm and a nominal mass composition of 90% aluminium and 10% silicon [29]. An example cross-sectional scanning electron microscope (SEM) image of the as-received coating is shown in figure 3. Small Si-aggregates are present in the Al-Si coating and a thin layer of Al_7Fe_2Si resulting from the Al-Si hot dipping process separates the base steel and coating. Samples were heated at set-points of 600°C, 700°C, 800°C, and 900°C. One sample for each furnace set-point was fitted with a k-type thermocouple to collect temperature data. For each test, the furnace was soaked for at least 5 minutes to ensure steady-state operation.

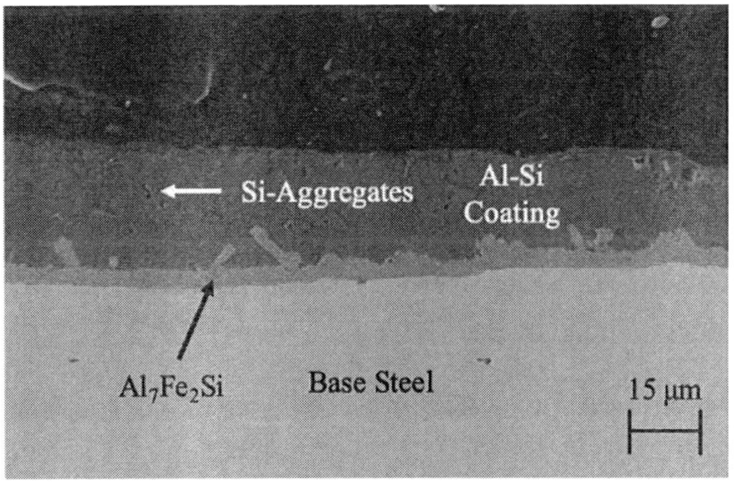

Figure 3. SEM image of cross-section of as-received Al-Si coated 22MnB5 steel at 750x magnification.

Ex situ surface roughness was characterized principally using a contact profilometer (Mitutoyo-SJ-400), which was calibrated against an optical profilometer (WYKO NT1100), which is considered to be more accurate. Both instruments are separately calibrated using their own calibration samples. The contact profilometer results are consistently lower than those of the optical profilometer by approximately 0.4 μm, as seen in figure 4, but show identical trends. The deviation could be attributed to differences in instrument resolution or the fact that the optical profilometer derives the roughness values from 2D maps whereas the contact profilometer relies on 1D line profiles. Since the discrepancy is reasonably constant, the proceeding analysis will be conducted with only the contact profilometer. Roughness profiles oriented at 0°, 45° and 90° with respect to the sample are recorded and averaged to produce the overall R_a value to account for any anisotropy at the surface of the specimens.

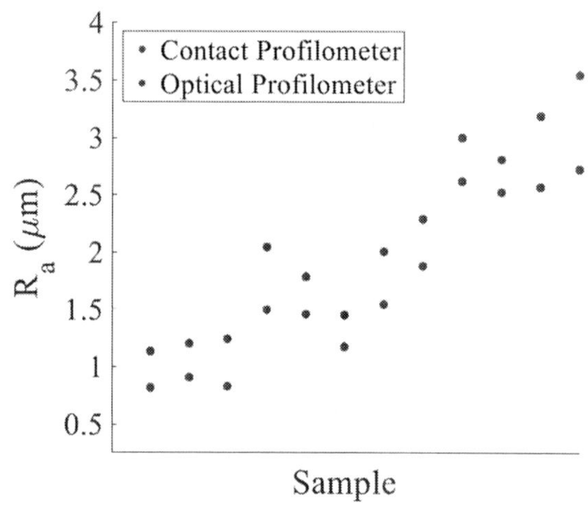

Figure 4. Comparison of contact and optical profilometer roughness values.

Finally, samples heated at the 600°C and 800°C set-points were removed from the furnace at intermediate heating times, immediately quenched with a stream of nitrogen gas, and then analysed using a SEM Zeiss Ultra Plus machine to assess intermetallic diffusion levels and chemical composition.

3. Results and Discussion

3.1 Speckle Pattern Characterization

To assess the validity of using speckle patterns to quantify the roughness of the blanks *in situ*, *ex situ* tests were conducted with various steel samples extracted at different furnace temperatures and dwell times to build a wide roughness sampling range. The $B\,/\,D$ ratio is sensitive to the cut-off parameter used to differentiate between the bright and dark pixels. We found that a threshold of 1.075 times the mean pixel intensity of the images yielded optimal results with respect to noise propagation for the *in situ* trials while giving a generally monotonic relationship between the $B\,/\,D$ ratio and surface roughness. An exponentially decreasing trend between the $B\,/\,D$ ratio and the roughness of the surface can be seen in figure 5. The best fit quadratic curve shown in figure 5 was then used to generate the *in situ* roughness profile in the following section.

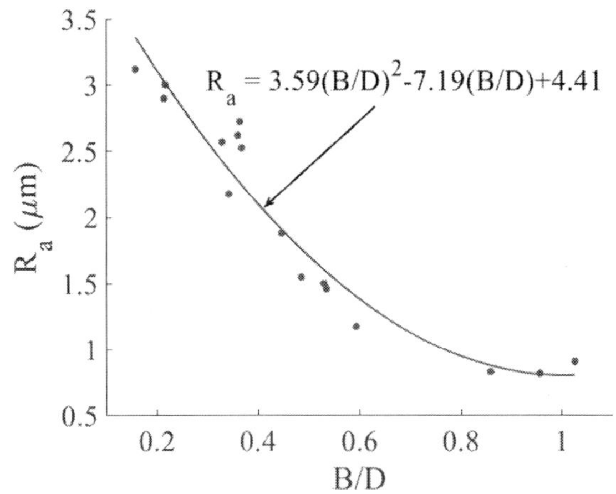

Figure 5. B / D ratio vs. R_a.

3.2 In Situ Analysis

The heating profiles for the different furnace set-points are displayed in figure 6a. Various inflection points in these curves can be detected, which is visualized in figure 6b when the heating rate is plotted alongside the temperature profile. The temperature derivative, calculated through first-order central finite differences, shows the melting point of the coating as well as the onset (T_{Ac1}) and conclusion (T_{Ac3}) of the austenitization process, aligning with the results of Jhajj et al. [30].

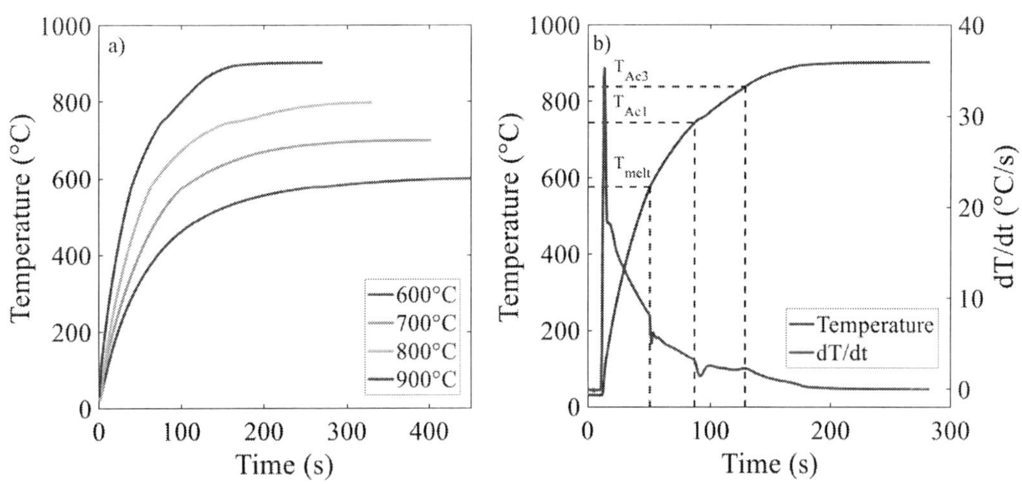

Figure 6. a) Heating profiles for various furnace set-points and b) temperature vs. dT/dt at the 900°C set-point.

An *in situ* B / D profile for a sample heated to 800°C is shown in figure 7. After approximately 20 seconds in the furnace, a distinct peak can be seen in the B / D plot, indicating a decrease in surface roughness and implying the coating has melted. Furthermore, a sharp decrease and subsequent increase can also be seen in the B / D signal shortly thereafter, before decreasing and finally plateauing. This closely resembles previously-reported *in situ* reflectivity measurements, which was hypothesized to be caused by an intermittent oxidation reaction between the two peaks [31]. Using *in situ* SEM images of the evolving surface state of hot stamping coupons, Barreau et al. hypothesized that an intermittent melting reaction occurs at approximately 655°C, potentially explaining the existence of the second peak [15].

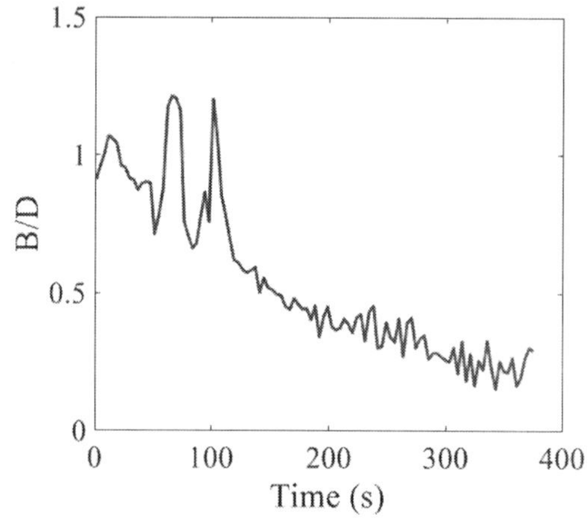

Figure 7. *In situ* B / D analysis for a sample heated to 800°C.

Notably, none of the samples heated to 600°C showed the two-peak behaviour, which echoes reflectance measurements from our previous study [31]. While the absence of the second peak was attributed to the existence of a "mushy" region on the Al-Si binary phase diagram, it is more likely that the samples simply do not reach the temperature necessary to incur the secondary melting reaction specified by Barreau et al. [15]. While the two-peak behaviour was detected for many of the individual trials heated at furnace set-points above 600°C, it was not present in every case. This could be because of the relatively low temporal resolution of the camera. To represent multiple trials, the mean of 5 B / D plots was taken to build an average curve for each temperature along with confidence intervals. An example of this process using the 700°C trials is shown in figure 8.

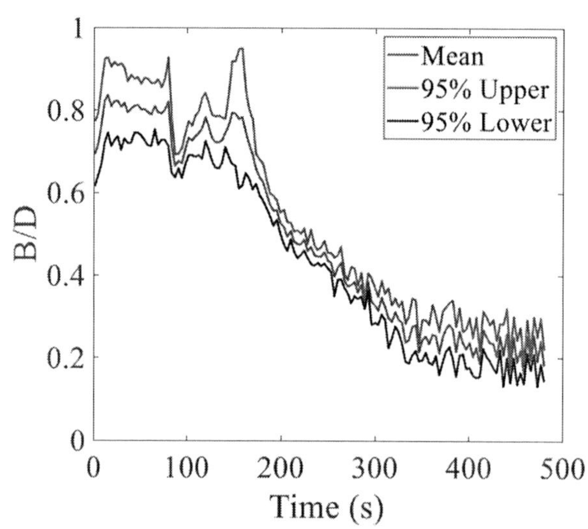

Figure 8. Mean B / D plot and confidence interval of 5 samples heated to 700°C.

Using the relation given in figure 6 and the average of 5 B / D curves, *in situ* roughness profiles were generated for samples heated at the different temperature set-points, as seen in figure 9. The different trials all exhibit an abrupt increase in roughness once the Al-Fe-Si compounds diffuse to the surface of the coating, and the final roughness increases with increasing furnace set-point. Samples were then extracted from the furnace and analyzed with the contact profilometer. The square data points in figure

IOP Conf. Series: Materials Science and Engineering **967** (2020) 012075 doi:10.1088/1757-899X/967/1/012075

9 show the average of four samples after a complete heating cycle at each of the four temperatures. In the case of samples heated to 900°C, the average roughness was 3.12 μm; accounting for the average 0.4 μm difference between the contact and optical profilometer, this result is consistent with roughness values of approximately 3.40 and 3.50 μm, reported by Shi et al. [9, 10] and Podor et al. [16] for coupons heated under similar conditions.

Furthermore, circular data points signify the average of two samples that were intermittently removed from the furnace at various dwell times and immediately quenched with a stream of nitrogen gas to preserve the surface microstructure. These samples were then measured with the profilometer and mostly show higher roughness values than what the B / D model predicts. However, this result is reasonable, given that any liquid constituent present at the surface of the samples would solidify when quenched, altering the overall roughness.

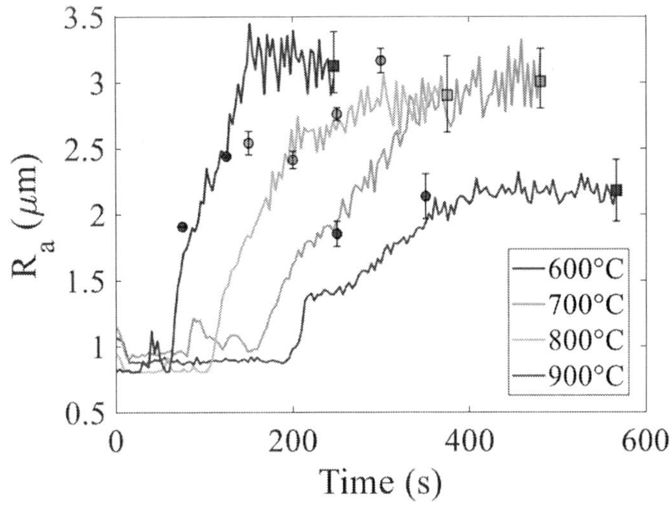

Figure 9. *In situ* roughness profiles for furnace set-points of 600°C, 700°C, 800°C and 900°C with *ex situ* contact profilometry measurements.

3.3 Ex situ SEM and EDS Analysis

The SEM and EDS analysis provided herein is from our previously published work [31]. The overall surface chemical composition measured using EDS over a 150 μm × 100 μm area for samples heated at set-points of 600°C and 800°C for various furnace dwell times can be seen in table 1. At both set-points, the increasing oxygen content implies the formation of an oxide layer. Additionally, rising iron levels indicate iron diffusion from the base steel to the outer layer. These results are consistent with the findings of Shi et al., as they detected alumina and iron oxides at longer heating durations [9, 10].

Backscattered SEM images of the sample surfaces can be seen in figure 10. The presented phases were approximated through the elemental compositions given by the EDS analysis. The propagation of $Al_{4.5}SiFe$ can be detected at both furnace set-points, aligning with the results of Grigorieva et al. [7]. Si-rich aggregates reported by Borsetto et al. [11] are present in the shorter dwell time samples but disappear when subjected to longer heating durations. Moreover, the surface also appears to become smoother in figure 10f compared to figure 10e, corresponding to the drop in the B / D signal between the two peaks, potentially caused by the intermittent melting reaction proposed by Barreau et al. [15].

Table 1. Surface chemical composition by furnace set-point and dwell time.

Furnace Set-point (°C)	Dwell Time (s)	Chemical Composition (at. %)			
		Al	Fe	Si	O
600	150	81.8	0.39	14.4	3.41
	210	86.8	1.03	8.37	3.76
	420	86.5	2.31	2.46	8.70
800	40	86.4	0.57	11.8	1.21
	50	84.9	1.01	11.8	2.28
	80	77.6	0.7	11.4	10.3
	100	80.3	0.75	11.4	7.61
	180	78.1	1.38	10.2	10.3

The evolution of the coating was further analysed using backscattered SEM cross-section images along with EDS line scans through the coating, as seen in figure 11. These images show progressive iron diffusion from the base metal to the outermost layer and a variety of Al-Fe-Si and Al-Fe intermetallics, including $Al_{4.5}FeSi$, $Al_{7.4}Fe_2Si$, and Al_5Fe_2. Iron diffusion occurs at a greater degree at the higher furnace set-point and the samples held at the longest dwell times show intermetallic layers reaching the surface, aligning with the final increase in surface roughness for the different trials seen in figure 9. Grigorieva et al. also identify $Al_{7.4}Fe_2Si$ and Al_5Fe_2 layers after the melting of the coating [7]. Additionally, Borsetto et al. show similar intermetallic compounds in samples heated to 600°C [11]. Heightened oxygen content at the surfaces of the samples heated to the longer dwell times further imply the formation of an oxide layer.

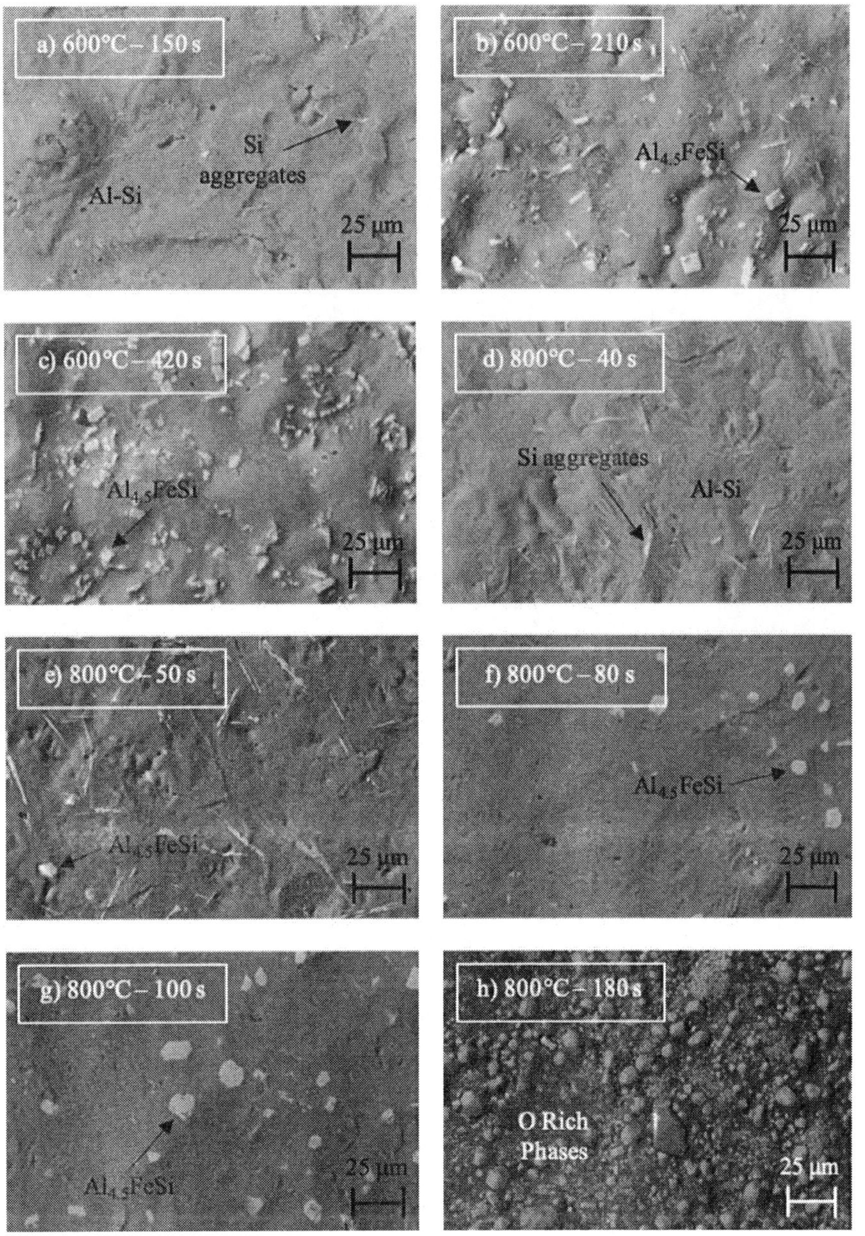

Figure 10. Backscattered SEM images depicting Al-Si coating evolution at the surface after heating at 600°C for: (a) 150 s, (b) 210 s and (c) 420 s; and at 800°C for (d) 40 s, (e) 50 s, (f) 80 s, (g) 100 s and (h) 180 s. (500x magnification).

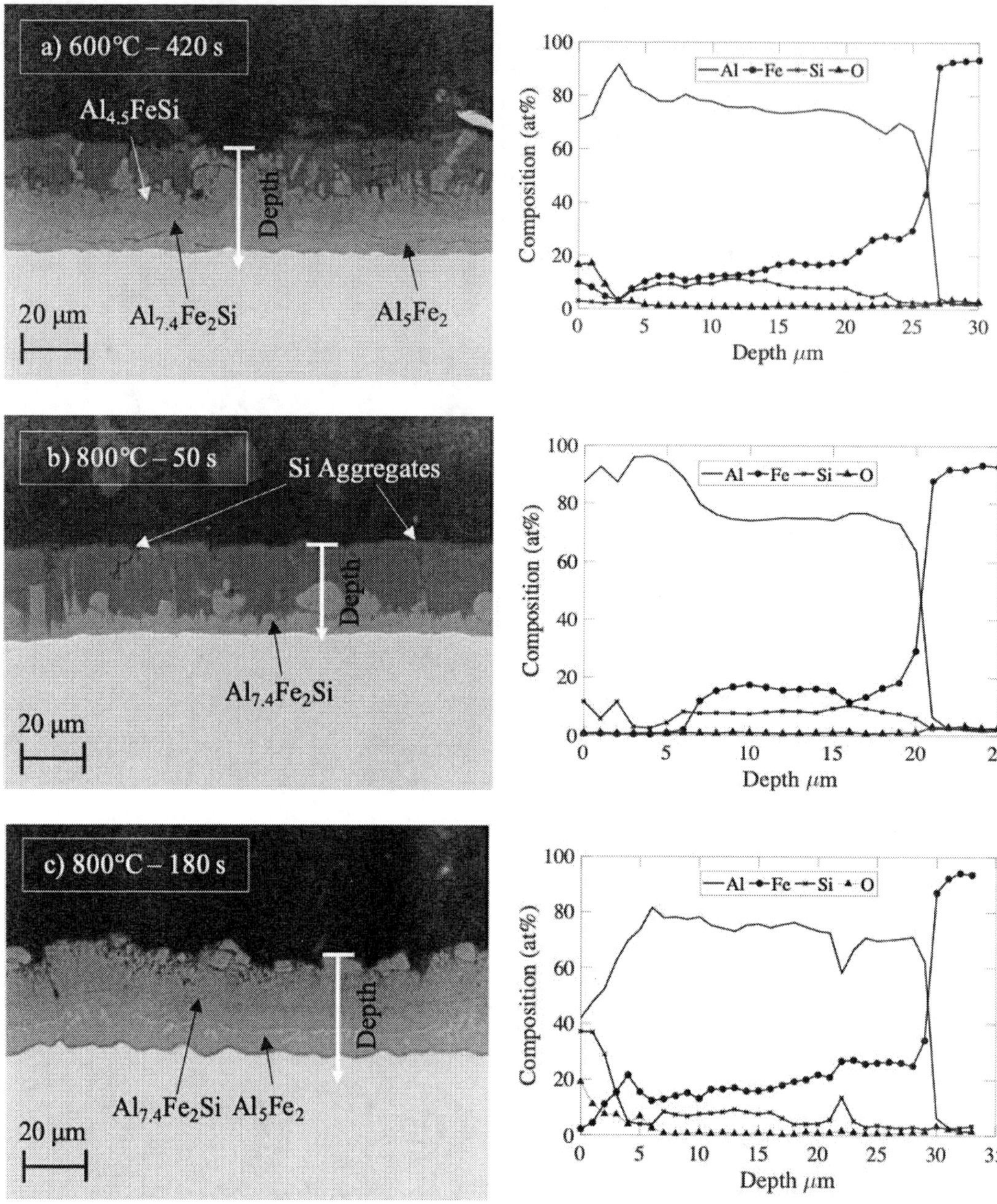

Figure 11. Back-scattered SEM cross-section images and EDS line scans showing coating evolution after heating at: (a) 600°C for 420 s; and 800°C for (b) 50 s and (c) 180 s (750x magnification).

4. Conclusions

The evolving surface roughness of Al-Si coated 22MnB5 steel significantly effects several hot stamping operations and must be understood to improve part quality and industrial processes. This work pioneers the application of speckle analysis for *in situ* roughness measurements on the intermetallic coating. These measurements are valuable for process control and optimization, and also provide important insight into the evolution of the coating within the furnace. Measurements conducted on samples heated above 600°C indicate abrupt changes in roughness proceeding the melting of the coating, which may be linked to the formation of an intermittent oxide layer and a secondary melting reaction. Longer hold times show a dramatic increase in the surface roughness followed by a stable value once the evolution of the coating was complete. Higher furnace set-point temperatures produced rougher surfaces due to the increased diffusion occurring within the coating. *Ex situ* scanning electron microscopy and energy dispersive X-ray spectroscopy confirmed that iron diffusion towards the surface occurred quicker and to a greater degree with longer heating durations and higher furnace set-points. These methods also indicate a distinct change in surface composition shortly after the melting of the coating and increased oxygen content, pointing to the formation of an oxide layer.

Future work will focus on adapting this technique and creating a diagnostic tool to assess surface roughness within industrial furnaces. Furthermore, the growth of the oxide layer and the reactions occurring within and at the surface of the coating will be characterized using *in situ* Fourier transform spectroscopy measurements and Raman microscopy, both of which can detect the presence of oxides and intermetallic phases based on their various bond energies. Finally, the effect of coating weight on the *in situ* radiative properties of the steel will be investigated.

5. References

[1] Fan D, Kim H, Birosca S and De Cooman B 2007 *Materials Science & Technology Conference* 99-110
[2] Karbasian H and Tekkaya A 2010 *J. Mater. Process. Technol.* **210** 2103-18
[3] Merklein M, Wieland M, Lechner M, Bruschi S and Ghiotti A 2016 *J. Mater. Process. Technol.* **228** 11-24
[4] Allély C, Dosdat L, Clauzeau O, Ogle K and Voloitch P 2014 *Surf. Coat. Technol.* **238** 188-96
[5] Murray J and McAlister A 1984 *Bulletin Alloy Phase Dia.* **5** 74-84
[6] Kolaříková M, Chotěborský R, Hromasová M and Linda M 2019 *Acta Polytech.* **59** 351-7
[7] Grigorieva R, Drillet P, Mataigne J-M and Redjaïmia A 2011 *Solid State Phenom.* **172-174** 784-90
[8] Jönsson M 2016 *International Conference on Advanced High Strength Steel and Press Hardening* 479-85
[9] Shi C, Daun K and Wells M 2016 *Metall. Mater. Trans.* B **47** 3301-09
[10] Shi C, Daun K and Wells M 2015 *Int. J. Heat Mass Transf.* **91** 818-28
[11] Borsetto F, Ghiotti A and Bruschi S 2009 *Key Eng. Mater.* **410-11** 289-96
[12] Liang W, Tao W, Zhu B and Zhang Y 2017 *Sci. China Technol. Sci.* **60** 1088-102
[13] Jenner F, Walter M, Iyengar R and Hughes R 2010 *Metall. Mater. Trans.* A **41A** 1554-63
[14] Ghiotti A, Bruschi S and Borsetto F 2011 *J. Mater. Process. Technol.* **211** 1694-1700
[15] Barreau M et al. 2020 *Mater. Charact.* **163** 1-13
[16] Podor R, Le Goff X, Lautru J, Brau H-P, Barreau M, Carrier X, Mendonça J, Nogues D and Candeias A 2020 *Microsc. and Microanal.* **26** 397-402
[17] Suehiro M, Maki J, Kusumi K, Ohgami M and Miyakoshi T, 2003 *International Body Engineering Conference & Exposition* 16-21
[18] Chang Y, Tang X, Zhao K, Hu P and Wu Y 2016 *J. Mater. Process. Technol.* **228** 25-33
[19] Ighodaro O, Biro E and Zhou Y 2016 *J. Mater. Process. Technol.* **236** 64-72
[20] Persson U 2006 *J. Mater. Process. Technol.* **180** 233-38
[21] Persson U 1992 *Opt. Laser. Eng.* **17** 61-7
[22] Fujii H and Asakura T 1977 *J. Opt. Soc. Am.* **67** 1171-76
[23] Rodríguez F, Cotto I, Dasilva D, Rey P and Van der Straeten K 2017 *Procedia Manuf.* **13** 519-25
[24] Kayahan E, Oktem H, Hacizade F, Nasibov H and Gundogdu O 2010 *Tribol. Int.* **43** 307-11
[25] Meireles J, da Silva L, Caetano D and Huguenin J 2012 *Opt. Laser. Eng.* **50** 1731-34
[26] Hamed A, El-Ghandoor H, El-Diasty F and Saudy M 2004 *Opt. Laser Technol.* **36** 249-53
[27] Xu D, Yang Q, Dong F and Krishnaswamy S 2018 *J. Eng.* **2018** 773-78
[28] Wong P and Li K 1999 *Opt. Laser Technol.* **31** 543-548
[29] ArcelorMittal 2019 automotive.arcelormittal.com/products/flat/PHS/usibor_ductibor
[30] Jhajj K, Slezak S and Daun K 2015 *Appl. Therm. Eng.* **83** 98-107
[31] Klassen C and Daun K 2020 *Surf. Coat. Technol.* **393** 1-9

Acknowledgements

This research was supported by the Natural Sciences and Engineering Research Council of Canada Discovery Grant (NSERC DG RGPIN-2018-03756). The authors would like to acknowledge Mr. Kaihsiang Lin for his technical assistance with the SEM and EDS analysis.

IOP Conf. Series: Materials Science and Engineering **967** (2020) 012076 doi:10.1088/1757-899X/967/1/012076

Evaluation of 22MnB5 Steel Austenitization Sub-Models for Simulating the Heating Phase of Hot Stamping

B Zhao[1,3], C Chiriac[2] and K J Daun[1]

[1]Department of Mechanical and Mechatronics Engineering, University of Waterloo, 200 University Ave. West, Waterloo, ON, N2L 3G1, Canada
[2]Ford Motor Company, Research and Innovation Center, 2101 Village Rd, Dearborn, Michigan USA
[3]Corresponding author: tom.zhao@uwaterloo.ca

Abstract. The heating phase of hot stamping is the most time- and energy- intensive part of the process, and therefore offers significant potential to improve process efficiency. This requires a robust thermometallurgical model that predicts the blank heating profile and the austenitization progress inside a roller hearth furnace. In this paper, an IM austenitization model for 22MnB5 steel is derived using dilatometry data and evaluated against a JMAK type model using temperature profiles of coupons heated within a laboratory-scale muffle furnace and blanks heated within a roller hearth furnace. The austenitization rates predicted by the two models are then compared to the inferred austenitization processes from those temperature profiles. The result shows that the IM model has potential to capture the trend of austenitization process, especially the rapid pearlite dissolution at the beginning of the transformation, whereas the JMAK type model is too restrictive and fails in this regard.

1. Introduction

Automakers increasingly rely on ultra-high strength steel (UHSS) parts made through hot stamping for automotive lightweighting. In the direct hot stamping process, blanks are first austenitized, usually in a roller hearth furnace, before they are transferred to a press to be formed and simultaneously quenched into fully martensitic components. Most often the parts are made of 22MnB5 alloy with an Al-Si coating to prevent oxidation and decarburization within the furnace. The coating melts at around 577 °C [1], and then transforms into an Al-Si-Fe intermetallic layer that provides long-term corrosion protection. The furnace parameters, consisting of zone temperatures, roller speed, and other ancillary parameters, must be adjusted to ensure full austenite transformation and proper coating transformation while minimizing cycle time, i.e. the time to austenitize one batch of parts. This is often done heuristically based on experience but considering that a typical roller hearth furnace contains more than 15 independently-set heating zones, in addition to the roller speed and batch layout, the outcome is rarely optimal. This is further complicated by the fact that the blank heating rate is largely determined by the radiative properties of the coating, which change dramatically as the coating evolves. Accordingly, there is a need for an accurate thermometallurgical model of the blank transformation within the furnace in order to optimize the heating stage of hot stamping.

Austenitization is a complex multi-stage phase transformation that is affected not only by the temperature of the blank but also its heating rate [2]. Verma et al. [3] recently developed a roller hearth furnace model that consists of a heat transfer sub-model coupled with an austenitization sub-model. This complete thermometallurgical model can predict both the blank temperature history as it travels through the furnace, and the gradual austenitization of the steel. Verma et al. compared two different austenitization sub-models: a first-order kinetics model proposed by Di Ciano et al. [4] that has two degrees-of-freedom; as well as a more detailed physical model proposed by Li et al. [5], which considers nucleation, growth and impingement and involves 15 degrees-of-freedom. In both cases, the model parameters were identified through nonlinear least-squares regression with dilatometry data. Verma et al. concluded that the model by Li et al. was not only susceptible to overturning but also limited to intercritical annealing as opposed to full annealing, which is the objective in hot stamping. This is also supported by Yan et al.'s work [6], which examined the Li et al. [5] model in more detail and concluded that it performs better at temperatures closer to onset temperature, T_{Ac1}, for shorter soaking times, and significantly underpredicts the austenite fraction, f_γ, at higher soak temperatures.

Zhu et al. [7] proposed a highly physical model using the 2D Cellular Automation method. This model simulates the nucleation and grain growth of austenite in pearlite, the growth of austenite in ferrite, and the grain coarsening of austenite. However, the calculations are computationally intensive to carry out, with the simulation of a 10-minute austenitization process taking over a full day to compute and is therefore unsuitable for furnace process optimization. Huang et al. [8] developed a non-isothermal austenitization model using the Internal State Variable (ISV) method, which includes an incubation period and a transition period. It is capable of predicting both the starting and finishing temperature of the austenitization process, as well as f_γ during the process.

Luo et al. [9] proposed a "model free" approach to predicting the austenitization kinetics for SA508 Gr.3 steel, using the Isoconversional Method (IM). "Model free" means the model does not presume a specific model parameterization. In particular, instead of deriving a set of constant model parameters, activation energy (E_a) and pre-exponential factor (A) for the entire process, these parameters are evaluated at different degrees of conversions, i.e., $E_a(f_\gamma)$, and $A(f_\gamma)$ based on dilatometry data measured for heating rates between 0.02 K/s and 500 K/s. An important advantage of such approach is that it can model complex conversion processes while avoiding oversimplified model assumption.

This paper evaluates the suitability of the IM approach for modelling austenitization of 22MnB5 steel in hot stamping. The model is first derived from a set of dilatometry measurements carried out on inductively heated 22MnB5 specimens. The predictive capability of the model is first assessed by comparing it to the dilatometry data set that was used in the derivation, after which, the IM model is then integrated into the roller hearth furnace model and evaluated using temperature measurements of coupons heated within a laboratory-scale muffle furnace and blanks heated within a roller hearth furnace.

2. Roller Hearth Furnace Model

The furnace model derived by Verma et al. [3] is applied to predict the heating rate of an unpatched, singe-gauge blank. Normally, ceramic rollers convey the blanks through the furnace. The blanks are heated using natural gas fired radiant tube burners, which may be located above and below the ceramic rollers. The furnace is separated into zones, each of which typically includes multiple burners and a single control thermocouple. The burners in each zone are simultaneously activated or deactivated via a hysteresis control system based on the temperature measured by the control thermocouple relative to the corresponding zone set-point temperature. Each zone is assumed to be large and isothermal relative to the blank, and the blank is assumed to be irradiated equally on both the top and bottom surface. Due to the small thickness, the blank temperature is assumed to be uniform. Since the contact area between rollers and the blank is small, conduction heat transfer from the rollers is excluded from the model. The air within the furnace is assumed to be quiescent [3], therefore the dominant convection mode is natural convection. With these assumptions, the blank heating curve is governed by Equation (1)

$$\rho c_p V \frac{dT}{dt} = Q_{rad} + Q_{conv} - \rho V \Delta h_\gamma \frac{df_\gamma}{dt} \tag{1}$$

where ρ and c_p denotes the density of the blank and specific heat, respectively [10], Δh_γ is the latent heat of austenitization, df_γ/dt is the instantaneous austenitization rate. Q_{conv} is the total convection obtained from

$$Q_{conv} = A\left[\bar{h}_{top}\left(T_{amb} - T_s\right) + \bar{h}_{bot}\left(T_{amb} - T_s\right)\right] \tag{2}$$

where T_{amb} is the ambient temperature, T_s is the blank surface temperature, i.e., the blank temperature, and the area A accounts for both the upper and lower blank surfaces. The average convection coefficients for the top ($\bar{h}_{top}$) and bottom ($\bar{h}_{bot}$) are calculated from natural convection correlations [11], and range approximately from 1 to 10 W/(m^2·K) for the top surface and 1 to 17 W/(m^2·K) for the bottom surface, becoming smaller as the blank temperature approaches that of the surrounding furnace air. The net radiative transfer to the blank, Q_{rad} is the energy received through radiation heat transfer and is given by

$$Q_{rad} = A\sigma\left[\alpha(T_s, T_{surr})T_{surr}^{\ 4} - \varepsilon(T_s)T_s^4\right] \tag{3}$$

where $\sigma = 5.67 \times 10^{-8}$ W/(m^2·K^4) is the Stefan-Boltzmann constant, and α and ε are the total absorptivity and emissivity. The spectral emissivity is plotted in Figure 1 (*a*) and the derived total absorptivity and emissivity by Jhajj et al. [12] is shown in Figure 1 (*b*) . Both the total emissivity and total absorptivity drop at 575 °C due to the melting of the Al-Si coating, and then rise as the intermetallic layer evolves and becomes progressively rougher. Changes in surface morphology have great influence on the radiative properties of the material, although the exact relationship remains unclear [1, 13, 14].

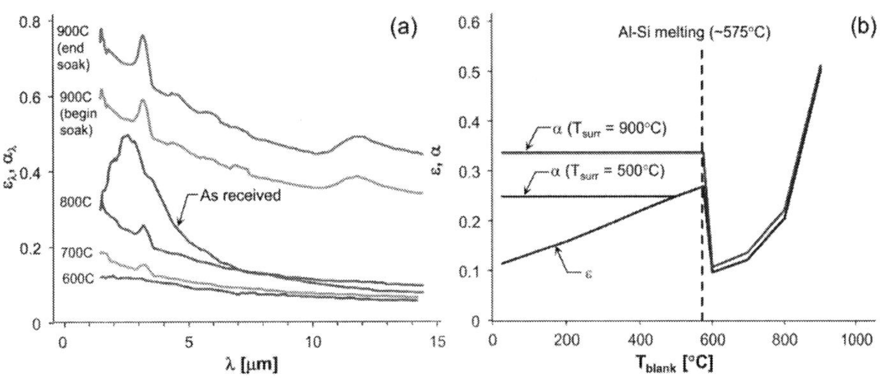

Figure 1. Spectral emissivity (a) [10] and total emissivity and absorptivity (b) [12] of 22MnB5.

The metallurgical sub-model is used to obtain the austentization rate, df_γ/dt, based on the blank temperature calculated from the heat transfer model. The metallurgical sub-model, in turn, is coupled with the heat transfer model through the latent heat of austenitization. The latent heat of austenitization for 22MnB5 has not been characterized in literature. Verma et al. estimated it to be 30 kJ/kg based on a rule of mixture using Δh_γ values for fully pearlitic and fully ferritic iron [3] and taking the as-received state of 22MnB5 being approximately 80%-ferrite/20%-pearlite. Equation (1) is solved simultaneously with a differential equation that governs the rate of austentization, df_γ/dt. In this work the explicit Euler technique is used to solve these equations.

3. Candidate Austenitization Kinetics Sub-models

3.1 F1 model

Khawam et al. [15] reviewed many solid-stated kinetics models, including the first-order kinetics model (F1), which is the Avrami (or Johnson–Mehl–Avrami–Kolmogorov, JMAK) equation, with $n = 1$. Di-Ciano et al. [4] modified the F1 model Equation to account for a non-isothermal heating rate

$$g\left(f_\gamma\right) = A\int_0^t \exp\left(\frac{E_a}{RT(t)}\right)dt \tag{4}$$

Taking the derivative of Equation (4) with respect to time yields

$$\frac{dg\left(f_\gamma\right)}{dt} = A\exp\left(\frac{E_a}{RT(t)}\right) \tag{5}$$

where R is the universal gas constant (8.314J/mol·K), $T(t)$ is the current temperature of the blank in K and $g(f_\gamma)$ is an intermediate variable

$$g(f_\gamma) = \ln(1 - f_\gamma) \tag{6}$$

The rate of austenitization is then given by

$$\frac{df_\gamma}{dt} = \begin{cases} 0, & T < T_{Ac1} \\ \dfrac{dg(f_\gamma)}{dt}\exp[-g(f_\gamma)] & T \geq T_{Ac1} \end{cases} \tag{7}$$

The two model parameters are the pre-exponential factor, A, and activation energy, E_a, which Di Ciano derived from dilatometry data to be 3.42×10^{18} s^{-1} and 402 kJ/(mol·K), respectively, for heating rates between 1-5 K/s.

3.2 Isoconversional Method (IM) model

Luo et al. [9] modelled the austenitization process of SA508 Gr.3 steel using an isoconversional method (IM), specifically the Friedman method, which is commonly used in thermal analysis but not for solid-state phase transformation kinetics. The general equation for a solid-state phase transformation is

$$\frac{df_\gamma}{dt} = A\exp\left(-\frac{E_a}{RT}\right)f(f_\gamma) \tag{8}$$

where f_γ is the austenitization fraction, and $f(f_\gamma)$ represents the specific austenitic transformation model, such as the F1 model, which is $f(f_\gamma) = 1 - f_\gamma$ according to Khawam et al. [15]. Instead of having a constant activation energy (E_a) and a pre-exponential factor (A) for the entire transformation process, in the IM approach those parameters are evaluated as the reaction progresses, i.e. at different specific f_γ values. At a specific f_γ, $f(f_\gamma)$ is constant, and Equation (8) can be rearranged into

$$\left.\frac{df_\gamma}{dt}\right|_{f_\gamma} = \left[A(f_\gamma)f(f_\gamma)\right]\exp\left[-\frac{E_a(f_\gamma)}{RT}\right] = A'(f_\gamma)\exp\left[-\frac{E_a(f_\gamma)}{RT}\right] \tag{9}$$

Since $A(f_\gamma)$ and $f(f_\gamma)$ are both constant at this specific f_γ value, they can be combined into a modified pre-exponential factor, $A'(f_\gamma)$, and after taking natural logarithm Equation (9) becomes

$$\ln\left.\frac{df_\gamma}{dt}\right|_{f_\gamma} = \ln\left(A'(f_\gamma)\right) - \frac{E_a(f_\gamma)}{R}\frac{1}{T} \tag{10}$$

Equation (10) is a linear equation of $1/T$ with a slope of $-[E_a(f_\gamma)/R]$ and an intercept of $\ln[A'(f_\gamma)]$. These parameters are found through linear regression with dilatometry data. Since heating rate is constant during the dilatometry test, the transformation rate df_γ/dt can be calculated using

$$\frac{df_\gamma}{dt} = \frac{df_\gamma}{dT}\cdot\frac{dT}{dt} = \frac{df_\gamma}{dT}\cdot\beta \tag{11}$$

where β represents the constant heating rate. A total of nine sets of dilatometry data with three repeats per heating rate (10, 100, 200 K/s) are used to derive the model parameters. The dilatometry data is measured with a 1.6 mm thick Al-Si coated 22MnB5 (90 g/m^2 coating weight) steel sample on a DIL805 A/D thermal-mechanical simulator. The sample composition is obtained through optical emission spectrometry (OES) and is listed in Table 1:

Table 1. Composition of 22MnB5 steel used in dilatometry measurement.

Elem	C	Mn	P	S	Si	Ni	Cr	V	Ti	Al	B	Ca
%	0.23	1.14	0.01	0.003	0.23	0.01	0.23	0.005	0.035	0.05	0.002	0.002

The austenite fraction f_γ is interpolated from the dilatometry curve using a lever rule procedure detailed in Huang et al.'s work [16] and is shown in Figure 2. The derivative of austenite fraction with respect to temperature, df_γ/dT, is calculated numerically from Figure 2 and the result is in Figure 3.

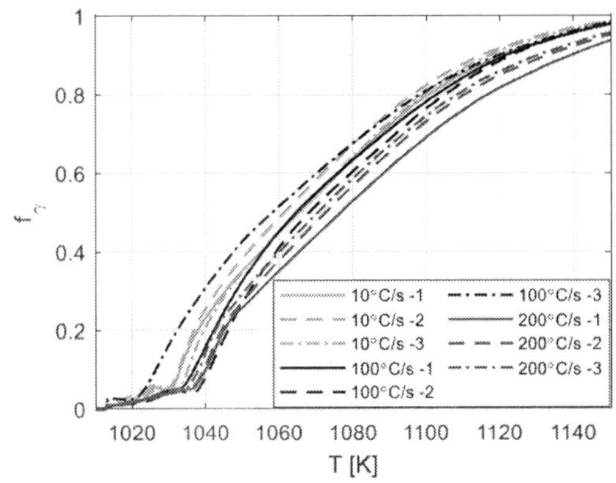

Figure 2. The austenite transformation fraction.

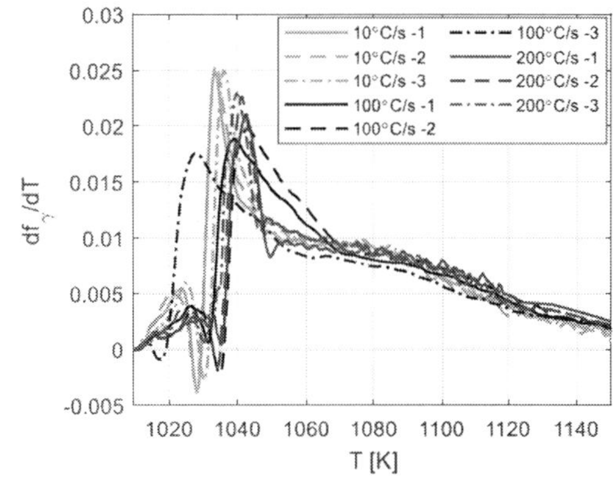

Figure 3. df_γ/dT vs T curve.

Figure 2 and Figure 3 show that the austenitization fraction appears to decrease before rising again at the beginning of the process, which is a measurement artifact caused by induction heating used by the dilatometer. As-received 22MnB5 is a mixture of ferrite and pearlite, which are magnetic phases, whereas austenite is nonmagnetic [17]. Austenitization happens quickly at the beginning stage with pearlite dissolution [16], causing the DIL805 controller to respond rapidly, which result in an overshoot of HF power for a short period of time until the controller stabilizes. In addition to the effect of temperature, the spike in HF power also triggers a magnetostriction effect that contributes to a minor change in length of the sample. Figure 2 and Figure 3 also show that the third dataset of 100 °C/s is an outlier and is not used to derive the model.

The process for obtaining $E_a(f_\gamma)$ and $A(f_\gamma)$ curves is summarized in the flow chart shown in Figure 4. Data pairs of $[\ln(df_\gamma/dt), 1/T]$ at this f_γ from each of the eight remaining datasets are plotted and a line is fitted through these points, as shown in Figure 5 for a subset of sampled f_γ values ($f_\gamma = 0.25$, 0.51, and 0.75). The actual fitting process follows a much more stringent sampling of f_γ values with different spacings at different ranges of f_γ, considering that the transformation rate is very rapid at the beginning and slows down afterwards, to have a reasonably evenly distributed sampled points. Figure 6 provides an example of how the 100 K/s heating rate dataset No.2 is sampled.

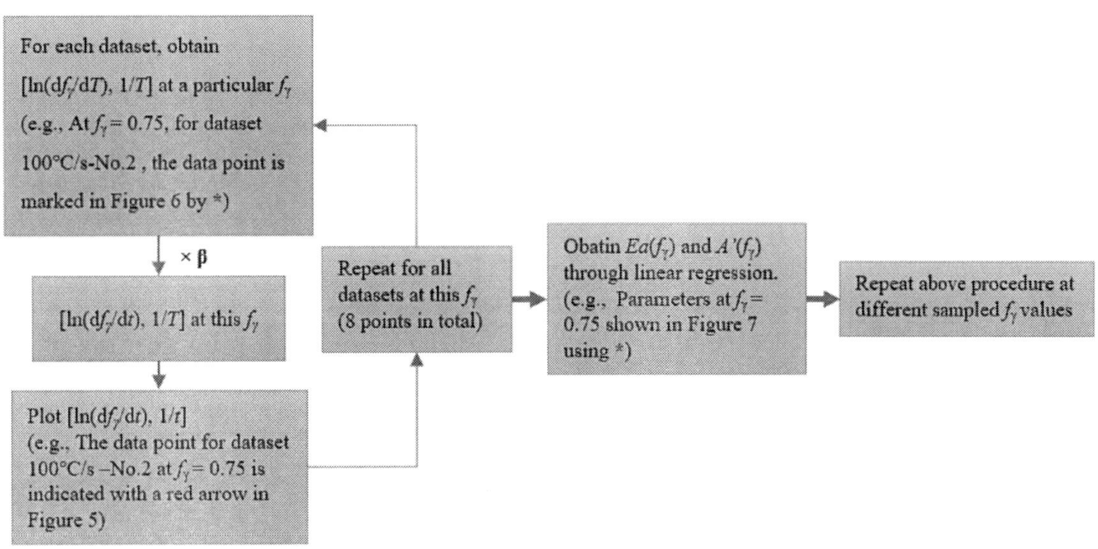

Figure 4. Flowchart illustrating the derivation process.

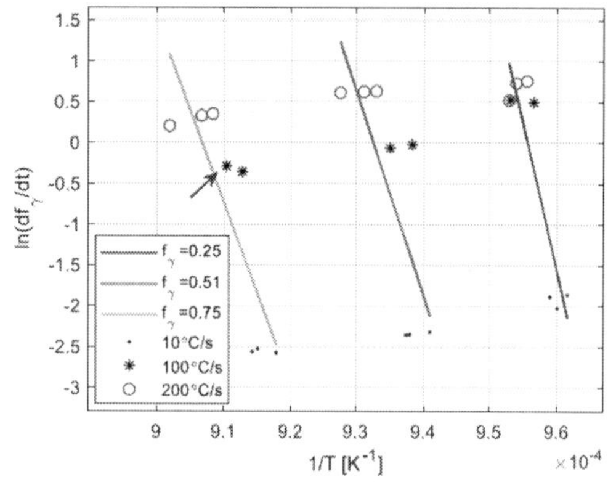

Figure 5. Fitting process to obtain the effective model parameters.

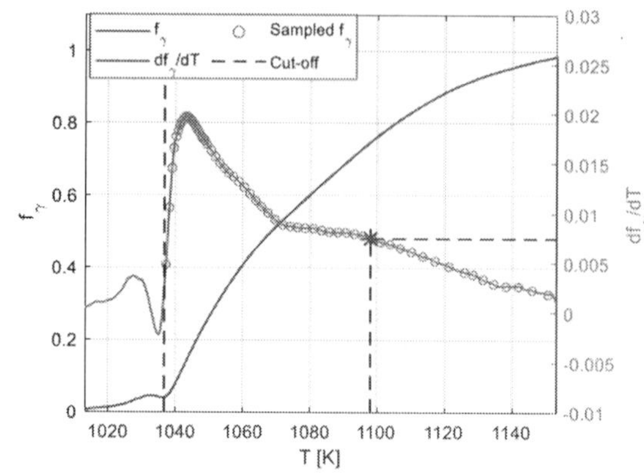

Figure 6. Spacing of sampled f_γ values and the workaround of measurement artifact, from dataset 100°C/s - No.2.

As a workaround for the measurement artifact, the sampling starts at lowest point of the "dip" caused by the artifact. It is marked with a vertical cut-off line in Figure 6. However, this workaround precludes values of f_γ below 0.05, which corresponds to the onset of rapid austenitization. Figure 7 shows the derived model parameters, which share the same order-of-magnitude as Luo et al.'s result [9] despite the fact that Luo et al. examined SA508 Gr.3 steel.

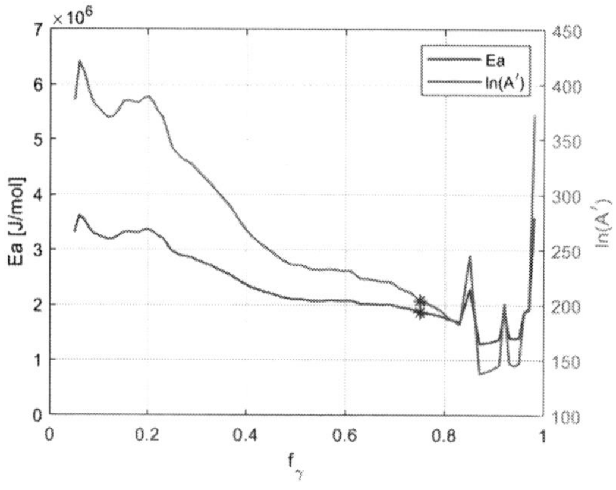

Figure 7. Derived model parameters.

With the derived model parameters $E_a(f_\gamma)$ and $A'(f_\gamma)$, the model prediction can be made by solving differential Equation (9) using the explicit Euler scheme. At each timestep, t_i, the austenite fraction $f_{\gamma,i}$ can be used to linearly interpolate its corresponding effective activation energy, $E_a(f_{\gamma,i})$, and pre-exponential factor, $A(f_{\gamma,i})$. Given a small enough change in austenite fraction, the two parameters are assumed to be constant within this timestep, and the transformation rate can be obtained through

$$\left.\frac{df}{dt}\right|_{f_{\gamma,i}} = A'\left(f_{\gamma,i}\right)\exp\left[-\frac{E_a\left(f_{\gamma,i}\right)}{RT_i}\right] \tag{12}$$

4.0 Discussion and Results

4.1 IM Sub-model Validation against Dilatometry Data

The temperature history of each dilatometry dataset is incorporated into the IM model to predict the austenitization process, and the model error is calculated by subtracting the data value at this temperature from the model prediction. Figure 8 shows an example of the model validated against the first run of 10°C/s dilatometry measurements; model prediction is in good agreement with the dilatometry data. This is further demonstrated in the error plot in Figure 9. The highest error is under 0.1 and tends to occur at the onset of austenitization. It may be due to the measurement artifact mentioned previously, which causes the model parameter to be missing for the first 5% of austenitization process. In practice, the austenitization fraction is set to $f_{\gamma,i} = 5\%$ when $f_{\gamma,i} \leq 5\%$ to retrieve the isoconversional values of the two model parameters.

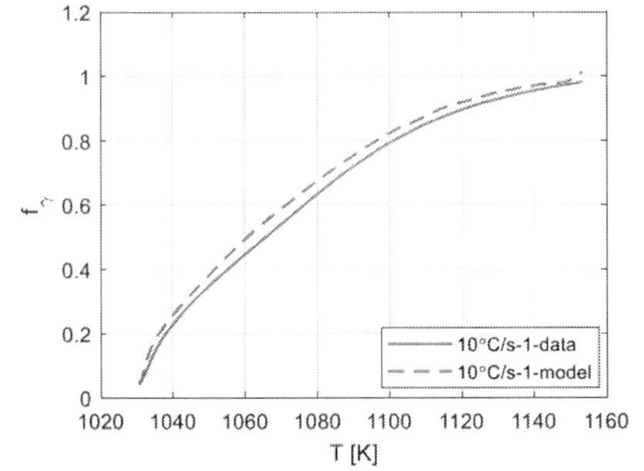

Figure 8. Validation against dilatometry data.

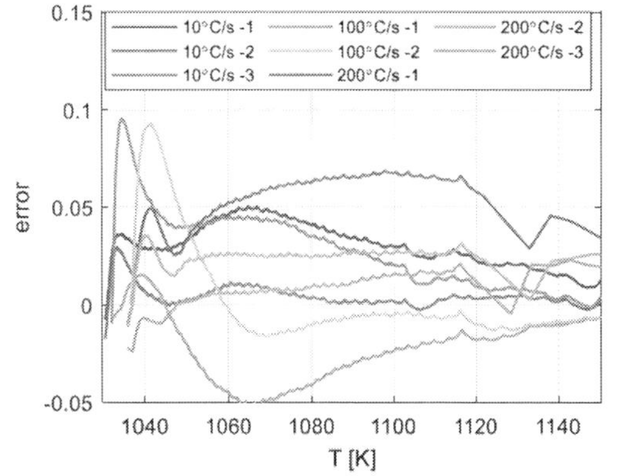

Figure 9. Validation error plot.

4.2 Comparison of F1 and IM sub-models for predicting austenitization in furnaces

The F1 and IM austenitization sub-models are combined with the heat transfer model and are used to predict the heating curves of coupons heated within a laboratory-scale muffle furnace and blanks heated in a roller hearth furnace. The heating curve predictions are compared with thermocouple measurements. The austenitization rate is also inferred from the thermocouple measurements by rearranging Equation (1) to solve for df_γ/dt,

$$\frac{df_\gamma}{dt} = \frac{(Q_{rad} + Q_{conv}) - \rho c_p V \dfrac{dT}{dt}}{\rho V \Delta h_\gamma} = \frac{Q_{net} - \Delta E_{sensible}}{\rho V \Delta h_\gamma} \tag{13}$$

which is then compared directly with result obtained from austenitization sub-models.

4.2.1. Muffle Furnace Measurements. A muffle furnace set to 910 °C is used heat Al-Si coated 22MnB5 coupons having dimension of 130 × 30 × 2mm, which is mounted in a ceramic holder as described in Ref [3] to minimize conduction heat transfer between the blank and the holder. A K-type thermocouple is welded to the centre of the blank and measured using NI - 9213® data acquisition unit. The roller hearth furnace model is adapted to this scenario by modeling the muffle furnace as a single heating zone in a roller hearth furnace. The F1 and IM models are activated at the same prescribed T_{Ac1}, which is apparent by the drop in heating rate. A comparison of the measured temperature profile and predicted temperature profiles using both the F1 and IM models is shown in Figure 10. Both kinetics sub-models generate temperature profiles that closely match the experimental measurements.

Figure 11 shows a more informative comparison between the predicted f_γ by both kinetics sub-models and the inferred f_γ obtained using equation (13). Both the IM and F1 models underpredict the austenite fraction. The austenitization transformation rate is initially very large due to rapid pearlite dissolution, as suggested by both Huang et al. [16] and Li et al. [5]. This is captured by the IM model but not the F1 model. Although inferred, F1 and IM austenite transformation are all set to start at the same onset temperature, there is a delayed start with IM model. This could be related to the missing model parameters due to the measurement artifact. Manually shifting the IM curve forward by 10 K shows a better agreement between it and the inferred result. With more data and a resolution to the measurement artifact, the IM model shows potential for providing an accurate instantaneous austenite phase fraction.

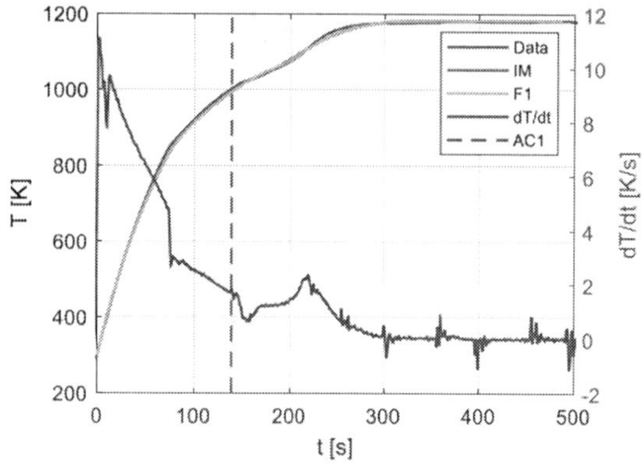

Figure 10. Validation against muffle furnace data.

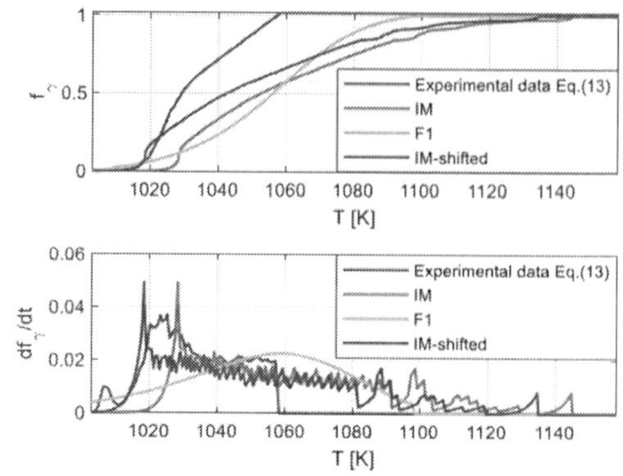

Figure 11. Validation against muffle furnace inferred austenitization result.

4.2.2. Roller Hearth Furnace Results. Next, the models are analyzed using temperature measurements made on instrumented blanks heated within a roller hearth furnace. A 625 × 297 × 1.6 mm 22MnB5 steel blank with 90 g/m^2 Al-Si coating weight is sent through the roller hearth furnace instrumented with K-type thermocouples and a Datapaq® unit. The measured temperature profile is compared to the F1 and IM model predictions in Figure 12. Qualitatively the model predictions match the experimental result, although with much less agreement than the muffle furnace results shown in Figure 10. This may be due to the idealization of a constant and uniform zone temperature, while in reality the zones are not isothermal and the average temperature fluctuates about a set-point [3]. Moreover, the radiative properties shown in Figure 1 change with the evolving surface state, which is a complex function of heating rate and is not yet completely understood [13, 14].

Inferring the austenite phase fraction using the thermocouple data via equation (13) initially produces a non-physical result shown in Figure 13, because the change in sensible energy inferred from the blank temperature, $\Delta E_{\text{sensible}}$, is larger than the heat transfer to the blank predicted by the model, Q_{net}. This issue likely arises from uncertainties in the radiative properties, which change abruptly over this temperature range as shown in Figure 1 due to coating evolution. In their muffle furnace experiments, Jhajj et al. [12] circumvent this problem by removing the Al-Si coating and applying a protective boron nitride (BN) coating, which remains inert at high temperatures, but this cannot be done for an industrial roller hearth furnace. Instead, the absorptivity value at 700 and 800 °C are heuristically increased by 10% and the modified result of net heat transfer is shown Figure 13.

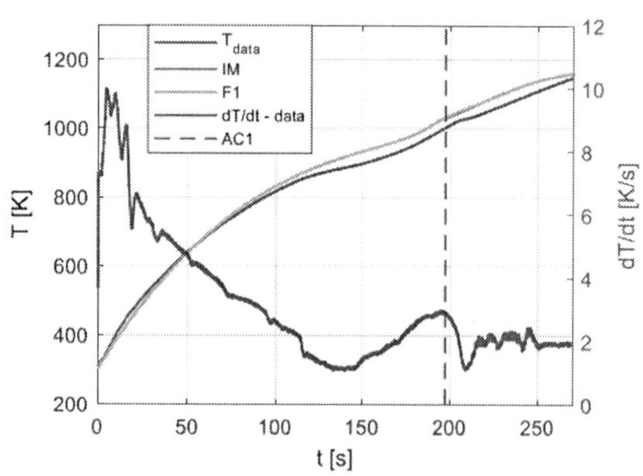

Figure 12. Validation against roller hearth furnace data.

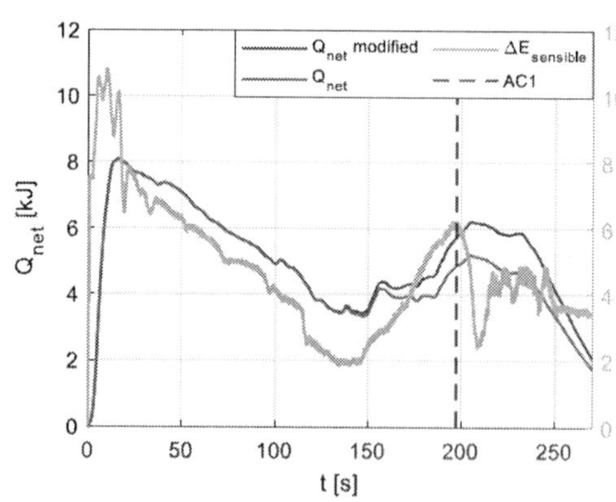

Figure 13. Comparison of change in sensible energy and total heat transfer.

With the modified absorptivity, the inferred f_γ is compared to the F1 and the IM prediction in Figure 14, and the rate of transformation comparison is shown in Figure 15. Similar to the muffle furnace result, the F1 model predicts a much lower austenitization rate close to T_{Ac1} compared to the IM model. The IM model, however, still shows a delayed start despite the model starts at the same austenitization onset temperature. Manually shifting the IM curve forward by 5 K shows a better agreement.

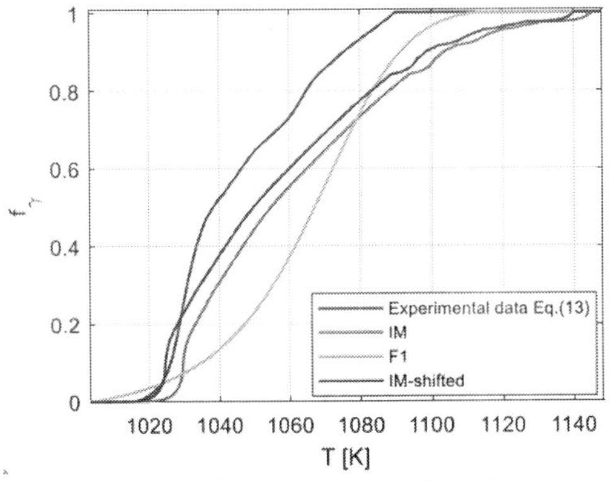

Figure 14. f_γ validation against datapaq inferred result.

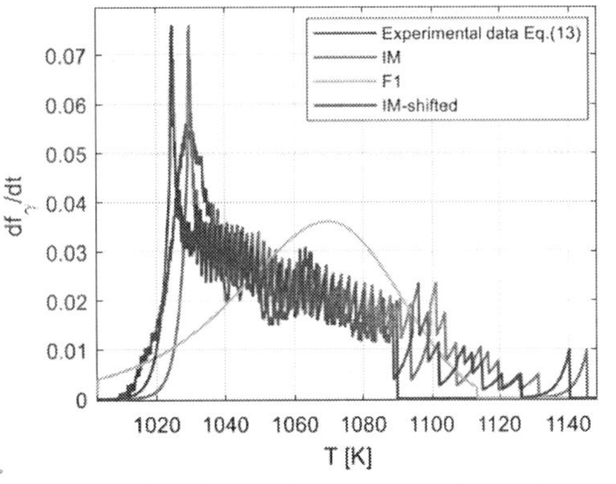

Figure 15. Rate of transformation validation against datapaq inferred result.

5.0 Conclusions

Maximizing the profitability of hot stamping operations involves identifying the optimal set of furnace parameters that minimize cycle time while achieving the functional requirements of the heating stage, i.e. austenitization and coating transformation. This, in turn, demands accurate models for both the heat transfer and metallurgical transformations undergone by the blank. This paper derives a new type of austenitization kinetics sub-model for 22MnB5 alloy, using the Isoconversional Method (IM), from dilatometry measurements. The model is incorporated into a thermometallurgical blank model, and the performance is compared with a first-order kinetics model using instrumented Al-Si coated 22MnB5 coupons heated in a muffle furnace and instrumented blanks heated within a roller hearth furnace.

Both the F1 and IM models under-predict the austenite fraction compared to values inferred from thermocouple measurements, although the inferred austenite phase fraction is subject to significant uncertainties due to the uncertain radiative properties of the evolving coating. However, the IM method has recreated the rapid austenitization associated with pearlite dissolution close to T_{Ac1}. This result shows that although F1 model is robust and easy to use, it does not have the necessary degrees-of-freedom to capture the physical transformation process. In contrast, the IM method captures the general trend better, although more dilatometry data with a wider range of heating rates and smaller intervals should be used to derive a more robust set of parameters. Lower heating rate data, in particular, would be more representative of conditions within a roller hearth furnace. Other candidate phase transformation n models, such as the ISV model, should also be evaluated as candidates for modeling austenitization within roller hearth furnaces.

Future work should also focus on reducing the uncertainty in radiative properties after the coating melts. The surface morphology and properties evolve as the coating transforms from its as-received Al-Si state to its final intermetallic form, which in turn drastically changes radiative properties. Experiments should be conducted where samples with coatings at different stages of the evolution are preserved and examined to uncover the relationship between coating evolution and changes in radiative properties.

6.0 References

[1] Klassen C and Daun K 2020 *Surf. Coat. Technol.* **393** 125795
[2] Caballero F, Capdevila C and García de Andrés C 2001 *Metall. Mater. Trans.* A **32** 1283-91
[3] Verma M, Yan H, Culham J, Di Ciano and M Daun K 2019 *J. Heat. Transfer.* **141**(6) 062101
[4] Di Ciano M, Field N, Wells M and Daun K 2018 *J. Mater. Eng. Perform.* **27**(4) 1792-1802
[5] Li N, Lin J, Balint D and Dean T 2016 *J. Mater. Process. Technol.* **237** 394-401
[6] Yan H, Di Ciano M, Verma M and Daun K 2018 Intercritical Annealing of 22MnB5 for Hot Forming Die Quenching in *IDDRG 37th Annual Conference* Waterloo ON Canada
[7] Zhu B, ZhangY, Wang C, Liu P, Liang W and Li J 2014 *Metall. Mater. Trans.* A **45** 3161-71
[8] Huang H, Wang B, Tang X and Li J 2017 *Metall. Mater. Trans.* A **48** 5799-5804
[9] Luo X, Han L and Gu J 2015 *Metals* **6**(1)
[10] ArcelorMittal Properties of Usibor 1500P
[11] Bergman T, Lavine A, and Dewitt D 2011 *Fundamentals of Heat and Mass Transfer 6th ed* Hoboken NJ Wiley
[12] Jhajj K, Slezak S and Daun K 2015 *Appl. Therm. Eng* **83** 98-107
[13] Barreau M et al. 2020 *Mater. Charact.* **163** 110266
[14] Shi C, Daun K and Wells M 2016 *Metall. Mater. Trans.* B **47** 3301-09
[15] Khawam A and Flanagan D 2006 *J. Phys. Chem.* B **110** 17315-28
[16] Huang J, Poole W and Militzer M 2004 *Metall. Mater. Trans.* A **35** 3364-75
[17] Güler M 2014 Magnetism and Microstructure Characterization of Phase Transitions in a Steel *Adv. Cond. Matter Phys.* **2014** 408607
[18] Karbasian H and Tekkaya A 2010 *J. Mater. Process. Technol* **210**(15) 2103-18
[19] Chiriac C and Sohmshetty R 2017 The Effects of the Heating Rate and the Incoming Microstructure on Phase Transformation Temperatures of 22MnB5 Steel in *CHS2*

Acknowledgement

This research was supported by the Natural Sciences and Engineering Research Council (NSERC) Canada CRDPJ 532078, Ford Motor Company, and Cosma International. The authors are grateful to Mr. Cameron Klassen for carrying out muffle furnace tests on instrumented coupons.

A simplified method for the evaluation of the layer compression test using one 3D digital image correlation system and considering the material anisotropy by the equibiaxial Lankford parameter

A A Camberg[1], T Tröster[1]

[1] Automotive Lightweight Design, Paderborn University, Mersinweg 7, 33100 Paderborn, Germany

E-mail: alan.camberg@uni-paderborn.de

Abstract. The layer compression test is an established method for the characterization of sheet metal behavior under equibiaxial tension. Due to its ability to achieve high strains without necking and the simple experimental setup, the test is a welcome alternative to the more complex Bulge test for flow curve determination. In addition, the layer compression test provides further data for the calibration of yield loci models. However, the evaluation of the test is not trivial, since a tactile extensometer measurement incorporates the error of closing gaps between single sheet layers. Moreover, the sheet metal anisotropy is neglected. To overcome the above-mentioned drawbacks, a measurement and evaluation methodology based on the optical strain measurement of two orthogonally positioned 3D digital image correlation (DIC) systems was introduced recently. The method delivers reliable results but increases the complexity and costs of the test. Here, a simplification to this method is proposed. By the usage of only one 3D DIC system, the major in-plane strain is acquired from an online measurement. The minor in-plane strain is calculated posteriori by the consideration of the biaxial Lankford parameter. In this way, the anisotropy and thus the correct flow behavior can be studied by only one 3D DIC system. The proposed method is evaluated and discussed.

1. Introduction

The finite element analysis (FEA) is an indispensable tool in the field of virtual product development. It allows to assess the formability of sheet metal parts far before prototyping. In this way, technical and economic risks can be identified and eliminated already in the digital phase of the manufacturing process. That keeps the development costs low and reduces the necessary tool reworking steps to a minimum. However, the prediction quality of forming simulations is determined by the quality of the underlying constitutive models [1]. Furthermore, the experimental characterization and model parameter identification has a significant influence on the expected reliability of the simulation models [2].

The yield locus as well as the hardening curve are necessary models to describe the plastic behavior of metals. In metal forming and other applications where high strains are expected, the flow curve gained from a uniaxial tensile test is insufficient to fully cover the relevant strain range. To characterize the materials flow behavior beyond the point of uniform elongation in the uniaxial tensile test, many tests have been proposed in the past as the Bulge test [3], different types of shear tests [4],[5] and the layer compression test [6]. The latter test has gained an increased interest over the last years as it provides further data for the calibration of higher-order yield loci models, such as the yield stress and the Lankford parameter under equibiaxial tension. However, the evaluation of the test is not trivial, since a tactile strain measurement incorporates the error of closing gaps between single sheet layers. The same applies for the measurement of the machine crosshead displacement which is additionally influenced by the machine stiffness. Furthermore, in both cases the sheet metal anisotropy is neglected.

Merklein and Kuppert [7] introduced a measurement and evaluation methodology based on the optical strain measurement of two orthogonally positioned 3D DIC systems to overcome the typical drawbacks of the layer compression test. The method presented in [7] delivers reliable results but increases the complexity and costs of the test at the same time. Besides that, many laboratories are not equipped with several DIC systems.

Within this contribution a simplification to this method is proposed. By the usage of only one 3D DIC system, the major strain is acquired from an online measurement. The minor strain is calculated posteriori by the consideration of the Lankford parameter under equibiaxial tension. The Lankford parameter itself is obtained by a measurement of the final diagonals of each sheet disc as proposed by Barlat et al. [8]. In this way, the anisotropy and thus the correct flow behaviour can be studied by only one 3D DIC system. The proposed method is evaluated and discussed using an exemplary aluminium alloy sheet material with a pronounced anisotropy.

2. The layer compression test and its evaluation methods

2.1. Basic test principle

The layer compression test has not yet been standardized, but in many cases it is based on the DIN 50106 [9] standard for compression tests of solid cylinder specimens. Initially introduced by Pawelski [6] in 1967, the layer compression test reassembles a cylinder geometry by single sheet metal discs stacked on each other as shown in figure 1.

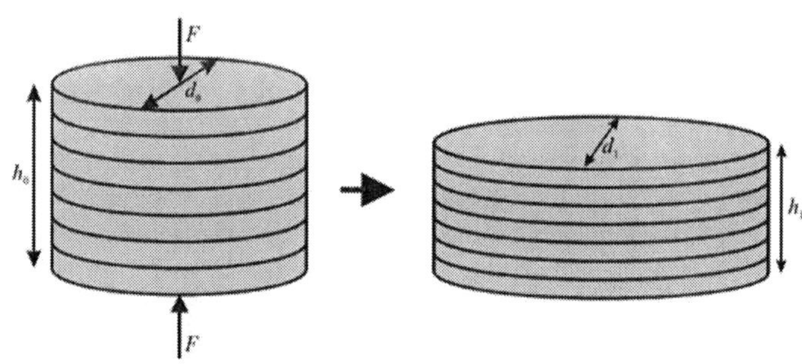

Figure 1. Layer compression specimen before and after compression. Assuming isotropic material behaviour, the initial diameter d_0 increases uniformly while the stack height h_0 decreases [5].

Under the assumption of isotropic material behavior, isochoric plastic flow and frictionless contact between the specimen surfaces and contact plates, the Cauchy stress tensor under uniaxial compression reads

$$\sigma_{uni.comp.} \mathrel{\widehat{=}} \begin{bmatrix} \sigma_x & \tau_{yx} & \tau_{zx} \\ \tau_{xy} & \sigma_y & \tau_{zy} \\ \tau_{xz} & \tau_{yz} & \sigma_z \end{bmatrix} = \begin{bmatrix} 0 & 0 & 0 \\ 0 & 0 & 0 \\ 0 & 0 & -k_f \end{bmatrix}, \tag{1}$$

whereas the Cauchy stress tensor for equibiaxial tension is equal to

$$\sigma_{biax.tens.} \mathrel{\widehat{=}} \begin{bmatrix} \sigma_x & \tau_{yx} & \tau_{zx} \\ \tau_{xy} & \sigma_y & \tau_{zy} \\ \tau_{xz} & \tau_{yz} & \sigma_z \end{bmatrix} = \begin{bmatrix} k_f & 0 & 0 \\ 0 & k_f & 0 \\ 0 & 0 & 0 \end{bmatrix}. \tag{2}$$

Since the hydrostatic part of the stress tensor has no influence on the plastic flow, the flow is controlled by the deviatoric part alone. An indifference between the stress deviator of both the uniaxial compression given by equation (1) and the equibiaxial tension given by equation (2) may be evidenced through an additive decomposition of both stress tensors as shown in equation (5) and equation (7).

$$\boldsymbol{\sigma} = \begin{bmatrix} \sigma_x & \tau_{yx} & \tau_{zx} \\ \tau_{xy} & \sigma_y & \tau_{zy} \\ \tau_{xz} & \tau_{yz} & \sigma_z \end{bmatrix} = \boldsymbol{\sigma}^{dev} + \boldsymbol{\sigma}^{vol} \tag{3}$$

$$\boldsymbol{\sigma}^{vol}_{uni.comp.} = \frac{1}{3}(\sigma_x + \sigma_y + \sigma_z)\bar{\bar{1}} = k_f \cdot \begin{bmatrix} -\frac{1}{3} & 0 & 0 \\ 0 & -\frac{1}{3} & 0 \\ 0 & 0 & -\frac{1}{3} \end{bmatrix} \tag{4}$$

$$\boldsymbol{\sigma}^{dev}_{uni.comp.} = \boldsymbol{\sigma}_{uni.comp.} - \boldsymbol{\sigma}^{vol}_{uni.comp.} = k_f \cdot \begin{bmatrix} 0-\left(-\frac{1}{3}\right) & 0 & 0 \\ 0 & 0-\left(-\frac{1}{3}\right) & 0 \\ 0 & 0 & -1-\left(-\frac{1}{3}\right) \end{bmatrix} = k_f \cdot \begin{bmatrix} \frac{1}{3} & 0 & 0 \\ 0 & \frac{1}{3} & 0 \\ 0 & 0 & -\frac{2}{3} \end{bmatrix} \tag{5}$$

$$\boldsymbol{\sigma}^{vol}_{biax.tens.} = \frac{1}{3}(\sigma_x + \sigma_y + \sigma_z)\bar{\bar{1}} = k_f \cdot \begin{bmatrix} \frac{2}{3} & 0 & 0 \\ 0 & \frac{2}{3} & 0 \\ 0 & 0 & \frac{2}{3} \end{bmatrix} \tag{6}$$

$$\boldsymbol{\sigma}^{dev}_{biax.tens.} = \boldsymbol{\sigma}_{biax.tens.} - \boldsymbol{\sigma}^{vol}_{biax.tens.} = k_f \cdot \begin{bmatrix} 1-\frac{2}{3} & 0 & 0 \\ 0 & 1-\frac{2}{3} & 0 \\ 0 & 0 & 0-\frac{2}{3} \end{bmatrix} = k_f \cdot \begin{bmatrix} \frac{1}{3} & 0 & 0 \\ 0 & \frac{1}{3} & 0 \\ 0 & 0 & -\frac{2}{3} \end{bmatrix} \tag{7}$$

$$\boldsymbol{\sigma}^{dev}_{biax.tens.} = \boldsymbol{\sigma}^{dev}_{uni.comp.} \,. \tag{8}$$

Under the assumption of isotropic and isochoric plastic flow the diameter of the stacked cylinder increases uniformly with the plastic flow. The true stress k_f and the thickness strain ε_3 follows with

$$k_f = \frac{F}{A} \Rightarrow k_f(h) = \frac{F}{A_0} \cdot \frac{h}{h_0} = \frac{F}{(d_0/2)^2\pi} \cdot \frac{h}{h_0} \quad \text{and} \tag{9}$$

$$\varepsilon_3(h) = ln\frac{h_0}{h} \,. \tag{10}$$

By making use of the relationships given above, the crosshead displacement of a universal testing machine is commonly used for the evaluation of stress and strain in the layer compression tests [10], [11]. Alternatively, as suggested by the DIN 50106, the strain can be determined by measuring either the displacement of the compression plates or the compression of the sample directly using a tactile extensometer. A major disadvantage of height-based measuring approaches is, however, that the setting behavior between the individual sheet metal discs is included in the measurement. That can lead to a significant error in the derived experimental results.

2.2. Anisotropic evaluation

Another way to determine the strain is a direct inline measurement of the sample diameter during the compression test. Thereby the measuring error caused by closing gaps between the individual sheet layers can be eliminated. In addition, both diagonals of the discs can be evaluated by using a second orthogonally aligned extensometer and thus a possible anisotropy in the plastic flow can be considered, see figure 2.

Such setup is used for example by Gese et al. [13]. By a system of four orthogonal tactile extensometer sets the surface increase of four discs within a stack are evaluated independently. For the strain and stress calculation the measurement of all four systems is averaged.

Merklein and Kuppert [7] introduced a measurement and evaluation methodology based on the optical strain measurement of two orthogonally positioned 3D digital image correlation (DIC) systems. By making use of the proportionality between the tangential strain measured on the specimen surface perpendicular to the rolling and transverse direction, ε_1 and ε_2 respectively, the diameter increase is calculated from equation (11) as introduced by the authors

$$A = \left(\frac{exp(\varepsilon_1) + exp(\varepsilon_2)}{2} \cdot \frac{d_0}{2}\right)^2 \cdot \pi \; . \tag{11}$$

The proportionality assumption between the tangential strain and the diameter increase is based on the elementary relationship between the circumference C and the diameter d of a circle $C/d = \pi$. If the circumferential strain is evaluated on a reasonably small section, the following applies $C_0/d_0 = C_i/d_i$. The method presented by the authors delivers reliable results but increases the complexity and costs of the test at the same time. Besides that, many laboratories are not equipped with several DIC systems, which further limits the dissemination of this effective but complex measurement method.

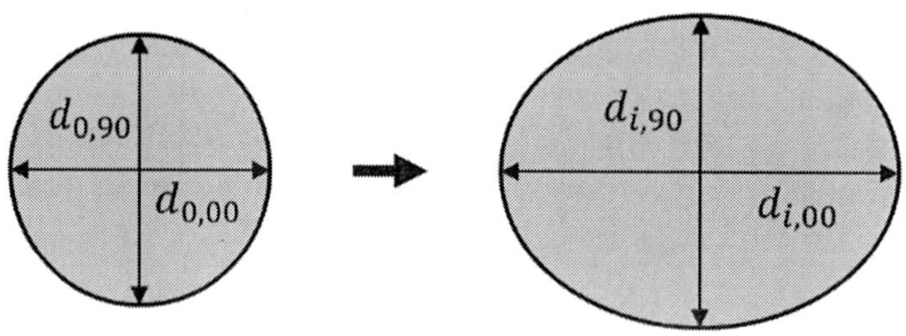

Figure 2. Layer compression specimen before and after compression. The non-uniform diameter increase is related to an anisotropy in the plastic flow.

2.3. Proposed method

As already thoroughly discussed in [7], the use of only one 3D DIC system for the evaluation of the layer compression test implies the assumption of an isotropic deformation, since no information about the orthogonal direction is available. This assumption may be true for materials with no pronounced anisotropy, but for highly anisotropic sheet materials such as aluminum, significant deviations may occur. The main issue here is the incorrect determination of the current cross-sectional area of the deformed specimen. Therefore, the calculated strain as well as the stress may show significant discrepancies from the actual values.

In order to enable a correct evaluation of anisotropic materials with only one 3D DIC system during a layer compression test, a simplification of the method proposed by Merklein and Kuppert [7] is introduced in this paper. By combining the strain data acquired from an inline DIC measurement perpendicular to the rolling direction, ε_1, the strain in the transvers direction, ε_2, is calculated from a relationship between both values $\varepsilon_2 = f(\varepsilon_1)$. The functional relationship is obtained from post-deformation measurements of single - now elliptic - sheet metal discs in analogy to the evaluation method of the biaxial Lankford parameter described by Barlat et al. [8]. With this, the actual shape of the cross section can be incorporated into the calculation of strains and stresses. The methodology is presented in figure 3.

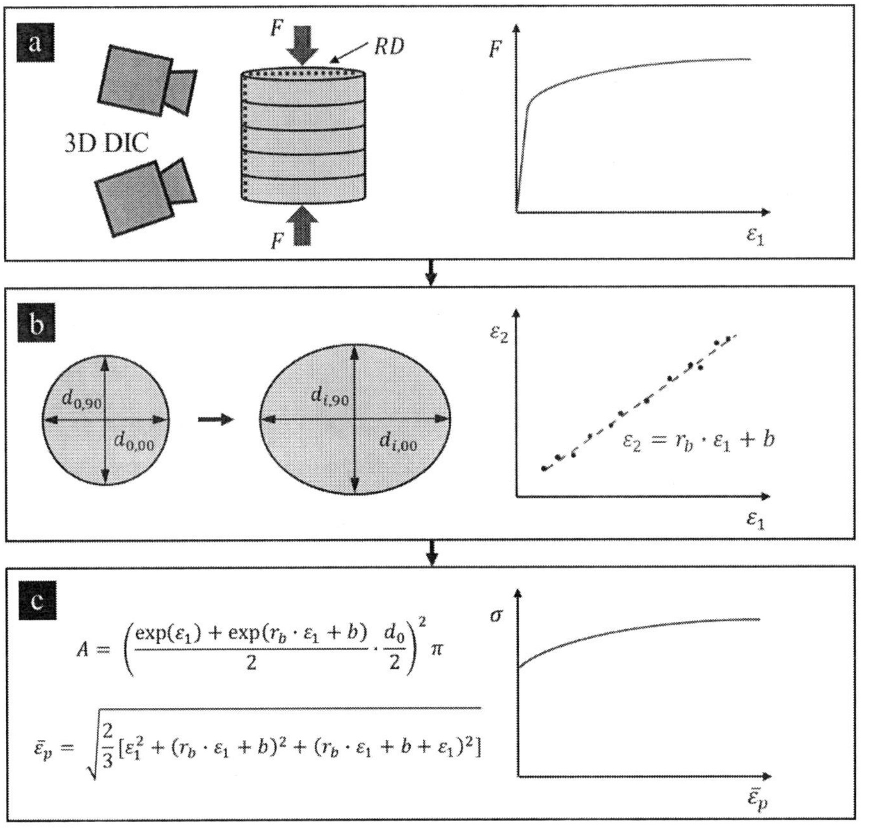

Figure 3. Proposed methodology for the evaluation of the layer compression test: a) Layer compression test with one 3D DIC system perpendicular to the rolling direction (RD), b) Posteriori evaluation of the functional relationship between ε_1 and ε_2, c) Calculation of the true stress from the elliptic cross-sectional area and the effective plastic stain under consideration of $\varepsilon_2 = f(\varepsilon_1)$.

By supplementing ε_2 in equation (11) with $\varepsilon_2 = f(\varepsilon_1) = r_b \cdot \varepsilon_1 + b$ equation (12) follows

$$A = \left(\frac{exp(\varepsilon_1) + exp(r_b \cdot \varepsilon_1 + b)}{2} \cdot \frac{d_0}{2}\right)^2 \cdot \pi, \tag{12}$$

whereby r_b represents the Lankford parameter under equibiaxial tension and b an offset factor. The stress is calculated from $\sigma = F/A$. Under the condition of isochoric flow, the effective von Mises plastic strain follows to

$$\bar{\varepsilon}_{p,vM} = \sqrt{\frac{2}{3}[\varepsilon_1^2 + \varepsilon_2^2 + \varepsilon_3^2]} = \sqrt{\frac{2}{3}[\varepsilon_1^2 + (r_b \cdot \varepsilon_1 + b)^2 + (r_b \cdot \varepsilon_1 + b + \varepsilon_1)^2]}. \tag{13}$$

3. Experimental procedures

3.1. Material and test setup

The studied material is a cold rolled AA6xxx aluminum alloy in W-temper with a nominal sheet thickness of 2.50 mm. The material is quenched between two cooled steel plates directly after solutionizing and tested within five minutes after quenching. In order to enable a quick conduction of the tests and to make the material easy to handle, the single discs of a diameter of 15 mm are pre-cut by laser and held in the sheet metal by a small rib. After quenching the single discs are removed from the sheet metal and layered to a stack of six discs with a height of 15 mm (d/h = 1). The discs are positioned uniformly with respect to the rolling direction and held in position by a cyanoacrylate adhesive applied between the single layers. The experiments are performed at 25 °C on an electromechanically driven

IOP Conf. Series: Materials Science and Engineering **967** (2020) 012077 doi:10.1088/1757-899X/967/1/012077

MTS Criterion universal tensile testing machine with a maximal load capacity of 100 kN. In order to reduce the effect of friction a grease (DuPont Molykote) is applied as lubricant between the specimen faces and the compression plates. The strain fields at the specimen's surface are accessed by a GOM Aramis 4M 3D DIC system positioned perpendicularly to the rolling direction. In addition, the displacements of both compression plates are tracked by three discrete points per plate. The force signal from the load cell is synchronized with the DIC system. The setup is presented in figure 4.

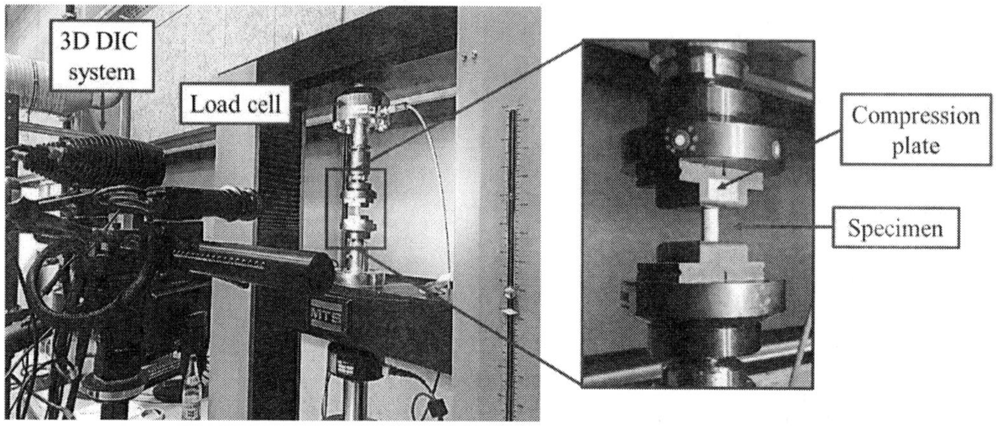

Figure 4. Experimental test setup.

3.2. Evaluation

For comparison purposes the layer compression test will be exemplary evaluated by four different approaches:

- Isotropic evaluation by the averaged compression plate displacement by using equations (9) and (10).
- Isotropic evaluation by a digital extensometer applied between the second and the fifth disc and by using equations (9) and (10).
- Isotropic evaluation by using the averaged tangential surface strain of two discs acquired from the 3D DIC system by using equation (11) and assuming $\varepsilon_1 = \varepsilon_2$.
- Anisotropic evaluation by using the averaged tangential surface strain of two discs acquired from the 3D DIC system and the determined relationship $\varepsilon_2 = f(\varepsilon_1)$ by using equations (12) and (13).

The different evaluation approaches are depicted in figure 5.

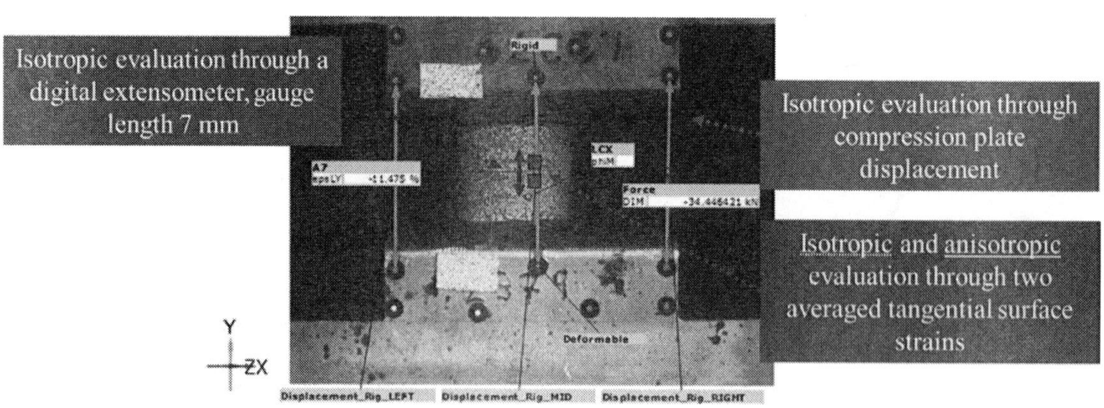

Figure 5. Different evaluation methods of the layer compression test used in this study.

4. Results

4.1. Lankford parameter at equibiaxial loading

Following the recommendations of Barlat et. al [9], the equibiaxial Lankford parameter r_b is obtained by a measurement of the final diagonals of several distinctly compressed discs. The slope of the linear fit describing the relationship between the longitudinal ε_1 (0° with respect to the rolling direction) and transversal strains ε_2 (90° with respect to the rolling direction) gives the r_b-value, which is found to be 1.1734. The offset b is found to be -0.0012. The dependency of ε_2 on ε_1 is depicted in figure 6 and given by equation (14)

$$\varepsilon_2 = 1.1734 \cdot \varepsilon_1 - 0.0012. \tag{14}$$

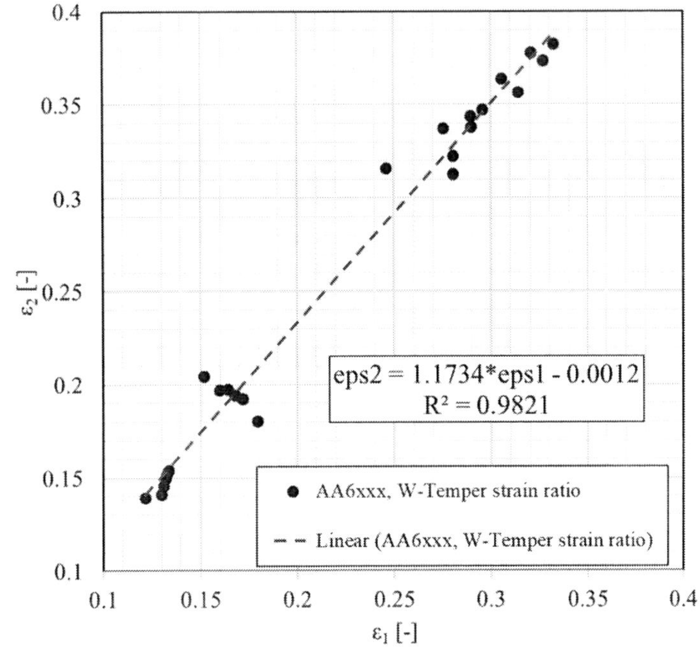

Figure 6. Identified r-value at equibiaxial loading and the functional relationship between ε_1 and ε_2 obtained from layer compression experiments.

4.2. Stress-strain curves

The results of the four different evaluation approaches described in 3.2 are shown in figure 7. For both height-based methods (ε_3), a high level of correspondence between the stress-strain curves can be observed until a strain of ≈ 0.45. Beyond this point both curves diverge and the extensometer-based result shifts progressively toward higher stresses. That indicates a stagnation of the locally measured strain and therefore leads, in accordance to equation (9), to higher stress values then those captured from the entire specimen height. In the case of results based on the tangential surface strain two main characteristics can be observed. First, the stiffness is higher compared to the height-based methods and the elastic regime corresponds to the Young's modulus of aluminum ($\approx$ 70 GPa). Secondly, the diameter-based methods give lower stresses whereby the isotropic evaluation converges at higher strains towards the solution based on the compression plate displacement. Both effects can be traced back to closing gaps between the individual layers with ongoing compression loading. The good correspondence of the isotropic diameter- (ε_1) and the isotropic height- (ε_3) evaluation at higher strains can be reasoned by a decreasing influence of the latter measurement by closing gaps at higher loads. The lower stress level of the anisotropic evaluation is caused by the correct incorporation of the higher cross-sectional area ($\varepsilon_1 < \varepsilon_2$) into the stress calculation. This in turn means, that in this case all isotropic evaluations ($\varepsilon_1 = \varepsilon_2$) overestimates the resulting stresses.

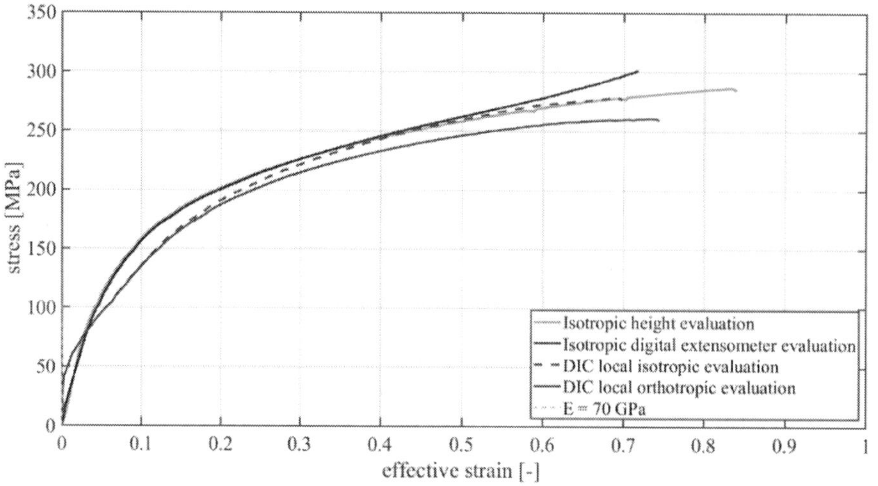

Figure 7. Comparison between different evaluation methods.

5. Conclusions

With the aim to increase the reliability of layer compression results evaluated by only one 3D DIC system, the standard evaluation method is enhanced by the materials anisotropic behavior. The consideration of the anisotropy is relying on the Lankford parameter which must be determined anyway for higher order yield loci models. Using the example of the investigated material, a non-negligible discrepancy between the isotropic and the anisotropic evaluation can be determined. This underlines the importance of the simple but nevertheless effective evaluation method proposed here. In addition, the yield stress at equibiaxial stress can be determined much more precisely than with a height-based evaluation method. In future work, the method should be compared with the results of an evaluation with two orthogonally positioned 3D DIC systems.

References

[1] Banabic D 2010 *Sheet metal forming processes: Constitutive modelling and numerical simulation* (Heidelberg und London: Springer)

[2] Merklein M 2006 *Charakterisierung von Blechwerkstoffen für den Leichtbau* (Bamberg: Meisenbach)

[3] Hill R 1950 *The London, Edinburgh, and Dublin Philosophical Magazine and Journal of Science* **41:322** 1133–42

[4] Marciniak Z 1961 Influence of the sign change of the load on the strain hardening curve of a copper test subject to torsion *Archiwum Mechaniki Stosowanej* **13** 743-51

[5] Yin Q 2014 *Verfestigungs- und Schädigungsverhalten von Blechwerkstoffen im ebenen Torsionsversuch* Dissertation, Technische Universität Dortmund

[6] Pawelski O 1967 *Archiv für das Eisenhüttenwesen* **38 (6)** 437–42

[7] Merklein M and Kuppert A 2009 *Int. J. Mater. Form.* **2 (1)** 483-6

[8] Barlat F, Brem J C, Yoon J W, Chung K, Dick R E, Lege D J, Pourboghart F, Choi S H and Chu E 2003 *Int. J. Plast.* **19** 1297-1319

[9] DIN 50106 Standard 2016 *Testing of metallic materials - Compression test at room temperature* (Berlin: Beuth Verlag)

[10] Tian H, Kang D and Lin J 2004 *Key Eng. Mat.* **274-276** 793-98

[11] Alves L M, Nielsen C V and Martins P A F 2011 *Exp. Mech.* **51 (9)** 1565-72

[12] Gese H, Keller S, Yeliseyev V and Dell H 2003 *Ermittlung von Fließwiderstandskurven bei großen Formänderungen für die Blechumformsimulation (Kennwertermittlung für die Praxis vol 7)* ed Frenz H and Wehrstedt A (Weinheim: Willey-VCH) pp 242-9

IOP Conf. Series: Materials Science and Engineering **967** (2020) 012078 doi:10.1088/1757-899X/967/1/012078

On the design of an air-cooling system of an IC engine block made from sheet metal

Prashant P Date[1*] and Shubham P Dhonde[2]

[1*] Department of Mechanical Engineering IIT Bombay, Powai, Mumbai 400076 (Corresponding Author)
[2] Department of Mechanical Engineering, IIT Guwahati, Guwahati, Assam, India

Abstract. The cast and machined automotive cylinder block and the cylinder head serve to locate and hold the combustion cylinders in place, provide for cooling channels for the combustion cylinders and damp out vibrations, besides housing allied components needed for combustion. In the present work, the authors attempt to redesign the cooling system of the engine block for manufacture from sheet metal, and reduce its weight. It is proposed to have an air-cooling system around each of the combustion cylinders using a finned cooling cylinder rotating concentric with respect to the combustion cylinder with a coolant occupying the annular gap between the combustion cylinder and the cooling cylinder. The peculiar design enables hotter air to enter the combustion cylinder thereby increasing the expected thermal efficiency and power output, besides enhancing the quality of air-fuel mixing. Manufacture of the engine block from sheet metal involves embossing flow channels into thin sheets for the coolant, or machining such channels in thick sheets, folding them over, brazing the flat surfaces between channels and brazing the fin to the periphery of the cooling cylinder

1. Introduction

A conventional IC engine involves manufacturing processes like casting and forging which are known to be environment unfriendly. Moreover, an engine block is a rather heavy casting providing great opportunity for light weighting. Besides, it needs to be machined subsequent to casting. Functionally, an engine block primarily houses the combustion cylinders, provides passages for circulation of the cooling fluid, and damps out the vibrations.

If a light weight structure constructed from alternate form of raw material using an alternate manufacturing route, serves the above-mentioned functionalities, then it might be possible to effectively reduce the weight of the engine. A light weight IC engine would also lead to benefits downstream, like lighter engine cradle and engine mounts.

This subject has not been touched upon by literature published in recent times, except for the ones by Date et. Al. [1-3] and Mahadevan [4], who proposed measures to manufacture the reciprocating components of the IC Engine (piston [1], connecting rod [3]), and transmission components like gears [2]. Water cooled light weight engine block for manufacture using sheet metal was designed by Basunathe [5]. Automotive cooling system component interactions have been documented by Walters [6], and studies on light weighting especially by material substitution (use of light-weight materials) for manufacture of automobile components were reported by Cole et. Al [7]. Materials and manufacturing processes used to make a cylinder block were reported by Nguyen [8]. Earlier, L. M. Taylor [9] had made attempts to make the engine block out of sheet metal. Taylor constructed the cylinder block from tubing and sheet metal, joining parts by oven brazing. However, one is unable to find research-based documentation on the design and manufacture of a light weight engine block using sheet metal. Of late, Dhonde [10] has made an attempt to design an IC engine block out of sheet metal.

The present work is about light weighting of an air-cooled engine block of an automobile, by redesigning the cooling system, for manufacture using sheet metal. Designing a cooling system compatible with multi-cylinder IC engines presents many challenges. The design challenges and how they can possibly be met using a novel design of an air-cooled engine block are discussed in the sections that follow.

2. Design of a sheet metal engine block

2.1. Engine parameters

The total power of a four-stroke engine generated by a four in-line cylinder configuration used in this study, was assumed to be about 100 kW. Each cylinder was considered to have an inner diameter of the liner equal to 80mm. Assuming a thermal efficiency of 34% and assuming that 33% of the total heat is carried away by the cooling system, the remaining 33% is carried away by the exhaust gases. Taking this into account, each cylinder would need a cooling system to dissipate an average of 7.5 kW of power. Heat transfer considerations apart from light weighting should therefore maximise the dissipation of heat for efficient cooling of the engine.

Consistent with the above, the peak cylinder liner temperature was estimated to be about 812 deg. C and the peak pressure of combustion was estimated to reach about 7 MPa. At these temperatures and pressures, the yield strength of the liner material should be at least 300 MPa at 812 deg. C. These values relate to a typical engine available in the laboratory.

2.2. Design requirements

In order to make the engine block manufacturable using sheet metal, it needs to be redesigned. Especially, the method of providing for cooling of the cylinders will have to change, since the current design is meant to be made by casting, and cannot be made from sheet metal, the way it is. To redesign the engine block, the functionalities of the conventional engine block, which is usually made as an aluminium casting, need to be achieved by design from sheet metal. *No changes in the combustion parameters are envisaged as a consequence of the engine block being re-designed.* Thus the process of generation of power in the engine cylinders remains unchanged, though the means of cooling and methods of performing the functions of the engine block would change as the engine block gets redesigned for manufacture using sheet metal. The major functions of the engine block are listed below :

a. The combustion cylinders need to be housed in the cylinder block, being held at a certain distance from each other.
b. Provision for cooling of the combustion cylinders should be made
c. Engine vibrations should be damped out
d. Adequate support for mounting allied parts should be available.

2.3. Design Concept

This may be achieved using sheet metal of different thicknesses, in the following manner :

a. The combustion cylinders are supported between two thick parallel plates. The cylinders are firmly held in position by the flanged plates.
b. Each combustion cylinder has around it a concentric cooling cylinder with fins. There is an annular gap between the combustion cylinder and the cooling cylinder. A certain quantity of thermal fluid (like Therminol XP) fills the annular gap. Cooling channels in the fins meant to dissipate heat open into the annular gap and the thermal fluid is circulated in these channels by natural convection (Figure 3).
c. The annular fluid also serves to damp the vibrations in each of the cylinders of the engine.

d. The chamber made from sheet metal that serves as the engine block would provide adequate opportunity to support / mounting of allied parts.

2.3.1. Thermal considerations for the engine block. Figure 1(a-c) shows the schematic arrangement of the engine block made from sheet metal. Figure 1(a) shows the arrangement of cylinders in a trapezoidal enclosure relative to the direction of movement of the vehicle, i.e., direction of entry of air into the engine, and the way the general features of the engine block would look from the top. The direction of reciprocation of the piston is perpendicular to the plane of paper in Figure 1(a). The trapezoidal shaped sheet metal chamber has two parts, the lower part that admits ambient air as the vehicle moves and the top part that has the inlet and outlet ports and corresponding ducts to and from these ports (Figure. 1a). The cylinders are closer to the larger side of the trapezium, and so are the inlet and outlet ports. The cylinders are arranged in line along the width of the vehicle, with the crank shaft running along the width of the vehicle (perpendicular to the plane of paper in Figure 1b). This figure also shows a schematic of engine action together with the engine combustion cylinder, cooling cylinder and the engine block. The path of air entering the lower half, acquiring heat from the cooling fins that dissipate heat from the combustion cylinders to the ambient air (Figure 1(b)). The heated air would try to expand, but is instead compressed by the tapering section (Figure 1(a)), thereby increasing its velocity

The incoming air will invariably carry dust particles along. As the air initially collides with the fins, part of the dust particles would get removed. Those smaller ones which enter the trapezoidal section of the sheet metal engine block would get removed as the hot air, at a high velocity, would collide with the wall at the back requiring the air to make a 180 degrees turn before entering into the inlet port of the combustion cylinder in the upper part of the cylinder block (Figure 1(b)). Thus the high velocity air entering the cylinders may be expected to generate a mild supercharging effect. This way, part of the heat lost from the engine may be returned to the combustion cylinders. The extra air that does not get into the cylinders is used for scavenging the exhaust gases preventing them from crossing over to the inlet port side

The engine cooling (and preheating of incoming air) is achieved by the finned cooling cylinders running concentric with the combustion cylinders (Figure 1(c)) The finned cylinders of a multiple in-line cylinder engine determine the power dissipated to the cooling air, and are made from sheet metal. Adequate number of fins and fin area must be available for efficient dissipation of heat as shown schematically in Figure 1(c). In this work, to account for adverse conditions, the maximum temperature of ambient air has been assumed to be 60 deg. C, which is higher than the temperatures seen by most Indian cities during peak summer. The objective here is to admit hotter air into the combustion cylinders thereby increasing the Carnot efficiency.

The loss of volumetric efficiency due to expansion of the hot air is compensated by the tapered geometry of the cylinder block. The angle of taper therefore is a very important design parameter. This way the velocity of air entering the cylinder may be expected to increase. Heated air would have a natural tendency to rise and will get into the upper part of the cylinder block, where a suction stroke would do the rest. In an engine operating at 6000 rpm, 3000 suction strokes per minute would be expected. Under these conditions, the cylinder liner would be at 812 deg C for total heat generation of about 25 KW. The temperature of the Therminol fluid in the fins would be expected to reach 107 deg C for an average air velocity of 10m/s (driving speed of about 40 kmph). In the worst case scenario, wherein the driving speeds of 10-15 km/hr at full power can be attained, the temperatures attained by the therminol fluid are of the order of 280 deg C, with the fluid circulating through the channels by natural convection. This is much below its maximum operating temperature of 315 deg C. These temperatures will influence the material selection for manufacturing, the design of the fins, and the manufacturing processes themselves.

The cooling system illustrated in Figure 1(a-c) is self-regulating. That is, when the vehicle is moving at a high speed, more heat is generated in the combustion cylinders and incoming air meets the fins at a greater velocity enhancing the rotational speed of the finned cooling cylinder. This would lead to turbulence in the coolant reservoir and enhance the heat transfer coefficient. Hence greater power generation is matched by more effective cooling from air flowing over the fins. In all the calculations it is assumed that only the fin area will dissipate heat to the incoming air. *The area of the engine block itself that might dissipate considerable amount of heat has not been taken into consideration at this time.* Hence the sheet metal engine block will further augment the heat carried by the air into the engine.

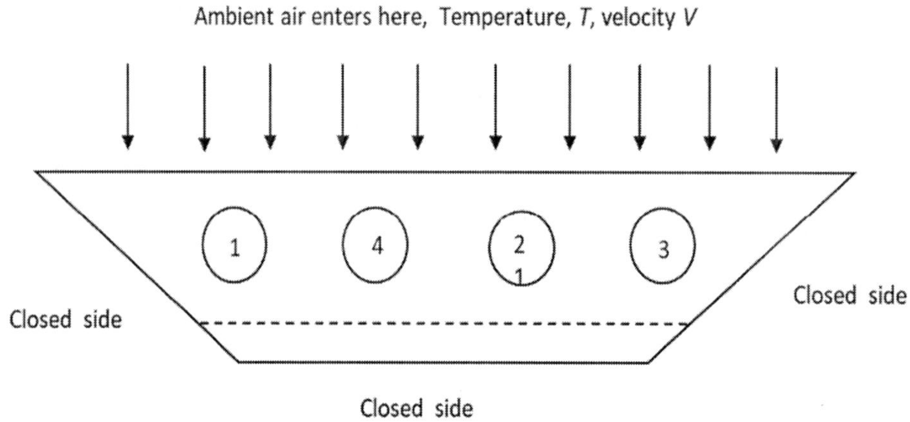

(a)

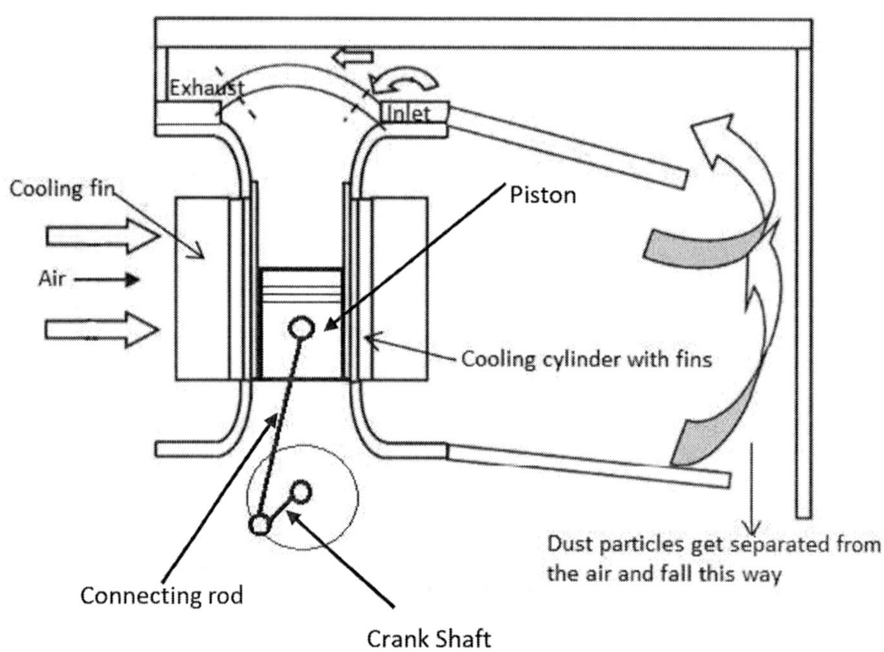

(b)

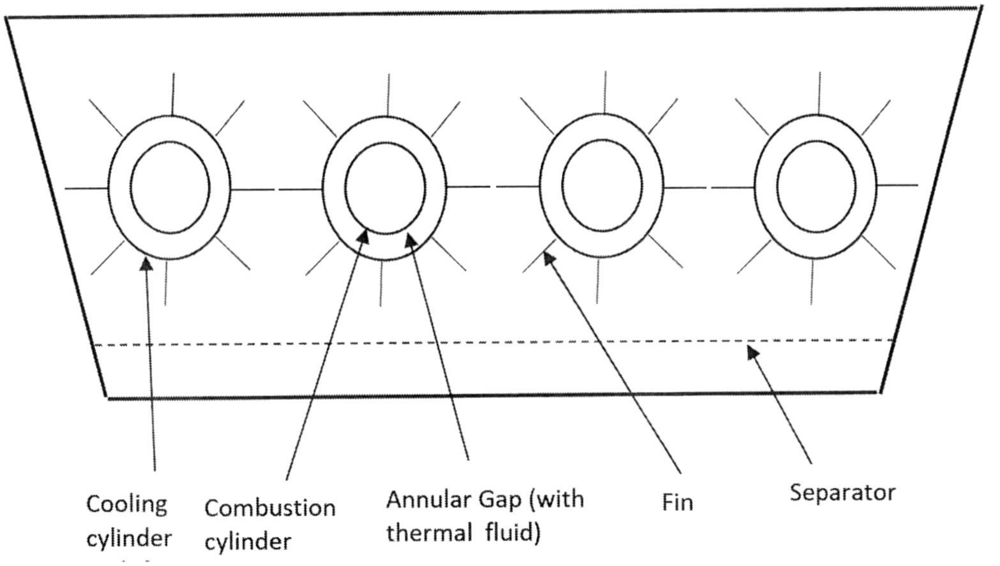

(c)

Figure 1. Schematic of combustion cylinders concentric with the finned cylinders (a) Positioning of finned cylinders (b) Side view showing the path of air. (c) Schematic of the engine block showing the finned cylinders around the combustion cylinders.

The air, after changing direction gets into the inlet of the cylinder, while some of it moves over to the exhaust side facilitating scavenging. The flue gases of all cylinders therefore enter a common line (perpendicular to the plane of paper in Figure 1(b)) and are discharged via an exhaust pipe.

The radius of curvature of the fin and the length of the fin (angle subtended by the fin at the centre of curvature, Figure 2), was also studied. Number of fins can go upto eight, exceeding which the cooling effect drops. Similarly, excessive curvature of the fin also restricts its utility. Hence the light curvature is applied through a subtended angle of 10 degrees (i.e., 'angle of curvature' = 10 deg.).

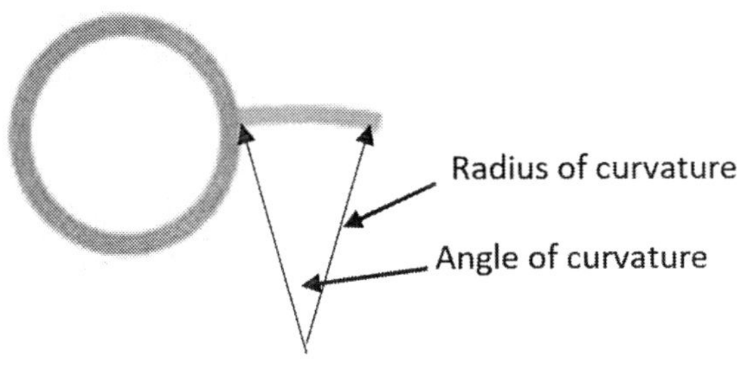

Figure 2. Curvature of a typical fin. Eight such fins are provided around the periphery of the cooling cylinder.

Based on simulations, eight fins were found to have the maximum dissipation of heat. Figure 3 shows a schematic view of a combustion cylinder with a concentric cooling cylinder on which one fin is shown to conceptually illustrate the features to enhance the cooling system. The fluid flows into the fin from a constant fluid filled annular gap between the combustion and cooling cylinders that serves as a reservoir of thermal fluid. Figure 3 shows the various features of the cooling fin with a network of horizontal and vertical cooling channels for flow of the thermal fluid.

The flow of the fluid through the channels made in the fins experiences inertia forces in terms of the centrifugal force (created by rotation of the cooling cylinder as the air enters the trapezoidal engine chamber), in addition to natural convection. At slow speeds natural convection would dominate, while at higher speeds, the flow of fluid would depend largely on the centrifugal force resulting from the speed of rotation of the cooling cylinder. Thus the effect of natural convection on circulation of the cooling fluid is relatively small at high speeds.

The new design of the cylinder block does not interfere with the combustion inside the automotive cylinders *per se*, but provides a framework to dissipate the heat to the incoming air and use hotter air for combustion in the combustion cylinders.

2.3.2. Aerodynamic loading on the sheet metal fins. The aerodynamic load on account of the air flow impinging on the fin is rather small, and does not cause any significant bending moment. A typical velocity is assumed to be 50 kmph for city driving conditions. For highway driving conditions, and no rotation of the finned cylinder, the bending stress on the fin corresponding to 120 kmph works out to about 132.5 kpa for a fin thickness of 10 mm. If the sheet thickness is reduced to 0.5 mm and cooling fluid is circulated through cooling channels embossed in the fin, the section modulus of the fin can be increased significantly. Even in the absence of any such stiffening features, the bending stress increases to 53 MPa.

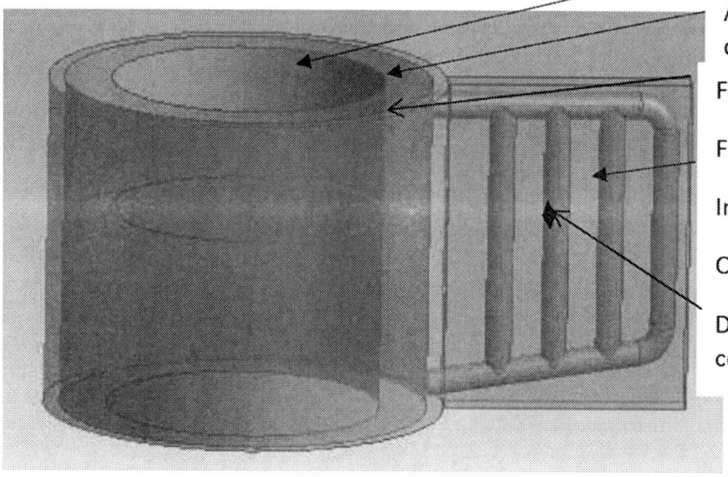

Figure 3. Schematic of the arrangement of a fin, attached to a finned cylinder concentric with the combustion cylinder (a) with parallel flow channels (b) with horizontal and vertical channels. The centre-to-centre distance between the vertical channels is about 17 mm.

3. Manufacture of the sheet metal Engine block

3.1. Material for manufacture

The fins would have to be made from conducting metallic sheets with the fluid passages embossed into the fins to facilitate circulation of the fluid, as well as impart a mild curvature as shown in Figure 2. The material for manufacture of the finned cylinder should have good thermal conductivity and weldability, together with moderate degree of ductility. Given the temperatures seen by the 4 wheeler engines, using aluminium alloys will be risky as they will soften at the temperatures they will get heated to. Hence a medium carbon steel will be advisable as it will have adequate strength at the operating temperature, good ductility, and good weldability from a manufacturing standpoint. The stress of 53 MPa is rather small for any significant deflection of the fin which may be made from Al 5083 alloy or Al 6061 alloy. Besides, both have very good weldability as well. However, aluminium alloys will be suitable at these stress levels only if the fins are effectively cooled by circulation of the thermal fluid.

3.2. Manufacture of the finned cylinder

The fins could be made by making the channels (by machining, as in the present case described in Figure 4) or by embossing if the sheet is thin enough, for the thermal fluid to flow through, bending of sheet metal at the two ends (Figure 4(a)) and folding them over at the line of bend to bring together the two halves of the fin (Figure 4(b)), which will involve a hairpin bend. In order to retain the overall shape of the fins, furnace brazing of the fins would be more practicable to prevent leakage of the fluid.

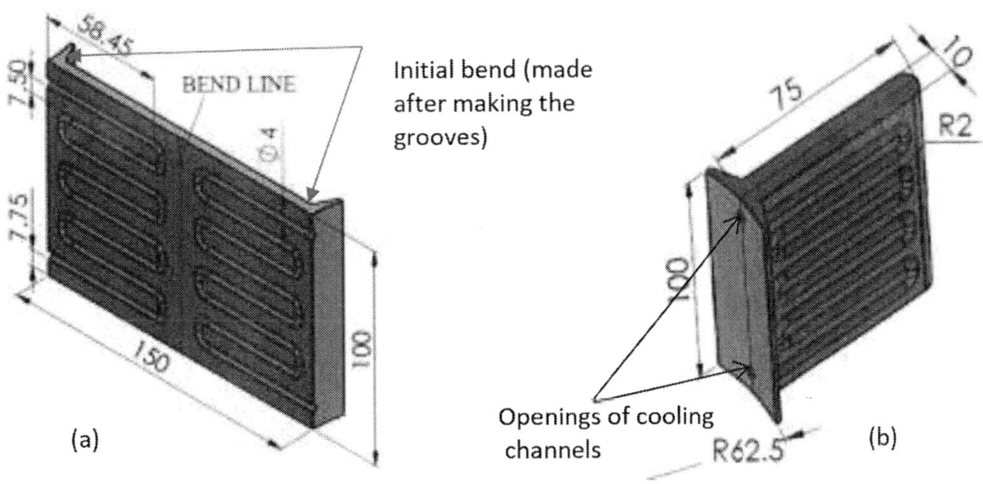

Figure 4. (a) Two halves of the fin embossed on a single sheet to be folded(bent at 180deg) along the bend line (b) The fin after bending, showing end openings of the cooling channel and the radiused (base) flange to enable brazing onto the finned cooling cylinder.

The base of the fin would be shaped as shown in Figure 4(b). This will enable assembly of the fin onto the outer surface of the finned cooling cylinder by brazing or similar technique. For better heat transfer, brazing the bottom of the fin with the outer surface of the cooling (finned) cylinder would be recommended. This will also strengthen the joint between the fin and the cylinder, making it leak proof. Pins may be inserted through the holes in the cylinder and those serving as openings for cooling channels to align these during brazing. The pins can be removed after the fins have been brazed to the cooling cylinder. In such an event, relatively smaller sheet thickness would be adequate. For instance, a thickness of 3mm would be enough for sufficient stiffness of the aluminium fin.

The cooling cylinder may be made from an aluminium sheet about 5-6 mm thick. The reservoir of thermal fluid would then be bounded by the outer surface of the combustion cylinder, the inner surface of the cooling cylinder and the thickness of the top and bottom plates, (which would be of the order of 10mm or so) and top and bottom seals.

3.3. Manufacture of trapezoidal engine chamber

The trapezoidal engine chamber is manufacturable by joining sheets/plates of suitable sizes. Two thick trapezoidal parallel plates, with flanged holes in one plate (equal to the number of cylinders) aligned with the corresponding flanged hole in the other plate, enable proper alignment of the combustion cylinder consisting of a liner inside it. The combustion cylinders could then be welded to the flanges in the parallel plates.

A negative (compressive) contact pressure can be generated between the combustion cylinder and the liner by subjecting the combination to high internal pressure (as in hydroforming) so as to cause a sufficiently high elastic contact pressure. At the combustion temperature the contact pressure could be partly relieved due to heat, but then the liner will develop only small tensile stresses.

Such a design holds promise due to a significant weight reduction (at least 20-30%) in addition to eliminating the radiator and the cooling circuit not being required any more.

4. Conclusions

A design and manufacturing scheme for manufacturing the engine block from sheet metal has been suggested. The functionalities of the cast engine block can be achieved by a sheet metal engine block re-designed for the purpose.

Following differences from the conventional engine block are noteworthy :

a. A separate coolant pump is not required. Coolant is circulated by natural convection, saving power needed to drive the coolant pump
b. A separate radiator is not required as the function is performed by the finned cooling cylinder made from sheet metal
c. No machining of the engine block is involved
d. The redesigned cooling system uses wrought elements which would have a greater strength/weight ratio compared to the cast elements
e. A number of expensive manufacturing processes typically associated with the conventional engine block would get eliminated.

References

[1] Nayak K C and Date P P 2018 Manufacture of Light automobile engine piston head using sheet metal, in Procedia Manufacturing, 17th International Conference on Metal Forming, Metal Forming 2018, 16-19 September 2018, Toyohashi, Japan, pp 940-48.

[2] Pandey Ankit Kr, Walunj B S and Date P P 2018 Simulation based approach for lightweighting of transmission components using tube hydroforming in *Procedia Manufacturing*, 17th International Conference on Metal Forming, Metal Forming 2018, 16-19 September 2018, Toyohashi, Japan, pp 915-22.

[3] Date P P, Kasture R M and Kore A S 2017 On the light weighting of automobile engine components : forming sheet metal connecting rod, IOP Conf. Series,Journal of Physics, (**896**) 2017.

[4] Mahadevan R 2017 Strategies for Better Fuel Economy, Light Weighting, and Friction Reduction in Cylinder Components of IC Engines In Bajpai R., Chandrasekhar U. (eds) Innovative Design and Development Practices in Aerospace and Automotive Engineering. Lecture Notes in Mechanical Engineering. Springer, Singapore.

[5] Basunathe S 2018 Design and Manufacturing of Engine Block Using Sheet Metal, *M. Tech Dissertation*, IIT Bombay, Mumbai, India, 2018.

[6] Walters J 2001 Automotive cooling system component interactions, Ph.D. thesis, Texas Tech. University.

[7] Cole G S and Sherman A M 1995 Light weight materials for automotive applications *Mater. Charact.* **35** pp. 3–9.

[8] Nguyen H 2005 Manufacturing Processes and Engineering Materials Used in Automotive Engine Block *Mater. Sci. Eng. Sect. B* **EGR250** pp. 11–23.

[9] Taylor L M 1944 Cylinder Block US Patent no. 2444963 June 1944.

[10] Dhonde S P 2017 Lightweighting of cooling system of IC Engines, Summer Term Project Report IIT Bombay Mumbai India.

IOP Conf. Series: Materials Science and Engineering **967** (2020) 012079 doi:10.1088/1757-899X/967/1/012079

Experimental issues in the instrumented 3 point bending VDA238-100 test

L. Wagner, P. Larour*, D. Dolzer, F. Leomann, C. Suppan

voestalpine Stahl GmbH, voestalpine-Straße 3, A-4020 Linz, Austria.

E-mail: patrick.larour@voestalpine.com

Abstract. The instrumented 3-point bending test according to VDA238-100 test standard is increasingly used within the steel and automotive industry. Originally developed for aluminum hemming characterization, this bending test has been shown to be also relevant for local formability and crash foldability assessment originally of press-hardened steel grades and more recently for newly developed advanced and ultra high strength AHSS/UHSS steels grades. This instrumented bending test delivers bending load vs. bending angle curves. It is commonly assumed that material failure shortly happens beyond maximum load after a 30N load drop. The bending angle at maximum force α_{Fmax} characterizes then the bendability of the investigated material. The assumption maximum force = bending crack initiation, while being true for press-hardened grades, is in too many case not valid for steel grades with tensile strength ≤1200MPa and cannot be universally trusted. An alternative approach is presented using passive acoustic emission sensors placed in the vicinity of the bending punch. The interpretation of such acoustic data is however quite subjective and still in trial status. Redundant crack detection systems based on load, acoustic as well as optical measurements may have to be considered together for increasing crack detection reliability within the VDA238-100 bending test specification.

1. Introduction

The instrumented 3 point bending test according to VDA238-100 has established itself in the last decade as a reference standard test for automotive sheet steel bendability characterization [1]. One major motivation is the correlation between bending angle at maximum load commonly assumed to describe crack initiation with axial & side impact crash ability for automotive PHS press-hardened steels ([2]-[4]) as well as for AHSS/UHSS advanced/ultra high strength steels ([2],[5]). Some correlation of bending angle at crack initiation (but not necessarily with bending angle at maximum load) has been found with material fracture toughness in [6]. The correlation with local formability from (notched) tensile test thickness fracture is however not that obvious [7]. There are some ongoing investigations for increasing VDA238 testing accuracy with regard to bending angle formula [8], test set up stiffness characterization and optimization [9]-[11], thickness dependency of bending angle at maximum load [12] or punch-sheet-liftoff and its prevention [8],[13]. The validity of maximum bending load as universal material bending failure criteria is questioned in [5],[6],[14]. Recently interest shifts therefore towards digital image analysis DIC assisted VDA238 bending test both for optical accurate crack detection as well as plane strain fracture strain determination for damage model calibration [14]-[19].

Some investigations deal with alternative acoustic measurement for damage evolution assessment during bending test [20]-[22]. The present investigation gives a critical assessment of the reliability of the bending angle at maximum load with regard to physical crack initiation detection in VDA238 bending test. The results shown in this contribution originate from R&D activities covering the last decade and are neither representative of the current material development and specification status nor of the actual test setup development stage.

2. Experimental procedure

Figure 1 shows the bending test setup and test parameters. In the following investigation and figures, VDA238-100 standard bending test conditions as defined in Figure 1 apply if not specifically mentioned differently. Sample failure is supposed to occur in the maximum load range. After reaching the maximum load, the test is however not interrupted but is performed up to a bending angle around 160°, which is dependent on the sheet thickness. The recorded bending load/displacement curve is used to calculate the bending load/bending angle curve. Longitudinal/transverse designation refer to a bending punch tool line 0°/90° to rolling direction RD according to VDA238 specification.

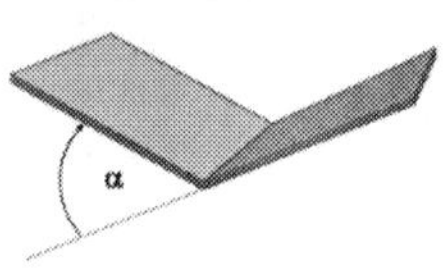

punch diameter: 0.8mm; roll diameter: 30mm
roll distance = 2xthickness+0.5mm; punch testing speed: 20mm/min
free rotating rolls, as delivered condition, no lubricant, no Teflon film
sample size: 60x60mm (alternatively 30x60mm).

Figure 1: VDA238-100 instrumented three-point bending and acoustic test set-up.

Figure 2a illustrates the load based bending properties. Additionally to the VDA238 bending angle at maximum load (α_{Fmax}), the average (absolute) postuniform slope between maximum force and inflection point (|PUBS-F|) has proved useful within own previous PHS material investigations [3]-[4]. Figure 2b delivers an insight on acoustic emission based properties. A passive piezoelectric acoustic sensor (Vallen ASCO P-AE type) is located in the vicinity of the bending punch (Figure 1). It delivers via peak holder some Acoustic Emission event amplitudes AE(i) which are then integrated (summed up) over time as a Cumulated Acoustic Emission signal (CAE, equation 1) and synchronized at 50Hz versus punch displacement and load. The cumulated acoustic emission CAE can be divided into a continuous friction-related and a crack-related acoustic event contribution (Figure 2b).

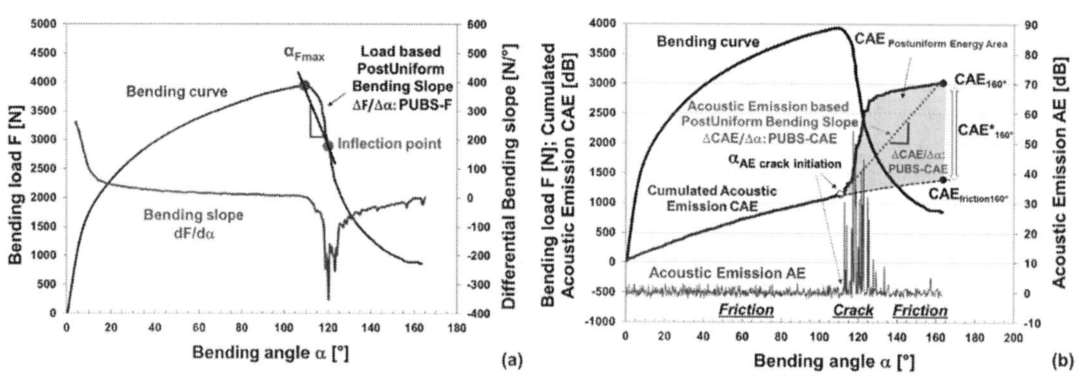

Figure 2: Bending curves with (a): Load based postuniform bending slope PUBS-F; (b): Cumulated Acoustic Emission CAE based postuniform bending slope PUBS-CAE.

The baseline friction contribution can be fitted as a 2nd order polynomial vs. bending angle (equation 2). A clear upper deviation of this baseline is observed for the CAE signal at onset of material failure. The crack-related CAE* contribution can then be derived by subtracting the CAE friction fit contribution from the total recorded CAE signal (equation 3). The postuniform bending slope is derived based on CAE signal between crack initiation and bending test end at 160° with or without the friction contribution (PUBS-CAE: equation 4, PUBS-CAE*: equation 5). The analogy based on bending force vs. bending angle would be the postuniform bending slope |PUBS-F| between maximum force and inflection point as defined in equation 6 and Figure 2a.

$$CAE(i) = \sum_{i=1}^{n} \mathrm{AE(i)} \tag{1}$$

$$CAE_{Crackfree\ Friction\ Fit} = -0.029555.\alpha^2 + 13.669319.\alpha \tag{2}$$

$$CAE^*_{friction\ free} = CAE_{Recorded} - CAE_{Crackfree\ Friction\ Fit} \tag{3}$$

$$PUBS\text{-}CAE = (CAE_{160°} - CAE_{Crack\ initiation\ AE})/(160° - \alpha_{Crack\ initiation\ AE}) \tag{4}$$

$$PUBS\text{-}CAE^*_{friction\ free} = CAE^*_{160°}/(160° - \alpha_{Crack\ initiation\ AE}) \tag{5}$$

$$|PUBS\text{-}F| = (F_{max} - F_{Inflection\ point})/(\alpha_{Inflection\ Point} - \alpha_{Fmax}) \tag{6}$$

The CAE signal is much smoother to analyze and can be considered as an integral acoustic energy of all acoustic events ever recorded (sample-punch first contact and friction, sample-rolls friction, roll bearing intrinsic friction). Tribological, hydraulic test setup as well as surrounding laboratory environment noise sources are therefore superposed on the actual material failure crack-related acoustic events of interest. The acoustic burst events range for the actual test set-up between 5 and ≈75dB depending on thickness and material strength and are often clearly audible for the operator (Figure 3).

Some samples break in a massive way (Figure 3a,b), other stepwise with individual acoustic peaks (Figure 3c), other sample only show some regular accumulated support roll friction noise and test artefacts which are not clearly linked to sample failure (Figure 3d). Isolated peaks before maximum force (Figure 3b-d) may be due to surrounding machine testing environment or maybe inner cracks on the compression side which do not affect the bending load curve. Slip-stick of bending rolls may (rarely) be an issue as well as possible lateral sample sliding under the punch (Figure 3d).

Due to the relative subjectivity of acoustic emission interpretation in the applied test setup, an automatic reliable test stop cannot be achieved. However it enables useful post-mortem analysis in order to check roughly within ±5° the bending crack initiation angle with regard to F_{max} location. If large discrepancies are detected between α_{Fmax} and $\alpha_{acoustic\ material\ failure}$ then this should be reported in the test result protocol. Run out bending tests without material failure at all can be rather easily identified as friction driven without significant crack-related acoustic burst events accumulation.

It is also recommended to let the test run until maximum achievable displacement (≈14mm) and bending angle (≈160°), in this way the postuniform bending crack behaviour can be also assessed, since fracture mechanic and local ductility as well as crash behaviour (crashindex) do not solely depend on α_{Fmax} level but also on how fast cracks propagate under bending loading [3]-[6].

A pre-load is advantageous to avoid initial impact vibration noise at punch-sample contact. The acoustic sensor should be ideally placed on an immobile tool to avoid dynamic inertial effects as well as hydraulic cylinder noise. It should be fixed somewhere on a flat surface on the punch grip as mostly found in the literature ([20]-[22]). The configuration immobile punch with moving rolls may be more suitable than moving punch and fixed rolls with regard to parasitic vibration noise. The less connecting parts and distance between AE sensor and the sample, the better for the signal quality.

A frequency range for the AE sensor (100kHz-400kHz) is suitable for metals. An application of the AE sensor directly on the sample (Figure 1) however does not bring significant advantages (as also seen in [20],[21]), moreover it could lead to damage of the sensor and is not practicable for mass scale testing. A mechanical damping of the testing frame should not to be necessary in usual laboratory conditions.

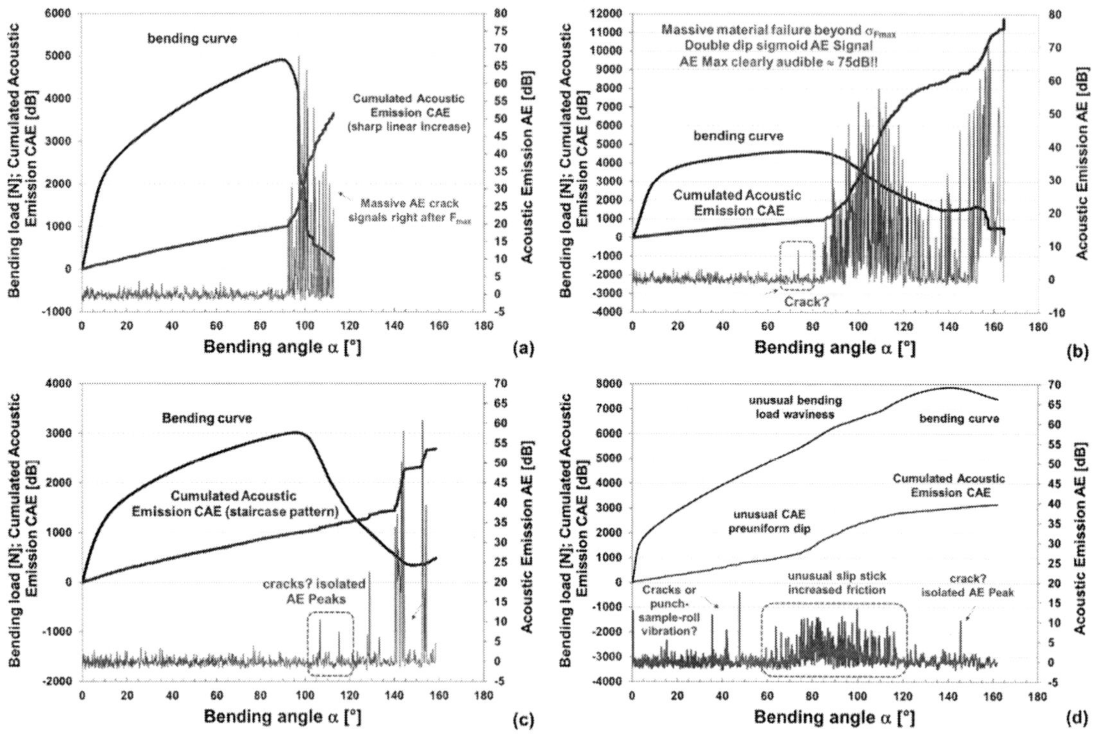

Figure 3: Typical bending curves with AE instrumentation (a): Massive AE-signal with sudden break; (b): Double dip AE-signal 75dB; (c): AE-Signal with single peaks; (d): AE-signal artefacts.

3. Experimental results of crack-free bending curves

Over the past years many examples have been collected with obviously no cracks at all observed on 160° fully bent samples, which actually compromises the whole VDA238 bending test significance. Exemplary bending curves with no visual sample failure are given for mild and high strength low alloy HSLA samples (Figure 4a) as well as for AHSS up to 1000MPa tensile strength (Figure 4b). Misleading α_{Fmax} VDA238 values between 100° and 140° can still be determined. Bending samples may not fail at all while a pronounced bending angle at maximum force α_{Fmax} is delivered as test result (Figure 5).

A mild steel in the as delivered and 15% uniaxial pre-strained condition showed obviously no cracks up to 160° (Figure 5a). α_{Fmax} even increases with increasing pre-straining level, which contradicts the assumption that pre-straining should decrease α_{Fmax}. This is rather linked to support roll transverse stiffness vs. sample thinning after pre-straining. Figure 5b shows bending curves with 20 replicates (crack-free at 160°) for HSLA grade HX340LAD, which unusually match together with atypical low scattering in α_{Fmax}. Figure 5c shows HX500LAD bending samples with crack-free metallographic sections neither at maximum force nor after 160° bending. The usual anisotropy between longitudinal and transverse direction for α_{Fmax} also unexpectedly disappear. Figure 5d shows a DP600 steel grade with varying thickness with no cracks up to 160° (only friction contribution below 5dB in AE-signal). A decrease of α_{Fmax} with increasing thickness is even suggested, which rather is a consequence of increasing lateral bending force triggering the maximum force sooner. Figure 5e,f show similarly results for DP800 and CP800 steel grade with α_{Fmax} between 100° and 160° without any cracks and low 5dB AE-signal. Thinner material often show a strictly monotonic increasing bending curve (α_{Fmax}=160°).

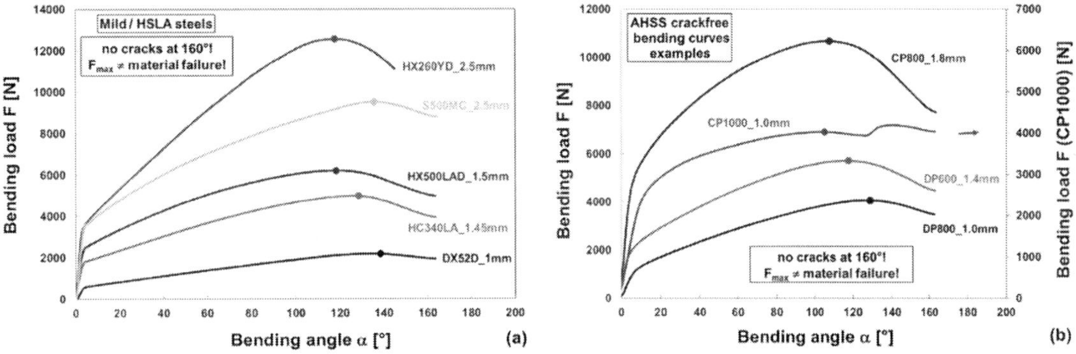

Figure 4: Bending curves with α_{Fmax} without failure (a): Mild/HSLA steels; (b): AHSS steels.

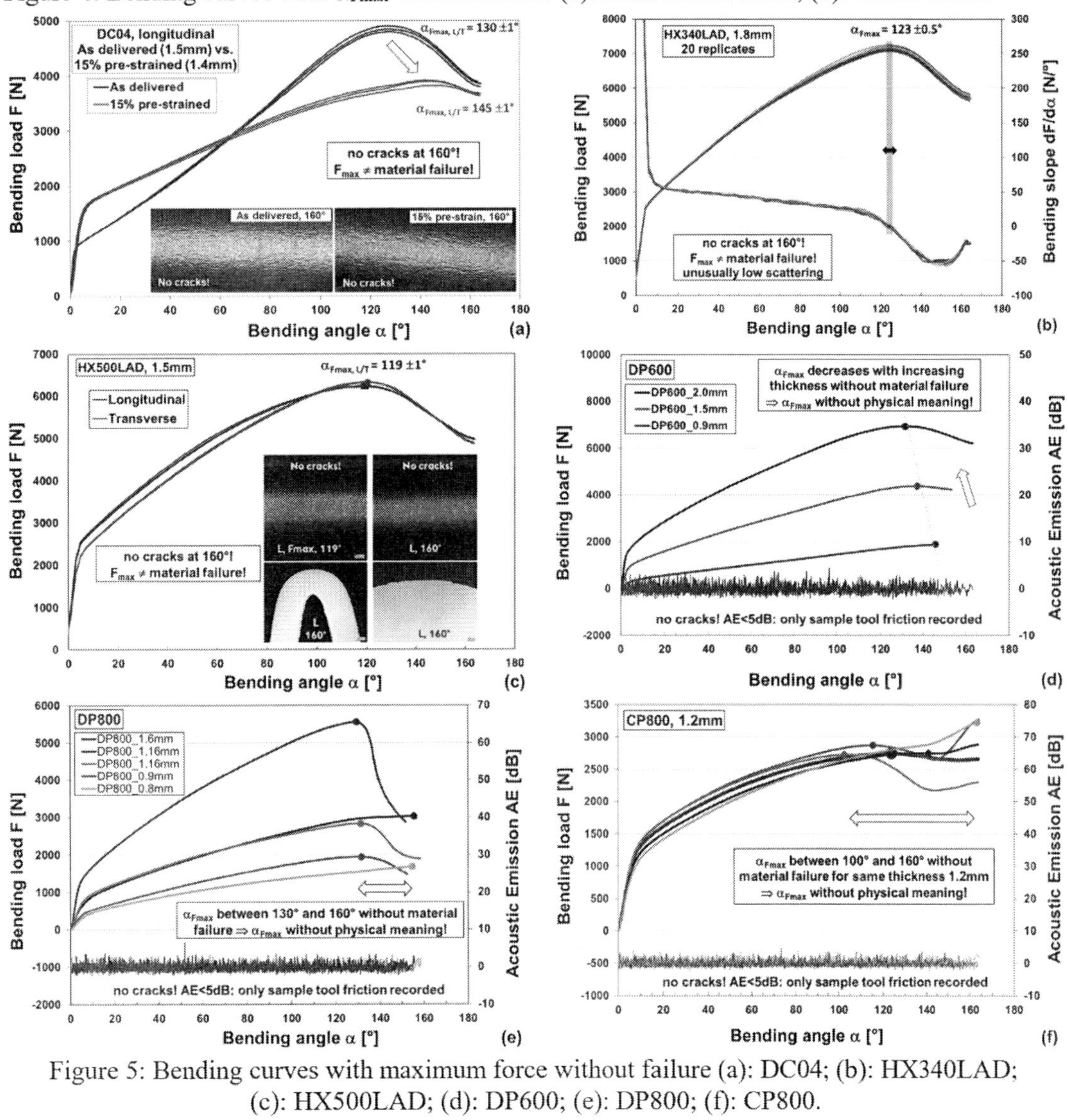

Figure 5: Bending curves with maximum force without failure (a): DC04; (b): HX340LAD; (c): HX500LAD; (d): DP600; (e): DP800; (f): CP800.

IOP Conf. Series: Materials Science and Engineering **967** (2020) 012079 doi:10.1088/1757-899X/967/1/012079

Some attention should be paid in minimizing the outward elastic lateral displacement of support rolls under loading conditions. There is some evidence that such a transverse stiffness issue do trigger maximum force arbitrarily depending on test setup design [9]-[11]. The maximum force if not due to material failure is rather a result of work hardening, strength, thickness, overall bending angle level and bending roll distance (2xthickness+0.5mm) interacting with lateral support roll stiffness.

As shown in Figure 6a with an older test setup [9], a 0.2mm support rolls elastic deformation triggers a maximum bending load regardless of the material investigated (no cracks at 160°). FE simulations (Abaqus explicit, CPE8R plane strain 2D shell elements, Hill48, no damage, no friction) show a high sensitivity of F_{max} location to bending roll transverse stiffness (fixing the rolls on a horizontal linear spring with stiffness K, Figure 6b). The horizontal compliance of the rolls produces a force maximum due to the altered geometric conditions during the bending test without any material failure occurrence.

The horizontal force acting on bending rolls assuming rigid tools in FEM Simulations reads [9]:

$$F_{\text{Roll,x,Rigid}} = 0.5 * F_{\text{Stamp,Rigid}} * \tan(\alpha_{\text{Rigid}}/2) \quad (7)$$

The horizontal displacements of rolls $\Delta X_{\text{Roll,X, Exp}}$ are taken from experiments in Figure 6a [9] in this case for DP600_2mm and HX340LAD_1.75mm. An average linear spring stiffness K of the bending roll support is determined from experimental data over all steel grades data (half of setup simulated):

$$K_{\text{Spring}} = F_{\text{Roll,x,Rigid}} / \Delta X_{\text{Roll,X,Exp}} / 2 \quad (8)$$

The simulation of compliant setups is then carried out fixing the rolls on a horizontal spring with stiffness K. Altering the horizontal compliance enables to arbitrarily shift the bending angle at maximum force. The higher the tool rigidity the higher the bending angle at maximum force until it disappears with the standard rigid tool configuration (Figure 6b). The test setup bending roll transverse stiffness must therefore be taken into account for reliable bending test FE simulation. A nonlinear spring might be able to better describe the horizontal deformation behaviour of the rolls [9]. The bending stiffness of the rolls should be ideally increased in such a way to minimize this issue. In any way the α_{Fmax} values are rather meaningless and even highly test setup dependent if no material failure occur.

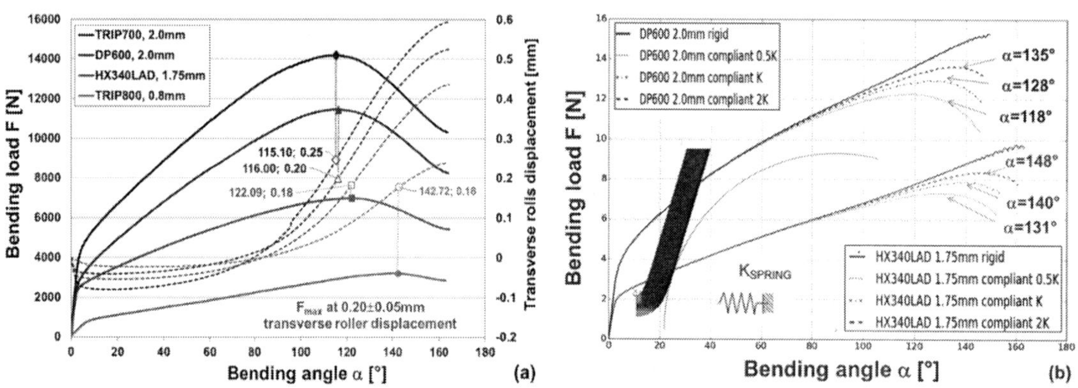

Figure 6: Bending force vs. bending angle vs. roll stiffness (a): Experiments [9]; (b) FE simulation.

4. Experimental results on mixed bending curves with/without cracks

A multitude of bending tests show some cracks with the bending punch in longitudinal direction but no cracks in transverse direction as for DP600 in Figure 7a (test interrupted at maximum load) or for DP800 in Figure 7b (test up to 160° but with roll distance 2xthickness instead of 2xthickness+0.5mm). Longitudinal/transverse directions in Figure 7 and Figure 8 mean that the bending punch line is longitudinal/transverse to rolling direction according to VDA238 nomenclature convention. Note in Figure 7b that when reducing the roll distance to 2xthickness (as prescribed in the older VDA238 specification draft before 2010) the bending curves show no maximum force and a continuous steady load increase in spite of crack initiation at some point during bending test.

This indicates that the maximum force VDA238 criteria is mainly dependent on the prescribed bending roll distance. Figure 8a shows a deviation in differential bending slope for DP800 20° after maximum load in longitudinal direction. Figure 8b confirms the usefulness of comparing longitudinal and transverse bending curves. Some deviation from ideal Finite Element simulated plasticity behaviour (Abaqus explicit, CPE8R plane strain 2D shell elements, Hill48, no damage model, no friction) is a strong indicator for damage initiation. Transverse curves follow longer the ideal plasticity path than the longitudinal curves (Figure 8b). Figure 9a and Figure 9b show examples of crack detection with Acoustic Emission respectively for DP1000 and CP1200 with cracks beginning 15 to 40° beyond maximum load. The thinner the material the larger the discrepancy between α_{Fmax} and $\alpha_{Acoustic\ crack\ initiation}$.

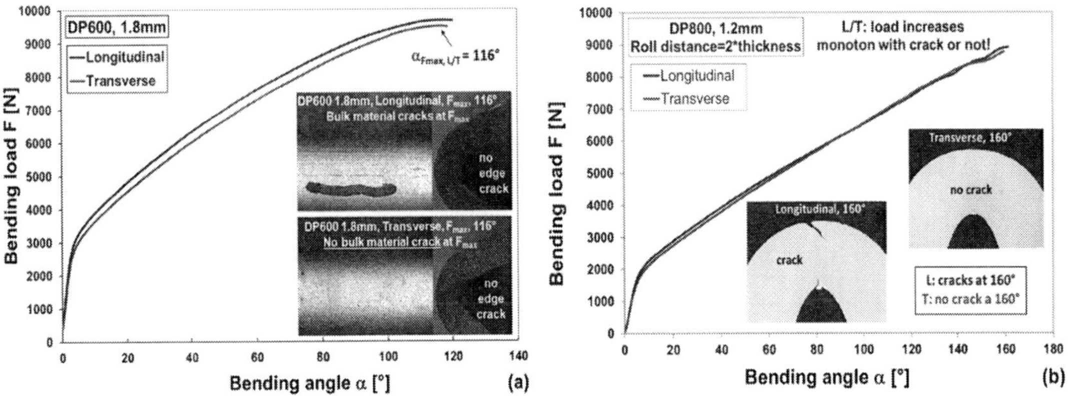

Figure 7: Bending curves longitudinal vs. transverse (a): DP600; (b): DP800 (d_{roll}=2xthickness).

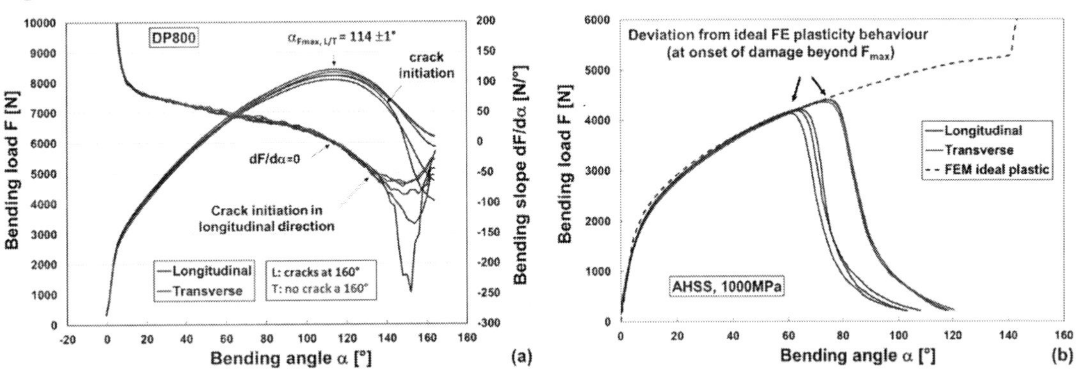

Figure 8: Bending curves longitudinal vs. transverse (a): DP800 crack initiation at bending slope deviation; (b): AHSS1000MPa crack initiation at deviation from ideal plastic FE simulation.

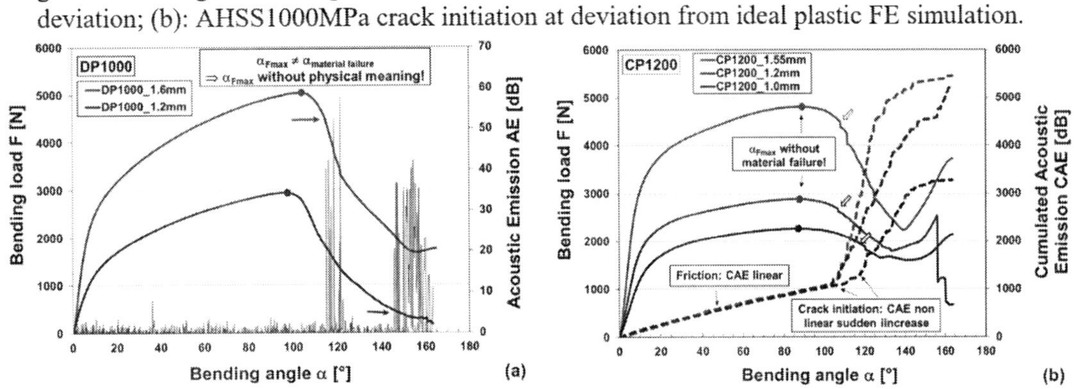

Figure 9: Bending curves with $\alpha_{crack,\ Acoustic}$ >> α_{Fmax} (a): DP1000; (b): CP1200.

The particular case of bending curves with sample-punch lift-off occurring with or without material failure should be considered. Punch lift off leads to a shift from 3 point to 4 point bending loading and a sudden secondary increase in bending load ("two humps" bending curves, Figure 5f & Figure 10a). As shown in Figure 10b the effect of roll distance on bending curve shape can also be simulated in FE simulations (Abaqus explicit, CPE8R plane strain 2D shell elements, Hill48, no damage model, no friction). An increase in thickness results in an increase in bending roll distance according to VDA238 specification (d_{roll}=2xthickness+0.5mm). The lower the bending roll distance, the more a continuous increase in bending load is seen without properly defined load maximum (α_{Fmax} =160°, Figure 11a) and a material failure which is hidden within the increasing bending curve. The bending load drop from material failure superpose on top of it, which leads to a rather unpredictable load curve depending on crack propagation rate thus preventing a proper α_{Fmax} analysis and even definition (Figure 11b). The 3 point bending loading state is also no more fulfilled when punch lift off occurs and cracks may be triggered prematurely, which is not representative of the true material bendability [13].

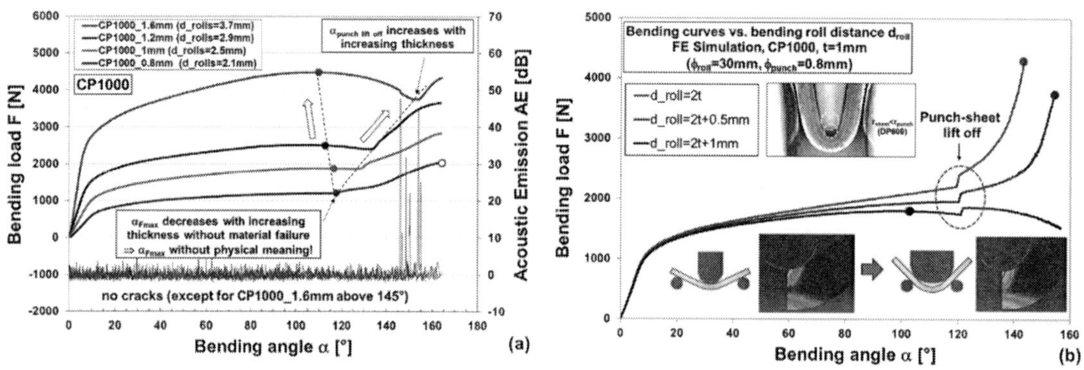

Figure 10: CP1000 bending curves (a): Experiment vs. thickness; (b): FE simulation vs. roll distance.

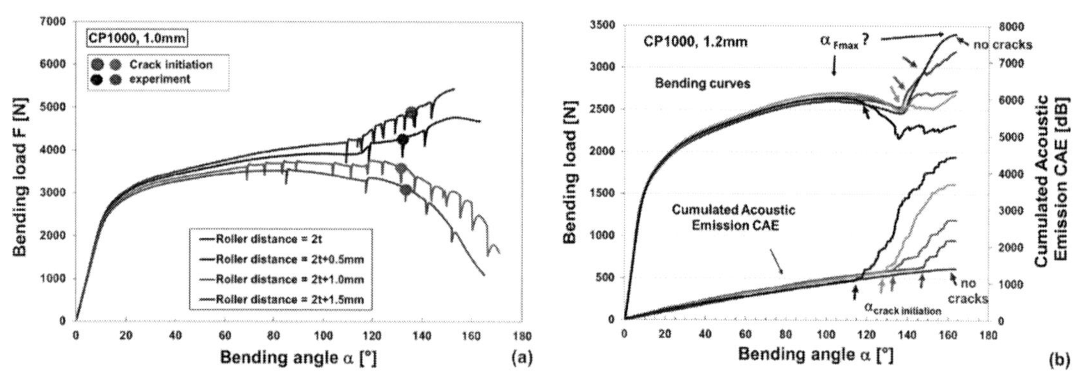

Figure 11: CP1000: (a) Crack initiation vs. roll distance; (b): Bending curves with/without cracks.

5. Acoustic fracture analysis on AHSS bending curves

A wide selection of Cumulated Acoustic Emission CAE testing is shown in the 1000MPa tensile strength range (Figure 12a) and more generally for typical cold rolled steels in the tensile strength range 600-1200MPa and thickness range 1-2mm (Figure 12b). The test setup characteristic crack-free base line of each CAE curve seems surprisingly to be quasi-independent on material strength, thickness or steel sheet coating with free bending rolls and can be fitted with a 2nd order polynomial vs. bending angle (equation 2, Figure 12b) and subtracted later on from total CAE signal (CAE*), which gives an opportunity for a crack detection method. A significant deviation from the known crack-free CAE behaviour will then give a strong indication on macroscopic failure start. How soon and how fast the friction-free CAE* amount grows is a valuable information with regard to the fracture toughness of the investigated microstructure in bending loading condition.

Some selected examples for accumulated acoustic emission CAE are shown with total experimental data in Figure 13a as well as corrected from the friction contribution in Figure 13b. The postuniform bending slopes with or without friction PUBS-CAE / PUBS-CAE* (equation 4&5) are also valuable physical crack propagation related properties. Such acoustic characteristics may be more reliable than load based PUBS-F postuniform bending slope (equation 6) for crack propagation speed analysis. While PUBS-CAE is based on accumulated acoustic energy, the PUBS-F postuniform bending slope may be only largely dependent on geometric effects and test setup compliance concealing the material-related information. Alternative plots based on the distribution of Acoustic Emission peaks in form of histograms may also be of interest to distinguish the damage behaviour of different microstructure in plane strain bending conditions.

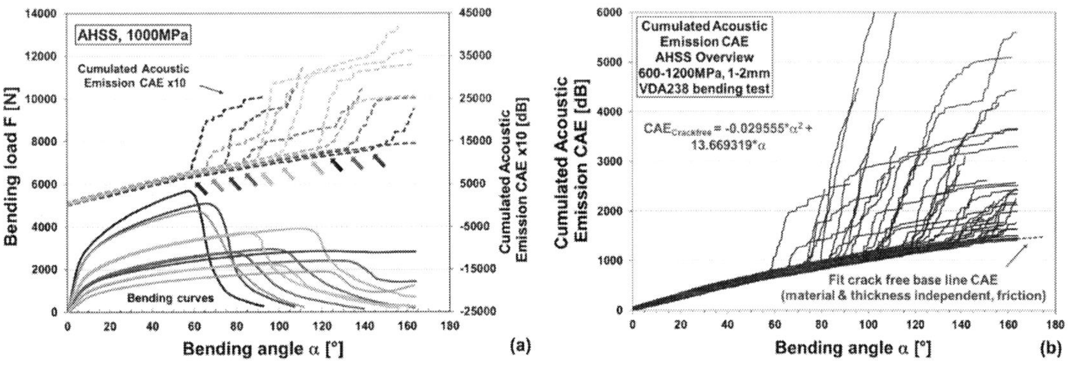

Figure 12: Acoustic emission: (a) 1000MPa strength; (b): Representative selection of AHSS/UHSS.

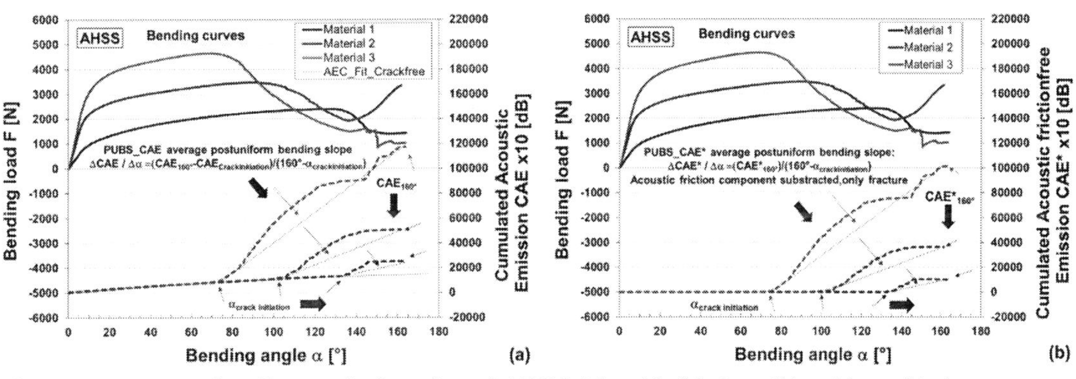

Figure 13: CAE vs. bending angle for selected AHSS (a): with friction; (b): without friction.

Table 1 shows the postuniform bending properties based on Figure 13. A much better differentiation can be reached with acoustic properties (CAE160, CAE160*, PUBS-CAE, PUBS-CAE*) in comparison to PUBS-F from bending curve postuniform slope between F_{max} and inflection point. Note the distinction between α_{Fmax} (at absolute load maximum) and $\alpha_{Fmax\,-30N}$ (from VDA238 specification 30N below F_{max} for cold rolled steels) which are often mixed up. In this case there is up to 6° difference between α_{Fmax} and $\alpha_{Fmax\,-30N}$. $\alpha_{Crack\ initiation\ AE}$ matches better with $\alpha_{Fmax\,-30N}$ than with α_{Fmax} for the selected materials.

Table 1: Postuniform bending properties from bending tests in Figure 13.

Material	α_{Fmax}	$\alpha_{Fmax-30N}$	$\alpha_{inflection}$	α_{Fmax} - $\alpha_{inflection}$	$\alpha_{Crack\ initiation\ AE}$	F_{max}	$F_{inflection}$	|PUBS-F|	$CAE_{160°}$	$CAE^*_{160°}$	PUBS-CAE	PUBS-CAE*
	[°]	[°]	[°]	[°]	[°]	[N]	[N]	[N/°]	[dB]	[dB]	[dB/°]	[dB/°]
Material 1	127	133	142	15	136	2397	1841	37	11718	10140	433	422
Material 2	90	96	122	32	100	3470	2490	31	5096	3656	67	61
Material 3	69	73	99	30	73	4648	3003	55	2519	1009	19	12

6. Conclusions

Whatever strength level considered up to 1200MPa there is always a counterexample to be found for which the VDA238 assumption maximum bending load F_{max} = crack initiation is contradicted. A non-negligible amount of tests show a crack initiation way after F_{max} or even no cracks at all (run out) in spite of a well-defined load maximum in the bending curve, thus α_{Fmax} largely underestimating the true bendability of the investigated steel grades. Transverse test are more affected than longitudinal ones (VDA nomenclature). The thinner the material, the lower the strength level and the lower the strain hardening, the more this problem of invalid test may also occur. This is a major obstacle for a widespread use of this bending testing method for AHSS-UHSS characterisation based on load measurement alone. Unless each sample is checked visually for crack occurrence at F_{max}-30N, the results cannot be trusted.

A significant improvement involves the use of online passive acoustic emission sensor for post-mortem crack detection analysis. This is however time and know how intensive and rather subjective since many test set up specific noise sources superpose to the actual crack events. Ideally some video instrumentation would complete the load and acoustic emission instrumentation to deliver a more robust assessment on crack initiation and propagation rate. Video instrumented crack detection is however also subjective depending on lightning contrast as well as spatial and temporal image resolution when focusing on a constantly moving punch bending line. Bending test instrumentation is therefore quite a challenging task with no ready to go commercial application.

More over the bending curve maximum force in the absence of material failure is directly dependent on the test set up support roll stiffness and would therefore vary arbitrarily from test lab to test lab, delivering actually misleading information on the bending behaviour. The sample-punch lift off topic for steels with low n-value challenges also the very definition of α_{Fmax} which is no more defined in case of a double hump bending curve.

Based solely on load instrumentation the VDA238 bending test therefore needs to be interpreted with special care taking into account the tested material, in order to avoid misleading results. Unless significant instrumentation improvement is made, the VDA238 bending test should remain in the field of martensitic press-hardened steels for which $F_{max} \approx$ crack is arguably fulfilled.

References

[1] VDA 238-100 (2010) Plate bending test for metallic materials. VDA, Berlin, Germany.
[2] Larour P, Pauli H, Kurz T and Hebesberger T (2010) Influence of post uniform tensile and bending properties on the crash behaviour of AHSS and press-hardening steel grades. Proc. IDDRG 2010, Graz, Austria.
[3] Larour P, Naito J, Pichler A, Kurz T and Murakami T (2015) Side impact crash behavior of presshardened steels - correlation with mechanical properties. Proc. CHS2 Int. Conf., 281-289, Toronto, Canada.
[4] Kurz T, Larour P, Lackner J, Steck T and Jesner G (2016) Press-hardening of zinc coated steel - characterization of a new material for a new process. *IOP Conf. Ser.: Mater. Sci. Eng.* **159** 012025.
[5] Wagner L, Larour P, Lackner J, Schauer H and Berger E (2019) Characterizing axial crash foldability of AHSS & UHSS sheets by means of L-profile compression tests. *IOP Conf. Ser.: Mater. Sci. Eng.* **651** 012014.
[6] Frómeta D, Lara A, Casas B and Casellas D (2019) Fracture toughness measurements to understand local ductility of advanced high strength steels. *IOP Conf. Ser.: Mater. Sci. Eng.* **651** 012071.
[7] Heibel S, Dettinger T, Nester W, Clausmeyer T and Tekkaya A E (2018) Damage Mechanisms and Mechanical Properties of High-Strength Multiphase Steels. *Materials* **2018**, 11, 761. DOI: 10.3390/ma11050761.
[8] Larour P, Hackl B, Leomann F and Benedyk K (2012) Bending angle calculation in the instrumented three-point bending test. Proc. IDDRG 2012, 203-215, Mumbai, India.

[9] Larour P, Hackl B and Leomann F (2013) Sensitivity analysis on the calculated bending angle in the instrumented bending test. Proc. IDDRG 2013, 317-322, Zurich, Switzerland.

[10] Aegerter J and Keller S (2015) Vorrichtung für die einfache und reproduzierbare Durchführung des Plättchen-Biegeversuches nach VDA 238-100. Proc. Werkstoffprüfung 2015, Bad Neuenahr, 291-296, 6-7 December, Germany.

[11] Aegerter J and Keller S (2015) Device for carrying out a bending test. WO2016102084A1 Patent.

[12] Durrenberger L and Dietsch P (2018) Bending angle correction regarding sheet thickness. *IOP Conf. Ser.: Mater. Sci. Eng.* **418** 012076.

[13] Wagner L, Schauer H, Pauli H and Hinterdorfer J (2019) Improved bendability characterization of UHSS sheets. *IOP Conf. Ser.: Mater. Sci. Eng.* **651** 012019.

[14] Arola A M, Kaijalainen A, Kesti V, Pokka A P and Larkiola J (2019) Digital image correlation and optical strain measuring in bendability assessment of ultra-high strength strctural steels. *Procedia Manuf.* **29** 398-405.

[15] Cheong K, Omer K, Butcher C, George R and Dykeman J (2017) Evaluation of the VDA 238-100 Tight Radius Bending Test using Digital Image Correlation Strain Measurement. *J. Phys.: Conf. Ser.* **896** 012075.

[16] Cheong K, Butcher C and Dykeman J (2018) The Influence of the Through-Thickness Strain Gradients on the Fracture Characterization of Advanced High Strength Steels. *SAE Int J Mater Manuf*, **11(4)**, 541-552. DOI: 10.4271/2018-01-0627.

[17] Noder J, Abedini A and Butcher C (2020) Evaluation of the VDA 238–100 Tight Radius Bend Test for Plane Strain Fracture Characterization of Automotive Sheet Metals. *Exp Mech* **60**, 787-800. DOI: 10.1007/s11340-020-00597-2.

[18] Sadhinoch M, Kampczyk M and Mulder R (2019) Statistics of Fracture in 3-Point Bending. *IOP Conf. Ser.: Mater. Sci. Eng.* **651** 012075.

[19] Dietsch P, Tihay K, Bui-Van A and Cornette D (2017) Methodology to assess fracture during crash simulation: fracture strain criteria and their calibration. *Metall. Res. Technol.* **114**, 607. DOI: 10.1051/metal/2016065.

[20] Boul I, Lang M, Mees G, Sponem F, Sturel T and Garat X (2007) Bending test and bendability of Ultra High Strength Steels. Proc. IDDRG 2007, Győr, Hungary.

[21] Labudde T and Bleck W (2011) Characterisation of ductility of ultra-high strength steels. *Steel Research Int. 2011, Special Edition*, 1090-1095.

[22] Seemuang N, Panich S and Slatter T (2017) Bendability evaluation of sheet metals in three-point bending test by using acoustic emission features. *J Appl Sci* **16 (2)**, 15-22. DOI: 10.14416/j.appsci.2017.12.001.

IOP Conf. Series: Materials Science and Engineering **967** (2020) 012080 doi:10.1088/1757-899X/967/1/012080

Edge crack sensitivity versus tensile local ductility of AHSS sheet steels

P Larour*, J Freudenthaler, M Kerschbaum, D Dolzer

voestalpine Stahl GmbH, voestalpine-Straße 3, A-4020 Linz, Austria.

E-mail: patrick.larour@voestalpine.com

Abstract. There is a growing interest in correlating usual tensile testing results with edge crack sensitivity testing from punched ISO16630 hole expansion ratio HER (10mm shear cut hole, 12% clearance, conical expansion tool). A new kind of tensile local ductility parameter has been developed lately based on broken sample surface of tensile specimens after testing. Reduction in area or thickness at fracture are more sensitive than conventional fracture elongation with a 50 to 80m gage length to characterize the local ductility potential of sheet steels. A representative amount (300 different sets of samples) of cold rolled sheet steels have been tested in the tensile strength range 600-1200MPa and thickness range 1-2mm with 3 replicates in the transverse and longitudinal direction with ISO 6892-1, type 2, A_{80} tensile samples. Correlation levels of ISO16630 HER values with conventional tensile mechanical properties such as uniform & fracture elongation, yield & tensile strength, n- & r-values or derivatives are disappointing low for the investigated AHSS grades. There is however a massive improvement in the empirical statistical correlation when using local ductility properties based on fracture area or thickness reduction measurements on broken tensile samples. Logarithmic local ductility strains correlate generally linearly with logarithmic hole expansion ratio. Logarithmic true local ductility values are proving more suitable than engineering strains for correlations. Transverse direction improves slightly the correlation quality vs. longitudinal direction. The correlation is also higher for thickness reduction in comparison to reduction of area based properties.

1. Introduction

There is some long lasting need to correlate tensile test mechanical properties with conical hole expansion ratio HER from ISO16630 ISO conical hole expansion tests HET [1], which are usually used for edge crack sensitivity characterization. A broad literature is available on the topic with mixed predictive correlation accuracy [2]-[9], the usual correlation method consisting in plotting individual basic properties such as sheet thickness, yield strength $R_{p0.2}$, tensile strength R_m, uniform elongation A_g, fracture elongation A_{xx}, strain hardening rate (n-value), plastic strain ratio (r-value) for a specific direction or averaged in 0/45/90° direction (n_m, r_m) [2]-[4],[7] or even strain rate sensitivity m [8],[9].

Combinations of such properties such as yield ratio $R_{p0.2}/R_m$, n*r [5], A_{xx}*r_m [2],[3], r/n [4] or postuniform strain A_{xx}-A_g [7]-[9] as well as logarithmic postuniform width strain [7] have also been proposed. The r-value should have a positive effect on hole expansion level by postponing local thinning. This may be especially true for milled condition for bulk material testing and thicker HSLA hot rolled steels [2],[4]. In punched condition however the random shear cut pre-damage along the hole edge may reduce the influence of the r-value alone.

Some integrated multifit hybrid models have been developed including thickness, tensile properties as well as material chemistry information such as sulphur content [3]. Different models have also been proposed for punched vs. machined milled hole edge condition [2]. Nonlinear dependencies instead of linear assumptions have also been proposed in [3],[6]. Such models need however to be updated constantly for AHSS-UHSS both for cold and hot rolled multiphase steel grades.

A recent decisive improvement may be brought by local area & thickness reduction measurement on flat sheet samples according to Hance test methodology and derivatives for local ductility assessment on broken surface of (milled) flat tensile test samples [10]-[16]. A strong correlation has been found between hole expansion ratio HER and reduction of area at fracture [11]. Alternatively the average percentage or logarithmic thickness reduction at fracture is also correlating well with the ISO16630 hole expansion ratio [12]. The logarithmic fracture thickness reduction at minimum thickness location has also proven valuable for local formability characterization and edge crack sensitivity prediction [14],[15]. The concept of postuniform log. thickness strain is also introduced in [14], for which the true uniform thickness strain is deduced from the total thinning at fracture. This may enable more accurate postuniform local ductility assessment and tensile test-HET correlations.

The hole expansion ratio may correlate even better with fracture toughness and essential work of fracture [8],[9],[16],[17]. There is an open debate whether tensile local ductility from bulk undamaged material can be correlated at all with the crack resistivity behavior of a shear cut pre-damaged hole expansion test [16],[17]. Punched tensile test may be an alternative option to clarify this issue [18],[19].

The present investigation delivers a correlation between conventional milled tensile test with punched ISO16630 hole expansion ratio on a much wider statistical basis than in previous investigations (300 correlation points in total for a wide variety of AHSS-UHSS cold rolled steel grades in the tensile strength range 600-1200MPa and thickness range 1mm to 2mm). The final aim is to reduce the amount of conical hole expansion testing with a deeper big data based correlation knowledge with tensile tests.

2. Experimental procedure

The reduction of area, thickness and width at fracture shown later on in this investigation have been determined based on the digitalized contour plot surrounding the fracture zone, the tensile sample broken surface being tilted flat under the microscope as previously described in [12],[13]. The area within the contour line has been integrated pixel by pixel, the thickness (average, mid & min location) as well as the width at fracture have then been derived as shown in Figure 1. The projected fracture area and width normal to tensile direction have been determined via the average fracture angle α (Figure 1).

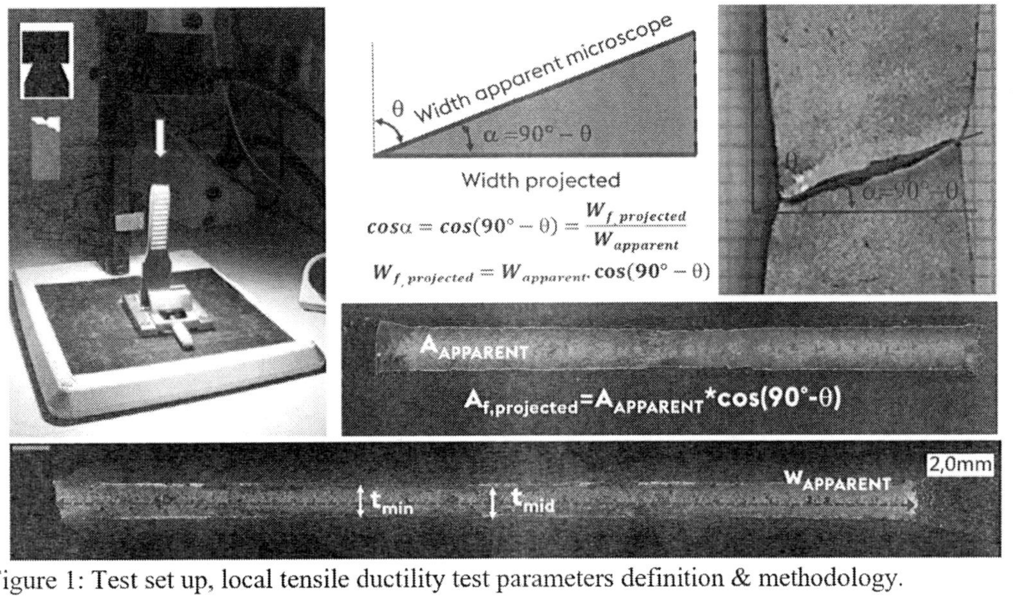

Figure 1: Test set up, local tensile ductility test parameters definition & methodology.

A set of equation formulas is given as follows for engineering and logarithmic local ductility parameters definition plotted later on in the diagrams with t_0, w_0 and A_0 (initial) and t_f, w_f and A_f (fracture) thickness, width and cross section area. W_f and A_f are projected in the plane normal to tensile direction according to the procedure described in Figure 1, [12],[13]. t_f is measured either as average thickness over the whole fracture area or at mid sample center or at min thickness location. This delivers the corresponding average/mid/min thickness reduction.

Following abbreviations are used in the following investigation: $R_{p0.2}$: yield strength, R_m: tensile strength; A_g: uniform elongation, A_{80} fracture elongation from tensile test (gage length L_0=80mm). r-value and n-value are determined at uniform elongation A_g. L/T stand for longitudinal and transverse tensile testing directions. The hole expansion ratio (HER) vs. local tensile ductility at fracture conversion from log-log to %-log or %-% is highly nonlinear as shown in equations (21)-(30). The conversion equations are identical for reduction of area in equations (21)-(25) as for thickness reduction parameters in equations (26)-(30). Equations (26)-(30) are valid for any kind of thickness reduction at fracture considered, calculated either from sample average, middle or min thickness location.

Width and thickness strain or logarithmic are defined in this investigation as positive (absolute) values. Postuniform logarithmic values are derived by subtracting the log. strain at uniform diffuse necking from the total local strain (only logarithmic strains are additive not the engineering strains).

n-value at uniform elongation n_{Ag}:	$n_{Ag} = \ln(1 + A_g)$	(1)
r-value at uniform elongation r_{Ag}:	$r_{Ag} = \varepsilon_{2,Ag}/\varepsilon_{3,Ag}$	(2)

% Reduction of area at fracture $Z_{\%}$:	$Z_{\%} = 100.(A_0 - A_f)/A_0$	(3)
Log. True Fracture Strain TFS_{log}:	$TFS_{log} = \varepsilon_{1,f,log} = ln(A_0/A_f)$	(4)
	$TFS_{log} = -ln(1 - Z\%/100)$	(5)
Log. postuniform True Fracture Strain $TFS_{PU,log}$	$TFS_{PU,log} = TFS_{log} - \ln(1 + A_{g\%}/100)$	(6)

% Fracture thickness reduction $e_{3f,\%}$ (>0):	$e_{3f,\%} = 100.(t_0 - t_f)/t_0$	(7)
Log. thickness reduction at fracture $\varepsilon_{3f,log}$ (>0):	$\varepsilon_{3f,log} = \ln(t_0/t_f)$	(8)
	$\varepsilon_{3f,log} = -ln(1 - e_{3f,\%}/100)$	(9)
Log. postunif. thickness reduc. at fracture $\varepsilon_{3f,PU,log}$ (>0)	$\varepsilon_{3f,PU,log} = \varepsilon_{3f,log} - \ln(1 + A_{g\%}/100)/(r_{Ag}+1)$	(10)

% Width reduction at fracture $e_{2f,\%}$ (>0):	$e_{2f,\%} = 100.(w_0 - w_f)/w_0$	(11)
Log. width reduction at fracture $\varepsilon_{2f,log}$ (>0):	$\varepsilon_{2f,log} = \ln(w_0/w_f)$	(12)
	$\varepsilon_{2f,log} = -ln(1 - e_{2,f,\%}/100)$	(13)
Log. postunif. width reduction at fracture $\varepsilon_{2f,PU,log}$ (>0):	$\varepsilon_{2f,PU,log} = \varepsilon_{2f,log} - r_{Ag}/(r_{Ag} + 1).\ln(1 + A_{g\%}/100)$	(14)

Log. width strain reduction at A_{80}, $\varepsilon_{2,A80,log}$ (>0):	$\varepsilon_{2,A80,log} = r_{Ag}/(r_{Ag} + 1).\ln(1 + A_{80\%}/100)$	(15)
Log. postuniform width strain at A_{80}, $\varepsilon_{2,PU,A80,log}$ (>0):	$\varepsilon_{2,PU,A80,log} = \varepsilon_{2,A80,log} - r_{Ag}/(r_{Ag} + 1).\ln(1 + A_{g\%}/100)$	(16)

$HER_{\%}$ to HER_{log}:	$HER_{log} = \ln(1 + HER_{\%}/100)$	(17)
HER_{log} to $HER_{\%}$:	$HER_{\%} = 100.\{exp(HER_{log}) - 1\}$	(18)
TFS_{log} to $Z_{\%}$:	$Z_{\%} = 100.\{1 - \exp(-TFS_{log})\}$	(19)
$\varepsilon_{3f,log}$ to $e_{3f,\%}$:	$\varepsilon_{3f,\%} = 100.\{1 - \exp(-\varepsilon_{3f,log})\}$	(20)

Fit HER_{log} vs. TFS_{log}:	$HER_{log} = K.TFS_{log} \approx 0.5.TFS_{log}$ with K≈0.5	(21)
Fit $HER_{\%}$ vs. TFS_{log}:	$HER_{\%} = 100.\{exp(K.TFS_{log}) - 1\} = 100.\{[exp(TFS_{log})]^K - 1\}$	(22)
	$HER_{\%} \approx 100.\left\{\sqrt{exp(TFS_{log})} - 1\right\}$ with K ≈ 0.5	(23)
Fit $HER_{\%}$ vs. $Z_{\%}$:	$HER_{\%} = 100.\{exp(-K.ln(1 - Z_{\%}/100)) - 1\} = 100.\{[1 - Z_{\%}/100]^{-K} - 1\}$	(24)
	$HER_{\%} \approx 100.\{1/\sqrt{1 - Z_{\%}/100} - 1\}$ with K ≈ 0.5	(25)

Fit HER_{log} vs. $\varepsilon_{3f,log}$:	$HER\ log = K.\varepsilon_{3f,log} \approx 0.5.\varepsilon_{3f,log}$ with K≈0.5	(26)
Fit $HER_{\%}$ vs. $\varepsilon_{3f,log}$:	$HER_{\%} = 100.\{exp(K.\varepsilon_{3f,log}) - 1\} = 100.\{[exp(\varepsilon_{3f,log})]^K - 1\}$	(27)
	$HER_{\%} \approx 100.\left\{\sqrt{exp(\varepsilon_{3f,log})} - 1\right\}$ with K ≈ 0.5	(28)
Fit $HER_{\%}$ vs. $e_{3f,\%}$:	$HER_{\%} = 100.\{exp(-K.ln(1 - e_{3f\%}/100)) - 1\} = 100.\{[1 - e_{3f\%}/100]^{-K} - 1\}$	(29)
	$HER_{\%} \approx 100.\{1/\sqrt{1 - e_{3f\%}/100} - 1\}$ with K ≈ 0.5	(30)

3. Experimental results

The correlation results with common tensile properties in longitudinal (L) and transverse (T) directions are shown in Figure 2 to Figure 7. The HER vs. new local ductility parameters on broken samples are shown later on in Figure 8 to Figure 15.

Only a poor up to non-existing correlation with hole expansion ratio HER can be achieved for yield or tensile strength (Figure 2a,b). A slight negative but not significant enough correlation is seen for uniform or tensile elongation (Figure 3a,b). Similarly a small weak negative tendency for HER vs. n-value can be observed in Figure 4a. The r-value on the contrary has rather a slightly positive effect on HER level (Figure 4b). The product n*r-value at uniform elongation does not bring any decisive input in the correlation (Figure 5a) while the ratio r/n of r-value to n-value at uniform elongation offers a clear positive significant correlation of $r^2 > 0.5$ (Figure 5b), as also seen in [4] for hot rolled AHSS CP-steels.

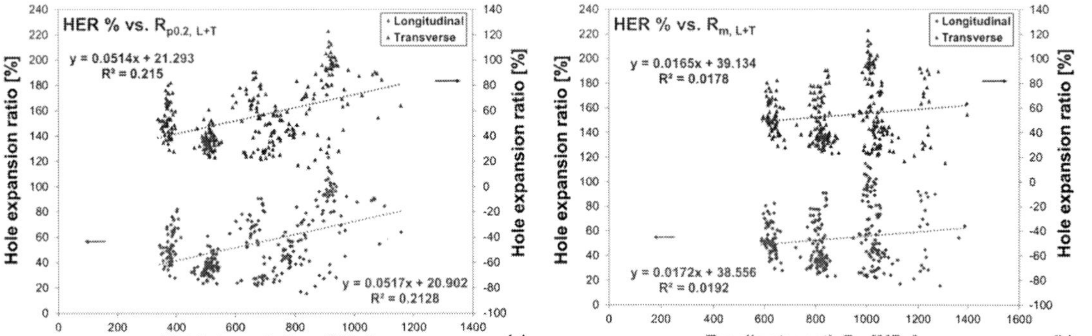

Figure 2: HER vs. (a) yield strength $R_{p0.2}$, L+T; (b): tensile strength R_m, L+T.

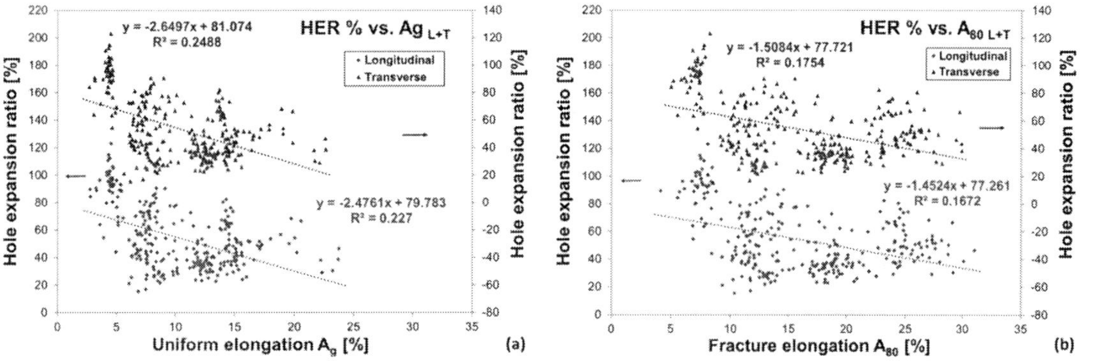

Figure 3: HER vs. (a) uniform elongation Ag, L+T; (b): fracture elongation A_{80}, L+T.

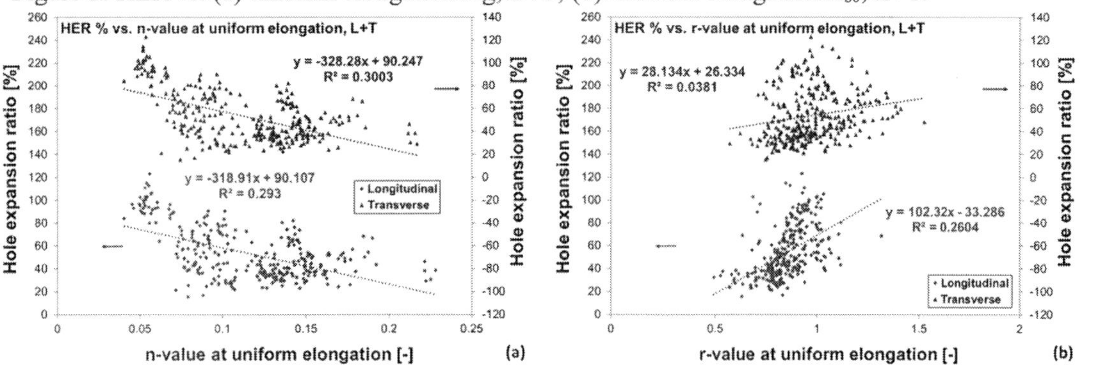

Figure 4: HER vs. (a) n-value at Ag, L+T; (b): r-value at Ag, L+T.

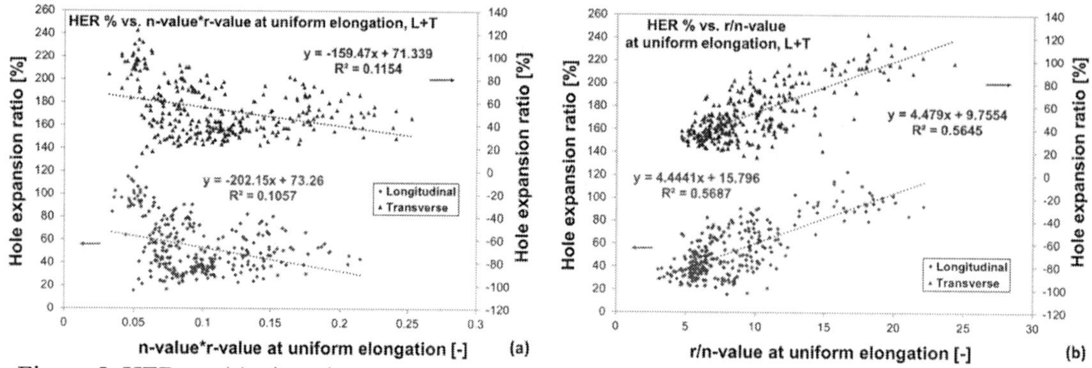

Figure 5: HER vs. (a) n*r-value at Ag, L+T; (b): r/n-value at Ag, L+T.

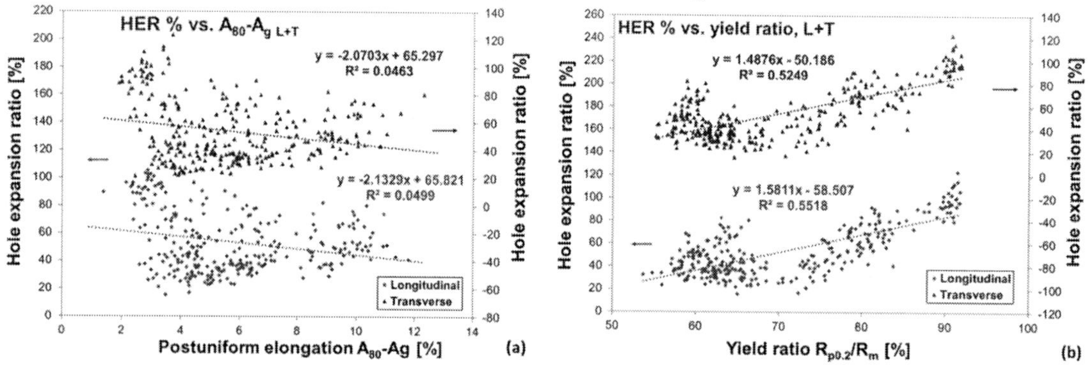

Figure 6: HER vs. (a) postuniform elongation A_{80}-A_g, L+T; (b): yield ratio $R_{p0.2}/R_m$, L+T.

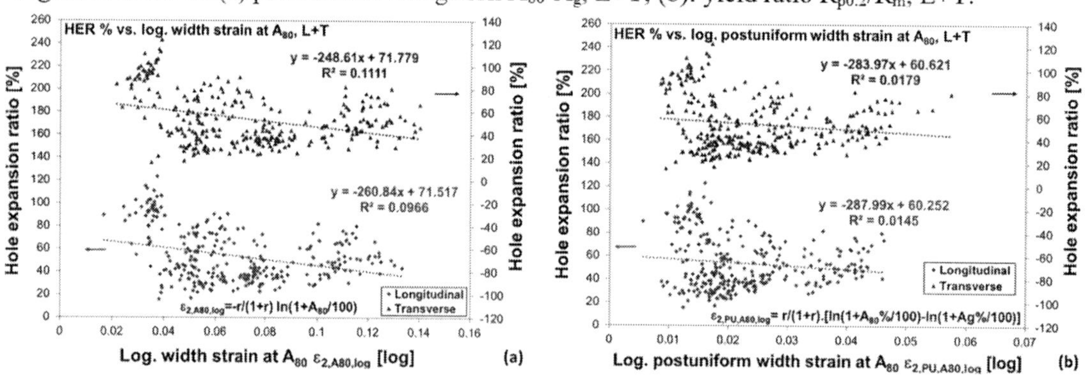

Figure 7: HER vs. (a) Log. width strain $\varepsilon_{2,A80,log}$, L+T; (b): Log. postunif. width strain $\varepsilon_{2,PU,A80,log}$, L+T.

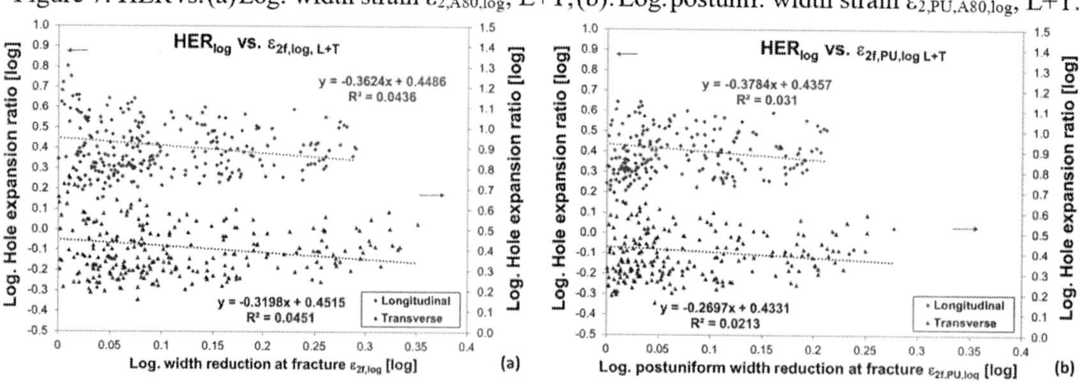

Figure 8: HER vs. (a) log. width strain $\varepsilon_{2f,log}$, L+T; (b): log. postuniform width strain $\varepsilon_{2f,PU,log}$, L+T.

The postuniform elongation A_{80}-Ag which is often mentioned in the literature does not deliver any significant correlation (Figure 6a). On the contrary the yield ratio $R_{p0.2}/R_m$ brings some significant correlation with $r^2>0.5$, the correlation seems particularly good for yield ratio values over 70% (Figure 6b). The correlation with the logarithmic width reduction up to fracture elongation according to equation (15)-(16) as proposed in [7] is almost non-existent (Figure 7a). The postuniform width reduction up to fracture elongation brings no improvement either (Figure 7b).

All in all only yield ratio and r/n-value ratio can be considered so far as helpful for HET-Tensile test correlations among conventional tensile properties. The correlation level r^2 around 50% is however not satisfying enough for reliable HER prediction.

No correlation for width strain measured on broken sample surfaces could be found with HER-values for total or postuniform log. fracture width strain (Figure 8a,b). This shows together with results in Figure 7 that the local necking in tensile width direction does not bring any additional information on the local formability behaviour as also argued in [14],[15].

Figure 9 shows exemplarily some randomly chosen HER-tensile correlation tests (50 sampling data sets in total in the 600-1200MPa tensile strength range). The correlation quality with HER is strikingly good for fracture % thickness reduction at min thickness location / average or Z-value (Figure 9a). The correlation with A_{80} fracture elongation is not clear and even show an opposite trend to HER values (Figure 9a). The correlation of logarithmic HER with logarithmic thickness reduction (at min thickness location or average) and True Fracture Strain TFS values is also obviously equally strong (Figure 9b).

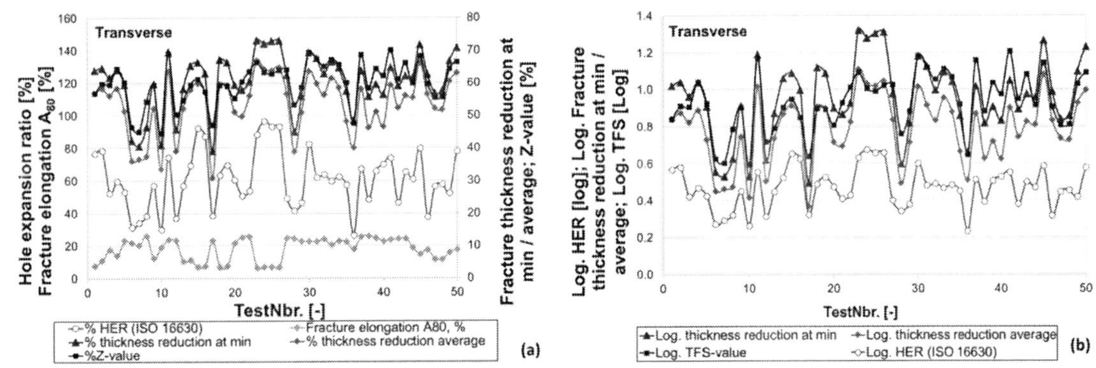

Figure 9: %HER vs. % thickness reduction at min thickness/average, %Z-value & $\%A_{80}$, transverse; (b): Log. HER vs. log. thickness reduction min thickness/average & log. TFS-value, transverse.

Figure 10a delivers the correlation results for all 300 data sets with the Tue Fracture Strain TFS_{log}. Figure 10b show a significant improvement of about 10% in correlation level r^2 when using instead postuniform $TFS_{log,PU}$ as defined in equations (5)-(6). Figure 11, Figure 12 and Figure 13 show the correlation results of log HER vs. log. thickness reduction average, middle or at min thickness location. As for TFS the correlations are quite linear and the postuniform local thickness reduction values deliver slightly better correlation. The correlation level with local parameters is higher ($r^2\approx70\%$) than with conventional mechanical properties ($r^2\approx50\%$). The transverse direction improves slightly the correlation level for any local ductility parameter. It is the most sensitive direction with regards to material defects stretched in rolling direction. The longitudinal direction as prescribed in VDA239 test standard is however also still suitable with acceptable fitting accuracy. TFS_{log} delivers a slightly lower correlation accuracy than with log. thickness reduction parameters. This can be compensated when only considering the postuniform strain $TFS_{PU\ log}$.

Correlation results for engineering strain values between HER% and % reduction of area Z (Figure 14a) or average % thickness reduction (Figure 14b) or % thickness reduction at sample middle (Figure 15a) as well as % thickness reduction at min thickness location (Figure 15b). The non-linearity of HER% vs. eng. % local ductility parameters is well captured through the chosen modelling approach.

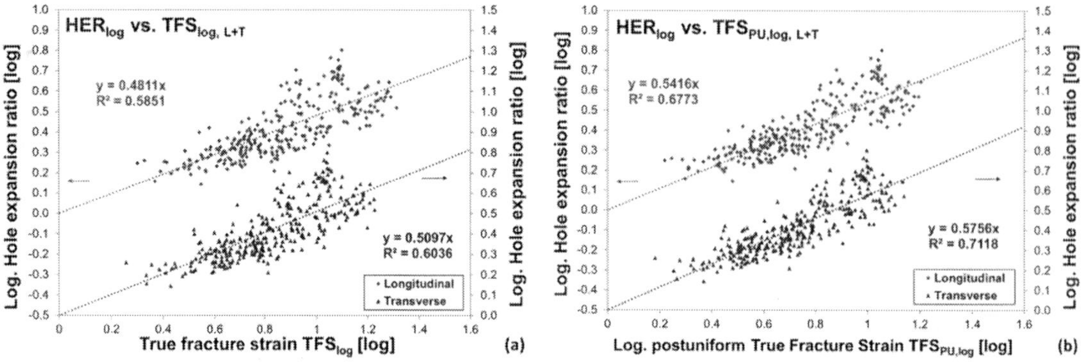

Figure 10: Log. HER vs. log. True Fracture Strain, L+T: (a): TFS_{log}; (b): $TFS_{PU,log}$.

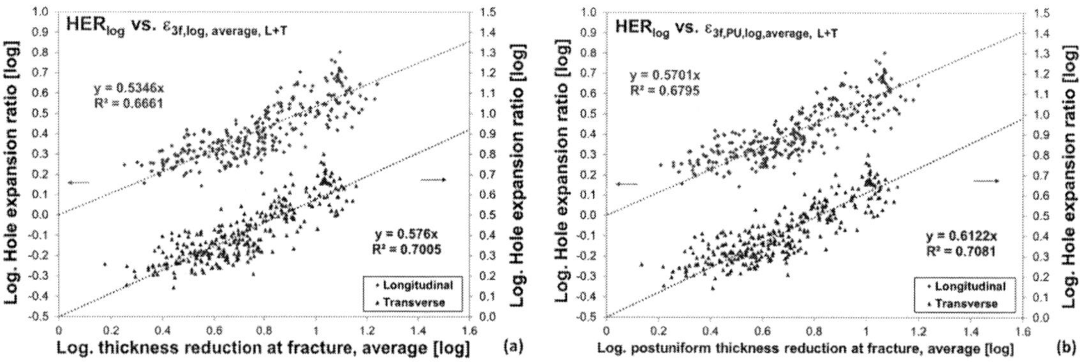

Figure 11: Log. HER vs. log. thickness reduction at fracture average, L+T: (a): $\varepsilon_{3f,log,average}$; (b): $\varepsilon_{3f,log,PU,average}$

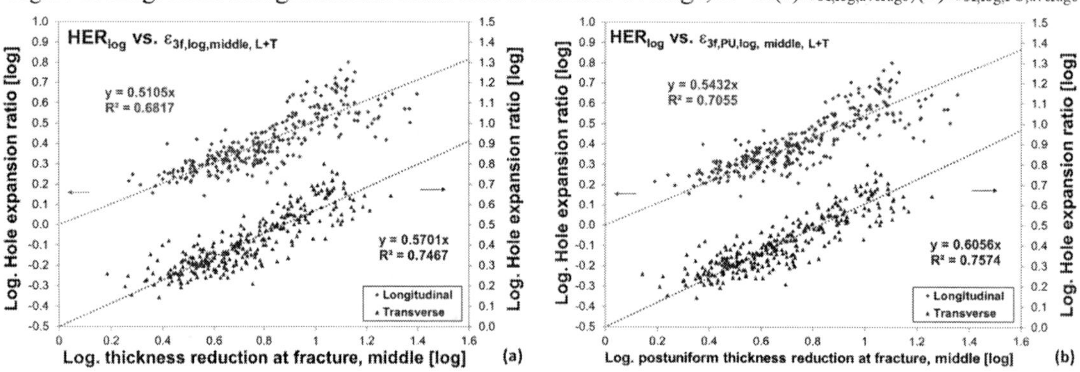

Figure 12: Log. HER vs. log. thickness reduction at fracture middle, L+T: (a): $\varepsilon_{3f,log,mid}$; (b): $\varepsilon_{3f,log,PU,mid}$

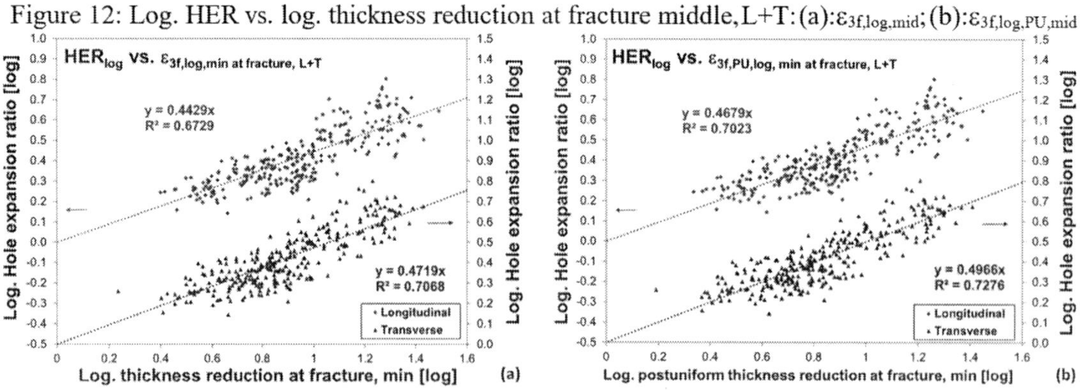

Figure 13: Log. HER vs. log. thickness reduction at fracture min, L+T: (a): $\varepsilon_{3f,log,min}$; (b): $\varepsilon_{3f,log,PU,min}$.

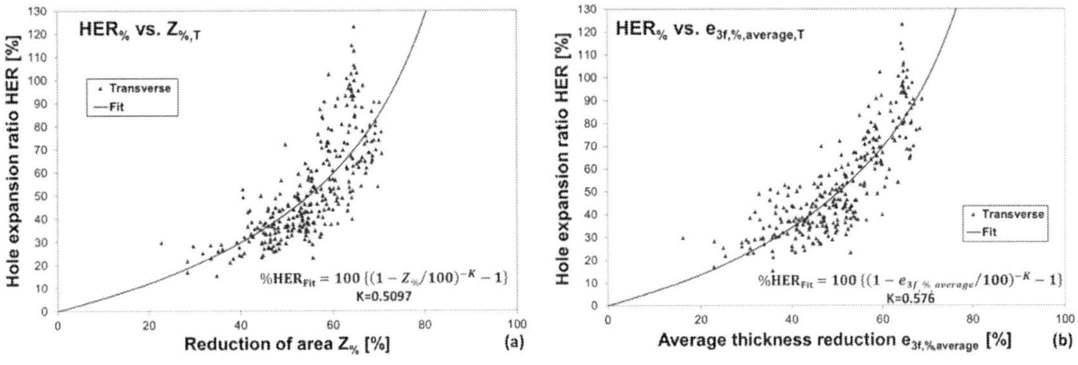

Figure 14: HER% experimental vs. Fit,T:(a): Z%;(b):$e_{3f,\%,average}$.

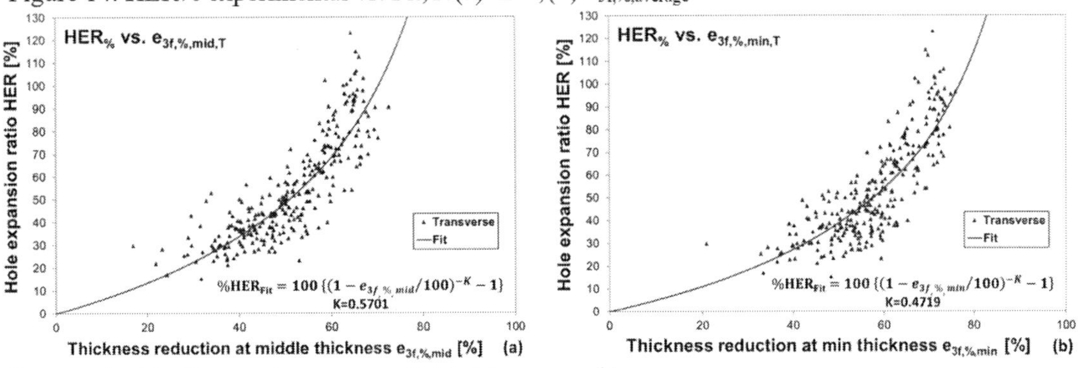

Figure 15: HER% experimental vs. Fit, T: (a):$e_{3f,\%,mid}$;(b):$e_{3f,\%,min}$.

4. Conclusions

Correlations from edge crack sensitivity (ISO16630 hole expansion ratio HER) with tensile testing (ISO 6892-1, type 2 A_{80} tensile samples) have been performed based on a wide thickness, tensile strength and microstructure range. Following conclusions can be drawn:

- No particular correlation can be achieved with yield or tensile strength, uniform or fracture elongation, n- or r-value alone.
- Yield ratio and r-value to n-value ratio deliver the most promising correlation results among conventional tensile test mechanical properties.
- Correlation with tensile test in transverse direction are slightly better than in longitudinal direction.
- Correlation with true log. values both for HER and local ductility properties are strongly linear.
- Engineering values bring an artificial strong non linearity in correlation results from the true logarithmic to engineering % strain conversion, which complicates analytical correlation models. Correlations should be done in true logarithmic strain modus and converted later on in engineering % strain formulation using the set of equations given if needed. Engineering strain formulation is actually no valid for such high strain levels found in local ductility.
- Tensile width strain does not correlate well with HER. The local necking in width direction does not bring additional information on local ductility but rather on global formability.
- Fracture area based local formability correlate slightly less with HER than thickness based parameters.
- Thickness strain measured on an average integral basis, or in the middle or at the min thickness location of broken samples surfaces actually comparatively all perform well in correlation with HER.
- Subtracting the postuniform component up to uniform diffuse necking improves the statistical correlation level with HER for all local ductility values, but especially for TFS.
- It actually does not matter which local formability tensile property is used, the correlation is still much better than with any conventional tensile properties or derivatives.

- 100% correlation will never be achieved. The correlation level (r^2 up to 70-75%) has been optimized in the order of the inherent scattering of both local tensile and hole expansion testing. Any correlation dealing with fracture strain behavior should be based on statistical regression with a significant data amount matching the confidence level expected. It should be also considered that the scattering of the mean of 3 replicates has been investigated both for tensile and HER values. Individual test values with higher scattering have not been considered.

References

[1] ISO 16630:2017(E) Metallic Materials—Sheet and Strip—Hole Expanding Test.

[2] Comstock R J, Scherrer D K and Adamczyk RD (2006) Hole expansion in a variety of sheet steels. Journal of Materials Engineering and Performance **15** 675-683.

[3] Comstock R J, Brown D and Scherrer D K (2011) Predicting Formability of Sheared Holes in a Variety of Sheet Steels. Materials Science and Technology (MS&T), Columbus, Ohio, USA.

[4] Goncalves J, Alibeigi S and Sarkar S (2019) Comprehensive understanding of effective parameters on edge cracking sensitivity of hot-rolled complex phase steels. *IOP Conf. Ser.: Mater. Sci. Eng.* **651** 012108.

[5] Narayanasamy R, Narayanan C S, Padmanabhan P at al. (2010) Effect of mechanical and fractographic properties on hole expandability of various automobile steels during hole expansion test. *Int J Adv Manuf Technol* **47** 365-380.

[6] Paul S K (2014) Non-linear Correlation Between Uniaxial Tensile Properties and Shear-Edge Hole Expansion Ratio. *Journal of Materials Engineering and Performance* **23** (10) 3610-3619.

[7] Kim J H, Kwon Y J, Lee T et al. (2018) Prediction of Hole Expansion Ratio for Various Steel Sheets Based on Uniaxial Tensile Properties. *Met. Mater. Int.* **24** (1), 187-194.

[8] Yoon J I, Jung J, Kim J G et al. (2017) Key factors of stretch-flangeability of sheet materials. Journal of Materials Science **52** 7808-7823.

[9] Yoon J I, Jung J, Joo S H et al. (2016) Correlation between fracture toughness and stretch-flangeability of advanced high strength steels. *Materials Letters* **180** 322-326.

[10] Hance B (2016) Advanced High Strength Steel: Deciphering Local and Global Formability. Int. Automotive Body Congress, Sept. 28-29, Dearborn, MI, USA.

[11] Hance B M (2017) Practical Application of the Hole Expansion Test.*SAE* 2017-01-0306.

[12] Larour P, Freudenthaler J and Weissböck T (2017) Reduction of cross section area at fracture in tensile test: measurement and applications for flat sheet steels. *J. Phys.: Conf. Ser.* **896** 012073.

[13] Larour P, Wagner L, Felbinger A and Angeli J (2019) Local fracture strain measurement in AHSS uniaxial flat tensile tests considering specimen geometry and fracture morphology. *IOP Conf. Ser.: Mater. Sci. Eng.* **651** 012016.

[14] Heibel S, Dettinger T, Nester W et al. (2018) Damage Mechanisms and Mechanical Properties of High-Strength Multiphase Steels. *Materials* **11** 761.

[15] Gruenbaum M, Aydin G, Dettinger T and Heibel S (2019) Local formability of AHSS: Measurement technique, specimen types and robustness.*IOP Conf.Ser.:Mater.Sci.Eng.* **651** 012056.

[16] Denks I A, Schneider M, Westhäuser S et al. (2019) On the Correlation between Suitable Material Parameters for the Prediction of Local Formability of AHSS. *Steel research int.* **90** (6) 1800460.

[17] Frómeta D, Parareda S, Lara A et al. (2020) Identification of fracture toughness parameters to understand the fracture resistance of AHSS steels. *Eng. Fracture Mechanics* **229** 106949.

[18] Yoon J I, Jung J, Ryu J H et al. (2017) Development of Methodology with Excellent Reproducibility for Evaluating Stretch-Flangeability Using a Sheared-Edge Tensile Test. *Exp Mech* **57**,1349-58.

[19] Cestoni H, Anderson D (2018) Effect of Edge Quality on Formability of an AA6xxx Aluminum Alloy. *IOP Conf. Series: Materials Science and Engineering* **418** 012065.

A new technique for characterising mechanical properties of materials under hot stamping conditions

R Zhang*, J Lin, Z Shi and Z Shao

Department of Mechanical Engineering, Imperial College London, London SW7 2AZ, UK

* r.zhang17@ic.ac.uk (R. Zhang)

Abstract. In order to characterise mechanical properties of materials (e.g. formability) under hot stamping conditions, significant efforts have been made to the development of the biaxial tensile testing method using cruciform specimens. However, no method for necking strain determination and no cruciform specimen design have been widely accepted. In this study, a new technique for characterising mechanical behaviour of materials under hot stamping conditions has been proposed. It includes two main parts: 1) a novel spatio-temporal method for determining necking and fracture strains, and 2) a cruciform specimen design for formability evaluation using biaxial testing method. In the first part, the theoretical base of the novel spatio-temporal method has been discussed, and the method has been validated by applying to uniaxial tensile tests on AA6082 specimens. The method has also been compared with several existing popular methods, in the determination of limit strain at onset of localised necking. It is found that the novel method has greater simplicity, stability and accuracy for the determination of localised necking strain. In the second part, a proposed cruciform specimen of AA5754 has been tested under the equi-biaxial tension, and both the necking initiation location and the strain path at the location where necking initiates, have been analysed. Furthermore, the novel spatio-temporal method has been applied to the biaxial tensile test for the determination of necking and fracture strains. The results show that the designed cruciform specimen enables to initiate fracture at the centre of the specimen and realisation of linear strain path under equi-biaxial tension.

1. Introduction

Hot stamping technologies have been widely applied to sheet metals for producing high strength complex-shaped engineering components [1,2]. Both application and optimisation of those technologies require a profound characterisation of mechanical properties (e.g. formability) under hot stamping conditions. Due to the complex temperature profile experienced by workpieces in hot stamping processes, the conventional tests, such as Nakajima and Marciniak tests, are particularly difficult to be used for formability evaluation under those extreme conditions [3]. It has been shown that biaxial tensile testing method using cruciform specimens has high potential to overcome this difficulty [4].

Formability evaluation using the biaxial tensile testing method usually requires three main parts: 1) a mathematical method for limit strain determination, 2) a biaxial tensile testing machine and 3) cruciform specimens with a suitable design. This study mainly focuses on the limit strain determination method and the cruciform specimen design. So far, many mathematical methods for limit strain determination have been developed and utilised. One of the most famous methods is the cross-section method provided in the standard ISO 12004-2:2008 [5]. Other methods include the linear best fit method [6] and the coefficient correlation method [7], etc. However, all those methods have their own limitations and disadvantages, which impedes their applications to biaxial tensile tests [8]. In respect to the cruciform specimen designs, the standard ISO 16842:2014 [9] provides a recommended cruciform test piece, but it is used for yield surface measurement rather than formability evaluation. Apart from this, many cruciform specimen designs have been proposed and tested [10], and a review of those existing cruciform specimen designs has been made in [11]. However, no one design has been widely accepted for formability evaluation so far.

This study aims to develop the biaxial tensile testing method for formability evaluation for materials under hot stamping conditions. A previously proposed spatio-temporal method for determination of necking and fracture strains, and cruciform specimen for formability evaluation, were analysed and discussed. The theoretical base of the spatio-temporal method was derived. Furthermore, the method validation was performed by applying to uniaxial tensile tests on AA6082 specimens. The spatio-temporal method was also compared with several popular existing methods, in respect of the values of the determined limit strain at onset of localised necking. Equi-biaxial tensile tests were carried out on AA5754 cruciform specimens, and both the necking initiation location and the corresponding strain path were analysed. Finally, the spatio-temporal method was applied to the biaxial tensile tests for the determination of necking and fracture strains.

2. Novel spatio-temporal method

2.1. Theoretical base

A novel spatio-temporal method has been proposed for determining necking and fracture strains by analysing average thickness reduction/major strain within two rectangular zones RZ and BZ [8]. In this study, average thickness strain is used for limit strain determination. Figure 1 shows the selection of the zones RZ and BZ around the necking initiation location, and their dimensions. The areas of the zone RZ, the zone BZ and the zone RZ excluding the zone BZ are designated as A_{RZ}, A_{BZ} and $A_{\mathrm{RZ-BZ}}$, and the corresponding thickness strain increments within these zones are $\Delta\varepsilon_3^{\mathrm{RZ}}$, $\Delta\varepsilon_3^{\mathrm{BZ}}$ and $\Delta\varepsilon_3^{\mathrm{RZ-BZ}}$, respectively.

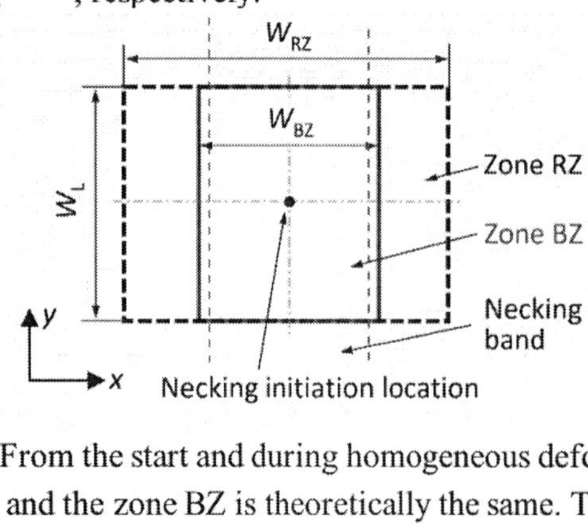

Figure 1. Schematic diagram showing selection of zones BZ and RZ for determination of necking and fracture strains using the spatio-temporal method. Note that zones RZ and BZ have same centre and dimension W_{L}, and zone RZ includes zone BZ.

From the start and during homogeneous deformation, the average thickness strain within the zone RZ and the zone BZ is theoretically the same. Therefore,

$$\Delta\varepsilon_3^{\mathrm{RZ}} = \Delta\varepsilon_3^{\mathrm{BZ}} \tag{1}$$

After the onset of localised necking, deformation concentrates in the necking band and $\Delta\varepsilon_3^{\mathrm{RZ-BZ}} = 0$. Thus,

$$\Delta\varepsilon_3^{\mathrm{RZ}} = \frac{\Delta\varepsilon_3^{\mathrm{RZ\text{-}BZ}} \times A_{\mathrm{RZ\text{-}BZ}} + \Delta\varepsilon_3^{\mathrm{BZ}} \times A_{\mathrm{BZ}}}{A_{\mathrm{RZ\text{-}BZ}} + A_{\mathrm{BZ}}} = \Delta\varepsilon_3^{\mathrm{BZ}} \times \frac{A_{\mathrm{BZ}}}{A_{\mathrm{RZ\text{-}BZ}} + A_{\mathrm{BZ}}} \tag{2}$$

Although the area A_{BZ} continues to increase slightly, the area $A_{\mathrm{RZ-BZ}}$ does not change because of the localised necking. Hence, the strains $\Delta\varepsilon_3^{\mathrm{RZ}}$ and $\Delta\varepsilon_3^{\mathrm{BZ}}$ also keep an almost linear relationship. This is the theoretical base of the spatio-temporal method to determine the onset of localised necking by fitting two straight lines in stable and instable deformation separately [8].

In order to enable that the line fitting is independent on testing speed and frame rate for strain measurement, the first straight line is fitted using the experimental points during the time range of $0 \sim t_S$, and the time range of $t_{IS} \sim t_F$ is used for fitting the second straight line, where t_F is time at fracture. The values of t_S and t_{IS} are recommended in Table 1, together with other recommended values for the dimensions in Ref. [8]. It should be noted that at least three experimental points are required for fitting the lines.

Table 1. Recommended values for the determination of necking and fracture strains using the spatio-temporal method, h and t_F are material thickness and fracture time, respectively

Items	Recommended values
W_{BZ}	$a \times h$, $a = 2 \sim 2.5$
W_{RZ}	$b \times W_{BZ}$, $b = 2 \sim 2.5$
W_L	$c \times h$, $c = 1.5 \sim 3$
t_S	$d \times t_F$, $d = 0.3 \sim 0.5$
t_{IS}	$e \times t_F$, $e = 0.990 \sim 0.998$

2.2. *Method validation*

The spatio-temporal method was validated by applying to uniaxial tensile tests on 1.5 mm thick aluminium alloy AA6082 sheet. Dog-bone shaped specimens, with a parallel length 34 mm and a width 6 mm in the reduced section, were cut along the rolling direction of the as-received sheet with T6 temper. The uniaxial tensile tests were carried out at room temperature with a constant tensile speed of 15 mm/min. The digital image correlation (DIC) technique was used for strain measurement, and a constant frame rate 125 fps was adopted for recording deformation. The commercial software GOM correlate 2018 was utilised for data post-processing, setting up with a facet size of 19 pixels and a point distance of 5 pixels.

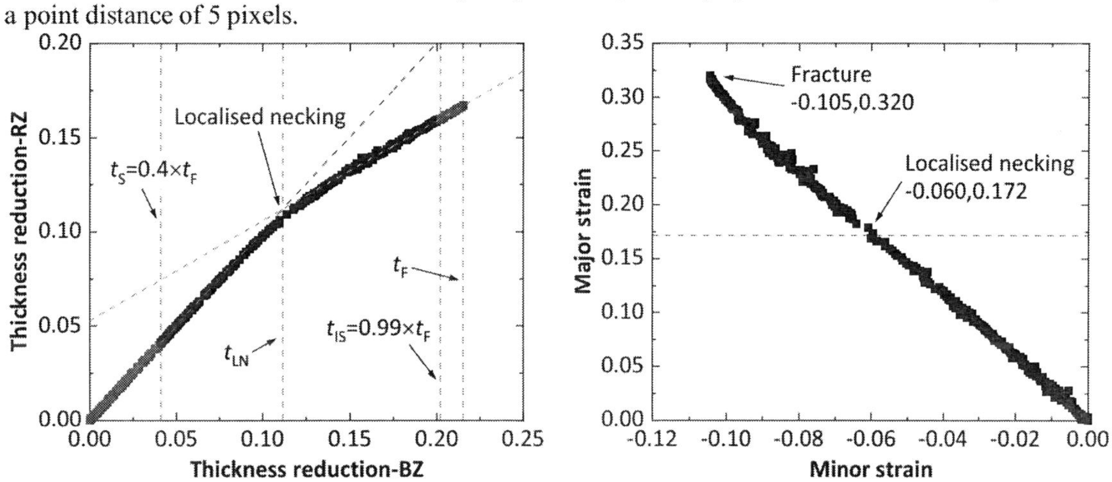

Figure 2. Validation of the spatio-temporal method for the determination of necking and fracture strains in uniaxial tensile tests on 1.5 mm thick AA6082 sheet, (a) determined onset of necking and fracture, (b) determined necking and fracture strains.

Figure 2(a) shows the determination of onset of localised necking and fracture using the spatio-temporal method, by adopting the following values as recommended in table 1, $a = 2$, $b = 2$, $c = 2$, $d = 0.4$, and $e = 0.99$. In the figure, the red points were selected for fitting the dashed green and red lines, representing stable and instable deformation, respectively. Indeed, different linear relationships exist between $\Delta\varepsilon_3^{RZ}$ and $\Delta\varepsilon_3^{BZ}$ before and after the onset of localised necking. Moreover, the recommended values of the time t_S and t_{IS} are reasonable for the line fitting. Based on the determined onset of necking and fracture, the corresponding limit strain was determined, as shown in Figure 2(b). The determined necking strain is (-0.06, 0.172) and the fracture strain is (-0.105, 0.32). In addition, the stain path is linear and is maintained throughout deformation. It can be observed that ratio of minor strain to major strain is about -0.34 rather than -0.5, which demonstrates the anisotropy of the material.

2.3. Comparison with existing methods

The spatio-temporal method was compared with several popular existing methods, in the determination of limit strain at onset of localised necking. Those existing methods include cross-section method (CS) [5], linear best fit method (LBF) [6], correlation coefficient method (CC) [7], gliding correlation coefficient method (GCC) [12], and gliding difference of mean to median method (GMM) [12]. By using a script implemented in the commercial software ARAMIS, the existing methods were successfully applied to limit strain determination in the uniaxial tensile tests on AA6082 specimens. Figure 3 shows a comparison of the determined necking strains by using the existing methods, and the necking and fracture strains by using the spatio-temporal method. For the CS method, five different sections were selected and thus, five different values of necking strain were determined. Furthermore, all the determined limit strains have a same ratio of minor strain to major strain, and thus, only the major strain was compared for the limit strain comparison. The determined major strain at onset of necking by using the spatio-temporal method is almost the same as that by using the CS method and the CC method, while it is much smaller than that determined by using the LBF method, the GCC method and the GDMM method. Specifically, the spatio-temporal method determined a major strain 0.172, which is 48.4% smaller than 0.333 by using the LBF method. It is noticed that the values of the major strain at necking, determined by using the LBF method, the GCC method and the GDMM method, are close to the value of the major strain at fracture, determined by using the spatio-temporal method.

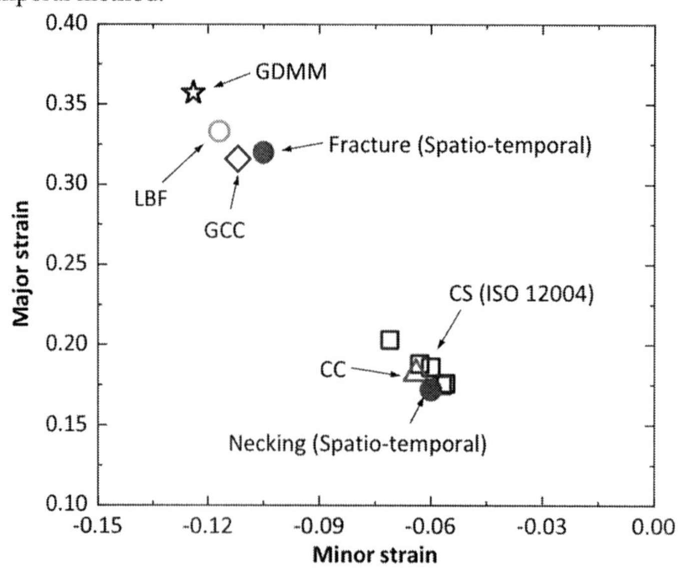

Figure 3. Comparison of the determined necking strains using the spatio-temporal method and the existing methods, CS: Cross-section method, LBF: Linear best fit method, CC: Correlation coefficient method, GCC: Gliding correlation coefficient method, GDMM: Gliding difference of mean to median method.

3. Cruciform specimen design

3.1. Geometry of cruciform specimen

To enable formability evaluation by using the biaxial testing method, two fundamental objectives must be achieved in designing cruciform specimen, i.e., 1) to initiate necking/fracture in gauge region, and 2) to achieve linear strain path with a designated strain ratio β, e.g. $\beta = 1$ for equi-biaxial tension. It has been known that compared to the uniaxial and plane-strain tensions, these two objectives are extremely difficult to realise in equi-biaxial tension. A type of cruciform specimen has been proposed to overcome this difficulty [11]. Figure 4 shows the geometry and dimensions of the proposed cruciform specimen, which includes features of notches between arms, slits in each arm, a circular thickness-reduced zone and importantly, a central thickness-reduced zone with arc-shaped profile through thickness. Except the dimensions of the thickness-reduced zones, other dimensions have been optimised to generate a relatively uniform strain distribution in the central region of the specimen [11].

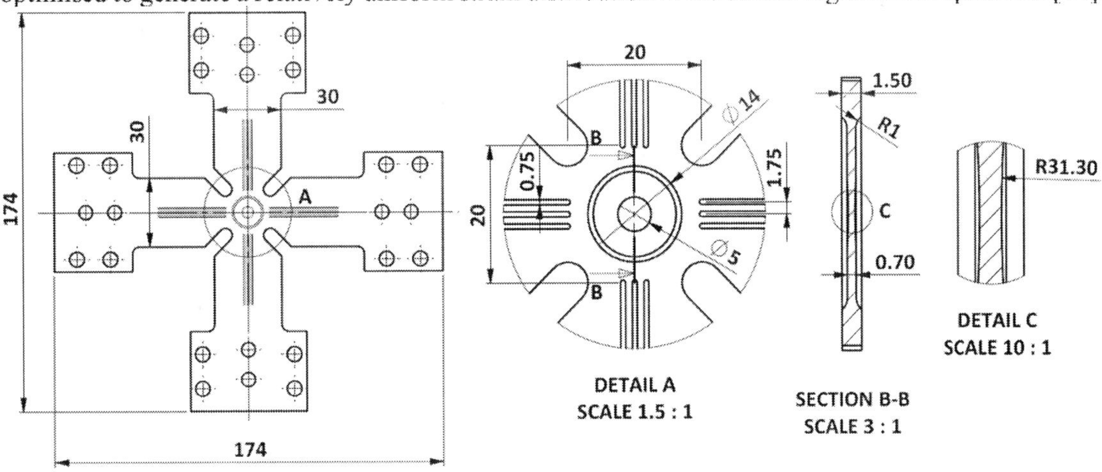

Figure 4. Geometry and dimensions (in millimetres) of crucifom specimen for formability evaluation.

3.2. *Deformation under equi-biaxial tension*

The AA5754 cruciform specimens were deformed under the equi-biaxial tension, by using a patented biaxial tensile system [13]. The relative grip speed in each arm is 15 mm/min. The DIC technique was adopted for strain measurement, with a constant frame rate of 125 fps. Figure 5 shows the measured distribution of major strain and thickness reduction in the circular thickness-reduced zone (diameter 14 mm), at different normalised times t/t_F. In the strain distribution at $t/t_F = 0.8$, deformation mainly concentrated in the central thickness-reduced zone due to smaller thickness, although it also occurred in other thickness-reduced zone. Both major strain and thickness reduction at the centre of the specimen is the highest. At $t/t_F = 1$, fracture started to initiate at the centre, and the highest major strain and thickness reduction were observed. This demonstrates that the first objective can be realised in the designed cruciform specimen under the equi-biaxial tension, i.e. localised necking/fracture initiates in the gauge region.

The strain distribution in Figure 5 was further analysed quantitatively. Figure 6(a) shows the thickness strain distribution along the path 1-1 perpendicular to the necking band, as indicated in Figure 5, at different normalised times. Indeed, the thickness reduction at the centre of the specimen is the highest throughout deformation. Figure 6(b) presents the strain path in the central area of the specimen where the localised necking initiates. Before $t/t_F = 0.9$, the strain path is almost linear, and corresponding strain ratio is calculated by $\beta = 1/k = 0.86$, where k is the slope of the fitted dashed line (red) for the experimental data using the least square method. This indicates that in the designed cruciform specimen under equi-biaxial tension, linear strain path with strain ratio β, which is very close to $\beta = 1$, can be achieved at the location where localised necking initiates. Therefore, the proposed cruciform specimen is able to realise the two objectives simultaneously, thus having high potential to be used for formability evaluation with the biaxial testing method. It should be noted that

IOP Conf. Series: Materials Science and Engineering **967** (2020) 012081 doi:10.1088/1757-899X/967/1/012081

the strain path abruptly changed at a time after $t/t_F = 0.9$. This is due to the increment of minor strain became zero and this phenomenon has also been observed in [14].

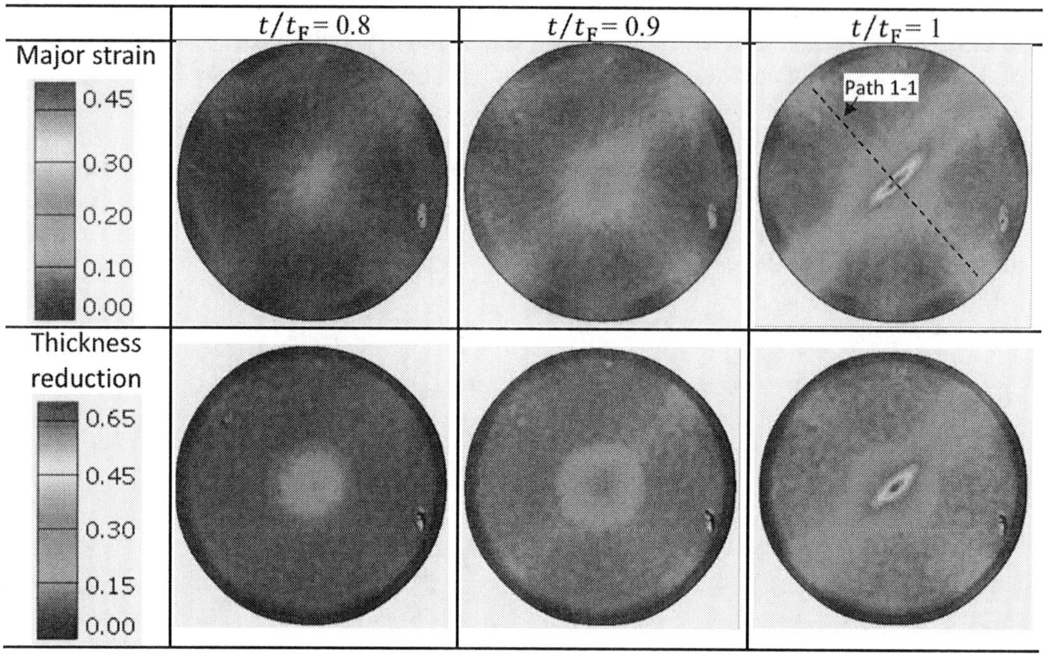

Figure 5. Strain distribution in the circular thickness-reduced zone in the cruciform specimen under equi-biaxial tension.

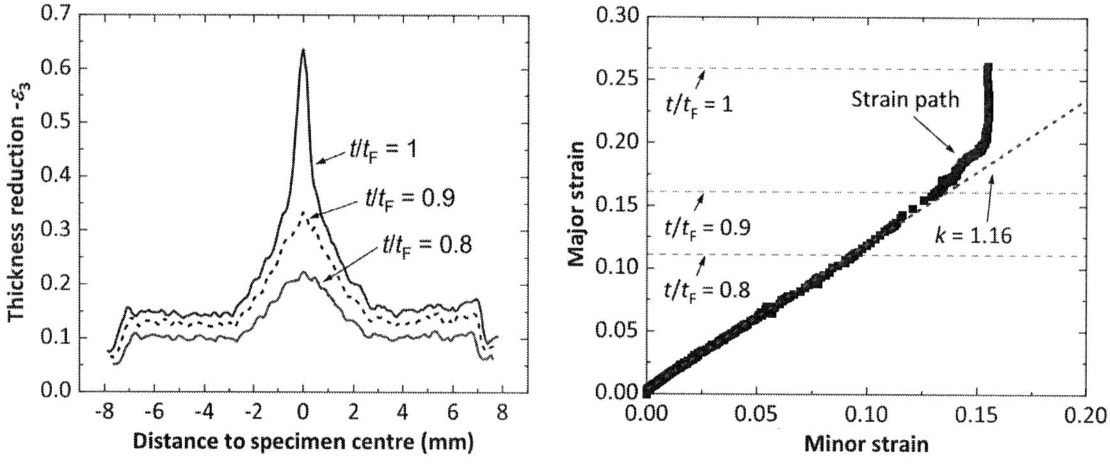

(a) (b)

Figure 6. Strain distribution along the path 1-1 and strain path at the location where necking initiates. Note that the path 1-1 is noted in Figure 5, (a) thickness strain distribution, (b) strain path at necking.

The spatio-temporal method was applied to the biaxial tensile test for the determination of the necking and fracture strains. Figure 7(a) shows the determined onset of localised necking and fracture in the space of thickness strain within the zones BZ and RZ, by adopting the following parameter: $a = 2$, $b = 2$, $c = 2$, $d = 0.4$, and $e = 0.99$. Indeed, the thickness strain within the zones BZ and RZ has two linear relationships for the starting and final stages, and the two fitted lines, using the least square method, are able to accurately express these linear relationships. Based on the determined onset of necking and fracture, the corresponding limit strain was further determined, as shown in Figure 7(b).

The limit strain at onset of localised necking is (0.133, 0.164), and the limit strain at fracture is (0.154, 0.262).

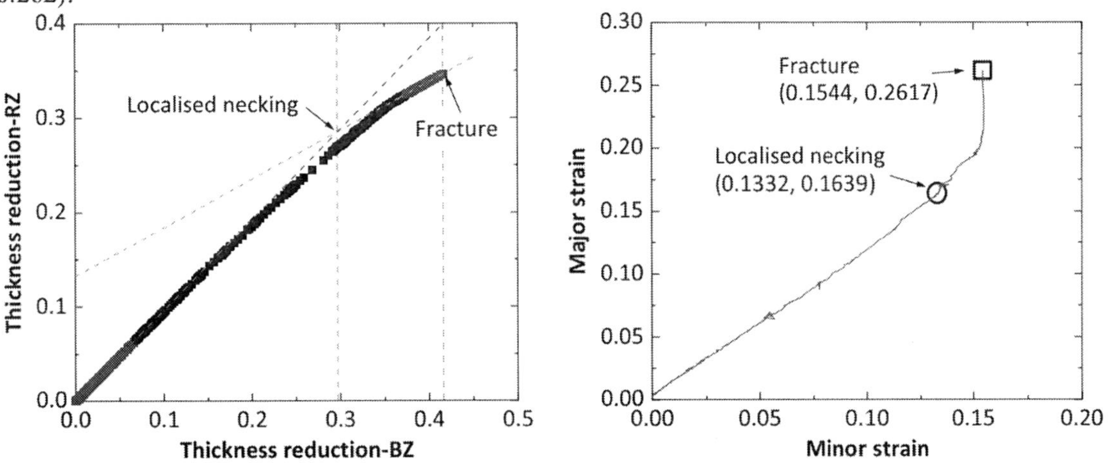

Figure 7. Determination of the onset of localised necking and fracture, and the corresponding limit strains, (a) onset of localised necking and fracture, (b) determined necking and fracture strains.

4. Conclusions

1) Compared with the existing methods, the novel spatio-temporal method has greater simplicity, stability and accuracy for the determination of localised necking strain.
2) The necking strain determined by the spatio-temporal method is almost the same as that by using the cross-section method in the standard ISO 12004-2:2008.
3) The designed cruciform specimen enables to initiate localised necking in the gauge region, and to achieve linear strain path with strain ratio β which is very close to 1 under the equi-biaxial tension.

Acknowledgements

The authors would like to thank financial support from EPSRC (Grant no. EP/R001715/1) on "Lightform: Embedding Materials Engineering in Manufacturing with Light Alloys". R. Zhang also appreciates financial support from the CSC-Imperial Scholarship (Grant no. 201700260069).

References

[1] Karbasian H and Tekkaya AE (2010) A review on hot stamping. Journal of Materials Processing Technology 210 (15):2103-2118. doi:10.1016/j.jmatprotec.2010.07.019

[2] Lin J, Balint D, Wang L, Dean TA and Foster AD (2013) Method of forming a component of complex shape from sheet material. United States Patent US 2013/0125606 A1, May 23, 2013

[3] Shao Z (2015) Development of a novel biaxial testing system for formability evaluation of sheet metals under hot stamping conditions. Imperial College London,

[4] Shao Z, Li N, Lin J and Dean T (2017) Formability evaluation for sheet metals under hot stamping conditions by a novel biaxial testing system and a new materials model. International Journal of Mechanical Sciences 120:149-158. doi:10.1016/j.ijmecsci.2016.11.022

[5] BS EN ISO 12004 (2008) Metallic materials — Sheet and strip — Determination of forming-limit curves Part 2: Determination of forminglimit curves in the laboratory. doi:10.3403/bseniso12004

[6] Volk W and Hora P (2011) New algorithm for a robust user-independent evaluation of beginning instability for the experimental FLC determination. International Journal of Material Forming 4 (3):339-346. doi:10.1007/s12289-010-1012-9

[7] Merklein M, Kuppert A and Geiger M (2010) Time dependent determination of forming limit diagrams. CIRP Annals - Manufacturing Technology 59 (1):295-298. doi:10.1016/j.cirp.2010.03.001

[8] Zhang R, Shi Z, Shao Z, Dean TA and Lin J (2020) A novel spatio-temporal method for determining necking and fracture strains of sheet metals. International Journal of Mechanical Sciences. doi:10.1016/j.ijmecsci.2020.105977

[9] BS ISO 16842 (2014) Metallic materials — Sheet and strip — Biaxial tensile testing method using a cruciform test piece. doi:10.3403/30246886

[10] Zhang R, Shao Z and Lin J (2019) A type of cruciform specimen applied to evaluate forming limits for boron steel under hot stamping conditions. IOP Conference Series: Materials Science and Engineering 651:012087. doi:10.1088/1757-899x/651/1/012087

[11] Zhang R, Shi Z, Shao Z, Dean TA and Lin J (2020) Effect of cruciform specimen design on strain paths and fracture location in equi-biaxial tension. Journal of Materials Processing Technology:Under review

[12] Hotz W, Merklein M, Kuppert A, Friebe H and Klein M Time dependent FLC determination comparison of different algorithms to detect the onset of unstable necking before fracture. In: Key Eng Mater, 2013. pp 397-404. doi:10.4028/www.scientific.net/kem.549.397

[13] Lin J, Li N, Shao Z and Qian B (2018) Planar test system. United States Patent US 20180238784 A1, Aug. 23, 2018

[14] Tasan C, Hoefnagels J, Ten Horn C and Geers M (2009) Experimental analysis of strain path dependent ductile damage mechanics and forming limits. Mechanics of Materials 41 (11):1264-1276. doi:10.1016/j.mechmat.2009.08.003

IOP Conf. Series: Materials Science and Engineering **967** (2020) 012082 doi:10.1088/1757-899X/967/1/012082

Interaction between forming and joint quality of the friction-assisted clinching

Zhiyong Li[1], Zhiyong Wang[1], Yong Li[2] and Shanling Han [2]*

[1] College of Transportation, Shandong University of Science and Technology, Qingdao, 266590, China

[2] College of Mechanical and Electronic Engineering, Shandong University of Science and Technology, Qingdao, 266590, China

* S.L.HAN@163.com

Abstract. In order to achieve the effective joining of aluminum and magnesium alloy sheets, a friction-assisted clinching is innovatively proposed to reduce material damage during the process and improve the joint quality. 2 mm thick Al5052 aluminum sheet and 2 mm thick AZ31 magnesium alloy sheet were selected as the research objects, the interaction between different forming parameters, i.e., die depth, rotational speed and rotation time, and the joint quality, i.e., the damage, neck and interlock were explored by DEFORM-3D. The influence of sheet temperature on damage and punch load is analysed, the rotational speed and rotation time have great influence on the forming joint, the die depth mainly affects the neck and interlock. Comparing the joint qualities under different forming parameter combinations, it is found that the joint quality is the best when the bottom die depth is 2.2 mm and the bottom punch is applied at 1000 rpm for 0.5 s. The research will provide a valuable reference for the realization of joining the aluminum and magnesium sheets.

1. Introduction

In recent years, a series of policies on energy saving and emission reduction of automobiles have been issued to cope with the increasing environmental pollution and energy crisis, has resulted in the increased demand for lightweight vehicles [1, 2]. Aluminum alloy and magnesium alloy have been widely used in automotive body-in-white due to higher specific strength [3]. The difference in chemical compositions and physical properties of these two alloys are large, the welding process can't realize an effective joining. As an advanced joining technology, clinching is developed in manufacturing of lightweight vehicle body structure.

The clinching process equipment is mainly composed of a punch, a holder, and a bottom die. The principle of clinching is that sheets are deformed locally under the pressure of the punch, and embedded in the bottom die due to the plastic flow characteristics of material, an S-shape interlock joint is finally formed. Due to it does not require any additional rivet or fastener, the clinching has an important advantage in automobile lightweight [4-6].

The deficiency of traditional clinching process is that it is impossible to realize the connection of materials with poor plasticity, the materials with poor plasticity are prone to damage and fracture under the pressure of punch. In view of this deficiency, scholars have done some research on it. In order to improve the formability of aluminum alloy, Lambiase et al. [7] proposed a friction-assisted clinching process. Aluminum alloy AA6061 was selected as the upper sheet, and the pre-holed CFRP was selected as the lower sheet. The results showed that under reasonable process parameters, friction-assisted crimping could significantly improve the formability of AA6061, and the forming effect of the joint was better than that of the traditional clinching process. Chen et al. [8, 9] combined clinching and hot stamping process, a hot stamping clinching process forming system with resistance heating system was developed, and the joint strength was tested to verify the feasibility of the new process. Kaczyński et al. [10] proposed a clinching process that preheated the bottom sheet to enhance the joint strength, and verified through experiments that the new process could increase the shear and tensile strength of the joint.

Based on the existing research, a new friction-assisted clinching process is proposed, the die depth, rotational speed and rotation time are selected as forming parameters, and the simulation study on the process is carried out to provide a valuable reference for the realization of joining the aluminum and magnesium sheets.

2. Modelling and finite element analysis

The new friction-assisted clinching process model is shown in Figure 1. It can be seen in Figure 1(c), the traditional fixed bottom die is replaced by a movable bottom die. The movable bottom die is composed of a base and a bottom punch. A through hole is machined in the base. The base and the bottom punch is used together, the bottom punch can rotate in the through hole. In the initial stage of the process, the top punch is pressed down, and the sheets move downward under the force of the top punch. When the lower sheet contacts the bottom punch, as shown in Figure 1(b), the bottom punch starts to rotate, and the lower sheet will be heated by the rotating friction of the bottom punch, and the top punch will continue to press down and stop moving when it reaches the specified stroke position, and the joint is formed. Two rotation time of 0.5 s and 2 s are selected, Figure 1(c) represents the process diagram when the bottom punch rotates for 0.5 s, that is, when the rotation time is 0.5 s, the rotation of bottom punch will stop at the position of Figure 1(c), and the 2 s rotation time represents the bottom punch rotation will be maintained to the end of the process, as shown in Figure 1(d). Because the die depth is different, the corresponding time in Figure 2(b) is also different, which is about h/2 s.

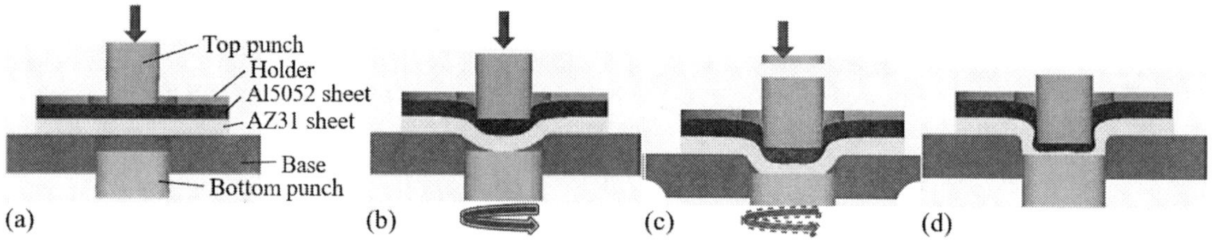

Figure 1. Schematic of friction-assisted clinching process. (a) Positioning, (b) punching and friction starting, (c) friction acting for 0.5 s, (d) joint formed.

2 mm thick Al5052 aluminum sheet and 2 mm thick AZ31 magnesium alloy sheet are selected as the research objects. The mechanical properties of Al5052 and AZ31 are shown in Table 1.

Table 1. Mechanical properties of Al5052 and AZ31.

Material	Density ρ (kg.m^{-3})	Elastic modulus E (GPa)	Yield strength σ_s (MPa)	Ultimate tensile strength σ_b (MPa)	Elongation δ (%)
Al5052	2680	71	195	230	10
AZ31	1780	45	169	297	8

The ductility of AZ31 magnesium alloy sheet is poor, if AZ31 sheet is selected as the upper sheet, the traditional friction-assisted clinching based on top punch rotating can indeed improve its ductility, and reduce the risk of fracture damage during the process. But the frictional heat will also make the sheet soft, it is difficult to press into the lower aluminum alloy sheet to form an effective joint. In view of this, this article selects Al5052 aluminum alloy sheet as the upper sheet and AZ31 magnesium alloy as the lower sheet. During the deformation of the AZ31 magnesium alloy, the temperature of the AZ31 magnesium alloy sheet is increased by the rotation friction of the bottom punch, so as to improve its ductility and reduce the fracture damage during the process, and obtain an effective joint.

The whole model was established by UG, the size of AZ31 sheet and Al5052 sheet are both 26 mm×26 mm×2 mm, the diameter of the top punch is 7 mm, the draft angle is 1°, the diameter of the bottom punch is 10 mm, the through hole diameter of the base is 10 mm. The established three-dimensional geometric model was exported to stl format and imported into DEFORM-3D software, and the tetrahedral mesh element was selected to mesh the upper and lower sheet, top punch and bottom punch. In order to improve the accuracy of the simulation results, the middle part of the upper and lower sheet were refined, and finally the finite element model was established as shown in Figure 2.

In order to observe the damage of sheet during the process, the Normalized Cockcroft & Latham damage model was used in the DEFORM-3D software, as shown in equation (1).

$$C = \int_0^{\bar{\varepsilon}_f} \frac{\sigma^*}{\bar{\sigma}} d\bar{\varepsilon} = \frac{\sigma^*}{\bar{\sigma}} \bar{\varepsilon}_f \tag{1}$$

Where C is the damage factor, $\bar{\varepsilon}_f$ is the equivalent strain at fracture, $\bar{\varepsilon}$ is the equivalent strain, σ^* is the maximum principal stress, $\bar{\sigma}$ is the equivalent stress, $\bar{\sigma} = \bar{\sigma}(\bar{\varepsilon}, \dot{\bar{\varepsilon}}, T)$, $\dot{\bar{\varepsilon}}$ is the equivalent strain rate, T is the temperature of sheet.

The die depth, the rotational speed and rotation time of the bottom punch are selected as variables to explore the influence of different parameters on the forming quality. The depth of the bottom die is determined by the stop position of the bottom punch. The pressing speed of the punch in all simulations is set to 2 mm/s, two rotation time of 0.5 s and 2 s are selected. The designed simulation parameters are shown in Table 2.

Table 2. Simulation parameters.

No.	Die depth (mm)	Rotational speed (rpm)	Rotation time (sec)
1	2.2	0	
2	2.2	1000	2.0
3	2.4	1000	2.0
4	2.0	1000	0.5
5	2.2	1000	0.5
6	2.4	1000	0.5
7	2.0	3000	0.5
8	2.2	3000	0.5
9	2.4	3000	0.5

Figure 2. Finite element model.

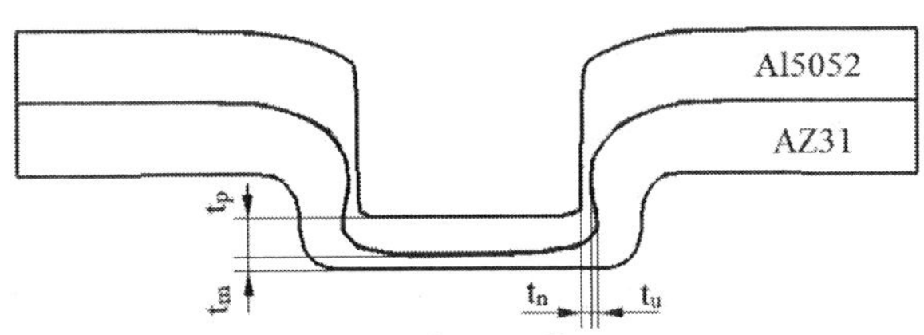

Figure 3. The joint quality parameters.

The joint quality parameters are shown in Figure 3. The main factors for evaluating the quality of the clinched joints are neck thickness t_n, interlock t_u, undercut thickness t_p of the upper sheet, and undercut thickness t_m of the lower sheet. The neck thickness mainly affects the shear strength of the joint. The thicker the neck, the higher the shear strength of the joint. The interlock value mainly affects the tensile strength of the joint. The greater the interlock value, the higher the joint tensile strength. The undercut thickness of the upper and lower sheets have an indirect effect on the neck thickness and interlock value. In theory, the lower the undercut of the sheet, the greater the neck thickness and interlock of the joint.

3. Results and discussion

It is an important goal of this article to reduce the fracture damage of the material during the forming process and improve the joint quality. First, the damage of AZ31 sheet of No.1, No.5 and No.8 during the simulation of the top punch stroke to 3 mm were extracted, as shown in Figure 4. When the top punch went up to 3 mm, AZ31 sheet and the bottom punch have come to contact, frictional heat had been applied to AZ31 sheet under the rotation of the bottom punch in simulation No.5 and No.8. The damage of AZ31 sheet at this time is more representative.

The Figure 4(a), 4(b) and 4(c) represent the side view of AZ31 sheet damage of No.1, No.5 and No.8, respectively, and the Figure 4(d), 4(e) and 4(f) represent the damage on the bottom of AZ31 sheet of No.1, No.5 and No.8, respectively. It can be seen from Figure 4 that the damage is mainly concentrated in the bottom of AZ31 sheet, this is the main part where AZ31 sheet is extruded by the bottom punch. The damage decreases step by step from the bottom to the top along the vertical direction. The damage degree of No.1 AZ31 sheet without frictional heat is the most serious. Frictional heat generated by rotation will soft AZ31 sheet, the ductility of AZ31 sheet is improved, and the damage is reduced. It can be seen that the AZ31 sheet damage value of 3000 rpm is lower than that of 1000 rpm. It shows that the higher the rotational speed is, the lower the sheet damage is.

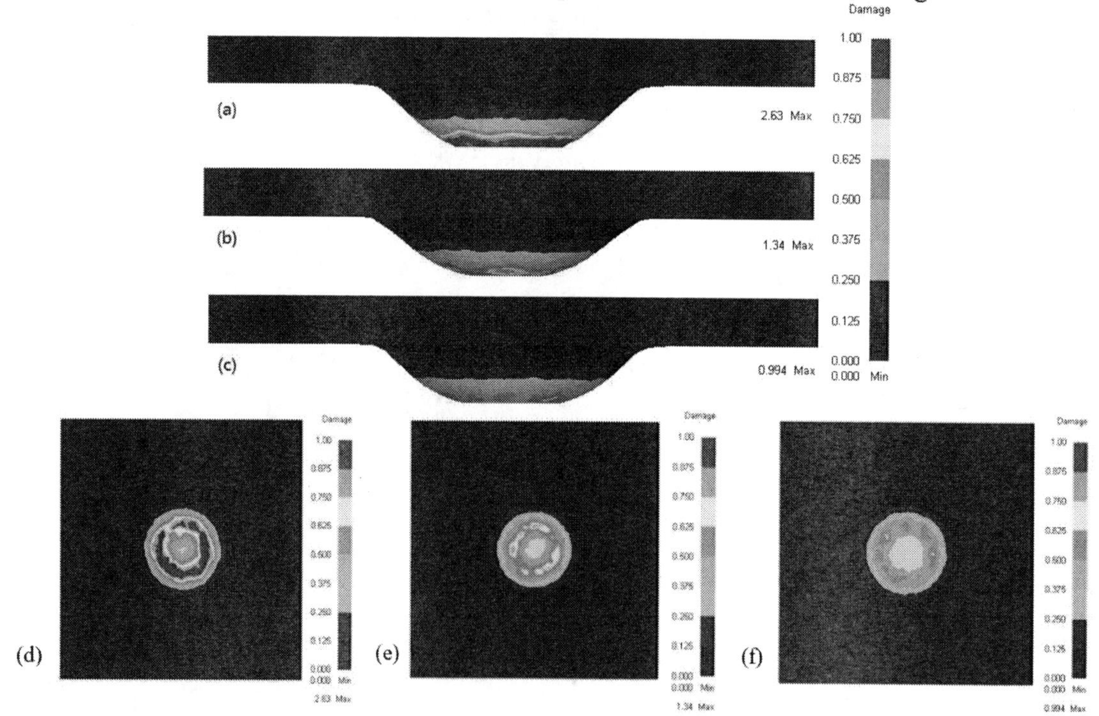

Figure 4. The side view of AZ31 sheet damage of (a) No.1, (b) No.5 and (c) No.8, the damage on the bottom of AZ31 sheet of (d) No.1, (e) No.5 and (f) No.8.

The simulation results of No.2 and No.3 are shown in Figure 5. It can be seen from Figure 5 that the neck thickness and interlock of No.2 and No.3 are basically 0, the undercut of the upper sheet is too large, and the undercut of the lower sheet is too small. The simulation results of No.2 and No.3 show that when the rotation time is 2 s, the die depth basically has no effect on the forming of the joint, and an effective joint cannot be formed during the process.

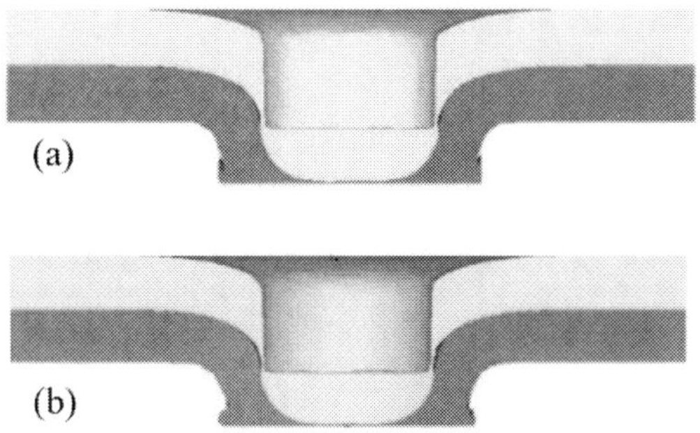

Figure 5. Simulation results of (a) No.2 and (b) No.3.

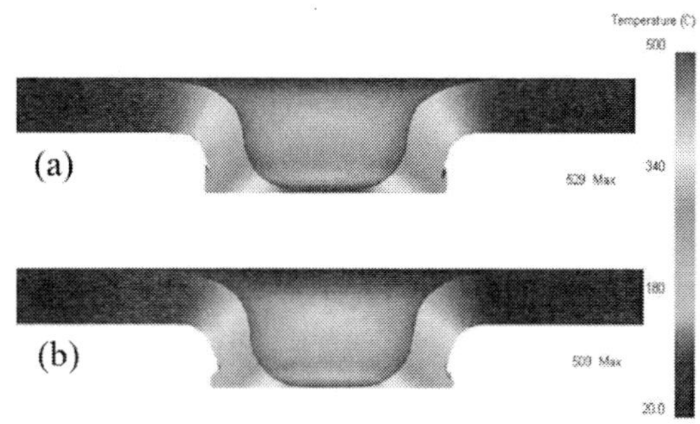

Figure 6. The temperature distributions of (a) No.2 and (b) No.3.

The temperature distributions of AZ31 sheet of No.2 and No.3 are shown in Figure 6. It can be seen that when the rotational speed continues to act on AZ31 sheet, the temperature of AZ31 sheet will continue to rise. The material of AZ31 tend to move around under the centrifugal force and the downward pressure of the top punch. The accumulation of the AZ31 material around the bottom die also causes much pressure to Al5052 sheet, and the S-shape interlock cannot be formed. When the undercut of AZ31 sheet is close to 0, under the action of heat generated by friction and plastic deformation, the bottom center temperature of AZ31 sheet in No.2 is as high as 529 ℃, and the bottom center temperature of AZ31 sheet in No.3 is as high as 509 ℃, which is close to the melting temperature of AZ31 magnesium alloy sheet, it will be pierced under the action of continuous pressure of the bottom punch. It is indicated that the 2 s of rotation time of the bottom punch cannot improve the joint quality. On the contrary, excess frictional heat generated by rotation make AZ31 sheet excessively softened, resulting in ineffective forming.

According to the above results, it can be seen that the main function of the bottom punch is to improve the ductility and reduce the damage by frictional heating the AZ31 magnesium alloy sheet when it is deformed. In order to reduce the adverse effects of excessive frictional heat, the rotation time of the bottom punch in the No.4-9 simulation is only 0.5 s.

The final joint of No.4-9 simulation results are shown in Figure 7, and the joint quality parameters are shown in Table 3.

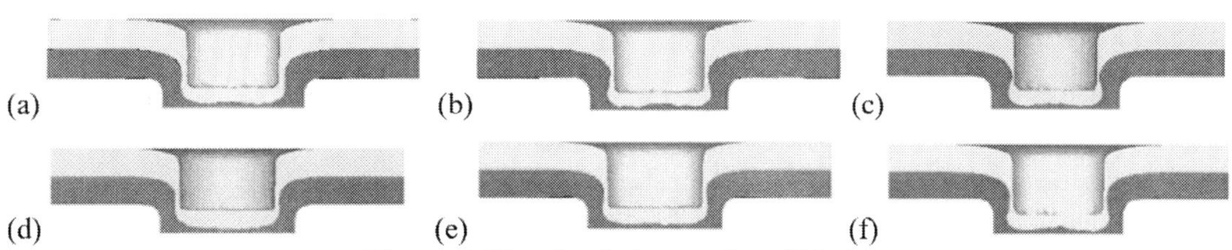

Figure 7. The simulation results of No.4-9.

By the simulation results in Figure 7 and joint quality parameters in table 3, it can be seen that in the case of the same rotational speed, as the die depth increases, the neck thickness gradually decreases and the interlock gradually increases. In the case of the same die depth, the joint quality parameters at the 1000 rpm rotational speed is better than those at 3000 rpm. In the case of 1000 rpm rotational speed, the total undercut is maintained at 1.4 mm by controlling the stop position of top punch stroke. At this time, the undercuts of the upper sheets are maintained between 1.0-1.1 mm and the undercuts of the lower sheet are maintained between 0.3-0.4 mm. There is little difference in joint quality parameters, indicating that the die depth has little effect on the undercut of the upper and lower sheets. The undercut of AZ31 sheet under 3000 rpm is thinner than that under 1000 rpm, the main reason is that the heat generated by friction and plastic deformation at 3000 rpm is too much, which causes the temperature of AZ31 sheet to be too high, the ability of AZ31 sheet to resist material deformation becomes worse. The results are similar to the No.2 and No.3.

Table 3. Joint quality parameters of No.4-9.

No.	Neck (mm)	Interlock (mm)	Undercut of upper sheet (mm)	Undercut of lower sheet (mm)	Total undercut (mm)
4	0.45	0	1.08	0.32	1.40
5	0.36	0.21	1.01	0.39	1.40
6	0.20	0.21	1.07	0.33	1.40
7	0.37	0	1.31	0.09	1.40
8	0.31	0.09	1.23	0.17	1.40
9	0.24	0.18	1.20	0.20	1.40

The simulation results of No.2, No.5, No.6, No.8, and No.9 are selected to extract the change of punch load with stroke as shown in Figure 8.

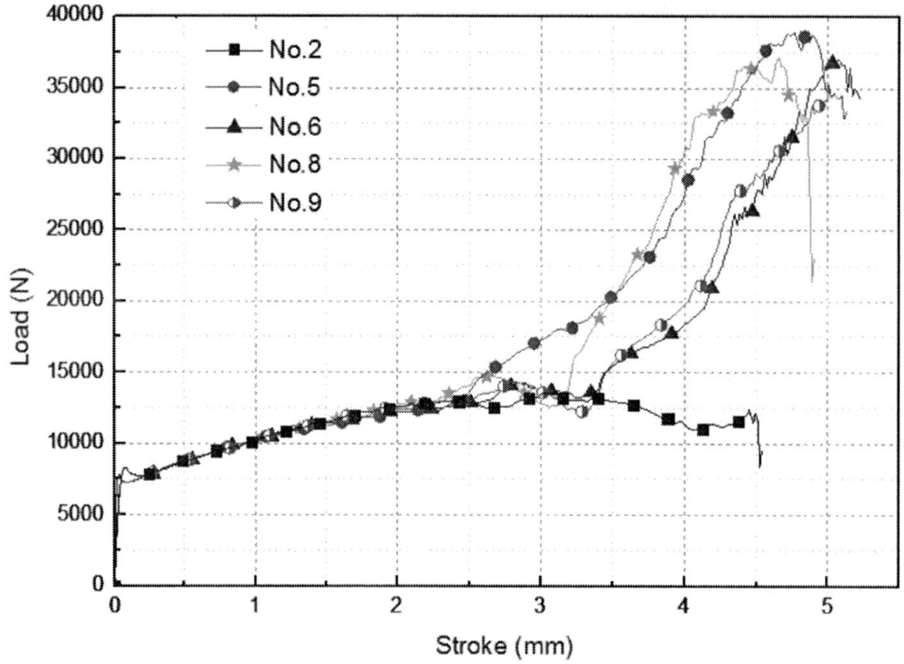

Figure 8. Load-stroke curves.

It can be seen from Figure 8 that the punch load curve is basically the same before AZ31 sheet contact to the bottom punch. The punch load begins to change after AZ31 sheet touches the bottom punch. The punch load of the No.2 changes smoothly with the stroke. Even when AZ31 sheet contact to the bottom punch, the punch load only slightly increases and then starts to drop. This is because after AZ31 sheet contact to the bottom punch, frictional heat is generated by rotation of the bottom punch. The rigidity of AZ31 sheet decreases with the increase of temperature. When AZ31 sheet is pierced, the punch load drops sharply. When the rotational speed of the bottom punch is the same and the punch descends to the same stroke, the punch load under the die depth of 2.2 mm is higher than that of 2.4 mm. The bottom die space of 2.2 mm is smaller than 2.4 mm, as the bottom die being filled with the material, the punch load will become larger and larger in the process, the punch load of No.4 and No.7 will be significantly higher than that of No.5 and No.8. When the die depth is same, the movement trend of the punch load is basically the same, indicating that the rotation time is shorter, and it will not have much impact on its forming quality, but it will have a certain effect on the load peak. The load peak value of 3000 rpm rotational speed is about 2000-3000N lower than that of 1000 rpm rotational speed, which is because the resistance of the higher temperature sheet is lower than that of the lower temperature sheet. The sudden load drop in the last stage of the curve in the Figure 8 is due to the moment when AZ31 sheet is pierced. By analyzing the load-stroke curve and comparing the forming quality of the joint, it can be concluded that a higher punch load is beneficial to the increase of the interlock, the interlock is increased, and the joint quality is improved.

4. Conclusion

An innovative friction-assisted clinching process for is proposed, which can increase the temperature of the lower sheet through friction, improve its ductility and reduce the damage in the clinching process. The continuous frictional heat generated by rotation of the bottom punch to the lower sheet will make the temperature of the AZ31 magnesium alloy sheet close to the melting temperature in the later stage of forming. Although the pressure of the top punch in the forming process is effectively reduced, the quality of the joint is poor and an effective joint cannot be formed.

When the lower sheet is in contact with the bottom punch, the damage of AZ31 sheet begun to increase. At this time, the damage degree of the bottom punch with the rotational speed of 1000 rpm is more serious than that of 3000 rpm, but the difference is small, and from the final joint quality, the rotational speed with 1000 rpm is a more suitable choice.

The die depth will affect the punch load and has a greater impact on the forming quality. As the die depth increases, the neck thickness will gradually decrease.

For the friction-assisted clinching of 2 mm thick Al5052 and 2 mm thick AZ31 dissimilar sheets, the joint quality is the best when the bottom die depth is 2.2 mm and the bottom punch is applied at 1000 rpm for 0.5 s.

References

[1] Li YB, Ma YW, Lou M, Zhang GT, Zhang QX, Qi L, and Deng L 2020 Advances in spot joining technologies of lightweight thin-walled structures. *Journal of Mechanical Engineering*, **56** 125-146.

[2] Wiesenmayer S, Graser M, and Merklein M 2020 Influence of the properties of the joining partners on the load-bearing capacity of shear-clinched joints. *Journal of Materials Processing Technology*, **283** 116696.

[3] Coppieters S, Zhang H, Xu F, Vandermeiren N, Breda A, and Debruyne D 2017 Process-induced bottom defects in clinch forming: simulation and effect on the structural integrity of single shear lap specimens. *Materials & Design*, **130** 336-348.

[4] Eshtayeh M, Hrairi M, and Mohiuddin A 2016 Clinching process for joining dissimilar materials: state of the art. *The International Journal of Advanced Manufacturing Technology*, **82** 179-195.

[5] Lin PC, Lo SM 2017 Friction stir clinching of alclad AA2024-T3 sheets. *The International Journal of Advanced Manufacturing Technology*, **92** 2425-37.

[6] Kumar S, Avinash G, and Vimal E 2020 Ductility effect on clinching joint strength in lap-shear configuration loading. *Materials Today: Proceedings*.

[7] Lambiase F, Paoletti A 2018 Friction-assisted clinching of Aluminum and CFRP sheets. *Journal of Manufacturing Processes*, **31** 812-822.

[8] Chen LW, Huang JM, and Hsu YC 2015 Investigation of the clinching process combines with hot stamping process for high-strength steel sheets. *MATEC Web of Conferences*, **21** 05004.

[9] Chen LW, Cai MJ 2018 Development of a hot stamping clinching tool. *Journal of Manufacturing Processes*, **34** 650-658.

[10] Kaczyński P, Skwarski M 2020 Partial heating as a new method for increasing the strength of clinch joints of thin-walled elements. *Thin-Walled Structures*, **148** 106610.

IOP Conf. Series: Materials Science and Engineering **967** (2020) 012083 doi:10.1088/1757-899X/967/1/012083

Numerical and Experimental Investigation on Hot Stamping of TWB B-Pillar

Cheng-Kai ChiuHuang, Ming-Feng Chiang and Ping-Kun Lee

China Steel Corporation, 1 Chung Kang Rd., Hsiao Kang, Kaohsiung 81233, Taiwan

185066@mail.csc.com.tw

Abstract. In order to achieve the targeted crashworthiness performance, the application of Tailor-Welded-Blank (TWB) on the crash-relevant components has become a trend in the automotive industry. This paper aims to investigate the mechanical property and structural performance of the hot stamped TWB B-pillars. After hot stamping, the strength grade for the hard zone is 1500 MPa while the soft zone's strength grade is 590 MPa. The CAE model shows that the intrusion distance is decreased for the TWB design compared to the monolithic material B-pillar. The three point bending performance could be adjusted by changing the range of the soft zone. It is observed that the weld line has a low crack sensitivity during hot stamping. The transition zone between the hard zone and soft zone is less than 1mm. The ATOS scanning results show that no significant effect on the part's size deviation for the TWB design compared to the monolithic material B-pillar.

1. Introduction

In recent years, hot stamping has become the most efficient way for automakers to reach the goal for body structure lightweight and safety performance improvement. During the traditional hot stamping process, a blank is first heated and soaked in the furnace at the temperature range 900-930 °C, then the blank is quickly transferred to the die for forming and in-die quenching. When the cooling rate is above 27 °C/s, the microstructure of the blank changes from the high temperature soft austenite phase to the hard martensitic phase with an ultimate tensile stress above 1450 MPa [1].

In order to enhance the crashworthiness and better protect the passenger's safety, the application of hot stamping parts has changed from the monolithic material to tailored property design [2]. The combination of tailor-welded blank (TWB) and hot stamping process can produce a body structural part which includes both the hard zone and soft zone. During the crash, the hard zone can prevent intrusion while the soft zone can absorb more kinetic energy [3]. Peister et al. investigated the mechanical property of TWB Usibor1500-AS/Ductibor500-AS via a hat shaped hot stamping die, and the results showed that the width of transition zone between hard/soft zones is smaller than 2 mm [4]. Peister et al. also discussed the axial crush energy absorption behavior of Usibor1500-AS/Ductibor500-AS TWB rails, and the weld seam failure was not observed during the crush test [5]. Choi et al. investigate the weld line's displacement of HPF1470/340Y TWB after hot-stamping via a lab-scale B-pillar, and the results show that the weld line's movement during forming is related to the range of soft zone [6]. The load–displacement curve under the three point bending test and part's size deviation for the monolithic material and TWB hot stamped parts are rarely discussed.

In this study, the numerical and experimental investigations on the CSC15B22/440W TWB hot stamped B-pillar were conducted. The width of transition zone between the hard zone and soft zone after hot stamping was measured. A simplified CAE model was used to predict the energy absorption performance for the three point bending tests. The effect of soft zone's range on the intrusion distance and energy absorption was discussed. Finally, the size deviation for the monolithic material and TWB hot stamped parts were compared.

2. Experimental method

2.1 TWBs sample preparation, die try-out and three point bending test

As shown in Figure 1, three different types of blanks (Monolithic material/TWB with the soft zone d=110 mm/TWB with the soft zone d=260 mm) are prepared for the B-pillar hot stamping die tryout. The material used for the hard zone is CSC15B22, and the material used for the soft zone is 440W. The thickness of 15B22 and 440W are both 1.8 mm. The main chemical composition of 15B22/440W are listed in Table 1.

(a) Monolithic (b)TWB_d=110 mm (c)TWB_d=260 mm

Figure 1. Preparation for the monolithic material blank and TWBs

Table 1. Main chemical composition of CSC15B22 and 440W

Material	C	Si	Mn	P	S	B
15B22	0.2-0.25	0.2-0.3	1.0-1.5	0.015	0.004	0.002
440W	0.05-0.1	0.04-0.08	1.0-1.5	0.015	0.0018	0.0002
Fe：balance						(wt%)

The B-pillar hot stamping die and three point bending test jig are shown in Figure 2. The monolithic material blank and TWBs are heated to 930 °C and soaked for 4 minutes in the furnace. Then the blank is manually transferred to the die. The transfer time is 8~10 seconds. Die quenching time is 15 seconds, and the die holding pressure is 550 ton. The die tryout parameters are listed in Table 2.

(a)TWB before die closing (b)After die quenching (c)Three-point-bending test jig

Figure 2. Hot stamping die tryout and three-point-bending test

Table 2. Die tryout parameters of monolithic and TWB B-pillars

Blank Temperature (°C)	Furnace Atmosphere	Soaking Time (sec)	Transfer Time (sec)	Die Holding Time (sec)	Die Holding Pressure (ton)
930	N_2	240	8~10	15	550

2.2 CAE model setup

In order to compare the energy absorption characteristic for the monolithic and TWB B-pillars, a three point bending CAE model is built by PAM-STAMP. Figure 3 shows the three point bending test jig model and the mesh of TWB B-pillar. Our test results show that the HAZ of the laser weld seam is very narrow (<1 mm), so the mechanical property of the weld line is not incorporated in the model, only material data for 15B22 and 440W are input. The weld line is built before meshing to insure the mesh has good connectivity between the hard zone and soft zone.

As shown in Figure 4, three different B-pillar designs are discussed in this study: Monolithic, TWB_soft zone 110 mm, and TWB_soft zone 260 mm. For the TWB_soft zone 110 mm design, the weld line is close to the B-pillar's bottom end and located outside the punch projection area. For the TWB_soft zone 260 mm design, the weld line is close to the projection of punch's center line.

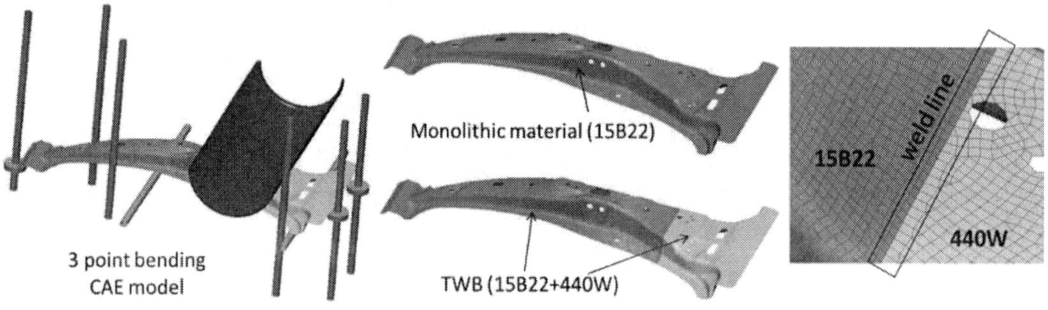

Figure 3. Schematic of three point bending CAE model

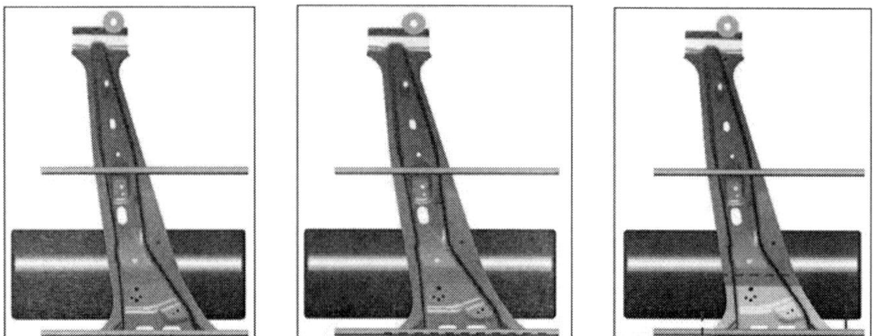

(a)Monolithic (b)TWB_soft zone 110 mm (c)TWB_soft zone 260 mm

Figure 4. Schematic of monolithic and soft zone's range for the TWB B-pillars

To further compare the intrusion distance for the monolithic and TWB B-pillar designs, a simplified fixed-end CAE model is used as shown in Figure 5. During the side impact, the intrusion displacement of the B-pillar is critical to the passenger's safety. Point A, which is close to the passenger's chest height, is selected as the observation point during the simulation. The intrusion displacements and section profiles for different B-pillar designs are compared under the same intrusion distance of the punch.

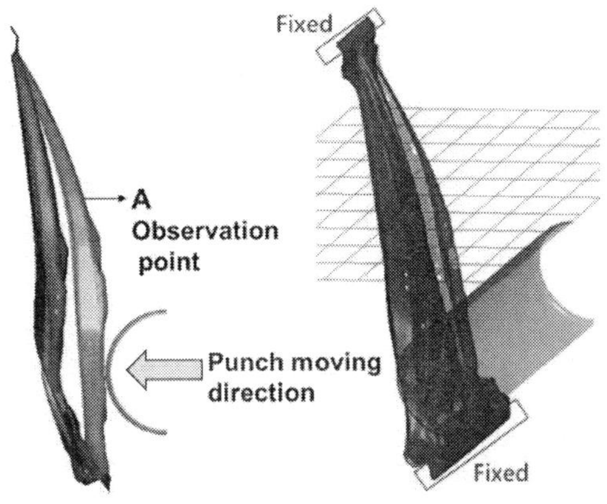

Figure 5. Schematic of intrusion CAE model

2.3 Mechanical property test and ATOS scanning

After die tryout, the tensile test specimen following ASTM E8 standard at different locations of 15B22 and 440W were cut from the hot-stamped B-pillars. The microstructure of the weld line is also examined by light optical microscope. Finally, ATOS 3D scanner (Compact Scan 8M) is used to measure the size deviation of hot-stamped B-pillars.

3. Results and Discussions

3.1 Hot stamping parts

Figure 6 shows the hot stamped TWB pillars with the soft zone 110 mm and 260 mm. A clear color difference is observed between the hard zone 15B22 and the soft zone 440W. During hot stamping, the blank holder first clamps the sheet's flange then the punch moves down to close the die. The weld line is under tangential and normal tension force during the drawing process. No crack is observed for the weld lines even on the corner and stepwise part profiles, indicating that the weld line has a low crack sensitivity during hot stamping. This is rational because when the blank is heated to 930 °C, the material of 15B22, 440W and weld line are all become austenite and have good formability at high temperature.

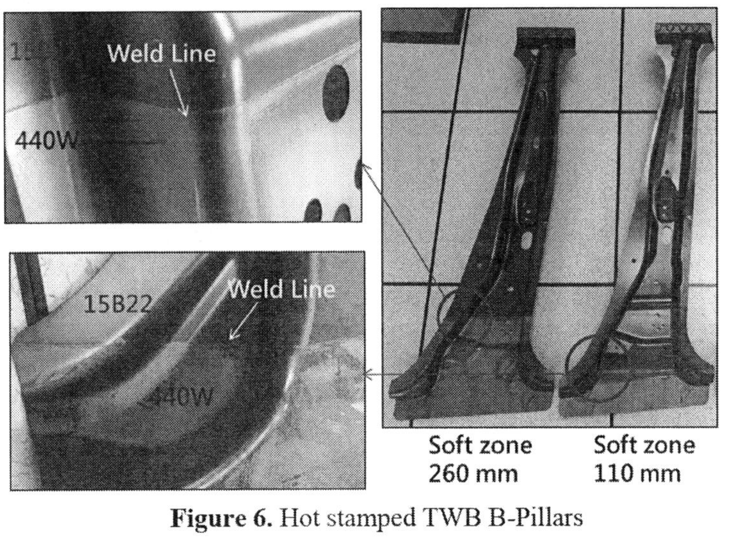

Figure 6. Hot stamped TWB B-Pillars

3.2 Material Strength

The tensile test specimen is cut from the hot stamped B-pillar as shown in Figure 7. Sample #1~#3 are located at the 440W soft zone. Sample #4~#5 are located at the flange area of hard zone 15B22, and sample #6~#7 are located on the side wall of hard zone 15B22. Two different types, laser cutting and CNC milling, of tensile test specimen are prepared. The tensile test results are listed in Table 3. For the laser cutting samples, the average tensile strength is 594 MPa for 440W, and 1511 MPa for 15B22. For the CNC milling samples, the average tensile strength is 590 MPa for 440W, and 1514 MPa for 15B22. No significant difference in strength is observed between the specimen prepared by laser cutting and CNC milling.

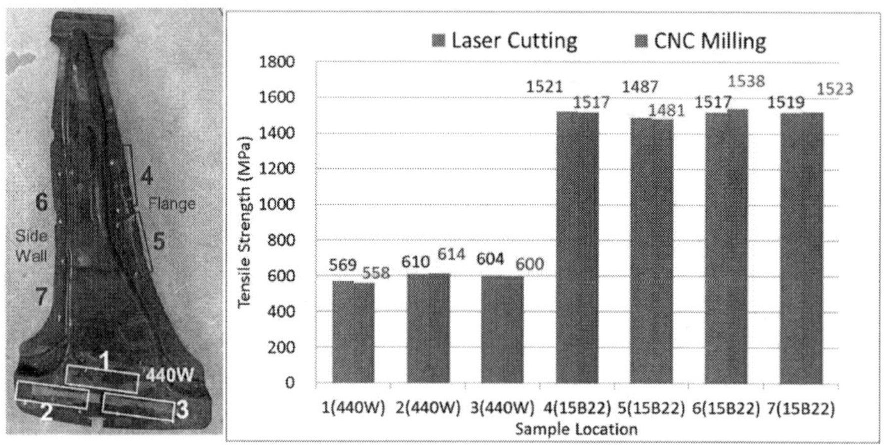

Figure 7. Tensile test results

Table 3. The average tensile strength of hot stamped 440W/15B22

Sample preparing method	Average Tensile Strength (MPa)	
	440W	15B22
Laser Cutting Material	594	1511
CNC Milling	590	1514

3.3 Three point bending CAE results

The experimental and simulation results for the three point bending test are shown in Figure 8. It can be observed that the punch's load-displacement curve is affected by the adoption of soft zone design. The monolithic, all hard zone 15B22 design, has the highest punch reaction force. For the TWB_soft zone 110 mm design, only a slightly drop in the reaction force observed because the punch is still in contact with the hard zone of the B-pillar under the punch movement. For the TWB_soft zone 260 mm design, a significant drop in the punch load is noticed because when the punch is pressing the B-pillar, the 440W soft zone will deform first due to its comparatively lower strength.

The discrepancy between the experimental and simulation results is noticed when the punch displacement exceeds 60 mm. A reasonable guess is that the B-pillar is not rigidly fixed in the jig during the test. Sliding between the B-pillar and test jig during the punch movement will affect the prediction of the B-pillar's deformation and thus affect the punch's load curve. Figure 9 shows the comparison of energy absorption capability between the simulation and experimental results under the punch movement range 0-60 mm. The CAE simulation shows a good agreement with the experimental results for all three types of B-pillar design.

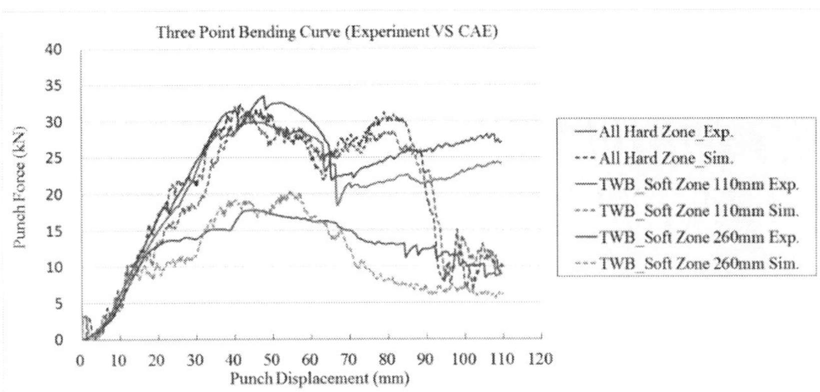

Figure 8. Experimental and CAE simulation results for the three point bending test

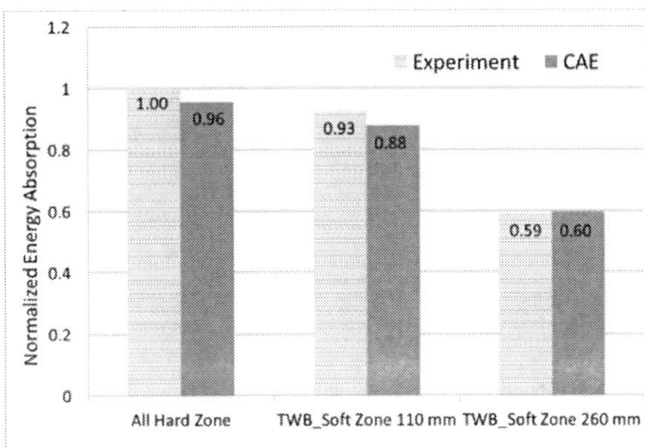

Figure 9. Comparisons for the energy absorption for the three point bending at the range of punch displacement 0-60mm

Figure 10 shows the intrusion distance at the observation point A for three different types of B-pillar design. It can be seen that the monolithic design has the maximum intrusion distance while the TWB_soft zone 260 mm design has the lowest intrusion value. As listed in Table 4, the intrusion distances for the monolithic, TWB_soft zone 110mm and TWB_soft zone 260 mm designs are 81.3, 79 and 71.9 mm respectively. A lower intrusion distance represents more safety room kept for the passenger during crash.

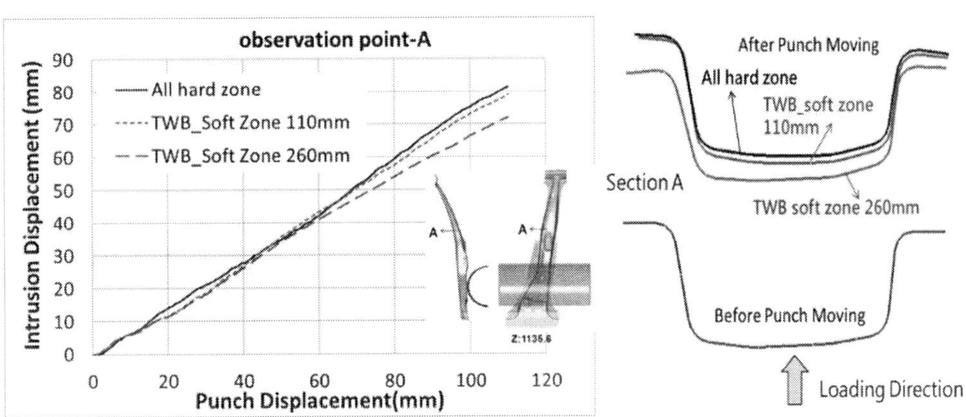

Figure 10. Simulation results for the intrusion distance and section profile at the observation point A.

IOP Conf. Series: Materials Science and Engineering **967** (2020) 012083 doi:10.1088/1757-899X/967/1/012083

Table 4. Intrusion distance of observation point A

Monolithic All hard zone (mm)	TWB Soft zone 110mm (mm)	TWB Soft zone 260mm (mm)
81.3	79	71.9

3.4 Microstructure and hardness distribution of the weld line

Figure 11 shows the microstructure of 440W and 15B22 specimen cut from the hot-stamped parts. The main microstructure of 440W is ferrite with a small portion of pearlite. The microstructure of 15B22 has become fully martensite due to its high hardenability after the die quenching process.

The micrograph and hardness distribution of the weld line is shown in Figure 12. The average hardness of 440W is 194 HV0.3, while the average hardness of 15B22 is 487 HV0.3. The width of transition zone between 440W and 15B22 is within 1 mm. Since the transition zone is very narrow, the neglecting of weld line's material property in the CAE model is acceptable. It can be noticed that the transition zone's hardness is roughly the average value of 440W and 15B22, and showing a continuous change in hardness between the base material and weld line.

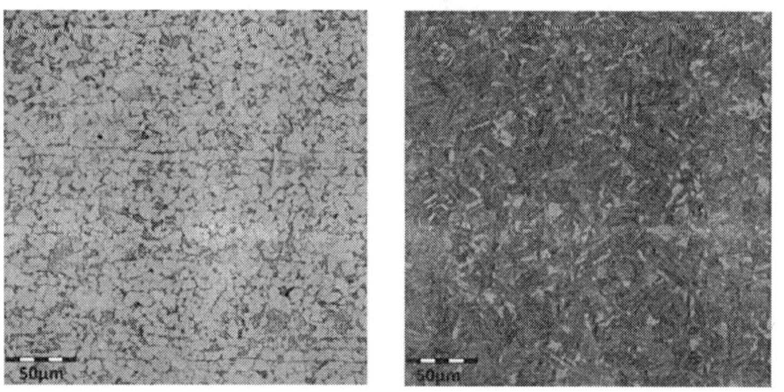

440W-mid-50X 15B22-mid-50X

Figure 11. Microstructures of hot-stamped 440W and 15B22

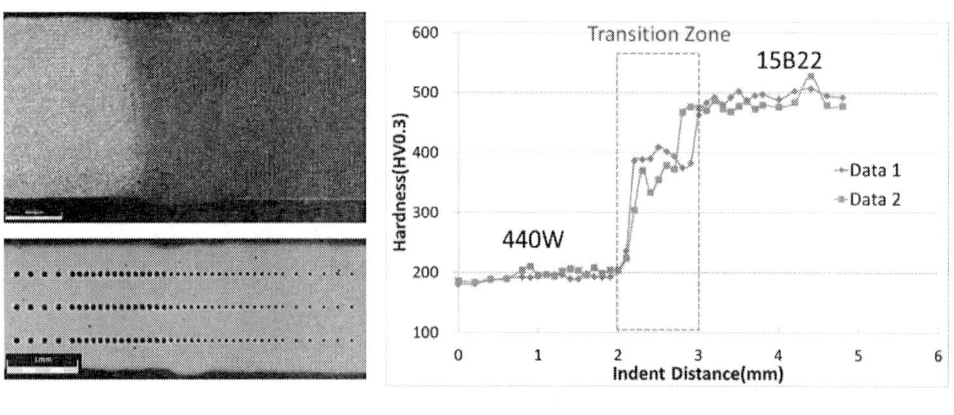

Figure 12. Micrograph of the laser weld line's cross section and hardness indent pattern. Hardness distribution profile of the weld line's cross section

3.5 ATOS 3D scanning results

Figure 13 shows the ATOS 3D scanning results for the monolithic, TWB_110 mm soft zone and TWB_260 mm soft zone's hot stamped B-pillars. The color shown in dark grey represents the area with the size deviation greater than +0.5 mm, while the color shown in light grey represents the area with the size deviation smaller than -0.5 mm.

For the monolithic B-pillar, the material is fully 15B22 so there is no boundary between the hard zone and soft zone. It can be noticed that only the upper left portion's geometry exceeds +0.5 mm size deviation. For the TWB_soft zone 110 mm design, the trend is quite similar. There is an insignificantly increase on the upper left area's size deviation (+1.08 mm). The size deviation of the top flange's middle area has a slightly increase, but still within the ±0.5 mm tolerance requirement. The 110mm soft zone is marked by the red dashed line. It is observed that no significant deterioration in the size deviation for the soft zone including the weld line's adjacent area.

For the TWB_soft zone 260 mm design, the size deviation of the soft zone is also within the ±0.5 mm tolerance requirement. No significant increase in the size deviation is observed. This is beneficial to the TWB's application, indicating that the adoption of TWB design does not noticeably deteriorate the part's dimension accuracy.

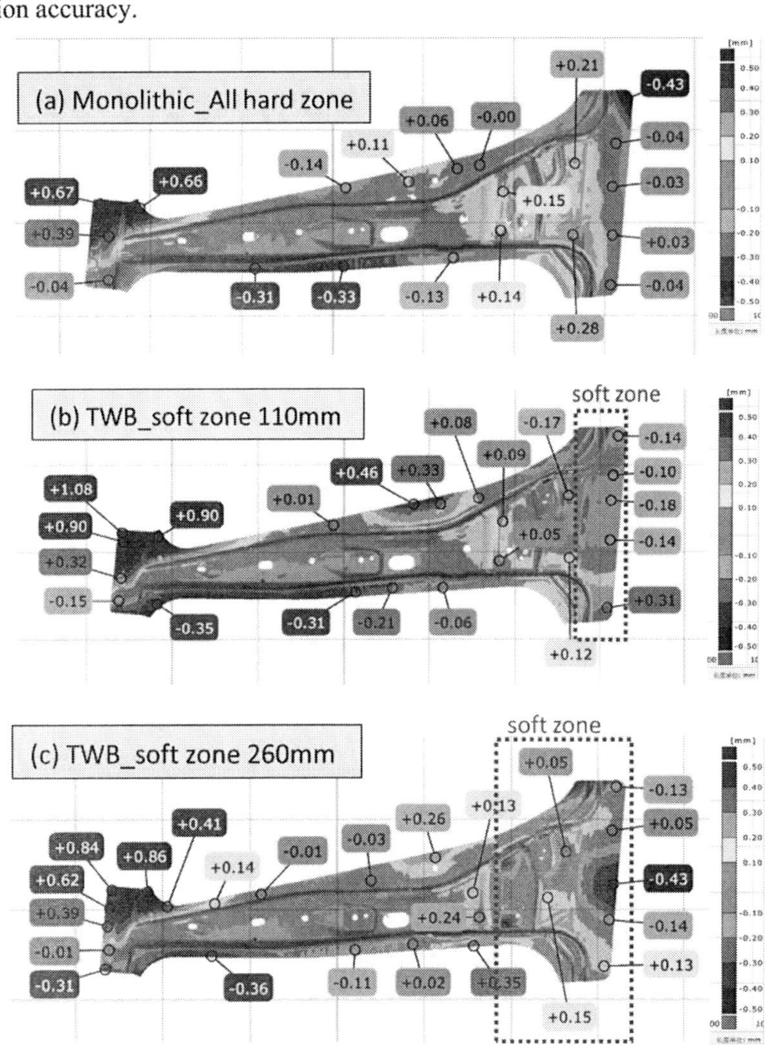

Figure 13. ATOS 3D scanning results for the (a)Monolithic material (b)TWB_soft zone 110 mm and (c)TWB_soft zone 260 mm B-pillars

4. Conclusion

Numerical and experimental investigations on the monolithic and TWB B-pillars are conducted in this study. The energy absorption behavior is strongly influenced by the adoption and the range of the soft zone. The TWB CAE model without incorporating the weld line's material property could effectively predict the punch's load-displacement curve for the three-point-bending test. The CAE results show that the adoption of TWB design decreases the intrusion distance at the passenger's chest height. The transition zone of the TWB is less than 1mm, and the die tryout shows that the weld line has a low crack sensitivity during hot stamping. The size deviation of TWB B-pillars show the same trend as the monolithic B-pillar design, and there is no significant increase in the size deviation observed in the soft zone. The adoption of TWB design does not noticeably deteriorate the part's dimension accuracy.

References

[1] Karbasian H and Tekkaya A E 2010 J. *Mater. Process. Technol.* A Review on Hot Stamping **210** p 2103–2118

[2] Taylor T and Clough A 2018 *Mater. Sci. Technol.* Critical Review of Automotive Hot-Stamped Sheet Steel from an Industrial Perspective **34** p 809–861

[3] Song Y L, Han Y, Hua L, Lu J and Yu C 2015 Optimal Design and Hot Stamping of B-Pillar Reinforcement Panel with Variable Strength based on Side Impact *Proc. of the 2nd Int. Conf. on Advanced High Strength Steel and Press Hardening,* ICHSU 2015 Changsha, China

[4] Peister C, George R, Omer K and Worswick M J 2017 Forming of an Axially Tailored Automotive Channel Section Through Hot Stamping of Tailor-Welded Blanks *Proc. of the 18th Int. Conf. on Experimental Mechanics,* ICEM 2018 Brussels, Belgium

[5] Peister C, O'Keeffe C, Imbert J and Butcher C 2017 Dynamic and Quasi-Static Testing and Modeling of Hot Stamped Tailor-Welded Axial Crush Rails *Proc. of the 36th Int. Conf. of International Deep Drawing Research Group,* IDDRG 2017 Munich, Germany

[6] Choi J W, Lee M G, Barlat F, Son H S and Nam J B 2012 *The Iron and Steel Institute of Japan Inter. J.* Hot Press Forming of Tailor Welded Blank: Experiments and FE Modeling **52** p 2059–2068

A full-field optimization approach for iterative definition of yielding for non-quadratic and free shape yield models in plane strain

Holger Hippke[a], **Bekim Berisha**[b] **and Pavel Hora**[a]
[a] Institute of Virtual Manufacturing, ETH Zurich, Tannenstrasse 3, 8092 Zurich, Switzerland
[b] inspire AG, inspire-ivp, Tannenstrasse 3, 8092 Zurich, Switzerland
E-mail: hhippke@ethz.ch

Abstract. An advanced strategy for iterative definition of initial yielding based on planar strain distribution is presented. It is shown that full-field DIC measurement of NTR5 samples provides information on initial yielding for plane strain. The interdependency of strain increment and yield locus under assumption of associated flow allows for definition of yield parameters using a non-linear optimization scheme with LS-OPT. Pivotal for research in direction of additional support points for definition of initial yielding was the discovery that definition of yielding based only on tensile and biaxial experiments is not sufficient for aluminum alloy. Special focus was placed on the area of generalized plane strain, which is the most critical stress state. Previous publications illustrated experimental options using cruciform tension and crystal plasticity as support points in generalized plane strain. This publication introduces an additional strategy to determine data for multiaxial stress states without need of additional experiments. The iterative strategy shows promising results for definition of yielding in generalized plane strain. Additionally, it is illustrated that common yield models such as non-quadratic YLD2000-2D and free-shape Vegter are sufficiently capable to describe yielding of aluminum alloy, if their full potential is exploited. The strategy is evaluated on the basis of Nakajima strain distributions and a conclusion is drawn on applicability and predictive capabilities.

1. Introduction

This publication is an extension of the work presented by Hippke et al. (2020) [1] by adding a full field optimization (FFO) based strategy for yielding definition. The underlying idea is to extend the scope of fitting strategies for yield models with the objective to increase model performance. The two experimental strategies previously published significantly improve model performance, but require large amounts of additional testing. FFO can be applied to standard geometries, reducing experimental cost to a minimum and providing an efficient alternative. Performance is evaluated using strain distributions of Nakajima specimens.
Iterative definition of parameters with FFO has become a regularly applied method in constitutive modeling. Güner at al. (2011) [2] pursued an iterative approach but was limited by computational power. Ilg et al. (2018,2019) [3, 4] focused on correct representation of the post necking range of tensile experiments using LS-Opt. A more complex application is discussed by Coppieters et al. (2018) [5], who applied FFO to a planar cruciform specimen. A great challenge of FFO is the definition of optimization problem. Especially both space and time discretization play a central role in definition of a well posed problem. Strong inhomogeneity of strain fields

may lead to a solution in local minimum, so more flexible optimization algorithms are needed. In order not be dependent on evaluation of derivatives, simulated annealing was chosen as optimization algorithm. The high number of function evaluations is of minor importance, as implicit FEM is used to reduce calculation time of an NTR5 sample to seconds. Belytschko-Tsay elements are used in simulation. The advantages and disadvantages of zero order and first order algorithms are described by Hao et al. (2019) [6].

2. Definition of reference configuration

In order to be able to evaluate performance of optimization based yield definition, a reference configuration is defined. This is achieved through tensile test and bulge experiment for YLD2000-2D, with an exponent of $m = 8$, as recommended by Barat et al. [7]. The fitted parameter set is provided in Table 2. To define a reference configuration of the Vegter yield criterion [8], additional plane strain and shear points need to be defined. This is achieved by NTR5 tensile tests and SH shear tests, using a geometry suggested by Roth and Mohr [9]. A challenge is, that only one stress is directly measured. The respective second stresses are derived using assumptions presented as part of the Vegter light model [10]: the stress ratio in plane strain is $\alpha_{ps} = f_{ps2}/f_{p12} = 0.5$ and the stress ratio in shear is $\alpha_{sh} = -1.0$. The measured stress in first principle direction is calculated as $\sigma_1 = F/A$, with the actual cross surface known through measurement of both in-plane plastic strains and volume consistency. The stress is evaluated at 1% of plastic strain and the initial yield stress in plane strain is calculated from tensile data using plastic work equivalence under the assumption of isotropic yielding. The shear stress is defined similarly as $\tau = F/A$, with the difference that the force is acting parallel to the plane of deformation. The rotation of the specimen is taken into account by measuring the angle between axis of major deformation and the loading axis. The resulting input parameters are identical with the parameters of the Vegter reference configuration in Table 3. Elastic constants are $E = 68GPa$ and $\nu = 0.33$. The hardening data in rolling direction is given by a Hockett-Sherby hardening law, as detailed in Table 1.

AA6014-T4	$A[MPa]$	$B[MPa]$	m	n	$\sigma_y = A - (A - B) \cdot e^{-m\epsilon_{pl}^n}$
$t = 1.014$mm	357.53	134.39	6.0	0.8338	

Table 1. Elastic constants and hardening parameters for AA6014-T4.

3. NTR5 samples in full-field optimization for plane strain

As optimization for the state of plane strain is desired at a minimal increase of required experimental testing, the NTR5 geometry is selected due to strain paths very close to perfect plane strain. Digital image correlation (DIC) is extracted to acquire 3D strain fields of the geometry. The software ARAMIS by GOM is used, as a direct import of DIC data is possible from ARAMIS into LS-Opt using an xml-database. In order to map DIC data to FE mesh, a set of alignment points is needed, as illustrated in Figure 1 (a) as red triangles. The error measure used in optimization is given in Equation (1). The resulting objective function is to minimize the error between measured $\bar{\epsilon}_i$ and simulated major and minor strain ϵ_i. Additionally, the forces are included and as the error values may change order of magnitude due to their respective physical definition, weighing factors w_i are introduced. The index j represents time progression and the index i represents element position. The error values are normalized with the maximum number of measured stages or times n and the number of spacial evaluation points m. In order to correlate respective strain data between simulation and measurement, a dynamic time warping (DTW) algorithm [11] is used. Great care is taken to define equivalent start and

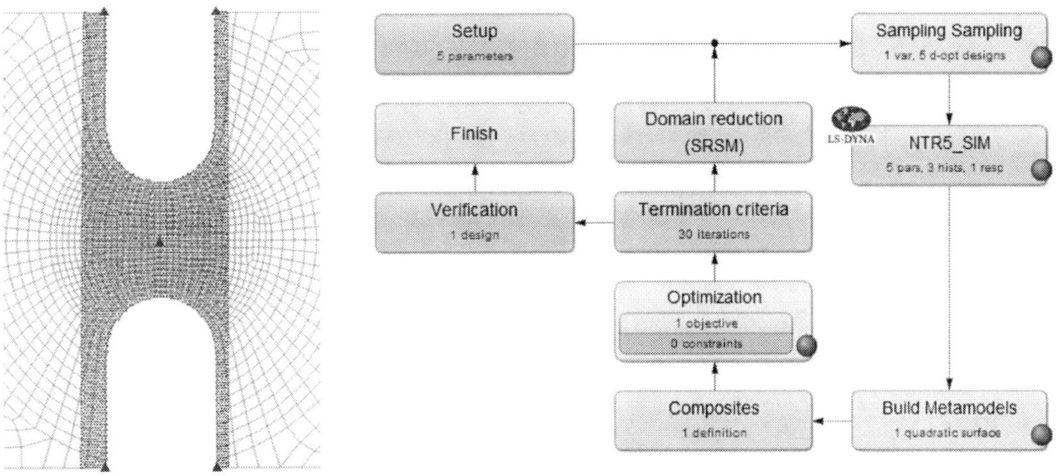

Figure 1. (a): Alignment of FE mesh and measurement data on the example of NTR5 in rolling direction. The points used for alignment are highlighted as red triangles. (b): Setup of full field optimization in LS-OPT. Strategy is similar for YLD2000 and Vegter. Only difference are additional constraints for optimization of Vegter yield criterion.

end of experiment. Both simulation and experiment are evaluated in steps of $\Delta t = 5s$ until a final time of $t_{end} = 115s$, which corresponds to a displacement of $1.35mm$.

$$f(x_i) = \frac{w_1}{n \cdot m} \sum_{j=1}^{n} \sum_{i=1}^{m} ||\epsilon_i(x) - \bar{\epsilon_i}||^2 + \frac{w_2}{n} \sum_{j=1}^{n} ||F_j(x) - \bar{F_j}||^2 \tag{1}$$

4. Optimization based definition of yielding

The mathematical formulation of an optimization problem is given in general form in Equation 2. Most commonly, an objective function $f(x)$ is formulated and then minimized, subject to additional inequality constraints $g_i(x)$ and equality constraints $h_j(x)$.

$$\begin{aligned} min(f(x)) \\ s.t. \\ g_i(x) \leq 0; i = 1, 2, ..., m \\ h_j(x) = 0; j = 1, 2, ..., l \end{aligned} \tag{2}$$

The general optimization problem given in Equation (2) is adjusted to fit the presented problem for YLD2000 as equivalent to the objective function given in Equation (1) with a constraint only for the lower and upper boundary of the exponent, $4 \leq m \leq 10$. The selected starting value is $m_0 = 6$.

For the Vegter yield criterion, the variable parameters change to the plane strain stresses in the respective direction. For the example of rolling direction, these are $f_{ps1,0}$ and $f_{ps2,0}$. Additional inequality constraints are defined in dependency of tensile and biaxial measurements. This set of constraints is necessary to guarantee stable definition of Bézier elements, as any plane strain point outside these boundaries return a non-convex yield surface. Constraints are given in Equation 3. $k \in [0, 45, 90]$ indicates the angle to rolling direction, β_k is the transformed R-value in the respective tensile direction with $\beta_k = -R_k/(1 + R_k)$. The constraints are defined in LS-Opt

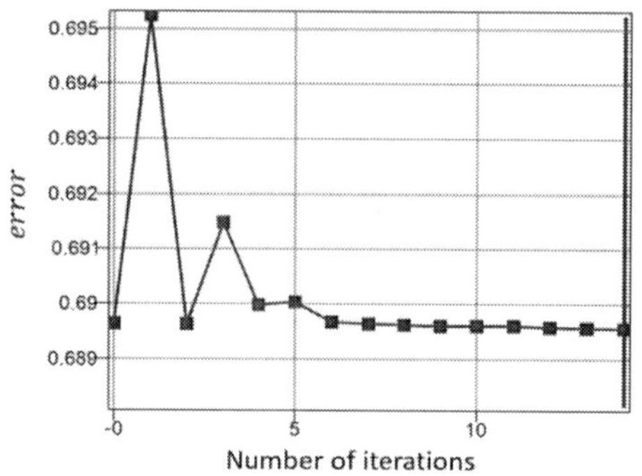

(a)

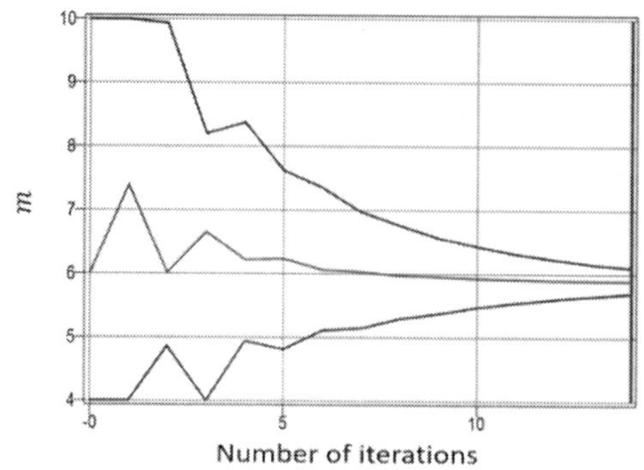

(b)

Figure 2. Convergence of full field optimization based on NTR5 sample for the error value (a) and the m-value (b).

as part of the sampling strategy.

$$\frac{f_{ps2,k}}{f_{ps1,k}-\sigma_k} \geq -\frac{\sigma_b}{|\sigma_b-\sigma_k|}$$

$$\frac{f_{ps2,k}}{\sigma_k - f_{ps1,k}} \leq \frac{1}{\beta_k} \quad (3)$$

$$\frac{f_{ps2,k}}{(R_b+1)\sigma_b - f_{ps1,k}} \leq \frac{1}{R_b} \quad (4)$$

$$f_{ps1,k} \geq \sigma_b.$$

A sequential response surface method (SRSM) is chosen with a quadratic polynomial metamodel and D-optimal sampling for optimization of YLD2000 m-value. For Vegter, the Kriging metamodel with space filling point selection performed more consistently, mostly due to the strong nonlinear interdependency of design variables and response. Adaptive simulated annealing (ASA) is used as optimization algorithm. Major and minor strain responses are compared simultaneously in agreement with Equation (1). The setup is shown in Figure 1 (b) as illustrated in LS-OPT.

4.1. Result for exponent m of YLD2000

The optimization strategy based on NTR5 specimen with the objective to minimize strain error by changing the exponent m resulted in a value of $m = 5.89$. The resulting yield locus is illustrated in Figure 5. The result is reasonable for an optimization in plane strain and illustrates that the methodology is successful. The optimization took 17 iterations to converge, successfully minimizing the resulting strain error. The evolution of the optimization is shown in Figure 2. As an additional control mechanism, the measured and simulated forces are compared for both YLD2000 and Vegter in Figure 4 (a). The difference between optimum and measured strain distributions is shown in Figure 3, illustrating a near perfect fit for major strain and only small deviations for minor strain.

4.2. Result for plane strain point of Vegter

It is necessary to evaluate the plane strain points of the Vegter yield criterion independently for each relevant angle to rolling direction. This is due to the piecewise definition of the Vegter

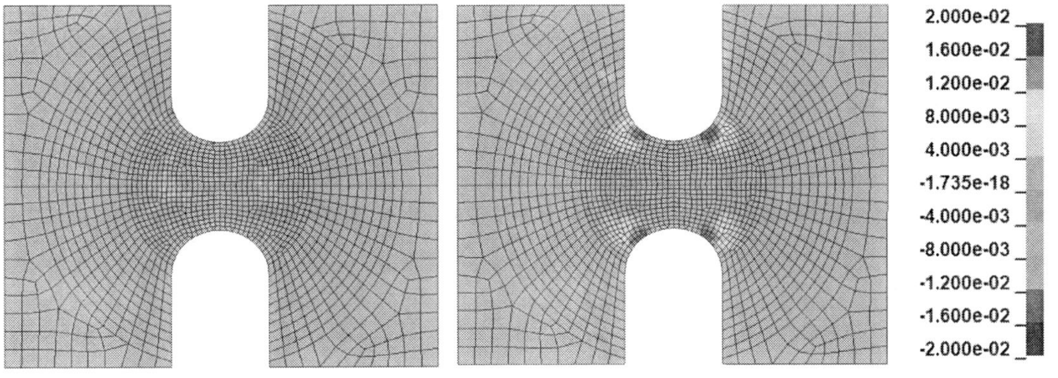

Figure 3. Resulting error distances of DTW algorithm: $\Delta\epsilon_{maj}$ (a) and $\Delta\epsilon_{min}$ (b) for YLD2000-2D optimization using NTR5 at the final state.

fit	m	α_1	α_2	α_3	α_4
		α_5	α_6	α_7	α_8
NTR5	5.89	0.9330	1.0670	0.8736	1.0500
		1.0240	1.0350	0.9173	1.2880
Reference	8	0.9218	1.1005	0.9024	1.0457
		1.0189	1.0261	0.9635	1.2346

Table 2. Parameters for YLD2000-2D yield locus fit of AA6014-T4 based on optimization strategies with NTR5 and the reference configuration.

	$f_{un,0}$	R_0	$f_{ps1,0}$	$f_{ps2,0}$	$f_{sh,0}$	f_{bi}
	$f_{un,45}$	R_{45}	$f_{ps1,45}$	$f_{ps2,45}$	$f_{sh,45}$	ρ_{bi}
	$f_{un,90}$	R_{90}	$f_{ps1,90}$	$f_{ps2,90}$	$f_{sh,90}$	
NTR5-based	1.0	0.83	1.105	0.379	0.5620	0.9900
	0.9408	0.44	1.141	0.740	0.5549	0.9860
	0.9350	0.76	1.066	0.316	0.5514	
Reference	1.0	0.83	1.0703	0.5352	0.5620	0.9900
	0.9408	0.44	1.0152	0.5076	0.5549	0.9860
	0.9350	0.76	1.0310	0.5155	0.5514	

Table 3. Parameters for Vegter yield locus fit of AA6014-T4 based on Reference case and NTR5 full field optimization. Optimization changed only plane strain parameters.

yield criterion with independent reference points for the Bézier elements in rolling direction (RD), diagonal direction (DD) and transverse direction (TD). In consequence, the optimization procedure outlined above is repeated for every direction. The optimization problem remains equivalent, with the objective function given in Equation (1) and constraints given in Equation (3). The resulting parameters are given in Table 3. To illustrate the optimization procedure, the constraints and error distribution are illustrated in Figure 4 (b) for rolling direction. The figure simultaneously displays the calculated optimum. The constraints are taken into account during sampling, successfully avoiding calculation of unreasonable combinations

IOP Conf. Series: Materials Science and Engineering **967** (2020) 012084 doi:10.1088/1757-899X/967/1/012084

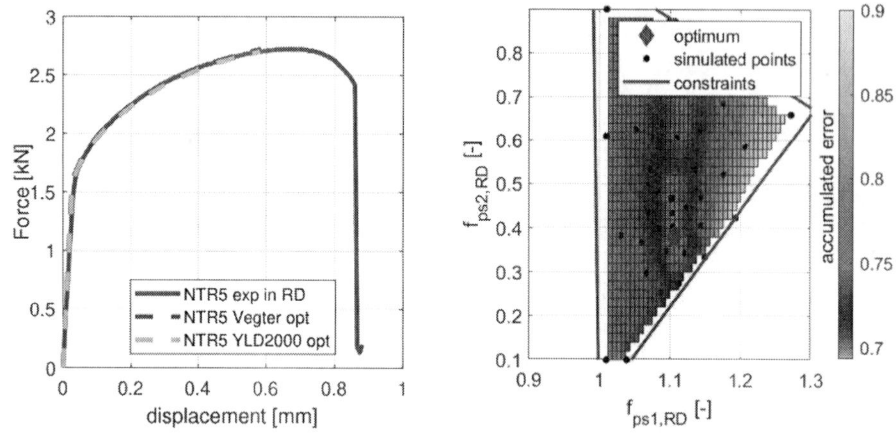

(a) (b)

Figure 4. (a) Resulting force against displacement for optimization result of YLD2000 and Vegter. Force is not part of the optimization and checked afterwards for first validation. Accuracy is sufficient for both optimization results. (b) Optimization of plane strain point with Vegter. Illustration of optimization constraints and accumulated error as well as resulting optimum in rolling direction.

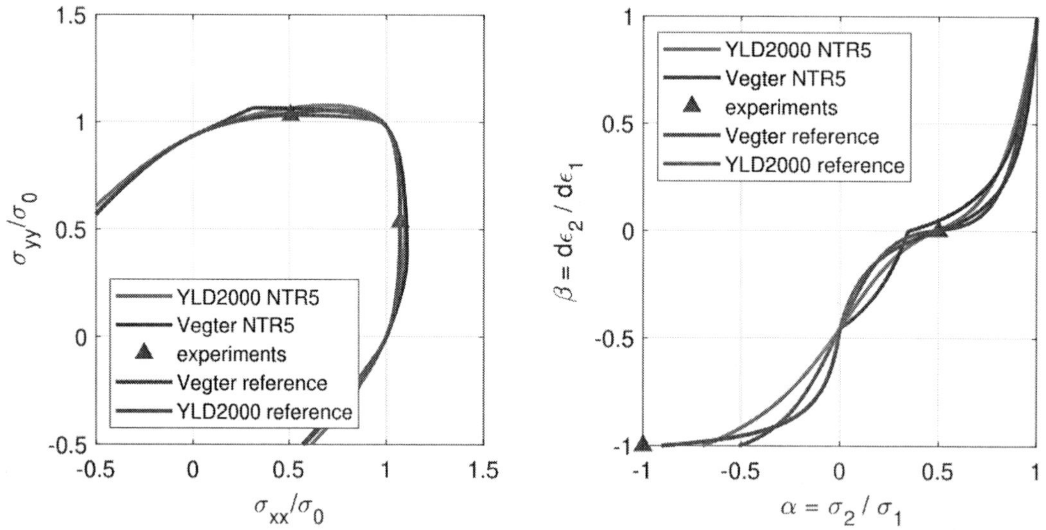

(a) (b)

Figure 5. Resulting yield locus (a) and derivative in rolling direction (b) for YLD2000 and Vegter based on NTR5 and both reference configurations.

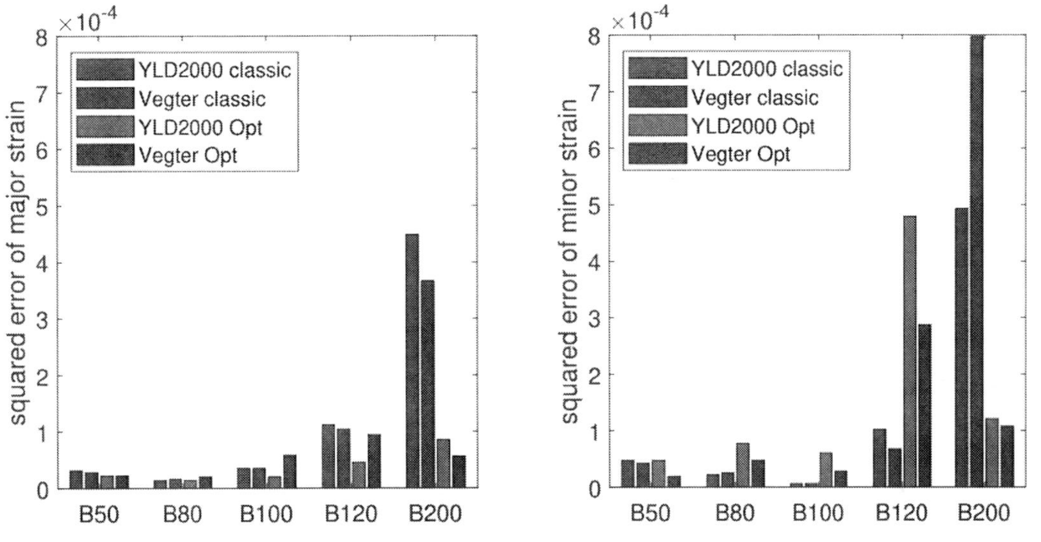

Figure 6. Calculated error of Nakajima sections for reference configurations and optimization based yielding divided into major strain (a) and minor strain (b).

5. Validation based on strain distributions of Nakajima specimen

Based on a database previously published in Hippke et al. (2020) [1], a comparison of strain distributions of Nakajima samples along central sections is made. Please refer to the cited publication for all details of the experiment. Both major and minor strain is compared in Figure 7 for B50, B80 and B100 and in Figure 8 for B120 and B200. The largest differences are apparent for B200 and the minor strain of B120. An average error is calculated to evaluate model performance. The error value is given in Equation (5), where n represent the number of measured points along a section and N represents the respective number of the evaluation height. In order not to bias the error calculation toward any geometry, $N = 3$ holds for all geometries.

$$err = \frac{1}{N \cdot n} \sum_{N} \sum_{n} (\epsilon_{sim} - \epsilon_{exp})^2 \quad (5)$$

The error is calculated separately for major strain and minor strain and is displayed in form of histograms in Figure 6. Additionally, the error values are given in Table 4, including the accumulated error under exclusion of geometry B120. This is chosen to illustrate the dominance of one geometry on the evaluated average.

6. Conclusion

Full field optimization in plane strain has been proven to provide a suitable addition to standard testing when calibrating yield surfaces. The validation on Nakajima specimen illustrates the advantage for a number of loading cases with an increase in predictive capability. The mean error of major and minor strain both show a strong advantage of the optimization based yield formulations. For minor strain, the advantage is slightly reduced by an increased error for B120 specimen. The advantage is independent of yield formulation as it holds for both YLD2000 and Vegter. In consequence, it is encouraged to include additional fitting strategies in standard procedures for definition of yielding. Two other approaches based on cruciform tension and

IOP Conf. Series: Materials Science and Engineering **967** (2020) 012084 doi:10.1088/1757-899X/967/1/012084

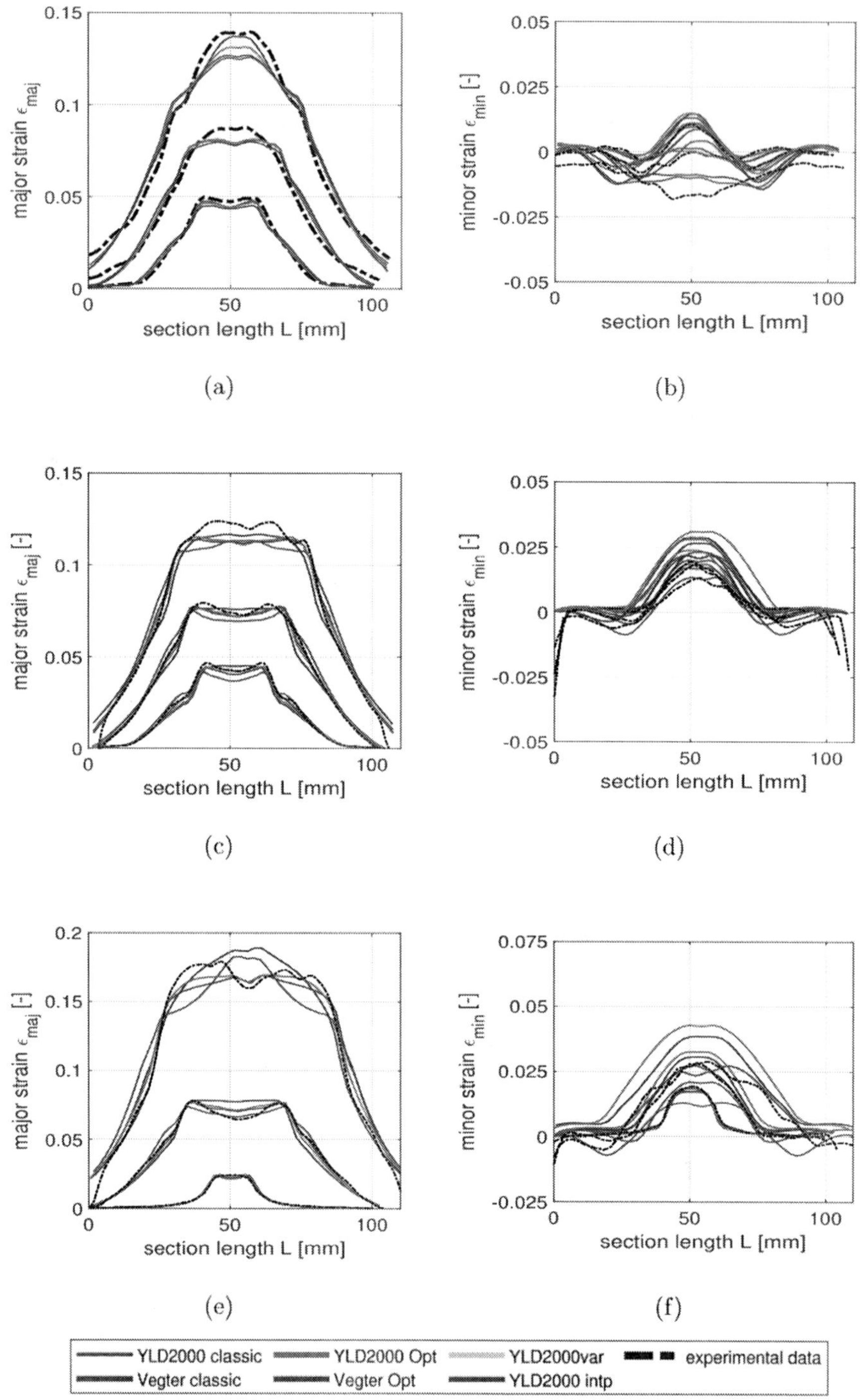

Figure 7. Resulting measurement and computation of major and minor strain distribution for the Nakajima samples B50, B80 and B100.

error [$\times 10^{-4}$]	B50	B80	B100	B120	B200	sum	sum*
of ϵ_{maj}							
YLD2000 classic	0.31	0.14	0.35	1.12	4.49	6.41	5.29
YLD2000 Opt	0.22	0.14	0.20	0.45	0.85	1.86	1.41
Vegter classic	0.28	0.16	0.35	1.12	4.49	5.50	4.38
Vegter Opt	0.22	0.20	0.58	0.94	0.56	2.50	1.56
of ϵ_{min}							
YLD2000 classic	0.47	0.22	0.06	1.02	4.93	6.70	5.68
YLD2000 Opt	0.47	0.77	0.60	4.79	1.21	7.84	3.05
Vegter classic	0.42	0.25	0.06	0.67	7.97	9.37	8.70
Vegter Opt	0.19	0.47	0.28	2.87	1.08	4.89	2.02

Table 4. Accumulated error per geometry and fitting strategy including total error per fitting strategy.

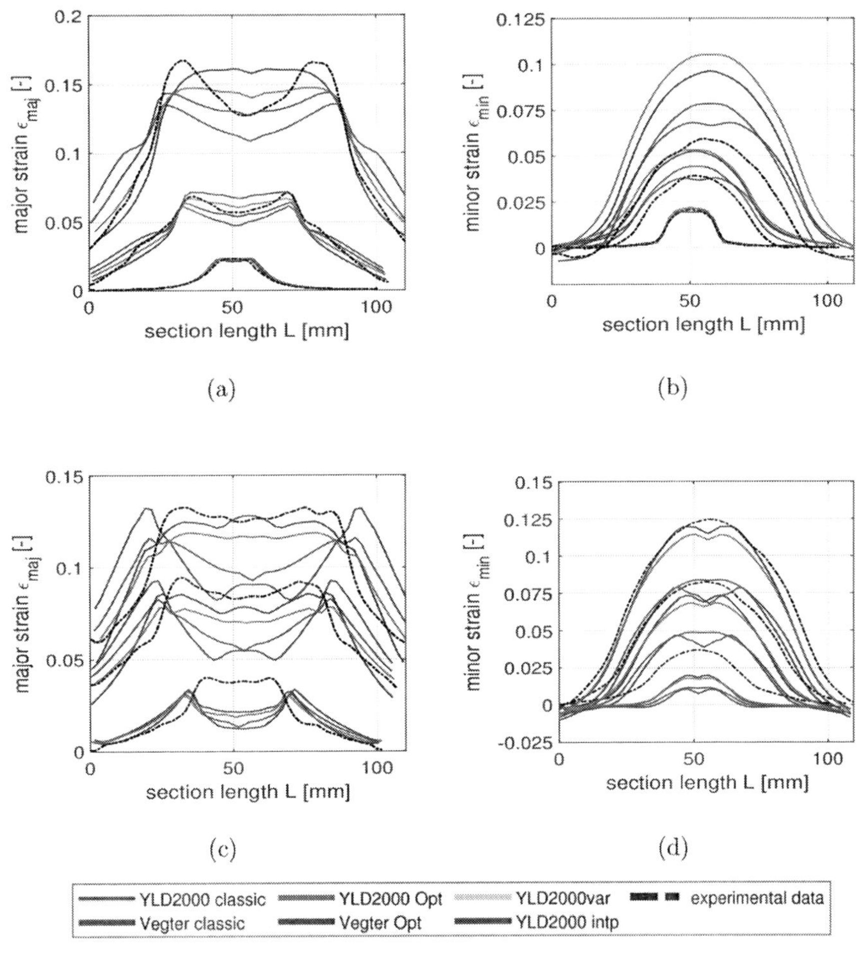

Figure 8. Resulting measurement and computation of major and minor strain distribution for the Nakajima samples B120 and B200.

crystal plasticity are discussed in Hippke et al. [1] with very similar results. This emphasises that model performance is more dependent on fitting strategy than on the complexity of the model itself. Additionally, the full field optimization strategy outlined here provides a methodology to increase precision at only very limited experimental cost. A short summary of the conclusion is given below:

- full field optimization based on NTR5 specimen provides sufficient information to significantly increase yield model performance,
- optimization strategy is suitable independent of type of yield model,
- the experimental cost is minimized when using full field optimization on standard geometries,
- precision of model prediction is increased significantly.

Future investigations address the strong influence of B120 geometry and include the shear geometry in full field optimization. Additionally, the known dependency of yield and fracture model provides an exciting field for future research.

Acknowledgements

This work was conducted within the scope of a Project (No.: 26515.1 PFIW-IW) with Innosuisse - Swiss Innovation Agency.

References

[1] Hippke H, Hirsiger S, Berisha B, Hora P. Optimized and validated prediction of plastic yielding supported by cruciform experiments and crystal plasticity. Int J Mater Form 2020;doi:10.1007/s12289-020-01569-6.

[2] Güner A, Yin Q, Soyarslan C, Brosius A, Tekkaya AE. Inverse method for identification of initial yield locus of sheet metals utilizing inhomogeneous deformation fields. Int J Mater Form 2011;4:121–128. doi:10.1007/s12289-010-1009-4.

[3] Ilg C, Haufe A, Koch D, Stander N, Witowski K, Svedin
, et al. Application of a full-field calibration concept for parameter identification of hs-steel with ls-opt. In: 15th international LS-Dyna users conference. 2018,.

[4] Ilg C, Witowski K, Koch D, Suanno P, Haufe A. Constitutive model parameter identification via full-field calibration. IOP Conference Series: Materials Science and Engineering 2019;651:012070. doi:10.1088/1757-899X/651/1/012070.

[5] Coppieters S, Hakoyama T, Debruyne D, Takahashi S, Kuwabara T. Inverse Yield Locus Identification of Sheet Metal Using a Complex Cruciform in Biaxial Tension and Digital Image Correlation. In: ICEM18; vol. 2. Proceedings; 2018, p. 382. doi:10.3390/ICEM18-05208.

[6] Hao Z, Coppieters S, Jiménez-Peña C, Debruyne D. Inverse identification of the post-necking work hardening behaviour of thick hss through full-field strain measurements during diffuse necking. Mechanics of Materials 2019;129:361 – 374. doi:10.1016/j.mechmat.2018.12.014.

[7] Barlat F, Brem J, Yoon JW, Chung K, Dick RE, Lege DJ, et al. Plane stress yield function for aluminum alloy sheets - part 1: theory. International Journal of Plasticity 2003;19:1297–1319.

[8] Vegter H, Van Den Boogaard AH. A plane stress yield function for anisotropic sheet material by interpolation of biaxial stress states. Int J Plasticity 2006;doi:10.1016/j.ijplas.2005.04.009.

[9] Roth CC, Mohr D. Determining the strain to fracture for simple shear for a wide range of sheet metals. International Journal of Mechanical Sciences 2018;149:224 – 240. doi:https://doi.org/10.1016/j.ijmecsci.2018.10.007.

[10] Vegter H, ten Horn C, Abspoel M. The vegter lite material model: simplifying advanced material modelling. Int J Mater Form 2011;4:85–92. doi:10.1007/s12289-010-1006-7.

[11] Stander N, Basudhar A, Roux W, Witowski K, Egglestone T, Goel T, et al. A design optimization and probabilistic analysis tool for the engineering analyst. Livermore Software Technology Corporation; 2019.

IOP Conf. Series: Materials Science and Engineering **967** (2020) 012085 doi:10.1088/1757-899X/967/1/012085

Study on the influence of the yield surface shape in the hole expansion test

M C Oliveira[1,*], D M Neto[1], J L Alves[2] and LF Menezes[1]

[1]CEMMPRE, Department of Mechanical Engineering, University of Coimbra, Polo II, Rua Luís Reis Santos, Pinhal de Marrocos, 3030-788 Coimbra, Portugal
[2]CMEMS, Microelectromechanical Systems Research Unit, University of Minho, Campus de Azurém, Guimarães, 4800-058, Portugal

*E-mail: marta.oliveira@dem.uc.pt

Abstract. The hole expansion test has become popular since it allows to study the formability of metallic sheets, in particular the onset of necking in stretch flanging areas. The accurate prediction of strain localization and consequent fracture requires a proper description of the plastic behavior, particularly the anisotropic yield function. In the context of strain localization prediction, the yield criterion adopted plays an important role, particularly when using an associated flow rule, since the direction of the plastic strain tensor is modelled by the normal to the yield surface. In this work, the parameters of an advanced yield criterion are calibrated considering a wide set of experimental data, which includes results from uniaxial and biaxial tension tests. This enables establishing yield surfaces with similar shape in the plane defined by the stress components in the rolling and transverse directions and a null shear component in the same plane. However, their shape changes slightly when considering non-null values for that shear component. The numerical simulations of the hole expansion test demonstrate the impact of these slight differences on the thickness strain distribution and, consequently, in the instant and location of the necking.

1. Introduction

The stress and strain distribution around a circular hole submitted to biaxial tensions has been the subject of several studies due to its relevance for industrial applications, but also because it poses challenges to the numerical prediction of the instant and location of necking occurrence [1]. In this context, the analytical solution derived for the extension of a circular hole in an infinite sheet under balanced biaxial tension [2] can be considered an interesting approximation to the stretch flanging process. This solution highlights the variation in the stress state from uniaxial tension, at the hole free edge, to plane strain in the surrounding region, attaining the balanced biaxial tension far from the edge. Although the free edge is stretched under uniaxial tension, it can generally accommodate larger deformation than the ones observed in the uniaxial tensile test. This is related with the compatibility between the edge and the surrounding material, which delays the geometrical instability observed under uniaxial tension. Moreover, the crack initiation is not always observed at the hole edge, and its occurrence inside the specimen can be related with the strong variation of the stress state in the radial direction [1]. The metallic sheets commonly used in stamping operations present orthotropic behavior as a result of the rolling operations involved in their production. This means that each radial direction around the circular hole is submitted to a stress state varying from uniaxial tension until balanced biaxial tension, with different mechanical behavior depending on the orientation to the rolling direction (RD). This indicates that the material located in the flat zone of the hole expansion tests covers a wide range of stress states located in the tension-tension quadrant of the yield surface. This contributes to the great sensitivity of the hole expansion test results from the constitutive model. Moreover, the experimental data commonly adopted to describe the material orthotropic behavior does not cover all this range of stress states, which can lead to some misinterpretations of the numerical results.

The process conditions considered in this work are the ones established for the "Benchmark 1 – Hole expansion of a high strength steel sheet" [3]. These will be described in the following section along with the finite element adopted to perform the numerical simulations. The yield criterion proposed by Cazacu et al. [4] considering two linear transformations [5], commonly designated by (CPB06ex2), is briefly presented as well as the anisotropy parameter calibration procedure adopted. In Section 3 the results are presented and discussed, in terms of stress and strains distributions.

2. Hole expansion test

The material considered is a dual phase steel (DP980) sheet (1.2 mm thickness). The circular blank has a diameter of 215 mm and a central hole with a diameter of 30 mm, as shown in Figure 1, which also presents the forming tools geometry. The interface between the blank and the punch head was lubricated, while no lubricant was applied to the interfaces between the blank and the upper/lower die. Moreover, the periphery of the blank was clamped using a draw-bead (see detail in Figure 1) and a blank-holding force of approximately 800 kN was applied [3,6].

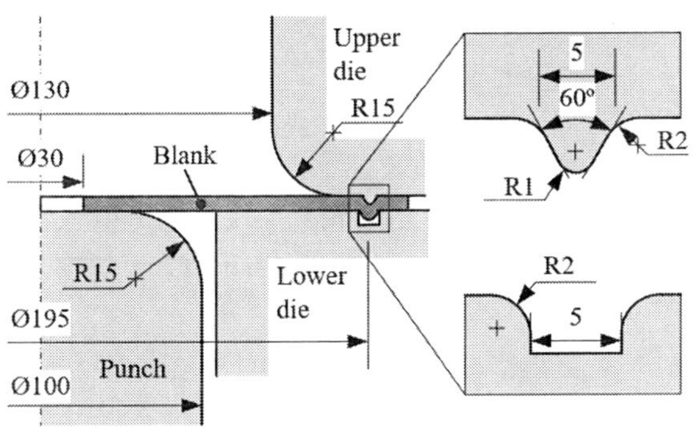

Figure 1. Schematic illustration of the tools geometry and specimen used in the hole expansion test. All dimensions are in millimeters.

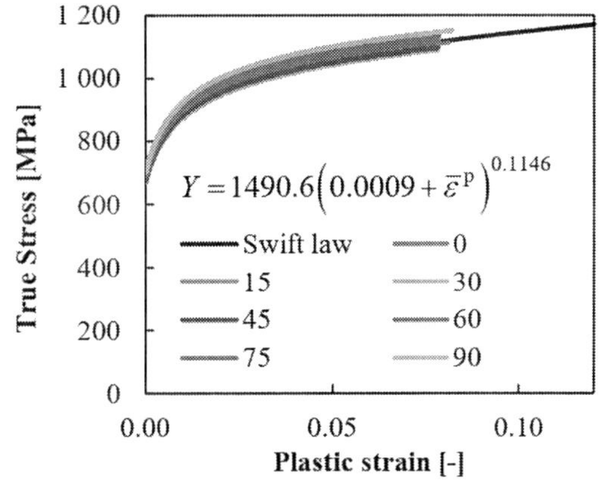

Figure 2. True stress–plastic strain curves from the uniaxial tensile tests and Swift law fitted with the result for the RD.

2.1. Finite element model

The forming tools are assumed as rigid and modelled with Nagata patches [7], including the real draw-bead geometry. Due to the previously described contact conditions, a null friction coefficient valued is adopted for the contact interface between the blank and the punch, while for the other contact zones a constant value of 0.15 was considered. A previous study indicates that the increase of the friction coefficient leads to a global decrease of the thickness strain in the flat region of the blank, which postpones the onset of necking but does not alter the location [8]. Only one-quarter model is modelled, allowing to perform the discretization of the blank with 64.800 hexahedral finite elements (3 layers through the thickness). All finite element simulations were performed with the in-house finite element code DD3IMP [9].

2.2. Mechanical Behavior

The Benchmark committee provided the uniaxial tension data for the DP980 steel, including the r-values and the yield stresses in-plane distribution, extracted from tests performed in every 15° to the

RD. Moreover, biaxial tension tests with cruciform specimens were performed to provide the directions of the plastic strain rates for the first quadrant of the yield loci, in the plane defined by the stress components in the RD and transverse direction (TD). The set of experimental data used in this work corresponds to the one reporter for a plastic work per unit volume of 0.01 [3].

The elastic behavior is considered isotropic and described by the generalized Hooke law with an elastic modulus of 210 GPa and a Poisson ratio of 0.3. The isotropic hardening behavior is described by the Swift law. The parameters of this law were determines based on the fitting of the stress–strain curve obtained from the uniaxial tensile test along RD, as shown in Figure 2, which also presents the comparison with the results for the other tensile tests. The elastoplastic model considers an associated flow rule with yielding described by the criterion briefly presented in the following section.

2.2.1. CPB06ex2 yield criterion. Cazacu et al. [4] proposed an isotropic yield criterion that allows the description of the strength-differential (SD) effect and extended it to orthotropy using the linear transformation of the deviatoric stress tensor $\boldsymbol{\sigma}'$, proposed by Barlat et al. [10], such that $\mathbf{s}=\mathbf{C}:\boldsymbol{\sigma}'$, where $\mathbf{C}$ is the constant 4th-order transformation tensor. To increase the number of anisotropy parameters in the formulation, instead of one linear transformation, two linear transformations were considered by Plunkett et al. [5], enabling an improved representation of the anisotropy yield surface. Thus, it also considers $\mathbf{s}'=\mathbf{C}':\boldsymbol{\sigma}'$, where $\mathbf{C}'$ is the constant 4th-order transformation tensor associated with the second linear transformation. The orthotropic form of this yield criterion is defined as

$$Y=B\begin{bmatrix}\left(|s_1|-k\,s_1\right)^a+\left(|s_2|-k\,s_2\right)^a+\left(|s_3|-k\,s_3\right)^a\\+\left(|s'_1|-k'\,s'_1\right)^a+\left(|s'_2|-k'\,s'_2\right)^a+\left(|s'_3|-k'\,s'_3\right)^a\end{bmatrix}^{\frac{1}{a}}, \tag{1}$$

where Y is the yield stress, the exponent a is a positive integer and k and k' are the material parameters that enables the description of the SD effect. In this equation, s_1, s_2 and s_3 correspond to the eigenvalues components of the tensor $\mathbf{s}$, while s'_1, s'_2 and s'_3 are the ones associated with $\mathbf{s}'$. For 3D stress conditions, assuming the Voigt notation, the constant transformation tensors involve each one 9 anisotropy coefficients and can be expressed in the principal axis of anisotropy as

$$\mathbf{C}=\begin{bmatrix}C_{11}&C_{12}&C_{13}&0&0&0\\C_{12}&C_{22}&C_{23}&0&0&0\\C_{13}&C_{23}&C_{33}&0&0&0\\0&0&0&C_{44}&0&0\\0&0&0&0&C_{55}&0\\0&0&0&0&0&C_{66}\end{bmatrix}\text{ and }\mathbf{C}'=\begin{bmatrix}C'_{11}&C'_{12}&C'_{13}&0&0&0\\C'_{12}&C'_{22}&C'_{23}&0&0&0\\C'_{13}&C'_{23}&C'_{33}&0&0&0\\0&0&0&C'_{44}&0&0\\0&0&0&0&C'_{55}&0\\0&0&0&0&0&C'_{66}\end{bmatrix}. \tag{2}$$

B is a constant defined such that Y reduces to the tensile yield stress in the RD, and is given by

$$B=\left[\frac{1}{f(\phi)+f(\phi')}\right]^{\frac{1}{a}}\text{ with }\begin{array}{l}f(\phi_i)=\left(|\phi_1|-k\,\phi_1\right)^a+\left(|\phi_2|-k\,\phi_2\right)^a+\left(|\phi_3|-k\,\phi_3\right)^a\\f(\phi'_i)=\left(|\phi'_1|-k'\,\phi'_1\right)^a+\left(|\phi'_2|-k'\,\phi'_2\right)^a+\left(|\phi'_3|-k'\,\phi'_3\right)^a\end{array}, \tag{3}$$

and

$$\begin{array}{l}\phi_i=(2/3)C_{i1}-(1/3)C_{i2}-(1/3)C_{i3}\\\phi'_i=(2/3)C'_{i1}-(1/3)C'_{i2}-(1/3)C'_{i3}\end{array}\text{ for } i=1,2,3, \tag{4}$$

The convexity is guaranteed for $a\geq 2$, $k\in[-1,1]$ and $k'\in[-1,1]$. The analysis of equation (2) would indicate a total of 18 anisotropy parameters. However, due to the homogeneity in stresses of the yield function given by equation (1) it is recommended to set $C_{11} = 1.0$. Therefore, for 3-D stress conditions this orthotropic yield criterion involves 17 anisotropy coefficients, the homogeneity parameter a, and 2 parameters associated with strength differential effects in plastic flow (k and k').

2.2.2. Identification procedure for the anisotropy parameters. Taking into account that there are no experimental results available to characterize any possible SD effect, k and k' are set equal to zero. Moreover, since it is difficult to measure the out-of-plane properties of the sheets, the parameters C_{44}, C_{55}, C'_{44} and C'_{55} are assumed to be isotropic, i.e. equal to 1.0. Therefore, it is necessary to identify 13 anisotropy parameters and the homogeneity parameter a. Taking into account the fact that the experimental database is constituted by 14 values extracted from uniaxial tensile tests and 9 values from the biaxial tension tests, the anisotropy parameters were determined using a downhill simplex method to minimize the following objective function:

$$F(\mathbf{A}) = w_{\sigma_\theta} \sum_\theta \left(\sigma_\theta \left(\mathbf{A}, \bar{\varepsilon}^{\mathrm{p}} \right) / \sigma_\theta \left(\bar{\varepsilon}^{\mathrm{p}} \right) - 1 \right)^2 + w_{r_\theta} \sum_\theta \left(r_\theta (\mathbf{A}) / r_\theta - 1 \right)^2 + w_{\beta_\varphi} \sum_\varphi \left(\beta_\varphi \left(\mathbf{A}, \bar{\varepsilon}^{\mathrm{p}} \right) / \beta_\varphi \left(\bar{\varepsilon}^{\mathrm{p}} \right) - 1 \right)^2 , \quad (5)$$

where $\mathbf{A}$ is the set of anisotropy parameters, σ_θ are the yield stresses and r_θ are the anisotropy coefficients, both obtained from uniaxial tensile tests, performed with the specimen oriented at an angle θ with RD, β_φ is the direction of the plastic strain rate in biaxial tension, for a loading direction φ, and w_i are weighting coefficients. This procedure was performed considering different even integer values for the homogeneity parameter a, trying to assure a similar description of the trends of the experimental values, which required the use of different sets of weighting coefficients. Unfortunately, this disables any comparison of the values attained for the objective function.

Figure 3 (a) presents the comparison between the experimental and predicted in-plane distribution of the r-values, showing that the maximum error occurs for the two lower values of a, at RD, attaining a maximum value of ~11% for a=4. For the other values of a, the maximum error is typically attained at TD and it is always lower than 3%. The comparison of the in-plane distribution of the yield stresses σ_θ normalized by the yield stress value for RD, σ_0, is presented in Figure 3 (b). Also in this case the maximum error occurs for the two lower values of a, with a=2 overestimating the yield stress for TD (~9%) while for a=4 this value is underestimated by ~3%. The yield stress for TD is also underestimated for all the other values of a. The experimental and predicted directions of the plastic strain rate β_φ, in function of the loading direction φ, are shown in Figure 4 (a). As previously, the maximum error occurs for the two lower values of a, with a=2 showing a particular different trend between the equibiaxial tension and plane strain along TD ($\varphi \in [50°;70°]$) with a maximum error of ~23%. For a=4, the maximum error reduces to ~12%. For all the other values of a, the differences seem marginal. This is highlighted in Figure 4 (b) that shows the projection of the yield surfaces, given by equation (1), for all the different values of a, in the biaxial plane with stress components along RD and TD ($\sigma_{\mathrm{RD}} - \sigma_{\mathrm{TD}}$ plane), assuming a null component for the shear component in the same plane (τ_{RDTD}). The stress components are also normalized with σ_0, to facilitate the comparison. Moreover, the fourth quadrant of the yield surface (tension-compression) is also shown to highlight that the differences in the yield surface shape are also marginal, except for the sets of parameters obtained with the two lower values of a. The β_φ values were optimized for the points represented with the diamonds. The points represented with the circles correspond to the yield stresses obtained from the uniaxial tensile tests. Therefore, the projection of the yield surfaces are also plotted assuming shear component τ_{RDTD} values equal to $0.25\sigma_0$, $0.4(3)\sigma_0$ and $0.5\sigma_0$, to enable the comparison with the tests performed at 15° and 75°, 30° and 60° and 45° with RD, respectively. These yield surfaces are not presented for a=2, since they are homothetic in relation to that obtained for a null shear components, i.e. all present the same evolution for the directions of the plastic strain rate β_φ. For the sets of parameters obtained with other values of a, the fact that the projections obtained for equal τ_{RDTD} values show slight differences indicates that the directions of the plastic strain rate β_φ will also differ when the stress state varies from uniaxial tension until balanced biaxial tension, along each orientation in relation to RD. In fact, they will only be equal to the ones presented in Figure 4 for a=2.

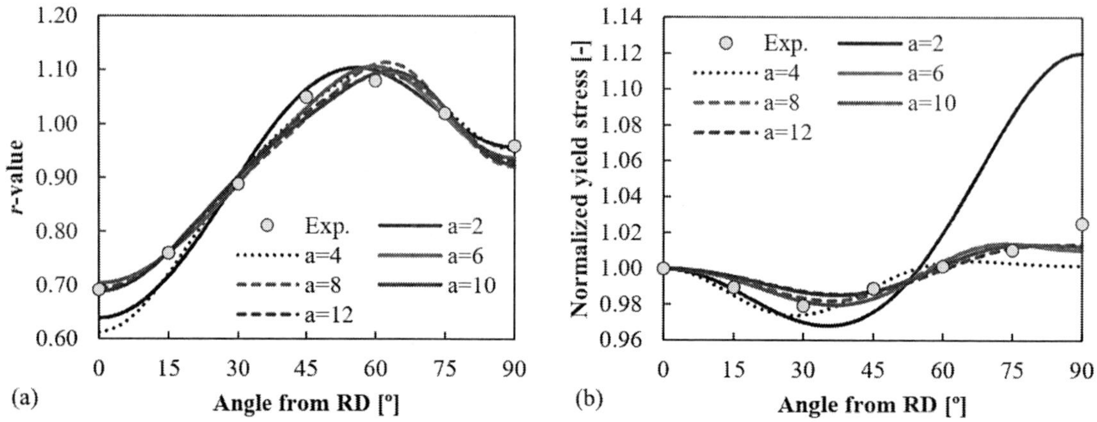

Figure 3. Experimental and predicted: (a) r_θ values and (b) σ_θ values.

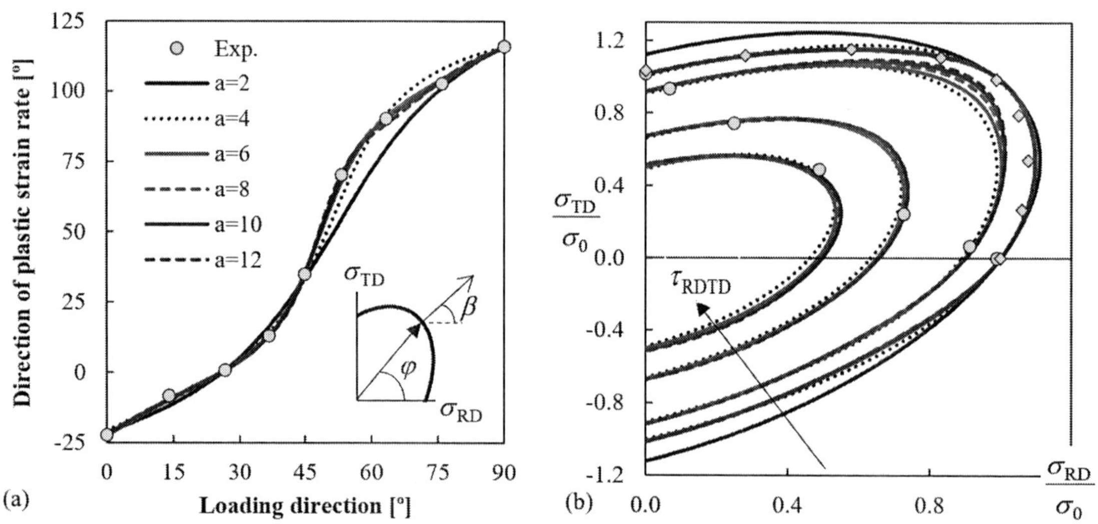

Figure 4. (a) Experimental and predicted β_φ values and (b) normalized yield surfaces in the $\sigma_{RD}-\sigma_{TD}$ plane, with null values for all other stress components, for all values of a. Except for a=2, the yield surfaces are also shown for τ_{RDTD} equal to $0.25\,\sigma_0$, $0.4(3)\ \sigma_0$ and $0.5\,\sigma_0$.

3. Results and discussion

The numerically predicted punch force evolution with its displacement is shown in Figure 5 (a), for the different sets of anisotropy parameters, showing that it mainly affects the instant the maximum force is attained. The comparison between the experimental [3] and numerical thickness strain evolutions was performed for a punch stroke of 15 mm. The results obtained along RD are shown in Figure 5 (b). For the direction at 45° with RD (labelled DD) the results are presented in Figure 6 (a). Finally, Figure 6 (b) plots the results along TD. The experimental results predict the highest decrease in thickness along the TD, followed by the RD and DD. Regarding the numerical prediction, except

IOP Conf. Series: Materials Science and Engineering **967** (2020) 012085 doi:10.1088/1757-899X/967/1/012085

for a=2, the DD is the one presenting the lowest decrease in thickness, while the RD presents the highest. In fact, for increasing values of a, the thickness strain along RD and TD tends to become more negative, while for the DD the trend is opposite. Therefore, although the trend for the distribution at 2 mm from the hole edge is globally well predicted (except for a=2), the variation is overestimated, as shown in Figure 7 (a). This seems to be related with the trend predicted for the radial displacement at 2 mm from the hole edge, as shown in Figure 7 (b), which is also clearly different for the set of parameters identified with the lower values of a. Note that the smaller numerical differences in the thickness strain distribution are observer for TD (see Figure 6 (b)), which mechanical behavior is dictated by the mechanical properties along the RD, due to the circumferential tensile stress component. Taking into account the similarities observed in section 2.2.2 between the yield surfaces for both RD and TD, particularly the ones obtained with $a \geq 6$, this indicates that the compatibility with the surrounding material also affects the strain distributions.

(a) 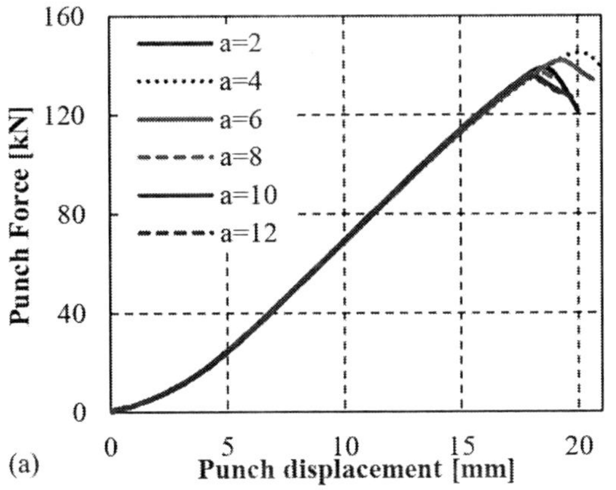

(b) 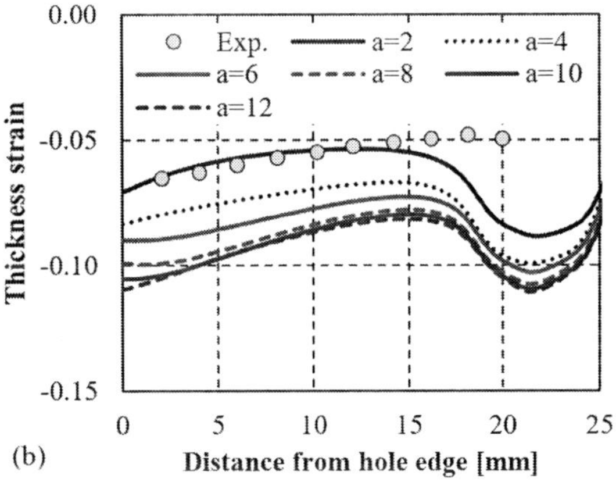

Figure 5. (a) Punch force evolution with its displacement and (b) Thickness strain distribution along the RD (stroke of 15 mm).

(a) 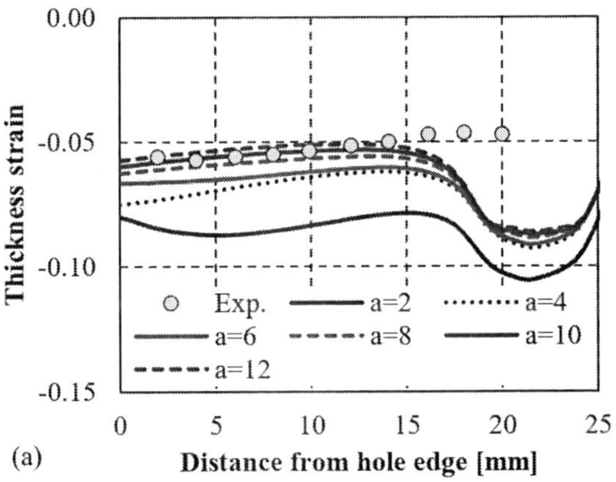

(b) 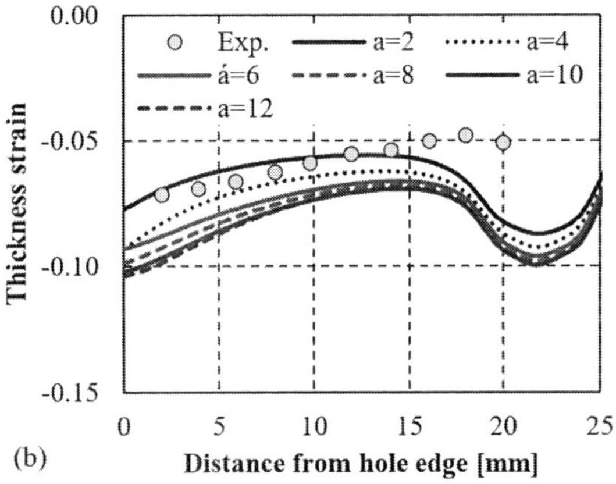

Figure 6. Thickness strain distribution along the: (a) DD and (b) TD (stroke of 15 mm).

Figure 8 (a) presents the distribution of the circumferential stress on the top surface of the blank, for a punch stroke of 15 mm. The radial stress component increases from the hole edge until the punch radius, leading to a similar behavior for the ratio between the radial and the circumferential stress components, as shown in Figure 8 (b). This ratio presents slight differences for each radial direction and evolves during the forming process. Since the Lode parameter attains the null value for a ratio of 0.5, the plane strain state is not located in the immediate surrounding to the hole edge. Although not shown here, for each point located in the flat zone, the ratio between the radial and the circumferential stress tends to decrease during the forming process. Thus, the location of the points submitted to plane strain tends to become closer to the hole edge. However, the strain localization occurs for points submitted to a stress state located between uniaxial tension and plane strain.

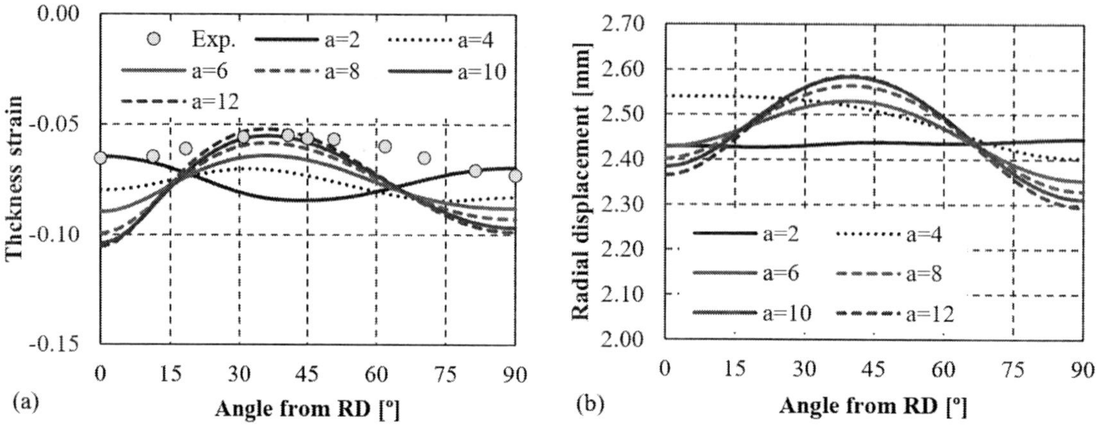

Figure 7. Distribution of: (a) thickness strain and (a) radial displacement at 2 mm from the hole edge (stroke of 15 mm).

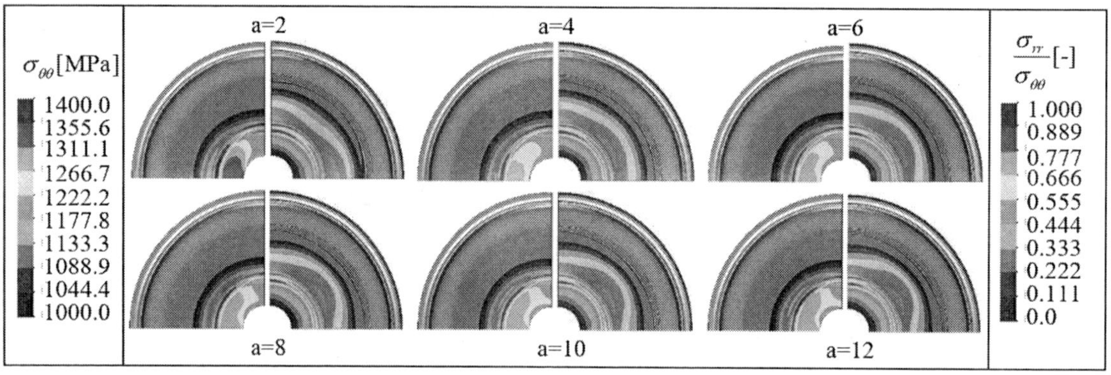

Figure 8. Distribution of the circumferential stress and of the ratio between the radial and the circumferential stress components, on the top surface of the blank (stroke of 15 mm).

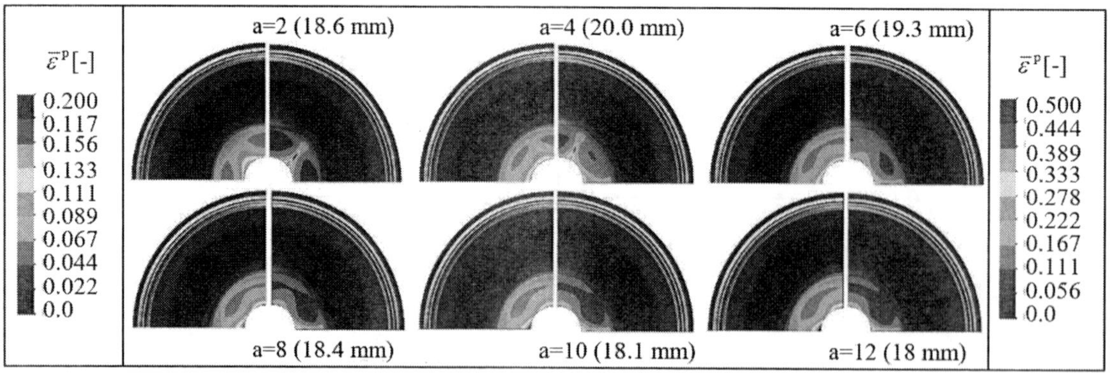

Figure 9. Equivalent plastic strain distribution on the top surface of the blank, for a punch stroke of 15 mm and close to the one corresponding to the maximum force, indicated between brackets.

Figure 9 shows the equivalent plastic strain distribution for a punch displacement of 15 mm and close to the one corresponding to the instant the maximum force is attained. For the stroke of 15 mm, the maximum values are always observed in the hole edge. However, for the set of parameters identified with the two lower values of a, the distribution presents a different trend from the others, which results in the different location for the onset of necking. For the sets of parameters with $a \geq 6$, the location is quite similar, although the stroke for which the maximum force is attained decreases with the increase of a. This is correlated with the differences observed for the thickness strain distribution for the stroke of 15 mm. The onset of necking was experimentally predicted for a stroke of 20 mm along the RD, at a distance from the hole edge of about 7.5 mm [3], which is similar to the one observed in Figure 9 for the sets of parameters with $a \geq 6$.

4. Conclusions

The analysis of the hole expansion test indicates that the flat zone is submitted to stress states from uniaxial tension at the hole free edge (circumferential direction), to balanced biaxial tension between the radial and the circumferential directions, close to the punch radius. The rotation of these stress tensors to the material frame, defined by the principal axis of anisotropy, shows that the flat zone of the hole expansion tests covers a wide range of stress states located in the tension-tension quadrant of the yield surface in the RD-TD plane, considering also different in-plane shear stress τ_{RDTD} components. Therefore, the accurate prediction of the strain distributions and, consequently, the onset of necking requires an accurate description of the yield surface shape for all this range.

Acknowledgments

The authors gratefully acknowledge the financial support of the Portuguese Foundation for Science and Technology (FCT) under projects with reference PTDC/EME-EME/30592/2017 and PTDC/EME-EME/31657/2017 and by European Regional Development Fund (ERDF) through the Portugal 2020 program and the Centro 2020 Regional Operational Programme (CENTRO-01-0145-FEDER-031657) under the project MATIS (CENTRO-01-0145-FEDER-000014) and UIDB/00285/2020.

References

[1] Lee J Y, Lee K J, Lee M G, Kuwabara T and Barlat F 2019 Numerical modeling for accurate prediction of strain localization in hole expansion of a steel sheet *Int. J. Solids Struct.* **156–157** 107–18

[2] Parmar A and Mellor P B 1978 Plastic expansion of a circular hole in sheet metal subjected to biaxial tensile stress *Int. J. Mech. Sci.* **20** 707–20

[3] Kuwabara T, Hakoyama T, Maeda T and Sekiguchi C 2018 Benchmark 1 - Hole expansion of a high strength steel sheet *The 11th International Conference and Workshop on Numerical Simulation of 3D Sheet Metal Forming Processes* ed T Kuwabara, T Hama, M Kuroda, S Takahashi and A Yamanaka (Tokyo: Nissei Eblo Inc.) pp 1–13

[4] Cazacu O, Plunkett B and Barlat F 2006 Orthotropic yield criterion for hexagonal closed packed metals *Int. J. Plast.* **22** 1171–94

[5] Plunkett B, Cazacu O and Barlat F 2008 Orthotropic yield criteria for description of the anisotropy in tension and compression of sheet metals *Int. J. Plast.* **24** 847–66

[6] Kuwabara T, Mori T, Asano M and Hakoyama T 2017 Material modeling of 6016-O and 6016-T4 aluminum alloy sheets and application to hole expansion forming simulation *Int. J. Plast.* **93** 164–86

[7] Neto D M, Oliveira M C and Menezes L F 2015 Surface Smoothing Procedures in Computational Contact Mechanics *Arch. Comput. Methods Eng.*

[8] Neto D M, Barros P D, Oliveira M C, Alves J L and Menezes L F 2018 Study of the frictional contact conditions in the hole expansion test *J. Phys. Conf. Ser.* **1063**

[9] Menezes L F and Teodosiu C 2000 Three-dimensional numerical simulation of the deep-drawing process using solid finite elements *J. Mater. Process. Technol.* **97** 100–6

[10] Barlat F, Maeda Y, Chung K, Yanagawa M, Brem J C, Hayashida Y, Lege D J, Matsui K, Murtha S J, Hattori S, Becker R C and Makosey S 1997 Yield function development for aluminum alloy sheets *J. Mech. Phys. Solids* **45** 1727–63

IOP Conf. Series: Materials Science and Engineering **967** (2020) 012086 doi:10.1088/1757-899X/967/1/012086

Heat generation when forming AHSS: experimental and numerical analysis of tensile and draw-bead tests

D M Neto[1], J R Barros[1], M C Oliveira[1], P V Antunes[1], A Ramalho[1], R L Amaral[2], A D Santos[2] and J L Alves[3] and L F Menezes[1]

[1] CEMMPRE, Department of Mechanical Engineering, University of Coimbra, Polo II, Rua Luís Reis Santos, 3030-788 Coimbra, Portugal
[2] Faculty of Engineering, University of Porto, Rua Dr. Roberto Frias,4200-465 Porto, Portugal
[3] CMEMS, Department of Mechanical Engineering, University of Minho, Campus de Azurém, 4800-058 Guimarães, Portugal

E-mail: diogo.neto@dem.uc.pt

Abstract. The increasing use of advanced high-strength steels in sheet metal forming processes requires improved knowledge concerning the thermal and contact conditions. This study presents the experimental and numerical analysis of tensile and draw-bead tests, considering a dual phase steel DP780. The temperature evolution was measured in both tests using an infrared thermal camera. The presented finite element model considers both the heat generated by plastic deformation and friction, as well as the heat loss to the environment by free convention and the contact conductance. The temperature rise in the uniaxial tensile test is accurately predicted by the numerical model. Regarding the draw-bead test, the pulling force is accurately predicted but the temperature variation is overestimated, requiring further investigation.

1. Introduction

Sheet metal forming is a manufacturing process widely used for making shaped components, from soda cans to automotive car bodies. Finished products have good quality, are geometrically accurate and are ready to be used. However, sheet metal forming processes are complex and require expensive tooling, which is economically feasible only for mass production [1]. Advanced High Strength Steels (AHSS) are replacing conventional steels in the production of sheet components, particularly in the automotive industry [2], due to their excellent weight to strength ratio. The high specific strength guarantees the safety and weight reduction requirements, assuring savings in the fuel consumption. However, AHSS present high yield strength and are susceptible to surface damage like adhesion [3], since their deformation involves higher contact pressures than those required by conventional steels.

The frictional conditions at the interface between the forming tools and the metal sheet may vary significantly during the deformation process. Besides, the frictional force depends on diverse parameters such as lubrication, normal pressure, surface roughness of the tool and sheet metal, sliding speed and temperature [4]. Thus, different tribological tests have been developed to model the friction conditions, which can be divided into two categories: (i) process tests and (ii) simulative tests [5]. Process tests comprise typical forming operations without changing the basic process kinematics, while simulative tests reproduce the tribological conditions of forming processes to study friction and/or lubrication. Most of these tests allow the indirect evaluation of the coefficient of friction based on the measurement of other physical parameters, such as strains or forces [6]. The most frequently applied simulative tests for sheet metal forming can be categorized according to the conditions arising in each region of the forming tool. The conditions in the flange region are usually represented by the strip drawing test, the draw-bead test, and the drawing with tangential compression.

The main function of the draw-beads is to increase the material flow resistance around the periphery of the part, changing the curvature of the metallic sheet passing the draw-bead. The sheet is subjected to multiple bending-unbending when passing through a draw-bead, generating a reverse tension-compression loading over the sheet thickness. The draw-bead test was developed by Nine [7] to model the friction conditions in draw-beads used in sheet metal forming processes. The draw-bead test enables to separate the deformation resistance from the frictional resistance, comparing the values of the pulling and the clamping forces when pulling the strip over fixed and rotating rollers (no frictional resistance). Thus, when the wrap angle of the middle roller is 180°, the coefficient of friction can be calculated through the difference in the drawing force obtained for the rotating and the fixed rollers.

This study explores the potential of using the draw-bead test to evaluate the mechanical and tribological behavior of an AHSS, including the heat generated by plastic deformation and friction. The equipment used was designed to enable the tests to be performed in a tensile test machine and it allows changing the draw-bead penetration depth and gap distance, as well as the pulling speed of the grip. The material considered is a dual phase steel DP780 steel with an initial thickness of 0.8 mm. The experimental tests were performed considering different values of pulling speed. The results include the evolution of the pulling force, as well as the specimen's temperature, which was acquired using an infrared thermal camera. The same procedure was used to evaluate the temperature evolution in the uniaxial tensile test. The thermomechanical models of both tests were built, taking into account the heat generated by plastic deformation and friction, as well as the heat losses to the environment by free convention and due to the contact conductance with the forming tools.

2. Experimental procedure

The uniaxial tensile test was performed along the rolling direction according to ASTM E8M [8] at room temperature with constant crosshead speed (1.3 mm/s), resulting in an initial strain rate of 1.6×10^{-2} s^{-1}. The specimen surface was coated with matt black paint (ensuring an emissivity close to 1) to improve the temperature field measurement with an infrared thermographic camera.

The draw-bead test is essentially composed by four rollers (see Figure 1), which are placed to induce plastic deformation on a sheet strip. The process is divided in three phases: (i) the strip is bended by the middle roller; (ii) the strip is pulled by a grip and (iii) the deformed strip springback. All rollers have 21 mm of diameter, while the dimensions of the sheet strip are 450×25×0.8 mm. Regarding the process parameters, the full penetration was considered (p=21.8 mm), the side clearance between rollers was c=1.55 mm and the total pulling displacement (at constant speed) was 200 mm. In order to maximize the heat generated by friction, the tests were carried out without lubrication. Figure 2 shows the equipment mounted in a tensile test apparatus. The evolution of the middle roller and grip forces was evaluated and the temperature field of the strip was measured with a thermographic infrared camera (FLIR A325), using an image resolution of 320×240 pixels matrix, at 60 frames/s.

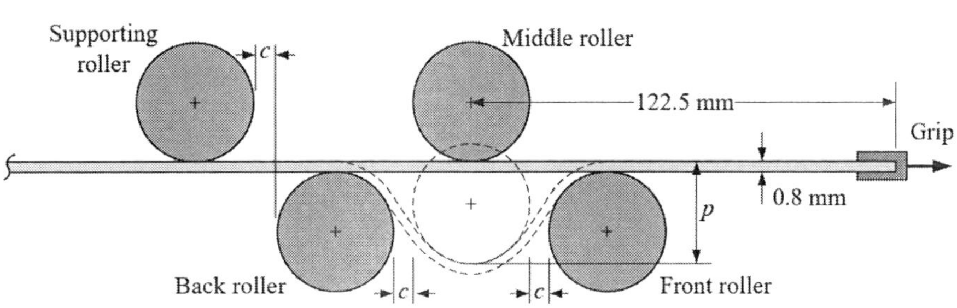

Figure 1. Scheme of the draw-bead test (initial position), composed by four rollers and the strip. The penetration is denoted by p, while the side clearance between rollers is represented by c.

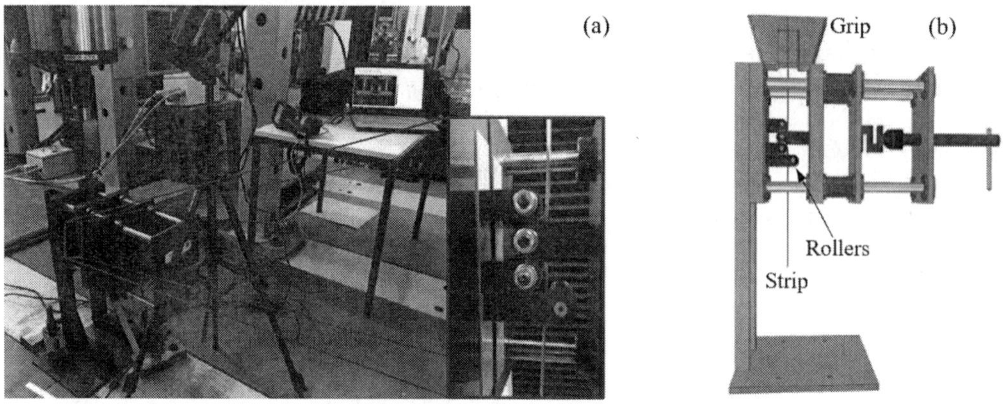

Figure 2. Setup of the draw-bead test: (a) draw-bead equipment in a uniaxial tensile apparatus for large specimen tests and infrared thermal camera; (b) 3D representation of the draw-bead equipment.

3. Numerical model

The transient thermo-mechanical analysis of both tests (uniaxial tensile and draw-bead) was carried out with the in-house implicit finite element code DD3IMP [9], which has been continuously developed for sheet metal forming simulation. The forming tools are assumed rigid and isothermal [10], while the elastoplastic behavior of the dual phase steel DP780 is assumed as temperature independent. The thermo-mechanical coupling was performed through a staggered algorithm [11], which is assumed unidirectional in this study, i.e. the mechanical behavior is independent of the temperature distribution.

The specimens are discretized with 3D linear hexahedral finite elements, using full integration in the thermal analysis and selective reduced integration in the mechanical problem. Three layers of finite elements are adopted in the thickness direction of the strip, to accurately capture the through-thickness gradients. The element size in the length direction is approximately 0.27 mm, i.e. the strip is discretized with 4167 elements. The friction between the strip and the rollers is defined by the Coulomb friction law, using a constant value for the friction coefficient, evaluated by fitting numerical and experimental force evolutions.

3.1. Thermal analysis

The differential equation that defines the thermal conduction within a solid is given by:

$$\rho c_{\mathrm{p}} \frac{\partial T}{\partial t} - k\left(\frac{\partial^2 T}{\partial x^2} + \frac{\partial^2 T}{\partial y^2} + \frac{\partial^2 T}{\partial z^2}\right) - \dot{q}_{\mathrm{p}} - \dot{Q}_{\mathrm{f}} = 0, \tag{1}$$

where ρ is the mass density, c_{p} is the specific heat capacity and k is the thermal conductivity. The thermal properties of the dual phase steel DP780 are: ρ=7900 Kg/m^3, c_{p}=450 J/(KgK) and k=50 W/(mK). The thermal power generated by plastic deformation is defined as the fraction of plastic power converted into heat, given by:

$$\dot{q}_{\mathrm{p}} = \beta \dot{w}^{\mathrm{p}} = \beta(\boldsymbol{\sigma} : \dot{\boldsymbol{\varepsilon}}^{\mathrm{p}}), \tag{2}$$

where β is the Taylor–Quinney factor, $\boldsymbol{\sigma}$ is the Cauchy stress tensor and $\dot{\boldsymbol{\varepsilon}}^{\mathrm{p}}$ is the plastic strain rate tensor. In this study, β is assumed constant and equal to 0.9 [12]. The thermal power generated by the friction forces can be expressed as:

$$\dot{Q}_{\mathrm{f}} = \eta(\mathbf{t}_{\mathrm{t}} \cdot \dot{\mathbf{g}}_{\mathrm{t}}), \tag{3}$$

where η defines the partition of heat dissipated to the strip, $\mathbf{t}_{\mathrm{t}}$ is the frictional force and $\dot{\mathbf{g}}_{\mathrm{t}}$ denotes the relative tangential slip velocity at the contact interface. In this study, the generated heat is equally partitioned between the two contacting bodies (strip and rollers), thus $\eta = 0.5$. The free convection and the contact conductance, both defined on the surface, are given by:

$$\dot{q}_{\text{conv}} = h_{\text{conv}}(T - T_{\infty}), \tag{4}$$

$$\dot{q}_{\text{c}} = h_{\text{c}}(T - T_{\text{roller}}) = h_{\text{sup}} \exp(-m\, g_{\text{n}})(T - T_{\text{roller}}), \tag{5}$$

where T_{∞} and T_{roller} are the environment and the rollers temperature, respectively. The heat transfer coefficient in free convection is denoted by h_{conv}, while the interfacial heat transfer coefficient h_{c} depends on the gap distance between the strip and the roller [13]. Thus, the h_{sup} is the upper threshold value (in contact) and m is a free parameter used to control the rate of decrease. Figure 3 presents the variation of the interfacial heat transfer coefficient with the gap distance, comparing three different values of h_{sup}. These conditions are analyzed since the heat transfer coefficient is strongly dependent on the contact pressure and, consequently, on the contact distance, as reported in several studies [14]. The heat transfer coefficient in free convection is 5 W/m²·K.

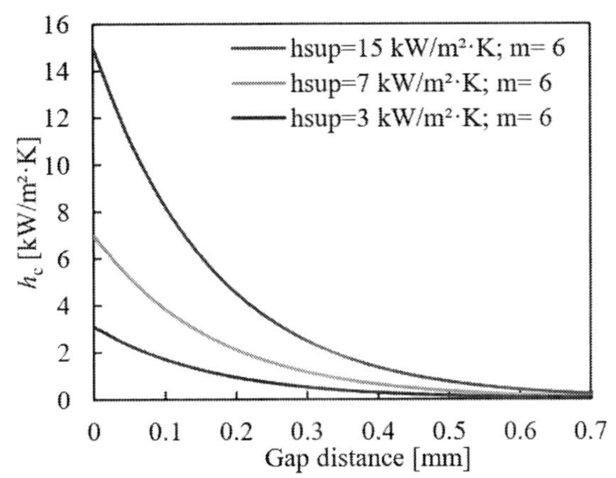

Figure 3. Variation of the interfacial heat transfer coefficient as a function of the normal distance to the roller surface, comparing three different values of upper threshold value.

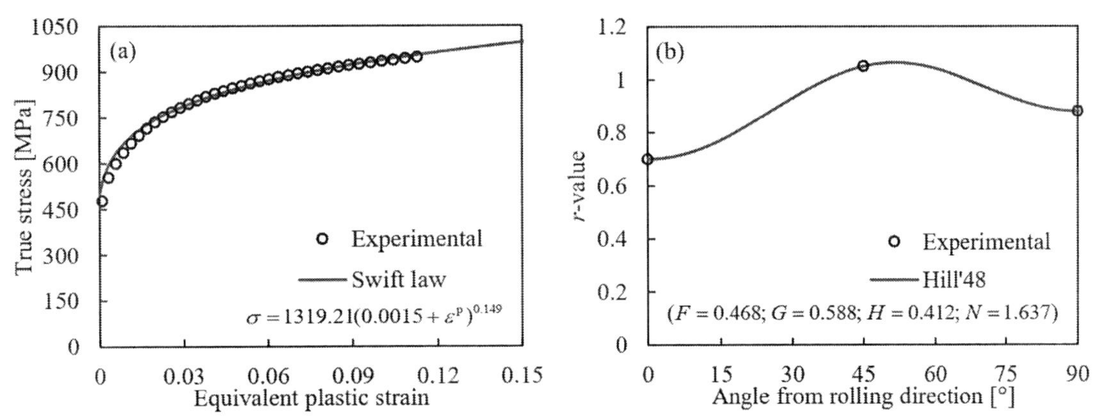

Figure 4. Mechanical behavior of the dual phase steel DP780, comparing the experimental data with the numerical model: (a) true stress–plastic strain curve from the uniaxial tensile test in the rolling direction; (b) evolution of the anisotropy coefficient (r-value) in the plane of the sheet.

3.2. Mechanical analysis

The mechanical behavior of the dual phase steel DP780 is described by an elastoplastic constitutive model. The elastic behavior is assumed isotropic, which is described by the Hooke's law, using 210 GPa for the Young's modulus and 0.30 for the Poisson's ratio. Regarding the plastic behavior, it is described by an isotropic work hardening law (Swift) and a yield criterion (Hill'48). Figure 4 presents the comparison between experimental and numerical results from uniaxial tensile tests. The parameters of the Swift law were calibrated using the stress–strain curve from the uniaxial tensile test with the specimen oriented along the rolling direction (see Figure 4 (a)). The anisotropy parameters of the Hill'48 yield criterion were calibrated using the experimental r-values, measured at every 45° from the rolling direction (see Figure 4 (b)).

4. Results and discussion

4.1. Uniaxial tensile test

The transient thermal analysis of the uniaxial tensile test was performed to validate the proposed model, namely the thermal power generated by plastic deformation and the thermal properties of the DP780 steel. The distribution of the predicted temperature variation in the specimen is presented in Figure 5 (a) for three different levels of engineering strain. The comparison between experimental and numerical evolution of the temperature variation in the midpoint of the specimen is presented in Figure 5 (b). The numerical prediction is in good agreement with the experimental measurement, presenting a linear rising up to the onset of necking. Considering the instant of onset of necking, the predicted and experimental temperature rise is about 25°C and 30°C, respectively. Although the onset of necking is numerically predicted for a higher value of strain, the rate of the temperature rise after necking is also accurately predicted (see Figure 5 (b)). Thus, the fraction of plastic power converted into heat (90%) and the heat transfer coefficient in free convection (5 W/m^2·K) were accurately selected for this example [12].

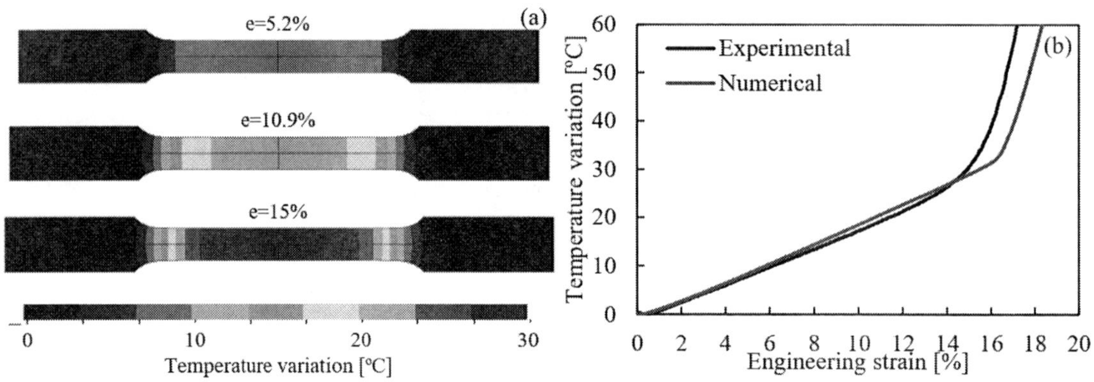

Figure 5. Temperature variation in the uniaxial tensile test considering 1.325 mm/s of crosshead speed: (a) distribution of the predicted temperature variation; (b) comparison between experimental and numerical temperature variation in the midpoint of the specimen.

4.2. Draw-bead test

The experimental tests were performed with different grades of AHSS steels to explore the potential of using the draw-bead test to evaluate the heat generated by plastic deformation and friction. However, due to some restrictions in setting the required laboratory conditions, the experiments were performed with a single sheet strip for each condition. The results are discussed here only for the DP780 considering all available data. The experimental evolution of the pulling force in the draw-bead test is presented in Figure 6 (a), comparing different values of pulling speed. After the initial sheet bending/unbending on the middle roller, the steady state of the pulling force is quickly achieved. There is no evidence that the pulling force is influenced by the pulling speed and even the 30 mm/s curve evolution lies in between the 15 and 60 mm/s curves, which indicates that the results dispersion corresponds to the level of uncertainty in the measurements. Accordingly, it is assumed that the experimental force evolution is approximately constant during the pulling operation, presenting a value of about 2.5 kN. Also, since the constitutive model adopted to describe the mechanical behavior of the dual phase steel DP780 is not strain rate sensitive (neither temperature sensitive), the predicted pulling force is independent from the pulling speed. Additionally, the influence of the friction coefficient on the numerical pulling force is presented in Figure 6 (b). Considering the frictionless condition (μ=0.0), the predicted pulling force is about 1.5 kN in the steady state regime. On the other hand, the predicted pulling force is close to 2.5 kN for μ=0.12, as highlighted in Figure 6 (b). Comparing experimental and numerical pulling force evolutions, the friction coefficient between the strip and the rollers can be accurately described by a constant value of 0.12.

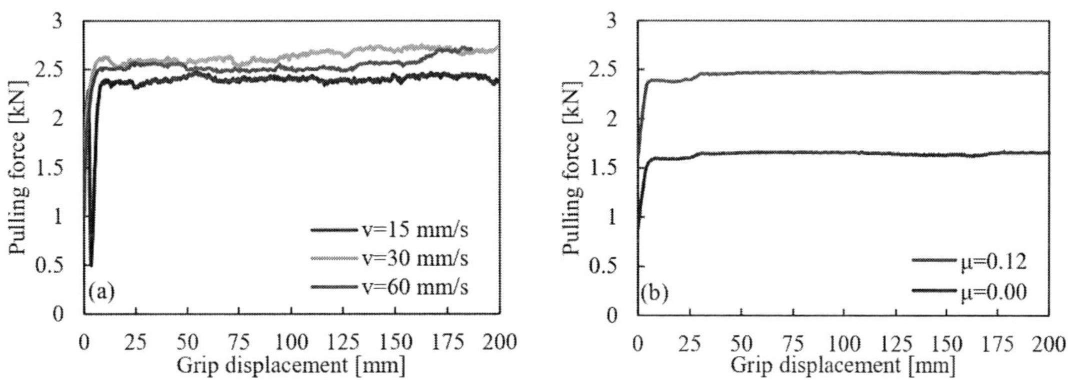

Figure 6. Evolution of the experimental and numerical pulling force: (a) influence of the pulling speed on the experimental force; (b) influence of the friction coefficient on the numerical force.

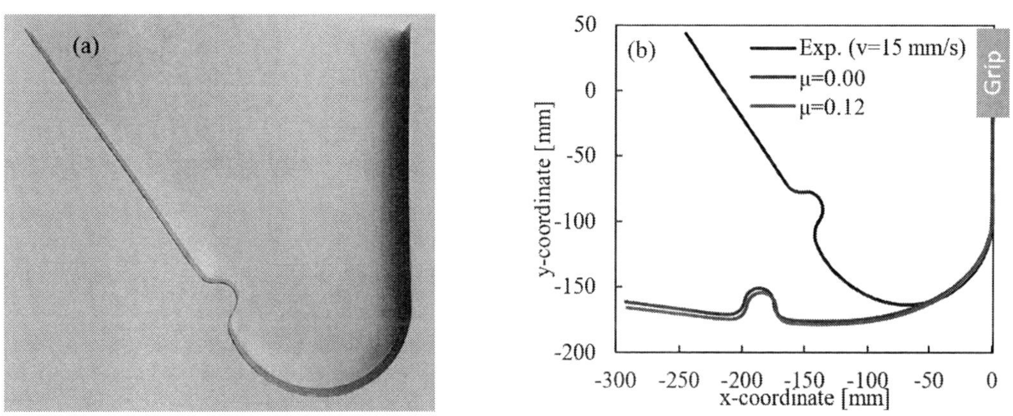

Figure 7. Profile of the strip after springback: (a) the experimental profile obtained for v=15 mm/s; (b) comparison between experimental and numerical profile.

Since the strain rate sensitive of the dual phase steel DP780 is neglected in the numerical model, the predicted final geometry is independent of the pulling velocity and therefore only the experimental condition with a pulling speed v=15 mm/s is considered. The experimental profile of the strip after springback is presented in Figure 7 (a). The curved region of the strip sheet was subjected to multiple bending-unbending induced by the draw-bead geometry. Thus, a decrease of the pulling distance leads to a reduction of the springback. Hence, the accurate assessment of the springback requires the distribution of the turning angle along the section profile [15]. The influence of the friction coefficient on the predicted profile of the strip after springback is presented in Figure 7 (b). Despite the large influence of the friction coefficient on the predicted pulling force (see Figure 6 (b)), the effect of the friction coefficient on the springback is negligible. Moreover, the springback predicted numerically is significantly lower than the one measured experimentally. This difference can be related with the adoption of an isotropic hardening law. The inclusion of the kinematic hardening can improve the springback prediction since the Bauschinger effect (reduction of the yield stress under reverse loading) is more evident in high strength steels, as well as the pre-strain dependency of the elastic modulus [16]. Indeed, the degradation of elastic stiffness due to plastic straining presents a strong impact on the predicted springback [17]. This may be the main reason why the springback is underestimated.

The thermal analysis of the draw-bead test is presented in Figure 8. The temperature variation at 75 mm ahead the middle roller is presented for three different values of pulling speed. The increase of the pulling speed leads to an increase of the temperature since the time available for the heat loss is lower. Considering the uniaxial tensile test, the influence of the interfacial heat transfer coefficient on the predicted temperature is negligible, since the temperature rise occurs far from the contact zone [12]. Nevertheless, in the draw-bead test, the heat generated by plastic deformation and frictional contact occurs near the contact zone. Thus, three different values of h_{sup} (see Figure 3) were adopted in the draw-bead simulation. The increase of the interfacial heat transfer coefficient leads to a decrease of the temperature rise, as shown in Figure 8 (a), for the lower values considered. The comparison between experimental and numerical temperature variation is presented in Figure 8 (b), considering h_{sup} =15 kW/m^2·K. Despite the large value for the interfacial heat transfer coefficient, the temperature variation is overpredicted by the numerical model, i.e. the numerical prediction of the temperature variation is at least twice the experimental value. The temperature rise is significantly larger in the uniaxial tensile test (Figure 5 (b)) than in the draw-bead test due to the heat lost by contact with the rollers. The frictionless simulation allows to conclude that most of the heat generated comes from plastic deformation.

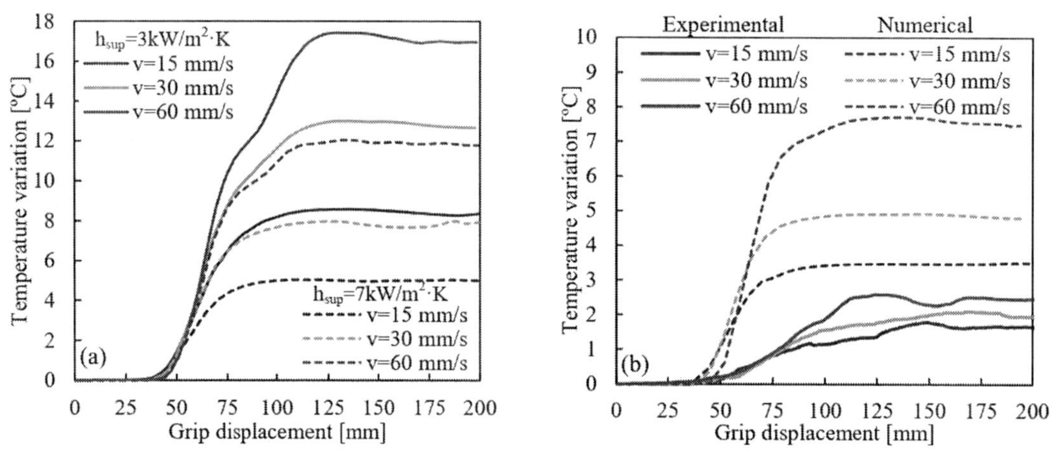

Figure 8. Temperature variation in the draw-bead test ahead the front roller considering different values of pulling speed: (a) influence of the interfacial heat transfer coefficient on the predicted temperature; (b) comparison between experimental and numerical (h_{sup}=15 kW/m^2·K) results.

5. Conclusions

This study presents the experimental and numerical analysis of tensile and draw-bead tests, using the DP780 steel with an initial thickness of 0.8 mm. In addition to the mechanical and tribological behaviour of this steel, the heat generated by plastic deformation and/or friction is evaluated experimentally and numerically. The predicted temperature rise in the uniaxial tensile test is in good agreement with the experimental measurement. Comparing the numerical and experimental results from the draw-bead test, the pulling force is accurately predicted by the numerical model, but the springback is underestimated while the temperature variation is overestimated. The temperature rise is predominantly influenced by the pulling speed, due to the time available for the occurrence of heat losses to the environment by free convention and the contact conductance with the rollers. Thus, the predicted temperature variation is significantly affected by the adopted interfacial heat transfer coefficient.

Acknowledgments
The authors gratefully acknowledge Prof. Joaquim G. Mendes for the availability of IR thermal equipment used on this research. This work was funded by the Portuguese Foundation for Science and Technology (FCT) under projects with reference PTDC/EME-EME/30592/2017 and PTDC/EME-EME/31657/2017 and by European Regional Development Fund through the Portugal 2020 program and the Centro 2020 Regional Operational Programme (CENTRO-01-0145-FEDER-031657) under the project MATIS (CENTRO-01-0145-FEDER-000014) and UIDB/00285/2020.

References
[1] Ingarao G, Di Lorenzo R and Micari F 2011 Sustainability issues in sheet metal forming processes: an overview *J. Clean. Prod.* **19** 337–47
[2] Lesch C, Kwiaton N and Klose F B 2017 Advanced High Strength Steels (AHSS) for Automotive Applications − Tailored Properties by Smart Microstructural Adjustments *steel Res. Int.* **88** 1700210
[3] Karlsson P, Gåård A, Krakhmalev P and Bergström J 2012 Galling resistance and wear mechanisms for cold-work tool steels in lubricated sliding against high strength stainless steel sheets *Wear* **286–287** 92–7
[4] Bay N, Azushima A, Groche P, Ishibashi I, Merklein M, Morishita M, Nakamura T, Schmid S and Yoshida M 2010 Environmentally benign tribo-systems for metal forming *CIRP Ann.* **59** 760–80
[5] Bay N, Olsson D D and Andreasen J L 2008 Lubricant test methods for sheet metal forming *Tribol. Int.* **41** 844–53
[6] Trzepiecinski T and Lemu H G 2019 Recent Developments and Trends in the Friction Testing for Conventional Sheet Metal Forming and Incremental Sheet Forming *Metals (Basel).* **10** 47
[7] Nine H D 1978 Drawbead Forces in Sheet Metal Forming *Mechanics of Sheet Metal Forming* (Boston, MA: Springer US) pp 179–211
[8] ASTM E8 / E8M 2016 Standard Test Methods for Tension Testing of Metallic Materials
[9] Menezes L F and Teodosiu C 2000 Three-dimensional numerical simulation of the deep-drawing process using solid finite elements *J. Mater. Process. Technol.* **97** 100–6
[10] Neto D M, Oliveira M C, Menezes L F and Alves J L 2014 Applying Nagata patches to smooth discretized surfaces used in 3D frictional contact problems *Comput. Methods Appl. Mech. Eng.* **271** 296–320
[11] Martins J M P, Neto D M, Alves J L, Oliveira M C, Laurent H, Andrade-Campos A and Menezes L F 2017 A new staggered algorithm for thermomechanical coupled problems *Int. J. Solids Struct.* **122–123** 42–58
[12] Neto D M, Simões V M, Oliveira M C, Alves J L, Laurent H, Oudriss A and Menezes L F 2020 Experimental and numerical analysis of the heat generated by plastic deformation in quasi-static uniaxial tensile tests *Mech. Mater.* **146** 103398
[13] Martins J M, Neto D M, Alves J L, Oliveira M C and Menezes L F 2016 Numerical modeling of the thermal contact in metal forming processes *Int. J. Adv. Manuf. Technol.* **87** 1797–811
[14] Omer K, Butcher C and Worswick M 2020 Characterization of heat transfer coefficient for non-isothermal elevated temperature forming of metal alloys *Int. J. Mater. Form.* **13** 177–201
[15] Neto D M, Oliveira M C, Santos A D, Alves J L and Menezes L F 2017 Influence of boundary conditions on the prediction of springback and wrinkling in sheet metal forming *Int. J. Mech. Sci.* **122** 244–54
[16] Chongthairungruang B, Uthaisangsuk V, Suranuntchai S and Jirathearanat S 2013 Springback prediction in sheet metal forming of high strength steels *Mater. Des.* **50** 253–66
[17] Eggertsen P-A and Mattiasson K 2010 On constitutive modeling for springback analysis *Int. J. Mech. Sci.* **52** 804–18

IOP Conf. Series: Materials Science and Engineering **967** (2020) 012087 doi:10.1088/1757-899X/967/1/012087

Forming limit of sheet metal in cylindrical deep drawing with a conical die

Luo Xin[1], Evsyukov C A[1] and Yu Zhongqi[2]

[1]Bauman Moscow State Technical University, ul. Baumanskaya 2-ya, 5/1, Moscow 105005, Russia
[2]Shanghai Key Lab of Digital Manufacturing for Thin-Wall Structure, Shanghai Jiao Tong University, Shanghai 200240, China

Abstract: It is well-known that the use of a conical die in cylindrical deep drawing can enhance the formability compared with conventional process. The most commonly observed failure mode in the process is fracture near the punch corner or wrinkling in the flange region. In order to obtain the limiting drawing ratio (LDR), a cylindrical deep drawing including only the conical die and flat punch is discussed based on numerical simulation in the paper. A finite element model for the deep drawing with a conical die is developed, and the accuracy of the model is verified by experiments results. The investigation shows that in the cylindrical deep drawing with a conical die, the types of failure above , which can obtain for thinner sheet materials, are obviously related to the sheet thickness and the conical angle, and the optimum conical angle is also found.

1. Introduction

Thin-walled cylindrical deep drawing is one of the most widely used process in sheet forming technology. The main method for producing thin-walled cylindrical cups is conventional deep drawing with a blank holder. In this case, the limit coefficient of drawing, according to Ref [1], does not exceed 2.0. By using a conical die without blank holder for stamping, the maximum drawing ratio can be slightly increased. Hence, the process is carried out without a blank holder, which greatly simplifies the structure of forming tools. The most commonly observed failure mode in the process is fracture near the punch corner or wrinkling in the flange region when drawing cylindrical cup through a conical die. But the deformation mechanism and the influence of processing parameters on the limit drawing ratio of this process have been little studied.

Soweby et al. [2] described an experimental study of deep drawing through conical dies without blankholder. Four common sheet metals with three different thicknesses are used and the wrinkling modes which developed during the different stages of deep drawing are explained. Narayanasamy et al [3] discussed an experimental study of deep-drawing of three different steels through a conical die using flat-bottomed and hemispherical-ended punches and obtained an relation between the limit drawing ratio and a non-dimensional parameter index involving the punch diameter, the blank thickness and the average plastic strain ratio. However, in all the two experimental studies only observed a conical die with a constant die angle. The effect of the die angle on the limit drawing ratio is not discussed in these works.

Mahadwie and Mellor [4] described an experimental study of the deep drawing of ten sheet metals through a conical die. A semi-empirical approach is adopted to demonstrate the relationships between the blank diameter, blank thickness, punch diameter, punch profile radius and anisotropy of the sheet metal. However, the influence of the die angle on the failure just for the same kind of material is not enough investigated.

In the present work, the influences of the die angle and the blank thickness on the formability of cylindrical cups through a conical die is studied based on numerical simulation. For varication of the finite element (FE) predictions, experiments for similar operation conditions are carried out. Finally, the types of failure of blanks and the optimum conical angle are found.

2 Experimental Procedure and Numerical Simulation

In this research, circular blanks of steel 08F (Russian standard) with 1 mm of thickness were used. The chemical compositions are given in table 1. The engineering stress-strain curve is obtained based on uniaxial tension test. The elastic-plastic material properties and the yield function are listed in table 2.

The forming test was performed on circular blanks of different diameters with a ram speed 1mm/s using a vertical press. The diameter of the punch is 36 mm and the diameter of the die is 40 mm. In Figure 1, the photo of the dies with 30°, 60° and 70° respectively used for forming test are shown. It is considered to be successful if a workpiece without any fractures and wrinkles was obtained during the forming process.

Figure 2 shows the geometries of the tools and a full FE-model in Software Dynaform. The forming tools consisting of punch and die were modelled as rigid bodies and the blank was modelled as deformable. Process was modelled with shell elements with average element sizes 0.3 mm ,0.5 mm and 0.7 mm. By comparing the result of simulation and the experiment (forming accuracy and time of simulation), 0.5 mm were used to discrete the blank in the simulation. According to Ref [5], anisotropy of the blank has little influence on wrinkling in process of deep drawing through a conical die, so during the simulation, the anisotropy of the blank was not considered.

The friction coefficient between the tools and the blank was obtained by comparison of experimental and FE simulation results about the maximum drawing force and the deformation work In this regard, initially approximate value (0.25, without lubricant) of the friction coefficient was determined by the data reported by [6]. The different values of friction coefficient around 0.25 were used in the FE simulations. The drawing load under the die angle of 60° and the blank diameter of 90 mm with different friction coefficients is shown in figure 3. According to figure 3, 0.23 were used as the friction coefficient between die and blank for the simulation.

In figure 4 and table 3, the comparison between experimental results and simulations are shown. The results of the simulation and experiment have a good consistency. The maximum error obtained from the die angle of 60° was 3.3 %. Therefore, the actuary of the simulation is verified.

Table 1. Chemical compositions of steel 08F.

C	Si	Mn	Ni	S	P	Cr	Cu	Ti
0.06	0.056	0.31	0.03	0.015	0.014	0.05	0.05	0.002

Table 2. Mechanical properties and work hardening equation.

Material	Mechanical properties				Work hardening equation (yield function) $\sigma = \sigma_s + k\varepsilon^n$	
	Yield stress $\sigma_s(MPa)$	Tensile strength $\sigma_b(MPa)$	Maximum plastic strain ε_{max}	Young's module E(GPa)	k	n
steel 08F	196.6	365	0.28	186	550	0.512

Figure 1 Dies with different die angles in deep drawing test.

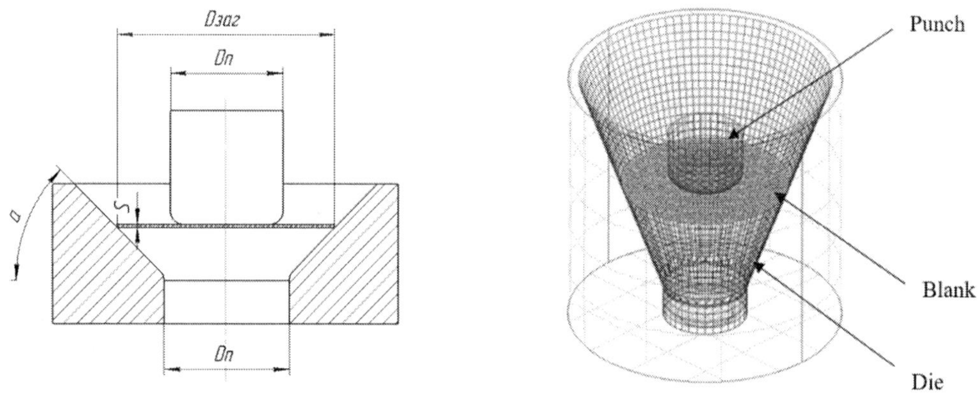

Figure 2 Tool geometry and FE-model in Dynaform.

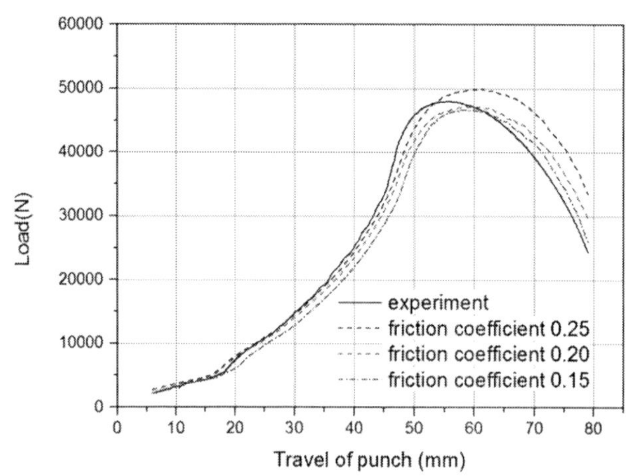

Figure 3. The drawing load under the die angle of 60° and the blank diameter of 90 mm with different friction coefficients

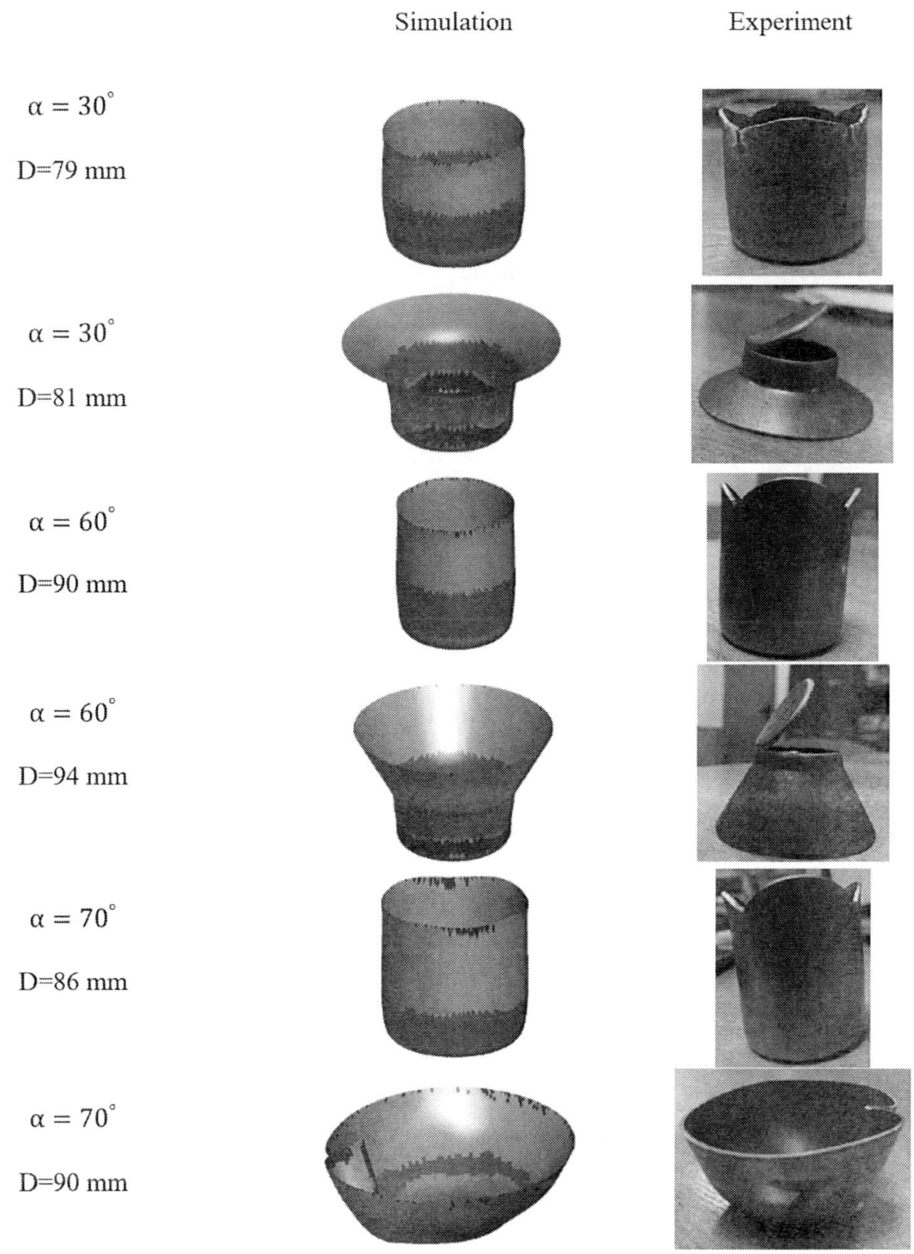

Figure 4. Comparison under different die angles with 1mm thickness of blank

Table 3. Results of experiment and simulation.

Die angle	Limit drawing ratio in the experiment	Limit drawing ratio in the simulation	Error between the experiment and simulation
30°	1.975	2.0	1.2%
60°	2.25	2.325	3.3%
70°	2.15	2.2	2.3%

IOP Conf. Series: Materials Science and Engineering **967** (2020) 012087 doi:10.1088/1757-899X/967/1/012087

3 Results and discussion

The influence of the die angle and blank thickness on the limit drawing angle is obtained through simulation using the verified FE model. Results of prediction of limit drawing ratio are shown in Figure 5 and Figure 6.

From figure 5, it can be observed that for the blank thickness of 0.5 mm, the main type of failure is wrinkling when the blank was drawing into a conical die without blank holder. The optimum conical angle is 40°. When the die angle is below the optimum die angle, with a decrease of conical angle, the process is more like the case of conventional deep drawing without a blank holder, so the limiting drawing ratio falls rapidly. When the angle is higher than the optimal die angle, wrinkles are more likely to occur during the deep drawing process, because the flange of blank will be further deformed without contact with the surface of the die, which contributes to the formation of wrinkles.

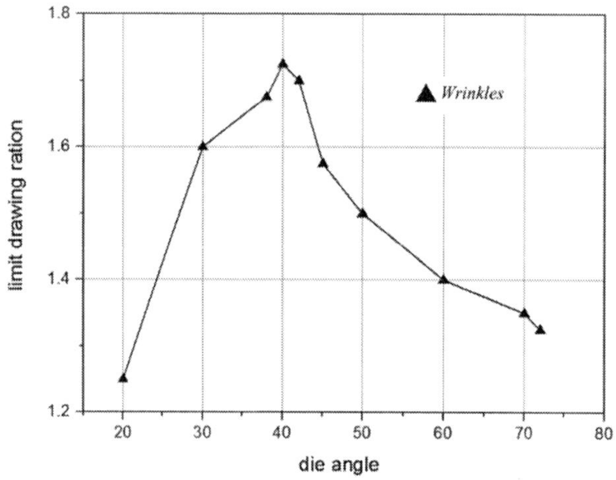

Figure 5. Variation of limiting drawing ratio with die angle for thickness 0.5 mm

Figure 6 show that with small die angels, the main type of failure is facture at the bottom of the cylindrical cup under the blank thicknesses of 1.0 mm and 1.5 mm, respectively. With an increase in the die angle, wrinkling at the edge of the flange becomes a limitation of the drawing process. There is a die angle at which the type of failure changes from factures to wrinkles, and at the same die angle, the maximum limiting drawing ratio is achieved

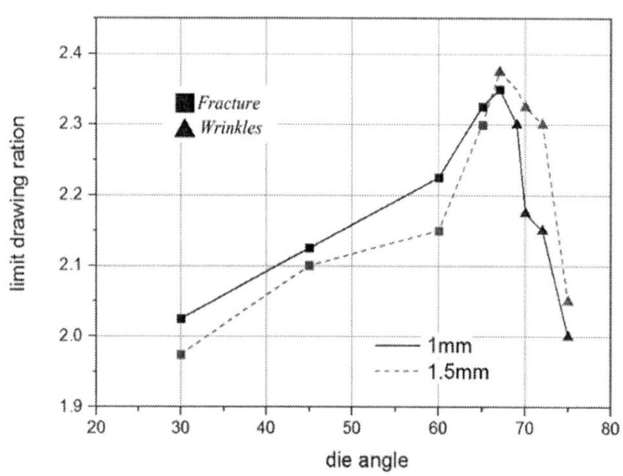

Figure 6. Variation of limiting drawing ratio with die angle for thickness 1.0 mm and 1.5 mm

The above results are good in agreement with the quoted results of Mahadwie and Mellor [4]. However, in the latter work the limit drawing ratio with several materials with diffident thickness under the same die angle was given. The present work draws attention to the influence of the die angle under the same material with different thickness.

4 Conclusion

The limit drawing ratio was obtained when forming through a conical die without blank holder using simulation method. The type of failure in the forming process depends on the blank thickness. For small blank thickness (blank thickness of 0.5 mm), the wrinkles at the edge of flange are more likely to occur during process. For thicker plates (blank thickness of 1 mm and 1.5 mm), the type of failure changes from factures to wrinkles with increase in die angle, and the maximum limiting drawing ratio is achieved at die angle, at which the change occurs.

References

[1] Eduard V and Theodor K 2001 *Thin plates and shells. Theory, analysis and applications.* (New York: Marcel Dekker) p 688
[2] Soweby Kaima M and Chakravrti P.C 1982 *Journal of Mechanical Working Technology* **6** p 35-50
[3] Narayanasamy R and Sowerby R 1993 Forming behavior of some sheet metal sheet material when drawn through a conical die. *Journal of Material Processing Technology*. **39** p. 43-53
[4] AL A H M and Mahdawle, Mellor P B 1986 The effect of work-hardening on deep-drawing through a conical die *Journal of Mechanical working technology* **13** p 317-24
[5] Zhang C and Yu T X 1991 An experimental investigation into the stamping of elastic-plastic circular plates. *Journal of Material Processing Technology*. 1991, **28** p 321-34
[6] Seireg A.A. 1998 *Friction and lubrication on mechanical design (Mechanical Engineering)* (New York: Marcel Dekker) p 568

IOP Conf. Series: Materials Science and Engineering **967** (2020) 012088 doi:10.1088/1757-899X/967/1/012088

New tool to evaluate the fracture resistance of thin high strength metal sheets

D Frómeta[1], A Lara[1] S Parareda[1], L. Grifé[1] and D Casellas[1,2]
[1] Eurecat, Centre Tecnològic de Catalunya, Unit of Metallic and Ceramic Materials, Plaça de la Ciència 2, 08243 Manresa, Spain
[2] Division of Mechanics of Solid Materials, Luleå University of Technology, 971 87 Luleå, Sweden

Abstract. Fracture toughness has become a key property to predict the fracture performance of high strength metal sheets (edge cracking resistance, crash failure behaviour, local formability, etc.). However, the measurement of the fracture toughness of thin sheets still being challenging, mainly because of complex, expensive and time-consuming specimen preparation. In this work, an innovative tool to readily assess the fracture resistance of thin advanced high strength metal sheets is presented. The device consists of a special cutting tool (punch and die) designed to introduce sharp notches in sheet specimens through a simple shearing process. This new method avoids the need for fatigue pre-cracking procedures and allows measuring the fracture toughness of thin metal sheets with easy and cheap specimen preparation. It has been used in this work to evaluate the crack propagation resistance of four different advanced high strength steel sheets. The obtained toughness values are in good agreement with those measured with fatigue pre-cracked specimens and they show to be suitable to predict edge formability of AHSS sheets.

1. Introduction

Recent developments in new high strength sheet materials for automotive lightweight construction has brought new challenges to the sheet manufacturing sector: the need for accurate characterization techniques and methods to avoid sheet cracking occurrence during forming and to predict in-service performance. The high strength and limited ductility of such alloys increase their cracking susceptibility and reduce their formability. Thus, end-use industries and their suppliers are currently facing serious productivity losses due to the occurrence of unexpected fractures during forming. Moreover, end-users are continuously asking for high-performance materials and reliable methods to predict crashworthiness in the transport sector and long-life performance in other engineering applications. In this regard, fracture toughness measured in the frame of fracture mechanics has shown to be a suitable material property to understand crack-related problems in high strength metal sheets [1-5]. Unfortunately, the experimental evaluation of fracture toughness in thin sheets is not straightforward. It involves tedious specimen preparation, including expensive and time-consuming fatigue pre-cracking operations, and complex crack growth measuring techniques. Furthermore, the thickness of these sheets (usually 1-3 mm), do not satisfy the size requirements described in fracture mechanics standard procedures [6,7]. As a consequence, the knowledge of the fracture properties of such high strength sheet materials is very limited.

The Essential Work of Fracture (EWF) methodology [8] has shown to be a suitable alternative to readily characterize the fracture toughness of such thin high strength metal sheets [2-5, 9]. The method is very simple and does not require the measurement of the current crack advance during the test, which supposes great advantage respect to traditional elastic-plastic fracture mechanics (EPFM) procedures. However, as it is well known in fracture mechanics testing, EWF measurements are strongly affected by the notch root radius [9]. Therefore, to obtain reliable notch-independent toughness values, it is necessary the use of fatigue pre-cracked specimens as recommended by standard procedures [6, 7]. This significantly increases the cost and time of the tests. Moreover, specialized equipment and skilled staff are required. To overcome such experimental difficulties in specimen preparation for fracture toughness characterization of thin metal sheets and to promote its application as routine testing for quality control and material selection, an innovative tool is presented in this work. The device, developed at the Unit of Metallic and Ceramic Materials of Eurecat and registered under European patent EP 3567364A1, consists of a two-pillar modular cutting die, equipped with a bevelled punch (Figure 1 left) designed to introduce crack-like sharp notches in the sheet specimens. The tool allows obtaining rectangular Double Edge Notched Tension (DENT) specimens (Figure 1 right) by means of a simple shearing process. It works with sheet blanks between 0.5 and 2 mm thickness approximately (its application to thicknesses above 2 mm is under study). In the present work, the new device has been used to evaluate the EWF of 4 different Advanced High Strength (AHSS) sheets of 1000-1200 MPa strength and ≈1.4 mm thickness. In order to validate the obtained fracture toughness results, they are compared to those obtained with conventional fatigue pre-cracked specimens.

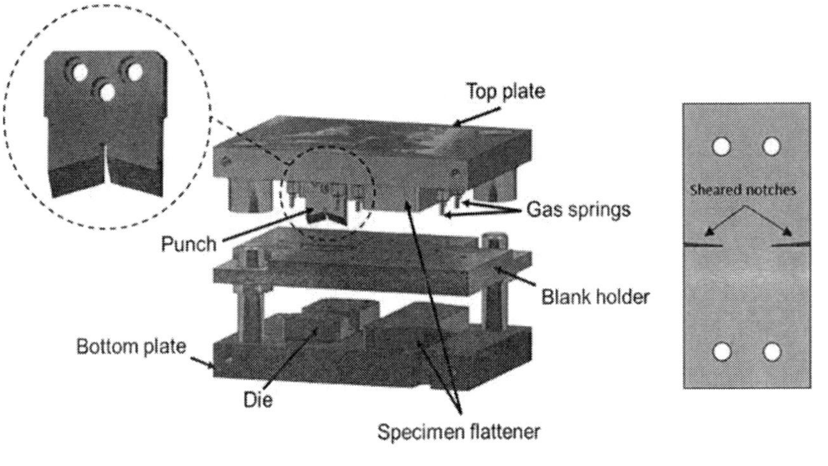

Figure 1. Left: Tool for introducing sharp notches in sheet metal specimens and detail of the bevelled punch. Right: DENT specimen with sheared sharp notches

2. Materials and methods

2.1. Materials

The materials investigated in this work are 4 AHSS grades in the form of 1.35-1.4 mm thickness sheets. The steels are a Complex Phase (CP) steel, two Dual Phase (DP) steels and a 3rd GEN Transformation Induced Plasticity (TRIP)-assisted steel. The mechanical properties of the studied steels for the transverse direction are indicated in Table 1. Sheet thickness is also given. CP steel has a homogeneous bainite/tempered martensite matrix and it is characterized by low strain hardening and high yield stress to tensile strength ratio. DP-A has a ferritic-bainitic matrix with some amount of tempered martensite and a lower amount of hard martensite islands. It shows higher strain hardening and greater elongation values than CP steel. DP-B has a more homogeneous ferrite-martensite distribution. It shows slightly lower elongation than DP-A and higher strain hardening than CP. The microstructure of the 3rd Gen TRIP-assisted steel consists of a bainitic matrix with islands of martensite/retained austenite. It has a considerable amount of retained austenite (≈15%). The strain-induced transformation of retained austenite to martensite (TRIP effect) enhances strain hardening and elongation values (both uniform and total).

Table 1. Mechanical properties of the investigated materials for the transverse direction

Material	Thickness, t [mm]	Yield strength, σ_{ys} [MPa]	Ultimate Tensile Strength, σ_{UTS} [MPa]	Uniform elongation, A_g [%]	Elongation at fracture, A_{80} [%]	Strain hardening exponent, $n_{2\text{-}4\%}$
CP	1.4	915	1008	4.8	8.8	0.05
DP-A	1.35	807	1057	6.6	9.6	0.13
DP-B	1.4	769	1040	5.3	8.7	0.09
3rd Gen	1.4	987	1216	9.2	12.6	0.11

2.2. Fracture toughness measurements: Essential Work of Fracture methodology

The Essential Work of Fracture (EWF) methodology was developed to quantify the ductile tearing resistance of thin ductile metal sheets [8]. The methodology permits to separate the total work of ductile fracture (W_f) in two energetic contributions: an essential work of fracture (w_e), developed in the fracture process zone and necessary to create new surfaces in the front of the crack tip and a non-essential plastic work (w_p) surrounding the fracture area. The first term is proportional to the fracture surface and the second to the plastic volume, according to:

$$W_f = w_e l_0 t_0 + w_p \beta {l_0}^2 t_0 \tag{1}$$

where l_0 is the ligament length (the non-fractured section between the two edge notches, see Figure 2), t_0 is the specimen thickness and β is a shape factor that depends on the shape of the plastic zone. Even though different specimen geometries can be used to obtain the EWF, the DENT specimen has shown to be the most suitable geometry for thin sheets since there is no buckling during the test. The experimental procedure for the determination of the EWF is schematized in Figure 2. W_f is obtained by testing a DENT specimen (Figure 2 left) up to fracture at a constant displacement rate and integrating the area under the load vs displacement curve. The specific work of fracture (w_f) is obtained by dividing W_f by the initial ligament area $l_0 t_0$. Thus, equation (1) can be rewritten as:

$$\frac{W_f}{l_0 t_0} = w_f = w_e + w_p \beta l_0 \tag{2}$$

If DENT specimens with different ligament lengths are tested and w_f is plotted against the ligament length l_0, a straight line with a positive intercept, which is the specific essential work of fracture (w_e), is obtained (Figure 2 right).

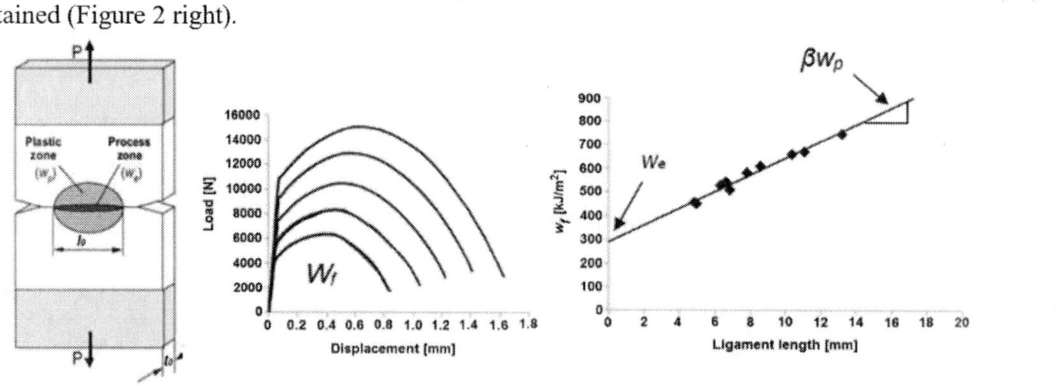

Figure 2. DENT specimen and experimental determination of the EWF: W_f for different ligament lengths (l_0) and plot of w_f against l_0. The y-intercept indicates the specific essential work of fracture, w_e [4].

2.3. Specimen preparation

In the present work, EWF tests have been performed with two different notch configurations: 1) fatigue pre-cracks according to standard procedures recommendations and 2) mechanically sheared notches obtained with the new tool. The experimental procedure for the preparation of both notch conditions is described in *Section 2.3.1* and *2.3.2.*

2.3.1. Fatigue pre-cracked specimens. For EWF tests with fatigue pre-cracked specimens, rectangular DENT specimens of 240 x 55 mm were used (Figure 3 left). Initial notches were machined by electrical discharging machining (EDM). Then, fatigue pre-cracks were nucleated at the notch root following the recommendations of the ASTM E1820 standard [7]. For fatigue pre-cracking, a resonance fatigue machine was used (Figure 3 right). The cracks were extended about 1-1.5 mm per side. All the specimens were machined at transverse orientation respect to the rolling direction. The final notch radius (ρ) at the crack tip is approximately 0.1 μm. Ligament lengths (l_0) from 5 to 15 mm were used and 3 specimens per ligament length were tested.

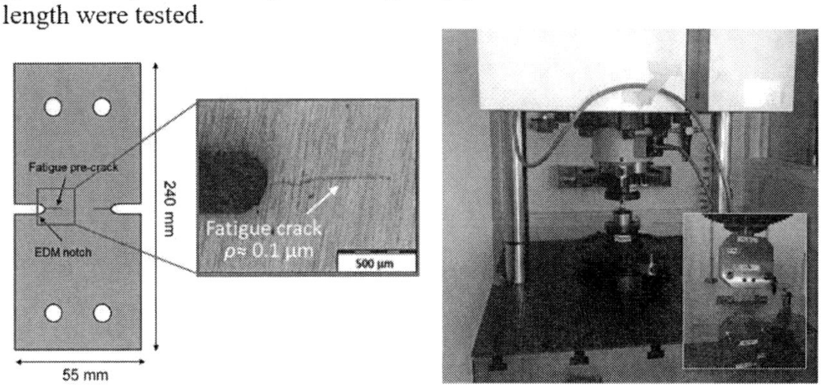

Figure 3. Left: DENT specimen and detail of the fatigue pre-crack. Right: Resonance fatigue machine.

2.3.2. Specimens with sheared notches. The new notching device (Figure 1) was used to prepare DENT specimens with different ligament lengths. Figure 4 shows some images of the experimental setup. The tool was mounted in a Zwick Roell 50 kN AllroundLine testing machine. The machine has two working areas, which permits to perform the notching and the subsequent tensile testing with a single testing equipment and easily switch from one operation to another (Figure 4a). The experimental procedure for specimen notching is schematized in Figure 5. The process is described as follows: first, a rectangular specimen of 200 x 55 mm (cut at transverse orientation respect to the rolling direction) is placed at the die and fixed using 2 pins (Figure 4c). This fixation system ensures the alignment of the specimen and that notches are always centred respect to the pinning holes. Then, the punch is moved downwards and, by means of a shearing process, two sharp notches (notch radius, ρ≈2 μm) are introduced in the specimen (notches symmetrical respect to the longitudinal axis of the specimen). The ligament length is modified by controlling the punch displacement, i.e. the greater the punch displacement the smaller the ligament between the two notches. After cutting, the punch returns to the initial position and the specimen can be extracted. Due to the shearing operation, the specimen is slightly bent at the end of the process. Therefore, a final flattening operation is performed before tensile testing. The specimen is placed in the base intended for that purpose (base for specimen flattening, Figure 4b) and pressed with the blank holder. This final step is optional and it does not affect the final result. However, it is highly recommendable since it facilitates the specimen manipulation and the fitting in the testing grips.

Due to the difficulty of measuring the length of the ligaments directly on the specimen, they were measured from the fracture surface after testing. Ligament lengths (l_0) from 8 to 20 mm were obtained and 2 specimens per ligament length were used. In total, 8 to 12 specimens per material were tested.

IOP Conf. Series: Materials Science and Engineering **967** (2020) 012088 doi:10.1088/1757-899X/967/1/012088

Figure 4. Images of the experimental setup for the notching process. a) Setup of the tool in the testing machine. b) Detail of the cutting tool. c) Specimen before (left) and after (right) the notching process.

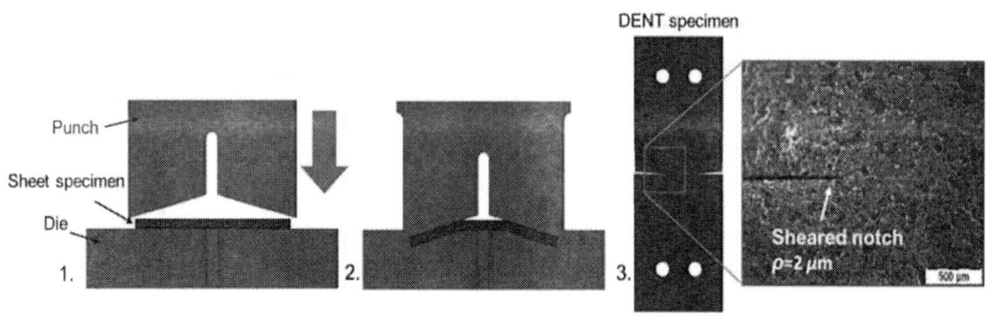

Figure 5. Schematization of the experimental procedure for the preparation of sheared notches in sheet specimens.

2.4. EWF tests

After notch preparation, the DENT specimens were tested up to fracture according to the European Structural Integrity Society (ESIS) protocol for EWF testing [10]. The tests were conducted at a constant displacement rate of 1 mm/min. The load-line displacement was measured by means of the video extensometer using initial extensometer marks separated 25 mm.

3. Results and discussion

3.1. EWF results

Figure 6 shows the load vs load-line displacement curves for the two specimen configurations. Figure 7 plots the values of w_f as a function of the ligament length. Numerical values of w_e and βw_p are given in Table 2 and Figure 8. As observed, for the same ligament length, both fatigue pre-cracked and sheared specimens show very similar load vs displacement curves (similar maximum load and displacement at fracture) in the four investigated materials. It explains the good agreement between w_f values for the two notch conditions (Figure 7) and the practically identical specific essential work of fracture, w_e and plastic work, βw_p (Figure 8). In general, very good repeatability is observed for sheared specimens, which

IOP Conf. Series: Materials Science and Engineering **967** (2020) 012088 doi:10.1088/1757-899X/967/1/012088

enhance the reliability of the obtained toughness values and confirms the robustness of the new process. It is worth noting that the similarity between specimens of the same ligament length is improved in sheared specimens. This is because the notch length is precisely defined by the punch displacement and, therefore, is easier to obtain multiple specimens with the same ligament. On the other hand, the ligament size in fatigue pre-cracked specimens is determined by the length of the fatigue cracks, which makes difficult to have two specimens with identical ligament.

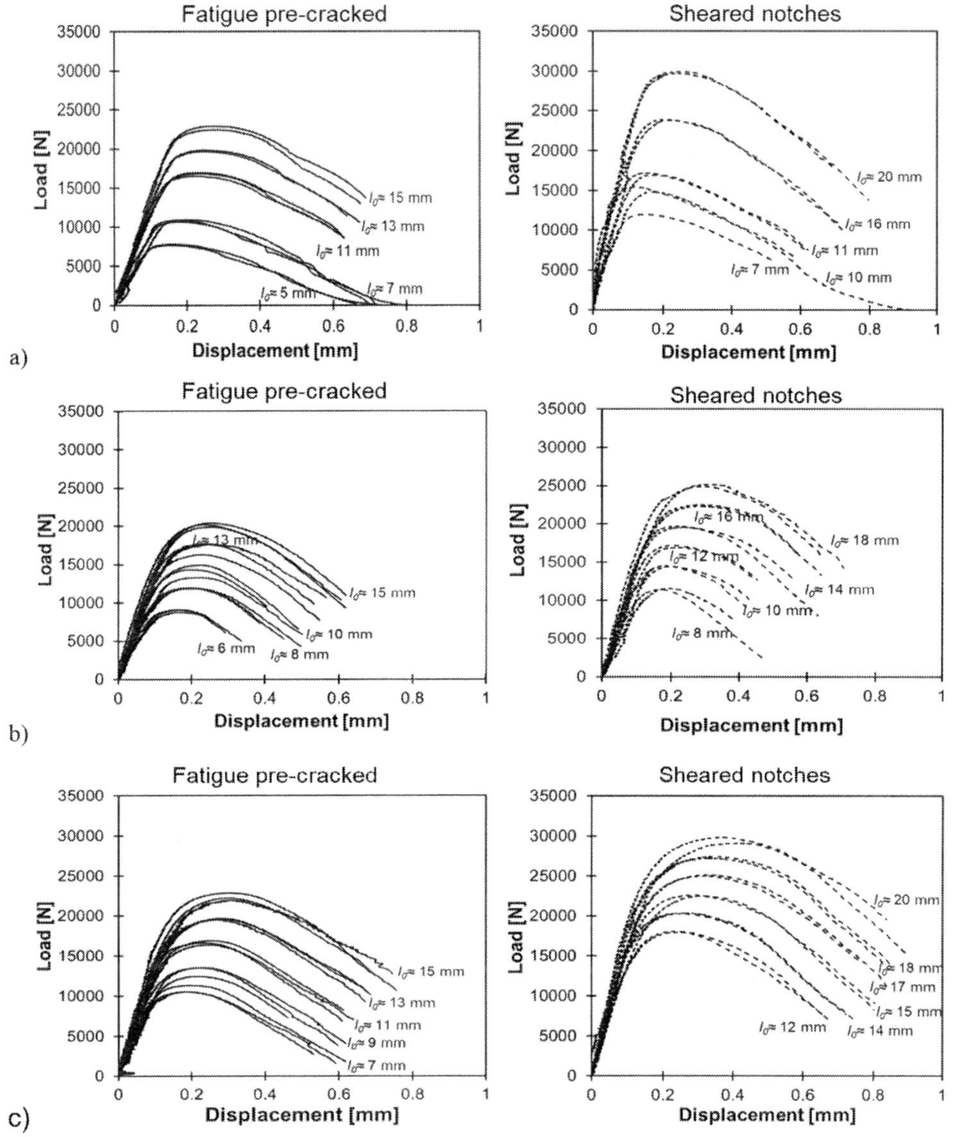

IOP Conf. Series: Materials Science and Engineering **967** (2020) 012088 doi:10.1088/1757-899X/967/1/012088

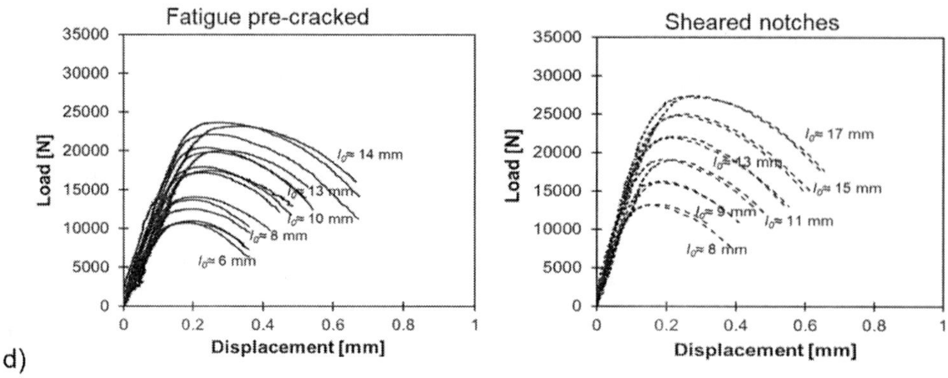

Figure 6. Load-displacement curves obtained from EWF tests with fatigue pre-cracked specimens (left) and specimens with sheared notches (right). a) CP, b) DP-A, c) DP-B and d) 3rd Gen AHSS.

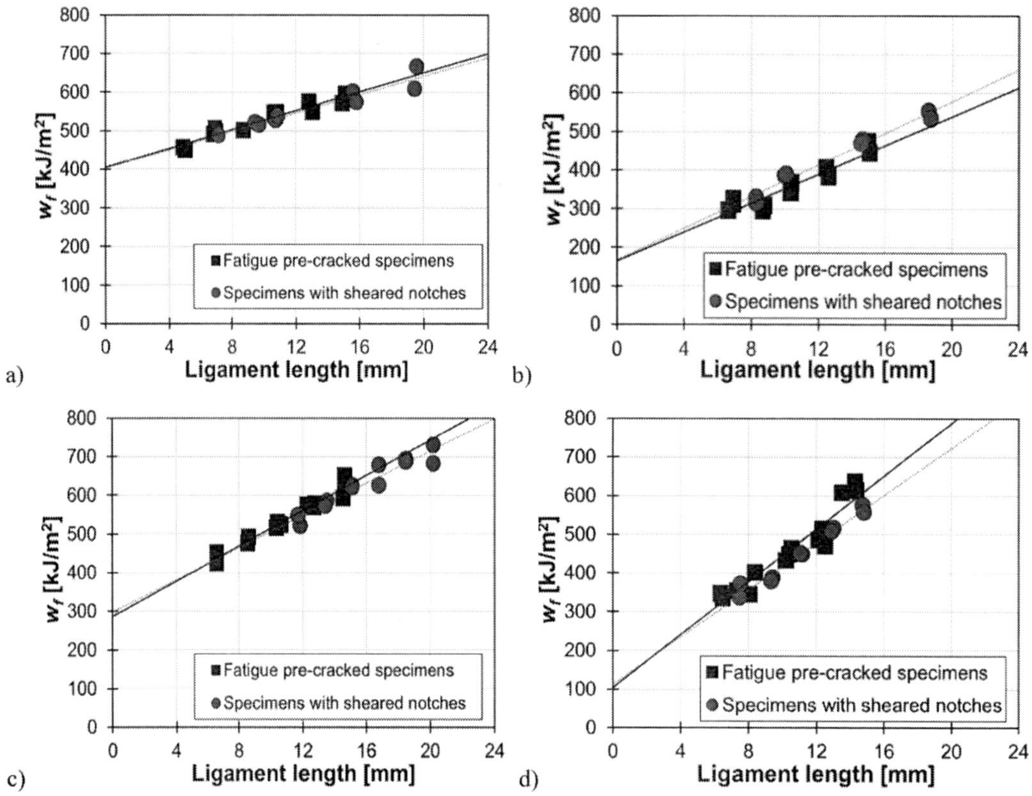

Figure 7. w_f values against ligament length for the two investigated specimen configurations. **a)** CP, **b)** DP-A, **c)** DP-B and **d)** 3rd Gen TRIP-assisted.

Table 2. EWF results obtained with fatigue pre-cracked specimens and specimens with sheared notches

	Fatigue pre-cracked		*Sheared notches*	
Material	w_e [kJ/m^2]	βw_p [MJ/m^3]	w_e [kJ/m^2]	βw_p [MJ/m^3]
CP	405 ± 11	12 ± 1	404 ± 19	12 ± 1
DP-A	149 ± 21	24 ± 2	163 ± 27	21 ± 2
DP-B	286 ± 17	23 ± 1	298 ± 32	21 ± 2
3rd GEN	104 ± 30	34 ± 3	113 ± 21	30 ± 2

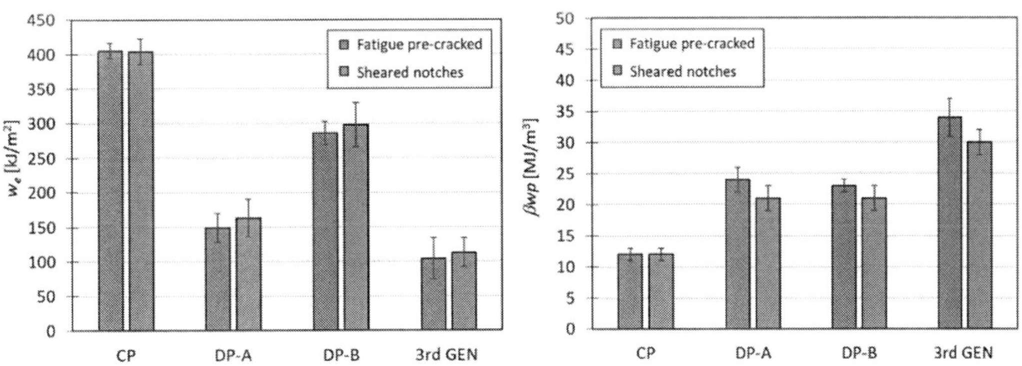

Figure 8. Results from EWF tests with fatigue pre-cracked (blue) and sheared (orange) specimens. Left: specific essential work of fracture, w_e. Right: non-essential plastic work, βw_p.

3.2. Fracture surface of DENT specimens

Figure 9 shows the fracture surfaces of different sheared and fatigue pre-cracked DENT specimens. It can be observed that for both specimen configurations the fracture aspect is quite similar and the ligament is well defined between the two notches. The major difference between the two notch types is the shape of the crack front (concave for the sheared notch and convex for the fatigue pre-crack). Overall, it was found that the morphology of the sheared notches is similar in the four investigated AHSS grades and independent of the ligament length.

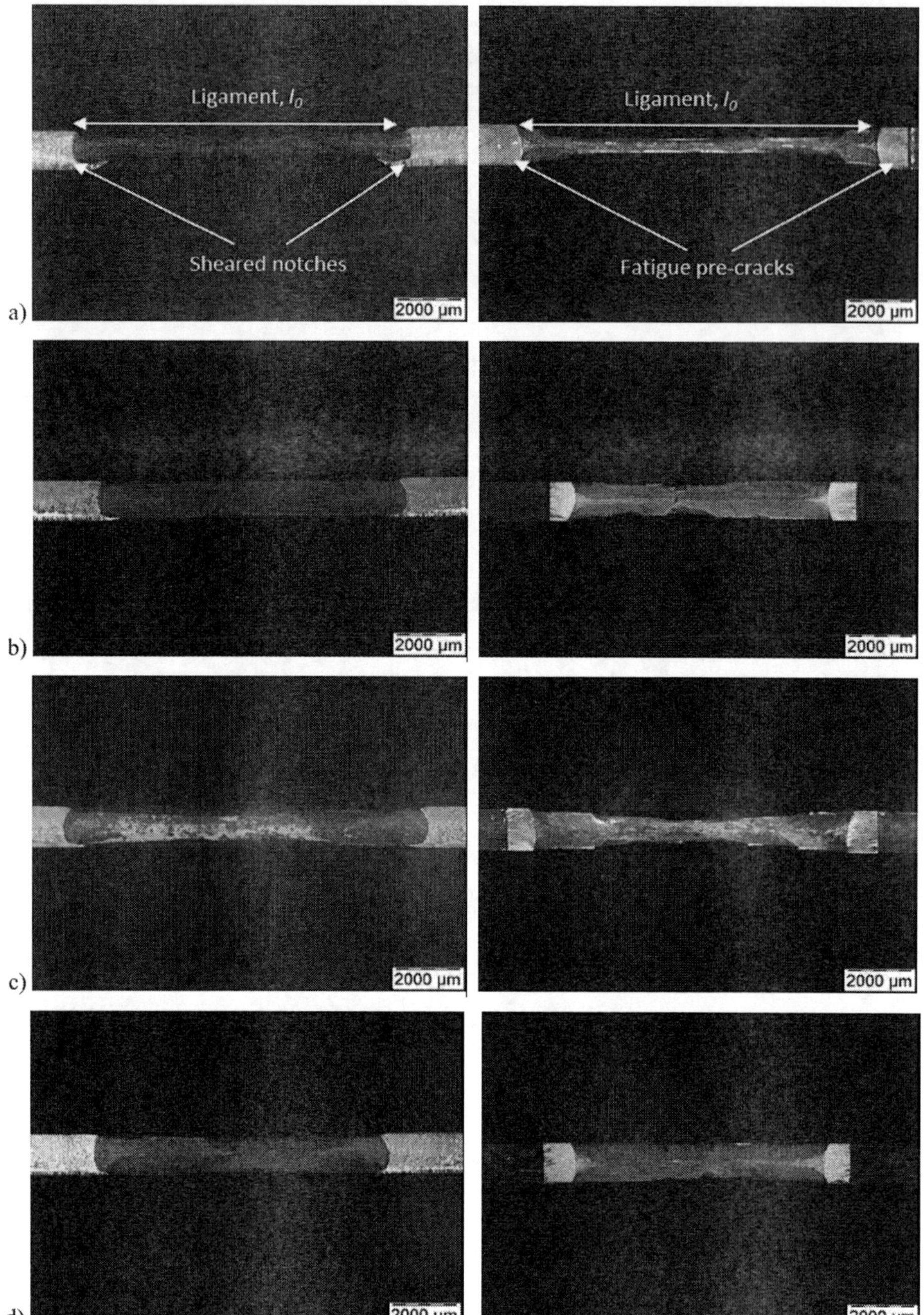

Figure 9. Fracture surfaces of DENT specimens with sheared notches (left) and fatigue pre-cracks (right). a) CP, b) DP-A, c) DP-B and d) 3rd GEN. The different areas (fractured ligament, sheared notch and fatigue pre-crack) are indicated in the first images.

4. Applications

As mentioned before, fracture toughness has shown to be a useful material property to predict crack-related problems in high strength metal sheets, such as edge fractures [1-3], crack formation during crash [4] or other fractures related to their local ductility [5]. It is illustrated in Figure 10, where w_e values of several AHSS grades are plotted against HER [2-3] and the maximum energy absorbed in axial crash tests [4]. As already discussed in [2-5], w_e shows a very good correlation with HER and crash resistance.

Therefore, w_e can be used to estimate the fracture resistance of AHSSs and predict their cracking behaviour during forming or crash. According to the linear data fittings depicted in Figure 10, expected values of HER and impact energy for the steel grades investigated in this work are plotted as a function of the obtained specific essential work of fracture values (results from EWF tests with sheared specimens). In the case of DP-A and 3rd GEN steel, the HER was experimentally evaluated according to ISO16630. As observed, the measured HER fits very well in the linear w_e vs HER data fitting, which validates the suitability of w_e for edge cracking resistance prediction. Therefore, the new testing procedure presented, can be used as a fast and cost-effective tool to readily assess the fracture performance of AHSS sheets.

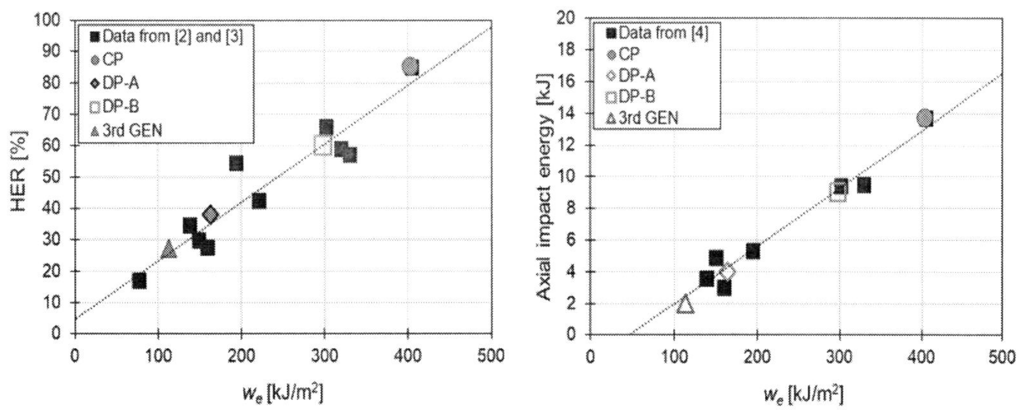

Figure 10. Correlation of w_e with HER (left) and axial impact energy (right). The data represented by black squares is extracted from references [2-4]. Blue symbols correspond to w_e values obtained in this work. Solid symbols represent experimentally evaluated values of HER and axial impact energy. Open symbols are expected values according to the linear data fitting (dashed line).

5. Conclusions

In the present work, an innovative device to prepare high strength metal sheet specimens for fracture toughness characterization has been presented. The tool can be easily mounted in a universal testing machine and offers an easy and cheap alternative to fatigue pre-cracking procedures. The new process has shown to be robust and reliable enough to evaluate the fracture toughness of four different AHSS sheets. It supposes a great time-saving in specimen preparation and it can be very useful to boost the use of fracture toughness measurements as routine testing for coil quality determination or for the selection of high strength sheet materials with enhanced cracking resistance.

6. References

[1] Yoon J I, Jung J, Joo S H, Song T J, Chin K G, Seo M H, Kim S J, Lee S and Kim H S Correlation between fracture toughness and stretch-flangeability of advanced high strength steels *Matter.Lett.* 180 (2016) 322-326.

[2] Casellas D, Lara A, Frómeta D, Gutiérrez D, Molas S, Pérez Ll, Rehrl J and Suppan C. Fracture Toughness to Understand Stretch-Flangeability and Edge Cracking Resistance in AHSS. *Metall. and Mat. Trans.* A 48 (2017) 86-94.

[3] Frómeta D, Lara A, Parareda S and Casellas D. Evaluation of Edge Formability in High Strength Sheets Through a Fracture Mechanics Approach. *AIP Conference Proceedings* 2113, 160007 (2019).

[4] Frómeta D, Lara A, Molas S, Casellas D, Rehrl J, Suppan C, Larour P and Calvo J. On the correlation between fracture toughness and crash resistance of advanced high strength steels. *Eng.*

Frac. Mech. 205 (2019) 319-332.

[5] Frómeta D, Lara A, Casas B, and Casellas D. Fracture toughness measurements to understand local ductility of advanced high strength steels. *IOP Conf. Ser.: Mater. Sci. Eng.* (2019) 651 012071

[6] ASTM E1820. Standard test method for measurement of fracture toughness. American Society for Testing and Materials

[7] ASTM E399. Standard test method for plane-strain fracture toughness of metallic materials. American Society for Testing and Materials

[8] Cotterell B and Reddel JK. The essential work of plane stress ductile fracture. *Int. J. Fract.* (1977) 267-277.

[9] Muñoz R, Lara A and Casellas D. Fracture toughness characterization of advanced high strength steels. *Int. Deep Drawing Research Group (IDDRG) Conference 2011* (Bilbao, Spain, June 5-8, 2011)

[10] Clutton E. Essential work of fracture. Moore DR, Pavan A, Williams JG, editors. Fracture mechanics testing methods for polymers, adhesives and composites, vol.28. ESIS Publ.; (2001) 177–95.

Acknowledgements
The present research work has been partially funded by the EU Horizon 2020 programme under grant agreement H2020-EU.2.1.3. – 814517 (FormPlanet).

IOP Conf. Series: Materials Science and Engineering **967** (2020) 012089 doi:10.1088/1757-899X/967/1/012089

Experimental Analysis of Warping Defects in Chain-die Forming Variable-width Profiles

Zhenye Liang[1], Tianxia Zou[1], Dayong Li[1*], Hua Xiao[2,3], Lei Shi[2,3], Shichao Ding[4] and Yinghong Peng[1]

[1] State Key Laboratory of Mechanical System and Vibration, Shanghai Jiao Tong University, Shanghai 200240, China

[2] Research Institute, Baoshan Iron & Steel Co., Ltd., Shanghai 201900, China

[3] State Key Laboratory of Development and Application Technology of Automotive Steels (BaoSteel), Shanghai 201900, China

[4] School of Mechanical and Mining Engineering, The University of Queensland, St Lucia, Brisbane, QLD 4072, Australia

* Corresponding author: dyli@sjtu.edu.cn

Abstract. Chain-Die forming is an emerging gradual forming technology suitable to fabricate advanced high strength steel (AHSS) products. In this paper, a variable-width product was designed, and the feasibilities of chain-die forming were investigated with experiments. The results show that chain-die forming is capable of manufacturing AHSS variable-width products with good precision. However, warping defects were found in the chain-die formed variable-width products. Meanwhile, web-warping and flange-warping show a similar trend for the designed variable-width profile, and the maximum deviation of warping was quantitatively analyzed. The occurrences of warping defects are ascribed to the insufficient tensile or compressive deformation, and both web area and flange area of hat profile tend to rotate to compensate for the geometrically necessary longitudinal strain.

1. Introduction

Chain-die forming is an emerging sheet metal forming technology [1], and figure 1 shows the schematic diagram of the chain-die forming process. The idea of chain-die forming is to enlarger the deformation zones by using multiple discrete die blocks, which approximate a roller with extreme larger radii [2]. Figure 2 shows a lab-scale chain-die forming machine and the discrete die blocks mounted to the chains. It has been proved that chain-die forming can achieve a limited residual longitudinal strain by enlarging forming length [3] compared with the roll forming process, thus AHSS products could be fabricated by

IOP Conf. Series: Materials Science and Engineering **967** (2020) 012089 doi:10.1088/1757-899X/967/1/012089

conserving the material's ductility [4].

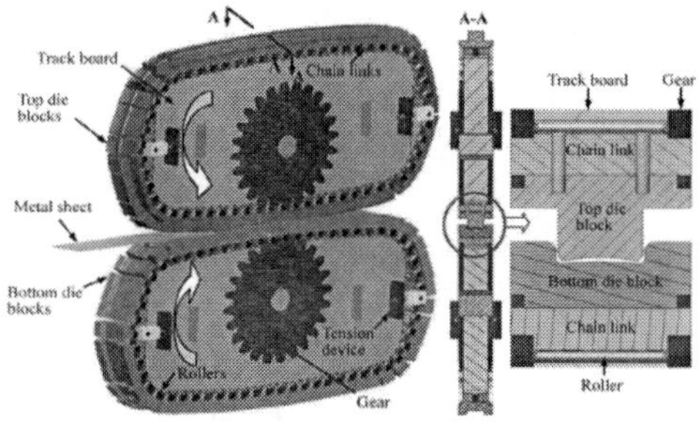

Figure 1. Schematic diagram of the chain-die forming process [4].

Figure 2. Lab-scale chain-die forming machine.

The complex material behaviors of AHSS are prone to induce forming defects in the forming process [5]. Warping is one of the major shape defects in chain-die forming and flexible roll forming of variable-width products, and it is defined as the height deviation between the formed samples and the horizontal line for web and flange area along the longitudinal direction. Many efforts have been made to investigate the mechanism of the web-warping defect in the forming processes. Jiao [6] established an analytical model to predict the web-warping of flexible roll-formed variable-width U-channel product by considering the relationship between longitudinal strain and web-warping. However, the analytical model is based on the geometrical analysis, the boundary condition and the material behavior were ignored. To take the material properties into account, a modified analytical model was further proposed by Jiao [7], which considers both the geometric and material parameters. It was found that the established model can provide a rapid initial design assessment for flexible roll-formed products. To reduce the web-warping error in cold roll-formed variable-width products, a new process called incremental counter forming (ICF) was proposed by Park [8]. Experimental results show that the web-warping error decreased when the longitudinal strain in the concave zone increases, however, if the longitudinal strain in the straight zone reaches a critical limit, the increase of longitudinal strain leads to a worsen warping defect. In the chain-die forming process plate is incrementally bended into the desired geometry, the deformation pattern is similar to the cold roll forming process, and warping will be introduced inescapability when chain-die forming a variable-width profile. Sun [9] investigated the relationship between longitudinal strain and web-warping defects in chain-die formed AHSS variable-width U-channel profile, and an analytical model was proposed based on Jiao's model [6].

In this study, a variable-width hat-channel product was designed, and chain-die forming experiments were conducted to investigate the feasibility of chain-die forming technology. Experimental results were analyzed to illustrate the warping behavior, including web-warping and flange-warping.

2. Chain-die forming experiments

Figure 3 shows the geometry and design parameters for the variable-width hat-channel profile. Meanwhile, varying H_2 were chosen to study the influence of geometric dimensions on warping, as listed in table 1. It is expected that a small value of H_2 will lead to an enhanced deformation at the transition area, which will further cause a worse warping defect. Note that the profile's length and height were assigned to be 100 millimeters and 30 millimeters respectively, as shown in Figure 3.

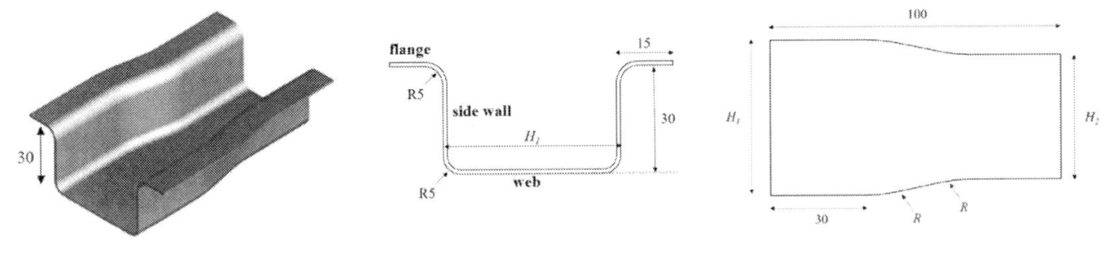

(a) (b) (c)

Figure 3. (a) Geometry, (b) sectional dimensions, and (c) critical dimensions of variable-width profile for the designed hat-channel product (Unit: mm).

Table 1. Design parameters for variable-width products (Unit: mm).

Configurations	R	H_1	H_2
Config 1	50	50	40
Config 2	50	50	34
Config 3	50	50	30

A chain-die forming machine usually consists of track board, rollers, gears, chains and die blocks mounted to the chains [3]. The chain-die forming machine used for the experiments has a 35-meter-radii trace board, and the die blocks have a length of 36 millimeters. Note that the die blocks are replaceable, which is one of the advantages of chain-die forming. QP1180 (QP refers to quenching and partitioning, and QP1180 is one of the third generation AHSS [4]) plates with a thickness of 1.2 mm are used in the chain-die forming experiments, and the basic mechanical properties are listed in table 2. The experiments are repeated at least three times to ensure repeatability. The ATOS Triple S scanner was used to acquires the geometry of experimental samples.

Table 2. Mechanical properties for QP1180 sheet metal [10].

t (mm)[a]	σ_y (MPa)[b]	σ_{UTS} (MPa)[c]	e_u (%)[d]	e_t (%)[e]	E (GPa)[f]
1.2	884	1403	16.2	18.2	204

[a] Thickness of the specimen.
[b] 0.2% offset yield strength.
[c] Ultimate tensile strength.
[d] Uniform elongation (engineer strain at maximum tensile load).
[e] Total elongation (engineer strain at fracture).
[f] Young's modulus.

3. Result and discussion

3.1. Chain-die forming experimental results

Table 3. shows the experimental results of the critical dimensions for chain-die formed variable-width products. It is found that the maximum deviation is 1.53mm and the maximum relative deviation is less than 5% without any mold compensation, which signifies that chain-die forming can be used to fabricate the AHSS variable-width hat-channel products.

Table 3. Experimental results for chain-die formed variable-width products.

	Configurations	Config 1	Config 2	Config 3
H_1	Designed (mm)	50.00	50.00	50.00
	Exp.(mm)	51.44	51.53	51.37
	Δ(mm)[a]	1.44	1.53	1.37
	σ(%)[b]	2.88	3.06	2.74
H_2	Designed (mm)	40.00	34.00	30.00
	Exp.(mm)	40.77	34.97	31.49
	Δ(mm)	0.77	0.97	1.49
	σ(%)	1.93	2.85	4.97

[a] Deviation Δ = designed value - experimental value.

[b] Relative deviation σ = Δ / designed value.

3.2 Analysis of warping in chain-die forming

Experimental results show that chain-die forming is applicable to manufacturing complex variable-width hat-channel profiles, however, warping defects occurred in the experimental samples. Previous studies have been focusing on the variable-width U-channel profiles for its simplicity compared with hat-channel profiles [9]. In this study, the variable width products have been assigned with a basic configuration of hat profile, and both web-warping and flange-warping were analyzed to provide a better understanding of the occurrences of warping defects in chain-die forming. Figure 4 depicts the variation of both web-warping and flange-warping with regard to longitudinal length for the designed variable-width geometry. Figure 4 shows that there was a combined upward and downward warping defects in the web area. It can also be found that there is a similarity between the variation of web-warping and flange-warping, and meanwhile, the locations of the warping peak are close to each other.

Figure 5 compares the maximum deviation of the warping ΔH (as a representative of the severity of warping defects, as depicted in figure 4) for the designed profiles with varying H_2. Note that the maximum deviation for flange-warping is averaged over the left and right flange. Figure 5 shows that with the decreasing of end width H_2, both web-warping and flange-warping are enhanced, however, the influence of H_2 on the flange-warping is relatively smaller than those on the web-warping.

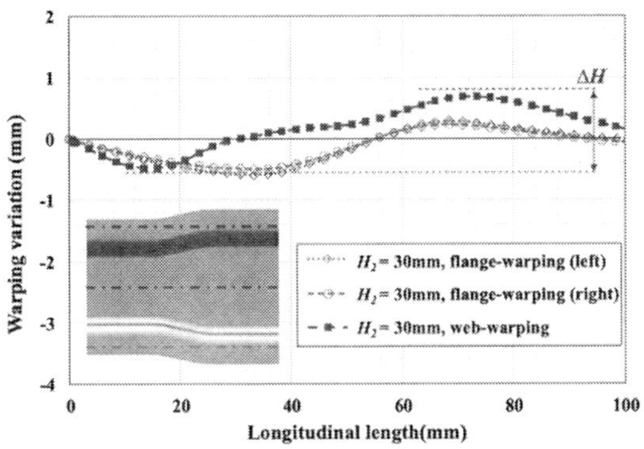

Figure 4. Variation of web-warping and flange-warping with regard to longitudinal length for chain-die formed variable-width profiles.

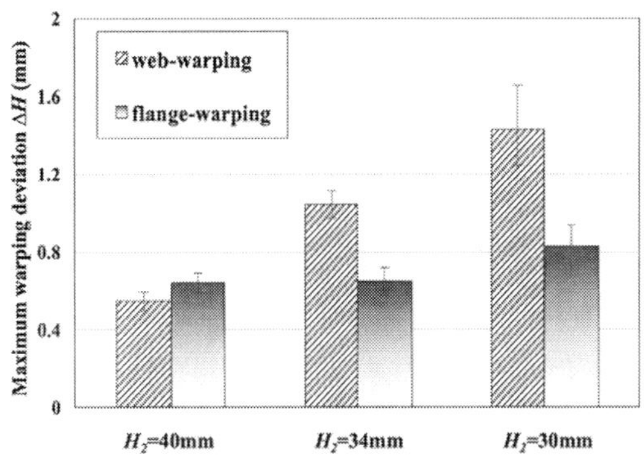

Figure 5. Comparison of maximum warping deviation for chain-die formed variable-width profiles with varying H_2.

Figure 6 shows the division of the designed variable-width profile. By Comparing the initial plate (figure 7(a)) with the target geometry, it can be found that when chain-die forming a variable-width product, longitudinal tensile or compressive strain is necessary and desired for the transition area. Taking the convex part as an example, $B'C'$ is geometrically needed to be stretched to equal to BC $(b'c')$, with an elongation of $B'b' + C'c'$. However, part of the elongation is compensated by the rotation of web and flange area represented with AB and $A'B'$ respectively, as shown in figure 7(b). It is believed that two factors cause insufficient elongation, the less constrained boundary condition during the chain-die forming process [11], and a limited elongation of QP1180. Thus, both web-warping and flange-warping in chain-die forming result from the contradiction between the geometric necessary of tensile (or compressive) deformation and the practical boundary condition and limited elongation of QP1180. Future work will focus on the elimination of warping defects, and pressure pad and mold surface compensation are expected to be efficient ways to reduce warping in the chain-die forming process.

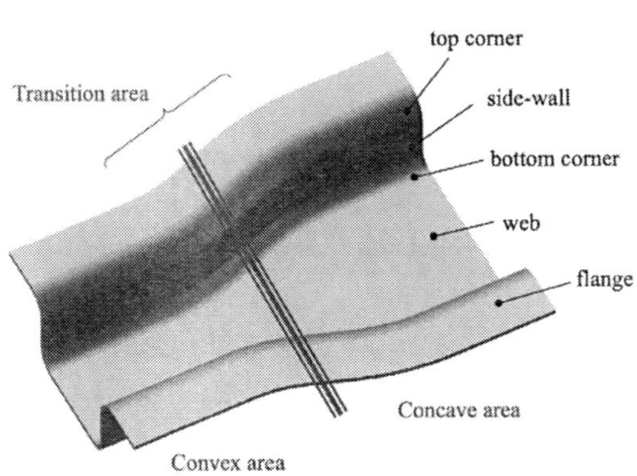

Figure 6. Illustration of the division for the designed variable-width profile.

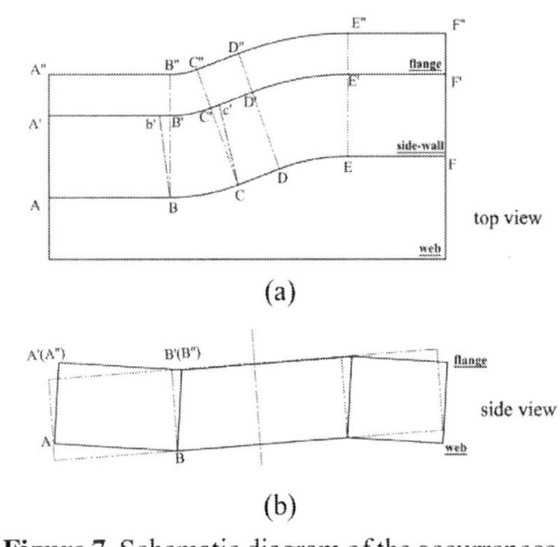

Figure 7. Schematic diagram of the occurrences of warping defects (a) the initial plate; (b) the rotation of web and flange.

IOP Conf. Series: Materials Science and Engineering **967** (2020) 012089 doi:10.1088/1757-899X/967/1/012089

4. Conclusions

A complex variable-width hat-channel product was designed, and the feasibility of chain-die forming was investigated with experiments. The experimental samples were digitalized and compared with the target geometry and warping defects were quantitatively analyzed. The flowing conclusions can be drawn.

1). Experimental results of variable-width hat-channel profiles indicate that chain-die forming is a promising technology for fabricating complex AHSS products.

2). Web-warping and flange-warping show a similar variation for the designed variable-width hat profile, and the cause of warping is ascribed to insufficient longitudinal tensile or compressive deformation.

References

[1] Ding S, Daniel W J T, Yuan J and Zhang Y 2011 *Yokohama tube and pipe 2011 joint symposium-innovative tube and pipe manufacturing and forming*. Osaka, Japan. p. 335-40.

[2] Qian Z, Sun Y, Li Y, Wang C, Meehan P A, Daniel, W J T and Ding S 2020 *J. Mater. Process. Tech.* **277**, 116484.

[3] Sun Y, Li Y, Daniel W J T, Meehan P A, Liu Z and Ding S 2017 *J. Mater. Process. Tech.* **243**, 322-34.

[4] Li Y, Sun Y, Xiao H, Bong H J, Shi L, Li S, Li D, Ding S and Wagoner, R H 2018 *Int. J. Mech. Sci.* **135**, 279-293.

[5] Lee, J-Y, Barlat F and Lee M-G 2015 *Int. J. Plasticity.* **71**, 113-135.

[6] Jiao J, Rolfe B, Mendiguren J and Weiss M 2015 *Int. J. Mech. Sci.* **90**, 228-238.

[7] Jiao J, Rolfe B, Mendiguren J and Weiss M 2016 *Int. J. Adv. Manuf. Tech.* **86**(5-8), 1541-1555.

[8] Park J-C, Yang D-Y, Cha M, Kim D and Nam J-B 2014 *Int. J. Mach. Tool. Manu.* **86**, 68-80.

[9] Sun Y, Qian Z, Daniel W J T, Meehan P A, Shi L and Ding S 2017 *Int. J. Adv. Manuf. Tech.* **92**(1-4), 1147-1164.

[10] Li Y, Liang Z, Zhang Z, Zou T, Li D, Ding S, Xiao H and Shi L 2019 *Int. J. Mech. Sci.* **159**, 195-212.

[11] Sun Y, Li Y, Li D, Meehan P A, Daniel W J T, Shi L, Xiao H, Zhang J and Ding S 2019 *J. Manuf. Process.* **39**, 208-225.

Acknowledgments

The authors would like to acknowledge the support of National Science Foundation of China (Grant No. U1860110). The authors would also like to give many thanks to BaoSteel Research Institute and Ningbo SaiRolf Metal Forming Co., Ltd. for providing experimental materials and facilities.

IOP Conf. Series: Materials Science and Engineering **967** (2020) 012090 doi:10.1088/1757-899X/967/1/012090

Deep drawing behavior of IN625 alloy under the influence of different process parameters

Ayush Morchhale[1], Nitin Kotkunde[1*] and Swadesh Kumar Singh[2]

[1]Department of Mechanical Engineerng, BITS-Pilani, Hyderabad, India
[2]Department of Mechanical Engineering, GRIET, Hyderabad, India

*nitink@hyderabad.bits-pilani.ac.in

Abstract The Ni-Cr based IN625 alloy has been extensively used in critical applications of aerospace and nuclear industries. Various components have been manufactured using sheet metal forming processes. In the present work, the deep drawability of material has been examined under the influence of different process parameters such as temperature, punch speed, lubrication, and blank holding pressure using the processing window. The deep drawability has been analyzed in three different zones, namely, safe, wrinkling, and fracture. The limit drawing ratio increased by 7.12 % on increasing the temperature from 300K to 673K. The thickness distribution across the deep-drawn component plays a crucial role in examining the quality and life of it. Therefore, in this context, a detailed analysis of different process parameters over the drawn cup's thickness distribution has been done. The minimum thickness was obtained near the punch corner or the cup junction region of the deep drawn cup. The uniformity in the thickness distribution and drawing height also increased with the forming temperature and decrease in punch speed. Additionally, two different yield criteria, namely, Hill 1948 and Barlat 1989, have been analyzed and further used for the numerical analysis using the user material subroutines in Abaqus software. The results predicted using Barlat 1989 yield criterion more closely followed the experimental results as the average absolute error obtained is well within the 5% of the acceptable limit.

1. Introduction

Inconel 625 (IN625) is a Ni-Cr-Fe based superalloy having a right blend of various mechanical properties, namely, high strength, ductility, and resistance towards corrosion, oxidation, creep and fatigue [1,2]. This Ni-based superalloy shows high-performance ability in extreme conditions and hence used in various critical applications of aerospace and nuclear industries [3]. Sheet metal forming (SMF) process is majorly used in different industries to deform material into the required shape using the appropriate stresses [4,5]. This process helps in forming the material into the required complex shape. This process is cost-effective and reduces the material wastage compared to the traditional machining and welding processes [6].

IN625 alloy is a very high strength metal and has limited workability, which makes it very difficult to form into the desired shape at the room temperature condition . Thus, researchers proposed that in the past, metal forming for such high strength metals at elevated temperature conditions [7–9]. Further, several popular metal forming processes are deep drawing, stretch forming, stamping, etc. Several process parameters play a very important role in determining the quality of a finally formed component. Researchers have adopted the optimization route of process parameters in order to get the best quality components. Determination of processing window is one such method for determining the safe drawing zone of the deep-drawn cups under the combined influence of different process parameters. In the last few decades, several studies have been devoted to designing the processing windows (PW) diagrams for various metal forming processes. Chu et al. [10] determined the theoretical PW of Al extrusion tubes using the hydroforming process. The PW has been designed on the basis of variation between the axial force (compressive) and internal pressure for predicting the bursting, wrinkling, and buckling phenomenon. Vollertsen et al. [11] determined PW's effect for a micro deep drawing process to investigate the effect of punch velocity. Finally, they concluded that the allowable blank holding pressure (BHP) increases with the punch velocity during the process. Prasad et al. [12] determined the deep drawing PW for the solution treated IN718 alloy to determine the safe drawing region for the material. Gao et al. investigated PW for the compression of the tube using the viscous pressure forming process by considering the variation of tube blank diameter and compression length. They found that the safe and wrinkling zone decreased with the increase in the tube compression length. Further, the wrinkle elimination zone increased, and the safe zone decreased with the tube blank diameter increase. Hashemi et al. [13] found PW for conical cup formation using the hydrodynamic deep drawing process using the Cu, steel, and Al sheets. It has been concluded that the sheets with less thickness and high strength displayed more uniformity in thickness hence better formability on the finally obtained product.

The hardening law and the yielding function play an important role in the accurate prediction of metal's deep drawing behavior. Hill 1948 yield criterion has first been proposed as a modification of von mises criterion in order to include the anisotropic behavior of the material. This criterion has four material constants [14]. The determination of yielding function involves simple calibrations; hence, it has been extensively used in the industries. Although, it has a drawback of not predicting the anisotropy value with high accuracy. Barlat and Lian later found a non-quadratic yielding function for the anisotropic effect in sheet metals and found the r-value accurately [15]. This proposed relation also has four material constants, which can be found using a large number of iterative computations [16].

In the present work, the PW has been determined under the influence of forming temperature and lubricating conditions. Further, the effect of different process parameters on the cup's thickness distribution has been analyzed individually. Additionally, the Hill 1948 and Barlat 1989 yielding function have been calibrated for analyzing the deep drawing behavior using the finite element analysis. The simulated results of thickness distribution and drawn heights have further been validated with the experimental results.

2. Material and Method

1 mm thick IN625 alloy sheet has been used in the present study for all analyses. The dimensions and design of the tensile test specimen have been considered according to the ASTM E08/E8M-11 standard. These specimens were cut using wire cut Electric discharge machining in five different directions, namely, 0°, 22.5°, 45°, 67.5°, and 90° with respect to the sheet's rolling direction for determining the anisotropic material properties. All the experiments have been carried out three times and the average material properties are reported below in Table 1.

Table 1. Average material properties of IN625 alloy

Temperature (K)	Yield Stress (MPa)	Ultimate Stress (MPa)	Elongation (%)	Anisotropic Coefficient r_0	r_{90}
300	812.07±8	978.55±9	39.65±0.3	0.851	0.749
673	609.51±6	805.18±4	46.57±0.5	1.086	0.741

The deep drawing experiments have been carried out on a 40-ton hydraulic press equipped with the induction coil heating setup to perform forming operations at elevated temperature conditions. The Molybdenum disulfide based Molykot lubricant has been used during the whole experimentation as it remains stable 1000K temperature. The punch speed, blank holding pressure, and temperature have been varied according to the experiments' requirement. The blank diameter of the circular blanks considered in the present study varied from 52mm to 62mm. The K-type contact thermocouple has been used in the present study to measure the temperature during all the high-temperature experimentations. The thickness and height measurement digital gauge has been used to measure the thickness and draw height accurately after the experiments. The experiments have been carried out at two different temperatures i.e., 300K and 673K. The BHP and punch speed were to 35bar, and 50mm/min, respectively, as these are the safe operating limits of the used hydraulic press in the present study.

3. Result and Discussion

3.1. Processing Window

The process windows obtained by performing deep drawing experiments at 300K and 673K under the influence of lubricating condition are shown in Figure 1(a & b) respectively. All the cups were drawn at different BHP within the range of 1bar to 35bar pressure and were characterized under different zones, namely, wrinkling, safe, and fracture. All the deep drawing operations were performed until 35bar pressure. The drawing ratio in the process window corresponds to the ratio of diameter for blank to that of punch. The different zones i.e., safe forming zone, fracture zone, and wrinkling zone, are marked in the process window at different conditions. It was observed that the processing temperature and BHP play a vital role in sheet metal forming industries. While performing the deep drawing operation, at lower BHP (up to 2bar) under all the processing conditions, the cups were difficult to form because of the formation of wrinkles, which in turn restricted the easy flow of material under all the considered processing conditions. Finally, this resulted in the cup fracture from the region of the punch corner. The increase in the formation of wrinkles has been observed with the increase in blank diameter, but it can also be compensated and suppressed by increasing the BHP. The cups were also fractured at higher BHP because of the hindrance in the flow of material and increase in the tensile stresses along the cup's walls above the critical limit [12].

The maximum diameter of 56mm blank size was observed to be successfully drawn at a BHP of 20bar at 300K. As a result, the LDR of 1.867 has been observed in Figure 1(a). At 673K, under the lubricated condition, the LDR has been increased to 2 and the successful cup has been drawn at 10bar BHP, as shown in Figure 8(b). Thus the LDR improved by approximately 7.12% as the forming temperature increased from 300K to 673K. This effective change in the LDR is due to the thermal softening, which made the material soft and helped in the easy forming of the material. The lubrication also helped and provided an added advantage for the easy drawing of material into the desired shape. Thus, the study of process window help in optimizing the process parameters for obtaining safe drawing limit of material.

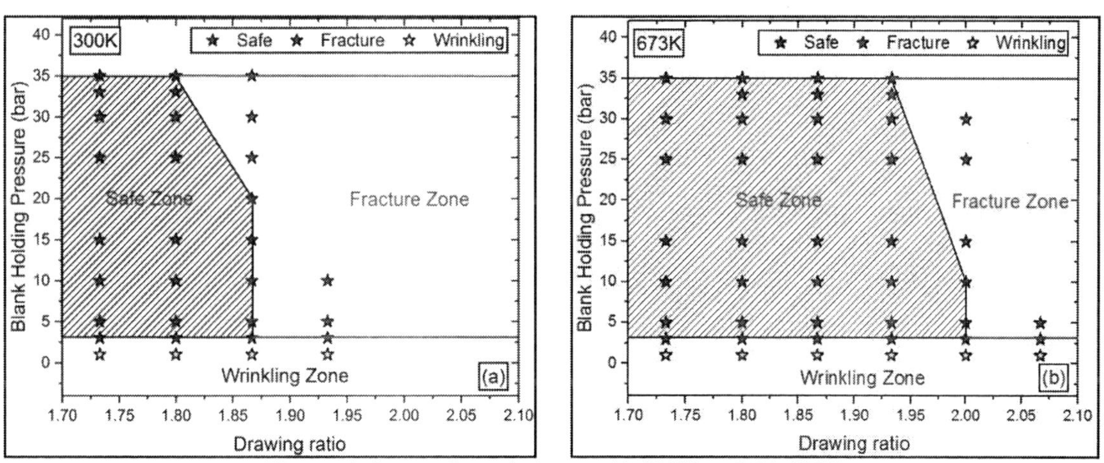

Figure 1. Processing window at (a) 300K and (b) 673K

3.2 Effect of Different Process Parameters over the Thickness Variation

The thickness distribution across the deep drawn cup is highly influenced by different process parameters. Maximum thinning rate (MTR) and thickness deviation (TD) are two different statistical parameters used for the quantitative evaluation of the thickness under the influence of different process parameters. In the formulation of TD and MTR, the terms t_i, t_m, t_{min} and N correspond to the initial thickness, average thickness, minimum thickness, and the total number of observations. The thickness has been measured at a fixed distance of 2mm in three different directions and the average values are reported in Figure 2.

$$\mathrm{TD} = \sqrt{\frac{\sum(t_i - t_m)^2}{N}} \times 100\ (\%) \tag{1}$$

$$\mathrm{MTR} = \frac{(t_i - t_{min})}{t_i} \times 100\ (\%) \tag{2}$$

The thickness distribution obtained from the successfully drawn cups with the highest drawing ratio at 300K and 673K are shown in Figure 2(a). All the cups drawn under different conditions displayed a similar trend for the thickness distribution. The thinning near to the base of the cup is less due to the increase in friction between the sheet metal and punch at this point of contact as stated by Prasad et al. [17]. Thus, the material near the base region of the punch will always be constrained, resulting in very less deformation than other regions of the deep-drawn cup. Additionally, the thickness near the cup junction is minimum because of large stretching and subsequently losing the contact with the punch corner. This is known as the critical region, and mostly deep-drawn cups fail from this region. More uniformity or less deviation in the thickness has been observed at higher forming temperatures. Further, the use of lubricant provides an added advantage in the deep drawing process as it helps in easy forming by reducing the forming load. In the present case, the cup formed at 673K under the lubricating condition was found to have the least variation in thickness, highest LDR, and drawn height.

Figure 2(b) shows a representative comparison of thickness for the case, 673K and 5mm/min punch speed, at three different BHP, namely, 3bar, 12bar, and 20bar for the blank size of 58mm. In the deep drawing process, at low BHP (below 3bar in the present case), the thickness of the sheet became closer to the outer periphery of circular blanks due to the formation of wrinkles. This ultimately led to the restricted flow of material followed by the fracture from the cup junction region. At high pressure, the blank experiences a rapid increase in the tensile stress along the cup's vertical walls. This results in the increase of stress concentration near the contact region of blank and punch, ultimately leading to the fracture of the cup. Thus, the cup's failure occurs when the generated tensile stress exceeds the material's load-bearing capacity at the cup junction. The blank holding pressure also seems to have an effective contribution in deciding the thickness distribution across the formed cup. With the increase in BHP, more uniformity in the thickness distribution has been observed. Further, the minimum thickness of the drawn cups also improved by 7.48% on increasing the BHP from 3bar to 20bar. Thus, the determination of optimum BHP is important in order to remove any failures in terms of wrinkling and fracture. The thickness distribution w.r.t the punch speed is shown in Figure 8(c). The punch speed seem to have significantly influence the drawn height of the cup. The TD seem to have reduced by 36.2% when deep drawing has been performed at 5mm/min. Hence, from the above discussion, it can be concluded that better quality deep drawn cups can be obtain at higher temperature, lower punch speed and nominal BHP.

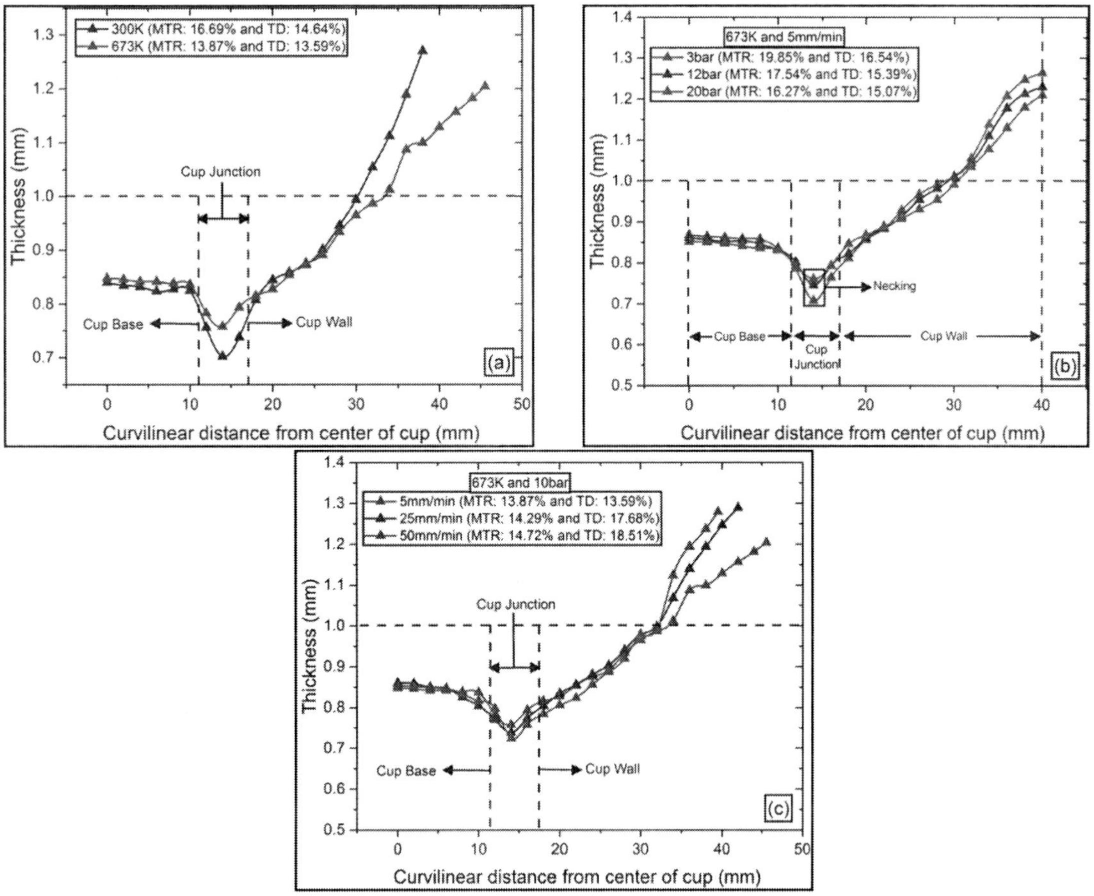

Figure 2. Influence of (a) temperature, (b) BHP and (c) punch speed on the thickness distribution of deep drawn cup

3.3. Numerical Analysis

The numerical analysis of any forming process is highly influenced by the used yielding function. Hence, in the present study two different yielding functions, namely, Hill 1948 and Barlat 1989 have been used for accurate prediction of deep drawing behavior of material. The material constants for both the yield criteria have been evaluated using the procedure mentioned by Pandre et al. [18]. Figure 3 shows the yield locus calibrated using these yield criteria. Barlat 1989 yield criterion seem to have closely following the experimentally normalized yield stresses at both the temperatures.

Table 2. Hill 1948 yield criterion material constants

Temperature	H	G	F	N
300 K	0.614015	0.540103	0.459897	1.694937
673 K	0.739487	0.47918	0.52082	1.415115

Table 3. Barlat 1989 yield criterion material constants

Temperature	a	c	h	p
300 K	0.887581	1.112419	1.036294	0.89
673 K	1.025793	0.974207	0.881727	0.94

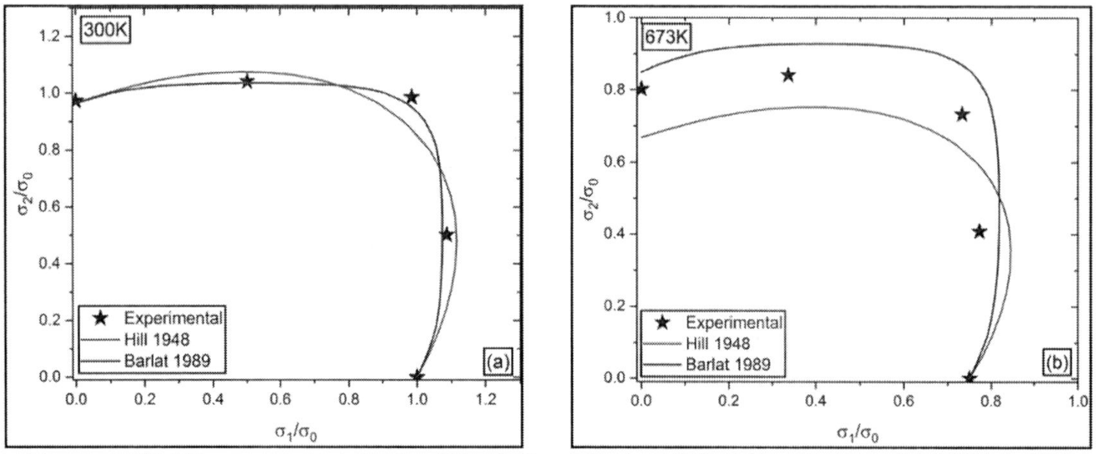

Figure 3. Yield Locus computed using Hill 1948 and Barlat 1989 yield criterion at (a) 300K and (b) 673K

The setup for performing deep-drawing operation consists of circular die, flat circular punch, blank holder plate, and circular blanks. The blanks are assigned with the properties of the deformable object. S4R planar elements have been used for the fast computation of results. Further, the die, punch, and blank holder plate are assigned with the rigid R3D4 elements as no results are desired over them. The user-defined material subroutine has been used for including the effect of yield criteria while doing the numerical simulation of the deep drawing process in ABAQUS software. The thickness distribution and drawn height have been obtained using both the above-calibrated yielding function and compared with the experimental results. The cup with height drawing ratio has been considered for the comparison with the numerically obtained results of thickness and drawn height, shown in Figures 4 (a and b) at 300K and 673K, respectively. Both the yield criteria helped in predicting the results with high accuracy. Though, Barlat 1989 criterion best predicted both the thickness distribution and drawn height among the two considered yielding functions. The average absolute error obtained is well within the 5% of the acceptable limit.

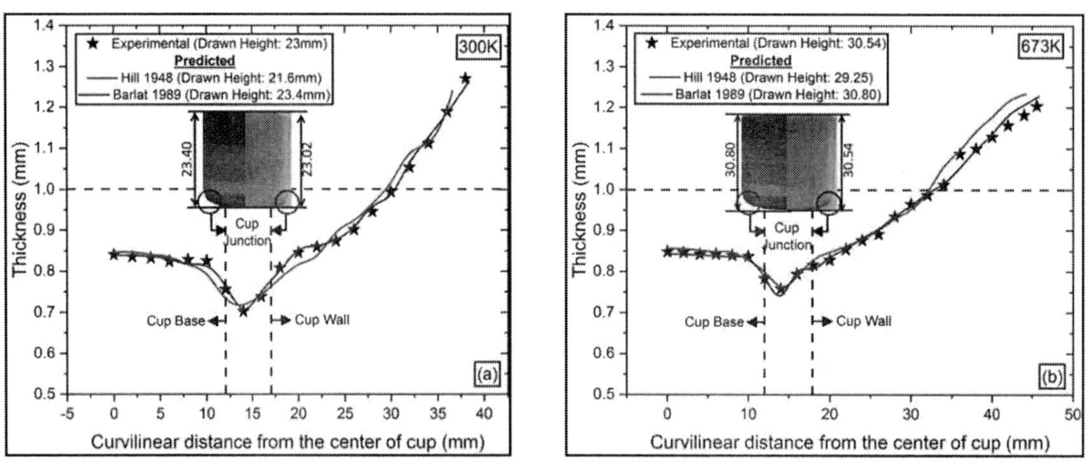

Figure 4. Experimental and predicted thickness distribution at (a) 300K and (b) 673K

4. Conclusions

The analysis of warm deep drawing process has been done in the present study. Some of the important results from the above study are stated below.

- The processing window has been obtained at 300K and 673K temperature. The safe drawing limit of the material improved with the forming temperature. LDR has been found to be increased by approximately 7.12%. The thickness distribution across the successfully drawn cup has been found to be greatly influenced by different process parameters. The minimum thickness in all the cases has been found to be near to punch corner region. The uniformity in thickness has been found to be increased with the increase in temperature and decrease in punch speed. The nominal BHP needs to be maintained during the whole process. The cup's wrinkling and fracture have been observed in the case of very low and high BHP, respectively.

- The Hill 1948 and Barlat 1989 anisotropic yielding function has been considered for the accurate prediction of the deep drawing behavior using numerical simulations. Barlat 1989 criterion helped in predicting the numerical results of thickness distribution and drawn height more accurately compared to Hill 1948 criterion.

Acknowledgement
The authors express their sincere gratitude towards Science and Engineering Research Board, Government of India (ECR/2016/001402) and BITS-Pilani, Hyderabad Campus for financially supporting the present work.

References

[1] Kotkunde N, Badrish A, Morchhale A, Takalkar P and Singh S K 2019 Warm deep drawing behavior of Inconel 625 alloy using constitutive modelling and anisotropic yield criteria *Int J Mater Form* **13** 355–69
[2] Badrish C A, Morchhale A, Kotkunde N and Kumar Singh S 2020 Prediction of flow stress using integrated JC-ZA constitutive model for Inconel 625 alloy *Materials Today: Proceedings*
[3] Badrish A, Morchhale A, Kotkunde N and Singh S K 2020 Influence of material modeling on warm forming behavior of nickel based super alloy *Int J Mater Form* **13** 445–65
[4] Mahalle G, Morchhale A, Kotkunde N, Gupta A K, Singh S K and Lin Y C 2020 Forming and fracture limits of IN718 alloy at elevated temperatures: Experimental and theoretical investigation *Journal of Manufacturing Processes* **56** 482–99
[5] Pandre S, Morchhale A, Kotkunde N and Singh S K 2020 Influence of processing temperature on formability of thin-rolled DP590 steel sheet *Materials and Manufacturing Processes* **35** 901–9
[6] Dharavath B, Morchhale A, Singh S K, Kotkunde N and Naik M T 2020 Experimental Determination and Theoretical Prediction of Limiting Strains for ASS 316L at Hot Forming Conditions *J. of Materi Eng and Perform*
[7] Badrish A, Morchhale A, Kotkunde N and Singh S K 2020 Parameter Optimization in the Thermo-mechanical V-Bending Process to Minimize Springback of Inconel 625 Alloy *Arab J Sci Eng* **45** 5295-5309
[8] Badrish C A, Morchhale A, Kotkunde N and Singh S K 2020 Experimental and finite element studies of springback using split-ring test for Inconel 625 alloy *Advances in Materials and Processing Technologies* **6** 540-46
[9] Pandre S, Kotkunde N, Takalkar P, Morchhale A, Sujith R and Singh S K 2019 Flow Stress Behavior, Constitutive Modeling, and Microstructural Characteristics of DP 590 Steel at Elevated Temperatures *J. of Materi Eng and Perform* **28** 7565–81
[10] Chu E and Xu Y 2004 Hydroforming of aluminum extrusion tubes for automotive applications. Part II: process window diagram *International Journal of Mechanical Sciences* **46** 285–97
[11] Vollertsen F and Hu Z 2010 Analysis of punch velocity dependent process window in micro deep drawing *Prod. Eng. Res. Devel.* **4** 553–9

[12] Prasad K S, Panda S K, Kar S K, Murty S V S N and Sharma S C 2018 Effect of solution treatment on deep drawability of IN718 sheets: Experimental analysis and metallurgical characterization *Materials Science and Engineering: A* **727** 97–112
[13] Hashemi A, Hoseinpour Gollo M and Seyedkashi S M H 2015 Process window diagram of conical cups in hydrodynamic deep drawing assisted by radial pressure Transactions of Nonferrous Metals Society of China **25** 3064–71
[14] Hill R 1950 The mathematical theory of plasticity vol 11 (Oxford university press)
[15] Barlat F and Lian K 1989 Plastic behavior and stretchability of sheet metals. Part I: A yield function for orthotropic sheets under plane stress conditions *International Journal of Plasticity* **5** 51–66
[16] Banabic D 2010 *Sheet Metal Forming Processes: Constitutive Modelling and Numerical Simulation* (Berlin Heidelberg: Springer-Verlag)
[17] Prasad K S, Panda S K, Kar S K, Murty S V S N and Sharma S C 2018 Prediction of fracture and deep drawing behavior of solution treated Inconel-718 sheets: Numerical modeling and experimental validation *Materials Science and Engineering: A* **733** 393–407
[18] Pandre S, Takalkar P, Morchhale A, Kotkunde N and Singh S K 2020 Prediction capability of anisotropic yielding behaviour for DP590 steel at elevated temperatures *Advances in Materials and Processing Technologies* **6** 476–84

Study on the Anti-friction properties of Chemically Etched surface texture and its synergistic Anti-friction properties with nano-lubricating oil in sheet metal deep drawing process

Tangjie Mei, Dongliang Zhang, Qi Zhang*

Xi'an Jiaotong University, Xi'an, 710049, China

*E-mail: henryzhang@xjtu.edu.cn

Abstract. In sheet metal deep drawing process, the sliding contact between mold and sheet metal will cause friction and wear between sheet metal and die, resulting in tensile breaking and die wear, which will not only shorten the service life of die seriously, but also lead to quality decline and the rising costs of the forming process. In this paper, an anti-friction method using chemically etched surface texture collocated with graphene nanoparticles added lubricating oil was proposed to reduce the COF between sheet metal and die during sheet metal deep drawing process. Several surface textures with different parameters were obtained by surface chemical etching treatment on the die. The surface texture was observed by Laser Confocal Microscope. The results show that the surface texture after corrosion consists of micron-scale convex and concave, in the process of friction, when the pressure is higher than 72 MPa, the concave in the surface texture forms a closed oil concave because of plastic deformation, lubricating oil provides static pressure bearing capacity in the closed oil concave, sheet metal was separated by the oil from die contact surface, which effectively reduced the COF. In addition, the anti-friction properties ware tested, which showed that when the mass fraction of graphene nanoparticles added is 0.3%, stable deposited film in the friction contact interface can be formed, and a good filling effect of the lubricating oil in the sealed oil concave can be guaranteed, which can preserve a better integrity of the surface texture.

1. Introduction

Sheet metal deep drawing process is a kind of metal plastic processing method as stamping, which is widely used in automobile manufacturing, rail transit, aviation and so on. But the friction and wear between sheet metal and die during the processing has become the most serious problem hinders the development of sheet metal drawing process[1-4]. Sulaimana et al.[5] studied the performance and physical properties of the DLC/TiALN-coated die surface. The results showed that the presence of a thin oil film reduces the sliding-originated surface tensile stresses of the DLC/TiALN coating. Costing models[6] investigated the tribological behavior of concentional and modified surfaces at the critical radii area with a strip bending rotation test to investigate the cause-effect relations as well as wear mechanisms. Flegler et al.[7] found out that a reduction of the roughness of a mill-finished sheet results in a significant improvement with regard to the friction and wear behaviour in the strip drawing test.

In this paper, the surface texture with different surface characteristics is prepared by chemical etching on C12. Through the friction experiment, the influence of different texture on the friction of die under different pressure is studied, and the nano-lubricating oil with graphene nanoparticles were used in the friction and wear tests to investigate the anti-friction properties in sheet metal deep drawing processing.

2. Chemical etching experiment

2.1. Sample Preparation

Cr12 die steel is selected as the test material, and prepared into Φ10 × 5 mm cylinder. The surface of all specimens was grinded and polished, and the initial surface was obtained after ultrasonic cleaning for 2 min (the surface roughness was about 0.08 μm). Samples were etched by a mixture of ferric chloride, nitric acid (a mass fraction of 65%) and distilled water, with a proportion of 20 g: 30 ml: 10 ml. The etching process was carried out in a 30° constant temperature water bath of 2.5 min, 5 min, 7.5 min, 10 min, 12.5 min, 15 min, 17.5 min and 20 min respectively.

2.2. Sample Characterization

In the experiment, ISO25178 were used to characterize the surface texture observed by laser confocal microscope (LSCM, model: OLS4000) under 50 times magnification.

2.3. Anslysis of Corrosion test results

As shown in Figure 1, it is a three-dimensional appearance when the etching time is 17.5 min. After etching, the surface appeared obvious gully appearance with concaves and convexes. At the same time, there formed many oil concaves (closed concaves) convenient for storing lubricating oil, which will squeeze the lubricating oil in the friction process, providing support force and reducing surface friction by hydrostatic effect.

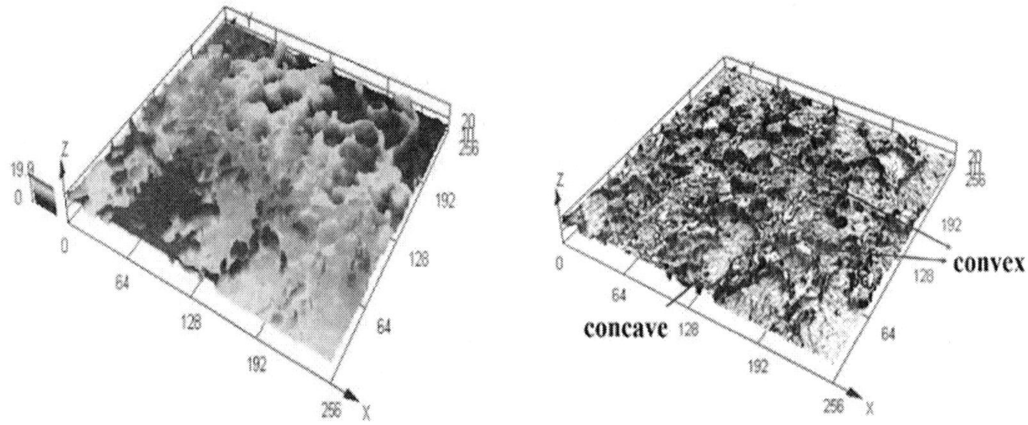

Figure 1. 3D topography of the surface texture

According to the following parameters to evaluate the quality of corrosion surface: S_a and $V_{vv} + V_{vc}$ as the main parameters to characterize the surface quality, S_{sk} S_{ku} as auxiliary parameters. The specific values are as follows:

As shown in Figure 2, following parameters selected to evaluate the quality of corrosion surface are S_a, V_{vv}, V_{vc}, S_{sk} and S_{ku}. S_a mainly represents the roughness of the surface, $V_{vv} + V_{vc}$ reflects the surface storage oil lubrication effect, S_{sk} and S_{ku} reflects the bearing capacity and undulation degree of rough surface. With the increase of the time of corrosion, S_a presents a wavy change, which is because the corrosion on the surface layer by layer during the time of corrosion. When the time of corrosion reached a certain time, the surface layer is gradually corroded completely, then the surface gradually tends to be smooth, and the next layer continues to be etched.

When the time of corrosion is 7.5 min, the surface is smoother than the other samples with the S_a of 0.67 μm; when the time of corrosion is 17.5 min, the surface is rough with the Sa of 2.546 μm. V_{vv}+V_{vc} showed the same change rule as S_a, and the consistency between them ware good. When the time of corrosion is 15 min and 17.5 min, the S_{sk} is lower than 0, the S_{ku} is lower than 3, the V_{vv}+V_{vc} is larger, which means there are more convexes than concaves on the surface after corrosion.

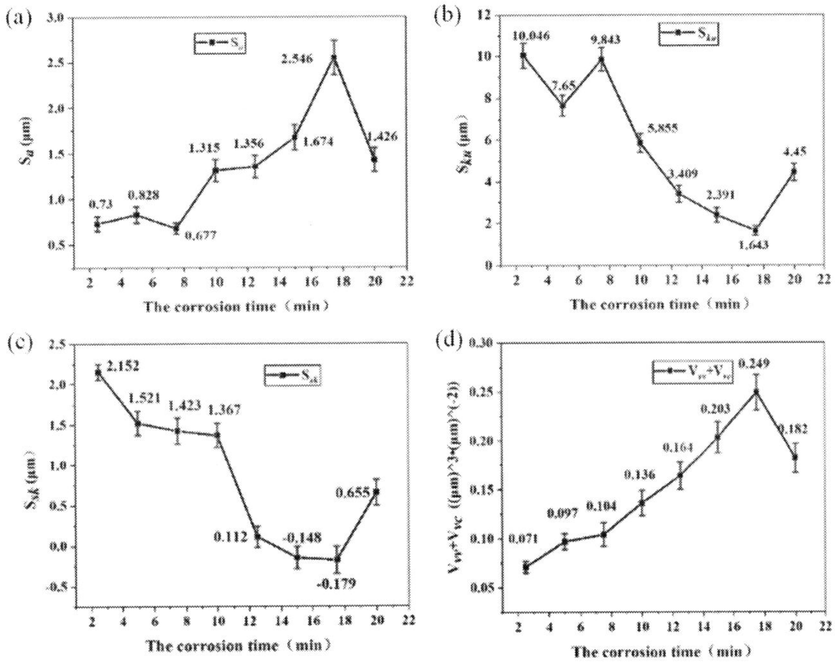

Figure 2. Surface morphology characterization parameters of the samples with surface texture: (a) S_a; (b) S_{ku}; (c) S_{sk}; (d) V_{vv}+V_{vc}.

3. Friction experiments

3.1. Test specimen

Five groups of etching samples with different gradients were selected as samples for friction experiment, which were recorded as samples 1, 2, 3, 4, 5(0 min, 7.5 min, 12.5 min, 15 min, 17.5 min).The lubricating oil used in this test is water-soluble emulsion oil.

3.2. Friction and wear experiments

The experiments verified the effect of surface texture on friction. The actual contact pressure is 9 MPa, 18 MPa, 36 MPa, 54 MPa, 72 MPa (706.5 N, 1413 N, 2826 N, 4239 N, 5652 N); the relative velocity rate is 10 mm / min.

3.3. Results of the friction and wear experiments

As shown in Figure 3, the results showed that after the run-in, the COF curve is stable and the data is efficacious.

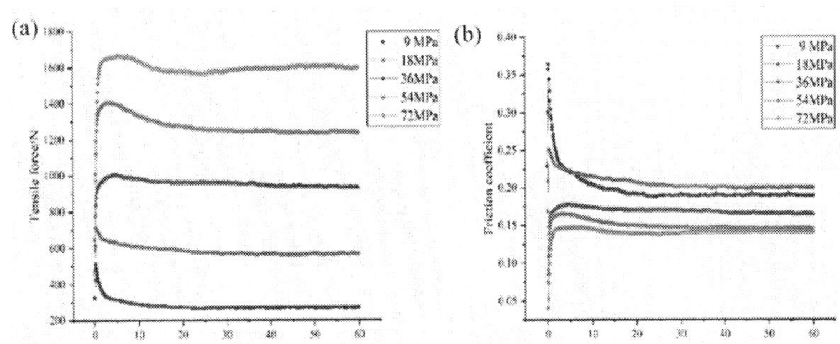

Figure 3. Friction experiment of samples 1~5: (a) tensile force curve; (b) COF curve.

As shown in Figure 4, the results showed that with the increase of pressure, the COF of sample 1 increases gradually while the COF of sample 2-5 decreases. At low pressure, the COF of etching sample is generally higher than that of non-etching sample, while at high pressure, the COF of etching sample is generally lower than that of non-etching sample. The lowest COF is 0.095 at the pressure of 72 MPa, which is 49.5% lower than that of 0.188, and the average value of COF of sample 2-4 is 0.644 times of sample 1.

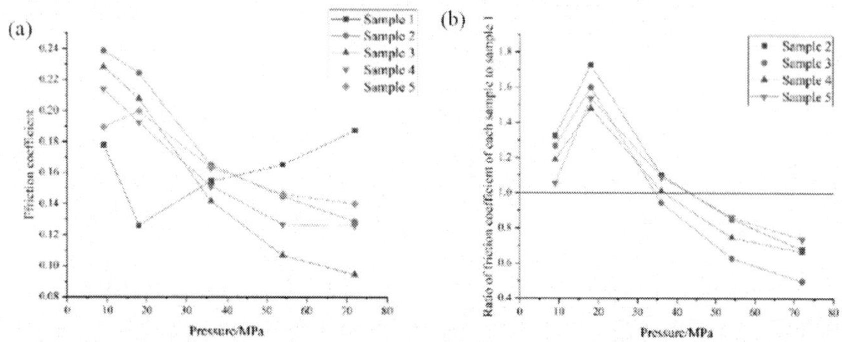

Figure 4. Friction experiments: (a) COF curve; (b) ratio of COF of each sample to sample 1.

4. Anti-friction properties of nano-lubricating oil

4.1. preparation of the nano-lubricationg oil

The nano-lubricating oil is prepared through the method of mechanical agitation and ultrasonic vibration. The graphene nano particles (as shown in Figure 5) ware mixed with water-soluble emulsion oil by the weight percentage of 0, 0.1, 0.3 and 0.5.

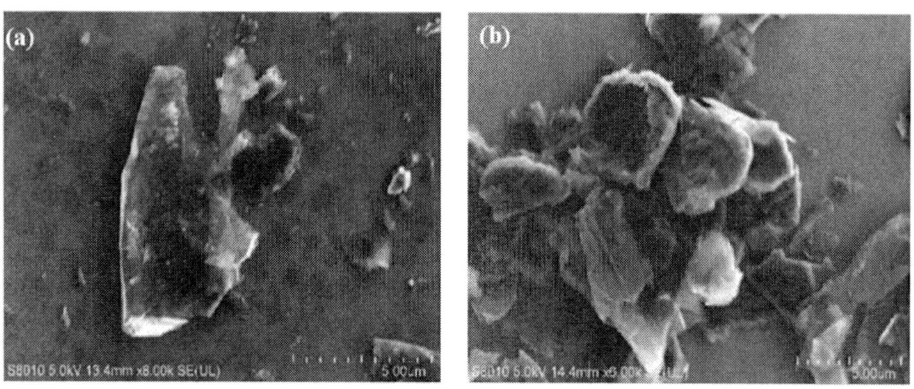

Figure 5. SEM graph of graphene nano-particles

4.2. Results

As shown in Figure 6, after selecting the average value of all the points (a total of 50 data points) in 5 s with stable friction, the average friction coefficient curve of each experiment was obtained, which first rapidly increased to the maximum value, then gradually decreased, and finally reached the stable value, that is, changed from static friction to sliding friction, and finally became stable gradually.

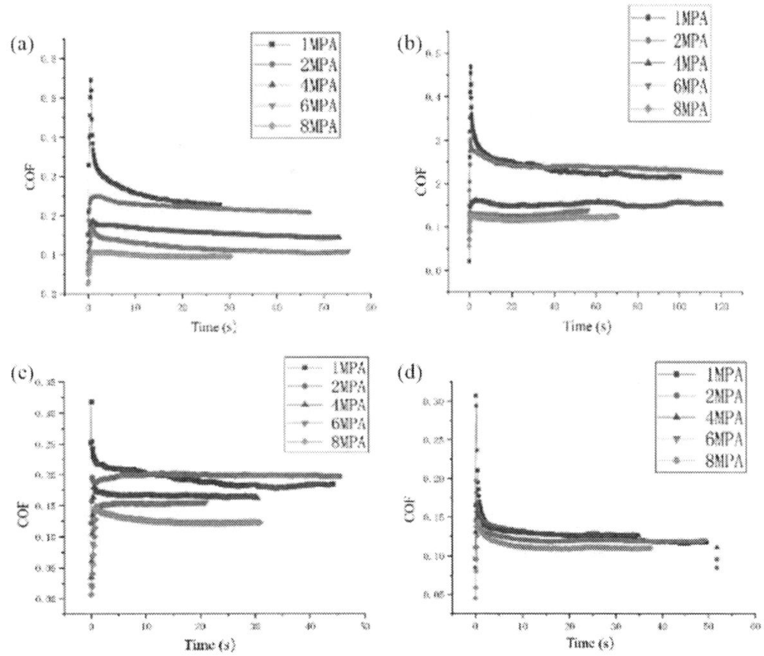

Figure 6. COF-time curve of nano-lubricating oil: a) 0%; b) 0.1%; c) 0.3%; d) 0.5.

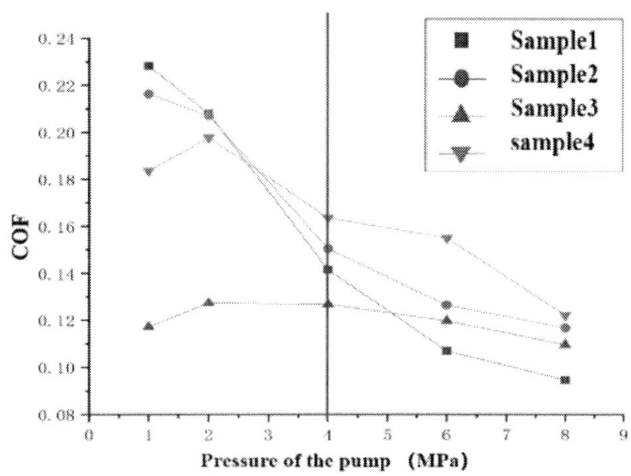

Figure 7. COF-pressure of the pump curve for each lubricating oil group.

The average COF curve of each lubricating oil group on different pressure ware shown in Figure 8. The results showed that, with the increasing of the pressure, the COF decrease for all oil samples. Meanwhile, when the percentage of weight achieve 0.3% (sample c), the COF (0.12) is lower than the other samples at the same pressure of pump. But when the pressure is more than 4 MPa, the COF of sample c is higher than pure lubricating oil. That's because when the low pressure is strong, graphene acts as a solid lubricant, and the oil is not completely extruded by the contact surface, and the combination can be effectively reduced. Because the fabric gap is filled with graphene particles, the lubricating oil is greatly squeezed by the contact surface, the liquid lubrication is greatly reduced, and the partial graphene reduces the effect of the friction effect, so the friction coefficient is still less than the friction coefficient of the friction coefficient of the oil sample c.

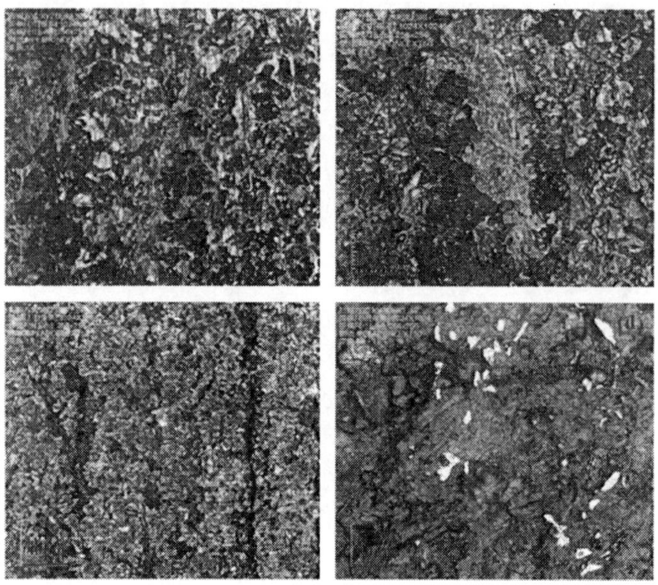

Figure 8. COF-time curve of nano-lubricating oil: (a) 0%; (b) 0.1%; (c) 0.3%; (d) 0.5%.

5. Conclusion

According to the results of friction and wear experiments, we can obtain the conclusion as below:
1) With the increase of the load, the COF between die and sheet metal with the surface texture prepared by chemical etching decrease gradually.
2) The surface texture prepared by chemical etching can effectively reduce the COF between die and sheet metal under high load condition (actual contact load ⩾ 72 MPa), with an average reduction of 35.6%;
3) When the S_a of the surface is about 1.3, the effect of reducing friction of the surface texture is the best, and the maximum reduction is 49.5% (actual contact load is 72 MPa);
4) The nano-lubricating oil with graphene nano particles show definite anti-friction and anti-wear performance during the friction and wear test against under the pure lubricating oil condition. When the graphene percentage is 0.3%, the anti-friction performance of the nano-lubricating oil is the best.

References

[1] Häfenrt T, Heberle J and Dobler M. Friction adjustment within dry deep drawing by locally laser textured tool surfaces. 2015, Key Engineering Materials, 57-64.
[2] Hazrati J, Stein P and Kramer P. Tool Texturing for Deep Drawing Applications; 2018, proceedings of the IOP Conference Series: Materials Science and Engineering, F.
[3] Meng B, Fu W and Wan M. Drawability and frictional behaviour of pure molybdenum sheet in deep-drawing process at elevated temperature. 2015, International Journal of Advanced Manufacturing Technology, 78(5-8): 1005-14.
[4] Recklin V, Dietrich F and Groche P. In-situ-measurement of the COF in the deep drawing process; Proceedings of the Journal of Physics: Conference Series, F, 2017.
[5] Ssulaiman H, Farahanar N and Bienk K. Effects of DLC/TiAlN-coated die on friction and wear in sheet-metal forming under dry and oil-lubricated conditions: Experimental and numerical studies. 2019, Wear, 438-439
[6] Krachenfels K, Rothammer B and Tremmel S. Experimental investigation of tool-sided surface modifications for dry deep drawing processes at the tool radii area; 2019, proceedings of the Procedia Manufacturing, F.
[7] Flegler F, Greche P and Abraham T. Dry Deep Drawing of Aluminium and the Influence of Sheet Metal Roughness. 2020, JOM, 72(7): 2511-6.

AUTHOR INDEX